Fundamentals of Electric Circuits

SEVENTH EDITION

DAVID BELL

OXFORD
UNIVERSITY PRESS

8 Sampson Mews, Suite 204, Don Mills, Ontario M3C 0H5
www.oupcanada.com

Oxford University Press is a department of the University of Oxford.
It furthers the University's objective of excellence in research, scholarship,
and education by publishing worldwide in

Oxford New York
Auckland Cape Town Dar es Salaam Hong Kong Karachi
Kuala Lumpur Madrid Melbourne Mexico City Nairobi
New Delhi Shanghai Taipei Toronto

With offices in
Argentina Austria Brazil Chile Czech Republic France Greece
Guatemala Hungary Italy Japan Poland Portugal Singapore
South Korea Switzerland Thailand Turkey Ukraine Vietnam

Oxford is a trade mark of Oxford University Press
in the UK and in certain other countries

Published in Canada by Oxford University Press

First Published 2009

Library and Archives Canada Cataloguing in Publication

Bell, David A., 1930–
Fundamentals of electric circuits / David A. Bell.—7th ed.
Includes index.
Early eds. published under title: Fundamentals of electric circuits.
ISBN 978-0-19-542524-6

1. Fundamentals of electric circuits—Textbooks. 2. Electric circuit analysis—
Data processing—Textbooks. I. Bell, David A., 1930– .
Fundamentals of electric circuits. II. Title.

TK454.B44 2009 621.319'2 C2008-903601-8

Cover image: NASA

This book is printed on permanent acid-free paper.

Printed and bound in Canada.

3 4 5 – 16 15 14

PREFACE

Since my earliest days as a student, I have often been shocked to find that many technical concepts that at first seemed impossibly complex were really quite simple once I understood them. Occasionally, I was almost ready to believe that *there was a conspiracy to deliberately make technical material difficult to understand*! In preparing this book for a two-semester course in electric circuit fundamentals I have strived to show that the subject matter is quite understandable to anyone interested.

Assuming no previous electrical knowledge, the coverage begins with explanations of basic electrical concepts, then progresses through simple resistive circuit calculations to complex ac network analysis techniques. How to correctly use basic electrical instruments (ammeters, voltmeters, and ohmmeters) is explained early in the book, and the use of an oscilloscope for ac waveform study is discussed in the first ac chapter. Many illustrations are provided to help students understand each new concept or analysis method as it is introduced. Explanatory captions are provided with most illustrations and each illustration is carefully discussed in the text.

Technical material is absorbed by the process of studying explanations, working through calculation examples, answering review questions, and solving problems. Numbered step-by-step circuit analysis procedures are listed in the text and many practical worked examples are offered. Practice problems are provided throughout, with answers given at the end of each chapter. Extensive sets of review questions and problems are also included at the chapter ends.

I am always grateful for suggestions that will improve my presentation of the material. Comments concerning this book would be very welcome.

David Bell

ORGANIZATION OF THE BOOK

Assuming that the reader has had no previous electrical instruction, the first chapter offers explanations for electrical phenomena and introduces the basic electrical units without getting into precise definitions. Unit definitions are covered in the appendices. Because the first laboratory experiments performed by students normally involve the use of ammeters, voltmeters, and ohmmeters, it is important that they quickly acquire knowledge of how to use these instruments. Methods of correctly connecting and reading such instruments, both digital and analog, are discussed in Chapter 2. Explanations of how the instruments operate are given later in the book.

Ohm's law is explained in Chapter 3, and many examples are provided showing calculations of current, voltage, resistance, and power dissipation. Conductors, insulators, and resistors are the subject of Chapter 4. Differences among the materials, insulator breakdown, conductor resistivity, thermal effects, etc., are treated in detail.

Chapters 5, 6, and 7 study resistive circuits as series circuits, parallel circuits, and series-parallel circuits. This leads to network analysis and network theorems, treated in Chapters 8 and 9. It is important to first acquire a thorough understanding of dc circuits and circuit analysis methods before commencing the study of alternating current circuits. So the circuit analysis topics are first covered as purely resistive circuits with direct voltage inputs. Later, the topics are again treated for ac impedance circuits.

Voltage cells, batteries, and dc power supplies are covered in Chapter 10. Treatment includes series-parallel connected cells and calculations involving lead-acid batteries. In treating magnetism and magnetic circuits (Chapters 11 and 12), every effort is again made to explain the basic phenomena in a simple, understandable, and accurate way. As always, appropriate illustrations and step-by-step worked examples are employed at each stage. Once an understanding of magnetism is achieved, the operation of dc measuring instruments is explained in Chapter 13, including basic explanations of digital and analog electronic instruments. Inductance and capacitance are introduced in Chapters 14 and 15, respectively, and the performance of inductive and capacitive components in dc circuits is the subject of Chapter 16.

The generation of an alternating voltage is explained in Chapter 17 in terms of a simple two-conductor generator. The sine wave instantaneous value, frequency, phase angle, rms value, etc., are explained using appropriate calculation examples. When the study of alternating voltage and current is commenced, students must become familiar with the use of oscilloscopes. To facilitate this, the front panels of typical analog and digital oscilloscopes are presented, oscilloscope controls are explained, and basic application to the study of waveforms is discussed. No attempt is made to show how oscilloscopes work.

Complex numbers and rectangular/polar conversions are gently introduced in Chapter 18. Then, in Chapter 19, the behavior of inductors and capacitors in ac circuits is investigated. This includes series and parallel *RLC* circuits, but resonance is left to a later chapter. Ac circuit analysis and power dissipation in ac circuits are studied in Chapters 20, 21, and 22.

The investigation of resonance (Chapter 23) commences with an explanation of the basic phenomenon and works through series and parallel resonance, resistance damping, and tuned coupled coils. With an understanding of *RLC* circuit analysis and resonance acquired, filter circuits can now be studied. Chapter 24 explains the operation of low-pass high-pass, band-pass, and notch filters, and (as always) offers appropriate calculation examples.

Transformers are investigated in Chapter 25, with the treatment including open-circuit and short-circuit tests, voltage regulation, efficiency, and referred (or reflected) quantities. Chapter 26 covers three-phase ac systems, progressing from a basic three-phase generator to three-phase circuit analysis, phase sequence measurement, and power factor correction. Chapter 27 presents harmonic analysis methods of investigating nonsinsuoidal waveforms.

The final chapter introduces the use of graphic analysis software for computer analysis of circuits. Examples presented include analysis of circuits from other chapters in the book.

SUPPLEMENTARY MATERIALS

Laboratory Manual containing practical experiments, suggested laboratory rules, and recommendations for report writing. Each experiment has the following:

- A *title*
- A one-paragraph *introduction*
- A numbered step-by-step *procedure*
- *Circuits* and *connection diagrams*
- An *analysis* section
- A unique *laboratory record sheet* that teaches students good data-recording methods

Instructor's Resource Manual containing:

- Solutions for all practice problems and end-of-chapter problems in the textbook
- A list of the laboratory equipment and components required to perform all experiments in the laboratory manual
- Model results for all experiments in the laboratory manual

Test Item File containing:

- Seventeen sets of double tests (Tests A and B) that cover the material in the textbook
- Model answers for all tests

Electronic Graphic Analysis File Containing:

- Circuit analysis examples using Mustisim, PSpice, and Micro-Cap

Power Point Transparencies

CONTENTS

CHAPTER 1
Basics of Electricity

CONTENTS

Objectives

You will be able to:

1. Explain electrification by friction and the concept of positive and negative electricity.
2. State the fundamental law of electrification by friction.
3. Describe the planetary atom and sketch a diagram to represent an atom and its component parts.
4. Describe what happens when an electric current flows.
5. Define a conductor, an insulator, and a resistance.
6. Explain potential difference and its relationship to current and resistance.
7. Sketch the basic construction of a voltage cell and discuss its operation.
8. Sketch the graphic symbols used to represent basic circuit components and draw simple circuit diagrams.
9. Explain conventional current direction and electron flow.
10. Explain the difference between direct current (dc) and alternating current (ac).
11. Discuss how electric shock occurs, and describe some of the situations in which electricity can be hazardous to humans.

INTRODUCTION

Human knowledge of electricity began with the study of the phenomenon known to experimenters as *electrification by friction*. The early concept of an electric fluid being transferred from one body to another is now understood as a motion of electrons that have become detached from their atoms. A flow of electrons constitutes an electric current, and the pressure that produces the electron flow is a potential difference between two bodies or between two terminals. A potential difference is produced by an excess of electrons on one body and/or a deficiency of electrons on the other. In the simplest source of electricity the excess and deficiency of electrons are the result of chemical action. When an electric current flows, heat is generated by the electron motion. In certain circumstances, light may also be produced—hence the electric lamp. The various components and interconnections of an electrical system comprise an electric circuit, and a circuit diagram is a graphic representation of an electric circuit. Because electricity can be dangerous, it is important to use it carefully and to take immediate action when someone suffers an electric shock.

1-1 ELECTRIFICATION BY FRICTION

The study of electricity began with investigations of *electrification by friction.* The ancient Greeks knew that when amber was rubbed with wool it could attract lightweight particles of other material, such as feathers or lint. The Greek word *elektron,* which means amber, is the origin of the word *electricity.*

Early in the seventeenth century it was discovered that amber was not the only material that had this property. Glass rubbed with silk and ebonite rubbed with fur were both found capable of attracting small particles of other materials. It was also discovered that when two silk-rubbed glass rods were brought close together they were repelled from each other [see Figure 1-1(a)]. Also, when a silk-rubbed glass rod and a fur-rubbed ebonite rod were brought close together there was a force of attraction between them [Figure 1-1(b)].

From these results it was concluded that there were two types of electricity and that materials *charged* with the same type of electricity repelled each other, while those charged with different types of electricity attracted each other. This conclusion produced a fundamental law:

Like charges repel; unlike charges attract.

Around the middle of the eighteenth century, Benjamin Franklin[1] suggested that when glass was rubbed with silk, some kind of *electric fluid* passed from the

[1]American statesman and scientist (1706–1790).

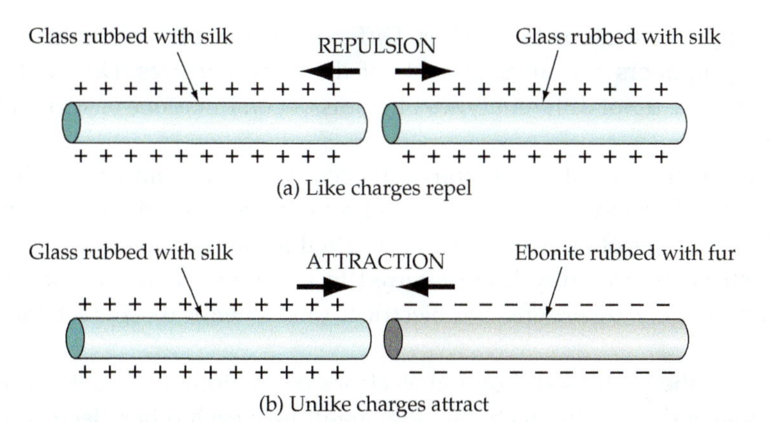

Figure 1-1 Glass and ebonite rods can be used to demonstrate electrification by friction. A silk-rubbed glass rod becomes positively charged. A fur-rubbed ebonite rod acquires a negative charge.

silk to the glass and that this gave the glass an increased amount of electric fluid, or *positive charge.* Conversely, when ebonite was rubbed with fur, the electric fluid passed from the ebonite to the fur, he argued. Thus, the ebonite acquired a reduced amount of electricity, or *negative charge.*

The modern explanation of electrification by friction utilizes our present day understanding that the atom consists of a central *nucleus* surrounded by orbiting *electrons.* The diagram in Figure 1-2(a) illustrates the concept of the *planetary atom.* The electrons have a negative charge, and relative to the nucleus they are extremely small particles. The nucleus [Figure 1-2(b)] is largely a cluster of two types of particles, *protons* and *neutrons,* each of which has a mass approximately *1800 times the mass of the electron.* Neutrons have no charge at all, and protons have a positive charge equal in magnitude to the negative charge on an electron. Consequently, the nucleus is positively charged. The three basic particles—*proton, neutron,* and *electron*—are

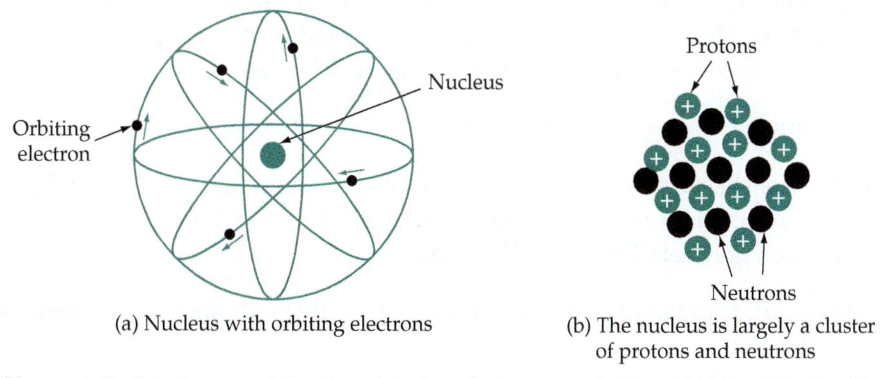

Figure 1-2 An atom consists of a central nucleus surrounded by orbiting electrons. The electrons have a negative charge and the protons in the nucleus have a positive charge.

similar from one atom to another. Differences between atoms are due to differing numbers and arrangements of the three particles. Different materials are made up of different types of atoms or combinations of several types of atoms.

It has been found that electrons can occupy only certain orbital rings, or *shells*, at fixed distances from the nucleus and that each shell can contain only a certain number of electrons. The outer shell is named the *valence shell*, and the electrons that occupy it are referred to as *valence electrons*. The valence electrons largely determine the electrical (and chemical) characteristics of an atom.

Because the protons and orbital electrons of an atom are equal in number and equal and opposite in charge, they neutralize each other electrically. So, each atom is normally electrically neutral; it exhibits neither a positive nor a negative charge. If an atom loses an electron it loses some negative charge, and so it exhibits a positive charge. In this case the atom is referred to as a *positive ion*. Similarly, an atom that gains an additional electron becomes negatively charged and is then termed a *negative ion*.

Benjamin Franklin's electric fluid can now be looked upon as being composed of electrons passing from one material to another. The material that loses electrons becomes positively charged as the result of the loss of negative charges. The material that gains electrons becomes negatively charged. So, glass rubbed with silk acquires its positive charge by losing some electrons to the silk, and ebonite rubbed with fur becomes negatively charged by gaining electrons from the fur.

1-2 VOLTAGE, CURRENT, AND RESISTANCE

When two oppositely charged bodies make contact, electrons flow from the negatively charged body to the positively charged body. This means that electrons move from a location with an excess of electrons to one with a deficiency of electrons. Electrons will also flow from a negatively charged body to an uncharged body or to one with a lower negative charge.

The movement of electrons constitutes a flow of electric current.

The flow of charge carriers (i.e., electrons) also occurs if the two charged bodies are connected by a piece of metallic material (Figure 1-3). The flow does not occur when nonmetallic material is employed.

Materials that readily allow electrons to pass through them are termed conductors.

Materials that do not permit electron flow are known as insulators.

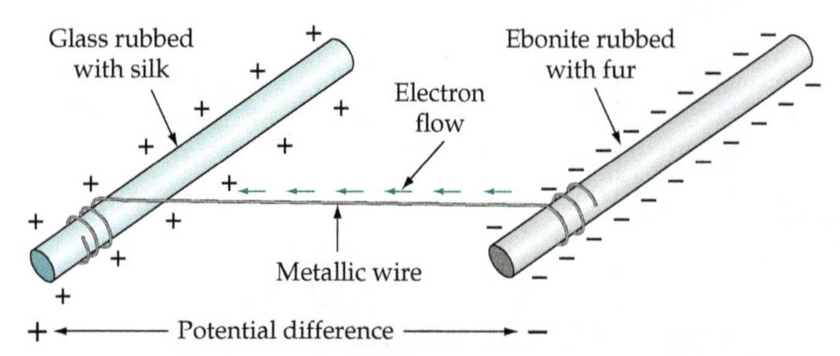

Figure 1-3 A potential difference (PD) exists between a negatively-charged body and a positively-charged body. When connected by a metallic wire, electrons flow from the negatively-charged body to the positively-charged body.

Because some materials are better conductors than others, it can be said that *they offer lower* **resistance** *to the flow of electric current.* Conductors obviously have much lower resistance than insulators. Electric current is measured in *amperes,* or *amps* (symbol A), and resistance is measured in *ohms* (symbol Ω—Greek letter omega).

The ability of two oppositely charged bodies to produce a flow of electricity between them may be thought of as a *potential* for the production of electric current. A positively charged body is said to have a *positive potential,* and a plus sign (+) is used to identify this potential on all diagrams. Similarly, a negatively charged body is described as having a *negative potential,* and a minus sign (−) is used for its identification. Two oppositely charged bodies are said to have a *potential difference* (PD) between them (see Figure 1-3). It is also possible for a potential difference to exist between two similarly charged bodies if one is charged to a higher potential than the other. Potential difference is measured in *volts* (symbol V), and the term *voltage* is commonly applied for potential difference. Another, more descriptive term for potential difference is *electromotive force* (emf).

Electromotive force is an electrical pressure that tends to cause current to flow when a suitable conducting path is provided.

Figure 1-4 compares the basic electrical quantities to the quantities in a water system. Water tank A in Figure 1-4(a) has a larger quantity of liquid than tank B. So, at the level of the connecting pipes, there is a pressure causing water to flow from tank A to tank B. Similarly, in Figure 1-4(b) body A has a larger quantity of electrons than body B. Consequently, there is an electrical pressure (electromotive force) causing a current of electrons to flow along the conductors from A to B. The large-diameter water pipe passes a greater water current than the small-diameter pipe, and the conductor with the large cross-sectional area passes a greater electrical current than the conductor with the small cross-sectional area. The large-diameter conductor offers less *resistance* to current flow than the small-diameter conductor.

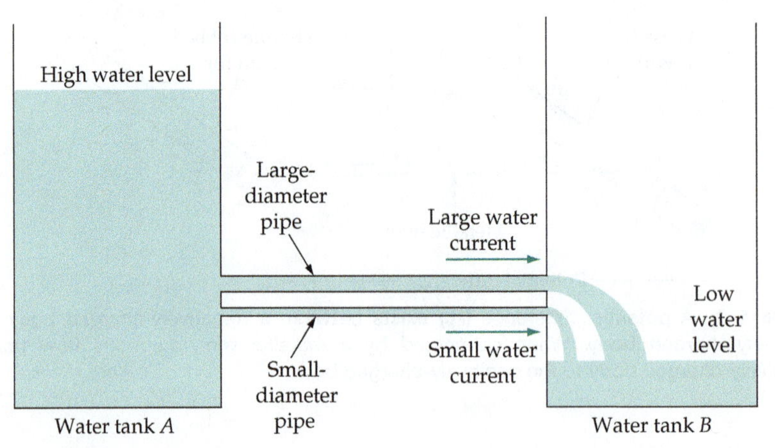

(a) Water pressure causes a current of water to flow between two tanks. The large-diameter pipe passes a larger current of water than the small-diameter pipe.

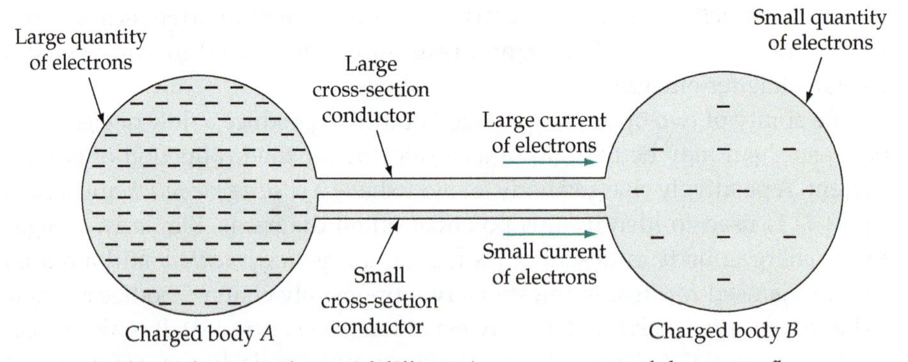

(b) Electrical pressure (potential difference) causes a current of electrons to flow. The large cross-section conductor passes a greater electrical current than the small cross-section conductor. The large cross-section conductor has a *lower resistance* than the small cross-section conductor.

Figure 1-4 Comparison of water system and electrical system. In both systems a pressure difference causes a current to flow along a suitable path.

Summarizing:

Voltage (emf, potential difference) is an electrical pressure that produces current flow (a movement of electrons) through a conductor. Resistance is opposition to current flow.

1-3 BASIC SOURCE OF ELECTRICITY

The simplest modern source of electricity is a *zinc-carbon voltage cell*. As illustrated in Figure 1-5, the cell consists basically of a zinc can containing certain chemicals and having a centrally located carbon rod (see Chapter 10). Chemical action removes electrons from the carbon and causes them to accumulate

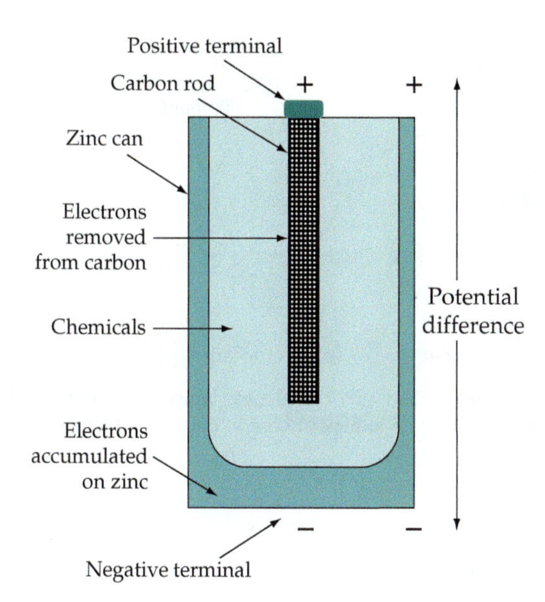

Figure 1-5 Cross-section of a zinc-carbon cell. Chemical action transfers electronic from the carbon to the zinc. The carbon rod is the positive terminal of the cell and the zinc can is the negative terminal.

on the zinc. Because the carbon has lost (negatively charged) electrons, it is positively charged. Also, the electron accumulation on the zinc makes the zinc negatively charged. The carbon rod is the *positive terminal* of the cell and the zinc can is the *negative terminal*. The outside of the zinc can is usually insulated and the bottom is left uncovered for connection purposes. A potential difference, or voltage, exists between the cell positive and negative terminals. When the terminals are connected via a suitable conducting path, electrons flow from the negative terminal to the positive.

1-4 ELECTRIC LAMP

Figure 1-6(a) illustrates the movement of electrons in a conductor when a potential difference is applied across the ends of the conductor. The electrons are repelled from the negative terminal and attracted toward the positive terminal. However, the atoms within the conductor impede the motion of the electrons, so that they cannot simply accelerate from the negative end to the positive end. Instead, each electron bounces about from one atom to another and merely drifts toward the positive end. Each time the electron strikes an atom, it dissipates energy in the form of heat. When a great many electrons are involved in a flow of current, the heat dissipation can be considerable. A simple electric lamp consists of a thin tungsten wire, or *filament*, contained in a glass bulb from which the air has been evacuated [see Figure 1-6(b)]. When a current of electrons is passed through the filament, many electron-atom collisions occur, and the resulting heat makes the filament glow white hot, so that it emits light.

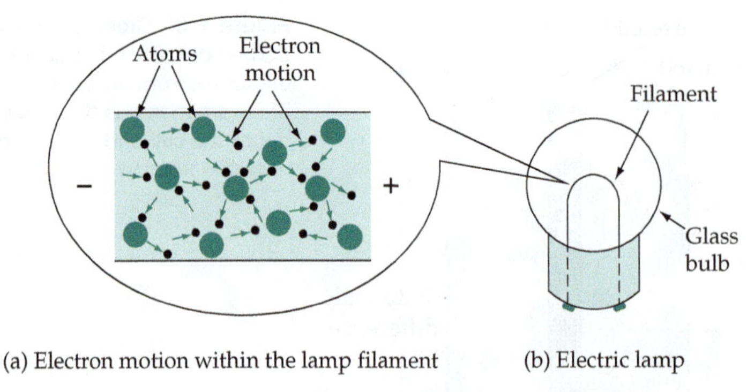

Figure 1-6 When a current of electrons flows in the filament of an electric lamp a great many electron-atom collisions occur. These collisions produce sufficient heat to cause the filament to glow.

1-5 ELECTRIC CIRCUIT

The flashlight circuit shown in Figure 1-7 is made up of a voltage cell, a switch, an electric lamp (or bulb), and connecting conductors. The voltage cell and lamp have already been discussed. In this case, the conductors are the spring at the bottom of the voltage cell, the metal tube that contains the cell, and the metal reflector which is connected to one terminal of the lamp. The switch is simply a thin sliding metal strip that can make or break contact between the tube/container and the reflector.

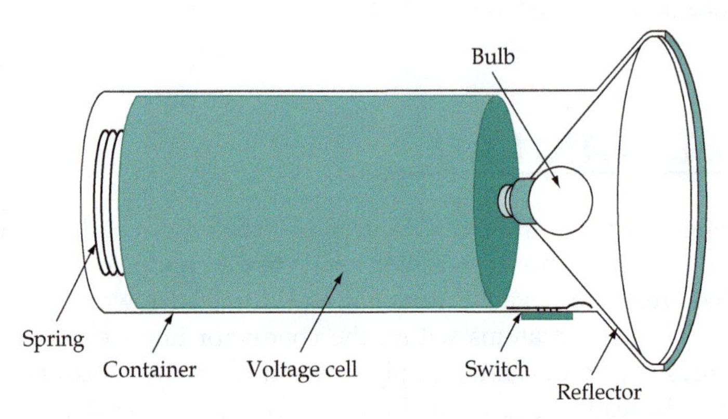

Figure 1-7 A flashlight consists of a battery, a light bulb, a switch, and connecting conductors.

Recall from Section 1-2 that a potential difference (voltage or emf) is an electrical pressure that causes current to flow when a suitable conducting path is provided. When the flashlight switch is open, the conducting path is interrupted so that no current flows. When the switch is closed, the potential difference between the terminals of the voltage cell causes a current of electrons to flow from the negative terminal through the conductors, the lamp filament, and the switch to the positive terminal of the cell.

The circuit in Figure 1-7 can be represented by a *circuit diagram* consisting of lines and graphic symbols. Figure 1-8(a) shows that a conductor is depicted by a straight line, and that the graphic symbol for a voltage cell is a long bar and a short bar side by side. The long bar represents the positive terminal and the short bar is the negative terminal. Lamp and switch symbols are also shown, and a *resistor* circuit symbol is illustrated. As will be explained later, a resistor is a component constructed to have a particular resistance value. Other circuit symbols are given in Appendix 1.

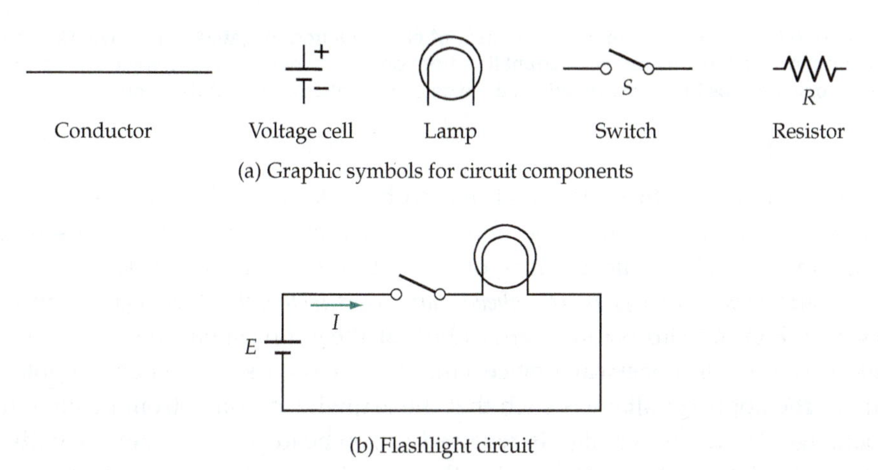

Figure 1-8 An electric circuit can be represented by a circuit diagram that uses graphic symbols for each component. The letter symbols *E*, *I*, and *R* identify voltage, current, and resistance respectively.

Figure 1-8(b) shows the complete diagram for the flashlight circuit in Figure 1-7. As illustrated on the diagram, the letter *E* is used to represent voltage (sometimes *V* is used instead of *E*), the letter *I* represents current, and *R* is used for resistance.

1-6 CURRENT DIRECTION

Early electrical researchers believed that a positive charge identified an increased quantity of electricity, and that a negative charge showed a reduced quantity. Consequently, they assumed that current flowed from positive to negative. This is a convention that remains in use today, even though current flow has long been accepted as a movement of electrons from negative to positive [see Figure 1-9(a)].

Current flow from positive to negative is known as the conventional current direction.

Electron flow from negative to positive is termed the direction of electron flow.

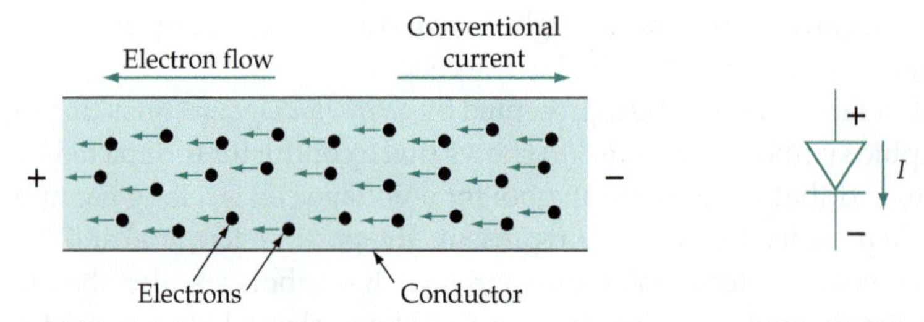

(a) Electron flow and conventional current direction

(b) Graphic symbols for electronic devices use arrowheads that point in the conventional current direction

Figure 1-9 **Electric current is a movement of electrons from negative to positive. Conventional current direction assumes current flow from positive to negative. All graphic symbols for electronic devices have arrowheads that indicate conventional current direction.**

It is important to understand and to be able to think in terms of both conventional current direction and electron flow. Electron flow is used to explain how electronic devices operate, but *the graphic symbol for every electronic device employs an arrowhead that points in the direction of conventional current.* Figure 1-9(b) shows the symbol for the simplest electronic device; a *diode.* A diode is a one-way device; current flow occurs only when the polarity of the applied voltage is such that the arrowhead points from positive to negative. When the voltage is reversed (arrowhead pointing from negative to positive) current cannot flow. So, the arrowhead indicates the direction of *conventional current* flow.

Electronic component and equipment manufacturers normally use conventional current direction when they indicate current on a circuit diagram. Conventional current direction is employed throughout this book. It should be noted that some books use the direction of electron motion as current direction.

1-7 DIRECT CURRENT AND ALTERNATING CURRENT

Current that flows continuously in one direction is termed *direct current* (abbreviated to dc). This is the kind of current supplied by a voltage cell. Because the voltage at the terminals of a voltage cell remains substantially constant, the current in a circuit supplied by the voltage cell also remains constant. A voltage cell can be described as a *direct voltage source.* Figure 1-10(a) shows the graph of direct current (I) plotted versus time (t); obviously, the current remains constant. The graphic symbol for a direct voltage source is also shown, and this is the voltage cell symbol already discussed.

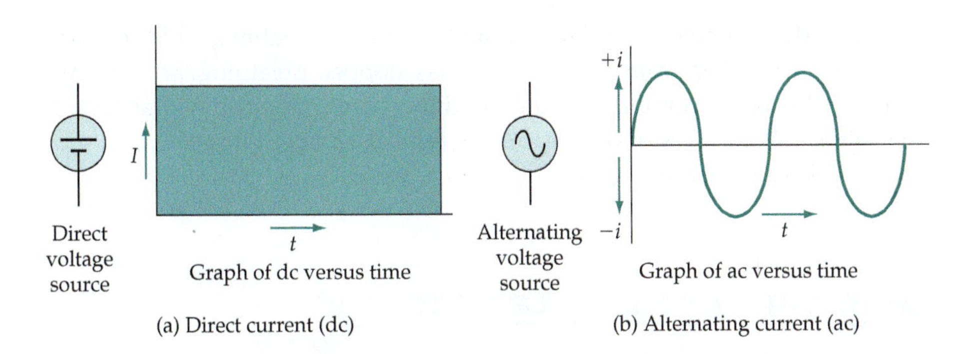

Figure 1-10 A graph of direct current (dc) plotted versus time shows that the current remains constant. A graph of alternating current (ac) versus time shows that the current reapedly reverses, alternately flowing in one direction and then in the other.

Alternating current (abbreviated to ac) is current that flows first in one direction for a time (usually a fraction of a second), and then reverses to flow in the opposite direction for a similar time. This is illustrated in Figure 1-10(b), where a graph of alternating current (i) is plotted versus time (t). Note that the current level increases from zero to a peak (positive or negative) value, decreases to zero again, and then increases once more in the opposite direction. This cycle repeats over and over again. The circuit symbol for an alternating voltage source is also illustrated.

1-8 ELECTRIC SHOCK

Electricity can kill, and so it is very important to treat electrical supplies and equipment with great care. The greatest domestic danger exists in bathrooms and kitchens, because of the presence of water and water taps. Water pipes are connected to other pipes that are buried in the ground, so they provide an electrical connection to ground. Normally, there is a high potential difference between at least one of the electric supply conductors and ground. Consequently, when a person holds a faulty appliance in one hand and a grounded water tap in the other, a path is provided for current to flow from the conductor to ground via the person. In this case the individual can experience a severe electric shock and may find it impossible to open either hand!

A similar danger exists when someone is using a portable appliance, such as an electric drill, if, for example, the supply cable is worn or has temporary taped splices. While holding the appliance in one hand and pulling the cable, the other hand might close around a bare wire. Here again it may be impossible to open either hand as the electricity flows through the body. An even more lethal situation exists when someone taking a bath reaches for an electrical appliance. The person's body is grounded via the water, and if the appliance is faulty *death is virtually inevitable*.

In a case of electric shock, it is most important to immediately get the victim away from the source of the shock. To avoid endangering a rescuer, the electricity supply must be switched off or disconnected before the victim is touched. Medical help should be summoned, and the victim should be made comfortable and kept calm. If the victim has stopped breathing, artificial respiration should be commenced and continued until medical help arrives. If the victim's heart has stopped beating, external heart compression should be applied by a trained individual.

Summary of Important Terms

- ***Potential difference*** (PD), ***voltage,*** and ***electromotive force*** (emf) are all names for the electrical pressure that tends to cause current flow.
- ***Electrical current*** is a movement of electrons from a negative potential to a positive potential.
- ***Electrical resistance*** is opposition to current flow.
- ***Conventional current direction*** is from positive to negative.
- ***Electron flow*** is electron movement from negative to positive.
- ***Direct current*** (dc) is current that constantly flows in a single direction.
- ***Alternating current*** (ac) is current that alternately flows in one direction and then in the opposite direction.

Review Questions

Section 1-1

1-1 Explain what is meant by electrification by friction. Also define positive and negative electricity as related to electrification by friction.

1-2 State the fundamental law that originated from the study of electrification by friction. Briefly explain the law.

1-3 Draw a diagram of the planetary atom. Identify the three basic particles that constitute an atom and define their relative quantities of charge and mass.

1-4 Define a valence electron and discuss the effect on an atom when it loses an electron and the effect when it gains an electron.

Section 1-2

1-5 Describe what occurs when an electric current flows between two charged bodies. Briefly define: conductor, insulator, resistance.

1-6 Explain what constitutes a potential difference and how it is related to electric current.

Section 1-3

1-7 Sketch the basic construction of a voltage cell and explain how it produces positive and negative terminals.

Section 1-4

1-8 Explain how electric current can produce light and briefly describe an electric lamp.

Section 1-5

1-9 Sketch a diagram for an electric circuit consisting of a voltage cell, a lamp, and a switch. Briefly explain the operation of the circuit.

1-10 Refer to the circuit symbols in Appendix 1. Sketch an electric circuit consisting of a generator, a switch, a resistor, and a lamp.

Section 1-6

1-11 Explain the difference between conventional current direction and direction of electron flow. Sketch the circuit diagram of a lamp supplied from a voltage cell, and identify the directions of conventional current flow and electron flow.

1-12 Sketch the circuit symbol for a diode showing the polarity of applied voltage for current flow. Show the direction of conventional current flow and the direction of electron movement.

Section 1-7

1-13 Define direct current, alternating current, emf, potential difference, and voltage.

1-14 Draw graphs to show the difference between direct current and alternating current. Briefly explain.

Section 1-8

1-15 Describe some of the most dangerous situations in which a person can suffer an electric shock. List the emergency actions that should be taken to aid a victim of electric shock.

CHAPTER 2
Measuring Current, Voltage, and Resistance

CONTENTS

Objectives

You will be able to:

1. Use metric prefixes and engineering notation to record electrical quantities and in making calculations.
2. Correctly connect multifunction meters for current, voltage, and resistance measurements.
3. Select appropriate meter switch positions for digital and analog multifunction instruments to operate as ammeters, voltmeters, and ohmmeters.
4. Estimate the accuracy of current, voltage, and resistance measurements.

INTRODUCTION

Very large and very small electrical quantities are most conveniently expressed using metric prefixes and/or engineering notation. The first laboratory investigations performed by students of electricity and electronics involve measurement of current, voltage, and resistance. This normally requires the use of digital or analog multifunction meters. It is very important that the correct meter terminals be used and that the instrument be correctly connected into the circuit. Appropriate selection of meter switch positions is also important. In the case of analog meters, the pointer position on the instrument scale must be read correctly.

2-1 METRIC PREFIXES AND ENGINEERING NOTATION

Scientific Notation

Most electrical and electronic calculations involve very large or very small numbers. These are most conveniently expressed using *scientific notation,* which means that they are written as numbers raised to the power of 10. Consider a resistance of 1000 Ω (1000 ohms) and a current of 0.001 A (0.001 amps), which are both quite common quantities.

$$1000\ \Omega = 1 \times 10 \times 10 \times 10\ \Omega$$

so, it can be written as $1 \times 10^3\ \Omega$

Similarly $$0.001\ \text{A} = \frac{1}{10 \times 10 \times 10}\ \text{A}$$

can be written as $1 \times 10^{-3}\ \text{A}$

Also, $$100\ \text{V} = 1 \times 10 \times 10\ \text{V}$$
$$= 1 \times 10^2\ \text{V}$$

and $$0.0001\ \text{A} = \frac{1}{10 \times 10 \times 10 \times 10}\ \text{A}$$
$$= 1 \times 10^{-4}\ \text{A}$$

Figure 2-1 shows a calculator displaying the number 8×10^{81}. Clearly, the calculator display could not show the number without using scientific notation. Note that in the SI system (or international system) of units listed in Appendix 2, spaces are used instead of commas when writing large numbers. Four-numeral numbers are an exception. One thousand is written as 1000 and ten thousand is 10 000.

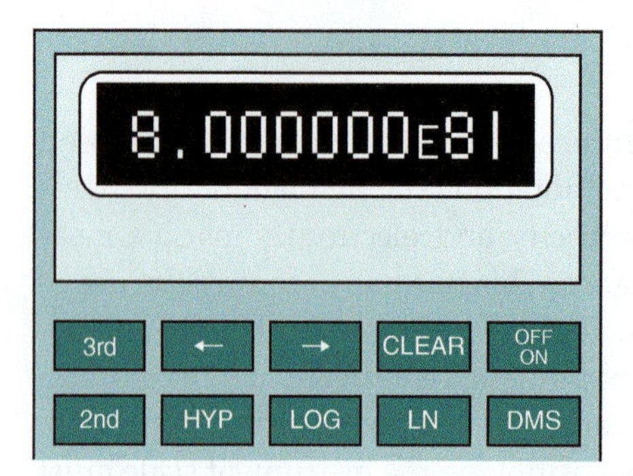

Figure 2-1 Scientific notation on a calculator. There are not sufficient numerals on the calculator display to show this number without scientific notation.

Metric Prefixes

Metric prefixes and the letter symbols for the various multiples and submultiples of 10 are listed in Table 2-1, with those most commonly used with electrical units shown bold. These prefixes are employed to simplify the writing of very large and very small quantities. Thus, 1000 Ω can be expressed as 1 kilohm, or 1 kΩ. Here *kilo* is the prefix that represents 1000 and *k* is the symbol for kilo. Similarly, 1×10^{-3} A can be written as 1 milliamp, or 1 mA.

TABLE 2-1 Scientific notation and metric prefixes

Value	Scientific Notation	Prefix	Symbol
1 000 000 000 000	10^{12}	tera	T
1 000 000 000	10^{9}	**giga**	**G**
1 000 000	10^{6}	**mega**	**M**
1 000	10^{3}	**kilo**	**k**
1 00	10^{2}	hecto	h
10	10	deka	da
0.1	10^{-1}	deci	d
0.01	10^{-2}	centi	c
0.001	10^{-3}	**milli**	**m**
0.000 001	10^{-6}	**micro**	**μ**
0.000 000 001	10^{-9}	**nano**	**n**
0.000 000 000 001	10^{-12}	**pico**	**p**

Engineering Notation

As already discussed, 1 kΩ, is 1×10^{3} Ω, and 1 mA is 1×10^{-3} A. Note also from Table 2-1 that 1×10^{6} Ω is expressed as 1 MΩ, and 1×10^{-6} A can be written as 1 μA. These quantities and most of the metric prefixes in Table 2-1 involve multiples of 10^{3} or 10^{-3}. Quantities that use 10^{3} or 10^{-3} are said to be written in *engineering notation*. A quantity such as $1 \times 10^4\ \Omega$ is more conveniently expressed as $10 \times 10^3\ \Omega$, or 10 kΩ. Also, 47×10^{-4} A is best written as 4.7×10^{-3} A, or 4.7 mA. For electrical calculations, engineering notation is more convenient than ordinary scientific notation.

Example 2-1

Write the following quantities using (a) scientific notation; (b) engineering notation; and (c) metric prefixes: 0.05 A, 10 000 Ω, 0.000 06 V.

Solution

(a)

$$0.05\ \text{A} = 5 \times 10^{-2}\ \text{A}$$

$$10\,000\ \Omega = 1 \times 10^{4}\ \Omega$$

$$0.000\,06\ \text{V} = 6 \times 10^{-5}\ \text{V}$$

(b)

$$0.05\ \text{A} = 50 \times 10^{-3}\ \text{A}$$

$$10\,000\ \Omega = 10 \times 10^{3}\ \Omega$$

$$0.000\,06\ \text{V} = 60 \times 10^{-6}\ \text{V}$$

(c)

$$0.05\ \text{A} = 50\ \text{mA}$$

$$10\,000\ \Omega = 10\ \text{k}\Omega$$

$$0.000\,06\ \text{V} = 60\ \mu\text{V}$$

Significant Figures

Consider the number 816, which is said to be accurate to three significant figures. If this number is rewritten as accurate to two significant figures, it becomes 820. This is because 816 is clearly closer to 820 than it is to 810. If the number 814 were rewritten accurate to two significant figures, it would be 810. In the case of 820 and 810, the zero is not a significant figure.

The number 2.03, written accurate to two significant figures, becomes 2.0. In this case the zero is a significant figure, and it obviously implies that the number is greater than 2.0 but less than 2.1. It also implies that the exact value is closer to 2.0 than to 2.1. Now consider the result of using an electronic calculator to resolve 5.4/2.3:

$$\frac{5.4}{2.3} = 2.347826087$$

Clearly, it does not make sense to have an answer containing ten figures when each of the original numbers has only two significant figures. The only reasonable approach is to use the same number of significant figures in the answer as in the original quantities. Consequently, the problem and answer become,

$$\frac{5.4}{2.3} = 2.3$$

Practice Problems

2-1.1 Express the following quantities using scientific notation: (a) 0.0003, (b) 300×257, (c) $(156 \times 10^4)/2000$.

2-1.2 Solve the following, and give the answers using engineering notation: (a) $(6 \times 10^{-6}) \times (3.3 \times 10^3)$, (b) $0.0047 \times 5 \times 10^5$, (c) $0.05/(2 \times 10^3)$.

2-2 CURRENT MEASUREMENT

Unit of Current

The ampere (or amp) is the unit of electric current.

The definition of the amp is given in Appendix 2. At this stage, the definition is less important than an understanding of typical current levels. Most electronic devices pass currents ranging from 1 mA to 50 mA. Some devices take currents as high as several amps, and many have input currents less than 1 μA. A typical electric lamp might use a current of approximately 1 A.

Ammeter

The instrument used to measure electric current is called an *ammeter* (shortened form of *amp-meter*). Different types of ammeters are available for measuring direct current (dc) and alternating current (ac). However, *multifunction instruments* can usually be employed to measure either type of current.

Ammeter Connection

An ammeter must be connected so that the current to be measured flows through the instrument. Consequently, *ammeters are always connected in series with the component in which current is to be measured.* An ammeter must have a *very low resistance* so that it does not affect the current level when connected into a circuit. At this stage of study it is helpful to simply think of an ammeter as equivalent to a piece of connecting wire. Figure 2-2 shows a circuit diagram of a voltage source and a resistor with an ammeter connected to measure the current flowing through the resistor. Note that the ammeter circuit symbol is a circle with an A in its center.

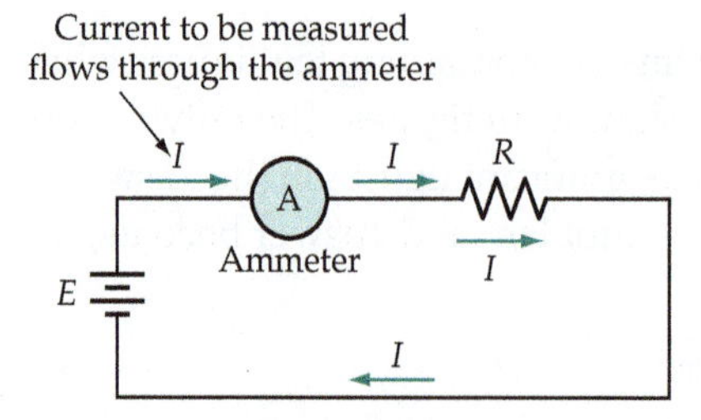

Figure 2-2 Circuit diagram showing an ammeter connected to measure the current passing through a resistor.

Digital Ammeter

Figure 2-3(a) shows a *digital* multifunction meter [also known as a digital *volt-ohm-milliammeter* (VOM)] connected as an ammeter to measure the current through a resistor. Note that the meter *function selector* switch is set for measurement of direct current. Note also that the instrument current terminals are identified as A (for *amps*) and C (for *common*). The measured current is displayed digitally, as illustrated, and the measured quantity (in this case mA) is also usually identified.

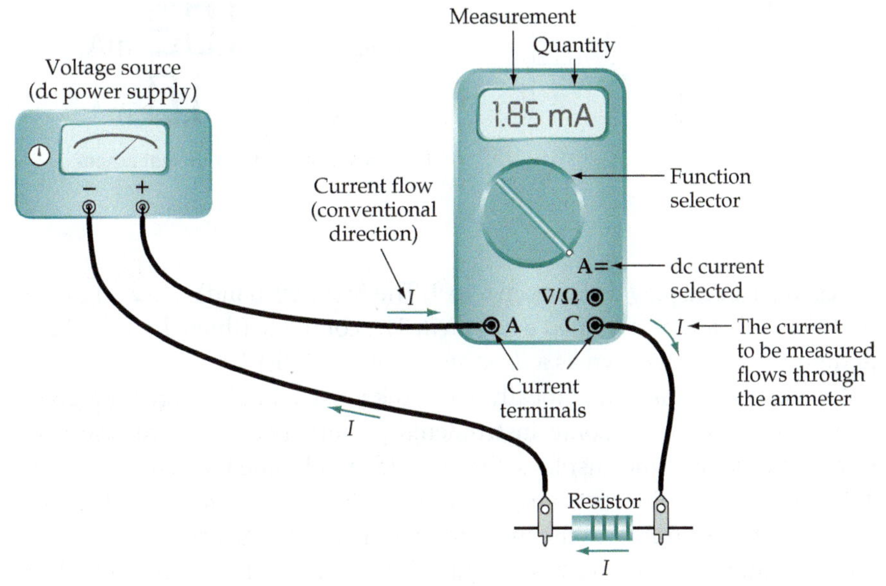

(a) Digital multifunction meter employed as an ammeter to measure the current passing through a resistor

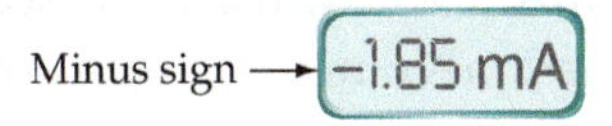

(b) A minus sign is displayed when the current direction is reversed

Figure 2-3 An ammeter must always be connected so that the current to be measured passes through the instrument. A digital volt-ohm-milliammeter (VOM) used as a dc ammeter must have its function selector switch set for dc current. The current terminal (A) and the common terminal (C) must be used to connect the instrument into the circuit.

With the instrument connected as shown, the (conventional direction) current enters the meter at the amps terminal (A) and leaves at the common terminal (C), giving a positive measurement. If the current direction is reversed, the measured current is a negative quantity, and a minus sign is displayed, as shown in Figure 2-3(b). Some instruments display a plus sign for a positive measurement, but most simply show no sign when the current direction is positive.

Each indicated numeral on a digital instrument is produced by means of a device known as a *seven-segment display* [see Figure 2-4(a)]. The segments are selectively energized to display any numeral from 0 to 9. For the 1.024 mA measurement in Figure 2-4(b), the three right-hand units are seven-segment

Figure 2-4 Digital instruments use seven-segment display devices to numerically indicate a measured quantity.

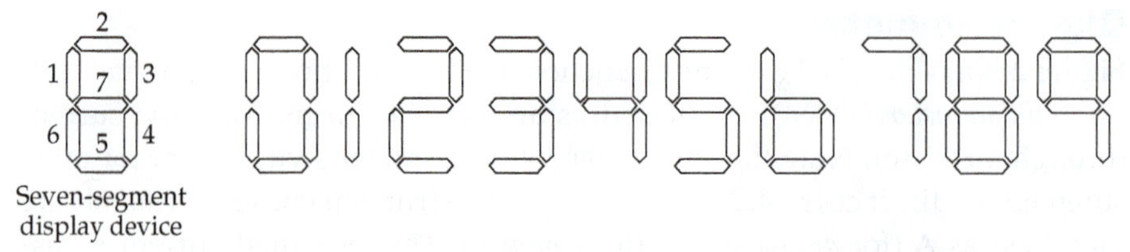

(a) Seven segment devices can be used to indicate any numeral from 0 to 9

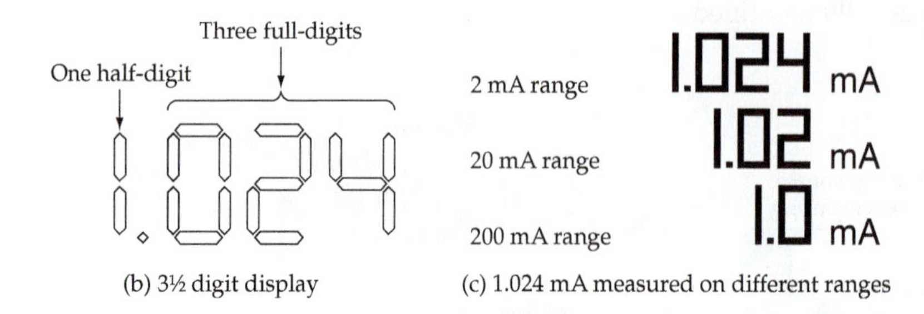

(b) 3½ digit display

(c) 1.024 mA measured on different ranges

devices that can display digits from 0 to 9. The first (left-hand) numeral can be only 1 or blank, and it is termed a *half-digit*. The complete (three digits and one half digit) display is known as a *three-and-a-half-digit* display.

Most digital *VOMs* automatically select the most suitable range for a given measurement; however, some instruments permit manual range selection. Figure 2-4(c) shows the displays that would be obtained when measuring 1.024 mA on three different ranges. On a 2 mA range the meter indicates a maximum of 1.999 mA, and the measurement is 1.024 mA. On a 20 mA range the maximum indicated measurement is 19.99 mA. The display has two decimal places and, consequently, 1.024 mA is displayed as 1.02 mA. On a 200 mA range the maximum measurement is 199.9 mA. With only one decimal place the measurement is 1.0 mA. Clearly, the most accurate measurement is the 1.024 mA displayed on the 2 mA range. *The most accurate measurement is always made on the lowest possible range.*

Because an ammeter has a very low resistance, it normally has no effect on the circuit current level. However, if the instrument is set to function as a voltmeter when connected as an ammeter, it introduces a very high resistance into the circuit and seriously alters the current level. This is a mistake often made when using a multifunction instrument.

Analog Ammeter

The *analog* (or *deflection-type*) VOM illustrated in Figure 2-5 has a pointer which is deflected over a calibrated scale to indicate the measurement. The instrument shown has a *function* switch and a *range* switch, but some analog VOMs use a single selector switch for range and function. The meter terminals are identified as + (*positive*) and – (*negative*) and it is important that the terminals

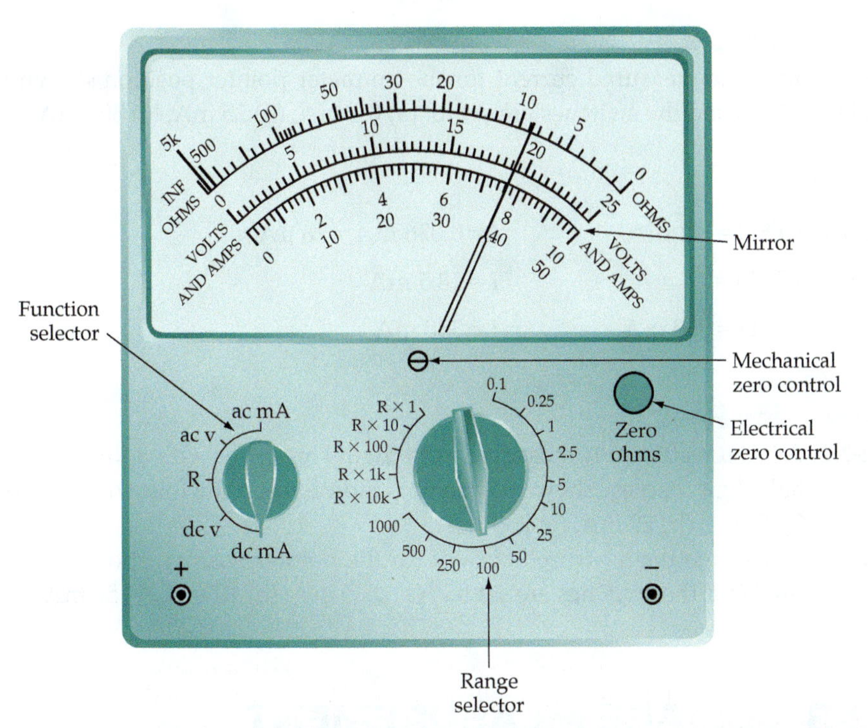

Figure 2-5 An analog VOM used as a dc ammeter must be connected so that the current flows into the positive (+) terminal and out of the negative (–) terminal. The switches should be first set to the highest dc current range. The range can then be switched down to give the greatest on-scale deflection.

be correctly connected. When the (conventional direction) current flows into the positive terminal and out of the negative terminal, the pointer deflects from left to right over the scale. When connected with the wrong polarity (for current flow into the negative terminal) the pointer attempts to deflect to the left of the scale zero mark. An analog instrument must be checked to ensure that it indicates zero before use, and the mechanical zero control must be adjusted if necessary (see Figure 2-5).

The ammeter illustrated in Figure 2-5 has its switches set for 100 dc mA. This means that when the pointer is at the right-hand side of the scale [full-scale deflection (FSD)] the measured current is 100 mA. Using the scale calibrations of 0 to 10, 10 represents 100 mA, and the pointer indication of 7.6 gives the measured current as 76 mA. When the range switch is set to 50 mA, the 0 to 50 scale is used. The 0 to 50 scale is used on the 5 mA, 50 mA, and 500 mA ranges, and the 0 to 10 scale is used with the 0.1 mA, 1 mA, 10 mA, 100 mA, and 1000 mA ranges. When the approximate level of current to be measured is unknown, the instrument should be set to its highest range before being connected into the circuit, in order to give the smallest possible deflection. The range can then be switched down until the greatest *on-scale* deflection is obtained.

Example 2-2

Determine the measured current for the ammeter pointer position shown in Figure 2-5 when the switches are set to: (a) 0.1 mA, (b) 25 mA, (c) 500 mA.

Solution

(a) For FSD = 0.1 mA, $I = 0.076$ mA = 76 μA

(b) For FSD = 25 mA, $I = 18.5$ mA

(c) For FSD = 500 mA, $I = 380$ mA

Practice Problems

2-2.1 A 598 μA current is measured on a digital ammeter with a three-and-a-half-digit display. Show the currents indicated on the following ranges: (a) 2 mA, (b) 20 mA, (c) 200 mA.

2-2.2 Determine the measured current for the pointer position shown in Figure 2-5 if the switches are set for: (a) 0.25 mA, (b) 10 mA, (c) 50 mA.

2-3 VOLTAGE MEASUREMENT

Unit of Voltage

The volt is the unit of emf, potential difference, or voltage.

The unit is defined in Appendix 2. Here again, as in the case of current, a knowledge of typical voltage levels is more important than the definition. Electronic circuits typically use supply voltages ranging from 5 V to 50 V, and within a circuit, potential differences as low as 10 mV might be measured. In North America the normal supply voltages are 115 V for lighting and 220 V for stoves, heating, and industrial machinery.

Voltmeter

The instrument used to measure the potential difference, or electrical pressure, between two points in a circuit is called a *voltmeter.* Multifunction instruments can normally be employed for measuring direct voltage (dc volts) or alternating voltage (ac volts).

Voltmeter Connection

A voltmeter must be connected so that the voltage to be measured appears at the voltmeter terminals. So, *voltmeters are always connected in parallel with the component that is to have its voltage measured.* A voltmeter must have a *very high resistance* so that it does not affect the circuit voltage level. Figure 2-6 shows a circuit diagram in which a voltmeter is connected to measure the voltage across a resistor. Note that the voltmeter circuit symbol is a circle with a V at its center.

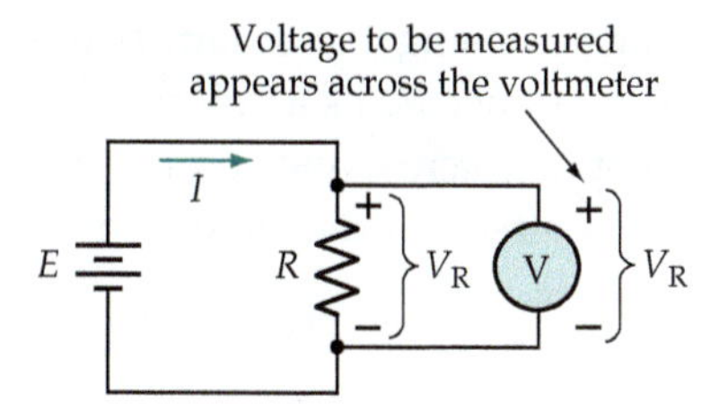

Figure 2-6 Circuit diagram showing a voltmeter connected to measure the voltage across a resistor.

Digital Voltmeter

Figure 2-7(a) shows a digital VOM connected as a voltmeter to measure the voltage across a resistor. Note that the meter function selector switch is set for measurement of dc voltage. Note also that the voltmeter terminal connections are V (for *volts*) and C (for *common*). The measured voltage is displayed digitally and the measured quantity (Vdc) is also identified. If the voltage is applied negative to V and positive to C, a minus sign is displayed to indicate the polarity reversal [Figure 2-7(b)].

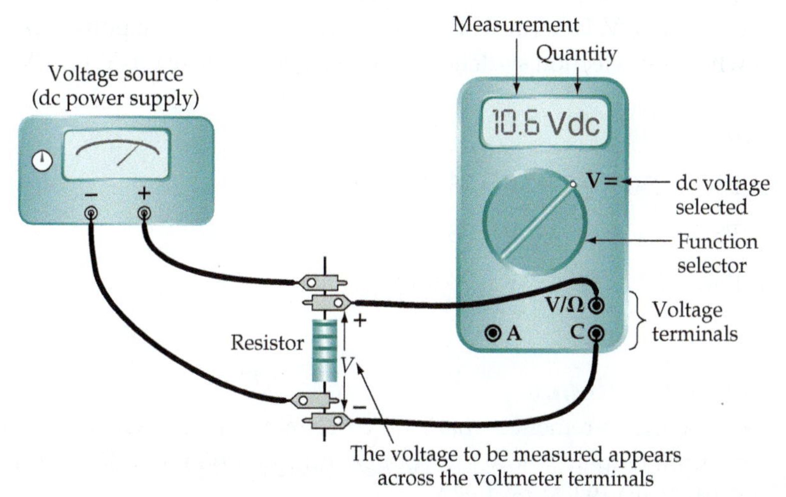

(a) Digital multimeter employed as a voltmeter to measure the voltage across the resistor

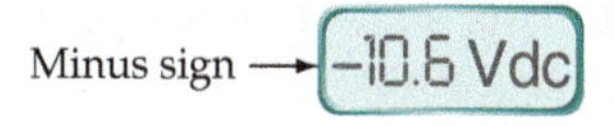

(b) A minus sign is displayed when the voltage polarity is reversed

Figure 2-7 A voltmeter must always be connected so that the voltage to be measured appears at the instrument terminals. A digital VOM used as a dc voltmeter must have its function selector switch set for dc volts. The voltage terminal (V) and common terminal (C) must be used to connect the instrument to the circuit.

As explained for the digital ammeter in Section 2-2, digital voltmeters usually have *autoranging* to automatically select the best range for an applied voltage. However, some digital instruments have manual range selection, and in this case the most accurate measurement is made on the lowest range that can display the voltage. Because a voltmeter has a very high resistance, it normally has no effect on the circuit in which the voltage is being measured. If the instrument is set to function as an ammeter when connected as a voltmeter, a very low resistance (a virtual short-circuit) is applied across the component under investigation. This can have a very serious affect on the circuit being tested.

Analog Voltmeter

When an analog VOM is used as a voltmeter, the instrument selector switchs must be set for a suitable dc volts range (see Figure 2-5), and the voltage to be measured must be connected to the meter terminals with the correct polarity to obtain a positive (left-to-right) deflection. As in the case of the analog ammeter, the measured voltage indicated by the pointer position depends upon the selected range.

Example 2-3

Refer again to the meter shown in Figure 2-5 and assume that the function switch is set to dc V. Determine the measured voltage for the pointer position shown when the selection switches are set for: (a) 250 V, (b) 50 V, (c) 1 V.

Solution

(a) For FSD = 250 V, $V = 185$ V

(b) For FSD = 50 V, $V = 38$ V

(c) For FSD = 1 V, $V = 0.76$ V

Practice Problems

2-3.1 A 9.7 V measurement is made on a digital voltmeter with a two-and-a-half digit display. Show the voltage indicated on the following ranges: (a) 199 V, (b) 19.9 V, (c) 1.99 V.

2-3.2 If the pointer in Figure 2-5 indicates 6 on the 10 V scale, determine the measured voltage for the same pointer position when the switches are set for: (a) 5 V, (b) 25 V, (c) 1000 V.

2-4 RESISTANCE MEASUREMENT

Unit of Resistance

The ohm is the unit of resistance.

As in the case of the other units, the ohm is defined in Appendix 2, and, once again, a knowledge of typical resistance values is more important than the definition. The resistance of a conductor is normally less than 0.01 Ω, and the resistance of an electric lamp filament might be approximately 1 Ω. In electronic circuits, resistances of 2.7 Ω to 22 MΩ are found, with values in the range of 100 Ω to 100 kΩ being the most common.

Ohmmeter

Resistance is measured by means of an *ohmmeter.* The ohmmeter has its own battery or internal power supply for the purpose of passing a current through the resistance to be measured. Any externally applied voltage may damage the instrument, or at least give an incorrect measurement. Thus, *an ohmmeter must never be connected to a circuit in which a supply is switched on.* A multifunction instrument (digital or analog) should never be left in its ohmmeter position, because if the terminals get short-circuited the battery will discharge.

Digital Ohmmeter

The use of a digital VOM as an ohmmeter is illustrated in Figure 2-8(a). The ohmmeter *ohms* (Ω) and *common* (C) terminals are connected directly to the terminals of the resistance to be measured. The polarity of the connection is unimportant. In the case of a component that is part of a circuit, the circuit supply must be switched off and one terminal of the component must be disconnected from the circuit [see Figure 2-8(b)]. The measurement is displayed digitally and the quantity is identified as Ω, kΩ, or MΩ.

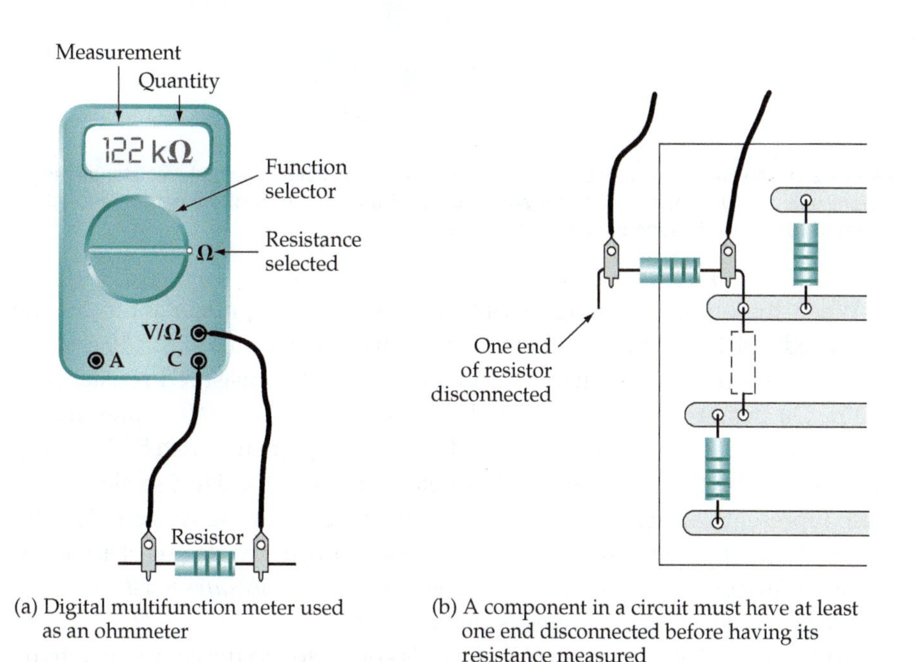

(a) Digital multifunction meter used as an ohmmeter

(b) A component in a circuit must have at least one end disconnected before having its resistance measured

Figure 2-8 Digital VOM used as an ohmmeter. The ohms (Ω) and common (C) terminals are used, and the function selector must be set to Ω.

Analog Ohmmeter

The resistance scale on an analog VOM (the top scale in Figure 2-9) is quite different from the current and voltage scales. The *0 ohms* position is on the right-hand side of the scale instead of the left. The left-hand end of the scale is marked *INF*, which means *infinity*. Note that the length of the scale representing a 10 Ω change from 0 Ω to 10 Ω is greater than the scale length from 20 Ω to 30 Ω, which also represents a 10 Ω change in resistance. Moving from right to left, the scale becomes more and more cramped so it is said to be *nonlinear*. The ammeter and voltmeter scales already discussed are termed *linear scales* because equal lengths on each scale represent equal current or voltage changes.

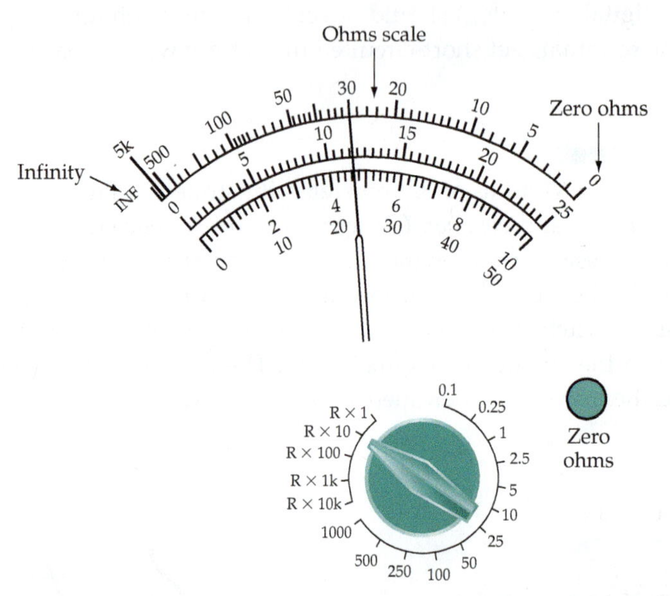

Figure 2-9 An analog VOM used as an ohmmeter must be set to a suitable resistance range. Before use, the terminals should be short-circuited and the zero ohms control should be adjusted for zero indication on the resistance scale.

When the ohmmeter range switch is at R × 1, the pointer indicates the measured resistance directly in ohms. With the range switch at R × 10 and the pointer indicating 30, as illustrated in Figure 2-9, the measured resistance is 300 Ω. On this range the small length of scale from 500 to INF represents all resistance values above 5 kΩ. Obviously, resistances greater than 5 kΩ cannot be measured on this range of the ohmmeter. Similarly, the 0 to 5 section of the scale represents all resistance values below 50 Ω when the range is R × 10. Again, these resistance values cannot be measured with any degree of accuracy. From this discussion it is seen that, *the analog ohmmeter measures resistances most accurately when its pointer position is around center scale.*

An analog VOM used as an ohmmeter is connected to the component to be measured in exactly the same way as a digital instrument, and the meter switches are set for a suitable resistance range. As always, the mechanical zero control should be adjusted to set the pointer exactly at the left-hand end of the scale before use. Before measuring a resistance, the ohmmeter terminals should be connected together (short-circuited), and the *Zero ohms* control (see Figure 2-9) should be adjusted until the pointer indicates exactly 0 ohms on the right-hand side of the scale.

Example 2-4

Determine the reading of the analog multimeter for the pointer position illustrated in Figure 2-9 when the range is selected as: (a) R × 1, (b) R × 10 k.

Solution

(a) At R × 1, $R = 30 \times 1\ \Omega = 30\ \Omega$

(b) At R × 10 k, $R = 30 \times 10\ \text{k}\Omega = 300\ \text{k}\Omega$

Practice Problem

2-4.1 Determine the measurements in Figure 2-9 when the pointer indicates 50 on the ohms scale and the meter switches are set for (a) 25 mA, (b) R × 1 kΩ.

2-5 INSTRUMENT ACCURACY AND PRECISION

Digital Instrument Accuracy

The accuracy of a digital VOM is typically stated by the manufacturer as ±0.5% Rdg ± 1d. This means ±(0.5% of the reading) ± 1 digit, where the 1 digit refers to a 1 in the right-hand numeral position of the display (see Figure 2-10). Example 2-5 demonstrates the measurement error calculation.

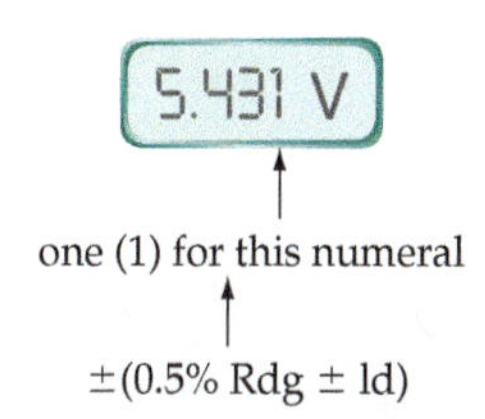

Figure 2-10 The accuracy of a digital display is typically expressed as a percentage of the reading plus 1 digit (±1% of Rdg + 1d).

Example 2-5

If the instrument accuracy for the measured quantity in Figure 2-10 is specified as ±0.1% Rdg ± 1d, calculate the upper and lower limits of the measured voltage.

Solution

$$\begin{aligned} \text{Error} &= \pm(0.1\% \text{ of } 5.431\ \text{V}) \pm 0.001\ \text{V} \\ &\approx \pm 0.005\ \text{V} \pm 0.001\ \text{V} \\ &\approx \pm 0.006\ \text{V} \end{aligned}$$

The measured voltage is:

$$\begin{aligned} E &= 5.431\ \text{V} \pm 0.006\ \text{V} \\ &= 5.425\ \text{V to } 5.437\ \text{V} \end{aligned}$$

Analog Instrument Accuracy

The accuracy of an analog instrument is specified by the manufacturer as a percentage of full-scale deflection (FSD). For example, a voltmeter with an FSD of 100 V might have its accuracy stated as ±2%. In this case, the maximum possible error *at all points on the scale* is ±2% of 100 V, or ±2 V. Thus, when the pointer is indicating exactly 100 V, as in Figure 2-11(a), the measured voltage is correctly stated as 100 V ± 2 V, or 98 V to 102 V. When the pointer of this voltmeter indicates 50 V on the 100 V range, as in Figure 2-11(b), the actual measured voltage must be taken as 50 V ± (2% of FSD), or 50 V ± 2 V. The ±2 V error is ±4% of 50 V. So, the measurement percentage error increases as the pointer moves further away from FSD. When an analog VOM is used as an ohmmeter the error at mid-scale can be shown to be four times the FSD error of the meter.

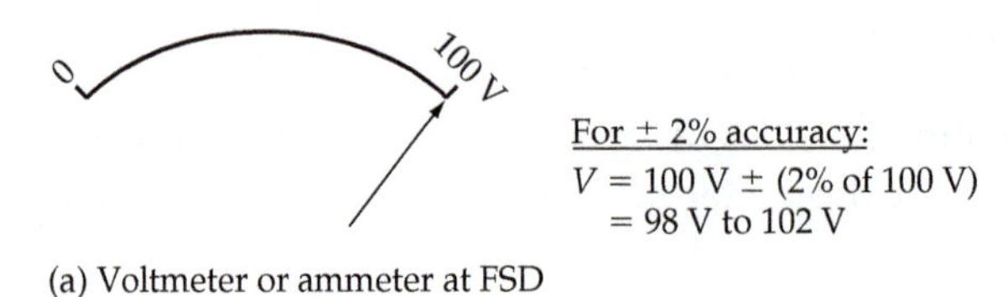

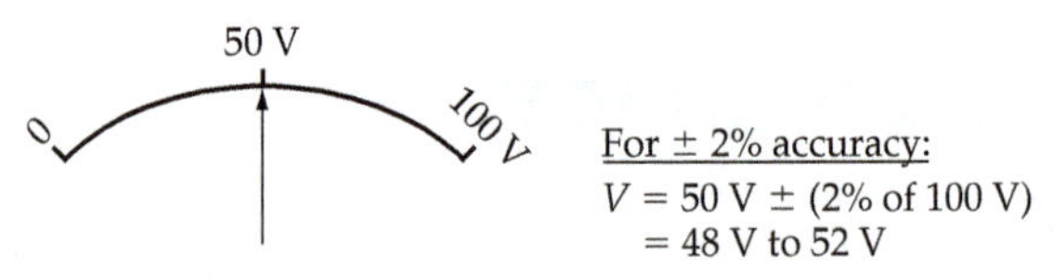

Figure 2-11 Errors in an analog instrument are related to full-scale deflection (FSD). A meter that has a possible ±2% error at FSD can have the same error (±2% of FSD) at all points on the scale.

Example 2-6

If the instrument illustrated in Figure 2-5 has an accuracy of ±1%, determine the upper and lower limits of the measurement for the pointer position shown when measuring on the 500 V range.

Solution

$$\text{Meter reading} = 380\text{ V}$$
$$\text{Full-scale deflection} = 500\text{ V}$$
$$\text{Error} = \pm 1\%\text{ of }500\text{ V} = \pm 5\text{ V}$$

The measured voltage is,

$$E = 380\text{ V} \pm 5\text{ V} = 375\text{ V to }385\text{ V}$$

Precision

The *precision* with which a measurement is made by an instrument is not the same as the accuracy of measurement, although accuracy and precision are related. The precision can be defined as the smallest change that can be detected in a measured quantity.

Consider the digital voltmeter indication shown in Figure 2-12(a). For the 7.60 V measured quantity, a 1 on the last (right side) numeral represents 0.01 V, or 10 mV. If the measured quantity increases or decreases by 10 mV, the reading becomes 7.61 V or 7.59 V, respectively. A voltage change smaller than 10 mV is unlikely to produce a change in the measured quantity. Therefore, 10 mV is the smallest voltage change that can be detected in this case, and it can be stated that the voltage is measured with a precision of 10 mV.

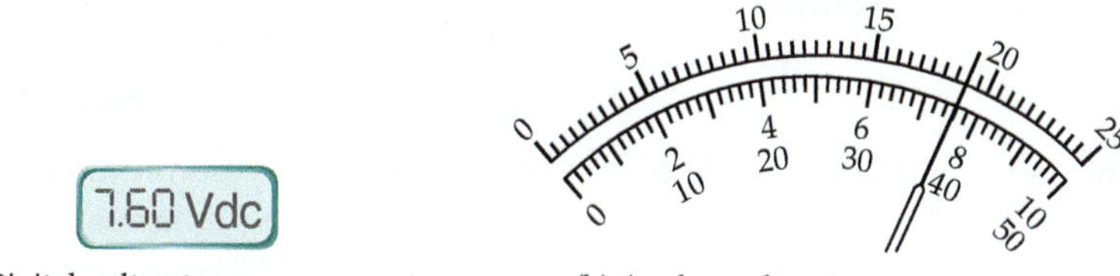

(a) Digital voltmeter measurement (b) Analog voltmeter measurement

Figure 2-12 The precision of a measurement is determined by the smallest detectable change in the measured quantity. The precision of the above digital reading is ±0.01 V. For the analog measurement the precision is ±(half of the smallest scale division).

If the digital instrument in Figure 2-12(a) has an accuracy of ±(0.5% + 1 d) the measured voltage is 7.60 V ±(0.5% + 10 mV), or 7.60 V ± 48 mV; which means that the actual voltage is somewhere between 7.648 V and 7.552 V. So, although the quantity is measured with a precision of 10 mV, the measurement accuracy is ±48 mV.

For the analog voltmeter measurement in Figure 2-12(b), the pointer position on the 10 V range can be read to within one-half of the smallest scale division. Since the smallest scale division represents 0.2 V (on the 10 V range), one-half of the scale division is 0.1 V. So 0.1 V is the smallest detectable change in this case, and the measurement precision of this instrument on the 10 V range is 0.1 V. If the instrument has an accuracy of ±2% of full scale (or ±2% of 10 V which is ±0.2 V), the measured quantity is 7.60 V ± 0.2 V; that is, 7.4 V to 7.8 V. In this case, the measurement is made to a precision of 0.1 V, but the measurement accuracy is ±0.2 V.

Practice Problems

2-5.1 If the accuracy of the digital measurement in Figure 2-4(b) is specified as ±0.3% Rdg ± 1d, determine the upper and lower limits of the indicated quantity.

2-5.2 For the analog measurement in Figure 2-9 determine the upper and lower limits of a 300 Ω measured resistance if the VOM is accurate to ±3% at full scale.

Summary of Instrument Characteristics

Instrument	Quantity measured	Instrument resistance	Instrument connection
Voltmeter	Electrical pressure in volts	Very high	In parallel with the points at which the voltage is to be measured
Ammeter	Electric current in amps	Very low	In series with the circuit or component in which the current is to be measured
Ohmmeter	Resistance in ohms	—	In parallel with the component that is to have its resistance measured

Review Questions

Section 2-1

2-1 List the names of the various metric prefixes and the corresponding symbols. Also, list the value represented by each prefix, both in the long form and in scientific notation.

2-2 When an electronic calculator is used to determine 7.2/3.1, the answer displayed is 2.322580645. How would you give the answer if 7.2 and 3.1 were both known to be accurate to two significant figures? Explain.

Section 2-2

2-3 Draw a circuit diagram for an ammeter connected to measure the current supplied to a resistor from a voltage source. Briefly explain.

2-4 Sketch a connection diagram showing a voltage source supplying current to a resistor and an ammeter connected to measure the current. Discuss the control positions and terminal polarity for digital and analog VOMs used as ammeters.

2-5 Describe seven-segment display devices as used in a digital instrument. Explain a three-and-a-half-digit display.

2-6 Explain what happens when a VOM with its controls set for voltage measurement is connected as an ammeter.

2-7 A digital ammeter with a three-and-a-half-digit display has 200 μA, 2 mA, 20 mA, and 200 mA ranges. Select the range that will give the most accurate measurement of (a) 295 μA, (b) 19.93 mA. Explain the effects of using a range that is too high and a range that is too low.

2-8 Discuss the purpose and use of the mechanical zero control on an analog instrument.

2-9 An analog ammeter has 1 mA, 2.5 mA, 50 mA, and 100 mA ranges. Select the range that will give the most accurate measurement of (a) 55 mA, (b) 0.9 mA. Explain the effects of using a range that is too high and a range that is too low.

Section 2-3

2-10 Draw a circuit diagram showing a voltmeter connected to measure the voltage applied to a resistor from a voltage source. Briefly explain.

2-11 Sketch a connection diagram showing a voltage source supplying current to a resistor and a voltmeter connected to measure the applied voltage. Discuss the control positions and terminal polarity for digital and analog VOMs used as voltmeters.

2-12 Explain what happens when a VOM with its controls set for current measurement is connected as a voltmeter.

Section 2-4

2-13 State the precautions necessary when using an ohmmeter to measure the resistance of a component connected in a circuit.

2-14 Discuss the use of the mechanical zero and zero ohms controls on an analog ohmmeter.

Section 2-5

2-15 Explain the accuracy specifications for (a) a digital instrument specified as accurate to $\pm(0.3\text{ Rdg} + 1\text{d})$, (b) an analog instrument specified as accurate to $\pm 1\%$.

2-16 Explain linear and nonlinear scales on analog instruments. Identify which part of a scale gives the most accurate measurements on (a) a voltmeter, (b) an ohmmeter, (c) an ammeter.

2-17 Sketch a typical analog ohmmeter scale suitable for measuring resistance values around 100 Ω. Show the zero, infinity, and 100 Ω positions.

Problems

Section 2-1

2-1 Express each of the following quantities in scientific notation: (a) 0.015, (b) 16 000, (c) 6260, (d) 0.0007, (e) 989 000.

2-2 Perform the following calculations, and give the answers in scientific notation: (a) $0.15 \times 1600/0.004$, (b) $87\,500/6.25$, (c) $0.5^2 \times 400$, (d) $2^{12}/\sqrt{256}$, (e) $0.003^3/1500$.

2-3 Express each of the following quantities in engineering notation: (a) 7800 Ω, (b) 0.000 06 A, (c) 0.019 A, (d) 91 500 Ω, (e) 0.05 A.

2-4 Perform the following calculations and give the answers in engineering notation: (a) 1290×620, (b) $0.000\,06/36$, (c) $27.9/0.009$, (d) $6.25/(125 \times 10^6)$.

Section 2-2

2-5 Show how 4.59 mA would be indicated on a three-digit display on (a) a 5 mA range, (b) a 50 mA range, (c) a 500 mA range. Identify the range that gives the greatest accuracy.

2-6 If the pointer in Figure 2-5 moves to 20 on the 0 to 50 scale, determine the measured quantities when the VOM controls are set for: (a) 1000 mA, (b) 25 mA, (c) 5 mA.

Section 2-3

2-7 A digital voltmeter has a three-and-a-half-digit display and manually selectable ranges of 2000 V, 200 V, 20 V, and 2 V. If the instrument is used to measure 13.65 V, show the voltage indicated on each range, and identify the range that gives the greatest accuracy.

2-8 For a new pointer position of 12 on the 0 to 25 scale in Figure 2-5, determine the measured quantities when the VOM controls are at: (a) 0.1 V, (b) 5 V, (c) 2.5 V.

Section 2-4

2-9 A digital instrument with a two-and-a-half-digit display has manually selectable resistance ranges of 200 Ω, 2 kΩ, and 20 kΩ. Show how a resistance of 139 Ω is displayed on each range, and identify the most suitable range.

2-10 For the ohmmeter scale shown in Figure 2-9 select the ranges that should be used when measuring resistances with the following approximate values: (a) 560 Ω, (b) 12 kΩ, (c) 1 MΩ, (d) 10 Ω.

2-11 Determine the measured resistance for the ohmmeter scale and pointer position in Figure 2-9 if the instrument range is set to $R \times 100$.

Section 2-5

2-12 A digital instrument has a specified accuracy of ±0.2% Rdg ± 1d. Calculate the maximum and minimum values of the voltage when the instrument indicates: (a) 9.418 V, (b) 1.290 V, (c) 3.067 V.

2-13 A digital instrument with an accuracy of ±0.3% Rdg ± 1d indicates the following current levels: (a) 7.240 mA, (b) 1.893 mA, (c) 4.235 mA. Calculate the maximum and minimum quantities in each case.

2-14 The accuracy of the resistance ranges on a digital multimeter is specified as ±0.6% Rdg ± 1d. Determine the upper and lower limits of the measured resistance when the instrument indication is (a) 4.663 kΩ, (b) 560.1 Ω, (c) 27.77 Ω.

2-15 Assuming a 10 V range and an instrument accuracy of ±1%, calculate the maximum and minimum levels of the voltage indicated in Figure 2-12(b).

2-16 Calculate the maximum and minimum values of the quantity indicated on the 1000 mA scale in Figure 2-5 if the VOM accuracy is specified as ±2%.

2-17 Assuming an R × 1 kΩ range and a VOM accuracy of ±3%, calculate the maximum and minimum resistances when the pointer indicates 25 on the resistance scale in Figure 2-9.

Practice Problem Answers

2-1.1 (a) 3×10^{-4}, (b) 7.71×10^{4}, (c) 7.8×10^{2}
2-1.2 (a) 19.8×10^{-3}, (b) 2.35×10^{3}, (c) 25×10^{-6}
2-2.1 .598 mA, 0.59 mA, 00.5 mA
2-2.2 185 μA, 6.8 mA, 38 mA
2-3.1 9 V, 9.7 V, (over-range)
2-3.2 3 V, 15 V, 600 V
2-4.1 8 mA, 50 kΩ
2-5.1 1.02 V to 1.028 V
2-5.2 264 Ω to 336 Ω

CHAPTER 3
Ohm's Law and Electrical Calculations

CONTENTS

Objectives

You will:

1. Gain an understanding of resistance, and be able to state Ohm's law and write the Ohm's law equation.
2. Given any two of resistance, current, or potential difference, calculate the third quantity.
3. Gain an understanding of conductance and be able to convert between resistance and conductance.
4. Understand electrical power dissipation, and be able to determine power dissipation in a circuit or component.
5. Understand and be able to calculate efficiency where it applies to electric circuits.
6. Understand electrical energy and be able to calculate electrical energy consumption.

INTRODUCTION

Current flow in an electric circuit is the result of an applied voltage. The opposition to current flow that exists in every circuit is termed *resistance,* and its converse is *conductance.* The relationship between voltage, current, and resistance is defined by the simple but extremely important *Ohm's law.* Electrical *power* is dissipated when current flows in a circuit or component. In many cases, this power is wasted as unwanted heat. Electrical *energy* is used when power is dissipated over a period of time.

3-1 RESISTANCE

The resistance of a conductor is a measure of its opposition to current flow.

When a potential difference is applied to the ends of a conductor, the quantity of current that flows depends on the applied voltage, the conductor material (the type of metal), the length of the conductor, and the cross-sectional area of the conductor. The type of metal is important because different metals offer different resistances to current flow. (This is investigated further in Chapter 4). The cross-sectional area is involved because, just as a large-diameter water pipe can carry a larger flow of water than a small-diameter pipe, a conductor with a large cross-sectional area passes more electric current than a small cross-section conductor of the same material [see Figures 3-1(a) and (b)].

How the conductor length affects its resistance is best understood by reconsidering the current flow process discussed in Section 1-4. Figure 1-6 shows that

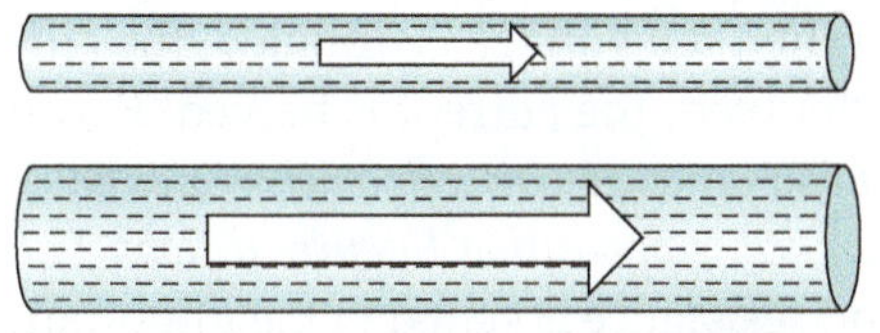

(a) A water pipe with a large cross-sectional area passes a larger current of water than one with a small cross-sectional area

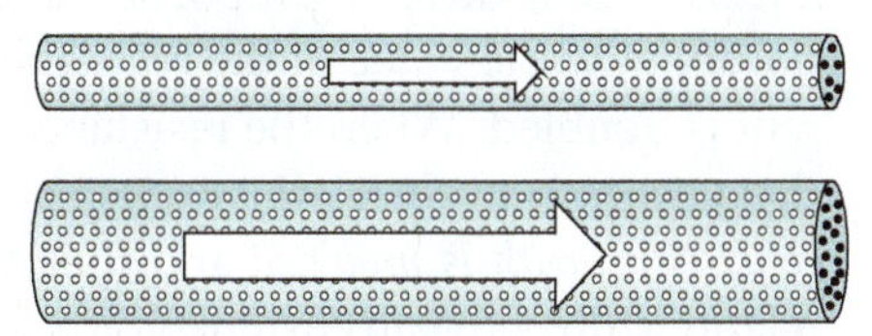

(b) A conductor with a large cross-sectional area passes a larger electrical current than one with a small cross-sectional area. The large cross-section conductor has a *lower resistance* than the smaller one.

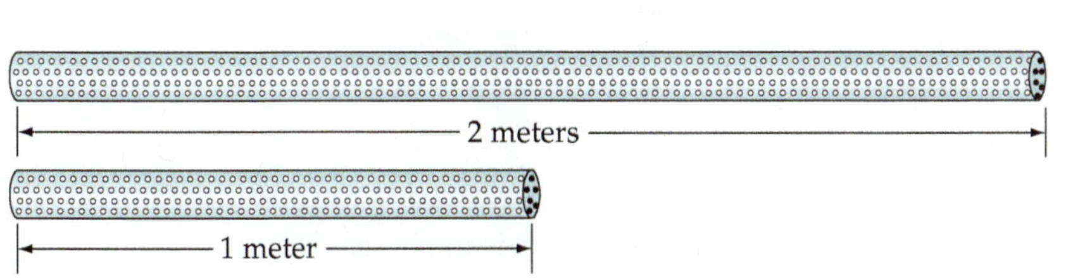

(c) Electrons flowing along a 2 m conductor have to pass twice as many atoms as those travelling through a similar 1 m conductor. The resistance of a 2 m conductor is twice that of a similar 1 m conductor.

Figure 3-1 A conductor with a large cross-sectional area offers a lower resistance to current flow than a conductor of similar length and material with a smaller cross-sectional area. A short conductor has a lower resistance than a similar longer conductor.

the electrons are continually colliding with the conductor atoms, and so the resistance of a conductor has a lot to do with its atoms. Figure 3-1(c) illustrates the fact that an electron moving through a 2 m conductor has to get past twice as many atoms than in the case of 1 m conductor (assuming the same cross-sectional area and material for each conductor). There are likely to be twice as many electron-atom collisions in the longer conductor. Consequently, it can be stated that a 2 m conductor has twice the resistance of a similar 1 m conductor.

As already mentioned in Section 2-4, the *ohm* (symbol Ω) is the unit of resistance, and *resistors* are components constructed to have a particular resistance. The letter symbol used to identify resistances is R; several resistors in a circuit might be numbered R_1, R_2, R_3, and so on.

3-2 OHM'S LAW

Consider Figure 3-2(a), in which an electromotive force (emf) E is shown producing a current I through a resistance R. If $E = 1$ V and $R = 1\ \Omega$, as illustrated, the current that flows is 1 A. This comes directly from the definition of the ohm in Appendix 2:

One ohm is that resistance which permits a current of 1 A to flow when a potential difference of 1 V is applied to the resistance.

Now consider the effect of doubling the applied voltage [Figure 3-2(b)]. If an electrical pressure of 1 V causes a 1 A current to flow through 1 Ω resistance, then doubling the pressure doubles the current to 2 A. Similarly, when E remains 1 V and the resistance is doubled, as in Figure 3-2(c), the current flow is halved (i.e., $I = 0.5$ A). This discussion shows that the current flow through a resistance is directly proportional to the applied voltage and inversely proportional to the value of the resistance. When the voltage is doubled, the current is doubled. When the resistance is doubled, the current is halved. Recall that *voltage is an electrical pressure that tends to cause current flow when a suitable conducting path is provided* and that *resistance is opposition to current flow.* The relationship between current, voltage, and resistance is stated in the important fundamental law of electricity known as *Ohm's law*:

Ohm's law:

$$\text{Current} = \frac{\text{voltage}}{\text{resistance}}$$

or,

$$I = \frac{E}{R} \tag{3-1}$$

where I is in amperes, E is in volts, and R is in ohms.

Equation 3-1 can be rewritten as:

$$E = IR, \text{ and } R = \frac{E}{I}$$

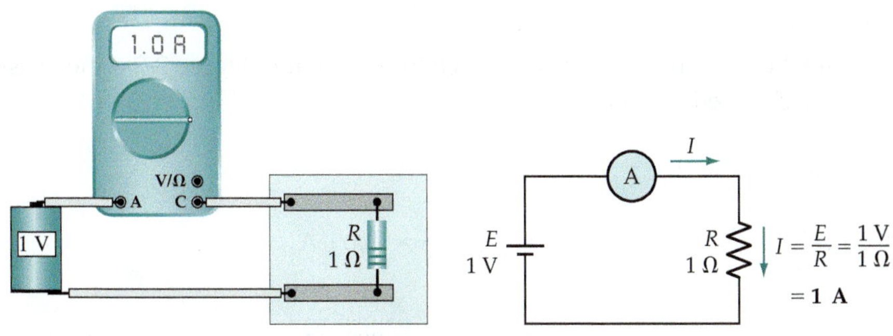

(a) Connection diagram and circuit diagram showing that a 1 V source produces a 1 A current through a 1 Ω resistance

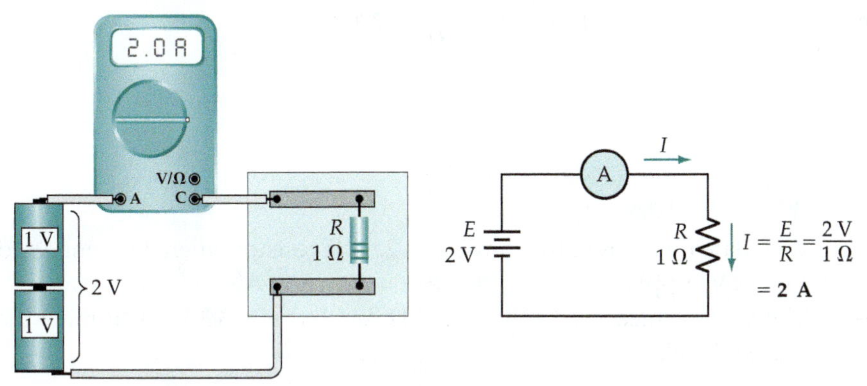

(b) A 2 V source produces a 2 A current in a 1 Ω resistance

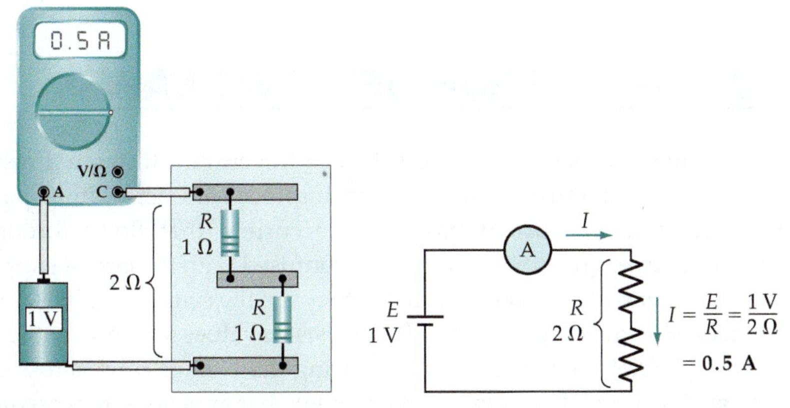

(c) A 1 V source produces 0.5 A in a 2 Ω resistance

Figure 3-2 Illustration of Ohm's law. Current (I) is directly proportional to voltage (E) and inversely proportional to resistance (R). Current = voltage/resistance ($I = E/R$). When a resistor voltage is doubled, the current is doubled. When the resistance is doubled, the current level is halved.

Example 3-1

Calculate the new current levels in the circuits shown in Figure 3-2 if the resistances are changed to 2.5 kΩ.

Solution

(a) Eq. 3-1, $$I = \frac{E}{R} = \frac{1\ \text{V}}{2.5\ \text{k}\Omega} = 400\ \mu\text{A}$$

(b) $$I = \frac{E}{R} = \frac{2\ \text{V}}{2.5\ \text{k}\Omega} = 800\ \mu\text{A}$$

(c) $$I = \frac{E}{R} = \frac{1\ \text{V}}{5\ \text{k}\Omega} = 200\ \mu\text{A}$$

Practice Problems

3-2.1 Calculate the current that flows in a 2.5 kΩ resistor when it is connected to a 12 V supply.

3-2.2 The voltage measured across a 22 Ω resistance is 3.3 V. Determine the current level.

3-3 APPLICATION OF OHM'S LAW

When Ohm's law is employed to calculate the current through a resistance, the voltage used must be the voltage that appears across that particular resistance, and the current must be the current that flows through that resistance. These quantities must not be confused with the circuit supply voltage and current, although in Figure 3-2 they are the same quantities. Consider the circuit shown in Figure 3-3(a). If the resistor values are known, the current levels can be determined by first measuring the voltage drop across each resistor, as illustrated. For example, the voltage that appears across resistor R_3 is V_3, and the current through R_3 is I_3. So, in Figure 3-3(b),

$$I_3 = \frac{V_3}{R_3}$$

Also, $$I_1 = \frac{V_1}{R_1} \quad \text{and} \quad I_2 = \frac{V_2}{R_2}$$

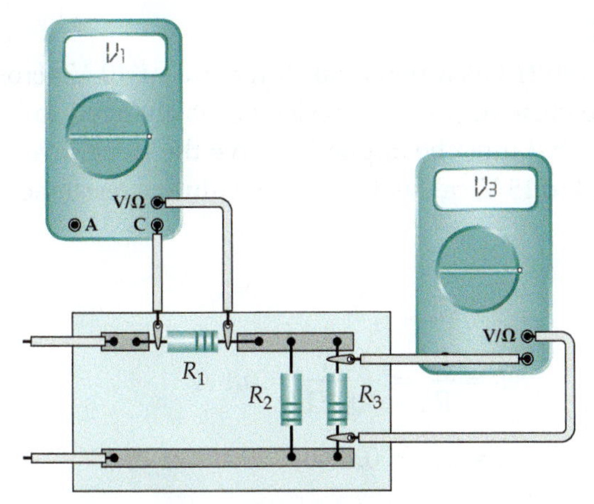

(a) Resistor current levels can be determined by measuring the voltage drops and applying Ohm's law

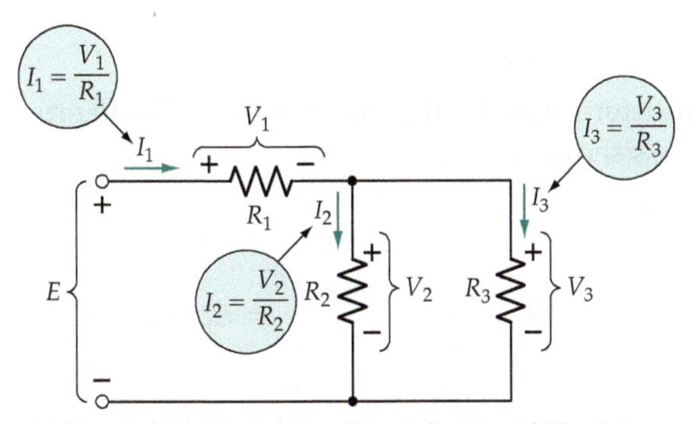

(b) Circuit diagram showing the application of Ohm's Law to determine individual resistor currents

Figure 3-3 When Ohm's law is used to calculate the current through a resistance, the voltage across the particular resistance and the resistance value must be used in the calculation.

Example 3-2

The measured voltages in the circuit shown in Figure 3-3 are: $V_1 = 9$ V, $V_2 = 10$ V, $V_3 = 10$ V. The resistance values are: $R_1 = 600\ \Omega$, $R_2 = 4\ \text{k}\Omega$, $R_3 = 800\ \Omega$. Calculate the individual resistor currents.

Solution

Eq. 3-1,

$$I_1 = \frac{V_1}{R_1} = \frac{9\ \text{V}}{600\ \Omega} = 15\ \text{mA}$$

$$I_2 = \frac{V_2}{R_2} = \frac{10\ \text{V}}{4\ \text{k}\Omega} = 2.5\ \text{mA}$$

$$I_3 = \frac{V_3}{R_3} = \frac{10\ \text{V}}{800\ \Omega} = 12.5\ \text{mA}$$

Example 3-3

A resistance of 100 Ω has a potential difference of 10 V across its terminals. (a) Calculate the current. (b) If the resistance is changed to 250 Ω, determine the new voltage that must be applied to give the same level of current. (c) If the emf is altered to 15 V, calculate the new value of resistance that will set the current at 300 mA.

Solution

(a) Eq. 3-1,
$$I = \frac{E}{R} = \frac{10\ \text{V}}{100\ \Omega} = 0.1\ \text{A}$$

(b)
$$E = IR = 0.1\ \text{A} \times 250\ \Omega = 25\ \text{V}$$

(c)
$$R = \frac{E}{I} = \frac{15\ \text{V}}{300\ \text{mA}} = 50\ \Omega$$

Example 3-4

Calculate the resistor values for the circuit in Figure 3-4 to produce the current and voltage levels shown.

Solution

From Eq. 3-1,
$$R_1 = \frac{V_1}{I_1} = \frac{9.3\ \text{V}}{150\ \mu\text{A}} = 62\ \text{k}\Omega$$

$$R_2 = \frac{V_2}{I_1} = \frac{5.7\ \text{V}}{150\ \mu\text{A}} = 38\ \text{k}\Omega$$

$$R_3 = \frac{V_3}{I_2} = \frac{10\ \text{V}}{2\ \text{mA}} = 5\ \text{k}\Omega$$

$$R_4 = \frac{V_4}{I_2} = \frac{5\ \text{V}}{2\ \text{mA}} = 2.5\ \text{k}\Omega$$

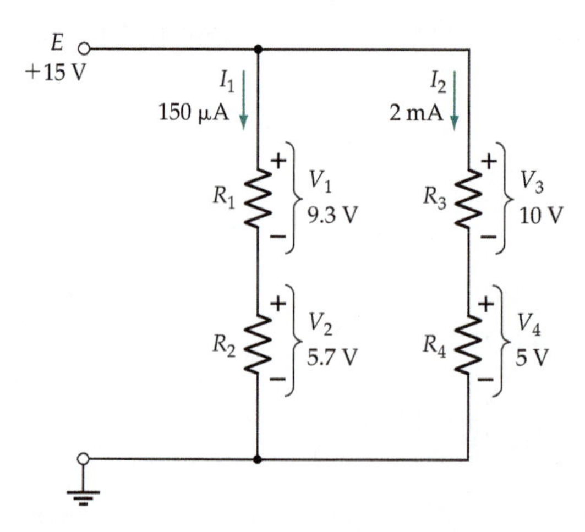

Figure 3-4 Illustration for Example 3-4. When the required current and voltage are known, the resistor values can be determined by applying Ohm's law.

Practice Problems

3-3.1 (a) Calculate the current through a 3.9 kΩ resistor when the voltage across it is measured as 22 V.

(b) If the current through the 3.9 kΩ resistor is to be exactly 4.5 mA, determine the required potential difference.

(c) If a 4.5 mA current is to flow through a resistance when 22 V is applied across it, determine the required resistance.

3-3.2 If the voltages in the circuit of Figure 3-3(b) are measured as $V_1 = 5$ V, $V_2 = 12$ V, and $V_3 = 12$ V, and the resistance values are $R_1 = 2.5$ kΩ, $R_2 = 10$ kΩ, and $R_3 = 15$ kΩ, calculate the currents levels.

3-4 CONDUCTANCE

In some circumstances it is convenient to use the reciprocal of resistance instead of the actual resistance value. The term *conductance* (symbol G) is applied to the reciprocal of resistance. Resistance is a measure of opposition to current flow, and so *conductance is a measure of the ease with which a conductor or resistor will pass a current.* The same graphic symbol is employed in circuit diagrams for conductance and resistance.

The *siemens* (symbol S) is the unit of conductance.

$$\text{Conductance} = \frac{1}{\text{resistance}}$$

or,

$$G = \frac{1}{R} \tag{3-2}$$

where G is in siemens and R is in ohms.

Ohm's law rewritten for conductance instead of resistance is,

$$I = EG$$

Example 3-5

Determine the conductance of a 100 Ω resistor, and the current that flows when a 10 V emf is applied across the resistor terminals.

Solution

Eq. 3-2,

$$G = \frac{1}{R} = \frac{1}{100\ \Omega}$$

$$= 10\text{ mS}$$

$$I = EG = 10\text{ V} \times 10\text{ mS}$$

$$= 100\text{ mA}$$

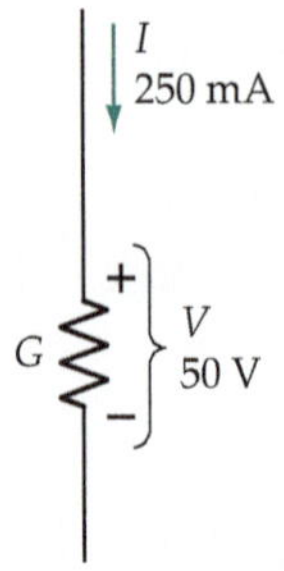

Figure 3-5 Conductance is determined as the inverse of resistance.

Example 3-6

Calculate the value of conductance that passes a 250 mA current when an emf of 50 V is applied across its terminals, as illustrated in Figure 3-5.

Solution

From Eq. 3-1,
$$R = \frac{E}{I} = \frac{50\ \text{V}}{250\ \text{mA}}$$
$$= 200\ \Omega$$

Eq. 3-2,
$$G = \frac{1}{R} = \frac{1}{200\ \Omega}$$
$$= 5\ \text{mS}$$

or
$$G = \frac{I}{E} = \frac{250\ \text{mA}}{50\ \text{V}}$$
$$= 5\ \text{mS}$$

Practice Problems

3-4.1 Convert 4.7 kΩ and 5.6 Ω into conductances.

3-4.2 Express 500 S and 33 μS as resistances.

3-5 ELECTRICAL POWER AND ENERGY

Power Dissipation

When an electric current flows in a conductor, heat is dissipated as a result of the work done in moving electrons past atoms (see Section 1-4). Work is also performed in a lamp filament when the electric current is converted into light. Similarly, work is done when a speaker converts electric current into sound and when an electric motor produces motion.

Power (P) is the time rate of doing work.

If a certain amount of work W is done in a time t, then $P = W/t$. It can be demonstrated that the power supplied to any electrical component is,

Power = current × voltage

or,
$$P = IE \qquad (3\text{-}3)$$

The unit of electric power is the watt (W) (see Appendix 2)

When an applied potential difference of 1 V causes a current of 1 A to flow in a circuit or device, the power supplied is

$$P = IE = 1\ \text{V} \times 1\ \text{A} = 1\ \text{W}$$

From Ohm's law $$E = IR, \text{ and } R = \frac{E}{I}$$

Substituting for I in Equation (3-3) gives

$$P = \frac{E}{R} \times E$$

or

$$P = \frac{E^2}{R} \tag{3-4}$$

And substituting for E in Equation (3-3) gives,

$$P = I \times IR$$

or

$$P = I^2R \tag{3-5}$$

From Equations 3-3, 3-4, and 3-5, it is seen that power can be calculated from a knowledge of any two of E, I, and R.

Electric heating elements are designed to convert electricity into heat for boiling water, space heating, and so on. As illustrated in Figures 3-6(a) and (b), power supplied to an electric lamp is converted into light, and a speaker converts input power into sound. The instrument employed for measuring power is called a wattmeter (see Section 13-8). Power dissipated in a conductor

(a) Power dissipated in a lamp is converted into light.

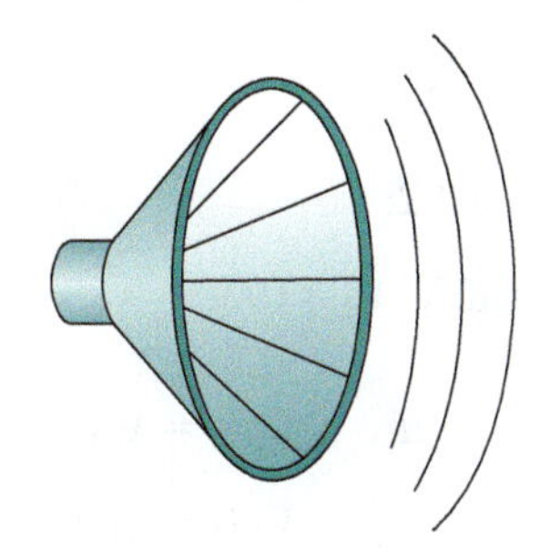

(b) Power supplied to a speaker is converted into sound.

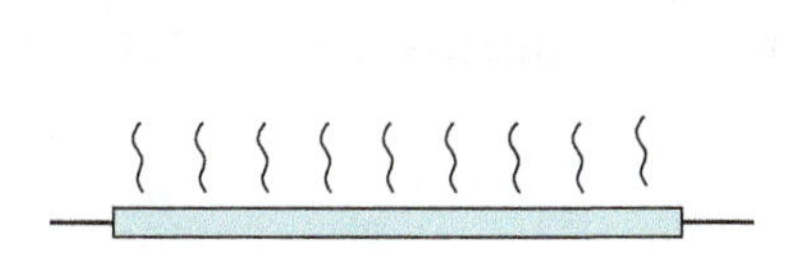

(c) Power dissipated in a conductor is wasted power.

(d) Power dissipated in an electronic device is wasted power.

Figure 3-6 Power is dissipated when an electric current flows. Power can be converted into light, heat, sound, and mechanical motion. Power dissipation in conductors and electronic devices is unwanted and wasteful.

is wasted power [Figure 3-6(c)]. Excessive power dissipation in a conductor can produce overheating which may damage or destroy the conductor insulation. Power dissipated in electronic devices is also wasted power [Figure 3-6(d)]. Here again, too much power dissipation can destroy a device, and so every device has a maximum power rating specified by the manufacturer.

To protect equipment and circuits from excessive power dissipation, a *fuse* may be connected into the circuit. A fuse is simply a short piece of thin copper wire or strip (see Figure 3-7). Power is dissipated in the fuse as the circuit current flows through it ($P = I^2R$). When the current flow exceeds the design limit, the power dissipated in the fuse causes it to melt and interrupt the current flow to the circuit.

Figure 3-7 A fuse is a thin strip of copper that melts when the current level in a circuit is too high. The melted fuse interrupts the current flow in the circuit.

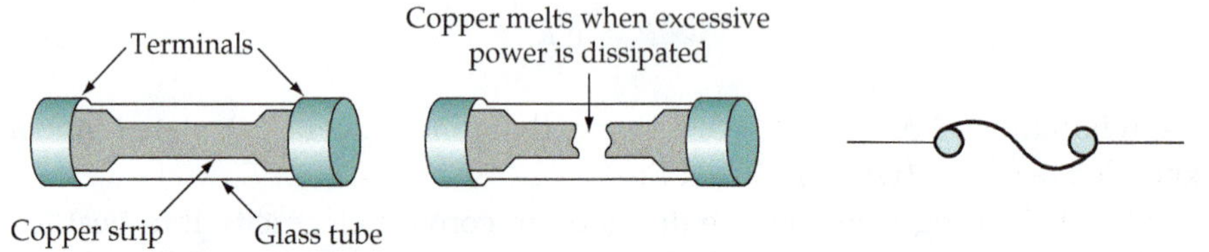

(a) A fuse is a thin strip of copper often contained in a glass tube (b) Circuit symbol for a fuse

Example 3-7

Determine the power dissipated in each of the resistors in Example 3-2.

Solution

From Example 3-2:

$V_1 = 9\text{ V}$, $V_2 = 10\text{ V}$, $V_3 = 10\text{ V}$, $I_1 = 15\text{ mA}$, $I_2 = 2.5\text{ mA}$, $I_3 = 12.5\text{ mA}$.

Eq. 3-3,

$$P_1 = V_1 I_1 = 9\text{ V} \times 15\text{ mA} = 135\text{ mW}$$

$$P_2 = V_2 I_2 = 10\text{ V} \times 2.5\text{ mA} = 25\text{ mW}$$

$$P_3 = V_3 I_3 = 10\text{ V} \times 12.5\text{ mA} = 125\text{ mW}$$

Example 3-8

A 120 Ω resistor has a specified maximum power dissipation of 1 W. Calculate the maximum current level.

Solution

Eq. 3-5

$$P = I^2R$$

so

$$I^2 = \frac{P}{R}$$

giving $$I = \sqrt{P/R} = \sqrt{1\ \text{W}/120\ \Omega}$$
$$= 91.3\ \text{mA}$$

Example 3-9

Calculate the current supplied to a 100 W lamp with a 115 V supply.

Solution

From Eq. 3-3 $$I = \frac{P}{E} = \frac{100\ \text{W}}{115\ \text{V}}$$
$$= 0.87\ \text{A}$$

Efficiency

The power supplied to electrical equipment is not all converted into useful power; usually, some power is wasted. For example, as illustrated in Figure 3-8, an electric motor produces heat as well as mechanical rotation, and an electric lamp gives off heat as well as light. In both cases, the useful output power is less than the input power.

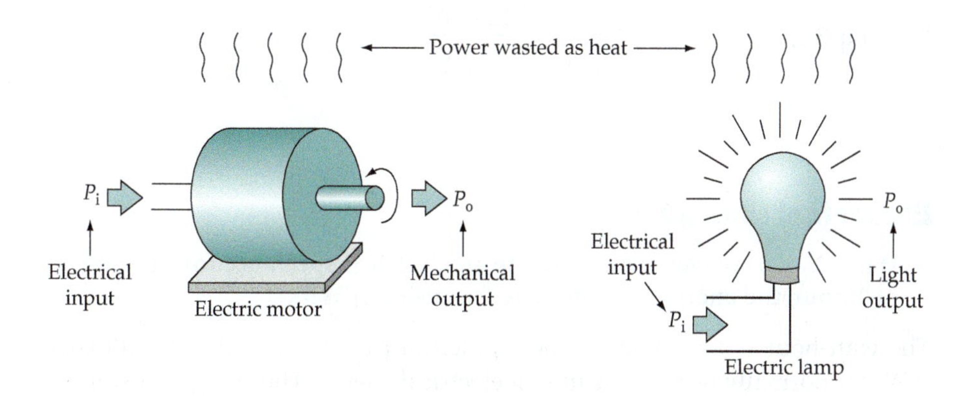

Figure 3-8 Some of the power delivered to an electric motor is wasted as heat. Similarly, some of the power supplied to an electric lamp is converted into heat instead of light. The efficiency of the equipment is the ratio of output power to input power expressed as a percentage.

The ability of an electric motor, electronic amplifier, or other component to convert power into a useful output is defined in terms of its *efficiency*. Thus, if 80% of the power supplied to an audio amplifier is delivered to a speaker and 20% is wasted as heat, then the amplifier has an efficiency of 80%.

The symbol for efficiency is η (Greek letter eta).

$$\text{Efficiency} = \frac{\text{power output}}{\text{power input}} \times 100\%$$

or

$$\eta = \frac{P_o}{P_i} \times 100\% \qquad (3\text{-}6)$$

Example 3-10

An electronic amplifier receives a supply of 1.2 W and delivers 700 mW of power to a speaker. Calculate its efficiency.

Solution

Eq. 3-6,
$$\eta = \frac{P_o}{P_i} \times 100\% = \frac{700 \text{ mW}}{1.2 \text{ W}} \times 100\%$$
$$= 58\%$$

Example 3-11

A 115 V electric motor produces an output of 750 W. Determine the input current if the motor efficiency is 60%.

Solution

From Eq. 3-6,
$$P_i = \frac{P_o}{\eta} \times 100\% = \frac{700 \text{ mW}}{60\%} \times 100\%$$
$$= 1250 \text{ W}$$

From Eq. 3-3,
$$I = \frac{P_i}{E} = \frac{1250 \text{ W}}{115 \text{ V}}$$
$$\approx 10.9 \text{ A}$$

Electrical Energy

When 1 W of power is delivered to an electrical appliance for a period of 1 hour, the energy consumed is 1 watt-hour (Wh).

The watt-hour is too small for most practical purposes, so the *kilowatt-hour* (kWh) is normally used as the unit of electrical energy. The energy consumed, or work done is,

$$\text{Energy consumed} = \text{power} \times \text{time}$$

or
$$W = P \times t \tag{3-7}$$

The quantity of energy consumed is calculated by multiplying the total power in kilowatts by the total consumption time in hours. This is demonstrated in Example 3-12. Energy consumption is measured by means of a *kilowatt-hour meter.*

It is explained in Appendix 2 that the *joule* is the unit of work and that 1 watt-second equals 1 joule. Thus, energy consumed can be expressed in *joules* (J), *kilojoules* (kJ), or *megajoules* (MJ).

Example 3-12

A small workshop has six 150 W electric lamps, three 1.5 kW electric motors, and a 3 kW electric oven. If everything operates for 8 hours per day for 5 days a week, calculate the energy consumed during 1 week. Express the energy in kilowatt-hours and in megajoules.

Solution

Total power $P = (6 \times 150\ \text{W}) + (3 \times 1.5\ \text{kW}) + 3\ \text{kW}$

$= 8.4\ \text{kW}$

Energy consumed per week:

Eq. 3-7 $W = Pt = 8.4\ \text{kW} \times 8\ \text{h} \times 5\ \text{days}$

$= 336\ \text{kWh}$

or $W = Pt = 8.4\ \text{kW} \times 8\ \text{h} \times 60\ \text{min} \times 60\ \text{s} \times 5\ \text{days}$

$= 1209.6\ \text{MJ}$

Practice Problems

3-5.1 An electric heater has a terminal voltage of 115 V and a current of 21.7 A. Calculate the power dissipated. Also, determine the required current and voltage if the heater is to dissipate 1 kW.

3-5.2 The current supplied to an electric lamp is measured as 20 A, and the terminal voltage as 11 V. If the lamp is supplied from a 12.5 V battery, determine the power dissipated in the lamp and the power wasted in the cables.

3-5.3 A 500 mA output current is taken from a 12 V electronic power supply. The power supply input is 115 V and 78 mA. Calculate the efficiency.

3-5.4 A house has three 100 W lamps, four 60 W lamps, and two 25 W lamps. If all of the lights are switched on for an average of three hours each day, determine the energy consumed in one week.

Summary of Formulas

- ***Ohm's law:***

$$\text{Current} = \frac{\text{voltage}}{\text{resistance}} \qquad I = \frac{E}{R}$$

- ***Conductance:***

$$\text{Conductance} = \frac{1}{\text{resistance}} \qquad G = \frac{1}{R}$$

- ***Power:***

$$\text{Power} = \frac{(\text{voltage})^2}{\text{resistance}} \qquad P = \frac{E^2}{R}$$

$$\text{Power} = (\text{current})^2 \times \text{resistance} \qquad P = I^2R$$

- ***Efficiency:***

$$\text{Efficiency} = \frac{\text{power output}}{\text{power input}} \times 100\% \qquad \eta = \frac{P_o}{P_i} \times 100\%$$

- ***Energy:***

$$\text{Energy consumed} = \text{power} \times \text{time} \qquad W = P \times t$$

Review Questions

Section 3-1

3-1 Define resistance and discuss the relative resistances of conductors with different lengths and cross-sectional areas.

Sections 3-2 and 3-3

3-2 Define the unit of resistance. Explain the relationship between current, voltage, and resistance.

3-3 State Ohm's law and write it in the forms required for calculating current, voltage, and resistance.

Section 3-4

3-4 Define conductance and the unit of conductance. State the relationship between resistance and conductance.

Section 3-5

3-5 Define power. Discuss the effects of power dissipation in conductors and electronic devices.

3-6 Write three forms of the equation for electrical power.

3-7 Explain what is meant by efficiency as applied to electrical equipment. Write the equation for efficiency.

3-8 Define electrical energy. Discuss units used in the measurement of energy.

Problems

Section 3-2

3-1 Determine the current that flows in each of the following resistances when a potential difference of 15 V is applied to the terminals of each component: 1 MΩ, 0.1 Ω, 1.5 kΩ, 270 Ω, 33 kΩ.

3-2 An emf of 100 V is applied across a resistance of 5 kΩ. (a) Calculate the current that flows in the resistance. (b) If the emf is changed to 110 V, determine the new value of resistance required to maintain the original current level.

3-3 Calculate the emf required to pass a current of 10 mA through a resistance of 3.3 kΩ.

3-4 A current of 3 A is to be passed through a 100 Ω resistor. Calculate the required voltage.

3-5 An electric lamp takes a current of 15 A when connected to a 12 V supply. Calculate the resistance of the lamp filament.

3-6 Determine the voltage required to pass a current of 100 mA through a resistance of 1 kΩ.

3-7 Determine the current that flows in a 2.2 kΩ resistor when it is connected to a 12 V battery.

Section 3-3

3-8 The circuit in Figure 3-9 has the following voltage and current levels: $V_1 = 12$ V, $V_2 = V_3 = 8$ V, $I_1 = 10$ mA, $I_2 = 3$ mA, $I_3 = 7$ mA. Calculate the resistor values.

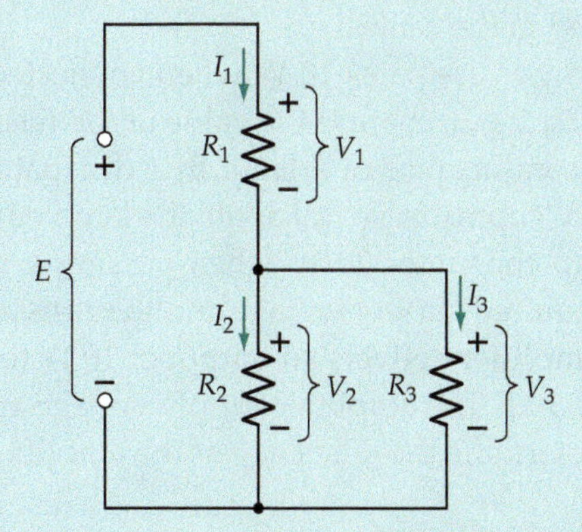

Figure 3-9

3-9 The voltage measured at the terminals of an electrical appliance is 113 V, the supply voltage is 115 V, and the current is 18 A. Calculate the resistance of the supply cables.

3-10 A 9 V battery supplies 100 mA of current to an electronic circuit, as illustrated in Figure 3-10. The circuit terminal voltage is 8.85 V. Calculate the resistance of the conductors connecting the battery to the circuit.

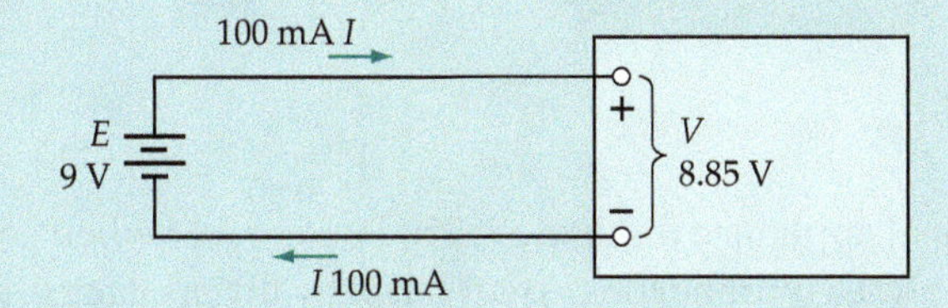

Figure 3-10

Section 3-4

3-11 Convert the following resistances into conductances: (a) 1 MΩ, (b) 0.1 Ω, (c) 1.5 kΩ, (d) 270 Ω (e) 33 kΩ

3-12 Express the following conductances as resistances: (a) 100 S, (b) 0.1 S, (c) 5 μS, (d) 750 S, (e) 2 mS.

3-13 Calculate the conductance of a heating element that takes a current of 20 A when connected to a 120 V supply.

3-14 Calculate the conductance of a circuit that takes 45 mA when its terminal voltage is 25 V. Determine the current that flows when the conductance is doubled.

Section 3-5

3-15 Determine the power dissipated in the lamp in Problem 3-5.

3-16 Calculate the power dissipated in the heater in Problem 3-13.

3-17 Calculate the power dissipated in the resistance in Problem 3-6.

3-18 Calculate the power dissipated in each resistance in Problem 3-8.

3-19 Calculate the current taken by a 100 W lamp from a 115 V supply. Also determine the resistance of the lamp filament.

3-20 For Problem 3-9, determine the power supplied to the appliance and the power dissipated in the cables.

3-21 A certain resistance dissipates 10 W when connected to a 50 V supply. Calculate the circuit current and the value of the resistance.

3-22 Determine the conductance of a lamp that dissipates 40 W when connected to a 120 V supply. Also, calculate the lamp current.

3-23 An electric lamp consumes 20 W when passing a current of 100 mA. Calculate the required supply voltage and the resistance of the lamp.

3-24 An electronic amplifier delivers an output of 10 W to a speaker, as illustrated in Figure 3-11. The amplifier supply voltage and current are 30 V and 470 mA. Determine the efficiency of the amplifier.

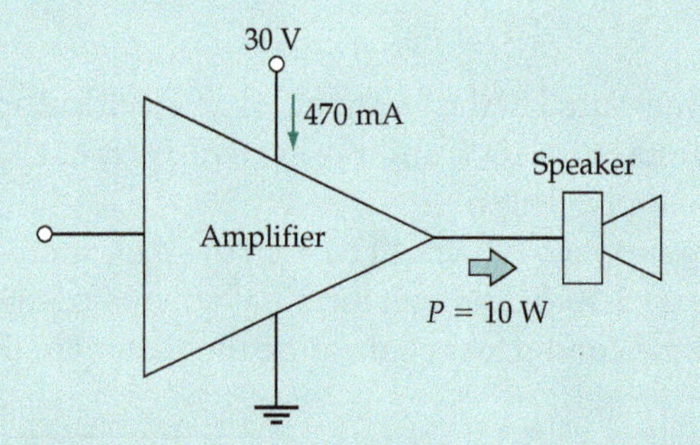

Figure 3-11

3-25 An electrical appliance dissipates 1250 W of power when connected to a 115 V supply. Determine the current taken by the appliance, and calculate the electric energy consumed over a period of 5 hours.

3-26 The appliance referred to in Problem 3-9 is used 8 hours each day for 365 days. Calculate the energy consumed (a) by the appliance, (b) by the cables.

3-27 A 230 V electrical appliance absorbs 2 kW and has an efficiency of 75%. Determine the supply current and the resistance offered by the appliance. Also, determine the energy employed usefully when the appliance is operated for 30 minutes.

3-28 A house with a 115 V supply has ten 100 W lamps, a 3.4 kW oven, a 300 W electric motor on the furnace, and a 200 W motor on the refrigerator. Calculate the supply current that flows when all the electrical equipment is in use at the same time. Determine the energy consumed daily if all of the equipment is in use for an average of 4 hours each day.

Practice Problem Answers

3-2.1 4.8 mA
3-2.2 150 mA
3-3.1 5.64 mA, 17.55 V, 4.9 kΩ
3-3.2 2 mA, 1.2 mA, 800 μA
3-4.1 213 μS, 179 mS
3-4.2 0.002 Ω, 30.3 kΩ
3-5.1 2.5 kW, 13.7 A, 72.8 V
3-5.2 220 W, 30 W
3-5.3 66.9%
3-5.4 12.39 kWh

CHAPTER 4
Conductors, Insulators, and Resistors

CONTENTS

Objectives

You will be able to:

1. Explain the relationship between insulator breakdown, insulator thickness, applied potential difference, and electric field strength. Solve problems involving these quantities.
2. Discuss the relationship between conductor resistance, conductor length, cross-sectional area, and specific resistance. Solve problems involving these quantities.
3. Solve problems involving conductor resistance and temperature.
4. Describe the various types of resistors and solve problems involving resistor power dissipation.
5. Solve problems involving resistor variation with temperature change.
6. Use the resistor color code to identify the resistor value and tolerance.

INTRODUCTION

Conductors carry electric current. Insulators protect conductors and protect people from conductors. Conductors can be destroyed if too much current is passed through them. The best conductor materials have the lowest resistance per cubic meter. Insulators can break down if subjected to excessive voltages. The best insulator materials have the highest breakdown voltage for a given thickness.

The resistance of any conductor can be calculated from a knowledge of its cross-sectional area, its length, and the resistance per cubic meter of the material. The conductor resistance and the current flowing in it can be used to calculate the conductor voltage drop and power dissipation.

A device constructed to have a certain value of resistance is known as a *resistor*. Large quantities of resistors are used in electronic circuits for providing voltage drops, and current limiting. These are manufactured in standard resistance values and with various ranges of accuracy and power dissipation capability. Small resistors use a color code to identify the resistance value and accuracy. The power rating of a resistor is important because it limits the maximum voltage that should be applied and the maximum current that should flow through the component. How a resistor value varies with temperature is defined by its temperature coefficient.

4-1 CONDUCTORS

Conducting Materials

Consider the atom of copper as illustrated by the two-dimensional diagrams in Figure 4-1. It is seen that the outer shell (or *valence shell*) of the copper atom contains only one electron. This electron is so weakly attached to the nucleus that it easily escapes and drifts off through the spaces between the atoms that make up a piece of copper. Because the free electrons can be given motion by the application of an electric field, current flow is easily achieved. Thus,

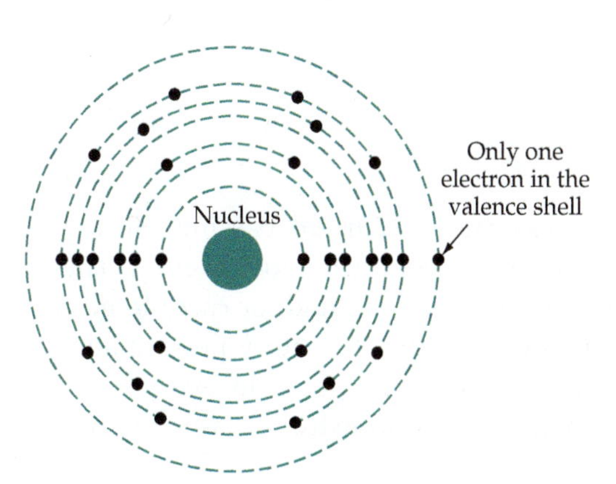

Figure 4-1 Two-dimensional diagram of copper atom. Copper has one electron in its valence (outer) shell.

copper is an excellent conductor of electricity. In general, all metals are good conductors, but some are better than others. Silver is the very best conductor (lowest resistance), followed closely by copper. Gold is also a relatively good conductor, as is aluminum.

Conductor Calculations

The function of a conductor is to conduct current from one point to another in an electric circuit. Electric cables usually consist of copper conductors sheathed with rubber or plastic insulating material. Cables that have to carry large currents must have relatively thick conductors (see Section 3-1). Where very small currents are involved, the conductor may be a thin strip of copper or an aluminum film. Between these two extremes, a wide range of conductors exist for various applications. Three different types of cables used in electronics equipment are illustrated in Figure 4-2. The one identified as a *coaxial cable* has a central insulated conductor and a circular plaited conducting screen as well as an outer insulating sheath. The other two are multiconductor cables, one circular and one flat.

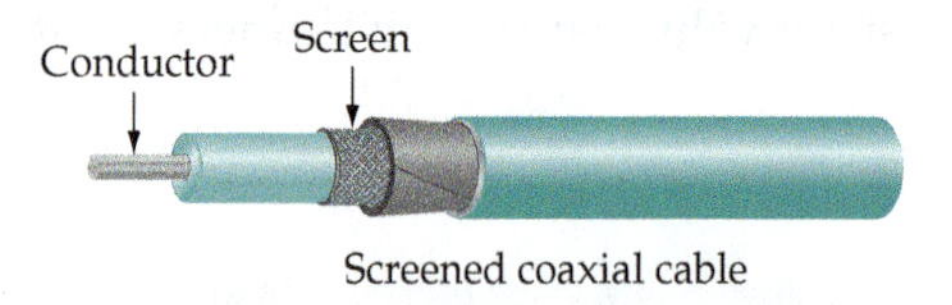

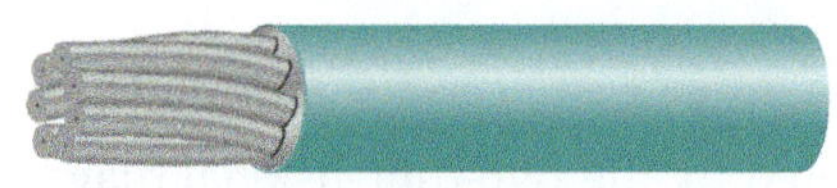

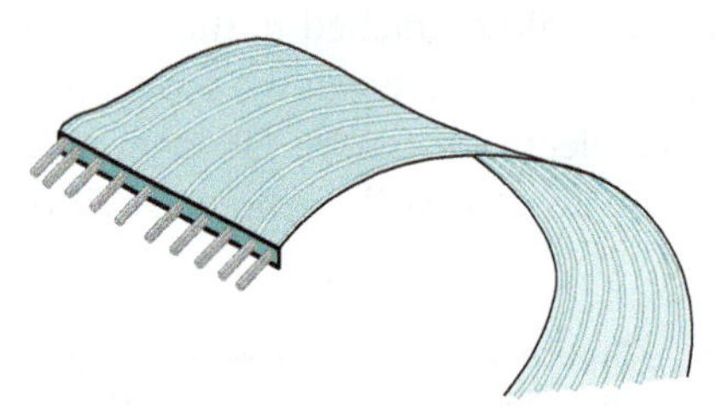

Figure 4-2 Some of the many different types of cables used with electronics equipment.

Because each conductor has a resistance, a current passing through it causes a voltage drop from one end of the conductor to the other. This is illustrated in Figure 4-3. When conductors are long and/or carry large currents, the conductor voltage drop may cause unsatisfactory performance of the equipment supplied. Power (I^2R) is also dissipated in every current-carrying conductor, and this is wasted power. In extreme cases the power

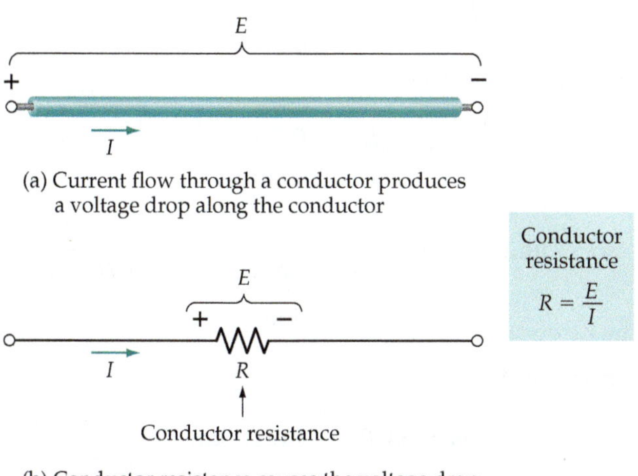

(a) Current flow through a conductor produces a voltage drop along the conductor

(b) Conductor resistance causes the voltage drop when a current flows

Conductor resistance

$$R = \frac{E}{I}$$

Figure 4-3 **Conductor resistance (*R*) is determined by using the voltage drop and current level with Ohm's law. The resistance per unit length (R/ℓ) is then used to select a suitable wire gauge.**

dissipated in conductors might generate sufficient heat to destroy the insulation or even melt the conductor. The conductor voltage drop and power dissipation are easily calculated when the resistance per unit length of the conductor is known.

The table in Appendix 4 gives standard *American Wire Gauge* (AWG) sizes, together with the metric equivalents. Diameters are given in millimeters and in *mils* (inches $\times 10^{-3}$), and resistances (for copper) are given in Ω/km and in Ω/1000 ft. Gauges #10, #12, and #14 are normally used for domestic wiring (in North America); #14 is used for lighting and other relatively low current applications, and #10 for electric dryers. Note that the resistance per unit length decreases by half for each gauge number decrease of 3 (for example from #16 to #13, and from #13 to #10).

To select a conductor for a given application, its resistance per unit length is first determined from the current level and the acceptable voltage drop along the cable. The AWG table is then used to find a wire gauge with the calculated resistance per kilometer (R/km) or with the next lower R/km value. Example 4-1 demonstrates the process, and Example 4-2 shows that the voltage drop along the *return* conductor must be taken into account as well as the voltage drop along the *line* (outgoing) conductor.

Example 4-1

A 50 cm long conductor is to carry a 500 mA current (see Figure 4-3). The voltage drop along the conductor is not to exceed 0.5 V. Select a suitable wire gauge from the table in Appendix 4.

Solution

Conductor (maximum) resistance

$$R = \frac{E}{I} = \frac{0.5\ \text{V}}{500\ \text{mA}} = 1\ \Omega$$

$$\text{Resistance/unit length} = \frac{R}{\text{length in km}} = \frac{1\ \Omega}{50\ \text{cm}/(1000 \times 100)}$$

$$= 2\ \text{k}\Omega/\text{km}$$

Referring to Appendix 4

$$\text{For \#38, } R = 2147\ \Omega/\text{km}$$

This is larger than the required 2 kΩ/km, so when $I = 500$ mA the conductor voltage drop would be larger than 0.5 V.

$$\text{\#37 has } R = 1715\ \Omega/\text{km}$$

When $I = 500$ mA, the voltage drop along this conductor (#37) will be less than the 0.5 V maximum, so #37 (or lower) is suitable.

Example 4-2

An electric motor is supplied with 1 A of current from a 115 V source (E_1), as shown in Figure 4-4(a). Each of the cables used is 40 m long and the resistance of the cables is 0.025 Ω/m. Determine the potential difference (E_2) at the motor terminals [see Figure 4-4(b)]. Calculate the power dissipated in the cables.

Solution

For 40 m of cable $\quad R = 40\ \text{m} \times 0.025\ \Omega/\text{m} = 1\ \Omega$

Voltage drop along each cable

$$V_1 = V_2 = IR = 1\ \text{A} \times 1\ \Omega$$

$$= 1\ \text{V}$$

Motor terminal voltage

$$E_2 = E_1 - V_1 - V_2 = 115\ \text{V} - 1\ \text{V} - 1\ \text{V}$$

$$= 113\ \text{V}$$

Power dissipated in the cables

$$P = (V_1 + V_2) \times I = 2\ \text{V} \times 1\ \text{A}$$

$$= 2\ \text{W}$$

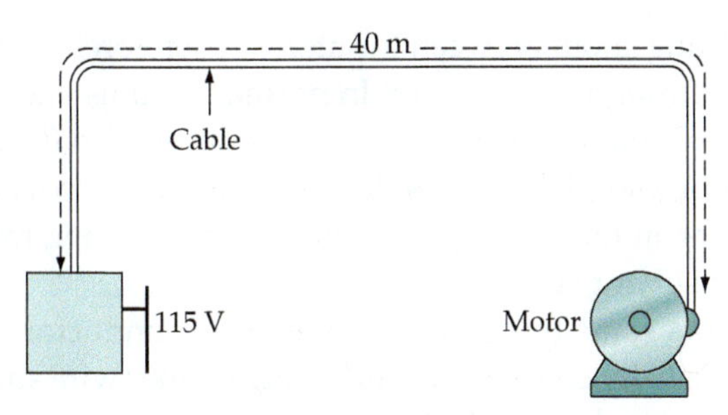

(a) 115 V supply connected to an electric motor via a 40 m two-conductor cable

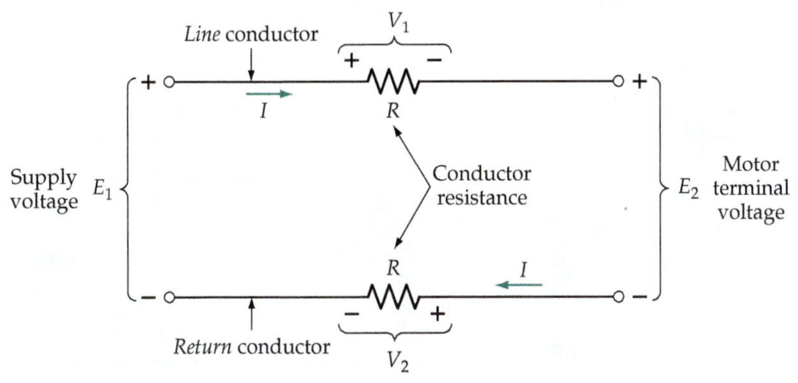

(b) Current flow through the cable causes voltage drops along the *line* and *return* conductors

Figure 4-4 Electric motor supplied from a 115 V source. The voltage drops along the connecting cables depend on the current level and the cable resistance. Power is also dissipated in the cables.

Practice Problems

4-1.1 A 2 kW electric heater is supplied from a 115 V source using 15 m of cable with two #14 copper wire conductors. Calculate the terminal voltage at the heater.

4-1.2 If the electric motor in Example 4-2 can operate satisfactorily with a terminal voltage of 110 V, select a suitable wire gauge for the supply cables.

4-1.3 Determine the power dissipated in the cables in (a) Problem 4-1.1 and (b) Problem 4-1.2.

4-2 INSULATORS

Insulating Materials

In materials that function as insulators, the valence shell electrons are so strongly attached to their atoms that there are no free electrons available for current flow. However, even in the best of insulating materials there are some free electrons drifting about between the atoms. Therefore, a very small electric current can flow through an insulator. In normal circumstances this current is so small that it is absolutely negligible. A current can also flow along the surface of an insulator, especially if it is dirty or wet. This is particularly evident with the insulators on high-voltage overhead cables during rain, when *arcing* occurs over the insulator surface.

Figure 4-5 shows some typical arrangements of conductors and insulators. An electric cable usually consists of conducting copper wire surrounded by an insulating sheath of rubber or plastic. Sometimes there is more than one conductor, and these are, of course, individually insulated. Other insulating materials are glass and porcelain.

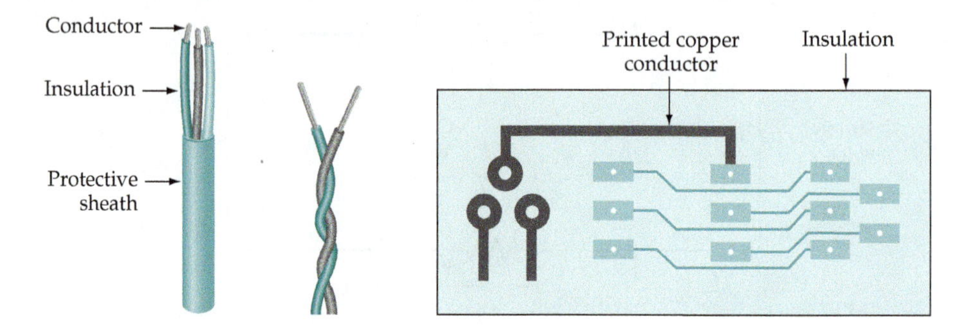

Figure 4-5 Conductors employed for industrial and domestic purposes normally have stranded copper wires with rubber or plastic insulation. In electronics equipment, flat cables of fine wires and thin printed circuit conductors are widely used.

Insulator Calculations

If a sufficiently high potential difference is applied across an insulator, electrons can be *pulled out* of the atoms, and substantial current flow might occur. This is known as *insulator breakdown*, and it normally results in the destruction of the insulator. To help avoid breakdown, all insulating materials (e.g., sheaths of electric cables) are rated according to the maximum voltage that may be safely applied across them. Table 4-1 lists typical *breakdown voltages* (or *dielectric strengths*) for various insulating materials. It is seen that air has the lowest breakdown voltage, at 30 kV/cm, and mica has the highest, at 2000 kV/cm. The figures given vary to some extent with the

TABLE 4-1 **Typical breakdown voltages of some insulating materials**

Material	**Breakdown Voltage (kV/cm)**
Air	30
Porcelain	70
Rubber	270
Glass	1200
Mica	2000

actual thickness of the insulating material. The *electric field strength* is a measure of the electric stress set up in an insulator. The field strength should always be substantially lower than that which might cause insulator breakdown. The field strength is simply the applied voltage divided by the insulator thickness.

$$\text{Electric field strength, } \xi = \frac{E}{d} \tag{4-1}$$

When E is in volts and d is in meters, the units of field strength are *volts/meter* (*V/m*). The situation is illustrated in Figure 4-6 and further explored in Example 4-3.

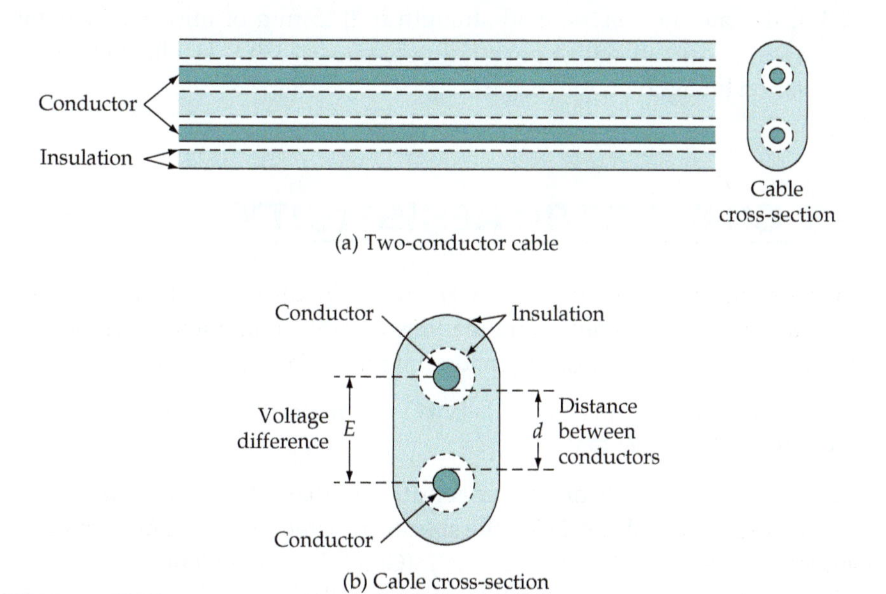

Figure 4-6 An electric field exists between two conductors separated by insulating material. Insulator breakdown can occur if the voltage difference is too large.

Example 4-3

Two electric conductors encased in rubber are exactly 0.25 cm apart.

(a) Calculate the electric field strength in the insulation when the potential difference between the conductors is 25 V.

(b) Determine the minimum potential difference between conductors that could cause insulator breakdown.

Solution

(a) Eq. 4-1,

$$\xi = \frac{E}{d} = \frac{25\ \text{V}}{0.25\ \text{cm}}$$

$$= 100\ \text{V/cm}$$

(b) From Table 4-1, insulation breakdown for rubber occurs at 270 kV/cm.

So $$E = d \times (\textit{breakdown voltage}) = 0.25\text{ cm} \times 270\text{ kV/cm}$$
$$= 67.5\text{ kV}$$

Practice Problems

4-2.1 Determine the electric field strength (in V/cm) in a 2 mm thick printed circuit board if a 45 V potential difference is applied between the conductors on opposite sides.

4-2.2 Calculate the voltage that may cause breakdown of the air between two terminals spaced 4 mm apart.

4-2.3 Calculate the electric field strength in 0.15 mm of mica that separates two conductors with a potential difference of 330 V. Is it likely that insulation breakdown will occur?

4-3 CONDUCTOR RESISTIVITY

The resistance per unit length of a given conductor is not always available. So, to allow for all possible combinations of cable length and cross-sectional area, the *specific resistance* of the conducting material is employed.

The specific resistance (or resistivity) of a material is the resistance of 1 meter cube of the material.

Because some metals are better conductors than others, each has its own specific resistance. Table 4-2 lists the specific resistances of various metals. The symbol used for specific resistance is ρ (Greek lowercase letter rho), and the units of ρ are ohm meters ($\Omega \cdot \text{m}$). The origin of the units will become apparent shortly.

TABLE 4-2 **Specific resistance for electrical conducting metals**

Metal	Specific Resistance at 20°C ($\Omega \cdot \text{m}$)
Silver	1.64×10^{-8}
Copper (annealed)	1.72×10^{-8}
Gold	2.45×10^{-8}
Aluminum	2.83×10^{-8}
Tungsten	5.5×10^{-8}
Nickel	7.8×10^{-8}
Constantan	49×10^{-8}

Consider the meter cube of copper illustrated in Figure 4-7(a). From Table 4-2, the specific resistance of copper (resistance of the cube) is $\rho = 1.72 \times 10^{-8}$ $\Omega \cdot$m. If the length of copper is doubled or tripled by placing blocks side by side as shown in Figure 4-7(b), the resistance is increased. Components connected together in this way are said to be *connected in series.* (Series-connected circuits are discussed in Chapter 5.) The resistance of two series-connected blocks is 2 $\rho\Omega$ and that of three is 3 $\rho\Omega$, and so on. From this it is derived that the total resistance of any length of copper with a cross-sectional area of 1 m^2 is,

$$R = \rho \times (\text{length in meters})$$
$$= 1.72 \times 10^{-8} \times (\text{length in meters})\ \Omega$$

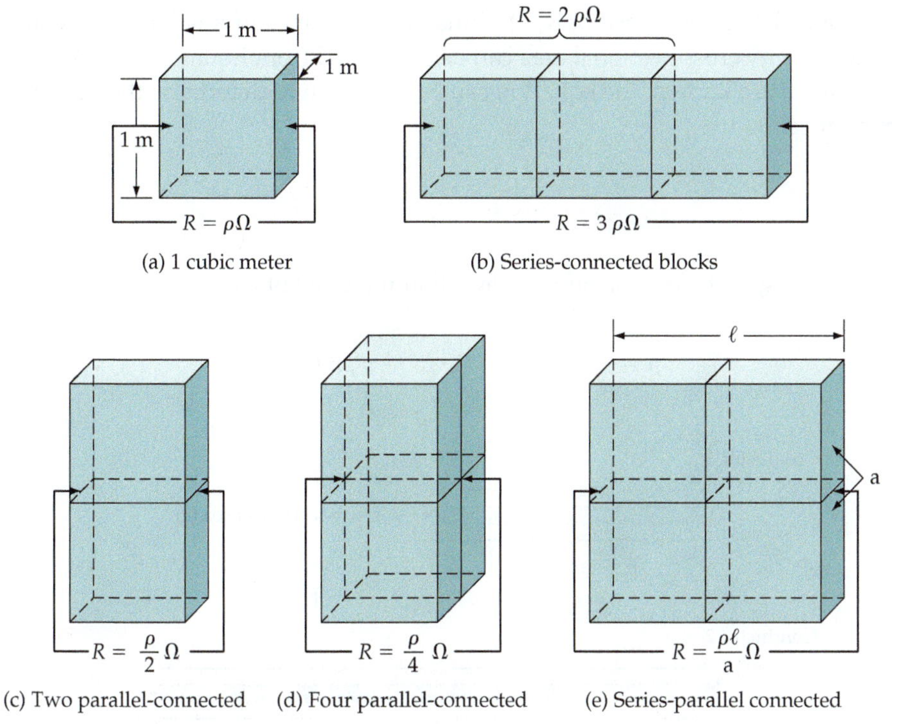

Figure 4-7 If a one meter cube of copper has a resistance of ρ Ω, three cubes in series will have a resistance of 3 $\rho\Omega$, and the resistance of two cubes in parallel will be $\rho/2$ Ω. This gives the equation: Resistance = (ρ × length)/(cross-sectional area).

Now consider the effect of doubling the cross-sectional area of the copper by placing two blocks as illustrated in Figure 4-7(c). Components connected together in this way are said to be *connected in parallel.* (Parallel-connected circuits are discussed in Chapter 6.) Because the cross-sectional area is doubled, the resistance is halved. If four blocks are parallel-connected [Figure 4-7(d)], the resistance is quartered. In the case of blocks connected in parallel, the resistance can be written as,

$$R = \frac{\rho}{\text{cross-sectional area in m}^2}$$

When blocks are put in series and parallel [Figure 4-7(e)], the two previous equations are combined to give,

$$R = \frac{\rho \ell}{a} \tag{4-2}$$

where R is the resistance in ohms, ρ is the specific resistance of the material, ℓ is the length in meters, and a is the cross-sectional area in m^2.

Cylindrical shaped conductors with different dimensions are illustrated in Figure 4-8. It is seen that the resistance of a conductor with length ℓ, and cross-section area a is calculated (from Equation 4-2) as $R_1 = \rho\ell/a$. A similar conductor with the same cross-sectional area and twice the length has a resistance of $2R_1$. Also, a conductor with the same length as the first but a cross-sectional area of $2a$ has a resistance of $R_1/2$. Thus, the resistance of any length of conductor of any cross-sectional area can be calculated from Equation 4-2. Where R, ℓ, and a are known, the equation can be rewritten to determine the specific resistance:

$$\rho = \frac{Ra}{\ell}$$

Substituting units for R, a, and ℓ shows that the units of ρ are,

$$\rho = \frac{\Omega \times \mathrm{m}^2}{\mathrm{m}} = \Omega \cdot \mathrm{m} = \textit{ohm meters}$$

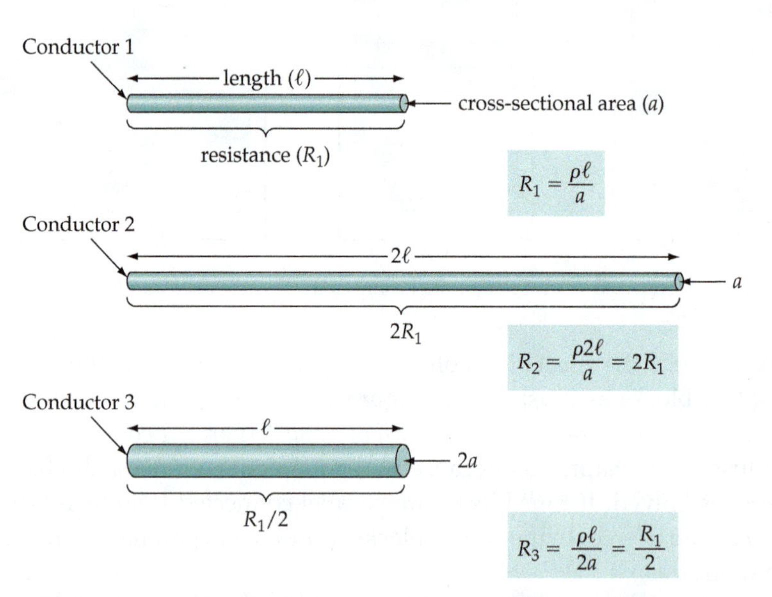

Figure 4-8 Conductor resistance is determined from the length, the cross-sectional area, and the specific resistance of the conductor material.

Example 4-4

Determine the resistance of a 20 m length of annealed copper wire with a 2 mm diameter.

Solution

Cross-sectional area, $a = \frac{\pi D^2}{4} = \frac{\pi \times (2 \times 10^{-3})^2}{4}$

$$= \pi \times 10^{-6}\ \text{m}^2$$

From Table 4-2, for copper

$$\rho = 1.72 \times 10^{-8}\ \Omega \cdot \text{m}$$

Eq. 4-2 $R = \frac{\rho \ell}{a} = \frac{1.72 \times 10^{-8} \times 20\ \text{m}}{\pi \times 10^{-6}\ \text{m}^2}$

$$= 0.109\ \Omega$$

Example 4-5

Calculate the resistance per meter length of aluminum wire 5 mm in diameter.

Solution

Cross-sectional area, $a = \frac{\pi D^2}{4} = \frac{\pi \times (5 \times 10^{-3})^2}{4}$

$$= 6.25\ \pi \times 10^{-6}\ \text{m}^2$$

From Table 4-2, for aluminum

$$\rho = 2.83 \times 10^{-8}$$

$$R = \frac{\rho \ell}{a} = \frac{2.83 \times 10^{-8} \times 1\ \text{m}}{6.25\pi \times 10^{-6}\ \text{m}^2}$$

$$= 1.44 \times 10^{-3}$$

Example 4-6

An electric heater draws a current of 15 A from a 115 V source. The two-conductor cable connecting the heater to the supply is 43 m long (see Figure 4-9). If the total voltage drop along the cables is not to exceed 12 V, determine the diameter of suitable copper wire and select a suitable wire size from Appendix 4.

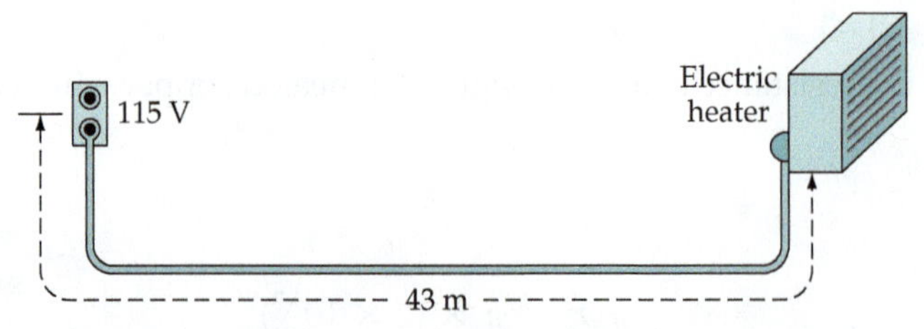

Figure 4-9 Illustration for Example 4-6. Conductor size can be determined by calculating the acceptable conductor resistance and then using the resistance to find the minimum cross-sectional area.

Solution

The maximum allowable resistance of the cable is,

$$R = \frac{E}{I} = \frac{\text{voltage drop along cable}}{I} = \frac{12\text{ V}}{15\text{ A}}$$

$$= 0.8\ \Omega$$

Total length of wire, $\ell = 2 \times 43\text{ m} = 86\text{ m}$

From Eq. 4-2,

$$a = \frac{\rho\ell}{R} = \frac{1.72 \times 10^{-8} \times 86\text{ m}}{0.8\ \Omega}$$

$$= 1.85 \times 10^{-6}\text{ m}^2$$

$$a = \frac{\pi}{4}D^2$$

so,

$$D = \sqrt{\frac{4a}{\pi}} = \sqrt{\frac{4 \times 1.849 \times 10^{-6}}{\pi}}$$

$$= 1.53\text{ mm}$$

The minimum wire diameter must be 1.53 mm. From Appendix 4, #14 has $D = 1.63$ mm, so #14 is suitable.

Practice Problems

4-3.1 Using the specific resistance of copper and the wire diameter, calculate the resistance per kilometer of #14 copper wire.

4-3.2 Determine the resistance of 1.25 m of nickel wire 1 mm in diameter.

4-3.3 A printed circuit board has 0.25 mm thick copper. When all unwanted copper is removed, the conducting strips are to carry 100 mA of current with a maximum voltage drop of 5 mV per meter. Determine the minimum width for the strips.

4-4 TEMPERATURE EFFECTS ON CONDUCTORS

The values of specific resistance listed in Table 4-2 refer only to conducting materials at a temperature of 20°C. So, in all resistance calculations using the specific resistance, the conductor temperature is assumed to remain constant at 20°C. Because the resistance of metals changes with temperature, it is necessary to be able to calculate the resistance values at higher or lower temperature levels.

The resistance of all pure metals tends to increase as the temperature of the metal rises. This may be explained in terms of the atoms actually vibrating at elevated temperatures and becoming greater obstructions in the path of moving electrons. Because the resistance increases with increasing temperatures, metals are said to have a *positive temperature coefficient* (PTC). Some materials, notably semiconductors, exhibit a decrease in resistance as their temperature rises. These have a *negative temperature coefficient* (NTC). Over the normal range of operating temperatures, all metals exhibit a nearly linear relationship between resistance and temperature.

In Figure 4-10(a) the variation in resistance of copper is plotted versus temperature. Note that the resistance of copper appears to go to zero at a temperature

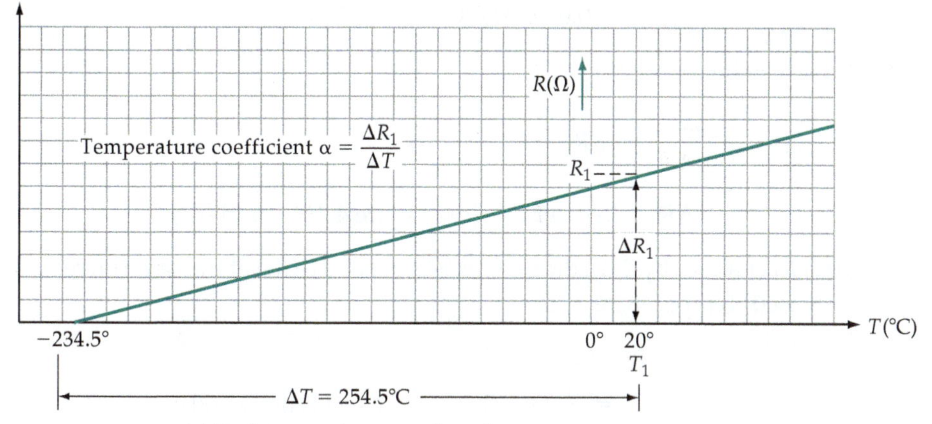

(a) Resistance of copper plotted versus temperature

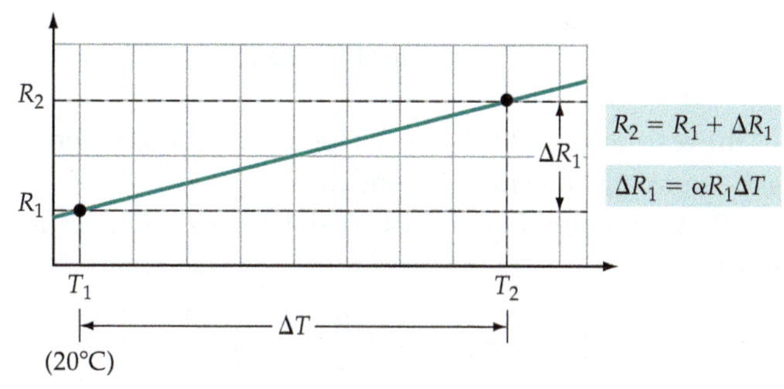

(b) The resistance (R_2) at a new temperature (T_2) can be determined from a knowledge of R_1 at T_1 and the temperature coefficient (α)

Figure 4-10 The resistance of a conductor at any temperature T_2 can be calculated from a knowledge of the resistance at a starting temperature T_1 and the temperature coefficient of the material.

of −234.5°C. (The actual graph is not as completely linear at the low temperature end as illustrated.) If the resistance (R_1) of a copper conductor at a temperature of 20°C is known, the new value of resistance at any other temperature can be calculated. Suppose that R_1 is 1 Ω at 20°C. If the temperature is reduced to −234.5°C, the temperature change is,

$$\Delta T = 20°\text{C} - (-234.5°\text{C})$$
$$= 254.5°\text{C}$$

and the resistance change per degree of temperature change is,

$$\alpha = \frac{\Delta R_1}{\Delta T} = \frac{1\ \Omega}{254.5°\text{C}}$$
$$= 0.003\ 93\ \Omega/°\text{C}$$

The quantity $\alpha = 0.003\ 93\ \Omega/°\text{C}$ is the *temperature coefficient for copper* at 20°C. For a resistance $R_x = 10\ \Omega$ at 20°C, the resistance change with temperature would be,

$$\frac{\Delta R_x}{\Delta T} = \alpha R_x = (0.003\ 93\ \Omega/°\text{C}) \times 10\ \Omega$$
$$= 0.0393\ \Omega/°\text{C}$$

So, for any value of resistance R_1,

$$\frac{\Delta R_1}{\Delta T} = \alpha R_1$$

or,

$$\Delta R_1 = \alpha R_1 \Delta T \qquad \text{[see Figure 4-10(a) and (b)]}$$

Because R_1 changes by ΔR_1, the new resistance value (R_2) at the new (increased or decreased) temperature T_2 is,

$$R_2 = R_1 + \Delta R_1$$
$$= R_1 + \alpha R_1 \Delta T \qquad \text{[see Figure 4-10(b)]}$$

This can be rewritten as

$$R_2 = R_1(1 + \alpha \Delta T) \tag{4-3}$$

where R_1 is the resistance at 20°C, α is the temperature coefficient of the material, and ΔT is the temperature change from 20°C.

If R_1 and R_2 are known, Equation 4-3 can be rearranged to determine ΔT or α as follows:

$$\Delta T = \frac{1}{\alpha}\left(\frac{R_2}{R_1} - 1\right)$$

and

$$\alpha = \frac{1}{\Delta T}\left(\frac{R_2}{R_1} - 1\right)$$

Table 4-3 lists the temperature coefficients for various metals at 20°C. Note that the alloy *constantan* has an extremely small temperature coefficient. This characteristic makes constantan useful in applications where the resistance is required to remain as constant as possible when the temperature changes. Temperature coefficients can also be determined at temperatures other than 20°C. Column 3 in Table 4-3 lists temperature coefficients at 0°C, which can be used instead of those at 20°C when the reference temperature is 0°C.

TABLE 4-3 **Temperature coefficients for metals**

Material	α at 20°C	α at 0°C
Silver	0.003 8	0.004 12
Copper	0.003 93	0.004 26
Gold	0.003 4	0.003 65
Aluminum	0.003 9	0.004 24
Tungsten	0.004 5	0.004 95
Nickel	0.006	
Constantan	0.000 008	

Example 4-7

The resistance of a length of copper cable is 5.8 Ω at 20°C. Determine the new resistance of the cable at 125°C (see Figure 4-11).

Solution

$$\Delta T = T_2 - T_1 = 125°\text{C} - 20°\text{C}$$

$$= 105°\text{C}$$

For copper at 20°C, $\alpha = 0.003\,93$ (see Table 4-3).

Eq. 4-3 $$R_2 = R_1(1 + \alpha\Delta T) = 5.8\,\Omega\,[1 + (0.003\,93 \times 105°\text{C})]$$

$$= 8.2\,\Omega$$

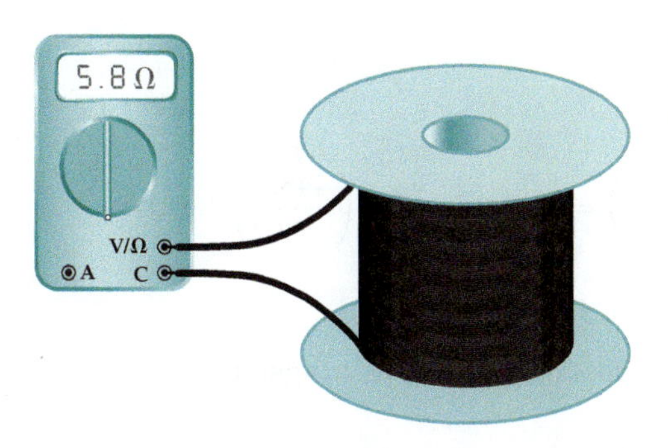

Figure 4-11 The resistance of a conductor increases when its temperature is increased.

Example 4-8

The resistance of a coil of nickel wire is 1 kΩ at 20°C. This falls to 880 Ω after the coil is submerged in a liquid for some time, as illustrated in Figure 4-12. Calculate the temperature of the liquid.

Solution

For nickel at 20°C $\quad \alpha = 0.006 \quad$ (see Table 4-3)

From Eq. 4-3,
$$\Delta T = \frac{1}{\alpha}\left(\frac{R_2}{R_1} - 1\right) = \frac{1}{0.006}\left(\frac{880\ \Omega}{1\ \text{k}\Omega} - 1\right)$$
$$= -20^\circ\text{C}$$

and
$$\Delta T = T_2 - T_1$$

or
$$T_2 = \Delta T + T_1 = -20^\circ\text{C} + 20^\circ\text{C}$$
$$= 0^\circ\text{C}$$

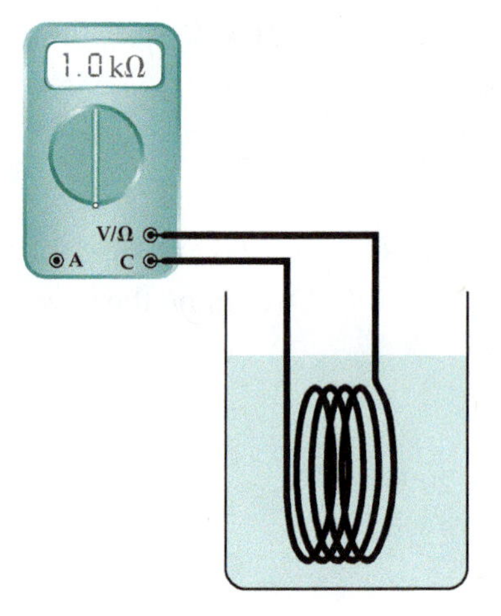

Figure 4-12 The temperature of a liquid can be determined from the resistance of a coil immersed in the liquid.

Example 4-9

A certain length of wire has its resistance measured as 330 Ω at 20°C and 448.8 Ω at 100°C. Calculate the temperature coefficient, and identify the material.

Solution

From Eq. 4-3
$$\alpha = \frac{1}{\Delta T}\left(\frac{R_2}{R_1} - 1\right) = \frac{1}{(100 - 20)^\circ\text{C}}\left(\frac{448.8\ \Omega}{330\ \Omega} - 1\right)$$
$$= 0.0045$$

From Table 4-3 $\quad \alpha = 0.0045$ identifies the material as tungsten.

Practice Problems

4-4.1 A tungsten filament lamp has a resistance of 11.8 Ω at 20°C. Determine its resistance at 2500°C.

4-4.2 A resistor is constructed of nickel wire to have a resistance of 1.25 kΩ at 20°C. Calculate its resistance at 55°C.

4-4.3 A fine wire has a resistance of 0.11 Ω at 20°C and 0.116 Ω at 36°C. Determine the temperature coefficient and identify the material.

4-5 RESISTOR CONSTRUCTION AND POWER RATING

Fixed-value Resistors

The function of a resistor is to offer a particular resistance to current flow. For a given current and known resistance, the voltage drop across the resistance can be predicted using Ohm's law. Similarly, for a given voltage applied to a resistor, the current that flows may be predetermined by selection of a suitable resistor value. The allowable power dissipation largely dictates the construction and physical size of a resistor.

Two common types of electronics resistors are *wire-wound* and *carbon composition*. A typical wire-wound resistor consists of a length of nickel wire wound on a ceramic tube and covered with porcelain [see Figure 4-13(a)]. Low-resistance connecting wires are provided, and the resistance value is usually printed on the side of the component. Carbon composition resistors are

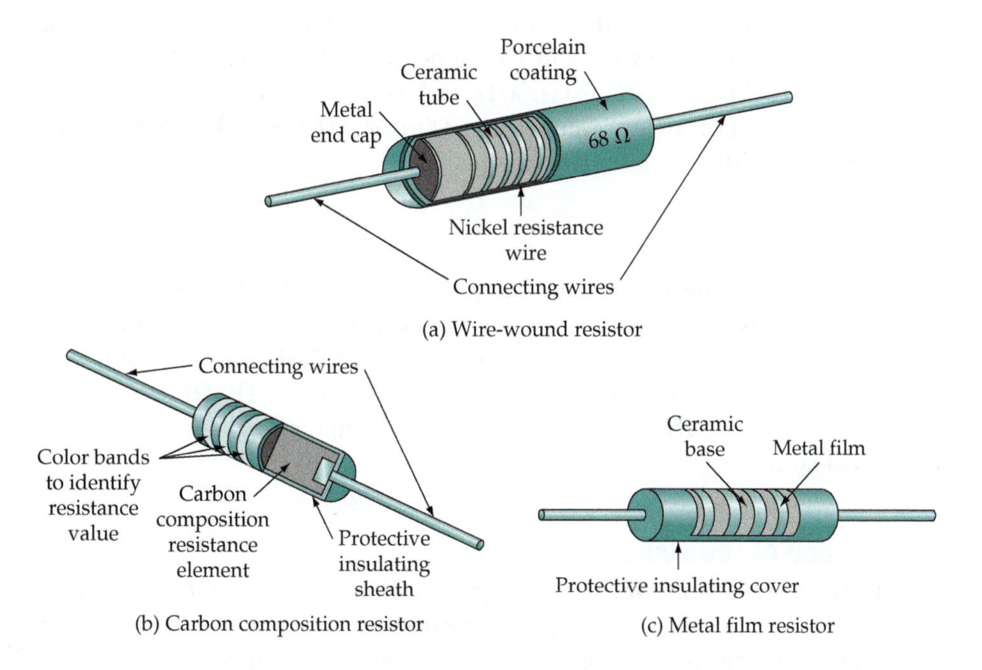

Figure 4-13 Individual resistors are typically wire-wound or carbon composition construction. Wire-wound resistors are used where high power dissipation is required. Carbon composition type is the least expensive. Metal film resistance values can be more accurate than carbon composition type.

constructed by molding mixtures of powdered carbon and insulating materials into a cylindrical shape [Figure 4-13(b)]. An outer sheath of insulating material affords mechanical and electrical protection, and copper connecting wires are provided at each end. Carbon composition resistors are smaller and less expensive than the wire-wound type. However, the wire-wound type is the more rugged of the two and is able to survive much larger power dissipations than the carbon composition type. Figure 4-13(c) shows a *metal film* resistor which has a thin metal film deposited on a porcelain cylinder and then etched to form a spiral. The thickness and width of the spiral determines the resistance. This method produces more accurate resistance values than carbon composition type.

Variable Resistors

Most resistors have standard fixed values, so they can be termed *fixed resistors. Variable resistors,* or *adjustable resistors,* are also used a great deal in electronics. The construction of a typical variable resistor is illustrated in Figure 4-14(a), and two frequently employed graphic symbols for a variable resistor are shown in Figure 4-14(b). The illustration in Figure 4-14(a) shows a coil of closely wound insulated resistance wire formed into a partial circle. The coil has a low-resistance terminal at each end, and a third terminal is connected to a movable contact with a shaft adjustment facility. The movable contact can be set to any point on a connecting track that extends over one (uninsulated) edge of the coil. Using the adjustable contact, the resistance from either end terminal to the center terminal may be adjusted from zero to the maximum coil resistance.

Another type of variable resistor, known as a *decade resistance box,* is shown in Figure 4-14(c). This is a laboratory component that contains precise values of switched series-connected resistors. As illustrated, the first switch from the right controls resistance values in 1 Ω steps from 0 Ω to 9 Ω, and the second switch selects values of 10 Ω, 20 Ω, 30 Ω, and so on. The decade box shown can be set to within ±1 Ω of any value from 0 Ω to 9999 Ω. Other decade boxes are available with different resistance ranges.

Resistor Tolerance

Standard (fixed-value) resistors normally range from 2.7 Ω to 22 MΩ (see Appendix 6). The resistance *tolerances* on these standard values are typically ±20%, ±10%, ±5%, or ±1%. A tolerance of ±10% on a 100 Ω resistor means that the actual resistance may be as high as 100 Ω + 10% (i.e., 110 Ω) or as low as 100 Ω − 10% (i.e., 90 Ω). Obviously, the resistors with the smallest tolerance are the most accurate.

Resistor Networks

Resistor networks are available in *integrated circuit* type *dual-in-line* packages, see Figure 4-15(a). One construction method uses a *thick film* technique in which conducting solutions are deposited in the required form.

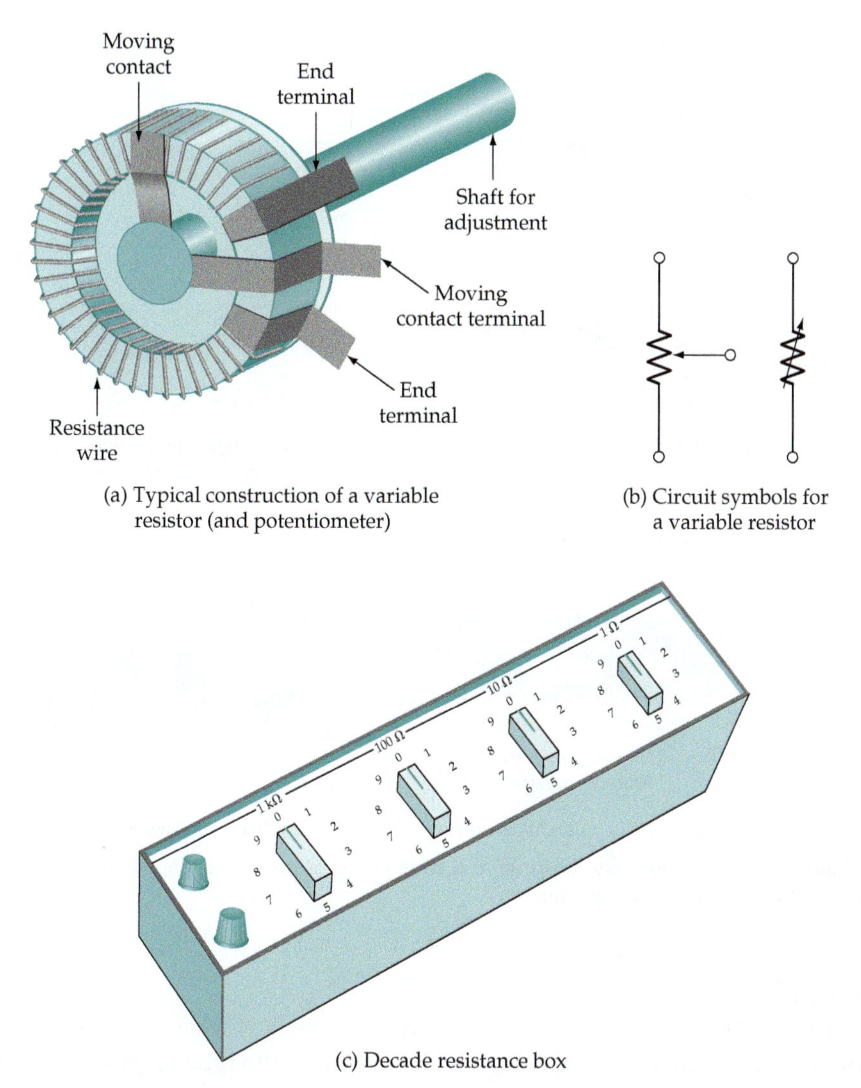

(a) Typical construction of a variable resistor (and potentiometer)

(b) Circuit symbols for a variable resistor

(c) Decade resistance box

Figure 4-14 Small variable resistors are used in electronic circuit construction. Large decade resistance boxes are employed in electronics laboratories.

Photoconductive Cell

The photoconductive cell is simply a resistor constructed of photoconductive material (cadmium selenide or cadmium sulfide) [Figure 4-15(b)]. When dark, the cell resistance is very high. When illuminated, the resistance decreases in proportion to the level of illumination.

Low-Power Variable Resistor

A small variable resistor suitable for mounting directly on a circuit board is shown in Figure 4-15(c). A threaded shaft, adjustable by a screwdriver, sets the position of the moving contact on a resistance wire.

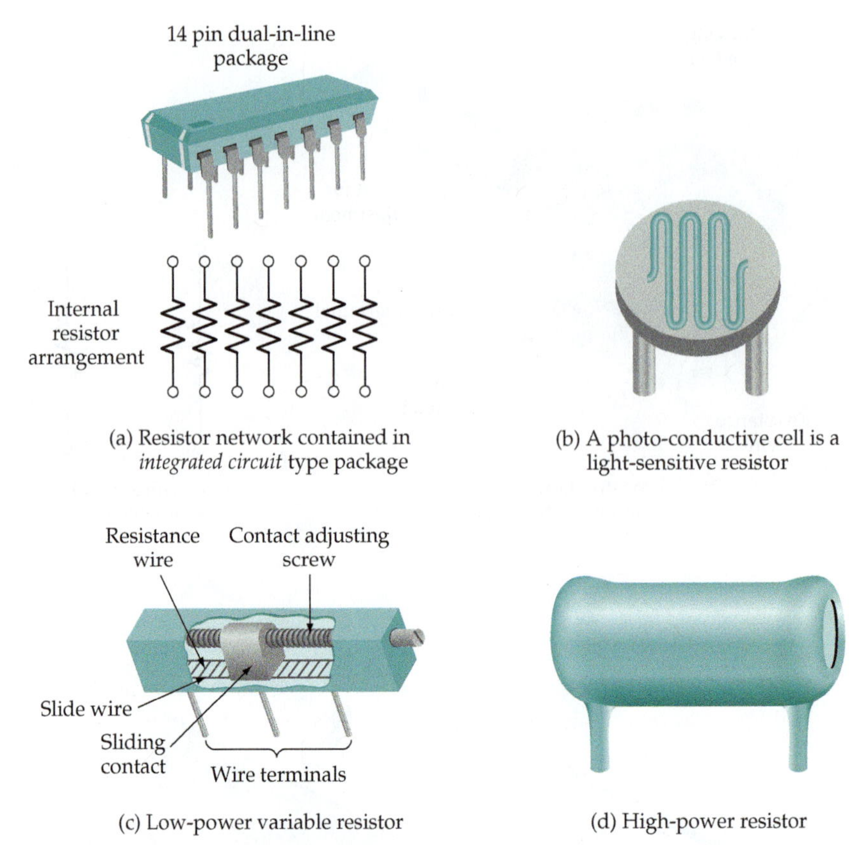

Figure 4-15 Resistors are available in a wide range of packages, depending upon their application and power dissipation.

High-Power Resistor

The type of high-power resistor shown in Figure 4-15(d) is usually wire-wound on the surface of a ceramic tube. Air flow through the tube helps to keep the resistor from overheating.

Resistor Power Ratings

Typical power ratings for wire-wound resistors start at 1 W and range to 10 W and much larger. Large surface areas are required to dissipate the heat generated, consequently, high power ratings result in physically large resistors. Carbon composition resistors are essentially low-power components; excessive power dissipation destroys them rapidly. The usual range of power ratings for carbon composition resistors is 1/8 W, 1/4 W, 1/2 W, 1 W, and 2 W (see Figure 4-16). The maximum current that may be permitted to flow through a resistor and the maximum voltage that may be applied across it are limited by the specified maximum power dissipation for the component. Using Equation 3-4 ($P = E^2/R$) and Equation 3-5 ($P = I^2R$) the maximum levels of current and voltage are easily calculated.

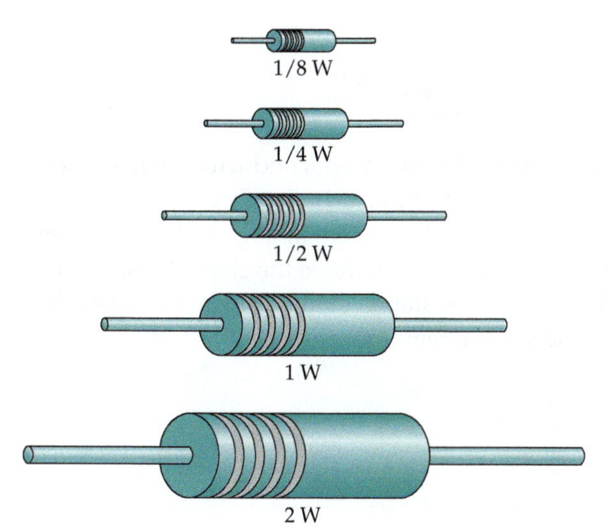

Figure 4-16 Relative size of carbon composition resistors with various power ratings.

Example 4-10

A 2.2 kΩ resistor has a specific maximum power dissipation of 1 W. Determine the maximum current that may be passed through the resistor and the maximum voltage that may be applied to its terminals.

Solution

Eq. 3-5 $$P = I^2R$$

so $$I_{(max)} = \sqrt{\frac{P_{(max)}}{R}} = \sqrt{\frac{1\ \text{W}}{2.2\ \text{k}\Omega}}$$

$$= 21.3\ \text{mA}$$

Eq. 3-4 $$P = \frac{E^2}{R}$$

so $$E_{(max)} = \sqrt{P_{(max)} \times R} = \sqrt{1\ \text{W} \times 2.2\ \text{k}\Omega}$$

$$= 46.9\ \text{V}$$

Practice Problems

4-5.1 Determine the maximum current that may be passed by 100 Ω and 100 kΩ, 1/8 W resistors.

4-5.2 A 6.8 Ω resistor is used in a circuit where it may have a maximum of 3.3 V across its terminals. Determine the minimum power rating for the resistor.

4-6 TEMPERATURE COEFFICIENT OF RESISTORS

Wire-wound resistors behave much the same as conductors when their temperature increases or decreases (see Section 4-4). So, they have the kind of resistance/temperature characteristic illustrated in Figure 4-10. Carbon composition resistors, on the other hand, usually have the characteristic illustrated in Figure 4-17. In this case the resistance tends to increase when the temperature varies above or below room temperature.

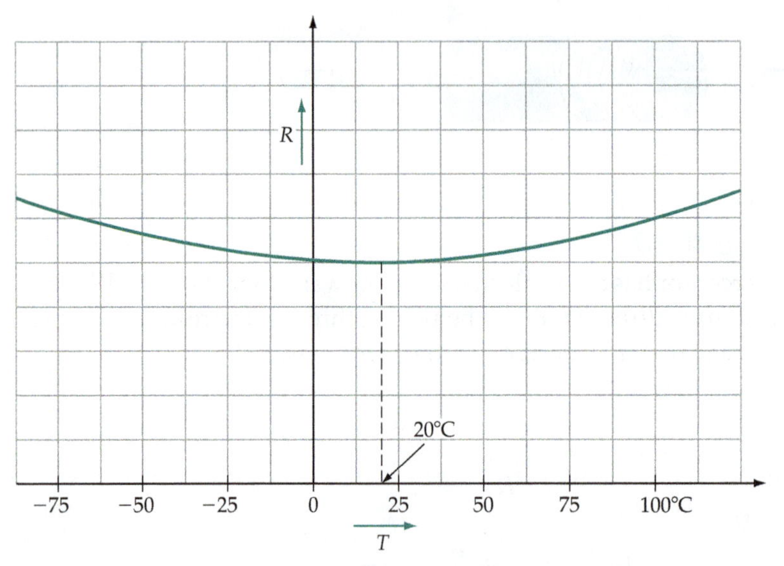

Figure 4-17 Typical variation of resistance with temperature change for carbon composition resistors. The temperature coefficient of resistors is usually expressed in parts per million per degree Celsius (ppm/°C).

The *temperature coefficient* (TC) of resistance is usually expressed in parts per million per degree Celsius (ppm/°C). To understand what this means, consider a resistor with a resistance of R Ω and a TC of ±100 ppm/°C. For every 1°C change in temperature, the maximum resistance change (ΔR) is a change of 100 parts in one million parts of R. Thus

$$\Delta R = \pm\frac{R}{1\,000\,000} \times \text{TC}$$

$$= \pm\frac{R}{1\,000\,000} \times (100 \text{ parts})$$

When the temperature change is ΔT °C, the resistance change is,

$$\Delta R = \pm\frac{R}{1\,000\,000} \times \text{TC} \times \Delta T \tag{4-4}$$

Example 4-11

Calculate the maximum change in resistance for a 470 Ω resistor with a TC of ±500 ppm/°C when its temperature increases from 20°C to 75°C.

Solution

$$\Delta T = 75°\text{C} - 20°\text{C} = 55°\text{C}$$

Eq. 4-4,

$$\Delta R = \pm\frac{R}{1\,000\,000} \times \text{TC} \times \Delta T$$

$$= \pm\frac{470\ \Omega}{1\,000\,000} \times 500 \times 55°\text{C}$$

$$\approx \pm 12.9\ \Omega$$

The resistance change calculated from the temperature coefficient is the maximum change in resistance that should occur over a given temperature range. With some resistors, the actual resistance change may be only a small fraction of the maximum, but other resistors of the same type may exhibit the maximum possible change There are circuit applications in which it is important that two (or more) resistors change by approximately the same proportion as the temperature increases or decreases. In this case it is said that the resistors are required to *track* each other. When resistors are constructed to fulfill this requirement, the manufacturer specifies the *matching* of the temperature coefficients. Sometimes the term *tracking temperature coefficient* is used. So, the TC for a certain type of resistor may be specified as:

Temperature coefficient = ±20 ppm/°C matched to ± 2 ppm/°C

or Tracking temperature coefficient = ±2 ppm/°C

Practice Problems

4-6.1 Two 2.2 kΩ resistors each have a temperature coefficient of ±200 ppm/°C. Determine the maximum resistance difference that could develop between the two due to a 35°C temperature change.

4-6.2 If the resistors in Problem 4-8.1 have a tracking temperature coefficient of ±10 ppm/°C, determine the maximum resistance difference that could develop due to the 35°C temperature change.

4-7 RESISTOR COLOR CODE

Because carbon composition and metal film resistors are physically small (some are less than 1 cm in length), it is not convenient to print the resistance value on the side. Instead, a *color code* in the form of colored bands is employed to identify the resistance value and tolerance. The color code (which uses the colors of the visible spectrum) is illustrated in Figure 4-18. Starting from one end of the resistor, the first two bands identify the first and second digits of the resistance value, and the third band indicates the number of zeros, as illustrated. An exception to this is when the third band is either silver or gold, which indicates a 0.01 or 0.1 multiplier, respectively. The fourth band is always either silver or gold, and in this position silver indicates a ±10% tolerance and gold indicates a ±5% tolerance. Where no fourth band is present, the resistor tolerance is ±20%.

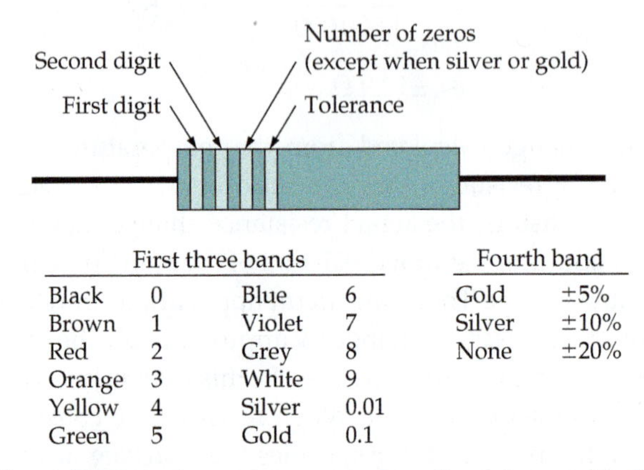

First three bands				Fourth band	
Black	0	Blue	6	Gold	±5%
Brown	1	Violet	7	Silver	±10%
Red	2	Grey	8	None	±20%
Orange	3	White	9		
Yellow	4	Silver	0.01		
Green	5	Gold	0.1		

Figure 4-18 The resistance value and accuracy of small resistors are identified by means of color-coded bands. The colors are numbered according to the visible spectrum.

Example 4-12

Identify the values of resistors with the following color codes: brown-black-red-silver, red-red-black-gold, and red-violet-gold-silver (see Figure 4-19).

Solution

Fig. 4-19(a)

$$\left.\begin{array}{cccc} \textit{brown} - & \textit{black} - & \textit{red} - & \textit{silver} \\ \downarrow & \downarrow & \downarrow & \downarrow \\ 1 & 0 & 00 & \pm 10\% \end{array}\right\} = 1000\ \Omega \pm 10\%$$

Fig. 4-19(b)

$$\left.\begin{array}{cccc} \textit{red} - & \textit{red} - & \textit{black} - & \textit{gold} \\ \downarrow & \downarrow & \downarrow & \downarrow \\ 2 & 2 & \overline{\textit{no zeros}} & \pm 5\% \end{array}\right\} = 22\ \Omega \pm 5\%$$

Fig. 4-19(c)

$$\left.\begin{array}{cccc} \textit{red} - & \textit{violet} - & \textit{gold} - & \textit{silver} \\ \downarrow & \downarrow & \downarrow & \downarrow \\ 2 & 7 & \overline{\times 0.1} & \pm 10\% \end{array}\right\} = 2.7\ \Omega \pm 10\%$$

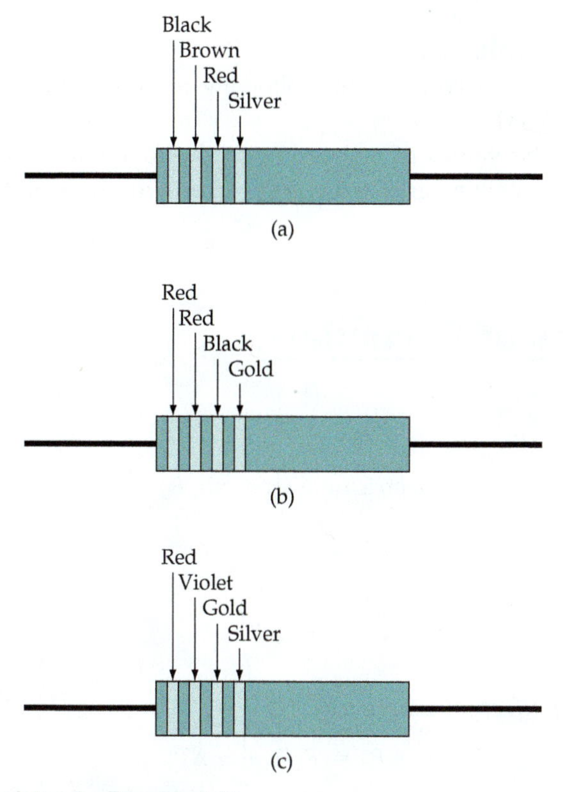

Figure 4-19 Resistors for Example 4-12.

COLOR CODE MEMORY AID

Black Bruins Relish Ornery Young Greenhorns.
Blue Violets Growing Wild Smell Good

Memory aid	Color	Number	Tolerance
Black	Black	0	
Bruins	Brown	1	
Relish	Red	2	
Ornery	Orange	3	
Young	Yellow	4	
Greenhorns	Green	5	
Blue	Blue	6	
Violets	Violet	7	
Growing	Gray	8	
Wild	White	9	
Smell	Silver	0.01	10%
Good	Gold	0.1	5%

Practice Problems

4-7.1 Write the color code for the following resistor values: 4.7 kΩ, 330 Ω, 1.5 MΩ, 5.6 Ω.

4-7.2 Identify the values of resistors with the following color codes: blue-gray-brown-silver, gray-red-red-silver, brown-gray-yellow-gold.

Summary of Formulas

- ***Insulator:***

$$\text{electric field strength, } \xi = \frac{E}{d}$$

- ***Resistance:***

$$R = \frac{\rho \ell}{a}$$

- ***Temperature effect on resistance:***

$$R_2 = R_1(1 + \alpha\, \Delta T)$$

- ***Resistance change with temperature:***

$$\Delta R = \pm \frac{R}{1\,000\,000} \times TC \times \Delta T$$

Review Questions

Section 4-1

4-1 Explain the function of a conductor and discuss conducting materials.

4-2 Write an equation for the voltage drop along a conductor with a given current and resistance/unit length.

Section 4-2

4-3 Explain the function of insulators and discuss insulating materials. Explain insulator breakdown.

4-4 Define electric field strength and write an equation for calculating electric field strength.

Section 4-3

4-5 Define specific resistance. Write an equation relating the resistance of a conductor to its length and cross-sectional area. Identify the units for each quantity in the equation.

Section 4-4

4-6 Using illustrations, explain how temperature affects the resistance of a metallic conductor.

4-7 Define temperature coefficient for conductors.

4-8 Write an equation relating conductor resistance at an elevated temperature to resistance at 20°C.

Section 4-5

4-9 Describe the two most common types of electronic resistor construction and discuss their relative advantages and disadvantages.

4-10 Sketch the construction of a variable resistor, identify each part of the device, and explain its operation.

4-11 Explain resistor tolerance.

4-12 Briefly explain the importance of using a resistor with the correct power rating.

4-13 Write equations for the maximum current and voltage that may be used with a resistor with a given power rating.

Section 4-6

4-14 Explain what is meant by TC = ±100 ppm/°C. Sketch the approximate shape of the resistance/temperature characteristic of a carbon composition resistor.

4-15 Define tracking temperature coefficient.

Section 4-7

4-16 Draw a sketch to show how resistors are color-coded. List the colors, and identify the significance of each.

Problems

Section 4-1

4-1 An electronic circuit that takes a current of 470 mA is connected to a power supply via two 10 m conductors. If the total voltage drop along the conductors is not to exceed 100 mV, determine the gauge of suitable copper wire conductors.

4-2 An electric heater is supplied with 5 A of current from a 120 V source. The cables used to connect the heater are each 50 m long and have conductor resistances of 0.01 Ω/m. Calculate the potential difference at the heater terminals and the power dissipated in the cables.

4-3 The conductors in the cables in Problem 4-2 are replaced with #14 copper wire. Calculate the new terminal voltage at the heater and the power dissipated in the cables.

4-4 An automobile starter motor is connected to a 12 V battery via the chassis and a 1 m long insulated copper conductor. The motor takes a current of

80 A, and its terminal voltage is to be not less than 11.5 V. Assuming that the chassis resistance is negligible, select a suitable wire gauge for the copper conductor.

4-5 An electric lamp has a terminal voltage of 113.6 V and a current of 1.13 A when connected to a 115 V supply via a 75 m twin-conductor cable. Determine the wire gauge of the conductor.

Section 4-2

4-6 Two conductors 0.5 cm apart in an electric cable have a potential difference of 500 V. Calculate the electric field strength in the insulation between the conductors.

4-7 Determine the minimum voltage that could cause breakdown in 4 mm of rubber insulation that separates two conductors.

4-8 The electric field strength in the insulation between the outer metal plate and the heating element in an electric smoothing iron is not to exceed 1000 kV/cm. Calculate the minimum insulation thickness if a 115 V supply is used.

4-9 Two conductors in a single cable have 0.9 cm of rubber insulation between them and a potential difference of 450 V. Calculate the electric field strength in the insulation, and determine the minimum voltage that might cause insulation breakdown.

4-10 Two terminals are placed 1.5 cm apart in air. Referring to Table 4-1, calculate the minimum potential difference that might produce arcing between the terminals.

4-11 Copper conecting bars that have a potential of 120 V with respect to ground are supported by porcelain insulators 2 cm thick. Determine the electric field strength in the porcelain, and calculate the minimum voltage that could cause breakdown (a) in the insulators and (b) in the surrounding air.

4-12 Two conductors are separated by 0.2 cm of rubber insulation. Determine the minimum potential difference between them that could cause insulation breakdown.

4-13 A porcelain insulator supports overhead electric cables 15 cm underneath horizontal bars on a grounded metal pylon. If the surrounding air sometimes breaks down allowing current to leak along the surface of the insulator, estimate the cable voltage with respect to ground.

4-14 The glass insulation between a heating element and a metal plate broke down when a 500 V potential difference was applied. Calculate the thickness of the glass.

Section 4-3

4-15 Calculate the resistance per meter of copper conductors with the following diameters: 1 mm, 2 mm, 3 mm, 4 mm, and 5 mm.

4-16 The resistance of a 65 cm length of wire 1 mm in diameter is measured as 0.0136 Ω. Determine the material of the wire.

4-17 Determine the resistance of copper and aluminum conductors that are 6 mm in diameter and 100 m long.

4-18 A strip of copper on a printed circuit board is 10 cm long and 0.3 cm wide. Determine the thickness of the copper strip if its resistance is measured as 0.0035 Ω.

4-19 A piece of electronic equipment takes 2.5 A from a 120 V supply. The cables connecting the supply to the equipment are 10 m long. If the terminal voltage at the equipment is to be not less than 115 V, determine the minimum diameter of suitable copper conductors.

4-20 Copper wire with a #13 gauge is to be replaced by aluminum wire having an equal or lower resistance. Select a suitable gauge for the aluminum wire.

4-21 The copper conductors on a printed circuit board are 0.05 mm thick and 2 mm wide. Determine the voltage drop per centimeter along the conductors when carrying a 300 mA current.

4-22 A 0.001 Ω ammeter shunt is to be constructed from a copper conductor which is 1 cm wide and 2 mm thick. Calculate the required length of conductor.

4-23 A 750 W electric motor is to be connected to a 130 V supply via two cables that are each 64 m in length. If the motor terminal voltage is to be not less than 125 V, determine the minimum thickness of the copper conductors in the cables.

4-24 A 330 Ω resistor is to be constructed of #40 gauge constantan wire. Determine the required length of wire.

Section 4-4

4-25 An aluminum conductor has a resistance of 10 Ω at 20°C. Determine the resistance of the conductor at 100°C.

4-26 The printed circuit board conductors described in Problem 4-21 are raised to a temperature of 65°C from the normal temperature level of 20°C. Determine the new voltage drop per centimeter when the current is 300 mA.

4-27 A coil of nickel wire is used to monitor the temperature of a liquid in which it is submerged. The coil resistance is 500 Ω at 20°C. Prepare a table of coil resistance values for liquid temperatures at 10° intervals ranging from 0°C to 100°C.

4-28 Two lengths of wire each have resistances of 1 kΩ at 20°C. At 75°C the resistance values are measured as $R_a = 1.209$ kΩ and $R_b = 1.187$ kΩ. Calculate the temperature coefficient of each wire and identify the materials.

4-29 Two conductors supplying current to a 250 V, 10 kW electric kiln normally operate at an average temperature of 150°C. If the conductors are each 9 m long and the total voltage drop along them is not to exceed 4 V, select a suitable wire gauge (from Appendix 4) for copper conductors.

4-30 A current of 50 A is applied to the starter motor of an automobile. The terminal voltage of the motor is to be not less than 11 V. Calculate the minimum thickness, and select a suitable wire gauge from Appendix 4, for two 3 m copper conductors that connect the starter to the 12 V battery. The maximum temperature of the conductors is 50°C.

Section 4-5

4-31 Calculate the maximum current and voltage limits for (a) a 1/4 W, 820 Ω resistor and (b) a 1/2 W, 1 MΩ resistor.

4-32 Calculate the minimum power rating for each of the following resistors, when they are used in a circuit in which they will have a 30 V potential diference applied across their terminals: 33 kΩ, 2.2 kΩ, 560 Ω, 1.2 kΩ, 180 Ω.

4-33 Calculate the minimum power rating for each of the following resistors, when they are used in a circuit in which they carry a maximum current of 33 mA: 270 Ω, 4.7 kΩ, 820 Ω, 5.6 kΩ, 22 Ω.

Section 4-6

4-34 A 560 Ω resistor with a TC of ±200 ppm/°C has its temperature raised from 20°C to 55°C. Determine the resistance change that occurs.

4-35 A 3.3 kΩ precision resistor with a TC of 10 ppm/°C is to operate over a temperature range of 20°C to 100°C. Calculate the maximum change in resistance that could occur over the temperature range.

4-36 Two 3.9 kΩ resistors are to track each other within 0.01% over a temperature change of 50°C. Calculate the required tracking temperature coefficient.

4-37 Calculate the new value of resistance for each of the resistors referred to in Problem 4-31 when its temperature changes from 20°C to 75°C. Take the TC of each resistor as ±150 ppm/°C.

Section 4-7

4-38 Identify the values of resistors with the following color codes: green-blue-orange, gray-red-brown-silver, and orange-white-green-gold.

4-39 State the identifying colors for the following 10% tolerance resistors: 47 kΩ, 22 Ω, 1 MΩ, 820 Ω, 330 Ω, 5.6 Ω.

Practice Problem Answers

4-1.1 110.8 V
4-1.2 #22
4-1.3 70.2 W, 4.2 W
4-2.1 225 V/cm
4-2.2 12 kV
4-2.3 22 kV/cm, no

4-3.1	8.24 Ω/km
4-3.2	0.124 Ω
4-3.3	1.4 mm
4-4.1	143.5 Ω
4-4.2	1.51 kΩ
4-4.3	0.0034 Ω/°C, gold
4-5.1	35.3 mA, 1.1 mA
4-5.2	1.6 W
4-6.1	30.8 Ω
4-6.2	0.77 Ω
4-7.1	Yellow-violet-red, orange-orange-brown, brown-green-green, green-blue-gold
4-7.2	680 Ω ± 10%, 8.2 kΩ ± 10%, 180 kΩ ± 5%

CHAPTER 5
Series Resistor Circuits

CONTENTS

Objectives

You will be able to:

1. Sketch circuit diagrams for series-connected resistors, series-connected lamps, and series-connected voltage sources.
2. Indicate current directions and voltage polarities throughout series circuits.
3. Write equations for total equivalent resistance, current level, and voltage drops in series resistive circuits.
4. State Kirchhoff's voltage law.
5. Solve problems involving equivalent resistance, current levels, voltage drops, and power dissipations in series resistive circuits.
6. Explain voltage dividers and potentiometers and solve problems involving voltage dividers and potentiometers.
7. Explain voltage-dropping resistors and current-limiting resistors and solve problems involving such components.
8. Discuss the effects of open-circuits and short-circuits in series resistive circuits and solve problems involving open-circuits and short-circuits.

INTRODUCTION

Resistors are said to be *in series* when they are connected in such a way that there is only one path through which current can flow. This means that the current in a series circuit is the same in all parts of the circuit. The voltage drop across each component in a series circuit depends on the current level and the component resistance. Two or more series-connected resistors can be used as a voltage divider, and a potentiometer is an adjustable resistor used as a variable voltage divider. The total power supplied to a series circuit is the sum of the powers dissipated in the individual components. Resistors may be connected in series with other components for the purpose of voltage dropping or current limiting.

5-1 SERIES RESISTOR CIRCUIT

Series-Connected Resistors

Three series-connected resistors are shown in Figure 5-1(a) with a power supply to provide electromotive force (emf). (DC power supplies are discussed in Chapter 10.) The resistors are mounted on a plug-in-type laboratory breadboard, in which the horizontal groups of sockets are connected together, as illustrated. The conductor from the positive terminal of the power supply is connected to the top of resistor R_1. The bottom of R_1 is connected to the top of R_2, the bottom of R_2 is connected to the top of R_3, and the bottom of R_3 has a conductor connecting it to the power supply negative terminal.

In Figure 5-1(b), three series-connected resistors are shown on a *printed circuit board*. In this case, the resistors are connected to thin strips of copper on the surface of the board, which is made of non-conducting material. Here again, the resistors are connected end-to-end, R_1 to R_2, R_2 to R_3.

Equivalent Resistance

The circuit diagram for three series-connected resistors and a voltage source is shown in Figure 5-1(c). The total resistance connected across the voltage source is,

$$R = R_1 + R_2 + R_3$$

The total is also called the *equivalent resistance* for the circuit. For any series circuit with a number of resistors (n), the equivalent resistance is

$$R_{eq} = R_1 + R_2 + R_3 + \cdots + R_n \quad (5\text{-}1)$$

For a circuit consisting of n equal-value resistors,

$$R_{eq} = n \times R_1$$

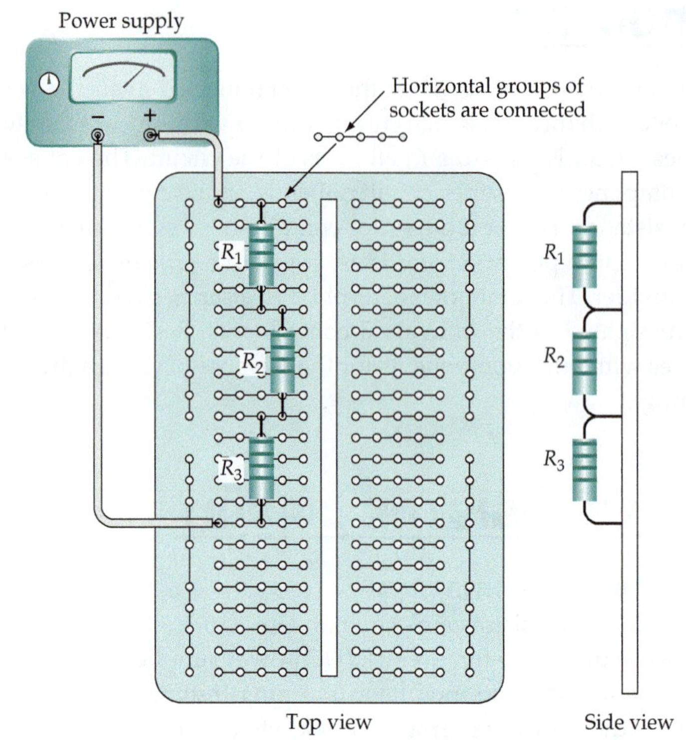

(a) Three series-connected resistors on a plug-in-type laboratory breadboard

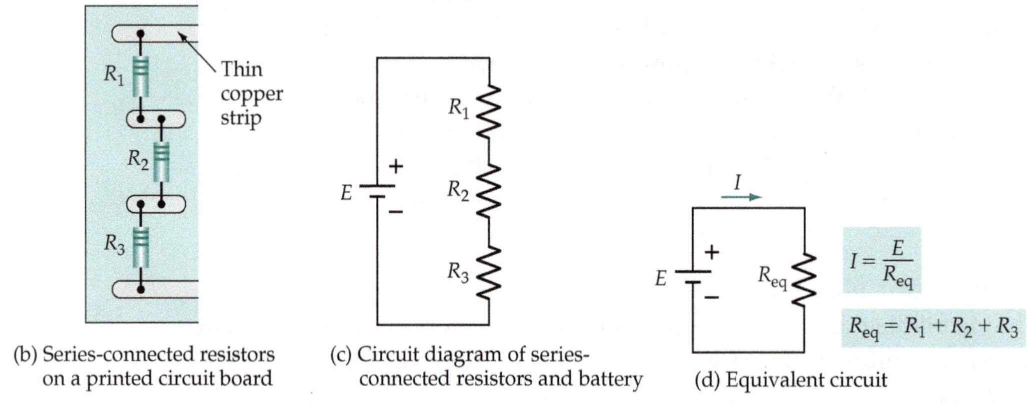

(b) Series-connected resistors on a printed circuit board

(c) Circuit diagram of series-connected resistors and battery

(d) Equivalent circuit

Figure 5-1 Resistors in series are connected end-to-end so that the total resistance (or equivalent resistance) is $R_{eq} = R_1 + R_2 + R_3 + \cdots R_n$.

For example, the equivalent resistance for ten 1 kΩ resistors connected in series would be 10 kΩ. Figure 5-1(d) shows the *equivalent circuit* for the series resistor circuit in Figure 5-1(c). The equivalent circuit simply consists of the voltage source and the equivalent resistance.

Current Levels

An ammeter connected to measure the current flowing in a series circuit is shown in Figure 5-2(a), and the circuit diagram is illustrated in Figure 5-2(b).

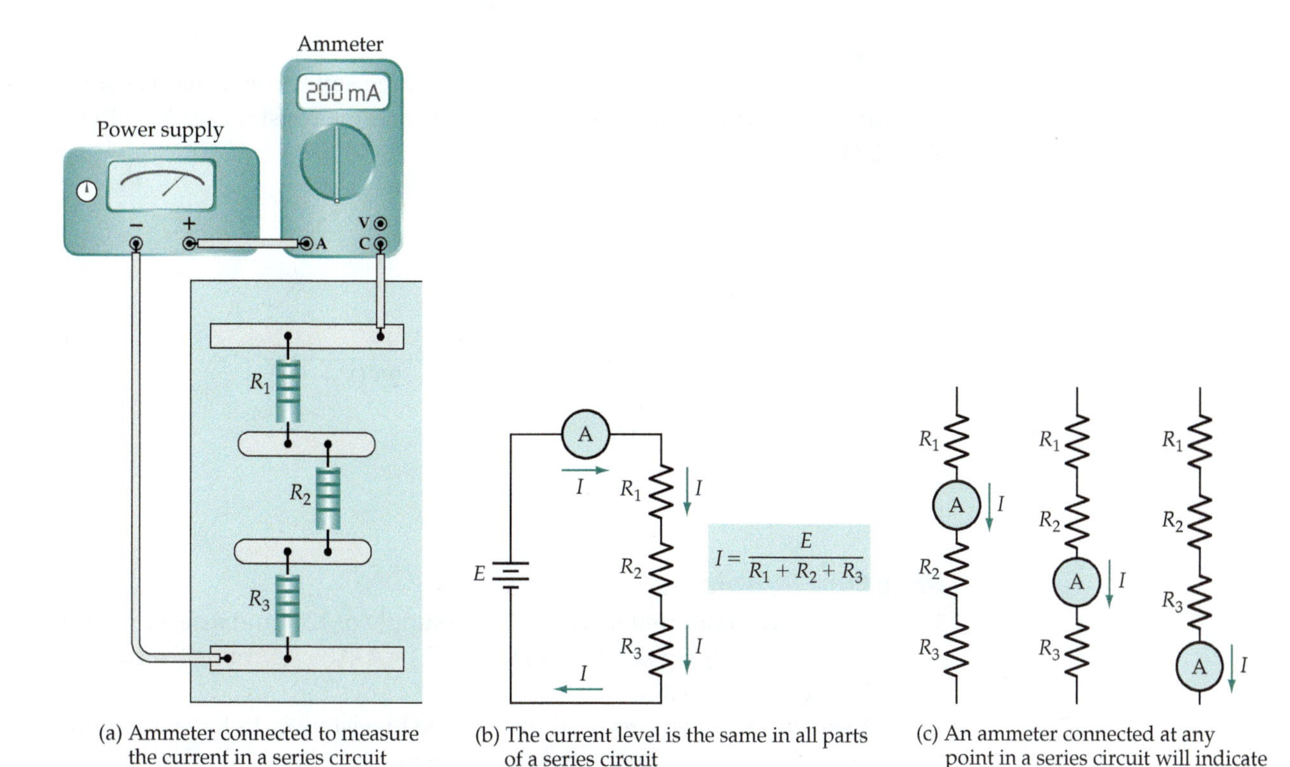

(a) Ammeter connected to measure the current in a series circuit

(b) The current level is the same in all parts of a series circuit

(c) An ammeter connected at any point in a series circuit will indicate the same level of current

Figure 5-2 There is only one path for current flow in a series circuit. Consequently, the current is the same in all parts of a series circuit.

Because the resistors are connected end-to-end, there is only one path for current flow in the circuit. Current flows from the positive terminal of the voltage source, through the ammeter, and into the top terminal of resistor R_1. Clearly, all of the current that flows into the one end of R_1 must flow out of the other end. So, the current flows out of the bottom terminal of R_1 into the top terminal of R_2, and from R_2 it moves through R_3 to the negative terminal of the voltage source.

Figure 5-2(c) shows that an ammeter connected in any part of the circuit would measure the same current level. It is seen that,

The current level is the same in all parts of a series circuit.

Using Ohm's law, the current through the series circuit is calculated as,

$$I = \frac{E}{R_1 + R_2 + R_3 + \cdots + R_n} \tag{5-2}$$

or

$$I = \frac{E}{R_{eq}}$$

Example 5-1

Calculate the current that flows through the three series-connected resistors in Figure 5-2(b). The supply voltage is $E = 9$ V, and the resistors are $R_1 = 15\ \Omega$, $R_2 = 25\ \Omega$, and $R_3 = 5\ \Omega$.

Solution

From Equation 5-2,

$$I = \frac{E}{R_1 + R_2 + R_3} = \frac{9\text{ V}}{15\ \Omega + 25\ \Omega + 5\ \Omega}$$

$$= 200\text{ mA}$$

Practice Problems

5-1.1 Four series-connected resistors have a supply of 12 V and resistor values of $R_1 = 7\text{ k}\Omega$, $R_2 = 5\text{ k}\Omega$, $R_3 = 8\text{ k}\Omega$, and $R_4 = 4\text{ k}\Omega$. Determine the current through each resistor.

5-1.2 A series-connected group of three 3.3 kΩ resistors is to have a current level of 2.7 mA. Calculate the required supply voltage.

5-2 VOLTAGES IN A SERIES CIRCUIT (Kirchhoff's Voltage Law)

Resistor Voltage Drops

The three-resistor series circuit is reproduced again in Figure 5-3(a) with the addition of a voltmeter to measure the voltage drop across R_1. The circuit diagram is redrawn in Figure 5-3(b). It is seen that the current flow causes a voltage drop, or potential difference, across each resistor. If there was no potential difference between the terminals of each resistor, there would be no current flow. Using Ohm's law, the voltage drops across each resistor are

$$V_1 = IR_1,\ V_2 = IR_2,\text{ and } V_3 = IR_3$$

Note that the polarity of the resistor voltage drops is always such that the (conventional) current direction is from positive to negative. So, for the circuit as shown, the polarity is plus (+) at the top of each resistor, minus (−) at the bottom. Also, note that the positive end of each resistor is the end nearest the positive terminal of the voltage source (via the current path). The negative end of each resistor is closest to the voltage source negative terminal. The sum of

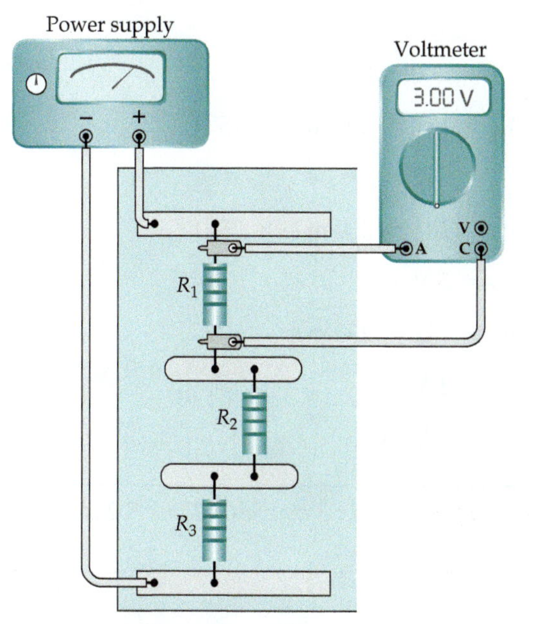

(a) Voltmeter connected to measure the voltage across R_1

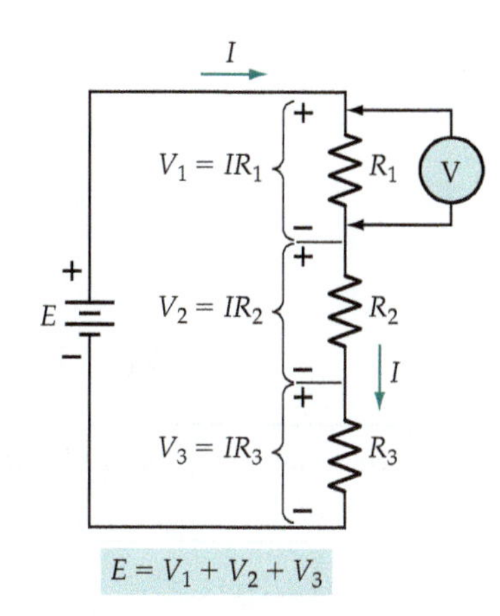

(b) The supply voltage equals the sum of the resistor voltage drops

Figure 5-3 The current in a series circuit produces voltage drops across each resistor: $V_1 = IR_1$, $V_2 = IR_2$, and $V_3 = IR_3$.

the resistor voltage drops is $V_1 + V_2 + V_3$, and, as shown in Figure 5-3(b), this must equal the applied emf E. For any series circuit with n resistors,

$$E = V_1 + V_2 + V_3 + \cdots + V_n$$

or

$$E = IR_1 + IR_2 + IR_3 + \cdots + IR_n \tag{5-3}$$

Therefore,

$$E = I(R_1 + R_2 + R_3 + \cdots + R_n) \tag{5-4}$$

Kirchhoff's Voltage Law

The relationship between the applied emf and the resistor voltage drops in a series circuit is defined by *Kirchhoff's voltage law*[1] (KVL):

> **In any closed electric circuit, the algebraic sum of the voltage drops must equal the algebraic sum of the applied emfs.**

Example 5-2

Using the component values given in Example 5-1, determine the voltage drops across each resistor in the circuit in Figure 5-3(b).

---

[1]Formulated by German physicist Gustav Kirchhoff (1824–1887).

Solution

From Example 5-1,

$$I = 200\text{ mA}$$

$$V_1 = IR_1 = 200\text{ mA} \times 15\ \Omega$$

$$= 3\text{ V}$$

$$V_2 = IR_2 = 200\text{ mA} \times 25\ \Omega$$

$$= 5\text{ V}$$

$$V_3 = IR_3 = 200\text{ mA} \times 5\ \Omega$$

$$= 1\text{ V}$$

$$V_1 + V_2 + V_3 = 9\text{ V} = E$$

Computer analysis of the kind of problem presented in Examples 5-1 and 5-2 is offered in Section 28-1.

Circuit Ground

Figure 5-4 shows a series resistor circuit with the negative supply terminal grounded. In this case, it is normal to measure the voltage at each junction in the circuit with respect to ground, as illustrated. The voltages measured are V_3, $V_2 + V_3$, and $V_1 + V_2 + V_3$, and the individual resistor voltages are found by subtraction $[V_2 = (V_2 + V_3) - V_3]$. This approach to voltage measurement is normally not necessary when a portable battery-operated voltmeter is used. However, some electronic instruments have one terminal grounded, and this can affect the current and voltage levels in a circuit under test.

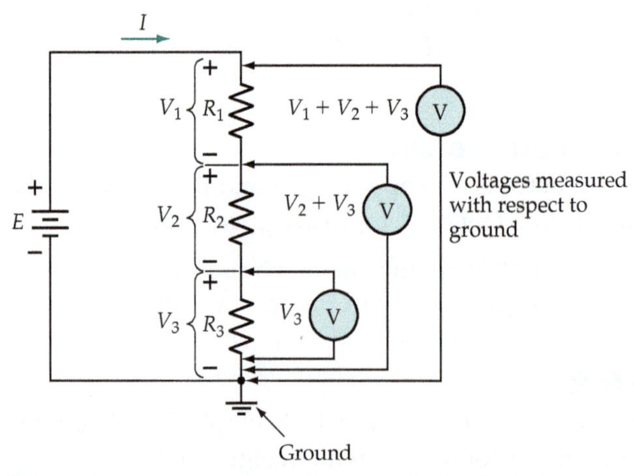

Figure 5-4 When a ground is present in a resistor circuit, all voltages are normally measured with respect to ground.

Series-Connected Voltage Sources

The three series-connected voltage cells in Figure 5-5(a) are arranged so that they all produce current in the same direction when applied to a circuit. The terminal voltages add together to give,

$$E = E_1 + E_2 + E_3$$

Because the voltage sources assist each other produce current, they are said to be connected in a *series-aiding* sequence.

In Figure 5-5(b) the bottom cell of the three has its negative terminal connected to the negative terminal of the middle cell. So, as illustrated by the circuit diagram for the cells, the total voltage is

$$E = E_1 + E_2 - E_3$$

In this case, because of its terminal polarity the bottom cell will attempt to produce current in the opposite direction to that from the other two cells. Consequently, the bottom cell is said to be connected in a *series-opposing* sequence with the top two cells.

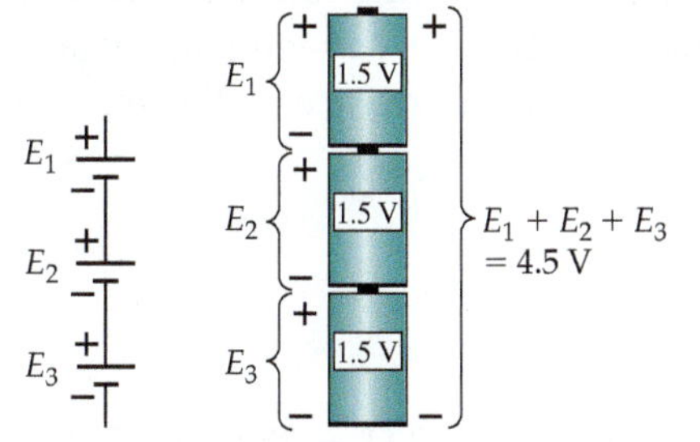

(a) Three voltage cells connected series-aiding

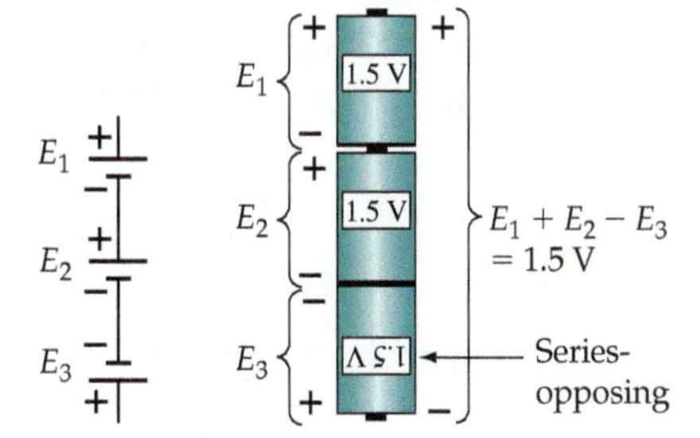

(b) Two series-aiding voltage cells and one series-opposing cell

Figure 5-5 When voltage sources are connected in series-aiding sequence, the terminal voltages add together. When one source is connected series-opposing, its terminal voltage subtracts from the other source terminal voltages.

Where more than one battery or other source of *emf* is used in a circuit, Kirchhoff's voltage law still applies. Consequently, for the circuit shown in Figure 5-6(a) and (b)

$$E_1 + E_2 = IR_1 + IR_2 + IR_3 + IR_4$$

The voltage equation for the circuit in Figure 5-6(c) is

$$E_1 - E_2 = IR_1 + IR_2 + IR_3 + IR_4$$

Analysis Procedure for a Series Resistor Circuit

1. *Determine the total applied voltage,* $E = E_1 + E_2 + \cdots$
2. *Calculate the total series resistance, Equation (5-1).*
3. *Calculate the circuit current, Equation (5-2).*
4. *Determine the voltage drop across each component,* $V_1 = IR_1$*, etc.*

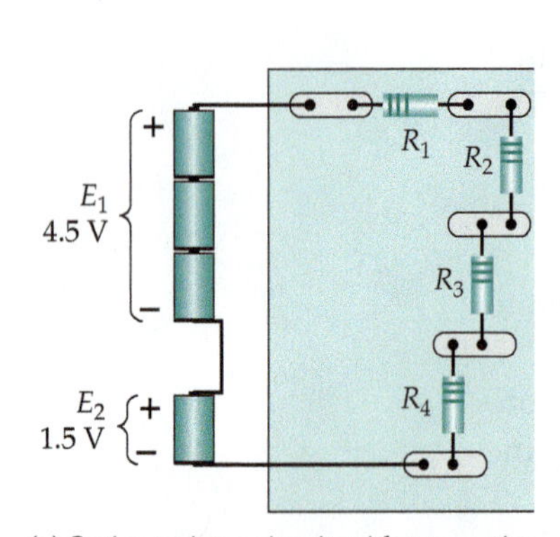

(a) Series resistor circuit with two series-aiding voltage sources

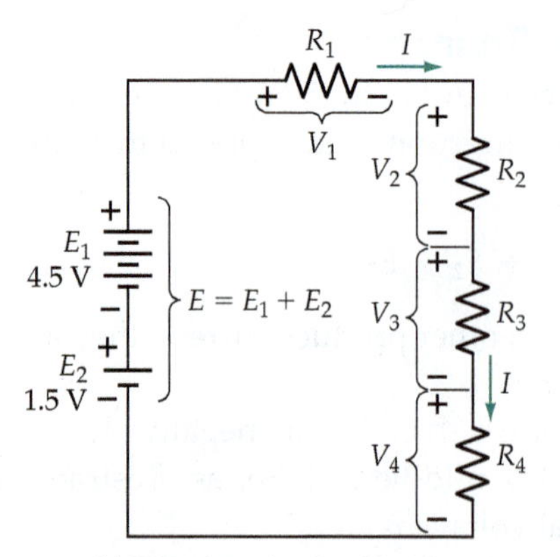

(b) Diagram of a circuit with series-aiding voltage sources

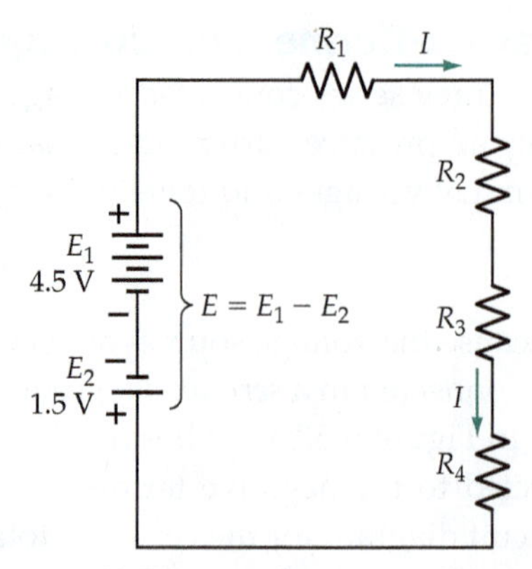

(c) Diagram of a circuit with series-opposing voltage sources

Figure 5-6 Series resistor circuit with two sources of emf (voltage sources). The sources may assist each other (series-aiding), as in (a) and (b), or oppose each other (series-opposing), as in (c).

Example 5-3

The four resistors in the circuit in Figure 5-6(b) have the following values: $R_1 = 5\ \Omega$, $R_2 = 13\ \Omega$, $R_3 = 25\ \Omega$, and $R_4 = 17\ \Omega$. The voltage sources are: $E_1 = 4.5$ V and $E_2 = 1.5$ V. Determine the circuit current and the resistor voltage drops.

Solution

$$E_1 + E_2 = 4.5\text{ V} + 1.5\text{ V} = 6\text{ V}$$

Eq. 5-1,

$$R_{eq} = R_1 + R_2 + R_3 + R_4 = 5\ \Omega + 13\ \Omega + 25\ \Omega + 17\ \Omega$$
$$= 60\ \Omega$$

Eq. 5.2,

$$I = \frac{E_1 + E_2}{R_{eq}} = \frac{6\text{ V}}{60\ \Omega}$$
$$= 0.1\text{ A}$$

$$V_1 = IR_1 = 0.1\text{ A} \times 5\ \Omega$$
$$= 0.5\text{ V}$$

$$V_2 = IR_2 = 0.1\text{ A} \times 13\ \Omega$$
$$= 1.3\text{ V}$$

$$V_3 = IR_3 = 0.1\text{ A} \times 25\ \Omega$$
$$= 2.5\text{ V}$$

$$V_4 = IR_4 = 0.1\text{ A} \times 17\ \Omega$$
$$= 1.7\text{ V}$$

Verifying Kirchhoff's voltage law:

$$E_1 + E_2 = V_1 + V_2 + V_3 + V_4$$

$$4.5\text{ V} + 1.5\text{ V} = 0.5\text{ V} + 1.3\text{ V} + 2.5\text{ V} + 1.7\text{ V}$$

$$6\text{ V} = 6\text{ V}$$

Example 5-4

Using the component values given in Example 5-3, determine the current and voltage levels when E_2 is reversed, as illustrated in Figure 5-6(c).

Solution

$$E_1 + E_2 = 4.5\text{ V} + (-1.5\text{ V}) = 3\text{ V}$$

$$R_{eq} = 60\ \Omega$$

From Eq. 5-2,

$$I = \frac{E_1 + E_2}{R_{eq}} = \frac{3\text{ V}}{60\ \Omega}$$

$$= 50\text{ mA}$$

$$V_1 = IR_1 = 50\text{ mA} \times 5\ \Omega$$

$$= 250\text{ mV}$$

$$V_2 = IR_2 = 50\text{ mA} \times 13\ \Omega$$

$$= 650\text{ mV}$$

$$V_3 = IR_3 = 50\text{ mA} \times 25\ \Omega$$

$$= 1.25\text{ V}$$

$$V_4 = IR_4 = 50\text{ mA} \times 17\ \Omega$$

$$= 850\text{ mV}$$

Practice Problems

5-2.1 Calculate the voltage drops across each resistor in Problem 5-1.1.

5-2.2 If a 4.5 V battery is connected in series-opposing sequence with the 12 V supply in Problem 5-1.1, determine the new current level and the voltage drop across each resistor.

5-3 VOLTAGE DIVIDER

Circuit and Equations

It has been shown that the voltage drops across a *string* of series-connected resistors add up to the value of the supply voltage E. Another way of looking at this is that the applied voltage is divided up between the resistors.

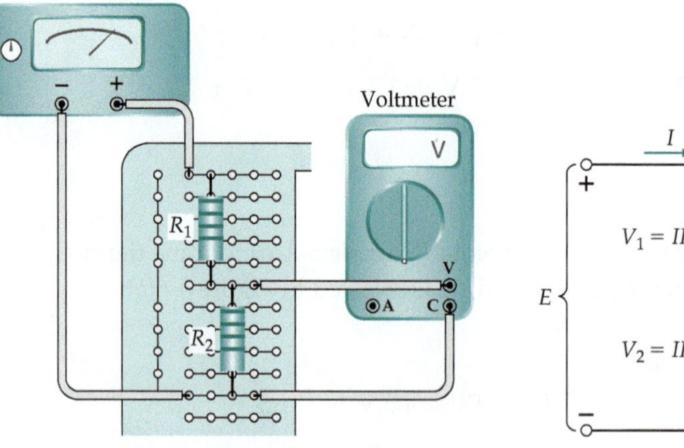

(a) Two resistors connected to function as a voltage divider

(b) Voltage divider circuit diagram

Figure 5-7 Two series-connected resistors can be used as a voltage divider. The circuit current is $I = E/(R_1 + R_2)$, and the resistor voltages are IR_1 and IR_2, or $V_1 = ER_1/(R_1 + R_2)$, and $V_2 = ER_2/(R_1 + R_2)$.

Figure 5-7 shows two series-connected resistors used as a *voltage divider* or *potential divider*. From previous studies

$$I = \frac{E}{R_1 + R_2}$$

Also
$$V_1 = IR_1$$

So
$$V_1 = \frac{E}{R_1 + R_2} \times R_1$$

Or
$$V_1 = E \times \frac{R_1}{R_1 + R_2} \quad \text{(5-5)}$$

And
$$V_2 = E \times \frac{R_2}{R_1 + R_2}$$

Also, if $R_1 = R_2$
$$V_1 = V_2 = \frac{E}{2}$$

Where more than two series-connected resistors are involved, the voltage drop across any one resistor R_n is

$$V_n = E \times \frac{R_n}{R_1 + R_2 + R_3 + \cdots + R_n} \quad \text{(5-6)}$$

When there are n equal-value resistors in series,

$$V_1 = V_2 = V_3 = \cdots \frac{E}{n}$$

Voltage-Divider Theorem

In a series resistor circuit, the portion of applied emf developed across any one resistor is the ratio of the resistor value to the total series resistance.

The voltage-divider theorem (illustrated by Equations 5-5 and 5-6) is important because it is applied over and over again in electronic circuits. A surprisingly large amount of electronic circuit design merely involves the selection of appropriate resistor values for voltage-divider networks.

Example 5-5

For the voltage-divider circuit in Figure 5-7, the applied *emf* is $E = 100$ V, and the resistor values are $R_1 = 22\ \Omega$ and $R_2 = 28\ \Omega$. Calculate the values of V_1 and V_2.

Solution

Eq. 5-5
$$V_1 = E \times \frac{R_1}{R_1 + R_2} = 100\text{ V} \times \frac{22\ \Omega}{28\ \Omega + 22\ \Omega}$$
$$= 44\text{ V}$$
$$V_2 = E \times \frac{R_2}{R_1 + R_2} = 100\text{ V} \times \frac{28\ \Omega}{28\ \Omega + 22\ \Omega}$$
$$= 56\text{ V}$$

Example 5-6

Apply the voltage-divider theorem to determine the voltages across each resistor in Figure 5-1, as reproduced in Figure 5-8. The circuit quantities are: $E = 9$ V, $R_1 = 15\ \Omega$, $R_2 = 25\ \Omega$, $R_3 = 5\ \Omega$.

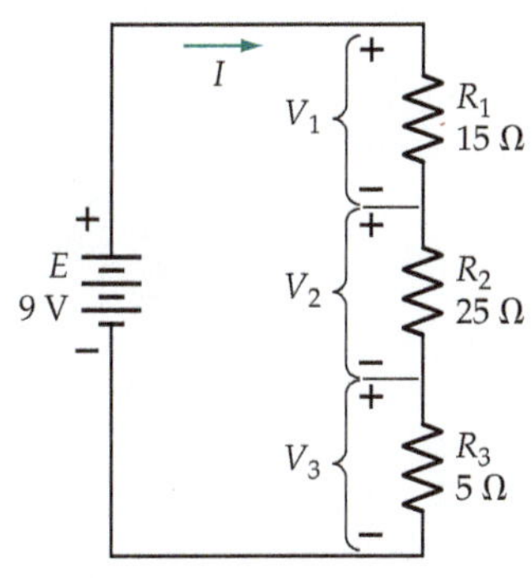

Figure 5-8 Series resistor circuit for Example 5-6.

Solution

Eq. 5-6
$$V_1 = E \times \frac{R_1}{R_1 + R_2 + R_3} = 9\text{ V} \times \frac{15\ \Omega}{15\ \Omega + 25\ \Omega + 5\ \Omega}$$
$$= 3\text{ V}$$
$$V_2 = E \times \frac{R_2}{R_1 + R_2 + R_3} = 9\text{ V} \times \frac{25\ \Omega}{15\ \Omega + 25\ \Omega + 5\ \Omega}$$
$$= 5\text{ V}$$
$$V_3 = E \times \frac{R_3}{R_1 + R_2 + R_3} = 9\text{ V} \times \frac{5\ \Omega}{15\ \Omega + 25\ \Omega + 5\ \Omega}$$
$$= 1\text{ V}$$

Compare these results to those obtained in Example 5-2.

Voltage-Divider Design

Equations 5-5 and 5-6 are correct only if there is no output (load) current from the junction of the resistors. This is because, as illustrated in Figure 5-9, output current I_o flows through resistor R_1, but not through R_2.

So $$V_{R1} = I_1R_1 = R_1(I_2 + I_o)$$

And $$V_{R2} = I_2R_2$$

If the resistor current is very much larger than the output current ($I_2 \gg I_o$), then $(I_2 + I_o) \approx I_2$, and Equations 5-5 and 5-6 are approximately correct.

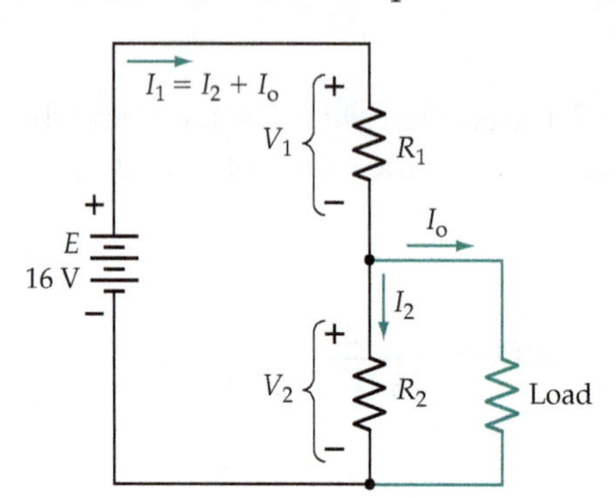

Figure 5-9 The resistor voltages in a voltage divider are affected by the presence of an output current I_o. The effect is minimized by making $I_2 \gg I_o$.

Design of a voltage divider circuit normally commences with selection of I_2 very much larger than I_o. A ratio of 100:1 is frequently used. The resistor values are then calculated using I_2 and the desired voltage drops. Finally, suitable standard value components are selected (see Appendix 6). Example 5-7 demonstrates the design procedure for a voltage divider circuit, and Example 5-8 shows how the output current can be taken into account when analyzing the circuit.

Example 5-7

A two-resistor voltage divider, as in Figure 5-9, has $E = 16$ V. The output voltage (V_2) is to be approximately 6 V, and the maximum output current (I_o) is 100 μA. Determine suitable resistor values.

Solution

$$I_2 \gg I_o$$

Select $$I_2 = 100\,I_o = 100 \times 100\ \mu\text{A}$$

$$= 10\ \text{mA}$$

$$R_2 = \frac{V_2}{I_2} = \frac{6\ \text{V}}{10\ \text{mA}}$$

$$= 600\ \Omega$$

$$R_1 \approx \frac{E - V_2}{I_2} = \frac{16\ \text{V} - 6\ \text{V}}{10\ \text{mA}}$$

$$\approx 1\ \text{k}\Omega$$

Example 5-8

Using the calculated resistance values, analyze the voltage divider circuit designed in Example 5-7 to determine the actual level of V_2.

Solution

$$V_2 = I_2R_2$$

And
$$V_1 = (I_2 + I_o)R_1$$

So
$$E = V_1 + V_2 = (I_2 + I_o)R_1 + I_2R_2$$

Giving
$$I_2 = \frac{E - I_oR_1}{R_1 + R_2} = \frac{16\text{ V} - (100\ \mu\text{A} \times 1\text{ k}\Omega)}{1\text{ k}\Omega + 600\ \Omega}$$

$$= 9.94\text{ mA}$$

$$V_2 = I_2R_2 = 9.94\text{ mA} \times 600\ \Omega$$

$$= 5.96\text{ V}$$

Practice Problems

5-3.1 A voltage divider as in Figure 5-7 is to be designed to produce 7.5 V from a 12 V supply. Using a 10 mA resistor current, calculate suitable resistor values.

5-3.2 A three-resistor voltage divider as shown in Figure 5-8 has $R_1 = 1.8\text{ k}\Omega$, $R_2 = 2.7\text{ k}\Omega$, $R_3 = 3.9\text{ k}\Omega$, and $E = 15$ V. Calculate the voltages V_1, V_2, and V_3.

5-3.3 A two-resistor voltage divider uses a 12 V battery supply and has $R_1 = 820\ \Omega$, and $R_2 = 330\ \Omega$. If the junction of R_1 and R_2 is grounded, determine the voltage at each battery terminal with respect to ground.

5-4 POTENTIOMETER

A variable resistor employed as a *potentiometer* is illustrated in Figure 5-10(a) and the circuit diagram is given in Figure 5-10(b). [The construction of such a resistor is shown in Figure 4-14(a) and discussed in Section 4-5.] A potentiometer is essentially a single resistor with terminals at each end and a movable contact that can be set to any point on the resistor. Terminals A and B in Figure 5-10 are the end terminals, and terminal C is the adjustable contact. The device is seen to function as an adjustable voltage divider. From Eq. 5-5, the output voltage is,

$$V_o = E \times \frac{R_2}{R_1 + R_2}$$

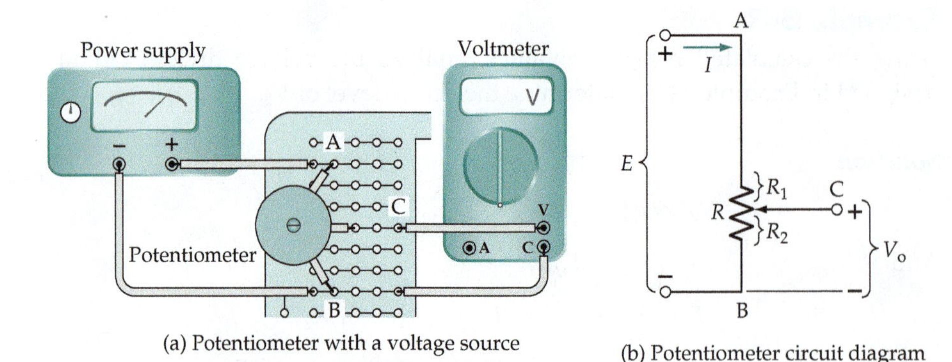

(a) Potentiometer with a voltage source

(b) Potentiometer circuit diagram

Figure 5-10 A potentiometer is a resistor with two fixed end terminals and one movable contact. When connected across a supply of E volts, the output is adjustable from $V_{o(min)} = 0$ to $V_{o(max)} = E$.

When the moving contact is exactly halfway between the two end terminals, the resistance from the moving contact to each end terminal is exactly half the total resistance of the potentiometer [see Figure 5-10(b)]. In this situation the output voltage between points C and B is,

$$V_o = E \times \frac{R/2}{R/2 + R/2}$$

$$= 0.5E$$

If the moving contact is adjusted until $R_2 = 0.25R$, then $R_1 = 0.75R$, and the output voltage becomes

$$V_o = E \times \frac{0.25R}{0.75R + 0.25R}$$

$$= 0.25E$$

When $R_2 = 0.75R$, the output voltage is

$$V_o = 0.75E$$

And when $R_2 = R$

$$V_o = E$$

It is seen that the potentiometer can be adjusted to give an output voltage ranging from 0 V to the supply voltage, E.

Ideally, the resistance between the moving contact on a potentiometer and either one of the end terminals should increase or decrease smoothly as the device is adjusted. Because of the construction of most potentiometers, however, this resistance change normally occurs in small jumps. How smoothly any given potentiometer changes its resistance when adjusted is defined in terms of the potentiometer *resolution*. The resolution is usually stated as a number of steps. So, the step change in resistance (Δr) of a 1 kΩ potentiometer with a resolution of 400 steps is,

$$\Delta r = \frac{1\ \text{k}\Omega}{400} = 2.5\ \Omega$$

A resolution of 400 steps (or 1 in 400) also means that the step change in the output voltage (Δv) is,

$$\Delta v = \frac{E}{400}$$

Figure 5-11 shows a potentiometer connected in series with two resistors. In this case the output voltage is adjustable from a minimum of V_3 to a maximum of $(V_2 + V_3)$. As demonstrated in Example 5-9, the voltage-divider equation can be used to calculate the extremes of the output voltage.

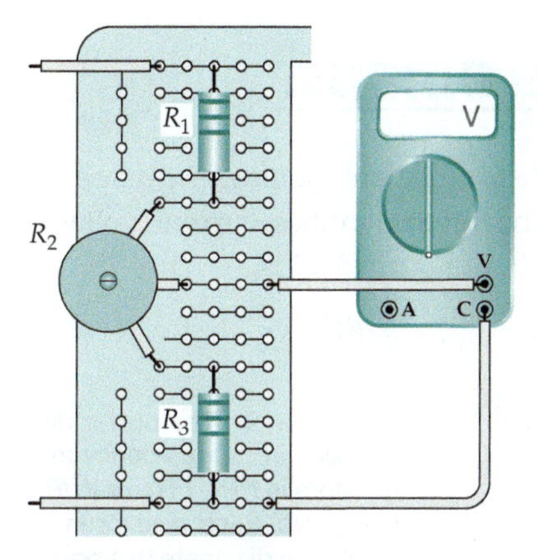

(a) Two resistors and a potentiometer connected in series

(b) Circuit diagram for series-connected potentiometer and resistors

Figure 5-11 A potentiometer R_2 connected in series with two resistors (R_1 and R_3) produces an output voltage that ranges from V_{R3} to $V_{R2} + V_{R3}$.

Example 5-9

Calculate the minimum and maximum values of V_o that can be obtained from the circuit in Figure 5-11.

Solution

With the moving contact at the lower end of R_2,

$$V_{o(min)} = E \times \frac{R_3}{R_1 + R_2 + R_3} = 30\text{ V} \times \frac{1\text{ k}\Omega}{3.5\text{ k}\Omega + 500\ \Omega + 1\text{ k}\Omega}$$

$$= 6\text{ V}$$

With the moving contact at the upper end of R_2,

$$V_{o(max)} = E \times \frac{R_2 + R_3}{R_1 + R_2 + R_3} = 30\text{ V} \times \frac{500\ \Omega + 1\text{ k}\Omega}{3.5\text{ k}\Omega + 500\ \Omega + 1\text{ k}\Omega}$$

$$= 9\text{ V}$$

The output current effects discussed in Section 5-3 apply also in the case of the potentiometer. Like the design procedure for a potential divider, calculation of the value of a potentiometer for a particular application commences with selection of a potentiometer current very much larger than any output (load) current.

Practice Problems

5-4.1 If the potentiometer circuit shown in Figure 5-11 has $E = 18$ V, $R_1 = 4.7$ kΩ, $R_2 = 2$ kΩ, and $R_3 = 5.6$ kΩ, calculate the maximum and minimum levels of V_o.

5-4.2 The circuit of Figure 5-11 is to be altered to make V_o adjustable from 2.5 V to 3.75 V. Determine the new resistance value for R_1 that will effect this change.

5-5 POWER IN A SERIES CIRCUIT

Using Equations 3-3, 3-4, or 3-5, it is possible to calculate the power dissipated in a resistor from a knowledge of any two of the three quantities: current, voltage, and resistance. So, the power dissipated in R_1 in Figure 5-12 is,

$$P_1 = V_1 I, \text{ or } P_1 = V_1^2/R_1, \text{ or } P_1 = I^2 R_1$$

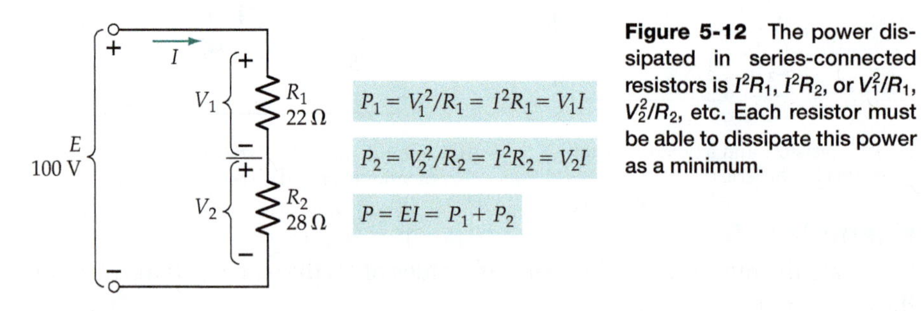

Figure 5-12 The power dissipated in series-connected resistors is I^2R_1, I^2R_2, or V_1^2/R_1, V_2^2/R_2, etc. Each resistor must be able to dissipate this power as a minimum.

The power dissipated in R_2 is calculated in exactly the same way as for R_1, and the total power dissipated in the circuit is the sum of the individual resistor power dissipations. For any series resistance circuit, the total power dissipated is

$$P = P_1 + P_2 + P_3 + \cdots + P_n \tag{5-7}$$

$$= V_1 I + V_2 I + V_3 I + \cdots + V_n I$$

$$= I(V_1 + V_2 + V_3 + \cdots + V_n)$$

so,

$$P = IE$$

The total power can also be calculated as

$$P = \frac{E^2}{R_{eq}} \tag{5-8}$$

Also

$$P = I^2 R_{eq} \tag{5-9}$$

Example 5-10

Determine the total power dissipation and the power dissipated in each resistor in Figure 5-12 when the component values are as specified in Example 5-5 ($E = 100$ V, $R_1 = 22\ \Omega$, $R_2 = 28\ \Omega$, $V_1 = 44$ V, $V_2 = 56$ V).

Solution

$$P_1 = \frac{V_1^2}{R_1} = \frac{(44\text{ V})^2}{22\ \Omega}$$

$$= 88\text{ W}$$

$$P_2 = \frac{V_2^2}{R_2} = \frac{(56\text{ V})^2}{28\ \Omega}$$

$$= 112\text{ W}$$

$$P = P_1 + P_2 = 88\text{ W} + 112\text{ W}$$

$$= 200\text{ W}$$

Or

$$P = \frac{E^2}{R_1 + R_2} = \frac{(100\text{ V})^2}{22\ \Omega + 28\ \Omega}$$

$$= 200\text{ W}$$

The physical size and type of construction of a resistor (see Section 4-5) determines the maximum power that it may dissipate. The maximum power that may be safely dissipated in any component is specified by the manufacturer and is referred to as its *power rating*. A wide range of resistors with various power ratings is available. Typical ratings for resistors usually employed in electronic circuits are 1/8 W, 1/4 W, 1/2 W, and 1 W. The power ratings for small potentiometers and variable resistors typically range from 1/2 W to 5 W. Every time the value of a resistor is calculated for a particular application, its power dissipation should also be determined. Where the component power dissipation exceeds the specified rating, the component is likely to *burn out*.

Example 5-11

A 2 kΩ potentiometer is connected across a 100 V supply. Determine the minimum power rating for the potentiometer.

Solution

Minimum power rating,

$$P = \frac{E^2}{R} = \frac{(100\text{ V})^2}{2\text{ k}\Omega}$$

$$= 5\text{ W}$$

Practice Problems

5-5.1 Calculate the power dissipation in each resistor in the circuit for Problem 5-4.1.

5-5.2 Calculate the minimum power rating for each resistor in the circuit shown in Figure 5-11(b).

5-6 VOLTAGE DROPPING AND CURRENT LIMITING

Sometimes a resistor is included in series with a circuit or electronic device to drop the supply voltage down to a desired level. In other circumstances this same kind of series resistor arrangement can be thought of as a current-limiting resistor. In the circuit shown in Figure 5-13(a), series resistor R_s limits the current to an electronic device that operates at a fixed voltage level lower than the source voltage. In Figure 5-13(b), R_s provides a voltage drop (V_s) to reduce the applied voltage to three series-connected lamps to ($E - V_s$). It can also be shown that R_s limits the current to the level required by the three lamps.

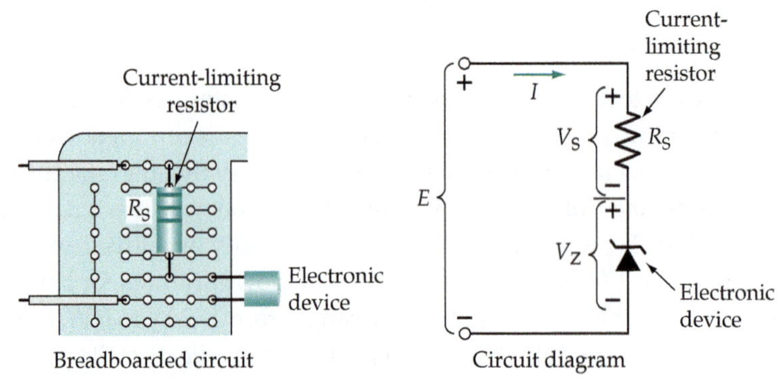

(a) Use of a current-limiting resistor

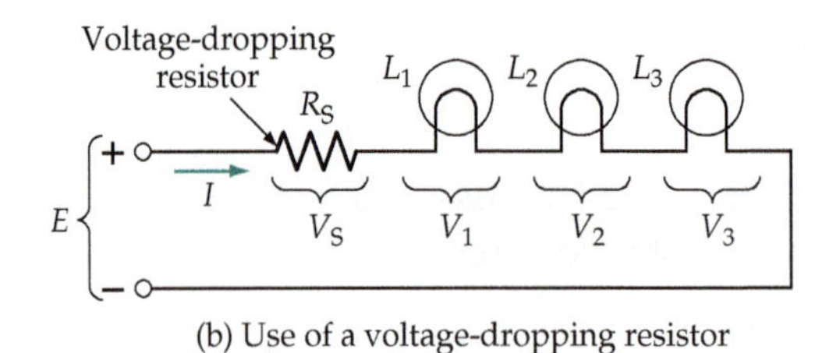

(b) Use of a voltage-dropping resistor

Figure 5-13 Resistors can be used for voltage-dropping or current-limiting. In each case, the resistor value is calculated as, $R_S = V_S/I$, and its power dissipation is I^2R_S.

Example 5-12

A certain electronic device, supplied via a series-connected resistor (R_S) as illustrated in Figure 5-13(a), uses a current of $I = 20$ mA and has a voltage drop of $V_Z = 6$ V. If the circuit supply voltage is 24 V, determine the required series resistance value. Also, calculate the power rating of the series resistor.

Solution

$$V_s = E - V_Z = 24\text{ V} - 6\text{ V}$$
$$= 18\text{ V}$$
$$R_1 = \frac{V_s}{I_s} = \frac{18\text{ V}}{20\text{ mA}}$$
$$= 900\ \Omega$$
$$P_s = I^2R_s = (20\text{ mA})^2 \times 900\ \Omega$$
$$= 0.36\text{ W}$$

Example 5-13

Three 6 V, 3 W lamps connected in series, as in Figure 5-13(b), are to be supplied from a 50 V source. Calculate the resistance value and power rating for the current-limiting resistor.

Solution

For each lamp,

$$P = VI$$

so,

$$I = \frac{P}{V} = \frac{3\text{ W}}{6\text{ V}}$$
$$= 0.5\text{ A}$$

The total voltage across the three lamps is

$$V = V_1 + V_2 + V_3 = 6\text{ V} + 6\text{ V} + 6\text{ V}$$
$$= 18\text{ V}$$
$$V_s = E - (V_1 + V_2 + V_3) = 50\text{ V} - 18\text{ V}$$
$$= 32\text{ V}$$
$$R_1 = \frac{V_s}{I_s} = \frac{32\text{ V}}{0.5\text{ A}}$$
$$= 64\ \Omega$$
$$P_s = V_sI_s = 32\text{ V} \times 0.5\text{ A}$$
$$= 16\text{ W}$$

Practice Problems

5-6.1 An electronic device operates at a voltage of 2.4 V and a current of 15 mA. If the device is to be connected to a 12 V supply, determine the resistance and power rating of a suitable current-limiting resistor.

5-6.2 The circuit shown in Figure 5-13(a) takes a constant current of 25 mA. If $E = 18$ V, and $R_s = 270\ \Omega$, calculate V_Z.

5-7 OPEN-CIRCUITS AND SHORT-CIRCUITS IN SERIES CIRCUITS

An *open-circuit* occurs in a series resistor circuit when one of the resistors becomes disconnected from an adjacent resistor (see Figure 5-14). An open-circuit can also occur when one of the resistors has been destroyed by excessive power dissipation. In the circuit of Figure 5-14, the open-circuit can be thought of as another resistance (R_{oc}) in series with R_1, R_2, R_3, and R_4. So, instead of the current being $I = E/(R_1 + R_2 + R_3 + R_4)$, it becomes,

$$I = \frac{E}{R_1 + R_2 + R_3 + R_4 + R_{oc}}$$

Suppose that $R_{oc} = 100\,000$ MΩ and $E = 100$ V; then, with $R_{oc} \gg (R_1 + R_2 + R_3 + R_4)$,

$$I \approx \frac{100\ \text{V}}{100\,000\ \text{M}\Omega}$$

$$= 1\ \text{nA}$$

Figure 5-14 An open-circuit in any part of a series circuit interrupts current flow throughout the circuit. There is no voltage drop across the resistors and the supply voltage appears at the open-circuit.

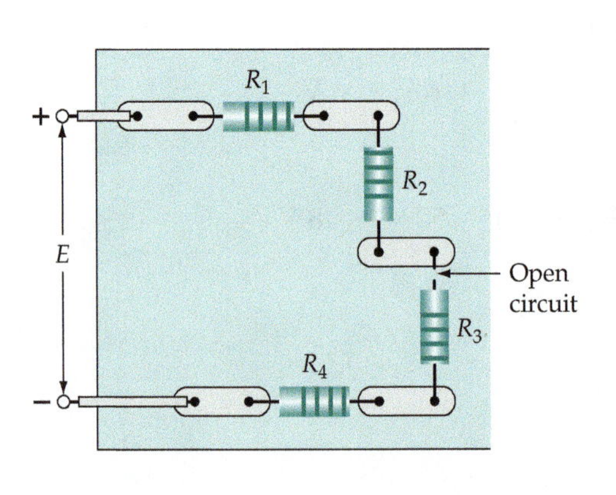

(a) Series resistor circuit with an open circuit

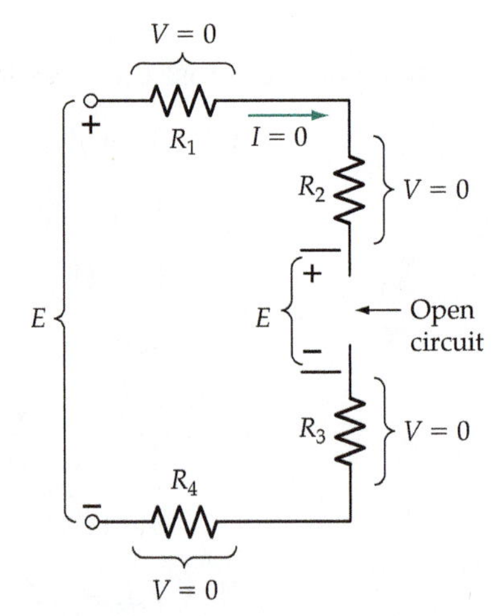

(b) Diagram of a series circuit with an open-circuited connection

This small current (1 nA) causes insignificant voltage drops across R_1, R_2, R_3, and R_4. With virtually zero voltage drop across the resistors, the voltage at the open circuit is,

$$V_{oc} = E$$

Figure 5-15 shows a series circuit with resistor R_3 *short-circuited.* In this case, the resistance between the terminals of R_3 is effectively zero. Consequently, instead of the current being $I = E/(R_1 + R_2 + R_3)$, it becomes,

$$I = \frac{E}{R_1 + R_2}$$

It is obvious that open-circuits and short-circuits seriously affect the current flow through a series resistor circuit.

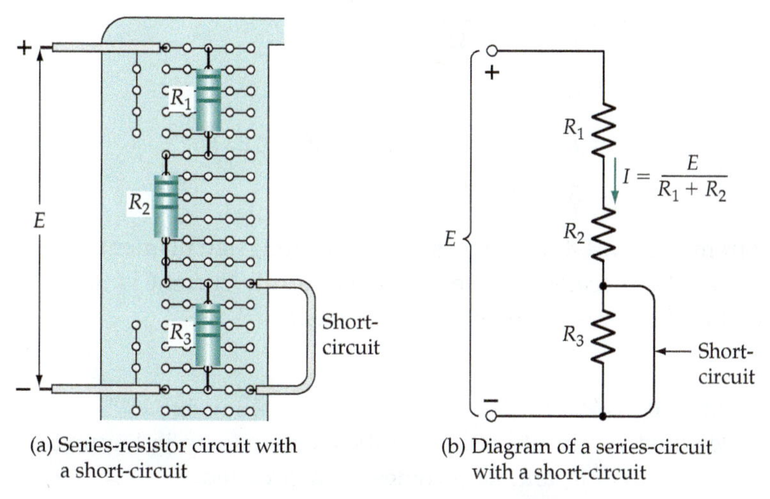

(a) Series-resistor circuit with a short-circuit

(b) Diagram of a series-circuit with a short-circuit

Figure 5-15 **A short-circuit increases the current level in a circuit. Consequently, the power dissipation is increased in those components that are not shorted.**

Example 5-14

The circuit of Figure 5-13(b) has three 6 V, 3 W lamps. Also, $E = 50$ V, and $R_s = 64\ \Omega$, as calculated in Example 5-13. (a) Determine the voltage at the terminals of lamp L_2 if L_2 becomes open-circuited. (b) If L_2 becomes short-circuited, calculate the current flow through the circuit and the power dissipated in each of the other two lamps.

Solution

(a) With L_2 open-circuited, no current flows and no voltage drop occurs across R_s, L_1, and L_3.

So $\qquad L_2 \text{ terminal voltage} = E = 50 \text{ V}$

(b) When the lamps are operating correctly, the power dissipated in each lamp is,

$$P_L = \frac{V_L^2}{R_L}$$

So the lamp resistance is,

$$R_L = \frac{V_L^2}{P_L} = \frac{(6\ \text{V})^2}{3\ \text{W}}$$

$$= 12\ \Omega$$

When L_2 is short-circuited,

$$R_{eq} = R_s + R_{L1} + R_{L3} = 64\ \Omega + 12\ \Omega + 12\ \Omega$$

$$= 88\ \Omega$$

$$I = \frac{E}{R_{eq}} = \frac{50\ \text{V}}{88\ \Omega}$$

$$= 0.568\ \text{A}$$

$$P_{L1} = P_{L2} = I^2 R_L = (0.568\ \text{A})^2 \times 12\ \Omega$$

$$= 3.87\ \text{W}$$

(Recall from Section 4-4 that the resistance of a tungsten filament changes with temperature. So, because the heating effect of the current is not taken into account, the above results are approximations.)

Practice Problems

5-7.1 Calculate the output voltage for the circuit shown in Figure 5-11 when the junction of R_2 and R_3 becomes open-circuited.

5-7.2 If R_3 in Figure 5-11 becomes short-circuited, determine the new levels of $V_{o(min)}$ and $V_{o(max)}$. Also, calculate the new power dissipation in each resistor.

Summary of Formulas

- ***Total series resistance:***

$$R_{eq} = R_1 + R_2 + R_3 + \cdots + R_n$$

- ***Current in series circuit:***

$$I = \frac{E}{R_1 + R_2 + R_3 + \cdots + R_n}$$

$$I = \frac{E}{R_{eq}}$$

- ***Total voltage:***

$$E = IR_1 + IR_2 + IR_3 + \cdots + IR_n$$

$$E = I(R_1 + R_2 + R_3 + \cdots + R_n)$$

- ***For a voltage divider:***

$$V_1 = E \times \frac{R_1}{R_1 + R_2}$$

$$V_n = E \times \frac{R_n}{R_1 + R_2 + R_3 + \cdots + R_n}$$

- ***Total power:***

$$P = P_1 + P_2 + P_3 + \cdots + P_n$$

$$P = \frac{E^2}{R_1 + R_2 + R_3 + \cdots + R_n}$$

$$P = I^2(R_1 + R_2 + R_3 + \cdots + R_n)$$

Review Questions

Section 5-1

5-1 Sketch the circuit diagram for four series-connected resistors supplied from a battery. Show the current direction, and indicate the polarity of all voltage drops. Define a series circuit in terms of the current.

5-2 Write the equations for total resistance and current in a series resistance circuit.

Section 5-2

5-3 Write the equations for the voltage drops across the resistors in a series circuit and for the total voltage. State Kirchhoff's voltage law.

5-4 Draw a circuit diagram for several series-connected resistors supplied from a voltage source that has one terminal grounded. Show how the circuit voltages should be measured. Briefly explain.

5-5 Sketch the circuit of three series-connected resistors supplied from two batteries that are connected series-opposing. Explain the effect that the series-opposing connection has on the circuit compared to the situation when the batteries are connected series-aiding.

Section 5-3

5-6 Sketch the circuit of two resistors employed as a voltage divider. Derive equations for the voltage drop across each resistor in terms of the supply voltage. State the voltage-divider theorem.

Section 5-4

5-7 Sketch the circuit diagram of a potentiometer supplied from a battery. Explain the principle of the potentiometer and discuss what is meant by the resolution of a potentiometer.

Section 5-5

5-8 Write equations for the total power supplied to a series resistor circuit in terms of, (a) the power dissipated in each component, (b) the supply voltage and the resistor values, (c) the current and the resistor values.

Section 5-6

5-9 Explain what is meant by a voltage-dropping resistor. Sketch the diagram of a circuit using a voltage-dropping resistor.

5-10 Five 12 V lamps are to be supplied from a 120 V source. Sketch a circuit diagram to show how the lamps should be connected. Briefly explain.

Section 5-7

5-11 Using circuit diagrams, explain open-circuits and short-circuits, and discuss the effect that each has on a series resistor circuit.

Problems

Section 5-1

5-1 Calculate current that flows through four series-connected resistors when $R_1 = 150\ \Omega$, $R_2 = 250\ \Omega$, $R_3 = 125\ \Omega$, $R_4 = 75\ \Omega$, and the battery voltage is $E = 12$ V.

5-2 When a new resistor is substituted in place of R_1 in Problem 5-1, the current level is measured as 25 mA. Calculate the new value of R_1.

5-3 A battery with an open-circuit voltage of 9 V and an internal resistance of 1 Ω has two resistors ($R_1 = 4.7\ \Omega$ and $R_2 = 3.3\ \Omega$) connected in series across its terminals. Determine the circuit current.

Section 5-2

5-4 A 4.7 kΩ resistor and a 2.2 kΩ resistor are connected in series to a 6 V supply. The supply is made up of four series-connected 1.5 V cells. Calculate the circuit current (a) when all of the cells are connected series-aiding, (b) when one cell is connected series-opposing.

5-5 Calculate the voltage drops across each resistor in Problem 5-1, and show that they add up to equal the supply voltage.

5-6 Five resistors are connected in series across a 30 V supply. $R_1 = 1.8$ kΩ, $R_2 = 680\ \Omega$, $R_3 = 1.2$ kΩ, $R_4 = 1.5$ kΩ, and $R_5 = 470\ \Omega$. Determine the circuit current and the voltage drop across each resistor.

5-7 Refer to the circuit of Figure 5-1(c). Determine E when $R_1 = 6.8$ kΩ, $R_2 = 12$ kΩ, $V_3 = 2$ V, and $I = 370\ \mu$A.

5-8 Calculate E_2 in the circuit of Figure 5-16.

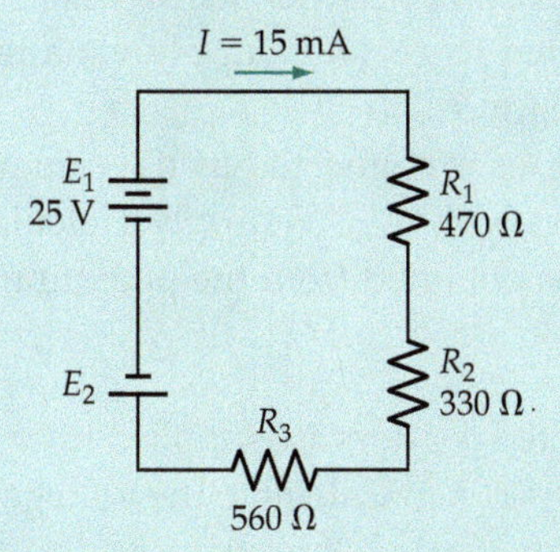

Figure 5-16

5-9 For the circuit shown in Figure 5-17, calculate the voltage at terminals A, B, and C with respect to ground.

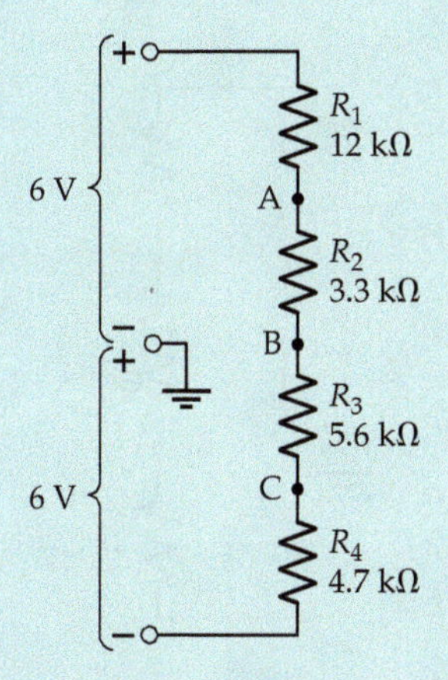

Figure 5-17

Section 5-3

5-10 Referring to Problem 5-4, calculate the voltage across each resistor for case (a) and case (b).

5-11 A voltage divider uses two resistors connected in series across a 75 V supply. If the resistor values are $R_1 = 37\ \Omega$ and $R_2 = 88\ \Omega$, calculate the voltage across each resistor.

5-12 A voltage divider is made up of two resistors R_1 and R_2 connected across a 15 V supply. R_2 has a resistance of 2.7 kΩ, and the voltage across it is to be 6.5 V. Determine the required resistance for R_1.

Section 5-4

5-13 A 100 kΩ potentiometer is connected in series with a resistance R_x. The two are supplied from a 15 V source. Determine the value of R_x if the maximum output voltage from the potentiometer is to be 12 V.

5-14 A 5 kΩ potentiometer (R_2) is to have a 3.3 V output when its moving contact is exactly halfway between its two fixed terminals. If the potentiometer is connected to an 18 V supply via a resistor R_1, calculate the required resistance for R_1.

5-15 A 3 kΩ potentiometer is connected as the center resistor in series with two 5.6 kΩ resistors with a ±6 V supply. Calculate the range of output voltage that may be obtained from the potentiometer.

Section 5-5

5-16 Three series-connected resistors are supplied from a 25 V source. $R_1 = 5$ kΩ, $R_2 = 13$ kΩ, and $R_3 = 11$ kΩ. Calculate the voltage drop across each resistor and the power dissipated in each.

5-17 The supply voltage for the circuit described in Problem 5-16 is to be adjusted to give $V_{R1} = 7$ V. Calculate the new supply voltage.

5-18 For the circuit in Figure 5-18, calculate the voltage levels at terminals A and B with respect to ground. Also determine the power dissipated in each resistor.

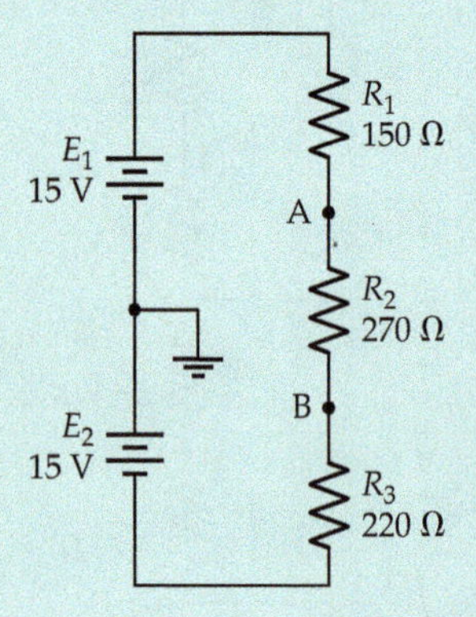

Figure 5-18

5-19 For the circuit of Figure 5-18, recalculate the voltage levels and power dissipated when E_2 becomes short-circuited.

5-20 Determine the voltage drop across L_1 in the circuit of Figure 5-19, and calculate the resistance of L_1 and the total power supplied to the circuit if $I = 117$ mA.

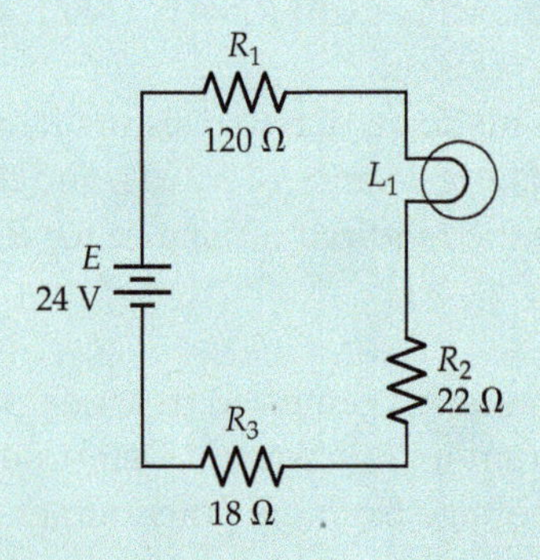

Figure 5-19

5-21 The terminal voltage of a 12 Ω electric heater is measured as 110 V. The heater is supplied via conductors from a 115 V source. Calculate the resistance of the conductors and the power dissipated in them.

5-22 Determine the maximum and minimum values of V_o in Figure 5-20. Also calculate the power dissipated in each resistor.

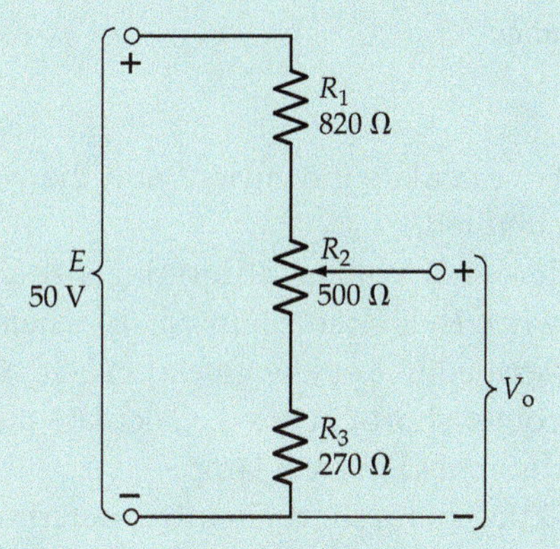

Figure 5-20

5-23 Calculate the minimum and maximum power dissipated in L_1 in the circuit in Figure 5-21.

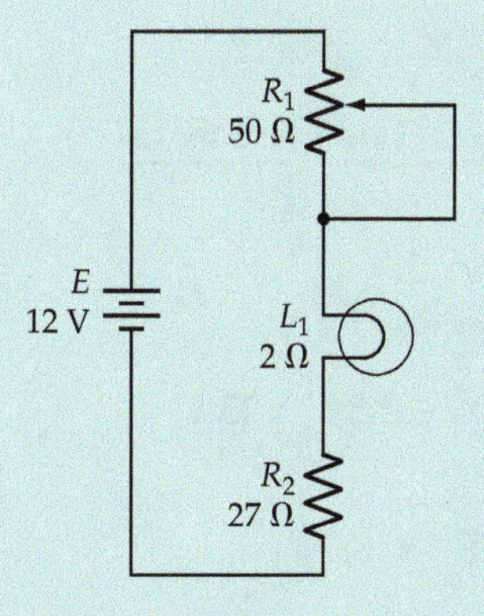

Figure 5-21

Section 5-6

5-24 For the circuit described in Review Question 5-5, assume $R_1 = 12\ \Omega$, $R_2 = 22\ \Omega$, $R_3 = 16\ \Omega$, $E_1 = 9$ V, and $E_2 = 6$ V. Determine the circuit current and resistor voltage drops when the batteries are connected (a) series-aiding and (b) series-opposing.

5-25 Calculate the power dissipated in each resistor in the circuit of Problem 5-24 for both case (a) and case (b) connections of the batteries. Calculate the total power dissipation in each case.

5-26 If the lamps described in Review Question 5-10 are each rated as 25 W, calculate the size and power rating of the required current-limiting resistor.

5-27 A small transistor radio normally takes 15 mA from a 9 V battery. Calculate the value of the resistor that must be connected in series with the radio if it is to be supplied from a 12 V battery. Calculate the power rating for the series resistor.

5-28 Twelve 6 V, 5 W series-connected lamps are to be supplied from a 230 V source. Calculate the resistance and power rating of the necessary voltage-dropping resistance.

Section 5-7

5-29 For Problem 5-6 recalculate the current and the voltage drops when R_3 becomes short-circuited.

5-30 Determine the levels of voltage at terminals A and B in the circuit of Figure 5-18, (a) when R_3 is open-circuited, (b) when R_3 is short-circuited.

5-31 In the circuit described in Review Question 5-10 and Problem 5-26, one of the lamps becomes short-circuited. Calculate the new level of current and the power dissipated in each lamp.

5-32 If R_2 in Figure 5-21 becomes short-circuited, determine the new minimum and maximum power dissipation in L_1.

5-33 For Problem 5-9 recalculate the voltage levels (a) when R_2 becomes short-circuited, (b) when R_2 is open-circuited.

5-34 For the circuit in Figure 5-20, calculate the maximum and minimum levels of V_o when R_3 becomes short-circuited.

Practice Problem Answers

5-1.1 0.5 mA
5-1.2 26.73 V
5-2.1 3.5 V, 2.5 V, 4 V, 2 V
5-2.2 312.5 μA, 2.2 V, 1.6 V, 2.5 V, 1.25 V
5-3.1 750 Ω, 450 Ω
5-3.2 3.2 V, 4.8 V, 7 V
5-3.3 +8.6 V, −3.4 V
5-4.1 8.2 V, 11.1 V
5-4.2 10.5 kΩ
5-5.1 10 mW, 4.3 mW, 12 mW
5-5.2 126 mW, 18 mW, 36 mW
5-6.1 640 Ω, 144 mW
5-6.2 11.25 V
5-7.1 30 V
5-7.2 0 V, 3.75 V, 197 mW, 28.1 mW

CHAPTER 6
Parallel Resistor Circuits

CONTENTS

Objectives

You will be able to:

1 Sketch circuit diagrams for parallel-connected resistors and parallel-connected lamps.
2 Indicate current directions and voltage polarities throughout parallel circuits.
3 Write equations for total equivalent resistance, current levels, and voltage drops in parallel resistor circuits.
4 State Kirchhoff's current law.
5 Solve problems involving equivalent resistance, current levels, voltage drops, and power dissipations in parallel resistor circuits.
6 Explain the current divider and solve problems involving current divisions.
7 Convert parallel resistor circuits into parallel conductance circuits and solve problems involving conductances in parallel.
8 Explain the effects of open-circuits and short-circuits in parallel resistor circuits.

INTRODUCTION

Resistors are said to be connected in parallel when the same voltage appears across every component. With different resistance values, different currents flow through each resistor. The total current taken from the supply is the sum of all the individual resistor currents. The equivalent resistance of a parallel resistor circuit is most easily calculated by using the reciprocal of each individual resistance value (i.e., using conductance values). Two resistors connected in parallel may be used as a current divider. In a parallel circuit, as in a series circuit, the total power supplied is the sum of the power dissipations in the individual components. Open-circuit and short-circuit condition in a parallel circuit have an effect on the total supply current.

6-1 PARALLEL RESISTOR CIRCUIT (Kirchhoff's Current Law)

Parallel-Connected Resistors

Resistors are connected in parallel when the circuit has two terminals that are common to every resistor. Figure 6-1(a) shows two resistors (R_1 and R_2) connected in parallel on a laboratory breadboard with a voltage applied from a power supply. In Figure 6-1(b), two parallel-connected resistors on a printed circuit board are supplied from a 9 V battery. In both circuits, the supply voltage appears across R_1 and across R_2. So it can be stated that

Resistors are connected in parallel when the same voltage is applied across each resistor.

Current Levels

The parallel resistor circuit diagram in Figure 6-1(c) shows that different currents flow in each parallel component. As illustrated, the current through each resistor is,

$$I_1 = \frac{E}{R_1} \quad \text{and} \quad I_2 = \frac{E}{R_2}$$

Now look at the current directions in the parallel resistor circuit diagram in Figure 6-1(c). I_1 (through R_1) is flowing away from junction A, and I_2 (through R_2) is also flowing away from A. The supply current I is flowing toward junction A. Currents I, I_1, and I_2 are the only currents entering or leaving the junction. The total current flowing into the junction must equal the total current flowing out of the junction. Consequently,

$$I = I_1 + I_2$$

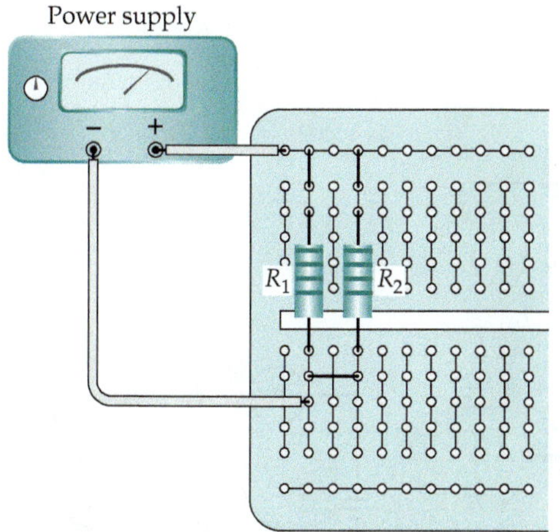

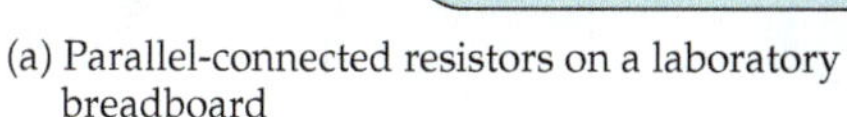
(a) Parallel-connected resistors on a laboratory breadboard

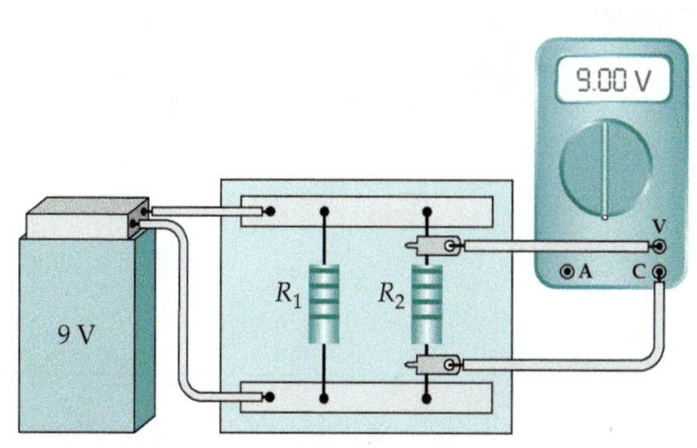

(b) Voltmeter measuring the voltage across parallel-connected resistors on a printed circuit board

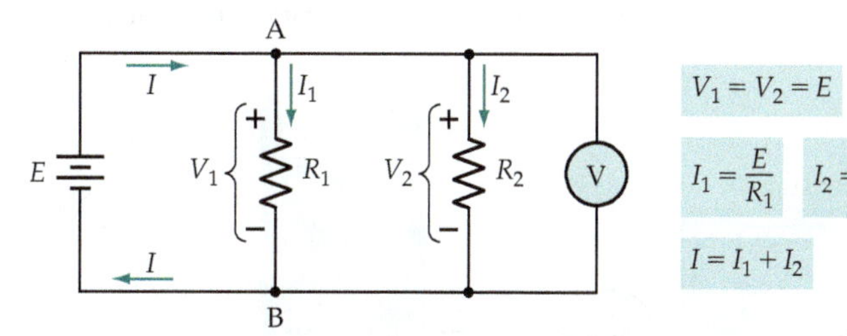

$V_1 = V_2 = E$

$I_1 = \frac{E}{R_1}$ $I_2 = \frac{E}{R_2}$

$I = I_1 + I_2$

(c) Circuit diagram for parallel-connected resistors

Figure 6-1 Resistors are connected in parallel when the same voltage appears across each resistor. The current through each component is $I_1 = E/R_1$, $I_2 = E/R_2$, etc.

The same reasoning applied to the currents at junction B (where I_1 and I_2 are entering B, and I is leaving B) also gives,

$$I = I_1 + I_2$$

In the situation where there are n resistors in parallel, the supply current is,

$$I = I_1 + I_2 + I_3 + \cdots + I_n \tag{6-1}$$

Kirchhoff's Current Law

The rule about currents entering and leaving a junction is defined in *Kirchhoff's current law* (KCL).

The algebraic sum of the currents entering a point in an electric circuit must equal the algebraic sum of the currents leaving that point.

Figure 6-2 further illustrates the currents in a parallel circuit. In Figure 6-2(a) an ammeter is shown connected in series with the conductor from the power supply to measure the supply current (I). The current I_2 (through resistor R_2) is measured by connecting the ammeter directly in series with R_2, as shown in Figure 6-2(b). Similarly, the current in resistor R_1 can be measured only by connecting the ammeter directly in series with R_1.

Figure 6-2 The supply current to a circuit is measured by connecting an ammeter in series with the conductor from the power supply. The current in resistor R_2 is measured by an ammeter connected directly in series with R_2.

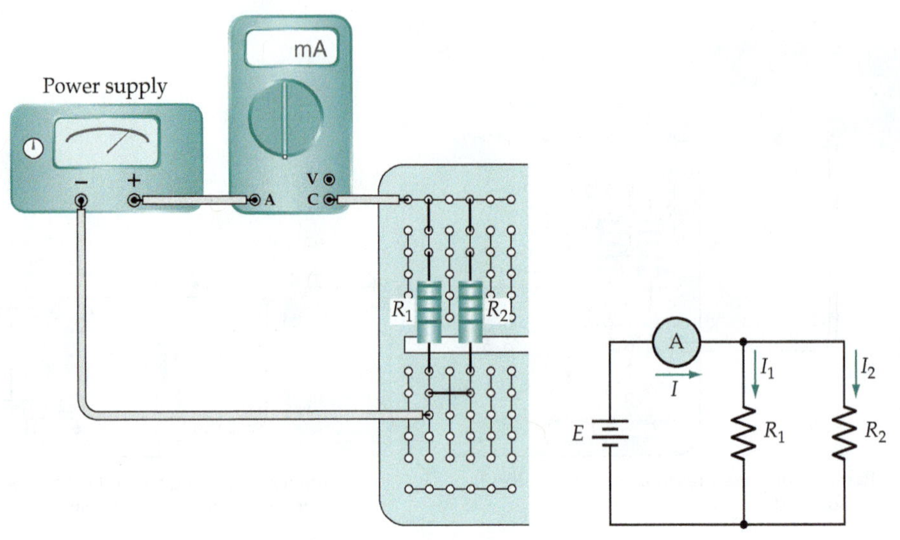

(a) Ammeter connected to measure supply current to parallel-connected resistors

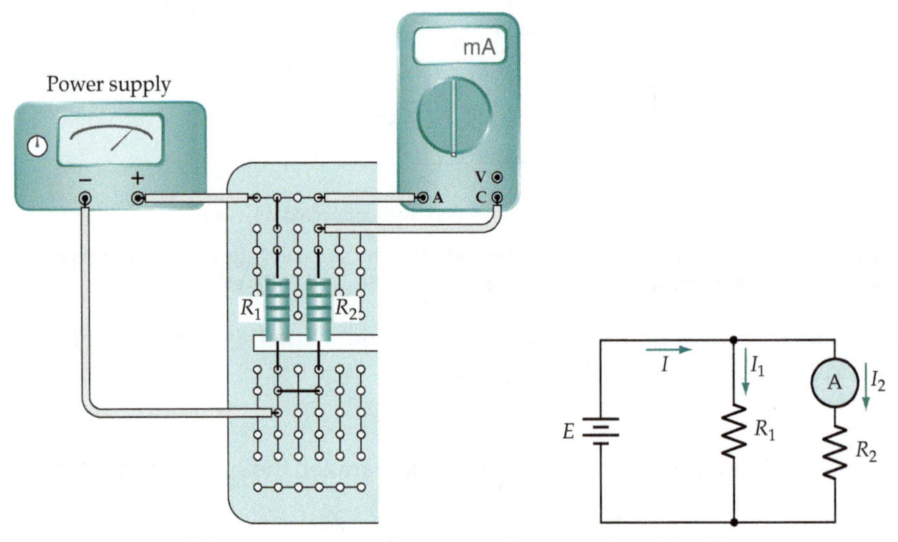

(b) Ammeter connected to measure the current in resistor R_2

Example 6-1

The resistors shown in Figure 6-2 have values of: $R_1 = 12\ \text{k}\Omega$, and $R_2 = 15\ \text{k}\Omega$. If the supply voltage is $E = 9$ V, calculate the current that flows through each resistor and the total current drawn from the battery.

Solution

$$I_1 = \frac{E}{R_1} = \frac{9\ \text{V}}{12\ \text{k}\Omega}$$
$$= 0.75\ \text{mA}$$
$$I_2 = \frac{E}{R_2} = \frac{9\ \text{V}}{15\ \text{k}\Omega}$$
$$= 0.6\ \text{mA}$$

$$I = I_1 + I_2 = 0.75 \text{ mA} + 0.6 \text{ mA}$$
$$= 1.35 \text{ mA}$$

Example 6-2

Calculate the individual resistor currents in the circuit shown in Figure 6-3. Also, calculate the total current that must be supplied by the voltage source.

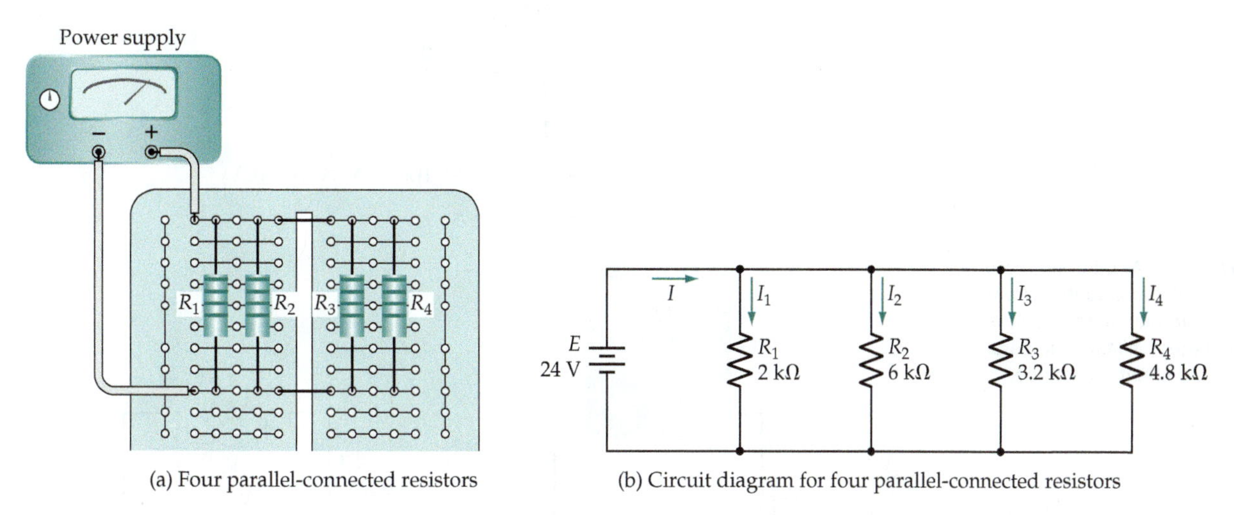

Figure 6-3 For four parallel-connected resistors: $I_1 = E/R_1$, $I_2 = E/R_2$, $I_3 = E/R_3$, and $I_4 = E/R_4$. The total supply current is $I = I_1 + I_2 + I_3 + I_4$.

Solution

$$I_1 = \frac{E}{R_1} = \frac{24 \text{ V}}{2 \text{ k}\Omega}$$
$$= 12 \text{ mA}$$

$$I_2 = \frac{E}{R_2} = \frac{24 \text{ V}}{6 \text{ k}\Omega}$$
$$= 4 \text{ mA}$$

$$I_3 = \frac{E}{R_3} = \frac{24 \text{ V}}{3.2 \text{ k}\Omega}$$
$$= 7.5 \text{ mA}$$

$$I_4 = \frac{E}{R_4} = \frac{24 \text{ V}}{4.8 \text{ k}\Omega}$$
$$= 5 \text{ mA}$$

$$I = I_1 + I_2 + I_3 + I_4$$
$$= 12 \text{ mA} + 4 \text{ mA} + 7.5 \text{ mA} + 5 \text{ mA}$$
$$= 28.5 \text{ mA}$$

Example 6-3

The microammeter (a very low current ammeter) shown in Figure 6-4 has a 1 Ω internal resistance and a parallel (shunting) resistance of 0.001 Ω. Calculate the level of the total current I if the meter current is 100 μA.

Solution

$$V_{\mathrm{m}} = I_{\mathrm{m}}R_{\mathrm{m}} = 100\ \mu\mathrm{A} \times 1\ \Omega$$
$$= 100\ \mu\mathrm{V}$$
$$I_{\mathrm{S}} = \frac{V_{\mathrm{m}}}{R_{\mathrm{S}}} = \frac{100\ \mathrm{mV}}{0.001\ \Omega}$$
$$= 100\ \mathrm{mA}$$
$$I = I_{\mathrm{S}} + I_{\mathrm{m}} = 100\ \mathrm{mA} \times 100\ \mu\mathrm{A}$$
$$= 100.1\ \mathrm{mA}$$

Figure 6-4 A microammeter with a parallel-connected shunting resistor may be represented by the meter resistance in parallel with the shunt resistance.

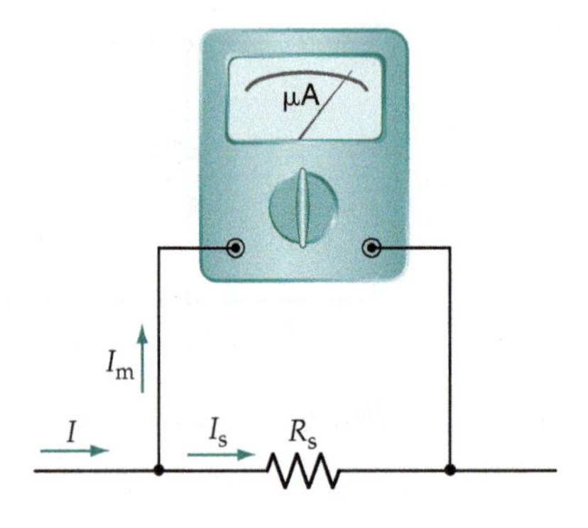

(a) Analog microammeter with a parallel (shunt) resistor

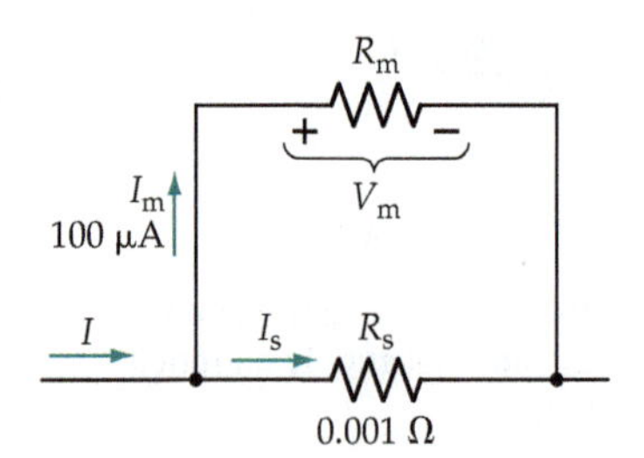

(b) The meter resistance is in parallel with the shunt

Practice Problem

6-1.1 Three parallel-connected resistors with a 5 V supply have values of $R_1 = 7.5\ \mathrm{k\Omega}$, $R_2 = 5\ \mathrm{k\Omega}$, and $R_3 = 8\ \mathrm{k\Omega}$. Determine the branch currents and the total supply current.

6-2 PARALLEL EQUIVALENT CIRCUIT

Equivalent Resistance

Consider the case of four resistors in parallel, as reproduced in Figure 6-5(a). From Equation (6-1), the battery current is,

$$I = I_1 + I_2 + I_3 + I_4$$

which can be rewritten as,

$$I = \frac{E}{R_1} + \frac{E}{R_2} + \frac{E}{R_3} + \frac{E}{R_4}$$

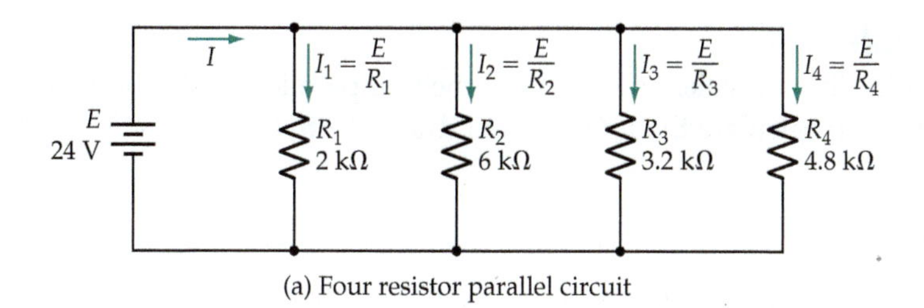

(a) Four resistor parallel circuit

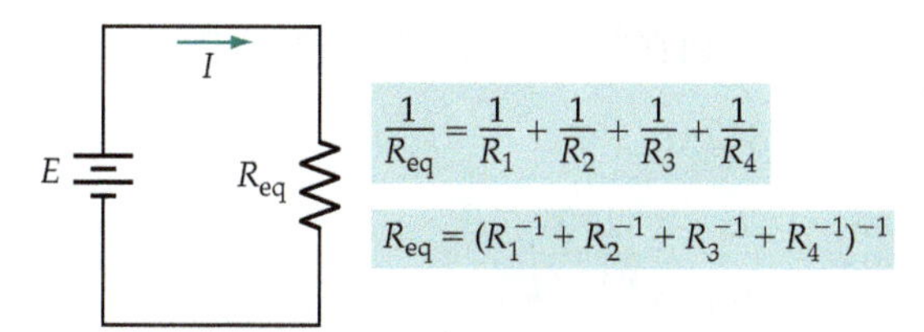

(b) Equivalent circuit

Figure 6-5 In the equivalent circuit for four parallel-connected resistors, the resistors are replaced by a single equivalent resistance: $R_{eq} = (R_1^{-1} + R_2^{-1} + R_3^{-1} + R_4^{-1})^{-1}$. The supply current is, $I = E/R_{eq}$.

or,

$$I = E\left[\frac{1}{R_1} + \frac{1}{R_2} + \frac{1}{R_3} + \frac{1}{R_4}\right]$$

For n resistors in parallel, this becomes,

$$I = E\left[\frac{1}{R_1} + \frac{1}{R_2} + \frac{1}{R_3} + \cdots + \frac{1}{R_n}\right] \tag{6-2}$$

If all the resistors in parallel could be replaced by just one (equivalent) resistance R_{eq} that would draw the same current from the battery, the equation for current would be written,

$$I = \frac{E}{R_{eq}}$$

where,

$$\frac{1}{R_{eq}} = \frac{1}{R_1} + \frac{1}{R_2} + \frac{1}{R_3} + \cdots + \frac{1}{R_n} \tag{6-3}$$

It is seen that,

The reciprocal of the equivalent resistance of several resistors in parallel is equal to the sum of the reciprocals of the individual resistances.

Equation 6-3 can be rearranged to give,

$$R_{eq} = 1\Big/\left[\frac{1}{R_1} + \frac{1}{R_2} + \frac{1}{R_3} + \cdots + \frac{1}{R_n}\right]$$

or,

$$R_{eq} = (R_1^{-1} + R_2^{-1} + R_3^{-1} + \cdots + R_n^{-1})^{-1} \tag{6-4}$$

The equivalent circuit of the parallel resistors and battery can now be drawn as illustrated in Figure 6-5(b).

Example 6-4

Determine the equivalent resistance of the four parallel resistors in Figure 6-5(a), and use it to calculate the total current drawn from the battery.

Solution

Eq. 6-4,

$$R_{eq} = (R_1^{-1} + R_2^{-1} + R_3^{-1} + \cdots + R_n^{-1})^{-1}$$

$$= [(2\ \text{k}\Omega)^{-1} + (6\ \text{k}\Omega)^{-1} + (3.2\ \text{k}\Omega)^{-1} + (4.8\ \text{k}\Omega)^{-1}]^{-1}$$

$$\approx 842\ \Omega$$

$$I = \frac{E}{R_{eq}} = \frac{24\ \text{V}}{842\ \Omega}$$

$$= 28.5\ \text{mA}$$

It should be noted that when two equal-value resistors are connected in parallel, the equivalent resistance is half the resistance value of one resistor. Also, the equivalent resistance for n parallel-connected equal-value resistors is,

$$R_{eq} = \frac{R_n}{n}$$

Two resistors in parallel can be represented as $R_1 \| R_2$, and any number of parallel-connected resistors can be represented as,

$$R_1 \| R_2 \| R_3 \| \cdots \| R_n$$

Analysis Procedure for Parallel Circuits

1. ***Calculate the current through each resistor in turn:***

$$I_1 = \frac{E}{R_1},\ I_2 = \frac{E}{R_2},\ \text{etc.},$$

2. ***Calculate the total supply current:***

$$I = I_1 + I_2 + \cdots + I_n$$

Alternatively:

1. ***Use Equation (6-4) to determine the equivalent resistance (R_{eq}) of all the resistors in parallel.***
2. ***Calculate the total supply current:***

$$I = \frac{E}{R_{eq}}$$

Practice Problems

6-2.1 Three resistors connected in parallel to a 14 V supply have values of: $R_1 = 68\ \text{k}\Omega$, $R_2 = 56\ \text{k}\Omega$, and $R_3 = 82\ \text{k}\Omega$. Calculate the equivalent resistance of the three, and determine the total supply current.

6-2.2 Determine the resistance of a fourth resistor in parallel with the three in Problem 6-2.1 if the supply current increases to 1 mA when the fourth resistor is connected.

6-2.3 If the three parallel-connected resistors in Problem 6-1.1 are to have a total supply current of 5 mA, determine the required supply voltage.

6-3 CONDUCTANCES IN PARALLEL

As discussed in Section 3-4, conductance is the reciprocal of resistance, and its unit is the *siemens* (S). In the case of parallel circuits, it is sometimes more convenient to use the conductance values of the resistors involved instead of the resistance values (see Example 6-5).

When Equation 6-3 is rewritten in terms of conductances, it becomes

$$G_{eq} = G_1 + G_2 + G_3 + \cdots + G_n \quad (6\text{-}5)$$

This is illustrated in Figure 6-6.

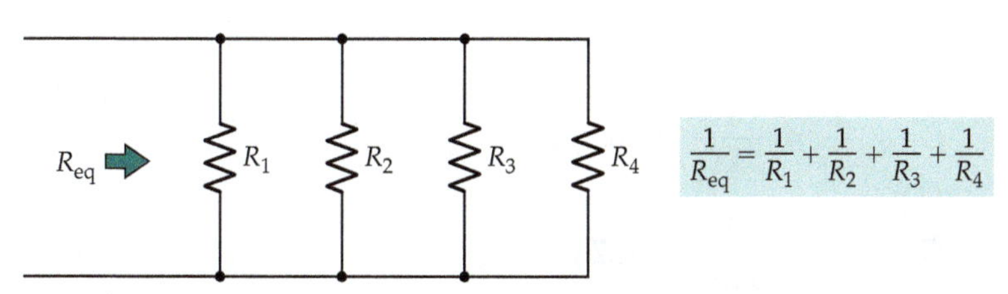

(a) The equivalent resistance for a parallel-resistor circuit is found by summing the reciprocals of the resistances

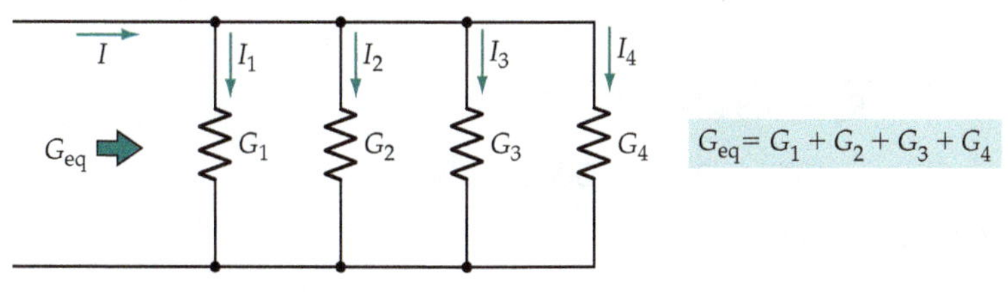

(b) The equivalent conductance for a parallel-conductance circuit is found by summing the individual conductances

Figure 6-6 In a parallel resistor circuit, resistors R_1, R_2, etc., can be expressed as conductances G_1, G_2, etc. The total conductance is $G_{eq} = G_1 + G_2 + G_3 + G_4$, and the supply current is $I = EG_{eq}$.

It will be recalled that Ohm's law as applied to conductances is changed from,

$$I = \frac{E}{R} \text{ to } I = EG$$

So, the current in each branch of the parallel circuit in Figure 6-6 can be calculated as

$$I_1 = EG_1,\ I_2 = EG_2, \text{ etc.,}$$

and the current drawn from the supply becomes,

$$I = EG_{eq}$$

Example 6-5

For the circuit shown in Figure 6-7 (reproduced from Figure 6-3), express each resistance as a conductance, and, using the conductance values, calculate the current through each resistor and the total current taken from the battery.

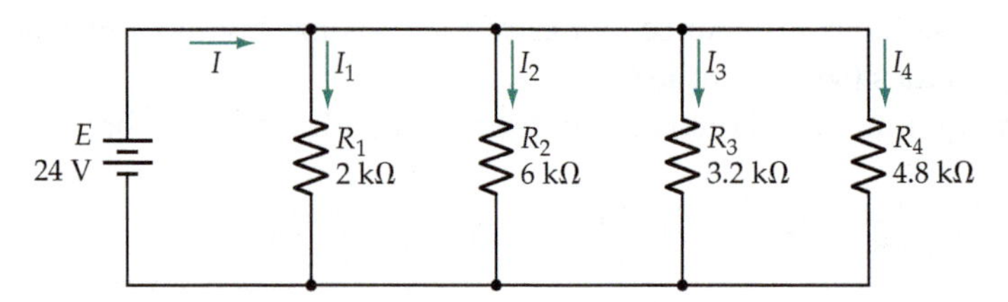

Figure 6-7 Four-resistor parallel circuit for Example 6-5.

Solution

$$G_1 = \frac{1}{R_1} = \frac{1}{2\ \text{k}\Omega} = 500\ \mu\text{S}$$

$$G_2 = \frac{1}{R_2} = \frac{1}{6\ \text{k}\Omega} = 167\ \mu\text{S}$$

$$G_3 = \frac{1}{R_3} = \frac{1}{3.2\ \text{k}\Omega} = 313\ \mu\text{S}$$

$$G_4 = \frac{1}{R_4} = \frac{1}{4.8\ \text{k}\Omega} = 208\ \mu\text{S}$$

$$I_1 = E \times G_1 = 24\ \text{V} \times 500\ \mu\text{S}$$
$$= 12\ \text{mA}$$

$$I_2 = E \times G_2 = 24\ \text{V} \times 167\ \mu\text{S}$$
$$= 4\ \text{mA}$$

$$I_3 = E \times G_3 = 24\ \text{V} \times 313\ \mu\text{S}$$
$$= 7.5\ \text{mA}$$

$$
\begin{aligned}
I_4 &= E \times G_4 = 24\ \text{V} \times 208\ \mu\text{S} \\
&= 5\ \text{mA} \\
G_{eq} &= G_1 + G_2 + G_3 + G_4 = (500 + 167 + 313 + 208)\ \mu\text{S} \\
&= 1.188\ \text{mS} \\
I &= E \times G_{eq} = 24\ \text{V} \times 1.188\ \text{mS} \\
&= 28.5\ \text{mA} = I_1 + I_2 + I_3 + I_4
\end{aligned}
$$

Practice Problems

6-3.1 Convert each of the resistances in Problem 6-2.1 into conductances, and determine the total conductance of the circuit.

6-3.2 Determine the conductance of each of the three resistors in Problem 6-1.1 and the total conductance of the three in parallel.

6-4 CURRENT DIVIDER

Refer to the two-resistor parallel circuit illustrated in Figure 6-8. Such a parallel combination of two resistors is sometimes termed a *current divider*, because the supply current is divided between the two *branches* (or parallel sections) of the circuit. The current levels are,

$$I_1 = \frac{E}{R_1} \quad \text{and} \quad I_2 = \frac{E}{R_2}$$

Also,

$$
\begin{aligned}
I &= I_1 + I_2 = \frac{E}{R_1} + \frac{E}{R_2} \\
&= E\left[\frac{1}{R_1} + \frac{1}{R_2}\right]
\end{aligned}
$$

This equation can be rewritten as,

$$I = E\left[\frac{R_1 + R_2}{R_1 \times R_2}\right] \tag{6-6}$$

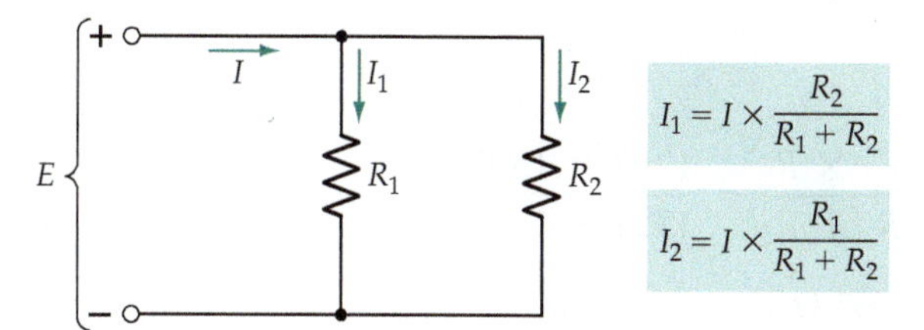

Figure 6-8 Two resistors can be connected in parallel to function as a *current divider*, giving $I_1 = IR_2/(R_1 + R_2)$ and $I_2 = IR_1/(R_1 + R_2)$.

or
$$I = \frac{E}{\left[\dfrac{R_1 \times R_2}{R_1 + R_2}\right]}$$

Also,
$$I = \frac{E}{R_{eq}}$$

where R_{eq} is the equivalent resistance of R_1 and R_2 in parallel ($R_1 \| R_2$). Therefore, for two resistors in parallel the equivalent resistance is,

$$R_{eq} = \frac{R_1 \times R_2}{R_1 + R_2} \tag{6-7}$$

From Eq. 6-6,
$$E = I\left[\frac{R_1 \times R_2}{R_1 + R_2}\right]$$

and substituting for E in $I_1 = E/R_1$ gives,

$$I_1 = \frac{I}{R_1}\left[\frac{R_1 \times R_2}{R_1 + R_2}\right]$$

Which produces,
$$I_1 = I\left[\frac{R_2}{R_1 + R_2}\right] \tag{6-8}$$

and
$$I_2 = I\left[\frac{R_1}{R_1 + R_2}\right] \tag{6-9}$$

Equations 6-8 and 6-9 can be used to determine how a known supply current divides into two individual resistor currents. Note that the equation for I_1 has R_2 in the numerator, and that for I_2 has R_1, in the numerator. When two equal-value resistors are connected in parallel, the supply current divides equally between the two. For two parallel-connected unequal resistors, the largest current is passed by the smallest value resistor.

Example 6-6

Calculate the equivalent resistance and branch currents for the circuit shown in Figure 6-8 when $R_1 = 12\ \Omega$, $R_2 = 15\ \Omega$, and $E = 9$ V.

Solution

From Eq. 6-7,
$$R_{eq} = \frac{R_1 \times R_2}{R_1 + R_2} = \frac{12\ \Omega \times 15\ \Omega}{12\ \Omega + 15\ \Omega}$$
$$\approx 6.67\ \Omega$$

$$I = \frac{E}{R_{eq}} = \frac{9\ \text{V}}{6.67\ \Omega}$$

$$= 1.35\ \text{A}$$

From Eq. 6-8, $$I_1 = I\left[\frac{R_2}{R_1 + R_2}\right] = 1.35\ \text{A} \times \left[\frac{15\ \Omega}{12\ \Omega + 15\ \Omega}\right]$$

$$= 0.75\ \text{A}$$

From Eq. 6-9, $$I_2 = I\left[\frac{R_1}{R_1 + R_2}\right] = 1.35\ \text{A} \times \left[\frac{12\ \Omega}{12\ \Omega + 15\ \Omega}\right]$$

$$= 0.6\ \text{A}$$

Compare these results to the results for Example 6-1.

It is important to note that Equations 6-8 and 6-9 refer only to circuits with two parallel branches. They *cannot be applied to circuits with more than two parallel branches.* However, similar equations can be derived for the current division in a multibranch parallel circuit.

Consider the circuit in Figure 6-9(a), which has four resistors connected in parallel. The total current (I) splits into four components, as illustrated. The parallel resistance for the whole circuit is R_{eq}, and the voltage drop across the parallel combination (and across each individual resistor) is,

$$E_R = IR_{eq}$$

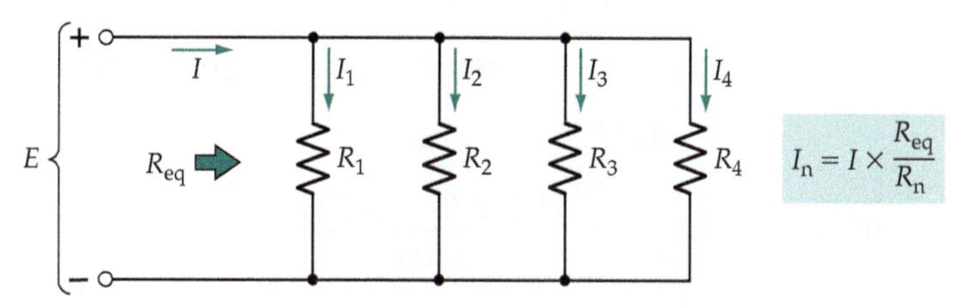

(a) The current in any resistor (R_n) can be calculated using I and R_{eq}

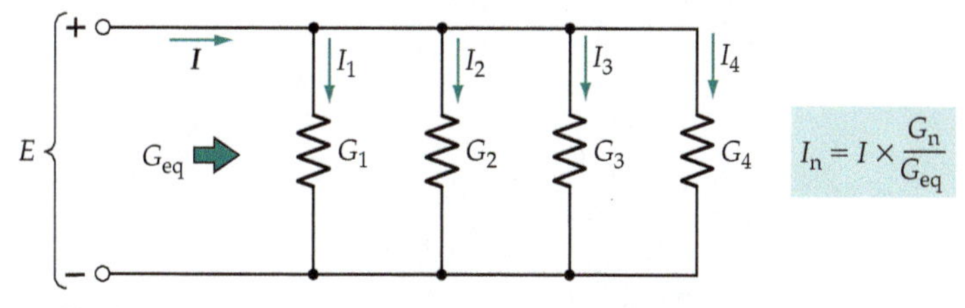

(b) The current in any conductance (G_n) can be calculated using I and G_{eq}

Figure 6-9 For a multiresistor current-divider circuit, the current through any resistor (R_n) can be calculated from $I_n = IR_{eq}/R_n$, where R_{eq} is the equivalent of all the resistors in parallel. Alternatively, $I_n = IG_n/G_{eq}$.

Also $$I_1 = \frac{E_R}{R_1}, I_2 = \frac{E_R}{R_2} \text{ etc.,}$$

So $$I_1 = \frac{IR_{eq}}{R_1}, I_2 = \frac{IR_{eq}}{R_2} \text{ etc.,}$$

For any multibranch parallel resistor circuit, the current in branch n is,

$$I_n = \frac{IR_{eq}}{R_n} \tag{6-10}$$

Equation 6-10 is known as the *current divider equation*. Because $R_{eq} = 1/G_{eq}$ and $R_n = 1/G_n$, the equation can be rewritten for conductances in parallel, as illustrated in Figure 6-9(b):

$$I_n = \frac{IG_n}{G_{eq}} \tag{6-11}$$

Current-divider Equation 6-11 is similar in form to voltage-divider Equation 5-6, except that conductances are involved instead of resistances.

Example 6-7

Use the current-divider equation to determine the branch currents in the circuit shown in Figure 6-9. The component values are $R_1 = 2\ \text{k}\Omega$, $R_2 = 6\ \text{k}\Omega$, $R_3 = 3.2\ \text{k}\Omega$, and $R_4 = 4.8\ \text{k}\Omega$. The supply current is $I = 28.5$ mA.

Solution

Eq. 6-4,
$$\begin{aligned} R_{eq} &= (R_1^{-1} + R_2^{-1} + R_3^{-1} + R_4^{-1})^{-1} \\ &= [(2\ \text{k}\Omega)^{-1} + (6\ \text{k}\Omega)^{-1} + (3.2\ \text{k}\Omega)^{-1} + (4.8\ \text{k}\Omega)^{-1}]^{-1} \\ &\approx 842\ \Omega \end{aligned}$$

From Eq. 6-10,
$$\begin{aligned} I_1 &= \frac{IR_{eq}}{R_1} = \frac{28.5\ \text{mA} \times 842\ \Omega}{2\ \text{k}\Omega} \\ &\approx 12\ \text{mA} \\ I_2 &= \frac{IR_{eq}}{R_2} = \frac{28.5\ \text{mA} \times 842\ \Omega}{6\ \text{k}\Omega} \\ &\approx 4\ \text{mA} \\ I_3 &= \frac{IR_{eq}}{R_3} = \frac{28.5\ \text{mA} \times 842\ \Omega}{3.2\ \text{k}\Omega} \\ &\approx 7.5\ \text{mA} \\ I_4 &= \frac{IR_{eq}}{R_4} = \frac{28.5\ \text{mA} \times 842\ \Omega}{4.8\ \text{k}\Omega} \\ &\approx 5\ \text{mA} \end{aligned}$$

These results are comparable to the current levels calculated in Example 6-2, which involves the same circuit as that shown in Figure 6-9.

Example 6-8

Three parallel-connected resistors that take a total supply current of 3 mA have resistance values of 10 kΩ, 12 kΩ, and 15 kΩ. Convert the resistance values into conductances, and determine the individual branch currents.

Solution

$$G_1 = \frac{1}{R_1} = \frac{1}{10\ \text{k}\Omega}$$

$$= 100\ \mu\text{S}$$

$$G_2 = \frac{1}{R_2} = \frac{1}{12\ \text{k}\Omega}$$

$$= 83.3\ \mu\text{S}$$

$$G_3 = \frac{1}{R_3} = \frac{1}{15\ \text{k}\Omega}$$

$$= 66.67\ \mu\text{S}$$

Eq. 6-5,
$$G_{eq} = G_1 + G_2 + G_3 = 100\ \mu\text{S} + 83.3\ \mu\text{S} + 66.67\ \mu\text{S}$$

$$= 250\ \mu\text{S}$$

Eq. 6-11,
$$I_1 = \frac{IG_1}{G_{eq}} = 3\ \text{mA} \times \frac{100\ \mu\text{S}}{250\ \mu\text{S}}$$

$$= 1.2\ \text{mA}$$

$$I_2 = \frac{IG_2}{G_{eq}} = 3\ \text{mA} \times \frac{83.3\ \mu\text{S}}{250\ \mu\text{S}}$$

$$= 1\ \text{mA}$$

$$I_3 = \frac{IG_3}{G_{eq}} = 3\ \text{mA} \times \frac{66.67\ \mu\text{S}}{250\ \mu\text{S}}$$

$$= 0.8\ \text{mA}$$

Practice Problems

6-4.1 A total current of 55 mA flows through two resistors in parallel. If the resistor values are $R_1 = 680\ \Omega$ and $R_2 = 560\ \Omega$, determine the individual resistor currents.

6-4.2 Use the current-divider equation to determine the individual resistor currents for a three-resistor parallel circuit with $R_1 = 33\ \text{k}\Omega$, $R_2 = 47\ \text{k}\Omega$, $R_3 = 27\ \text{k}\Omega$, and $I = 2.13$ mA.

6-5 POWER IN PARALLEL CIRCUITS

Whether a resistor is connected in a series circuit or a parallel circuit, the power dissipated in the resistor is,

$$P = \text{(resistor current)} \times \text{(resistor voltage)}$$

For R_1 in the circuit of Figure 6-10,

$$P_1 = EI_1 \quad \text{or, } P_1 = E^2/R_1 \quad \text{or, } P_1 = I_1^2 R_1$$

The power dissipated in resistor R_2 is calculated in a similar way. The total power output from the supply is, of course,

$$P = EI = E(I_1 + I_2 + I_n)$$

$$= EI_1 + EI_2 + EI_n$$

or $$P = P_1 + P_2 + P_n$$

For any parallel (or series) combination of n resistors, Equation 5-7 applies:

$$P = P_1 + P_2 + P_3 + \cdots + P_n$$

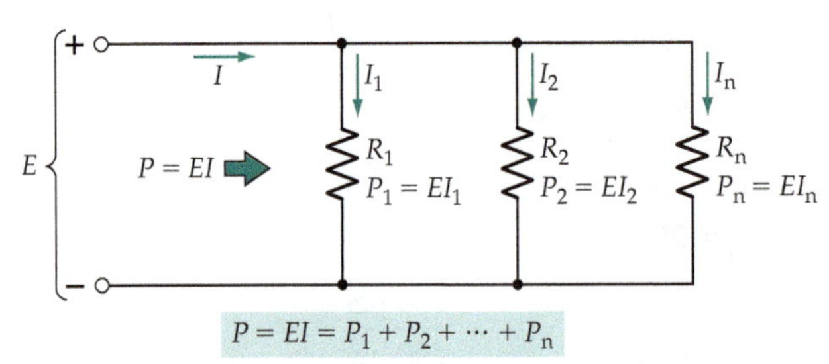

Figure 6-10 The power dissipated in a parallel resistor circuit is: $P = EI$, or $P = P_1 + P_2 + \cdots + P_n$, where $P_1 = EI_1$, $P_2 = EI_2$, etc.

Example 6-9

For the circuit described in Example 6-6, calculate the power dissipations in R_1 and R_2 and the total power supplied to the circuit.

Solution

$$P_1 = I_1^2 R_1 = (0.75\text{ A})^2 \times 12\ \Omega$$

$$= 6.75\text{ W}$$

$$P_2 = I_2^2 R_2 = (0.6\text{ A})^2 \times 15\ \Omega$$

$$= 5.4\text{ W}$$

$$P = P_1 + P_2 = 6.75\text{ W} + 5.4\text{ W}$$

$$= 12.15\text{ W}$$

Also, $$P = EI = 9\text{ V} \times 1.35\text{ A}$$

$$= 12.15\text{ W}$$

Example 6-10

Three 12 V lamps are connected in parallel to a 12 V supply, as illustrated in Figure 6-11. The lamp ratings are: $P_1 = 3$ W, $P_2 = 5$ W, and $P_3 = 10$ W. Determine the current that flows through each lamp and the total power delivered by the supply.

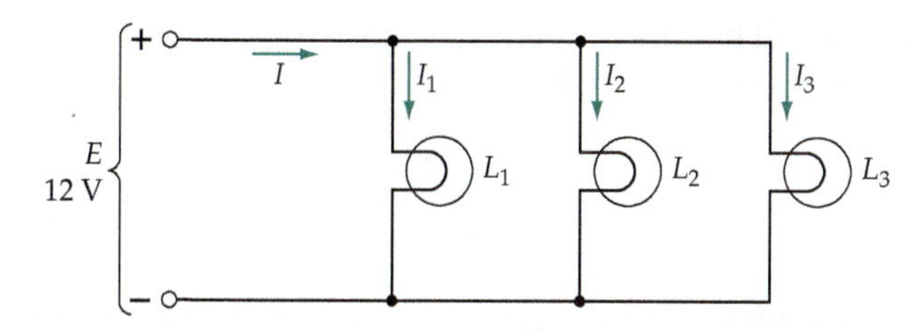

Figure 6-11 Parallel-connected lamps can be treated exactly like parallel-connected resistors.

Solution

$$P_1 = EI_1$$

giving,

$$I_1 = \frac{P_1}{E} = \frac{3\ \text{W}}{12\ \text{V}}$$

$$= 0.25\ \text{A}$$

$$I_2 = \frac{P_2}{E} = \frac{5\ \text{W}}{12\ \text{V}}$$

$$\approx 0.42\ \text{A}$$

$$I_3 = \frac{P_3}{E} = \frac{10\ \text{W}}{12\ \text{V}}$$

$$\approx 0.83\ \text{A}$$

Total supply power,

$$P = P_1 + P_2 + P_3 = 3\ \text{W} + 5\ \text{W} + 10\ \text{W}$$

$$= 18\ \text{W}$$

Practice Problems

6-5.1 Calculate the power dissipated in each resistor in Problem 6-2.1 and the total power taken from the supply.

6-5.2 A third resistor connected in parallel with the two resistors in Problem 6-4.1 causes the power supplied to the circuit to increase to 1.5 W. Determine the power dissipated in each resistor.

6-6 OPEN-CIRCUITS AND SHORT-CIRCUITS IN PARALLEL CIRCUITS

When one of the components in a parallel resistor circuit is open-circuited, as illustrated in Figure 6-12(a), no current flows through that branch of the circuit. The other branch currents are not affected by such an open circuit, because each of the other resistors still has the full supply voltage applied to its terminals. When I_1 goes to zero, the total current drawn from the supply is reduced from,

$$I = I_1 + I_2 + I_3$$

to

$$I = I_2 + I_3$$

Because there is no voltage drop across a resistor with zero current through it, the voltage measured at the open-circuit is the supply voltage (E). An open-circuit can also occur in the supply line to the parallel resistor combination, as shown in Figure 6-12(b). In this case the open-circuit is effectively an infinite resistance in series with the voltage source and resistors. The result is that no supply current can flow, and consequently I_1, I_2, and I_3 are all zero. Once again, the supply voltage (E) appears at the open-circuit terminals.

Figure 6-12(c) shows a short-circuit across resistor R_3. This has the same effect whether it is across R_1, R_2, or R_3, or across the voltage source terminals. In this case, the current that flows through each resistor is effectively zero. Also, the supply has a short-circuit across its terminals, and consequently, the supply short-circuit current flows:

$$I_{sc} = \frac{E}{r_i} \tag{6-12}$$

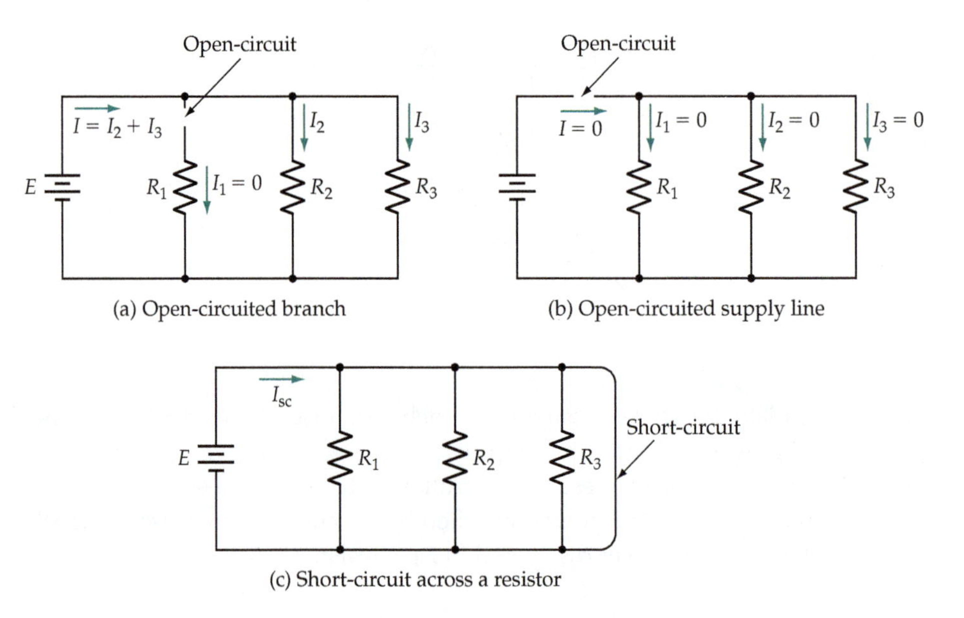

Figure 6-12 An open-circuit in one branch of a parallel circuit interrupts the current flow in that branch. An open-circuit in the supply line interrupts the supply current. A short-circuit in any branch of a parallel circuit shorts the supply.

where r_i is the battery internal resistance (see Section 10-2). In this situation, an abnormally large supply current flows and the battery could be seriously damaged.

Example 6-11

Determine the supply current for the circuit analyzed in Example 6-2 (reproduced in Figure 6-13) when (a) R_1 is open-circuited, (b) R_3 and R_4 are open-circuited, (c) R_4 is short-circuited. Assume that the supply source resistance is $r_i = 0.1\ \Omega$.

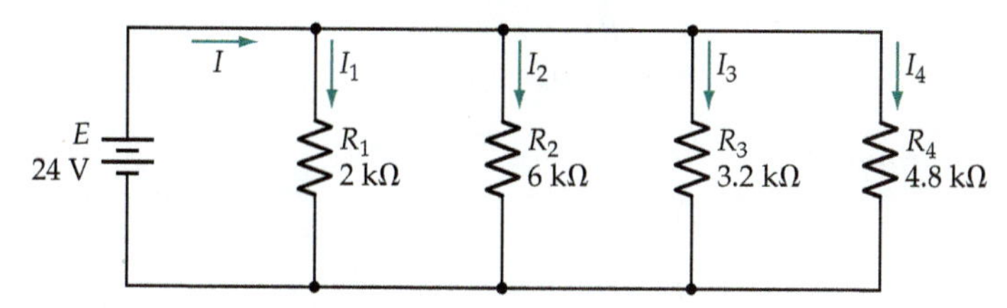

Figure 6-13 Parallel resistor circuit for Example 6-11.

Solution

(a)
$$I = I_2 + I_3 + I_4 = 4\text{ mA} + 7.5\text{ mA} + 5\text{ mA}$$
$$= 16.5\text{ mA}$$

(b)
$$I = I_1 + I_2 = 12\text{ mA} + 4\text{ mA}$$
$$= 16\text{ mA}$$

(c) Eq. 6-12,
$$I_{sc} = E/r_i = 24\text{ V}/0.1\ \Omega$$
$$= 240\text{ A}$$

Practice Problem

6-6.1 For the circuit described in Problem 6-4.2, determine the supply voltage, the total current taken from the supply when R_1 is open-circuited, and the voltage measured at the open-circuit.

Summary of Formulas

- ***Total supply current to parallel circuit:***

$$I = I_1 + I_2 + I_3 + \cdots + I_n$$

$$I = E\left[\frac{1}{R_1} + \frac{1}{R_2} + \frac{1}{R_3} + \cdots + \frac{1}{R_n}\right]$$

- ***Total parallel resistance:***

$$R_{eq} = (R_1^{-1} + R_2^{-1} + R_3^{-1} + \cdots + R_n^{-1})^{-1}$$

- ***Total parallel conductance:***

$$G_{eq} = G_1 + G_2 + G_3 + \cdots + G_n$$

- ***Two resistors in parallel:***

$$R_{eq} = \frac{R_1 \times R_2}{R_1 + R_2}$$

- ***Current through R_1:***

$$I_1 = I\left[\frac{R_2}{R_1 + R_2}\right]$$

- ***Current through R_n:***

$$I_n = \frac{IR_{eq}}{R_n}$$

- ***Total power supplied:***

$$P = P_1 + P_2 + P_3 + \cdots + P_n$$

Review Questions

Section 6-1

6-1 Sketch the circuit diagram for two parallel-connected resistors supplied by a battery. Show all voltage polarities and current directions, and write expressions for the current through each resistor and for the total current drawn from the battery.

6-2 State Kirchhoff's current law. Also, explain the component voltage relationships in a parallel circuit.

Section 6-2

6-3 Derive an equation for the equivalent resistance of n resistors connected in parallel. Repeat the derivation using conductances instead of resistances.

Section 6-3

6-4 Show that for conductances connected in parallel:

$$G_{eq} = G_1 + G_2 + G_3 + \cdots + G_n$$

Section 6-4

6-5 Derive equations for the current through each of two parallel connected resistors, in terms of the total supply current. Also, show that for two parallel resistors,

$$R_1 \| R_2 = \frac{R_1 \times R_2}{R_1 + R_2}$$

6-6 For two resistors R_1 and R_2 connected in parallel, derive the equation for I_1 in terms of supply current and resistor values.

Section 6-5

6-7 Write equations for the total power dissipated in a parallel circuit in terms of (a) the power dissipated in each component, (b) the supply voltage and equivalent resistance of the circuit, and (c) the supply current and circuit equivalent resistance.

Section 6-6

6-8 Discuss the effects that open-circuit and short-circuit conditions can have on parallel resistance circuits.

Problems

Section 6-1

6-1 For the circuit shown in Figure 6-14, determine the individual resistor currents and the current taken from the battery.

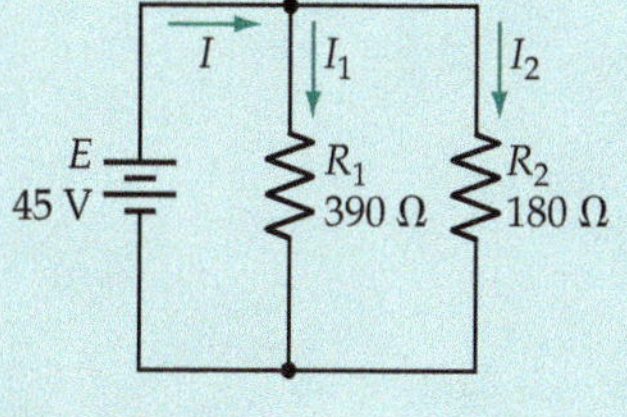

Figure 6-14

6-2 Four lamps are connected in parallel to a 120 V supply. Three of the lamp currents are: $I_{L1} = 0.8$ A, $I_{L2} = 0.5$ A, and $I_{L3} = 0.2$ A. If the total supply current is 2.75 A, calculate the resistance of each of the four lamps.

6-3 Calculate the current through each resistor and the total battery current for the circuit shown in Figure 6-15.

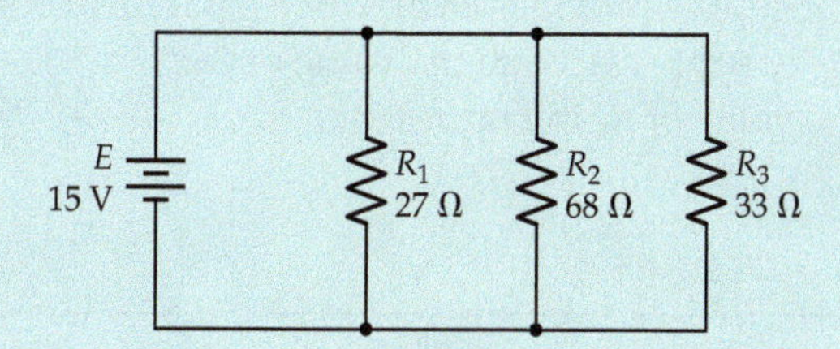

Figure 6-15

Section 6-2

6-4 Four parallel-connected resistors are supplied from a 25 V battery. Three of the resistors have values of $R_1 = 1$ kΩ, $R_2 = 12$ kΩ, and $R_3 = 8.2$ kΩ. If the battery current is measured as 36.5 mA, determine the value of the fourth resistor.

6-5 Determine the equivalent resistance for the two parallel-connected resistors shown in Figure 6-14. Also draw the equivalent circuit, and use it to calculate the total supply current.

6-6 Determine the equivalent resistance for the three parallel-connected resistors shown in Figure 6-15. Also draw the equivalent circuit, and use it to calculate the total supply current.

6-7 Determine the voltage E_2 for the circuit shown in Figure 6-16.

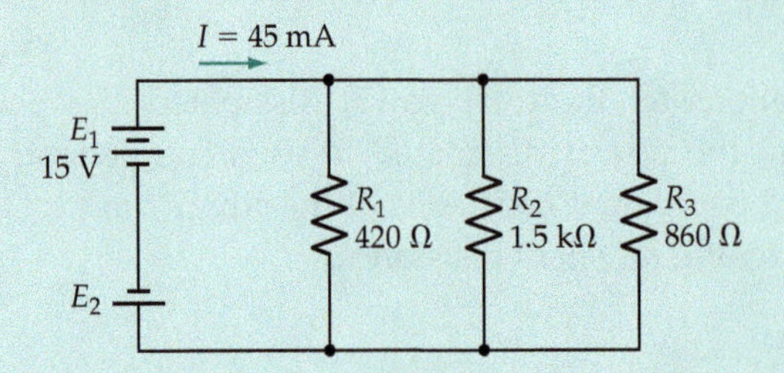

Figure 6-16

6-8 If the battery voltage drops to 14.2 V in the circuit of Figure 6-15, calculate the new resistance value for R_3 that will return the battery current to the level at $E = 15$ V.

Section 6-3

6-9 For the circuit in Figure 6-14, convert each resistance value into a conductance. Use the conductance values to calculate all current levels.

6-10 For the circuit in Figure 6-15, convert each resistance value into a conductance, and use the conductance values to determine the ciruit current levels.

6-11 A 175 Ω resistor is connected in parallel with a 265 Ω resistor. Determine the value of a third parallel resistor if the total conductance of the three in parallel is to be 13.44 mS.

6-12 A 30 V power supply provides 88 mA to two parallel-connected conductances. If $G_1 = 0.6$ mS, calculate the value of G_2.

6-13 Calculate the value of R_4 in Figure 6-17.

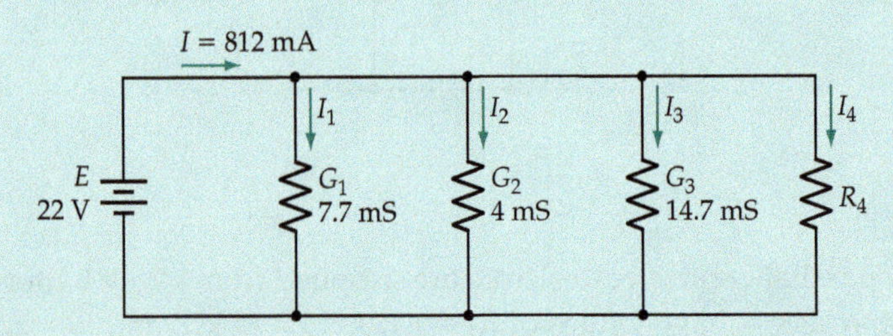

Figure 6-17

Section 6-4

6-14 Two conductances connected in parallel take a total current of 750 mA. If $G_1 = 14$ mS and $G_2 = 16$ mS, calculate the current through each conductance.

6-15 Use the current-divider rule to determine each branch current in the circuit of Figure 6-14.

6-16 For the circuit of Figure 6-15, use the current-divider rule to determine the branch currents.

6-17 Apply the current-divider rule to the circuit in Figure 6-16 to determine each branch curent.

6-18 Calculate the equivalent resistance and total supply current for the circuit in Figure 6-18, and use the current-divider rule to determine each branch current.

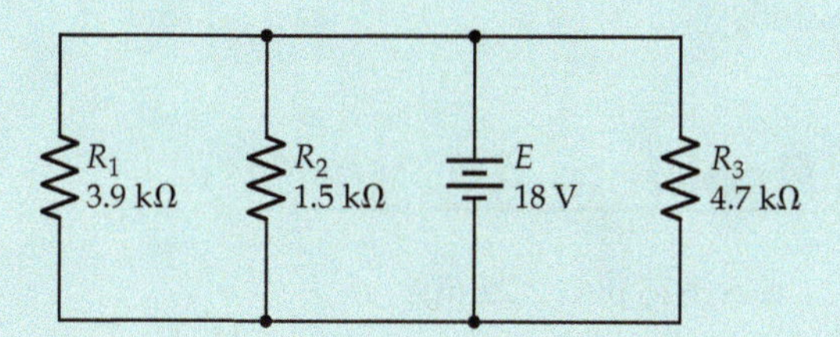

Figure 6-18

Section 6-5

6-19 For the circuit shown in Figure 6-14, calculate the power dissipated in each resistor and the total power supplied by the battery.

6-20 Calculate the power dissipated in each resistor and the total circuit power for the circuit in Figure 6-15.

6-21 Four 115 V lamps are connected in parallel to a 115 V supply. The lamp ratings are $P_1 = 100$ W, $P_2 = 40$ W, $P_3 = 60$ W, and $P_4 = 25$ W. Determine the current that flows through each lamp and the total power delivered from the supply.

6-22 Three parallel-connected resistors, $R_1 = 1.2$ kΩ, $R_2 = 3.9$ kΩ, and $R_3 = 4.7$ kΩ, are supplied from a 5 V source. Calculate the equivalent resistance of the three in parallel. Also determine the individual resistor currents and the total power dissipated in the circuit.

6-23 Two parallel-connected resistors ($R_1 = 620$ Ω, and $R_2 = 880$ Ω) take a total current of 75 mA from the supply. Calculate I_1, I_2, and the level of the supply voltage. Also, determine the power dissipated in each resistor.

6-24 The following parallel-connected lamps have a 120 V supply: $L_1 = 60$ W, $L_2 = 40$ W, and $L_3 = 100$ W. If the supply current is not to exceed 3 A, determine the power of one additional lamp that may be connected.

6-25 For the circuit in Figure 6-16, calculate the power dissipated in each resistor and the total power supplied by the battery.

6-26 Calculate the power dissipated in each resistor in the circuit in Figure 6-17. Also determine the total power supplied by the battery.

Section 6-6

6-27 If E_2 in Figure 6-16 becomes short-circuited, calculate the value of the additional parallel resistance required to give $I = 100$ mA.

6-28 Repeat Problem 6-22 when R_2 is open-circuited.

6-29 If R_4 in Figure 6-17 becomes open-circuited, determine the new level of supply voltage required to keep the total current at 812 mA.

6-30 If the voltage source in Figure 6-18 has a resistance of 0.9 Ω, calculate the source current when R_2 becomes short-circuited.

6-31 Determine each branch current in Figure 6-17 when one of the connections between G_3 and R_4 is open-circuited.

6-32 Calculate each resistor current in Figure 6-18 when the connection between the battery positive terminal and the junction of R_1 and R_2 is open-circuited.

Practice Problem Answers

6-1.1 667 μA, 1 mA, 625 μA, 2.29 mA
6-2.1 22.34 kΩ, 627 μA
6-2.2 37.5 kΩ
6-2.3 10.9 V
6-3.1 14.7 μS, 17.8 μS, 12.2 μS, 44.7 μS
6-3.2 133 μS, 200 μS, 125 μS, 458 μS
6-4.1 24.8 mA, 30.2 mA
6-4.2 728 μA, 511 μA, 890 μA
6-5.1 2.88 mW, 3.5 mW, 2.39 mW, 8.77 mW
6-5.2 418 mW, 511 mW, 571 mW
6-6.1 24 V, 1.4 mA, 24 V

CHAPTER 7
Series-Parallel Resistor Circuits

CONTENTS

Objectives

You will be able to:

1 Indicate current directions and voltage polarities throughout series-parallel resistor circuits.

2 Reduce a series-parallel resistor circuit to a simple series or parallel equivalent circuit.

3 Solve problems involving equivalent resistance, current levels, voltage drops, and power dissipations in series-parallel resistor circuits.

4 Troubleshoot series-parallel resistor circuits for open-circuit and short-circuit conditions.

INTRODUCTION

Not all resistor circuits are simple series or parallel arrangements. Many are combinations of parallel resistors connected in series with other resistors or combined with other parallel resistor groups. These are described as series-parallel circuits. The simplest approach to analyzing a series-parallel circuit is to resolve each series-connected group of resistors into its equivalent resistance, and each parallel-connected group into its equivalent resistance. The process is repeated as many times as necessary. As in all types of circuits, open-circuit and short-circuit conditions affect the currents and voltage drops throughout a series-parallel circuit.

7-1 SERIES-PARALLEL RESISTOR CIRCUITS

Simple Series-Parallel Circuit

Series-parallel resistor circuits consist of combinations of series-connected and parallel-connected resistors. Figure 7-1(a) shows a very simple three-resistor series-parallel circuit on a printed circuit board, with its circuit diagram illustrated in Figure 7-1(b). Resistors R_2 and R_3 are seen to be connected in parallel, and resistor R_1 is in series with the parallel combination of R_2 and R_3. Different current levels flow in each branch, and the component voltage drops depend on the branch currents and the component resistances. The supply current depends on the supply voltage and on the circuit resistance offered to the voltage source. Kirchhoff's voltage and current laws are applied for analyzing series-parallel circuits.

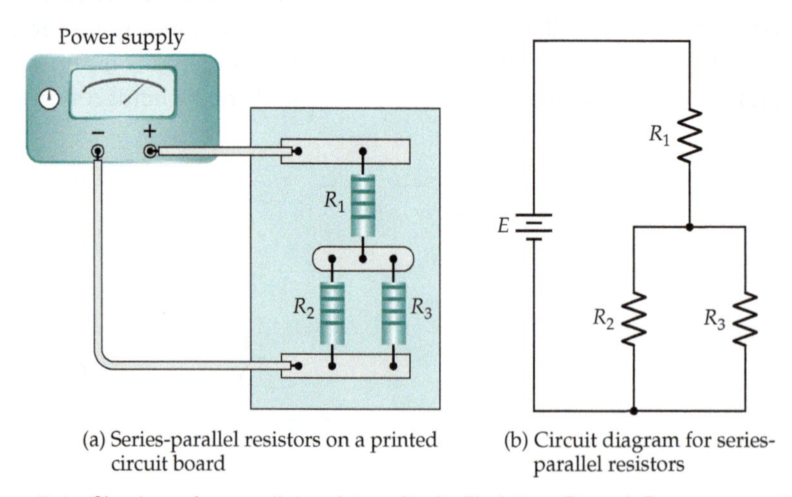

(a) Series-parallel resistors on a printed circuit board

(b) Circuit diagram for series-parallel resistors

Figure 7-1 Simple series-parallel resistor circuit. Resistors R_2 and R_3 are connected in parallel and resistor R_1 is in series with $R_2 \| R_3$.

Series-Parallel Equivalent Circuits

In the circuit shown in Figure 7-2(a) [similar to Figure 7-1(b)], resistors R_2 and R_3 are in parallel, and together they are in series with R_1. The total supply current is easily calculated if R_2 and R_3 are first replaced with their equivalent resistance ($R_{eq} = R_2 \| R_3$) as illustrated in Figure 7-2(b). The circuit now becomes a simple two-resistor series circuit.

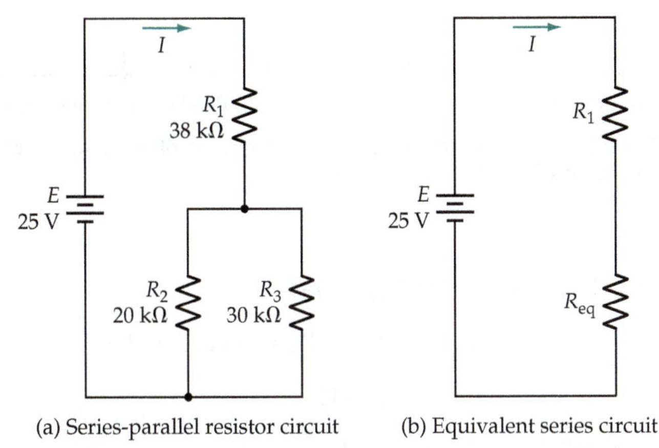

(a) Series-parallel resistor circuit (b) Equivalent series circuit

Figure 7-2 A series-parallel resistor circuit can be simplified by replacing each group of parallel-connected resistors with its equivalent resistance.

Example 7-1

Calculate the supply current for the circuit shown in Figure 7-2(a).

Solution

Draw the equivalent circuit as shown in Figure 7-2(b),

$$R_{eq} = R_2 \| R_3$$

Eq. 6-7
$$R_{eq} = \frac{R_2 \times R_3}{R_2 + R_3} = \frac{20\ \text{k}\Omega \times 30\ \text{k}\Omega}{20\ \text{k}\Omega + 30\ \text{k}\Omega}$$

$$= 12\ \text{k}\Omega$$

$$I = \frac{E}{R_1 + R_{eq}} = \frac{25\ \text{V}}{38\ \text{k}\Omega + 12\ \text{k}\Omega}$$

$$= 0.5\ \text{mA}$$

Another series-parallel resistor combination is shown in Figure 7-3(a). In this case the circuit is reduced to a simple parallel circuit when R_2 and R_3 are replaced by their equivalent resistance ($R_{eq} = R_2 + R_3$) [see Figure 7-3(b)].

Example 7-2

Determine the supply current for the circuit in Figure 7-3(a).

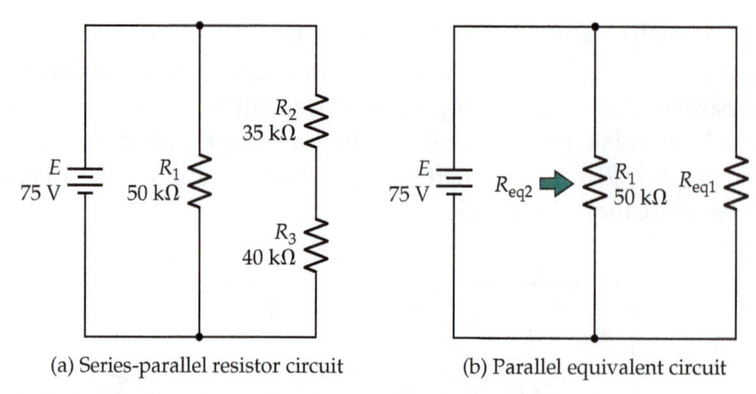

Figure 7-3 To simplify some series-parallel circuits, groups of series-connected resistors should first be replaced with their equivalent resistances.

Solution

Draw the equivalent circuit as in Figure 7-3(b).

$$R_{eq1} = R_2 + R_3 = 35\text{ k}\Omega + 40\text{ k}\Omega$$

$$= 75\text{ k}\Omega$$

R_1 and R_{eq1} in parallel

$$R_{eq2} = R_1 \| R_{eq1}$$

Eq. 6-7

$$R_{eq2} = \frac{R_1 \times R_{eq1}}{R_1 + R_{eq1}} = \frac{50\text{ k}\Omega \times 75\text{ k}\Omega}{50\text{ k}\Omega + 75\text{ k}\Omega}$$

$$= 30\text{ k}\Omega$$

$$I = \frac{E}{R_{eq2}} = \frac{75\text{ V}}{30\text{ k}\Omega}$$

$$= 2.5\text{ mA}$$

Practice Problems

7-1.1 A series-parallel circuit as in Figure 7-2(a) has a 5 V supply and resistor values of $R_1 = 7.5$ kΩ, $R_2 = 5$ kΩ, and $R_3 = 8$ kΩ. Calculate the supply current.

7-1.2 In the circuit for Problem 7-1.1, an additional resistor (R_4) is connected in parallel with R_1. If $R_4 = 3.3$ kΩ, determine the new level of supply current.

7-2 CURRENTS IN A SERIES-PARALLEL CIRCUIT

The circuit of Figure 7-2(a) is reproduced in Figure 7-4 with the branch currents and voltages identified. It is seen that the supply current (I) flows through resistor R_1 and that it splits up into I_2 and I_3 to flow through R_2 and R_3,

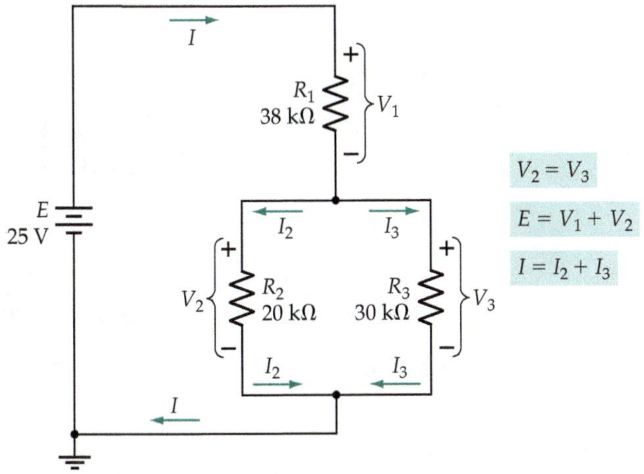

Figure 7-4 Currents and voltages in the series-parallel circuit of Figure 7-2(a). Supply current I flows through R_1, then splits into I_2 and I_3 to pass through R_2 and R_3, respectively.

respectively. Returning to the supply negative terminal, the current is once again I. It is seen that

$$I = I_2 + I_3$$

Similarly, the supply current splits up between the resistors in Figure 7-5, which is a reproduction of the circuit shown in Figure 7-3(a). Here, I_1 flows through R_1, and I_2 flows through R_2 and R_3. The supply current is

$$I = I_1 + I_2$$

In each of these cases (Figures 7-4 and 7-5), the current through the individual resistors can be calculated using the current-divider rule.

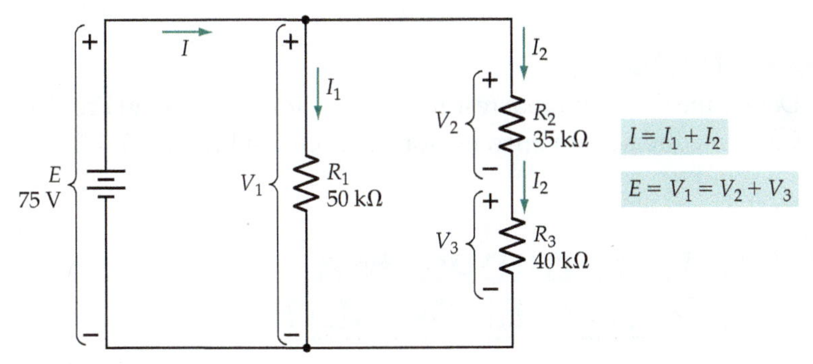

Figure 7-5 Currents and voltages in the series-parallel circuit of Figure 7-3(a). Supply current I splits into I_1 and I_2 to pass through R_1 and $(R_2 + R_3)$, respectively.

Example 7-3

Calculate the branch currents in the circuit of Figure 7-4, using the information available from Example 7-1.

Solution

From Example 7-1, $I = 0.5 \text{ mA}$

From Eq. 6-8, $I_2 = I\left[\dfrac{R_3}{R_2 + R_3}\right] = 0.5 \text{ mA}\left[\dfrac{30 \text{ k}\Omega}{20 \text{ k}\Omega + 30 \text{ k}\Omega}\right]$

$= 0.3 \text{ mA}$

and, $I_3 = I\left[\dfrac{R_2}{R_2 + R_3}\right] = 0.5 \text{ mA}\left[\dfrac{20 \text{ k}\Omega}{20 \text{ k}\Omega + 30 \text{ k}\Omega}\right]$

$= 0.2 \text{ mA}$

Example 7-4

Find the individual branch currents for the circuit in Figure 7-5, using the information given in Example 7-2.

Solution

From Example 7-2, $I = 2.5 \text{ mA}$

and, $R_{eq1} = 75 \text{ k}\Omega$

From Eq. 6-8 $I_1 = I\left[\dfrac{R_{eq1}}{R_{eq1} + R_1}\right] = 2.5 \text{ mA}\left[\dfrac{75 \text{ k}\Omega}{75 \text{ k}\Omega + 50 \text{ k}\Omega}\right]$

$= 1.5 \text{ mA}$

and $I_2 = I\left[\dfrac{R_1}{R_{eq1} + R_1}\right] = 2.5 \text{ mA}\left[\dfrac{50 \text{ k}\Omega}{75 \text{ k}\Omega + 50 \text{ k}\Omega}\right]$

$= 1 \text{ mA}$

Practice Problems

7-2.1 Determine the individual resistor currents for the circuit in Problem 7-1.1.

7-2.2 Calculate the resistor currents for the circuit in Problem 7-1.2.

7-3 VOLTAGE DROPS IN A SERIES-PARALLEL CIRCUIT

As always, the voltage drop across any resistor is the product of the resistance value and the current through the resistor.

In Figure 7-4 $V_1 = IR_1$

And $V_2 = I_2R_2 = V_3 = I_3R_3$

Also $E = V_1 + V_2$

In Figure 7-5 $V_1 = I_1R_1, V_2 = I_2R_2, V_3 = I_2R_3,$

And $E = V_1 = V_2 + V_3$

Once the branch currents are known, the voltages across each resistor can be readily calculated, and the component power dissipations can be determined. In some circumstances it might be more convenient to calculate the resistor voltages first, then use these voltages to determine the branch currents.

Example 7-5

Calculate the voltage drop across each resistor in Figure 7-6(a), and the voltages V_A and V_B that would be measured with respect to ground.

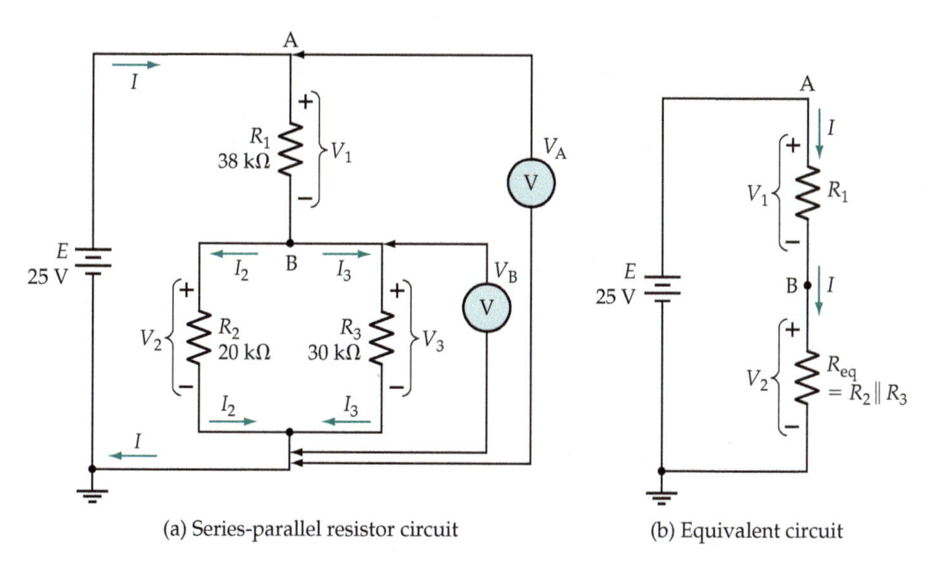

Figure 7-6 Series-parallel resistor circuit reproduced from Figure 7-4.

Solution

From Examples 7-1 and 7-3,

$$I = 0.5 \text{ mA}, I_2 = 0.3 \text{ mA, and } I_3 = 0.2 \text{ mA}$$

$$V_1 = I_1R_1 = 0.5 \text{ mA} \times 38 \text{ k}\Omega$$
$$= 19 \text{ V}$$

$$V_2 = I_2R_2 = 0.3 \text{ mA} \times 20 \text{ k}\Omega$$
$$= 6 \text{ V}$$

$$V_3 = I_3R_3 = 0.2 \text{ mA} \times 30 \text{ k}\Omega$$
$$= 6 \text{ V}$$

$$V_B = V_3 = 6 \text{ V}$$

$$V_A = V_1 + V_2 = E$$
$$= 25 \text{ V}$$

Computer analysis of the type of circuit referred to in Examples 7-1, 7-3, and 7-5 is offered in Section 28-2.

Example 7-6

Using the voltage-divider rule, analyze the circuit shown in Figure 7-6(a) to determine the resistor voltage drops and branch currents.

Solution

In Figure 7-6(b), $R_{eq} = R_2 \| R_3 = 12\ \text{k}\Omega$ (from Example 7-1)

For voltage divider R_1 and R_{eq}

$$V_2 = E\frac{R_{eq}}{R_1 + R_{eq}} = 25\ \text{V} \times \frac{12\ \text{k}\Omega}{38\ \text{k}\Omega + 12\ \text{k}\Omega}$$

$$= 6\ \text{V}$$

$$V_1 = E\frac{R_1}{R_1 + R_{eq}} = 25\ \text{V} \times \frac{38\ \text{k}\Omega}{38\ \text{k}\Omega + 12\ \text{k}\Omega}$$

$$= 19\ \text{V}$$

$$I_1 = \frac{V_1}{R_1} = \frac{19\ \text{V}}{38\ \text{k}\Omega}$$

$$= 0.5\ \text{mA}$$

$$I_2 = \frac{V_2}{R_2} = \frac{6\ \text{V}}{20\ \text{k}\Omega}$$

$$= 0.3\ \text{mA}$$

$$I_3 = \frac{V_3}{R_3} = \frac{6\ \text{V}}{30\ \text{k}\Omega}$$

$$= 0.2\ \text{mA}$$

Example 7-7

Analyze the circuit shown in Figure 7-7 to determine the resistor voltage drops and branch currents.

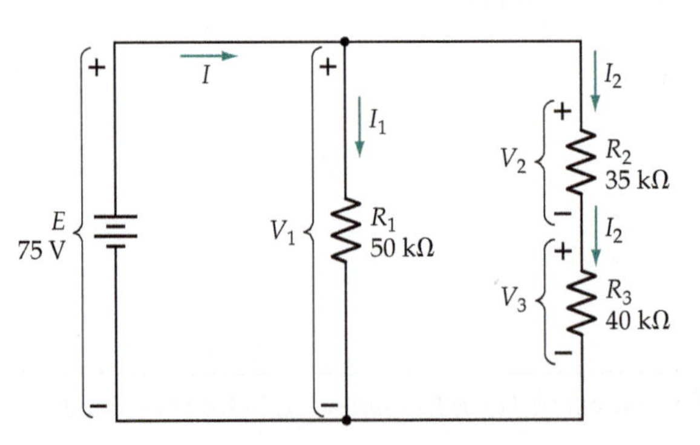

Figure 7-7 Series-parallel circuit reproduced from Figure 7-5.

Solution

$$V_1 = E = 75 \text{ V}$$

$$I_1 = \frac{V_1}{R_1} = \frac{75 \text{ V}}{50 \text{ k}\Omega}$$

$$= 1.5 \text{ mA}$$

For voltage divider R_2 and R_3

$$V_3 = E\frac{R_3}{R_2 + R_3} = 75 \text{ V} \times \frac{40 \text{ k}\Omega}{40 \text{ k}\Omega + 35 \text{ k}\Omega}$$

$$= 40 \text{ V}$$

$$V_2 = E\frac{R_2}{R_2 + R_3} = 75 \text{ V} \times \frac{35 \text{ k}\Omega}{40 \text{ k}\Omega + 35 \text{ k}\Omega}$$

$$= 35 \text{ V}$$

$$I_2 = \frac{V_2}{R_2} = \frac{35 \text{ V}}{35 \text{ k}\Omega}$$

$$= 1 \text{ mA}$$

$$I_3 = \frac{V_3}{R_3} = \frac{40 \text{ V}}{40 \text{ k}\Omega}$$

$$= 1 \text{ mA}$$

Practice Problems

7-3.1 Determine each resistor voltage drop in the circuit for Problem 7-1.1.

7-3.2 An additional resistor (R_4) is connected in series with R_3 in the circuit for Problem 7-1.1. If $R_4 = 3.3 \text{ k}\Omega$, determine the voltage drops across each resistor.

7-4 LADDER NETWORKS

As the name suggests, a *ladder network* is a ladder-shaped resistor network, and it is a special case of a series-parallel circuit. Like all series-parallel circuits, a ladder network can be analyzed by replacing series-connected resistors and parallel-connected resistors with their equivalent resistance values. Figure 7-8(a) shows a six-resistor ladder network, and Figures 7-8(b) and (c) illustrate the process of circuit simplification.

Example 7-8

Determine the voltage across resistor R_6 in the ladder network shown in Figure 7-8(a).

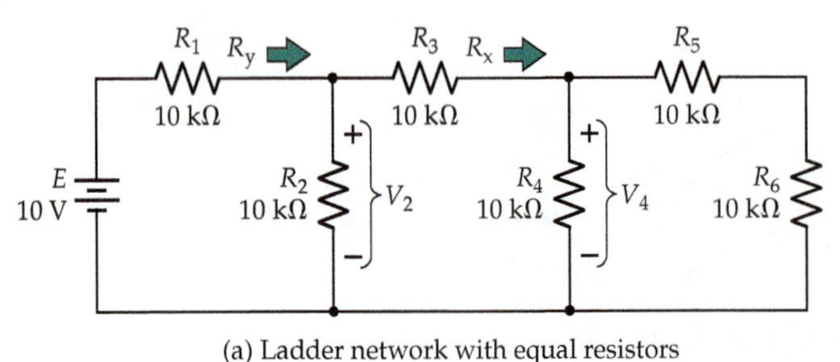

(a) Ladder network with equal resistors

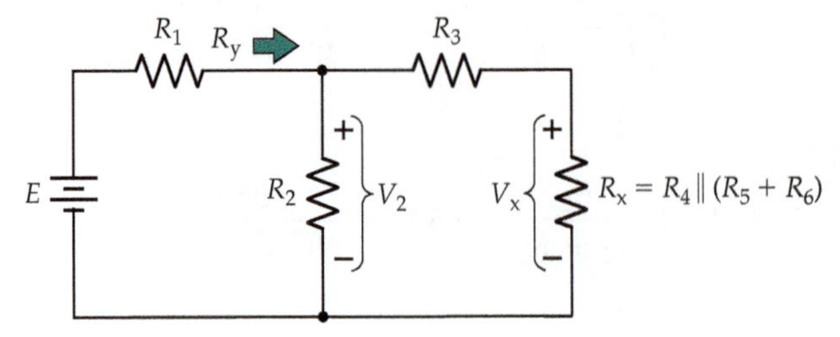

(b) Circuit simplification by replacement of $R_4 \| (R_5 + R_6)$

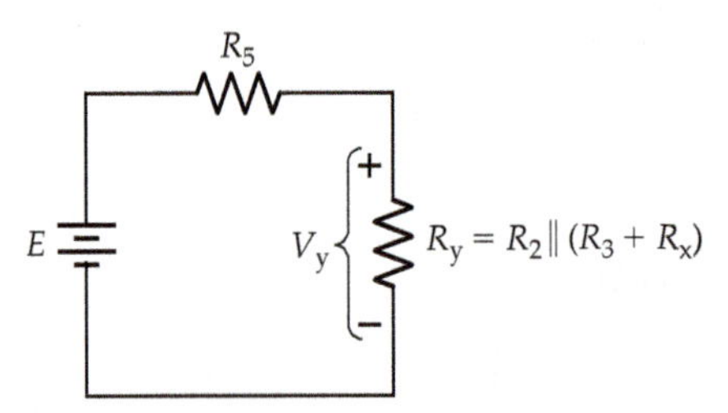

(c) Further simplification by replacement of $R_2 \| (R_3 + R_x)$

Figure 7-8 A ladder network is a series-parallel circuit. The circuit can be analyzed by the usual process of replacing series-connected and parallel-connected components with their respective equivalent resistances.

Solution

For Figure 7-8(b),

$$R_x = R_4 \| (R_5 + R_6) = 10\ \text{k}\Omega \| (10\ \text{k}\Omega + 10\ \text{k}\Omega)$$

$$= [(10\ \text{k}\Omega)^{-1} + (10\ \text{k}\Omega + 10\ \text{k}\Omega)^{-1}]^{-1}$$

$$= 6.67\ \text{k}\Omega$$

For Figure 7-8(c),

$$R_y = R_2 \| (R_3 + R_x) = 10\ \text{k}\Omega \| (10\ \text{k}\Omega + 6.67\ \text{k}\Omega)$$

$$= [(10\ \text{k}\Omega)^{-1} + (10\ \text{k}\Omega + 6.67\ \text{k}\Omega)^{-1}]^{-1}$$

$$= 6.25\ \text{k}\Omega$$

$$V_y = E\frac{R_y}{R_1 + R_y} = 10\ \text{V} \times \frac{6.25\ \text{k}\Omega}{10\ \text{k}\Omega + 6.25\ \text{k}\Omega}$$

$$= 3.75\ \text{V}$$

$$V_2 = V_y = 3.75\ \text{V}$$

$$V_x = V_2 \frac{R_x}{R_3 + R_x} = 3.75\ \text{V} \times \frac{6.67\ \text{k}\Omega}{10\ \text{k}\Omega + 6.67\ \text{k}\Omega}$$
$$= 1.5\ \text{V}$$
$$V_4 = V_x = 1.5\ \text{V}$$
$$V_6 = V_4 \frac{R_6}{R_5 + R_6} = 1.5\ \text{V} \times \frac{10\ \text{k}\Omega}{10\ \text{k}\Omega + 10\ \text{k}\Omega}$$
$$= 0.75\ \text{V}$$

Some ladder networks have a special relationship between component values that allows them to be analyzed quite quickly. Two such circuits are shown in Figure 7-9. For both circuits,

$$R_5 + R_6 = R_4$$

Also, as demonstrated in Example 7-9,

$$R_3 + R_x = R_2$$

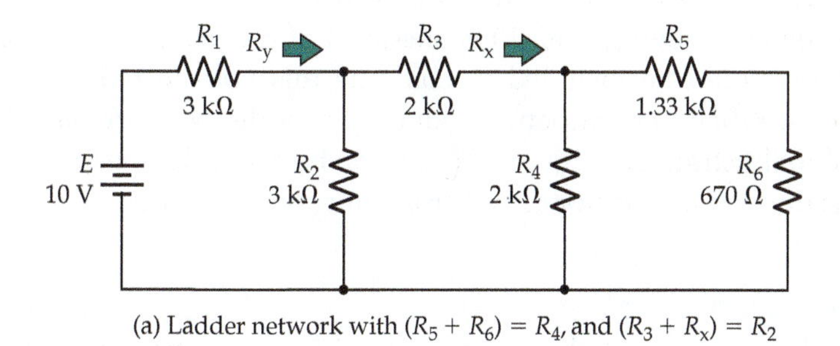

(a) Ladder network with $(R_5 + R_6) = R_4$, and $(R_3 + R_x) = R_2$

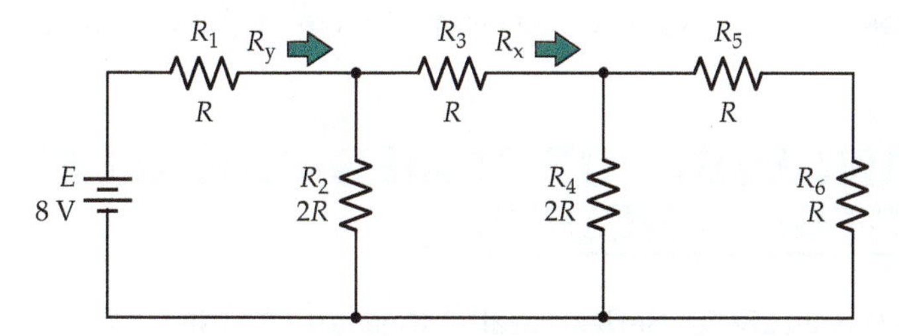

(b) Ladder network with R and $2R$ resistor values

Figure 7-9 Two ladder networks with special relationships between the resistance values.

Example 7-9

Analyze the circuit in Figure 7-9(a) to determine the voltage across resistor R_6.

Solution

$$R_x = R_4 \| (R_5 + R_6) = 2\ \text{k}\Omega \| (1.33\ \text{k}\Omega + 670\ \Omega)$$
$$= [(2\ \text{k}\Omega)^{-1} + (1.33\ \text{k}\Omega + 670\ \text{k}\Omega)^{-1}]^{-1}$$
$$= 1\ \text{k}\Omega$$

$$
\begin{aligned}
R_y &= R_2 \| (R_3 + R_x) = 3\text{ k}\Omega \| (2\text{ k}\Omega + 1\text{ k}\Omega) \\
&= [(3\text{ k}\Omega)^{-1} + (2\text{ k}\Omega + 1\text{ k}\Omega)^{-1}]^{-1} \\
&= 1.5\text{ k}\Omega \\
V_2 = V_y &= E\frac{R_y}{R_1 + R_y} = E \times \frac{1.5\text{ k}\Omega}{3\text{ k}\Omega + 1.5\text{ k}\Omega} \\
&= E/3 \\
V_4 = V_x &= V_2\frac{R_x}{R_3 + R_x} = (E/3) \times \frac{1\text{ k}\Omega}{2\text{ k}\Omega + 1\text{ k}\Omega} \\
&= E/9 \\
V_6 &= V_4\frac{R_6}{R_5 + R_6} = (E/9) \times \frac{670\ \Omega}{1.33\text{ k}\Omega + 670\ \Omega} \\
&= E/27
\end{aligned}
$$

Example 7-9 shows that for the circuit in Figure 7-9(a), each subsequent junction voltage (moving from E) is one-third of the previous voltage. Additional stages could be included to continue this voltage reduction to new lower levels. When the component values of a ladder network have the R:2R relationship illustrated in Figure 7-9(b), it can be easily demonstrated that the voltage is divided by a factor of two from junction to junction.

Practice Problem

7-4.1 Analyze the circuit shown in Figure 7-9(b) to determine the voltages across resistors R_2, R_4, and R_6 in relation to the input voltage E.

7-5 ANALYSIS OF SERIES-PARALLEL RESISTOR CIRCUITS

Analysis Procedure for Series-Parallel Resistor Circuits:

1. ***Draw a circuit diagram identifying all components by number and showing all currents and resistor voltage drops.***
2. ***Convert all series branches of two or more resistors into a single equivalent resistance.***
3. ***Convert all parallel combinations of two or more resistors into a single equivalent resistance.***
4. ***Repeat procedures 2 and 3 until the desired level of simplification is achieved.***

When the above procedure is followed, the final circuit should be a straightforward series or parallel circuit, which can be analyzed in the usual way.

Once the current through each equivalent resistance, or the voltage across it, is known, the original circuit can be used to determine individual resistor currents and voltages.

Example 7-10

Analyze the series-parallel circuit in Figure 7-10(a) to calculate all resistor currents and voltages. Determine the voltage that would be measured at points A, B, and C with respect to ground.

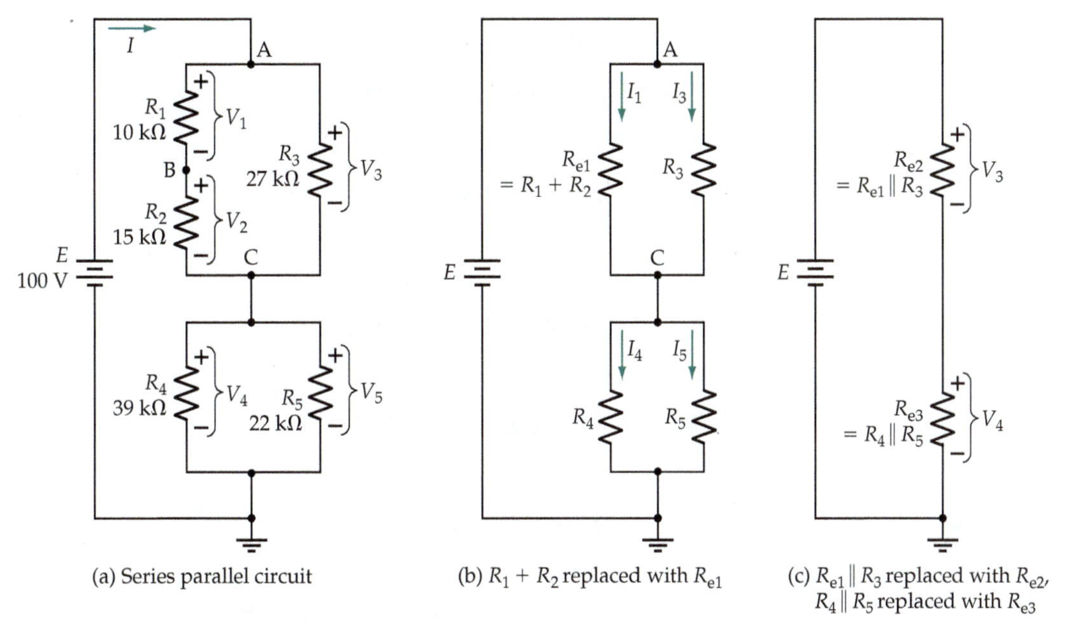

(a) Series parallel circuit (b) $R_1 + R_2$ replaced with R_{e1} (c) $R_{e1} \| R_3$ replaced with R_{e2}, $R_4 \| R_5$ replaced with R_{e3}

Figure 7-10 To analyze a series-parallel circuit, each group of series-connected resistors is first replaced with its equivalent resistance, and each parallel-connected group is replaced with its equivalent resistance. The process is repeated as many times as necessary.

Solution

$$R_{e1} = R_1 + R_2 = 10\ \text{k}\Omega + 15\ \text{k}\Omega$$

$$= 25\ \text{k}\Omega$$

The circuit is now modified as shown in Figure 7-10(b), with R_{e1} replacing R_1 and R_2.

$$R_{e2} = R_{e1} \| R_3$$

$$= \frac{R_{e1} \times R_3}{R_{e1} + R_3} = \frac{25\ \text{k}\Omega \times 27\ \text{k}\Omega}{25\ \text{k}\Omega + 27\ \text{k}\Omega}$$

$$= 12.98\ \text{k}\Omega$$

$$R_{e3} = R_4 \| R_5$$

$$= \frac{R_4 \times R_5}{R_4 + R_5} = \frac{39\ \text{k}\Omega \times 22\ \text{k}\Omega}{39\ \text{k}\Omega + 22\ \text{k}\Omega}$$

$$= 14.07\ \text{k}\Omega$$

The circuit is now further modified as shown in Figure 7-10(c), with R_{e2} replacing $R_{e1} \| R_3$ and R_{e3} replacing $R_4 \| R_5$. So,

$$I = \frac{E}{R_{e2} + R_{e3}} = \frac{100\ \text{V}}{12.98\ \text{k}\Omega + 14.07\ \text{k}\Omega}$$

$$\approx 3.7\ \text{mA}$$

$$V_3 = IR_{e2} = 3.7\ \text{mA} \times 12.98\ \text{k}\Omega$$

$$\approx 48\ \text{V}$$

$$V_4 = IR_{e3} = 3.7\ \text{mA} \times 14.07\ \text{k}\Omega$$

$$\approx 52\ \text{V}$$

$$I_1 = \frac{V_3}{R_1 + R_2} = \frac{48\ \text{V}}{10\ \text{k}\Omega + 15\ \text{k}\Omega}$$

$$\approx 1.92\ \text{mA}$$

$$V_1 = I_1 R_1 = 1.92\ \text{mA} \times 10\ \text{k}\Omega$$

$$\approx 19.2\ \text{V}$$

$$V_2 = I_1 R_2 = 1.92\ \text{mA} \times 15\ \text{k}\Omega$$

$$\approx 28.8\ \text{V}$$

$$I_3 = \frac{V_3}{R_3} = \frac{48\ \text{V}}{27\ \text{k}\Omega}$$

$$\approx 1.78\ \text{mA}$$

$$I_4 = \frac{V_4}{R_4} = \frac{52\ \text{V}}{39\ \text{k}\Omega}$$

$$\approx 1.33\ \text{mA}$$

$$I_5 = \frac{V_4}{R_5} = \frac{52\ \text{V}}{22\ \text{k}\Omega}$$

$$\approx 2.36\ \text{mA}$$

$$V_C = V_4 = 52\ \text{V}$$

$$V_B = V_C + V_2 = 52\ \text{V} + 28.8\ \text{V}$$

$$= 80.8\ \text{V}$$

$$V_A = E = 100\ \text{V}$$

Example 7-11

Determine the current through R_4 in the circuit shown in Figure 7-11. The supply voltage is $E = 15$ V and the resistors values are: $R_1 = 2.7$ kΩ, $R_2 = 820$ Ω, $R_3 = 1.2$ kΩ, and $R_4 = 3.9$ kΩ.

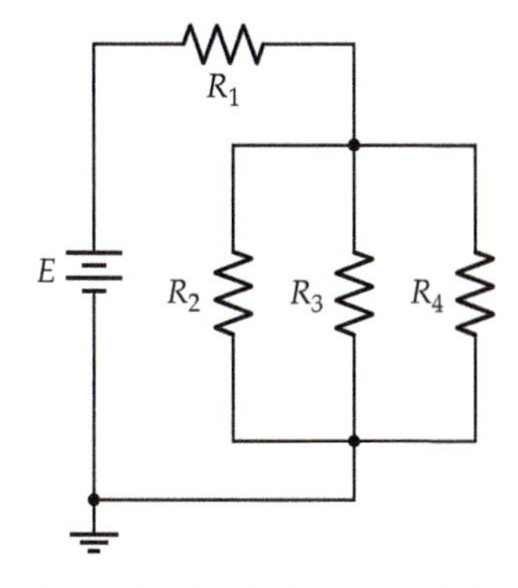

Figure 7-11 Series-parallel circuit for Example 7-11.

Solution

$$R_{eq} = R_2 \| R_3 \| R_4 = 820\ \Omega \| 1.2\ \text{k}\Omega \| 3.9\ \text{k}\Omega$$

Eq. 6-4,

$$R_{eq} = \left(R_2^{-1} + R_3^{-1} + R_4^{-1}\right)^{-1}$$
$$= (820\ \Omega^{-1} + 1.2\ \text{k}\Omega^{-1} + 3.9\ \text{k}\Omega^{-1})^{-1}$$
$$= 433\ \Omega$$

$$I_1 = \frac{E}{R_1 + R_{eq}} = \frac{15\ \text{V}}{2.7\ \text{k}\Omega + 433\ \Omega}$$
$$= 4.79\ \text{mA}$$

$$V_2 = V_3 = V_4 = I_1 R_{eq} = 4.79\ \text{mA} \times 433\ \Omega$$
$$= 2.07\ \text{V}$$

$$I_4 = \frac{V_4}{R_4} = \frac{2.07\ \text{V}}{3.9\ \text{k}\Omega}$$
$$= 0.53\ \text{mA}$$

Practice Problem

7-5.1 A series-parallel circuit consists of three resistors $R_1 \| R_2 \| R_3$ connected in series with resistors $R_4 \| R_5$. The supply is $E = 45$ V and the resistor values are: $R_1 = 2.2$ kΩ, $R_2 = 3.9$ kΩ, $R_3 = 1.2$ kΩ, $R_4 = 1.5$ kΩ, $R_5 = 1.8$ kΩ. Analyze the circuit to determine all resistor voltages and currents.

7-6 TROUBLE-SHOOTING SERIES-PARALLEL CIRCUITS

Open-Circuits and Short-Circuits

The effect of an open-circuit or short-circuit condition on a series-parallel circuit depends on just where in the circuit the fault occurs. Consider Figure 7-12(a), where an open-circuit is shown at one end of R_1. This has the same effect as an open-circuit in the supply line, so that all current levels are zero. Also, because the currents are zero, there are no voltage drops across the resistors, and consequently all of the supply voltage (E) appears across the open-circuit.

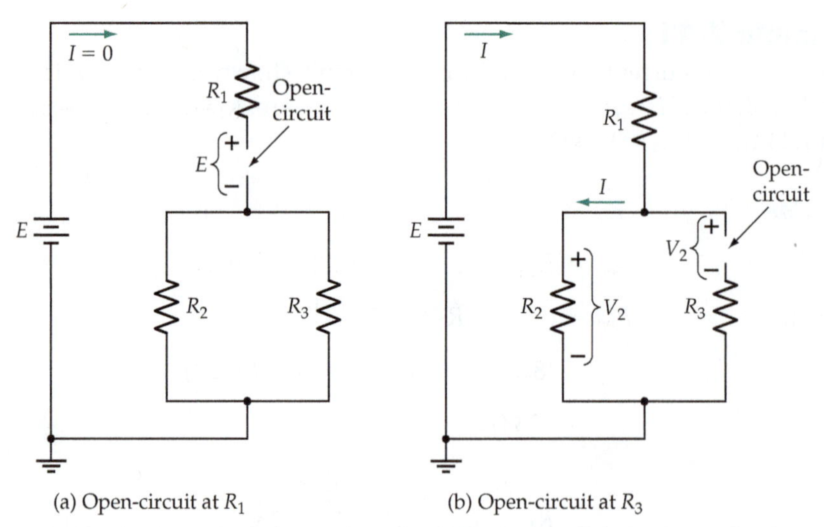

Figure 7-12 An open-circuit in one branch of a series-parallel circuit usually alters the current and voltage levels in several branches of the circuit.

In the case of an open-circuit at one end of one of the parallel-connected resistors, as for R_3 in Figure 7-12(b), I_3 goes to zero. The current through R_1 and R_2 is now equal to the supply current and is calculated as

$$I = \frac{E}{R_1 + R_2}$$

Also, because there is no current through R_3, there is no voltage drop across it and the voltage at the open circuit is equal to V_2.

For the short-circuit condition shown in Figure 7-13(a), the resistance between the terminals of R_1 is effectively zero. Therefore, the supply voltage appears across R_2 and R_3 in parallel. This gives a supply current of

$$I = \frac{E}{R_2 \| R_3}$$

and the branch currents are

$$I_2 = \frac{E}{R_2}, \text{ and } I_3 = \frac{E}{R_3}$$

It is seen that the levels of current through R_2 and R_3 have been increased from the normal condition (before the short-circuit). This could cause excessive power dissipation in the components if they have previously been operating near their maximum rating.

The short-circuit condition illustrated in Figure 7-13(b) effectively reduces I_2 and I_3 to zero and increases the supply current to,

$$I = \frac{E}{R_1}$$

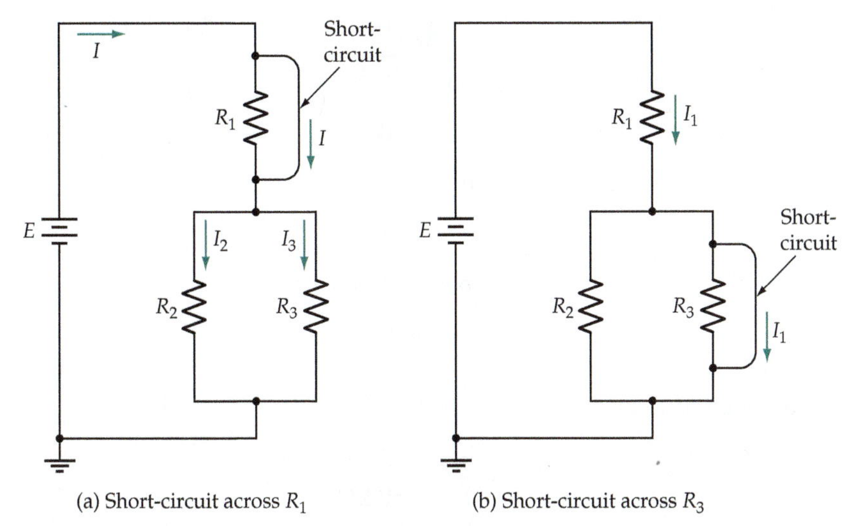

Figure 7-13 A short-circuit in one branch of a series-parallel circuit usually alters the current and voltage levels in several branches of the circuit.

Obviously, the current through R_1 is now greater than normal, and again the component power dissipation might present a problem.

Example 7-12

Calculate the supply current and resistor power dissipations for the circuit in Figure 7-14 (reproduced from Figure 7-7) for the following conditions: (a) R_2 short-circuited, (b) R_3 open-circuited.

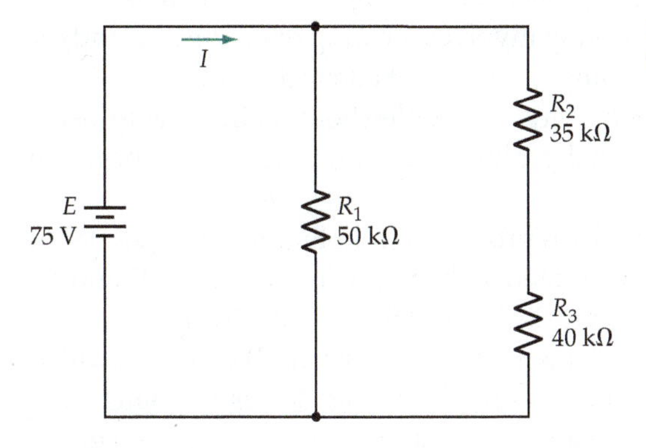

Figure 7-14 Parallel resistor circuit for Example 7-12.

Solution

(a) R_2 shorted,

$$R_{eq} = R_1 \| R_3 = 50\ \text{k}\Omega \| 40\ \text{k}\Omega$$

$$= 22.2\ \text{k}\Omega$$

$$I = \frac{E}{R_{eq}} = \frac{75\ \text{V}}{22.2\ \text{k}\Omega}$$

$$= 3.38\ \text{mA}$$

$$P_1 = \frac{E^2}{R_1} = \frac{(75\ \text{V})^2}{50\ \text{k}\Omega}$$

$$= 112.5\ \text{mW}$$

$$P_2 = 0$$

$$P_3 = \frac{E^2}{R_3} = \frac{(75\ \text{V})^2}{40\ \text{k}\Omega}$$

$$= 141\ \text{mW}$$

(b) R_3 open-circuited,

$$I = \frac{E}{R_1} = \frac{75\ \text{V}}{50\ \text{k}\Omega}$$

$$= 1.5\ \text{mA}$$

$$P_1 = \frac{E^2}{R_1} = \frac{(75\ \text{V})^2}{50\ \text{k}\Omega}$$

$$= 112.5\ \text{mW}$$

$$P_2 = 0,\ P_3 = 0$$

Trouble-Shooting

Regardless of whether a series, parallel, or series-parallel circuit is under investigation, trouble-shooting usually involves locating open-circuits and/or short-circuits. This involves searching for incorrect connections or burned-out components. The most convenient way to trouble-shoot a circuit is to use a voltmeter with its common terminal grounded or connected to the negative terminal of the supply source.

Consider the circuit in Figure 7-15, reproduced from Figure 7-10(a). Refer to Example 7-10 to note that the calculated voltage levels for points B and C, measured with respect to ground, are 80.8 V and 52 V, respectively. Also note that the voltage at point A (measured with respect to ground) equals E, which is 100 V. So, when investigating this circuit, the voltage levels at points A, B, and C should be measured, and any incorrect results should be used as a guide to possible faults. The following are example conditions and possible faults for the circuit in Figure 7-15:

Measured Voltages	Possible Fault
$V_A = 100$ V, $V_B = 100$ V, $V_C < V_B$	R_1 shorted or R_2 opened
$V_A = 100$ V, $V_B = V_C < 100$ V	R_1 opened or R_2 shorted
$V_A = 100$ V, $V_B = V_C = 100$ V	R_4 and R_5 opened or R_3 shorted
$V_A = 100$ V, $V_C = 0$ V, $V_B > 0$	R_4 or R_5 shorted

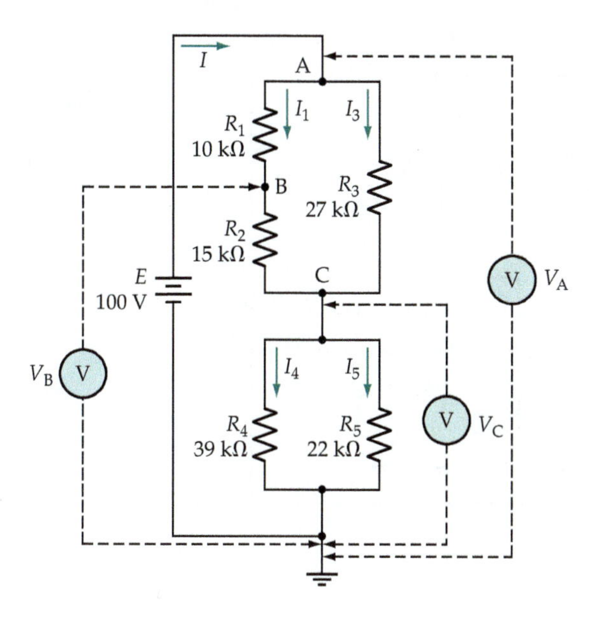

Figure 7-15 All junction voltages should be measured with respect to ground when troubleshooting a resistor circuit.

Practice Problems

7-6.1 If the circuit in Figure 7-2(a) has R_3 open-circuited, as illustrated in Figure 7-12(b), determine the new levels of resistor currents.

7-6.2 Determine the power dissipations in each resistor in the circuit of Figure 7-2(a), then calculate the new power dissipations when R_1 is short-circuited as shown in Figure 7-13(a).

Problems

Section 7-1

7-1 The circuit shown in Figure 7-2(a) has its components changed to $R_1 = 750\ \Omega$, $R_2 = 330\ \Omega$, $R_3 = 560\ \Omega$, and its supply voltage changed to 80 V. Determine the circuit equivalent resistance and the total supply current.

7-2 The series-parallel circuit in Figure 7-3(a) is changed as follows: $R_1 = 2.7\ \text{k}\Omega$, $R_2 = 8.2\ \text{k}\Omega$, $R_3 = 15\ \text{k}\Omega$, and $E = 50$ V. Determine the circuit equivalent resistance and the total supply current.

7-3 Determine the equivalent resistance and the total supply current for the circuit in Figure 7-11 if $E = 9$ V, $R_1 = 3.3\ \text{k}\Omega$, $R_2 = 12\ \text{k}\Omega$, $R_3 = 6.8\ \text{k}\Omega$, and $R_4 = 18\ \text{k}\Omega$.

7-4 Analyze the circuit in Figure 7-16 to determine the equivalent resistance and total supply current.

7-5 Determine the equivalent resistance and the total supply current for the circuit in Figure 7-17.

7-6 For the circuit in Figure 7-18, determine the value of R_1 that will give a supply current of $I = 200$ mA.

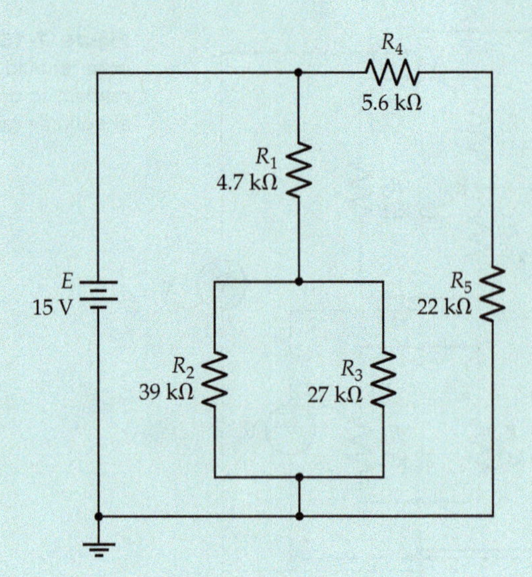

Figure 7-16

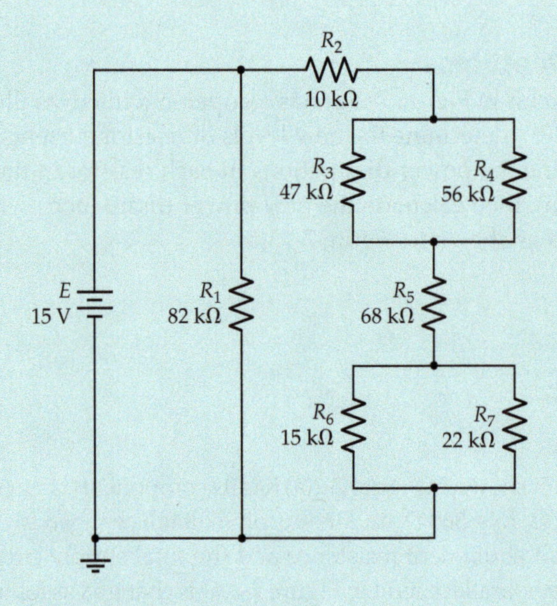

Figure 7-17

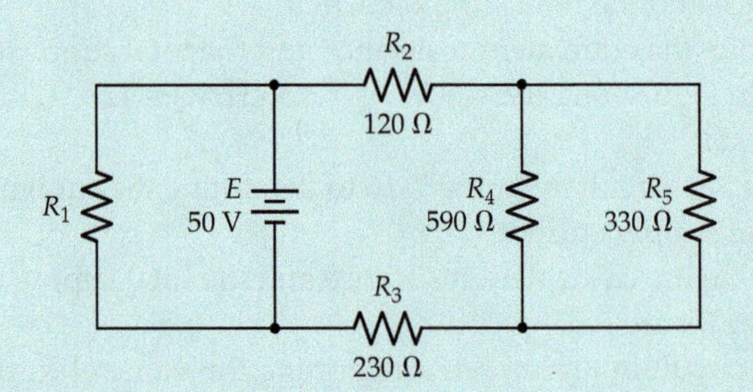

Figure 7-18

Section 7-2

7-7 Calculate the branch currents for the circuit in Problem 7-1.

7-8 Calculate all branch currents for the circuit in Problem 7-2.

7-9 Determine the branch currents for the circuit described in Problem 7-3.

7-10 Analyze the circuit in Figure 7-16 to determine all branch currents.

7-11 Determine the branch currents for the circuit in Figure 7-17.

7-12 Calculate the branch currents for the final circuit analyzed in Problem 7-6.

Section 7-3

7-13 Calculate the branch voltages for the circuit in Problem 7-1.

7-14 Calculate all branch voltages for the circuit in Problem 7-2.

7-15 Determine the branch voltages for the circuit described in Problem 7-3.

7-16 Analyze the circuit in Figure 7-16 to determine all branch voltages.

7-17 Determine the branch voltages for the circuit in Figure 7-17.

7-18 Calculate the branch voltages for the final circuit analyzed in Problem 7-6.

Section 7-4

7-19 A ladder network as illustrated in Figure 7-9(a) uses a 50 V supply and has the following resistor values: $R_1 = 10$ kΩ, $R_2 = 5$ kΩ, $R_3 = 4$ kΩ, $R_4 = 5$ kΩ, $R_5 = 1$ kΩ, and $R_6 = 250$ Ω. Determine the voltages across R_2, R_4, and R_6.

7-20 The $R/2R$ ladder network in Figure 7-9(b) has an additional stage ($R_7 = R$ and $R_8 = R$) connected to R_6. The value of R_6 is changed to $2R$, and the input is $E = 20$ V. Determine the voltages across resistors R_2, R_4, R_6, and R_8.

7-21 A ladder network as in Figure 7-9(a) has the following resistor values: $R_1 = 2$ kΩ, $R_2 = 14$ kΩ, $R_3 = 4$ kΩ, $R_4 = 20$ kΩ, $R_5 = 8$ kΩ, and $R_6 = 12$ kΩ. If the input is 36 V, calculate the voltages across R_1, R_3, and R_5.

Section 7-5

7-22 Determine all resistor currents and voltages for the circuit shown in Figure 7-19.

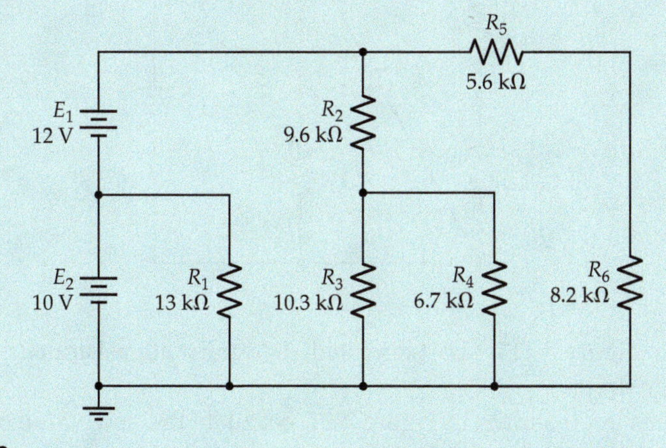

Figure 7-19

7-23 Repeat Problem 7-22 with the E_2 polarity reversed.

7-24 For the circuit in Figure 7-20, calculate the voltages at terminals A, B, and C with respect to ground.

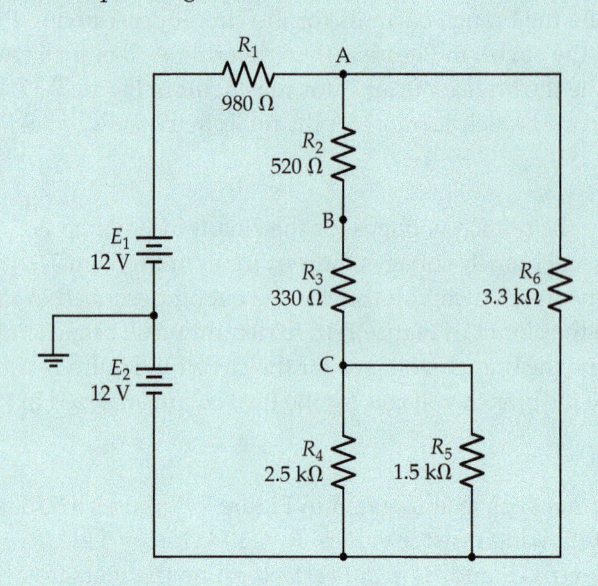

Figure 7-20

7-25 Repeat problem 7-24 with the polarity of E_1 reversed.

7-26 Calculate the voltages at terminals A, B, and C with respect to ground for the circuit in Figure 7-20 when R_4 is short-circuited.

7-27 If R_6 in Figure 7-20 is open-circuited, calculate the voltages at terminals A, B, and C with respect to ground.

7-28 Calculate all resistor currents and voltages in the circuit shown in Figure 7-21.

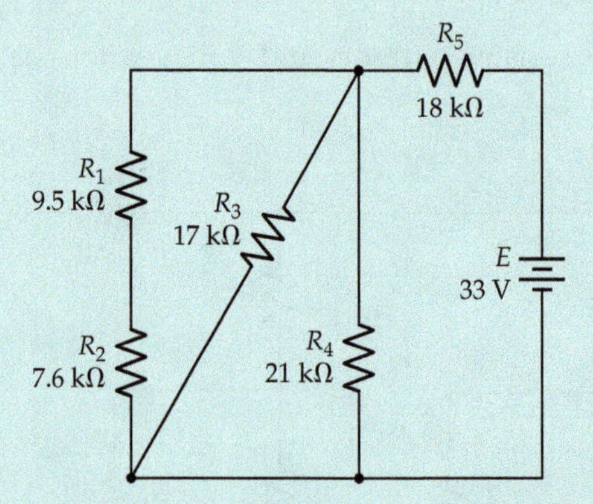

Figure 7-21

7-29 If R_2 in Figure 7-21 is short-circuited, determine all resistor currents and voltage drops.

7-30 For the circuit shown in Figure 7-21, calculate the new value of supply voltage that will cause 2.2 mW to be dissipated in R_2.

7-31 For the circuit in Figure 7-22, determine the R_2 value that will give a supply current of $I = 168\ \mu A$.

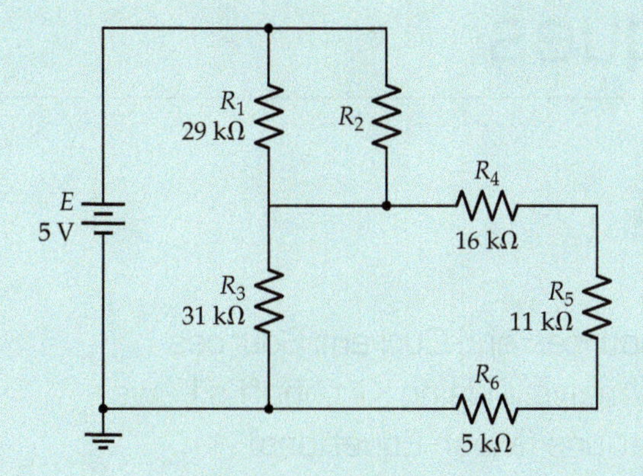

Figure 7-22

7-32 For Problem 7-31, determine the new levels of resistor currents when R_3 becomes open-circuited.

Section 7-6

7-33 If resistor R_3 in Figure 7-17 becomes short-circuited, calculate the new levels of current and voltage throughout the circuit.

7-34 If resistor R_1 in Figure 7-17 becomes open-circuited, determine the new levels of current and voltage throughout the circuit.

7-35 Determine the new level of supply current for the circuit in Figure 7-16 when R_2 is short-circuited.

7-36 Calculate the supply current for the circuit in Figure 7-16 when R_3 becomes open-circuited.

7-37 Calculate the new value of R_1 for the circuit in Figure 7-18 to produce a 200 mA supply current when R_5 is short-circuited.

Practice Problem Answers

7-1.1 473 μA
7-1.2 931 μA
7-2.1 473 μA, 291 μA, 182 μA
7-2.2 284 μA, 573 μA, 358 μA, 647 μA
7-3.1 3.55 V, 1.45 V, 1.45 V
7-3.2 3.42 V, 1.58 V, 1.12 V, 0.46 V
7-4.1 4 V, 2 V, 1 V
7-5.1 19.9 V, 19.9 V, 19.9 V, 25.1 V, 25.1 V, 9 mA, 5.1 mA, 16.6 mA, 16.7 mA, 13.9 mA
7-6.1 431 μA, 431 μA
7-6.2 9.5 mW, 1.8 mW, 1.2 mW, 0, 31.25 mW, 20.8 mW

CHAPTER 8
Network Analysis Techniques

CONTENTS

Objectives

You will be able to:

1. Sketch the circuits of voltage sources and current sources.
2. Convert current sources into voltage sources and vice versa.
3. Solve problems involving parallel and series operation of current and voltage sources.
4. Analyze resistor networks using Kirchhoff's laws.
5. Use loop equations to analyze resistor networks.
6. Solve resistor network problems using nodal analysis.
7. Apply delta-wye (Δ-Y) transformations to resistor networks.
8. Analyze resistor circuits involving dependent voltage and current sources.

INTRODUCTION

Some resistor networks are so complex that they cannot be analyzed by the simple series-parallel techniques described in Chapter 7. An example is a circuit that has more than one voltage source. Also, some circuits have current sources as well as voltage sources. A network can sometimes be simplified by converting voltage sources into current sources or vice versa. By direct application of Kirchhoff's voltage and current laws, equations can be derived for the analysis of any circuit. Other useful techniques for determining the voltage and current levels in complex circuits are loop equations, nodal analysis, and delta-wye (Δ-Y) transformations.

8-1 VOLTAGE SOURCES AND CURRENT SOURCES

Constant-Voltage Source

Voltage cells and batteries can be represented by an *ideal cell* in series with the internal resistance of the cell or battery (see Section 10-2). Because an ideal cell is assumed to have a voltage that does not change, it is termed a *constant-voltage source*. However, the output current produces a voltage drop across the internal resistance, and this varies with changes in the current level. Figure 8-1(a) shows a voltage source with constant voltage E, a source resistance R_S, and an external load resistance R_L. Sometimes, instead of a cell symbol, the symbol for a *dc* voltage generator is used to represent a constant-voltage source [Figure 8-1(b)]. Using the voltage-divider rule (see Section 5-3), the output voltage developed across R_L from the constant-voltage source can be quickly determined.

$$V_L = \frac{ER_L}{R_S + R_L} \tag{8-1}$$

If $R_S \ll R_L$, $\quad V_L \approx E$

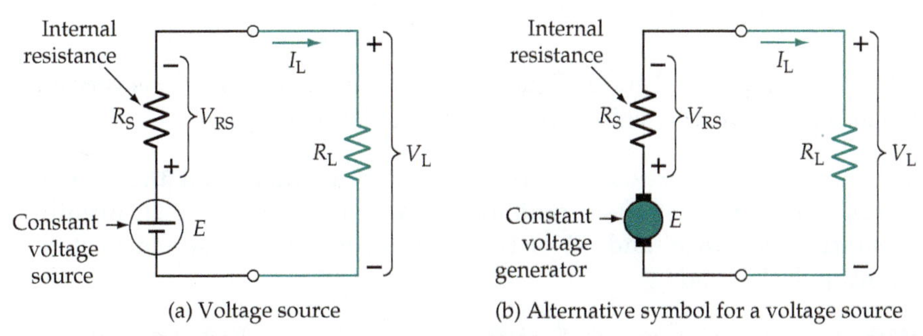

Figure 8-1 A voltage source is represented by a constant-voltage cell or generator in series with its source resistance.

From Equation 8-1 it is seen that, if R_L is very much larger than R_S, virtually all of the source voltage appears across R_L.

Where the load resistance is very much larger than the source resistance, the constant voltage source is assumed to have zero source resistance, and all of the source voltage is assumed to be applied to the load.

In this case (with $R_S \ll R_L$), the source is termed an *ideal voltage source;* one that has (effectively) a zero source resistance.

Constant-Current Source

Certain electronic devices can produce a current that tends to remain constant regardless of how the load resistance varies. So, it is possible to have a *constant-current source.* Figure 8-2 shows that a constant current source consists of a constant current generator in parallel with a source resistance R_S. Note that load resistance R_L, is also in parallel with the current generator. It is seen that some of the source current flows through R_S and some flows through R_L. Using the current-divider rule (see Section 6-4), the output current (or load current) from a constant-current source can be determined in terms of R_L and R_S.

$$I_L = \frac{IR_S}{R_L + R_S} \tag{8-2}$$

If $R_L \ll R_S$, $\qquad I_L \approx I$

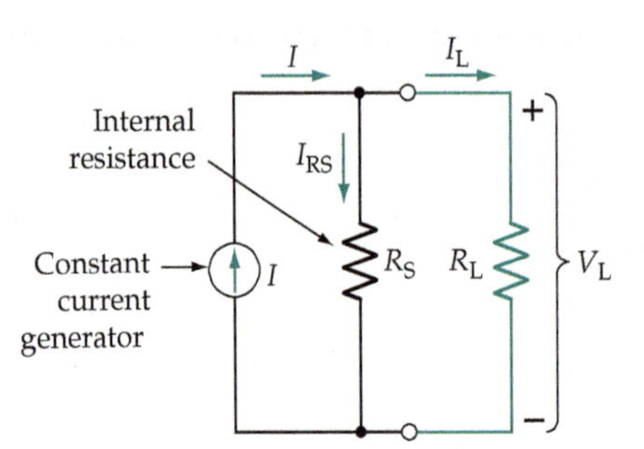

Figure 8-2 A current source is represented by a constant-current generator in parallel with its source resistance.

Referring to Figure 8-2 and to Equation 8-2, it is obvious that if R_S is very much larger than R_L then I_{RS} will be very much smaller than I_L.

Where the load resistance is very much smaller than the source resistance, a constant-current source is assumed to have an infinite source resistance, and all of the source current is assumed to flow through the load.

With $R_L \ll R_S$ the term ideal source is once again applied. An *ideal current source* is one that has (effectively) an infinitely high source resistance.

Source Conversions

For the voltage source shown in Figure 8-3(a), the load current is calculated as,

$$I_L = \frac{E}{R_L + R_S}$$

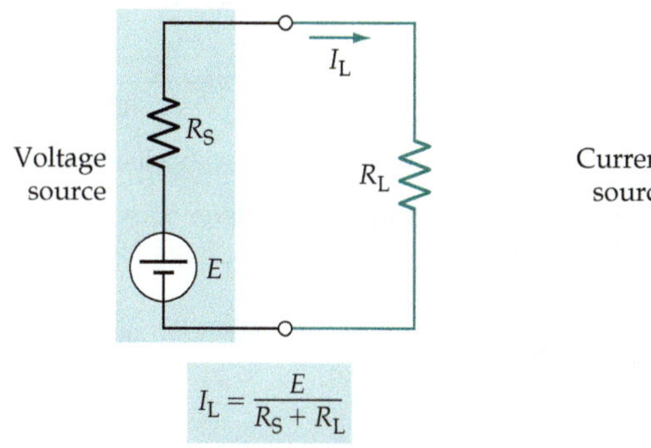

(a) Voltage source with a load resistor

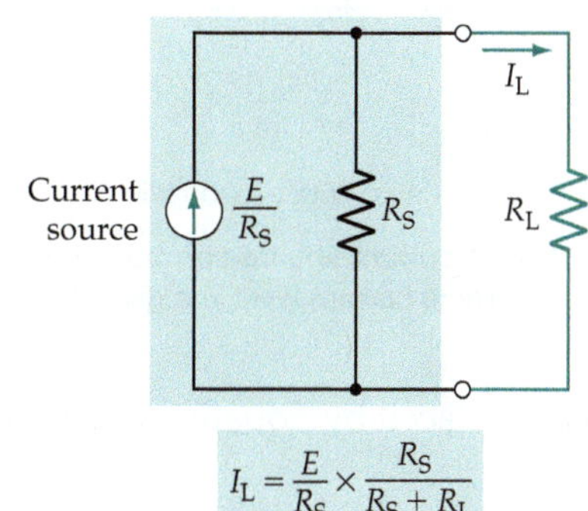

(b) Current source with a load resistor

Figure 8-3 A voltage source can be converted into an equivalent current source that will produce the same current level in a given load resistor.

If E in this equation is multiplied by R_S/R_S, nothing is changed because this is the same as multiplying by 1:

$$I_L = \frac{E(R_S/R_S)}{R_L + R_S}$$

which can be rewritten as,

$$I_L = \frac{E}{R_S}\left(\frac{R_S}{R_L + R_S}\right) \tag{8-3}$$

In Equation 8-3, E/R_S is a current, and $R_S/(R_S + R_L)$ is exactly the same as in the current-divider equation, (Equation 8-2). So, although it is derived from a voltage source, Equation 8-3 represents the output from a current source. As illustrated in Figure 8-3(b), a current source with $I = E/R_S$ and source resistance R_S gives the same current through R_L as the voltage source in Figure 8-3(a).

A voltage source with a voltage E and a series source resistance R_S can be replaced by a current source with a current E/R_S and a parallel source resistance R_S.

For the circuit shown in Figure 8-4(a), the load voltage is,

$$V_L = I_L R_L$$

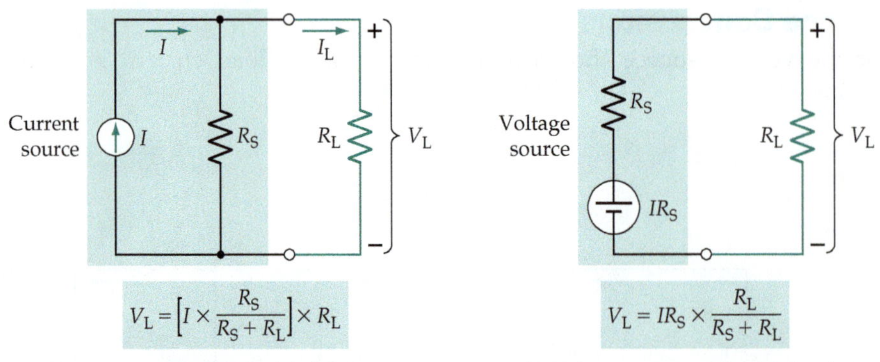

(a) Current source with a load resistor (b) Voltage source with a load resistor

Figure 8-4 A current source can be converted into an equivalent voltage source that will produce the same current level in a given load resistor.

I_L can be replaced by Equation 8-2, giving,

$$V_L = \left(\frac{IR_S}{R_L + R_S}\right)R_L$$

or

$$V_L = IR_S\left(\frac{R_L}{R_L + R_S}\right) \quad \textbf{(8-4)}$$

In Equation 8-4, IR_S is a voltage, and $R_L/(R_L + R_S)$ is exactly the same as in the voltage-divider equation (Equation 8-1). Consequently, although it is derived from a current source, Equation 8-4 represents the output from a voltage source. Figure 8-4(b) illustrates the fact that a voltage source with a voltage $E = IR_S$ and resistance R_S has the same effect on a load resistor as the current source in Figure 8-4(a).

A current source with a current I and a parallel source resistance R_S can be replaced by a voltage source with a voltage IR_S and a series source resistance R_S.

Example 8-1

The voltage source in Figure 8-3(a) has $E = 10$ V, $R_S = 1\ \Omega$, and $R_L = 10\ \Omega$. Determine I and R_S for the equivalent current source. Then calculate V_L and I_L for each type of source.

Solution

For the current source,

$$I = \frac{E}{R_S} = \frac{10\text{ V}}{1\ \Omega} = 10\text{ A}$$

$$\text{Current source } R_S = \text{voltage source } R_S = 1\ \Omega$$

Using the voltage source,

$$V_L = \frac{ER_L}{R_S + R_L} = \frac{10\text{ V} \times 10\ \Omega}{1\ \Omega + 10\ \Omega}$$

$$= 9.09\text{ V}$$

$$I_L = \frac{E}{R_S + R_L} = \frac{10\text{ V}}{1\ \Omega + 10\ \Omega}$$

$$= 0.909\text{ A}$$

Using the current source,

$$I_L = \frac{IR_S}{R_L + R_S} = \frac{10\text{ A} \times 1\ \Omega}{10\ \Omega + 1\ \Omega}$$

$$= 0.909\text{ A}$$

and

$$V_L = I_L \times R_L = 0.909\text{ A} \times 10\ \Omega$$

$$= 9.09\text{ V}$$

Parallel and Series Operation

Voltage sources can be operated in series without difficulty (see Sections 5-2 and 10-4). Figure 8-5(a) shows two series-connected voltage sources, and Figure 8-5(b) shows the equivalent circuit. The equivalent circuit voltage is the sum of the individual source voltages, and the equivalent circuit resistance is the sum of the source resistances.

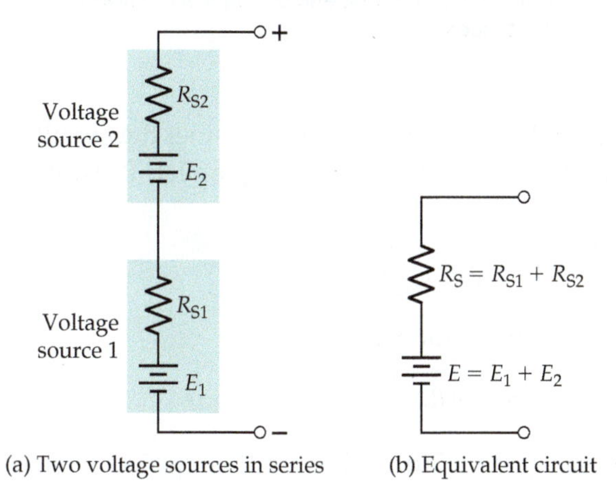

Figure 8-5 Voltage sources can be operated in series without difficulty. The equivalent output voltage is the sum of the individual source voltages, and the equivalent source resistance is the sum of the individual source resistances.

Voltage sources can be operated in parallel only if the source voltages are equal. As illustrated in Figure 8-6(a) and (b), the equivalent circuit source resistance is the parallel combination of the individual source resistances, and the equivalent circuit voltage is, of course, equal to the source voltages. Figure 8-6(c) shows that when sources with unequal voltages are connected in parallel, the lower-voltage source will tend to discharge the higher-voltage source.

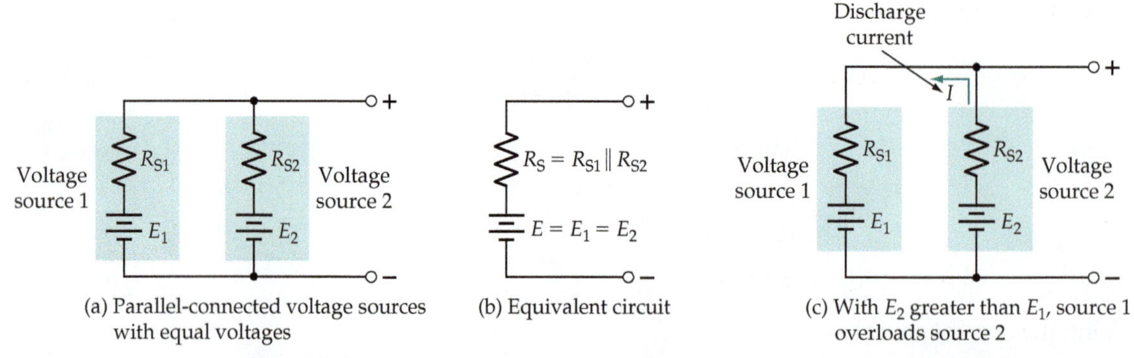

Figure 8-6 Voltage sources with equal voltages can be operated in parallel. Voltage sources with unequal voltages should not be connected in parallel.

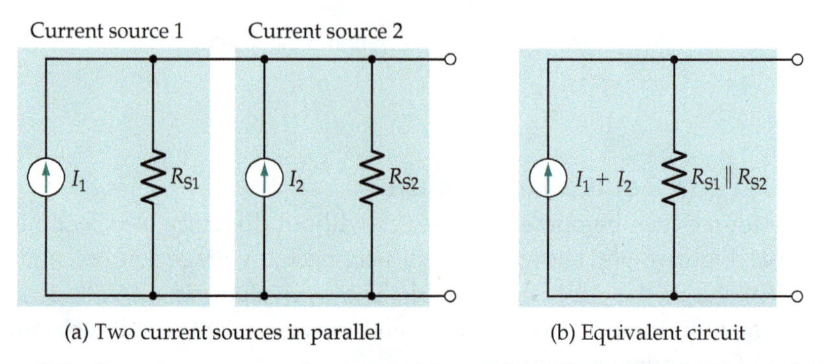

Figure 8-7 Current sources can be operated in parallel without difficulty. The equivalent source current is the sum of the individual source currents, and the equivalent resistance is the parallel combination of the source resistances.

Similar, but converse, rules apply to current sources. Consider the circuit in Figure 8-7(a) which has two current sources connected in parallel. The output current from the two is

$$I = I_1 + I_2$$

and the internal resistance of the combination is,

$$R = R_{S1} \| R_{S2}$$

giving the equivalent circuit in Figure 8-7(b). It is seen that current sources may be operated in parallel without difficulty.

Now consider two current sources connected in series (Figure 8-8). If they both generate exactly the same level of current, there is no problem. But suppose that one generates an output of 1 A and the other has an output of 2 A. The 2 A current would flow through the 1 A source, thus changing the 1 A output level. Alternatively, the resistance of the 1 A source may alter the level of the 2 A current. Consequently, series operation of current sources with differing output currents is not possible.

A current source may be connected in series or in parallel with a voltage source, but the rules discussed still apply. When a conversion is made to have two voltage sources, parallel operation is possible only when the two voltages are equal. When the conversion creates two current sources, series operation is possible only when the two currents are equal.

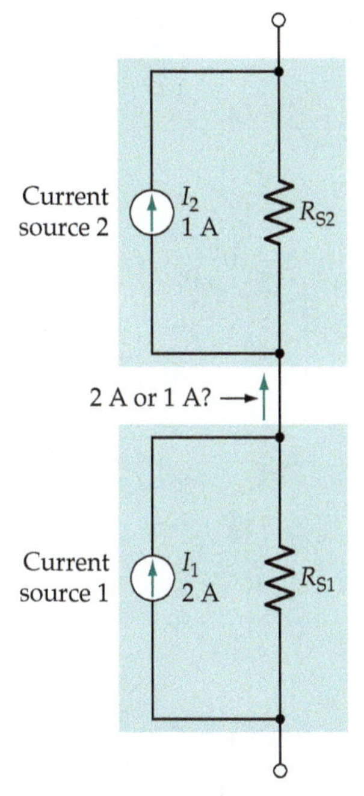

Figure 8-8 Current sources cannot be operated in series, except when the individual source currents are equal.

Practice Problems

8-1.1 A voltage source has $E = 500$ mV, $R_S = 1$ kΩ, and $R_L = 82$ kΩ. Convert the voltage source into its equivalent current source, and calculate the load current.

8-1.2 A constant current device has $I = 7.5$ mA, $R_S = 20$ kΩ, and $R_L = 2.7$ kΩ. Derive the equivalent voltage source, and calculate the load voltage.

8-2 NETWORK ANALYSIS USING KIRCHHOFF'S LAWS

Kirchhoff's voltage law (KVL) (Section 5-2) and Kirchhoff's current law (KCL) (Section 6-1) readily lend themselves to the derivation of equations for solving complex circuits. Consider the simple series-parallel circuit in Figure 8-9(a). The first step in analyzing the circuit is to identify the junctions as A, B, C, and so on, as shown in Figure 8-9(b). Then, the branch current directions are identified and numbered [Figure 8-9(c)], and the resistor voltage drops and polarities are labeled. Applying Kirchhoff's voltage law to the closed path starting at point A and going through points B, E, and F and back to A again gives,

$$E = V_1 + V_2$$

and because $V_1 = I_1R_1$ and $V_2 = I_2R_2$,

$$E = I_1R_1 + I_2R_2 \quad \textbf{(1)}$$

Summing the voltage drops around path BCDEB,

$$V_2 = V_3$$

or

$$I_2R_2 = I_3R_3 \quad \textbf{(2)}$$

For the path ABCDEFA,

$$E = V_1 + V_3$$

So

$$E = I_1R_1 + I_3R_3 \quad \textbf{(3)}$$

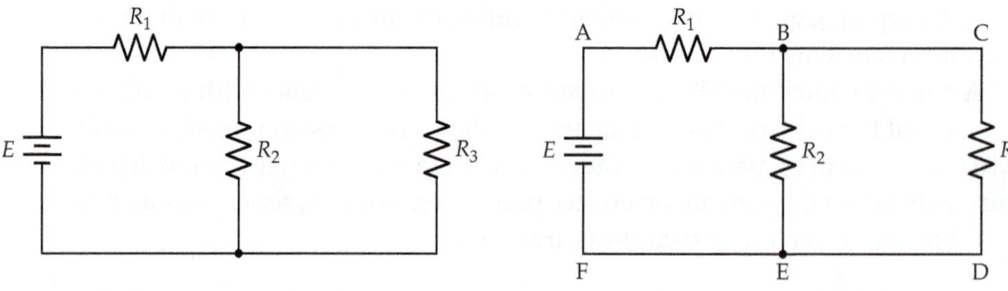

(a) Circuit to be analyzed by Kirchhoff's laws

(b) Circuit with junctions identified

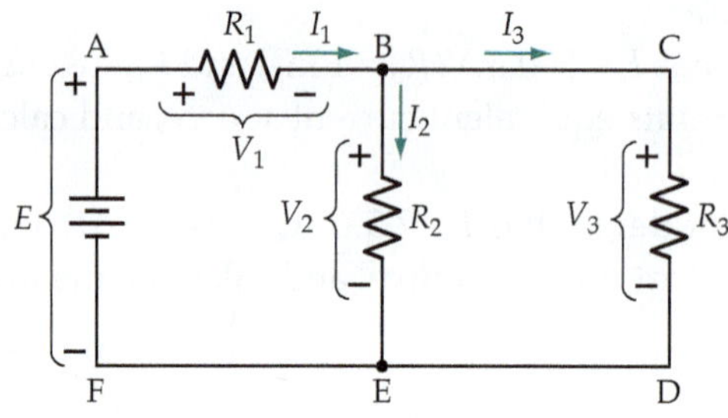

(c) Branch currents and voltage drops identified

Figure 8-9 **Kirchhoff's laws may be used to analyze a resistance network. For the circuit shown Kirchhoff's voltage law gives $E = V_1 + V_2$, and Kirchhoff's current law gives $I_1 = I_2 + I_3$.**

and from Kirchhoff's current law, at point B,

$$I_1 = I_2 + I_3 \tag{4}$$

Substituting the expression for I_1 from Eq. 4 into Eq. 1,

$$E = (I_2 + I_3)R_3 + I_2R_2$$

or

$$E = I_2R_1 + I_3R_1 + I_2R_2 \tag{5}$$

From Eq. 2

$$I_3 = \frac{I_2R_2}{R_3} \tag{6}$$

Substituting for I_3 from Eq. 6 into Eq. 5,

$$E = I_2R_1 + \frac{I_2R_2}{R_3}R_1 + I_2R_2$$

or

$$E = I_2\left(R_1 + \frac{R_1R_2}{R_3} + I_2R_2\right)$$

So

$$I_2 = \frac{E}{R_1 + R_2 + (R_2R_1/R_3)} \tag{7}$$

From Eq. 7, the level of current I_2 can be calculated. Then I_2 can be substituted into Eq. 6 to find I_3, and I_1 can be determined by substituting I_2 and I_3 into Eq. 4. Once all the current levels are known, the voltage levels are easily calculated.

Network Analysis Procedure Using Kirchhoff's Laws

1. *Convert all current sources to voltage sources.*
2. *Lebel all junctions on the network* **A, B, C,** *etc.*
3. *Identify current directions and voltage polarities, and number them according to the resistor involved.*
4. *Identify each current path according to the labeled junctions, and applying Kirchhoff's voltage law, write the voltage equations for the current paths.*
5. *Applying Kirchhoff's current law, write the equations for the currents entering and leaving all junctions where more than one current is involved.*
6. *Solve the equations by substitution to find the unknown currents.*

Note that in some circumstances currents and voltage polarities will turn out to be negative when the circuit is analyzed. This simply means that the assumed current directions and/or voltage polarities were incorrect.

Example 8-2

Using Kirchhoff's laws, analyze the circuit in Figure 8-10 to determine I_1, I_2, and I_3.

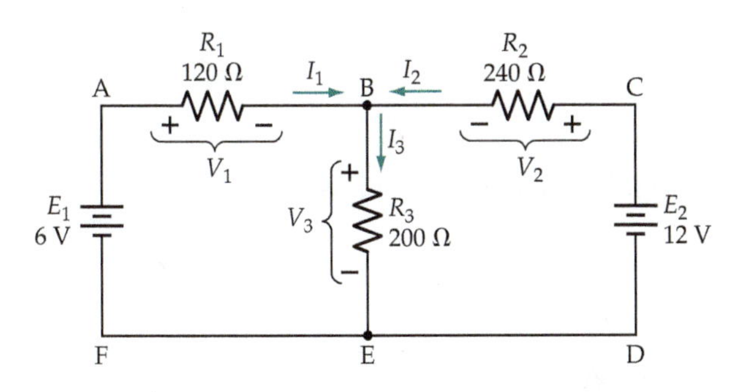

Figure 8-10 A resistor network with two voltage sources can be analyzed by use of Kirchhoff's laws. For the circuit shown, $E_1 = V_1 + V_3$, $E_2 = V_2 + V_3$ and $I_3 = I_1 + I_2$.

Solution

For the path ABEFA,

$$E_1 = V_1 + V_3$$
$$= I_1R_1 + I_3R_3$$

So

$$6\text{ V} = 120I_1 + 200I_3 \qquad (1)$$

For CBEDC,

$$E_2 = V_2 + V_3$$
$$= I_2R_2 + I_3R_3$$

giving

$$12\text{ V} = 240I_2 + 200I_3 \qquad (2)$$

For ABCDEFA,

$$E_1 = V_1 - V_2 + E_2$$
$$E_1 - E_2 = I_1R_1 - I_2R_2$$
$$6\text{ V} - 12\text{ V} = 120I_1 - 240I_2$$
$$-6\text{ V} = 120I_1 - 240I_2 \qquad (3)$$

For point B,

$$I_3 = I_1 + I_2 \qquad (4)$$

Substituting for I_3 from Eq. 4 into Eq. 1,

$$6\text{ V} = 120I_1 + 200I_1 + 200I_2$$

or

$$6\text{ V} = 320I_1 + 200I_2 \tag{5}$$

From Eq. 3

$$240I_2 = 120I_1 + 6\text{ V}$$

So

$$I_2 = \frac{120I_1}{240} + \frac{6\text{ V}}{240}$$

Giving

$$I_2 = \frac{I_1}{2} + \frac{1\text{ V}}{40} \tag{6}$$

Substituting for I_2 from Eq. 6 into Eq. 5,

$$6\text{ V} = 320I_1 + \frac{200I_1}{2} + \frac{200\text{ V}}{40}$$

or

$$6\text{ V} = 420I_1 + 5\text{ V}$$

giving

$$I_1 = \frac{6\text{ V} - 5\text{ V}}{420\ \Omega}$$

$$\approx 2.38\text{ mA}$$

Substituting for I_1 in Eq. 1,

$$6\text{ V} = (120 \times 2.38\text{ mA}) + 200I_3$$

which gives

$$I_3 \approx 28.57\text{ mA}$$

Substituting for I_1 and I_3 in Eq. 4,

$$28.57\text{ mA} = 2.38\text{ mA} + I_2$$

giving

$$I_2 \approx 26.19\text{ mA}$$

Practice Problems

8-2.1 Use Kirchhoff's laws to determine the current through resistor R_2 in the circuit of Figure 8-11.

8-2.2 Apply Kirchhoff's laws to the circuit in Figure 8-12 to calculate the current in resistor R_1.

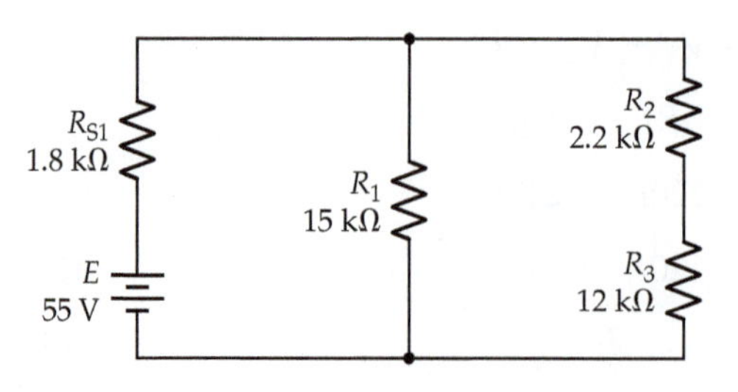

Figure 8-11 Circuit for Practice Problem 8-2.1.

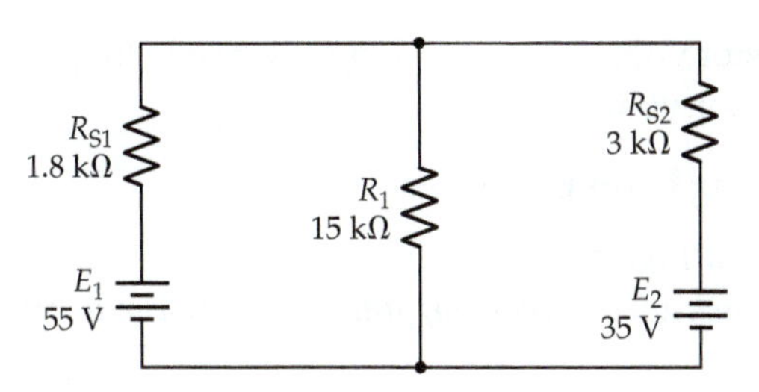

Figure 8-12 Circuit for Practice Problem 8-2.2.

8-3 LOOP EQUATIONS (MESH EQUATIONS)

By the use of *loop equations,* also termed *mesh equations,* complex circuits can be solved with greater ease than by direct application of Kirchhoff's laws. The circuit of Figure 8-10 is reproduced in Figure 8-13, and *loop currents* are shown in a clockwise direction. The loop current is simply the current that circulates in the closed current path. In Figure 8-13, loop current I_1 is exactly the same as the branch current that flows through R_1. But I_1 is *not* the branch current in R_3 because loop current I_2 also flows through R_3 in the opposite direction to I_1. The branch current in R_3 is $I_3 = (I_1 - I_2)$. Loop current I_2 is the same as the branch current flowing in resistor R_2. A third loop could be drawn through E_1, R_1, R_2, E_2, and back to E_1. However, because all of the circuit voltage drops are included in the first two loops, the third loop is redundant.

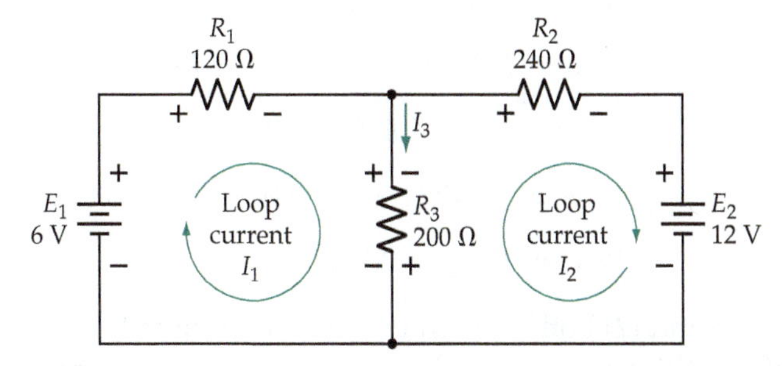

Figure 8-13 In the loop equation (or mesh equation) method of circuit analysis, the sum of the voltage drops around each circuit loop is equated to zero. For loop 1, $0 = I_1R_1 + I_1R_3 - I_2R_3 - E_1$.

The loop currents could actually be assigned either direction, but assigning a clockwise direction simplifies the process of writing equations for the voltage drops around each loop. When the circuit analysis is complete, those branch currents that are in the same direction as the (clockwise) loop currents have a plus sign. Those that are in the opposite direction are given a minus sign. In Figure 8-13, note that the voltage drop I_1R_3 is positive (+) at the top of R_3, while I_2R_3 is positive (+) at the bottom of the resistor. Also note that the polarity of E_1 is opposite to the resistor voltage drops in loop 1, while in loop 2, E_2 has the same polarity as the resistor voltage drops.

DC Circuit Analysis Procedure Using Loop Equations

1. *Convert all current sources into voltage sources.*
2. *Draw all loop currents in a clockwise direction, and identify them by number.*
3. *Identify all resistor voltage drops as + to – in the direction of the loop current.*
4. *Identify all voltage sources according to their correct polarity.*
5. *Write the equations for the voltage drops around each loop in turn, by equating the sum of the voltage drops to zero.*
6. *Solve the equations to find the unknown currents.*

Example 8-3

Using loop equations, solve the resistor network shown in Figure 8-13 to determine the current through R_3.

Solution

For loop 1,
$$0 = I_1R_1 + I_1R_3 - I_2R_3 - E_1$$

or,
$$E_1 = I_1(R_1 + R_3) - I_2R_3$$

$$6\text{ V} = I_1(120\ \Omega + 200\ \Omega) - (I_2 \times 200\ \Omega)$$

$$6\text{ V} = 320I_1 - 200I_2 \qquad \textbf{(1)}$$

For loop 2,
$$0 = I_2R_2 + E_2 + I_2R_3 - I_1R_3$$

Therefore,
$$-E_2 = I_2(R_2 + R_3) - I_1R_3$$

$$-12\text{ V} = I_2(240\ \Omega + 200\ \Omega) - (I_1 \times 200\ \Omega)$$

$$-12\text{ V} = 440I_2 - 200I_1 \qquad \textbf{(2)}$$

Equations (1) and (2) both contain two unknown quantities, I_1 and I_2. One of these two quantities must be eliminated before the other can be determined. Examining Equations (1) and (2) it is seen that if Equation (1) is multiplied by 440/200, then the multiple of I_2 becomes 440, as in Equation (2). This facilitates elimination of I_2.

Eq. 1 × 440/200,
$$13.2\text{ V} = 704I_1 - 440I_2$$

Eq. 2,
$$-12\text{ V} = -200I_1 + 440I_2$$

Adding
$$1.2\text{ V} = 504I_1 + 0$$

So
$$I_1 = \frac{1.2\text{ V}}{504} \approx 2.38\text{ mA}$$

Substituting for I_1 in Eq. 2,

$$-12\text{ V} = 440I_2 - (200 \times 2.38\text{ mA})$$

$$I_2 = \frac{-12\text{ V} + (200 \times 2.38\text{ mA})}{440}$$

$$\approx -26.19\text{ mA}$$

The negative sign indicates that the actual current direction for I_2 is opposite to the (clockwise) loop current.

$$I_3 = I_1 - I_2 \approx 2.38\text{ mA} - (-26.19\text{ mA})$$

$$\approx 28.57\text{ mA}$$

Compare these current levels to the answers obtained in Example 8-2.

Practice Problems

8-3.1 Using loop equations, analyze the circuit in Figure 8-14 to determine the current in resistor R_3.

8-3.2 Calculate the current in resistor R_3 in Figure 8-15 by the use of loop equations.

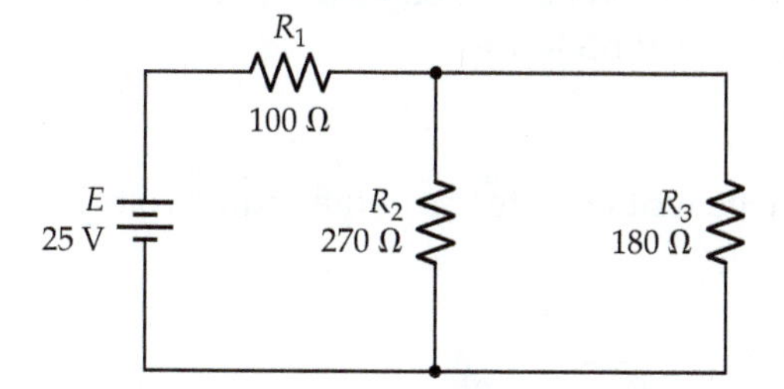

Figure 8-14 Circuit for Practice Problem 8-3.1.

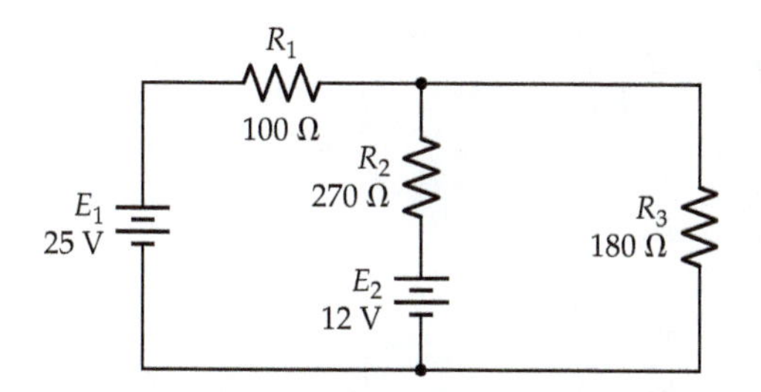

Figure 8-15 Circuit for Practice Problem 8-3.2.

8-4 NODAL ANALYSIS

A *voltage node* is a junction in an electrical circuit at which a voltage can be measured with respect to another (reference) node. If one point in the circuit is grounded [see Figure 8-16(a)], that point is usually selected as the reference

Figure 8-16 For nodal analysis of a circuit, the voltage at each node (or junction in the circuit) is identified with respect to ground (node zero). Then circuit equations are written to determine the node voltages.

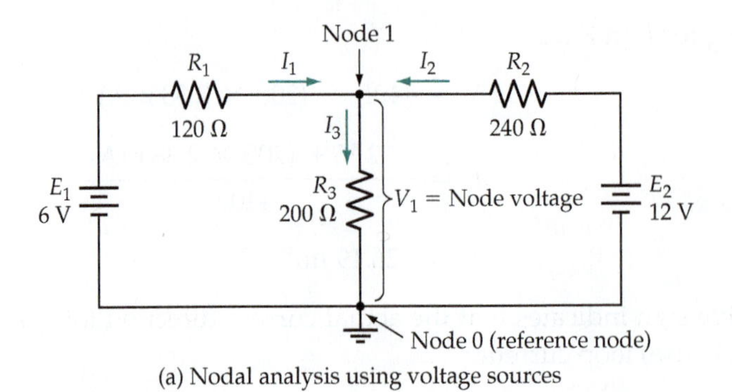

(a) Nodal analysis using voltage sources

(b) Nodal analysis using current sources

node. Otherwise, any convenient junction can be treated as a reference node. From Figure 8-16(a), the current equation at node 1 is,

$$I_3 = I_1 + I_2 \tag{1}$$

An equation for each current can also be written in terms of the source voltages and the node voltage:

$$I_1 = \frac{E_1 - V_1}{R_1} \tag{2}$$

$$I_2 = \frac{E_2 - V_1}{R_2} \tag{3}$$

$$I_3 = \frac{V_1}{R_3} \tag{4}$$

Substitution of Eq. 2, 3 and 4 into Eq. 1 gives,

$$\frac{V_1}{R_3} = \frac{E_2 - V_1}{R_2} + \frac{E_1 - V_1}{R_1}$$

$$= \frac{E_2}{R_2} - \frac{V_1}{R_2} + \frac{E_1}{R_1} - \frac{V_1}{R_1}$$

$$V_1\left(\frac{1}{R_1} + \frac{1}{R_2} + \frac{1}{R_3}\right) = \frac{E_1}{R_1} + \frac{E_2}{R_2}$$

Giving $$V_1 = \frac{(E_1/R_1) + (E_2/R_2)}{1/R_1 + 1/R_2 + 1/R_3} \quad (5)$$

Because V_1 is the only unknown quantity in Eq. 5, the node voltage can be quickly determined. Then, using Eq. 2, 3, and 4, the branch currents are readily calculated.

This (nodal analysis) method is obviously simpler than either the Kirchhoff's law or the loop equation approach to network analysis. However, it can be simplified even further if the voltage sources and their series resistances are first converted to current sources, as illustrated in Figure 8-16(b).

Nodal Analysis Procedure

1. ***Convert all voltage sources into current sources and redraw the circuit diagram.***
2. ***Identify all nodes and choose a reference node.***
3. ***Write the equations for the currents flowing into and out of each node, with the exception of the reference node.***
4. ***Solve the equations to determine the node voltage and the required branch currents.***

Example 8-4

For the circuit shown in Figure 8-16(a), convert the voltage sources into current sources, and then use nodal analysis to determine the current through R_3.

Solution

The circuit with voltage sources converted into current sources is shown in Figure 8-16(b). Voltage source E_1 and series resistance R_1 convert to current source 1.

From Section 8-1, $$I_1 = \frac{E_1}{R_1} = \frac{6\text{ V}}{120\ \Omega}$$

And $$R_{S1} = R_1 = 120\ \Omega$$

For voltage source E_2 and series resistance R_2, current source 2 is,

$$I_2 = \frac{E_2}{R_2} = \frac{12\text{ V}}{240\ \Omega}$$

And $$R_{S2} = R_2 = 240\ \Omega$$

The modified circuit now shows three current paths, through R_{S1}, R_3, and R_{S2}. Writing the current equation for node 1,

$$I_1 + I_2 = I_3 + I_4 + I_5 \quad (1)$$

where $$I_3 = \frac{V_1}{R_3},\ I_4 = \frac{V_1}{R_{S1}},\ I_5 = \frac{V_1}{R_{S2}}$$

Substituting into Eq. 1,

$$\frac{E_1}{R_1} + \frac{E_2}{R_2} = \frac{V_1}{R_3} + \frac{V_1}{R_{S1}} + \frac{V_1}{R_{S2}} \tag{2}$$

$$\frac{E_1}{R_1} + \frac{E_2}{R_2} = V_1\left(\frac{1}{R_3} + \frac{1}{R_{S1}} + \frac{1}{R_{S2}}\right)$$

$$V_1 = \frac{(E_1/R_1) + (E_2/R_2)}{1/R_3 + 1/R_{S1} + 1/R_{S2}}$$

$$= \frac{(6\text{ V}/120\ \Omega) + (12\text{ V}/240\ \Omega)}{(1/200\ \Omega) + (1/120\ \Omega) + (1/240\ \Omega)}$$

$$\approx 5.71\text{ V}$$

$$I_3 = \frac{V_1}{R_3} \approx \frac{5.71\text{ V}}{200\ \Omega}$$

$$\approx 28.6\text{ mA}$$

Example 8-5

Using nodal analysis, determine the current that flows through resistor R_3 in the circuit shown in Figure 8-17(a).

Solution

Voltage source E_1 with series resistance R_1 becomes current source 1 in Figure 8-17(b).

$$I_1 = \frac{E_1}{R_1} \quad \text{and} \quad R_1 = R_S$$

Voltage source E_2 with series resistance R_2 becomes current source 2 in Figure 8-17(b).

$$I_4 = \frac{E_2}{R_2} \quad \text{and} \quad R_2 = R_S$$

The nodes are identified and numbered as 0, 1, and 2 [see Figure 8-17(b)]. The circuit is redrawn in Figure 8-17(c) to more easily identify which currents flow into the nodes and which flow out.

At node 1,

$$I_1 + I_4 = I_2 + I_3 + I_5 \tag{1}$$

So

$$\frac{E_1}{R_1} + \frac{E_2}{R_2} = \frac{V_1}{R_1} + \frac{V_1}{R_2} + \frac{V_1 - V_2}{R_3}$$

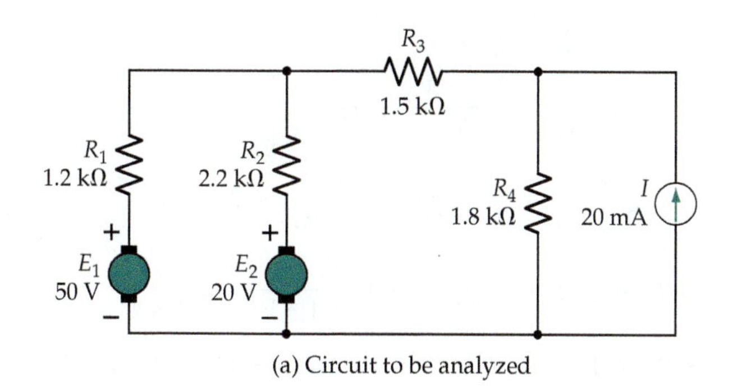

(a) Circuit to be analyzed

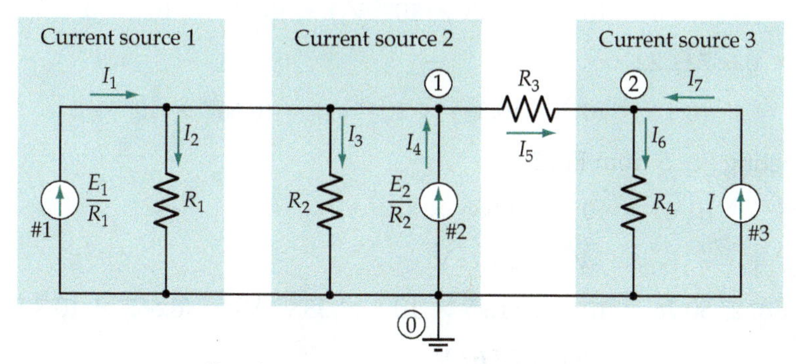

(b) Voltage sources replaced by current sources

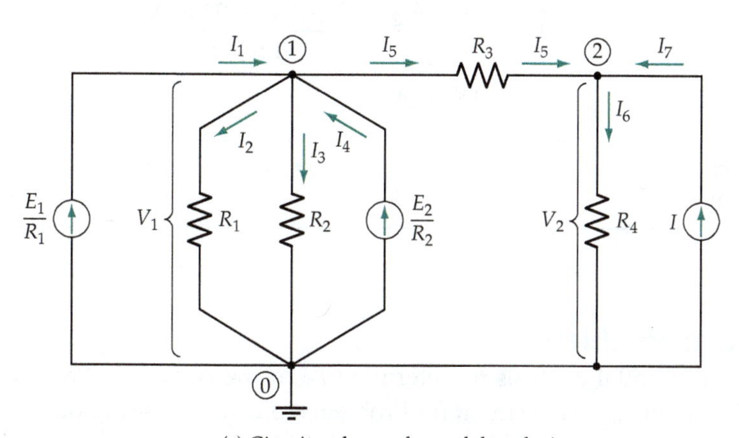

(c) Circuit redrawn for nodal analysis

Figure 8-17 Nodal analysis of a circuit with two voltage sources and one current source. The voltage sources shown in (a) are first converted to current sources, as in (b). Then, the circuit is redrawn in (c) to clearly show that there are only two nodes besides the reference node.

or

$$\frac{E_1}{R_1} + \frac{E_2}{R_2} = \frac{V_1}{R_1} + \frac{V_1}{R_2} + \frac{V_1}{R_3} - \frac{V_2}{R_3}$$

$$\frac{50\text{ V}}{1.2\text{ k}\Omega} + \frac{20\text{ V}}{2.2\text{ k}\Omega} = V_1\left(\frac{1}{1.2\text{ k}\Omega} + \frac{1}{2.2\text{ k}\Omega} + \frac{1}{1.5\text{ k}\Omega}\right) - \frac{V_2}{1.5\text{ k}\Omega}$$

$$50.76 \times 10^{-3} = (1.95 \times 10^{-3}\, V_1) - (0.667 \times 10^{-3}\, V_2) \qquad \textbf{(2)}$$

At node 2,

$$I_6 = I_5 + I_7 \qquad (3)$$

$$\frac{V_2}{R_6} = \frac{V_1 - V_2}{R_3} + I_7$$

$$\frac{V_2}{1.8\ \text{k}\Omega} = \frac{V_1}{1.5\ \text{k}\Omega} - \frac{V_2}{1.5\ \text{k}\Omega} + 20 \times 10^{-3}$$

$$-20 \times 10^{-3} = \frac{V_1}{1.5\ \text{k}\Omega} - V_2\left(\frac{1}{1.5\ \text{k}\Omega} + \frac{1}{1.8\ \text{k}\Omega}\right)$$

$$-20 \times 10^{-3} = (0.667 \times 10^{-3}\ V_1) - (1.22 \times 10^{-3}\ V_2) \qquad (4)$$

Eq. 2 $\quad 50.76 \times 10^{-3} = (1.95 \times 10^{-3}\ V_1) - (0.667 \times 10^{-3}\ V_2)$

Eq. 4 $\times$ 0.667/1.22

$$-10.93 \times 10^{-3} = (0.365 \times 10^{-3}\ V_1) - (0.667 \times 10^{-3}\ V_2) \qquad (5)$$

Subtracting Eq. 5 from Eq. 2,

$$61.69 \times 10^{-3} = 1.585 \times 10^{-3}\ V_1$$

giving $\quad V_1 \approx 38.92\ \text{V}$

From Eq. 2, $50.76 \times 10^{-3} = (1.95 \times 10^{-3} \times 38.92\ \text{V}) - (0.667 \times 10^{-3}\ V_2)$

$$V_2 \approx 37.68\ \text{V}$$

$$I_S = \frac{V_1 - V_2}{R_3} = \frac{38.92\ \text{V} - 37.68\ \text{V}}{1.5\ \text{k}\Omega}$$

$$\approx 0.83\ \text{mA}$$

Computer analysis of the type of circuit referred to in Example 8-5 is offered in Section 28-3.

Practice Problems

8-4.1 Apply nodal analysis to determine I_{a2} in the circuit for Problem 8-3.1.
8-4.2 Calculate I_{R1} in the circuit for Problem 8-3.2 by the use of nodal analysis.

8-5 DELTA-WYE TRANSFORMATIONS

Consider the circuit shown in Figure 8-18(a), in which resistors R_1, R_2, and R_3 are in the form of a *delta* (Δ) network. The circuit, as shown, could be analyzed using loop equations or nodal analysis. However, the analysis can be considerably simplified if the Δ network is replaced by the Y network shown in Figure 8-18(b). For the Δ network to be replaced by the *Y* network, both networks must have exactly the same effect on the rest of the circuit. Consequently, the resistance measured between any two of terminals A, B, and C must be exactly

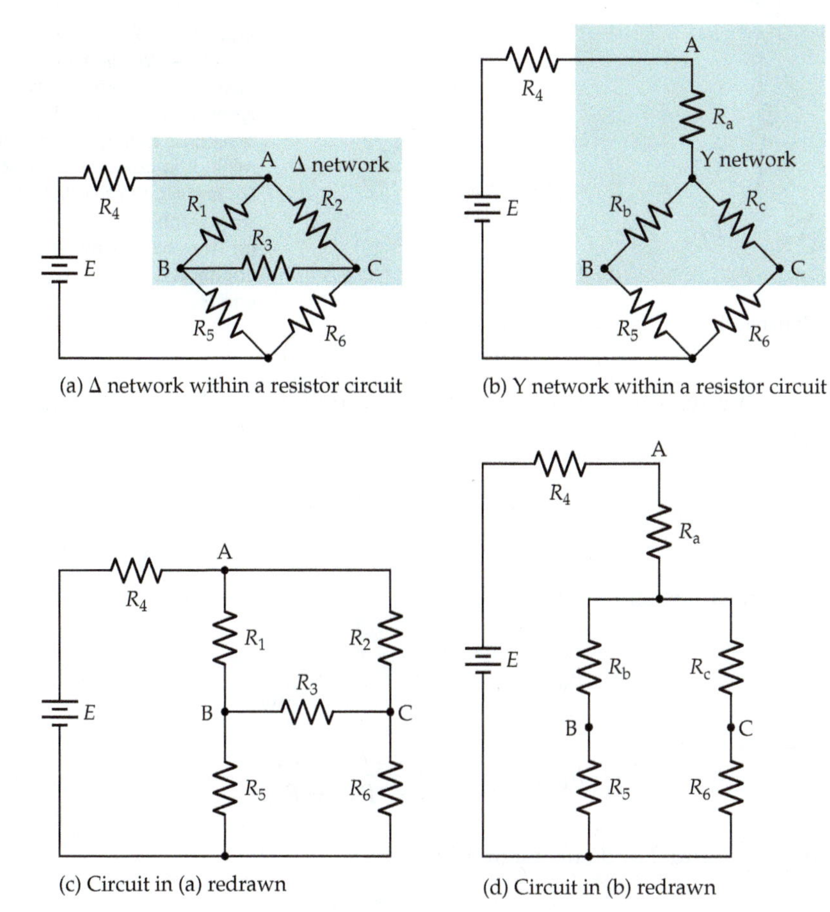

Figure 8-18 **Circuit simplification by Δ-Y transformation. A delta network can be converted into an equivalent Y network, and vice versa, to facilitate circuit analysis.**

the same in each case. Figures 8-18(c) and (d) represent exactly the same circuits as in Figures 8-18(a) and (b), respectively.

For Δ-Y and Y-Δ transformation, equations must be derived to relate the resistors in the two networks. Consider Figures 8-19(a) and (b). For the Y network, the resistance between terminals A and B is,

$$R_{AB} = R_a + R_b$$

and for the Δ network,

$$R_{AB} = R_{ab} \| (R_{ac} + R_{bc})$$

Therefore,

$$R_{AB} = \frac{R_{ab}(R_{ac} + R_{bc})}{R_{ab} + R_{ac} + R_{bc}}$$

which gives

$$R_{AB} = \frac{R_{ab}R_{ac} + R_{ab}R_{bc}}{R_{ab} + R_{ac} + R_{bc}}$$

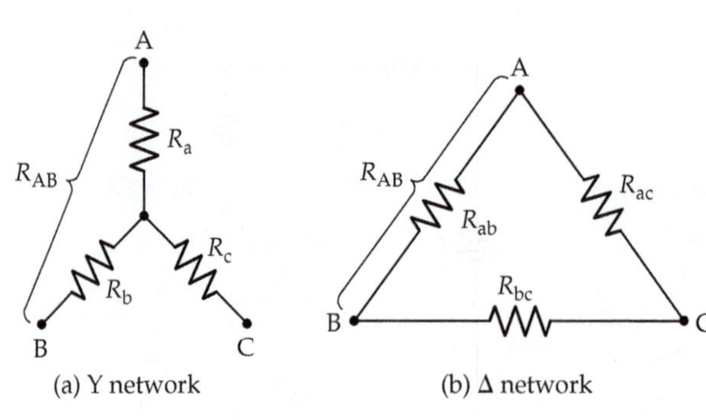

Figure 8-19 Y and delta networks identified for convenience in converting between types. In the Y networks each resistor is named for the terminal it is connected to; R_a, R_b, and R_c. For the delta network each resistor is named for the two terminals it is connected across; R_{ab}, R_{bc}, and R_{ac}.

For the same resistance value between terminals A and B on both Y and Δ networks,

$$R_a + R_b = \frac{R_{ab}R_{ac} + R_{ab}R_{bc}}{R_{ab} + R_{ac} + R_{bc}} \quad (1)$$

Similarly, it can be shown that

$$R_a + R_c = \frac{R_{ab}R_{ac} + R_{ac}R_{bc}}{R_{ab} + R_{ac} + R_{bc}} \quad (2)$$

and

$$R_b + R_c = \frac{R_{ab}R_{bc} + R_{ac}R_{bc}}{R_{ab} + R_{ac} + R_{bc}} \quad (3)$$

Adding Eq. 1 and 2 and subtracting Eq. 3 gives,

$$R_a = \frac{R_{ab}R_{ac}}{R_{ab} + R_{ac} + R_{bc}} \quad (8\text{-}5)$$

Also, Eq. 1 + Eq. 3 − Eq. 2 gives,

$$R_b = \frac{R_{ab}R_{bc}}{R_{ab} + R_{ac} + R_{bc}} \quad (8\text{-}6)$$

and, Eq. 2 + Eq. 3 − Eq. 1 gives,

$$R_c = \frac{R_{ac}R_{bc}}{R_{ab} + R_{ac} + R_{bc}} \quad (8\text{-}7)$$

Equations 8-5, 8-6, and 8-7 can be used to convert a network from Δ to Y configuration. An examination of Equation 8-5 shows that *to obtain the* Y *network resistor* (R_a) *connected to terminal* A, *the two* Δ *network resistors connected to* A *must be multiplied together, and the product divided by the sum of the three* Δ *network resistors.* A similar procedure applies for obtaining the Y network resistors connected to terminals B and C. By further manipulation of Equations 1, 2 and 3, equations for conversion from Y to Δ can be obtained:

$$R_{ab} = \frac{R_aR_b + R_aR_c + R_bR_c}{R_c} \tag{8-8}$$

$$R_{ac} = \frac{R_aR_b + R_aR_c + R_bR_c}{R_b} \tag{8-9}$$

$$R_{bc} = \frac{R_aR_b + R_aR_c + R_bR_c}{R_a} \tag{8-10}$$

Once again, examination of the three equations reveals a definite pattern. *To obtain the Δ resistor connected between any two terminals, divide the Y resistor connected to the terminal opposite those two (terminal C for the resistor between A and B) into the sum of the products of each pair of Y resistors.*

The Y network is also known as a *star network* or a *T network.* Figures 8-20(a) and (c) show Y networks redrawn in a slightly different form. Delta networks are also termed *mesh networks* or *π networks.* Figures 8-20(b) and (d) show delta networks redrawn in a different form.

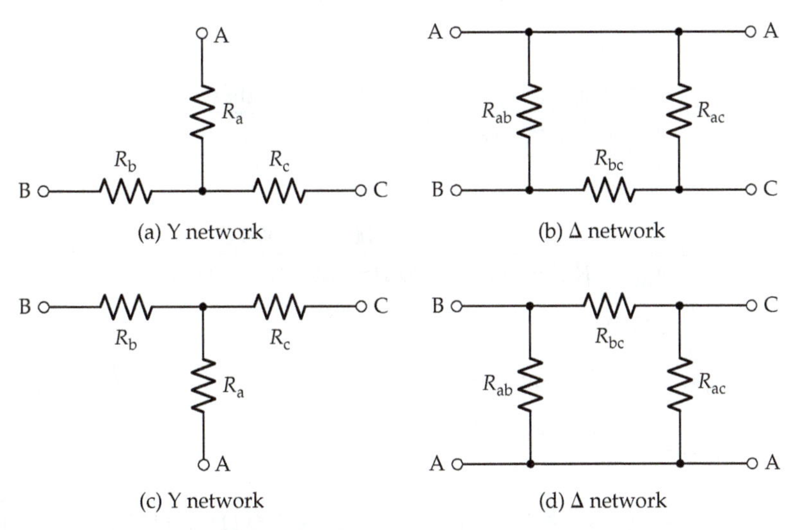

Figure 8-20 Y networks are also known as *T* or star networks, and Δ networks are sometimes referred to as *π* or mesh networks.

Δ-Y and Y-Δ Transformation Procedure:

1. ***When starting with a Δ network, draw a Y network.***
   When starting with a Y network, draw a Δ network.
2. ***Identify the three corresponding terminals on each network as A, B, and C.***

3. *Identify the resistors on the* **Δ** *network as follows:*
   Resistor between terminals **A** *and* **B** *as* R_{ab}
   Resistor between terminals **A** *and* **C** *as* R_{ac}
   Resistor between terminals **B** *and* **C** *as* R_{bc}
4. *Identify the resistors on the* **Y** *network as follows:*
   Resistor connected to terminal **A** *as* R_a
   Resistor connected to terminal **B** *as* R_b
   Resistor connected to terminal **C** *as* R_c
5. *For* **Δ** *to* **Y** *transformation, substitute the* **Δ** *network resistor values into Equations 8-5, 8-6, and 8-7 to obtain the* **Y** *network resistor values.*
6. *For* **Y** *to* **Δ** *transformation, substitute the* **Y** *network resistor values into Equations 8-8, 8-9, and 8-10 to obtain the* **Δ** *network resistor values.*

Example 8-6

Convert the Δ network shown in Figure 8-19(b) to a Y network. Then convert back again to prove the formula. Take $R_{ab} = 500\ \Omega$, $R_{ac} = 400\ \Omega$, and $R_{bc} = 300\ \Omega$.

Solution

Converting from Δ to Y:

Eq. 8-5 $$R_a = \frac{R_{ab}R_{ac}}{R_{ab} + R_{ac} + R_{bc}} = \frac{500\ \Omega \times 400\ \Omega}{500\ \Omega + 400\ \Omega + 300\ \Omega}$$

$$= 166.7\ \Omega$$

Eq. 8-6 $$R_b = \frac{R_{ab}R_{bc}}{R_{ab} + R_{ac} + R_{bc}} = \frac{500\ \Omega \times 300\ \Omega}{500\ \Omega + 400\ \Omega + 300\ \Omega}$$

$$= 125\ \Omega$$

Eq. 8-7 $$R_c = \frac{R_{ac}R_{bc}}{R_{ab} + R_{ac} + R_{bc}} = \frac{400\ \Omega \times 300\ \Omega}{500\ \Omega + 400\ \Omega + 300\ \Omega}$$

$$= 100\ \Omega$$

Converting back from Y to Δ:

Eq. 8-8 $$R_{ab} = \frac{R_aR_b + R_aR_c + R_bR_c}{R_c}$$

$$= \frac{(166.6\ \Omega \times 125\ \Omega) + (166.6\ \Omega \times 100\ \Omega) + (125\ \Omega \times 100\ \Omega)}{100\ \Omega}$$

$$= \frac{50\ 000}{100} = 500\ \Omega$$

Eq. 8-9 $$R_{ac} = \frac{R_aR_b + R_aR_c + R_bR_c}{R_b} = \frac{50\ 000}{125}$$

$$= 400\ \Omega$$

$$\text{Eq. 8-10} \quad R_{bc} = \frac{R_aR_b + R_aR_c + R_bR_c}{R_a} = \frac{50\,000}{166.7}$$

$$= 300\ \Omega$$

Practice Problems

8-5.1 The resistor values for the Y network in Figure 8-19(a) are $R_a = 1.8\ \text{k}\Omega$, $R_b = 3.9\ \text{k}\Omega$, and $R_c = 2.7\ \text{k}\Omega$. Calculate the equivalent Δ network component values.

8-5.2 The Δ network in Figure 8-20(b) has $R_{ab} = 47\ \Omega$, $R_{bc} = 39\ \Omega$, and $R_{ac} = 56\ \Omega$. Calculate the equivalent Y network component values.

8-6 DEPENDENT VOLTAGE AND CURRENT SOURCES

Externally Dependent Sources

All of the voltage and current sources already used in circuit analysis in the preceding pages can be classified as *independent sources*, because they have fixed quantities unaffected by any other voltage or current. *Dependent sources* (also termed *controlled sources*) are dependent on a separate voltage or current; they are controlled by an external quantity. Consider Figure 8-21(a) which shows a voltage source (E_2) with the source voltage identified as KV_1. This means that

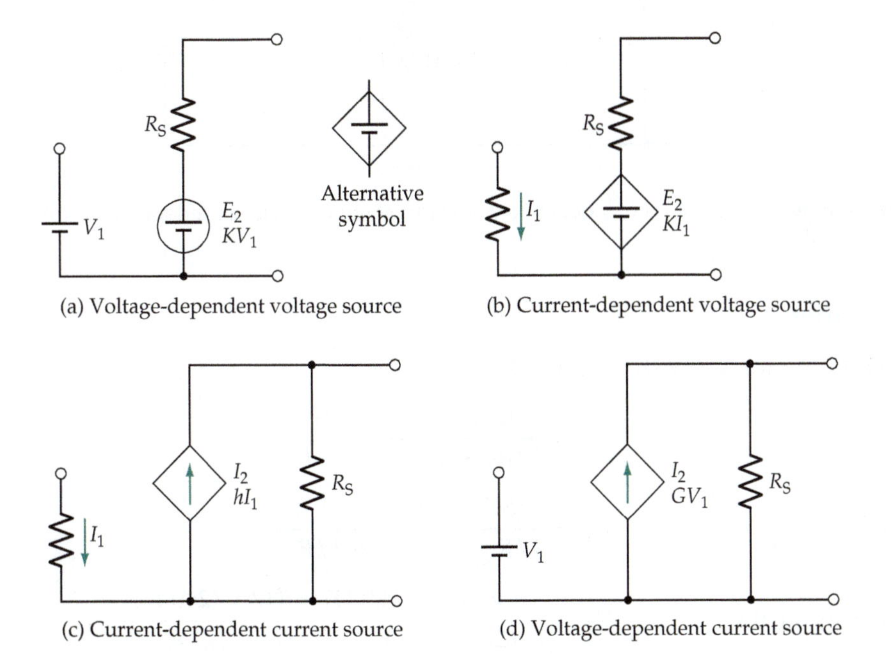

Figure 8-21 Dependent voltage and current sources. Each source is controlled by an external quantity.

if V_1 is 1 V and K is 5, then $E_2 = 5$ V. Note that instead of a circle a diamond shape is sometimes used to identify a dependent source. Figure 8-21(b) shows that the controlling quantity in a dependent voltage source could be a current instead of a voltage. In this case with $K = 10 \times 10^3$ and $I_1 = 1$ mA, $E_2 = 10$ V.

Two types of dependent current sources are illustrated in Figure 8-21(c) and (d). The current source in Figure 8-21(c) is dependent on a separate controlling current; $I_2 = hI_1$, where h is a constant. The current source in part (d) of Figure 8-21 is dependent on a controlling voltage, and in this case the constant might be defined as A/V, mA/V, or μA/V, which could also be stated as S (Siemens), mS, or μS. If V_1 is 1 V and G = 4 mA/V, $I_2 = 4$ mA.

Once the source voltage or current is calculated for a dependent source, it is used exactly like an independent source in any method of circuit analysis. One difference with a dependent source is that circuit analysis might be required for several different levels of controlling voltage or current. This is explored in Example 8-7.

Example 8-7

Analyze the circuit in Figure 8-22 to determine V_1 when $V_X = 0.9$ V and when $V_X = 0.33$ V. Note that Figure 8-22 is a modified version of the circuit in Figure 8-16(a) which is analyzed in Example 8-4.

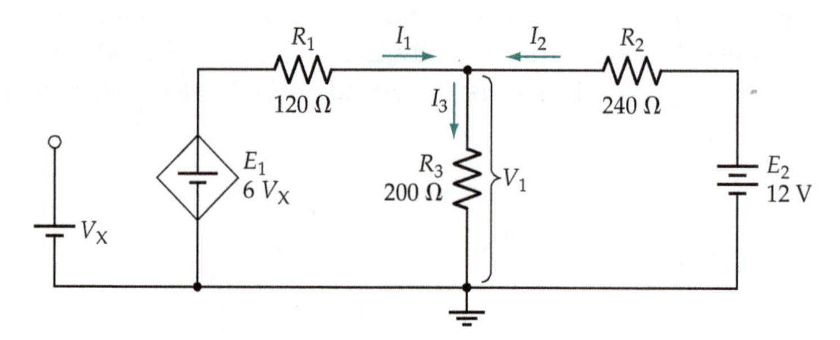

Figure 8-22 Circuit with a dependent voltage source E_1 and an independent voltage source E_2.

Solution

From the analysis in Example 8-4,

$$V_1 = \frac{(E_1/R_1) + (E_2/R_2)}{1/R_3 + 1/R_{S1} + 1/R_{S2}}$$

or

$$V_1 = \frac{(6V_X/R_1) + (E_2/R_2)}{1/R_3 + 1/R_{S1} + 1/R_{S2}}$$

For $V_X = 0.9$ V

$$V_1 = \frac{(6 \times 0.9\ \text{V}/120\ \Omega) + (12\ \text{V}/240\ \Omega)}{(1/200\ \Omega) + (1/120\ \Omega) + (1/240\ \Omega)}$$

$$= 5.4\ \text{V}$$

For $V_X = 0.3$ V
$$V_1 = \frac{(6 \times 0.33\ \text{V}/120\ \Omega) + (12\ \text{V}/240\ \Omega)}{(1/200\ \Omega) + (1/120\ \Omega) + (1/240\ \Omega)}$$
$$= 3.8\ \text{V}$$

Circuit-Dependent Sources

The dependent sources discussed above have controlling voltages or currents that are independent of the circuits in which the sources operate. Some other sources depend on a current or voltage level within the circuit. These can be termed *circuit-dependent sources* to distinguish them from externally dependent sources. Figure 8-23(a) shows a circuit with a current source I_2 that depends on the level of current I_1. This current (I_1) is obviously dependent on V_C which is affected by both I_1 and I_2. When analyzing this circuit, I_2 must be expressed as hI_1, and I_1 must be defined as V_C/R_2. Note that the source resistance is not shown in Figure 8-23(a). The current source is assumed to be an ideal source; one in which $R_S \approx \infty$.

The dependent current source in Figure 8-23(b) has $I_2 = GV_2$, where V_2 is the voltage across resistor R_2. Here again, the controlling quantity must be defined in terms of the circuit conditions in order to analyze the circuit.

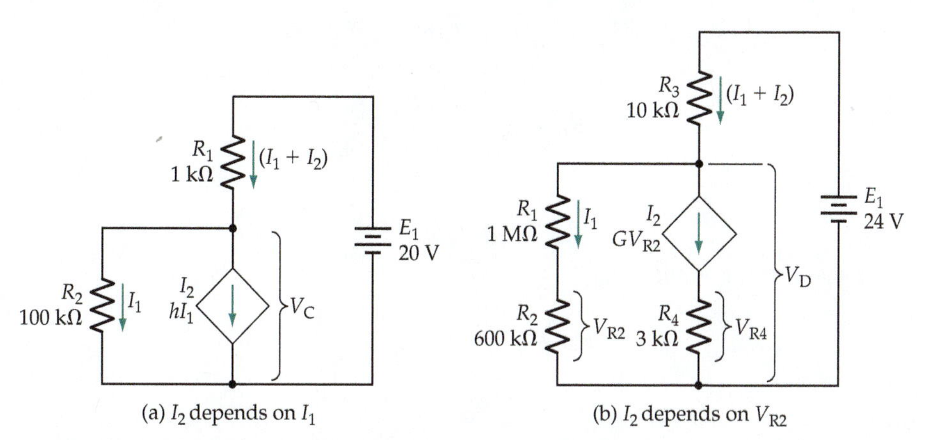

Figure 8-23 Circuits with current sources that depend on a current or voltage in another part of the same circuit.

Example 8-8

Calculate V_C, I_1, and I_2 for the circuit shown in Figure 8-23(a) if $h = 50$.

Solution

$$V_C = I_1 R_2$$

and
$$I_2 = hI_1 = 50I_1$$
$$E_1 = V_C + (I_1 + I_2)R_1$$
$$E_1 = I_1R_2 + (I_1 + hI_1)R_1$$

Giving,
$$I_1 = \frac{E_1}{R_2 + R_1(1 + h)} = \frac{20\ \text{V}}{100\ \text{k}\Omega + 1\ \text{k}\Omega\ (1 + 50)}$$
$$= 132\ \mu\text{A}$$
$$I_2 = hI_1 = 50 \times 132\ \mu\text{A}$$
$$= 6.6\ \text{mA}$$
$$V_C = I_1R_2 = 123\ \mu\text{A} \times 100\ \text{k}\Omega$$
$$= 13.2\ \text{V}$$

Practice Problem

8-6.1 Taking $G = 0.17$ mA/V, analyze the circuit in Figure 8-23(b) to determine V_D and V_{R4}. For simplicity, assume $I_2 \gg I_1$.

Summary of Formulas

- ***Δ-to-Y conversion:***

$$R_a = \frac{R_{ab}R_{ac}}{R_{ab} + R_{ac} + R_{bc}}$$
$$R_b = \frac{R_{ab}R_{bc}}{R_{ab} + R_{ac} + R_{bc}}$$
$$R_c = \frac{R_{ac}R_{bc}}{R_{ab} + R_{ac} + R_{bc}}$$

- ***Y-to-Δ conversion:***

$$R_{ab} = \frac{R_aR_b + R_aR_c + R_bR_c}{R_c}$$
$$R_{ac} = \frac{R_aR_b + R_aR_c + R_bR_c}{R_b}$$
$$R_{bc} = \frac{R_aR_b + R_aR_c + R_bR_c}{R_a}$$

Review Questions

Section 8-1

8-1 Sketch circuit diagrams for voltage and current sources. Show all voltage polarities and current directions.

8-2 Write load current and load voltage equations for current and voltage sources.

8-3 Discuss the effects of series and parallel operation of each type of source.

Section 8-2

8-4 Write Kirchhoff's voltage and current laws, and list the procedure for analyzing a complex network by use of Kirchhoff's laws.

Section 8-3

8-5 Sketch a resistor circuit with two or more current loops, and list the procedure for network analysis using loop equations.

Section 8-4

8-6 Define a voltage node, and list the procedure for nodal analysis of resistor circuits.

Section 8-5

8-7 Draw Δ and Y resistor networks, and write equations for the resistance between each pair of terminals for both networks.

Section 8-6

8-8 Sketch the circuit of a voltage source that is dependent on (a) an external voltage and (b) an external current. Briefly explain.

8-9 Sketch the circuit of a current source that is dependent on (a) an external current and (b) an external voltage. Briefly explain.

8-10 Sketch a current source that is dependent on (a) a current within the circuit and (b) a voltage within the circuit. Explain the difference between externally-dependent sources and circuit-dependent sources.

Problems

Section 8-1

8-1 A current source has $I = 15$ A and $R_S = 5\ \Omega$. Determine E and R_S for the equivalent voltage source. If $R_L = 25\ \Omega$, calculate the output voltage and current in each case.

8-2 For the circuit of Figure 8-24, convert the voltage sources (E_1, R_1) and (E_2, R_2) into their equivalent current sources.

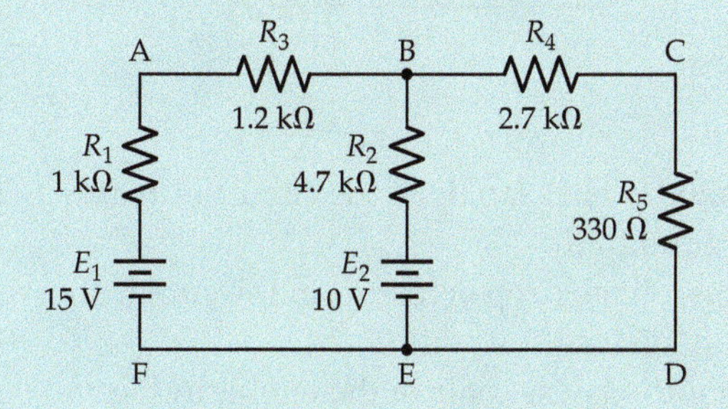

Figure 8-24

8-3 For the circuit shown in Figure 8-25, convert the voltage source (E_1, E_2, R_1) into an equivalent current source.

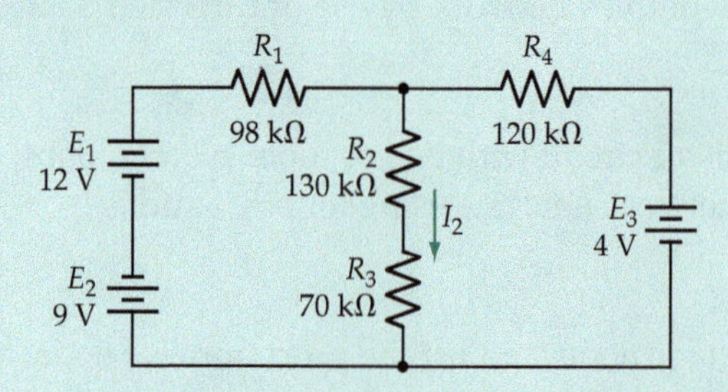

Figure 8-25

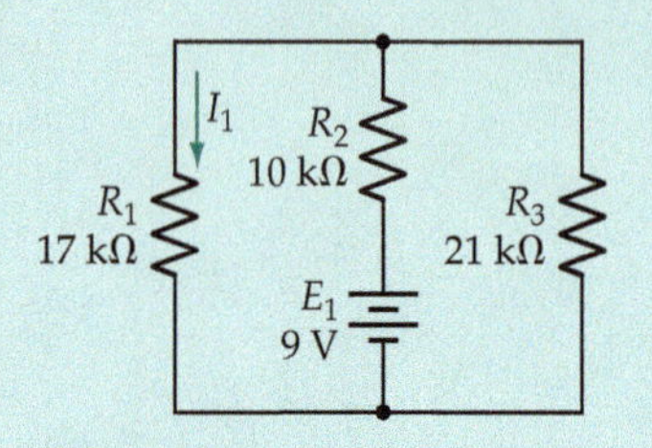

Figure 8-26

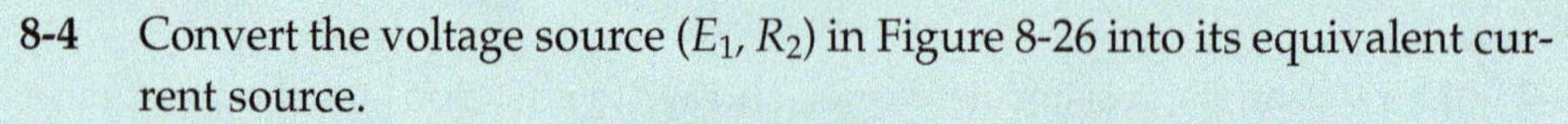

8-4 Convert the voltage source (E_1, R_2) in Figure 8-26 into its equivalent current source.

8-5 Refer to the circuit shown in Figure 8-31. Convert the current source (I_1, R_1) into its equivalent voltage source.

8-6 Convert the current source (I_1, R_1) in the circuit of Figure 8-32 into its equivalent voltage source.

8-7 For the circuit in Figure 8-33, convert the current source (I_1, R_3) into the equivalent voltage source.

Section 8-2

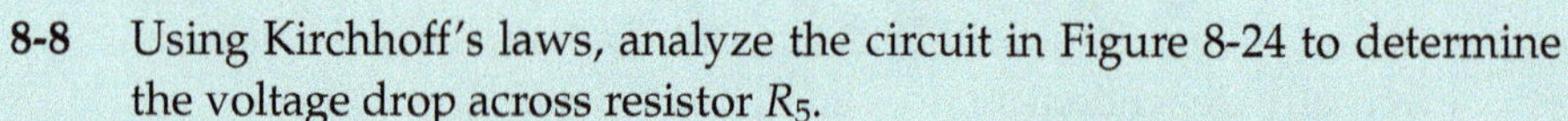

8-8 Using Kirchhoff's laws, analyze the circuit in Figure 8-24 to determine the voltage drop across resistor R_5.

8-9 Using Kirchhoff's laws, determine I_2 in the circuit shown in Figure 8-25.

8-10 Analyze the circuit in Figure 8-26 by use of Kirchhoff's laws to determine the current I_1.

8-11 Apply Kirchhoff's laws to the circuit in Figure 8-27 to calculate I_3.

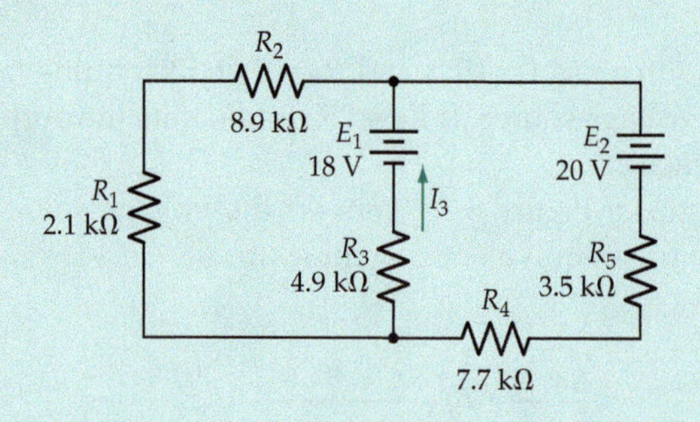

Figure 8-27

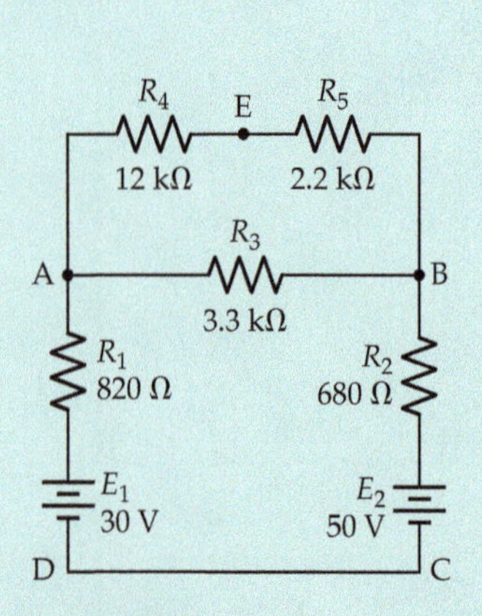

Figure 8-28

8-12 Using Kirchhoff's laws, analyze the circuit in Figure 8-28 to determine the current through R_5.

8-13 Calculate the voltage drop across R_5 in the circuit shown in Figure 8-29. Use Kirchhoff's laws.

8-14 Using Kirchhoff's laws, analyze the circuit in Figure 8-30 to determine the current through R_3.

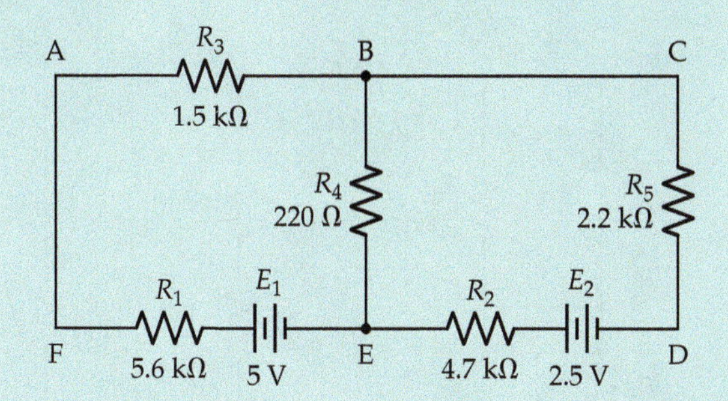

Figure 8-29

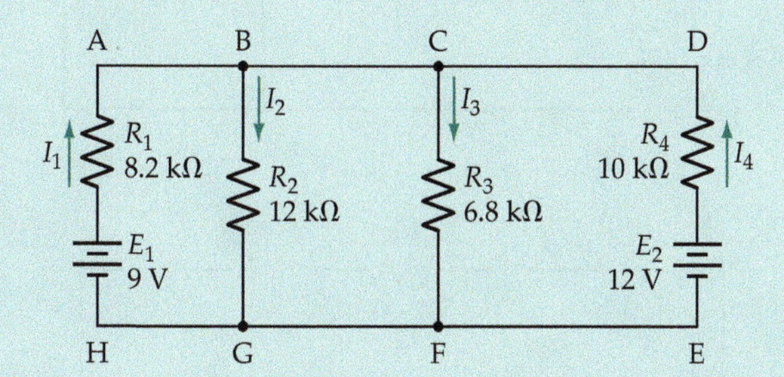

Figure 8-30

8-15 Apply Kirchhoff's laws to the circuit in Figure 8-31 to calculate the current through resistor R_3.

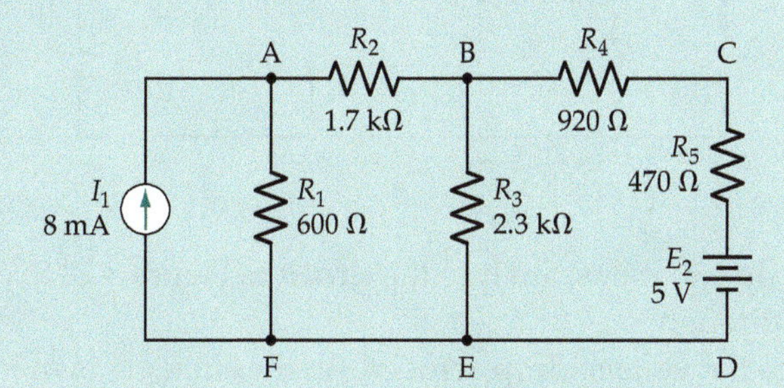

Figure 8-31

8-16 Calculate the current through resistor R_5 for the circuit in Figure 8-32. Use Kirchhoff's laws.

8-17 Determine the current through resistor R_4 for the circuit in Figure 8-33. Use of Kirchhoff's laws.

Section 8-3

8-18 Using loop equations, analyze the circuit in Figure 8-24 to determine the voltage drop across resistor R_5.

8-19 Using loop equations, determine I_2 in the circuit in Figure 8-25.

8-20 Analyze the circuit in Figure 8-26 by use of loop equations to calculate the current I_1.

8-21 Apply the loop equation method to the circuit in Figure 8-27 to calculate I_3.

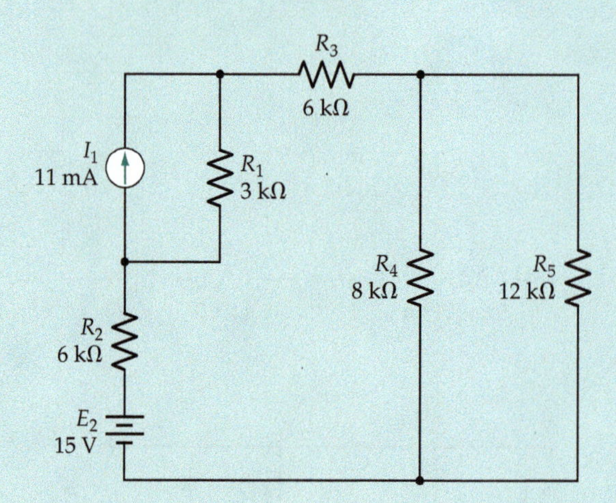

Figure 8-32

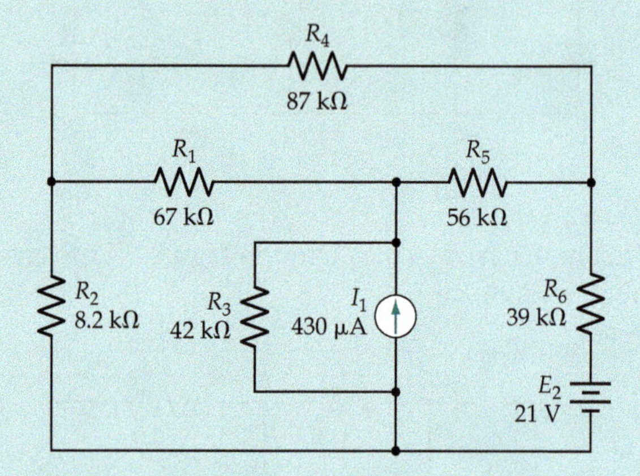

Figure 8-33

8-22 Using loop equations, analyze the circuit in Figure 8-28 to determine the current through R_5.

8-23 Calculate the voltage drop across R_5 in the circuit shown in Figure 8-29. Use loop equations.

8-24 Using loop equations, analyze the circuit in Figure 8-30 to determine the current through R_3.

8-25 Apply loop equations to the circuit in Figure 8-31 to calculate the current through resistor R_3.

8-26 Calculate the current through resistor R_5 for the circuit in Figure 8-32. Use loop equations.

8-27 Using loop equations, determine the current through resistor R_4 for the circuit in Figure 8-33.

Section 8-4

8-28 Apply nodal analysis to the circuit in Figure 8-24 to determine the voltage drop across resistor R_5.

8-29 Using nodal analysis, determine I_2 in the circuit shown in Figure 8-25.

8-30 Analyze the circuit in Figure 8-26 by use of nodal analysis to calculate the current I_1.

8-31 Apply the nodal analysis method to the circuit in Figure 8-27 to calculate I_3.

8-32 Use nodal analysis with the circuit in Figure 8-28 to determine the current through R_5.

8-33 Calculate the voltage drop across R_5 in the circuit shown in Figure 8-29. Use nodal analysis.

8-34 Using nodal analysis, determine the current through R_3 in the circuit in Figure 8-30.

8-35 Apply nodal analysis to the circuit in Figure 8-31 to calculate the current through resistor R_3.

8-36 Calculate the current through resistor R_5 for the circuit in Figure 8-32. Use nodal analysis.

8-37 Using nodal analysis, determine the current through resistor R_4 for the circuit in Figure 8-33.

Section 8-5

8-38 Apply Y-Δ transformation to resistors R_1, R_2, and R_3 in the circuit shown in Figure 8-16(a), then determine the current taken from E_1.

8-39 Modify the circuit in Figure 8-25 to replace resistors R_1, R_2, and R_4 with the equivalent delta network.

8-40 Redraw the circuit in Figure 8-33, replacing resistors R_1, R_4, and R_5 with the equivalent Y network.

Section 8-6

8-41 The voltage sources shown in Figure 8-30 are replaced with dependent sources, $E_1 = 8V_X$ and $E_2 = 7V_Y$. Analyze the circuit to determine I_3 (a) when $V_X = 1$ V and $V_Y = 0.8$ V, (b) when $V_X = 0.8$ V and $V_Y = 1$ V.

8-42 Analyze the circuit in Figure 8-34 to determine I_4 when $I_1 = 300$ μA.

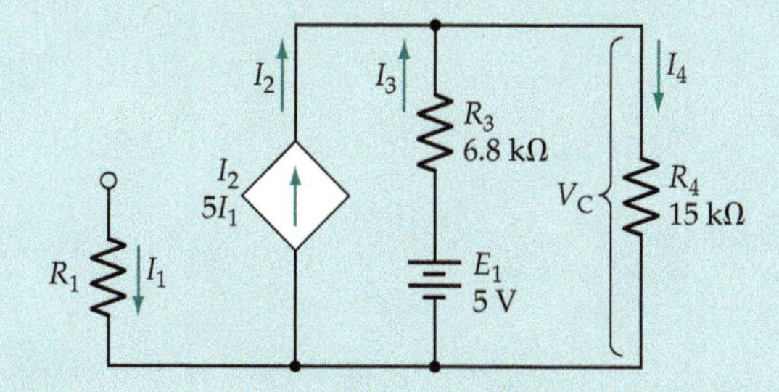

Figure 8-34

8-43 Calculate I_3 for the circuit in Figure 8-35 (a) for $h = 20$, (b) for $h = 50$.

8-44 Determine I_3 in the circuit shown in Figure 8-36 when $K = 5$.

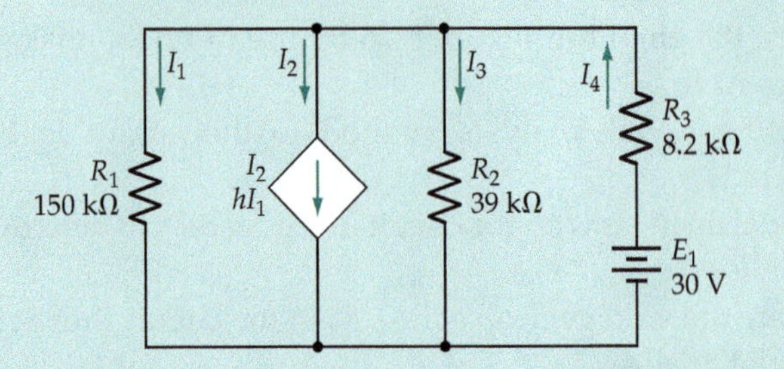

Figure 8-35

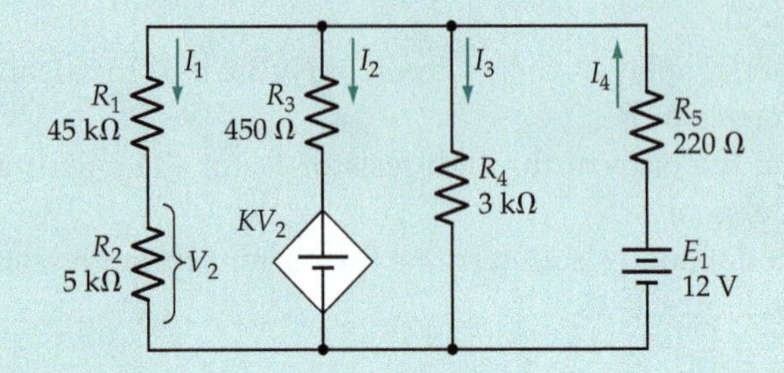

Figure 8-36

Practice Problem Answers

8-1.1	500 μA, 1 kΩ, 6.02 μA
8-1.2	150 V, 20 kΩ, 17.84 V
8-2.1	3.11 mA
8-2.2	2.95 mA
8-3.1	72 mA
8-3.2	84.9 mA
8-4.1	48.07 mA
8-4.2	97.1 mA
8-5.1	8.3 kΩ, 5.75 kΩ, 12.45 kΩ
8-5.2	18.54 kΩ, 12.91 kΩ, 15.38 kΩ
8-6.1	14.6 V, 2.82 V

CHAPTER 9
Network Theorems

CONTENTS

Objectives

You will be able to:

1. State the superposition theorem and apply it to simplify the analysis of complex resistive networks.
2. State Thévenin's theorem and use it in the analysis of complex resistive networks.
3. State Norton's theorem and apply it to solve problems involving complex resistive networks.
4. State Millman's theorem and apply it to problems involving complex resistive networks.
5. State the maximum power transfer theorem and apply it to solve problems involving variations in load on a given source.

INTRODUCTION

Network analysis can be simplified by the use of *network theorems*, which state certain rules that may be applied in particular circumstances. The *superposition theorem*, for example, enables a circuit with several voltage sources and/or current sources to be treated as several circuits that each have only one source. *Thévenin's theorem* permits complex networks to be reduced to a single voltage source in series with a resistance. Thus simplified, the load current and voltage can be very easily determined for many different values of load resistance. *Norton's theorem* is just as powerful as Thévenin's theorem. In this case the network is reduced to a single current source and parallel resistance. How a circuit consisting of many sources in parallel may be treated as just one source is defined by *Millman's theorem*, and the *maximum power transfer theorem* predicts optimum load conditions.

9-1 THE SUPERPOSITION THEOREM

In a network containing more than one source of voltage or current, the current through any branch is the algebraic sum of the currents produced by each source acting independently.

The above statement simply means that the current flowing in a given branch of a circuit can be calculated for each voltage or current source acting alone, then all of the currents are added to determine the actual branch current. The process is illustrated in Figure 9-1. For the circuit shown in Figure 9-1(a), source voltage E_2 is first taken as zero to give the circuit in Figure 9-1(b). This is a

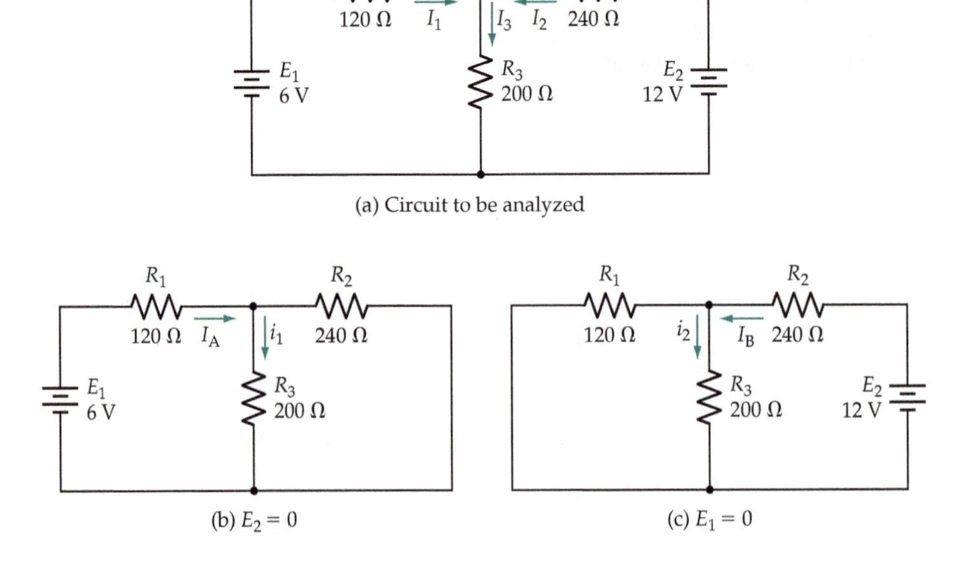

Figure 9-1 Application of the superposition theorem to determine the current I_3 through resistor R_3 in circuit (a). The current i_1 produced by source E_1 when $E_2 = 0$ is first determined from circuit (b). Then the current i_2 due to E_2 when $E_1 = 0$ is determined from circuit (c). The result is $I_3 = i_1 + i_2$.

simple series-parallel circuit that is easily analyzed to determine current i_1 through resistor R_3. E_1 is then taken as zero to give the circuit shown in Figure 9-1(c), which can be quickly analyzed to determine current i_2 through R_3. Currents i_1 and i_2 are added to give the resultant current that flows through R_3 when both voltage sources are present in the circuit. Application of the superposition theorem simplifies the analysis of a network that has more than one source.

Procedure for Applying the Superposition Theorem

1. ***Select one source, and replace all other sources with their internal resistances.***
2. ***Determine the level and direction of the current that flows through the desired branch as a result of the single source acting alone.***
3. ***Repeat steps 1 and 2 using each source in turn until the branch current components have been calculated for all sources.***
4. ***Algebraically sum the component currents to obtain the actual branch current.***

Example 9-1

Use the superposition theorem to calculate the current I_3 in the circuit shown in Figure 9-1(a) (reproduced from Figure 8-10).

Solution

In Figure 9-1(b) voltage source E_2 is replaced with its internal resistance (R_2), and in Figure 9-1(c) voltage source E_1 is replaced with its internal resistance (R_1).

For Figure 9-1(b), equivalent resistance:

$$R_{eq1} = R_1 + R_2 \| R_3 = 120\ \Omega + 240\ \Omega \| 200\ \Omega$$
$$\approx 229.09\ \Omega$$

and

$$I_A = \frac{E_1}{R_{eq1}} \approx \frac{6\ \text{V}}{229.09\ \Omega}$$
$$\approx 26.19\ \text{mA}$$

Using the current-divider rule (see Section 6-4),

$$i_1 = I_A \times \frac{R_2}{R_3 + R_2} \approx 26.19\ \text{mA} \times \frac{240\ \Omega}{200\ \Omega + 240\ \Omega}$$
$$\approx 14.29\ \text{mA}$$

For Figure 9-1(c),

$$R_{eq2} = R_2 + R_1 \| R_3 = 240\ \Omega + 120\ \Omega \| 200\ \Omega$$
$$= 315\ \Omega$$

$$I_B = \frac{E_2}{R_{eq2}} \approx \frac{12\ \text{V}}{315\ \Omega}$$
$$\approx 38.1\ \text{mA}$$

Using the current-divider rule,

$$i_2 = I_B \times \frac{R_1}{R_1 + R_3} \approx 38.1 \text{ mA} \times \frac{120\ \Omega}{120\ \Omega + 200\ \Omega}$$

$$\approx 14.29 \text{ mA}$$

Total current, $I_3 = i_1 + i_2 \approx 14.29 \text{ mA} + 14.29 \text{ mA}$

$$\approx 28.58 \text{ mA}$$

Note: The fact that i_1 and i_2 happen to be equal is peculiar to this particular circuit.

Example 9-2

For the circuit in Figure 9-2(a) [reproduced from Figure 8-17(a)], use the superposition theorem to determine the current through resistor R_3.

Solution

The circuits for each individual source acting independently are shown Figure 9-2(b), (c), and (d).

Figure 9-2 Superposition theorem applied to determine the current I_3 in circuit (a). With $E_2 = 0$, and $I_4 = 0$, i_1 is calculated from circuit (b). Current i_2 is next found from circuit (c), in which $E_1 = 0$ and $I_4 = 0$. Finally, when E_1 and E_2 are both zero, i_3 is determined from circuit (d). Then $I_3 = i_1 + i_2 - i_3$.

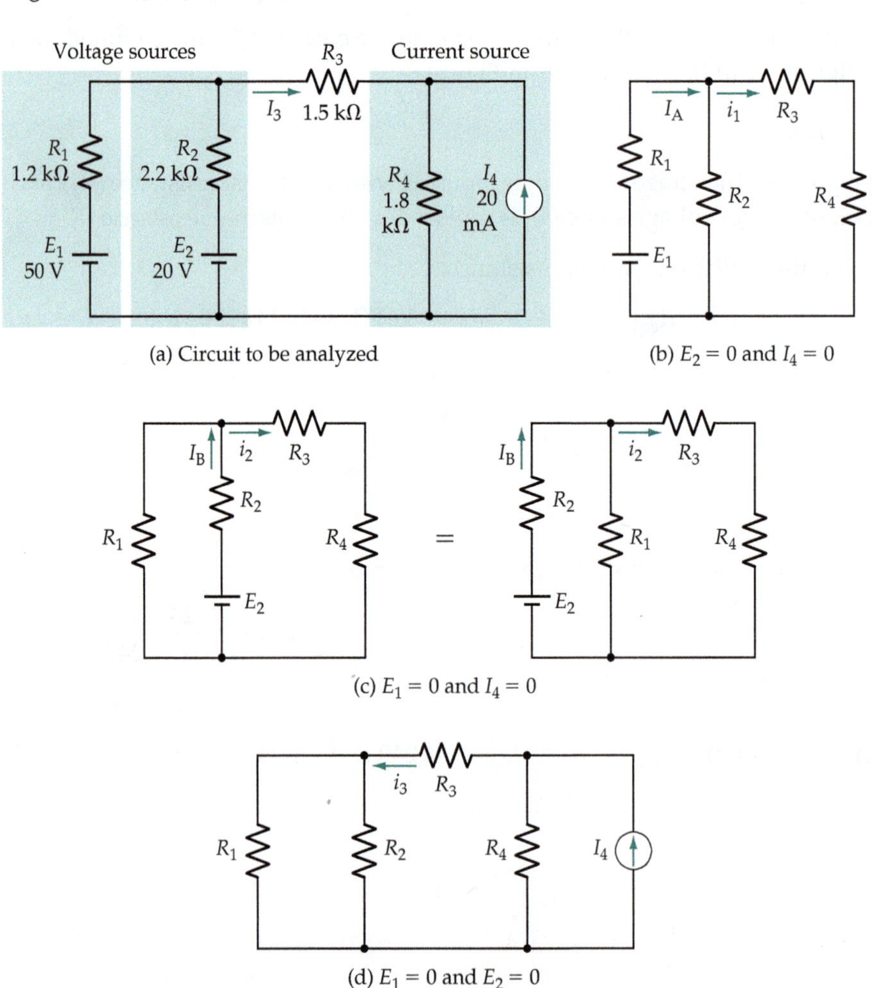

In Figure 9-2(b),

$$I_A = \frac{E_1}{R_1 + R_2 \| (R_3 + R_4)}$$

$$= \frac{50\text{ V}}{1.2\text{ k}\Omega + 2.2\text{ k}\Omega \| (1.5\text{ k}\Omega + 1.8\text{ k}\Omega)}$$

$$\approx 19.84\text{ mA}$$

and using the current-divider rule,

$$i_1 = I_A \times \frac{R_2}{R_2 + (R_3 + R_4)}$$

$$\approx 19.84\text{ mA} \times \frac{2.2\text{ k}\Omega}{2.2\text{ k}\Omega + (1.5\text{ k}\Omega + 1.8\text{ k}\Omega)}$$

$$\approx 7.94\text{ mA}$$

For Figure 9-2(c),

$$I_B = \frac{E_2}{R_2 + R_1 \| (R_3 + R_4)}$$

$$= \frac{20\text{ V}}{2.2\text{ k}\Omega + 1.2\text{ k}\Omega \| (1.5\text{ k}\Omega + 1.8\text{ k}\Omega)}$$

$$\approx 6.49\text{ mA}$$

Using the current-divider rule,

$$i_2 = I_B \times \frac{R_1}{R_1 + (R_3 + R_4)}$$

$$\approx 6.49\text{ mA} \times \frac{1.2\text{ k}\Omega}{1.2\text{ k}\Omega + (1.5\text{ k}\Omega + 1.8\text{ k}\Omega)}$$

$$\approx 1.73\text{ mA}$$

For Figure 9-2(d), using the current-divider rule,

$$i_3 = I_4 \times \frac{R_4}{R_4 + R_3 + (R_1 \| R_2)}$$

$$= 20\text{ mA} \times \frac{1.8\text{ k}\Omega}{1.8\text{ k}\Omega + 1.5\text{ k}\Omega + (1.2\text{ k}\Omega \| 2.2\text{ k}\Omega)}$$

$$\approx 8.83\text{ mA}$$

Current through R_3, $I_3 = i_1 + i_2 - i_3 \approx 7.94\text{ mA} + 1.73\text{ mA} - 8.83\text{ mA}$

$$\approx 0.84\text{ mA}$$

This compares with the result of Example 8-5.

It should be noted that power dissipation in a branch of a circuit must be determined using the *total calculated current* for that branch. *The superposition theorem cannot be applied directly to power dissipation calculations.*

Practice Problems

9-1.1 Apply the superposition theorem to the circuit shown in Figure 8-24 to determine the current through resistor R_3.

9-1.2 Apply the superposition theorem to the circuit of Figure 8-25 to determine the current through resistor R_1.

9-2 THÉVENIN'S THEOREM

Any two-terminal network containing resistances and voltage sources and/or current sources may be replaced by a single voltage source in series with a single resistance. The emf of the voltage source is the open-circuit emf at the network terminals, and the series resistance is the resistance between the network terminals when all sources are replaced with their internal resistances.

By means of Thévenin's theorem, any one resistor in a network can be isolated. The entire remaining portion of the network can be replaced by a single source of emf and a single resistor. Then the current through the isolated resistor may be easily calculated for any value of resistance.

Consider the resistor network shown in Figure 9-3(a). Resistor R_L is variable. Consequently, a complete circuit analysis would seem to be necessary for each value of R_L. The circuit is simplified by replacing all of the network to the left of terminals A and B by its *Thévenin equivalent circuit.* E_{TH} is determined as the open-circuit output voltage at load terminals A and B [see Figure 9-3(b)]. R_{TH} is calculated as the resistance 'seen' when 'looking into' the load terminals with the source voltage taken as zero [Figure 9-3(c)]. This results in the circuit of Figure 9-3(d), which affords easy calculation of the load current for any number of values of R_L. Determining the Thévenin equivalent circuit for a network is sometimes termed *Thévenizing the circuit.*

Thévenizing Procedure

1. ***Calculate the open-circuit voltage (E_{TH}) at the network terminals.***
2. ***Redraw the network with each source replaced with its internal resistance.*** *(Note that when no source resistances are shown, voltage sources should be short-circuited and current sources should be open-circuited.)*
3. ***Calculate the resistance (R_{TH}) of the redrawn network as seen from the output terminals.***

Example 9-3

Use Thévenin's theorem to analyze the circuit shown in Figure 9-3(a) to determine the load current for $R_L = 7.5\ \text{k}\Omega$.

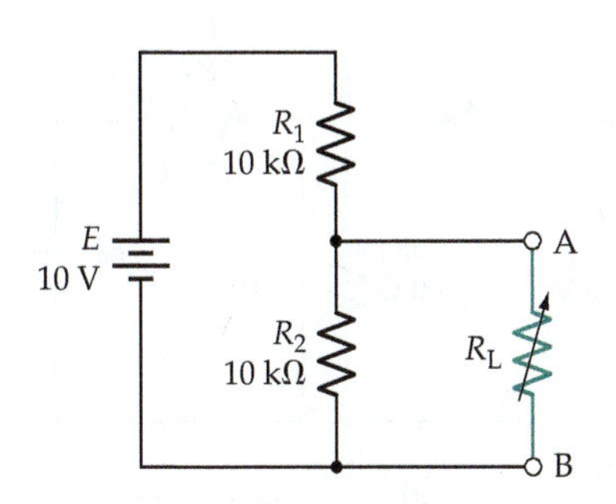

(a) Circuit with variable load

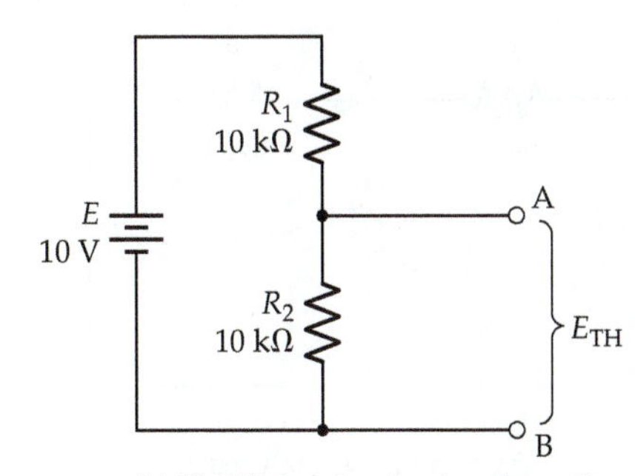

(b) The Thévenin circuit voltage E_{TH} is the open-circuit load voltage

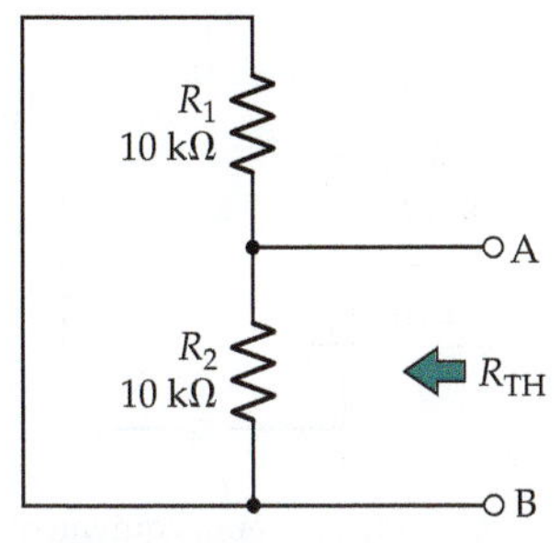

(c) The Thévenin circuit resistance R_{TH} is the measured resistance at the output terminals when $E = 0$

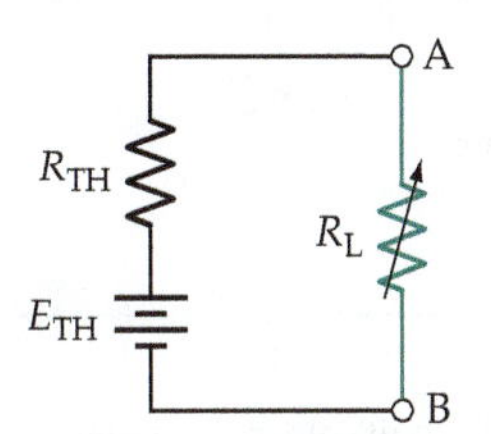

(d) The Thévenin equivalent circuit with a variable load

Figure 9-3 Application of Thévenin's theorem. The network to the left of terminals *A* and *B* in circuit (a) is replaced by its open-circuit output voltage E_{TH} and internal resistance R_{TH} to give the Thévenin equivalent circuit in (d).

Solution

From Figure 9-3(b),

$$E_{TH} = \frac{E \times R_2}{R_1 + R_2} = \frac{10\text{ V} \times 10\text{ k}\Omega}{10\text{ k}\Omega + 10\text{ k}\Omega}$$

$$= 5\text{ V}$$

From Figure 9-3(c),

$$R_{TH} = R_1 \| R_2 = 10\text{ k}\Omega \| 10\text{ k}\Omega$$

$$= 5\text{ k}\Omega$$

From Figure 9-3(b),

$$I_L = \frac{E_{TH}}{R_{TH} + R_L} = \frac{5\text{ V}}{5\text{ k}\Omega + 7.5\text{ k}\Omega}$$

$$= 400\ \mu\text{A}$$

Example 9-4

For the circuit in Figure 9-4(a) (reproduced from Figure 8-10), an external load R_L is to be connected across resistor R_3, as illustrated. Determine the Thévenin equivalent circuit for the network, and calculate the load current when R_L is 330 Ω.

Figure 9-4 Application of Thévenin's theorem to determine the current through load R_L connected across terminals B and E in circuit (a). The open-circuit output voltage across B and E is first calculated to give E_{TH} [circuit (b)]. Then, the internal resistance R_{TH} is determined with E_1 and E_2 both equal zero, as in circuit (c).

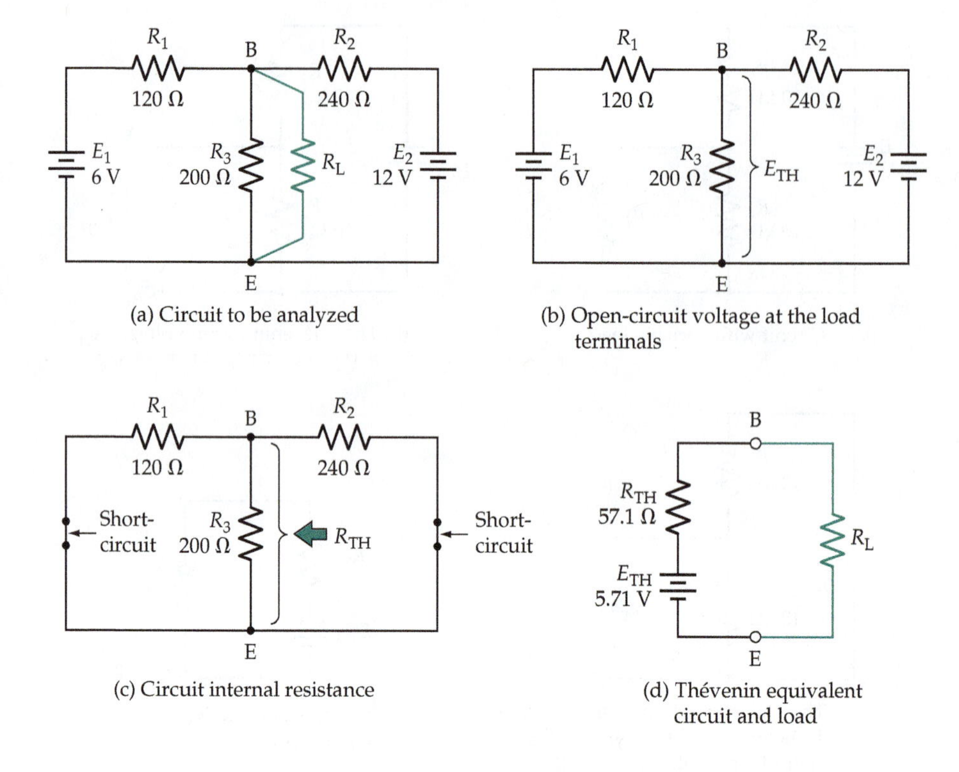

Solution

The open-circuit voltage at terminals B and E must first be calculated, [see Figure 9-4(b)].

From Example 8-2, $I_3 \approx 28.57$ mA

$$E_{TH} = V_{BE} = I_3 R_3 \approx 28.57 \text{ mA} \times 200\ \Omega$$
$$\approx 5.71 \text{ V}$$

Replacing E_1 and E_2 with short-circuits gives the circuit in Figure 9-4(c).
The resistance seen looking into terminals B and E is,

$$R_{TH} = R_3 \| R_1 \| R_2 = 200\ \Omega \| 120\ \Omega \| 240\ \Omega$$
$$= 57.1\ \Omega$$

The Thévenin equivalent circuit is now as shown in Figure 9-4(d).

$$I_L = \frac{E_{TH}}{R_{TH} + R_L} = \frac{5.71 \text{ V}}{57.1\ \Omega + 300\ \Omega}$$
$$\approx 14.75 \text{ mA}$$

Obviously, I_L can easily be recalculated for any number of values of R_L.

Example 9-5

Derive the Thévenin equivalent circuit for the network shown in Figure 9-5(a), and calculate the output voltage for $R_{L1} = 12$ kΩ and for $R_{L2} = 5.6$ kΩ.

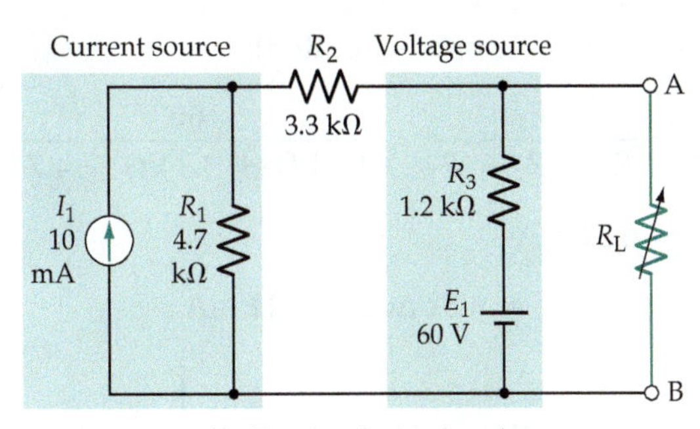

(a) Circuit to be analyzed

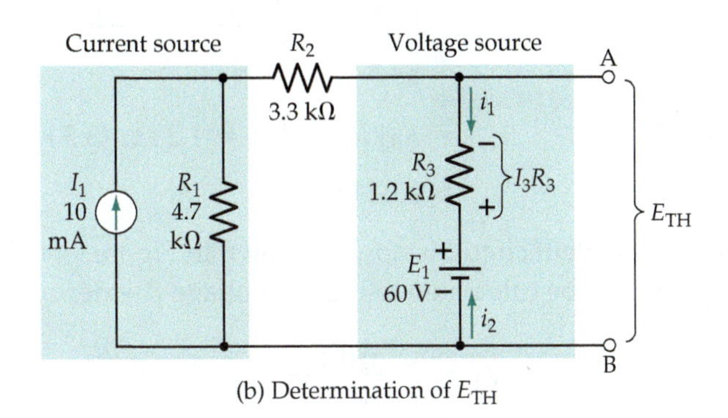

(b) Determination of E_{TH}

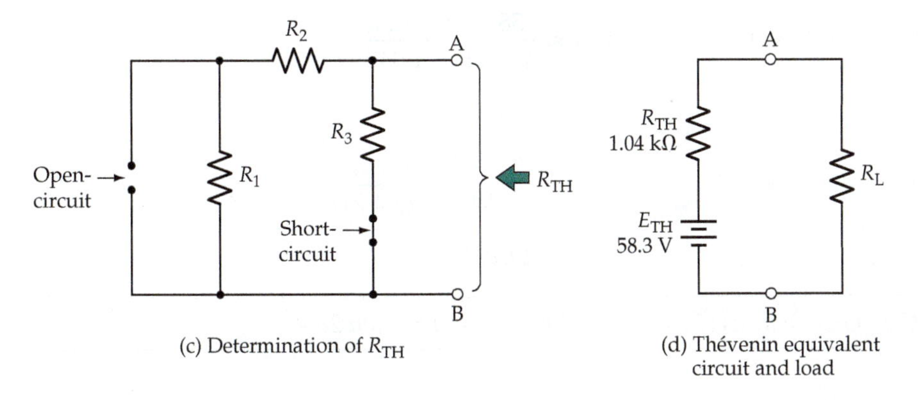

(c) Determination of R_{TH}

(d) Thévenin equivalent circuit and load

Figure 9-5 Application of Thévenin's theorem to the circuit of Figure 9-5(a). The open-circuit output voltage across *A* and *B* is first calculated to give E_{TH}. Then the internal resistance R_{TH} is determined with I_1 open-circuited and E_1 short-circuited.

Solution

The open-circuit voltage at terminals A and B can be calculated by any of the methods already discussed [Figure 9-5(b)]. Using the superposition theorem, a current i_1 flows through R_3 due to current source I_1. Also, a current i_2 flows in the opposite direction due to voltage source E_1.

For I_1 acting alone (voltage source E_1 replaced by R_3), using the current-divider rule:

$$i_1 = I_1 \times \frac{R_1}{R_1 + (R_2 + R_3)}$$

$$= 10 \text{ mA} \times \frac{4.7 \text{ k}\Omega}{4.7 \text{ k}\Omega + 3.3 \text{ k}\Omega + 1.2 \text{ k}\Omega}$$

$$\approx 5.11 \text{ mA}$$

For E_1 acting alone (I_1 source open-circuited):

$$i_2 = \frac{E_1}{R_3 + R_2 + R_1} = \frac{60\text{ V}}{1.2\text{ k}\Omega + 3.3\text{ k}\Omega + 4.7\text{ k}\Omega}$$

$$\approx 6.52\text{ mA}$$

$$I_3 = i_2 - i_1 \approx 6.52\text{ mA} - 5.11\text{ mA}$$

$$\approx 1.41\text{ mA}$$

From Figure 9-5(b),

$$E_{TH} = V_{AB} = E_1 - I_3R_3$$

$$\approx 60\text{ V} - (1.41\text{ mA} \times 1.2\text{ k}\Omega)$$

$$\approx 58.3\text{ V}$$

From Figure 9-5(c),

$$R_{TH} = R_3 \| (R_2 + R_1) = 1.2\text{ k}\Omega \| (3.3\text{ k}\Omega + 4.7\text{ k}\Omega)$$

$$\approx 1.04\text{ k}\Omega$$

The Thévenin equivalent circuit is now as shown in Figure 9-5(d). The output voltage across R_L can be calculated using the voltage-divider rule:

$$V_L = \frac{E_{TH}R_L}{R_L + R_{TH}}$$

For $R_L = 12\text{ k}\Omega$,

$$V_{L1} \approx \frac{58.3\text{ V} \times 12\text{ k}\Omega}{12\text{ k}\Omega + 1.04\text{ k}\Omega}$$

$$\approx 53.65\text{ V}$$

For $R_L = 5.6\text{ k}\Omega$,

$$V_{L2} \approx \frac{58.3\text{ V} \times 5.6\text{ k}\Omega}{5.6\text{ k}\Omega + 1.04\text{ k}\Omega}$$

$$\approx 49.17\text{ V}$$

Computer analysis of Example 9-5 is offered in Section 28-4.

Practice Problems

9-2.1 Use Thévenin's theorem to determine current I_3 in the circuit of Figure 7-2(a).

9-2.2 With R_1 open-circuited in Figure 7-18 apply Thévenin's theorem to calculate the voltage across resistor R_5.

9-3 NORTON'S THEOREM

In Section 8-1 it was shown that voltage sources can be converted to current sources, and vice versa. Therefore, a Thévenin equivalent circuit, which is a voltage source, can be converted into a current source. Alternatively, instead of deriving the Thévenin voltage equivalent circuit for a complex network, a current equivalent circuit can be derived directly for the network. In this case the equivalent circuit is termed a *Norton equivalent circuit,* and the theory for its application is stated in *Norton's theorem:*

Norton's Theorem

> **Any two-terminal network containing resistances and voltage sources and/or current sources may be replaced by a single current source in parallel with a single resistance. The output from the current source is the short-circuit current at the network terminals, and the parallel resistance is the resistance between the network terminals when all sources are replaced by their internal resistances.**

As with Thevenizing, the determination of the Norton equivalent circuit is termed *Nortonizing*. Figure 9-6 illustrates the Nortonizing procedure. I_N is determined as the short-circuit output current at load terminals A and B [see Figure 9-6(b)].

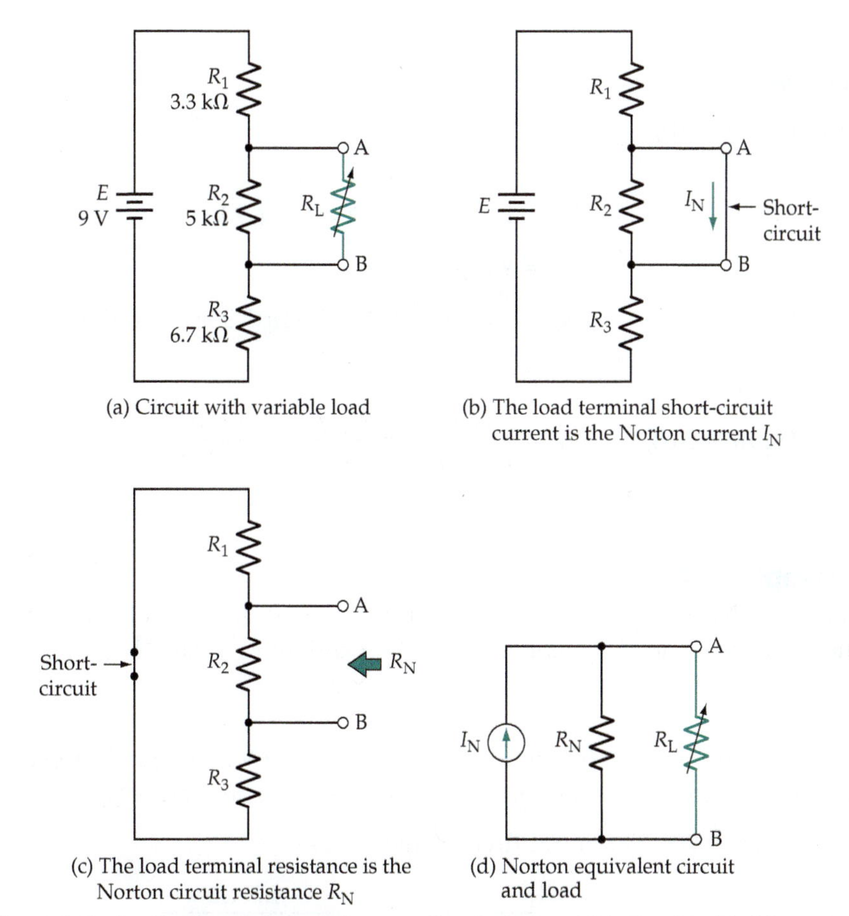

Figure 9-6 Application of Norton's theorem. Terminals *A* and *B* in the circuit in (a) are short-circuited so that I_N can be calculated [part (b)]. *E* is taken as zero, and R_N is calculated [part (c)]. The Norton circuit is then constructed in (d).

R_N is calculated as the resistance 'seen' when 'looking into' the load terminals with the source voltage taken as zero [Figure 9-6(c)]. The Norton equivalent circuit is now shown in Figure 9-6(d).

Nortonizing Procedure

1. ***Calculate the short-circuit current (I_N) at the network terminals.***
2. ***Redraw the network with each source replaced by its internal resistance.*** *(Note that when no source resistances are shown, voltage sources should be short-circuited and current sources should be open-circuited.)*
3. ***Calculate the resistance (R_N) of the redrawn network as seen from the output terminals.***

As with the Thévenin theorem, application of the Norton theorem can save a lot of calculations when several values of load resistor are involved.

Example 9-6

Use Norton's theorem to analyze the circuit in Figure 9-6(a) to determine the load current for $R_L = 5.6\ \text{k}\Omega$.

Solution

From Figure 9-6(b),

$$I_N = \frac{E}{R_1 + R_3} = \frac{9\ \text{V}}{3.3\ \text{k}\Omega + 6.7\ \text{k}\Omega}$$

$$= 900\ \mu\text{A}$$

From Figure 9-6(c),

$$R_N = R_2 \| (R_1 + R_3) = 5\ \text{k}\Omega \| (3.3\ \text{k}\Omega + 6.7\ \text{k}\Omega)$$

$$= 3.3\ \text{k}\Omega$$

From Figure 9-6(d),

$$I_L = \frac{I_N R_N}{R_L + R_N} = \frac{900\ \mu\text{A} \times 3.3\ \text{k}\Omega}{5.6\ \text{k}\Omega + 3.3\ \text{k}\Omega}$$

$$= 333\ \mu\text{A}$$

Example 9-7

Derive the Norton equivalent circuit for the network shown in Figure 9-5(a), and calculate the load current for $R_L = 12\ \text{k}\Omega$ and for $R_L = 5.6\ \text{k}\Omega$.

Solution

The circuit of Figure 9-5(a) is reproduced in Figure 9-7(a) with R_L replaced with a short-circuit. Use the superposition theorem to determine I_{sc}.

From Figure 9-7(b) by current-divider rule,

$$i_1 = I_1 \times \frac{R_1}{R_1 + R_2} = 10\ \text{mA} \times \frac{4.7\ \text{k}\Omega}{4.7\ \text{k}\Omega + 3.3\ \text{k}\Omega}$$

$$\approx 5.88\ \text{mA}$$

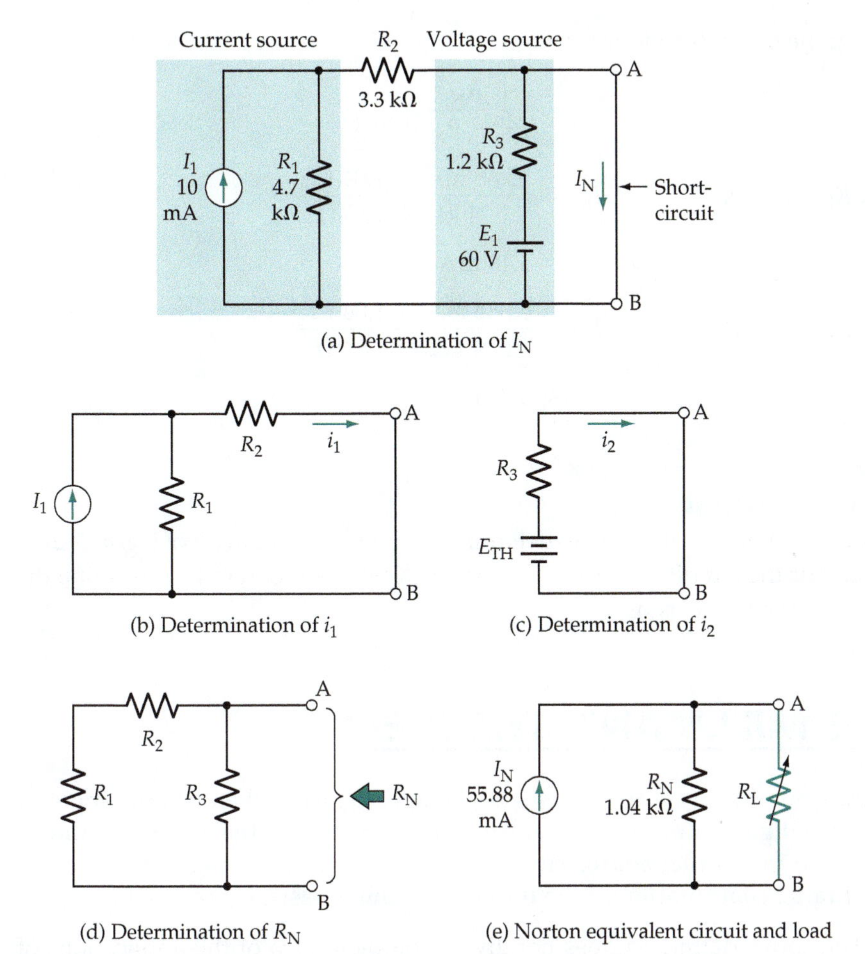

Figure 9-7 Application of Norton's theorem to the circuit of Figure 9-5(a). The output current is first calculated with *A* and *B* short-circuited to give I_N. Then the internal resistance R_N determined with I_1 open-circuited and E_2 short-circuited.

From Figure 9-7(c),

$$I_2 = \frac{E_1}{R_3} = \frac{60\ \text{V}}{1.2\ \text{k}\Omega}$$

$$= 50\ \text{mA}$$

$$I_N = I_{sc} = i_1 + i_2 = 5.88\ \text{mA} + 50\ \text{mA}$$

$$= 55.88\ \text{mA}$$

In Figure 9-7(d), the current source and voltage source are replaced with their internal resistances.

$$R_N = R_3 \| (R_2 + R_1) = 1.2\ \text{k}\Omega \| (3.3\ \text{k}\Omega + 4.7\ \text{k}\Omega)$$

$$\approx 1.04\ \text{k}\Omega$$

The Norton equivalent circuit is now as shown in Figure 9-7(e).

Using the current-divider rule,

$$I_L = \frac{I_N R_N}{R_L + R_N}$$

For $R_L = 12\ \text{k}\Omega$,

$$I_{L1} = \frac{55.88\ \text{mA} \times 1.04\ \text{k}\Omega}{1.04\ \text{k}\Omega + 12\ \text{k}\Omega} \approx 4.46\ \text{mA}$$

For $R_L = 5.6\ \text{k}\Omega$,

$$I_{L2} = \frac{55.88\ \text{mA} \times 1.04\ \text{k}\Omega}{1.04\ \text{k}\Omega + 5.6\ \text{k}\Omega} \approx 8.75\ \text{mA}$$

Practice Problems

9-3.1 Use Norton's theorem to determine current I_2 in the circuit of Figure 7-2(a).

9-3.2 For the circuit in Figure 7-17, apply Norton's theorem to determine the current in resistor R_5.

9-4 MILLMAN'S THEOREM

Any number of current sources in parallel can be represented by a single current generator in which the generator current is the algebraic sum of the individual source currents and the generator resistance is the parallel combination of the individual source resistances.

The above definition does not give a complete idea of the applications of Millman's theorem, because it refers only to current sources. In Section 8-1 it is shown that voltage sources can be converted into current sources, and vice versa. Consequently, Millman's theorem can also be applied to voltage sources in parallel or to a combination of voltage and current sources. Also, instead of ending up with a single current generator, the complete network can be represented as a single voltage generator.

One very important point is that *this theorem applies only to sources connected directly in parallel*; it does not apply where there are resistors between the sources. For example, Millman's theorem could not be applied directly to the two-generator circuit shown in Figure 9-5(a) because of the presence of resistor R_2.

Procedure for Application of Millman's Theorem

1. *Convert all voltage sources into current sources.*
2. *Algebraically add all source currents to obtain the final generator current.*
3. *Determine the resistance value of the source resistances in parallel to obtain the final generator resistance.*
4. *If required, convert the current generator into a voltage generator.*

Example 9-8

Apply Millman's theorem to the circuit shown in Figure 9-8(a) to obtain the equivalent current generator circuit and the equivalent voltage generator circuit.

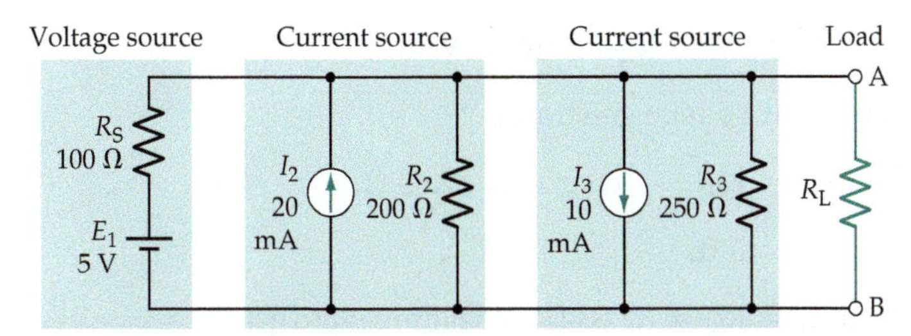

(a) Circuit that can be analyzed by the Millman Theorem (all sources in parallel)

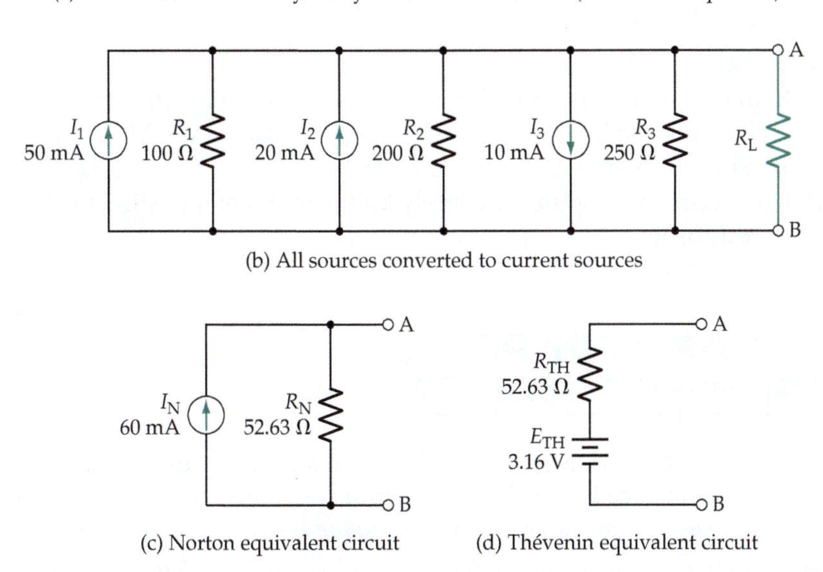

(b) All sources converted to current sources

(c) Norton equivalent circuit (d) Thévenin equivalent circuit

Figure 9-8 Application of Millman's theorem. All sources are first converted to current sources, and the total source current is summed to give I_N. Then the internal resistance R_N is determined as the equivalent of all source resistances in parallel.

Solution

Converting the voltage generator,

$$I_1 = \frac{E_1}{R_S} = \frac{5\text{ V}}{100\ \Omega}$$

$$= 50\text{ mA [see Figure 9-8(b)]}$$

Current source resistance,

$$R_1 = R_S = 100\ \Omega\text{ [see Figure 9-8(b)]}$$

For the final equivalent current generator,

$$I_N = I_1 + I_2 + I_3 = 50 \text{ mA} + 20 \text{ mA} + (-10 \text{ mA})$$
$$= 60 \text{ mA}$$
$$R_N = R_1 \| R_2 \| R_3 = 100 \ \Omega \| 200 \ \Omega \| 250 \ \Omega$$
$$\approx 52.63 \ \Omega \text{ [see Figure 9-8(c)]}$$

For the equivalent voltage generator,

$$E_{TH} = I_N R_N = 60 \text{ mA} \times 52.63 \ \Omega$$
$$\approx 3.16 \text{ V}$$

and $$R_{TH} = R_N \approx 52.63 \ \Omega \text{ [see Figure 9-8(d)]}$$

Practice Problems

9-4.1 In the circuit of Figure 8-17(a), resistor R_3 is shorted, and a new resistor R_5 is connected in parallel with R_4. If $R_5 = 3.3 \text{ k}\Omega$, apply Millman's theorem to determine the current through R_5.

9-4.2 For the circuit in Figure 8-24, apply Millman's theorem to find the current in resistor R_4.

9-5 MAXIMUM POWER TRANSFER THEOREM

Maximum output power is obtained from a network or source when the load resistance is equal to the output resistance of the network or source as seen from the terminals of the load.

The truth of the maximum power transfer theorem is easily tested by considering the circuit in Figure 9-9 and calculating the power output for various values of load resistance. It is also instructive to calculate the load current and voltage as well as the input power.

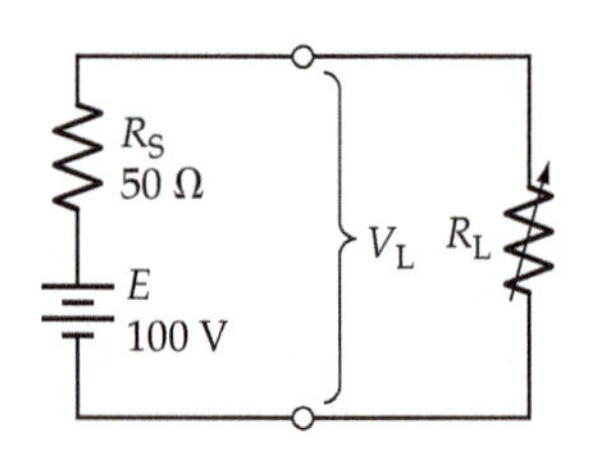

Figure 9-9 Maximum transfer of power from a source to a load occurs when the load resistance equals the source resistance.

Example 9-9

Calculate the output (load) power for the circuit in Figure 9-9 for the following load resistor values: 5 Ω, 10 Ω, 25 Ω, 50 Ω, 100 Ω, 250 Ω, and 500 Ω.

Solution

For $R_L = 50 \ \Omega$,

$$I_L = \frac{E}{R_S + R_L} = \frac{100 \text{ V}}{50 \ \Omega + 50 \ \Omega}$$
$$= 1 \text{ A}$$

$$V_L = I_L R_L = 1\text{ A} \times 50\ \Omega$$
$$= 50\text{ V}$$
$$P_L = I_L V_L = 50\text{ V} \times 1\text{ A}$$
$$= 50\text{ W}$$
$$P_i = E \times I_L = 100\text{ V} \times 1\text{ A}$$
$$= 100\text{ W}$$

The following table gives the various quantities for each load resistance value:

$R_L(\Omega)$	5	10	25	50	100	250	500
V_L(V)	9.09	16.7	33.3	50	66.7	83.3	90.9
I_L(A)	1.82	1.67	1.33	1	0.67	0.33	0.18
P_L(W)	16.5	27.8	44.4	50	44.4	27.8	16.5
P_i (W)	182	167	133	100	67	33	18

The load resistance values used for power calculations in Example 9-9 are all multiples of each other: ($10\ \Omega = 2 \times 5\ \Omega$), ($25\ \Omega = 5 \times 5\ \Omega$), and so on. The reason for this becomes apparent when the graph of V_L, I_L, and P_L are plotted versus R_L (see Figure 9-10). The resistance scale is *logarithmic* rather than *linear*, and this gives the particular shape to the graph of P_L. Clearly, the output power is at a peak when $R_L = R_S$. Also, note that the output current is largest when R_L is very small, and the output voltage is largest when R_L is very large. When the efficiency is calculated from P_o and P_i, it is found to be 50% when R_L equals R_S, and it increases progressively as R_L becomes larger.

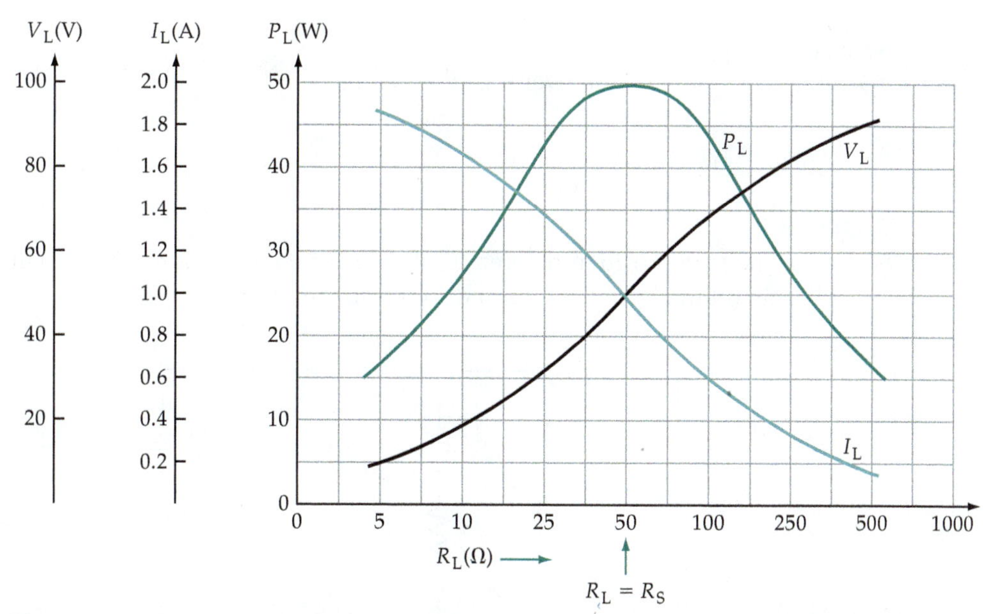

Figure 9-10 Plot of load voltage, current, and power for various values of load resistance. Maximum output power occurs when $R_L = R_S$.

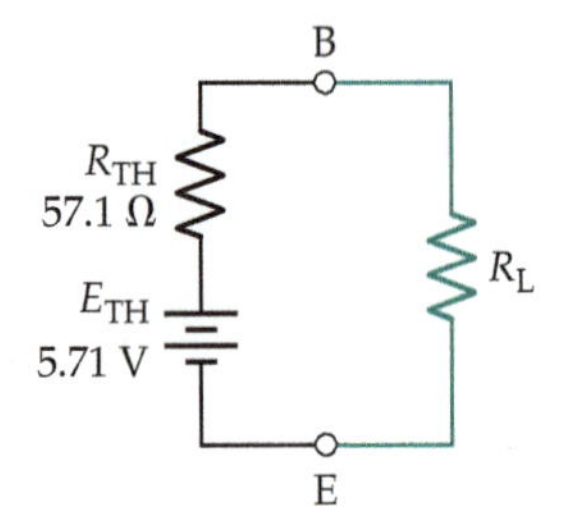

Figure 9-11 Circuit for Problem 9-5.1.

Practice Problem

9-5.1 Calculate the output power for the circuit in Figure 9-11 [reproduced from Figure 9-4(d)] for load resistor values of: 0.1 R_{TH}, 0.2 R_{TH}, 0.5 R_{TH}, R_{TH}, 2 R_{TH}, 5 R_{TH}, 10 R_{TH}. Plot a graph of output power versus load resistance.

Review Questions

Section 9-1

9-1 State the superposition theorem and list the steps involved in applying it to the analysis of a resistor network.

Section 9-2

9-2 State Thévenin's theorem and list the procedure for Thévenizing a circuit. Explain the major advantages of using this theorem.

Section 9-3

9-3 State Norton's theorem and list the steps for Nortonizing a circuit. Compare the Norton equivalent circuit to the Thévenin equivalent circuit.

Section 9-4

9-4 State Millman's theorem and list the procedure for its application to circuit analysis.

Section 9-5

9-5 State the maximum power transfer theorem.

Problems

Section 9-1

9-1 Use the superposition theorem to determine the current through R_1 in the circuit in Figure 9-12 (reproduced from Figure 8-27).

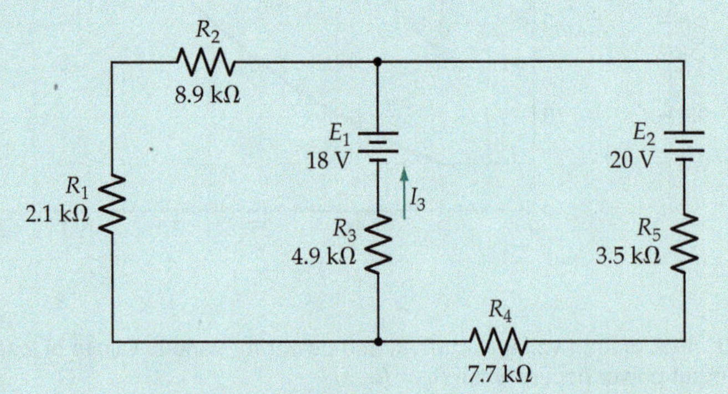

Figure 9-12

9-2 Apply the superposition theorem to analyze the circuit in Figure 9-13 (reproduced from Figure 8-28) to determine the current through R_5.

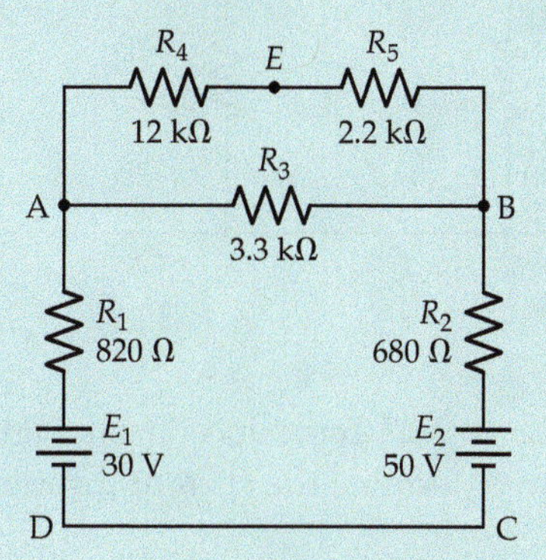

Figure 9-13

9-3 Using the superposition theorem, analyze the circuit in Figure 9-14 (reproduced from Figure 8-29) to determine the current through R_5.

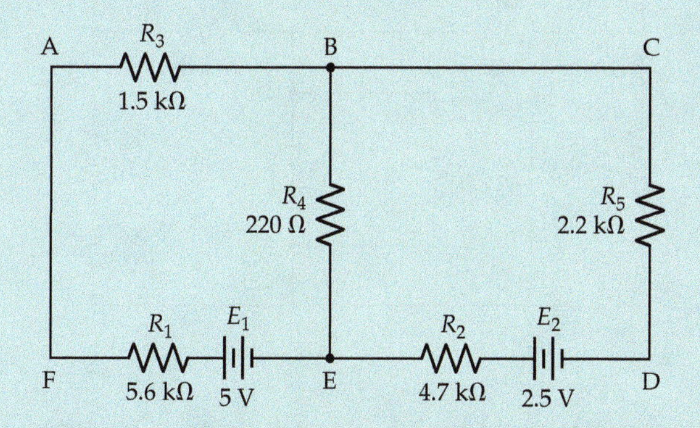

Figure 9-14

9-4 Determine the current through resistor R_3 in Figure 9-15 (reproduced from Figure 8-30). Apply the superposition theorem.

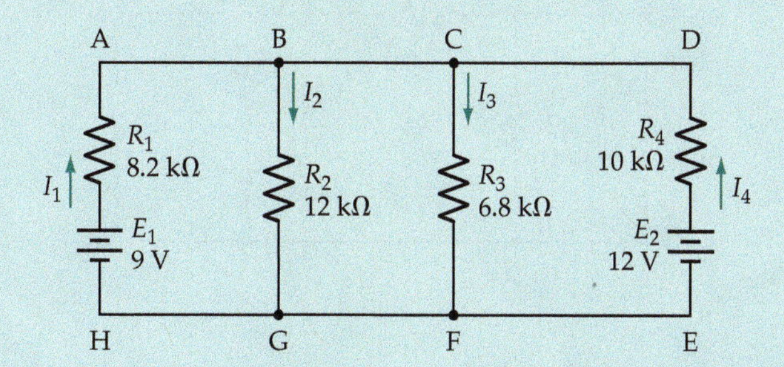

Figure 9-15

9-5 Analyze the circuit in Figure 9-16 (reproduced from Figure 8-31) by using the superposition theorem, to determine the current through R_3.

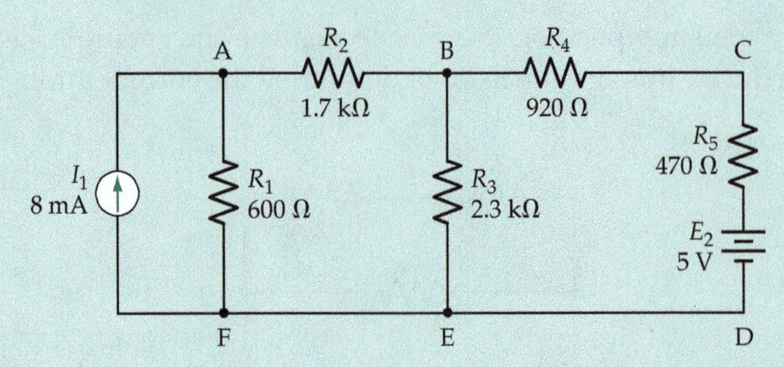

Figure 9-16

9-6 For the circuit in Figure 9-17 (reproduced from Figure 8-33) use the superposition theorem to determine the current through R_2 when R_4 is open-circuited.

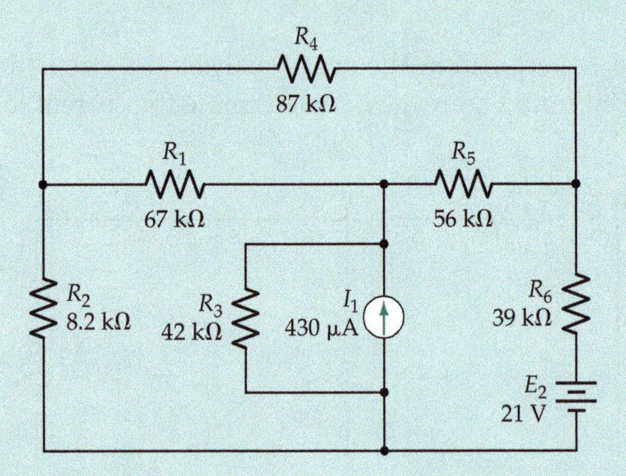

Figure 9-17

9-7 Calculate the current through R_2 when R_L is open-circuited in the circuit in Figure 9-18 [reproduced from Figure 9-5(a)]. Use the superposition theorem.

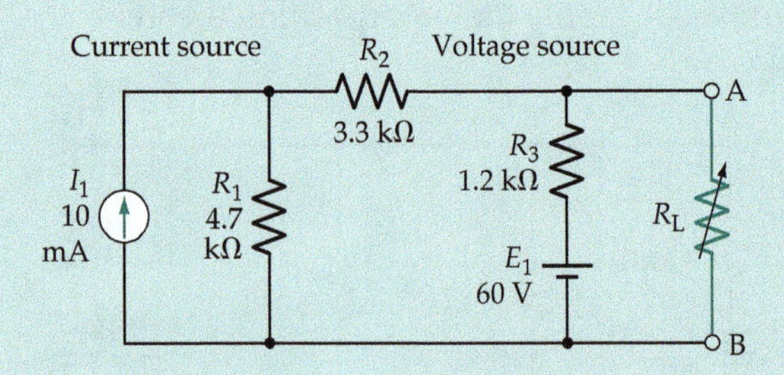

Figure 9-18

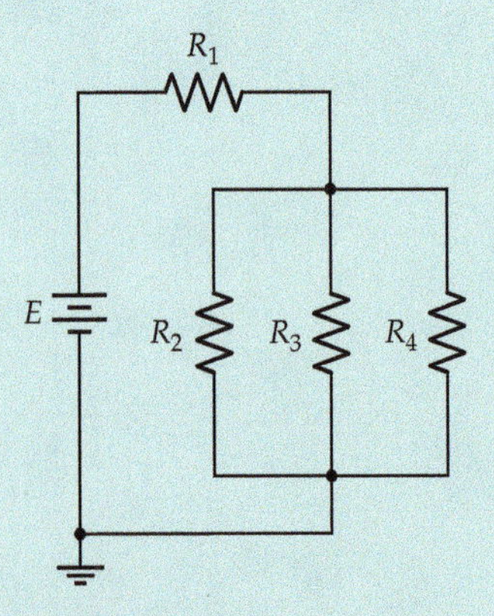

Figure 9-19

Section 9-2

9-8 Apply Thévenin's theorem to the circuit in Figure 9-19 (reproduced from Figure 7-11) to determine the current through resistor R_2.

9-9 For the circuit in Figure 9-20 (reproduced from Figure 7-16) use Thévenin's theorem to determine the current through resistor R_5.

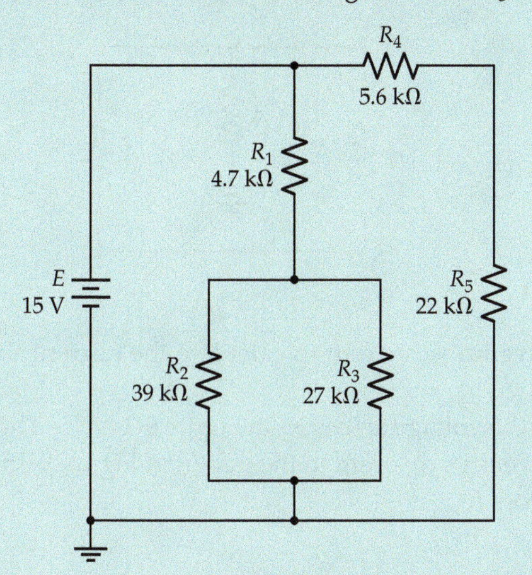

Figure 9-20

9-10 Using Thévenin's theorem, determine the current through R_4 in Figure 9-21 (reproduced from Figure 7-18) when R_1 is open-circuited.

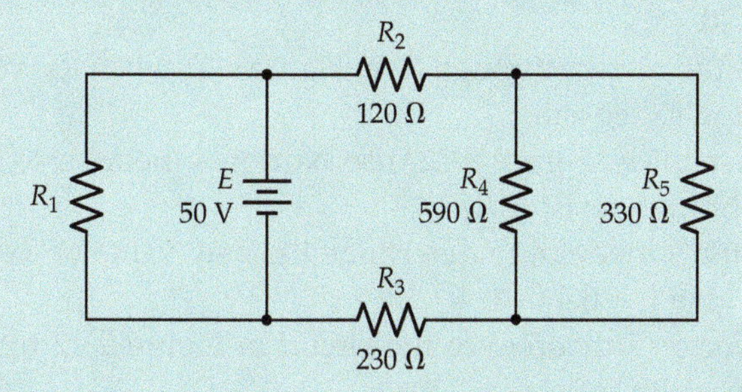

Figure 9-21

9-11 For the circuit in Figure 9-22 (reproduced from Figure 7-21) use Thévenin's theorem to determine the current through R_3.

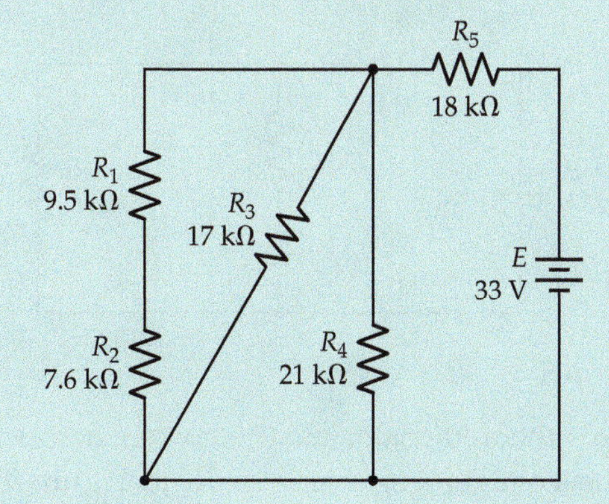

Figure 9-22

9-12 Calculate the current through R_3 in Figure 9-23 (reproduced from Figure 8-26) by the use of Thévenin's theorem.

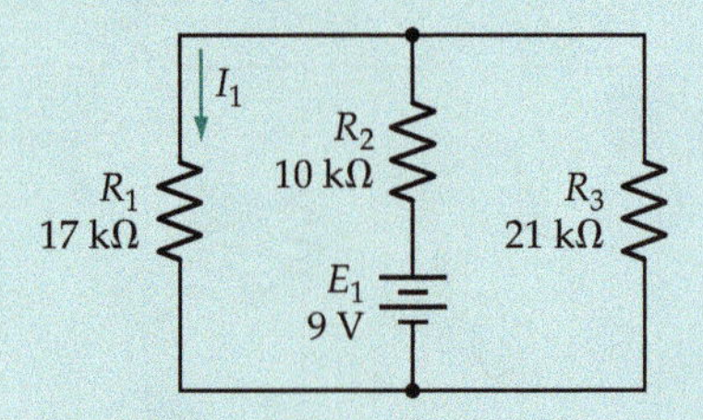

Figure 9-23

9-13 Apply Thévenin's theorem to calculate the current through R_5 in Figure 9-13.

9-14 Determine the voltage across R_5 in Figure 9-14. Use Thévenin's theorem.

9-15 Apply Thévenin's theorem to the circuit in Figure 9-15 to determine the voltage across R_3.

Section 9-3

9-16 Apply Norton's theorem to the circuit in Figure 9-19 to determine the current through resistor R_4.

9-17 Using Norton's theorem, determine the voltage across R_4 in the circuit in Figure 9-20.

9-18 Calculate the current through R_5 in Figure 9-21 when R_1 is open-circuited. Use Norton's theorem.

9-19 For the circuit in Figure 9-22, use Norton's theorem to determine the current through resistor R_4.

9-20 Use Norton's theorem to determine the new value of R_1 in Figure 9-23 that will give $I_1 = 0.4$ mA.

9-21 Apply Norton's theorem to the circuit in Figure 9-12 to determine the current through resistor R_1.

9-22 For the circuit in Figure 9-24 (reproduced from Figure 8-15) apply Norton's theorem to determine the current through resistor R_3.

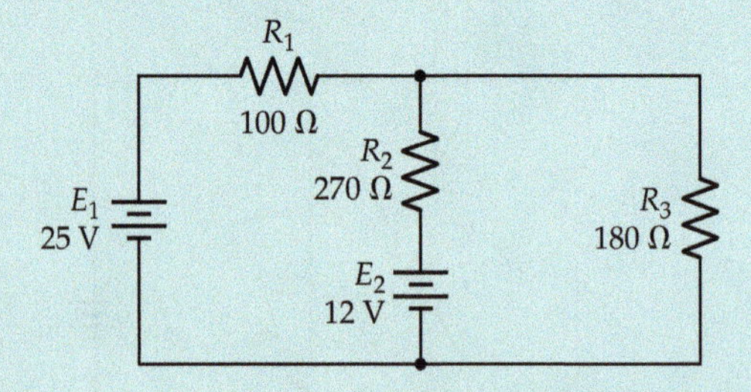

Figure 9-24

9-23 Using Norton's theorem, calculate the voltage across R_3 in Figure 9-14.

9-24 Determine the voltage across resistor R_2 in Figure 9-15 by the use of Norton's theorem.

Section 9-4

9-25 Use Millman's theorem to determine the current through resistor R_3 in Figure 9-15.

9-26 Apply Millman's theorem to the circuit in Figure 9-25 (reproduced from Figure 8-10) to determine the voltage across resistor R_3.

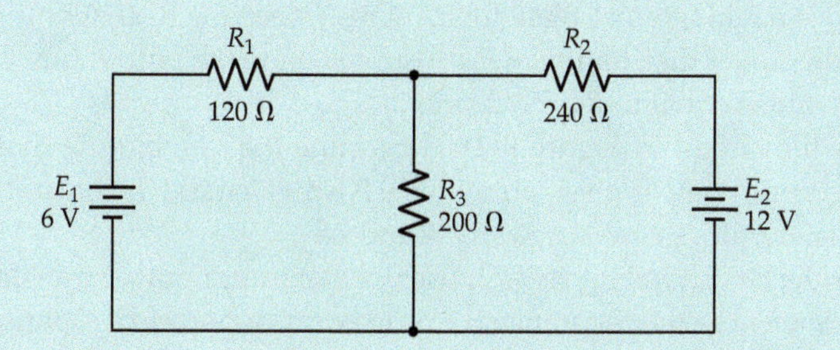

Figure 9-25

9-27 Use Millman's theorem to reduce the voltage and current generators in Figure 9-26 to a single current generator. Then calculate the load current for $R_L = 3.3\ \text{k}\Omega$ and for $R_L = 4.7\ \text{k}\Omega$.

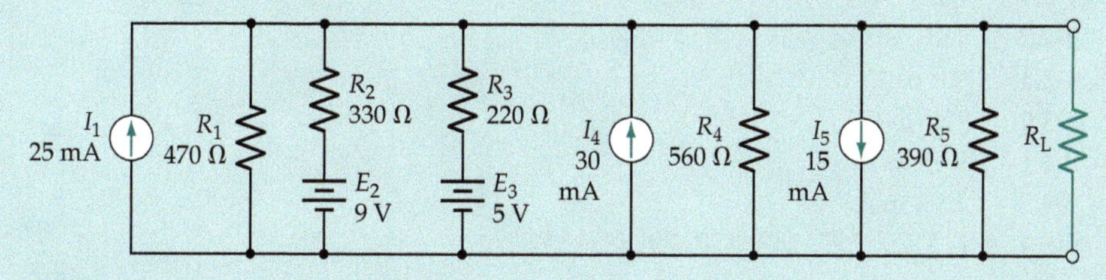

Figure 9-26

9-28 Apply Millman's theorem to the circuit in Figure 9-27 (reproduced from Figure 8-25) to calculate I_2.

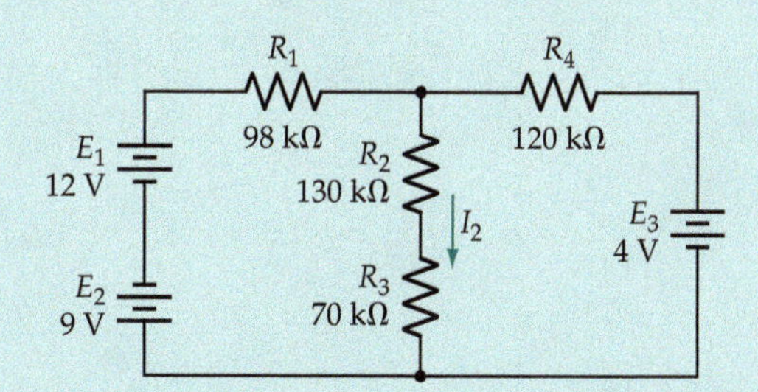

Figure 9-27

9-29 Use Millman's theorem to determine the current through resistor R_1 in the circuit shown in Figure 9-12.

Section 9-5

9-30 Using a current generator with $I = 44.5$ mA and $R = 71\ \Omega$, demonstrate the maximum power transfer theorem by calculating the load voltage, current, and power for various values of load resistance. Also, plot a graph of the quantities versus R_L.

9-31 Use the Thévenin circuit obtained in Problem 9-12 ($E_{TH} = 5.67$ V, $R_{TH} = 6.3\ \text{k}\Omega$) to calculate the new resistance for R_3 that will give maximum power dissipation in R_3. Also, calculate the output power for resistor values equal to R_3, $R_3/2$, and $2R_3$.

9-32 For the circuit in Figure 9-21, determine the value of R_5 that will give maximum power dissipation in R_5. Also, calculate the output power for resistor values equal to R_5, $R_5/2$ and $2R_5$.

9-33 For the circuit in Figure 9-22, use the maximum power transfer theorem to determine the resistance R_4 for maximum power dissipation.

Practice Problem Answers

9-1.1 2.73 mA
9-1.2 2.9 μA
9-2.1 0.2 mA
9-2.2 18.8 V
9-3.1 0.3 mA
9-3.2 133 μA
9-4.1 9.99 mA
9-4.2 2.96 mA
9-5.1 (47, 79, 127, 143, 127, 79, 47) mW

CHAPTER 10
Voltage Cells, Batteries, and DC Power Supplies

CONTENTS

Objectives

You will be able to:

1. Sketch the basic construction and explain the operation of a simple voltage cell.
2. Discuss the performance of voltage cells and solve problems involving internal resistance, ampere-hour rating, no-load output voltage, and performance under load.
3. Describe the construction, performance, and applications of the following voltage cells: zinc-carbon, manganese-alkaline, mercury, nickel-cadmium cells, solar.
4. Show how cells may be connected in series, parallel, and series-parallel. Solve problems involving batteries of cells.
5. Describe the construction and operation of a lead-acid battery. Discuss its performance and maintenance, and solve problems involving battery charge and discharge.
6. Discuss the performance and application of dc power supplies.

INTRODUCTION

A *voltage cell* consists basically of two different metal plates immersed in an acid solution. The action of the acid removes electrons from one plate and accumulates them on the other plate. In this way a potential difference is produced between the two plates, and a current can be made to flow through an external circuit.

Different types of voltage cells exist for different applications. Some can be recharged and reused many times; others can be used only once and are then discarded. One of the most important of the rechargeable cells is the lead-acid cell, which is the basic unit of the automobile battery.

Batteries of voltage cells can be operated in series or in parallel, or in series-parallel combinations. The performance of a voltage cell or battery may be defined in terms of the maximum voltage and current it can supply, and its ampere-hour rating.

10-1 SIMPLE VOLTAGE CELL

The simple voltage cell illustrated in Figure 10-1(a) is seen to consist of copper and zinc *electrodes* partially immersed in a chemical solution known as the *electrolyte*. Diluted sulfuric acid is frequently used as the electrolyte, but any one of several other chemical compounds may be employed. Sulfuric acid is a combination of *hydrogen* and *sulfate* in which each hydrogen atom

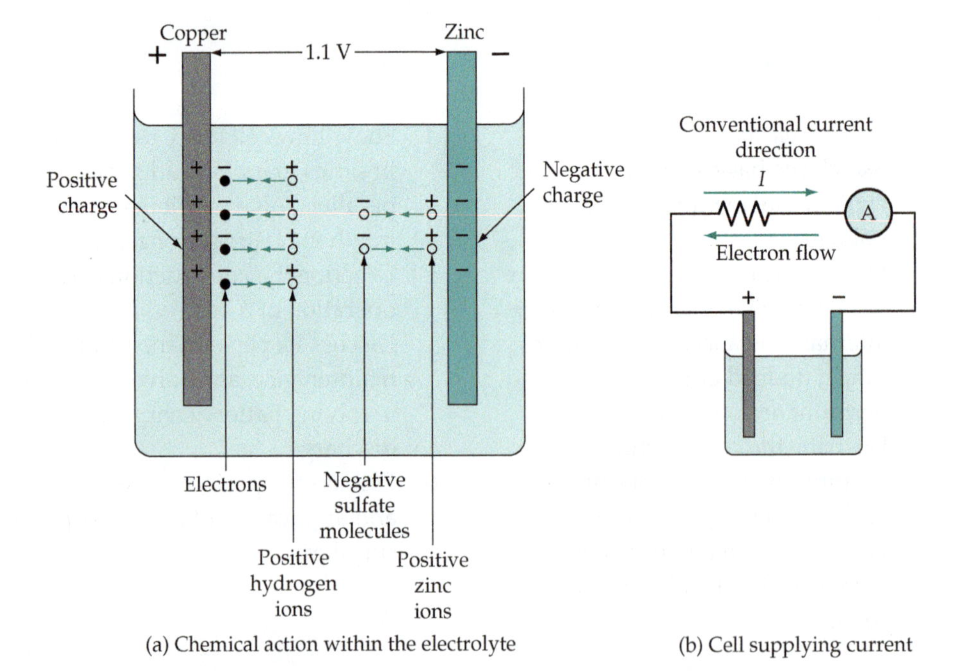

Figure 10-1 A simple voltage cell consists of copper and zinc rods partially immersed in a liquid electrolyte. The electrolyte causes positive ions to be removed from the zinc, creating a negative charge, and electrons to be taken from the copper, leaving it positively charged. When a circuit is connected between the terminals, electrons flow from the negative terminal to the positive terminal.

has given up an electron to the sulfate molecules. The solution remains chemically stable until the electrodes are introduced. Then the surface of the zinc electrode readily dissolves in the sulfuric acid, and the zinc atoms combine with the sulfate to form *zinc sulfate.* In this process the zinc atoms leave electrons behind them on the zinc electrode [see Figure 10-1(a)]. This is because the sulfate already has an excess of electrons acquired from the hydrogen. As the zinc sulfate forms, hydrogen ions are released, and each hydrogen ion is an atom that is short of one electron (i.e., a positive ion). These ions travel to the copper electrode, where they acquire electrons, as illustrated.

In giving up electrons, the copper electrode becomes positively charged. Similarly, the zinc electrode becomes negatively charged as it accumulates excess electrons. Consequently, a potential difference is created between the copper and zinc electrodes. If no current is drawn from the cell, the chemical action eventually ceases when the zinc becomes so negative that it repels the (negative) sulfate molecules, and the positive charge on the copper repels the (positive) hydrogen ions. In this condition the potential difference at the cell terminals is typically 1.1 V.

An external electrical circuit connected to the cell terminals provides a path for electrons to flow from the (negative) zinc electrode to the (positive) copper electrode [Figure 10-1(b)]. So, the zinc electrode loses negative charges and the copper electrode gains negative charges (i.e., becomes less positive). Once the electrodes begin to lose some of the accumulated charges, the chemical action resumes, and the electrolyte continues to provide electrons to supply the output current from the cell.

10-2 VOLTAGE CELL CHARACTERISTICS

Polarization

The hydrogen released in the simple voltage cell frequently remains clinging to the copper electrode in the form of bubbles. When large areas of the electrode become coated with hydrogen bubbles, the active surface area is reduced and the cell's ability to supply current is diminished. The effect is known as *polarization,* and it also occurs in more complex cells. Materials employed to combat polarization are termed *depolarizers.*

Output Voltage and Current

The potential difference produced at the terminals of a cell depends on the electrode materials. For the zinc and copper electrodes of the simple voltage cell, the terminal voltage is approximately 1.1 V. For most other types of cells the terminal voltage is typically 1.5 V. This potential difference is produced regardless of the physical size of the electrodes or the quantity of the electrolyte.

The output current supplied by a cell depends on the output voltage and the external load resistance. As always,

$$I = \frac{E}{R}$$

where I is the output current, E is the cell terminal voltage, and R is the external load resistance. However, there is a limit to the maximum current that a given cell can supply, and when this limit is approached, the cell's terminal voltage begins to fall, so that the output current cannot be increased. The maximum current that can be drawn from a cell is directly related to the physical size of the cell's component parts. Obviously, electrodes with large surface areas can release more atoms and generate more free electrons than is possible with small surface areas. So, a physically large voltage cell can produce a larger output current than a small cell.

Ampere-Hour Rating

The amount of energy that can be supplied by any cell (or battery of cells) is defined in terms of its rating in *ampere-hours* (Ah).

$$\text{Ah rating} = I \times t \qquad \textbf{(10-1)}$$

If a cell can supply a current of 5 A for a maximum time of 1 h,

$$\text{Ah rating} = I \times t = 5\text{ A} \times 1\text{ h}$$
$$= 5\text{ Ah}$$

When the same cell is supplying a current of only 1 A, it can be expected to sustain this current for a time of,

$$t = \frac{\text{Ah rating}}{I} = \frac{5\text{ Ah}}{1\text{ A}}$$
$$= 5\text{ h}$$

For a time of 50 h, a cell with a rating of 5 Ah can be expected to supply a continuous current of,

$$I = \frac{\text{Ah rating}}{t} = \frac{5\text{ Ah}}{50\text{ h}}$$
$$= 100\text{ mA}$$

The above current and time quantities are illustrated by the graph of I versus t in Figure 10-2.

Usually the Ah rating for a voltage cell applies only to a certain range of load currents. For a given cell, the Ah rating when supplying a high current is not as great as that at a low current level. One reason for this is that polarization

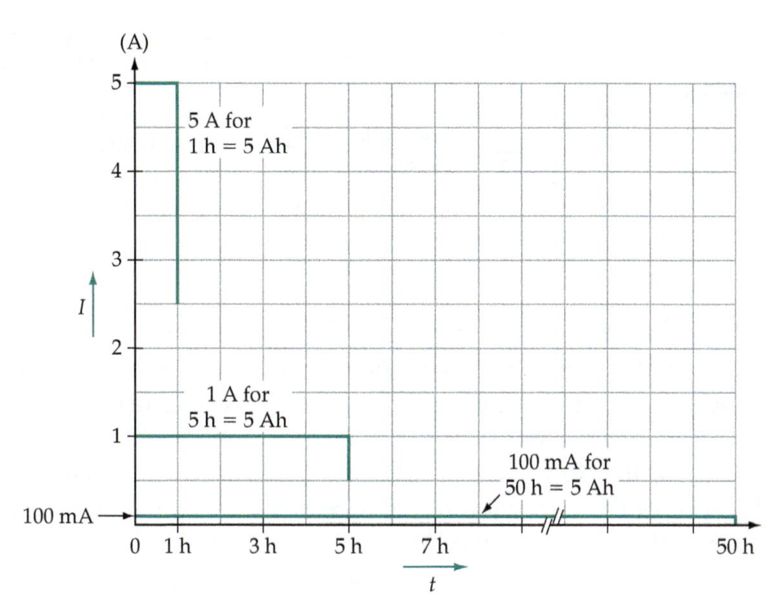

Figure 10-2 A voltage cell with an ampere-hour rating of 5 Ah might be expected to supply 5 A for 1 h, 1 A for 5 h, or 100 mA for 50 h.

occurs more rapidly when the load current is high. Cells can also be classified according to their Ah rating per unit volume or per unit weight, both of which can be important for cells used in portable equipment.

Example 10-1

A voltage cell supplies a load current of 0.5 A for a period of 20 h until its terminal voltage falls to an unacceptable level, (see Figure 10-3). Calculate the Ah rating of the cell and determine how long it could be expected to supply a current of 200 mA.

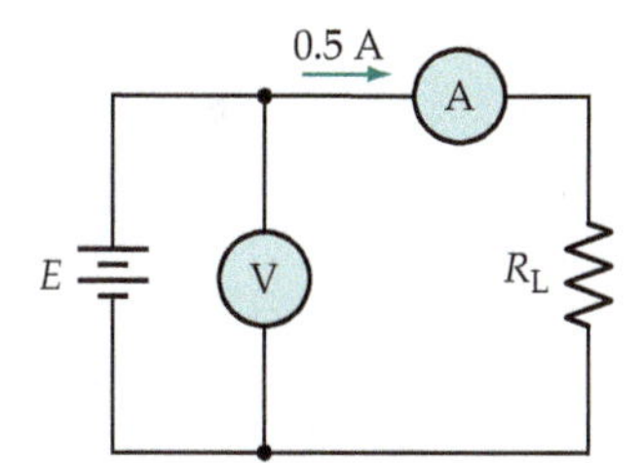

Figure 10-3 Circuit of a voltage cell supplying a constant load current. The ampere-hour rating of a voltage cell defines the time for which the cell can sustain a given current level.

Solution

From Eq. 10-1,

$$\text{Ah rating} = I \times t = 0.5\text{ A} \times 20\text{ h}$$

$$= 10\text{ Ah}$$

$$t = \frac{\text{Ah rating}}{I} = \frac{10\text{ Ah}}{200\text{ mA}}$$

$$= 50\text{ h}$$

Primary and Secondary Cells

The simple voltage cell already discussed is classified as a *primary cell.* Primary cells can supply current until the negative electrode is completely dissolved, the electrolyte is exhausted, or perhaps until polarization renders the cell useless.

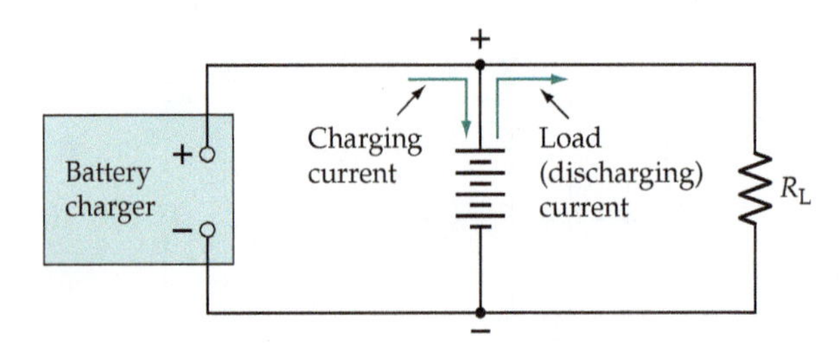

Figure 10-4 Secondary cells are rechargeable. The charging current flows into the cell in the opposite direction to the load current.

A *secondary cell,* on the other hand, can be *recharged.* That is, the electrode surface can be reformed and the electrolyte returned to its original condition. This is done by passing a current through the cell in an opposite direction to the current that normally flows from the cell (see Figure 10-4). The *lead-acid cell* discussed in Section 10-7 is a secondary cell because it is rechargeable. Most *dry cells,* such as the *zinc-carbon cells* (see Section 10-3) cannot be recharged; consequently, they are primary cells.

Cell Equivalent Circuit

As already explained, when no current is being drawn from a voltage cell, charges accumulate on the electrodes, and the chemical action ceases within the electrolyte. At this time, the potential difference between the cell terminals is known as the *open-circuit output voltage* or the *no-load output voltage.*

When current flows from the cell, electrons are transferred externally from the negative electrode to the positive electrode. The potential difference between the electrodes falls, and the chemical action within the electrolyte commences to replace the lost charges. If the load current is very small, the terminal voltage under load may not be noticeably different from the open-circuit output voltage. When the load current is large, the terminal voltage falls below the open-circuit voltage level. If the load current is made too large, the cell terminal voltage falls to near zero, because the chemical action cannot occur fast enough to replace the charges removed from the electrodes. The maximum current that may be taken from a voltage cell without a substantial drop in output voltage is proportional to the surface area of the electrodes and to the composition of the electrolyte.

To account for the terminal voltage drop of a cell when supplying current, each cell is said to have a certain *internal resistance.* The cell can then be represented by an equivalent circuit consisting of its internal resistance together with an *ideal cell* that has a terminal voltage equal to the actual cell's open-circuit terminal voltage. An ideal cell is a theoretical cell with a terminal voltage that is assumed to remain constant. The equivalent circuit of a voltage cell (discussed in Section 8-1) is illustrated in Figure 10-5.

When no current is flowing from the equivalent circuit, there is no voltage drop across the internal resistance (R_S), and the terminal voltage is equal to the ideal cell voltage. That is, the output voltage is the open circuit voltage E_C of

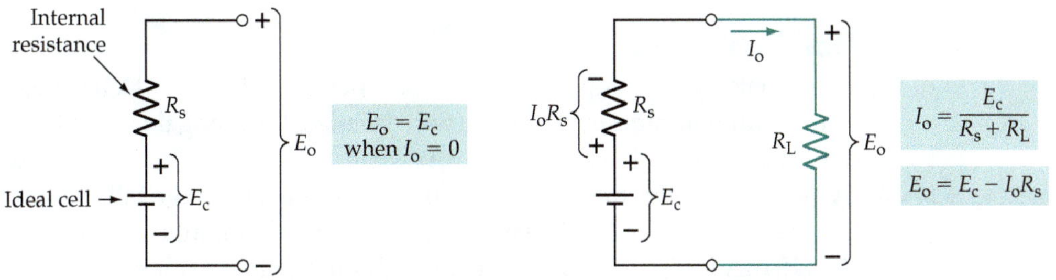

(a) Cell equivalent circuit when $I_o = 0$ (b) Cell equivalent circuit when supplying a load current

Figure 10-5 **Equivalent circuit of a voltage cell under no-load and load conditions. When a load current flows, the terminal voltage is reduced by the cell internal voltage drop.**

the cell [see Figure 10-5(a)]. When a load resistance (R_L) is connected to the cell terminals [Figure 10-5(b)], the current that flows is,

$$I_o = \frac{E_c}{R_s + R_L} \qquad \textbf{(10-2)}$$

When the cell is supplying current, a voltage drop occurs across R_s, and the output voltage becomes,

$$E_o = E_c - I_o R_s \qquad \textbf{(10-3)}$$

Obviously, a cell that has a low internal resistance is capable of supplying a larger current than one that has a relatively high internal resistance.

Example 10-2

A certain voltage cell has an open-circuit terminal voltage of $E_c = 2$ V and an internal resistance of $R_s = 0.1\ \Omega$. Determine the output voltage of the cell when the load current is 2 A. Also, determine the cell current that flows if the terminals are short-circuited.

Solution

When $I_o = 2$ A:

From Eq. 10-3 $\quad E_o = E_c - I_o R_s = 2\text{ V} - (2\text{ A} \times 0.1\ \Omega)$

$$= 1.8\text{ V}$$

When the cell is short-circuited $R_L = 0$

Eq. 10-2 $\quad I_o = \dfrac{E_c}{R_s + R_L} = \dfrac{2\text{ V}}{0.1\ \Omega + 0}$

$$= 20\text{ A}$$

Practice Problems

10-2.1 A 9 V battery is to supply 100 mA for a period of 5 hours. Determine the required minimum Ah rating, and calculate how long the same battery could be used to supply 35 mA.

10-2.2 A battery supplies the following currents before its voltage falls below a useful level: 10 mA for 30 minutes, 50 mA for 10 minutes, 25 mA for 4 h, 18 mA for 7 h. Calculate the total Ah rating of the battery.

10-2.3 The terminal voltage of a cell is measured as 1.55 V on open-circuit and as 1.47 V when supplying 50 mA. Calculate the cell internal resistance and short-circuit current.

10-2.4 The terminal voltage of a 2.2 V cell is to drop by not more than 5% when supplying a load current of 20 A. Determine the required maximum internal resistance of the cell.

10-3 MISCELLANEOUS VOLTAGE CELLS

Zinc-Carbon Cell

The *zinc-carbon* cell is a primary cell (cannot be recharged) commonly used in electric flashlights and other portable equipment. As illustrated in Figure 10-6(a), this cell basically consists of a zinc can containing electrolyte with a carbon rod at the center. The zinc can is the negative electrode of the cell, and the carbon rod is the positive electrode. The electrolyte is in the form of a moist paste, so that the *dry cell* is, in fact, not dry at all. The cell becomes useless when the electrolyte dries out because the necessary chemical reactions cannot occur.

The operation of the zinc-carbon cell is similar to that of the zinc-copper cell described in Section 10-1. Electrons accumulate on the zinc electrode, and

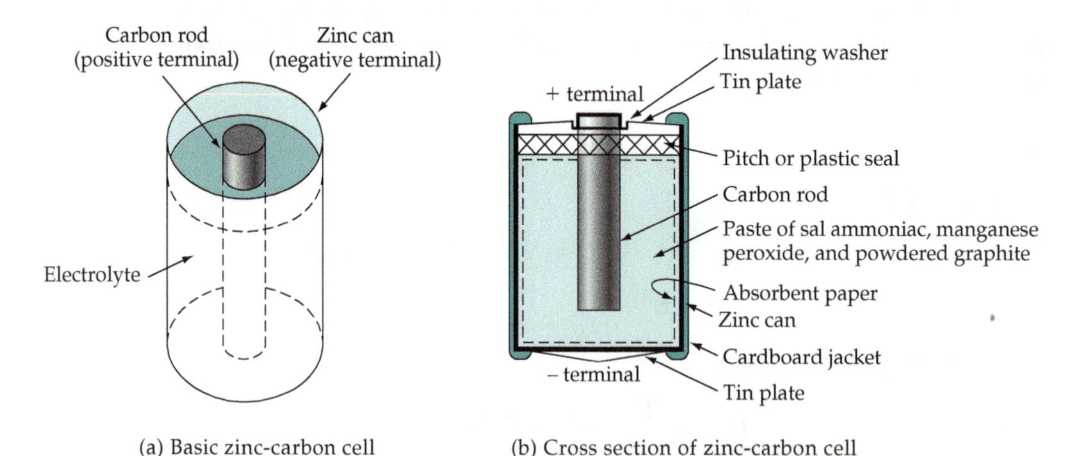

(a) Basic zinc-carbon cell (b) Cross section of zinc-carbon cell

Figure 10-6 Construction of a zinc-carbon cell. The zinc can is the negative terminal and the carbon rod is the positive terminal. The electrolyte is a paste of sal ammoniac, manganese peroxide, and powdered graphite.

electrons are removed from the carbon electrode as zinc atoms combine with the electrolyte. This creates a potential difference between the electrodes, with the zinc being negative and the carbon positive.

Figure 10-6(b) shows the construction of the zinc-carbon cell in more detail. The electrolyte consists of *sal ammoniac* (ammonium chloride) mixed with *manganese peroxide* and powdered graphite. The manganese peroxide acts as a depolarizer by combining with the hydrogen, which appears at the positive electrode. The combination results in *manganous oxide* and water, which eliminates the polarizing effect of the hydrogen and helps to keep the electrolyte moist.

The electrolyte must be hermetically sealed in the can to prevent it from drying out. A layer of plastic at the top of the can is the usual type of seal employed. The inside surface of the zinc is protected by a layer of absorbent paper [see Figure 10-6(b)] to help to prevent the zinc from reacting directly with the carbon and other impurities in the electrolyte. Such reactions tend to set up little voltage cells at the surface of the zinc, and this is referred to as *local action.* Local action reduces the output voltage of the cell and shortens its life by using up the zinc. To combat local action due to impurities in the zinc, the zinc surface is usually coated with mercury, or else mercury is mixed with the zinc during the manufacturing process.

Most dry cells have a surrounding outer jacket of cardboard or tin plate. Also, tin-plate discs are used at the top and bottom, as illustrated. As well as providing protection for the cell, the outer jacket helps to protect battery-powered equipment from the corrosive action that occurs on the outside of the zinc can when dry cells are unused for a long time.

The typical output voltage from a zinc-carbon cell is 1.5 V, but it could range from 1.4 V to 1.6 V in a good cell. This terminal voltage is available regardless of the physical size of the cell. The very small *C cell* and the much larger *No. 6 cell* illustrated in Figure 10-7 each have a terminal voltage of approximately 1.5 V. For larger output voltages, several cells must be connected in series, (see Section 10-4). In continuous use, the terminal voltage of the zinc-carbon cell tends to fall off rapidly unless the current is very small. So, this type of cell is best applied where it will be used intermittently.

The output current that may be drawn from a zinc-carbon cell very much depends on the physical size of the cell. The No. 6 cell can supply as much as 20 A for a short time. But, at this rate it would tend to discharge very quickly, and 1 A would be a more appropriate maximum load. The D cell, which is commonly used in flashlights, cannot supply more than about 5 A. Again, this would be an excessive load, and a maximum of a few hundred milliamps is normal. The C cells supply even smaller maximum currents, corresponding to their smaller size.

Manganese-Alkaline Cell

A manganese-alkaline cell can sustain a constant terminal voltage for heavy loads over a much longer period of time than the less expensive zinc-carbon cell.

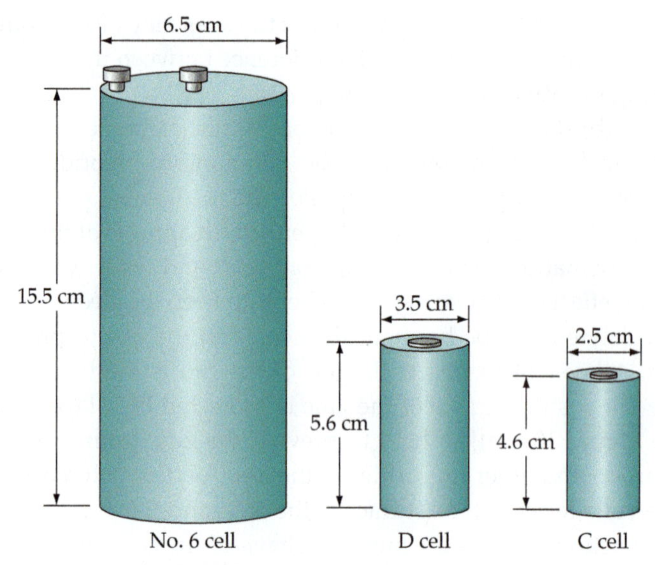

Figure 10-7 All zinc-carbon cells have a typical terminal voltage of 1.5 V, regardless of size. The largest cells are able to deliver the greatest currents. Alternatively, they may be used to supply a small current for a longer time than would be possible with a small cell.

The electrolyte is *potassium hydroxide,* the positive electrode is a layer of *manganese dioxide,* and the negative electrode is a *zinc-mercury* amalgam. The cell is enclosed in a steel container, and typically has 1.5 V open-circuit terminal voltage.

Mercury Cell

The mercury cell is a modification of the manganese-alkaline cell, in which *mercuric oxide* is used as the positive electrode (i.e., instead of manganese dioxide). The open-circuit terminal voltage of the mercury cell is usually around 1.4 V, falling to about 1.3 V under load conditions. The major advantage of this type of cell is that its terminal voltage is maintained constant for much longer periods than either the manganese-alkaline or zinc-carbon cells.

Comparison of Primary Cells

Figure 10-8 shows graphs of cell terminal voltage plotted versus time for different types of *D* cells for 200 mA average current levels. It is obvious from the graph that the terminal voltage of the zinc-carbon cell falls off relatively rapidly when supplying a constant load. The peformance of the manganese-alkaline cell is substantially better than that of the zinc-carbon cell, but the mercury cell gives by far the best performance.

Rechargeable Cells

The initial cost of a set of rechargeable batteries and a charger is much greater than the cost of primary cell batteries. However, because the rechargeable

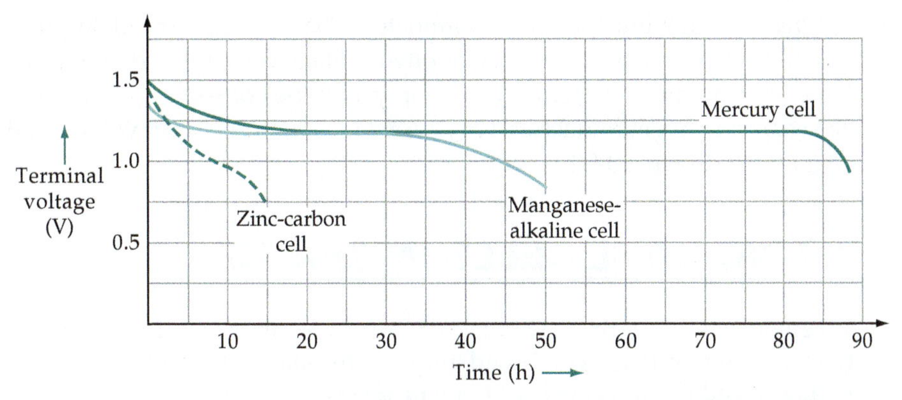

Figure 10-8 Comparison of primary D cells. Graphs of cell terminal voltage plotted versus time for zinc-carbon, manganese-alkaline, and mercury cells supplying 200 mA load currents.

batteries can last for many years, they are by far the least expensive in the long run. *Nickel-cadmium* cells have been extensively used for rechargeable batteries in portable electronics equipment, and they are largely being replaced in new equipment by batteries of *lithium-ion* cells which have approximately twice the energy storage capacity for a given weight. Lithium-ion cells typically have a 3.6 V terminal voltage compared to 1.3 V for Nickel-cadmium cells.

Solar Cell

The solar cell is an electronic device that exhibits a potential difference between its two terminals when illuminated. In this case it is the light energy, instead of a chemical action, that produces the terminal voltage. The cross-section of a solar cell illustrated in Figure 10-9(a) shows that the device is a junction of two-types of semiconductor material, (*n*-type and *p*-type). The incident light causes electrons to be released within the upper (*p*-type) layer, and to cross the junction to accumulate on the lower (*n*-type) layer. This makes the lower layer more negative than the upper layer, thus creating a potential difference between the terminals. Current flow through an external circuit connected to the cell terminals

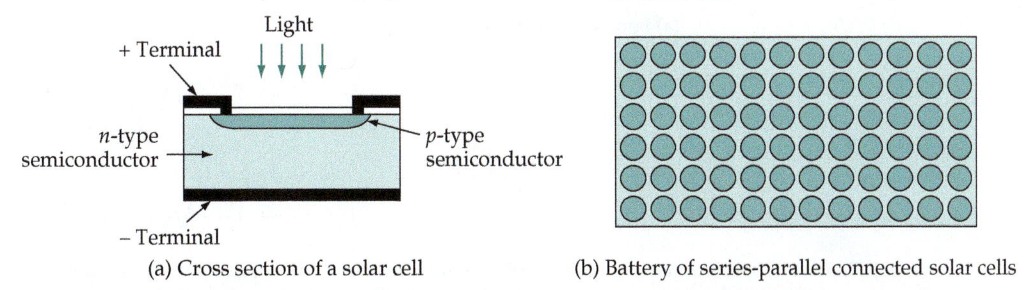

Figure 10-9 Cross-section of a solar cell. The light energy causes electrons to accumulate on the lower (*n*-type) semiconductor layer, making it more negative than the upper (*p*-type) layer. Batteries of solar cells can be constructed to produce a desired output voltage and load current.

can be maintained while the cell is illuminated. The typical solar cell output is 0.5 V and 50 mA, and as in the case of other voltage cells, the size of the cell dictates the amount of current it can supply. Batteries of series-parallel connected solar cells are manufactured to give a wide range of output voltage and current levels, [Figure 10-9(b)].

10-4 VOLTAGE CELLS IN SERIES

When the positive terminal of one voltage cell is connected to the negative of another cell, the combined output is the sum of the cell terminal voltages, and the cells are said to be *in series*.

Output Voltage and Current

Figure 10-10(a) shows a battery of four cells connected in series. The negative terminal of each cell is connected to the positive terminal of the previous cell. With this arrangement, the cell voltages add together to give a relatively large battery terminal voltage.

$$E = E_1 + E_2 + E_3 + E_4 + \cdots \qquad \textbf{(10-4)}$$

If each cell in Figure 10-10(a) has an *emf* of 1.5 V, the total output voltage is,

$$E = 1.5\text{ V} + 1.5\text{ V} + 1.5\text{ V} + 1.5\text{ V}$$
$$= 6\text{ V}$$

Obviously, if the cell emfs are not all 1.5 V, the output voltage from the battery will still be the sum of the individual cell voltages.

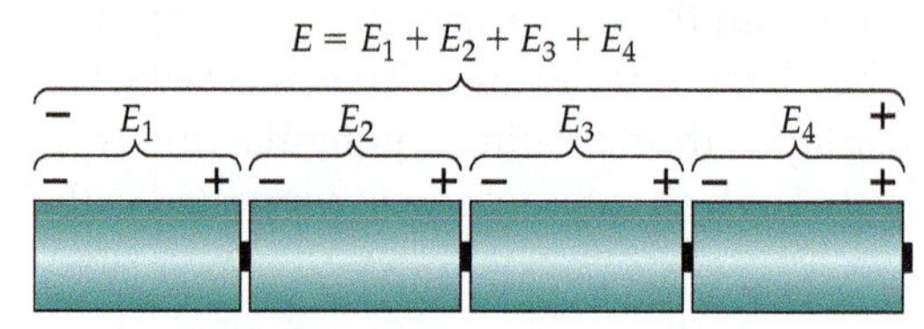

(a) Four series-connected voltage cells

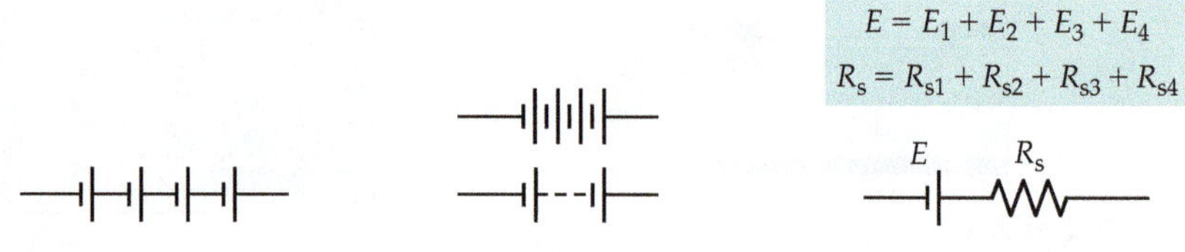

(b) Circuit symbol for four series-connected cells

(c) Alternative circuit symbols for series-connected cells

(d) Equivalent circuit for series-connected cells

Figure 10-10 Series-connected cells with circuit diagrams and equivalent circuits. The output voltage is the sum of the individual cell voltages, and the battery internal resistance is the sum of the cell internal resistances.

Figure 10-10(b) shows the circuit diagram for the four cells in series. Alternative ways of graphically representing a battery of series-connected cells are shown in Figure 10-10(c), and the battery equivalent circuit is illustrated in Figure 10-10(d). The total internal resistance of the battery of series-connected cells is, of course, the sum of the internal resistances of the individual cells.

$$R_s = R_{s1} + R_{s2} + R_{s3} + \cdots \qquad (10\text{-}5)$$

When an external load resistance (R_L) is connected to a group of four series-connected cells, the output current is,

$$I = \frac{E_1 + E_2 + E_3 + E_4}{R_L + R_{s1} + R_{s2} + R_{s3} + R_{s4}} \qquad (10\text{-}6)$$

and, $I = I_1 = I_2 = I_3 = I_4$

Note from the circuit that the output current flows through each individual cell. So, for example, if each cell is capable of supplying a maximum current of 1 A, the output current from the group of cells cannot exceed 1 A. The open-circuit output voltage, and the output current from the battery, can be calculated from Equations 10-4 and 10-6, respectively. The terminal voltage of the battery when supplying a load current is determined by use of Equation 10-3. Voltage cells that are not identical can be connected in series, however, the maximum current that can be supplied by the battery of cells is limited to the maximum output of the lowest current cell.

Series-connected cells produce an output voltage equal to the sum of the individual cell voltages, and supply a maximum current equal to the maximum that can he taken from the lowest current cell.

Example 10-3

The battery shown in Figure 10-11(a) is made up of four individual cells each of which has an open-circuit terminal voltage of $E = 2$ V and an internal resistance of 0.1 Ω. Determine the terminal voltage of the battery for no-load and

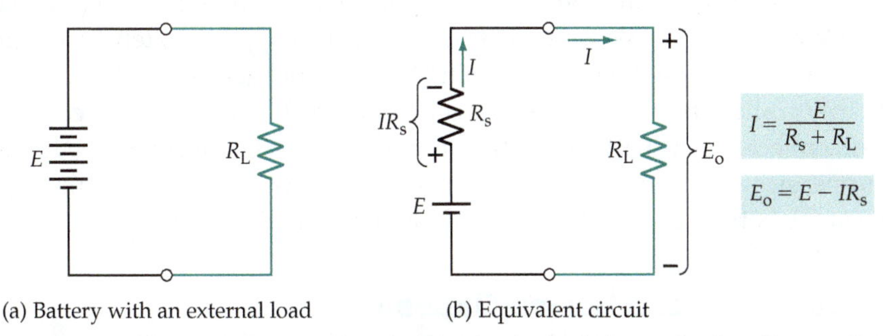

(a) Battery with an external load (b) Equivalent circuit

Figure 10-11 Circuit diagram and equivalent circuit of a battery of cells with an external load. The load current is $I = E/(R_s + R_L)$, and the battery terminal voltage is $E_o = E - (IR_s)$.

for $R_L = 9.6\ \Omega$. Also, determine the current that flows when the battery terminals are short-circuited.

Solution

$$E_c = E_1 + E_2 + E_3 + E_4 = 4 \times 2\ \text{V}$$
$$= 8\ \text{V}$$
$$R_s = R_{s1} + R_{s2} + R_{s3} + R_{s4} = 4 \times 0.1\ \Omega$$
$$= 0.4\ \Omega$$

For no-load, $$I_o = 0$$

and, $$E_o = E - I_o R_s = 8\ \text{V} - 0$$
$$= 8\ \text{V}$$

For $R_L = 9.6\ \Omega$, $$I_o = \frac{E_c}{R_s + R_L} = \frac{8\ \text{V}}{0.4\ \Omega + 9.6\ \Omega}$$
$$= 0.8\ \text{A}$$
$$E_o = E_c - I_1 R_s = 8\ \text{V} - (0.8\ \text{A} \times 0.4\ \Omega)$$
$$= 7.68\ \text{V}$$

For short-circuit, $$R_L = 0$$
$$I_o = \frac{E_c}{R_s + R_L} = \frac{8\ \text{V}}{0.4\ \Omega + 0}$$
$$= 20\ \text{A}$$

Note that short-circuiting a battery or cell terminals is to be avoided because it rapidly discharges the cells. Also, the large current that flows could permanently damage rechargeable cells.

9 V Battery

The construction of a commonly-used 9 V battery is illustrated in Figure 10-12. Six 1.5 V zinc-carbon cells are used, each of which is approximately rectangular prism shaped. The (negative terminal) bottom of each cell makes contact with the (positive terminal) top of the cell underneath. So the stacked cells are series-connected. The battery positive terminal connects directly to the top of the uppermost cell, and a flat insulated conductor makes contact from the negative of the bottom cell to the battery negative terminal.

Series-Aiding and Series-Opposing

The series cells illustrated in Figure 10-13(a) each have their positive terminal connected to the negative terminal of the adjacent cell. With this arrangement

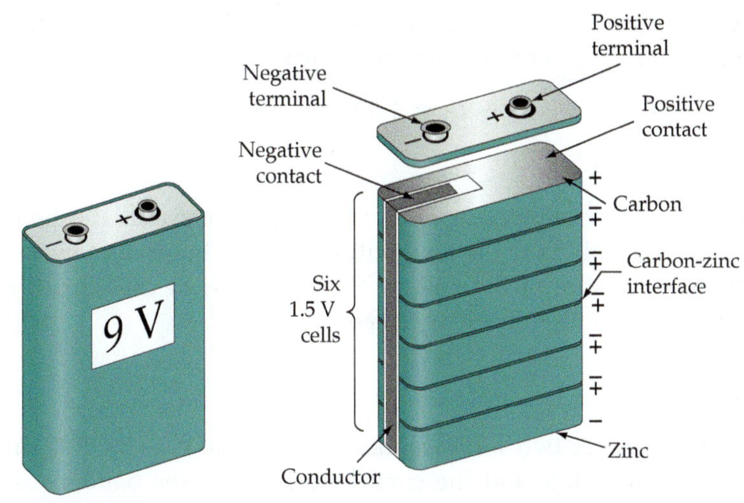

Figure 10-12 A 9 V battery consists of six 1.5 V cells connected in series.

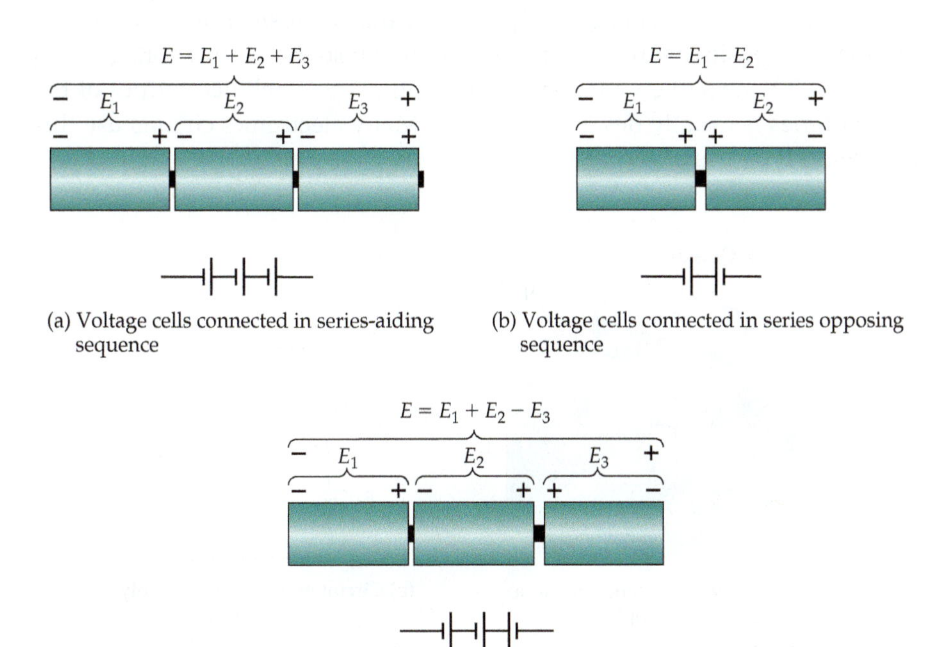

(a) Voltage cells connected in series-aiding sequence

(b) Voltage cells connected in series opposing sequence

(c) One cell connected in series opposing sequence

Figure 10-13 When voltage cells are connected in series, they are always connected series-aiding so that their voltages add together. Cells connected series-opposing give a reduced output voltage.

they assist each other to produce a current flow. The connection method is termed *series-aiding*. When cells are series-connected to produce currents in opposite directions, as shown in Figure 10-13(b), the connection is referred to as *series-opposing*. Series-opposing connections are normally avoided and usually occur only as a result of error. The total output voltage is still the sum of the individual cell voltages, but the polarity of the voltages must be considered. As illustrated in Figure 10-13(c), when one of three 1.5 V cells is connected series-opposing, the resultant terminal voltage of the battery is,

$$E = E_1 + E_2 + E_3 = 1.5\text{ V} + 1.5\text{ V} + (-1.5\text{ V})$$
$$= 1.5\text{ V}$$

If a discharged cell is connected in series with fully charged cells, the discharged cell adds little or nothing to the total output voltage, but it does increase the total internal resistance of the battery. So, a discharged cell should *not* be connected in series with fully-charged cells.

Plus/Minus Supply

Figure 10-14(a) shows two series-connected 4.5 V batteries with their common terminal grounded, and the circuit diagram for the batteries is illustrated in Figure 10-14(b). Because the negative terminal of battery B_1 is grounded, its output voltage is $E_1 = 4.5$ V, that is, *positive* with respect to ground. The positive terminal of B_2 is grounded; so the output voltage of B_2 is $E_2 = -4.5$ V, negative with respect to ground. The combined output of the two batteries is usually designated ±4.5 V. Many electronics circuits use this *plus/minus* type of supply.

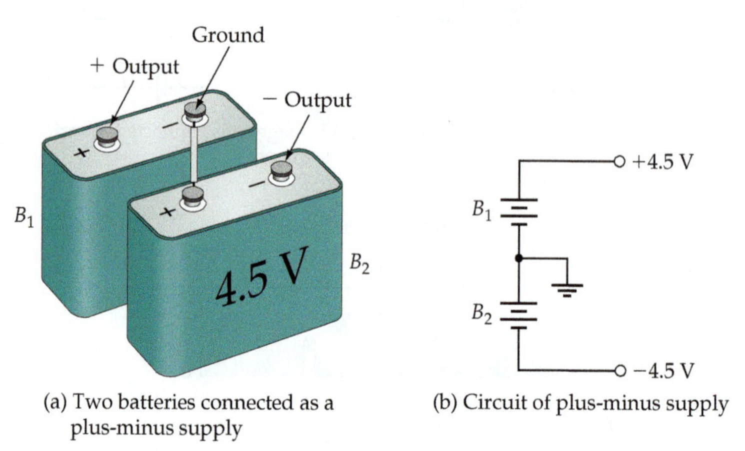

(a) Two batteries connected as a plus-minus supply

(b) Circuit of plus-minus supply

Figure 10-14 Two 4.5 V batteries may be connected in series with their common terminal grounded to provide a ±4.5 supply.

Practice Problems

10-4.1 A battery consists of six series-connected 1.5 V cells that each have an internal resistance of 0.17 Ω. Determine the battery terminal voltage when supplying a 200 mA load.

10-4.2 A battery made up of three series-connected 2.2 V cells has a potential difference of 6 V at its terminals when supplying a 0.33 Ω load. Calculate its short-circuit current.

10-5 VOLTAGE CELLS IN PARALLEL

When the positive terminal of one voltage cell is connected to the positive of another cell, and the two negative terminals are also connected together, the cells are said to be *in parallel*.

Consider two voltage cells connected in parallel as illustrated in Figure 10-15(a). If each cell has a terminal voltage of exactly 1.5 V, the output voltage from the parallel combination is also 1.5 V.

$$E_o = E_1 = E_2 = E_3 = \cdots$$

If each cell is capable of supplying a maximum current of 1 A, the total output current that can be drawn from the two-cell parallel combination is 2 A. Similarly, for three such cells in parallel the maximum output current is three times the maximum current per cell. Of course, the actual output current still depends on the output voltage and the load resistance. As shown by Figure 10-15(b), neglecting cell internal resistances, the output current is,

$$I_o = \frac{E}{R_L}$$

But the output current I_o is the sum of the individual currents from each cell. Therefore,

$$I_o = I_1 + I_2 + I_3 + \cdots + I_n \qquad \textbf{(10-7)}$$

For the case of two identical 1.5 V cells in parallel supplying a load resistor of 5 Ω (neglecting the internal resistances),

$$I_o = \frac{E_c}{R_L} = \frac{1.5\text{ V}}{5\ \Omega} = 0.3\text{ A}$$

and

$$I_1 = I_2 = 0.15\text{ A}$$

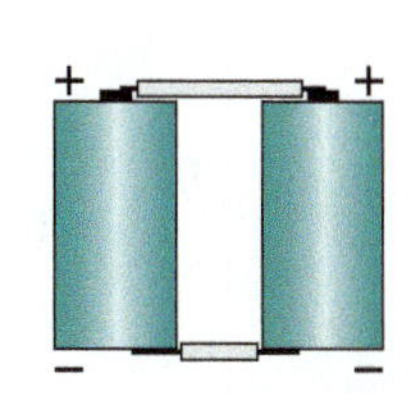

(a) Two voltage cells connected in parallel

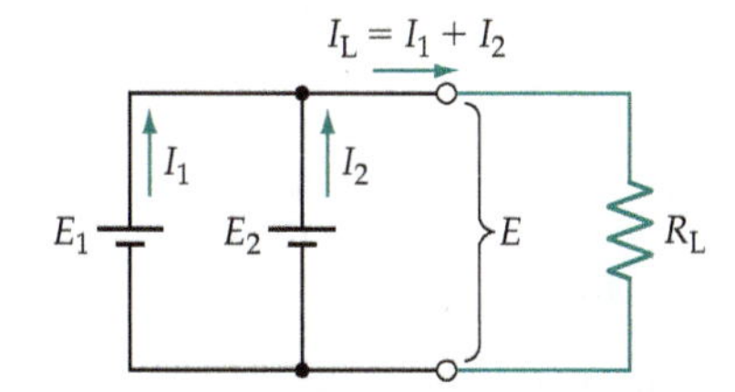

(b) Circuit diagram for two parallel-connected cells with a load resistor

Figure 10-15 Parallel-connected cells give an output voltage equal to the terminal voltage of one cell, and a maximum output current equal to the sum of the cell maximum current levels.

Example 10-4

A 0.2 Ω resistance is to be supplied from a parallel-connected battery of 2 V cells. If each cell can supply a maximum current of 0.5 A, determine the number of cells that should be connected in parallel. Assume the cell internal resistances to be negligible.

Solution

$$I_L = \frac{E_c}{R_L} = \frac{2\text{ V}}{0.2\ \Omega}$$

$$= 10\text{ A}$$

Number of parallel cells, $$N_p = \frac{I_L}{\text{current per cell}} = \frac{10\text{ A}}{0.5\text{ A}}$$

$$= 20$$

In a group of parallel-connected cells, each cell can be represented by its equivalent circuit (E_1R_{s1}, E_2R_{s2}, etc.,), as shown in Figure 10-16(a). The complete battery of cells can be represented by an equivalent circuit with a single source voltage (E) and a single source resistance (R_s), [Figure 10-16(b)]. E is the terminal voltage of each cell, and R_s equals the parallel combination of the cell internal resistances. For the case of two 1.5 V cells, each having an internal resistance of 0.1 Ω, the equivalent circuit has,

$$E = E_1 = E_2 = 1.5\text{ V}$$

and, $$R_s = R_{s1} \| R_{s2} = (0.1\ \Omega)/2$$

$$= 0.05\ \Omega$$

For three identical cells in parallel, the internal resistance of one cell would be divided by three to find the internal resistance of the equivalent circuit,

$$R_s = R_{s1} \| R_{s2} \| R_{s3} \tag{10-8}$$

$$= R_{s1}/3$$

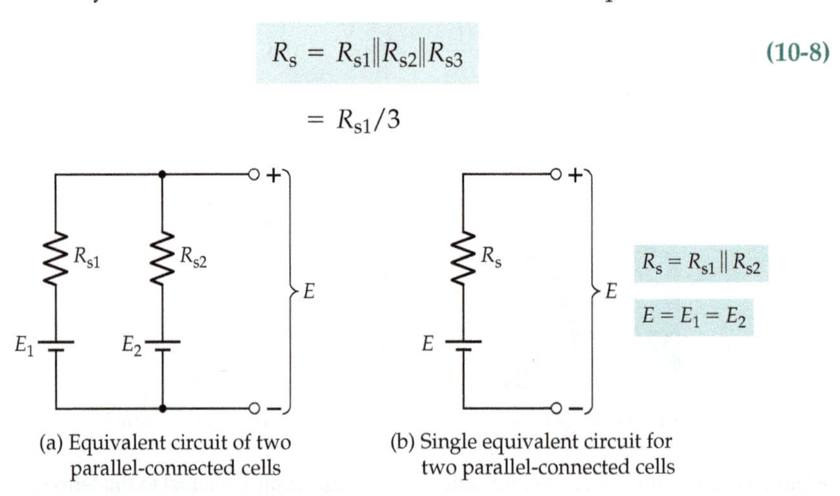

Figure 10-16 **The equivalent circuit for two parallel-connected cells shows that the internal resistance of the battery is $R_s = R_{s1} \| R_{s2}$.**

Taking the cell internal resistance into consideration, the output current from a group of parallel cells supplying a load resistance R_L is,

$$I_o = \frac{E}{R_s + R_L} \tag{10-9}$$

In some cases cells connected in parallel may not have voltages that are exactly equal. For example, in the case of the two cells in Figure 10-15, E_1 might be 1.54 V while E_2 is 1.49 V. When this occurs, the cell with the largest terminal voltage will tend to discharge through the one with the smaller terminal voltage, until both have equal voltages. If secondary (i.e., rechargeable) cells are used, one cell is charged from the other. *Cells that do not have closely equal terminal voltages should not be connected in parallel.*

Example 10-5

Four cells connected in parallel each have a terminal voltage of $E = 2$ V and internal resistances of $R_s = 0.1\ \Omega$. Each cell can normally supply a maximum load current of 500 mA for 4 h. Determine the open-circuit output voltage of the parallel combination and the maximum load current that can be supplied. Also, calculate the Ah rating and the internal resistance for the battery of cells.

Solution

$$E = E_1 = E_2 = E_3 = E_4$$

$$= 2\text{ V}$$

From Eq. 10-7,

$$I_{o(max)} = I_1 + I_2 + I_3 + I_4 = 4 \times 500\text{ mA}$$

$$= 2\text{ A}$$

$$\text{Ah rating of battery} = (4\text{ cells}) \times (\text{Ah rating per cell})$$

$$= 4 \times (500\text{ mA} \times 4\text{ h})$$

$$= 8\text{ Ah}$$

From Eq. 10-8,

$$R_s = R_{s1} \| R_{s2} \| R_{s3} \| R_{s4} = 0.1\ \Omega \| 0.1\ \Omega \| 0.1\ \Omega \| 0.1\ \Omega$$

$$= 0.025\ \Omega$$

Parallel-connected (similar) cells produce an output voltage equal to the terminal voltage of one cell and can supply a maximum output current equal to the sum of the maximum currents from each cell.

It is very important to ensure that parallel cells are correctly connected. Correct and incorrect ways of connecting cells in parallel are illustrated in Figure 10-17. All positive terminals should be connected together, and all negative terminals should be connected together, as shown in Figure 10-17(a).

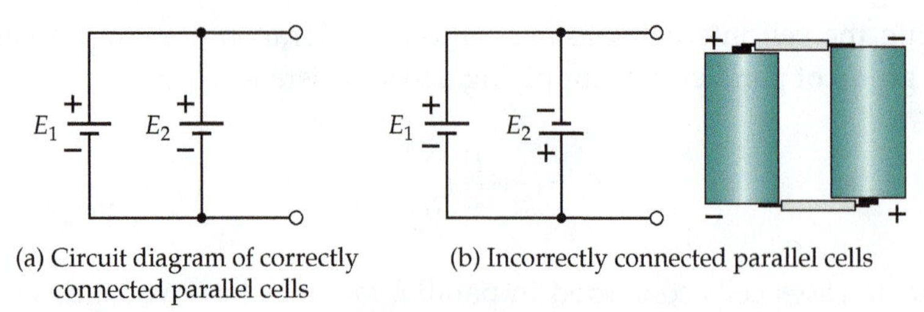

Figure 10-17 Parallel-connected cells should always be connected positive-to-positive and negative-to-negative. Incorrectly connected cells discharge rapidly.

Where positive and negative terminals are connected together as in Figure 10-17(b), *the cells short-circuit each other and discharge rapidly.*

Practice Problems

10-5.1 Each cell in a battery of 100 parallel-connected cells has a 1.45 V terminal voltage when supplying a maximum current of 60 mA. Determine the minimum load resistance that can be connected to the battery.

10-5.2 A battery is made up of 10 cells connected in parallel. Each cell has a 2.2 V open-circuit terminal voltage and an internal resistance of 0.015 Ω. Calculate the output voltage of the battery when supplying a 100 A load current.

10-5.3 A certain load connected to the battery in Problem 10-5.2 causes the terminal voltage to drop to 2 V. If this load can be supplied for a maximum of 3 hours, determine the Ah rating of the battery.

10-6 VOLTAGE CELLS IN SERIES-PARALLEL

Two parallel-connected groups of cells are shown in Figure 10-18(a), with each group consisting of three cells in series. The circuit diagram for a series-parallel group of cells is illustrated in Figure 10-18(b). If the terminal voltage of each cell is 1.5 V, then the terminal voltage of each group of series cells is,

$$E = E_1 + E_2 + E_3 = 1.5\text{ V} + 1.5\text{ V} + 1.5\text{ V}$$
$$= 4.5\text{ V}$$

and the terminal voltage of the entire battery is 4.5 V.

The currents flowing from each series group of cells in Figure 10-18(b) are identified as I_1 and I_2. If each cell is capable of supplying a maximum of 1 A, the greatest output current that can be taken from each series group is also 1 A. For the parallel combination (or entire battery of cells) the maximum output current is,

$$I = I_1 + I_2 = 1\text{ A} + 1\text{ A}$$
$$= 2\text{ A}$$

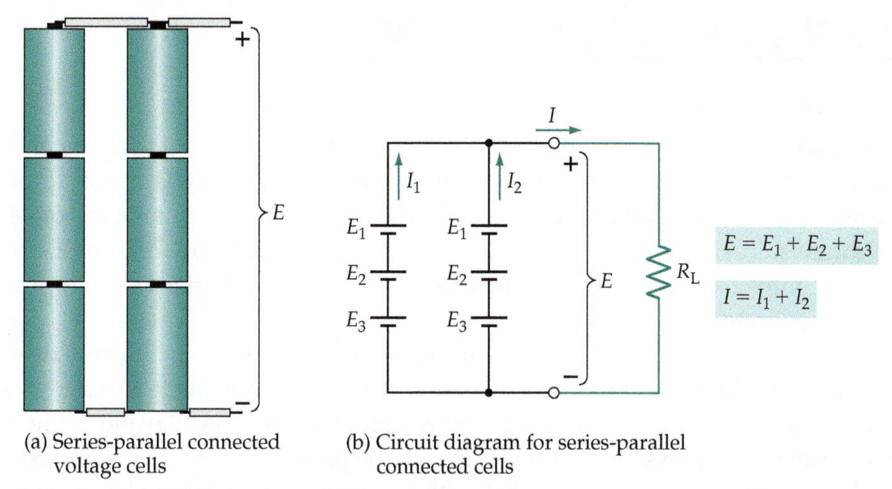

(a) Series-parallel connected voltage cells

(b) Circuit diagram for series-parallel connected cells

Figure 10-18 **Voltage cells can be connected in parallel combinations of series-connected groups. This arrangement gives a terminal voltage equal to that of each series-connected group, and a maximum load current equal to the sum of the group load currents.**

This kind of series-parallel arrangement of cells can be employed to give any desired combination of output voltage and maximum current. Where a greater output voltage is required, more cells are added to each series group. For larger maximum output currents, the number of parallel groups is increased.

Example 10-6

A 30 V, 500 mA supply is to be created from 1.5 V cells that can each produce 100 mA. Determine the total number of cells required.

Solution

Number of series cells, $$N_s = \frac{\text{output voltage}}{\text{cell voltage}} = \frac{30\text{ V}}{1.5\text{ V}}$$

$$= 20$$

Number of parallel cells, $$N_p = \frac{\text{load current}}{\text{cell current}} = \frac{500\text{ mA}}{100\text{ mA}}$$

$$= 5$$

$$\text{Total number of cells} = N_p \times N_s = 20 \times 5$$

$$= 100$$

Practice Problem

10-6.1 A battery of solar cells consists of 50 parallel-connected groups of 20 series-connected cells. Each cell has a 0.45 V terminal voltage when supplying a current of 50 mA. Calculate the maximum output power of the battery.

10-7 LEAD-ACID BATTERY

Construction

The lead-acid battery is the most commonly used type of storage battery and is well known for its application in automobiles. The battery is made up of several cells, each of which consists of lead plates immersed in an electrolyte of dilute sulfuric acid. The voltage per cell is typically 2 V to 2.2 V. For a 6 V battery, three cells are connected in series, and for a 12 V battery six cells are series-connected.

The construction of a lead-acid automobile type battery is illustrated in Figure 10-19. The electrodes are lead-antimony alloy plates with a pattern of recesses, so that they are in the form of grids [Figure 10-19(a)]. Lead oxide (termed *active material*) is pressed into the recesses of the plates. Each electrode consists of several plates connected in parallel, with porous rubber separators in between, as illustrated in Figure 10-19(b). This arrangement and the shape of the plates

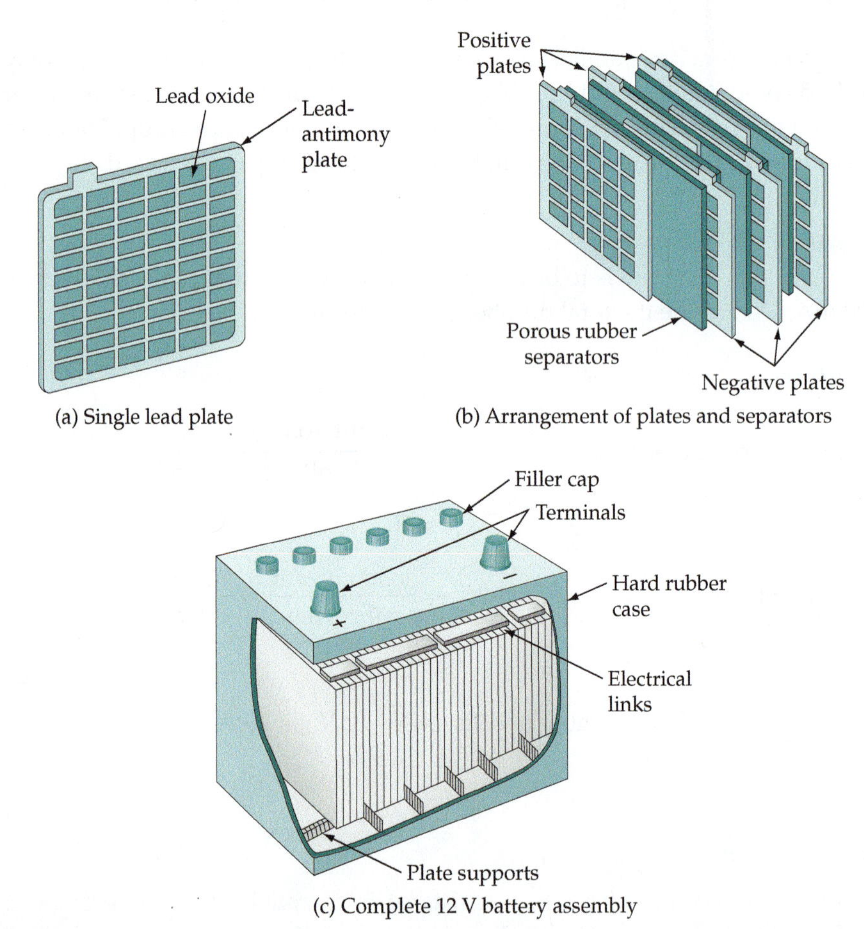

Figure 10-19 A lead-acid battery consists of several series-connected lead-acid cells in a single container. Three cells are used for a 6 V battery, six cells are employed for a 12 V battery.

give the largest possible electrode surface area within the size limitations of the battery.

The complete 12 V battery, shown in Figure 10-19(c), has a hard rubber case divided into six sections for the six separate cells. Projections are provided on the inside at the bottom of the case to support the plates. These projections ensure that the lower edges of the plates are normally well above the level of any active material that falls to the bottom of a cell. Such material can short out the positive and negative plates and render a cell useless. Every cell has a threaded filler cap with a small hole in its center. The filler caps provide access for adding electrolyte, and the holes allow gases to be vented to the atmosphere.

Low-resistance electrical links connect the cells in series, and hefty battery terminals are provided. Because the battery may be required to supply a very heavy current, it is important that the resistance of all electrical connections be very low to minimize voltage drops. A current of 250 A is not unusual for a battery driving an automobile starter.

Operation

When a lead-acid cell is charged, the lead oxide on the positive plates changes to *lead peroxide,* and that on the negative plates becomes *spongy* or *porous* lead. In this condition, the positive plates are brown in color, and the negative plates are gray.

When the battery is discharging (i.e., supplying a current), atoms from the spongy lead on the negative plates combine with sulfate molecules to form *lead sulfate* and *hydrogen.* As always, electrons are left behind on the negative plates so that they maintain a negative potential. The hydrogen released in the electrolyte combines with the lead peroxide on the positive plate, removing electrons from the plate to keep a positive potential. The combination of lead peroxide and hydrogen at the positive electrode produces water and lead sulfate. The water dilutes the electrolyte, making it a weaker solution, and the lead sulfate that is produced at both positive and negative plates tends to fill the pores of the active material. Both these effects (dilution of the electrolyte and formation of lead sulfate) render each cell less efficient and eventually cause the battery output voltage to fall.

When the battery is recharged, a current (conventional direction) is made to flow into the positive electrode of each cell. This current causes the lead sulfate at the negative electrode to recombine with hydrogen ions, re-forming sulfuric acid in the electrolyte and spongy lead on the negative plates. Also, the lead sulfate on the positive electrodes recombines with water to regenerate lead peroxide on the positive plates and sulfuric acid in the electrolyte. The final result of charging the cell is that the electrodes are re-formed and the electrolyte is returned to its original strength. With proper care a lead-acid battery is capable of sustaining a great many cycles of charge and discharge, giving satisfactory service for several years.

Ampere-Hour Rating of Lead-Acid Battery

Typical ampere-hour ratings for 12 V lead-acid automobile batteries range from 100 Ah to 300 Ah. This is usually specified for an 8 h discharge time, and it defines the amount of energy that can be drawn from the battery until the voltage drops to about 1.7 V per cell.

For a 240 Ah rating, the battery could be expected to supply 30 A for an 8 h period (see Figure 10-20). With greater load currents, the discharge time is obviously shorter. However, the ampere-hour rating is also likely to be reduced for a shorter discharge time because the battery is less efficient when supplying larger currents. Another method of rating a lead-acid battery is to define what its terminal voltage will be after about five seconds of supplying perhaps 250 A. This corresponds to the kind of load that a battery experiences in starting an automobile. It is important to avoid battery overloads that may demand excessive currents. Drawing a larger current than the battery is designed to supply may cause severe damage.

The rating of a battery is typically stated for temperatures around 25°C, and this must be revised for operation at lower temperatures. Because the chemical reactions occur more slowly at reduced temperatures, the available output current and voltage are less than at 25°C. Around −18°C a fully charged battery may be capable of delivering only 60% of its normal ampere-hour rating. As the cell is discharged and the electrolyte becomes weaker, freezing of the electrolyte becomes more likely. A fully charged cell is less susceptible to freezing, but even a fully charged cell may fail when its temperature falls to about −21°C.

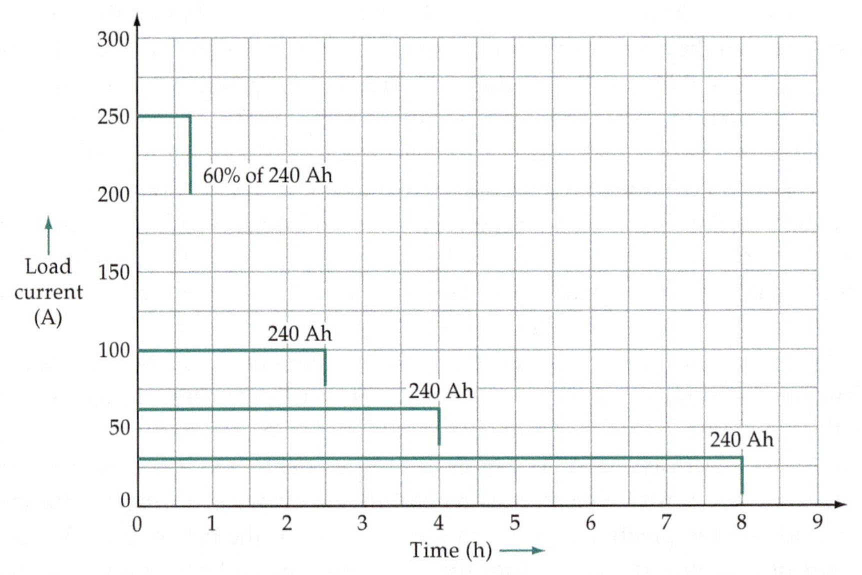

Figure 10-20 The ampere-hour rating for a lead-acid battery is typically 240 Ah. For high current levels and for extremely low temperatures the ampere-hour rating must typically he reduced to 60% of the normal rating.

Example 10-7

A lead-acid battery has a rating of 300 Ah. (a) Determine how long the battery might be employed to supply 25 A. (b) If the battery rating is reduced to 100 Ah when supplying large currents, calculate how long it could be expected to supply 250 A. (c) Under very cold conditions the battery supplies only 60% of its normal rating. Find the length of time that it might continue to supply 250 A to the starter motor of an automobile.

Solution

(a) Eq. 10-1:

$$t = \frac{\text{Ah rating}}{I} = \frac{300\ \text{Ah}}{25\ \text{A}}$$

$$= 12\ \text{h}$$

(b)

$$t = \frac{\text{Ah rating}}{I} = \frac{100\ \text{Ah}}{250\ \text{A}} = 0.4\ \text{h}$$

$$= 24\ \text{minutes}$$

(c)

$$\text{Ah rating} = 60\%\ \text{of}\ 100\ \text{Ah}$$

$$= 60\ \text{Ah}$$

$$t = \frac{60\ \text{Ah}}{250\ \text{A}} = 0.24\ \text{h}$$

$$= 14.4\ \text{minutes}$$

Charging

When a battery is to be charged, a dc charging voltage must be applied to its terminals. The polarity of the charging voltage must be such that it causes current to flow into the battery in opposition to the normal direction of discharge current. This means that the positive output terminal of the battery charger must be connected to the positive terminal of the battery, and the charger negative terminal must be connected to the battery negative terminal. The arrangement is shown in Figure 10-21. The battery charger normally has a voltmeter and ammeter to monitor the charging voltage and current, and a control to adjust the rate of charge.

The output voltage of a battery charger must be greater than the battery voltage in order to cause current to flow into the battery positive terminal. The charging current depends on the difference between the battery voltage and the charging voltage and on the internal resistances of the battery. A very large charging current is to be avoided because it could cause the battery to overheat, possibly resulting in warping of the lead plates. The maximum safe charging current is frequently taken as the maximum output current from the battery when discharging at its 8 h rate.

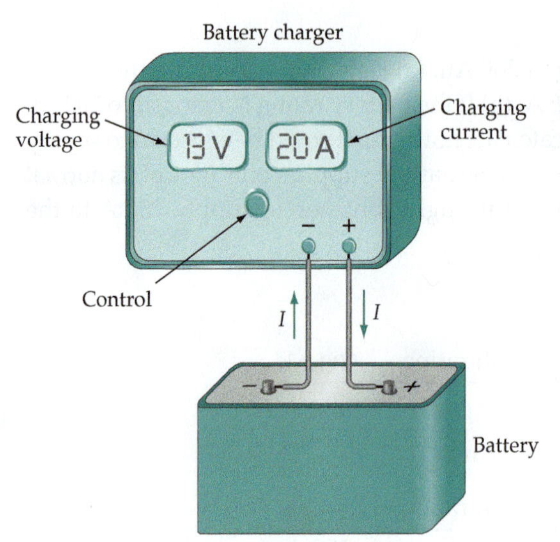

Figure 10-21 When a battery is being charged, (conventional direction) current flows from the positive terminal of the charger into the positive terminal of the battery and from the battery negative terminal into the charger negative terminal.

Example 10-8

A battery with a rating of 300 Ah is to be charged. (a) Determine a safe maximum charging current. (b) If the internal resistance of the battery is 0.008 Ω and its (discharged) terminal voltage is 11.5 V, calculate the initial output voltage level for the battery charger.

Solution

(a) Safe charging current at the 8 h rate,

$$I = \frac{\text{Ah rating}}{t} = \frac{300\ \text{Ah}}{8\ \text{h}}$$

$$= 37.5\ \text{A}$$

(b) Charging voltage,

$$I = \frac{(\text{charger voltage}) - (\text{battery voltage})}{\text{battery internal resistance}} = \frac{E_{\text{Ch}} - E_{\text{Bat}}}{r_{\text{i}}}$$

Therefore, $E_{\text{Ch}} = (I \times r_{\text{i}}) + E_{\text{Bat}} = (37.5\ \text{A} \times 0.008\ \Omega) + 11.5\ \text{V}$

$$= 11.8\ \text{V}$$

Specific Gravity

When a lead-acid battery is in a nearly discharged condition, the electrolyte is in its weakest state. Conversely, the electrolyte is at its strongest (or greatest density) when the battery is fully charged. The density of electrolyte related to the density of water is termed its *specific gravity*. The specific gravity of the electrolyte (measured by means of a *hydrometer*) is used as an indication of the state of charge of a lead-acid battery. Electrolyte with a specific gravity of 1100 to 1150 is 1.1 to 1.15 times as dense as water. At 1100 to 1150 the cell is completely discharged. When the specific gravity is 1280 to 1300, the cell may be assumed to be fully charged.

Care of Lead-Acid Batteries

- The level of the electrolyte in each cell should be checked regularly, and distilled water added as necessary to keep the top of the plates covered by about 1 cm of liquid.
- Battery terminals should be kept clean and lightly coated with petroleum jelly to avoid corrosion.
- In cold weather, batteries should always be maintained in a nearly fully charged condition to avoid freezing.
- Lead-acid batteries should never be allowed to remain for long periods in a discharged state because lead sulfate could harden and permanently clog the pores of the electrodes.
- Before storing it for a long time, the battery should be completely charged, and the electrolyte should be drained so that the battery is stored dry.

Practice Problems

10-7.1 A 12 V lead-acid battery with a 400 Ah rating is to supply a 0.6 Ω load. Determine the maximum time period that the load may be connected. If the battery rating is reduced to 70% of normal at low temperatures, calculate the length of time that a 0.8 Ω load may be connected.

10-7.2 An automobile battery has an 80 Ah rating when supplying a 200 A starter current. If the rating is reduced to 65% at low temperatures, calculate the maximum time period that it may keep the starter going.

10-7.3 A 400 Ah battery has an internal resistance of 0.01 Ω and a terminal voltage of 11.6 V when discharged. Determine the maximum safe charging current for the battery and the initial output voltage of the charger.

10-8 DC POWER SUPPLIES

Laboratory Power Supplies

Most electronics equipment uses a dc power supply rather than a battery. The power supply converts the usual domestic or industrial alternating voltage supply (see Chapter 17) into a direct voltage. In electronics laboratories dc power supplies are extensively used to supply experimental circuitry. Each laboratory-type power supply is designed to provide a particular output voltage, or range of voltages, and a specified maximum output current. When selecting a dc power supply for a particular application, the required load voltage and current should both be known.

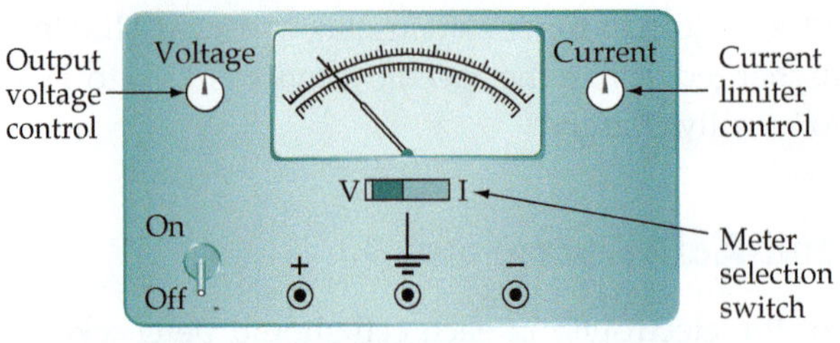

(a) Single output voltage dc power supply

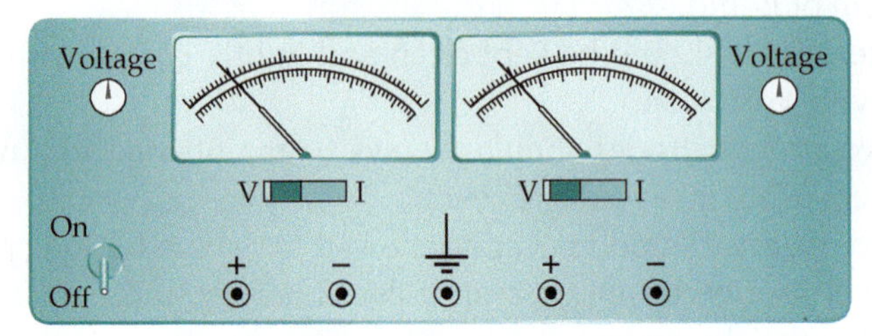

(b) Dual output voltage dc power supply

Figure 10-22 **The front panel of a laboratory dc power supply usually has an on/off switch, one or two meters, voltage and current controls, and a set of output terminals.**

Figure 10-22(a) shows the front panel of a typical laboratory-type dc power supply. As illustrated, there is an on/off switch for the ac input. A meter is provided for indicating output voltage or current, as selected by the *V/I* switch. Voltage and current controls are included for setting the output voltage and maximum output current to the desired level. *Positive* (+), *negative* (−), and *ground* terminals are available for connection purposes.

The output voltage may be taken from the + and − terminals and the ground terminal left unconnected. This gives a *floating* (i.e., ungrounded) output. Alternatively, either the − or + terminals may be connected to the ground terminal to give a voltage that is either positive or negative with respect to ground, as desired.

The voltage control is adjusted to set the output to the required voltage. Where there is no possibility of overload, the current control may be set to its maximum position. To limit the output current to a particular level, the output terminals should be short-circuited and the current control adjusted from minimum to the desired maximum current level.

Figure 10-22(b) shows the terminals of a dual output voltage supply with each output independently adjustable. The outputs may be series-connected to provide a greater voltage than can be obtained from one supply. Parallel connection for greater output current is also possible, but care must be taken to ensure that the load current is equally distributed between the two outputs. This type of supply is most frequently used with one + terminal and one − terminal grounded to give a *plus/minus* output voltage.

Portable DC Power Supplies

Small portable dc supplies are widely used in such applications as, providing power to telephone answering machines, and charging the batteries in cellular telephones, digital cameras, etc. Typical outputs are: (5.9 V, 350 mA), (9 V, 300 mA), (11 V, 1 A). It is important that the power supply designed for a particular application is used for that purpose. Use of the wrong power supply could result in equipment damage.

Summary of Formulas

- *Cell output current:*

$$I_o = \frac{E_c}{R_s + R_L}$$

- *Cell output voltage:*

$$E_o = E_c - I_o R_s$$

- *For cells in series:*

$$E = E_1 + E_2 + E_3 + \cdots$$

$$R_s = R_{s1} + R_{s2} + R_{s3} + \cdots$$

$$I_o = \frac{E_1 + E_2 + \cdots}{R_s + R_L}$$

$$I = I_1 = I_2 = I_3 = I_4$$

- *For cells in parallel:*

$$E = E_1 = E_2 = E_3 = \cdots$$

$$R_s = R_{s1} \| R_{s2} \| R_{s3} \| \cdots$$

Review Questions

Section 10-1

10-1 Draw a sketch to show the basic construction of a simple voltage cell. Explain the operation of the cell.

Section 10-2

10-2 Discuss the output voltage and current that can be supplied by a simple voltage cell and define ampere-hour rating.

10-3 Define primary cell, secondary cell, open-circuit output voltage, no-load output voltage, internal resistance, and ideal cell.

10-4 Sketch the equivalent circuit for a voltage cell with an external load resistance (R_L). Write equations for the cell terminal voltage and output current. Explain briefly.

Section 10-3

10-5 Draw a sketch to show the basic construction of a typical zinc-carbon dry cell. Explain the operation of the cell, and discuss its ampere-hour rating.

10-6 Discuss the construction, terminal voltages, ampere-hour ratings, and application for the following: manganese-alkaline cell, mercury cell, nickel-cadmium cell, lithium-ion cell, solar cell.

Section 10-4

10-7 Sketch the circuit diagram for four voltage cells connected in series. Draw the equivalent circuit for the series combination of cells, and write the equations for output voltage and current.

10-8 Draw circuit diagrams to show series-aiding and series-opposing connected cells and briefly explain. Also, show how a $\pm$ 10 V supply could be constructed using 2 V cells.

Section 10-5

10-9 Sketch the circuit diagram for four voltage cells connected in parallel. Draw the equivalent circuit for the parallel combination of cells, and write the equations for output voltage and current.

10-10 Explain what occurs when cells having different terminal voltages are connected in parallel. Also, discuss the result of incorrectly connecting voltage cells in parallel.

Section 10-6

10-11 Sketch a circuit diagram showing three parallel-connected groups of four series-connected voltage cells. Draw the equivalent circuit for the battery of cells, and write the equations for output voltage and current.

Section 10-7

10-12 Describe the construction of a 12 V lead-acid battery and explain the chemical process involved in charge and discharge within the individual cells.

10-13 Discuss typical ampere-hour ratings of a lead-acid battery with respect to supplying small currents, large currents, and low-temperature performance.

10-14 Explain how a lead-acid battery should be connected to a battery charger, and discuss charging voltage and current levels.

10-15 Define specific gravity and state typical specific-gravity values for discharged and fully charged lead-acid batteries.

10-16 Discuss the care of lead-acid batteries.

Section 10-8

10-17 Describe typical laboratory and portable-type dc power supplies.

Problems

Section 10-2

10-1 A voltage cell supplies a current of 0.75 A for 10 h. Then its terminal voltage drops to a low level. Calculate the Ah rating of the cell, and determine how long it might be expected to supply a current of 250 mA.

10-2 An electronic circuit draws a constant current of 8 mA from a 9 V battery. If the battery has to be replaced every 30 days, calculate its ampere-hour rating.

10-3 A 3 V battery supplies current continuously to a 2 Ω resistance. If the battery has a rating of 8 Ah, how long will it take to discharge?

10-4 A 1.5 V cell with a rating of 5 Ah is used to drive an electric clock. The cell has to be replaced approximately every 12 months. Determine the load resistance of the clock mechanism.

10-5 The open-circuit voltage from a certain voltage cell is 1.65 V, and its internal resistance is 0.05 Ω. Calculate the terminal voltage of the cell when the output current is 5 A. Determine the cell current when the terminals are short-circuited.

10-6 The cell in Problem 10-4 has a terminal voltage of 1.55 V when open-circuited and 1.5 V when connected to the clock. Calculate the internal resistance and short-circuit current for the cell.

10-7 A battery with an internal resistance of 3 Ω has an open-circuit voltage of 9.35 V. Calculate the minimum load resistance that can be connected to the battery if its terminal voltage is not to fall below 9 V, (see Figure 10-23).

10-8 A voltage cell has an open-circuit terminal voltage of 2.2 V and a short-circuit current of 1.47 mA. Determine the terminal voltage of the cell when supplying 100 μA.

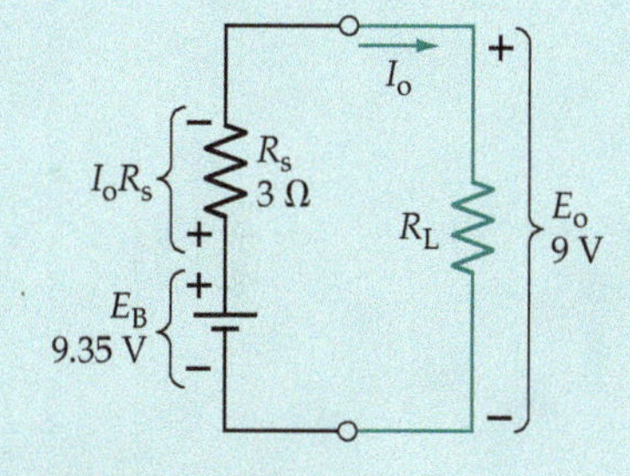

Figure 10-23

Section 10-4

10-9 A battery is made up of five series-connected voltage cells, each of which has an open-circuit terminal voltage of 1.6 V and an internal resistance of 0.08 Ω. Calculate the battery terminal voltages for no-load condition and for $R_L = 6\ \Omega$.

10-10 The terminal voltage of a battery consisting of 6 series-connected cells is measured as 9.5 V on open-circuit and 8.9 V when supplying 180 mA. Calculate the average internal resistance of one cell.

10-11 A 12 V supply is to be constructed using 1.5 V cells with $r_i = 0.05\ \Omega$. Determine the number of cells required and calculate the internal resistance of the battery (see Figure 10-24). Also, calculate the battery terminal voltage when the load resistance is 10 Ω.

10-12 A plus-minus power supply constructed of eight 1.5 V cells has outputs of +6.5 V and −6.4 V on open-circuit. When supplying 500 mA, the

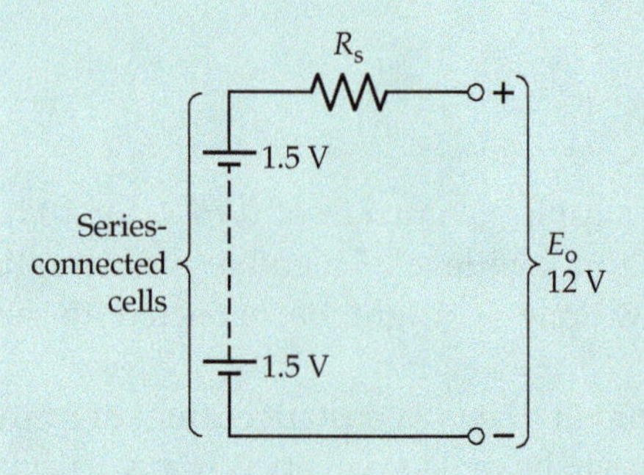

Figure 10-24

outputs are +6.2 V and −6.15 V. Calculate the internal resistance of each half of the power supply.

10-13 A ±9 V supply is to be constructed using zinc-carbon *D* cells. Determine the total number of cells required, and sketch the circuit diagrams for connecting the cells. If one of the cells is accidentally connected in reverse, what will be the output voltages from the supply?

10-14 Ten series-connected solar cells each have an open-circuit terminal voltage of 0.45 V and an internal resistance of 0.8 Ω. Determine the maximum current that can be drawn from the battery of cells if the terminal voltage is not to fall below 4 V.

Section 10-5

10-15 A parallel-connected battery of 1.5 V cells is to supply current to a 0.15 Ω resistor. If each cell can supply a maximum current of 1 A, determine the number of cells required. Neglect the internal resistances of each cell.

10-16 A 9 V power supply capable of sustaining a 1 A load current is to be constructed by parallel-connecting several of the type of batteries described in Problem 10-7, (see Figure 10-25). Determine the required number of batteries if the terminal voltage of the supply is not to fall below 9 V. Also, calculate the internal resistance and short-circuit current for the power supply.

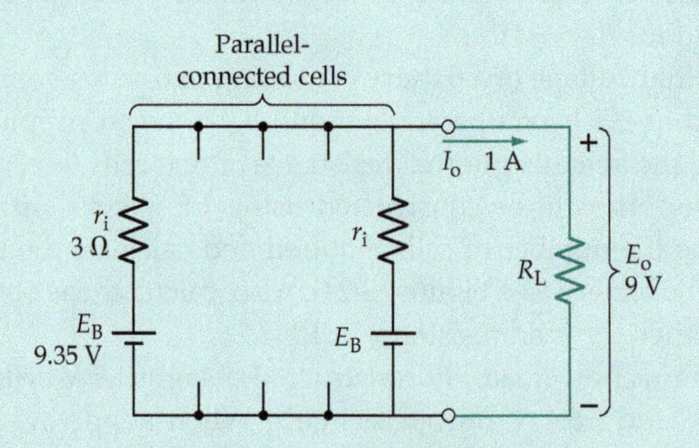

Figure 10-25

10-17 Eight voltage cells connected in parallel each have an open-circuit terminal voltage of 1.7 V and an internal resistance of 0.07 Ω. Each cell can normally supply a maximum load current of 750 mA for a time period of 12 h. Determine the output voltage of the battery of cells and the maximum load current that can be supplied. Also, calculate the Ah rating and internal resistance for the parallel combination of cells.

10-18 If the 10 solar cells described in Problem 10-14 are connected in parallel, determine the internal resistance and short-circuit current for the combination.

10-19 A 10 kW electric motor drives a vehicle at an average speed of 50 km/h. Ten parallel-connected 12 V, 100 Ah batteries supply the motor. Determine the maximum distance that the vehicle may travel before the batteries must be recharged.

10-20 A power supply made up of parallel-connected 1.5 V cells is to sustain a constant load current of 39 mA. The cells, which each have a rating of 5 Ah, are to be replaced every 14 days. Determine the number of cells required.

10-21 For the vehicle referred to in Problem 10-19, calculate the total Ah rating for batteries that will provide a range of 400 km.

Section 10-6

10-22 A 4.5 V power supply constructed of series-parallel-connected 1.5 V cells is to provide a constant current of 575 mA to an electronic circuit. The cells are each capable of supplying 100 mA without any serious drop in terminal voltage. Determine the number of cells required. If each cell has a 6 Ah rating, calculate the ampere-hour rating for the power supply.

10-23 A power supply to be constructed from voltage cells is required to supply a maximum current of 20 A with a terminal voltage of approximately 6 V. Several voltage cells are available, each of which has a terminal voltage of 1.5 V and each of which can supply a maximum current of 5 A. Determine the number of cells required and draw a circuit diagram to show how they should be connected.

10-24 A 100 m^2 roof area of a house is covered with solar cells which each occupy 5 cm^2. The terminal voltage of each cell is 0.46 V when supplying a current of 120 mA. Determine how the cells should be arranged to produce a 115 V dc supply, (see Figure 10-26). Calculate the maximum output current.

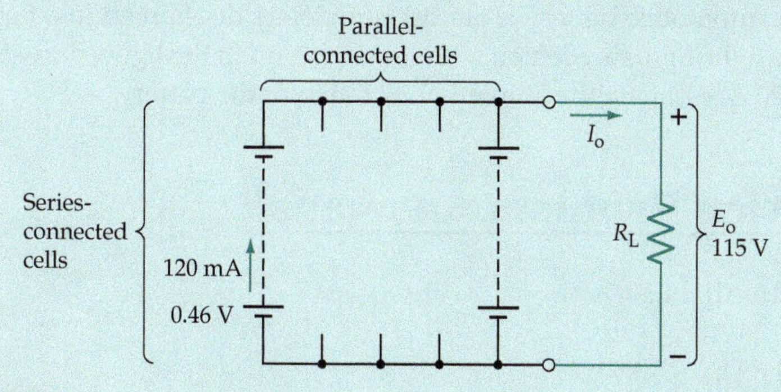

Figure 10-26

10-25 A solar cell supply with an output of approximately 13 V is to be constructed to recharge the 12 V batteries of a remote radio transmitter. The charging current is to be approximately 400 mA, and the available solar cells each have an output of 0.45 V and 65 mA. Determine the total number of cells required.

10-26 A 120 V dc power supply is to be constructed using 12 V batteries. Each battery can produce a continuous current of 10 A without a serious drop in terminal voltage. Determine the number of batteries required to produce an output current of 30 A. Also, determine the ampere-hour rating of the power supply if each battery is rated as 400 Ah.

Section 10-7

10-27 An automobile battery is to supply the starter motor with 150 A of current for a minimum time of 20 minutes. If the battery supplies only 65% of its normal rating under extremely cold conditions, calculate the minimum normal Ah rating for the battery.

10-28 The terminal voltage of a 12 V battery drops to 11.9 V when the load current is 10 A. Determine the short-circuit current for the battery. If the battery has a 300 Ah rating, which is reduced by 50% when overloaded, how long is the battery likely to sustain the short-circuit current?

10-29 A certain lead-acid battery has a rating of 400 Ah. (a) Calculate how long the battery should be able to supply a current of 12 A. (b) When supplying large currents the normal battery rating is reduced to 150 Ah. Determine how long it can be expected to supply 240 A to an automobile starter motor. (c) If the battery rating is reduced to 60% of normal when very cold, calculate how long it could supply 240 A.

10-30 A 12 V battery becomes fully charged with a constant current of 20 A over a time period of 10 h. If the battery rating is reduced to 60% of normal when supplying high current levels, calculate how long is it likely to sustain an 80 A load.

10-31 A 400 Ah lead-acid battery is to be charged. (a) Calculate a maximum safe level of charging current. (b) The internal resistance of the battery is 0.007 Ω, and the terminal voltage when discharged is 11.4 V. Determine the initial output voltage level for the battery charger.

10-32 An automobile battery is almost completely discharged in a time of 8 h when the lights are left on with the engine off. If the lights draw a current of 20 A, estimate the ampere-hour rating of the battery.

Practice Problem Answers

10-2.1 0.5 Ah, 14.3 h
10-2.2 0.24 Ah
10-2.3 1.6 Ω, 0.97 A

10-2.4 0.0055 Ω
10-4.1 8.8 V
10-4.2 20 A
10-5.1 0.24 Ω
10-5.2 2.05 V
10-5.3 400 Ah
10-6.1 22.5 W
10-7.1 20 h, 18.7 h
10-7.2 15.6 min
10-7.3 50 A, 12.1 V

CHAPTER 11
Magnetism

CONTENTS

Objectives

You will be able to:

1. Discuss the behavior of bar magnets and sketch the magnetic force fields around bar magnets in various situations.
2. State the fundamental law of magnetism.
3. Describe magnetic lines of force and list their characteristics.
4. Define magnetic flux and flux density and solve problems involving these quantities in relationship to bar magnets.
5. Sketch the magnetic fields that occur around current-carrying conductors and relate the direction of the fields to the current directions.
6. Sketch the magnetic fields that occur around various shaped coils.
7. Solve problems involving the flux and flux density of the magnetic fields around coils.
8. Explain electromagnetic induction.
9. Explain the (atomic) theory of magnetism and classify various magnetic and nonmagnetic materials.
10. Define magnetomotive force and magnetic field strength and solve problems involving these quantities.
11. Calculate the force exerted on a current-carrying conductor situated in a magnetic field and define the direction of the force.
12. Calculate the torque exerted on a coil pivoted in a magnetic field and show how it is produced.

INTRODUCTION

Some 2,000 years ago, Chinese mariners used a suspended bar of mineral known as *lodestone* as a crude compass. The mineral had magnetic properties, so its ends tended to point to the earth's magnetic poles. Today's magnetic compasses are much more sensitive than the lodestone, but they fulfill essentially the same purpose. As well as direction finding, there are a great many other applications of magnetism.

The magnetic force field around a bar magnet is easily plotted by means of iron filings or by use of compasses. It is found that the magnetic field passes through nonmagnetic material including air and vacuum. However, iron and steel are by far the best magnetic conductors, and in the presence of a magnetic field, pieces of iron and steel become magnetized.

A magnetic field also exists around a current-carrying conductor, and a current-carrying coil generates a field similar to that of a bar magnet.

11-1 MAGNETIC FIELD

North and South Poles

A bar magnet suspended on thread tends to align itself in relation to the earth in a north-south direction. This is also the case with the magnetic compass, which is simply a pivoted magnetized needle. The two ends of the bar magnet and the compass needle are identified as north pole (N) and south pole (S), in accordance with the direction that they tend to point.

Consider the effect of bringing the N pole of a bar magnet close to the N pole of a suspended magnet. As illustrated in Figure 11-1(a), the result is repulsion between the two magnets. A similar result occurs when the S pole of the handheld magnet is brought close to the S pole of the suspended magnet. However, when N and S poles are brought together, the magnets are attracted to each other, [see Figure 11-1(b)]. The same effects are demonstrated when one pole of a bar magnet is brought close to a magnetic compass, [Figure 11-1(c)]. The needle of the compass always turns so that its S pole points to the N pole of the magnet, or so that its N pole points to the magnet's S pole. These results give rise to the fundamental law of magnetism:

Like poles repel; unlike poles attract.

This is similar to the fundamental law of electrification by friction, (Section 1-1).

Because unlike poles attract, the pole of the magnetic compass that points to the earth's north pole must be *unlike* the earth's north pole, [Figure 11-1(d)]. Consequently, the earth's north pole must actually be a south magnetic pole! Alternatively, it can be stated that the north pole of a compass is actually a *north-seeking pole* with respect to the earth.

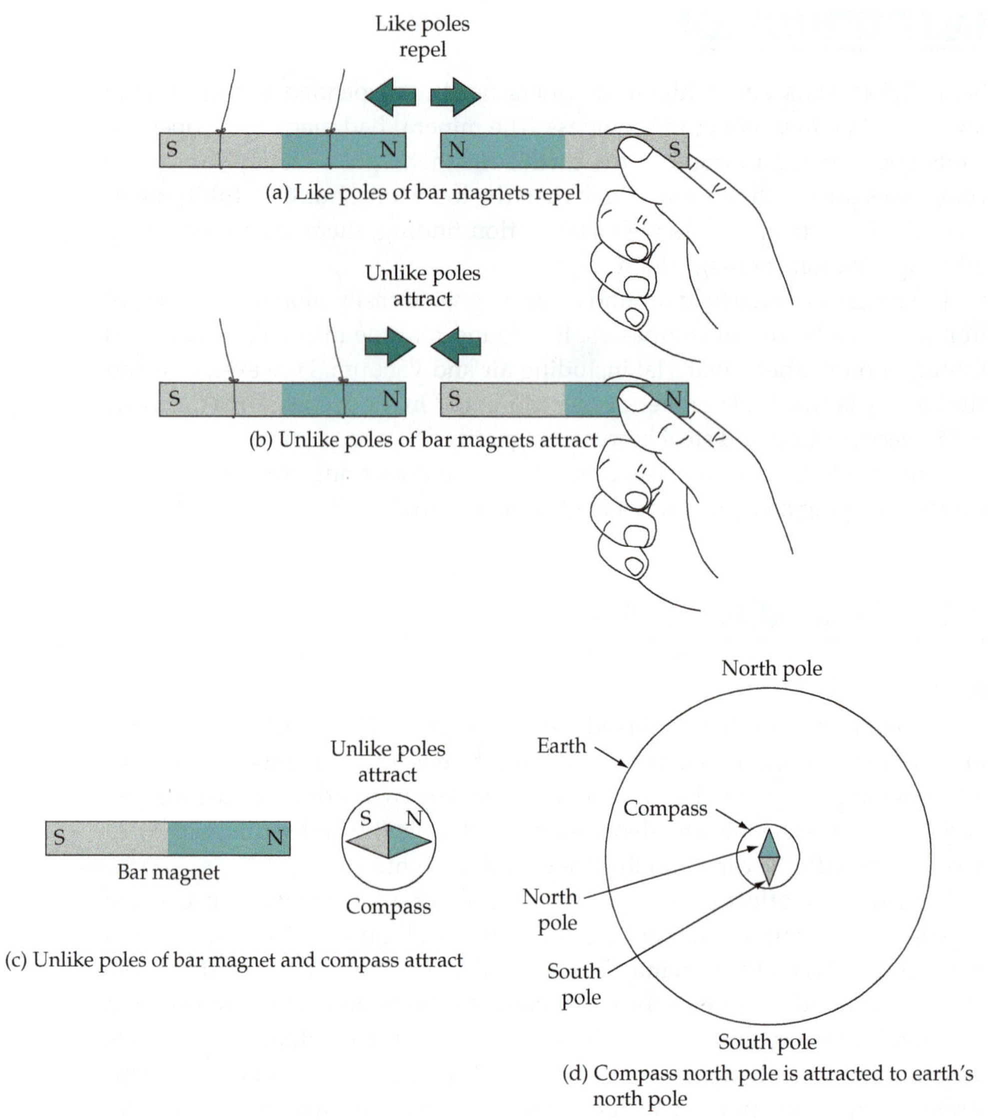

Figure 11-1 Like magnetic poles repel; unlike poles attract. Two north poles repel, two south poles repel, and a north and south pole attract. A compass north pole points to the earth's north pole, which must be unlike the compass north pole!

Magnetic Lines of Force

The fact that attraction or repulsion occurs when the poles of two magnets are brought together demonstrates that a *force field* exists around the magnets, and that the field tends to be concentrated at the magnet's poles. This force field around a magnet is termed a *magnetic field,* and it can be investigated further by means of iron filings sprinkled on a piece of cardboard placed on top of a magnet. When the cardboard is gently tapped, the iron filings align themselves into chains, as shown in Figure 11-2(a). The lines plotted by the pattern of iron filings stretch from one pole to the other and are termed *magnetic lines of force.* Collectively, the magnetic lines of force are referred to as the *magnetic flux.*

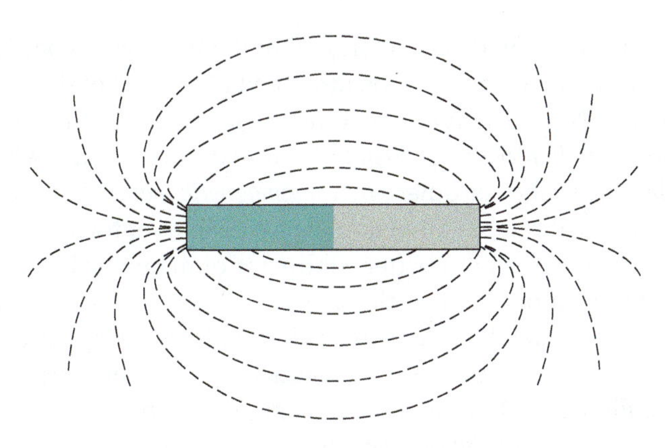

(a) Force field around a magnet, plotted by iron filings

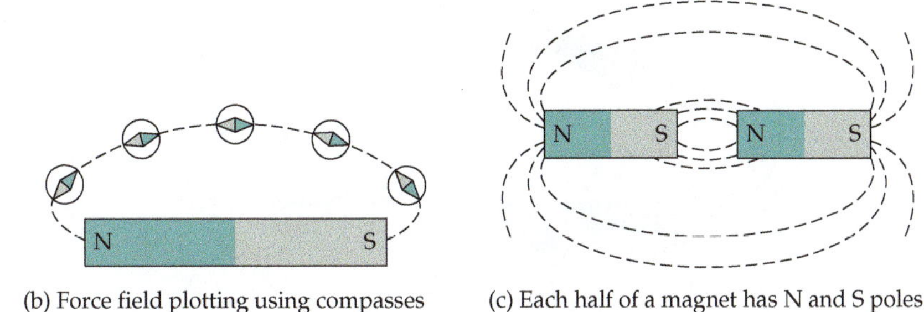

(b) Force field plotting using compasses (c) Each half of a magnet has N and S poles

Figure 11-2 **The magnetic force field that exists around a bar magnet can be plotted by the use of iron filings or by compasses.**

The concentration of iron filings at the poles of the magnet shows that the magnetic force field has its greatest strength close to the magnet's poles. The field is seen to become weaker at points farther away from the poles as it spreads out.

The magnetic lines of force can also be plotted by the use of several compasses, as illustrated in Figure 11-2(b). The compass needles point along the lines of force in a direction from the magnet's north pole to its south pole. Because of this, it is assumed that the magnetic lines of force exit from the north pole of the magnet and enter the magnet at its south pole. The direction of the lines of force is defined as: *the direction in which an isolated north pole would move if it were placed in the magnetic field.* In fact, an isolated north pole could not exist because all magnets have north and south poles. However, it can be imagined that if one did exist, an isolated north pole would be repelled from the magnet's N pole and move along a line of force toward the S pole of the magnet.

When a bar magnet is cut into two pieces, as shown in Figure 11-2(c), it is found that each half is still a magnet with N and S poles. Also, the lines of force cross the gap between the two pieces of the magnet, as illustrated, demonstrating that the magnetic lines of force pass through the magnet itself. So, it is seen that *magnetic lines of force form closed loops.*

When a piece of soft iron is placed in a magnetic field, it is found that the lines of force bend away from their usual paths in order to pass through the iron, [see Figure 11-3(a)]. It is also found that the iron has become magnetized temporarily and that a S pole can be identified at the end of the soft iron nearest the magnet's N pole, and a N pole can be identified at the other end. This effect is known as *induced magnetism*. No similar effect occurs when pieces of wood or glass or rubber or other such *nonmagnetic materials* are placed in the magnetic field. So, it can be stated that magnetic lines of force pass most easily through iron; that is, iron is a better conductor of magnetic lines of force than such materials as air or wood or glass.

It can now be understood that the individual iron filings used to plot the magnetic lines of force must become magnetized, so that each has its own N and S poles. Figure 11-3(b) illustrates the situation and shows that the poles of each little magnet or group of magnets are likely to be adjacent to like poles of

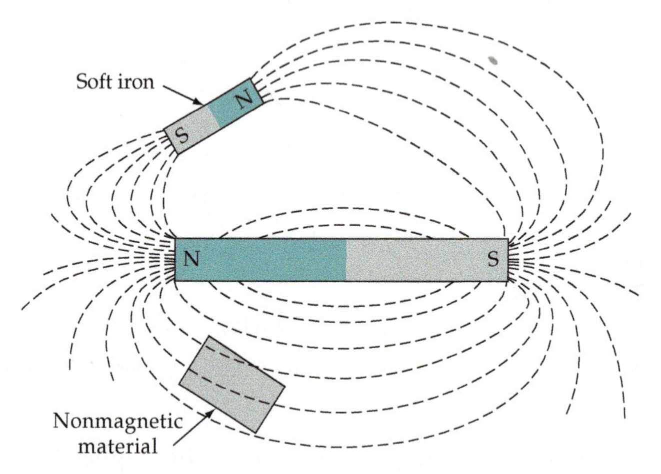

(a) Effects of magnetic and nonmagnetic material on the shape of magnetic force field

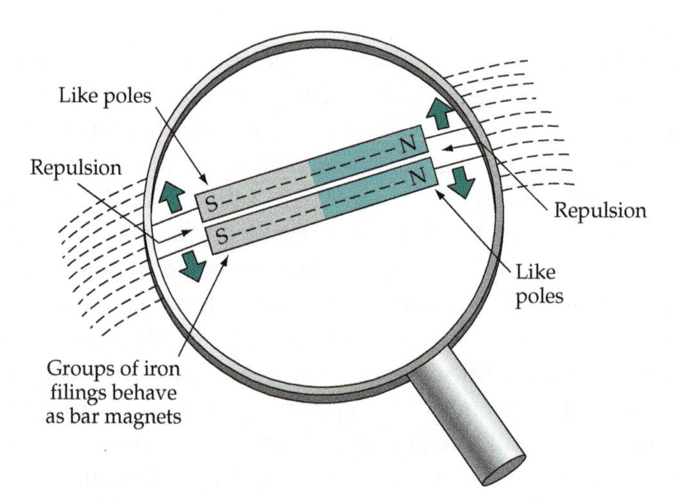

(b) 'Magnified' view of lines of force plotted by iron filings

Figure 11-3 **Magnetic lines of force form closed loops and distort to pass through iron. Magnetic lines of force tend to repel each other, and two lines of force never intersect.**

other magnetized iron filings alongside them. So, repulsion occurs, and this demonstrates another piece of information about the magnetic lines of force; that is, *magnetic lines of force tend to repel each other.* This fact is also shown by the field pattern at the poles of the magnet in Figure 11-2(a). Because lines of force repel each other, *two lines of force cannot intersect.*

Figure 11-4(a) shows the pattern of the magnetic field when two N poles (or two S poles) are brought close together. It is clearly seen that the lines of force from each pole repel one another. Conversely, when a N and S pole are placed close together [Figure 11-4(b)], the lines of force run out of one pole into the other. In Figure 11-4(c), the field pattern at the poles of a horseshoe-shaped magnet is shown. Note that the magnetic lines of force take the shortest path between the poles, and this shows that *the lines of force are always in a state of tension.*

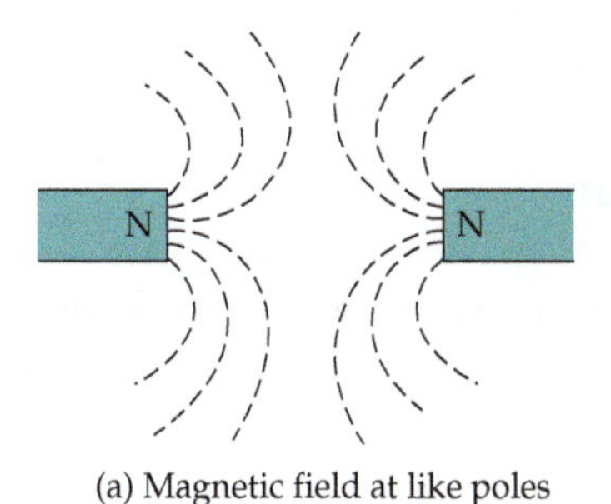

(a) Magnetic field at like poles

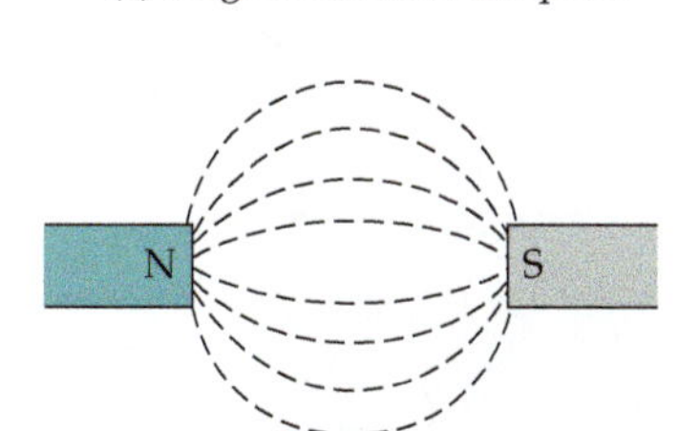

(b) Magnetic field at unlike poles

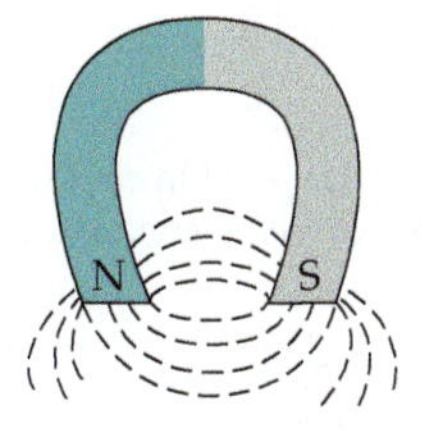

(c) Lines of force take the shortest path

Figure 11-4 The magnetic force field plotted at two like poles shows repulsion, and the field at two unlike poles demonstrates attraction.

Summarizing:

- Every magnet has two poles identified as a north pole (N) and a south pole (S). The N pole is actually a *north-seeking pole,* and the S pole is a *south-seeking pole,* with respect to the earth.
- The force field that exists around a magnet is concentrated at its poles.
- Like poles repel; unlike poles attract.

- Magnetic lines of force in a field around a magnet can be plotted by the use of iron filings.
- Lines of force form closed loops.
- Lines of force exit from a magnet's N pole and enter at its S pole.
- Lines of force travel most easily through soft iron.
- Lines of force repel each other.
- Lines of force cannot intersect.
- Lines of force from two like poles repel each other.
- Lines of force from two unlike poles run into each other.
- Lines of force are always in a state of tension.
- A piece of soft iron placed in a magnetic field is temporarily magnetized by induction.

11-2 ELECTROMAGNETISM

Magnetic Field Around a Conductor

When an electric current flows in a conductor a magnetic field is set up around the conductor. This is easily demonstrated by again using iron filings, as illustrated in Figure 11-5(a). A conductor is passed vertically through a hole in a

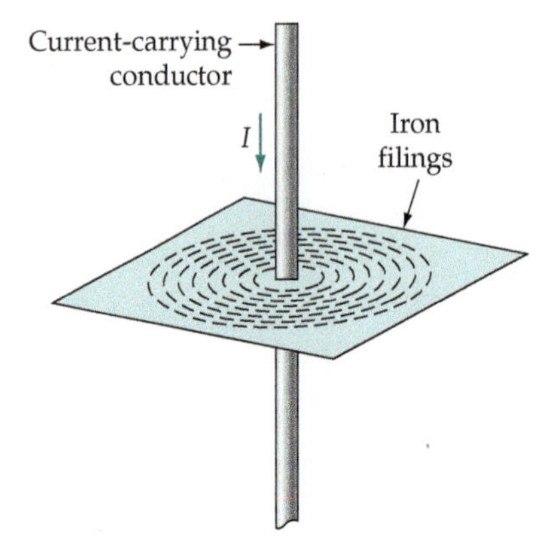

(a) Iron filings form concentric rings around a current-carrying conductor

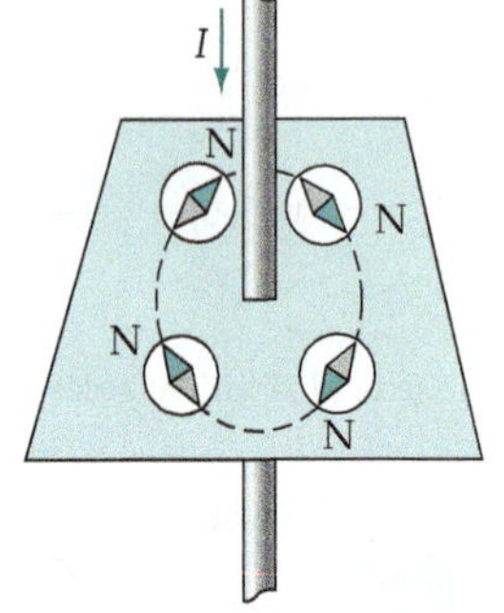

(b) Compasses point in circle around a current-carrying conductor

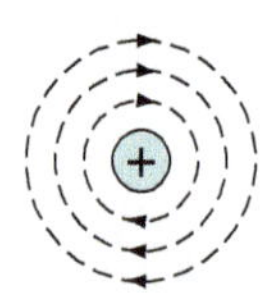

(c) Lines of force are in a clockwise direction around a conductor which is carrying current away from the viewer

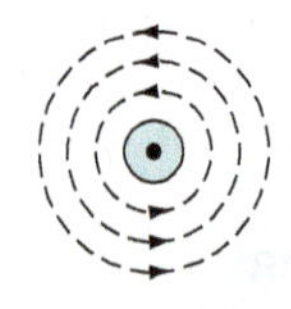

(d) Lines of force are in a counter-clockwise direction around a conductor which is carrying current toward the viewer

Figure 11-5 A magnetic force field in the form of concentric rings can be shown to exist around a current-carrying conductor.

horizontal piece of cardboard. Iron filings are sprinkled on the cardboard, and when the current through the conductor is zero there is no evidence of a magnetic field. However, when a current flows through the conductor, gentle tapping of the cardboard causes the iron filings to settle in concentric rings around the conductor.

As with other magnetic fields, the field around a current-carrying conductor can also be investigated by the use of magnetic compasses. Figure 11-5(b) shows the result of placing four compasses on the horizontal cardboard. For a current flowing in a (conventional) direction down through the cardboard, the compass needles point in a clockwise direction around the conductor. Because the direction of magnetic lines of force are defined in terms of the direction in which a free north pole would move, it can be stated that for the conditions illustrated, the magnetic lines of force are in a clockwise direction. When the current direction is reversed so that it flows up through the cardboard, the compass needles reverse and point in a counterclockwise direction.

Figure 11-5(c) and (d) further illustrate the direction of a magnetic field around a current-carrying conductor. The conductor is assumed to be directly opposite the viewer, so only its cross-sectional area is shown. In Figure 11-5(c), current is assumed to be flowing away from the viewer (i.e., into the page). This is indicated by the + sign, which represents the tail of an arrow. In this case the magnetic lines of force around the conductor have a clockwise direction. In Figure 11-5(d), the dot at the center of the conductor cross section represents the point of an arrow. Therefore, current is flowing toward the viewer (i.e., out of the page). The magnetic lines of force now have a counterclockwise direction.

Two memory aids for determining the direction of the magnetic flux around a current-carrying conductor are shown in Figure 11-6. The *right-hand-screw rule* as illustrated in Figure 11-6(a) shows a wood screw being turned clockwise and progressing into a piece of wood. The horizontal direction of the screw is analogous to the direction of current in a conductor, and the circular motion of the screw shows the direction of magnetic flux around the conductor. In the *right-hand rule,* illustrated in Figure 11-6(b), a right hand is closed around a conductor with the thumb pointing in the (conventional) direction of current flow. The fingers point in the direction of the magnetic lines of force around the conductor.

Because a current-carrying conductor has a magnetic field around it, when two current-carrying conductors are brought close together there will be interaction between the fields. Figure 11-7(a) shows the effect on the fields when two conductors carrying currents in opposite directions are adjacent. The directions of the magnetic fields round the conductors are in opposition, and this is similar to the situation when two like magnetic poles are brought close together. The fields exert a force of repulsion on each other, tending to push the conductors apart. When the adjacent conductors have currents flowing in the same direction [Figures 11-7(b) and (c)], the magnetic fields assist each other.

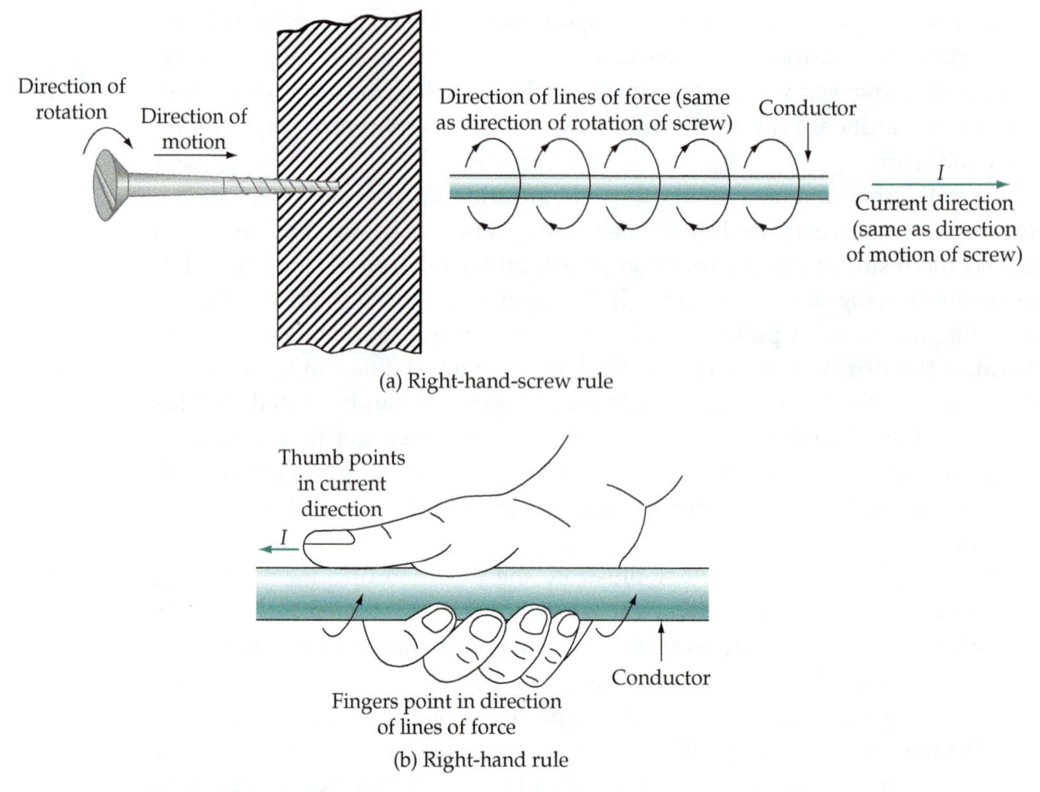

Figure 11-6 The right-hand-screw rule and the right-hand rule can be used for determining the direction of the magnetic lines of force around a current-carrying conductor.

Because the lines of force are always in tension, they are always trying to find the shortest path. Consequently, the fields exert a force that tends to pull the conductors together, as illustrated.

Magnetic Field Around a Coil

Now consider the effect of passing a current through a one-turn coil of wire. Figures 11-8(a) and (b) show that all the magnetic flux generated by the electric current passes through the center of the coil. Therefore, the one-turn coil acts like a little magnet and has a magnetic field with an identifiable N pole and S pole. Instead of a single turn, the coil may have many turns, as illustrated in Figure 11-8(c). In this case the flux generated by each of the individual current-carrying turns tends to link up and pass out of one end of the coil and back into the other end. This type of coil, known as a *solenoid,* obviously has a magnetic field pattern very similar to that of a bar magnet.

The right-hand rule for determining the direction of flux from a solenoid is illustrated in Figure 11-8(d). When the solenoid is gripped with the right hand so that the fingers are pointing in the direction of current flow in the coils, the thumb points in the direction of the flux (i.e., toward the N-pole end of the solenoid).

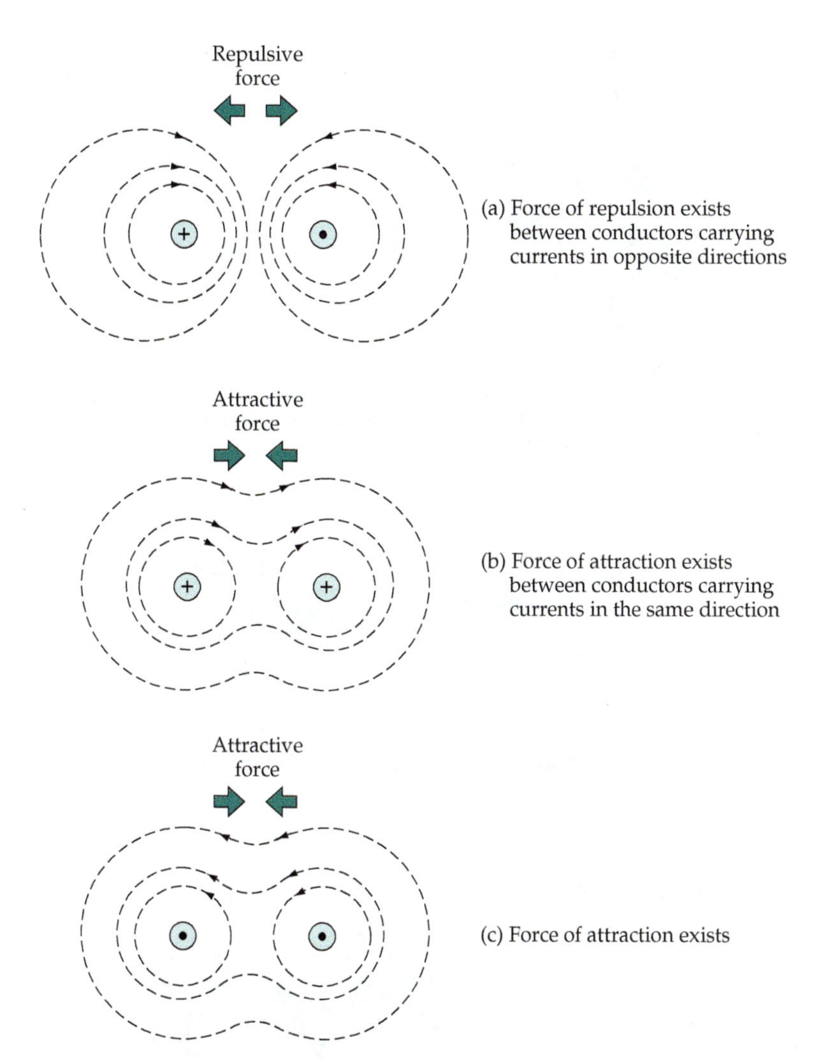

Figure 11-7 The magnetic field around two parallel current-carrying conductors shows that a force of repulsion occurs when the currents are in opposite directions and that an attractive force exists when the currents are in the same direction.

Electromagnetic Induction

It has been demonstrated that a magnetic flux is generated by an electric current flowing in a conductor. The converse is also possible; that is, a magnetic flux can produce a current flow in a conductor. Consider Figure 11-9(a), in which a handheld bar magnet is shown being brought close to a coil of wire. As the bar magnet approaches the coil, the flux from the magnet *brushes across* the coil conductors, or *cuts* the conductors. This produces a current flow in the conductors proportional to the total flux that *cuts* the coil. A current flows when the coil circuit is closed by a resistor, as shown in the figure. Whether or not the circuit is closed, an *electromotive force (emf)* can be measured at the coil terminals. This effect is known as *electromagnetic induction.*

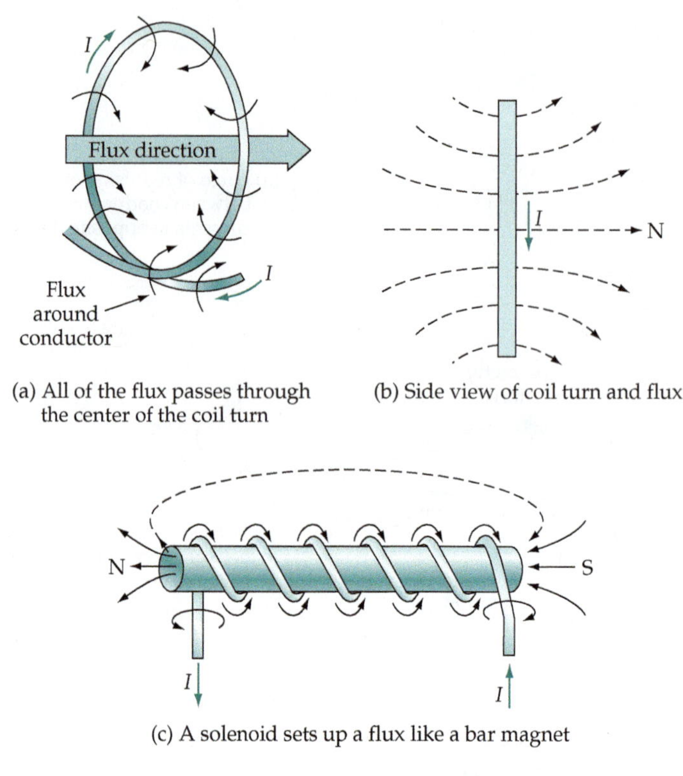

(a) All of the flux passes through the center of the coil turn

(b) Side view of coil turn and flux

(c) A solenoid sets up a flux like a bar magnet

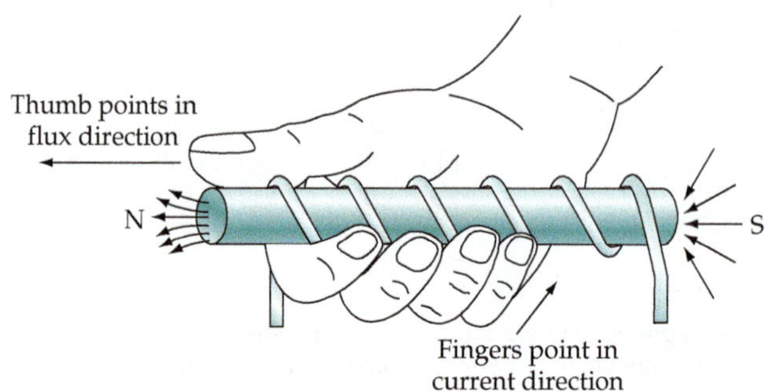

(d) Right-hand rule for solenoid flux direction

Figure 11-8 In current-carrying coils, the magnetic lines of force around the conductors all pass through the center of the coil.

It is important to note that the emf within the coil (which produces the current flow) is generated only when the magnetic field is in motion with respect to the coil. When both the field and the coil are stationary, no emf is produced.

Now consider Figure 11-9(b), which shows a solenoid placed close to another coil. Both the solenoid and the other coil are stationary. The solenoid also has a battery and a switch that can be closed to pass current through the turns of the solenoid.

When the switch is open, there is obviously no current flow, and no flux to generate an emf in the second coil. However, with the switch closed, current

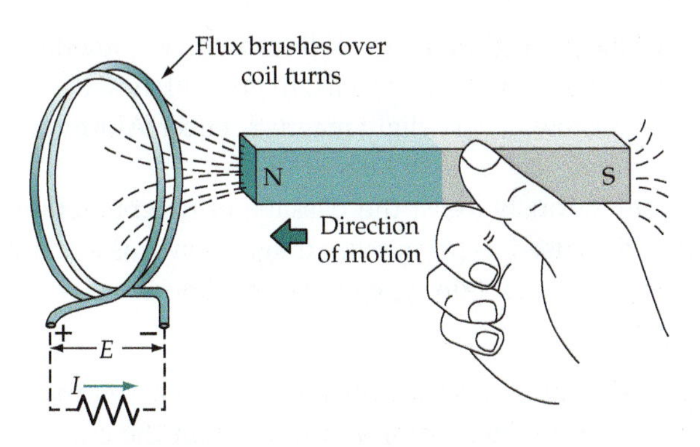

(a) An emf induced in a coil by the motion of the flux from the bar magnet

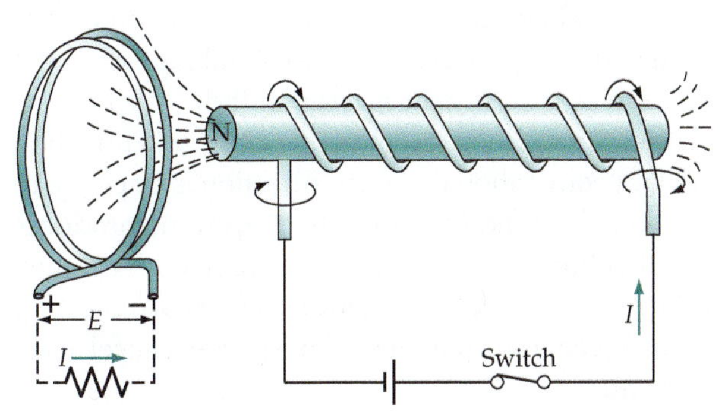

(b) An emf induced in a coil by the motion of the flux from a solenoid when the current is switched on or off

Figure 11-9 An electromotive force (emf) is induced in a coil when the coil is brushed by a magnetic field. The magnetic field may be from a bar magnet or from a current-carrying coil.

flows through the solenoid, and the flux grows from zero to its maximum level. During the time that the flux is growing, it is moving out from the solenoid and brushing over the coil. Consequently, an emf is generated in the coil. When the flux has reached its maximum level, it becomes stationary and is no longer able to generate the emf in the coil. If the switch is now opened, the flux falls to zero again, and in doing so it once more brushes over the conductors of the coil. Again, an emf is generated in the coil while the flux is in motion.

11-3 THEORY OF MAGNETISM

In relation to magnetism, all materials can be categorized into four groups:

- *Nonmagnetic* materials have no more effect on a magnetic field than air or vacuum. This group includes such materials as wood and rubber.
- *Diamagnetic* materials exhibit a very slight opposition to magnetic lines of force. They tend to be repelled from both poles of a magnet, and if a bar of diamagnetic substance is suspended in a magnetic field, it tends to align itself at right angles to the field. The effect is so slight that these materials are usually classified as nonmagnetic. Copper and silver are diamagnetic.

- *Paramagnetic* materials assist the passage of magnetic lines of force. Sometimes this term is used to include ferromagnetic materials. In general, however, it is applied only to material that shows a very slight magnetic effect. Aluminum and platinum are paramagnetic.
- *Ferromagnetic* materials tremendously assist the passage of magnetic lines of force. These are the materials used as permanent magnets and as electromagnets. Iron, nickel, cobalt, and a certain type of ceramic known as *ferrite* are all ferromagnetic materials.

It was shown in Section 11-2 that a current flowing through a conductor produces a magnetic field around the conductor and that when the current-carrying conductor is bent in a circle, a magnetic flux with a definite direction is generated. Now recall that current flow is in fact electron motion. So, it follows that an electron in orbit around an atom can be thought of as a circular current flow and that a magnetic flux is produced by the orbital motion of the electrons. This effect is referred to as the *magnetic moment* of the electron.

Figure 11-10(a) shows two electrons orbiting in the same plane and direction around adjacent atoms. Note that conventional current direction is opposite to that of the electron motion. The orbital motion of the two electrons produces magnetic fluxes in the same direction, and these fluxes reinforce each other. Now look at Figure 11-10(b). In this case the two electrons are shown orbiting in opposite directions. So, they generate magnetic fluxes that cancel each other, and the net result is zero flux.

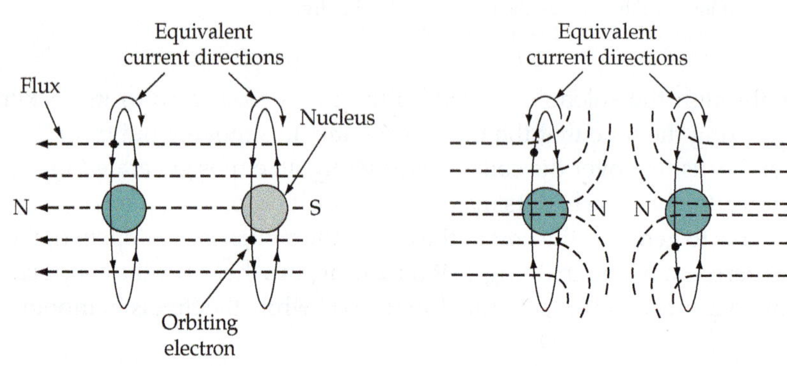

Figure 11-10 The magnetic moment of electrons in adjacent atoms may assist each other to establish a magnetic flux or oppose and cancel each other's magnetic effect.

Within nonmagnetic material, it can be assumed that all the magnetic effects of orbiting electrons add up to zero and that the condition is unchanged by the influence of a magnetic field. Diamagnetic materials tend to be affected by a magnetic field to the extent that they set up their own very weak magnetic flux that opposes the external magnetic field. The effect is almost undetectable, but it is fair to assume that the external magnetic field causes an alignment of some orbital electrons and that the combined effect of these produces the opposing flux. With paramagnetic materials that exhibit very weak magnetization, it may be assumed that some orbiting electrons are aligned in such a way that they produce a magnetic flux in the same direction as the external flux.

In ferromagnetic materials, groups of atoms seem to act together. Within these groups, large numbers of orbital electrons are aligned in such a way that their magnetic fields reinforce each other. The groups of atoms are termed *magnetic domains*, and while the material is unmagnetized, the net effect of the magnetic domains is to cancel each other completely. When an external magnetic field is applied to ferromagnetic material, the resulting alignment of the magnetic domains produces a tremendous increase in the magnetic flux. In the case of soft iron, the flux reduces almost to zero when the external field is removed. So, the magnetic domains appear to return (almost) to their original state. With hard steel and certain alloys of iron and nickel, the magnetic domains apparently remain realigned, and the steel continues to produce a magnetic flux even when the external field is removed. In this case the metal has become a *permanent magnet*.

11-4 MAGNETIC FLUX AND FLUX DENSITY

The total lines of force in a magnetic field are referred to as the *magnetic flux* (symbol Φ), and the flux per unit cross-sectional area of the field is termed *the flux density* (symbol B), (see Figure 11-11).

The *weber (Wb)* is the SI unit of magnetic flux. The weber is defined as the magnetic flux that, linking a single-turn coil, produces an emf of 1 V when the flux is reduced to zero at a constant rate in 1 s.

The *tesla (T)* is the SI unit of magnetic flux density. The tesla is the flux density in a magnetic field when I Wb of flux occurs in a plane of 1 m^2; that is, the tesla can be described a 1 Wb/m^2.

The relationship between flux and flux density is stated by the equation

$$B = \frac{\Phi}{A} \qquad (11\text{-}1)$$

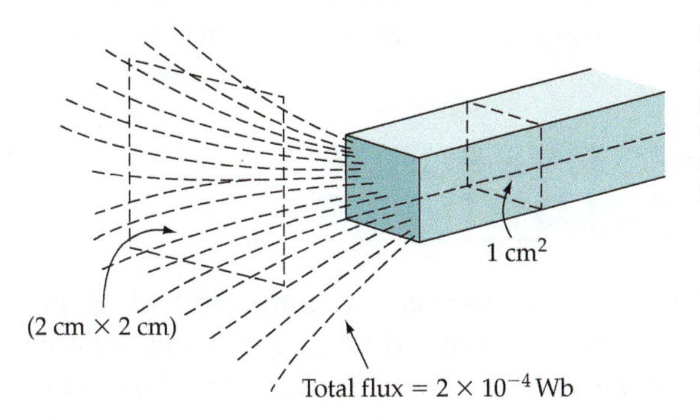

Figure 11-11 Because the magnetic flux spreads over a larger area as it is emitted from a pole of a bar magnet, the external flux density is less than that within the magnet.

In Equation 11-1, B is the flux density in *teslas* (T), when Φ is the total flux in webers, and A is the cross-sectional area in m^2.

The flux density is frequently different from one point to another in the same magnetic field. For example, in the case of the bar magnet in Figure 11-11, the flux density is obviously greatest at points close to the poles of the magnet, or within the metal.

Example 11-1

The total flux emitted from the pole of a bar magnet is 2×10^{-4} Wb, (see Figure 11-11).

(a) If the magnet has a cross-sectional area of 1 cm^2, determine the flux density within the metal.

(b) If the flux spreads out so that at a certain distance from a pole it is distributed over an area of 2 cm by 2 cm, find the flux density at that point.

Solution

(a) Flux density within the metal:

Eq. 11-1,

$$B = \frac{\Phi}{A} = \frac{2 \times 10^{-4}\ \text{Wb}}{1 \times 10^{-4}\ \text{m}^2}$$

$$= 2\ \text{T}$$

(b) Flux density away from the pole:

$$B = \frac{\Phi}{A} = \frac{2 \times 10^{-4}\ \text{Wb}}{(2 \times 10^{-2})^2\ \text{m}^2}$$

$$= 0.5\ \text{T}$$

Practice Problems

11-4.1 Determine the flux density within a coil with a cross-sectional area of 9 cm^2 if the total flux is 18 μWb. Calculate the new flux density if the coil diameter is doubled and the total flux is unchanged.

11-4.2 The flux density of 1.7 T measured in the air gap between two magnetic poles is found to be constant over an area of 3 cm^2. If the poles each have a cross-sectional area 2.2 cm, determine the flux density within the poles.

11-5 MAGNETOMOTIVE FORCE AND MAGNETIC FIELD STRENGTH

Just as an electric current is the result of an electromotive force (emf) acting on the electric circuit, so a magnetic flux is produced by a *magnetomotive force* (mmf), symbol F_m, acting on the magnetic circuit. In the case of a solenoid or any other current-carrying coil, the magnetomotive force is the product of the

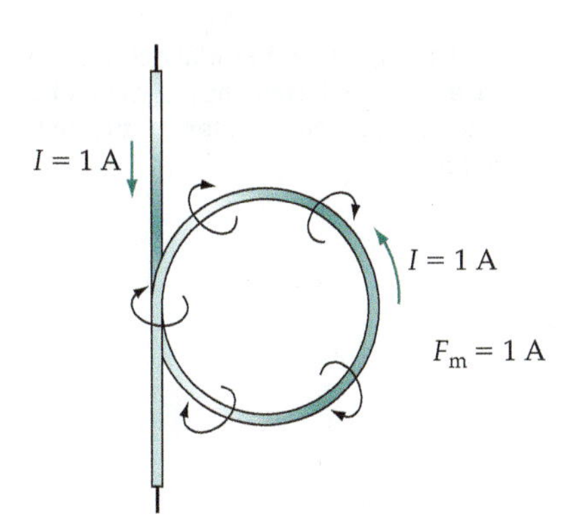

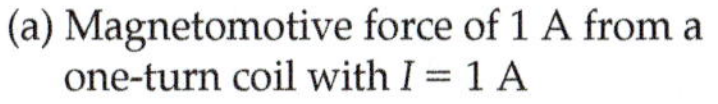
(a) Magnetomotive force of 1 A from a one-turn coil with $I = 1$ A

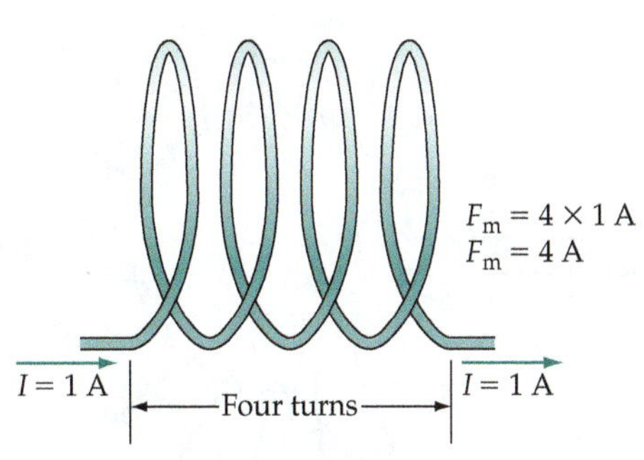

(b) Magnetomotive force of 4 A from a four-turn coil with $I = 1$ A

Figure 11-12 The magnetomotive force (mmf) that generates a magnetic flux depends on coil current and number of coil turns.

current and the number of turns on the coil. Figure 11-12(a) shows a one-turn coil carrying a current of 1 A. All of the flux set up by the 1 A passes through the center of the coil. Therefore, the magnetomotive force is,

$$\begin{aligned} F_m &= 1\text{ A} \times 1\text{ turn} \\ &= 1\text{ A} \end{aligned}$$

In the case of the four-turn coil with a current of 1 A [Figure 11-12(b)], the effect of the current is multiplied four times, because each turn of the coil generates a flux that passes through the center of the coil. Therefore, the magnetomotive force is,

$$\begin{aligned} F_m &= 1\text{ A} \times 4\text{ turns} \\ &= 4\text{ A} \end{aligned}$$

In general, for a coil of N turns carrying a current of I amperes, the magnetomotive force is,

$$F_m = NI\text{ amperes} \tag{11-2}$$

The more descriptive term, *ampere-turns*, is sometimes used as the units of mmf. However, the approved SI unit of mmf is the ampere.

The quantity of flux that can be set up in a magnetic circuit is, of course, proportional to the mmf. It is also inversely proportional to the length of the magnetic circuit. Consider the *toroidal* coil shown in Figure 11-13. The length of the magnetic path is l, as illustrated. If l is very large, the mmf has to act over a long distance, and if l is very small, the mmf acts over a short distance. Obviously, the greatest intensity of magnetic flux occurs for the shortest magnetic path. This gives rise to the expression for *magnetic field strength:*

$$H = \frac{F_m}{l}$$

or

$$H = \frac{NI}{l} \tag{11-3}$$

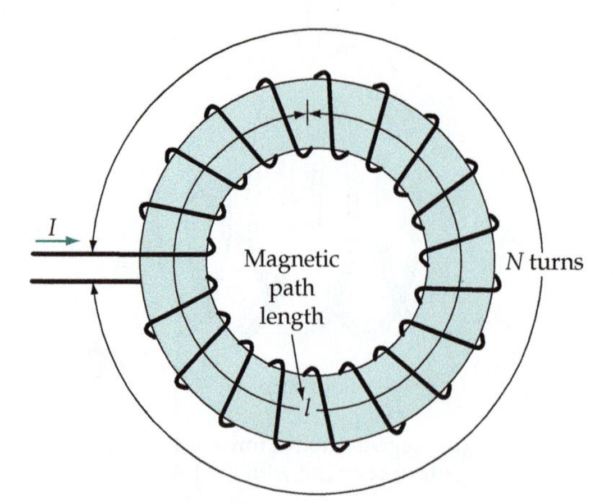

Figure 11-13 A toroidal coil wound on a circular former has a magnetic path through the center of the ring material.

When I is in amperes, N is number of turns, and l is in meters, the units of magnetic field strength are *amperes per meter* (A/m). Equation 11-3 for magnetic field strength is analogous to the equation for electric field strength, (Equation 4-1).

Example 11-2

The torodial coil shown in Figure 11-13 has 100 turns and carries a current of 0.5 A. If the length of the magnetic circuit is 10 cm, determine the mmf and the magnetic field strength.

Solution

Eq. 11-2,

$$F_m = NI = 100 \times 0.5\text{ A}$$
$$= 50\text{ A}$$

Eq. 11-3,

$$H = \frac{NI}{l} = \frac{100 \times 0.5\text{ A}}{10 \times 10^{-2}\text{ m}}$$
$$= 500\text{ A/m}$$

Practice Problems

11-5.1 A toroidal coil with 190 turns has a magnetic path length of 3.8 cm. Determine the current required to set up a magnetic field strength of 175 A/m.

11-5.2 A 10 mA current is to be passed through a toroidal coil that has a 4.5 cm magnetic path length. Calculate the number of coil turns required if the magnetic field strength is to be 120 A/m.

11-6 FORCE ON CURRENT-CARRYING CONDUCTORS

Force on a Conductor

Figure 11-14(a) shows an end view of a current-carrying conductor in a magnetic field. With the current direction as indicated by the + sign (i.e., away from the viewer), the flux set up by the current is in a clockwise direction around the conductor. When the field flux has the vertically down direction indicated, the flux from the conductor tends to assist the field flux on the right-hand side of the conductor. The conductor flux also opposes (or weakens) the field flux on the left-hand side of the conductor. Because the magnetic lines of force always tend to be in tension, those lines of force on the right-hand side of the conductor tend to straighten out. The effect of this is to produce a force that pushes the conductor to the left.

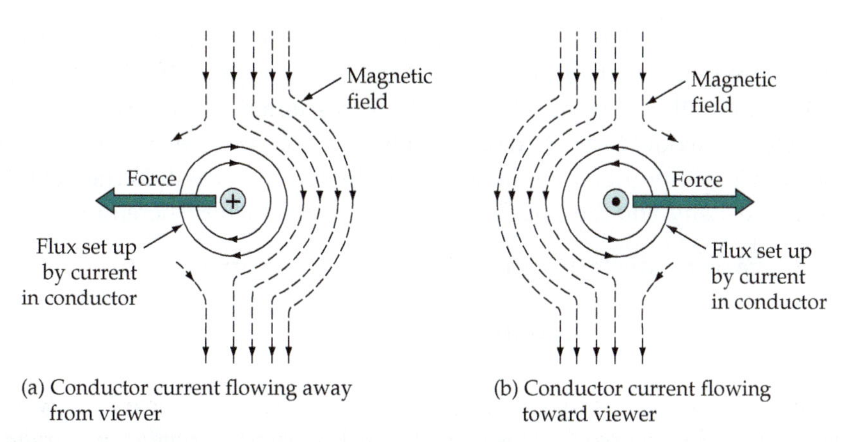

Figure 11-14 A current-carrying conductor situated in a magnetic field experiences a force due to the interaction of the conductor flux and the field flux. The conductor flux reinforces the field flux on one side of the conductor and weakens it on the other side.

When the current direction is reversed, as illustrated in Figure 11-14(b), the flux around the conductor is in a counterclockwise direction. So, the field flux is assisted on the left-hand side of the conductor and weakened on its right side. The result is a force that pushes the conductor to the right.

The effect is further illustrated in Figure 11-15, where the field flux is shown vertically downward. The current direction is such that the field set up around the conductor weakens the main field in front of the conductor and strengthens it behind the conductor. The result, as illustrated, is that a force acts to push the conductor forward. *Note that the conductor must be at right angles to the direction of the magnetic flux; otherwise, maximum force is not generated.*

The magnitude of the force acting on a current-carrying conductor in a magnetic field is proportional to the flux density of the field and to the density of the flux set up by the conductor. Because the flux set up around the conductor is

Figure 11-15 The force on a current-carrying conductor situated in a magnetic field is proportional to the current, the flux density of the field, and the length of the conductor within the field.

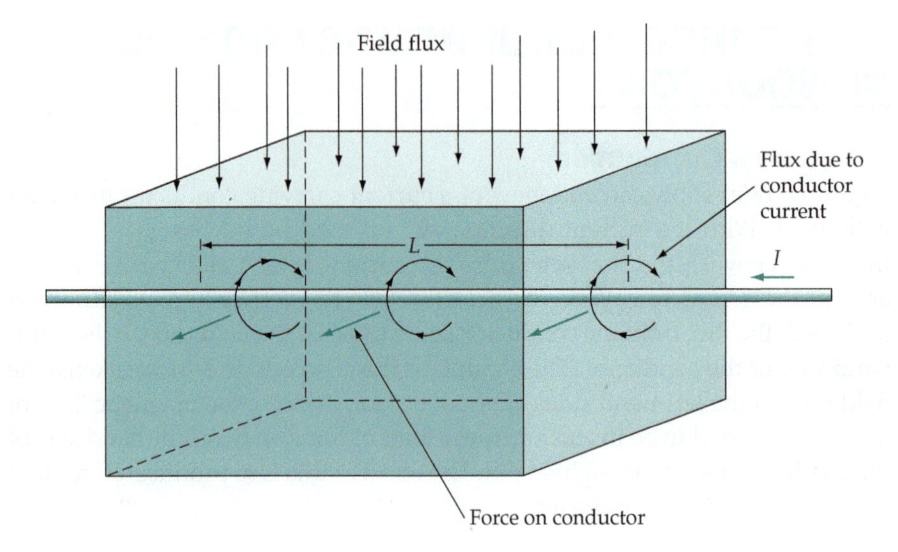

directly proportional to the current, the force on the conductor is proportional to the current and to the flux density of the magnetic field. If a 2 m length of conductor is within the magnetic field, the force exerted on it is obviously twice the force that would be exerted on a 1 m length. So, the force on the conductor is seen to be proportional to the length of the conductor within the field, the field flux density, and the current carried by the conductor. Therefore,

$$\text{(force on conductor)} = \text{(field flux density)} \times \text{(current)} \times \text{(length)}$$

or

$$F = BI\ell \qquad \textbf{(11-4)}$$

When B is in teslas, I is in amperes, and ℓ is in meters, the force F is given in *newtons* (N). So, *a force of 1 N is exerted on a 1 m length of conductor carrying a current of 1 A and situated in a magnetic field having a flux density of 1 T.*

Example 11-3

A conductor carrying a current of 15 A is situated at right angles to a magnetic field with a flux density of 0.7 T, (see Figure 11-15). If the length of the conductor within the field is 20 cm, determine the force on the conductor, Also calculate the new current level required to increase the force to 25 N.

Solution

Eq. 11-4

$$F = BI\ell = 0.7\text{ T} \times 15\text{ A} \times 20 \times 10^{-2}\text{ m}$$

$$= 2.1\text{ N}$$

$$I = \frac{F}{B\ell} = \frac{25\text{ N}}{0.7\text{ T} \times 20 \times 10^{-2}}$$

$$= 179\text{ A}$$

Force on a Coil

In Figure 11-16(a) a single-turn coil is shown pivoted in the magnetic field set up by N and S poles. Consideration of the current direction through the coil shows that there is a downward force is on its left-hand side and an upward force on its right-hand side [see Figure 11-16(b)]. The coil length within the magnetic field is ℓ meters, and (from Equation 11-4) the force on each side of the coil is,

$$F = BI\ell$$

For a coil with N turns, $$F = BI\ell N$$

Because the force acts at a radius (r), there is a *torque* on each side of the coil.

$$\text{torque on each side} = F \times r$$
$$= BI\ell Nr$$

With B is in teslas, I in amperes, ℓ in meters, and r in meters, the torque is expressed in *joules* (J). For the two sides of the coil the total torque is,

$$\text{Total torque} = 2\,BI\ell Nr$$

The coil diameter D equals 2 r.

So $$\text{Torque} = B\ell IND \qquad \textbf{(11-5)}$$

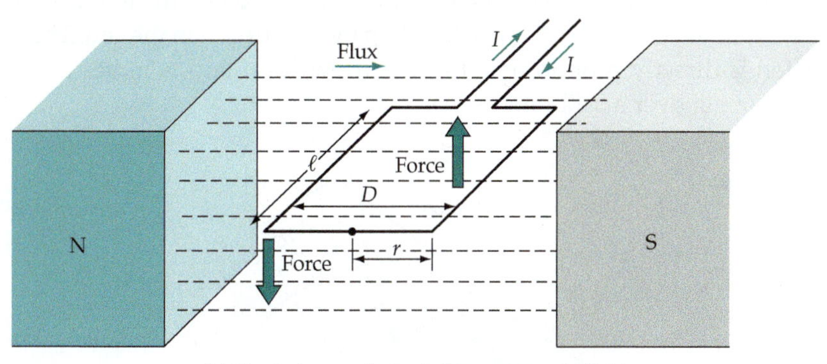

(a) Single-turn coil pivoted in a magnetic field

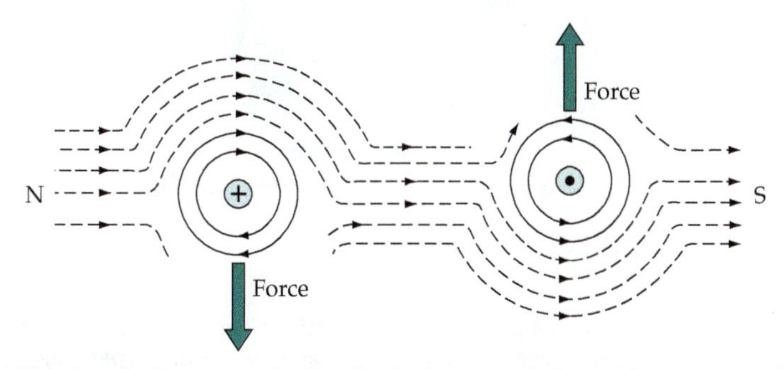

(b) Showing the force on each side of a single-turn coil pivoted in a magnetic field

Figure 11-16 A force is exerted on each side of a current-carrying coil pivoted in a magnetic field. This force tends to cause the coil to rotate.

Example 11-4

Calculate the torque acting on a 100-turn coil pivoted in a magnetic field having a flux density of 0.5 T. The current through the coil is 100 mA, its radius is 1 cm, and its axial length is 2 cm.

Solution

Eq. 11-5

$$\text{Torque} = B\ell IND$$

$$= 0.5\ \text{T} \times 2 \times 10^{-2} \times 100\ \text{mA} \times 100 \times 2 \times 10^{-2}$$

$$= 2 \times 10^{-3}\ \text{J}$$

PMMC Instruments

Figure 11-17 shows the basic construction of a *permanent magnet moving-coil* (PMMC) instrument as used in electromechanical ammeters, voltmeters, and ohmmeters. The instrument consists basically of a small lightweight coil pivoted within the field of a permanent magnet. A cylindrical soft iron core is included between the poles of the magnet so that the coil rotates in the narrow air gap between the pole surfaces and the core. When a current is passed through the coil windings, a torque is exerted on the coil by the interaction of the field from the magnet and the field set up by the current in the coil. The coil deflection is directly proportional to the current level and is indicated by a pointer moving over a calibrated scale.

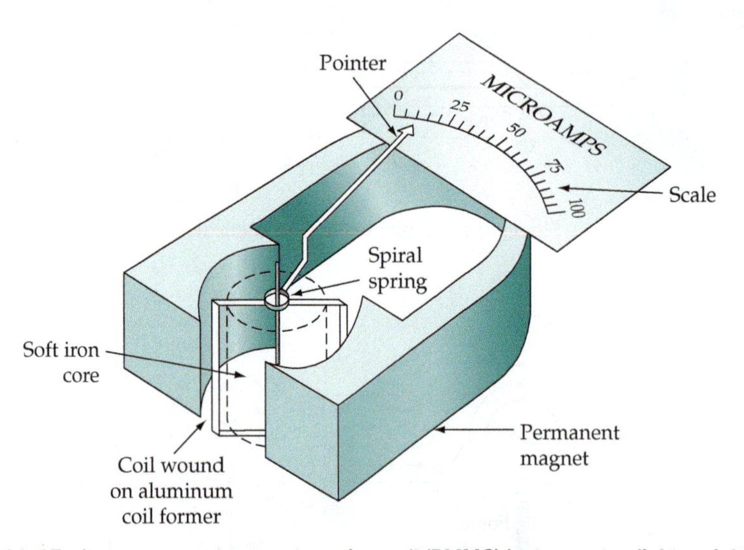

Figure 11-17 In a permanent magnet moving-coil (PMMC) instrument, a light-weight coil is pivoted to move in the air gap between a soft iron core and the poles of a permanent magnet. Current flowing in the coil produces the deflecting force, and spiral springs provide a controlling force. A pointer attached to the coil indicates the amount of deflection.

In addition to a *deflecting force* provided by the coil current and the field from the permanent magnet, a *controlling force* is needed. This is the force that returns the coil and pointer to the zero position when no current is flowing through the coil. The controlling force also balances the deflecting force, so that the pointer remains stationary for any constant level of current through the coil. The controlling force is usually provided by spiral springs (see Figure 11-17) which also function as connecting leads for conducting current through the coil.

Another force, known as a *damping force,* is required for correct operation of a deflection type instrument. When no damping force is present, the pointer swings above and below its final position on the scale for some time before settling down. In the case of the PMMC instrument, the damping force uses *eddy currents,* (see Section 12-7). The coil is wound on an aluminum frame or coil former in which eddy currents are generated by any rapid movement of the coil in the magnetic field. The eddy currents set up a magnetic flux that opposes the original movement that generates them, and this damps the pointer oscillations.

The simple arrangement illustrated in Figure 11-16 is also the principle behind the electric motor, as illustrated in Figure 11-18.

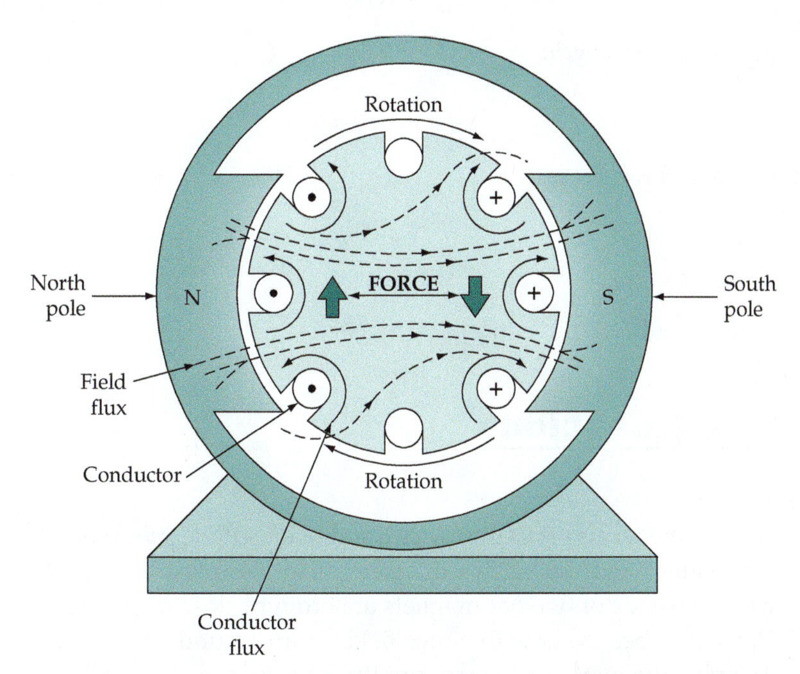

Figure 11-18 Cross section of a basic electric motor. The outer shell of the motor is a ring-shaped magnet with inward-pointing N and S poles. A cylindrical soft iron core is pivoted between the poles, and conductors are located in slots on the periphery of the core. Current flows through the conductors in one direction when passing the N pole, and in the reverse direction when passing the S pole. This produces an upward force on left side of the pivoted core, and a downward force on right side, resulting in a clockwise rotation of the core.

Practice Problems

11-6.1 A conductor situated at right angles to a magnetic field experiences a force of 0.083 N. If the current is 286 mA and the conductor length within the field is 5.8 cm, calculate the flux density of the field.

11-6.2 A torque of 3.3 mJ is to be exerted on a 75-turn coil pivoted in a magnetic field that has a flux density of 1.5 T. If the axial length of the coil is 1.65 cm and its diameter is 2 cm, determine the required current.

Summary of Formulas

- ***Flux density:***

$$B = \frac{\Phi}{A}$$

- ***Magnetomotive force:***

$$F_{m} = NI$$

- ***Magnetic field strength:***

$$H = \frac{NI}{\ell}$$

- ***Mechanical force:***

$$F = BI\ell$$

- ***Mechanical torque:***

$$\text{Torque} = B\ell IND$$

Review Questions

Section 11-1

11-1 Discuss the terms *north-seeking pole* and *south-seeking pole* in relation to a bar magnet and the earth's magnetic field. Also, explain what occurs when the ends of two bar magnets are brought close together.

11-2 Draw sketches to show the force field set up around a bar magnet, and describe two methods of mapping the force field.

11-3 Define magnetic lines of force and magnetic flux, and list the characteristics of magnetic lines of force. Also, explain how their direction is defined.

11-4 Sketch the magnetic fields that occur at the following arrangements: (a) two adjacent N poles of bar magnets, (b) two adjacent S poles, (c) adjacent N and S poles, (d) A bar magnet with a piece of soft iron in its field.

Section 11-2

11-5 Sketch the shape of the magnetic field around a current-carrying conductor, and show how the direction of the field is related to the direction of the current in the conductor.

11-6 Sketch the magnetic field that occurs around two adjacent current-carrying conductors: (a) for currents in opposite directions, (b) for currents in the same direction. Explain how a force occurs between the conductors, and indicate the direction of the force in each case.

11-7 Draw sketches to show the form of magnetic field that occurs with (a) a single-turn coil with current flowing through it, and (b) a long current-carrying solenoid. Sketch a toroid-shaped coil, and show what its magnetic field should look like.

11-8 Discuss electromagnetic induction, and draw sketches to explain the principle.

Section 11-3

11-9 Define the following classifications of materials: nonmagnetic, diamagnetic, paramagnetic, and ferromagnetic.

11-10 Using illustrations, explain the theory of magnetism.

Section 11-4

11-11 Define the weber and the tesla, and state their relationship.

Section 11-5

11-12 Define magnetomotive force and magnetic field strength, and explain how the two are related.

Section 11-6

11-13 Draw illustrations to show that a force is exerted on a current-carrying conductor situated in a magnetic field. Show how the direction of the force is related to the directions of the current and the magnetic field.

11-14 Discuss the relationships between the force on a current-carrying conductor in a magnetic field and the factors responsible for the force. Also, derive the equation for the force, and state the units for each factor.

11-15 Derive the expression for the torque on a pivoted coil situated within a magnetic field.

11-16 Draw a sketch to illustrate the construction of a permanent magnet moving coil instrument. Define the three forces involved in the instrument operation and briefly explain.

Problems

Section 11-4

11-1 The total flux emitted from the pole of a magnet is 0.5 μWb. (a) If the magnet has a cross-sectional area of 1.5 cm^2, determine the flux density within the metal. (b) Calculate the flux density a short distance from the pole if all of the flux is contained within an area 5 cm by 5 cm.

11-2 A circular magnetic pole with a diameter of 5 cm emits a total flux of 120 μWb. Calculate the flux density within the pole. Also, calculate the flux density a short distance from the pole if all of the flux is retained within an area 10% larger than the pole area.

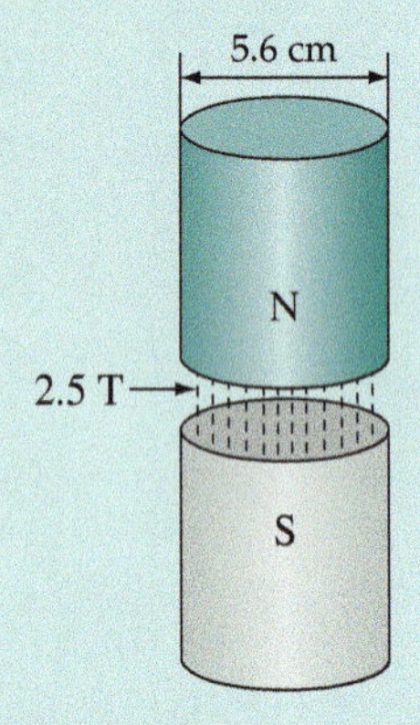

Figure 11-19

11-3 The flux density in an air gap between two N and S poles is 2.5 T. The poles are circular with a diameter of 5.6 cm (see Figure 11-19). Calculate the total flux crossing the air gap.

11-4 In Problem 11-3, allow for the flux spreading over an area 10% larger than the pole area. Recalculate the total flux and the flux density within the poles.

11-5 The cross-sectional area of a coil is 15 cm^2 and the flux density within it is 30 mT. Calculate the total flux.

11-6 If the diameter of the coil in Problem 11-5 is reduced by one third, determine the new level of flux density required to keep the total flux unchanged.

11-7 The flux density within a coil is 45 mT, and the total flux is 13 μWb. If the coil has a square cross section, calculate its dimensions.

11-8 The iron ring in Figure 11-20 has two air gaps, *a* and *b*. At gap *a*, the metal has a circular cross-section 1 cm in diameter. The cross-section at gap *b* is square with 1 cm sides. The flux density at gap *a* is measured as 20 mT. Calculate the flux density at gap *b*.

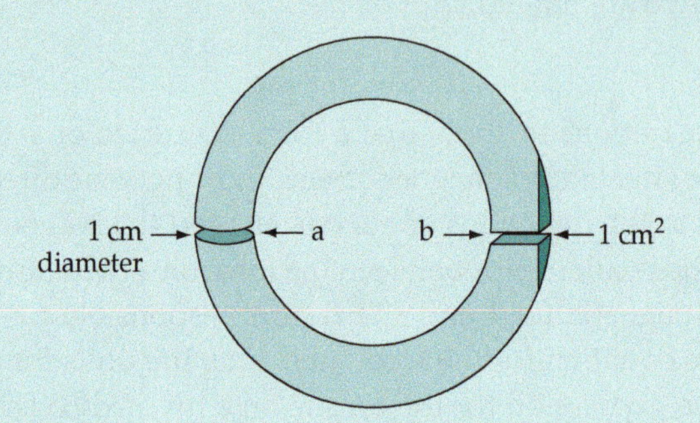

Figure 11-20

11-9 A magnetic pole with a cross-sectional area of 9 cm^2 has a flux density of 2 T. Calculate the total flux emitted from the pole.

11-10 For Problem 11-8, recalculate the flux density at gap *b* if the flux spreads out at each gap over an area 10% larger than the pole face area.

11-11 A magnetic pole with a cross-sectional area 3 cm by 3 cm has a flux density of 1.11 mT. Calculate the flux density in an area 4 cm by 4 cm a short distance from the pole if 90% of the total flux passes through this area.

11-12 The flux density within a long coil is measured as 0.4 μT at a point close to one end and as 0.48 μT at the middle of the coil, (see Figure 11-21). If the diameter of the coil is exactly 2.5 cm at the end, determine its diameter at the middle.

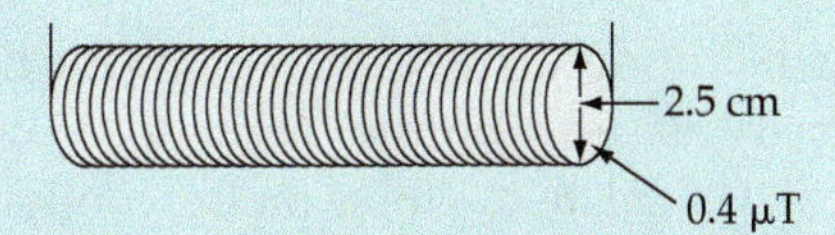

Figure 11-21

Section 11-5

11-13 A toroid-shaped coil with 280 turns and a magnetic path length of 50 cm has a current of 7.5 A. Calculate the mmf and magnetic field strength.

11-14 A 500-turn toroidal coil with a current of 33 mA has a magnetic field strength of 110 A/m. Calculate the magnetic path length.

11-15 A toroidal coil has a magnetic path length of 33 cm and a magnetic field strength of 650 A/m. The coil current is 250 mA. Determine the total number of coil turns.

11-16 The coil current in Problem 11-15 is reduced to 180 mA. Calculate the new magnetic field strength.

11-17 Two coils wound on a single ring-shapted core (as in Figure 11-22) produce mmf in the same direction over a common magnetic path length of 45 cm. Coil *a* has 1500 turns and a current of 80 mA. Coil *b* has 1000 turns. If the magnetic field strength is 400 A/m, determine the current flowing in coil *b*.

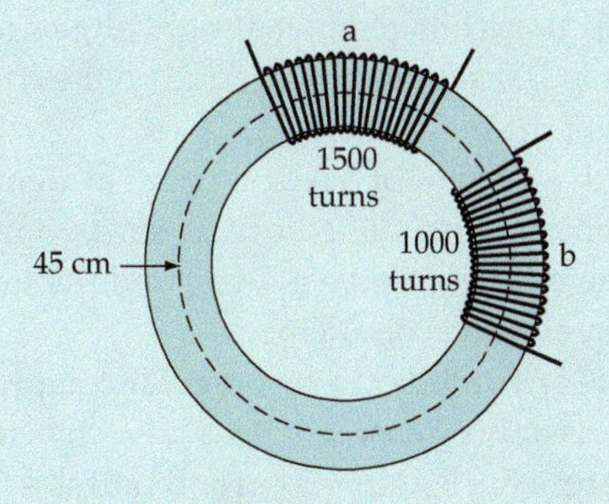

Figure 11-22

11-18 For Problem 11-17, determine the new magnetic field strength if the current in coil *a* is reversed.

11-19 For Problem 11-17 with the current in coil *a* reversed, determine the level of coil *a* current that will completely cancel the effects of the current in coil *b*.

11-20 A toroidal coil with a 9 cm magnetic path length has a center-tapped coil with a total of 350 turns. One half of the coil has a current of 15 mA, and the other half has a 25 mA current. Both currents are in the same direction. Calculate the magnetic field strength.

Section 11-6

11-21 A conductor situated at right angles to a magnetic flux has a current of 5 A. The flux density is 1.5 T, and the conductor length within the magnetic field is 12 cm. Calculate the force on the conductor.

11-22 For Problem 11-21, determine the new flux density required to maintain the force unchanged when the current is reduced to 500 mA.

11-23 Two 25 cm conductors are situated at right angles to a magnetic flux with a flux density of 500 mT. Conductor a has a current of 100 A, and the current in conductor b is 39 A. Calculate the force exerted on each conductor.

11-24 For Problem 11-23, the length of conductor b within the magnetic field is increased to 33 cm. Calculate the new level of current that must flow in conductor a if the total force on the two conductors is to remain unchanged.

11-25 A 600-turn coil is pivoted within a magnetic field with a flux density of 2.2 T. The coil current is 50 μA, its radius is 0.75 cm, and its axial length is 1 cm. Calculate the torque acting on the coil.

11-26 For the coil in Problem 11-25, determine the level of coil current to produce a torque of 6 μJ.

11-27 A 320-turn coil with a radius of 1.5 cm and an axial length of 3 cm is situated in a magnetic field with a flux density of 5 T. If the torque on the coil is to be 1.5 mJ, determine the required coil current.

11-28 A 500-turn coil with a radius of 1 cm and an axial length of 2.5 cm is pivoted between the poles of a magnet at right angles to the magnetic flux. The coil current is 100 μA, and the torque on the coil is 50 μJ. Calculate the flux density of the magnetic field.

11-29 A conductor with a current of 8 A is situated at right angles to the magnetic flux crossing a short air gap between two cylindrical magnetic poles. The diameter of each pole is 7 cm, and the total flux is 125 μWb. Calculate the force on the conductor.

11-30 A torque of 33 mJ is exerted on a rectangular coil pivoted within a magnetic field that has a flux density of 0.9 T. The coil current is 25 mA, and its dimensions are: axial length = 3 cm and radius = 2 cm. Determine the total number of coil turns.

11-31 If the coil in Problem 11-27 has 75 of its turns wound in the opposite direction to the other turns, calculate the new level of coil current that must flow for the conditions described.

11-32 Two insulated conductors (a and b) are contained within a single electric cable. Conductor a has a current of 25 A, and the current in conductor b is 9.5 A. The cable is 15 cm in length, and is situated in a magnetic field with a flux density of 0.85 T. Calculate the force on the cable: (a) when the conductor currents are in the same direction, (b) when the currents are in opposite directions.

Practice Problem Answers

11-4.1 20 mT, 5 mT
11-4.2 2.32 T
11-5.1 35 mA
11-5.2 540 turns
11-6.1 5 T
11-6.2 88.9 mA

CHAPTER 12
Magnetic Circuits

CONTENTS

Objectives

You will be able to:

1. Define reluctance and permeability and explain the terms: permeability of free space, and relative permeability.
2. Calculate the magnetic field strength around a conductor when a current flows in the conductor.
3. Calculate the field strength, flux density, and total flux in an air-cored toroid.
4. Determine the magnetomotive force required to set up a given flux within an air gap.
5. Determine the field strength, flux density, and total flux in an air-cored solenoid and explain the result of introducing an iron core into the solenoid.
6. For various-shapes of magnetic cores, determine the coil current required to set up a given flux within the core or within an air gap in the core.
7. Calculate the energy stored in an air gap and determine the current that must flow in the coil of a given electromagnet in order to lift iron bars and plates.
8. Sketch and explain typical magnetization curves and hysteresis loops for ferromagnetic materials.

INTRODUCTION

Magnetic circuits are in many ways analogous to electrical circuits. Magnetomotive force, magnetic flux, and reluctance are the counterparts of electromotiveforce, current, and resistance, respectively. Electrical conductivity also has its magnetic analog in permeability. For all nonmagnetic materials, the permeability has a very small fixed value, known as the permeability of free space. Ferromagnetic materials have relative permeability values that relate their magnetic properties to the pemeability of free space.

Magnetic circuits made up of cores with different cross-sectional areas, and perhaps an air gap, are termed composite magnetic circuits. The magnetomotive force required to set up a given flux is determined for each section of the circuit and then the quantities are added together to find the total mmf for the circuit. Each type of ferromagnetic material has its own particular magnetization curve. Because of the hysteresis effect, the magnetization curves change to loops when the flux within a sample of ferromagnetic material is reversed several times. This gives rise to a loss of energy in magnetic cores.

12-1 RELUCTANCE AND PERMEABILITY

Recall from Section 4-3 that the resistance of a conductor can be calculated using Eq. 4-2:

$$R = \frac{\rho \ell}{A}$$

where ρ is the resistivity of the conducting material, ℓ is the length of the conductor, and A is its cross-sectional area [see Figure 12-1(a)].

The magnetic circuit analog of the resistance of electrical circuits is termed *reluctance*. Therefore, *reluctance is a measure of the opposition offered by a magnetic circuit to the setting up of flux*, just as resistance is opposition to current flow in an electrical circuit. The corresponding equation for reluctance is,

$$R_m = \frac{1}{\mu} \times \frac{l}{A} \tag{12-1}$$

In Equation 12-1, l and A are length and cross-sectional area, respectively, of the magnetic circuit, and μ is the *permeability* of the material in the magnetic circuit [Figure 12-1(a)]. The reciprocal of permeability corresponds to the resistivity (ρ) of the electrical circuit. This means that magnetic permeability is the analog of electrical conductivity. Just as ρ is the resistance of a cubic meter of electrical material, so $1/\mu$ can be termed the *reluctance of a cubic meter of magnetic material.*

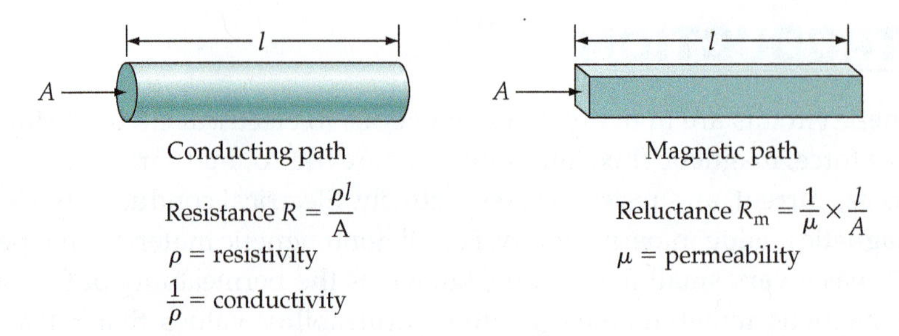

(a) Comparison of electrical resistance and magnetic reluctance

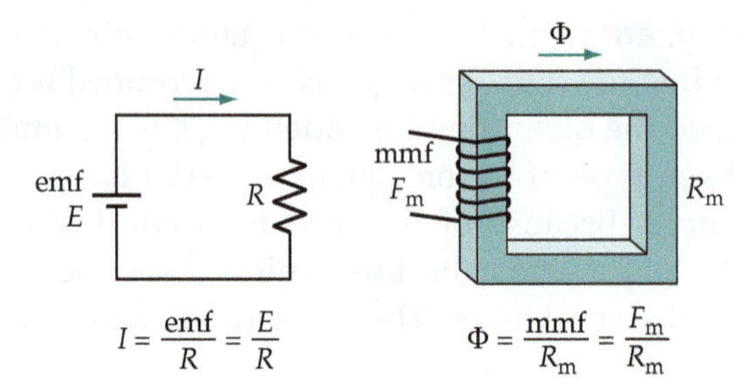

(b) Comparison of electrical and magnetic circuits

Figure 12-1 Magnetic reluctance compared to electrical resistance, and a magnetic circuit compared to an electrical circuit.

From Ohm's law, the electrical current is,

$$I = \frac{E}{R}$$

Similarly, the magnetic flux equation is,

$$\Phi = \frac{F_m}{R_m} \tag{12-2}$$

This is illustrated in Figure 12-1(b).

Rewriting the equation for reluctance,

$$R_m = \frac{F_m}{\Phi}$$

The units of reluctance are *amperes per weber* (A/Wb).

The magnetic circuit analog of conductance is *permeance,* and since conductance is the reciprocal of resistance, so permeance is the reciprocal of reluctance,

$$\text{Permanance} = \frac{1}{R_m} = \frac{\Phi}{F_m} \tag{12-3}$$

Substituting from Eqs. 11-1, 11-2 and 11-3 into the above equations gives,

$$\mu = \frac{B}{H} \tag{12-4}$$

$$\text{Permeability} = \frac{\text{Flux density}}{\text{Magnetic field strength}}$$

The SI unit for permeability is *henrys/meter* (H/m). The *henry* is the unit of inductance and is discussed in Chapter 14. However, as will be seen, μ is invariably used as a ratio.

12-2 PERMEABILITY OF FREE SPACE

Two long current-carrying conductors are illustrated in Figures 12-2(a) and (b). Conductor A is shown in Figure 12-2(b) with the magnetic flux around it in concentric circles. Conductor B is situated within the magnetic field generated

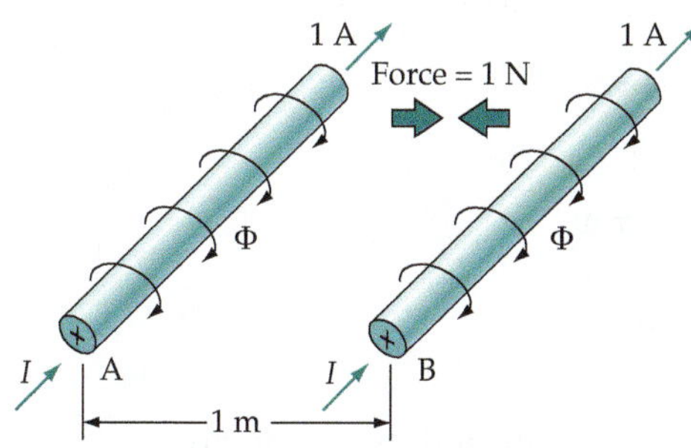

(a) Two long current-carrying conductors

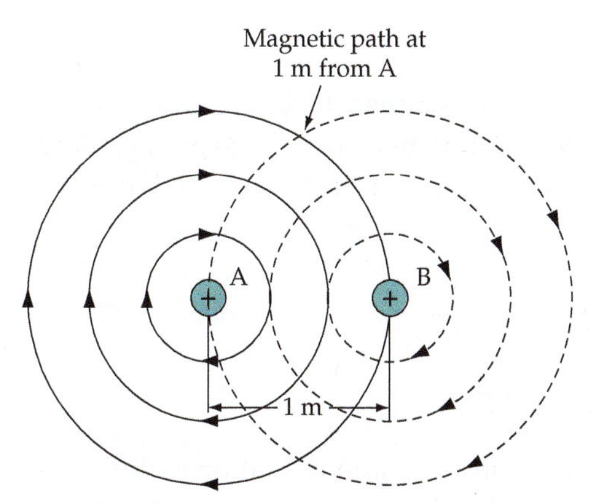

(b) Cross section of two long, current-carrying conductors

Figure 12-2 Two long conductors situated 1 m apart in air with currents of 1 A produce a (repulsive or attractive) force of 2×10^{-7} N.

by A, and therefore a force is exerted upon conductor B. The magnitude of the force can be determined using Equation 11-4:

$$F = BI\ell$$

Giving
$$B = \frac{F}{I\ell}$$

The definition of the ampere (see Appendix 2) states that the force per meter length exerted between two conductors 1 m apart when they are each carrying a current of 1 A is 2×10^{-7} N. Therefore, substituting $F = 2 \times 10^{-7}$ N, $I = 1$ A, and $\ell = 1$ m,

$$B = 2 \times 10^{-7}\ \text{T}$$

It is seen that the flux density in air (or any nonmagnetic material) at 1 m from a long conductor carrying 1 A is 2×10^{-7} T. Returning to Figure 12-2(b), the length of the magnetic path at 1 m from conductor A is the circumference of a circle with a radius of 1 m. Therefore, the length of the magnetic path is 2π meters, and the mmf is

$$F_{\text{m}} = I \times N = 1\ \text{A} \times (1\ \text{turn})$$
$$= 1\ \text{A (or 1 ampere-turn)}$$

The magnetic field strength is calculated as,

$$H = \frac{F_{\text{m}}}{l} = \frac{1\ \text{A}}{2\pi\ \text{m}}$$
$$= \frac{1}{2\pi}\ \text{A/m}$$

From Eq. 12-4, the permeability of the material around the conductor is,

$$\mu = \frac{B}{H}$$

The material around the conductor was assumed to be air; however, it could be a vacuum or any nonmagnetic material. In this case, the permeability is a constant that is designated μ_{o} and is referred to as the *permeability of free space.* Substituting for B and H gives

$$\mu_{\text{o}} = \frac{2 \times 10^{-7}\ \text{T}}{(1/2\pi)\ \text{A/m}}$$

or,
$$\mu_{\text{o}} = 4\pi \times 10^{-7}$$

The permeability of free space is employed in calculations relating to flux density and magnetic field strength in air-cored coils and in air gaps in iron cores (see Figure 12-3).

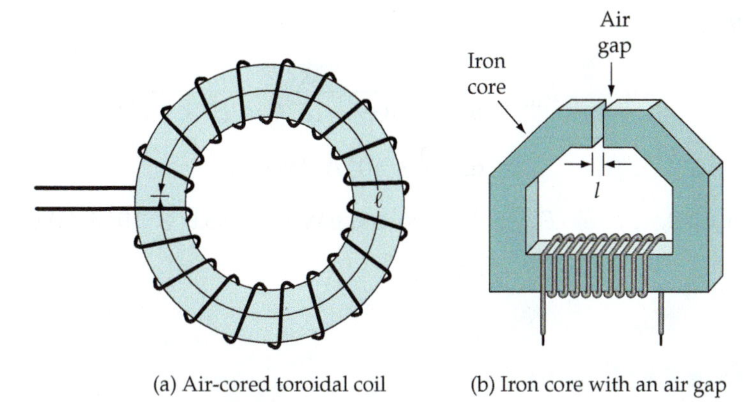

(a) Air-cored toroidal coil (b) Iron core with an air gap

Figure 12-3 The permeability of free space (μ_o) is used in calculations involving the flux density in air-cored coils, and in air gaps in iron cores.

Example 12-1

An air-cored toroidal coil of the type shown in Figure 12-3(a) has 3000 turns and carries a current of 0.1 A. The cross-sectional area of the coil is 4 cm^2, and the length of the magnetic circuit is 15 cm. Determine the magnetic field strength, the flux density, the total flux within the coil, and the reluctance of the magnetic circuit.

Solution

Eq. 11-3,
$$H = \frac{NI}{l} = \frac{3000 \times 0.1\text{ A}}{15 \times 10^{-2}\text{ m}}$$
$$= 2000\text{ A/m}$$

From Eq. 12-4,
$$B = \mu_o H = 4\pi \times 10^{-7} \times 2000\text{ A/m}$$
$$= 2.5 \times 10^{-3}\text{ T}$$

From Eq. 11-1,
$$\Phi = BA = 2.5 \times 10^{-3}\text{ T} \times 4 \times 10^{-4}\text{ m}^2$$
$$= 1\ \mu\text{Wb}$$

Eq. 12-1,
$$R_m = \frac{l}{\mu_o A} = \frac{15 \times 10^{-2}}{4\pi \times 10^{-7} \times 4 \times 10^{-4}}$$
$$= 298.4 \times 10^6\text{ A/Wb}$$

Example 12-2

Determine the mmf required to generate a total flux of 100 μWb in an air gap 0.2 cm long. The cross-sectional area of the air gap is 25 cm^2 [see Figure 12-3(b)].

Solution

Eq. 12-1, $$R_m = \frac{l}{\mu_o A} = \frac{0.2 \times 10^{-3}}{4\pi \times 10^{-7} \times 25 \times 10^{-4}}$$

$$= 636.6 \times 10^3 \text{ A/Wb}$$

From Eq. 12-2, $$F_m = \Phi R_m = 100\ \mu\text{Wb} \times 636.6 \times 10^3 \text{ A/Wb}$$

$$= 63.7 \text{ A}$$

Practice Problems

12-2.1 Calculate the current required in a 263 turn coil to establish a flux density of 0.033 T in a 0.01 cm air gap.

12-2.2 A 120 turn toroidal coil has a 1 mm diameter at its cross section and a coil length of 2 cm. Calculate the magnetic field strength, flux density, and the total flux within the coil when a current of 1.5 A flows in the coil.

12-3 SOLENOID

A *solenoid* is a long, thin, air-cored coil. When a current is passed through the coil, a magnetic field is set up, as shown in Figure 12-4(a). The path of the magnetic lines of flux is made up of two components: the length of the path within the

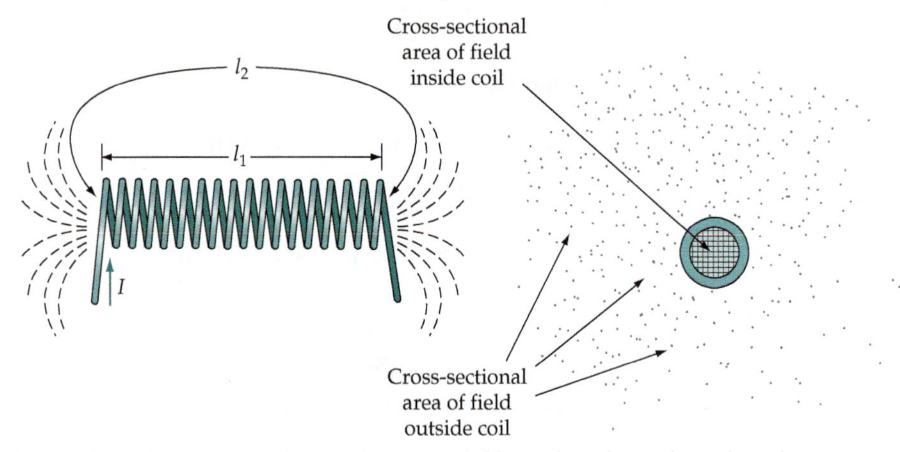

(a) Lengths and cross-sectional areas of magnetic fields inside and outside a solenoid

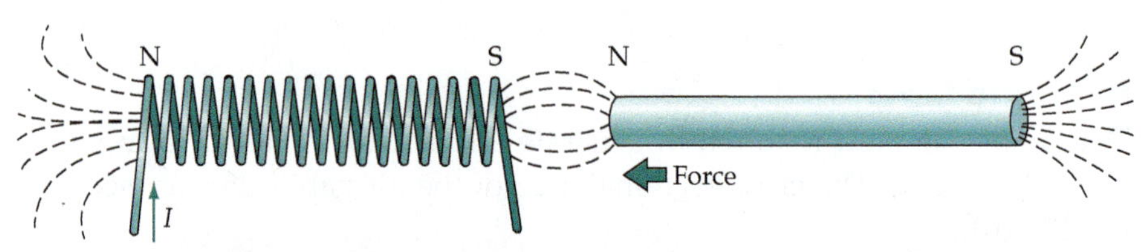

(b) Effect of bringing a soft iron core close to a solenoid

Figure 12-4 The magnetic field of an air-cored solenoid passes through the solenoid and through the surrounding air. An iron bar located close to the solenoid will become magnetized and will experience a force tending to pull it into the solenoid.

coil (l_1), and the length of the path followed by the flux outside the coil (l_2). The length l_2 is longer than l_1, but not very much longer. Also, as illustrated in the figure, the cross-sectional area of the path outside the coil is very much larger than that inside the coil. Consequently, because they are both of the same material (i.e., air), the reluctance of path l_2 is very much smaller than the reluctance of path l_1. So, the mmf required to set up a given total flux along path l_1 is very much larger than that required to set up the same flux along l_2. The total mmf required for the solenoid is the sum of the two components:

$$\text{Total mmf} = (\text{mmf for } l_1) + (\text{mmf for } l_2)$$

But,

$$(\text{mmf for } l_1) \gg (\text{mmf for } l_2)$$

Therefore,

$$\text{total mmf} \approx \text{mmf for } l_1$$

In Chapter 11 it was shown that magnetic lines of force are always in a state of tension and always seeking to flow through the path of least reluctance. It was also seen that the reluctance of ferromagnetic materials is very much smaller than that of air. When a bar of ferromagnetic material is brought close to one end of a solenoid, the bar becomes magnetized by induction, [Figure 12-4(b)]. The induced magnetism produces poles as illustrated, so that the bar is attracted to the solenoid. If the bar is of suitable dimensions, it will be sucked right into the solenoid and remain there. When this happens, the magnetic path inside the (iron-cored) solenoid has a very small reluctance, and it is the air path outside the solenoid that requires the greatest mmf to set up a given total flux.

Example 12-3

An air-cored solenoid has length $l = 15$ cm and inside diameter $D = 1.5$ cm. If the coil has 900 turns, determine the total flux within the solenoid when the coil current is 100 mA.

Solution

$$F_m = NI = 900 \times 100 \text{ mA}$$

$$= 90 \text{ A}$$

$$A = \pi\left(\frac{D}{2}\right)^2 = \pi\left(\frac{1.5 \times 10^{-2}}{2}\right)^2$$

$$= 176.7 \times 10^{-6} \text{ m}^2$$

Eq. 12-1,

$$R_m = \frac{l}{\mu_o A} = \frac{15 \times 10^{-2}}{4\pi \times 10^{-7} \times 176.7 \times 10^{-6}}$$

$$= 675.5 \times 10^6 \text{ A/Wb}$$

From Eq. 12-2,

$$\Phi = \frac{F_m}{R_m} = \frac{90 \text{ A}}{675.5 \times 10^6}$$

$$= 0.133\ \mu\text{Wb}$$

Practice Problems

12-3.1 The flux density within an air-cored solenoid is measured as 7.9 mT. If the solenoid has a length of 8 cm, a diameter of 1.25 cm, and 390 turns, calculate the coil current.

12-3.2 If a current of 0.85 A flows in the turns of the solenoid in Problem 12-3.1, determine the total flux.

12-4 RELATIVE PERMEABILITY

When iron or steel is inserted into a current-carrying coil, it is found that the total magnetic flux through the coil is increased a great many times, (see Figure 12-5). This occurs even though the mmf and magnetic field strength are not altered in any way. The reason (already discussed) is that iron and steel are much better conductors of magnetic flux than any other materials. Because permeability is the magnetic analog of conductivity, it is obvious that iron and steel have very much greater permeabilities than nonmagnetic materials. Clearly, magnetic circuits that use iron or steel have much smaller reluctances than magnetic circuits with nonmagnetic materials.

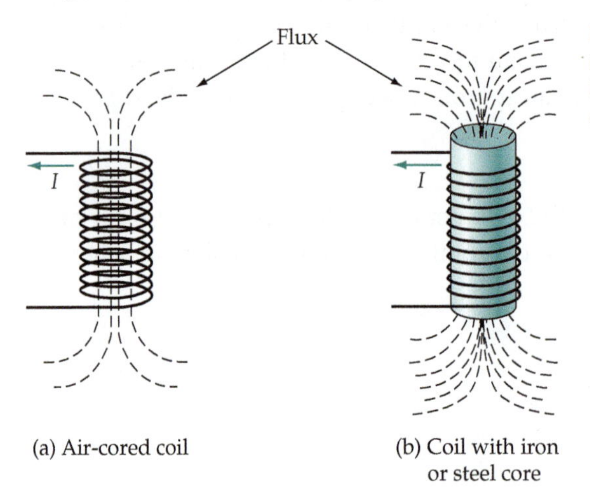

Figure 12-5 Magnetic flux is tremendously increased when an iron or steel core is inserted into a current-carrying coil.

The improvement in total flux and flux density when iron or steel is involved in a circuit is taken into account by assigning each material a *relative permeability*, (symbol μ_r). This necessitates modification of Eq. 12-4,

$$\mu_r \mu_o = \frac{B}{H} \tag{12-5}$$

Equation 12-1 is also modified to,

$$R_m = \frac{l}{\mu_r \mu_o A} \tag{12-6}$$

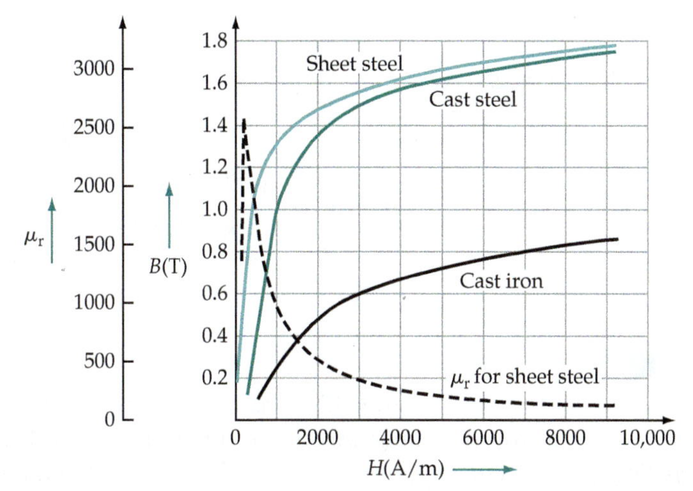

Figure 12-6 Magnetization curves (flux density *B* versus magnetic field strength *H*) for cast iron, cast steel, and sheet steel, and the relative permeability curve for sheet steel.

In the case of air and other nonmagnetic materials, $\mu_r = 1$. Depending on the particular type of iron or steel, the relative permeability can range from 400 to perhaps 2500.

Figure 12-6 shows how the flux density in magnetic materials changes with the level of magnetic field strength. The solid lines in Figure 12-6 show graphs of flux density (B) plotted versus magnetic field strength (H) for various specimens of iron and steel. The measurements were made on ring-shaped specimens, so the magnetic circuit was a closed iron circuit with no air gaps, allowing maximum possible flux densities to be achieved. The magnetization curves can be used for direct determinations of flux density for a given value of magnetic field strength. Alternatively, the relative permeabilities of each specimen can be calculated and plotted versus H. The broken line in Figure 12-6 shows the plot of μ_r for sheet steel. This is derived simply by taking corresponding values of B and H from the permeability curve for sheet steel, and substituting them into Equation 12-5 to calculate the values of μ_r. Note that the relative permeability is by no means a constant quantity; instead, it very much depends on the magnetic field strength H. From the plot of μ_r versus H for sheet steel, the greatest permeability is obviously achieved at a magnetic field strength of approximately 250 A/m.

It should be noted that all the B/H curves shown in Figure 12-6 are typical, and that for any given specimen of magnetic material the actual B/H relationships may be different from those illustrated. This is because slight differences in the manufacturing process can significantly affect the material's magnetic properties. The shape of the B/H curve is considered further in Section 12-7.

Example 12-4

Calculate the relative permeability of sheet steel at a magnetic field strength of 250 A/m.

Solution

From Figure 12-6, at $H = 250\ \text{A/m}$,

$$B \approx 0.8\ \text{T}$$

From Eq. 12-5,

$$\mu_r = \frac{B}{\mu_o H} = \frac{0.8\ \text{T}}{4\pi \times 10^{-7} \times 250\ \text{A/m}}$$

$$= 2546$$

Example 12-5

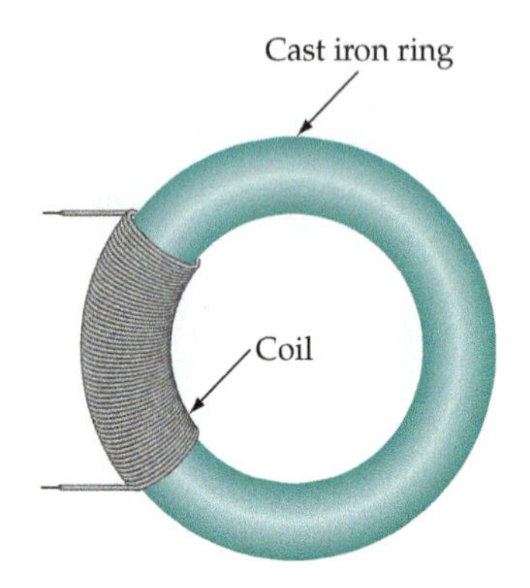

Figure 12-7 Cast iron ring for Example 12-5.

The cast iron ring illustrated in Figure 12-7 has a 3000 turn coil that carries a current of 0.1 A. The cross-sectional area of the ring is 4 cm^2 and the length of the magnetic path is 15 cm. Determine the flux density and the total flux in the ring.

Solution

Eq. 11-3,

$$H = \frac{NI}{l} = \frac{3000 \times 0.1\ \text{A}}{15 \times 10^{-2}\ \text{m}}$$

$$= 2000\ \text{A/m}$$

From Figure 12-6, for cast iron at $H = 2000\ \text{A/m}$,

$$B \approx 0.5\ \text{T}$$

From Eq. 11-1,

$$\Phi = BA \approx 0.5\ \text{T} \times 4 \times 10^{-4}\ \text{m}^2$$

$$\approx 200\ \mu\text{Wb}$$

Compare this to the total flux for the similar air-cored toroid in Example 12-1.

Example 12-6

The cast steel core shown in Figure 12-8 has a relative permeability of 800, and the coil has 700 turns. If the total flux in the core is to be 4×10^{-4} Wb, determine the required current through the coil.

Solution

The length of the magnetic circuit is l in Figure 12-8.

where

$$l = 2(3\ \text{cm} + 8\ \text{cm})$$

$$= 22\ \text{cm}$$

Eq. 12-6,

$$R_m = \frac{l}{\mu_r \mu_o A} = \frac{22 \times 10^{-2}}{800 \times 4\pi \times 10^{-7} \times 4 \times 10^{-4}}$$

$$= 547.1 \times 10^3\ \text{A/Wb}$$

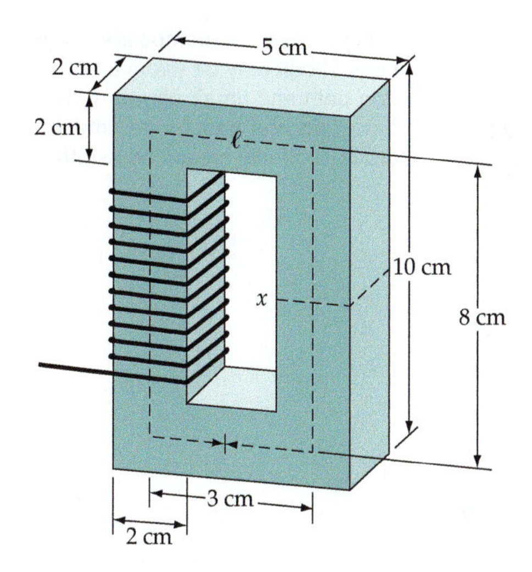

Figure 12-8 Cast steel core for Examples 12-6 and 12-7.

From Eq. 12-2,

$$F_m = \Phi R_m = 4 \times 10^{-4}\ \text{Wb} \times 547.1 \times 10^3\ \text{A/Wb}$$

$$= 219\ \text{A}$$

$$I = \frac{F_m}{N} = \frac{219\ \text{A}}{700}$$

$$= 313\ \text{mA}$$

Practice Problems

12-4.1 Determine the relative permeability of sheet steel at 3500 A/m, cast steel at 4000 A/m, and cast iron at 8000 A/m.

12-4.2 A sheet steel magnetic core with a cross-sectional area of 3.3 cm^2 and a magnetic path length of 12 cm has a 2500-turn coil. If the coil current is 75 mA, calculate the total flux in the core.

12-4.3 If the flux density in the core described in Problem 12-4.2 is to be altered to 1.15 T, determine the required coil current.

12-5 COMPOSITE MAGNETIC CIRCUITS

Where a magnetic circuit has an air gap in its path, as illustrated in Figure 12-9(a), the air gap is in series with the rest of the magnetic circuit. The air gap and magnetic path reluctances must be calculated independently, and then the two should be added to determine the total reluctance. The same procedure applies when the cross-sectional area of a core is not constant along its entire length [Figure 12-9(b)].

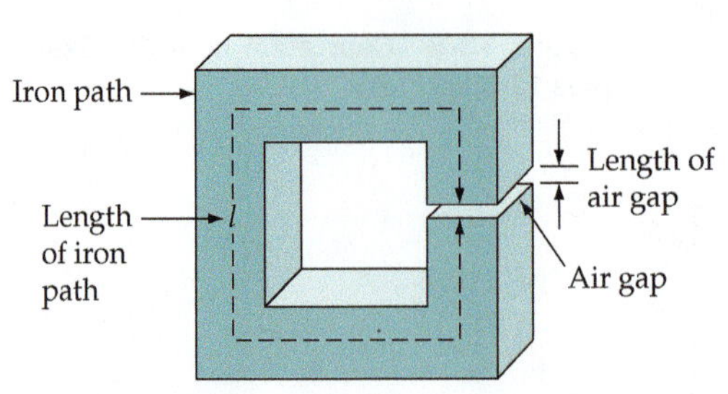

(a) Magnetic circuit consisting of an iron path and an air gap in series

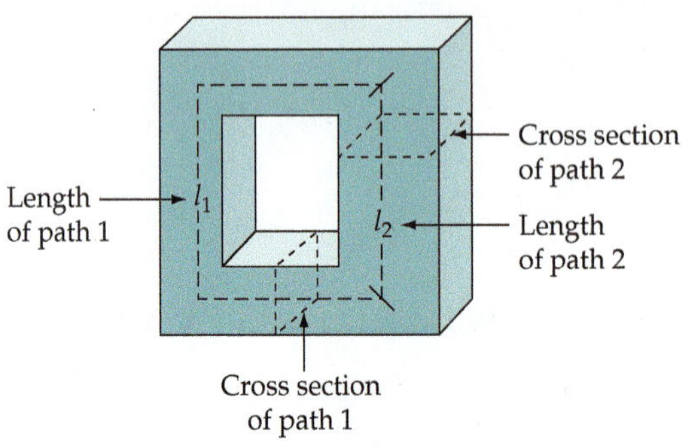

(b) Magnetic circuit made up of two iron paths with different cross-sectional areas and lengths

Figure 12-9 A composite magnetic circuit may be made up of an iron path and an air gap, or it may have two iron paths with different cross-sectional areas and lengths.

Example 12-7

The core shown in Figure 12-8 has a 1 mm air gap at point x. $\mu_r = 800$, and $N = 700$, as described for Example 12-6. Calculate the new level of current through the coil to give a total core flux of 4×10^{-4} Wb.

Solution

From Example 12-6, for the iron path,

$$R_{m1} = 547.1 \times 10^3 \text{ A/Wb}$$

For the air gap,

Eq. 12-6,
$$R_{m2} = \frac{l}{\mu_r \mu_o A} = \frac{1 \times 10^{-3}}{4\pi \times 10^{-7} \times 4 \times 10^{-4}}$$
$$= 1.99 \times 10^6 \text{ A/Wb}$$
$$R_m = R_{m1} + R_{m2} = 547.1 \times 10^3 \text{ A/Wb} + 1.99 \times 10^6 \text{ A/Wb}$$
$$= 2.54 \times 10^6 \text{ A/Wb}$$
$$F_m = \Phi R_m = 4 \times 10^{-4} \text{ Wb} \times 2.54 \times 10^6 \text{ A/Wb}$$
$$= 1016 \text{ A}$$
$$I = \frac{F_m}{N} = \frac{1016 \text{ A}}{700}$$
$$= 145 \text{ A}$$

Note, from Examples 12-6 and 12-7, the significant increase in coil current required for a given flux density when a very short air gap is introduced into the core.

Example 12-8

The uneven ring-shaped core shown in Figure 12-10 has a relative permeability of 1000, and the flux density in the section with the largest cross-sectional area is to be 0.75 T. If the current through a coil wound on the core is to be 500 mA, determine the number of coil turns required.

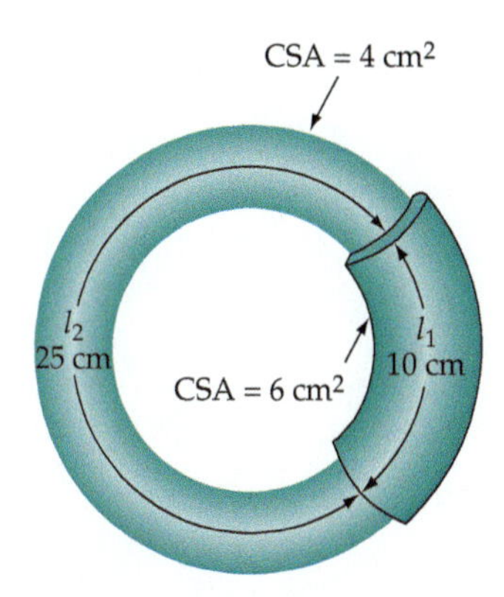

Figure 12-10 Composite magnetic circuit for Example 12-8.

Solution

For the section with the largest cross-sectional area,

Eq. 12-6, $$R_{m1} = \frac{l_1}{\mu_r \mu_o A_1} = \frac{10 \times 10^{-2}}{1000 \times 4\pi \times 10^{-7} \times 6 \times 10^{-4}}$$

$$= 133 \times 10^3 \text{ A/Wb}$$

For the section with the smaller cross-sectional area,

Eq. 12-6, $$R_{m2} = \frac{l_2}{\mu_r \mu_o A_2} = \frac{25 \times 10^{-2}}{1000 \times 4\pi \times 10^{-7} \times 4 \times 10^{-4}}$$

$$= 497 \times 10^3 \text{ A/Wb}$$

$$R_m = R_{ml} + R_{m2} = 133 \times 10^3 \text{ A/Wb} + 497 \times 10^3 \text{ A/Wb}$$

$$= 630 \times 10^3 \text{ A/Wb}$$

$$F_m = BAR_m = 0.75 \times 6 \times 10^{-4} \text{ Wb} \times 630 \times 10^3 \text{ A}$$

$$= 283.5 \text{ A}$$

$$N = \frac{F_m}{I} = \frac{283.5 \text{ A}}{500 \text{ mA}}$$

$$= 567 \text{ turns}$$

When crossing an air gap, the magnetic lines of force tend to bulge out as shown in Figure 12-11(a). This is because lines of force repel each other when passing through nonmagnetic material. The effect of this bulging, or *fringing* as it is termed, is to increase the cross-sectional area of the magnetic field at the air gap and consequently decrease its flux density. In a short air gap with a large cross-sectional area, the fringing may be insignificant. In other situations, 10% is typically added to the air gap cross-sectional area to allow for fringing.

In Figure 12-11(b), an electromagnet is shown lifting an iron bar. In this case the uneven surfaces create air gaps at the junction between the magnet and the iron bar, and fringing occurs as illustrated. Some lines of force actually cross the gap between the magnet's poles without passing through the iron bar. This is termed *leakage flux,* and unlike the fringing flux, it does not assist in holding

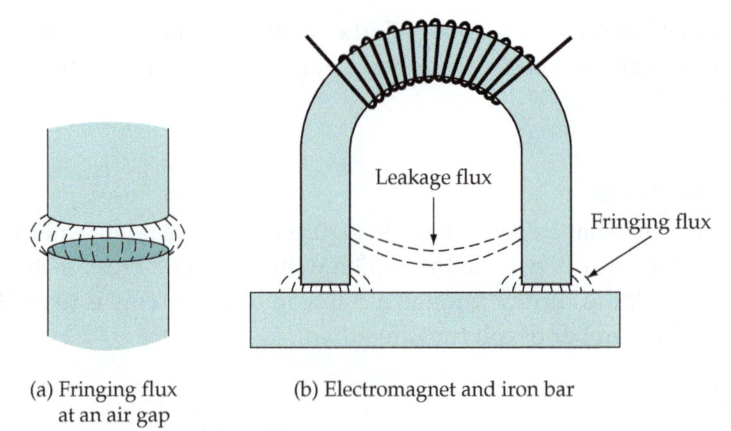

Figure 12-11 Fringing (bulging of magnetic fields) occurs at air gaps. Leakage flux may bypass part of a magnetic path at adjacent poles.

the iron bar to the magnet. So, the leakage flux is wasted. However, the iron path around the magnetic circuit has a much lower reluctance than the air path taken by the leakage flux; consequently, leakage flux is usually small enough to be neglected.

Example 12-9

The magnetic core shown in Figure 12-12 has the following dimensions: $l_1 = 10$ cm, $l_2 = l_3 = 18$ cm, cross-sectional area of l_1 path $= 6.25 \times 10^{-4}$ m^2, cross-sectional area of l_2 and l_3 paths $= 3 \times 10^{-4}$ m^2, length of air gap (l_4) = 2 mm. Determine the current that must be passed through the 600 turn coil to produce a total flux of 100 μWb in the air gap. Assume that the metal has a permeability of 800.

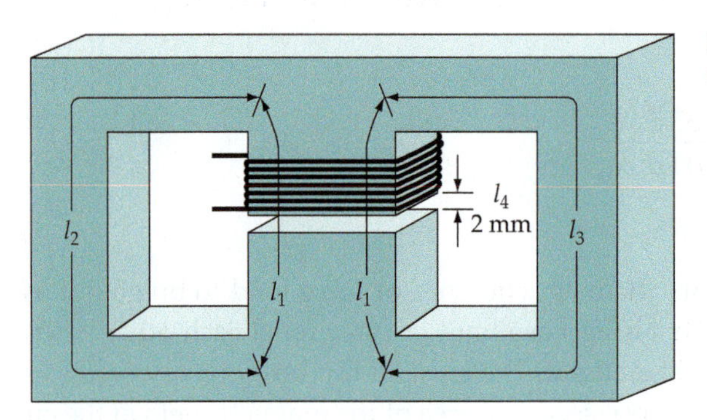

Figure 12-12 Composite magnetic core for Example 12-9.

Solution

For the air gap,

Eq. 12-1,

$$R_{m4} = \frac{l_4}{\mu_o A_4} = \frac{2 \times 10^{-3}}{4\pi \times 10^{-7} \times 6.25 \times 10^{-4}}$$

$$= 2.55 \times 10^6 \text{ A/Wb}$$

For path l_1,

$$\text{Eq. 12-6,} \quad R_{m1} = \frac{l_1}{\mu_r \mu_o A_1} = \frac{10 \times 10^{-2}}{800 \times 4\pi \times 10^{-7} \times 6.25 \times 10^{-4}}$$

$$= 0.16 \times 10^6 \text{ A/Wb}$$

Because l_2 and l_3 are in parallel, and each has a cross-sectional area of 3×10^{-4} m^2, they can be treated as a single path with a cross-sectional area of 6×10^{-4} m^2.

$$\text{Eq. 12-6,} \quad R_{m2,3} = \frac{l_{2,3}}{\mu_r \mu_o A_{2,3}} = \frac{18 \times 10^{-2}}{800 \times 4\pi \times 10^{-7} \times 6 \times 10^{-4}}$$

$$= 0.3 \times 10^6 \text{ A/Wb}$$

$$R_m = R_{m4} + R_{m1} + R_{m2,3} = (2.55 + 0.16 + 0.3) \times 10^6 \text{ A/Wb}$$

$$= 3 \times 10^6 \text{ A/Wb}$$

$$F_m = \Phi R_m = 100\ \mu\text{Wb} \times 3 \times 10^6 \text{ A/Wb}$$

$$= 300 \text{ A}$$

$$I = \frac{F_m}{N} = \frac{300 \text{ A}}{600}$$

$$\approx 500 \text{ mA}$$

Practice Problems

12-5.1 For Problem 12-4.2, calculate the required coil current to maintain the total core flux if two 0.1 mm air gaps are cut in the core.

12-5.2 A magnetic core with a relative permeability of 1800 has a cross-sectional area of 2.5 cm^2, a magnetic path length of 9 cm, and a 3900 turn coil. A 0.7 mm air gap in the core is to have a flux density of 0.9 T. Allowing 10% for fringing, calculate the required coil current.

12-6 FORCE BETWEEN TWO MAGNETIC SURFACES

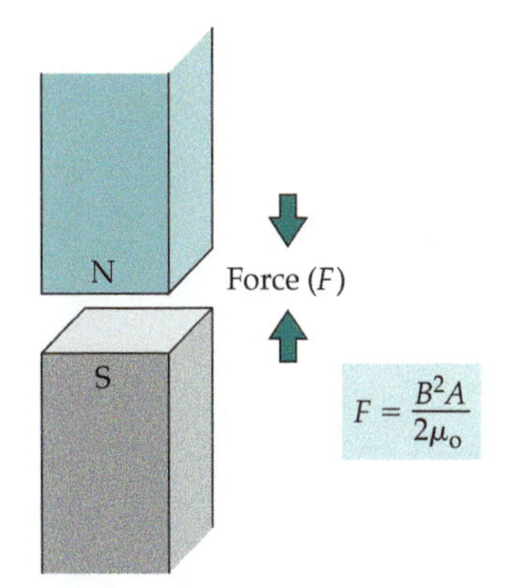

Figure 12-13 The force between two magnetic surfaces may be calculated from the air gap flux density and cross-sectional area.

The air gap in a magnetic circuit can be thought of as two magnetic poles of opposite polarity separated by air. Because opposite poles attract, there is a force of attraction tending to pull the two poles together. The magnitude of this force can be calculated in terms of the flux density and the cross-sectional area of the air gap. This is illustrated in Figure 12-13. If the magnetic field is the result of current flowing in a coil wound on the core, energy must be supplied to the coil in order to set up the flux in the air gap. This energy input becomes energy stored in the air gap. Consider a ferromagnetic core with an air gap. The magnetic field strength in the air gap is

Eq. 11-3, $$H = \frac{IN}{\ell}$$

Giving, $$I = \frac{Hl}{N} \tag{1}$$

If I is increased from zero to its maximum level during the time Δt, the core flux changes from zero to a maximum of Φ. This flux change links with the coil and induces a voltage (V) at the coil terminals. From the definition of the weber [that a flux change of 1 Wb in a time of 1 s induces 1 V emf in a one-turn coil (see Section 11-4)] the voltage induced in each turn of the coil is,

$$V = \frac{\Phi}{\Delta t}$$

For a coil of N turns, the total induced voltage is

$$V = \frac{\Phi N}{\Delta t} \tag{2}$$

The coil supply voltage is (almost) equal in magnitude to the induced voltage, and the power input to the coil can be calculated as,

$$P = VI$$

However, P increased from zero to its maximum level during time Δt. Assuming that the power increases linearly, the average power input can be shown to be,

$$P_{av} = \frac{VI}{2}$$

Also, the energy (in joules) supplied to the coil in a time of Δt is,

$$W = P_{av}\Delta t = \frac{V \times I \times \Delta t}{2}$$

Substituting from Equations (1) and (2) for V and I,

$$W = \frac{\Phi H l}{2}$$

and $\Phi = BA$, and $H = B/\mu_o$

So, the energy supplied is, $$W = \frac{BA(B/\mu_o)l}{2}$$

Or energy stored in an air gap is,

$$W = \frac{B^2 A l}{2\mu_o} \tag{12-7}$$

When two magnetic surfaces are separated by a short distance, the mechanical energy involved in pulling them apart is,

$$W = F \times d$$

Assuming that the surfaces are still attracted to each other, the electrical energy supplied to the air gap is equal to the mechanical energy supplied to pull the surfaces apart. Therefore,

$$Fd = \frac{B^2Al}{2\mu_o}$$

In this case d and l both represent the thickness of the air gap, and they cancel out, giving the force between the surfaces as,

$$F = \frac{B^2A}{2\mu_o} \quad \text{(12-8)}$$

This expression includes the cross-sectional area (A) of the air gap but does not include its length. When B is in teslas and A is in square meters, the force F is in newtons. Equation 12-8 can now be used to calculate the mechanical pull exerted by an electromagnet (see Figure 12-14).

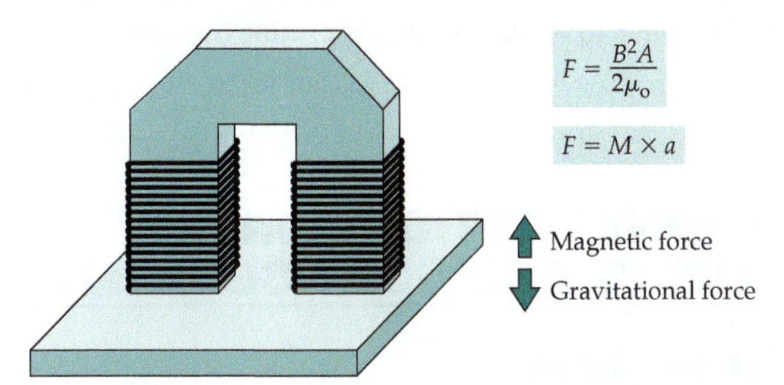

Figure 12-14 The weight that can be lifted by an electromagnet can be determined from knowledge of the contact area and the flux density at the poles of the magnet.

Example 12-10

The electromagnet shown in Figure 12-14 has pole pieces that each have a cross-sectional area of 25 cm^2. The total flux crossing each pole is 250 μWb. Determine the maximum weight of iron plate that can be lifted by the magnet.

Solution

$$B = \frac{\Phi}{A} = \frac{250\ \mu\text{Wb}}{25 \times 10^{-4}\ \text{m}^2}$$

$$= 0.1\ \text{T}$$

Total cross-sectional area,

$$A = 2 \times 25 \times 10^{-4}\ \text{m}^2$$
$$= 50 \times 10^{-4}\ \text{m}^2$$

Eq. 12-8,
$$F = \frac{B^2 A}{2\mu_o} = \frac{(0.1\ \text{T})^2 \times 50 \times 10^{-4}\ \text{m}^2}{2 \times 4\pi \times 10^{-7}}\ \text{newtons}$$
$$\approx 19.9\ \text{N}$$
$$F = m \times a$$

With a being the acceleration due to gravity (see Appendix 2), the mass (m) that can be lifted is,

$$m = \frac{F}{a} \approx \frac{19.9\ \text{N}}{9.81\ \text{m/s}^2}$$
$$\approx 2.03\ \text{kg}$$

Practice Problem

12-6.1 A magnet with two poles, each with a cross-sectional area of 1.6 cm^2, is required to lift 1 kg bars of soft iron that completely cover the pole faces. Allowing 10% for fringing, determine the required flux density in the core.

12-7 MAGNETIZATION CURVES, HYSTERESIS, AND EDDY CURRENTS

Magnetization Curves

A typical B/H graph (or *curve*) is shown in Figure 12-15, with a slightly smaller scale than the graphs in Figure 12-6. Examination of the curve shows that the material is initially not magnetized because at zero magnetic field strength the flux density is zero. As the field strength is increased from zero, the flux density increases very slowly at first, giving a gradual slope, section o to a of the curve. Further increase in H causes B to increase progressively more rapidly, until there is a near-linear relationship between B and H over section a to b of the curve. With continued increase in H past point b, changes in B become less and less until the curve flattens out again to a very gradual slope beyond point c.

The shape of the curve can be explained in terms of the magnetic domains in ferromagnetic material (see Section 11-3). During portion o to a of the curve, the magnetic field strength is too weak to produce much realignment of the magnetic domains. Consequently, the increase in flux density is relatively small for these levels of field strength. Over portion a to b, the increasing field

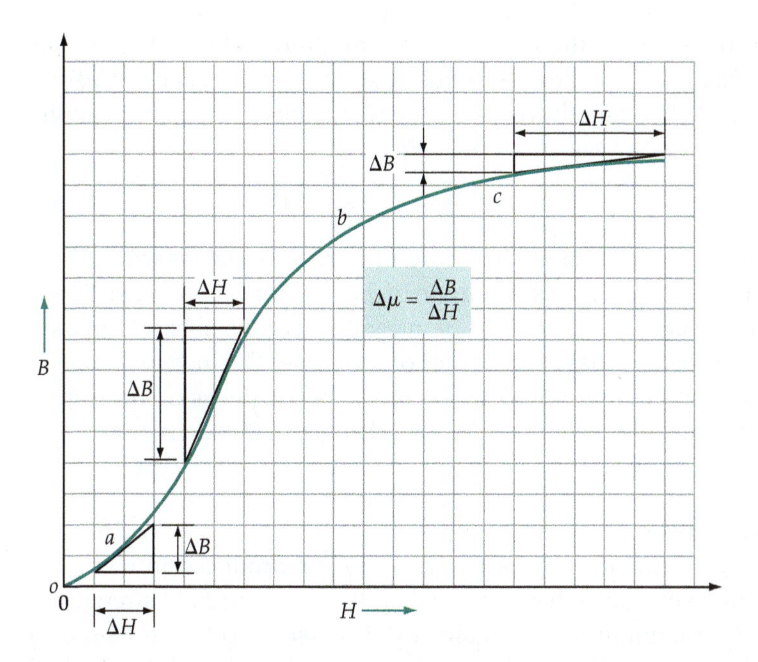

Figure 12-15 Permeability is the ratio *B/H*. When the flux density and magnetic field strength in a core vary by small amounts (Δ*B* and Δ*H*), the incremental permeability Δ*B*/Δ*H* can be important.

strength is obviously causing more and more of the domains to be aligned in the same direction. So, *B* increases in a near-linear relationship to *H*. From point *b* to *c*, all the domains are becoming completely aligned in the required direction for maximum magnetic effect. So, during the *b* to *c* portion of the curve, there are fewer and fewer magnetic domains available to be aligned. Therefore, the increases in flux density in relation to increases in *H* are now becoming smaller. Finally, from point *c* onwards, it can be assumed that effectively all available magnetic domains are completely aligned in the direction enforced by the magnetic field strength. The magnetic material is said to be *saturated* at this point. The slope of the *B/H* curve beyond point *c* shows the same order of increase in *B* with respect to *H* as would be obtained with non-magnetic material.

From Equation 12-5, the relationship between *B* and *H* is the permeability,

$$\mu_r\mu_o = \frac{B}{H}$$

Ideally, it is desirable to have the largest possible value of permeability, in order to achieve the greatest flux density with the smallest possible magnetic field strength. The plot of μ_r for sheet steel (Figure 12-6) shows that the largest possible value of μ_r is achieved around the center of the linear portion of the *B/H* curve. This direct relationship between constant values of *B* and *H* can be referred to as the *normal permeability* of the material. Another important relationship is that between *changing values* of *B* and *H*. Referring to Figure 12-15 again, the slope of the B/H curve at any point is known as the *incremental permeability.* This parameter defines the flux density change (ΔB) for a given magnetic field strength change (ΔH). So, the incremental permeability is,

$$\Delta\mu = \frac{\Delta B}{\Delta H}$$

Reference to the B/H curve in Figure 12-15 reveals that (as with μ_r) $\Delta\mu$ is greatest at the center of the linear portion of the magnetization curve. Both the initial and final values of incremental permeability (see the illustration) are considerably smaller than that at the center of the curve. The incremental permeability is important in applications where only small changes occur in the magnetic field strength and the largest possible changes in flux density are desired.

Hysteresis

The magnetization curve of Figure 12-15 is reproduced in Figure 12-16(a) in order to investigate what occurs when the magnetic field strength is reduced from its maximum level. As already discussed, portion *o-a-b-c* of the curve occurs when the material is initially unmagnetized and the field strength is increased from zero. Commencing at point *c,* when the field strength is decreased the flux density does not move down the *c-b-a-o* curve, as might be expected. Instead, it falls along the curve *c-d,* as illustrated. The effect can be explained in terms of a kind of friction-force resisting movement of the magnetic domains. So, instead of returning to their original unmagnetized state, the magnetic domains remain partially aligned and the material retains some magnetic flux, even though the field strength has been reduced to zero. The material has, in fact, been magnetized, and the retained flux density (*o-d*) is referred to as *remanence,* or *residual magnetism.* The ability of a given ferromagnetic material to retain residual magnetism is termed its *retentivity.*

To reduce the remanence to zero, a negative or reverse magnetic field strength must be applied (i.e., the current in the magnetizing coil must be reversed). As the reversed value of H is increased, the flux density moves down the *d-e* portion of the curve to a zero level of B. The magnetic field strength (*o-e*) required to reduce the remanence to zero is termed the *coercive force.*

When the reversed value of H is increased beyond the level needed to reduce B to zero, it is found that the flux density now increases in a negative direction. A new magnetization curve is generated, giving the curve from *e* to *f* in Figure 12-16(b). At point *f*, the reversed flux density has reached its saturation point, and this corresponds to point *c* on the positive half of the curve. Once again, when the reversed field strength is reduced to zero, the reversed flux density does not become zero. Instead, portion *f-g* of the B/H graph is generated, and *o-g* represents the retained flux density. Section *o-g* is exactly equal to *o-d* although in the opposite direction. To reduce the flux density to zero once more, the field strength must be increased in a positive direction, giving portion *g* to *h* of the graph. This again represents a coercive force, and *o-h* is

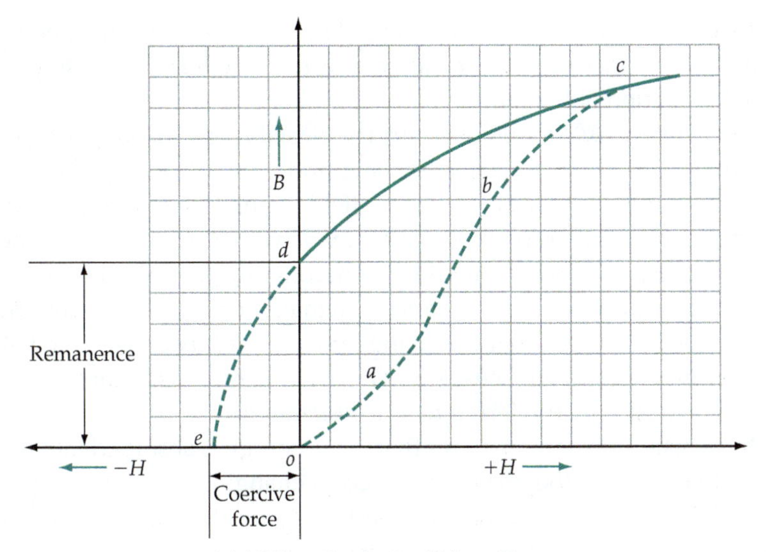

(a) Effect of reducing H from H_{max}

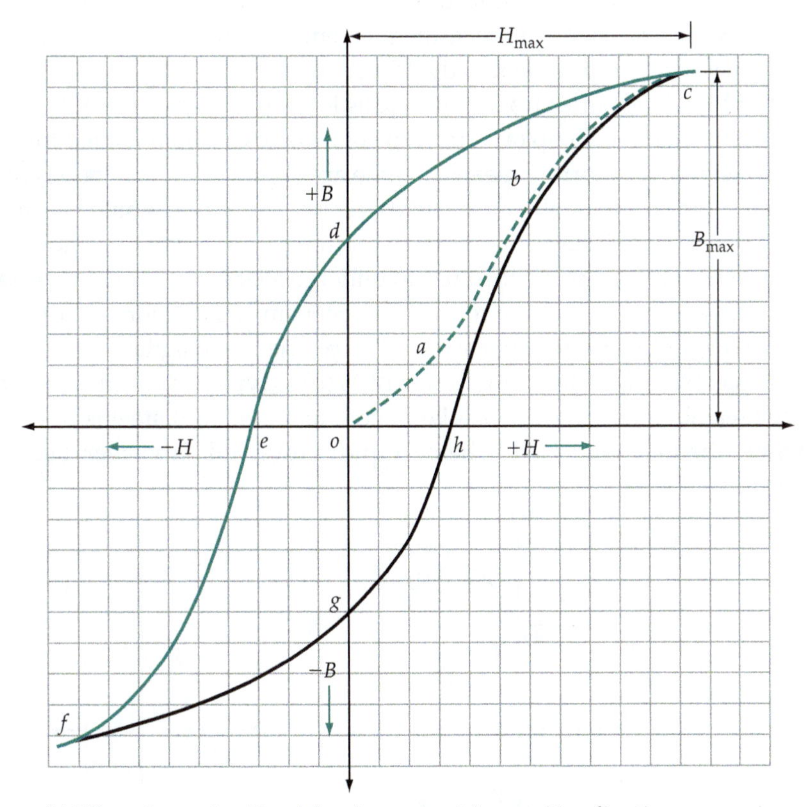

(b) Effect of reversing H and then increasing it in a positive direction once again

Figure 12-16 When the magnetic field strength H in a sample of iron is decreased to zero, some flux density B may remain. If H is reversed, the residual magnetism is also reversed. Instead of giving a one line graph, a plot of B versus H is likely to produce a hysteresis loop.

found to be equal in magnitude to *o-e*. Finally, when the magnetic field strength is further increased in a positive direction, the *B/H* relationship traces out the *h* to *c* section of the graph.

It is seen that the *B/H* graph now forms a closed loop. The loop is symmetrical, with $+B_{(max)}$ and $-B_{(max)}$ equal in magnitude, and $+H_{max}$ equal in magnitude to $-H_{max}$. When a positive magnetic field strength (*o-h*) is applied to the specimen to which the graph refers, the flux density *B* is zero. Only when *H* is increased to H_{max} does the flux density become B_{max}. Similarly, when *H* is reduced from a positive level to zero, *B* remains equal to *o-d*. Continuing around the loop, it is seen that the changing levels of flux density lag behind the changes in magnetic field strength. This lagging effect is termed *hysteresis*, and the *B/H* graph is then referred to as a *hysteresis loop*.

The hysteresis loop demonstrates that some energy is absorbed into a magnetic core to overcome the *friction* involved in changing the alignment of the magnetic domains. So, a core that is subjected to repeated and rapid reversals of the magnetic field (as in the case of alternating currents) may absorb a lot of energy in this way. This energy results in heating of the core, and it is obviously wasted or lost energy. In fact, the area enclosed by the hysteresis loop can be shown to be proportional to the lost energy.

Figure 12-17 shows three typical hysteresis loops for three different types of ferromagnetic material. Loop (a) is the type of hysteresis loop obtained for soft iron. The fact that it is narrow means that the area enclosed by the loop is relatively small, and consequently the hysteresis losses in the core would be a minimum. For greatest efficiency, therefore, a soft iron core should be employed in situations where the magnetic field has to undergo a large number of reversals each second. Also, note that the residual magnetism is small for the soft iron core; consequently, soft iron is not suitable for permanent magnets.

The hysteresis loop shown in Figure 12-17(b) is typical of hard steel, and its large area obviously indicates a large core loss. So, hard steel is not suitable where the magnetic field is being rapidly reversed. However, because

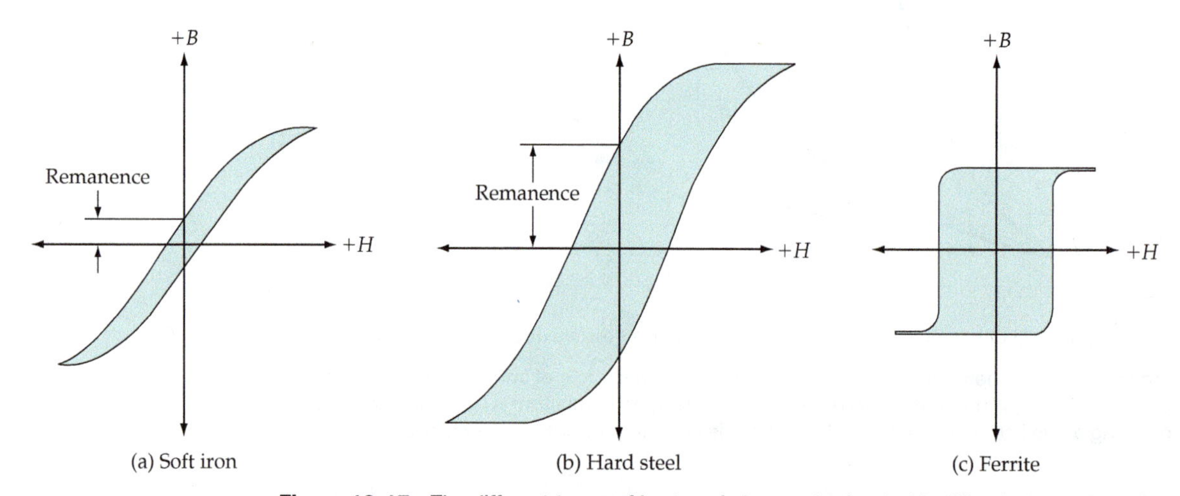

Figure 12-17 The different types of hysteresis loops obtained with different ferromagnetic materials influence the application of the materials.

its residual magnetism is large, hard steel is very suitable for permanent magnets.

The third hysteresis loop, Figure 12-17(c), is that of a material known as *ferrite*. This is a *ceramic core* made up of iron oxides. The shape of the loop suggests a large hysteresis loss. However, it also shows that B tends to remain constant in one direction until the value of H is increased almost to its maximum level in the opposite direction; then B rapidly reverses. This is exactly the characteristic required for a *magnetic memory*, and consequently, memories are one area of ferrite application.

Eddy Currents

In Section 11-2 it is explained that a changing magnetic field induces a voltage in a conductor situated within the field. When the current direction in a coil is continually reversing, a ferromagnetic core in that coil constitutes a conductor in a changing magnetic field. Consequently, voltages are induced in the core, and circulating currents (or *eddy currents*) are caused to flow, as illustrated in Figure 12-18(a).

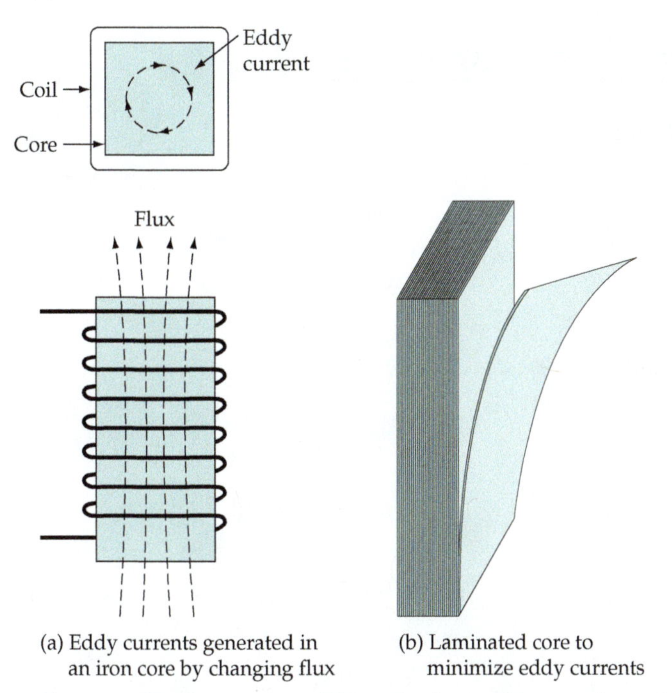

(a) Eddy currents generated in an iron core by changing flux

(b) Laminated core to minimize eddy currents

Figure 12-18 Eddy currents induced in a solid magnetic core by a continuously reversing coil current can overheat the core. The effect is minimized by constructing the core of sheets, called laminations.

Eddy currents cause heating of the core and add considerably to the total core losses. Even when the core is nonmagnetic, eddy currents are generated if the material is an electrical conductor. To combat the eddy current losses, magnetic cores used with alternating current are always made up of thin sheets, termed *laminations* [see Figure 12-18(b)]. The surfaces of the laminations are varnished or otherwise thinly insulated on either side so that they offer a high resistance to the flow of circulating eddy currents. By this method the eddy current core losses are rendered negligible without affecting the magnetic performance of the core. Note that the orientation of the laminations is such that they allow the setting up of the magnetic lines of force without difficulty. The varnished surface interface between laminations does not constitute an air gap in the magnetic path. The single negative effect of laminated cores is that the total cross-sectional area of the magnetic material is reduced by the total thickness of the insulation. This is usually taken into account by allowing an approximately 10% reduction in the thickness of the core when making the magnetic calculations.

Practice Problem

12-7.1 If the core in Example 12-6 is laminated, calculate the new current level required in the coil.

Summary of Formulas

- *Reluctance:*

$$R_m = \frac{1}{\mu} \times \frac{l}{A}$$

$$R_m = \frac{F_m}{\Phi}$$

- *Permeability:*

$$\mu = \frac{B}{H}$$

$$\mu_o = 4\pi \times 10^{-7}$$

$$\mu_r \mu_o = \frac{B}{H}$$

$$\Delta\mu = \frac{\Delta B}{\Delta H}$$

- *Energy stored in air gap:*

$$W = \frac{B^2 A l}{2\mu_o}$$

- *Force between two magnetic surfaces:*

$$F = \frac{B^2 A}{2\mu_o}$$

Review Questions

Section 12-1

12-1 Define reluctance and permeability. Derive the equation relating the two quantities.

Section 12-2

12-2 Show that the permeability of free space is $4\pi \times 10^{-7}$.

Section 12-3

12-3 Using illustrations, explain why the length of a solenoid can be taken as the total length of the magnetic path in calculations of mmf, flux density, etc. Discuss what occurs when a thin cylindrical iron core is brought close to one end of the solenoid.

Section 12-4

12-4 Explain the term relative permeability, and write the equation relating relative permeability to flux density and magnetic field strength.

Section 12-5

12-5 Define fringing and leakage flux, and discuss their effects on magnetic calculations.

Section 12-6

12-6 Derive equations for energy stored in an air gap and for the force between two magnetic surfaces.

Section 12-7

12-7 Sketch a typical magnetization curve for an initially unmagnetized sample of ferromagnetic material, and explain its shape. Also, show how the incremental permeability can be determined from the curve, and discuss its importance in some applications.

12-8 Sketch a typical hysteresis loop for a sample of ferromagnetic material, and explain its shape. Also, identify and define remanence and coercive force.

12-9 Sketch typical hysteresis loops for (a) soft iron, (b) hard steel, and (c) ferrite. In each case, state the applications that the material is most suitable for and explain why.

12-10 Explain the origin of eddy currents in a magnetic core; discuss the effects of eddy currents and the method employed to combat the effects.

Problems

Section 12-2

12-1 A toroidal coil (as in Figure 12-19) has 1750 turns and carries a 5 A current. The cross-sectional area of the coil is 2.25 cm^2, and the length of the magnetic path is 11 cm. Calculate the magnetic field strength, total flux, and flux density within the coil.

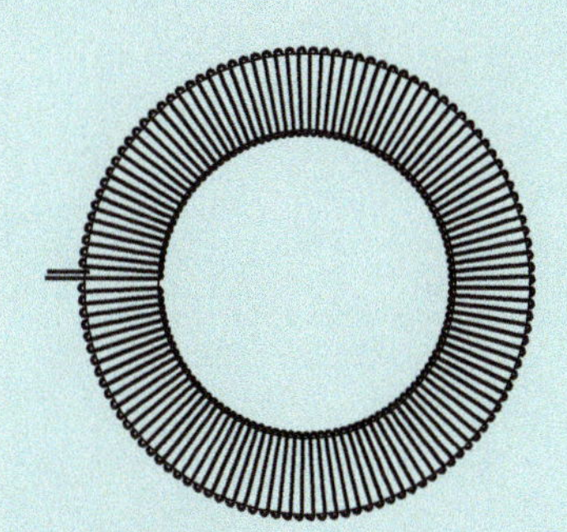

Figure 12-19

12-2 If the coil described in Example 11-2 has an air core with a cross-sectional area of 3.2 cm^2, calculate the total flux generated.

12-3 If the coil described in Problem 12-1 is to have a total flux of 0.2 μWb, determine the new level of current that must flow through the coil.

12-4 All of the mmf generated by the coil in Problem 12-1 is applied to produce a flux in a 1 mm air gap that has a cross-sectional area of 2.25 cm^2. Calculate the magnetic field strength, flux density, and total flux in the air gap.

12-5 A 0.5 mm air gap has a cross-sectional area of 7 cm^2. Calculate the mmf required to generate a total flux of 50 μWb in the air gap.

12-6 A coil with a magnetic path length of 28 cm and a cross-sectional area of 15 cm^2 produces a total flux of 45 μWb. If the coil current is to be 1.5 A, calculate the required number of turns.

Section 12-3

12-7 An air-cored solenoid is 18 cm long and has an inside diameter of 1.7 cm, (see Figure 12-20). The coil has 1400 turns. Calculate the current that must flow to give a total flux of 0.1 μWb within the coil.

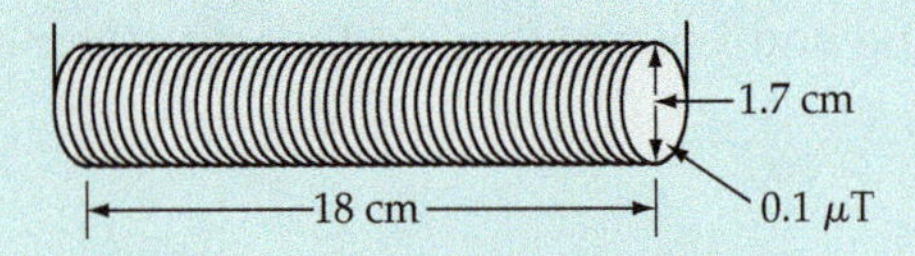

Figure 12-20

12-8 Determine the total flux within the solenoid described in Problem 12-7 when the coil current is 150 mA.

12-9 A short coil with a diameter of 1 cm is situated at the center of a 3300 turn solenoid. If the solenoid is 27 cm long and has a 1.6 cm diameter, calculate the solenoid current that will give a total flux of 0.9 μWb within the small coil.

12-10 An air-cored solenoid is to be 25 cm long and 2.2 cm in diameter. Calculate the required number of coil turns for a 100 mA current to produce 0.3 μWb total flux.

Section 12-4

12-11 Referring to the magnetization curves shown in Figure 12-6, calculate the relative permeability of (a) cast iron at a magnetic field strength of 2000 A/m, and (b) cast steel at 1000 A/m.

12-12 Determine the relative permeability of sheet steel, cast steel, and cast iron when all three have a flux density of 0.7 T.

12-13 If the coil described in Problem 12-1 has a cast steel core, determine the value of total flux within the core when the current is 500 mA.

12-14 If the coil described in Problem 12-6 has a cast iron core, calculate the flux density when the coil current is 100 mA.

12-15 A cast iron ring has a cross-sectional area of 9 cm^2, and the length of the magnetic path is 25 cm, (see Figure 12-21). A coil of 5000 turns is wound on the ring and a current of 75 mA is passed through the coil. Calculate the magnetic field strength, the total flux, and the flux density within the core.

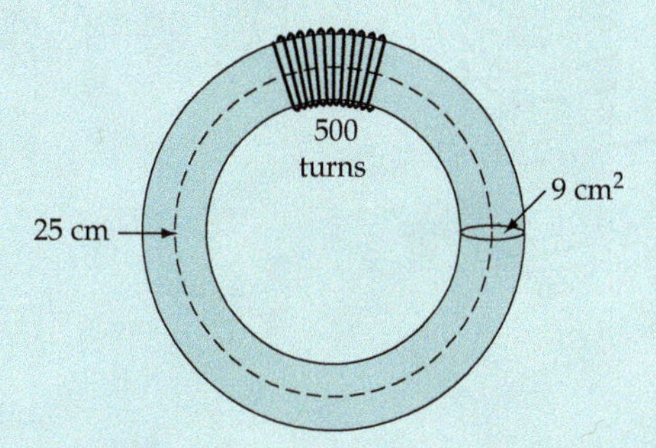

Figure 12-21

12-16 Recalculate the total core flux for Problem 12-15 when a sheet steel core is used.

Section 12-5

12-17 If the cast iron ring described in Problem 12-15 has a 2 mm air gap, calculate the new level of coil current required to establish the previously calculated level of flux density in the core.

12-18 The horseshoe-shaped magnet shown in Figure 12-11(b) has a magnetic path length of 50 cm and a cross-sectional area of 25 cm^2. The bar attached to the magnet's poles has a cross-sectional area of 35 cm^2, and the magnetic path through it is 15 cm long. Each air gap is approximately 5 mm thick. The magnet is cast iron and the bar is cast steel. Determine the current that must be passed through the 3000-turn coil to establish a total flux of 1500 μWb around the circuit.

12-19 The magnetic core shown in Figure 12-8 has four 0.5 mm air gaps. If the core material has $\mu_r = 800$ and the coil has 1400 turns, determine the current required to give a total core flux of 700 μWb.

12-20 Assume that the magnetic core shown in Figure 12-10 has a 1 mm air gap halfway along the section with the largest cross-sectinal area. The relative permeability of the material is 1000 and the flux density in the air gap is to be 0.75 T. Determine the number of coil turns required if the coil current is to be 500 mA.

12-21 In Problem 12-17 fringing at the 2 mm air gap tends to increase the cross-sectional area of the air gap by 10%. Recalculate the coil current required to establish the previously calculated level of flux density in the air gap.

12-22 The magnetic core shown in Figure 12-22 is constructed of sheet steel, and the coil has 900 turns. Calculate the coil current required to produce a flux of 330 μWb in the air gap.

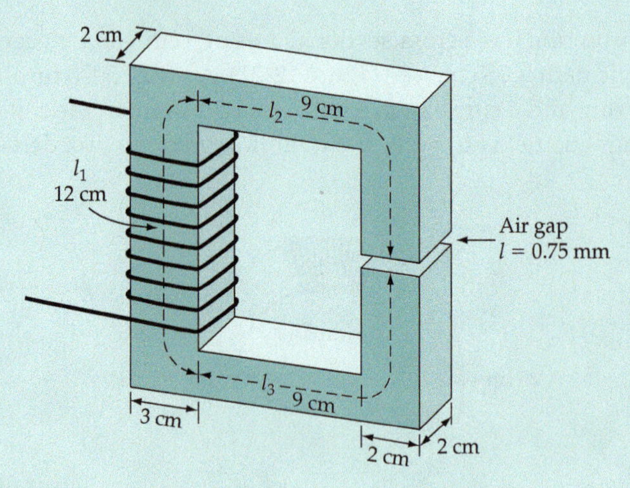

Figure 12-22

12-23 The magnetic core shown in Figure 12-12 is made of sheet steel and has the following dimensions: $l_1 = 8$ cm, $l_2 = l_3 = 15$ cm, cross-sectional area of l_1 path is 5×10^{-4} m^2, cross-sectional area of l_2 and l_3 paths is 2×10^{-4} m^2, and length of air gap is 1.75 mm. Calculate the required number of coil turns for a 5 mA current to produce a 1.2 T flux density in the air gap.

Section 12-6

12-24 An electromagnetic switch with a 1500 turn coil has a U-shaped sheet steel core with a cross-sectional area of 3 cm^2 and a magnetic path length of 10 cm, as illustrated in Figure 12-23. The sheet steel cross-bar lifted by the core has a magnetic path length of 4 cm, a cross-sectional area of 3 cm^2, and a weight of 80 gm. The two air gaps are 0.5 cm thick. Taking the relative permeability of sheet steel as 230, calculate the required coil current.

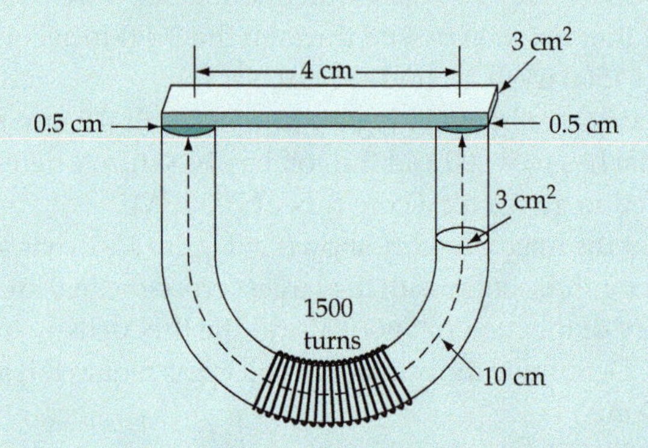

Figure 12-23

12-25 An electromagnet has two poles, each with a cross-sectional area of 35 cm^2. When the magnet is lifting a soft iron plate that completely covers the pole faces, the total flux around the magnetic circuit is 620 μWb. Determine the maximum weight of the iron plate that can be lifted.

12-26 For Problem 12-22, calculate the energy stored in the air gap.

12-27 For the electromagnet and steel bar described in Problem 12-18, calculate the maximum weight of steel bar that can be lifted.

Section 12-7

12-28 For Problem 12-24, recalculate the required coil current if the core and crossbar are laminated so that the cross-sectional area is reduced by 10%.

12-29 For Problem 12-22, assume that the core is constructed of laminated sheet steel so that the 2 cm thickness of the core actually contains only 1.8 cm of metal. Also assume that the fringing at the air gap increases its cross-sectional area by 10%. Recalculate the coil current required to produce the same flux density in the air gap.

12-30 For Problem 12-19, apply the same considerations (for a laminated core and air gap, fringing) described in Problem 12-29. Recalculate the coil current.

12-31 An electromagnet used for lifting steel plates has two poles each of which has cross-sectional areas of 100 cm^2. The magnet is constructed of laminated sheet steel so that the actual cross-sectional area of the magnetic path is 0.9 times the measured area. The magnetic path length through the core is 35 cm. The path length through the sheet steel being lifted is 20 cm, and the cross-sectional area of the sheets is 0.02 m^2. The weight of each sheet is 15 kg, and the air gaps are each 2 mm thick. Calculate the current that must flow in the 7000 turn coil. Take $\mu_r = 250$ for the core and for the steel plates.

Practice Problem Answers

12-2.1	10 mA
12-2.2	9000 A, 11.3 mT, 8.9 nWb
12-3.1	1.29 A
12-3.2	0.64 μWb
12-4.1	364, 312, 83
12-4.2	479 μWb
12-4.3	24 mA
12-5.1	167 mA
12-5.2	139 mA
12-6.1	0.29 T
12-9.1	348 mA

CHAPTER 13
DC Measuring Instruments

CONTENTS

Objectives

You will be able to:

1 Sketch and explain circuit diagrams for analog dc ammeters, voltmeters, and ohmmeters.

2 Calculate ammeter shunt resistances for various current ranges, voltmeter multiplier resistances for various voltage ranges, and the resistance measured at various points on analog ohmmeter scales.

3 Sketch and explain basic circuit/block diagrams for analog and digital electronic voltmeters, ammeters, and ohmmeters.

4 Solve problems involving ammeter and voltmeter resistances and the accuracy of resistance measurements by ammeter and voltmeter.

5 Explain megohmmeters and discuss their use.

6 Sketch the basic construction of an electrodynamic instrument, and explain its operation as a wattmeter.

7 Draw the circuit diagram of a Wheatstone bridge, derive its balance equations, and calculate the measured resistance.

INTRODUCTION

Electronic volt-ohm-milliammeters (VOMs) are partially based on electromechanical analog voltmeter, ohmmeter, and ammeter circuits. Although these instruments have been largely replaced by the very much more convenient digital instruments, knowledge of their operating principles can be useful. Electronic voltmeters use electronic amplifiers to present high input resistances and amplify low-level voltages. Digital voltmeters use analog-to-digital converters (ADCs) to convert measured voltages for digital display. Electronic ammeters and ohmmeters determine current and resistance by measuring the voltage drop across a resistance. The instrument used for measuring very high resistances is known as a *megohmmeter.* An *electrodynamic instrument* is used for power measurement.

13-1 BASIC AMMETERS

Analog DC Ammeter Circuit

Because the deflection of a PMMC instrument (see Section 11-6) is directly proportional to the current through its coil, the instrument is essentially an *ammeter.* However, it can be employed directly as an ammeter only for the very small current levels that flow in the moving coil. Modifications must be made where it is desired to have an ammeter that measures larger currents.

In the dc ammeter illustrated in Figure 13-1(a) a resistor known as a *shunt* is shown connected in parallel with the PMMC instrument. Figure 13-1(b) shows the equivalent circuit of the complete instrument, including the coil resistance R_m and the shunt resistance R_s. It is seen that a small portion of the current to be measured passes through the coil, and the rest passes through the shunt. From knowledge of the coil resistance and the meter *full-scale deflection* (FSD) current, the resistance of the shunt can be determined for any desired level of current to be measured. When the ammeter is

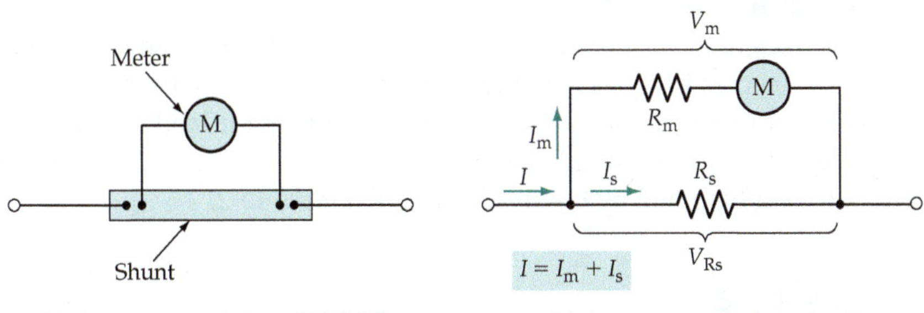

(a) Ammeter consisting of PMMC instrument and shunt

(b) Ammeter equivalent circuit

Figure 13-1 In an ammeter a low-resistance shunt causes most of the circuit current to be bypassed around the low-current PMMC instrument. The instrument measures a portion of the total current and indicates total current on its scale.

designed to indicate a current of (for example) 100 A, the meter scale is recalibrated to read 100 A at FSD and proportional levels at other points. The procedure for determining the shunt resistance value is demonstrated by Example 13-1.

Example 13-1

A PMMC instrument has a coil resistance of 200 Ω and produces full-scale deflection when the coil current is 750 μA. Determine the value of shunt resistance required for the instrument to function as an ammeter with a FSD = 1 A.

Solution

From Figure 13-1(b), the meter voltage is,

$$V_m = I_m \times r_m$$

At FSD,

$$V_m = I_{FSD} \times r_m = 750\ \mu A \times 200\ \Omega$$
$$= 150\ mV$$

and the shunt voltage is,

$$V_s = V_m = 150\ mV$$

shunt current,

$$I_s = I - I_m = 1\ A - 750\ \mu A$$
$$= 999.25\ mA$$

and the shunt resistance is,

$$R_s = \frac{V_m}{I_s} = \frac{150\ mV}{999.25\ mA}$$
$$\approx 0.15\ \Omega$$

Ammeter Resistance

It is very important for an ammeter to have a low resistance, because the ammeter is always connected in series with the load that is to have its current measured. If the ammeter resistance is not very much smaller than the load resistance, the load current can be substantially altered by the inclusion of the ammeter in the circuit. This is illustrated by Figure 13-2 and Example 13-2.

Example 13-2

An ammeter with a resistance of 1 Ω is to be used to measure the current supplied to a 4 Ω resistor from a 100 V source. Calculate the current through the resistor before the ammeter is connected and after it is included in the circuit.

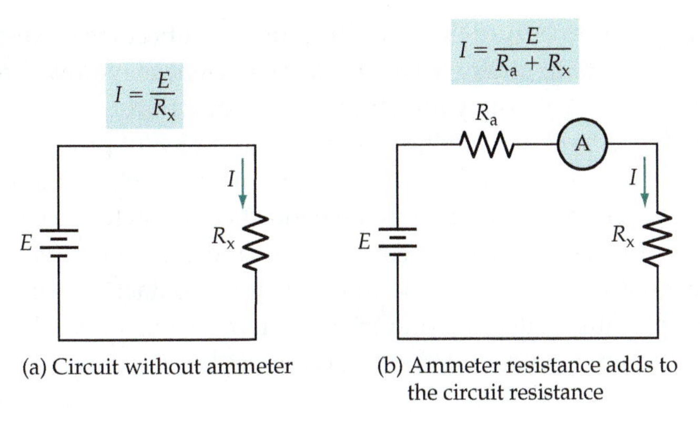

(a) Circuit without ammeter

(b) Ammeter resistance adds to the circuit resistance

Figure 13-2 An ammeter should have a very low resistance so that it does not add significantly to the total resistance in a circuit and introduce an error in the measurement.

Solution

Without the ammeter,

$$I = \frac{E}{R_x} = \frac{100\ \text{V}}{4\ \Omega}$$

$$= 25\ \text{A}$$

With the ammeter in circuit,

$$I = \frac{E}{R_x + R_a} = \frac{100\ \text{V}}{4\ \Omega + 1\ \Omega}$$

$$= 20\ \text{A}$$

Multirange Ammeters

A multirange ammeter can be constructed simply by using several values of shunt resistor with a rotary switch to select the desired range. Figure 13-3(a) shows the circuit arrangement. When an instrument is used in this fashion,

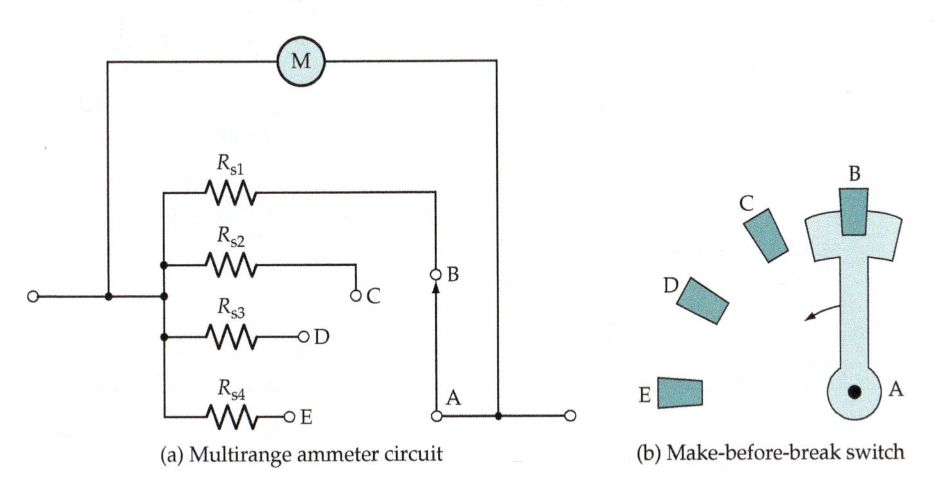

(a) Multirange ammeter circuit

(b) Make-before-break switch

Figure 13-3 In a multirange ammeter any one of several shunts can be selected. A make-before-break switch must be used to keep the low-current PMMC instrument shunted at all times.

care must be taken to ensure that the shunt does not become open-circuited, even for a brief instant; otherwise a very large current may flow through the coil of the instrument, possibly resulting in its destruction.

The *make-before-break* switch illustrated in Figure 13-3(b) protects an instrument from the possibility of the shunts becoming open-circuited in a multi-range ammeter. The wide-ended moving contact connects to the terminal it is being switched-to before it loses contact with the previous terminal. So, it *makes* contact with the next terminal before it *breaks* contact with the previous terminal. During the switching time there are two shunts in parallel with the instrument, and an open-circuited shunt is avoided.

Ayrton Shunt

Because ammeter shunts have very low resistance values, the switch contact resistances in parallel with the meter in the circuit shown in Figure 13-3 could introduce errors. The *Ayrton shunt* shown in Figure 13-4(a) avoids switch contact resistance in parallel with the meter, and protects the moving coil meter from the possibility of excessive current flow. When the moving contact of the switch is connected to terminal B, the resistance of the shunt in parallel with the meter is $(R_1 + R_2 + R_3)$. This is illustrated in Figure 13-4(b). When the moving contact is switched to terminal C, the shunt becomes $(R_1 + R_2)$, and resistor R_3 is now in series with the meter [Figure 13-4(c)]. Finally, with the moving contact at terminal D, the shunt is resistor R_1, and $(R_2 + R_3)$ is in series with the meter, [Figure 13-4(d)]. Note that there is a shunt in parallel with the meter at all times, and that the switch contact resistance is not part of the meter/shunt

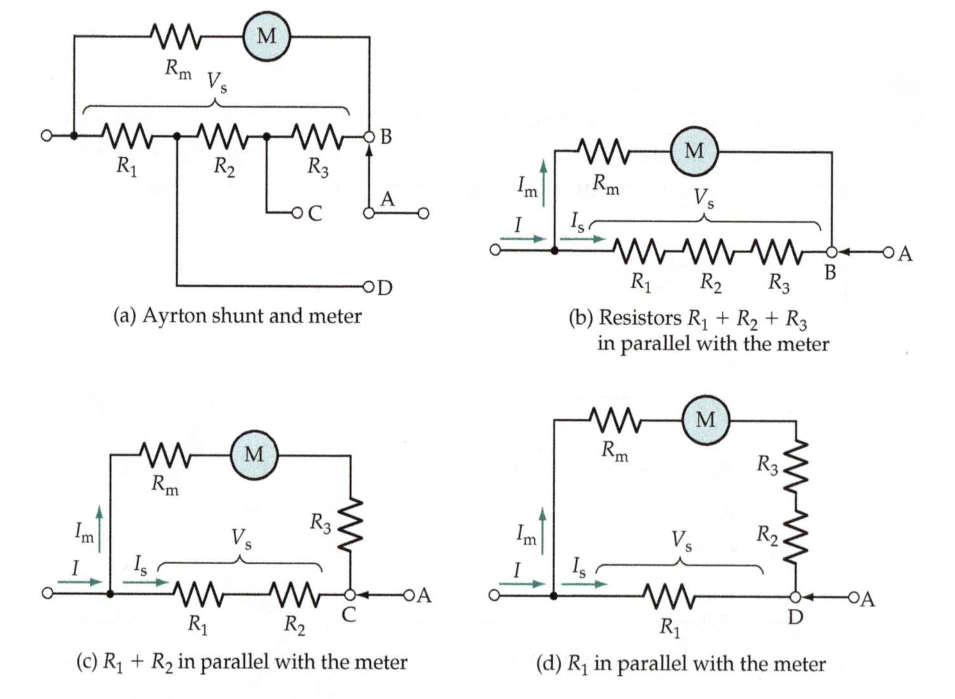

Figure 13-4 An Ayrton shunt has several series-connected shunts in parallel with the PMMC instrument. By selection of a connecting point at one of the junctions of the shunts, one or more shunts can be placed in parallel with the PMMC circuit while the others are in series with the meter.

parallel combination. The shunt resistances are normally so much smaller than the low-current meter resistance that they have virtually no effect when they are directly in series with the meter.

Practice Problems

13-1.1 The ammeter described in Example 13-1 is to have its range changed to 100 mA and 10 mA. Calculate the new shunt resistance values.

13-1.2 The ammeter in Example 13-1 is used to measure the current supplied to a 1 Ω load. If the supply voltage is 1 V, determine the measured current and the measurement error.

13-1.3 An ammeter using an Ayrton shunt, as illustrated in Figure 13-4, has $R_1 = R_2 = R_3 = 0.0334\ \Omega$. The meter resistance is 50 Ω, and the FSD current is 200 μA. Determine the ammeter full-scale range for each of the switched positions.

13-2 BASIC VOLTMETERS

Analog DC Voltmeter Circuit

The deflection of a PMMC instrument is proportional to the current through its coil, and the current is, of course, proportional to the voltage across the coil resistance; so the scale of the instrument could be calibrated to indicate the applied voltage. Because the coil resistance is usually quite small, this would produce a voltmeter capable of measuring only very low voltage levels. To increase the voltage range, the voltmeter resistance must be increased by connecting a resistor (R) in series with the instrument [see Figure 13-5(a)]. The series resistor increases the range of the voltmeter, so it is termed a *multiplier resistor*. As in the case of the ammeter, all calculations involving the voltmeter must include the meter resistance R_m [Figure 13-5(b)].

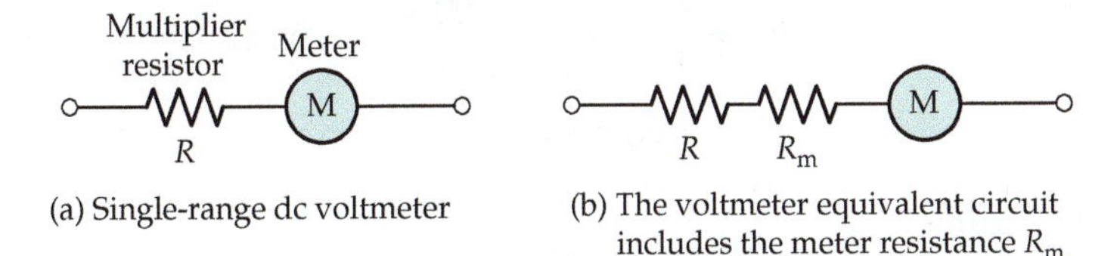

Figure 13-5 A dc voltmeter is constructed by connecting a resistor in series with a PMMC instrument. The pointer position is directly proportional to the applied voltage.

Multirange Voltmeters

Two multirange voltmeter circuits are illustrated in Figure 13-6. In Figure 13-6(a) any one of several multiplier resistors is selected by means of a rotary switch, as illustrated. Unlike the case of the ammeter, the rotary switch used with the voltmeter should be a *break-before-make* type; that is, the moving contact should

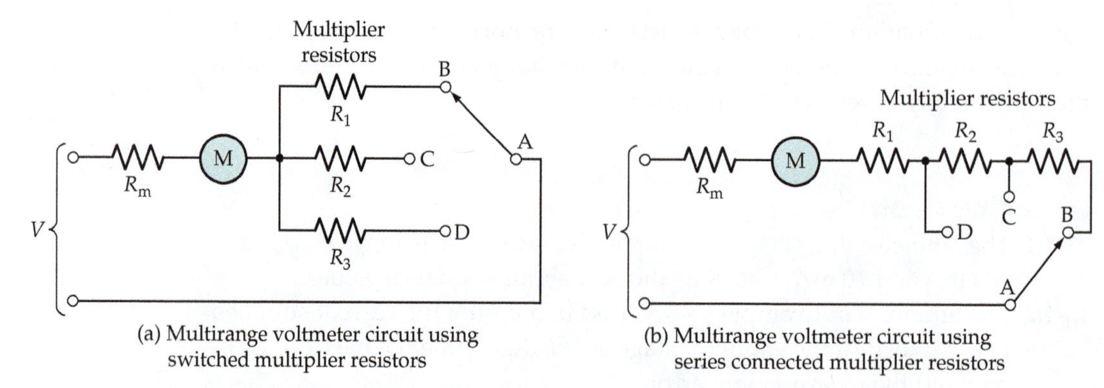

Figure 13-6 Multirange dc voltmeters use several standard value multiplier resistors and a range selector switch.

disconnect from one terminal before connecting to the next terminal. This is to keep the voltmeter resistance as high as possible. For the circuit shown in Figure 13-6(b), the multiplier resistors are in series, and each junction is connected to one of the switch terminals. The multiplier resistors are R_1, $(R_1 + R_2)$, or $(R_1 + R_2 + R_3)$.

Example 13-3

A PMMC meter with a coil resistance of 100 Ω and an FSD current of 50 μA is to be used in the voltmeter circuit in Figure 13-6(a). The voltmeter ranges are to be 100 V, 50 V, and 10 V. Determine the required values of multiplier resistor for each range.

Solution

$$I = \frac{E}{R_1 + R_m}$$

For the 100 V range,

$$R_1 = \frac{E}{I} - R_m = \frac{100\ \text{V}}{50\ \mu\text{A}} - 100\ \Omega$$

$$= 1.9999\ \text{M}\Omega$$

For the 50 V range,

$$R_1 = \frac{E}{I} - R_m = \frac{50\ \text{V}}{50\ \mu\text{A}} - 100\ \Omega$$

$$= 0.9999\ \text{M}\Omega$$

For the 10 V range,

$$R_1 = \frac{E}{I} - R_m = \frac{10\ \text{V}}{50\ \mu\text{A}} - 100\ \Omega$$

$$= 199.9\ \text{k}\Omega$$

Voltmeter Resistance

In Example 13-3, the total voltmeter resistance is

$$R_v = R \text{ of multiplier} + R_m$$

For the 100 V range this adds up to 2 MΩ. Dividing the 2 MΩ resistance by the 100 V range gives the *resistance per volt*, or *sensitivity*, of the voltmeter.

$$\text{Sensitivity} = \frac{R_v}{\text{range}} = \frac{2\ \text{M}\Omega}{100\ \text{V}}$$
$$= 20\ \text{k}\Omega/\text{V}$$

Calculating for the 50 V and 10 V ranges gives the same 20 kΩ/V constant. The voltmeter sensitivity is an important constant and is usually printed on the face of all analog voltmeters. To find the total voltmeter resistance, the resistance per volt is multiplied by the voltmeter range. For example, on a 50 V range with a sensitivity of 20 kΩ/V, the total voltmeter resistance is,

$$R_v = (20\ \text{k}\Omega/\text{V}) \times (50\ \text{V})$$
$$= 1\ \text{M}\Omega$$

A *voltmeter should have a very high resistance,* because it is normally connected *in parallel* with the circuit where the voltage is to be measured. If the voltmeter resistance is too low it can alter the circuit conditions; an effect termed *voltmeter loading*. To minimize voltmeter loading the voltmeter operating current should be very small, (its resistance should be high). Figure 13-7 and Example 13-4 illustrate the situation where a low-resistance voltmeter could substantially affect a circuit under investigation.

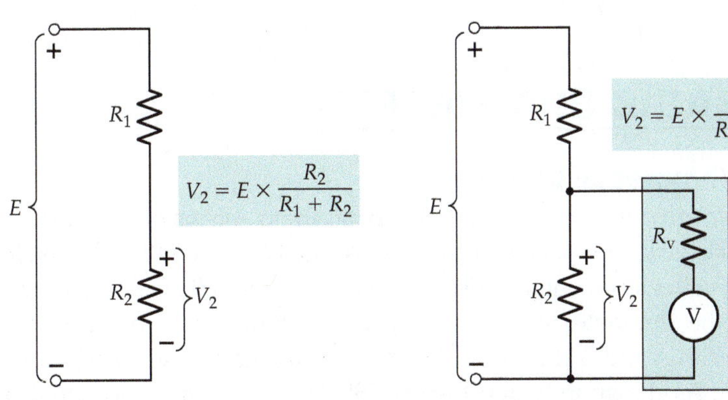

Figure 13-7 A voltmeter should have a very high resistance so that its current is extremely small. Voltmeter loading effect can change the conditions in a circuit and give a measurement error.

Example 13-4

For the circuit shown in Figure 13-7, $E = 100$ V, $R_1 = 100$ kΩ, and $R_2 = 1$ kΩ. The voltmeter range is 2 V and its sensitivity is 1 kΩ/V. Determine the voltage across resistor R_2, (a) without the voltmeter in the circuit, (b) with the voltmeter connected.

Solution

(a) Without the voltmeter:

$$V_2 = \frac{ER_2}{R_1 + R_2} = \frac{100\ \text{V} \times 1\ \text{k}\Omega}{100\ \text{k}\Omega + 1\ \text{k}\Omega}$$

$$\approx 0.99\ \text{V}$$

(b) With the voltmeter:

Voltmeter resistance, $R_v = (\text{range}) \times (\text{sensitivity}) = (2\ \text{V}) \times (1\ \text{k}\Omega/\text{V})$

$$= 2\ \text{k}\Omega$$

$$R_v \| R_2 = 2\ \text{k}\Omega \| 1\ \text{k}\Omega$$

$$\approx 666.7\ \Omega$$

$$V_2 = \frac{E(R_v \| R_2)}{R_1 + (R_v \| R_2)} \approx \frac{100\ \text{V} \times 666.7\ \Omega}{100\ \text{k}\Omega + 666.7\ \Omega}$$

$$\approx 0.66\ \text{V}$$

Practice Problems

13-2.1 The voltmeter in Example 13-3 is to have additional ranges of 100 mV, 30 V, and 250 V. Calculate the resistance of the multiplier resistor for each range.

13-2.2 A voltmeter on a 10 V range is used to measure the output voltage from the potentiometer in Figure 5-11. If the voltmeter sensitivity is 2 kΩ/V, determine the maximum and minimum measured voltages.

13-3 BASIC OHMMETERS

Simple Ohmmeter Circuit

The *ohmmeter* provides a quick, but not very accurate, means of resistance measurement. The basic *series ohmmeter* circuit illustrated in Figure 13-8(a) is made up of a moving-coil instrument in series with a battery and adjustable resistor (R_1). A and B are the ohmmeter terminals where the resistance to be measured (R_x) is connected. With A and B short-circuited ($R_x = 0$), resistor R_1 is adjusted for FSD on the meter. As shown in Figure 13-8(b), FSD is taken as an indication of zero ohms. When terminals A and B are open-circuited, the pointer should indicate infinity. So, the zero deflection point on the dial is marked as infinite resistance.

When an unknown resistance (R_x) is connected to terminals A and B, some current flows through the meter, giving a reading between zero and infinity. As will be seen, for a given value of R_x the pointer position on the dial depends on the resistance of R_1.

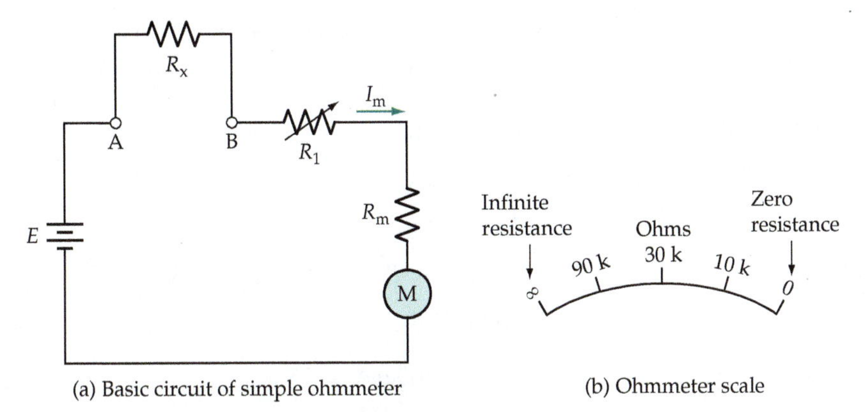

(a) Basic circuit of simple ohmmeter (b) Ohmmeter scale

Figure 13-8 **A simple ohmmeter consists of a PMMC instrument, a series resistor R_1, and a battery. When $R_x = 0$ the pointer deflects to full scale. When R_x is open-circuited the pointer deflection is zero. At $R_x = R_1$ the pointer indicates half-scale deflection.**

From Figure 13-8(a) it is seen that the meter current is,

$$I_m = \frac{E}{R_1 + R_x + R_m} \quad \textbf{(13-1)}$$

where R_m is the coil resistance of the instrument. If $R_m \ll R_1$, the equation is simplified to

$$I_m \approx \frac{E}{R_1 + R_x} \quad \textbf{(13-2)}$$

Example 13-5

The series ohmmeter shown in Figure 13-8(a) is made up of a 3 V battery, a 100 μA (FSD) meter, and a resistance R_1 of 30 kΩ. Determine the unknown resistance R_x when the pointer indicates: (a) 0.5 FSD, (b) 0.25 FSD, and (c) 0.75 FSD.

Solution

When $R_x = 0$:

Eq. 13-2, $$I_m \approx \frac{E}{R_1 + R_x} = \frac{3\text{ V}}{30\text{ k}\Omega + 0}$$

$$= 100\ \mu\text{A} = \text{FSD}$$

(a) At 0.5 FSD, $$I_m = 0.5 \times 100\ \mu\text{A} = 50\ \mu\text{A}$$

From Eq. 13-2, $$R_1 + R_x = \frac{E}{I_m} = \frac{3\text{ V}}{50\ \mu\text{A}}$$

$$= 60\text{ k}\Omega$$

and $$R_x = 60\ \text{k}\Omega - R_1 = 60\ \text{k}\Omega - 30\ \text{k}\Omega$$
$$= 30\ \text{k}\Omega$$

(b) At 0.25 FSD, $$I_m = 0.25 \times 100\ \mu\text{A} = 25\ \mu\text{A}$$

$$R_1 + R_x = \frac{E}{I_m} = \frac{3\ \text{V}}{25\ \mu\text{A}} = 120\ \text{k}\Omega$$

and, $$R_x = 120\ \text{k}\Omega - R_1 = 120\ \text{k}\Omega - 30\ \text{k}\Omega$$
$$= 90\ \text{k}\Omega$$

(c) At 0.75 FSD, $$I_m = 0.75 \times 100\ \mu\text{A} = 75\ \mu\text{A}$$

$$R_1 + R_x = \frac{E}{I_m} = \frac{3\ \text{V}}{75\ \mu\text{A}} = 40\ \text{k}\Omega$$

and, $$R_x = 40\ \text{k}\Omega - R_1 = 40\ \text{k}\Omega - 30\ \text{k}\Omega$$
$$= 10\ \text{k}\Omega$$

The ohmmeter scale is now as marked in Figure 13-8(b).

The results of Example 13-5 demonstrate that the ohmmeter scale is nonlinear. Also, the portion of the scale from 0.75 FSD to FSD includes all resistance measurements from 10 kΩ to zero, and the portion from zero deflection to 0.25 FSD includes all values from 90 kΩ to infinity. This shows that the useful range of the ohmmeter scale is approximately from 0.25 FSD to 0.75 FSD. The actual resistance values marked on the scale depend on the value of R_1, which (instead of being variable) should be a fixed-value precision resistor. Where $R_1 = 30\ \text{k}\Omega$, as in Example 13-5, the measured resistance at the center of the scale is 30 kΩ.

Ohmmeter with Zero Adjust

The circuit of Figure 13-8(a) relies on the battery voltage remaining absolutely constant. When the battery terminal voltage falls (as it does with use), the instrument scale is no longer accurate. So some means of adjusting for battery voltage variations must be built into the circuit. Variable resistor R_2 shown in the circuit of Figure 13-9 provides the necessary adjustment. To measure an unknown resistance (R_x), the ohmmeter terminals are first short-circuited, and

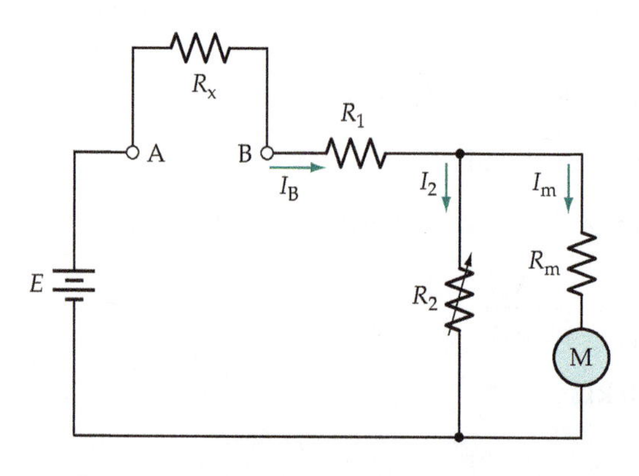

Figure 13-9 Series ohmmeter circuit with zero adjustment. To use the ohmmeter, the terminals are first short-circuited and R_2 is adjusted for zero ohms indication. Then the unknown resistor (R_x) is connected and the indicated resistance is read from the meter.

R_2 is adjusted to give FSD (zero ohms) on the meter. Then the terminals are open-circuited, and R_x is connected directly to the ohmmeter terminals. The parallel combination of R_2 and R_m is usually very much smaller than R_1, so that normally $R_x = R_1$ when the meter indicates half-scale. The presence of R_2 also allows for new values of standard resistor R_1, to be switched into the circuit to change the ohmmeter range.

Example 13-6

The ohmmeter in Figure 13-9 has $E = 3$ V, $R_1 = 30$ kΩ, $R_2 = 20$ Ω, $R_m = 20$ Ω, and the meter current is 50 μA at full scale. Determine the meter current at $R_x = 0$ and at $R_x = 30$ kΩ.

Solution

When $R_x = 0$:

Eq. 13-2,
$$I_B \approx \frac{E}{R_1 + R_x} \approx \frac{3\text{ V}}{30\text{ k}\Omega + 0}$$
$$= 100\ \mu\text{A}$$

Because $R_2 = R_m$, I_B divides equally between R_2 and R_m.

$$I_2 = I_m = 0.5 I_B = 0.5 \times 100\ \mu\text{A}$$
$$= 50\ \mu\text{A} = \text{FSD}$$

When $R_x = 30$ kΩ:

Eq. 13-2,
$$I_B \approx \frac{E}{R_1 + R_x} \approx \frac{3\text{ V}}{30\text{ k}\Omega + 30\text{ k}\Omega}$$
$$= 50\ \mu\text{A}$$
$$I_2 = I_m = 0.5 I_B = 0.5 \times 50\ \mu\text{A}$$
$$= 25\ \mu\text{A} = 0.5\ \text{FSD}$$

Note that the scale for the ohmmeter in Example 13-6 is exactly the same as that for the circuit Example 13-5, [Figure 13-8(b)]. This is because the circuit in Example 13-6 has $R_x = R_m$ and a meter with FSD = 50 μA instead of 100 μA. Practice Problem 13-3.2 offers an opportunity to show that resistor R_2 can be adjusted to deal with a falling battery voltage.

Practice Problems

13-3.1 Determine the value of internal resistor R_1, for an analog ohmmeter that is to indicate 10 kΩ, at half-scale. Also, calculate the measured resistance when the ohmmeter indicates 0.33 FSD and 0.67 FSD.

13-3.2 For the ohmmeter in Example 13-6, calculate the new resistance that R_2 must be adjusted to when E falls to 2.5 V.

13-4 ELECTRONIC VOLTMETERS (ANALOG AND DIGITAL)

Analog Electronic Voltmeter

An analog electronic voltmeter uses an electronic amplifier to improve on the performance of an electromechanical voltmeter. The amplifier gives the instrument a very high input resistance, so that voltmeter loading effect (Section 13-2) is considerably reduced. Also, voltage levels that are normally too small for measuring with an electromechanical voltmeter can be amplified to measurable levels in an electronic instrument.

The basic circuit of one type of electronic voltmeter is illustrated in Figure 13-10(a). This particular circuit is made up of three stages: an input attenuator,

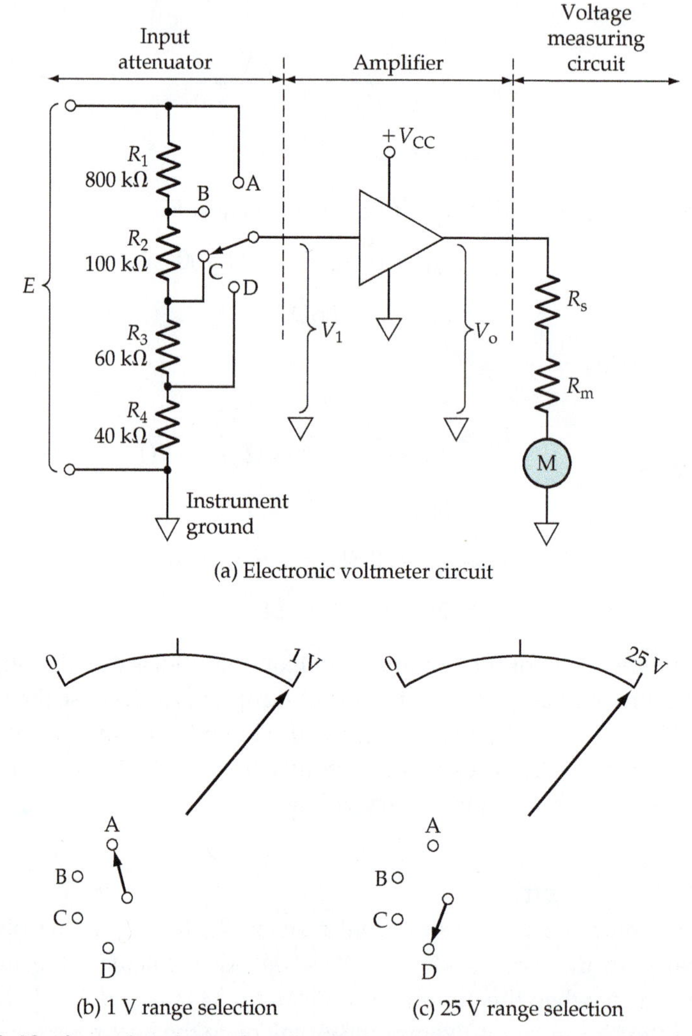

Figure 13-10 An analog electronic voltmeter basically consists of an input attenuator, an electronic amplifier, and a voltage measuring stage. The attenuator selection switch is the voltmeter range selector.

an electronic amplifier, and a voltage measuring stage. Note the large triangular graphic symbol normally used to represent an amplifier. Also note the small triangular symbol representing the instrument ground.

The input attenuator is simply a voltage divider (see Section 5-3) which divides (or attenuates) high input voltages to measurable levels. The amplifier has a very high input resistance, so that there is virtually no loading effect on the attenuator resistors. It also has a low output resistance to supply the current required by the voltage measuring stage. The amplifier voltage gain (or amplification) is normally 1, which means that a 1 V input produces a 1 V output. Its function in the circuit shown is solely to offer a high input resistance and a low output resistance. In this situation it is said to be a buffer between the attenuator and voltage-measuring stages. So, it is referred to as a *buffer amplifier.* A dc supply voltage (V_{CC}) must be provided for the amplifier, and this may be derived from a battery or power supply contained within the instrument. The internal operation of the amplifier cannot be understood until electronic devices are studied.

The voltage measuring stage shown in Figure 13-10 is typically designed to give meter FSD for an amplifier output of 1 V. The attenuator switch is the voltmeter range-selection switch. With the switch at position A, a maximum input of 1 V is passed to the voltage measuring-stage to give FSD. So, position A of the range-selection switch is identified at the 1 V range position [see Figure 13-10(b)].

When the selection switch is at position D, the voltage divider theorem gives the attenuator output as,

$$V_i = \frac{ER_4}{R_1 + R_2 + R_3 + R_4}$$

or,

$$E_{max} = \frac{V_i(R_1 + R_2 + R_3 + R_4)}{R_4}$$

For FSD, $V_i = 1$ V.

So, the maximum input voltage to the attenuator is,

$$E_{max} = \frac{1\text{ V }(800\text{ k}\Omega + 100\text{ k}\Omega + 60\text{ k}\Omega + 40\text{ k}\Omega)}{40\text{ k}\Omega}$$

$$= 25\text{ V}$$

As illustrated in Figure 13-10(c), position D of the switch gives a 25 V voltmeter range.

Example 13-7

Calculate the input resistance of the voltmeter in Figure 13-10(a) if the amplifier input resistance has no effect on the attenuator. Also, determine the voltmeter range at positions B and C of the range-selection switch.

Solution

$$
\begin{aligned}
R_i &= R_1 + R_2 + R_3 + R_4 \\
&= 800\ \text{k}\Omega + 100\ \text{k}\Omega + 60\ \text{k}\Omega + 40\ \text{k}\Omega \\
&= 1\ \text{M}\Omega
\end{aligned}
$$

At position B,

$$
\begin{aligned}
E_{max} &= \frac{V_i(R_1 + R_2 + R_3 + R_4)}{R_2 + R_3 + R_4} \\
&= \frac{1\ \text{V}\ (800\ \text{k}\Omega + 100\ \text{k}\Omega + 60\ \text{k}\Omega + 40\ \text{k}\Omega)}{100\ \text{k}\Omega + 60\ \text{k}\Omega + 40\ \text{k}\Omega} \\
&= 5\ \text{V}
\end{aligned}
$$

At position C,

$$
\begin{aligned}
E_{max} &= \frac{V_i(R_1 + R_2 + R_3 + R_4)}{R_3 + R_4} \\
&= \frac{1\ \text{V}\ (800\ \text{k}\Omega + 100\ \text{k}\Omega + 60\ \text{k}\Omega + 40\ \text{k}\Omega)}{60\ \text{k}\Omega + 40\ \text{k}\Omega} \\
&= 10\ \text{V}
\end{aligned}
$$

An electronic voltmeter circuit can be modified to measure low voltage levels by arranging for the amplifier to have a voltage gain greater than 1. For example, if the amplifier has a precise gain of 10, an input of 100 mV to the attenuator produces a 1 V output. So the instrument scale can be calibrated for an FSD of 100 mV.

Digital Voltmeter

Digital instruments use a circuit known as an *analog-to-digital converter* (ADC). As with other electronic circuits, the operation of an ADC cannot be understood until electronic devices are studied; however, its function can be explained. The quantity to be measured is first divided into discrete levels. For example, as illustrated in Figure 13-11(a), if 1 V is divided into 1000 equal levels, each level represents 1 mV. In this way, the input voltage is identified as a number. (For example, 0.5 V is 500 discrete 1 mV levels, so 0.5 V can be assigned the number 500.) This number is then represented by a code at the ADC output terminals [see Figure 13-11(b)] in which voltages are either *high* or *low,* and the code is expressed digitally in zeros and ones. Because the ADC illustrated has five output terminals, it is said to have a *5-bit output.*

The number of output bits for the ADC depends on the desired *resolution* of the measurement. A voltage that is divided into 1000 equal levels for measurement has a resolution of 1 in 1000. In this case, if the maximum measured voltage is 1 V it can only be measured in 1 mV steps. So, 94.3 mV would be indicated

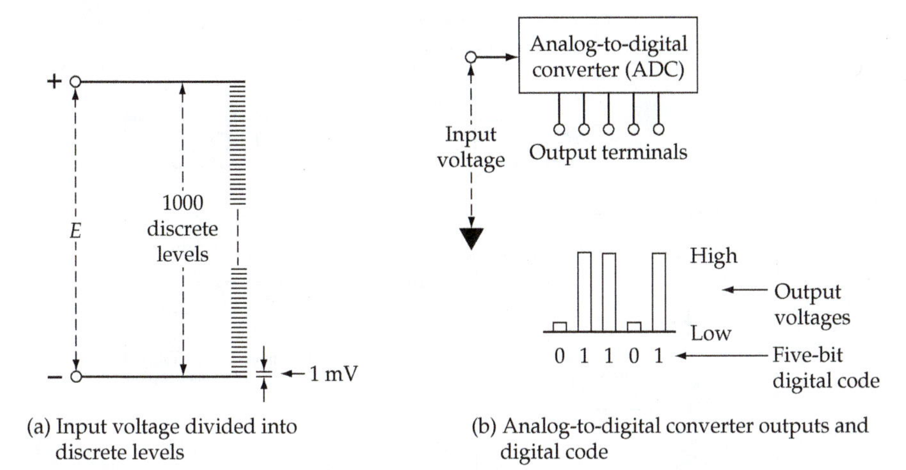

(a) Input voltage divided into discrete levels

(b) Analog-to-digital converter outputs and digital code

Figure 13-11 An analog-to-digital converter (ADC) used in digital instruments produces an output in the form of a digital code that represents the input voltage.

as 94 mV. If 1 V is divided into 100 equal (10 mV) steps, 94.3 mV would be measured as 90 mV. It can be shown that,

$$\text{Measurement resolution} = 2^n$$

where n is the number of bits in the ADC output.

Example 13-8

Determine the measurement resolution obtained with 5-bit and 10-bit ADCs.

Solution

5-bit ADC,

$$\text{resolution} = 2^n = 2^5 = 32$$

10-bit ADC,

$$\text{resolution} = 2^n = 2^{10} = 1024$$

The basic digital voltmeter circuit shown in Figure 13-12 is quite similar to the analog meter circuit in Figure 13-10. The instrument has an input attenuator and a similar amplifier to provide a high input resistance and a low output resistance. The amplifier may also have a precise voltage gain for low voltage measurements. In the case of the digital voltmeter, the amplifier output is converted into digital form (*digitized*) by the ADC.

The ADC output is further processed by the *display driver,* which selects the appropriate display segments to indicate the measured quantity, (see Section 2-2). Note that in Figure 13-12 the decimal point on the display is selected by the position of the range switch, which is made up of two rotary switches acting together. These are known as *ganged switches.* The decimal point between 1 and 9 is energized on the 2 V range, so that the meter indicates a maximum

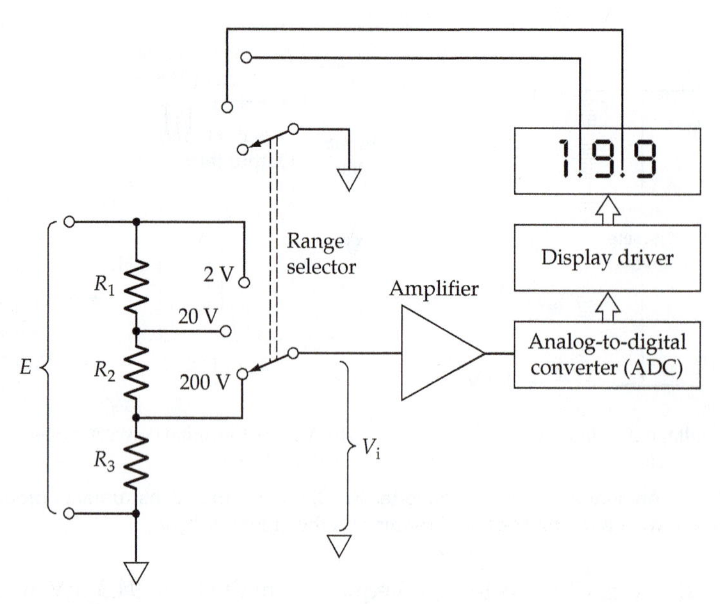

Figure 13-12 A digital voltmeter is similar to an analog electronic voltmeter except that the voltage measuring stage uses an analog-to-digital converter, a display driver, and a digital display.

of 1.99 V. On the 20 V range, the decimal point between the two nines is selected, to display a maximum of 19.9 V. Most digital meters have *autoranging* circuits, which select the appropriate range and decimal point position.

Practice Problems

13-4.1 Calculate the new resistance values for R_1 to R_4 in the attenuator in Figure 13-10(a) to give voltage ranges of 1 V, 10 V, 100 V, and 1000 V. The input resistance is to remain 1 MΩ.

13-4.2 Determine the measurement resolution of a digital voltmeter using an ADC with (a) a 4-bit output, and (b) an 8-bit output.

13-4.3 Calculate the required resistance values for R_1, R_2, and R_3 in the circuit in Figure 13-12 if the meter input resistance is to be 1 MΩ.

13-5 ELECTRONIC AMMETERS AND OHMMETERS

Electronic Ammeter

Current measurement by an electronic instrument is made simply by using an electronic voltmeter circuit to measure the voltage drop across a low resistance shunt, as shown in Figure 13-13. If the current to be measured (I) is 1 A and the shunt resistance is 0.1 Ω, as illustrated, the shunt voltage drop is 100 mV. In this case, an amplifier with a voltage gain of 10 produces a 1 V output to the

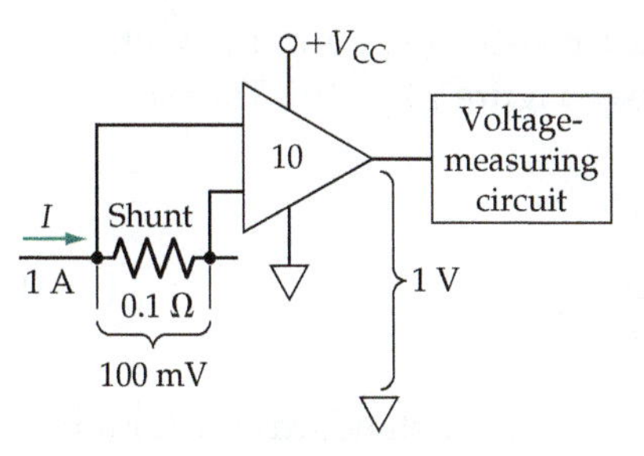

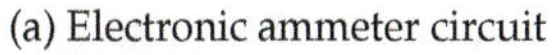

(a) Electronic ammeter circuit

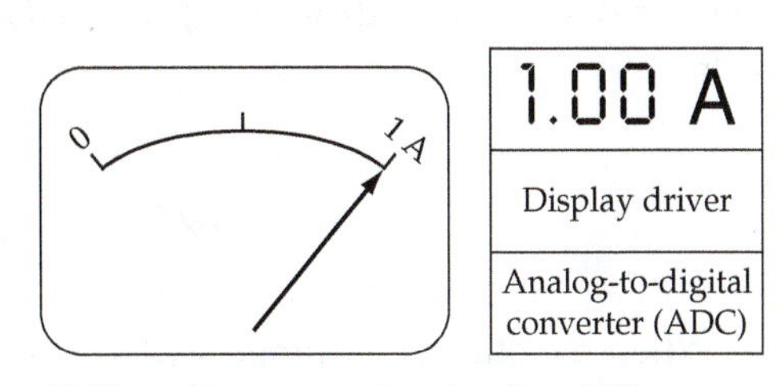

(b) The voltage-measuring circuit could be either analog or digital

Figure 13-13 Current measurement by means of an electronic instrument. The voltmeter circuit is used to measure the voltage drop across the shunt resistance, and the instrument display is calibrated to indicate current.

voltage measuring stage. The instrument display (analog or digital) can be designed to indicate 1 A [Figure 13-13(b)].

As explained in Section 13-1, an ammeter must have a very low resistance so that it does not affect the current to be measured. Instead of listing the ammeter resistance, the voltage drop across the terminals of electronic ammeters is usually specified for a given current level. This is termed the *burden voltage.* For the circuit in Figure 13-13(a), the ammeter resistance is the resistance of the shunt (0.1 Ω), and the burden voltage for the 1 A current level is 100 mV.

Electronic Ohmmeter

One type of ohmmeter circuit used with an electronic instrument is illustrated in Figure 13-14. The circuit consists of a 1.5 V battery (E_B), several standard resistors (R_1, R_2, R_3), and an electronic voltmeter circuit (analog or digital) set on a 1.5 V range. The resistance to be measured (R_x) is connected in parallel with the voltmeter (at terminals A and B), as illustrated. The battery voltage is

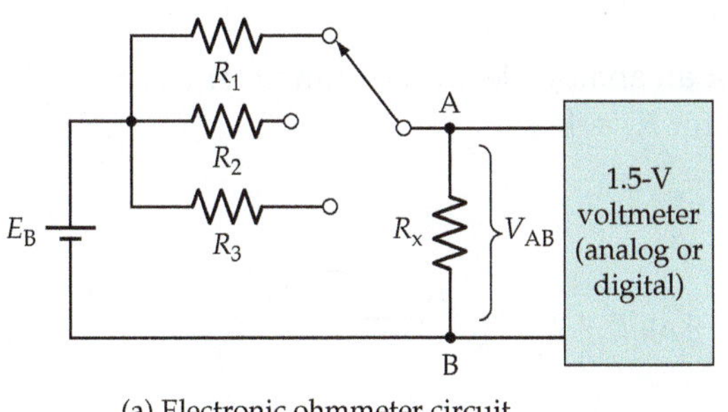

(a) Electronic ohmmeter circuit

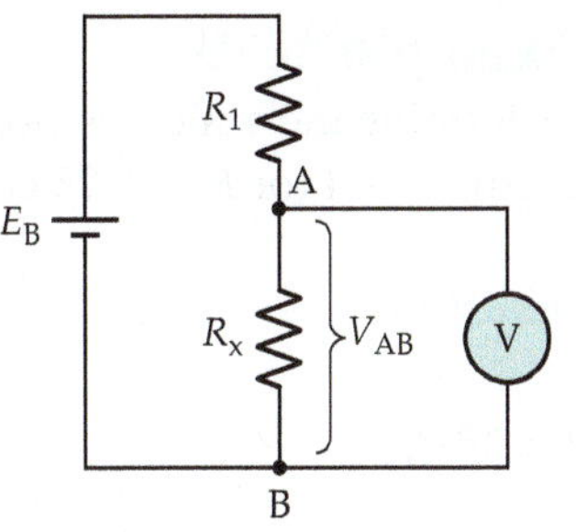

(b) E_B is divided across R_1 and R_x

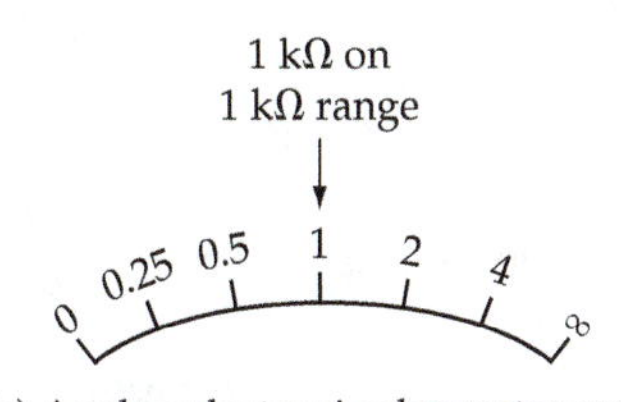

(c) Analog electronic ohmmeter scale

Figure 13-14 In a series ohmmeter used in an electronic instrument, a battery voltage is divided across a standard resistor (R_1) and the resistance to be measured (R_x). The voltmeter measures the voltage drop across R_x and indicates the resistance.

divided across R_x and the selected standard resistor (R_1), and the voltmeter measures the voltage drop (V_{AB}) across R_x [see Figure 13-14(b)]. The measured voltage is,

$$V_{AB} = E_B \times \frac{R_x}{R_1 + R_x} \qquad \textbf{(13-3)}$$

When the ohmmeter range switch is set to the 1 kΩ standard resistor and a 1 kΩ (unknown) resistor is connected at terminals A and B, the voltage across R_x is,

$$V_{AB} = 0.5E_B = 0.5 \times 1.5 \text{ V}$$
$$= 0.75 \text{ V}$$

This is the center point on the scale of the 1.5 V analog instrument, and it represents a 1 kΩ measured resistance, [Figure 13-14(c)]. The analog meter will always indicate half-scale when R_x equals the standard resistor on any selected range.

When terminals A and B are short-circuited, V_{AB} is zero, and the analog voltmeter pointer indicates zero voltage on the scale. So, the zero voltage point is also the zero resistance point on the resistance scale. When A and B are open-circuited, V_{AB} equals E_B, and the pointer indicates FSD, (1.5 V) Consequently, FSD represents an R_x of infinity (∞). (Note that this is the opposite of the basic ohmmeter scale discussed in Section 13-3.) An adjustment method can be included to deal with variations in the battery voltage.

Where a digital display is used, the digital meter can be designed to produce a 1 kΩ reading for 0.75 V across R_x, a 0 Ω reading for $V_{AB} = 0$ V, and appropriate resistance readings for other V_{AB} levels.

Example 13-9

Determine the scale positions for an analog electronic ohmmeter as shown in Figure 13-14 for $R_x = 0.25R_1$ and for $R_x = 4R_1$.

Solution

Eq. 13-3,

$$V_{AB} = E_B \times \frac{R_x}{R_1 + R_x}$$

or,

$$\text{Deflection} = \text{FSD} \times \frac{R_x}{R_1 + R_x}$$

For $R_x = 0.25R_1$,

$$\text{Deflection} = \text{FSD} \times \frac{0.25R_1}{R_1 + (0.25R_1)}$$
$$= 0.2 \text{ FSD}$$

For $R_x = 4R_1$,

$$\text{Deflection} = \text{FSD} \times \frac{4R_1}{R_1 + (4R_1)}$$

$$= 0.8\ \text{FSD}$$

Practice Problems

13-5.1 An electronic ammeter has a 300 mV burden voltage when measuring 100 mA. Calculate the resistance of the ammeter shunt.

13-5.2 Calculate the resistances measured by the ohmmeter in Example 13-9 when $R_1 = 1\ \text{k}\Omega$ and the analog meter indicates 0.33 and 0.66 of full scale.

13-6 MEASURING RESISTANCE BY AMMETER AND VOLTMETER

If the voltage across a resistor and the current flowing through it are measured, the resistance value can be calculated by applying Ohm's law. However, an error occurs depending on how the ammeter and voltmeter are connected, and this error may be insignificantly small or large enough to be important.

Consider the arrangement shown in Figure 13-15(a). Because the voltmeter is connected directly across R_x, it measures the actual resistor voltage. However, the ammeter measures the resistor current I_x and the current I_V that flows through the voltmeter.

$$\text{Measured resistance} = \frac{\text{Voltmeter reading}}{\text{Ammeter reading}}$$

$$R = \frac{V_x}{I_x + I_V} \qquad (13\text{-}4)$$

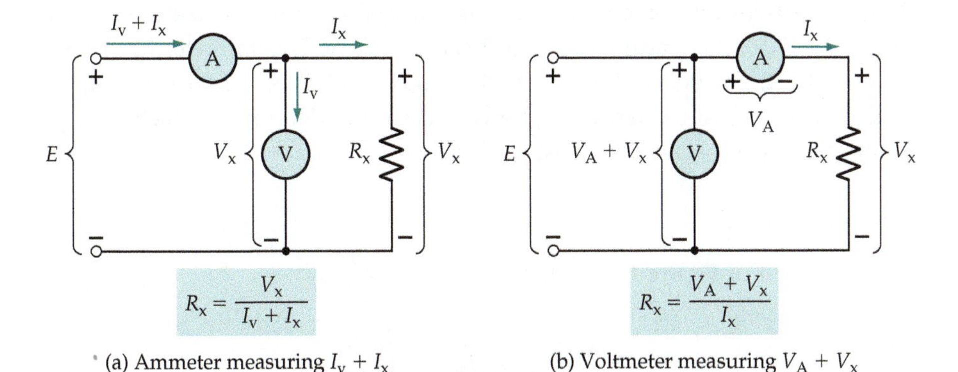

(a) Ammeter measuring $I_V + I_X$ (b) Voltmeter measuring $V_A + V_X$

Figure 13-15 Errors can occur when a resistance is measured by use of an ammeter and voltmeter. To minimize errors, the voltmeter should be connected directly across R_x when R_x is low, and the ammeter should be directly in series with R_x when R_x is high.

The actual resistance of R_X is $R_X = V_X/I_X$, so the presence of I_V in the equation introduces an error into the result. If I_V is very much smaller than I_X, the error may be insignificant. Obviously, this requires that I_X be a large current, which is the case when R_X is a small resistance.

Now consider the arrangement illustrated in Figure 13-15(b), where R_X and the ammeter are in series and the voltmeter is connected in parallel with the two of them. In this case, the ammeter measures the actual current through R_X, but the voltmeter measures the voltage across R_X plus the voltage across the ammeter. So,

$$R = \frac{V_X + V_A}{I_X} \tag{13-5}$$

Again, the actual resistance of R_X is $R_X = V_X/I_X$, and consequently, the presence of V_A in Equation 13-5 produces an error in the result. If V_A is very much smaller than V_X, the error could be insignificant. This requires that V_X be a large voltage, which means that R_X should be a large resistance.

It is seen that for the greatest accuracy when R_X is a small resistance, the voltmeter should be connected directly across R_X. For greatest accuracy when R_X is a large resistance, the ammeter should be connected directly in series with R_X. The correct arrangement can be easily determined by first connecting the ammeter in series with R_X, then observing the ammeter reading with the voltmeter temporarily connected directly across R_X [i.e., connected as in Figure 13-15(a)]. If the ammeter reading is not noticeably altered when the voltmeter is connected, the readings will give an accurate result. When the ammeter reading is noticeably changed by connecting the voltmeter, the voltmeter should be moved to the other side of the ammeter.

Example 13-10

An ammeter and voltmeter connected as in Figure 13-15(a) indicate 10 A and 99 V, respectively. When the voltmeter is changed to reconnect the circuit as in Figure 13-15(b), the readings become 10 A and 100 V. The ammeter has a resistance of 0.1 Ω. The voltmeter is on its 100 V range and its sensitivity is 20 kΩ/V. Calculate the measured resistance for each case, and determine which arrangements give the most accurate result.

Solution

For the circuit of Figure 13-15(a):

$$R_X = \frac{V_X}{I_X + I_V} = \frac{99\text{ V}}{10\text{ A}}$$

$$= 9.9\ \Omega$$

and the voltmeter resistance is,

$$R_V = 20\ \text{k}\Omega/\text{V} \times (\text{voltmeter range})$$
$$= 20\ \text{k}\Omega/\text{V} \times 100\ \text{V}$$
$$= 2\ \text{M}\Omega$$

$$I_V = \frac{V_X}{R_V} = \frac{99\ \text{V}}{2\ \text{M}\Omega}$$
$$= 49.5\ \mu\text{A}$$

So, $$I_V \ll I_X$$

For the circuit of Figure 13-15(b):

$$R_X = \frac{V_X + V_A}{I_X} = \frac{100\ \text{V}}{10\ \text{A}}$$
$$= 10\ \Omega$$

and, $$V_A = I_A R_A = 10\ \Omega \times 0.1\ \Omega$$
$$= 1\ \text{V}$$

In this case V_A is not very much smaller than V_X, and so the connection shown in Figure 13-15(a) gives the most accurate result.

Practice Problem

13-6.1 The insulation resistance of an electrical cable is to be measured by use of a 1000 V source, a 1500 V voltmeter with a sensitivity of 10 kΩ/V, and a 500 μA ammeter with a resistance of 100 Ω. If the insulation resistance is 5 MΩ, determine the measured resistance using each of the two methods of connecting the instruments.

13-7 THE MEGOHMMETER

The *megohmmeter* (or *megger*) is an instrument designed for measuring very high resistances, such as the insulation resistance of electrical cables. A high voltage source is required to pass a measurable current through such resistances. So, the megger is essentially an ohmmeter with a very low current instrument and a high voltage source. As illustrated in Figure 13-16(a), the voltage is usually produced by a hand-cranked generator. The generated voltage may range from 100 V to 2.5 kV.

As in the case of a low-resistance analog ohmmeter, the scale of the megger indicates *infinity* (∞) when measuring an open-circuit, zero on a short-circuit, and half-scale when the unknown resistance equals a standard resistor inside

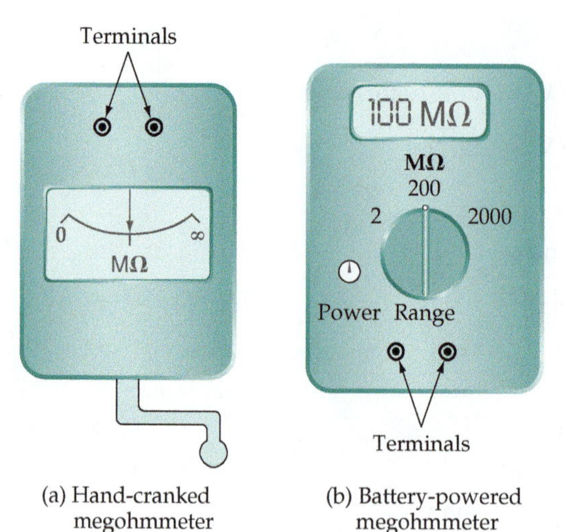

Figure 13-16 A megohmmeter (megger) is essentially an ohmmeter for measuring very high resistances. A high-voltage source and a low-current meter are required.

the megohmmeter. At other points on the scale, the deflection is proportional to the ratio of the unknown and standard resistors. The range of the instrument can be altered by switching different values of standard resistor into the circuit.

Battery-powered megohmmeters are also available, and these are essentially very high resistance ohmmeters. One such instrument is illustrated in Figure 13-16(b). The battery voltage is typically increased (by electronic circuitry) to a level of 1000 V in order to produce a measurable current through the unknown resistance. The measurement is made when the power button is pressed and held briefly. This arrangement minimizes the current drain on the battery.

13-8 POWER MEASUREMENT

The power supplied to a load in a dc circuit can be determined by measuring the load voltage and current and multiplying them together: $P = EI$. However, it is much more convenient to have an instrument that indicates power directly. The meter used for this purpose is called a *wattmeter,* and the analog instrument applied as a wattmeter is known as a *dynamometer,* or as an *electrodynamic instrument.* The construction of an electrodynamic instrument to some extent resembles the PMMC instrument (see Section 11-6); the major difference being that the permanent magnet is replaced by two *field coils,* as illustrated in Figure 13-17(a). (Compare this to Figure 11-17.) The magnetic field in which the lightweight moving coil is situated is generated by passing current through the stationary field coils. Then, when a current is passed through the moving coil, the moving coil and the meter pointer are deflected.

The deflection of the pointer of a electrodynamic instrument is proportional to the current through the moving coil; it is also proportional to the flux

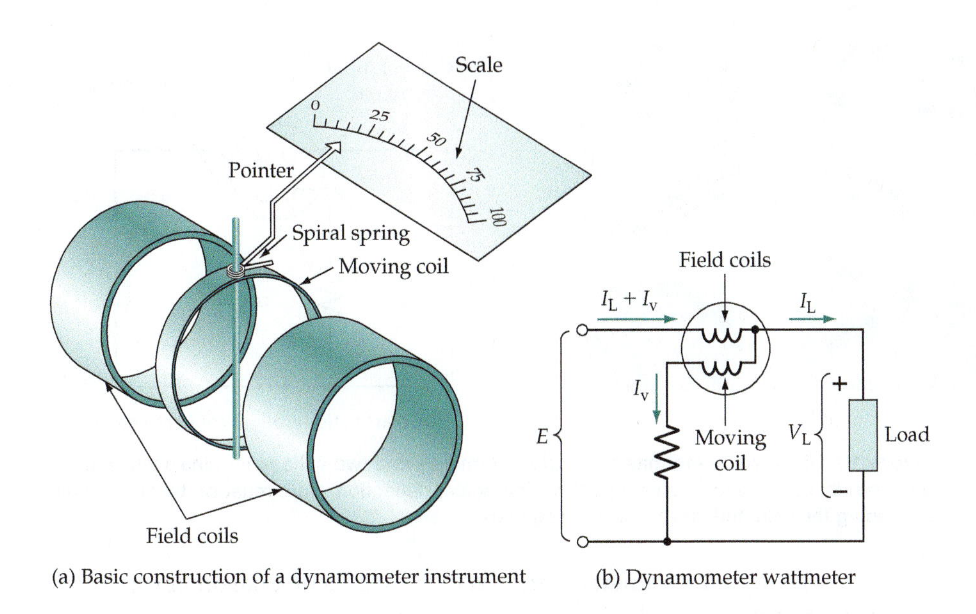

(a) Basic construction of a dynamometer instrument

(b) Dynamometer wattmeter

Figure 13-17 An electrodynamic instrument is similar to a PMMC instrument, except that the magnet is replaced by two field coils. When a load current is passed through the field coils and the load voltage is applied to the moving coil circuit the instrument indicates load power.

density of the magnetic field set up by the stationary coils. This means, of course, that the deflection is also proportional to the current through the field coils. Consider the arrangement shown in Figure 13-17(b). The moving coil of the instrument has a series resistor and is connected in parallel with the load. Consequently, current I_V through the moving coil is directly proportional to the load voltage. The field coils are connected in series with the load, so the current flowing through them is $(I_V + I_L)$. If I_V is very much smaller than I_L, I_V can be neglected and the field coil current assumed to be approximately equal to I_L. Because the meter deflection is proportional to the field coil current and to the moving-coil current,

$$\text{Deflection} \propto V_L \times I_L$$

or,

$$\text{Deflection} \propto P$$

The scale of the instrument can be calibrated to indicate watts, which makes it an *electrodynamic wattmeter.*

The scale of the wattmeter illustrated in Figure 13-18(a) is calibrated for a maximum deflection of 120 W. So, the scale is read directly only when the range switches are set to 1 A and 120 V. When set to 0.5 A and 60 V, FSD represents: $(0.5\ \text{A} \times 60\ \text{V}) = 30\ \text{W}$. It is important to avoid excessive current flow in the voltage and current coils of a wattmeter. When set to the 60 V range, the applied voltage should not exceed 60 V even when the pointer indication is less than full scale. The input terminal of each coil of the wattmeter shown in Figure 13-18 is identified with an arrow. Sometimes an asterisk (*) or a plus/minus sign (±) is used. The input terminal of the current coil should be connected to the supply, and the voltage coil input terminal should be connected to the current coil [see the circuit in Figure 13-18(b)]. Incorrect connection could result in pointer deflection to the left of zero.

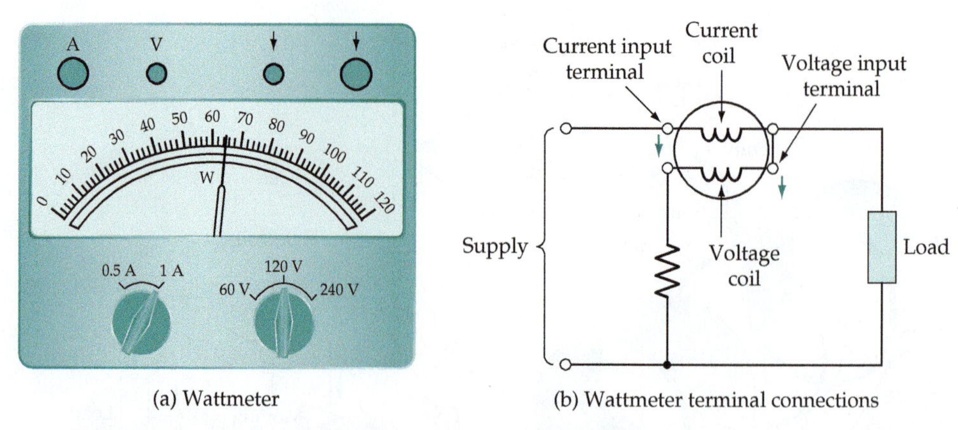

(a) Wattmeter (b) Wattmeter terminal connections

Figure 13-18 A wattmeter has two current terminals and two voltage terminals which must be correctly connected to give a positive (on-scale) deflection. Care must be taken to avoid exceeding the selected current and voltage ranges.

Figure 13-19 Portable digital wattmeter.

A portable digital wattmeter is shown in Figure 13-19. Like other digital instruments this meter used ADCs to process the analog input voltage and current quantities and produce a digital read-out of power in watts. Note that, as well as a watts (W) measurement, the meter displays measurements of VA and VAR. These are *ac* quantities that are explained in Chapter 21.

13-9 THE WHEATSTONE BRIDGE

Like the ohmmeter, the *Wheatstone bridge* can be employed for measurement of resistance. However, unlike the ohmmeter, the Wheatstone bridge measures resistance with a high degree of accuracy.

The circuit of a Wheatstone bridge is shown in Figure 13-20(a). R_1 and R_3 are precision resistors, R_4 is an adjustable precision resistor, R_2 is the unknown resistance to be measured, and G is a *galvanometer.* A galvanometer is essentially a PMMC instrument that has its zero at the center of the scale. It is also an extremely sensitive instrument; so very small currents through its coil cause it to deflect either to the left or to the right of zero, depending on the current direction. As applied in the Wheatstone bridge, the galvanometer is used to detect the zero-current condition; so it is usually termed a *null detector.*

In Figure 13-20(b), the Wheatstone bridge circuit is drawn in a slightly different form. This is the way the circuit is most frequently represented, however, although it may look different, it does not differ in any way from the circuit shown in Figure 13-20(a). Referring again to Figure 13-20(a), when the galvanometer indicates a null condition, the voltages on each side of the galvanometer must be equal. Also, because there is no current flowing through the galvanometer, current I_1 flows through resistors R_1 and R_2, and current I_2 flows through R_3 and R_4. So,

$$V_{R2} = V_{R4}$$

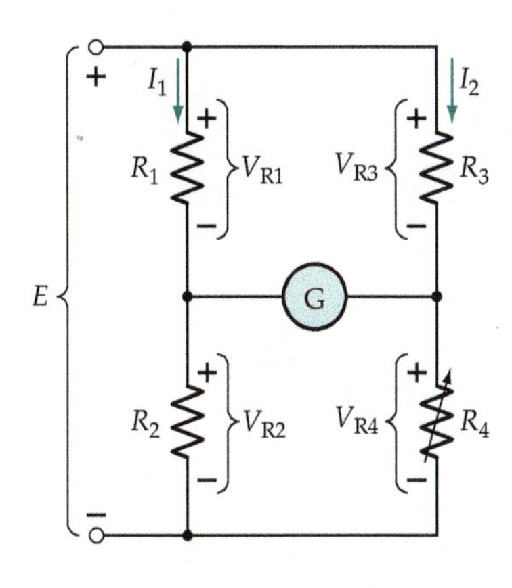

(a) Wheatstone bridge circuit

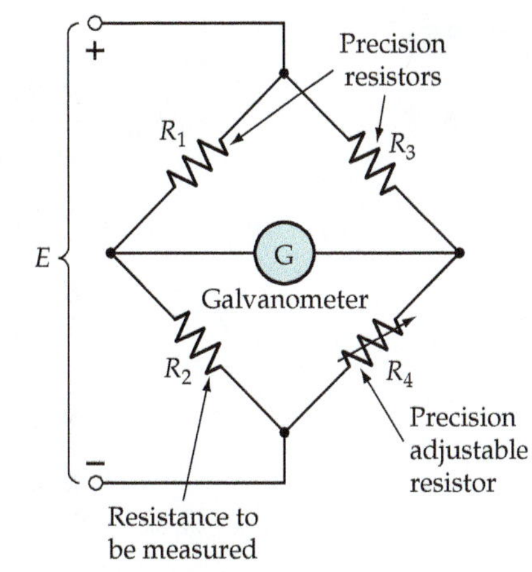

(b) Usual way to show a Wheatstone bridge

Figure 13-20 In a Wheatstone bridge a sensitive galvanometer is connected between the junctions of two potential dividers (R_1R_2) and (R_3R_4). When R_4 is adjusted to give zero deflection on the galvanometer, $R_2/R_1 = R_4/R_3$.

or,
$$I_1R_2 = I_2R_4 \tag{1}$$

Also,
$$V_{R1} = V_{R3}$$

or,
$$I_1R_1 = I_2R_3 \tag{2}$$

Dividing Equation (1) by Equation (2) gives

$$\frac{I_1R_2}{I_1R_1} = \frac{I_2R_4}{I_2R_3}$$

or,
$$\frac{R_2}{R_1} = \frac{R_4}{R_3}$$

So, the measured resistance is,

$$R_2 = \frac{R_1R_4}{R_3} \tag{13-6}$$

Because the precise values of R_1, R_3, and R_4, are known, the resistance R_2 can be accurately determined.

Example 13-11

A Wheatstone bridge has $R_1 = 5\ \text{k}\Omega$, $R_3 = 10\ \text{k}\Omega$, and $R_4 = 1994\ \Omega$ when null condition is indicated on the galvanometer. Determine the value of the unknown resistance R_2. Also, calculate the range of measured resistance if R_4 is adjustable from 100 Ω to 5 kΩ.

Solution

Eq. 13-6,
$$R_2 = \frac{R_1 R_4}{R_3} = \frac{5\ \text{k}\Omega \times 1994\ \Omega}{10\ \text{k}\Omega}$$
$$= 997\ \Omega$$
$$R_{2(\text{min})} = \frac{R_1 R_{4(\text{min})}}{R_3} = \frac{5\ \text{k}\Omega \times 100\ \Omega}{10\ \text{k}\Omega}$$
$$= 50\ \Omega$$
$$R_{2(\text{max})} = \frac{R_1 R_{4(\text{max})}}{R_3} = \frac{5\ \text{k}\Omega \times 5\ \text{k}\Omega}{10\ \text{k}\Omega}$$
$$= 2.5\ \text{k}\Omega$$

Practice Problem

13-9.1 A Wheatstone bridge is to measure resistances ranging from 10 Ω to 100 Ω. If $R_1 = 250\ \Omega$ and $R_3 = 200\ \Omega$, calculate the required maximum and minimum resistances for R_4.

Summary of Formulas

- ***For measuring resistance by ammeter and voltmeter:***

$$R = \frac{V_\text{L}}{I_\text{L} + I_\text{V}} \quad \text{or} \quad R = \frac{V_\text{L} + V_\text{A}}{I_\text{L}}$$

- ***Ohmmeter circuit:***

$$I_\text{m} = \frac{E}{R_1 + R_\text{x} + r_\text{m}}$$

$$I_\text{m} \approx \frac{E}{R_1 + R_\text{x}}$$

- ***Wheatstone bridge circuit:***

$$R_2 = \frac{R_1 R_4}{R_3}$$

Review Questions

Section 13-1

13-1 Sketch the circuit of a basic analog ammeter and briefly explain its operation. Comment on the resistance of an ammeter.

13-2 Draw a sketch to show the circuit of a multirange ammeter, and discuss the type of switch required for range changing. Also, sketch the circuit of an ammeter using an Ayrton shunt, and briefly explain its operation.

Section 13-2

13-3 Sketch two circuits for an electromechanical voltmeter, and briefly explain their operation. Discuss any special requirement for the range changing switch, and comment on the resistance of a voltmeter in relation to its usual applications.

Section 13-3

13-4 Sketch the basic circuit of an electromechanical series ohmmeter, and explain its operation. Also, show how adjustments may be made for battery voltage variations.

13-5 Sketch the typical scale for an ohmmeter, and explain the various pointer positions.

Section 13-4

13-6 Draw a basic circuit diagram for a multirange analog electronic voltmeter. Explain its operation, discuss its input resistance, and show how the circuit should be modified for measuring low voltage levels.

13-7 Show by sketch what occurs when a voltage is divided into discrete levels for digitizing. Show the type of voltages produced at the output of an ADC and the digital code that represents the outputs.

13-8 Draw the basic circuit/block diagram of a digital voltmeter, and discuss its operation.

Section 13-5

13-9 Show how current measurements are made by an electronic instrument. Explain.

13-10 Draw a circuit diagram for an electronic ohmmeter. Explain its operation, and sketch a typical scale for an analog electronic ohmmeter.

Section 13-6

13-11 Sketch circuits to show the two possible arrangements for measuring resistance by use of an ammeter and voltmeter. Discuss the error sources with each arrangement.

Section 13-7

13-12 Explain the basic operation and applications of electromechanical and electronic megohmmeters.

Section 13-8

13-13 Draw a sketch to show the basic construction of a electrodynamic instrument, and discuss its operation. Show how this instrument can be employed as a wattmeter, and explain how it can give a direct indication of power.

Section 13-9

13-14 Sketch the circuit of a Wheatstone bridge, explain its operation, and derive an equation from which the unknown resistance can be calculated.

Problems

Section 13-1

13-1 A PMMC instrument has a coil resistance of 270 Ω and gives FSD for a current of 100 μA. Determine the value of shunt resistance required to convert the instrument into a 100 mA ammeter.

13-2 An ammeter consists of a 37.5 μA meter in parallel with a 0.0018 Ω shunt. The meter coil resistance is 1.2 kΩ. Calculate the measured current at full scale.

13-3 Determine the new values of shunt resistor for the ammeter described in Problem 13-1 to change its range to (a) 1 A, and (b) 10 A.

13-4 An ammeter that indicates FSD at 300 mA has a 50 μA meter with a coil resistance of 900 Ω. Calculate the shunt resistance.

13-5 An ammeter with a resistance of 0.5 Ω is connected in series with a 20 V supply and a load that normally takes a current of 20 A from the supply. Calculate the current indicated by the ammeter.

13-6 The ammeter in Problem 13-2 indicates 15 A when connected in series with a load and a 5 V source. Calculate the current that flows when the ammeter is not connected in the circuit.

13-7 A PMMC instrument has a resistance of 100 Ω and FSD for a current of 100 μA. An Ayrton shunt is connected to the instrument to convert it into an ammeter. The Ayrton shunt has four resistors, each of which is 0.001 Ω. Determine the various ranges to which the ammeter may be switched.

13-8 The ammeter in Problem 13-7 is set on its lowest range and connected to measure the current flowing in a precise 0.9 Ω resistor. If the supply voltage is exactly 2 V, determine the indicated current and the error introduced by the ammeter.

Section 13-2

13-9 A PMMC instrument with a resistance of 75 Ω and FSD current of 100 μA is to be used as a voltmeter with 250 V, 100 V, and 50 V ranges. Determine the required values of multiplier resistor for each range.

13-10 The voltmeter in Problem 13-9 is used to measure the voltage drop across each of two resistors connected as a potential divider as in Figure 13-7. The supply voltage is 50 V, and the resistor values are $R_1 = 56$ kΩ and $R_2 = 68$ kΩ. If the voltmeter is set on its 50 V range, calculate the measured levels of V_{R1} and V_{R2}.

13-11 A voltmeter consists of a 150 μA meter with a 100 Ω coil resistance in series with three series-connected resistors, as in Figure 13-6(b). The resistor values are $R_1 = 166.6$ kΩ, $R_2 = 333.3$ kΩ, and $R_3 = 500$ kΩ. Determine the voltmeter range when the multiplier resistor is (a) R_1 alone, (b) $R_1 + R_2$, and (c) $R_1 + R_2 + R_3$.

13-12 A voltmeter is to be constructed using a 37.5 μA meter with a resistance of 900 Ω, and three series-connected resistors as in Problem 13-11. The instrument ranges are to be 3 V, 30 V, and 300 V. Calculate the resistor values.

13-13 Two resistors are connected in series across a 250 V supply. The resistor values are $R_1 = 330$ kΩ and $R_2 = 220$ kΩ. The voltmeter described in Problem 13-9 is used to measure the voltage across each resistor. Determine the voltage indicated in each case.

13-14 Calculate the sensitivity of each of the voltmeters referred to in Problems 13-9, 13-11, and 13-12.

13-15 The voltmeter in Problem 13-9 is used on its lowest range to measure the voltages across R_4 and R_5 in the circuit shown in Figure 7-16. Determine the measured voltages.

13-16 The voltmeter in Problem 13-12 is used to measure the output of a potential divider circuit as in Figure 5-11. The supply voltage is 12 V, and the resistor values are $R_1 = 27$ kΩ, $R_2 = 5$ kΩ, and $R_3 = 18$ kΩ. Calculate the measured maximum and minimum levels of V_o.

Section 13-3

13-17 A series ohmmeter as in Figure 13-8 uses a 200 μA meter and a 20 kΩ precision resistor. The supply is a battery with voltage $E = 4$ V. Determine the value of resistance measured by the ohmmeter at pointer deflections of 1/3, 1/2, and 2/3 of FSD. Draw a sketch of the instrument scale, showing current levels and resistance values.

13-18 A 50 μA meter with $R_m = 50$ Ω is to be used to construct a series ohmmeter with a centre scale indication of 15 kΩ. A shunt resistor R_2 is to be included across the meter, as in Figure 13-9. Determine suitable resistance values for R_1 and R_2 and a suitable battery voltage.

13-19 A series ohmmeter has a meter with 100 μA FSD and a resistance of $R_m = 30$ Ω. The supply battery has a terminal voltage of $E = 4$ V, and the series resistor is 10 kΩ. An adjustable shunt resistor connected across the meter has a value of $R_2 = 30\ \Omega$. Determine the resistance measured at 1/4, 1/2, and 3/4 of FSD.

13-20 For the ohmmeter in Problem 13-19, determine the new value that R_2 must be adjusted to when the battery voltage falls to 3 V.

13-21 For the ohmmeter in Problem 13-18, calculate the measured resistance at 1/3 and 2/3 of FSD.

13-22 For the ohmmeter in Problem 13-18, determine the new value that R_2 must be adjusted to when the battery voltage falls by 10%.

Section 13-4

13-23 The input resistance for the electronic voltmeter in Figure 13-10(a) is to be 2 MΩ, and the voltage ranges are to be 1 V, 30 V, 100 V, and 300 V. Determine suitable resistance values for R_1 through R_4.

13-24 The voltmeter in Problem 13-23 is to use a 100 μA meter with a 1.5 kΩ coil resistance. Determine a suitable resistance value for the resistor in series with the meter (R_s in Figure 13-10).

13-25 An electronic voltmeter as in Figure 13-10(a) is to have ranges of 50 mV, 500 mV, 5 V, and 50 V. Determine suitable resistance values for R_1 through R_4 and a suitable voltage gain for the amplifier if V_o is to be 1 V.

13-26 The electronic voltmeter in Problem 13-25 is to use an ADC and digital display. Determine a suitable bit number for the ADC if the voltage is to be measured with a 1 mV resolution.

13-27 Determine a suitable bit number for the ADC in the digital voltmeter circuit in Figure 13-12.

13-28 A digital voltmeter circuit as in Figure 13-12 is to have a 1 MΩ input resistance and ranges of 300 mV, 3 V, and 30 V. A display that can indicate a maximum of 299 is used. Calculate suitable resistance values for R_1, R_2, and R_3. Also, determine a suitable bit number for the ADC.

Section 13-5

13-29 An electronic ammeter as in Figure 13-13 is to measure current levels up to 500 mA. If the amplifier has a gain of 1 and a maximum output of 1 V, calculate the shunt resistance and the burden voltage.

13-30 The maximum current level for the ammeter in Problem 13-29 is to be measured with a resolution of 1 mA. Determine a suitable ADC bit number for the digital voltage-measuring stage.

13-31 Calculate the burden voltage for the electronic ammeter in Figure 13-13 when measuring 30 mA and 300 mA.

13-32 The electronic ohmmeter in Figure 13-14 is set to $R_2 = 10$ kΩ. Calculate the measured resistances when the pointer position is 0.4 FSD and 0.25 FSD.

13-33 If the ohmmeter in Figure 13-14(a) uses a 1.5 V digital voltmeter, determine the required ADC bit number to measure the 1 kΩ resistance with a 10 Ω resolution.

Section 13-6

13-34 An ammeter and voltmeter employed to measure resistance give readings of 196 μA and 240 V, respectively, when the voltmeter is connected directly in parallel with the resistor to be measured. When the ammeter is connected directly in series with the resistor, the readings are 100 μA and 240 V. The ammeter has a resistance of 50 Ω, and the voltmeter sensitivity is 10 kΩ/V. If the voltmeter is on a 250 V range, calculate the value of measured resistance for each case and determine which of the two gives the most accurate result.

13-35 The ammeter described in Problem 13-2 is used together with the voltmeter in Problem 13-12 to measure a resistance R_L. With the voltmeter on its lowest range connected directly in parallel with R_L, the meter readings are 2.75 V and 22 A. Determine the meter readings when the ammeter is directly in series with R_L.

13-36 A resistance R_x measured by a voltmeter and an ammeter gives readings of 500 V and 150 μA when the ammeter is directly in series with R_x. When the voltmeter is directly in parallel with R_x, the readings are 500 V and 200 μA. Determine the resistance of R_x and the sensitivity of the voltmeter.

13-37 Two voltmeters (V_1 and V_2), each having a sensitivity of 20 kΩ/V, are employed to measure the insulation resistance of an electrical installation. V_1 is connected in parallel with the supply, and V_2 is in series with the supply and the resistance to be measured. Both voltmeters are on the 300 V range. If V_1 indicates 250 V and V_2 indicates 62.5 V, calculate the insulation resistance.

Section 13-9

13-38 A Wheatstone bridge has a 100 Ω precision resistor (R_1) connected in series with the unknown resistor (R_2). Another precision resistor ($R_3 = 150$ Ω) is connected in series with the variable resistor (R_4). When the galvanometer indicates null, the resistance of R_4 is found to be 119.25 Ω. Calculate the unknown resistance.

13-39 A Wheatstone bridge as in Figure 13-20 is to be constructed to measure resistances in the range of 300 Ω to 30 kΩ. Precision resistors with values of 1 kΩ and 5 kΩ are available for use as R_1 and R_2, respectively. Calculate the required range of adjustment for R_4.

13-40 In a Wheatstone bridge circuit as in Figure 13-20 $R_3 = 1$ kΩ, R_4 is adjustable from 100 Ω to 10 kΩ, and R_1 may be selected as 1 kΩ or 10 kΩ. Determine the two measurement ranges of the bridge.

Practice Problem Answers

13-1.1 1.51 Ω, 16.22 Ω
13-1.2 870 mA, 13%
13-1.3 100 mA, 150 mA, 300 mA
13-2.1 1.9 kΩ, 599.9 kΩ, 5 MΩ
13-2.2 5.77 V, 8.55 V
13-3.1 10 kΩ, 20 kΩ, 5 kΩ
13-3.2 30 Ω
13-4.1 1 kΩ, 9 kΩ, 90 kΩ, 900 kΩ
13-4.2 16, 256
13-4.3 900 kΩ, 90 kΩ, 10 kΩ
13-5.1 3 Ω
13-5.2 493 Ω, 1.94 kΩ
13-6.1 3.75 MΩ, 5 MΩ
13-9.1 80 Ω, 8 Ω

CHAPTER 14
Inductance

CONTENTS

Objectives

You will be able to:

1. Draw sketches to illustrate the generation of current in a conductor moving through a magnetic field, and explain the process of electromagnetic induction.
2. State Lenz's law and Faraday's law.
3. Solve problems involving the emf produced in the secondary of two inductively coupled windings.
4. Using sketches explain self-inductance, and solve problems involving the self-inductance of air-cored and ferromagnetic-cored coils.
5. Explain mutual inductance and solve problems involving the mutual inductance between two coils having a single core.
6. Describe various types of inductors and their characteristics, and calculate the energy stored in an inductive circuit.
7. Calculate the equivalent inductance of several inductors connected in series or parallel.

INTRODUCTION

Electromagnetic induction occurs when a magnetic flux is in motion with respect to a single conductor, or when a coil induces an emf in the conductor or coil. Because the growth or decline of current through a coil generates a changing flux, an emf is induced in the coil by its own current change. The same effect can induce an emf in an adjacent coil. The level of emf induced in each case depends on the self-inductance of the coil, or on the mutual inductance between the two coils. In all cases, the polarity of the induced emf is such that it opposes the original change that induced the emf.

Components called inductors or chokes are constructed to have specified values of inductance. Inductors can be operated in series or in parallel. Even the shortest of conductors has an inductance. This is usually an unwanted quantity and is termed stray inductance.

14-1 ELECTROMAGNETIC INDUCTION

In Section 11-2 it is shown that an emf is generated within a conductor or a coil when a magnetic flux cuts the conductor or coil. This effect is termed *electromagnetic induction.* It is important to note that the emf is generated only when the flux is in motion with respect to the conductor or coil. When the flux and the conductor are stationary, no emf is generated.

Figure 14-1 illustrates electromagnetic induction in a single conductor in motion across a magnetic field. A center-zero galvanometer (see Section 13-9) is connected in series with the conductor, and it is found that the galvanometer deflects to the left when the conductor is moved up through the field, [Figure 14-1(a)]. This indicates that the current induced in the conductor is flowing through the galvanometer from right to left, as illustrated. It also shows that the polarity of the induced emf in the conductor is positive (+) on the right-hand side of the conductor and negative (−) on the left-hand side, as shown in the figure. When the direction of the current induced in the conductor is considered, it is seen that the current sets up its own flux in a clockwise direction around the conductor [see Figure 14-1(a)]. The effect of this (conductor) flux is to strengthen the magnetic field above the conductor and to weaken it below the conductor, as shown in Figure 14-1(b). The strengthening of the magnetic field above the conductor and weakening of the field below the conductor make it more difficult for the conductor to move upward. So, it is seen that the magnetic flux set up by the current induced in the conductor opposes the direction of motion of the conductor.

A similar effect is produced when the conductor is moved down through the field, as illustrated in Figures 14-1(c) and (d). In this case, the induced emf is positive on the conductor's left-hand side and negative on its right-hand

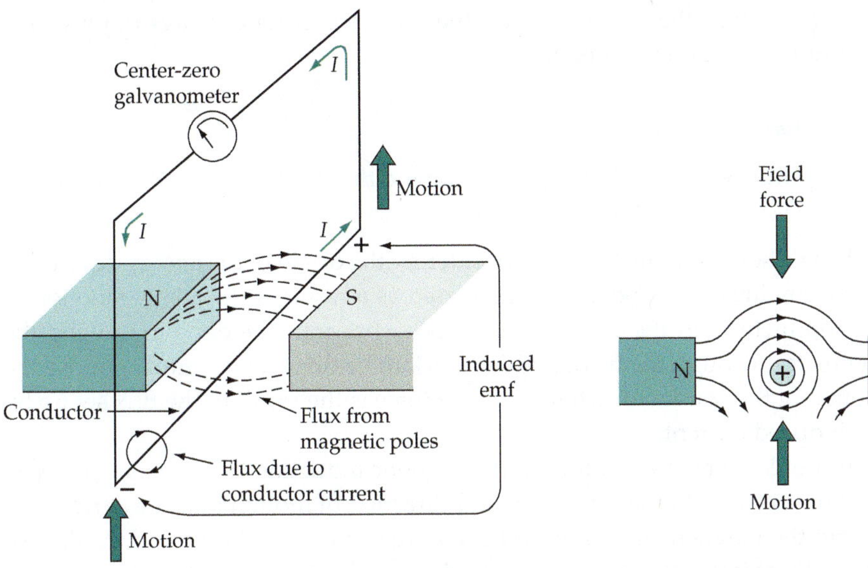

(a) Conductor moving up through magnetic field

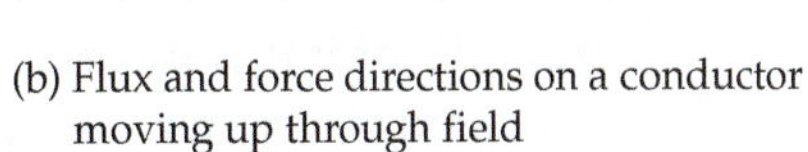

(b) Flux and force directions on a conductor moving up through field

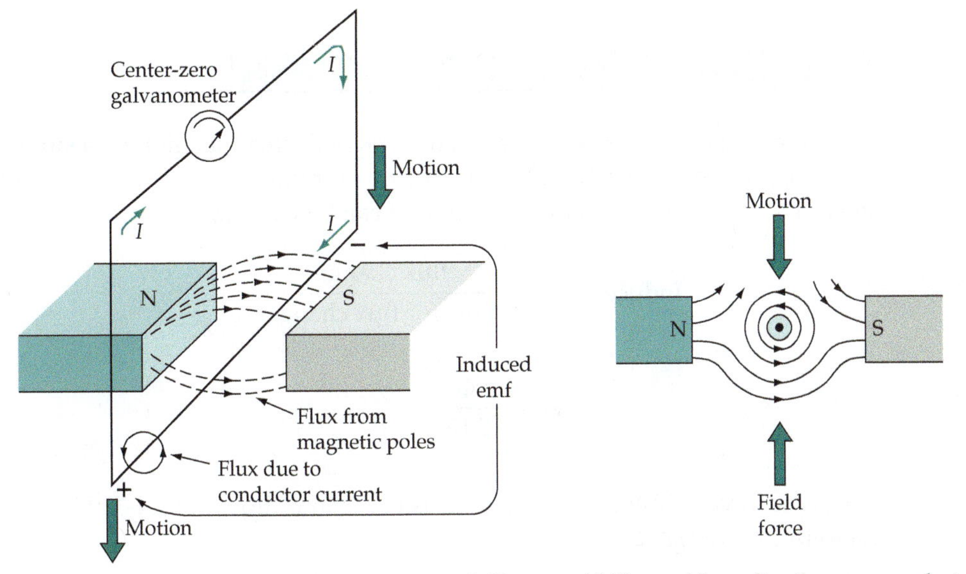

(c) Conductor moving down through magnetic field

(d) Flux and force directions on a conductor moving down through field

Figure 14-1 Electromagnetic induction. An emf is generated in a conductor in motion through a magnetic field. A current flows when an external circuit is connected to the conductor.

side. The galvanometer deflection is to the right, and the current direction is as shown in the illustration. Figure 14-1(d) shows that the induced current in the conductor now generates a flux that strengthens the magnetic field below the conductor and weakens it above the conductor. Again, the field set up by the induced current opposes the conductor's direction of motion.

The fact that the flux set up by the induced current always opposes the direction of motion is stated in *Lenz's law*[1].

Lenz's Law:

The induced current always develops a flux that opposes the motion or change producing the current.

Lenz's law can be justified by considering the origin of the energy developed in the conductor. It is obviously generated as a result of the relative motion of the conductor and the magnetic field. So, work must be done in moving the conductor through the field, and for work to be done a force must oppose the motion of the conductor. This opposing force is the result of the flux set up by the induced current.

It should be noted that for electromagnetic induction to occur as illustrated in Figure 14-1, both the conductor and direction of motion must be at right angles to the magnetic field flux. If the conductor were to be moved axially, for example, or if it were oriented in a horizontal direction from the N pole to the S pole, no electromagnetic induction effect would occur.

14-2 INDUCED EMF AND CURRENT

In Section 11-4 the weber is defined as the magnetic flux that, linking a single-turn coil, produces an emf of 1 V when the flux is reduced to zero at a constant rate in 1 s. So, the equation for induced emf can be written,

$$\text{Induced emf} = \frac{\text{flux change}}{\text{time for flux change}}$$

$$e_L = \frac{\Delta\Phi}{\Delta t} \tag{14-1}$$

Here, e_L is in volts, $\Delta\Phi$ is in Wb, and Δt is in seconds. Equation 14-1 also originates from *Faraday's law*[2].

Faraday's Law:

The emf induced in an electric circuit is proportional to the rate of change of flux linking the circuit.

If the total field flux in Figure 14-1 were 1 Wb, and if the conductor were moved through the field in exactly 1 s, the emf measured at the conductor

[1]Formulated by Russian physicist Heinrich Lenz (1804–1865)

[2]Formulated by English chemist and physicist Michael Faraday (1791–1867)

terminals would be exactly 1 V. If the conductor were to be moved through the field in 0.5 s, the voltage generated would be,

$$e_L = \frac{1 \text{ Wb}}{0.5 \text{ s}} = 2 \text{ V}$$

Similarly, if the flux were doubled the generated voltage would be doubled.

Figure 14-2 shows that an emf is induced in a coil when a changing magnetic flux links with the coil. The emf is induced when the flux is increasing or decreasing, but not while the flux is a constant quantity.

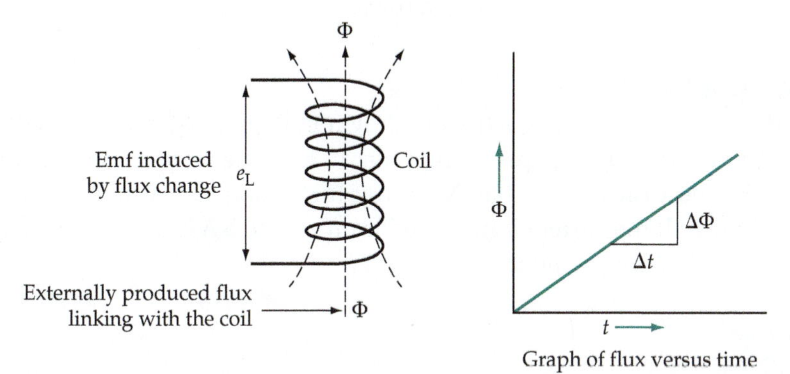

Figure 14-2 An emf is induced in a coil when a changing magnetic flux links with the coil.

Now consider Figure 14-3, which shows two coils on a ring-shaped Iron core, Flux generated by a current flowing in the left-hand coil will pass through the iron core and link with the right-hand coil. Because the left-hand coil is the *input* coil and the right-hand coil is the *output*, the coils are identified as *primary* and *secondary*, respectively. While the current in the primary is constant, the core flux will not change and no emf will be induced in the secondary winding. When the primary current is increased or decreased (by adjustment of R), the core flux grows or declines, and in doing so it induces an emf in the secondary winding.

If N is the number of turns in the secondary winding, the induced emf is,

$$e_L = \frac{\Delta \Phi N}{\Delta t} \tag{14-2}$$

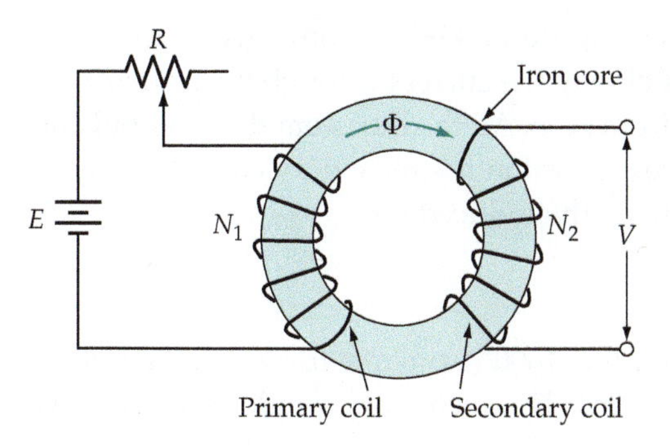

Figure 14-3 An iron core can be used to channel magnetic flux from one coil into another. An emf is induced in the secondary coil when the flux level changes in the primary.

Example 14-1

In the circuit shown in Figure 14-3, the core flux is increased from zero to 50 mWb in a time of 4 s. Calculate the number of secondary turns required if the induced voltage is to be 1.5 V.

Solution

From Eq. 14-2,
$$N = \frac{e_L \, \Delta t}{\Delta \Phi} = \frac{1.5 \text{ V} \times 4 \text{ s}}{50 \text{ mWb}}$$
$$= 120 \text{ turns}$$

Example 14-2

The dimensions of the magnetic core shown in Figure 14-3 are cross-sectional area $A = 3 \text{ cm}^2$, magnetic path length $l = 10$ cm. The relative permeability is $\mu_r = 250$, the primary coil has $N_p = 100$ turns, and the secondary coil has $N_s = 75$ turns. If the current is increased from zero to 5 A in 0.1 s, determine the emf induced in the secondary.

Solution

Eq. 11-2,
$$F_m = I \times N_p = 5\text{A} \times 100$$
$$= 500 \text{ A}$$

Eq. 12-6,
$$R_m = \frac{l}{\mu_r \mu_o A} = \frac{10 \times 10^{-2}}{250 \times 4\pi \times 10^{-7} \times 3 \times 10^{-4}}$$
$$= 1.06 \times 10^6 \text{ A/Wb}$$

Eq. 12-2,
$$\Phi = \frac{F_m}{R_m} = \frac{500 \text{ A}}{1.06 \times 10^6 \text{ A/Wb}}$$
$$= 471 \text{ µWb}$$

Induced emf,
$$e_L = \frac{\Delta \Phi N_s}{\Delta t} = \frac{471 \text{ µWb} \times 75}{0.1 \text{ s}}$$
$$= 0.35 \text{ V}$$

An air-cored solenoid is shown in Figure 14-4 with another (secondary) coil wound on top. When the switch is closed, the current through the primary coil causes the magnetic flux to grow from zero. All the flux from the solenoid cuts the secondary winding and induces an emf in the secondary coil. Once again, Equation 14-2 can be used to calculate the induced emf.

Example 14-3

The solenoid shown in Figure 14-4 has 1000 (primary) turns, is 30 cm long, and has a cross-sectional area of 5 cm². The secondary winding has 5 turns.

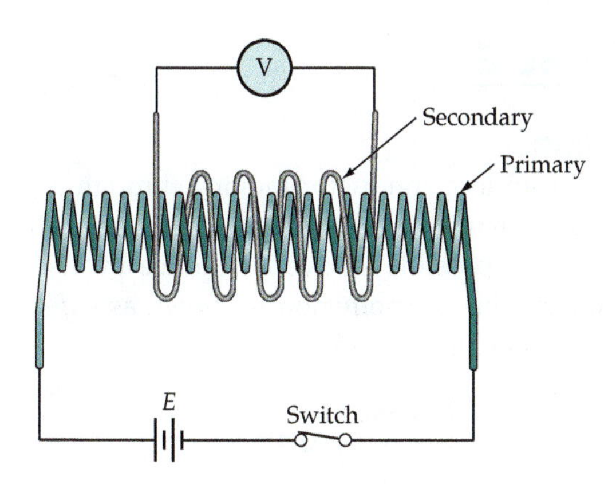

Figure 14-4 A secondary winding on top of an air-cored solenoid has an emf induced in it when a current increases or decreases in the solenoid.

Determine the emf induced in the secondary coil when the solenoid current is increased from zero to 20 A in a time period of 10 ms.

Solution

Eq. 11-2,
$$F_m = N_P I_P = 1000 \times 20\text{ A}$$
$$= 20\,000\text{ A}$$

Eq. 12-1,
$$R_m = \frac{l}{\mu_o A} = \frac{30 \times 10^{-2}}{4\pi \times 10^{-7} \times 5 \times 10^{-4}}$$
$$= 477.5 \times 10^6\text{ A/Wb}$$

Eq. 12-2,
$$\Phi = \frac{F_m}{R_m} = \frac{20\,000\text{ A}}{477.5 \times 10^6\text{ A/Wb}}$$
$$= 41.9\ \mu\text{Wb}$$

Induced emf,
$$e_L = \frac{\Delta\Phi N_s}{\Delta t} = \frac{41.9\ \mu\text{Wb} \times 5}{10\text{ ms}}$$
$$\simeq 21\text{ mV}$$

Practice Problems

14-2.1 Calculate the voltage induced in a 250-turn coil when the magnetic flux through the coil changes from 0.3 μWb to 0.9 μWb in a time period of 4 ms.

14-2.2 A magnetic core with the same dimensions and number of coil turns as in Example 14-2 produces a 420 mV secondary emf when the primary current is increased from zero to 1 A in 50 ms. Determine the relative permeability of the core.

14-2.3 A 20 cm long solenoid has an internal diameter of 2 cm and a 100 turn secondary winding. A 33 mV secondary voltage is measured when the primary current changes by 500 mA in 2.5 ms. Determine the number of primary turns.

14-3 SELF-INDUCTANCE

Coil and Conductor Inductance

It has been shown that an emf is induced in a conductor moving through a magnetic field and that the growth of current in a coil can induce an emf in another magnetically coupled coil. It is also possible for a coil to induce a voltage in itself as its current level changes. This phenomenon is known as *self-inductance,* and the principle is illustrated in Figure 14-5.

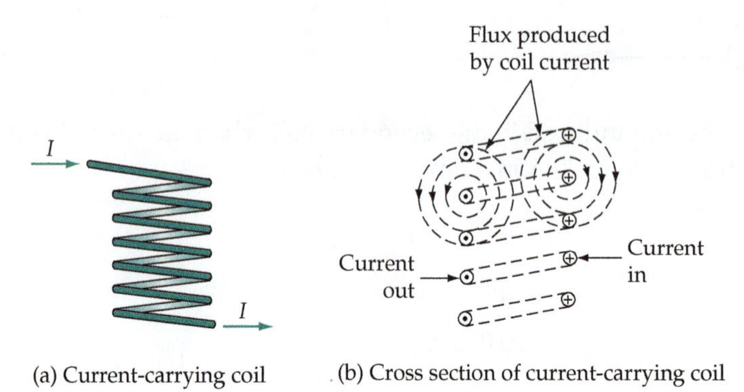

(a) Current-carrying coil (b) Cross section of current-carrying coil

Figure 14-5 Magnetic flux growing outwards around the turns of a coil cuts (or brushes over) the other coil turns and induces an emf in the coil.

A coil and its cross-sectional area are shown in Figure 14-5, with arrow tails and points indicating the current directions in each turn. Every turn of the coil has a flux around it produced by the current flowing through the coil. However, for convenience, the illustration shows the growth of flux around only one turn on the coil. It is seen that as the current grows, the flux expands outward and cuts (or brushes over) the other turns. This causes currents to be induced in the other turns, and the direction of the induced currents is such that they set up a flux that opposes the flux inducing them. Remembering that the current through the coil causes the flux to grow around all turns at once, it is seen that the flux from every turn induces a current that opposes it in every other turn.

To set up opposing fluxes, the induced current in a coil must be in opposition to the current flowing through the coil from the external source of supply. The induced current is, of course, the result of an induced emf. So, it is seen that the self-inductance of a coil sets up an induced emf that opposes the external emf that is driving current through the coil. Because this induced emf is in opposition to the supply voltage, it is usually termed the *counter-emf,* or *back-emf.* The counter-emf occurs only when the coil current is growing or declining. When the current has reached a constant level, the flux is no longer changing and no counter-emf is generated.

Even a single conductor has self-inductance. Figure 14-6 shows that when current is increasing in a conductor, flux may grow outward from the center of the conductor. This flux cuts other portions of the conductor and induces a counter-emf.

Flux within conductor

Conductor cross section

Figure 14-6 The growth of current within a conductor induces emfs in other portions of the conductor.

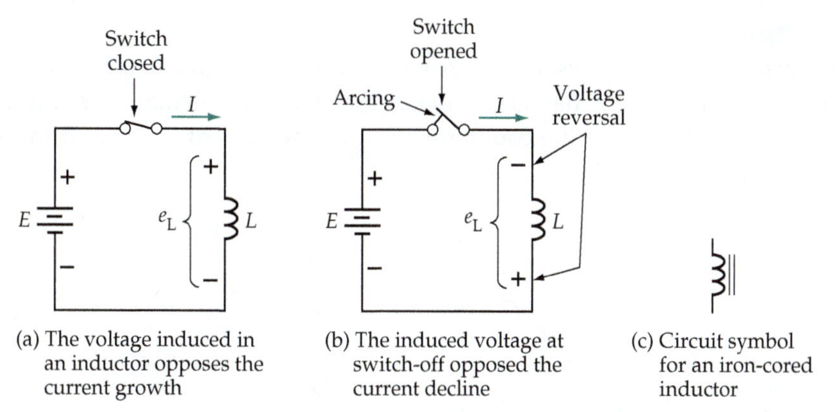

Figure 14-7 **The counter-emf induced in a coil always opposes the growth or decline of the current.**

Figure 14-7 illustrates the polarity of the counter-emf induced in a coil for a given supply voltage polarity. In Figure 14-7(a), the switch is closed and current (I) commences to grow from zero. The polarity of the counter-emf (e_L) is such that it opposes the growth of I, so it is *series-opposing* with the supply voltage. When the switch is opened the current tends to fall to zero, [Figure 14-7(b)]. But now the polarity of e_L is such that it opposes the decline of I. It is *series-aiding* with the supply voltage. In fact, e_L may cause arcing at the switch terminals, as illustrated, as it attempts to maintain the flow of current.

The amplitude of the counter-emf induced in a coil by a given rate of change in current depends on the coil's inductance.

The SI unit of inductance is the henry[3] (H).

The inductance of a circuit is *one henry* (1 H), when an emf of 1 V is induced by the current changing at the rate of 1 A/s.

The relationship between inductance, induced voltage, and rate-of-change of current is,

$$L = \frac{e_L}{\Delta i / \Delta t} \tag{14-3}$$

where L is the inductance in henrys, e_L is the induced counter-emf in volts, and ($\Delta i / \Delta t$) is the rate of change of current in amperes/second. A negative sign is sometimes included in front of e_L to show that the induced emf is in opposition to the applied emf. When $e_L = 1$ V and $\Delta i / \Delta t = 1$ A/s, $L = 1$ H.

A coil constructed to have a certain inductance is usually referred to as an *inductor* or a *choke*. Note the graphic symbols for an inductor shown in Figure 14-7.

---

[3]Named for the American physicist Joseph Henry (1797–1887).

Example 14-4

The current in a coil grows linearly from 0 A to 10 A in a time of 0.25 s [see Figure 14-8(a)]. If the coil has an inductance of 0.75 H, calculate the induced counter-emf. Determine the counter-emf when the current is reduced to zero in 1 ms [Figure 14-8(b)].

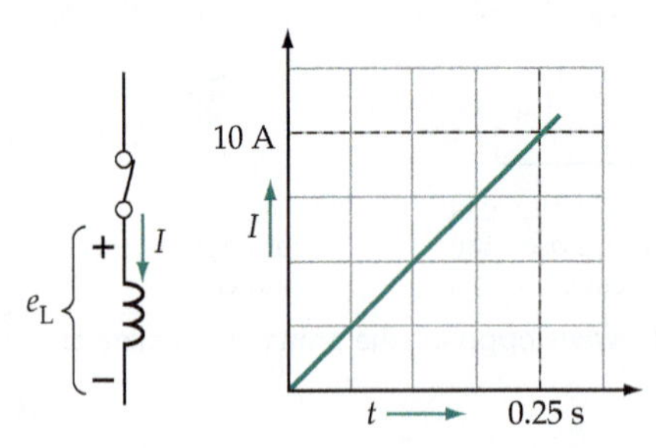

(a) A constant rate of current growth induces a constant opposing voltage

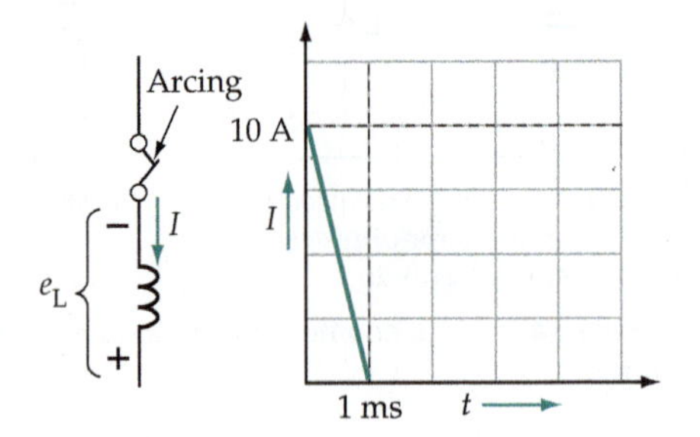

(b) Rapid current decling induces a high voltage which opposes the decline

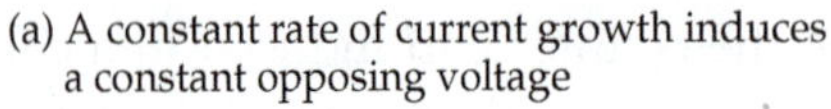

Figure 14-8 Illustration for Example 14-4.

Solution

Current increasing:

From Eq. 14-3,

$$e_L = L\left(\frac{\Delta i}{\Delta t}\right) = 0.75\text{ H} \times \left(\frac{10\text{ A} - 0}{0.25\text{ s}}\right)$$

$$= 30\text{ V}$$

Current decreasing:

$$e_L = L\left(\frac{\Delta i}{\Delta t}\right) = 0.75\text{ H} \times \left(\frac{10\text{ A} - 0}{1\text{ ms}}\right)$$

$$= 7500\text{ V}$$

Calculation of Self-Inductance

An expression for inductance can be derived involving the coil dimensions and the number of turns [see Figure 14-9]. From Eq. 14-2,

$$e_L = \frac{\Delta\Phi N}{\Delta t}$$

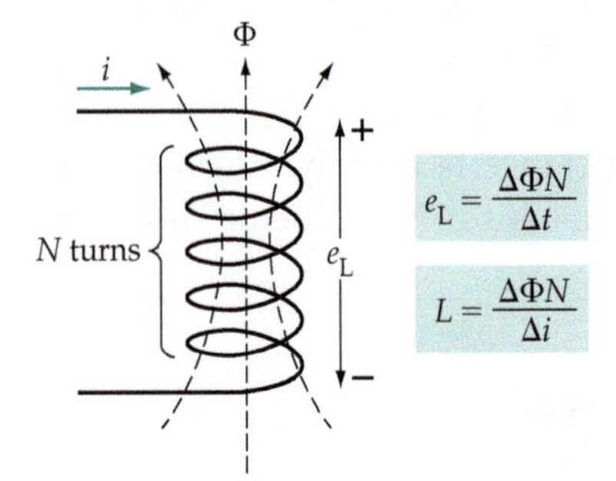

Figure 14-9 The inductance of a coil depends on the number of turns and the flux changes.

Substituting for e_L into Eq. 14-3 gives,

$$L = \frac{\Delta\Phi N/\Delta t}{\Delta i/\Delta t}$$

or,

$$L = \frac{\Delta\Phi N}{\Delta i} \quad \text{(14-4)}$$

Also, $\Phi = BA, B = \mu_r\mu_o H, \text{and } H = IN/l$

So, $\Phi = \mu_r\mu_o IN\frac{A}{l}$

Because I is a maximum current level, it also represents the change in current (Δi) from zero to the maximum level. So, change in flux is,

$$\Delta\Phi = \mu_r\mu_o \Delta i N\frac{A}{l} \quad \text{(14-5)}$$

Substituting for $\Delta\Phi$ in Eq. 14-4 gives,

$$L = \mu_r\mu_o N^2\frac{A}{l} \quad \text{(14-6)}$$

Note that, as illustrated in Figure 14-10, the inductance is proportional to the cross-sectional area of a coil and to the square of the number of turns. It is also inversely proportional to the coil length. So, maximum inductance is obtained with a short coil that has a large cross-sectional area and a large number of turns.

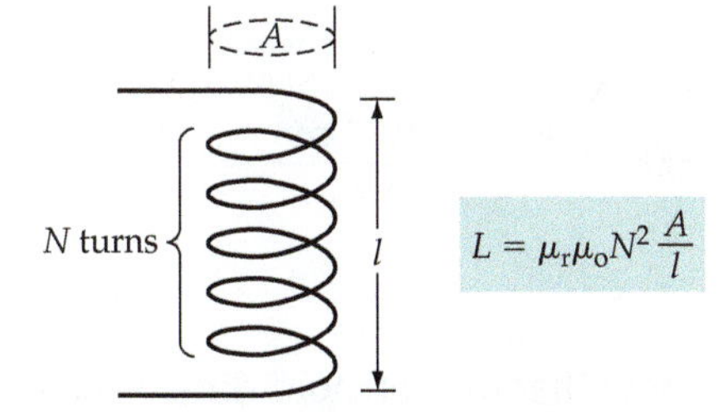

Figure 14-10 Coil inductance can be calculated from its dimensions and its core permeability.

Equation 14-6 now offers a means of calculating the inductance of a coil of known dimensions. Alternatively, it can be used to determine the required dimensions for a coil to have a given inductance. However, it is not so easily applied to iron-cored coils, because the permeability of ferromagnetic material changes when the flux density changes (see Figure 12-6). Consequently, the inductance of an iron-cored coil is constantly changing as the coil current increases and decreases.

Example 14-5

A solenoid with 900 turns has a total flux of 1.33×10^{-7} Wb through its air core when the coil current is 100 mA. If the flux takes 75 ms to grow from zero to its maximum level, calculate the inductance of the coil. Also, determine the counter-emf induced in the coil during the flux growth.

Solution

$$\Delta\Phi = 1.33 \times 10^{-7}\ \text{Wb},\ \Delta i = 100\ \text{mA, and}\ \Delta t = 75\ \text{ms}$$

Eq. 14-4,

$$L = \frac{\Delta\Phi N}{\Delta i} = \frac{1.33 \times 10^{-7}\ \text{Wb} \times 900}{100\ \text{mA}}$$

$$\approx 1.2\ \text{mH}$$

Eq. 14-2,

$$e_L = \frac{\Delta\Phi N}{\Delta t} = \frac{1.33 \times 10^{-7}\ \text{Wb} \times 900}{75\ \text{ms}}$$

$$\approx 1.6\ \text{mV}$$

Example 14-6

The air-cored solenoid described in Example 14-5 has $l = 15$ cm and inside diameter $D = 1.5$ cm. Recalculate its inductance using the coil dimensions.

Solution

$$A = \pi\left(\frac{D}{2}\right)^2 = \pi \times \left(\frac{1.5 \times 10^{-2}\ \text{m}}{2}\right)^2$$

$$\approx 1.77 \times 10^{-7}\ \text{m}^2$$

Eq. 14-6,

$$L = \mu_r\mu_o N^2 \frac{A}{l}$$

$$\approx 1 \times 4\pi \times 10^{-7} \times 900^2 \times \frac{1.77 \times 10^{-4}\ \text{m}^2}{15 \times 10^{-2}\ \text{m}}$$

$$\approx 1.2\ \text{mH}$$

This matches the answer to Example 14-5. Note that the solenoid dimensions employed are those used in Example 12-3.

Example 14-7

An air-cored coil is to be 2.5 cm long and to have an average cross-sectional area of 2 cm^2. Determine the number of turns required if the coil is to have an inductance of 100 μH.

Solution

From Eq. 14-6,
$$N = \sqrt{\frac{LI}{\mu_r\mu_o A}} = \sqrt{\frac{100 \times 10^{-6}\,\text{H} \times 2.5 \times 10^{-2}\,\text{m}}{4\pi \times 10^{-7} \times 2 \times 10^{-4}\,\text{m}^2}}$$
$$\approx 100 \text{ turns}$$

Example 14-8

A cast steel ring as in Figure 14-11 has a cross-sectional area of 4 cm^2 and a magnetic path length of 15 cm. A 300 turn coil is wound on the ring. Determine the coil inductance when the current level is (a) 2 A, and (b) 0.5 A.

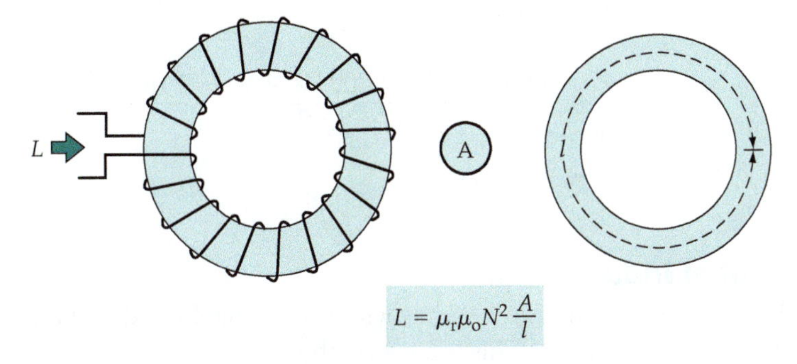

Figure 14-11 Inductance can be calculated from core permeability, number of coil turns, and coil dimensions.

Solution

(a) For $I = 2\,A$,

$$H_1 = \frac{NI}{l} = \frac{300 \times 2\,\text{A}}{15 \times 10^{-2}\,\text{m}}$$
$$= 4000\,\text{A/m}$$

From Figure 12-6, at $H = 4000\,\text{A/m}$, $B \approx 1.57$ T for cast steel,

$$\mu_r\mu_o = \frac{B}{H} = \frac{1.57\,\text{T}}{4000\,\text{A/m}}$$
$$\approx 3.93 \times 10^{-4}$$

Eq. 14-6,
$$L = \mu_r\mu_o N^2 \frac{A}{l}$$
$$\approx \frac{3.93 \times 10^{-4} \times 300^2 \times 4 \times 10^{-4}\,\text{m}^2}{15 \times 10^{-2}\,\text{m}}$$
$$\approx 94.3\,\text{mH}$$

(b) For $I = 0.5$ A,

$$H_2 = \frac{NI}{l} = \frac{300 \times 0.5 \text{ A}}{15 \times 10^{-2} \text{ m}}$$

$$= 1000 \text{ A/m}$$

From Figure 12-6, at $H = 1000$ A/m, $B \approx 1.05$ T for cast steel,

$$\mu_r \mu_o = \frac{B}{H} = \frac{1.05 \text{ T}}{1000 \text{ A/m}}$$

$$\approx 1.05 \times 10^{-3}$$

Eq. 14-6,

$$L = \mu_r \mu_o N^2 \frac{A}{l}$$

$$\approx \frac{1.05 \times 10^{-3} \times 300^2 \times 4 \times 10^{-4} \text{ m}^2}{15 \times 10^{-2} \text{ m}}$$

$$\approx 252 \text{ mH}$$

Incremental Inductance

The steel-cored coil in Example 14-8 has less inductance for a direct current of 2 A than for a current of 0.5 A. This is because the 2 A current tends to drive the core into magnetic saturation, while with 0.5 A the core is operating on the steepest (near linear) portion of the *B/H* curve (Figure 12-6). So, when inductance is specified for a coil with a ferromagnetic core, the coil direct current should also be specified. Recall from Section 12-7 that the incremental permeability of ferromagnetic material varies along the length of the *B/H* curve. When the incremental permeability is used, the calculated inductance is known as the *incremental inductance.* This quantity applies in the case of an inductor in which the current level fluctuates while remaining substantially constant [see Figure 14-12(a)]; for example, a current of 10 mA increasing and decreasing by

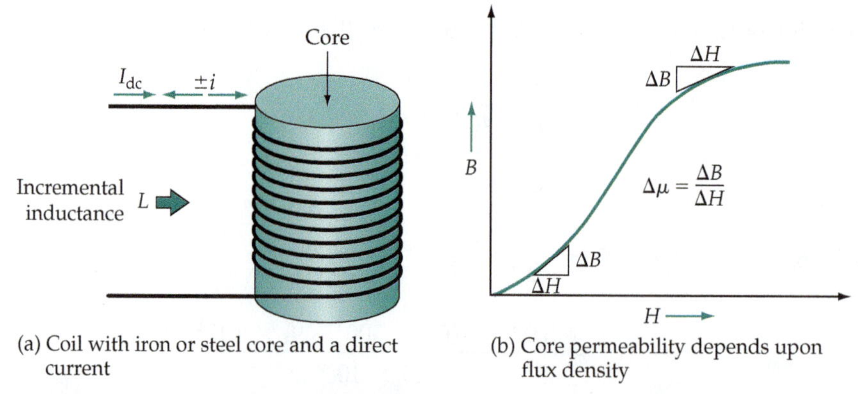

(a) Coil with iron or steel core and a direct current

(b) Core permeability depends upon flux density

Figure 14-12 The inductance of a coil with an iron or steel core depends upon the core permeability, which can be altered by the level of direct current through the coil.

1 mA. As explained for incremental permeability in Section 12-7, a constant level of bias current is frequently employed to bias an iron core to the center of the steepest part of its *B/H* characteristic [Figure 14-12(b)]. This gives the largest possible incremental inductance.

Noninductive Coil

In many cases it is desirable to have a noninductive coil; for example, precision resistors are usually noninductive. To construct such a coil, the winding is made of two side-by-side conductors, as illustrated in Figure 14-13. Every coil turn has an adjacent turn carrying current in the opposite direction. The magnetic fields generated by side-by-side turns cancel each other out. In this way, no counter-emf is generated, and the coil is noninductive.

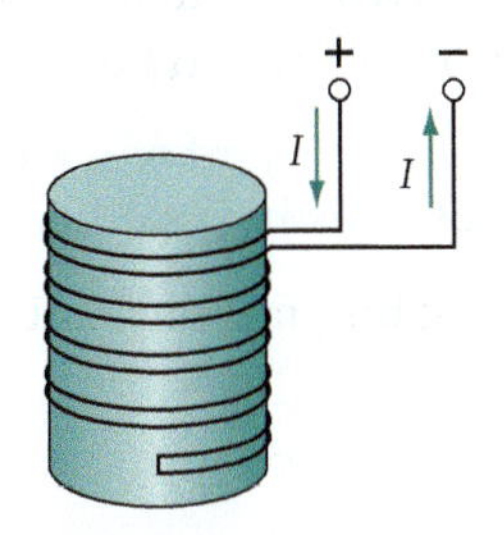

(a) Noninductive coil with adjacent conductors carrying currents in opposite directions

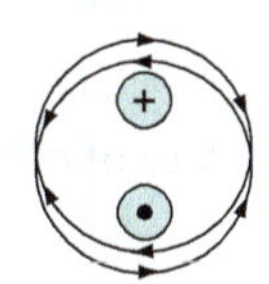

(b) Magnetic fields cancel when adjacent conductors carry currents in opposite directions

Figure 14-13 A noninductive coil is constructed by winding a double set of coil turns and ensuring that adjacent turns carry currents in opposite directions.

Practice Problems

14-3.1 Calculate the inductance of a 600 turn air-cored coil 5 cm long and 2.5 cm in diameter. Determine the counter-emf generated when the coil current collapses from 100 mA to zero in 5 ms.

14-3.2 Determine the number of turns required to create a 1.5 mH inductor if the dimensions of the coil are to be 4 cm long and 1.5 cm in diameter.

14-3.3 A ring made of sheet steel has a magnetic path length of 6.9 cm, a cross-sectional area of 2.7 cm^2, and a 160 turn coil. Determine the coil inductance when the current level is 300 mA.

14-4 MUTUAL INDUCTANCE

When the flux from one coil cuts another adjacent (magnetically coupled) coil, an emf is induced in the second coil. This was shown in Section 14-2. Following Lenz's law, the emf induced in the second coil sets up a flux that opposes the original flux from the first coil. So, the induced emf is again a counter-emf,

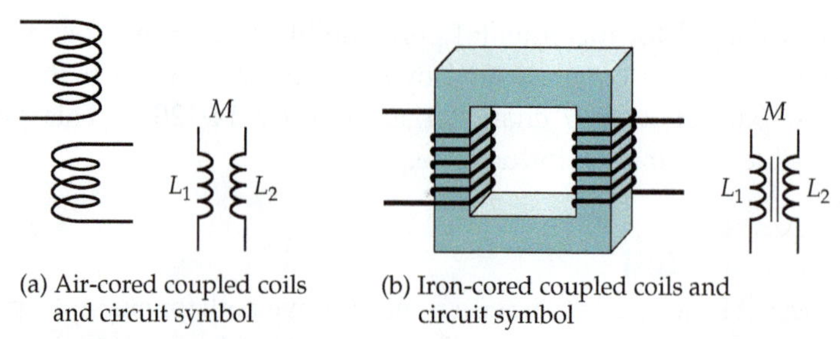

(a) Air-cored coupled coils and circuit symbol

(b) Iron-cored coupled coils and circuit symbol

Figure 14-14 Coupled coils and the circuit symbols.

and in this case the inductive effect is referred to as *mutual inductance.* Figure 14-14 shows coils with mutual inductance; termed *coupled coils.* Note that the graphic symbol for air-cored coupled coils is two side-by-side coil symbols, while that for coils with an iron-cored has two lines to represent the core.

Like self-inductance, mutual inductance is measured in henrys (H):

Two coils have a mutual inductance of 1 H when an emf of 1 V is induced in one coil by current changing at the rate of 1 A/s in the other coil.

This definition gives rise to the equation relating mutual inductance to induced voltage and rate of change of current:

$$M = \frac{e_L}{\Delta i/\Delta t} \tag{14-7}$$

where M is the mutual inductance in henrys, e_L is the *emf* in volts induced in the secondary coil, and $(\Delta i/\Delta t)$ is the rate of change of current in the primary coil in amperes/second.

Recall from Section 14-2 that the coil which has a current passed through it from an external source is termed the primary, and the coil that has an induced emf is referred to as the secondary. An equation for the emf induced in the secondary coil can be written from Eq. 14-2,

$$e_L = \frac{\Delta\Phi N_s}{\Delta t} \tag{14-8}$$

In Equation 14-8 $\Delta\Phi$ is the total change in flux linking with the secondary winding, N_s is the number of turns in the secondary winding, and Δt is the time required for the flux change. Substituting for e_L from Equation 14-8 into Equation 14-7 gives,

$$M = \frac{\Delta\Phi N_s}{\Delta i} \tag{14-9}$$

Figure 14-15(a) illustrates the fact that when the two coils are wound on a single ferromagnetic core, effectively all of the flux generated by the primary coil links with the secondary coil. However, when the coils are air-cored as in

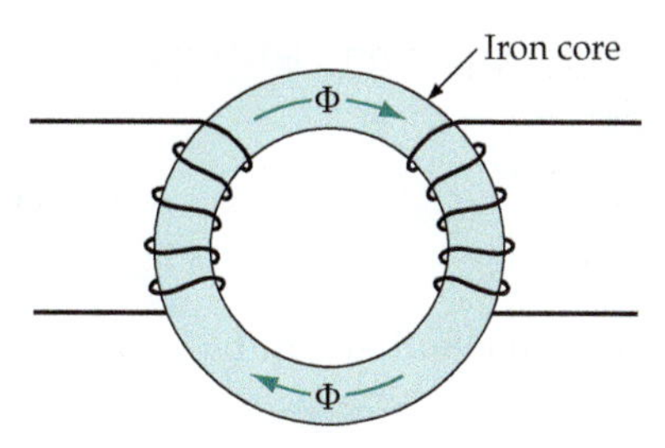

(a) Showing that with an iron core virtually all of the flux passes from one coil to the other

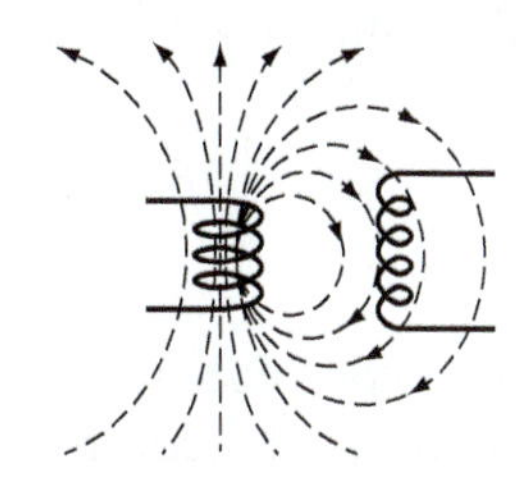

(b) Depending upon how closely air-cored coils are coupled, only a portion of the flux from the primary links with the secondary

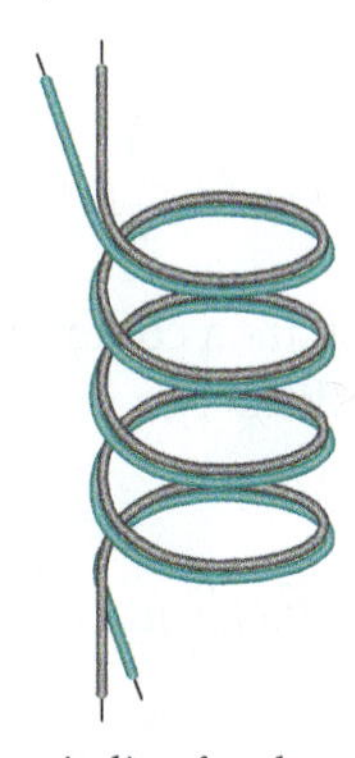

(c) Bifilar winding for close coupling

Figure 14-15 The amount of flux from a primary winding that links with a secondary depends on how closely the coils are coupled. The coefficient of coupling defines the linkage.

Figure 14-15(b), only a portion of the flux from the primary links with the secondary. Depending on how much of the primary flux cuts the secondary, the coils may be classified as *loosely coupled* or *tightly coupled.* One way to ensure tight coupling is shown in Figure 14-15(c), where each turn of the secondary winding is side-by-side with one turn of the primary winding. Coils wound in this fashion are said to be *bifilar.*

The amount of flux linkage from primary to secondary is also defined in terms of a *coefficient of coupling* (k). If all the primary flux links with the secondary, the coefficient of coupling is 1. When only 50% of the primary flux links with the secondary coil, the coefficient of coupling is 0.5. So,

$$k = \frac{\text{flux linkages between primary and secondary}}{\text{total flux produced by primary}}$$

Returning to Equation 14-9, when $\Delta\Phi$ is the total flux change in the primary coil, the flux linking with the secondary is $k\ \Delta\Phi$. So, the equation for M becomes,

$$M = k\frac{\Delta\Phi N_s}{\Delta t} \qquad \textbf{(14-10)}$$

Also, substituting for $\Delta\Phi$ from Equation 14-5 into Equation 14-10 gives,

$$M = kN_{\mathrm{p}}N_{\mathrm{s}}\mu_{\mathrm{r}}\mu_{\mathrm{o}}\frac{A}{l} \qquad \text{(14-11)}$$

Each winding considered alone has a self-inductance that can be calculated from Equation 14-6. So, for the primary coil,

$$L_1 = \mu_{\mathrm{r}}\mu_{\mathrm{o}}N_{\mathrm{p}}^2\frac{A}{l}$$

and for the secondary,

$$L_2 = \mu_{\mathrm{r}}\mu_{\mathrm{o}}N_{\mathrm{s}}^2\frac{A}{l}$$

$M = k\sqrt{L_1L_2}$

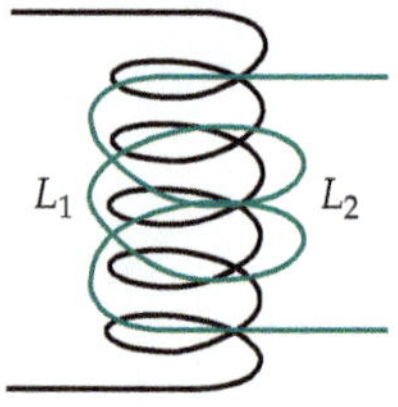

L_1 L_2

M

Figure 14-16 The mutual inductance between two coupled coils can be determined from the individual coil inductances and the coefficient of coupling (k).

Assuming that the two windings share a common core (magnetic or nonmagnetic as in Figure 14-16), the only difference in the expressions for L_1 and L_2 is the number of the coil turns. So,

$$L_1L_2 = N_{\mathrm{p}}^2N_{\mathrm{s}}^2\left(\mu_{\mathrm{r}}\mu_{\mathrm{o}}\frac{A}{l}\right)^2$$

Giving,

$$\sqrt{L_1L_2} = N_{\mathrm{p}}N_{\mathrm{s}}\mu_{\mathrm{r}}\mu_{\mathrm{o}}\frac{A}{l} \qquad \text{(14-12)}$$

Comparing Equations 14-11 and 14-12, it is seen that,

$$M = k\sqrt{L_1L_2} \qquad \text{(14-13)}$$

Example 14-9

Two identical coils are wound on a ring-shaped iron core that has a relative permeability of 500. Each coil has 100 turns, and the core dimensions are: cross-sectional area $A = 3\ \mathrm{cm}^2$, and magnetic path length $l = 20$ cm. Calculate the inductance of each coil and the mutual inductance between the coils.

Solution

From Eq. 14-6,

$$L_1 = L_2 = \mu_{\mathrm{r}}\mu_{\mathrm{o}}N^2\frac{A}{l}$$

$$= 500 \times 4\pi \times 10^{-7} \times 100^2 \times \frac{3 \times 10^{-4}\ \mathrm{m}^2}{20 \times 10^{-2}\ \mathrm{m}}$$

$$\approx 94.2\ \mathrm{mH}$$

Because the coils are wound on the same iron core, $k = 1$.

Eq. 14-13, $$M = k\sqrt{L_1 L_2} = \sqrt{9.42 \text{ mH} \times 9.42 \text{ mH}}$$

$$\approx 94.2 \text{ mH}$$

Example 14-10

Two 100 turn end-to-end solenoids each have $l = 20$ cm and cross-sectional area = 3 cm^2. Calculate their coefficient of coupling when the mutual inductance between them is measured as 0.62 μH.

Solution

$$L_1 = L_2 = \mu_r \mu_o N^2 \frac{A}{l}$$

$$= 1 \times 4\pi \times 10^{-7} \times 100^2 \times \frac{3 \times 10^{-4} \text{ m}^2}{20 \times 10^{-2} \text{ m}}$$

$$\approx 18.8 \text{ μH}$$

From Eq. 14-13, $$k = \frac{M}{\sqrt{L_1 L_2}} \approx \frac{0.62 \text{ μH}}{\sqrt{(18.8 \text{ μH})^2}}$$

$$\approx 0.033$$

Practice Problems

14-4.1 An iron ring with a relative permeability of 600 has a 4 cm magnetic path length, 1 cm^2 cross-sectional area, an 80 turn coil, and a 190 turn coil. Calculate the mutual inductance between the coils.

14-4.2 A 50 turn secondary coil is wound directly on top of a 600 turn solenoid that is 10 cm long and 1.25 cm in diameter. The secondary is 10 cm long and 1.8 cm in diameter. Calculate the mutual inductance between the coils.

14-5 TYPES OF INDUCTORS

There are many different types of inductors, ranging from large high-current iron-cored chokes to tiny resistor-style low-current coils.

Power Supply Inductor

Figure 14-17 shows the circuit of an inductor used with a dc power supply (see Section 10-8), and Figure 14-18 shows the appearance of a typical power supply inductor. In this application, the inductor is usually required to pass a direct

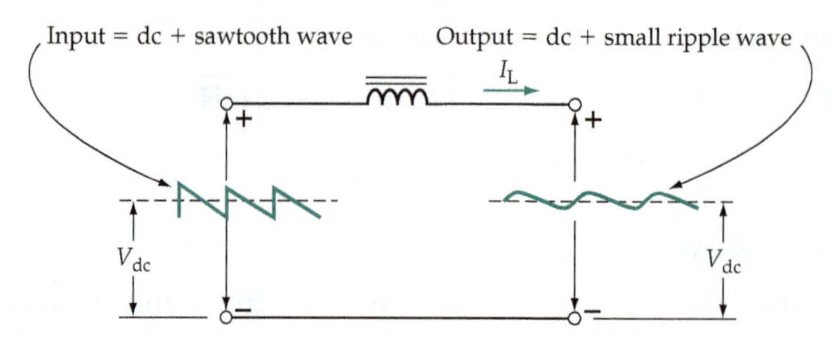

Figure 14-17 An inductor used with a dc power supply helps to smooth the increases and decreases in current level.

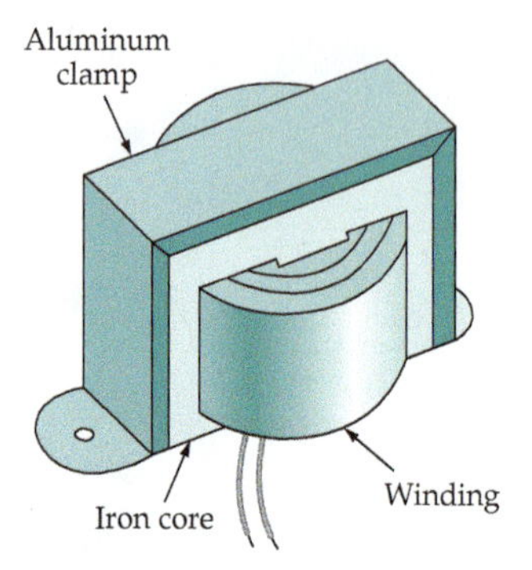

Figure 14-18 Typical power supply inductor.

current that has a fluctuating level. Because the inductor opposes any change in the current level through its windings, it tends to smooth out the fluctuations, (see the input and output voltages in Figure 14-17). This is exactly why the inductor is employed in this particular application. The presence of the direct current through the windings makes the incremental inductance of this component important. So, the inductance must be specified at a given level of direct current. Typical values for such inductors range from 50 mH to 20 H, with direct currents up to about 10 A and insulation voltage ratings up to 1000 V.

High-Frequency Inductor

A low-current high-frequency type of inductor is illustrated in Figure 14-19. The core in this case is ferrite material (see Section 12-7) in two mating sections known as a *pot core*. As well as increasing the coil's inductance, the pot core screens the coil to protect adjacent components against flux leakage and to protect the coil from external magnetic fields. The coil is wound on a bobbin, so the number of turns is easily modified.

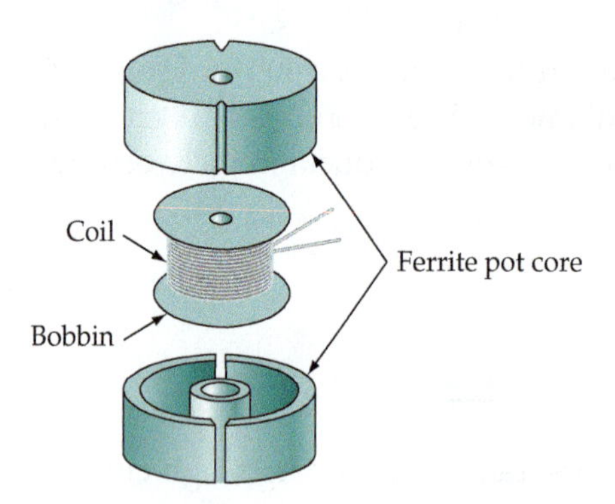

Figure 14-19 Some low-current, high-frequency inductors are wound on bobbins contained in a ferrite pot core. The ferrite core increases the winding inductance and screens the inductor.

Three different types of low-current inductors are illustrated in Figure 14-20. Figure 14-20(a) shows a type that is available either as an air-cored inductor or with a ferromagnetic core. With an air core, the inductance values range typically from 2.4 μH to 100 μH. With a ferromagnetic core, inductance values up to about 10 mH can be obtained. Depending on the thickness of the wire used and the physical size of the inductor, the maximum current can range from

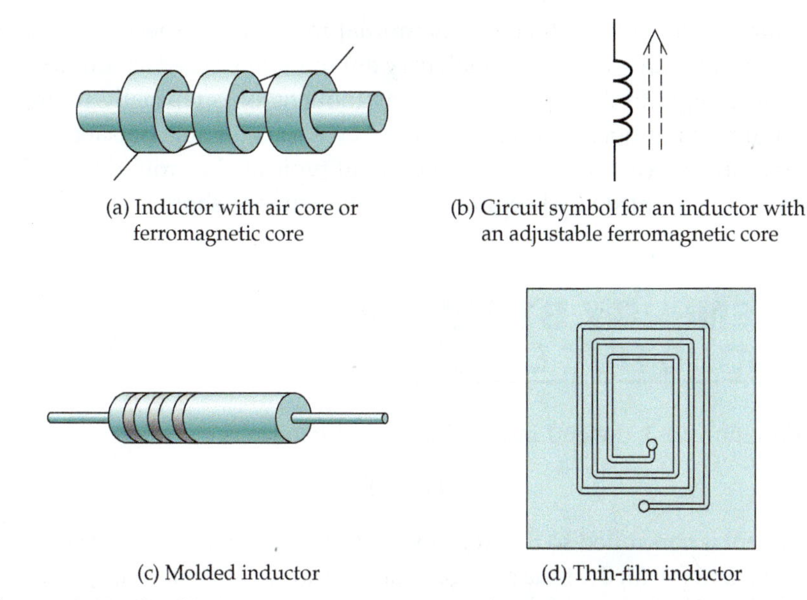

(a) Inductor with air core or ferromagnetic core

(b) Circuit symbol for an inductor with an adjustable ferromagnetic core

(c) Molded inductor

(d) Thin-film inductor

Figure 14-20 Small inductors may be wound on an insulating tube with an adjustable ferrite core, molded like small resistors, or deposited as a conducting film on an insulating material.

about 50 mA to 1 A. The core in such an inductor may be made adjustable so that it can be screwed into or partially out of the coil, making the inductance variable. Note the graphic symbol for an inductor with an adjustable core [Figure 14-20(b)].

Molded Inductors

A small *molded inductor* is shown in Figure 14-20(c). Typical available values for this type range from 1.2 μH to 10 mH, with maximum currents of about 70 mA. The values of molded inductors are identified by a color code, similar to molded resistors. Figure 14-20(d) shows a tiny *thin-film inductor* used in certain types of electronic circuits. In this case the inductor is simply a thin metal film deposited in the form of a spiral on a ceramic base.

Laboratory Inductors

Laboratory-type variable inductors can be constructed in *decade box* format [see Figure 4-14(c)], in which precision inductors are switched into or out of a circuit by means of rotary switches. Alternatively, two coupled coils can be employed as a variable inductor. The coils may be connected in series or in parallel, and the total inductance is controlled by adjusting the position of one coil relative to the other.

Stray Inductance

Inductance is (change in flux linkages)/(change in current). So, every current-carrying conductor has some self-inductance; typically less than 0.1 μH. Also every pair of adjacent conductors has mutual inductance. These *stray inductances* are usually unwanted, although they are sometimes used as components in a circuit design. In dc applications, stray inductance is normally unimportant, but in radio-frequency ac circuits it can be a considerable nuisance. Stray inductance in a piece of connecting wire might typically be around 0.1 μH, so it is normally minimized by keeping connecting wires as short as possible.

14-6 ENERGY STORED IN AN INDUCTIVE CIRCUIT

From Equation 14-3, the emf induced in an inductance is,

$$e_L = L(\Delta i/\Delta t)$$

If the current is controlled so that it grows linearly from zero to I in a time t, the average current is $0.5I$, the time change is $\Delta t = t$, and the current change is $\Delta i = I$. So, the induced voltage e_L is a constant quantity.

$$e_L = L(I/t)$$

From Equations 3-3 and 3-7, the electrical energy input is,

$$W = (\text{average voltage}) \times (\text{average current}) \times t$$

So, the energy supplied to the inductive circuit is,

$$W = e_L \times 0.5I \times t$$

Substituting for e_L, the energy stored is,

$$W = 0.5LI^2 \quad \textbf{(14-14)}$$

When L is in henrys and I is in amperes, W is given in joules. It is important to note that energy is stored in an inductor only when a current is flowing in the coil. When the current is reduced to zero, there is no energy stored [Figure 14-21(a)].

Substituting from Equations 12-5 and 14-6 into Equation 14-14 gives,

$$W = \frac{B^2 A l}{2\mu_r \mu_o} \text{ joules} \quad \textbf{(14-15)}$$

Figure 14-21(b) illustrates how the energy stored in an inductance can be calculated from the inductor dimensions etc. Equation 14-15 contains flux density (B), which is established by the coil current. So, once again note that energy is stored in an inductor only when a current flows in its coil. When μ_r in Equation 14-15 is replaced by 1 for an air core, the equation is exactly the same as Equation 12-7, which represents the energy stored in an air gap between two magnetic surfaces.

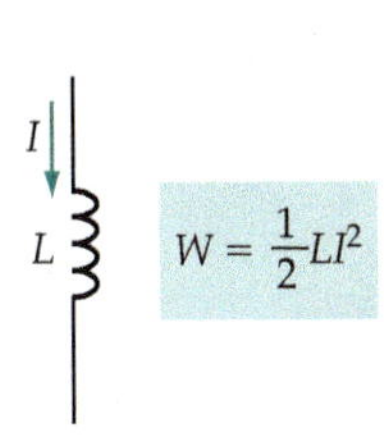

(a) Energy stored is dependent on the inductance and on the current

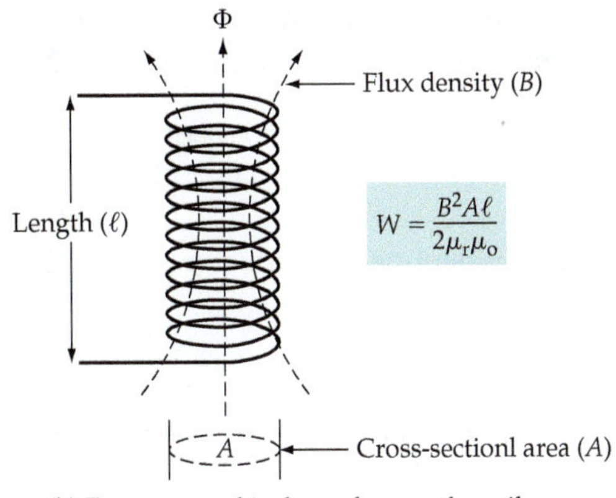

(b) Energy stored is dependent on the coil dimensions and on the flux density

Figure 14-21 Energy is stored in an inductor that has a constant coil current. The stored energy is reduced to zero when the current level is zero.

Example 14-11

A 20 H choke with a coil resistance of 180 Ω has a 300 V supply. Calculate the energy stored.

Solution

The steady-state current is,

$$I = \frac{E}{R} = \frac{300\text{ V}}{180\ \Omega}$$

$$\approx 1.67\text{ A}$$

Eq. 14-14,

$$W = 0.5LI^2 \approx 0.5 \times 20\text{ H} \times (1.67\text{ A})^2$$

$$\approx 27.9\text{ J}$$

Practice Problems

14-6.1 Calculate the energy stored in the solenoid in Problem 14-3.1 when a constant current of 100 mA flows in the coil.

14-6.2 Determine the energy stored in the magnetic core in Example 12-6.

14-7 INDUCTORS IN SERIES AND IN PARALLEL

Series-Connected Inductors

When inductors are connected in series or parallel and there is no mutual induction between them, they can be treated exactly like resistors. For series-connected inductors, the same current flows through all inductors, and the

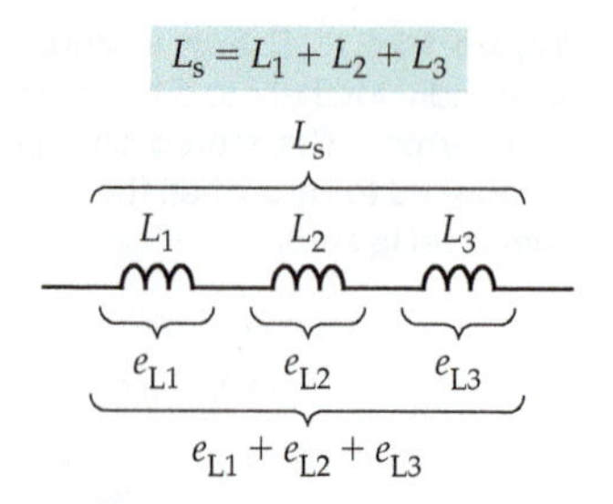

Figure 14-22 Series-connected inductors behave similarly to resistors in series. Total series inductance is $L_s = L_1 + L_2 + L_3 + \cdots$

counter-emf generated in each inductor is proportional to the rate of change of current. So, as illustrated in Figure 14-22, the total counter-emf is,

$$e_{L(total)} = e_{L1} + e_{L2} + e_{L3} + \cdots$$
$$= L_S(\Delta i/\Delta t)$$

For series-connected inductors the total inductance is,

$$L_S = L_1 + L_2 + L_3 + \cdots \tag{14-16}$$

Note that Equation 14-16 is correct only when there is no mutual induction between the inductors.

Parallel-Connected Inductors

For parallel-connected inductors, the same voltage appears across each inductor; consequently, the counter-emfs generated in each inductor must be equal.

$$e_L = e_{L1} = e_{L2} = e_{L3} = \cdots$$

As illustrated in Figure 14-23, the total current change is,

$$\Delta i_{total} = \Delta i_1 + \Delta i_2 + \Delta i_3 + \cdots$$

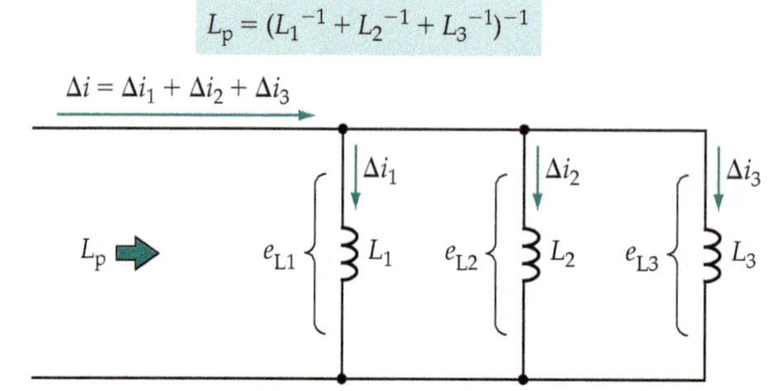

Figure 14-23 Parallel-connected inductors behave similarly to resistors in parallel. Total parallel inductance is $L_p = (L_1^{-1} + L_2^{-1} + L_3^{-1})^{-1}$.

From Eq. 14-3,

$$\frac{1}{L_{total}} = \frac{\Delta i_{total}}{e_L \Delta t} = \frac{\Delta i_1 + \Delta i_2 + \Delta i_3 + \cdots}{e_L \Delta t}$$
$$= \frac{\Delta i_1}{e_L \Delta t} + \frac{\Delta i_2}{e_L \Delta t} + \frac{\Delta i_3}{e_L \Delta t} + \cdots$$
$$= \frac{1}{L_1} + \frac{1}{L_2} + \frac{1}{L_3} + \cdots$$

So, for parallel-connected inductors the total inductance is determined from,

$$\frac{1}{L_p} = \frac{1}{L_1} + \frac{1}{L_2} + \frac{1}{L_3} + \cdots$$

or,

$$L_p = \left(L_1^{-1} + L_2^{-1} + L_3^{-1}\right)^{-1} \tag{14-17}$$

Note that Equation 14-17 is correct only when there is no mutual induction between the inductors.

Example 14-12

Three inductors have values of $L_1 = 10$ mH, $L_2 = 100$ μH, and $L_3 = 500$ μH. Determine the total inductance of the three when connected (a) in series, and (b) in parallel. Assume that there is no mutual induction between the inductors.

Solution

(a) In series:

Eq. 14-16,
$$L_s = L_1 + L_2 + L_3 = 10\text{ mH} + 100\ \mu\text{H} + 500\ \mu\text{H}$$
$$= 10.6\text{ mH}$$

(b) In parallel:

Eq. 14-17,
$$L_p = (L_1^{-1} + L_2^{-1} + L_3^{-1})^{-1}$$
$$= [(10\text{ mH})^{-1} + (100\ \mu\text{H})^{-1} + (500\ \mu\text{H})^{-1}]^{-1}$$
$$\approx 82.6\ \mu\text{H}$$

Inductors with Mutual Inductance

Figure 14-24 illustrates the situation when mutual induction exists between two series-connected coils. The coil connection in Figure 14-24(a) is such that their fluxes assist each other, and the coils are said to be connected *series-aiding*. Figure 14-24(b) shows the graphic symbols used in circuit diagrams for mutually coupled coils that are connected series-aiding. The dots at the same ends of each coil indicate that the coils have fluxes in the same direction when a current flows.

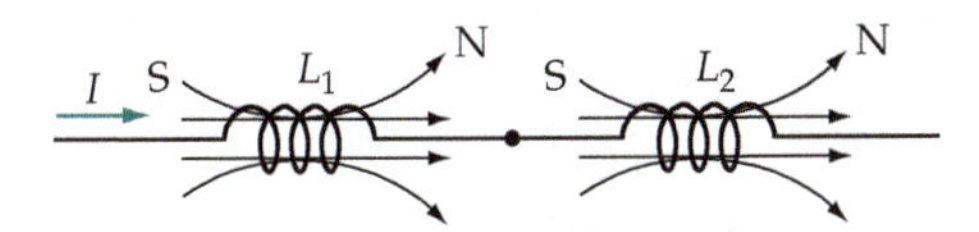

(a) Mutually coupled coils connected series-aiding

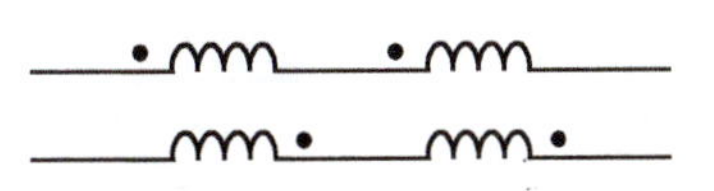

(b) Circuit symbols for mutually coupled coils connected series-aiding

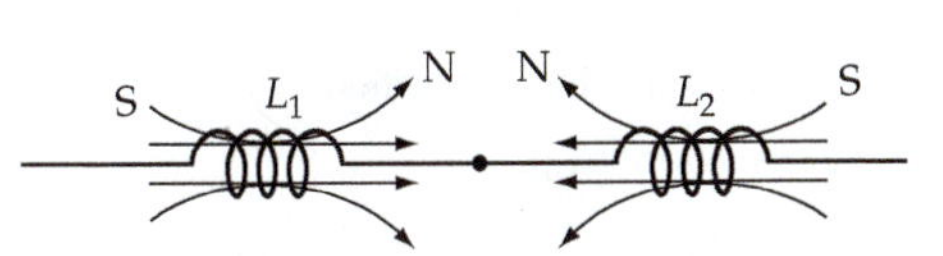

(c) Mutually coupled coils connected series-opposing

(d) Circuit symbols for mutually coupled coils connected series-opposing

Figure 14-24 As in the case of voltage cells, inductors can be connected series-aiding or series-opposing. Where the coils are mutally coupled, the total inductance is $L = L_1 + L_2 + 2M$ for a series-aiding connection, or $L = L_1 + L_2 - 2M$ for a series-opposing connection.

The two coils shown in Figure 14-24(c) have their fluxes in opposite directions, and so they are said to be connected *series-opposing*. In Figure 14-24(d) the graphic symbols for series-opposing coils are illustrated. In this case the dots at opposite ends of each coil indicate that their fluxes are in opposition when a current flows.

For series-aiding connections, the total emf induced in L_1 is the sum of the emf due to the coil's self-inductance and the emf induced from L_2. So,

$$e_{L1} = L_1\frac{\Delta i}{\Delta t} + M\frac{\Delta i}{\Delta t}$$

$$= \frac{\Delta i}{\Delta t}(L_1 + M)$$

Similarly,

$$e_{L2} = \frac{\Delta i}{\Delta t}(L_2 + M)$$

so,

$$e_{L(total)} = \frac{\Delta i}{\Delta t}(L_1 + M + L_2 + M)$$

Giving,

$$L_{total} = L_1 + M + L_2 + M$$

So, for series-aiding connections [Figure 14-25(a)],

$$L = L_1 + L_2 + 2M \tag{14-18}$$

For series-opposing connection [Figure 14-25(b)] the emf induced in L_1 from L_2 is in opposition to the self-induced emf. So, using the same procedure as above, the total inductance for series-opposing connections is,

$$L = L_1 + L_2 - 2M \tag{14-19}$$

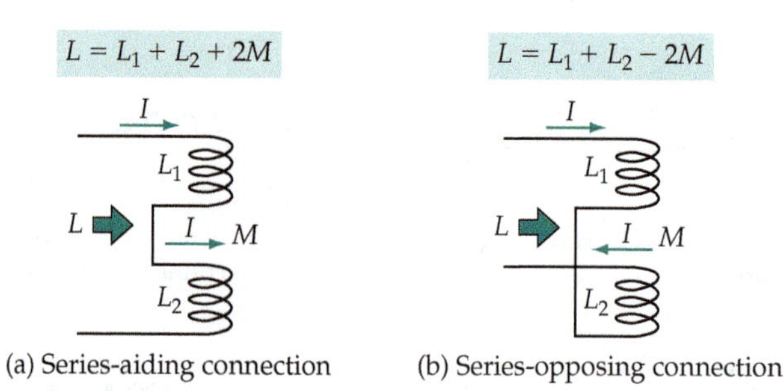

(a) Series-aiding connection (b) Series-opposing connection

Figure 14-25 Coils connected series-aiding and series-opposing.

Example 14-13

Two 500 μH coils have a mutual inductance of 200 μH. Determine the total inductance of the two coils when they are connected (a) series-aiding, and (b) series-opposing.

Solution

(a) Series-aiding:

Eq. 14-18, $L = L_1 + L_2 + 2M$

$= 500\ \mu H + 500\ \mu H + 2(200\ \mu H)$

$= 1.4\ mH$

(b) Series-opposing:

Eq. 14-19, $L = L_1 + L_2 - 2M$

$= 500\ \mu H + 500\ \mu H - 2(200\ \mu H)$

$= 600\ \mu H$

If the mutual inductance between two adjacent coils is not known, it can be determined by measuring the total inductance of the coils in series-aiding and series-opposing connections.

For series-aiding, $L_a = L_1 + L_2 + 2M$

For series-opposing, $L_b = L_1 + L_2 - 2M$

Subtracting, $L_a - L_b = 4M$

So,

$$M = \frac{L_a - L_b}{4} \quad \textbf{(14-20)}$$

Recall that the mutual inductance between two coils is given by Equation 14-13 as, $M = k\sqrt{L_1 L_2}$

From these two equations, the coefficient of coupling of the two coils can be determined.

Example 14-14

Use the results of Example 14-13 to determine the mutual inductance and coefficient of coupling for the two coils.

Solution

From Example 14-13, $L_a = 1.4\ mH$ and $L_b = 600\ \mu H$

Eq. 14-20, $M = \frac{L_a - L_b}{4} = \frac{1.4\ mH - 600\ \mu H}{4}$

$= 200\ \mu H$

From Eq. 14-13, $k = \frac{M}{\sqrt{L_1 L_2}} = \frac{200\ \mu H}{\sqrt{500\ \mu H \times 500\ \mu H}}$

$= 0.4$

Practice Problems

14-7.1 Calculate the inductance that must be connected in parallel with a 100 μH inductor to give a total inductance of 70 μH. Assume no mutual inductance between the two.

14-7.2 Determine the total inductance of the two coils in Problem 14-4.2 when connected (a) series-aiding, and (b) series-opposing.

14-7.3 The two inductors in Problem 14-7.1 are connected in series, and the total inductance is measured as 220 μH. Calculate the coefficient of coupling between the two.

Summary of Formulas

- ***Induced emf:***

$$e_L = \frac{\Delta\Phi}{\Delta t} \qquad e_L = \frac{\Delta\Phi N}{\Delta t}$$

- ***Inductance:***

$$L = \frac{e_L}{\Delta i/\Delta t} \qquad L = \frac{\Delta\Phi N}{\Delta i}$$

- ***Flux change:***

$$\Delta\Phi = \mu_r\mu_o\Delta i\, N\frac{A}{l}$$

- ***Self-inductance:***

$$L = \mu_r\mu_o N^2\frac{A}{l}$$

- ***Induced emf:***

$$e_L = \frac{\Delta\Phi N_s}{\Delta t}$$

- ***Mutual inductance:***

$$M = \frac{e_L}{\Delta i/\Delta t} \qquad M = \frac{\Delta\Phi N_s}{\Delta i}$$

$$M = k\frac{\Delta\Phi N_s}{\Delta i} \qquad M = kN_pN_s\mu_r\mu_o\frac{A}{l}$$

$$M = k\sqrt{L_1L_2} \qquad M = \frac{L_a - L_b}{4}$$

- ***Energy stored:***

$$W = 0.5LI^2$$

$$W = \frac{B^2Al}{2\mu_r\mu_o}$$

- ***Inductances in series:***

$$L_s = L_1 + L_2 + L_3 + \cdots$$

- ***Inductances in parallel:***

$$L_p = (L_1^{-1} + L_2^{-1} + L_3^{-1})^{-1}$$

- ***Total inductance (series-aiding):***

$$L = L_1 + L_2 + 2M$$

- ***Total inductance (series-opposing):***

$$L = L_1 + L_2 - 2M$$

Review Questions

Section 14-1

14-1 Draw sketches to show the direction of the induced current in a conductor in motion through a magnetic field. Also, show the field generated by the conductor. Briefly explain.

14-2 State Lenz's law, and explain it in relation to the induced emf and current in a conductor moving through a magnetic field.

Section 14-2

14-3 Draw a sketches to show how a changing magnetic flux through a coil induces an emf in the coil.

14-4 State Faraday's law, and define the weber. Use the definition of the weber to write an equation for induced emf in a coil with a changing magnetic flux.

Section 14-3

14-5 Define electromagnetic induction, primary, secondary, self-inductance, counter-emf, and back-emf.

14-6 Draw sketches to show how the current through a coil induces a counter-emf in the coil. Explain.

14-7 Define the henry and use the definition to write an equation for the inductance of a coil.

14-8 Derive an equation for inductance L in terms of flux linkages and current change. Also, write the equation that relates L to the dimensions of a core.

14-9 Discuss the difficulties that arise with ferromagnetic cored inductors in relation to coil current. Define incremental inductance.

14-10 Show how a coil can be wound to be noninductive.

Section 14-4

14-11 Explain mutual inductance and define the relationship between two coils when they have a mutual inductance of 1 H. Using the definition, write an equation for the mutual inductance between two coils.

14-12 Derive an equation for mutual inductance in terms of flux linkages and current change. Also explain coefficient of coupling, and show how it affects the equation for mutual inductance.

14-13 Derive the equation relating mutual inductance to the dimensions of a core. Write the equation relating mutual inductance to the individual coil inductances.

Section 14-5

14-14 Briefly discuss power supply inductors, high frequency inductors, and stray inductance.

Section 14-6

14-15 Write an equation for the energy stored in an inductive circuit in terms of the inductance value and the current through the coil. Also, write the equation for the energy stored in terms of the flux density and core dimensions.

Section 14-7

14-16 Derive the equations for the inductance of several inductors when connected (a) in series, (b) in parallel. Assume that there is no mutual induction between the inductors.

14-17 Using illustrations, explain what occurs when mutual inductance exists between two series-connected inductors when they are connected (a) series-aiding, and (b) series-opposing. Derive equations for the total inductance in each case.

14-18 Explain how the mutual inductance between two coils can be determined experimentally, and derive the appropriate equation.

Problems

Section 14-2

14-1 The flux linking a 100 turn coil increases from zero to 0.12 Wb in a time of 0.3 s. Calculate the induced voltage.

14-2 A 300-turn solenoid has a secondary winding with 250 turns. The solenoid is 20 cm long and 3 cm in diameter. Calculate the emf induced in the secondary when the solenoid current is increased from zero to 500 mA in a time of 5 ms.

14-3 A cast iron core with two coils has a closed magnetic path 20 cm long and a cross-sectional area of 5 cm^2, as illustrated in Figure 14-26. The primary winding has $N_p = 400$ turns, and the secondary winding has $N_s = 250$ turns. If the current through the primary increases from zero to 1 A in a time of 10 ms, determine the emf induced in the secondary.

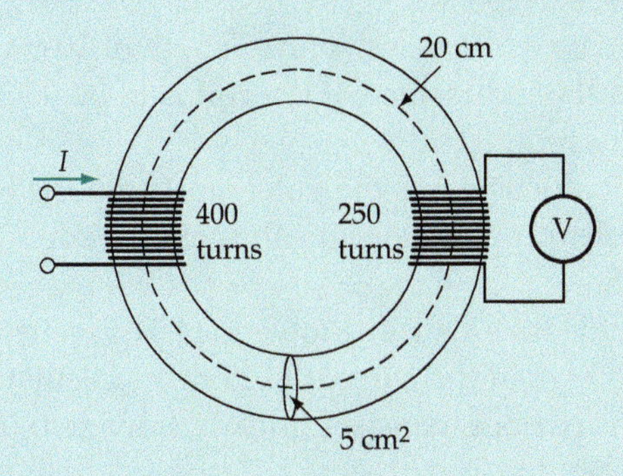

Figures 14-26

14-4 Another secondary winding having 200 turns is added on the core in Problem 14-3. Calculate the emf induced in each secondary when the primary current is increased from zero to 1 A in a time of 3.3 ms.

14-5 Two coils are wound on a ring-shaped cast steel core. The secondary of the two has $N_s = 250$ turns, and the output voltage from it is to be 25 V when the primary current increases from zero to its maximum level in a time of 8 ms. If the primary winding has 200 turns, determine the maximum level of the current that must flow through the primary. The core cross-sectional area is 10 cm^2 and the magnetic path length is 30 cm.

14-6 An additional 100 turns are added to the primary winding in Problem 14-5. Recalculate the required primary current.

14-7 A 5000 turn air-cored solenoid is 28 cm long and has an inside diameter of 2 cm. A secondary winding wound on top of the solenoid is to have 1 V induced in it when the solenoid current increases from zero to 15 A in a period of 2.5 ms (see Figure 14-27). Determine the number of turns required on the secondary.

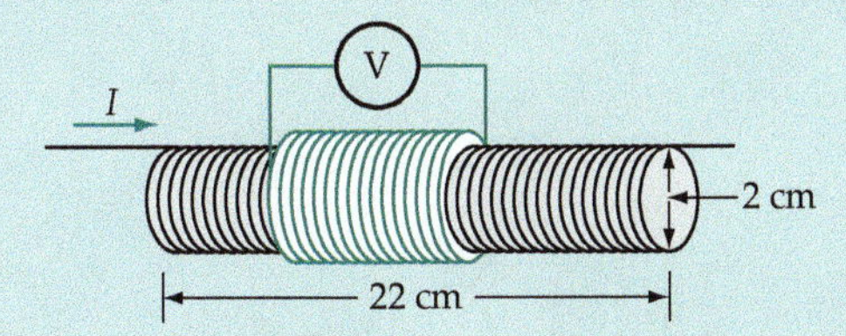

Figures 14-27

14-8 The secondary output voltage for the coil arrangement Problem 14-7 is to be increased to 1.8 V without altering the primary current or the number of coil turns. How can the increase be achieved?

Section 14-3

14-9 A coil current grows linearly from zero to a maximum of 30 A in a time period of 15 ms. If the coil has an inductance of 5 mH, determine the level of the counter-emf.

14-10 Calculate the new maximum current level for Problem 14-9 if the counter-emf is not to exceed 100 mV.

14-11 A solenoid is to be constructed with a length of 10 cm and inside diameter of 1 cm. If the inductance of the coil is to be 500 μH, calculate the number of turns required.

14-12 Calculate the inductance of the solenoid (primary winding only) described in Problem 14-2. Determine the counter-emf induced during the current growth.

14-13 For the solenoid described in Problem 14-11, determine the total flux through the coil when the current is 25 mA. Calculate the counter-emf generated if the current increases linearly from zero in a time period of 50 ms.

14-14 Calculate the inductance of the solenoid (primary winding only) described in Problem 14-7. Determine the counter-emf induced during the current growth.

14-15 Calculate the inductance of a 2000 turn coil that is 12 cm in length and has an average inside diameter of 1.2 cm.

14-16 For Problem 14-15, recalculate the coil inductance (a) when its length is doubted, and (b) when its diameter is doubled.

14-17 A cast iron ring has a cross-sectional area of 1 cm^2 and a magnetic path length of 10 cm. A 100 turn coil is wound on the ring. Determine the coil inductance when the current through it is (a) 1 A, and (b) 5 A.

14-18 For Problem 14-17, recalculate the coil inductance for a 1 A current (a) when the core length is doubled, and (b) when its diameter is doubled.

Section 14-4

14-19 Two 75 turn coils are wound on an iron core that has a closed magnetic path, as shown in Figure 14-28. The core dimensions are magnetic path length = 33 cm, and cross-sectional area = 9 cm^2. If the core has a relative permeability of 900, calculate the inductance of each coil and the mutual inductance between the coils.

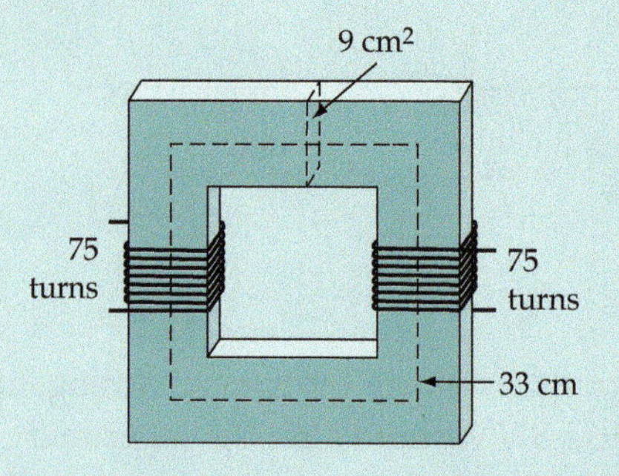

Figures 14-28

14-20 Calculate the inductance of the secondary winding in Problem 14-2. Determine the mutual inductance between the two windings.

14-21 Two 150 turn solenoids each have a length of 22 cm and a cross-sectional area of 3.3 cm^2. The mutual inductance between the two is measured as 2 μH. Calculate the coefficient of coupling.

14-22 Calculate the inductance of the secondary winding in Problem 14-7. Determine the mutual inductance between the two windings.

14-23 Calculate the self-inductance of each winding in Problem 14-3, and determine the mutual inductance between the two windings. Assume that the core has a relative permeability of 500.

14-24 If the two coils in Problem 14-2 are separated so that their coefficient of coupling is 0.4, calculate the mutual inductance between them.

Section 14-6

14-25 Calculate the energy stored in a 5 H choke that has a resistance of 105 Ω. The applied voltage is 250 V.

14-26 Calculate the energy stored in the solenoid (primary winding only) in Problem 14-2 when the current remains constant at 500 mA.

14-27 When 500 V is applied to a choke with a resistance of 120 Ω, it is found that the energy stored is 33 J. Calculate the inductance of the choke.

14-28 Calculate the energy stored in the primary winding in Problem 14-5 when the current remains constant at 1 A.

14-29 Calculate the energy stored in the coil in Problem 14-9 when a constant current of 15 A flows.

14-30 For the coil in Problem 14-17, determine the energy stored when the current is constant at (a) 1 A, and (b) 5 A.

Section 14-7

14-31 Four inductors have values of $L_1 = 100$ mH, $L_2 = 50$ mH, $L_3 = 750$ μH, and $L_4 = 1$ H. Determine the total inductance of the four when connected (a) in series, and (b) in parallel. Assume that there is no mutual induction between the inductors.

14-32 Two inductors that have values of 500 μH and 800 μH are connected in parallel. Calculate the value of a third inductor that, when connected in parallel with the other two, will give a total inductance of 250 μH.

14-33 A 600 μH coil and a 400 μH coil have a mutual inductance of 100 μH. Determine the total inductance of the two when they are connected (a) series-aiding, and (b) series-opposing.

14-34 Use the results of Problem 14-33 to determine the coefficient of coupling between the two coils.

14-35 Two identical coils when connected series-aiding give a total inductance of 850 μH. When connected series-opposing, their total inductance is 250 μH. Determine the mutual inductance between the two, the individual inductances, and the coefficient of coupling.

Practice Problem Answers

14-2.1 37.5 mV
14-2.2 742
14-2.3 836
14-3.1 44.4 mH, 88.8 mV
14-3.2 519
14-3.3 173 mH
14-4.1 28.6 mH
14-4.2 66.6 μH
14-6.1 22.2 μJ
14-6.2 43.9 mJ
14-7.1 233 μH
14-7.2 696 μH, 430 μH
14-7.3 0.37

CHAPTER 15
Capacitance

CONTENTS

Objectives

You will be able to:

1. Draw sketches to illustrate the process of electronic charge storage in two metal plates separated by a layer of insulating material.
2. Solve problems involving electric flux, electric flux density, and electric field strength in the dielectric of a capacitor.
3. Calculate the capacitance of a capacitor with given dimensions and given type of dielectric.
4. Describe the various types of capacitors and list the characteristics of each.
5. Calculate the equivalent capacitance of several capacitors connected in series or in parallel.
6. Calculate the energy stored in a charged capacitor.

INTRODUCTION

A capacitor consists of a layer of insulating material sandwiched between two metal plates. The electric field in a capacitor has many parallels to a magnetic field, including a parameter known as the permittivity, which is analogous to magnetic permeability. The capacitance value of a capacitor is a measure of the amount of electric charge that can be stored in the device. The capacitance can be calculated from knowledge of the capacitor dimensions and the permittivity of the insulating material. There are many different types of capacitors, each with its own particular characteristics and application. Parallel-connected and series-connected capacitors cannot be treated in the same way as parallel and series resistance circuits.

15-1 ELECTRIC CHARGE STORAGE

Figure 15-1 shows a device that consists of a layer of insulating material sandwiched between two metal plates. This device is known as a *capacitor* and, as will be shown, it has the ability to store an electric charge. Figure 15-2(a) shows a battery and switch (S_1) for connecting the battery to the plates of the capacitor. Initially, there is no voltage difference between the plates, but when the switch is closed the plates assume the same voltages as the battery terminals. So, with S_1 closed, a voltage difference exists between the capacitor plates.

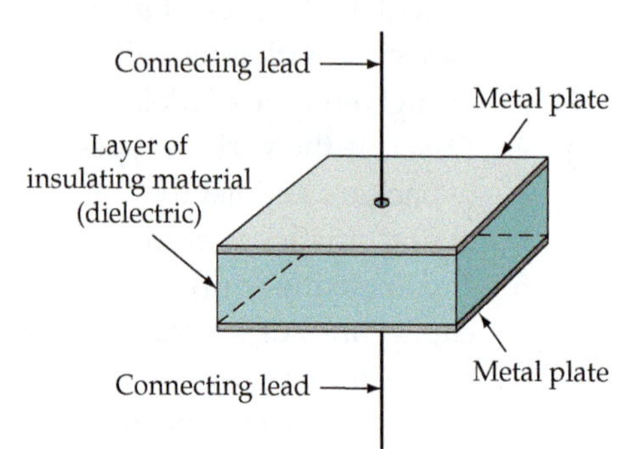

Figure 15-1 A capacitor basically consists of a layer of insulating material sandwiched between two metal plates.

Recall that the positive terminal of a battery is positive because many of its (negative) electrons have been removed. Also, the battery's negative terminal has an accumulation of electrons. For the capacitor plates in the illustration to have the same voltage difference as the battery terminals, electrons must be removed from the top plate to make it positive, and the removed electrons must be delivered to the bottom plate to make it negative. This means that electrons flow out of the top plate and into the bottom plate, as illustrated in Figure 15-2(a). The flow of electrons constitutes a (conventional direction) current flow into the top plate and out of the bottom plate. The current flow lasts

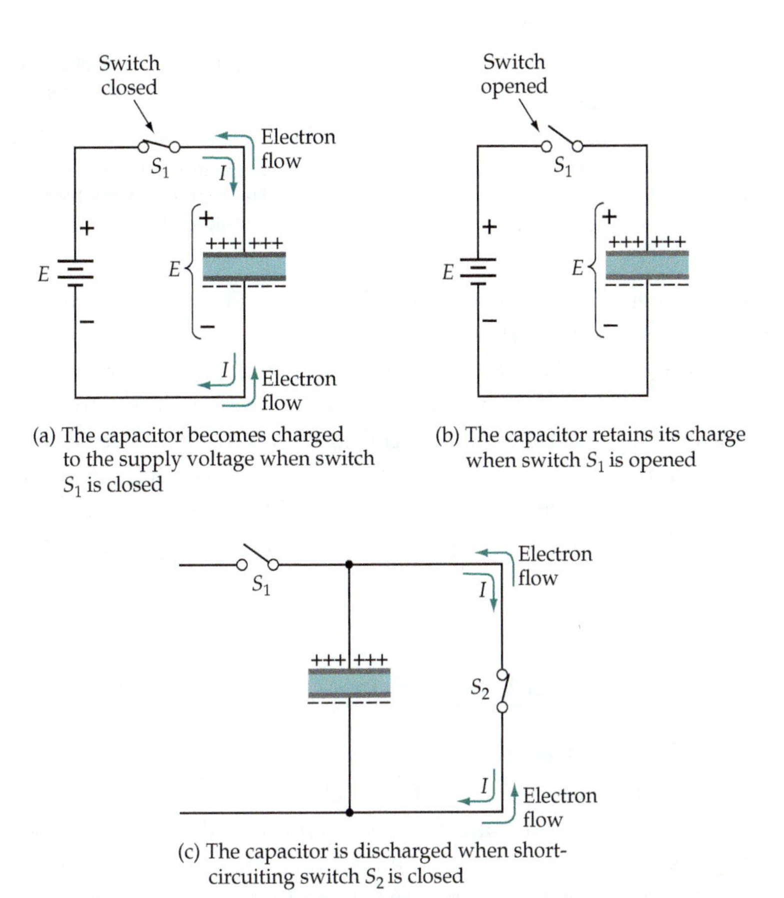

(a) The capacitor becomes charged to the supply voltage when switch S_1 is closed

(b) The capacitor retains its charge when switch S_1 is opened

(c) The capacitor is discharged when short-circuiting switch S_2 is closed

Figure 15-2 **A capacitor connected to a voltage source assumes the same terminal voltage as the source, and the capacitor terminal voltage remains constant when the source is disconnected. The capacitor can be discharged by short-circuiting its terminals.**

only a brief time (less than milliseconds) and ceases just as soon as the plates are at the same voltages as the battery terminals.

If the switch is now opened, the capacitor plates are found to retain their voltage difference, [see Figure 15-2(b)]. It can be shown that the capacitor has, in fact, stored an electric charge, and that the quantity of charge stored depends on the dimension of the plates and the insulating material, as well as on the battery voltage. The capacitor was *charged* by a brief pulse of current that flowed when switch S_1 was closed. If short-circuiting switch S_2 in Figure 15-2(c) is now closed, the surplus of electrons on the bottom plate returns to the top plate. Once again, this can be described as a conventional direction current flow from the positive plate to the negative plate. It is seen that, closing S_2 *discharges* the capacitor and returns the plate voltage difference to zero.

The layer of insulating material between the capacitor plates is known as the *dielectric*. Typical dielectric materials are rubber, mica, paper, and air. The capacitor plates may be any conducting material. Graphic symbols for fixed and variable capacitors are illustrated in Figure 15-3. The straight line identifies the terminal that should always be made positive whenever there is a voltage

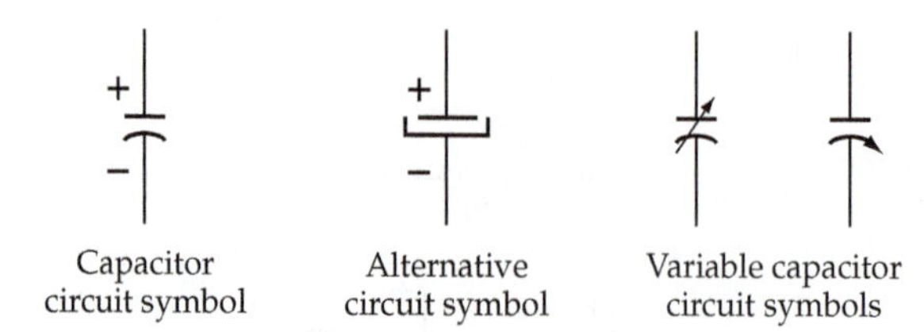

Figure 15-3 Circuit symbols for fixed and variable capacitors. In situations where there is a dc voltage difference between the capacitor terminals, the straight line represents the terminal that should be the most positive.

difference between the terminals. As will be explained, the correct connection can be very important for certain types of capacitors.

Because a capacitor stores a charge and a voltage cell also stores electricity, the two are comparable. However, when a load current is drawn from a voltage cell, the chemical action within the cell continues to create free electrons to supply the current. In the case of the capacitor there is no such chemical action. The current flow from the capacitor rapidly declines when the surplus electrons are transferred between the plates. So, a capacitor cannot be treated as a voltage source.

15-2 ELECTRIC FIELD

Electric Lines of Force

In Chapter 1 it was explained that a force exists between electrically charged bodies, and that bodies with like charges repel each other, while those with unlike charges experience a force of attraction. These forces are similar in many ways to the forces that occur between magnetic poles. As with magnetism, a *force field* exists around electrically charged bodies. *Electric lines of force* are spoken of, although their existence cannot be demonstrated as easily as in the case of magnetism. The *electric field strength, electric flux,* and *electric flux density* can all be calculated.

In Figure 15-4(a), the bottom plate of a capacitor is shown grounded, and the top plate has a voltage (or potential) of $+E$ with respect to ground. In the space between the two plates, the voltage at the halfway point becomes $+0.5E$. Similarly, at $0.25d$ and $0.75d$, the voltages are $+0.25E$ and $+0.75E$, respectively. At each of these points a line exists along which the voltage is a constant quantity. Appropriately, the lines are termed *equipotential lines.* The equipotential lines are found to be horizontal in the space between the plates, and at the edges of the plates they curve outward, as illustrated in Figure 15-4(b).

The electric lines of force (similar to magnetic lines of force) are found to always cross the equipotential lines at right angles. So, the lines of force are vertical in the space between the plates and bulge outward at the plate edges, as shown in the figure.

The direction of the electric lines of force is assumed to be the direction in which a free, positively charged particle would move if it were placed in the electric field, Such a particle would be repelled from the positive plate of the capacitor and attracted to the negative plate. So, the electric lines of force are said to flow from positive

(a) Equipotential lines in the dielectric of a capacitor

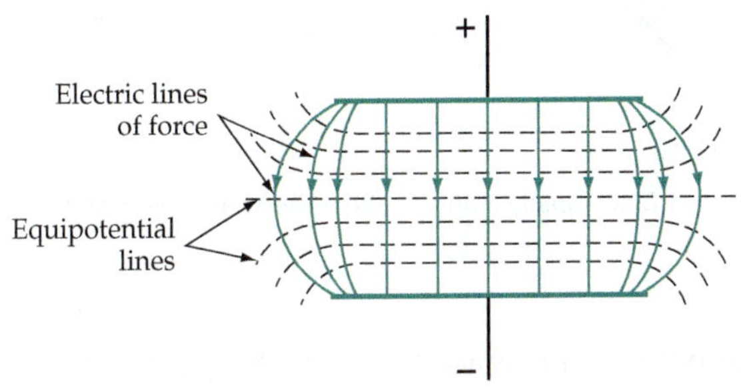

(b) Electric lines of force and equipotential lines between the plates of a capacitor

Figure 15-4 An electric force field exists around charged bodies. Electric lines of force and equipotential lines can be plotted. Electric field strength and flux density can be calculated.

to negative. As with magnetic lines of force, it is found that *electric lines of force repel each other, never intersect, and are always in a state of tension.*

Electric Charge, Flux, and Flux Density

Electric charge is measured in coulombs (see Appendix 2), and the coulomb is also the unit of electric flux. So, a body that is charged to Q coulombs emits a total electric flux of Q coulombs. Similarly, a capacitor that has Q coulombs of charge has a total electric flux of Q coulombs between its plates.

electric flux,
$$\Psi = Q \tag{15-1}$$

where Ψ is the electric flux in coulombs, and Q is the charge in coulombs. The electric flux density is simply the flux per unit area, [see Figure 15-5(a)],

electric flux density,
$$D = \frac{Q}{A} \tag{15-2}$$

When Q is total charge in coulombs and A is the area in m^2, D is electric flux density in C/m^2.

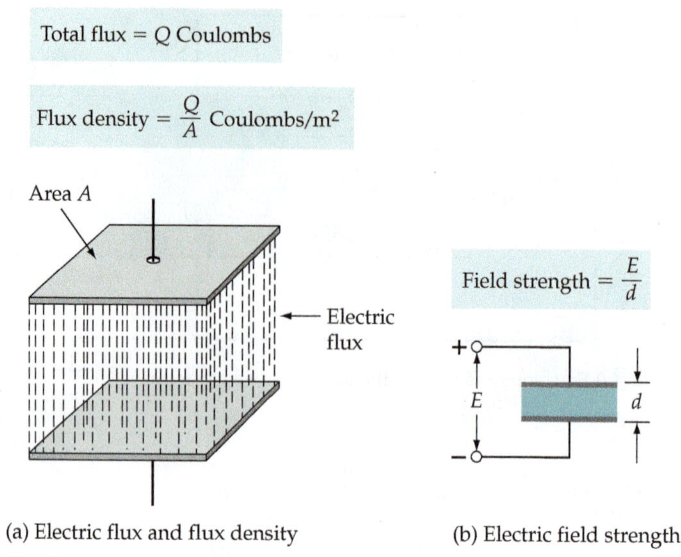

(a) Electric flux and flux density (b) Electric field strength

Figure 15-5 Electric flux and flux density can be calculated, and field strength can be determined.

The electric field strength in the space between the plates of a capacitor depends on the applied voltage and on the distance between the plates:

electric field strength,
$$\xi = \frac{E}{d} \tag{15-3}$$

Where E is the voltage difference between the plates in volts and d is the distance between the plates in meters, the electric field strength is in V/m, [see Figure 15-5(b) and Section 4-2].

Dielectric Effects

The insulating dielectric material between the plates of a capacitor is subjected to electric stress; that is, the applied voltage (electric pressure) is pressing for current flow. Where the voltage between the plates is low, the electric stress on the dielectric is low. The stress is greater when the voltage is high, and if the stress becomes large enough the material may break down. As discussed in Section 4-2, the stress on the dielectric is measured in volts/meter, and it is, in fact, the electric field strength. The *dielectric strength* of a given material is the voltage per unit thickness at which the dielectric may break down, see Table 15-1, (reproduced from Table 4-1).

Because the dielectric is an insulating material, the electrons within the dielectric are never detached from their atoms (except in a breakdown situation). However, a certain amount of distortion of the atoms occurs, as illustrated in Figure 15-6. When the capacitor plates are at the same voltage (zero voltage difference), there is no electric stress on the dielectric, and the electrons

TABLE 15-1 Breakdown voltage of various capacitor dielectric materials

Dielectric	Dielectric Strength (kV/cm)
Air	30
Ceramic	30
Porcelain	70
Paper	500
Mica	2000

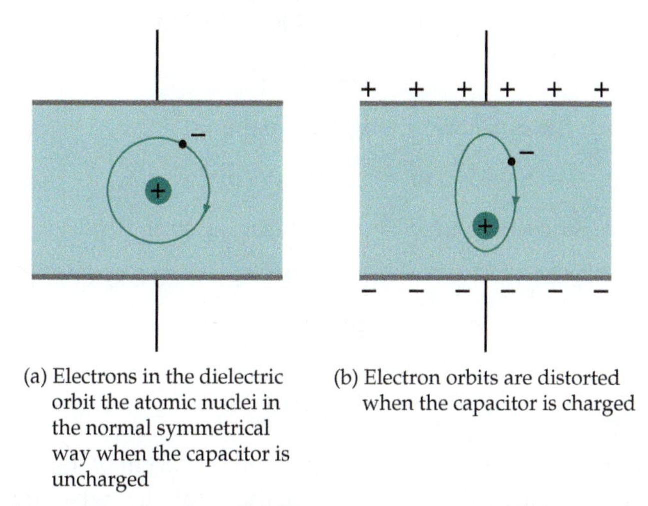

(a) Electrons in the dielectric orbit the atomic nuclei in the normal symmetrical way when the capacitor is uncharged

(b) Electron orbits are distorted when the capacitor is charged

Figure 15-6 **Electric stress occurs on the dielectric of charged capacitors. This distorts the orbits of electrons within the dielectric.**

in the dielectric atoms are orbiting normally, [Figure 15-6(a)]. When a voltage difference exists between the plates, the orbiting electrons experience a force attracting them toward the positive plate, while the atomic nucleus is attracted toward the negative plate. The result is that the electrons tend to go into a distorted orbit around the nucleus, as illustrated in Figure 15-6(b). In this situation the atoms are said to be *polarized.*

When a capacitor is discharged by short-circuiting its plates, the polarized atoms may be expected to return to their normal state. With some dielectric materials, it is found that the polarized atoms do not return completely to their normal state. Consequently, when the short circuit is removed from the capacitor plates, a small voltage can again be measured across the plates. The plate voltage is, of course, the result of residual polarization within the dielectric. This phenomenon of energy being retained by the dielectric is referred to as *dielectric absorption.*

Example 15-1

A capacitor has a plate area of 400 cm^2 and a dielectric thickness of 0.2 mm. If the capacitor has a charge of 10×10^{-3} C when the voltage between its plates is 25 V, determine the electric field strength and electric flux density within the dielectric. Calculate the capacitor breakdown voltage if the dielectric is paper.

Solution

Eq. 15-3,
$$\xi = \frac{E}{d} = \frac{25\ \text{V}}{0.2 \times 10^{-3}\ \text{m}}$$
$$= 125\ \text{kV/m}$$

Eq. 15-2,
$$D = \frac{Q}{A} = \frac{10 \times 10^{-3}\ \text{C}}{400 \times 10^{-4}\ \text{m}^2}$$
$$= 0.25\ \text{C/m}^2$$

From Table 15-1, the dielectric strength for paper is,

$$500\ \text{kV/cm} = 50\ 000\ \text{kV/m}$$

$$\text{Breakdown voltage} = d \times (\text{dielectric strength})$$
$$= 0.2 \times 10^{-3}\ \text{m} \times 50\ 000\ \text{kV/m}$$
$$= 10\ \text{kV}$$

Practice Problems

15-2.1 Two conductors in a cable are separated by 0.3 cm of insulation. Calculate the electric field strength within the insulation when the conductor voltage difference is 120 V. Also, determine the voltage at which the insulation may break down if its dielectric strength is 270 kV/cm.

15-2.2 A capacitor to contain a charge of 40 μC is to have a maximum electric flux density of 1.6 mC/m^2. Calculate the minimum plate area for the capacitor.

15-3 CAPACITANCE AND CAPACITOR DIMENSIONS

Plate Area and Dielectric Thickness

The quantity of charge that can be stored by a capacitor with a given terminal voltage is termed its *capacitance.* The capacitance of a capacitor has a definite relationship to the area of the plates and to the thickness of the dielectric.

Refer to Figure 15-7(a), and recall that electrons are attracted to a positive voltage. The presence of the positive voltage on the top plate causes electrons to be attracted (from the negative terminal of the battery providing the voltage)

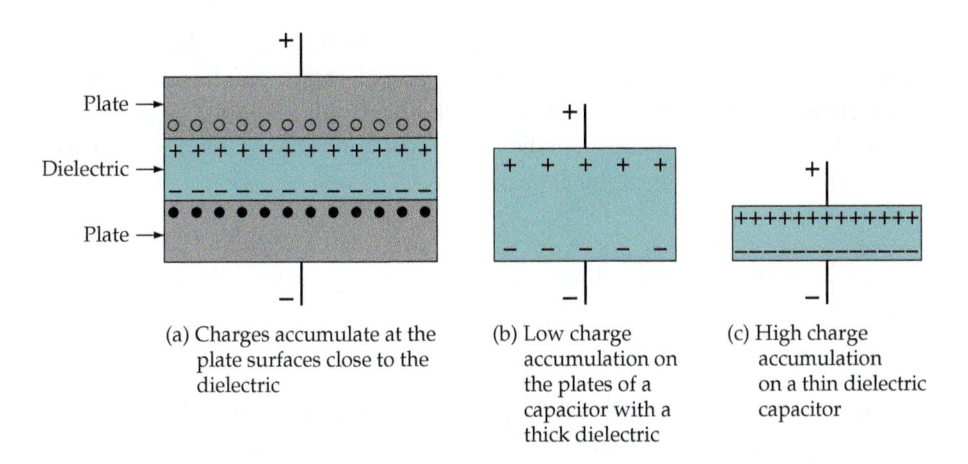

Figure 15-7 A capacitor with a thin dielectric (plates close together) can store more charge than a similar capacitor with a thick dielectric (plates far apart).

into the (nearby) bottom plate of the capacitor. The electrons cannot pass through the insulating dielectric to the positive top plate. Instead, they get as close to the top plate as possible by accumulating at the surface of the bottom plate close to the dielectric. Similarly, the negative voltage on the bottom plate causes electrons to be driven out of the adjacent top plate. Consequently, a positive charge (i.e., a lack of electrons) is accumulated at the surface of the top plate close to the dielectric.

Now look at Figure 15-7(b), which shows a capacitor consisting of two metal plates with a very thick dielectric between them. With such a thick piece of insulating material between the two plates, the voltage on each plate has very little influence on the electrons in the other plate. Consequently, only a relatively small charge can be stored. In Figure 15-7(c), a capacitor with a very thin dielectric is shown. In this case the voltage on each plate has a strong influence on the very close adjacent plate. The result is, of course, that a relatively large number of electrons is accumulated on the bottom plate, and an equally large number of electrons is driven out of the top plate, constituting a relatively large charge on the capacitor. This suggests (and it can be experimentally demonstrated) that the amount of charge stored in a capacitor is inversely proportional to the thickness of the dielectric.

Figure 15-8 shows two identical capacitors connected in parallel. If each capacitor has a charge of Q coulombs, the total charge stored by the two is $2Q$ coulombs. Paralleling the two similar capacitors means that the plate area has been doubled. So, a capacitor with twice the plate area would also store twice the charge. Now it can be written,

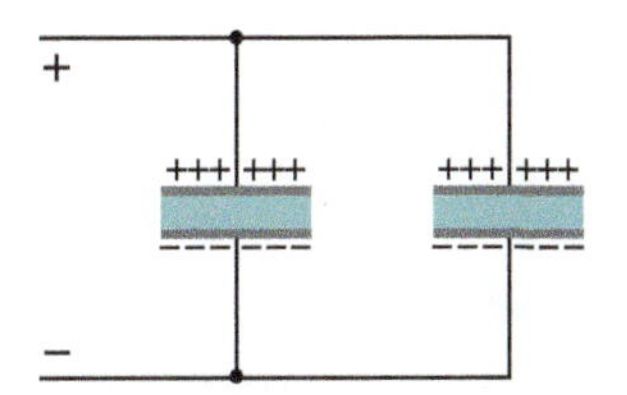

Figure 15-8 Connecting two capacitors in parallel increases the total plate area and consequently increases the capacitance.

$$\text{Capacitance} \propto \frac{\text{plate area}}{\text{distance between plates}}$$

or,

$$C \propto \frac{A}{d}$$

Refer once more to Figure 15-7. It can be shown that for a given applied voltage, a definite number of electrons is attracted into the bottom plate and that the same number of electrons is driven out of the top plate of the capacitor. If the applied voltage is doubled, twice as many electrons are displaced from the positive plate to the negative plate, and the accumulated charge is doubled. So, the charge on a capacitor is directly proportional to the applied voltage as well as to the capacitance of the capacitor:

$$Q \propto CE$$

The *farad*[1] (F) is the SI unit of capacitance.

The *farad* is the capacitance of a capacitor that contains a charge of 1 coulomb when the potential difference between its terminals is 1 volt.

It is found that the farad is a very large unit; consequently, microfarads (μF), nanofarads (nF), and picofarads (pF) are most frequently used for expressing capacitance.

From the definition of the farad, the equation for charge becomes,

$$Q = CE \tag{15-4}$$

Where C is capacitance in farads and E is voltage difference between the plates in volts, Q is charge in coulombs.

Permittivity

Before proceeding further, another factor, known as the *permittivity* must be introduced. Electric permittivity is analogous to magnetic permeability. It specifies the ease with which electric flux is permitted to pass through a given dielectric material. Like permeability, permittivity is subdivided into the *permittivity of free space* (ϵ_o), and the *relative permittivity* (ϵ_r), which is also known as the *dielectric constant.* Table 15-2 lists the relative permittivity for various dielectric materials. Note that the permittivity of air is 1.0006, and so it is usually taken as 1, (i.e., the same as for vacuum).

Permittivity of free space,

$$\epsilon_o = \frac{1}{36\pi \times 10^9}$$

or,

$$\epsilon_o \approx 8.84 \times 10^{-12}$$

The dielectric permittivity has a similar relationship to field strength and flux density as does the magnetic permeability:

$$\epsilon_r \epsilon_o = \frac{D}{\xi} \tag{15-5}$$

ξ is the electric field strength in V/m, D is electric flux density in C/m^2, and $\epsilon_r \epsilon_o$ is a ratio.

---

[1]Named for the English chemist and physicist Michael Faraday (1791–1867).

TABLE 15-2 Dielectric constants for various materials

Material	Typical Dielectric Constant
Vacuum	1
Air	1.0006
Ceramic (low loss)	6 to 20
Ceramic (high)	>1000
Glass	5 to 100
Mica	3 to 7
Mylar	3
Oxide film	5 to 25
Paper	4 to 6
Polystyrene	2.5
Teflon	2

It is interesting to note that[2]:

$$\frac{1}{\sqrt{\begin{bmatrix}\text{permeability}\\ \text{of free space}\end{bmatrix} \times \begin{bmatrix}\text{permittivity}\\ \text{of free space}\end{bmatrix}}} = \frac{1}{\sqrt{\mu_o \epsilon_o}}$$

$$= \frac{1}{\sqrt{4\pi \times 10^{-7} \times \dfrac{1}{36\pi \times 10^{9}}}}$$

$$= 3 \times 10^{8}$$

or,

$$\frac{1}{\sqrt{\mu_o \epsilon_o}} = \begin{cases}\text{velocity of light and other}\\ \text{electromagnetic waves in}\\ \text{meters per second}\end{cases}$$

Capacitance Equation

Substituting for D and ξ from Equations 15-2 and 15-3 into Equation 15-5,

$$\epsilon_r \epsilon_o = \frac{Q}{A} \times \frac{d}{E}$$

And substituting for Q from Equation 15-4,

$$\epsilon_r \epsilon_o = \frac{CE}{A} \times \frac{d}{E}$$

or,

$$\epsilon_r \epsilon_o = \frac{Cd}{A}$$

[2]This relationship was discovered in 1865 by Scottish physicist James C. Maxwell (1831–1879). From the relationship, he was able to predict the existence of electromagnetic waves many years before they were demonstrated experimentally.

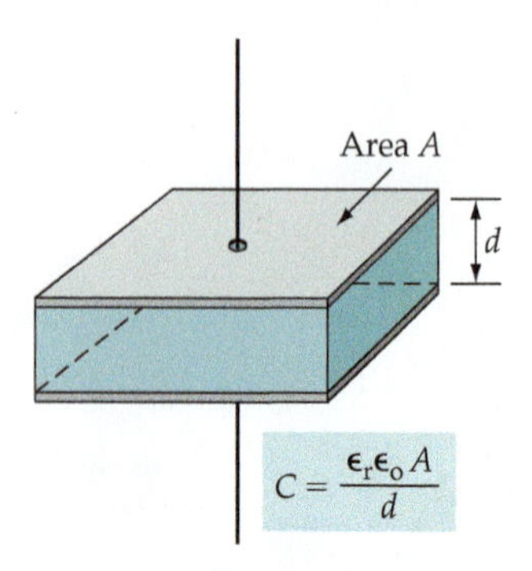

Figure 15-9 Capacitance can be calculated from the plate area (A), dielectric thickness (d), and the relative permittivity of the dielectric.

Giving,

$$C = \frac{\epsilon_r \epsilon_o A}{d} \qquad (15\text{-}6)$$

When A is plate area in m^2 and d is dielectric thickness in meters, C is the capacitance in farads, (see Figure 15-9).

Example 15-2

Calculate the capacitance of a capacitor with a plate area of 400 cm^2 and a dielectric thickness of 0.1 mm: (a) when the dielectric is air, (b) when the dielectric is mica with a relative permittivity of 5. Determine the charge on the capacitor in both cases above when the applied voltage is 25 V.

Solution

(a) Air dielectric:

Eq. 15-6, $$C = \frac{\epsilon_r \epsilon_o A}{d} = \frac{8.84 \times 10^{-12} \times 400 \times 10^{-4}\,\text{m}^2}{0.1 \times 10^{-3}\,\text{m}}$$

$$= 3536\ \text{pF}$$

(b) Mica dielectric:

Eq. 15-6, $$C = \frac{\epsilon_r \epsilon_o A}{d} = \frac{5 \times 8.84 \times 10^{-12} \times 400 \times 10^{-4}\,\text{m}^2}{0.1 \times 10^{-3}\,\text{m}}$$

$$= 17.68\ \text{nF}$$

Capacitor charge:

For air dielectric,

Eq. 15-4, $$Q = CE = 3536\ \text{pF} \times 25\ \text{V}$$

$$= 88.4 \times 10^{-9}\ \text{C}$$

For mica dielectric,

Eq. 15-4, $$Q = CE = 17.68\ \text{nF} \times 25\ \text{V}$$

$$= 442 \times 10^{-9}\ \text{C}$$

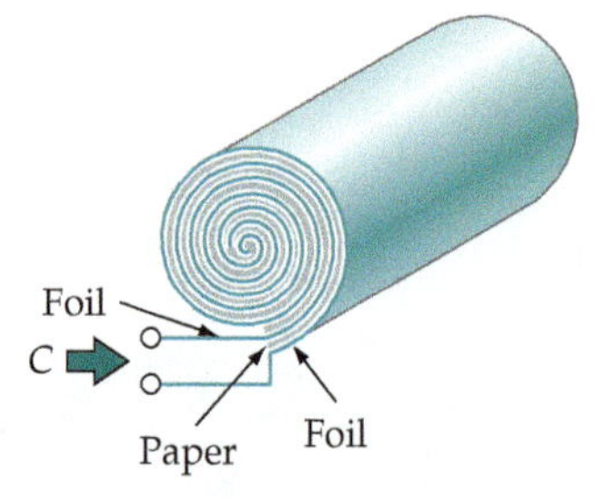

Figure 15-10 Illustration for Example 15-3.

Example 15-3

A 1 μF capacitor is to be constructed from rolled-up sheets of aluminum foil separated by layers of paper 0.1 mm thick, (see Figure 15-10). Calculate the required area for the foil if the relative permittivity of the paper is 6.

Solution

From Eq. 15-6,

$$A = \frac{Cd}{\epsilon_r \epsilon_o} = \frac{1\ \mu\text{F} \times 0.1 \times 10^{-3}\ \text{m}}{6 \times 8.84 \times 10^{-12}}$$

$$\approx 1.89\ \text{m}^2$$

Example 15-4

If the dielectric strength in Example 15-3 is 30 000 kV/m, calculate the maximum charge if the applied voltage is not to exceed 10% of the capacitor breakdown voltage.

Solution

$$\text{Breakdown voltage} = d \times (\text{dielectric strength})$$

$$= 0.1 \times 10^{-3}\ \text{m} \times 30\,000\ \text{kV/m}$$

$$= 3\ \text{kV}$$

$$\text{Maximum voltage} = 10\%\ \text{of}\ 3\ \text{kV} = 300\ \text{V}$$

Eq. 15-4,

$$Q = CE = 1\ \mu\text{F} \times 300\ \text{V}$$

$$= 300\ \mu\text{C}$$

Practice Problems

15-3.1 Two conductors in a cable are 0.5 cm in diameter, and the average thickness of the insulation between the conductors is 0.25 cm. Calculate the cable capacitance per meter if the insulation has a dielectric constant of 2.2.

15-3.2 Determine the charge contained on 100 m of the cable in Problem 15-3.1 when the conductor voltage difference is 120 V.

15-3.3 Two copper films on opposite sides of a printed circuit board are separated by 2 mm of insulation with a dielectric constant of 2. If each plate is 2 cm by 1 cm, calculate the capacitance.

15-4 CAPACITOR TYPES AND CHARACTERISTICS

In addition to the actual value of capacitance, there are many other important characteristics that must be considered in selecting a capacitor for a particular application. These include the maximum voltage that the capacitor has to be able to survive, the capacitance value accuracy, and terminal polarization.

Working Voltage

The maximum voltage that may be safely applied to a capacitor is usually expressed in terms of its *dc working voltage.* A *maximum rms ac voltage* (see Chapter 17) may also be specified, and this is usually little more than half the maximum dc voltage. These limits should never be exceeded; otherwise, dielectric breakdown may occur [Figure 15-11(a)].

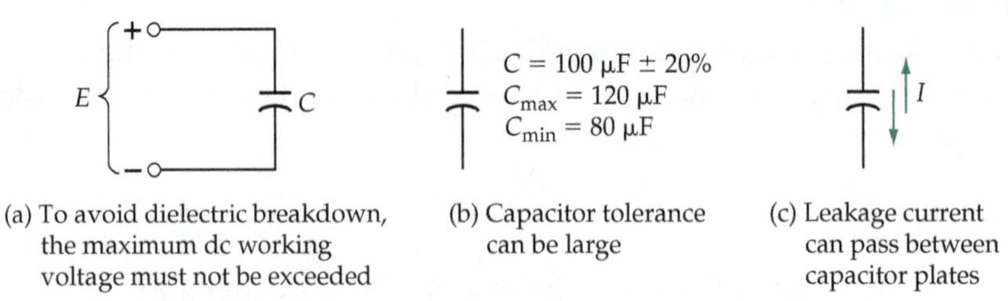

(a) To avoid dielectric breakdown, the maximum dc working voltage must not be exceeded

(b) Capacitor tolerance can be large

(c) Leakage current can pass between capacitor plates

Figure 15-11 Capacitor working voltage, tolerance, and leakage current are all important considerations in the use of a capacitor.

Capacitor Tolerance

This is simply the accuracy with which the capacitance value is specified. Tolerances of ±5% to ±20% are normal for most small-value capacitors, but more precise capacitors can be purchased at increased cost. In the case of larger-value capacitors, the capacitance has to be a definite minimum value in many applications. Consequently, manufacturers tend to specify the tolerance as −10% + 150%. This means, for example, that a 100 μF capacitor could have a value as low as 90 μF or as high as 250 μF [Figure 15-11(b)].

Temperature Effects

Every capacitor has an operating temperature range specified by the manufacturer. Typical ranges are −20°C to +65°C, −40°C to +65°C, and −55°C to +125°C. Obviously, no capacitor should be employed in an environment where the temperatures may be beyond its specified range. The capacitance value is also likely to change slightly over the temperature range. The maximum change that can occur is specified by the manufacturer either in *parts per million per degree celsius* (ppm/°C), or as a percentage change for the temperature extremes. Sometimes a graph is provided of capacitance percentage change versus temperature.

Leakage Current

Despite the fact that the dielectric is an insulator, small leakage currents flow between the plates of a capacitor. The actual level of leakage current depends on the insulation resistance of the dielectric. *Plastic film* capacitors, for example, may have insulation resistances higher than 100 000 MΩ. At the other extreme, an *electrolytic* capacitor may have a microamp (or more) of leakage current, with only 10 V applied to its terminals, [Figure 15-11(c)].

Capacitor Equivalent Circuit

An *ideal capacitor* has a dielectric that has an infinite resistance and plates that have zero resistance. However, an ideal capacitor does not exist because all dielectrics have some leakage current and all capacitor plates have some resistance. The complete equivalent circuit for a capacitor, as shown in Figure 15-12(a), consists of an ideal capacitor C in series with a resistance R_D representing the resistance of the plates, and in parallel with a resistance R_L representing the leakage resistance of the dielectric. Usually, the plate resistance can be completely neglected, and the equivalent circuit becomes that shown in Figure 15-12(b). With capacitors that have a very high leakage resistance (e.g., mica and plastic film capacitors), the parallel resistor is frequently omitted in the equivalent circuit, and the capacitor is then treated as an ideal capacitor. This cannot normally be done for electrolytic capacitors, for example, which have relatively low leakage resistances. The parallel RC circuit in Figure 15-12(b) can be shown to have an equivalent series RC circuit, as in Figure 15-12(c). This is explained in Section 20-6.

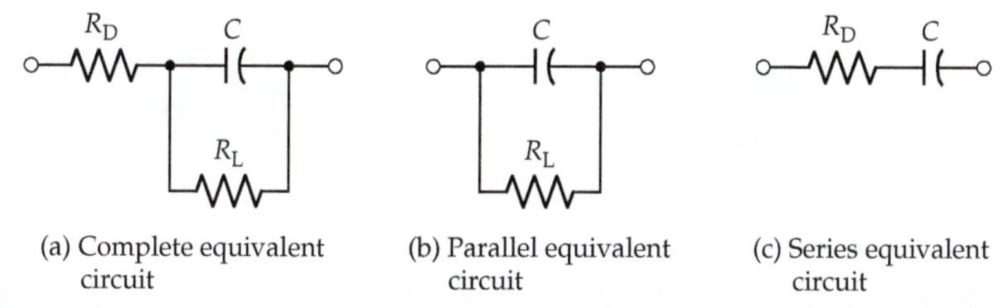

Figure 15-12 A capacitor equivalent circuit consists of the capacitance C, the leakage resistance R_L in parallel with C, and the plate resistance R_D in series with C and R_L.

Because the dielectric is largely responsible for determining its most important characteristics, capacitors are usually identified by the type of dielectric used.

Air Capacitors

A typical capacitor using air as a dielectric is illustrated in Figure 15-13. The capacitance is variable, as is the case with virtually all air capacitors. There are two sets of metal plates, one set fixed and one movable. The movable plates

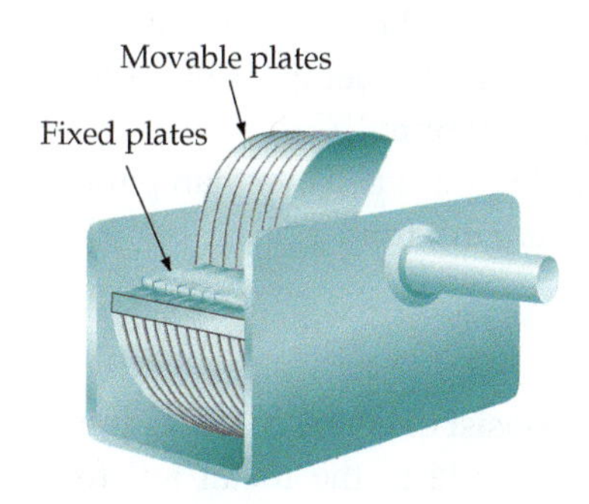

Figure 15-13 A variable air capacitor is made up of a set of movable plates and a set of fixed plates separated by air.

can be adjusted into or out of the spaces between the fixed plates by means of the rotatable shaft. In this way, the area of the plates opposite each other is increased or decreased, and the capacitance value is altered.

Paper Capacitors

In its simplest form, a paper capacitor consists of a layer of paper between two layers of metal foil. The metal foil and paper are rolled up, as illustrated in Figure 15-14(a); external connections are brought out from the foil layers, and the complete assembly is dipped in wax or plastic. A variation of this is the *metalized paper* construction, in which the foil is replaced by thin films of metal deposited on the surface of the paper. One end of the capacitor sometimes has a band around it, [Figure 15-14(b)]. This does not mean that the device is polarized but simply identifies the terminal that connects to the outside metal film, so that it can be grounded to avoid pickup of unwanted signals.

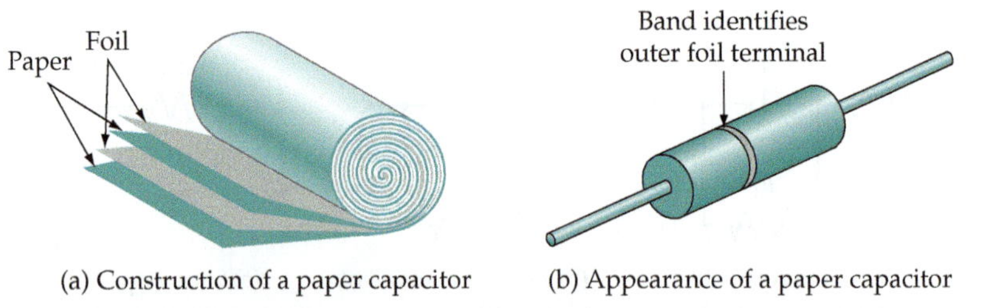

Figure 15-14 In a paper capacitor, two sheets of metal foil separated by a sheet of paper are rolled up together. External connections are made to the foil sheets.

Paper capacitors are available in values ranging from about 500 pF to 50 μF, and with dc working voltages up to about 600 V. They are among the lowest-cost capacitors for a given capacitance value, but are physically larger than several other types having the same capacitance value.

Plastic Film Capacitors

The construction of plastic film capacitors is similar to that of paper capacitors, except that the paper is replaced by a thin film that is typically polystyrene or Mylar. This type of dielectric gives insulation resistances greater than 100 000 MΩ. Working voltages are as high as 600 V, with the dielectric capable of surviving 1500 V surges for a brief period. Capacitance tolerances of ±2.5% are typical, as are temperature coefficients of 60 to 150 ppm/°C.

Plastic film capacitors are physically smaller but more expensive than paper capacitors. They are typically available in values ranging from 5 pF to 0.47 μF.

Mica Capacitors

As illustrated in Figure 15-15(a), mica capacitors consist of layers of mica alternated with layers of metal foil. Connections are made to the metal foil for

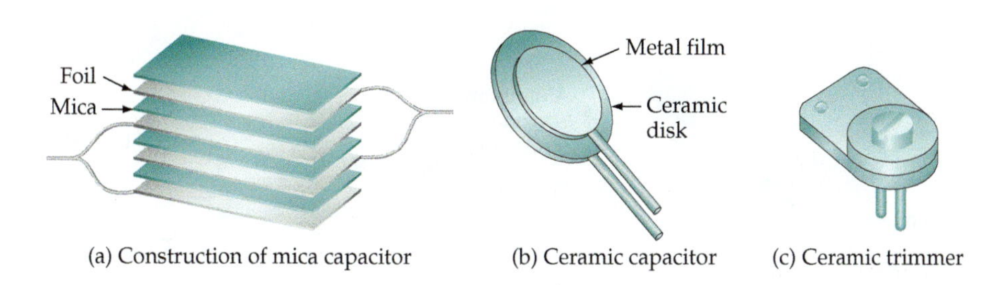

Figure 15-15 Mica capacitors consist of sheets of mica interleaved with foil. A ceramic disc silvered on each side makes a ceramic capacitor; in a ceramic trimmer, the plate area is screwdriver adjustable.

capacitor leads, and the entire assembly is dipped in plastic or encapsulated in a molded plastic jacket. Typical capacitance values range from 1 pF to 0.1 μF, and voltage ratings as high as 35 000 V are possible. Precise capacitance values and wide operating temperatures are obtainable with mica capacitors. In a variation of the process, *silvered mica* capacitors use films of silver deposited on the mica layers instead of metal foil.

Ceramic Capacitors

The construction of a typical ceramic capacitor is illustrated in Figure 15-15(b). Films of metal are deposited on each side of a thin ceramic disc, and copper wire terminals are connected to the metal. The entire unit is then encapsulated in a protective coating of plastic. Two different types of ceramic are used, one of which has extremely high relative permittivity. This gives capacitors that are much smaller than paper or mica capacitors having the same capacitance value. One disadvantage of this particular ceramic dielectric is that its leakage resistance is not as high as with other types. Another type of ceramic gives leakage resistances on the order of 7500 MΩ, but because of its lower permittivity, it produces capacitors that are relatively large for a given value of capacitance. The range of capacitance values available with ceramic capacitors is typically 1 pF to 0.1 μF, with dc working voltages up to 1000 V.

Figure 15-15(c) shows a variable ceramic capacitor known as a *trimmer.* The capacitance value of the trimmer can be (screwdriver) adjusted by altering the plate area on each side of the dielectric. Typical ranges available are 1.5 pF to 3 pF and 7 pF to 45 pF.

Electrolytic Capacitors

The most important feature of electrolytic capacitors is that they can have a very large capacitance in a physically small container. For example, a capacitance of 5000 μF can be obtained in a cylindrical package approximately 5 cm long by 2 cm in diameter. In this case the dc working voltage is only 10 V. Similarly, a 1 F (one farad) capacitor is available in a 22 cm by 7.5 cm cylinder, with a working voltage of only 3 V. Typical values for electrolytic capacitors range from 1 μF through 100 000 μF.

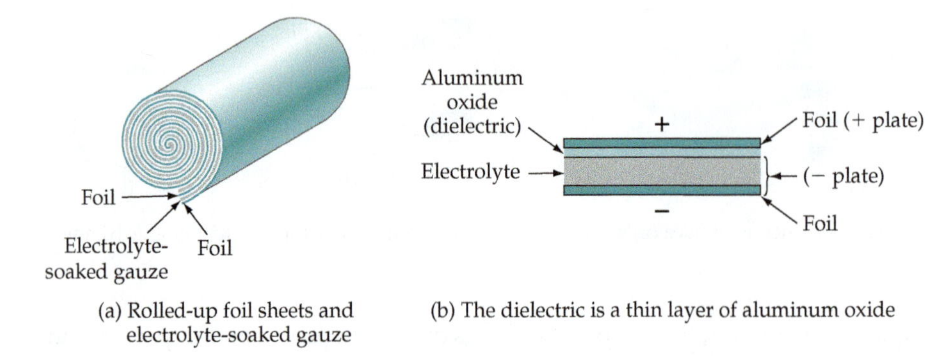

Figure 15-16 An electrolytic capacitor is constructed of rolled up foil sheets separated by electrolyte-soaked gauze, the dielectric is a layer of aluminum oxide at the positive plate.

The construction of an electrolytic capacitor is similar to that of a paper capacitor, [see Figure 15-16(a)]. Two sheets of aluminum foil separated by a fine gauze soaked in electrolyte are rolled up and encased in an aluminum cylinder for protection. A direct voltage is applied to the capacitor terminals after it is assembled, and this causes a thin layer of aluminum oxide to form on the surface of the positive plate next to the electrolyte, [Figure 15-16(b)]. The aluminum oxide is the dielectric, and the electrolyte and positive sheet of foil are the capacitor plates. The extremely thin oxide dielectric results in very large capacitance values.

Polarization

Electrolytic capacitors normally have one terminal identified as the most positive connection. Consequently, they are said to *be polarized.* This usually limits their application to situations where the polarity of the applied voltage will not change.

It is very important that electrolytic capacitors be connected with the correct polarity. When incorrectly connected, gas forms within the electrolyte and ***the capacitor may explode!*** Such an explosion could have tragic consequences for the eyes of an experimenter who happens to be closely examining the circuit. The terminal designated as positive must be connected to the most positive of the two points in the circuit where the capacitor is to be installed. Figure 15-17 illustrates some circuit situations where the capacitor must be correctly connected. Nonpolarized electrolytic capacitors can be obtained. They consist essentially of two capacitors in one package connected *back to back,* so one of the oxide films is always correctly biased.

Electrolytic capacitors are available with dc working voltages greater than 400 V, but in this case capacitance values do not exceed 100 μF. In addition to their low working voltage and polarized operation, another disadvantage of electrolytic capacitors is their relatively high leakage current.

Tantalum Capacitors

This is another type of electrolytic capacitor. Powdered tantalum is *sintered* (or baked) typically into a cylindrical shape. The resulting solid is quite porous,

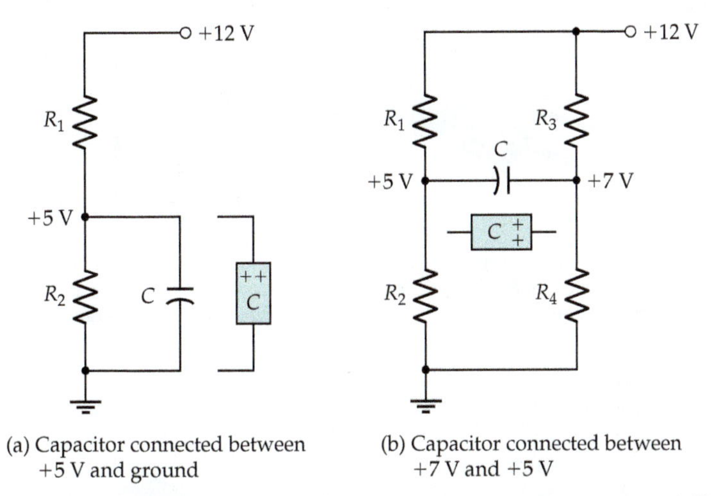

Figure 15-17 It is very important that polarized capacitors be correctly connected. The capacitor positive terminal voltage must be more positive than the voltage at the negative terminal.

so that when immersed in a container of electrolyte, the electrolyte is absorbed into the tantalum. The tantalum then has a large surface area in contact with the electrolyte, (see Figure 15-18). When a dc *forming voltage* is applied, a thin oxide film is formed throughout the electrolyte-tantalum contact area. The result is a large capacitance value in a small volume.

A typical tantalum capacitor in a cylindrical shape 2 cm by 1 cm might have a capacitance of 100 μF and a dc working voltage of 20 V. Other types are available with a working voltage up to 630 V, but with capacitance values on the order of 3.5 μF. Like aluminum-foil electrolytic capacitors, tantalum capacitors must be connected with the correct polarity.

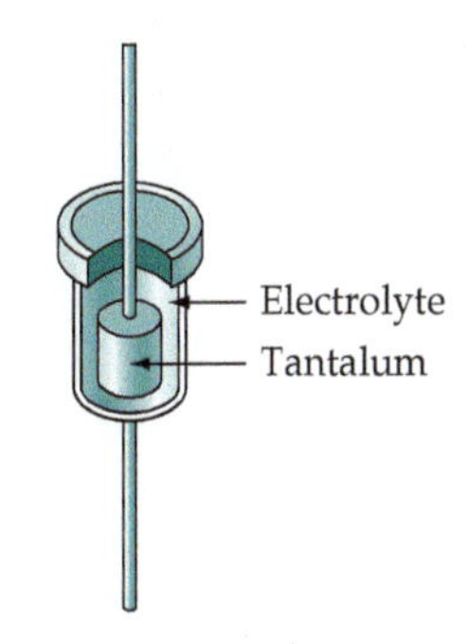

Figure 15-18 A tantalum capacitor has a relatively large capacitance in a small volume.

Stray Capacitance

Capacitance exists anywhere conductors are separated by an insulator. Every conductor has a capacitance to ground, all pairs of adjacent conductors have capacitance between them, and coils have capacitance between turns, termed *distributed capacitance.* Unwanted capacitance is referred to as *stray capacitance,* and because capacitance is inversely proportional to dielectric thickness, stray capacitance is most easily minimized by keeping conductors as far apart as possible. In dc applications, stray capacitance may be absolutely negligible, but in alternating current circuits, particularly at radio frequencies, the stray capacitance (like stray inductance) can be a serious problem. Stray capacitance within a circuit might typically be 10 pF, and the input capacitance of some instruments is around 30 pF.

Capacitor Identification Codes

Capacitors that are physically large enough usually have the capacitance value, tolerance, and maximum dc working voltage printed on the side of the device, [see Figure 15-19(a)]. Small capacitors normally have a printed number

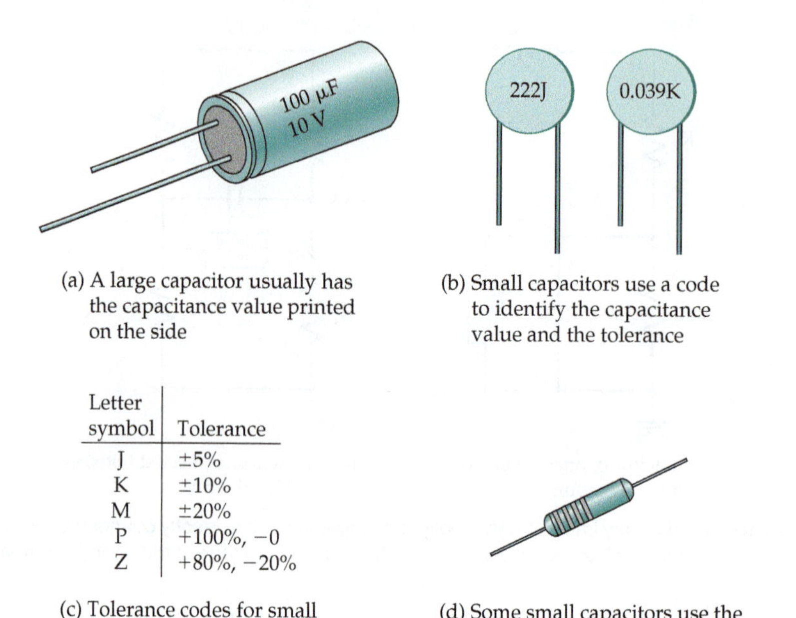

(a) A large capacitor usually has the capacitance value printed on the side

(b) Small capacitors use a code to identify the capacitance value and the tolerance

Letter symbol	Tolerance
J	±5%
K	±10%
M	±20%
P	+100%, −0
Z	+80%, −20%

(c) Tolerance codes for small capacitors

(d) Some small capacitors use the same color code as resistors

Figure 15-19 Capacitor values are printed directly on the side of large capacitors. Smaller capacitors use a code to identify their capacitance values and tolerance.

that gives the capacitance value in picofarads (pF) if a whole number is used, and in microfarads (μF) when a decimal quantity is shown.

Consider the two capacitors illustrated in Figure 15-19(b). For the capacitor with the whole number 222, the first two numerals (from the left) give the capacitance value in pF, and the third numeral is a multiplier. So, the capacitor has a value (in pF) of 22 followed by an additional two zeros, (2200 pF). The device with the printed decimal quantity 0.039, has a capacitance of 0.039 μF. For both capacitors, the letter alongside the numerical quantity identifies the capacitor tolerance. The *J* on the 2200 pF capacitor indicates a ±5% tolerance, and the *K* on the 0.039 μF device identifies its tolerance as ±10%. The tolerance codes are shown in the table in Figure 15-19(c).

Some small capacitors use a color code similar to the code used for small resistors, [Figure 15-19(d)]. The colors represent the same numerical quantities as used with resistors, (see Section 4-7).

15-5 CAPACITORS IN SERIES AND IN PARALLEL

Parallel-Connected Capacitors

When capacitors are connected in parallel, the result is the same as increasing the total plate area. From Equation 15-6, the capacitance is given by,

$$C = \frac{\epsilon_r \epsilon_0 A}{d}$$

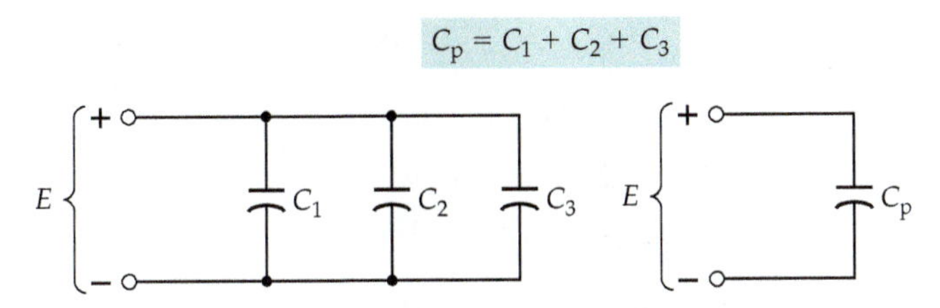

Figure 15-20 Parallel-connected capacitors have a total capacitance of $C_p = (C_1 + C_2 + C_3 + \cdots)$. The supply voltage appears across each capacitor.

When three capacitors with plate areas of A_1, A_2, and A_3, and dielectric thickness d are connected in parallel as shown in Figure 15-20, the total capacitance is,

$$C_p = \frac{\epsilon_r \epsilon_o}{d}(A_1 + A_2 + A_3)$$

$$= \frac{\epsilon_r \epsilon_o A_1}{d} + \frac{\epsilon_r \epsilon_o A_2}{d} + \frac{\epsilon_r \epsilon_o A_3}{d}$$

or,

$$C_p = C_1 + C_2 + C_3 + \cdots \tag{15-7}$$

It is seen that the total capacitance of capacitors in parallel is the sum of the individual capacitance values.

For the parallel-connected capacitors in Figure 15-20, the applied voltage is clearly common to all three. From Equation 15-4, the charge on each capacitor is,

$$Q_1 = C_1 E,\ Q_2 = C_2 E, \text{ and } Q_3 = C_3 E$$

So that the total charge is,

$$Q = Q_1 + Q_2 + Q_3$$

$$= C_1 E + C_2 E + C_3 E$$

or,

$$Q = C_p E$$

Example 15-5

Three capacitors have values $C_1 = 1\ \mu F$, $C_2 = 2\ \mu F$, and $C_3 = 3\ \mu F$. Determine the total capacitance and the charge on each capacitor when the three are connected in parallel across 100 V supply.

Solution

Eq. 15-7,

$$C_p = C_1 + C_2 + C_3 = 1\ \mu F + 2\ \mu F + 3\ \mu F$$

$$= 6\ \mu F$$

From Eq. 15-4,

$$Q_1 = C_1E = 1\ \mu\text{F} \times 100\ \text{V}$$
$$= 100\ \mu\text{C}$$
$$Q_2 = 2\ \mu\text{F} \times 100\ \text{V}$$
$$= 200\ \mu\text{C}$$
$$Q_3 = 3\ \mu\text{F} \times 100\ \text{V}$$
$$= 300\ \mu\text{C}$$

Series-Connected Capacitors

Considering the three series-connected capacitors in Figure 15-21, it is seen that connecting capacitors in series amounts to increasing the total thickness of the dielectric between the two outer plates. From Equation 15-6, the capacitance is given by,

$$C = \frac{\epsilon_r \epsilon_o A}{d}$$

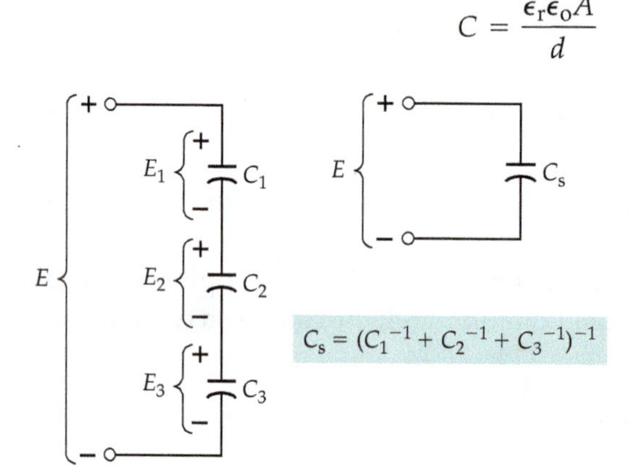

Figure 15-21 Series-connected capacitors have a total capacitance of $C_s = (C_1^{-1} + C_2^{-1} + C_3^{-1} + \cdots)^{-1}$. The supply voltage divides across the capacitors.

So, when three capacitors with similar plate areas and with dielectric thickness of d_1, d_2, and d_3 are connected in series, the total capacitance is,

$$C_s = \frac{\epsilon_r \epsilon_o A}{d_1 + d_2 + d_3}$$

Giving,

$$\frac{1}{C_s} = \frac{d_1 + d_2 + d_3}{\epsilon_r \epsilon_o A}$$

$$= \frac{d_1}{\epsilon_r \epsilon_o A} + \frac{d_2}{\epsilon_r \epsilon_o A} + \frac{d_3}{\epsilon_r \epsilon_o A}$$

or,

$$\frac{1}{C_s} = \frac{1}{C_1} + \frac{1}{C_2} + \frac{1}{C_3} + \cdots$$

This can be rearranged as,

$$C_s = (C_1^{-1} + C_2^{-1} + C_3^{-1} + \cdots)^{-1} \tag{15-8}$$

The total capacitance of capacitors connected in series is the reciprocal of the sum of the reciprocals of the individual capacitances.

Note that the equations for total capacitance with parallel-connected and series-connected capacitors are the reverse of the equations for similar resistor circuits, and for similar inductor circuits.

For the three series-connected capacitors in Figure 15-21, the charging current must flow through all three capacitors. If a charging current I flows for a time t, then *the charge supplied to each capacitor is equal to the charge supplied to all three capacitors:*

$$Q = Q_1 = Q_2 = Q_3$$

and,

$$Q = It$$

From Eq. 15-4,

$$E = \frac{Q}{C}$$

So, the voltage across the individual capacitors can be calculated as,

$$E_1 = \frac{Q}{C_1}, E_2 = \frac{Q}{C_2}, \text{ and } E_3 = \frac{Q}{C_3}$$

Note that the largest capacitor has the smallest terminal voltage. The total applied voltage is the sum of the individual capacitor voltages, (see Figure 15-21):

$$E = E_1 + E_2 + E_3$$

Example 15-6

If the three capacitors in Example 15-5 are connected in series across the 100 V supply, determine the total capacitance and the voltage across each capacitor.

Solution

Eq. 15-8,

$$C_s = (C_1^{-1} + C_2^{-1} + C_3^{-1})^{-1}$$
$$= [(1\ \mu F)^{-1} + (2\ \mu F)^{-1} + (3\ \mu F)^{-1}]^{-1}$$
$$\approx 0.545\ \mu F$$

Eq. 15-4,

$$Q = CE = 0.545\ \mu F \times 100\ V$$
$$\approx 54.5\ \mu C$$

and,

$$Q = Q_1 = Q_2 = Q_3$$

$$E_1 = \frac{Q}{C_1} \approx \frac{54.5\ \mu C}{1\ \mu F}$$
$$\approx 54.5\ V$$

$$E_2 = \frac{Q}{C_2} \approx \frac{54.5\ \mu C}{2\ \mu F}$$
$$\approx 27.3\ V$$

$$E_3 = \frac{Q}{C_3} \approx \frac{54.5\ \mu\text{C}}{3\ \mu\text{F}}$$

$$\approx 18.2\ \text{V}$$

$$E = E_1 + E_2 + E_3 = 54.5\ \text{V} + 27.3\ \text{V} + 18.2\ \text{V}$$

$$= 100\ \text{V}$$

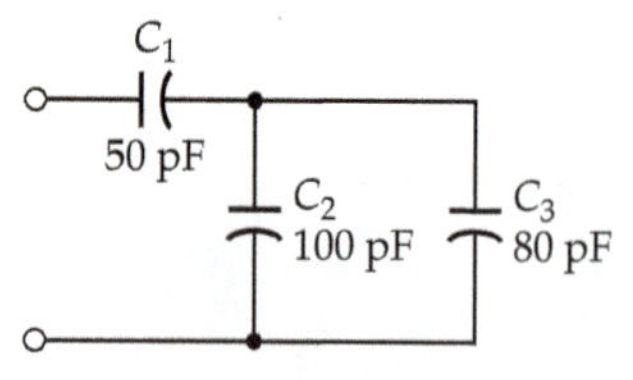

Figure 15-22

Example 15-7

A 50 pF capacitor is connected in series with two parallel-connected capacitors with values of 100 pF and 80 pF, as in Figure 15-22. Calculate the total capacitance.

Solution

$$C_{2,3} = C_2 \| C_2 = 100\ \text{pF} + 80\ \text{pF}$$

$$= 180\ \text{pF}$$

$$C_{eq} = (C_1^{-1} + C_{2,3}^{-1})^{-1} = (50\ \text{pF}^{-1} + 180\ \text{pF}^{-1})^{-1}$$

$$= 39.13\ \text{pF}$$

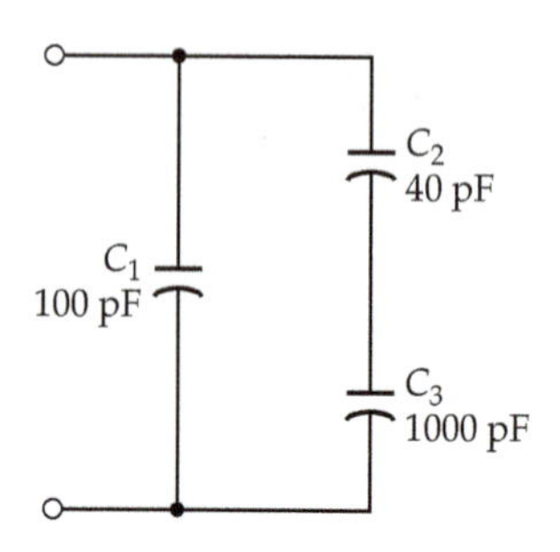

Figure 15-23

Practice Problems

15-5.1 Calculate the total capacitance of a capacitor $C_1 = 100$ pF in parallel with two series-connected capacitors, $C_2 = 40$ pF and $C_3 = 1000$ pF, as in Figure 15-23.

15-5.2 If a 15 V supply is connected to the capacitor circuit in Problem 15-5.1, calculate the total charge and the voltage across each capacitor.

15-6 ENERGY STORED IN A CHARGED CAPACITOR

From Equation 15-4,

$$C = \frac{Q}{E}$$

also,

$$Q = It$$

so,

$$C = \frac{It}{E}$$

or,

$$I = \frac{CE}{t} \tag{15-9}$$

When a capacitor is charged from a constant current source for a time t seconds, the voltage across it grows linearly from zero to E volts. The constant level of input current is given by Equation 15-9, and the average input voltage is $0.5E$.

From Equations (3-3) and (3-7), the electrical energy is,

$$W = (\text{average voltage}) \times (\text{average current}) \times t$$

The energy supplied to the capacitive circuit is,

$$W = 0.5E \times I \times t$$

Substituting for I from Equation 15-9, the energy stored is,

$$W = 0.5CE^2 \qquad (15\text{-}10)$$

When C is in farads and E is in volts, W is given in joules, (see Figure 15-24). Note the similarity between Equations 15-10 and 14-14.

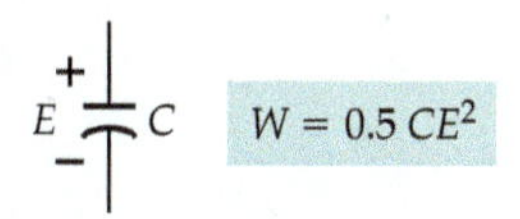

Figure 15-24 The energy stored in a capacitor depends on the capacitance and the terminal voltage.

Example 15-8

Calculate the energy stored in each of the three series-connected capacitors referred to in Example 15-6. Calculate the total energy stored.

Solution

For C_1, $W = 0.5C_1E^2 = 0.5 \times 1\ \mu\text{F} \times (54.5\ \text{V})^2$

$\approx 1.49\ \text{mJ}$

For C_2, $W = 0.5C_2E^2 = 0.5 \times 2\ \mu\text{F} \times (27.3\ \text{V})^2$

$\approx 745\ \mu\text{J}$

For C_3, $W = 0.5C_3E^2 = 0.5 \times 3\ \mu\text{F} \times (18.2\ \text{V})^2$

$\approx 497\ \mu\text{J}$

Total energy stored, $W = 0.5C_sE^2 = 0.5 \times 0.545\ \mu\text{F} \times (100\ \text{V})^2$

$\approx 2.73\ \text{mJ}$

Also, total energy stored is,

$$W = W_1 + W_2 + W_3 = 1.49\ \text{mJ} + 745\ \mu\text{J} + 497\ \mu\text{J}$$

$$\approx 2.73\ \text{mJ}$$

Practice Problems

15-6.1 Determine the energy stored in each of the capacitors in Problem 15-5.2.

15-6.2 Calculate the voltage that must be applied to a 100 μF capacitor to store 33 mJ of energy.

Summary of Formulas

- ***Electric flux density:***

$$D = \frac{Q}{A}$$

- ***Electric field strength:***

$$\xi = \frac{E}{d}$$

- ***Electric charge:***

$$Q = CE$$

- ***Permittivity:***

$$\epsilon_o \approx 8.84 \times 10^{-12}$$

$$\epsilon_r \epsilon_o = \frac{D}{\xi}$$

- ***Capacitance:***

$$C = \frac{\epsilon_r \epsilon_o A}{d}$$

- ***Capacitors in parallel:***
  Total capacitance: $C_p = C_1 + C_2 + C_3 + \cdots$
- ***Capacitors in series:***
  Total capacitance: $C_S = (C_1^{-1} + C_2^{-1} + C_3^{-1})^{-1}$
- ***Energy stored:***

$$W = 0.5CE^2$$

Review Questions

Section 15-1

15-1 Using illustrations, explain the process of electric charge storage in two metal plates separated by a thin layer of insulating material.

Section 15-2

15-2 Discuss the electric field, comparing it to a magnetic field. Write the equations for electric field strength and electric flux density, and explain each.

15-3 Explain electric lines of force, equipotential lines, dielectric strength, dielectric absorption, permittivity, and relative permittivity.

Section 15-3

15-4 State the definition of the farad, and from the definition write an equation for the charge stored in a capacitor.

15-5 Explain permittivity, relative permittivity, and the permittivity of free space.

15-6 Write the equation for the permittivity of free space, and the equation for permittivity in terms of flux density and field strength.

15-7 Explain why the capacitance of a capacitor is proportional to the plate area and inversely proportional to dielectric thickness.

15-8 Write an equation for the capacitance of a parallel-plate capacitor.

Section 15-4

15-9 Discuss the following characteristics of capacitors: working voltage, tolerance, temperature effects, leakage current, polarization.

15-10 Sketch the equivalent circuit of a capacitor, and briefly explain.

15-11 Describe the construction of air capacitors, paper capacitors, plastic film capacitors, and mica capacitors. Also, explain the most important parameter for each type of capacitor.

15-12 Repeat Review Question 15-11 for ceramic, electrolytic, and tantalum capacitors.

15-13 Discuss stray capacitance, and explain the problems associated with stray capacitance.

Section 15-5

15-14 Write equations for total capacitance of capacitors connected in parallel, and for the total charge on parallel-connected capacitors.

15-15 Write equations for the total capacitance of capacitors connected in series, and for the terminal voltage on each capacitor.

Section 15-6

15-16 Write an equation for the energy stored in a charged capacitor.

Problems

Section 15-2

15-1 A capacitor has a plate area of 50 cm^2 and a dielectric thickness of 0.5 mm, as in Figure 15-25. The capacitor charge is 10×10^{-9} C when the applied voltage is 20 V. Calculate the electric field strength and flux density.

15-2 Calculate the plate area for a capacitor that has to store a charge of 750 μC with an electric flux density of 0.3 C/m^2.

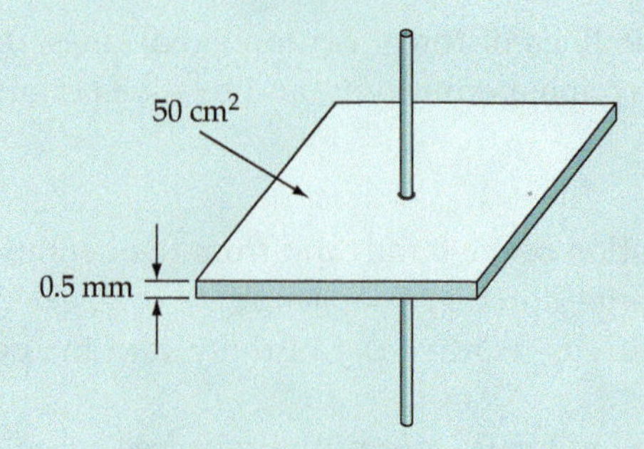

Figure 15-25

15-3 If the dielectric strength for the capacitor described in Problem 15-1 is 100 000 V/m, calculate the applied voltage at which the dielectric is likely to break down.

15-4 Calculate the electric field strength in a capacitor with 60 V applied to its terminals: (a) when the dielectric thickness is 0.3 mm, (b) when the dielectric is 0.025 mm thick.

Section 15-3

15-5 Determine the capacitance of a capacitor with a plate area of 50 cm^2 and a dielectric thickness of 0.5 mm: (a) When the dielectric is air, (b) When the dielectric is glass with a relative permittivity of 75.

15-6 For Problem 15-5, determine the necessary applied voltage in each case to store a charge of 3.3 nC.

15-7 A 20 μF capacitor is to be constructed using metal foil with a plastic film dielectric. If the dielectric is 0.05 mm thick and has a relative permittivity of 3, calculate the area of the metal foil plates.

15-8 Two conductors which are 2 mm in diameter are separated by an insulation with an average thickness of 0.3 cm, as in Figure 15-26. The dielectric constant of the insulation is 3.5. Calculate the capacitance per meter of the cable.

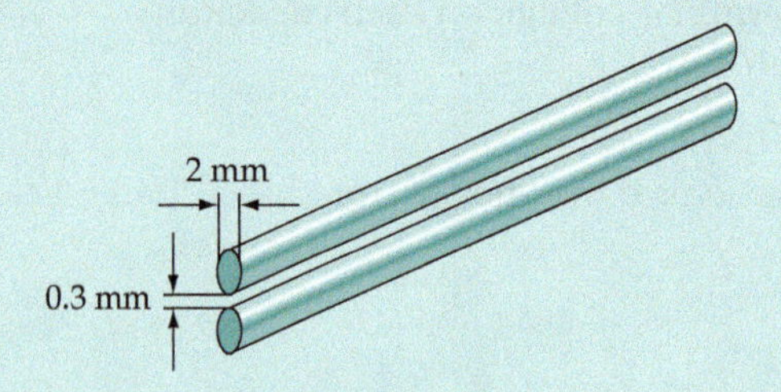

Figure 15-26

15-9 A 50 μF electrolytic capacitor has aluminum foil plates that are 10 cm × 75 cm. If the oxide dielectric has a relative permittivity of 20, calculate the thickness of the dielectric.

15-10 Calculate the capacitance of two 0.5 m^2 sheets of foil separated by 0.2 mm of paper with a dielectric constant of 4.7.

15-11 A 10 km electric cable has two conductors, each of which is 1 cm thick. The average distance between the conductors is 0.5 cm, and the insulation material is oil-soaked paper with a relative permittivity of 5. The voltage difference between the conductors is 500 V. Determine the capacitance of the cable.

15-12 Two copper conductors each of which are 10 cm long and 3 mm wide are printed on opposite sides of a circuit board. The board material is 1.5 mm thick and has a dielectric constant of 1.8. Calculate the capacitance.

15-13 Two sheets of aluminum foil 1 m in width and 15 m long are rolled up with two sheets of plastic. The plastic is 0.5 mm thick and has a relative permittivity of 3. Calculate the capacitance of the roll. Also, calculate its charge when the terminal voltage is 12 V.

15-14 Calculate the charge on the printed circuit conductors in Problem 15-12 if their voltage difference is: (a) 33 V, (b) 5 V.

15-15 A printed circuit board has a total conductor area of 25 cm^2. The opposite side of the board is completely coated with copper. The board is made of teflon which is 1 mm thick with a dielectric constant of 2. Determine the total capacitance.

15-16 Calculate the applied voltage for the capacitor in Problem 15-10 to store a charge of: (a) 15 μC, (b) 1.5 nC.

15-17 The charge taken by a capacitor is 0.13 μC when its terminal voltage is 50 V. The plate area of the is 39 cm^2, and the dielectric thickness is 0.9 mm. Calculate the relative permittivity of the dielectric.

15-18 The dielectric in the capacitor in Problem 15-17 is replaced by paper with a dielectric constant of 5.5. Calculate the new capacitance value and the new level of charge when the terminal voltage is 50 V.

15-19 Two circular metal plates 30 cm in diameter are separated by air, as illustrated in Figure 15-27. The plates have a charge of 0.01 μC when the voltage difference between them is 10 V. Calculate the distance between the plates.

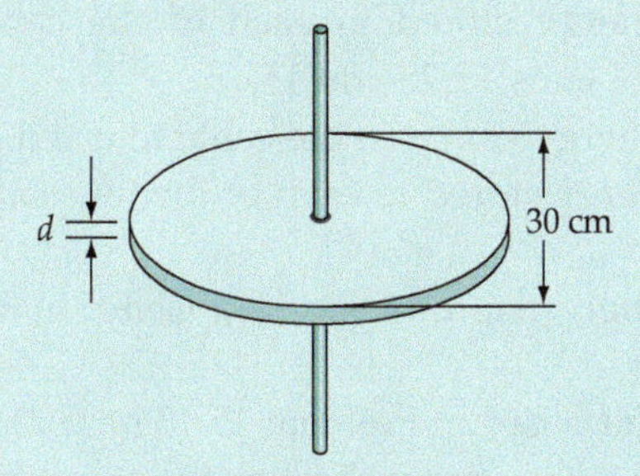

Figure 15-27

15-20 Determine the capacitance of the two plates in Problem 15-19 when the distance between them is: (a) 0.3 mm, (b) 1 mm.

Section 15-5

15-21 Determine the dielectric constant for the capacitor described in Problem 15-1, and list several possible dielectric materials.

15-22 A 5 μF capacitor and a 3 μF capacitor are connected in parallel, and the parallel combination is connected in series with a 7 μF capacitor and a 10 μF capacitor, see Figure 15-28. Calculate the total capacitance, the voltage across each capacitor, and the charge on each capacitor when 55 V is applied to the circuit terminals.

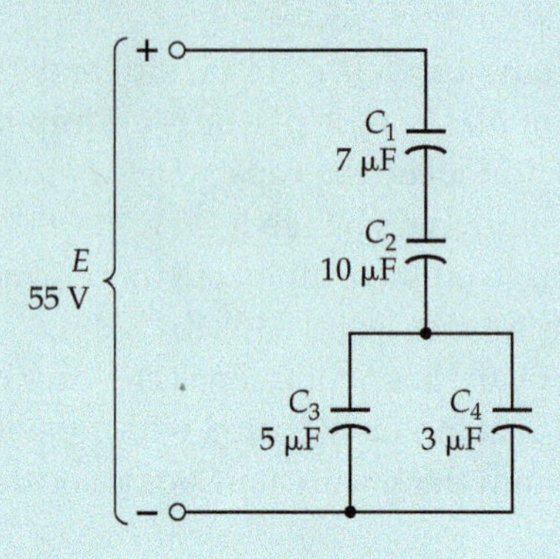

Figure 15-28

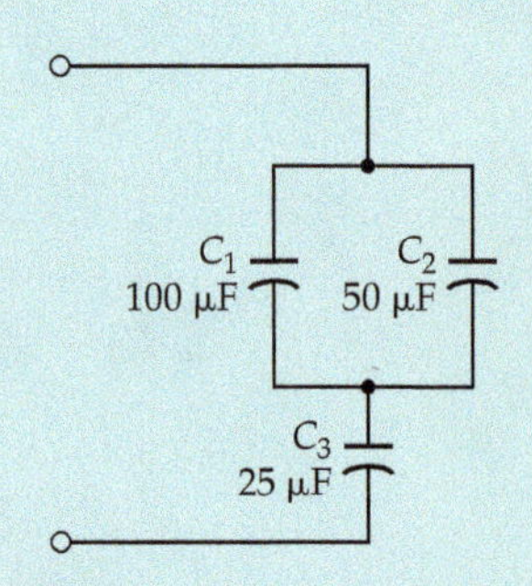

Figure 15-29

15-23 Four capacitors, with values of 100 μF, 50 μF, 25 μF, and 10 μF, are connected in series to a 25 V supply. Calculate the total capacitance, the voltage across each capacitor, and the charge on each capacitor.

15-24 Repeat the calculation in Problem 15-23 for the case of all capacitors connected in parallel.

15-25 A 100 μF capacitor is connected in parallel with a 50 μF capacitor, then the two are connected in series with a 25 μF capacitor, as shown in Figure 15-29. Calculate the total capacitance.

15-26 For Problem 15-25, calculate the voltage across each capacitor when a 25 V supply is applied to the terminals of the circuit.

Section 15-6

15-27 Calculate the energy stored in each of the capacitors in the circuit referred to in Problems 15-25 and 15-26.

15-28 Determine the energy stored in the cable referred to in Problem 15-11.

15-29 Calculate the energy stored in each of the four capacitors referred to in Problem 15-23.

15-30 Determine the energy stored in each capacitor in the circuit described in Problem 15-24.

15-31 The printed circuit board in Problem 15-15 has half the conductor area at 12 V with respect to the copper sheet on the other side, and half at 3 V with respect to the copper sheet. Calculate the energy stored in each half.

15-32 Calculate the voltage that must be applied to two series-connected capacitors to store 700 μJ of energy if the capacitances are $C_1 = 33$ μF and $C_2 = 15$ μF. Also, determine the energy stored in each capacitor.

Practice Problem Answers

15-2.1 40 kV/m, 81 kV
15-2.2 $25 \times 10^{-3}\ m^2$
15-3.1 38.9 pF
15-3.2 0.47 μC
15-3.3 1.8 pF
15-5.1 138.5 pF
15-5.2 2 nC, 15 V, 14.4 V, 0.6 V
15-6.1 11.25 nJ, 4.15 nJ, 180 pJ
15-6.2 25.7 V

CHAPTER 16
Inductance and Capacitance in DC Circuits

CONTENTS

Objectives

You will be able to:

1 Sketch a graph showing the growth of current with time in an inductive-resistive circuit.
2 Calculate the instantaneous levels of current and voltage in an *RL* circuit at any given time.
3 Define the time constant of an *RL* circuit and explain its special relationship to the performance of the circuit.
4 Solve problems involving open circuiting inductors in which energy is stored.
5 Sketch the various inductor and resistor voltage waveforms that result when a square wave is applied to an *RL* circuit.
6 Sketch a graph showing the growth of a capacitor voltage with time in a capacitive-resistive circuit.
7 Calculate the instantaneous levels of voltage and current in an *RC* circuit at any given time.
8 Define the time constant of an *RC* circuit and explain its special relationship to the performance of the circuit.
9 Solve problems involving discharge of charged capacitors.
10 Sketch the various capacitor and resistor voltage waveforms that result when a square wave is applied to an *RC* circuit.

INTRODUCTION

When an inductive-resistive series circuit has its supply voltage switched *on*, the inductor produces an initial maximum level of counter-emf that gradually falls to zero. The circuit current is zero initially and grows gradually to its maximum level. The behavior of an *RL* circuit is most easily understood by plotting the graphs of instantaneous current and instantaneous inductor voltage versus time. Care must be taken in open-circuiting *RL* circuits to avoid high induced voltages.

In the case of a capacitive-resistive series circuit, when the supply is first switched on the charging current is initially at its maximum level, then it gradually falls to zero. The capacitor voltage is zero at first and grows gradually to its maximum level. As with the *RL* circuit, the behavior of an *RC* circuit can be represented graphically by plotting instantaneous current and voltage versus time. Because energy is stored in a charged capacitor, a large current can flow when the capacitor terminals are short-circuited.

16-1 *RL* CIRCUIT OPERATION

When switch S_1 is closed in the simple battery and resistor circuit shown in Figure 16-1(a), the current tends to jump instantaneously to its maximum level of $I = E/R$. This is illustrated by the graph of i versus t in Figure 16-1(b), and it assumes that R is purely resistive. (Actually, there can be no such thing as *pure resistance,* and as will be seen later, there is always some *rise time* involved in the current going from zero to its maximum level.)

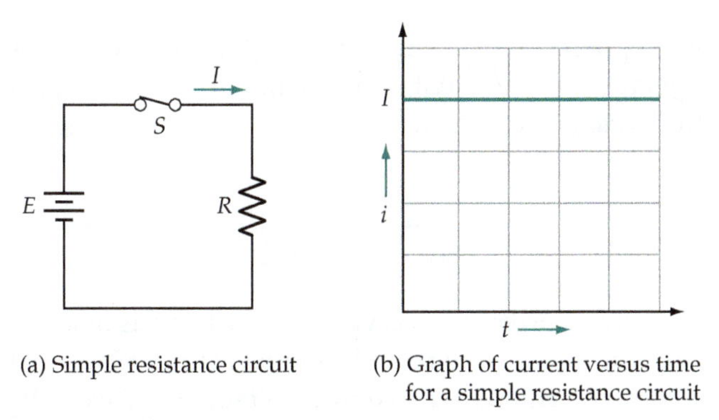

(a) Simple resistance circuit

(b) Graph of current versus time for a simple resistance circuit

Figure 16-1 In a simple resistance circuit, the current tends to jump to its final level and remain constant.

Now consider the circuit shown in Figure 16-2(a), and assume that L is a pure inductance. Recall from Equation 14-3 that when the current changes through an inductor, a counter-emf is generated that has the value,

$$e_L = L\frac{\Delta i}{\Delta t} \text{ volts}$$

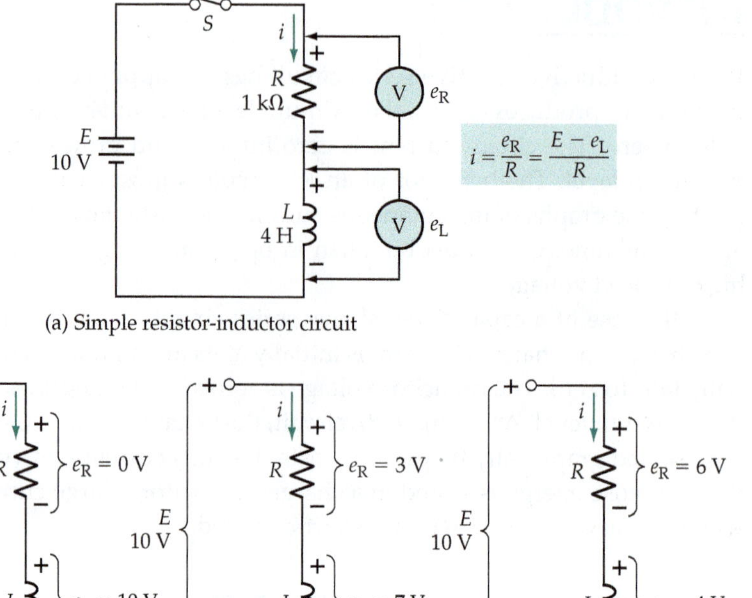

(a) Simple resistor-inductor circuit

$e_R = 0$ V, $e_L = 10$ V; $e_R = 3$ V, $e_L = 7$ V; $e_R = 6$ V, $e_L = 4$ V

$i_o = \frac{e_R}{R} = \frac{0\text{ V}}{1\text{ k}\Omega}$ $\qquad i_1 = \frac{e_R}{R} = \frac{3\text{ V}}{1\text{ k}\Omega}$ $\qquad i_2 = \frac{e_R}{R} = \frac{6\text{ V}}{1\text{ k}\Omega}$

(b) Voltages at $t = 0$ (c) Voltages at $t = t_1$ (d) Voltages at $t = t_2$

Figure 16-2 The counter-emf generated in an inductor when the current level changes causes the current in an *RL* circuit to grow slowly to its final level when the supply voltage is switched on.

The counter-emf opposes the current change that generates it, so, as shown in the circuit, its polarity is such that it opposes the current (i). Using Kirchhoff's voltage law, the instantaneous current level is,

$$i = \frac{E - e_L}{R} \tag{16-1}$$

Although the actual current level is very small when S is first closed, its *rate of growth* is large. The initial rapid rate of growth of current causes the counter-emf (e_L) to be virtually equal to the supply voltage (E). This is illustrated in Figure 16-2(b). Putting $e_L = E$ into Equation 16-1 and using the voltage and component values shown in the circuit, the initial current level is,

$$i_o = \frac{10\text{ V} - 10\text{ V}}{1\text{ k}\Omega} = 0$$

So, on the graph of i versus t, i is zero at $t = 0$; point 1 in Figure 16-3. At this instant the rate of change of current ($\Delta i_o/\Delta t_o$ on the graph) is at its maximum value. Because $e_L = L\,\Delta i/\Delta t$, e_L also has its maximum level at $t = 0$.

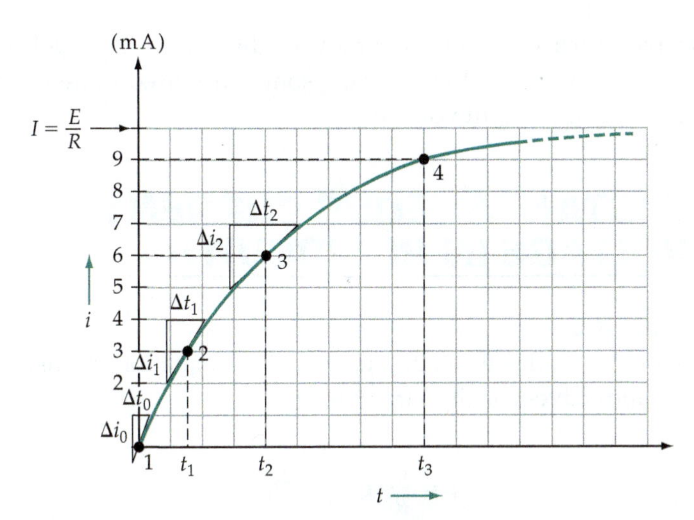

Figure 16-3 Graph showing the typical growth of current plotted versus time for a series *RL* circuit, starting from the instant of supply voltage switch-on.

Some time t_1 after S is closed, the rate of change of current has decreased to a value ($\Delta i_1/\Delta t_1$ on the graph in Figure 16-3) that gives $e_L = 7$ V. So, from Figure 16-2(c), at time t_1 the instantaneous current level is,

$$i_o = \frac{10\text{ V} - 7\text{ V}}{1\text{ k}\Omega} = 3\text{ mA}$$

Current $i_1 = 3$ mA is plotted at time t_1 on the graph of i versus t; point 2 in Figure 16-3.

Because e_L is falling, the current level is increasing. As the current level grows, it gets closer to its maximum level of $I = E/R$, and in doing so the rate of change of current $\Delta i/\Delta t$ continues to decrease. This decrease in $\Delta i/\Delta t$ is the cause of the decreasing level of counter-emf, ($e_L = L\,\Delta i/\Delta t$).

After some time t_2, $\Delta i/\Delta t$ has fallen to a level that gives $e_L = 4$ V. So, from Figure 16-2(d),

$$i_2 = \frac{10\text{ V} - 4\text{ V}}{1\text{ k}\Omega} = 6\text{ mA}$$

Current $i_2 = 6$ mA is plotted at time t_2 in Figure 16-3; point 3. At time t_3, $\Delta i/\Delta t$ is such that $e_L = 1$ V, giving,

$$i_3 = \frac{10\text{ V} - 1\text{ V}}{1\text{ k}\Omega} = 9\text{ mA}$$

The value of i_3 is plotted on the graph at time t_3; point 4. The final current level when e_L becomes zero is, of course, $I = E/R$.

Because the counter-emf is initially large and falls off smoothly to zero, the current grows continuously from zero to its maximum level, as illustrated in Figure 16-3. As the current approaches maximum, the rate of change of current

$(\Delta i/\Delta t)$ keeps decreasing, and this produces the decreasing level of e_L. The terms *step response* and *forced response* are sometimes used to describe the *RL* circuit response to a dc input voltage.

16-2 INSTANTANEOUS CURRENT AND VOLTAGE IN *RL* CIRCUITS

Current Equation

By application of calculus it can be shown that the equation for instantaneous current in an inductive-resistive circuit is,

$$i = \frac{E}{R}(1 - \epsilon^{-tR/L}) \tag{16-2}$$

where,

i = instantaneous current, in amperes, at time t
E = supply voltage
R = series resistance, in ohms (including the winding resistance of the inductor)
ϵ = exponential constant = 2.718
t = time, in seconds, from current commencement
L = inductance of the inductor, in henrys

Using Equation 16-2, the instantaneous current levels can be calculated for several different time intervals from $t = 0$ for a given circuit. The corresponding values obtained in this way can be used to plot an accurate graph of i versus t for the circuit. Equation 16-2 can also be manipulated to determine the value of t, R, or L for a given current level.

$$\epsilon^{-tR/L} = \frac{E}{E - iR}$$

Taking the natural logarithm of both sides,

$$t\frac{R}{L} = \ln\left(\frac{E}{E - iR}\right) \tag{16-3}$$

Example 16-1

For the circuit in Figure 16-2(a), calculate the times t_1, t_2, and t_3 for plotting the inductor current versus time graph in Figure 16-3.

Solution

From Eq. 16-3, $$t = \frac{L}{R}\ln\left(\frac{E}{E - iR}\right)$$

At $i = 3\text{ mA}$, $t_1 = \dfrac{4\text{ H}}{1\text{ k}\Omega} \ln\left[\dfrac{10\text{ V}}{10\text{ V} - (3\text{ mA} \times 1\text{ k}\Omega)}\right]$

$= 1.45\text{ ms}$

At $i = 6\text{ mA}$, $t_2 = \dfrac{4\text{ H}}{1\text{ k}\Omega} \ln\left[\dfrac{10\text{ V}}{10\text{ V} - (6\text{ mA} \times 1\text{ k}\Omega)}\right]$

$= 3.7\text{ ms}$

At $i = 9\text{ mA}$, $t_3 = \dfrac{4\text{ H}}{1\text{ k}\Omega} \ln\left[\dfrac{10\text{ V}}{10\text{ V} - (9\text{ mA} \times 1\text{ k}\Omega)}\right]$

$= 9.2\text{ ms}$

Example 16-2

Determine the instantaneous levels of current at 2 ms intervals from $t = 0$ for the circuit of Figure 16-2(a).

Solution

Eq. 16-2, $i = \dfrac{E}{R}(1 - \epsilon^{-tR/L})$

At $t = 2\text{ ms}$, $i = \dfrac{10\text{ V}}{1\text{ k}\Omega}(1 - \epsilon^{-2\text{ ms} \times 1\text{ k}\Omega/4\text{H}}) \approx 3.93\text{ mA}$ Point 2

At $t = 4\text{ ms}$, $i = \dfrac{10\text{ V}}{1\text{ k}\Omega}(1 - \epsilon^{-4\text{ ms} \times 1\text{ k}\Omega/4\text{H}}) \approx 6.32\text{ mA}$ Point 3

At $t = 6\text{ ms}$, $i = 7.77\text{ mA}$ Point 4

At $t = 8\text{ ms}$, $i = 8.65\text{ mA}$ Point 5

At $t = 10\text{ ms}$, $i = 9.18\text{ mA}$ Point 6

At $t = 12\text{ ms}$, $i = 9.5\text{ mA}$ Point 7

At $t = 14\text{ ms}$, $i = 9.7\text{ mA}$ Point 8

At $t = 16\text{ ms}$, $i = 9.82\text{ mA}$ Point 9

At $t = \infty$, $i = 10\text{ mA}$ = maximum current level

The corresponding values of t and i are plotted in Figure 16-4.

Time Constant

Referring to the graph of i versus t plotted in Figure 16-4, it is seen that when $t = 4\text{ ms}$, i is 6.32 mA. Note that 6.32 mA is 63.2% of the maximum current level (10 mA). Also,

$$\frac{L}{R} = \frac{4\text{ H}}{1\text{ k}\Omega} = 4\text{ ms} = t$$

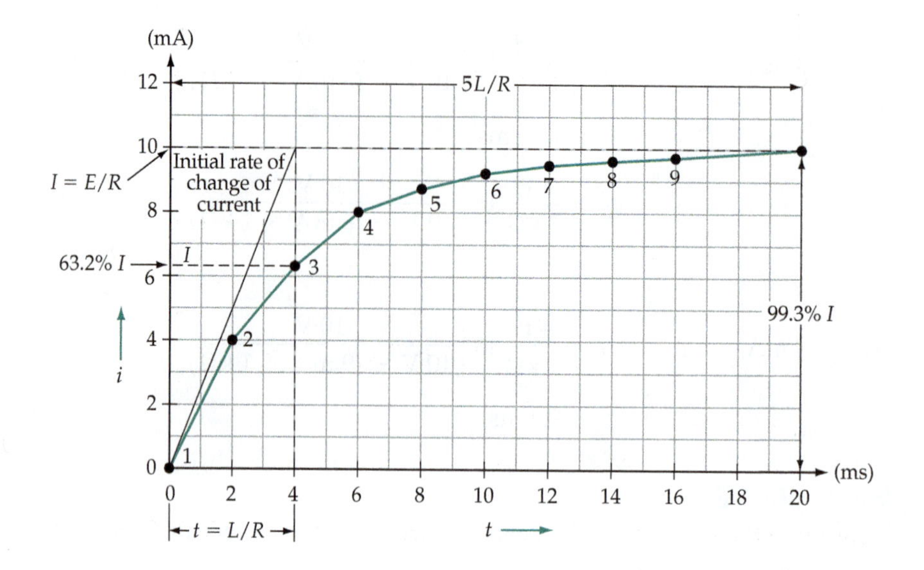

Figure 16-4 Graph of current (i) versus time (t) for a series RL circuit, as determined in Example 16-2. The current increases to 63.2% of its maximum level at $t = L/R$ and to 99.3% of its maximum at $t = 5L/R$.

So, when $t = L/R$, the instantaneous current level is always 63.2% of E/R. The quantity L/R is termed the *time constant* of an inductive-resistive circuit, and the time constant is very important in determining the behavior of the circuit. Sometimes the Greek letter τ is used as the symbol for the time constant. It can be shown that after a time of $t = 5L/R$, the current is 99.3% of its maximum level, (see Figure 16-4). By drawing a straight line from the origin at a tangent to the graph of i versus t, it is seen that if the initial rate of change of current were maintained, the current would reach its maximum level in a time of $t = L/R$.

Counter-emf

Equation 16-2 can be manipulated to derive an equation for the instantaneous counter-emf at any time.

Eq. 16-2, $$i = \frac{E}{R}(1 - \epsilon^{-tR/L})$$

So, $$iR = E - E\epsilon^{-tR/L}$$

From Eq. 16-1, $$iR = E - e_L$$

So, $$E - e_L = E - E\epsilon^{-tR/L}$$

Giving, $$e_L = E\epsilon^{-tR/L} \qquad \textbf{(16-4)}$$

Alternatively, when the instantaneous current levels have been calculated, the corresponding emf values can be determined from Equation 16-1,

$$e_L = E - iR$$

Example 16-3

Calculate the level of the counter-emf for the circuit in Figure 16-2(a) at $t = L/R$ and at $t = 5L/R$.

Solution

At $t = L/R$,

Eq. 16-4, $e_L = E\epsilon^{-tR/L} = 10\epsilon^{-1}$

≈ 3.68 V Point 1 in Figure 16-5

At $t = 5L/R$, $e_L = E\epsilon^{-tR/L} = 10\epsilon^{-5}$

≈ 0.067 V Point 2 in Figure 16-5

The graph of e_L versus t in Figure 16-5 shows how the counter-emf (e_L) changes with time. At $t = 0$, e_L is equal to the supply voltage (E). At $t = L/R$, e_L has fallen by 63.2% of E, and at $t = 5L/R$, e_L has fallen through 99.3% of its initial level. Comparing the graphs of i versus t and e_L versus t, it is seen that inductor terminal voltage drops as the current increases.

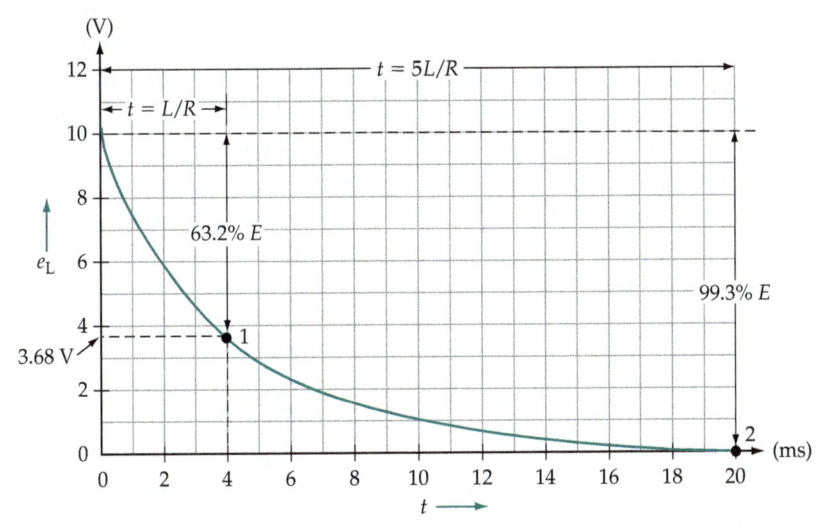

Figure 16-5 Graph of inductor voltage (e_L) versus time (t) for a series RL circuit. The voltage falls by 63.2% of its maximum level at $t = L/R$ and by 99.3% of its maximum at $t = 5L/R$.

Practice Problems

16-2.1 A series RL circuit has $L = 33$ mH, $R = 3.3$ kΩ, and $E = 12$ V. Determine the instantaneous current levels at five evenly spaced time intervals between the instant the supply is switched on and the time the current arrives at its maximum level.

16-2.2 An inductor is included in series with a 100 Ω resistor to limit the rate of current growth when the supply is switched on. If the current is to reach 80% of its maximum level in a time of 3 ms, determine the required inductance.

16-3 OPEN-CIRCUITING AN INDUCTIVE CIRCUIT

Counter-emf

From Equation 14-14, the energy stored in an inductive circuit is,

$$W = 0.5LI^2 \text{ joules}$$

The equation shows that energy is stored in an inductor only while a steady-state current is flowing through the inductor. When the current is reduced to zero, there is obviously no energy stored.

Now consider what occurs when the switch is opened in an inductive circuit, such as that shown in Figure 16-6. Before the switch is opened, a constant current is flowing through the inductor, and energy is stored [Figure 16-6(a)]. When the switch is opened, the current must go to zero; consequently, the energy contained in the inductor must also go to zero. This means that the energy must somehow flow out of the inductor at the instant that the switch is opened. Now recall that the counter-emf generated in an inductor opposes the current change that produces the counter-emf. The current begins to fall when the switch is opened, so the counter-emf generated tends to oppose the fall in current; that is, its polarity is such that it tries to maintain the current flow. This is illustrated in Figure 16-6(b), where the counter-emf is positive at the bottom of the inductor and negative at the top. This polarity is opposite to that of the counter-emf when the switch was first closed [see Figure 16-2(b)].

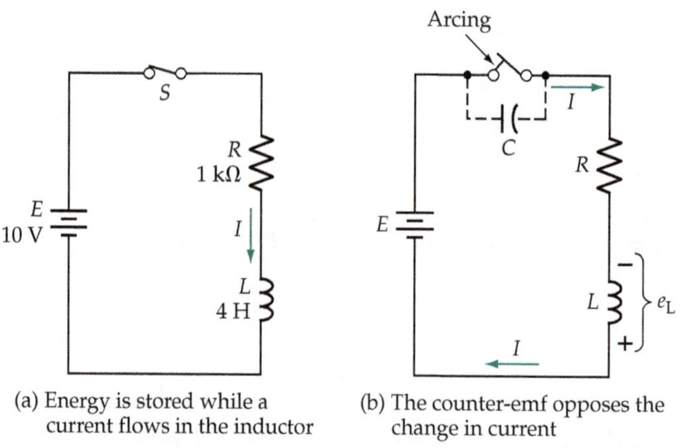

Figure 16-6 When an inductive circuit is open-circuited, the energy stored in the inductor tends to keep current flowing in the circuit; arcing may occur at a switch.

The counter-emf causes a spark to jump across the open switch contacts. The arcing across the contacts continues for the brief period necessary to discharge the energy contained in the inductor. To produce such an arc across open contacts requires a very high voltage. The high voltage is, of course, the counter-emf. It is not surprising that a very high voltage is generated when it is remembered that the equation for counter-emf is,

$$e_L = L\frac{\Delta i}{\Delta t}$$

Because the open-circuit attempts to reduce the current to zero instantaneously, the rate of change of current $\Delta i/\Delta t$ is a very large quantity, and so $L\,(\Delta i/\Delta t)$ is also very large.

To get some idea of the voltage generated when the switch is opened in an inductive circuit, consider again the circuit in Figure 16-6(a). The steady-state current level while the switch is closed is,

$$I = \frac{E}{R} = \frac{10\ \text{V}}{1\ \text{k}\Omega} = 10\ \text{mA}$$

At the instant the switch is opened, the counter-emf is such that for a brief time period, the 10 mA continues to flow across the air gap between the switch contacts. The resistance of the air gap is likely to be much greater than 1 MΩ. Thus, voltage across the air gap will be in excess of,

$$\begin{aligned} e &= IR = 10\ \text{mA} \times 1\ \text{M}\Omega \\ &= 10\,000\ \text{V} \end{aligned}$$

This 10 000 V is the generated counter-emf.

Controlling the Counter-emf

It is seen that when a switch is opened in an inductive circuit, the high level of counter-emf might damage the circuit insulation. Obviously, some means of discharging the energy in the inductor must be provided. A capacitor connected across the switch, as shown in Figure 16-6(b), is sometimes employed for this purpose. The capacitor initially behaves as a short-circuit when the switch is opened, and this allows current flow to continue until the capacitor is charged to the level of the supply voltage.

Another method of protecting an inductive circuit against high levels of counter-emf is shown in Figure 16-7. Resistor R_2, in parallel with R_1 and L, has

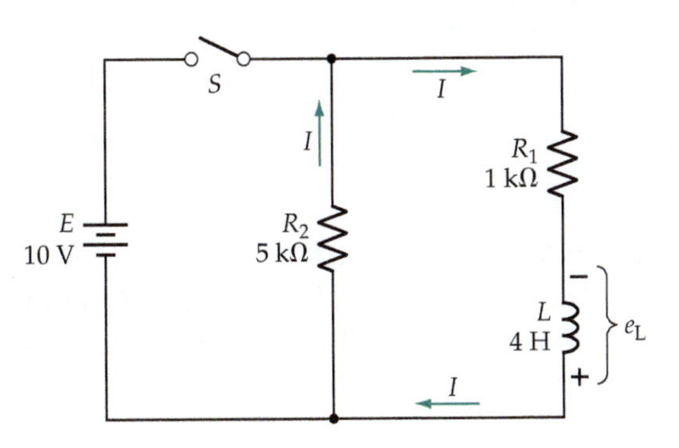

Figure 16-7 One method of protecting an inductive circuit from the effects of high counter-emf levels when the switch is open-circuited. The energy stored in the inductor is discharged via resistor R_2.

no effect on the steady-state current through R_1 and L while the switch remains closed. When the switch is opened, the total resistance in series with L is $(R_1 + R_2)$. So, the maximum level of counter-emf that must be generated to maintain the initial level of I is,

$$e_L = I(R_1 + R_2)$$

For $R_1 = 1\ \text{k}\Omega$ and $R_2 = 5\ \text{k}\Omega$ in the circuit of Figure 16-7, the counter-emf is

$$e_L = 10\ \text{mA}\ (5\ \text{k}\Omega + 1\ \text{k}\Omega)$$
$$= 60\ \text{V}$$

With the source disconnected and the inductor discharging via resistors, the RL circuit is said to be operating *source-free.*

Example 16-4

A circuit connected as shown in Figure 16-7 has $R_1 = 500\ \Omega$, $L = 500\ \text{mH}$, and $E = 100\ \text{V}$. Determine how long it takes the inductor current to grow to 1.7 mA after the switch is closed. Calculate the required value of R_2 if the counter-emf is not to exceed 300 V when the switch is opened.

Solution

Eq. 16-3,

$$t = \frac{L}{R_1} \ln\left(\frac{E}{E - iR_1}\right)$$
$$= \left(\frac{500\ \text{mH}}{500\ \Omega}\right) \ln\left(\frac{100\ \text{V}}{100\ \text{V} - (1.7\ \text{mA} \times 500\ \Omega)}\right)$$
$$= 8.5\ \mu\text{s}$$

Maximum current,

$$I = \frac{E}{R_1} = \frac{100\ \text{V}}{500\ \Omega}$$
$$= 200\ \text{mA}$$

Counter-emf,

$$e_L = I(R_1 + R_2)$$

So,

$$R_2 = \frac{e_L}{I} - R_1 = \frac{300\ \text{V}}{200\ \text{mA}} - 500\ \Omega$$
$$= 1\ \text{k}\Omega$$

Practice Problem

16-3.1 For a series RL circuit with $E = 12\ \text{V}$, $R = 3.3\ \text{k}\Omega$, and $L = 33\ \text{mH}$, as in Problem 16-2.1, determine the required parallel resistance to limit the counter-emf to 18 V at switch-off. The circuit is to be arranged as in Figure 16-7.

16-4 *RL* CIRCUIT WAVEFORMS

Consider the *RL* circuit once again, as illustrated in Figure 16-8(a). Suppose that switch S_1 is closed for an accurately measured time t_1, then opened again for an exactly equal time t_2, and that the sequence is repeated over and over. The resulting voltage applied to the circuit would be as shown in Figure 16-8(b). For obvious reasons the shape is termed a *square wave* or *pulse wave*. The time t_1 is referred to as the *pulse width*, and t_2 is termed the *space width*. It would not really be practical to attempt to manually open and close a switch in order to generate a square wave; suitable electronic instruments are available for the purpose. However, going along with the assumption that a square wave is applied as illustrated, the resulting

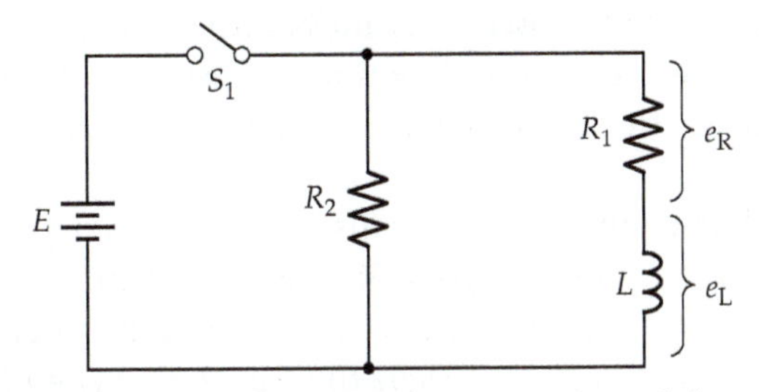

(a) Circuit for generating the waveforms shown below

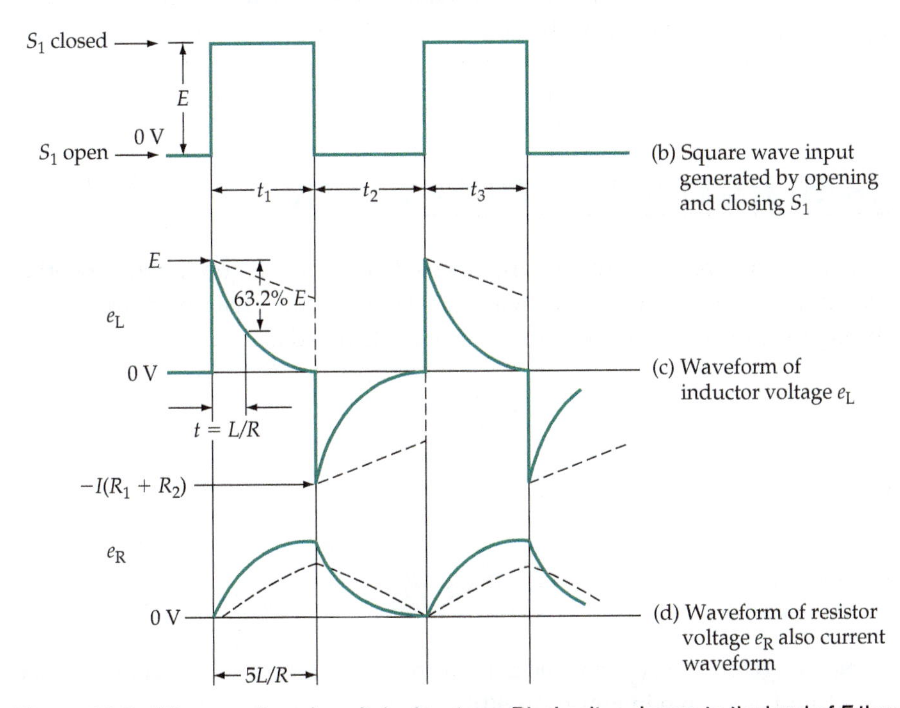

(b) Square wave input generated by opening and closing S_1

(c) Waveform of inductor voltage e_L

(d) Waveform of resistor voltage e_R also current waveform

Figure 16-8 When a voltage is switched on to an *RL* circuit, e_L jumps to the level of *E* then decreases exponentially while *I* increases exponentially from zero. When the voltage is switched off, e_L jumps to $-I(R_1 + R_2)$ then rises exponentially to zero while *I* decreases exponentially from its maximum level.

voltages across L and R can be investigated. Note that during t_1, the circuit is responding to a *step input*, and during t_2 it is said to be operating *source free*.

When the supply voltage (E) is switched *on* to the circuit, the counter-emf (e_L) immediately jumps to the level of the applied voltage, as shown in Figure 16-8(c). Then, as the current grows e_L slowly falls to zero, as illustrated. The counter-emf remains at zero until the switch is opened. As already discussed, at this instant the counter-emf immediately jumps (negatively) to $-I(R_1 + R_2)$, from which it again slowly returns to zero. The positive and negative step and decline of e_L are repeated over and over as the switch is closed and opened, giving the *spike waveform* illustrated in Figure 16-8(c).

The current through R_1, and L is zero initially when S_1 is closed and rises to its maximum level of $I = E/R_1$ when e_L has fallen to zero. When S_1 is opened, the current falls slowly as e_L moves from $-I(R_1 + R_2)$ back to zero. The resulting current waveform is shown in Figure 16-8(d). Because the voltage across R_1 is simply IR_1, the waveform of e_R is exactly the same as the current waveform.

Recalling that counter-emf e_L falls by 99.3% of its maximum level in a time $t = 5(L/R)$, it is seen from Figure 16-8(c) that $t_1 = t_2 \approx 5(L/R)$. So, the time constant (L/R) is short by comparison to the pulse width (t_1 and t_2), and the circuit is said to have a *short time constant*. When L/R is greater than t_1 and t_2, the typical shape of the waveforms are as shown by the dashed lines in Figure 16-8(c) and (d). In this case, the circuit is said to have a *long time constant*.

16-5 *RC* CIRCUIT OPERATION

A capacitor and resistor are shown connected in series in Figure 16-9(a), together with a supply voltage (E) and a switch (S). The capacitor charging current flows through the resistor. So, the current can be calculated as,

$$i = \frac{e_R}{R}$$

and

$$e_R = E - e_C$$

Giving,

$$i = \frac{E - e_C}{R} \tag{16-5}$$

If the charge on the capacitor is zero at the instant the switch is closed, then $e_C = 0$, and as shown in Figure 16-9(b),

$$i = \frac{10\text{ V} - 0}{1\text{ k}\Omega} = 10\text{ mA}$$

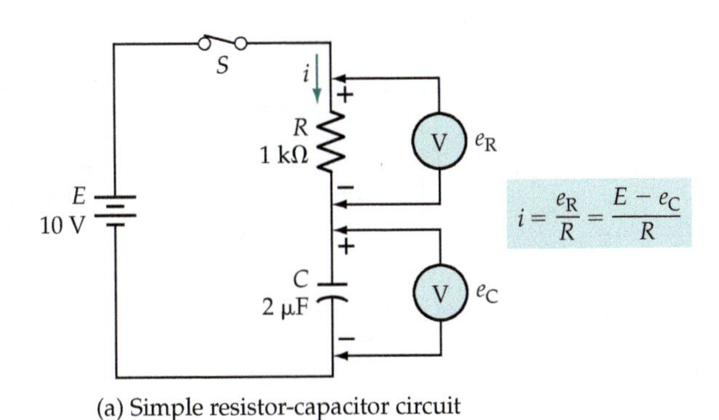

(a) Simple resistor-capacitor circuit

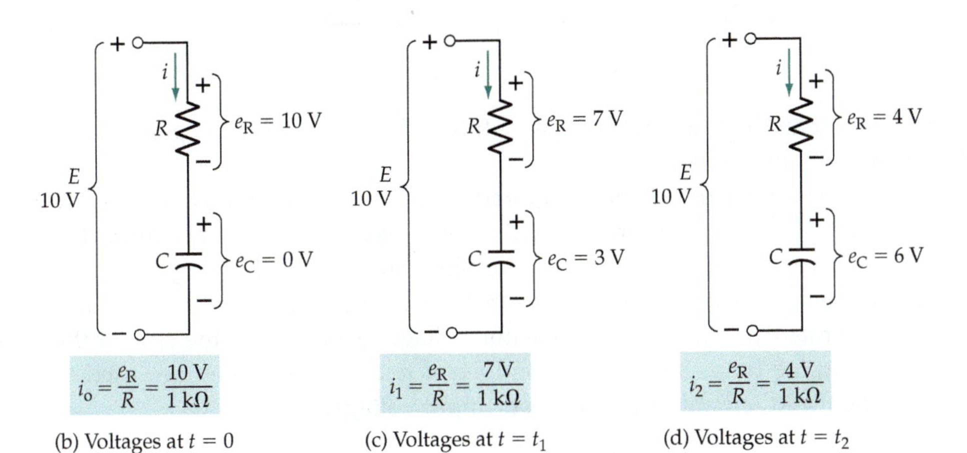

(b) Voltages at $t = 0$ (c) Voltages at $t = t_1$ (d) Voltages at $t = t_2$

Figure 16-9 The capacitor voltage in a series *CR* circuit tends to grow slowly from zero to its final level when the supply voltage is first switched on.

The current flow causes the capacitor to charge with the polarity illustrated. After a time t_1 the capacitor voltage might be 3 V [see Figure 16-9(c)]. Then the charging current becomes,

$$i = \frac{10\text{ V} - 3\text{ V}}{1\text{ k}\Omega} = 7\text{ mA}$$

It is seen that, because C has accumulated some charge, the voltage across R is reduced, and consequently the charging current is reduced from 10 mA to 7 mA. Because the charging current has been reduced, the capacitor voltage is now growing at a slower rate than before. The instantaneous levels of e_C at $t=0$ and $t=t_1$ can now be plotted on the graph of e_C versus time; points 1 and 2 in Figure 16-10.

After time t_2, the capacitor voltage has grown to 6 V [Figure 16-9(d)]. The charging current now becomes,

$$i = \frac{10\text{ V} - 6\text{ V}}{1\text{ k}\Omega} = 4\text{ mA}$$

The charging current has been further reduced (from 7 mA to 4 mA), so the capacitor is charging at an even slower rate than before. Because the charging current has been decreasing, the time for the capacitor to charge from 3 V to 6 V

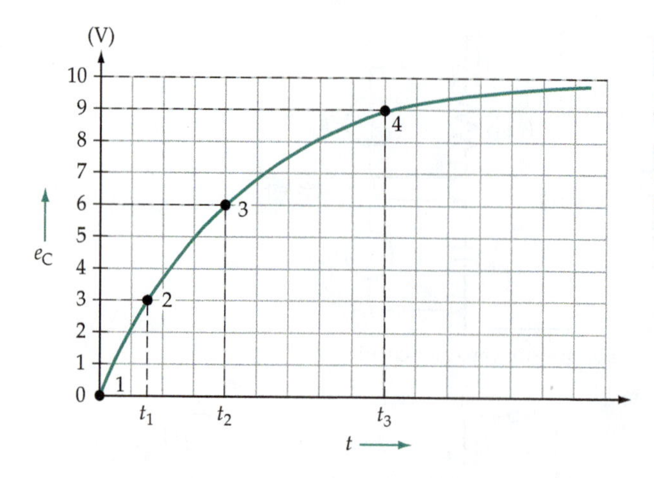

Figure 16-10 Graph showing the typical growth of capacitor voltage plotted versus time for a series *RC* circuit, starting from the instant of supply voltage switch-on.

is longer than the time for it to charge from 0 V to 3 V. Point 3 is plotted at t_2 and $e_C = 6$ V in Figure 16-10.

An even longer time is now required for the capacitor voltage to grow by another 3 V (point 4 in Figure 16-10). Because e_C is continuously increasing, the voltage across *R* is continuously decreasing, and so the charging current is continuously decreasing. This means that *C* is charged at a rapid rate initially, then the rate decreases as the capacitor voltage grows. As in the case of the *RL* circuit, the terms *step response* and *forced response* are sometimes used to describe the *RC* circuit response to a dc input voltage.

16-6 INSTANTANEOUS CURRENT AND VOLTAGE IN *RC* CIRCUITS

Voltage Equation

The equation for the instantaneous voltage on a capacitor in a resistive-capacitive circuit can be derived by differential calculus:

$$e_C = E - (E - E_o)\epsilon^{-t/(CR)} \qquad \textbf{(16-6)}$$

where.

e_C = capacitor voltage at time t
E = supply voltage
E_o = initial level of capacitor voltage
ϵ = exponential constant = 2.718
t = time, in seconds, from commencement of charge
C = capacitance value, in farads
R = charging resistance, in ohms

Using Equation 16-6, the instantaneous levels of capacitor voltage can be calculated for several different time intervals from $t = 0$ for a given circuit. The corresponding values of e_C and t can then be plotted to give an accurate graph of e_C versus t for the circuit. The equation can also be manipulated to obtain expressions for t, C, and R for a given capacitor voltage level.

When the capacitor is initially uncharged,

$$E_o = 0$$

and

$$e_C = E - \left(E\epsilon^{-t/(CR)}\right)$$

or,

$$e_C = E\left(1 - \epsilon^{-t/(CR)}\right) \qquad (16\text{-}7)$$

Also, from Equation 16-6,

$$\epsilon^{t/(CR)} = \frac{E - E_o}{E - e_C}$$

Taking the natural logarithm of both sides,

$$\frac{t}{CR} = \ln\left(\frac{E - E_o}{E - e_C}\right) \qquad (16\text{-}8)$$

Equation 16-8 can be further simplified if the capacitor starting voltage (E_o) is assumed to be zero:

$$\frac{t}{CR} = \ln\left(\frac{E}{E - e_C}\right) \qquad (16\text{-}9)$$

Equation 16-9 can be used to determine t, C, or R when the other quantities are known.

Example 16-5

For the circuit in Figure 16-9(a), calculate the times t_1, t_2, and t_3 for plotting the capacitor current versus time graph in Figure 16-10.

Solution

From Eq. 16-9,

$$t = CR \ln\left(\frac{E}{E - e_C}\right)$$

At $e_C = 3$ V,

$$t_1 = 2\ \mu\text{F} \times 1\ \text{k}\Omega \times \ln\left(\frac{10\ \text{V}}{10\ \text{V} - 3\ \text{V}}\right)$$

$$= 0.7\ \text{ms}$$

At $e_C = 6$ V,

$$t_2 = 2\ \mu\text{F} \times 1\ \text{k}\Omega \times \ln\left(\frac{10\ \text{V}}{10\ \text{V} - 6\ \text{V}}\right)$$

$$= 1.8\ \text{ms}$$

At $e_C = 9$ V,

$$t_3 = 2\ \mu\text{F} \times 1\ \text{k}\Omega \times \ln\left(\frac{10\ \text{V}}{10\ \text{V} - 9\ \text{V}}\right)$$

$$= 4.6\ \text{ms}$$

Example 16-6

Determine the instantaneous values of capacitor voltage at 1 ms intervals from $t = 0$ for the circuit in Figure 16-9(a).

Solution

Assuming that the initial level of capacitor voltage is zero, Equation 16-7 can be used:

$$e_C = E(1 - \epsilon^{-t/(CR)})$$

At $t = 0$, $e_C = 0\ \text{V}$ — Point 1 in Figure 16-11

At $t = 1$ ms, $e_C = 10\ \text{V}(1 - \epsilon^{-1\ \text{ms}/(2\ \mu\text{F} \times 1\ \text{k}\Omega)})$

$= 3.93\ \text{V}$ — Point 2

At $t = 2$ ms, $e_C \approx 6.32\ \text{V}$ — Point 3

At $t = 3$ ms, $e_C \approx 7.77\ \text{V}$ — Point 4

At $t = 4$ ms, $e_C \approx 8.65\ \text{V}$ — Point 5

At $t = 5$ ms, $e_C \approx 9.18\ \text{V}$ — Point 6

At $t = 6$ ms, $e_C \approx 9.5\ \text{V}$ — Point 7

At $t = 7$ ms, $e_C \approx 9.7\ \text{V}$ — Point 8

At $t = 8$ ms, $e_C \approx 9.82\ \text{V}$ — Point 9

At $t = 10$ ms, $e_C = 9.93\ \text{V}$ — Point 10

At $t = \infty$, $e_C = 10\ \text{V} =$ maximum voltage level

A computer programs for solving the type of problem presented in Example 16-6 is offered in Section 28-5.

Time Constant

Referring to the graph of e_C versus t plotted in Figure 16-11, it is seen that when $t = 2$ ms, e_C is 6.32 V. Note that 6.32 V is 63.2 percent of the maximum voltage level (10 V).

Also, $$t = CR = 2\ \mu\text{F} \times 1\ \text{k}\Omega = 2\ \text{ms}$$

So, when $t = CR$, the instantaneous capacitor voltage level is always 63.2% of E. The quantity CR is the *time constant* (τ) of a resistive-capacitive circuit, and, as in the case of an RL circuit, the time constant largely determines the behavior of the circuit. After a time period of 5 RC, the capacitor voltage is 99.3% of its maximum level. By drawing a straight line at a tangent to the graph of e_C/t, it can be shown that if the initial rate of charge were maintained, the capacitor voltage would reach its maximum level in a time of $t = CR$ (see Figure 16-11).

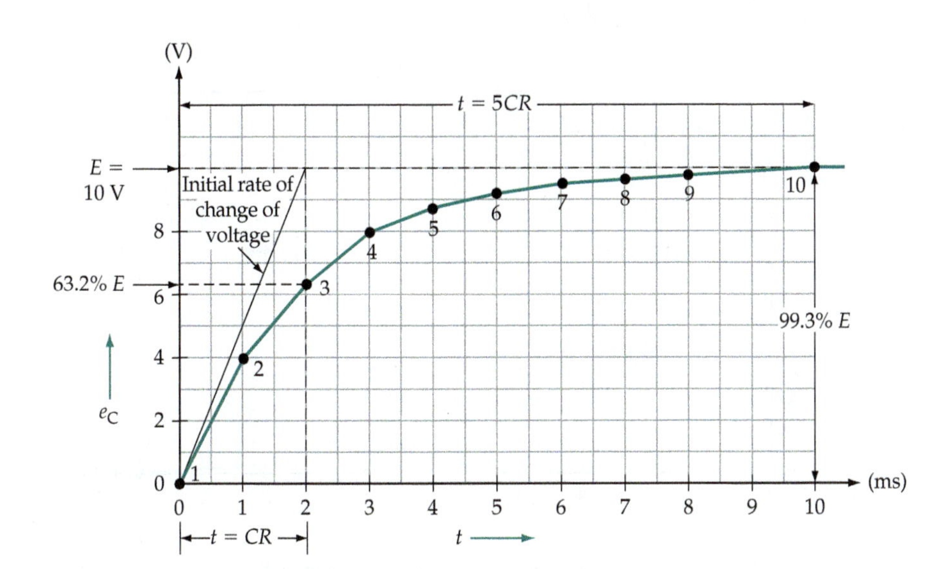

Figure 16-11 Graph of capacitor voltage (e_C) versus time (t) for a series CR circuit. The voltage increases to 63.2% of its maximum level at $t = CR$, and to 99.3% of its maximum at $t = 5CR$.

Charging Current

Equation 16-7 can be manipulated to determine an equation for the instantaneous charging current at any time:

Eq. 16-7, $$e_C = E\left(1 - \epsilon^{-t/(CR)}\right)$$

Eq. 16-5, $$i = \frac{E - e_C}{R}$$

So, $$i = \frac{E - E\left(1 - \epsilon^{-t/(CR)}\right)}{R}$$

Giving, $$i = \frac{E\epsilon^{-t/(CR)}}{R} \qquad \textbf{(16-10)}$$

Alternatively, when the instantaneous capacitor voltages are known, the corresponding current levels can be determined from Equation 16-5.

$$i = \frac{E - e_C}{R}$$

The graph of i_C versus t in Figure 16-12 shows how the charging current changes with time. At $t = 0$, $i_C = E/R$. At $t = CR$, i_C has fallen by 63.2% of E/R. And at $t = 5CR$, i_C has fallen through 99.3% of its initial level.

Example 16-7

Calculate the level of capacitor charging current for the circuit in Figure 16-9(a) at $t = CR$ and $t = 5CR$.

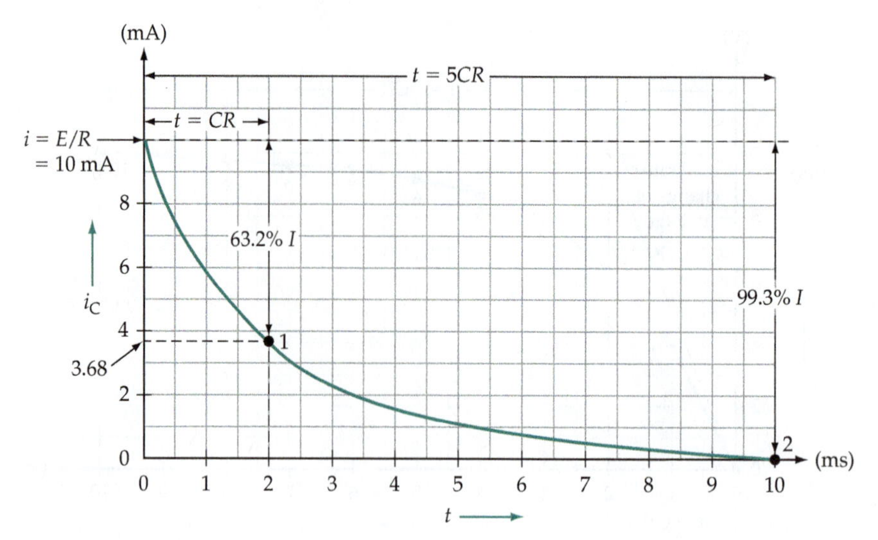

Figure 16-12 Graph of capacitor charging current (i_C) versus time (t) for a series *CR* circuit. The current falls by 63.2% of its maximum level at $t = CR$, and by 99.3% of its maximum at $t = 5CR$.

Solution

At $t = CR$,
$$i = \frac{E\epsilon^{-t/(CR)}}{R} = \frac{10\text{ V} \times \epsilon^{-1}}{1\text{ k}\Omega}$$

$$\approx 3.68\text{ mA}$$ Point 1 in Figure 16-12

At $t = 5CR$,
$$i = \frac{10\text{ V} \times \epsilon^{-5}}{1\text{ k}\Omega}$$

$$\approx 67\ \mu\text{A}$$ Point 2 in Figure 16-12

Practice Problems

16-6.1 A series *RC* circuit has $C = 10\ \mu\text{F}$, $R = 3.3\text{ k}\Omega$, and $E = 12\text{ V}$. Determine the instantaneous current levels at five evenly spaced time intervals from the instant the supply is switched *on* until the current approaches zero.

16-6.2 Two series-connected resistors $R_1 = 4.7\text{ k}\Omega$ and $R_2 = 22\text{ k}\Omega$ are connected via a switch to a 5 V supply. A 100 pF capacitor is connected in parallel with R_1. Determine the circuit current level at switch-on, and the time taken for the capacitor to discharge to zero when the supply is open-circuited.

16-7 DISCHARGING A CAPACITOR

Capacitor Voltage

From Equation 15-10, the energy stored in a charged capacitor is,

$$W = 0.5CE^2 \text{ joules}$$

From the equation, it is seen that energy is stored in a capacitor when a steady-state voltage exists across the capacitor terminals.

When the switch is opened in a series *CR* circuit such as that shown in Figure 16-13(a), there is no effect on the charge stored in the capacitor. The capacitor is able to hold its charge as a static quantity, and its terminal voltage is constant. This contrasts with the case of an inductor, where the charge is a dynamic quantity that cannot be maintained when the current is interrupted. The capacitor can hold its charge indefinitely, and if it is charged to a high voltage ***it can be dangerous*** to someone working on a circuit believed to be safely disconnected from the supply.

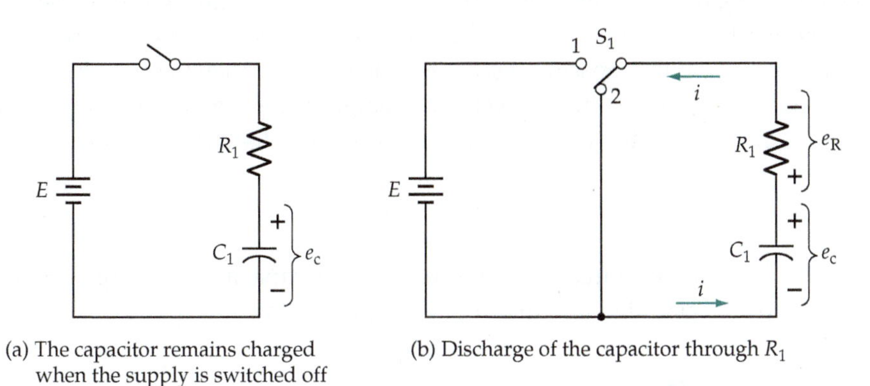

(a) The capacitor remains charged when the supply is switched off

(b) Discharge of the capacitor through R_1

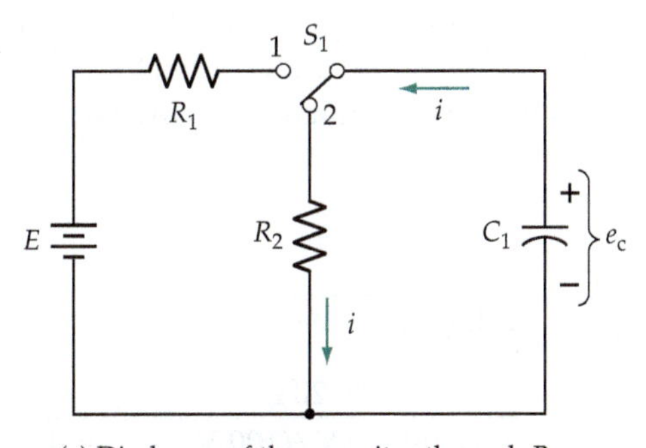

(c) Discharge of the capacitor through R_2

Figure 16-13 **Method of discharging a charged capacitor. When the supply is switched *off*, C_1 discharges through R_1 or R_2.**

Safe Discharging

Sometimes the switch employed in an *RC* circuit has two contacts, as illustrated in Figure 16-13(b). Then, when S_1 is switched to position 2, after being in position 1 for some time, C_1 is discharged via R_1. In this case, the capacitor has already been charged up to *E* volts with the polarity shown. When S_1 is in position 2, the capacitor acts like a voltage source and causes current to flow through R_1, as illustrated. This current discharges the capacitor after a time approximately equal to 5*CR*. Note that the direction of the discharging current through R_1 is opposite to the charging current direction when S_1 was in position 1. Consequently, the voltage drop (e_R) across R_1 has reversed polarity.

During discharge, the capacitor voltage at any instant can be determined from Equation 16-6. With S_1 in position 2 in the circuit of Figure 16-13(b), the supply voltage to the *CR* circuit is zero. So, Equation 16-6 can be simplified by putting $E = 0$. This gives,

$$e_C = E_o \epsilon^{-t/(CR)} \quad \textbf{(16-11)}$$

Figure 16-13(c) shows a slightly different discharge circuit arrangement. When S_1 is in position 1, C_1 is charged from the battery via R_1, as before. When S_1 is switched to position 2, R_1 is taken out of the circuit and the capacitor is discharged via resistor R_2. By selecting R_2 larger than R_1 the discharge time for the capacitor may be made longer than the charge time. Alternatively, R_2 may be selected smaller than R_1 so the capacitor discharge time is shorter than the charging time. In this circumstance, the initial level of discharge current can be very large. For example, if e_C is 10 V and $R_2 = 0.1\ \Omega$, the initial level of discharge current is

$$i = \frac{10\ \text{V}}{0.1\ \Omega} = 100\ \text{A}$$

With the voltage source disconnected and the capacitor discharging via a resistor, the *RC* circuit is said to be operating *source-free*.

Example 16-8

A circuit connected as shown in Figure 16-13(c) has $E = 100$ V and $C_1 = 1\ \mu$F. Determine the value of R_1, if C_1 is to become charged to 50 V in 20 ms. Also, calculate the value of R_2 that will limit the maximum discharge current to 1 mA.

Solution

From Eq. 16-9,

$$R_1 = \frac{t}{C \ln[E/(E - e_C)]}$$

$$= \frac{20\ \text{ms}}{1\ \mu\text{F} \ln[100\ \text{V}/(100\ \text{V} - 50\ \text{V})]}$$

$$\approx 29.9\ \text{k}\Omega$$

The maximum level of e_C is,

$$e_C = E = 100\ \text{V}$$

The maximum discharge current is,

$$i = \frac{E}{R_2}$$

So,

$$R_2 = \frac{E}{i} = \frac{100\ \text{V}}{1\ \text{mA}}$$

$$= 100\ \text{k}\Omega$$

Practice Problem

16-7.1 The circuit in Figure 16-13(b) has $R_1 = 470\ \Omega$, $C_1 = 2.7\ \mu F$, and $E = 18$ V. Calculate the time for the capacitor to reach 80% of maximum charge, and determine the initial level of discharge current at switch-off.

16-8 *RC* CIRCUIT WAVEFORMS

As explained in Section 16-4, a *square wave* can theoretically be generated by opening and closing a switch connected between a circuit and the supply battery. Refer to the circuit of Figure 16-14(a), and assume that switch S_1 is held in position 1 for an accurately timed period t_1, and then moved to position 2 for

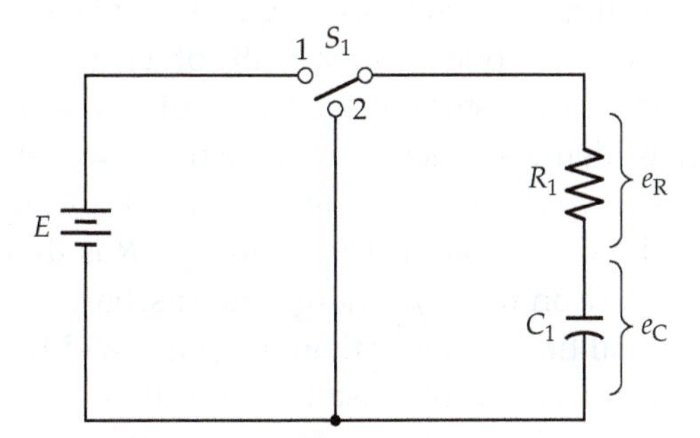

(a) Circuit for generating the waveforms shown below

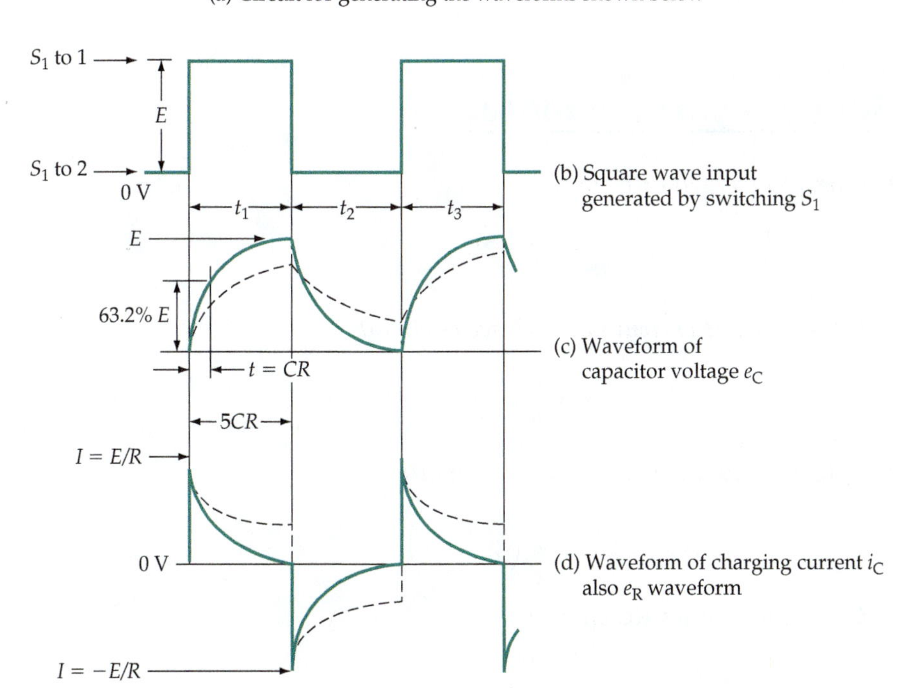

Figure 16-14 When a voltage is switched on to a series *CR* circuit, e_C increases exponentially toward the level of *E* over a time of 5*CR*, while the current at first jumps to $I = E/R$ then decreases exponentially to zero. When the voltage is switched off, e_C decreases exponentially from its maximum level, while *I* jumps to $-E/R$ then rises exponentially to zero.

an equal time t_2. During t_1 capacitor C_1 is charged from the battery via R_1, and during t_2 the capacitor discharges via R_1. The input waveform to the circuit is as shown in Figure 16-14(b).

When the switch is first set to position 1, the capacitor voltage starts to grow slowly from 0 V toward E, as illustrated in Figure 16-14(c). When e_C finally reaches E, it remains constant until S_1 is moved to position 2. Then e_C immediately begins to fall slowly toward zero, at which level it remains until the switch is moved over again. This growth and decline of e_C are repeated over and over as the switch is moved from one position to the other, resulting in the waveform shown.

When S_1 is switched to position 1 and e_C is zero, the charging current is a maximum of $I = E/R$. Then as the capacitor voltage grows, the current falls off to zero. This is illustrated in Figure 16-14(d). Similarly, when S_1 is moved to position 2, the current becomes a discharge current, and it has an initial maximum level of $I = -E/R$. So, as shown in Figure 16-14(d), the current is a negative quantity falling off to zero during t_2. Once again, the sequence is repeated as S_1 is moved from one position to the other, and the illustrated *spike waveform* is generated. The resistor voltage (e_R) is, of course, iR_1. Consequently, the waveform of e_R is identical in shape to the current waveform.

The waveforms shown in Figure 16-14 are those typically obtained when $t_1 = t_2 = 5CR$, [see Figure 16-14(c)]. The quantity CR is thought of as a *short time constant* by comparison with the charge and discharge times of the capacitor. When CR is several times greater than the pulse width, the circuit is said to have a *long time constant.* In this case the typical waveforms would be as shown by the broken lines in Figures 16-14(c) and (d).

Summary of Formulas

- ***Current in a series RL circuit:***

$$i = \frac{E - e_L}{R}$$

- ***Instantaneous current in an RL series circuit:***

$$i = \frac{E}{R}(1 - \epsilon^{-tR/L})$$

- ***R, L, and t values in an RL series circuit:***

$$t\frac{R}{L} = \ln\left(\frac{E}{E - iR}\right)$$

- ***Current in a series RC circuit:***

$$i = \frac{E - e_C}{R}$$

- ***Instantaneous voltage in a series RC circuit:***

$$e_C = E - (E - E_o)\epsilon^{-t/(CR)}$$

- ***When $E_o = 0$:***

$$e_C = E\left(1 - \epsilon^{-t/(CR)}\right)$$

- ***R, C, and t values in a series RC circuit:***

$$\frac{t}{CR} = \ln\left(\frac{E}{E - e_C}\right)$$

- ***Instantaneous discharging voltage in a series RC circuit:***

$$e_C = E_o\epsilon^{-t/(CR)}$$

Review Questions

Section 16-1

16-1 Sketch an approximate representation of the current growth in a series inductive-resistive circuit from the instant of supply switch-on. Carefully explain how the current and the counter-emf vary with time.

Section 16-2

16-2 Define the time constant of an *RL* circuit, and explain the special relationships between the time constant and the time at which the inductor's instantaneous current and voltage reach certain levels.

Section 16-3

16-3 Explain what occurs when the supply switch in a series *RL* circuit is opened. Discuss the danger to the circuit that occurs, and methods used to avoid damage.

Section 16-4

16-4 Sketch the voltage and current waveforms that result when a square wave is applied to a series *RL* circuit: (a) For a circuit with a long time constant; (b) for a circuit with a short time constant. Explain briefly.

Section 16-5

16-5 Sketch an approximate representation of the voltage growth across a capacitor being charged via a series resistor. Carefully explain how the charging current and voltage vary with time.

Section 16-6

16-6 Define the time constant of an *RC* circuit, and explain the special relationship between the time constant and the time at which the capacitor's instantaneous voltage and charging current reach certain levels.

Section 16-7

16-7 Discuss what occurs when the supply to an *RC* circuit is interrupted, and when a fully charged capacitor is short-circuited. Explain any danger that results and the methods used to avoid damage.

Section 16-8

16-8 Sketch the voltage and current waveforms that result when a square wave is applied to a series *CR* circuit: (a) For a circuit with long time constant; (b) for a circuit with short time constant. Explain briefly.

Problems

Section 16-2

16-1 A circuit consisting of a 100 mH inductor in series with a 500 Ω resistor has a 50 V supply. Calculate the instantaneous levels of circuit current at 0.2 ms intervals from switch-on to $t = 1$ ms, and plot the graph of i versus t.

16-2 A 500 mH inductor with a coil resistance of 280 Ω is to be connected to a 12 V supply via a resistor. The inductor current is to be approximately 63% of its maximum level 0.39 ms after switch-on, (see Figure 16-15). Select a suitable resistor value, and determine the time taken for the current to reach its approximate maximum level.

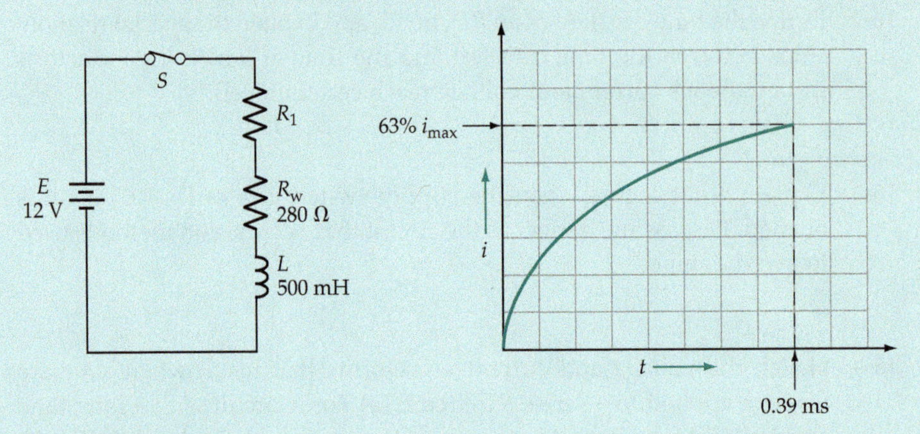

Figure 16-15

16-3 For the circuit described in Problem 16-1, calculate the instantaneous levels of inductor voltage at 0.25 ms intervals, and plot the graph of e_L versus t.

16-4 For the circuit in Problem 16-2, determine the inductor current and voltage levels at 0.1 ms, 0.2 ms, and 0.3 ms from the instant of supply switch-on.

16-5 A series *RL* circuit with a 35 V supply is to have an instantaneous current of 5 mA at a time of 22 ms from switch-on. If $R = 4.7$ kΩ, determine the required value of L.

16-6 For the circuit in Problem 16-5, calculate the approximate times from switch-on for the inductor voltage to reach (a) 13 V, and (b) zero.

16-7 A 600 mH inductor with a winding resistance of 250 Ω is connected in series with a 330 Ω resistor and a 24 V supply, as in Figure 16-16. Determine the times from the instant the supply is switched on until the inductor voltage reaches (a) 5 V, and (b) 17 V.

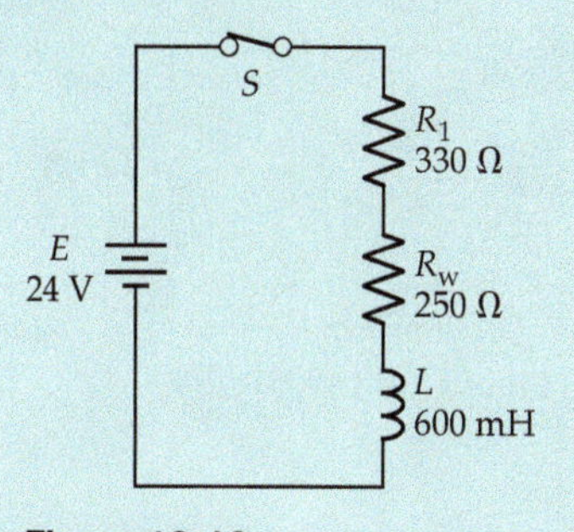

Figure 16-16

16-8 An inductor is included in series with an electric lamp filament to limit the current growth at switch-on. The filament has a resistance of 1 Ω, and the current is to be approximately 30% of maximum 20 ms after switch-on. Determine a suitable inductance value.

16-9 A 24 V supply is connected to an inductor via a 1.2 kΩ resistor. A time of 270 μs is to occur from the instant of switch-on until the resistor voltage reaches 5 V. Determine the required value of inductance.

16-10 For the circuit described in Problem 16-9, calculate the times from switch-on for the resistor voltage to reach: 3 V, 6 V, 12 V, and approximately 24 V.

Section 16-3

16-11 For the series circuit described in Problem 16-7, show how an additional resistor should be connected to limit the counter-emf when the supply is switched off. Determine the required resistor value to limit the counter-emf to a maximum of 75 V.

16-12 Calculate the counter-emf generated in the inductor in Problem 16-5 when the supply is replaced with a short circuit. Determine the possible counter-emf generated when the supply is switched off, if the open-circuited switch resistance is 10 MΩ.

16-13 A 500 mH inductor is connected to a 30 V supply via a 12 kΩ resistor, as in Figure 16-17. If the switch is open-circuited, and the resistance between its terminals is 5 MΩ, calculate the counter-emf generated by the inductor. Also, determine a suitable value of resistor to be connected in parallel with the inductor to limit its counter-emf to a maximum of 60 V.

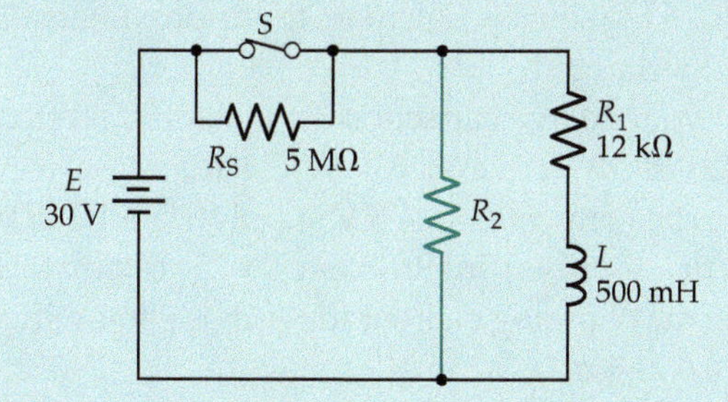

Figure 16-17

16-14 The circuit in Figure 16-7 has $R_1 = 1.5$ kΩ, $R_2 = 2.7$ kΩ, $L = 300$ mH, and $E = 15$ V. Calculate the time required from switch-on for the inductor voltage to fall to 3 V. Also determine the counter-emf generated at switch-off.

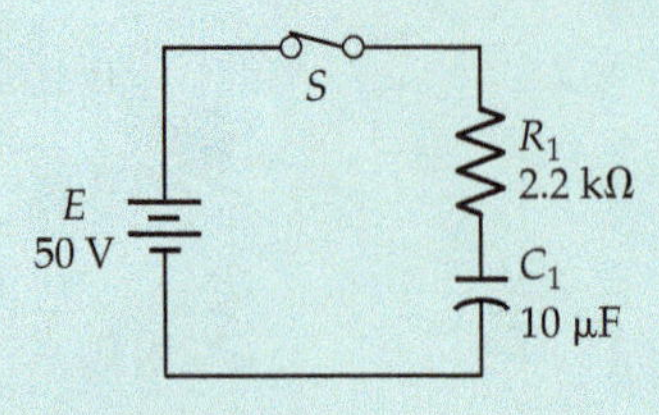

Figure 16-18

Section 16-6

16-15 A series circuit consisting of a 10 μF capacitor and a 2.2 kΩ resistor is connected to a 50 V supply, (see Figure 16-18). Calculate the instantaneous levels of capacitor voltage at 15 ms intervals from supply switch-on up to $t = 105$ ms. Plot the graph of e_C versus t.

16-16 A 200 μF capacitor is to be connected to a 15 V supply via a resistor. The capacitor voltage is to be approximately 63% of maximum 0.78 s after switch-on. Select a suitable resistor value and calculate the approximate time for the capacitor to become fully charged.

16-17 For the circuit described in Problem 16-15, calculate the instantaneous charging current levels at 20 ms intervals from switch-on, and plot the graph of i versus t.

16-18 Calculate the charging current and capacitor voltage for the circuit in Problem 16-16 at time intervals of 0.2 s, 0.4 s, 0.8 s, and 1.6 s from switch-on.

16-19 A series *CR* circuit with a 30 V supply is to have an instantaneous capacitor voltage of 22 V at a time of 65 μs from supply switch-on. If $C = 0.1$ μF, determine the required resistance of R.

16-20 For the circuit described in Problem 16-19, calculate the times from switch-on for the charging current to reach (a) 30 mA, and (b) approximately zero.

16-21 A 47 μF capacitor is connected in series with a 4.7 kΩ resistor and a 5 V supply. Determine how long it takes from the instant of switch-on for the capacitor voltage to reach 2.2 V.

16-22 A printed circuit board has a 150 pF capacitance between conductors and ground. If a dc voltage is applied to the circuit board conductors via a 33 kΩ resistance, how long will it take for the conductor voltage to reach 90% of maximum?

16-23 A 1500 μF capacitor is charged from a 200 V supply and then connected in parallel with an uncharged 1200 μF capacitor. Determine the terminal voltage of the two capacitors in parallel.

16-24 A 120 km cable with a capacitance of 50 pF per meter has a dc voltage applied from a 6 kΩ source. Calculate the approximate time for the cable to become fully charged.

16-25 The terminal voltage of a capacitor is to take 750 μs from the instant of supply switch-on to arrive at 3.3 V. Calculate the required capacitor value if it is to be connected to a 5 V supply via a 1.5 kΩ resistor.

16-26 For the circuit described in Problem 16-25, calculate the times from switch-on for the charging current to reach the folowing levels: 200 μA, 400 μA, 800 μA, 1.6 mA, 3.2 mA.

Section 16-7

16-27 The 10 km cable described in Problem 15-11 has an insulation resistance of 10 MΩ between the two conductors. Calculate the time required for the potential difference between the conductors to fall from 500 V to 10 V when the supply is disconnected.

16-28 Calculate the time required for the voltage of the fully charged capacitor in Problem 16-21 to fall to 3 V when the supply is replaced with a short-circuit.

16-29 Sketch the series circuit described in Problem 16-21, and show how an additional resistor and switch should be included for safely discharging the capacitor. Determine the value of the additional resistor if the capacitor discharge current is not to exceed 1 mA. Also, calculate approximately how long it will take for the capacitor to become completely discharged.

16-30 Determine the time required for a 100 μF capacitor to discharge from 120 V to 20 V if the discharge occurs through a 100 MΩ insulation resistance, as in Figure 16-19. Calculate the value of a resistor that will safely discharge the capacitor with a maximum current of 10 mA, and determine the new time required for discharge from 120 V to 20 V.

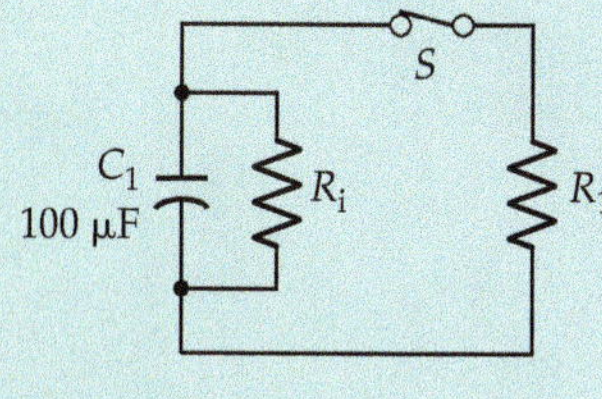

Figure 16-19

Practice Problem Answers

16-2.1 2.3 mA, 3.15 mA, 3.46 mA, 3.57 mA, 3.62 mA
16-2.2 186 mH
16-3.1 1.65 kΩ
16-6.1 1.34 mA, 0.49 mA, 181 μA, 66.7 μA, 25 μA
16-6.2 227 μA, 2.35 μs
16-7.1 2.04 ms, 38.3 mA

CHAPTER 17
Alternating Current and Voltage

CONTENTS

Objectives

You will be able to:

1. Draw a sketch of a simple alternating current generator, explain how the generator produces an ac output, and sketch the output waveform.
2. Calculate the instantaneous output voltage from an ac generator at any given time.
3. For a given sinusoidal waveform, determine peak value, rms value, average value, and frequency.
4. Calculate the time period and wavelength of a given sine wave.
5. Determine the phase relationship between two sine waves.
6. Calculate the instantaneous, peak, average, and rms levels of current in a resistor when a sinusoidal voltage is applied. Also, determine the power dissipated in the resistor.
7. Identify the most important controls on the front panel of analog and digital oscilloscopes and explain the function of each control.
8. Use oscilloscopes to measure voltage, frequency, and phase difference.
9. Sketch and explain the circuits of ac voltmeters using rectifier circuits, and solve problems involving rectifier voltmeter circuits.
10. Sketch and explain the basic circuits of analog and digital electronic voltmeters.
11. Discuss the range of operating frequency of typical ac voltmeters.

INTRODUCTION

Alternating current (ac) is current that is continuously reversing direction; alternately flowing in one direction and then in the other. Alternating voltage can be similarly described. Usually ac voltages and currents are sinusoidal, and definite relationships exist among the peak, average, and effective values. The effective value, or rms value, of an ac quantity is the normally quoted value and is the value normally applied in Ohm's law and power calculations. The frequency and, phase angle of an ac quantity can also be important, and the wavelength can be calculated from the frequency and the velocity of the wave. The oscilloscope is the instrument used for the study of alternating waveforms.

17-1 GENERATION OF ALTERNATING VOLTAGE

Electromagnetic induction has been discussed in Sections 11-2 and 14-1. In Section 14-1 it was shown how an emf is induced in a conductor in motion through a magnetic field. Now consider two series-connected conductors rotating in a magnetic field, as illustrated in Figure 17-1. Conductors 1 and 2 form a loop situated within the field set up between the two magnetic poles. The conductors are mechanically fastened to an axle equipped with a crank for hand turning. Each conductor is connected to a *slip ring,* and the slip rings have *brushes* for electrically connecting to the external terminals (A and B). When the loop is rotated, the conductors cut (or *brush through*) the magnetic flux, causing an emf to be generated in each conductor.

Figure 17-2 illustrates the conditions that exist at each of several instants while the loop is being rotated in the magnetic field. These can be explained as follows.

- **Figure 17-2(a):** The loop is horizontal, and conductor 1 is moving up while conductor 2 is moving down. Because the motion of each conductor is parallel to the direction of the magnetic field at this instant, no flux is being cut by the conductors and so no emf is generated. As illustrated by point 'a' on the graph of voltage (e) versus time (t), the voltage at the output terminals (A and B in Figure 17-1) is zero.
- **Figure 17-2(b):** The conductors are now moving at an angle across the field, so each is cutting some flux, and an emf is induced in each conductor. According to Lenz's law, the direction of the induced emf must be such that it opposes the motion that causes the emf. So, the emf induced in conductor 1 causes a current to flow toward the viewer, as illustrated. This current produces a counterclockwise magnetic flux around the conductor, which strengthens the magnetic field in front of the conductor and weakens it behind (i.e., it opposes the motion of the conductor). Similarly, the induced

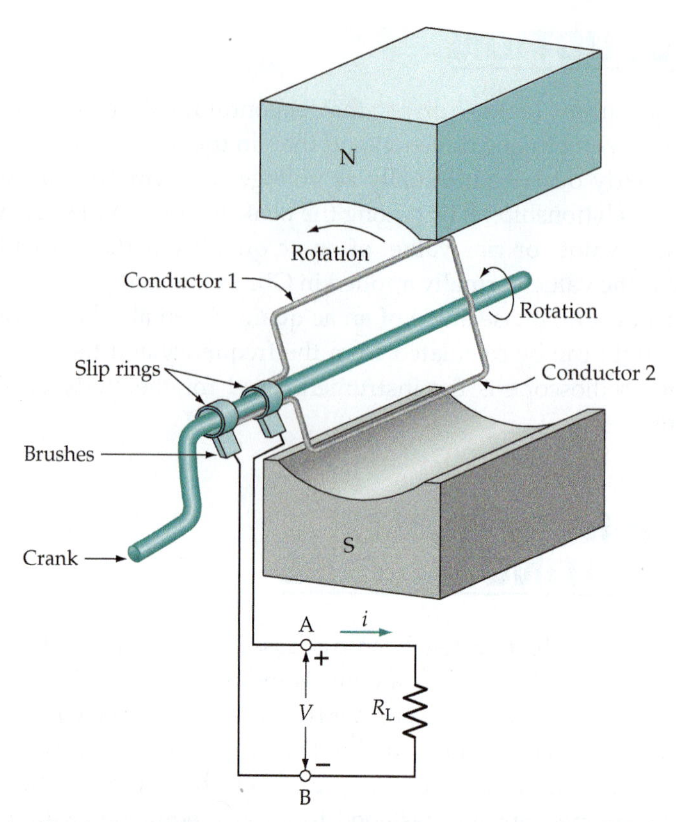

Figure 17-1 Simple alternating voltage generator consisting of a conducting loop rotated in a magnetic field. An emf is induced in each conductor (and produced at the output terminals) as the conductors are swept past the magnetic poles. The emf reverses polarity when the conductors move from one pole to the other.

emf in conductor 2 causes a current in the direction shown, away from the viewer. A clockwise flux is set up around conductor 2, and this again weakens the flux behind conductor 2 and strengthens it ahead of the conductor, again opposing the motion of the conductor. It is seen that the current directions in the two conductors are such that they assist each other [i.e., the current is flowing around the loop (in Figure 17-1) in a single direction]. Consequently, the output voltage is positive at terminal A and negative at terminal B. (Conductor 1 is connected to A, and conductor 2 is connected to B.) A current flows in the direction shown (from A to B) through the external load R_L. When the growth of the terminal voltage e is plotted versus time t, the result is as illustrated by the graph (or waveform) in Figure 17-2(i). Point 'a' represents the zero output voltage obtained for the conditions in Figure 17-2(a), and point 'b' is the output voltage some time later when the conductor positions are as shown in Figure 17-2(b).

- **Figure 17-2(c):** The conductors are now moving perpendicular to the magnetic field. So, they are cutting maximum flux and generating maximum emf. Again, the current directions assist each other, so that current flows around the loop, and terminal A remains positive while terminal B is negative. The

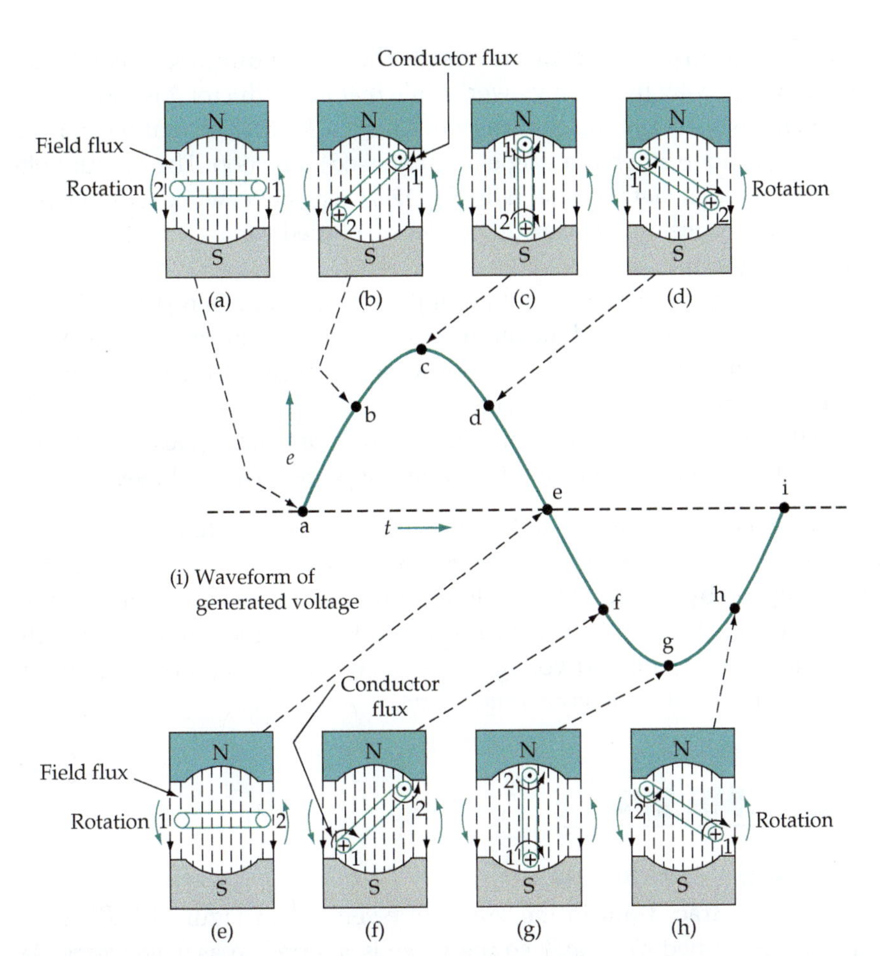

Figure 17-2 A sinusoidal voltage waveform is generated at the terminals of a conducting loop rotated in a magnetic field. Peak output voltage is produced when the conductors are moving perpendicular to the field [see (c) and (g)]. Zero output is generated at the instants when the conductors are moving parallel to the field [see (a) and (e)].

output current through R_L is now greater than before, and as shown on the graph of *e*-versus-*t*, the output voltage has reached a peak value, point 'c'.

- **Figure 17-2(d):** Now the conductors are again moving at an angle less than 90° with respect to the direction of the magnetic field. So, the amount of flux being cut is reduced, and consequently the generated emf is less than for the conditions in Figure 17-2(c). The current direction and output voltage polarity are the same as before. However, the output voltage has now dropped below the peak level, as shown at point 'd' on the *e*-versus-*t* graph.
- **Figure 17-2(e):** At this time the conductors are again moving parallel to the direction of the magnetic field. No flux is being cut, so no emf is generated. The output voltage is zero and is plotted at point 'e' on the voltage-versus-time graph.
- **Figure 17-2(f):** The conductors are once more moving at an angle with respect to the direction of the magnetic field. However, both conductors now have

the induced current directions reversed. The current direction in conductor 1 is flowing away from the viewer, while that in conductor 2 is toward the viewer. It is seen that the currents still assist each other, so that the current around the loop still has a single (but reversed) direction. The output voltage is now such that terminal B (in Fig. 17-1) is positive, while terminal A is negative. The instantaneous output voltage plotted versus time gives point 'f' on the graph.

- **Figure 17-2(g):** The motion of the conductors is once again perpendicular to the magnetic field, and maximum flux is being cut. So, maximum emf is again generated. The output voltage is still a negative quantity and has reached its peak value, plotted at point 'g' on the *e*-versus-*t* graph. It is easily seen that as the loop rotation continues the output voltage goes to zero once more, then reverses again, and the cycle is repeated over and over.

Because the voltage generated by the rotating loop is alternately positive and negative, it is referred to as an *alternating voltage.* The current produced in a load supplied by an alternating voltage flows first in one direction and then in the other. So, it is an *alternating current (ac).* The designation *ac* is normally applied to both current and voltage. The voltage/time graph constructed in Figure 17-2(i) is spoken of as an *ac waveform.*

17-2 SINE WAVE

Sinusoidal Waveform

The waveform traced out in the graph of *e*-versus-*t* in Figure 17-2(i) has a shape that is termed *sinusoidal;* so the wave is referred to as a *sine wave.* As shown below, these terms arise because the instantaneous voltage is found to depend on the trigonometrical sine of an angle.

In order to derive an equation for the instantaneous emf induced in the rotating loop, recall from Section 14-2 that the induced emf depends on the amount of flux cut per second. From Equation 14-1, the emf induced in a single conductor is,

$$e = \frac{\Delta\Phi}{\Delta t}$$

where $\Delta\Phi$ is the total flux cut by the conductor during time Δt. Because $\Phi = B \times A$, the equation for e can be rewritten as,

$$e = B \times \frac{\Delta A}{\Delta t}$$

where B is the flux density between the magnetic poles, and ΔA is the area swept by one conductor during time Δt.

Consider a conductor moving past a pole, as illustrated in Figure 17-3. If the length of each conductor within the magnetic field is $\ell/2$, and if the conductors

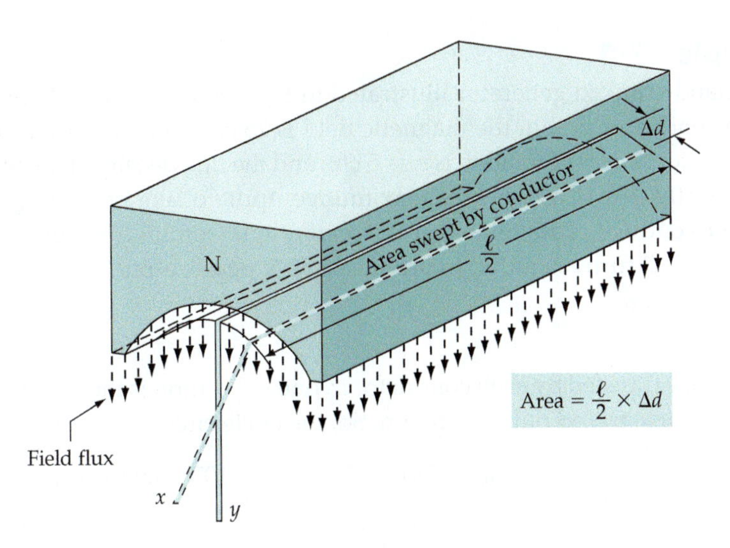

Figure 17-3 A conductor with a length of $\ell/2$ moving through a distance Δd sweeps an area of flux of $\ell/2 \times \Delta d$.

move through a distance Δd during time Δt, then the area swept by the conductor during Δt is,

$$\text{Area} = \frac{\ell}{2} \times \Delta d$$

For two conductors, the total area swept is,

$$\Delta A = 2\left(\frac{\ell}{2} \times \Delta d\right)$$

or,

$$\Delta A = \ell \times \Delta d$$

and the equation for emf induced in the loop becomes,

$$e = B\ell \frac{\Delta d}{\Delta t} \qquad \textbf{(17-1)}$$

Because maximum flux is being cut at the instant shown in Figure 17-3, the peak or maximum value of emf (E_m) is being generated. Also, because $\Delta d/\Delta t$ is in meters/second (i.e., a velocity v), the equation can be written as,

$$E_m = B\ell v \qquad \textbf{(17-2)}$$

where E_m is in volts, B is in teslas, ℓ is in meters, and v is in m/s. In this case v is the actual linear velocity of the conductors. It can also be described as the *peripheral velocity* with respect to the circle traced out by each conductor. It is important to note that Equation 17-2 is an expression for the maximum or *peak value* (E_m) of the voltage generated, (i.e., the voltage represented by point 'c' on the *e/t* graph in Figure 17-2).

Example 17-1

In the hand-cranked generator illustrated in Figures 17-1 and 17-2, the length of each conductor within the magnetic field is 25 cm. The distance from each conductor to the axis of rotation is $r = 5$ cm, and the flux density of the magnetic field is $B = 0.1$ T. (a) Calculate the maximum output voltage from the generator when the conductors are rotated at 100 revolutions/minute, (b) Determine the number of coil turns required to produce a 50 V peak output.

Solution

(a) Distance traveled by each conductor in one revolution is the circumference of the circle traced out by rotation. So, the circle circumference is,

$$d = 2\pi r = 2\pi \times 5 \times 10^{-2} \text{ metres (m)}$$

$$= 0.1\pi \text{ m}$$

$$\text{distance traveled per minute} = d \times 100 \text{ rpm} = 0.1\pi \text{ m} \times 100 \text{ rpm}$$

$$= 10\pi \text{ m}$$

So, the conductor peripheral velocity is,

$$v = \frac{10\pi}{60} \text{ meters/sec (m/s)}$$

$$\approx 0.524 \text{ m/s}$$

From Equation 17-2, the peak *emf* generated in the loop is,

$$E_m = B\ell v$$

where $\ell = 2 \times$ (length of each conductor). So,

$$E_m \approx 0.1 \text{ T} \times 2 \times 25 \times 10^{-2} \times 0.524 \text{ m/s}$$

$$\approx 26.2 \text{ mV}$$

(b) Number of coil turns,

$$N = \frac{E_{peak}}{E_m \text{ for one turn}} = \frac{50 \text{ V}}{26.2 \text{ mV}}$$

$$= 1908$$

Instantaneous Voltage

Now consider Figure 17-4, which illustrates the movement of conductor 1 from its position in Figure 17-2(a) toward that in Figure 17-2(b). It is seen that as the conductor moves through a distance Δd, the distance that it moves directly across the flux is $\Delta d'$, and

$$\Delta d' = \Delta d \sin \alpha$$

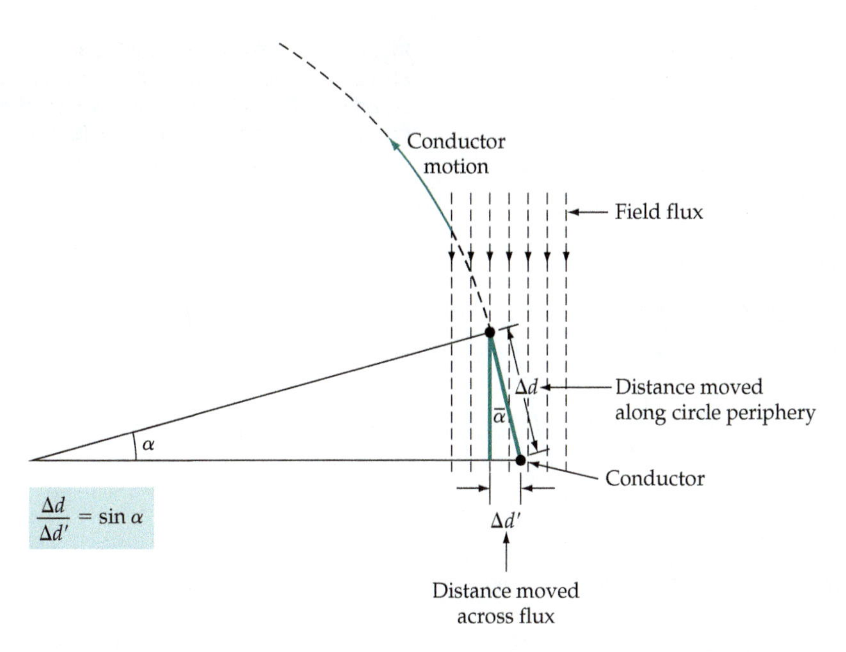

Figure 17-4 A conductor moving through a distance Δd at an angle α with respect to a magnetic field, moves $\Delta d'$ across the field, where $\Delta d' = \Delta d \sin \alpha$. The emf generated in the conductor is directly proportional to $\sin \alpha$.

where α is the angle moved through by the conductor in circular motion. Substituting in Equation 17-1, the instantaneous emf is,

$$e = B\ell \frac{\Delta d'}{\Delta t}$$

$$= B\ell \frac{\Delta d \sin \alpha}{\Delta t}$$

giving,

$$e = B\ell v \sin \alpha \tag{17-3}$$

$B\ell v$ is the expression obtained in Equation 17-2 for the peak or maximum value of the voltage generated (E_m). So,

$$e = E_m \sin \alpha \tag{17-4}$$

In Equation 17-4, α is usually termed the phase angle of the instantaneous voltage (e). E_m is, of course, the peak value, but it is also referred to as the *amplitude* of the waveform.

It is seen that [as shown in Figure 17-2(i)] the generated voltage has a sinusoidal waveform. This is further illustrated in Figure 17-5, which shows that one complete waveform is produced when the conductor loop rotates through 360°, or one complete revolution. The instantaneous voltage level at any angle a can be determined from Equation 17-4.

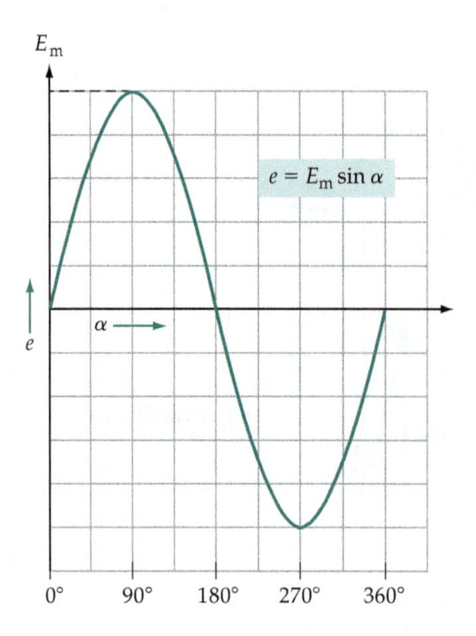

Figure 17-5 The output voltage from a simple generator (as in Figure 17-1) is a sinusoidal waveform. The voltage (e) at any instant can be calculated from the peak value (E_m) and the sin of the angle (α).

Example 17-2

For the generator described in Example 17-1, calculate the instantaneous output voltage at points 'b', 'c', 'd', and 'f' on the graph of e/t in Figure 17-2. Assume that the phase angles of the conductors at these points with respect to the flux direction are 45°, 90°, 135°, and 225°, respectively.

Solution

From Equation 17-4, the instantaneous emf generated is,

$$e = E_m \sin \alpha$$

At point 'b' $\quad e = 26.2 \text{ mV} \sin 45° \approx 18.5 \text{ mV}$

At point 'c' $\quad e = 26.2 \text{ mV} \sin 90° = 26.2 \text{ mV}$

At point 'd' $\quad e = 26.2 \text{ mV} \sin 135° \approx 18.5 \text{ mV}$

At point 'f' $\quad e = 26.2 \text{ mV} \sin 225° \approx -18.5 \text{ mV}$

Practice Problems

17-2.1 A single-loop generator as in Figure 17-1 is to produce a peak output of 100 mV. The conductor dimensions are $\ell = 6$ cm and $r = 2$ cm, and the magnetic flux density is 0.5 T. Determine the required rate of rotation.

17-2.2 For the generator in Problem 17-2.1, calculate the angle of the conductors with respect to the field flux for instantaneous output voltages of 66 mV, 70.7 mV, and 85 mV.

17-3 FREQUENCY, PHASE ANGLE, AND WAVELENGTH

Frequency

Consider Figure 17-6, which once again illustrates the sine-wave output from an *ac* generator. It has been shown that when the conductors in the simple generator are rotated through one complete revolution, the output voltage goes through one complete cycle of sinusoidal change. The time taken for one cycle of change (from zero volts through $+E_m$, zero, and $-E_m$ back to zero again) is referred to as the *time period* of the waveform and is designated T (see Figure 17-6). If $T = 1$ s, the waveform has *a frequency* (f) of 1 *cycle per second*, or 1 *hertz* (Hz).

A waveform has a frequency of *1 hertz*[1] *(Hz)* when it goes through one complete cycle of change in a period of 1 second.

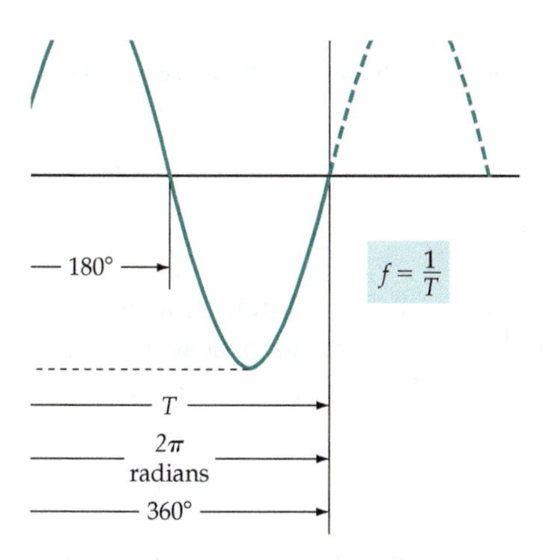

Figure 17-6 One cycle of a sinusoidal voltage waveform. The instantaneous voltage (e) increases from zero to $+E_m$, then decreases through zero to $-E_m$, and returns again to zero.

When two complete cycles of the waveform occur in one second, the frequency is two cycles per second, or 2 Hz. In this case, the time period of one cycle is $T = 0.5$ s, and,

$$f = 1/0.5\ \text{s} = 2\ \text{Hz} \qquad \text{[see Figure 17-7(a)]}$$

Similarly, if the waveform has a time period of 2 s, then its frequency is one cycle per 2 s. So,

$$f = 1/2\ \text{s} = 0.5\ \text{Hz} \qquad \text{[see Figure 17-7(b)]}$$

[1]Named for the German physicist Heinrich R. Hertz (1857–1894).

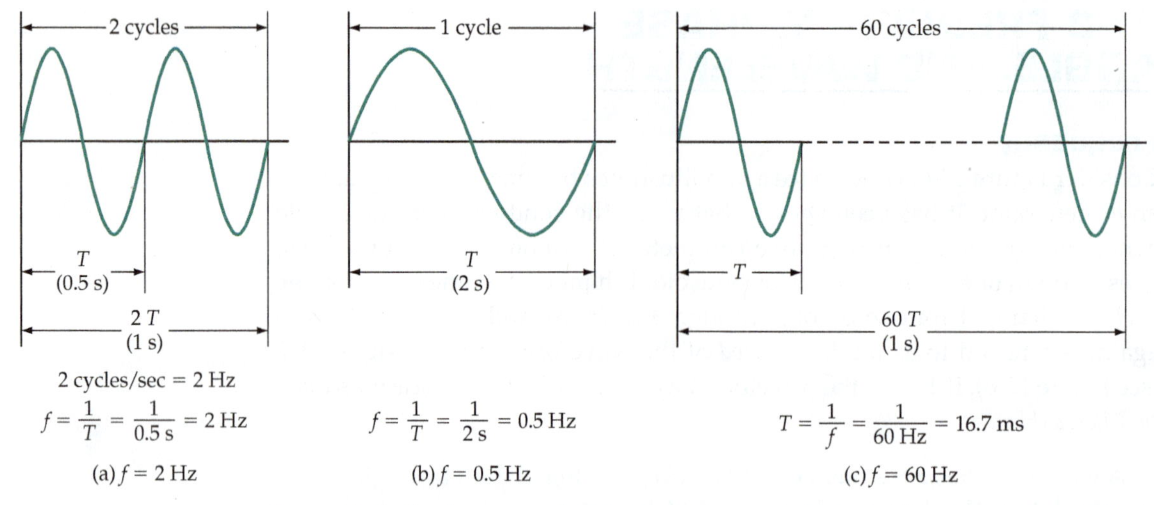

Figure 17-7 The frequency of a waveform is measured in *hertz* (*Hz*), or *cycles per second*.

The frequency of the waveform is always found by inverting the time period.

$$f = \frac{1}{T} \tag{17-5}$$

where f is in hertz and T is in seconds.

The usual (North American) frequency for domestic and industrial ac supplies is 60 Hz [see Figure 17-7(c)]. In electronics, the frequencies of ac waveforms can range from zero (i.e., dc) through kilohertz (kHz), megahertz (MHz), and gigahertz (GHz).

Phase Angle

For one complete revolution, the conductors in the simple generator obviously rotate through 360°. If the angle of rotation of the conductors is measured in *radians* instead of degrees, one complete revolution represents 2π radians. As T is the time period in seconds (or time for one complete revolution), the angle through which the conductors move in 1 s is given by,

$$\text{Angular velocity} = \frac{1}{T}\ \text{revolutions/second}$$

$$= \frac{360}{T}\ \text{degrees/second}$$

or,

$$\text{Angular velocity} = \frac{2\pi}{T}\ \text{radians/second}$$

and the phase angle (θ) (or α in Eq. 17-3 and 17-4) of the generated voltage at any instant $t_1, t_2, t_3, \cdots$ measured from $t = 0$ is,

$$\theta = \frac{2\pi}{T} \times t\ \text{radians}$$

Also, $$\theta = \frac{360}{T} \times t \text{ degrees}$$

Because $f = 1/T$, the phase angle equations can be rewritten,

$$\theta = 360\,ft \text{ degrees}$$

and, $$\theta = 2\pi ft \text{ radians}$$

or, $$\theta = \omega t \text{ radians}$$

where ω is the angular velocity in radians/s (i.e., $\omega = 2\pi f$).

Equation 17-4 can now be written in the form normally employed to represent ac waveforms:

$$e = E_m \sin \omega t \tag{17-6}$$

also, $$e = E_m \sin (360\,ft)^\circ$$

Figure 17-8 illustrates the use of Equation 17-6.

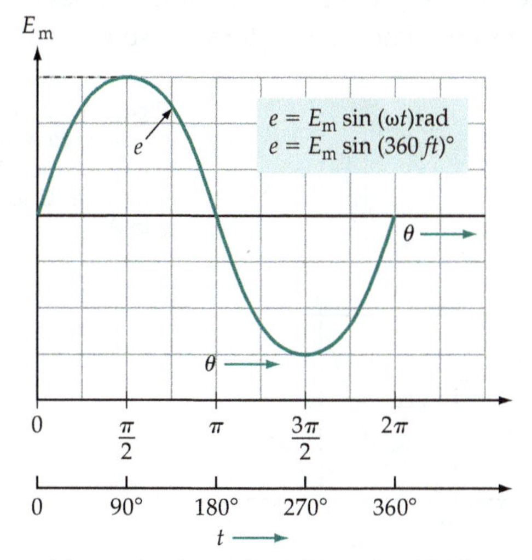

(a) e can be determined from ωt or $360\,ft$

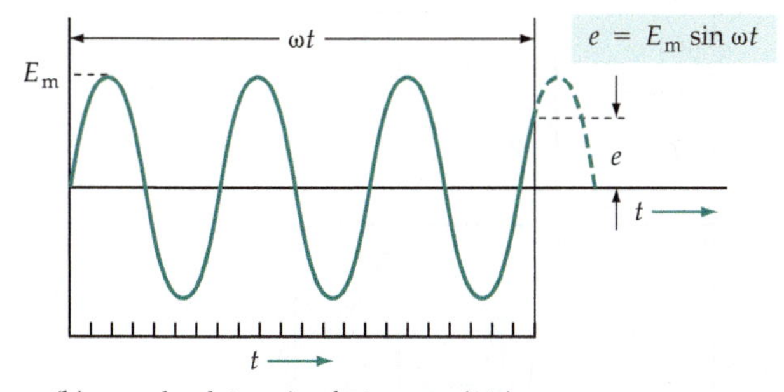

(b) e can be determined at any point in a waveform containing many cycles

Figure 17-8 The phase angle of a waveform can be expressed in degrees or radians, and either can be used in the calculation of instantaneous level.

Example 17-3

An ac waveform with a frequency of 1.5 kHz has a peak value of 3.3 V. Calculate the instantaneous levels of voltage at $t_1 = 0.65\ \mu s$ and at $t_2 = 1.2$ ms.

Solution

Eq. 17-6, $e = E_m \sin(2\pi ft)$

At t_1, $e = 3.3\text{ V} \sin(2\pi \times 1.5\text{ kHz} \times 0.65\ \mu\text{s})$ rad

≈ 20.2 mV

At t_2, $e = 3.3\text{ V} \sin(2\pi \times 1.5\text{ kHz} \times 1.2\text{ ms})$ rad

≈ -3.1 V

Phase Difference

Four different waveforms are illustrated in Figure 17-9. Waveforms B, C, and D have smaller amplitudes than waveform A. Waveform B goes through its cycle of change exactly at the same time as waveform A, so the two are said to be

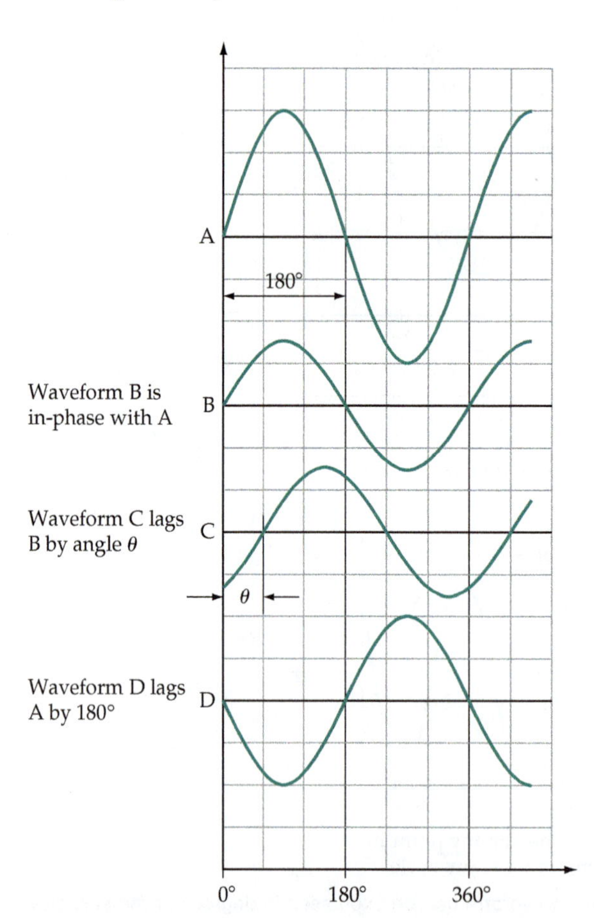

Figure 17-9 In-phase and out-of-phase sine waves. Waveform B is in phase with waveform A. Waveform C lags waveforms A and B.

in phase with each other. Waveforms C and D are *out of phase* with waveforms A and B. Because waveforms A and B commence their cycles ahead of C by an angle θ (degrees or radians), A and B are said to *lead C by* θ. Alternatively, waveform C may be said to *lag* waveforms A and B by θ. The angle θ may also be referred to as the *phase difference* between the waveforms. Waveform D lags A and B by 180° and is said to be in *antiphase* to A and B because it is at its negative peak when A and B are at their positive peak levels, and vice versa.

Wavelength

The *wavelength* of an ac waveform depends on its velocity. In the case of radio waves, the velocity is the speed of light, which is,

$$c = 3 \times 10^8 \text{ m/s}$$

For voltage waves moving along widely spaced conductors, the transmission speed is also 3×10^8 m/s. When the conductors are close together, the velocity depends on the insulation employed. Figure 17-10(a) shows a pair of long conductors connected to a generator, and parts (b), (c), (d), and (e) show the instantaneous voltage levels at points on the conductor as the terminal voltage changes. In Figure 17-10(b), e_1, represents the initial or zero level of a voltage waveform at the beginning of a long conductor. A fraction of a second later, in Figure 17-10(c), e_1 has traveled some distance along the conductor, and the instantaneous level of the voltage at the generator terminals has changed to e_2. Both e_1 and e_2 are traveling along the conductor at a velocity of 3×10^8 m/s. In Figure 17-10(d) the voltage at the generator terminals has changed to e_3. By the time the generator terminal voltage has gone through one complete cycle [Figure 17-10(e)], e_1 has been moving along the conductor for the time period T. So, during one complete cycle, the distance travelled by e is,

$$\begin{aligned} \text{distance} &= c \times T \text{ meters} \\ &= (3 \times 10^8 \text{ m/s} \times T) \text{ m} \end{aligned}$$

This distance represents the *wavelength* (λ) of the alternating voltage. The equation for wavelength is,

$$\lambda = c \times T$$

or,

$$\lambda = \frac{c}{f} \qquad (17\text{-}7)$$

where λ is wavelength in meters, c is velocity of light in m/s, and f is the frequency of the waveform in hertz.

Example 17-4

Determine the wavelengths of 60 Hz and 1 MHz alternating waves. Calculate the frequency of a voltage that has a wavelength of 33 m.

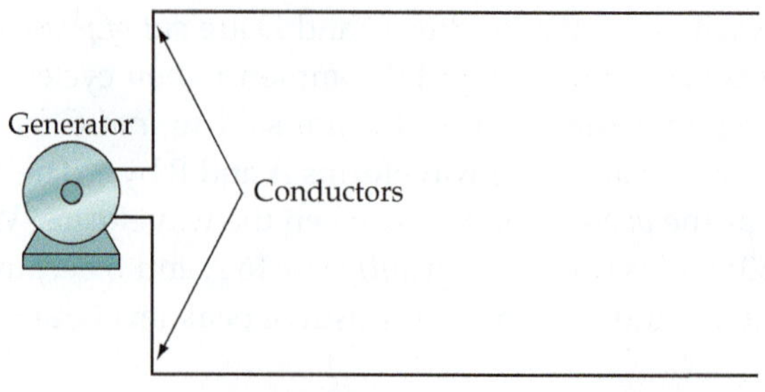

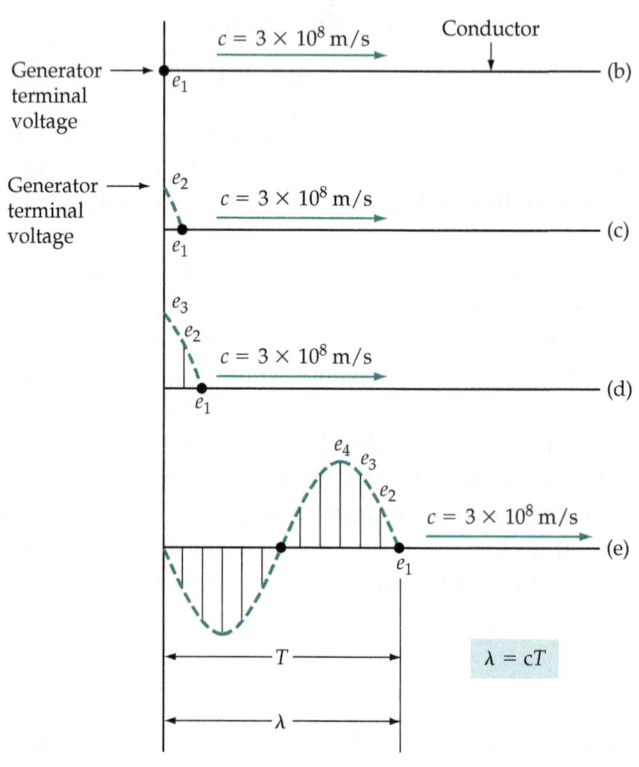

Figure 17-10 An alternating voltage wave moves along widely spaced conductors at the speed of light, ($c = 3 \times 10^8$ m/s). The time period of one cycle is $T = 1/f$, and the wavelength is $\lambda = c/f$.

Solution

For 60 Hz,

Eq. 17-7,
$$\lambda = \frac{c}{f} = \frac{3 \times 10^8 \text{ m/s}}{60 \text{ Hz}}$$
$$= 5 \times 10^6 \text{ m}$$

For 1 MHz,
$$\lambda = \frac{c}{f} = \frac{3 \times 10^8 \text{ m/s}}{1 \text{ MHz}}$$
$$= 300 \text{ m}$$

From Eq. 17-7,

$$f = \frac{c}{\lambda} = \frac{3 \times 10^8 \text{ m/s}}{33 \text{ m}}$$

$$= 9.09 \text{ MHz}$$

Practice Problems

17-3.1 A voltage waveform has a peak value of 60 V and a frequency of 50 kHz. Calculate the instantaneous amplitudes at $t_1 = 3$ μs and $t_2 = 17$ μs.

17-3.2 Calculate the wavelength of the 740 kHz and 1200 kHz radio broadcast frequencies.

17-4 RESISTIVE LOAD WITH AC SUPPLY

Current Level

Figure 17-11(a) shows a resistor (R) connected to an ac voltage source (e). Note the graphic symbol for the ac voltage source. Unlike the case of a dc circuit, a continuous current direction cannot be shown, because the current through R will change direction each time the polarity of e reverses. However, it is usually convenient to show an instantaneous current direction (i). The instantaneous current level can be found by Ohm's law:

$$i = \frac{e}{R}$$

or,

$$i = \frac{E_m}{R} \sin \omega t \tag{17-8}$$

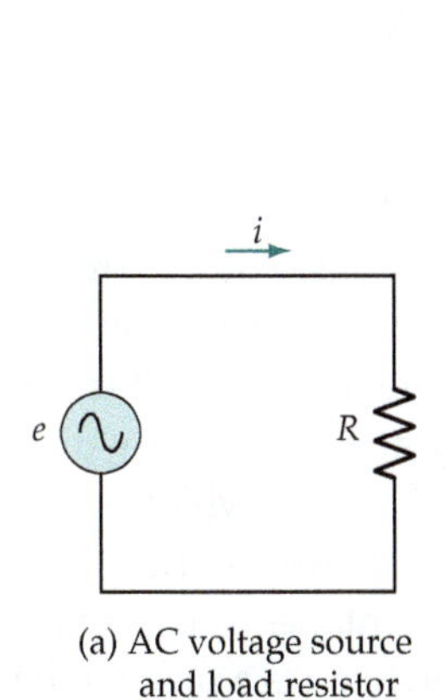

(a) AC voltage source and load resistor

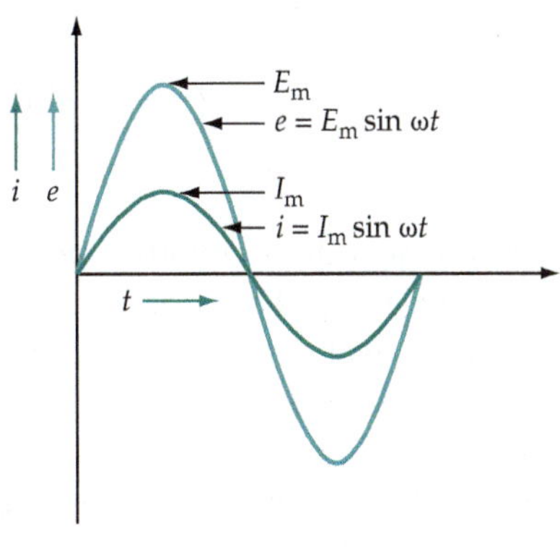

(b) Voltage and current waveforms

Figure 17-11 The alternating current in a purely resistive circuit is in phase with the alternating voltage. As the voltage increases or decreases, the current also increases or decreases.

The peak value of the current occurs at the peak value of the instantaneous voltage (i.e., when $e = E_m$). So, the maximum or peak current level is,

$$I_m = \frac{E_m}{R}$$

and,

$$i = I_m \sin \omega t \tag{17-9}$$

Because e is a sinusoidal quantity and i is directly proportional to e, the current in the circuit of Figure 17-7(a) is also a sinusoidal quantity. Furthermore, when e is zero, i must be zero, and when e is a maximum, i is at its peak value. This means that i is *in phase with e*, and the two waveforms can be plotted as shown in Figure 17-11(b).

Power Dissipation

From Equation 3-5, the power dissipated in a resistor is,

$$P = I^2R$$

and the instantaneous power is,

$$p = i^2R = (I_m \sin \omega t)^2 R$$

or,

$$p = I_m^2 R \sin^2 \omega t \tag{17-10}$$

The maximum instantaneous power dissipation that occurs is,

$$P_m = I_m^2 R$$

So,

$$p = P_m \sin^2 \omega t \tag{17-11}$$

The instantaneous power dissipated in the load resistance at any instant can easily be calculated using the above equations. Note that when I_m is a negative quantity, Equation 17-10 gives,

$$p = (-I_m)^2 R \sin^2 \omega t$$

which becomes,

$$p = I_m^2 R \sin^2 \omega t$$

which is a positive quantity. Obviously, power is dissipated by a current flowing in either direction through a resistor. The fact that p is always a positive quantity means that all the power supplied is dissipated in the resistor R. If the instantaneous power were to become a negative quantity at any time, it would imply that the resistor was supplying power to the signal source, which is something a resistor cannot do. The fact that the instantaneous power is always a positive quantity also explains the shape of the graph of instantaneous power plotted in Figure 17-12. Peak power dissipation occurs both when i is a positive peak and when i is a negative peak. Zero power dissipation occurs when i is zero.

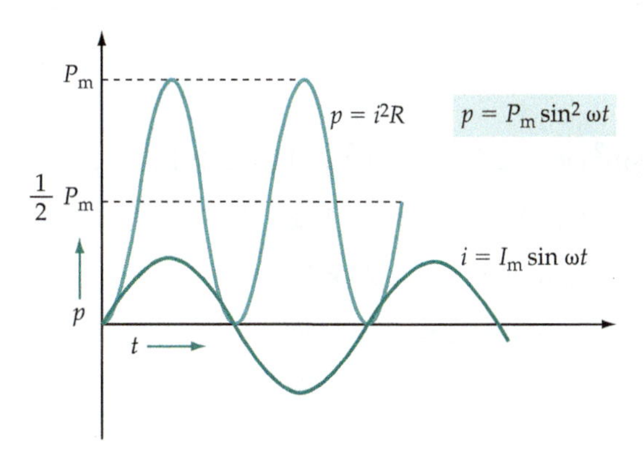

Figure 17-12 The instantaneous power dissipated in a purely resistive circuit is zero when the current is zero. The power increases to a peak as the current increases to a positive or negative peak level.

Examination of the waveform of instantaneous power dissipation in Figure 17-12 reveals that it is a perfect sine wave symmetrical about the level $0.5P_m$. So, $0.5P_m$ is the average value of the power dissipated in the resistor; or, average power dissipated over one complete cycle is half the peak power:

$$P = 0.5\,P_m \tag{17-12}$$

Example 17-5

The circuit shown in Figure 17-11(a) has a 60 Hz supply voltage with a maximum value of 160 V. If $R = 10\ \Omega$, calculate the values of instantaneous current and power at phase angles of $\pi/4$, $\pi/2$, $5\pi/4$, and $3\pi/2$ radians.

Solution

$$I_m = \frac{E_m}{R} = \frac{160\ \text{V}}{10\ \Omega}$$

$$= 16\ \text{A}$$

$$P_m = I_m^2 R = (16\ \text{A})^2 \times 10\ \Omega$$

$$= 2.56\ \text{kW}$$

Current calculations:

For $\omega t = \pi/4$,

Eq. 17-9, $\quad i = I_m \sin \omega t = 16\ \text{A} \sin \pi/4 \approx 11.3\ \text{A}$

For $\omega t = \pi/2$, $\quad i = 16\ \text{A} \sin \pi/2 \approx 16\ \text{A}$

For $\omega t = 5\pi/4$, $\quad i = 16\ \text{A} \sin 5\pi/4 \approx -11.3\ \text{A}$

For $\omega t = 3\pi/2$, $\quad i = 16\ \text{A} \sin 3\pi/2 \approx -16\ \text{A}$

Power calculations:

For $\omega t = \pi/4$,

Eq. 17-11, $p = P_m \sin^2 \omega t = 2.56\text{ kW} \sin^2 \pi/4 \approx 1.28\text{ kW}$

or, $p = i^2R = (11.3\text{ A})^2 \times 10\ \Omega = 1.28\text{ kW}$

For $\omega t = 5\pi/4$, $p = 2.56\text{ kW} \sin^2 5\pi/4 \approx 1.28\text{ kW}$

For $\omega t = 3\pi/2$, $p = 2.56\text{ kW} \sin^2 3\pi/2 \approx 2.56\text{ kW}$

Practice Problem

17-4.1 An instantaneous peak power of 30 mW is dissipated in a 1.2 kΩ resistor with an ac voltage source. Calculate the instantaneous resistor voltages at phase angles of $\pi/3$ and $2\pi/3$ radians.

17-5 PEAK, AVERAGE, AND RMS VALUES OF SINE WAVES

Peak and Peak-to-Peak Values

As discussed, the peak value of a sinusoidal waveform is the maximum value; E_m for a voltage wave, I_m in the case of a current waveform. The *peak-to-peak* value is $2E_m$ or $2I_m$, as illustrated in Figure 17-13. For sinusoidal waveforms, there are direct relationships between peak, average, and rms values.

Average Value

The average value of a waveform can be determined by differential calculus, or by the more laborious process of taking the average value of a number of equally spaced instantaneous levels. The process is illustrated in Fig. 17-13, where each 180° half (or half cycle) of the current waveform is divided into nine 20° sections. The instantaneous values of the waveform at the center of each section is measured and the average value calculated as,

$$I_{av} = \frac{i_1 + i_2 + i_3 + \cdots + i_9}{9}$$

This process can be applied to waveforms of any shape; in fact, it may be the only way of determining the average value of some irregular waveforms. For the sine wave, however, it is not necessary to measure each instantaneous value, because they can be calculated from the maximum value and the sine of each angle. So, the average value of the *positive half-cycle* of the current wave is,

$$I_{av} = \frac{I_m \sin 10^\circ + I_m \sin 30^\circ + I_m \sin 50^\circ + \cdots + I_m \sin 170^\circ}{9}$$

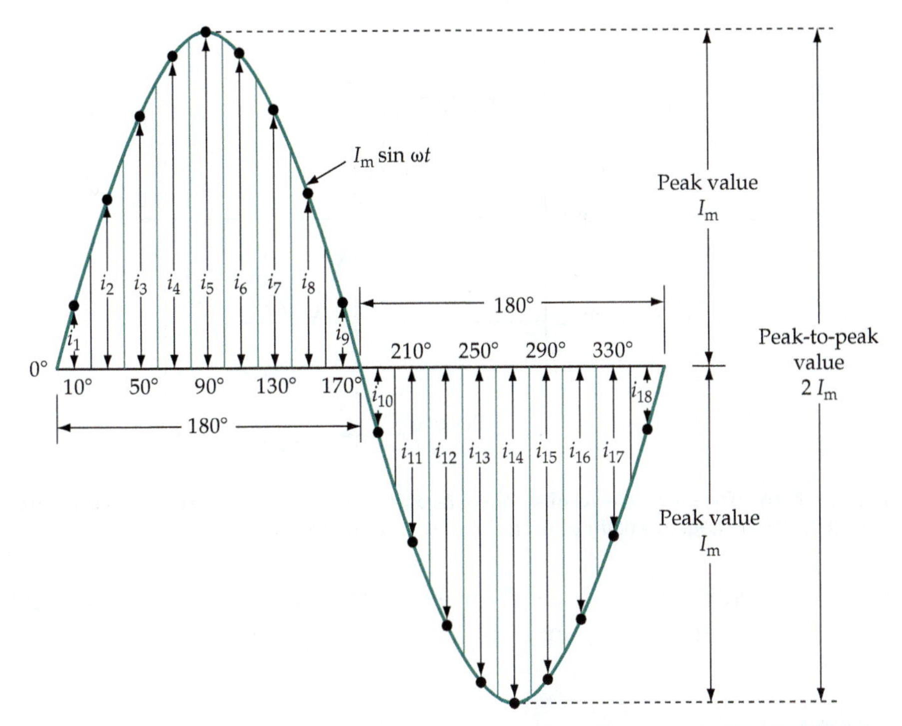

Figure 17-13 The average value of a waveform can be determined by averaging a number of equally spaced instantaneous amplitudes. The root mean squared (rms) value can also be calculated from a number of instantaneous amplitudes. Differential calculus offers simpler techniques.

Substituting the appropriate values (or using calculus), it is found that, as illustrated in Figure 17-14,

$$I_{av} = \frac{2}{\pi} I_m \approx 0.636 I_m \qquad \textbf{(17-13)}$$

Similarly, the average value of *the positive half-cycle* of a sinusoidal voltage wave is,

$$E_{av} = \frac{2}{\pi} E_m \approx 0.636 E_m \qquad \textbf{(17-14)}$$

Returning to Figure 17-13, the average value of the second 180° (or *negative half-cycle)* of the waveform is found to be,

$$I_{av} = -0.636 I_m$$

For the complete cycle,

$$\begin{aligned} I_{av} &= 0.636\, I_m - 0.636 I_m \\ &= 0 \end{aligned}$$

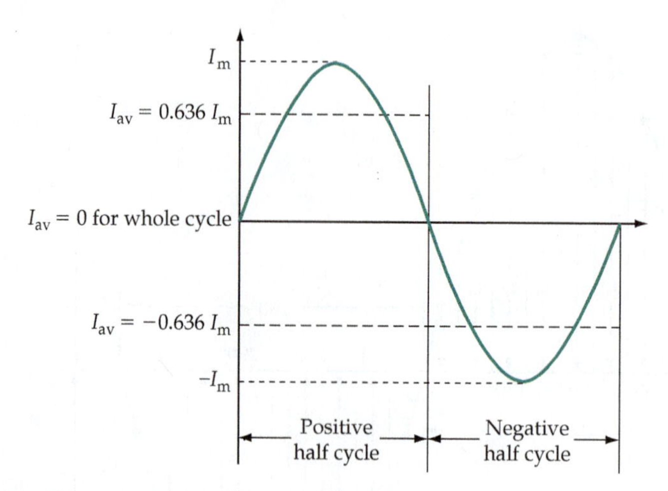

Figure 17-14 The average level of a sinusoidal current is 0.636 I_m for the positive half-cycle, −0.636 I_m for the negative half-cycle, and zero for the whole cycle.

So, over the entire 360° sinusoidal waveform, the average value of the current (or voltage) is zero, (see Figure 17-14).

RMS Value

Taking each instantaneous current value in Figure 17-13 in turn, the instantaneous power dissipated in a resistor through which the current is flowing is,

$$p_1 = i_1^2R,\ p_2 = i_2^2R,\ p_3 = i_3^2R,\ \cdots\ p_n = i_n^2R$$

So, the average power dissipated in the resistor is,

$$P = \frac{i_1^2R + i_2^2R + \cdots + i_n^2R}{n}$$

As explained in Section 17-4, the negative instantaneous current values also give a positive power dissipation in the load.

When a direct current is flowing through a resistor, the power dissipated is given by Equation 3-5 [see Figure 17-15(a)]:

$$P = I^2R$$

To find the equivalent alternating current that dissipates the same amount of power as a direct current I flowing through the resistor, the dc and ac power dissipation expressions are equated.

$$I^2R = \frac{i_1^2R + i_2^2R + \cdots + i_n^2R}{n}$$

which gives,

$$I^2 = \frac{i_1^2 + i_2^2 + \cdots + i_n^2}{n}$$

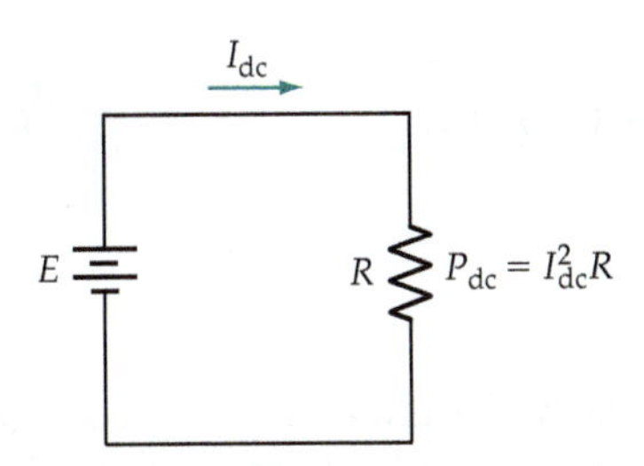

(a) DC power dissipation in a resistor

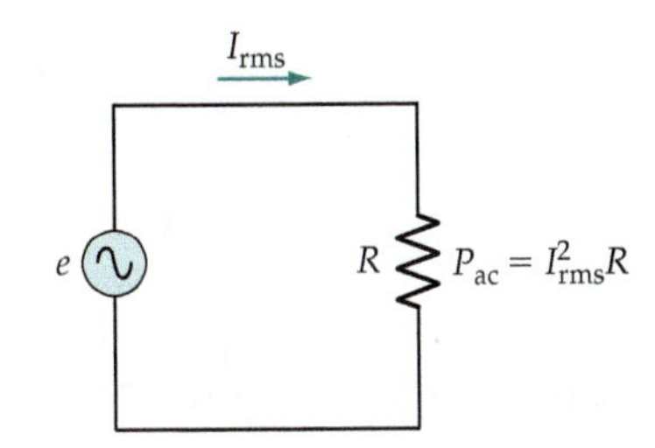

(b) AC power dissipation in a resistor

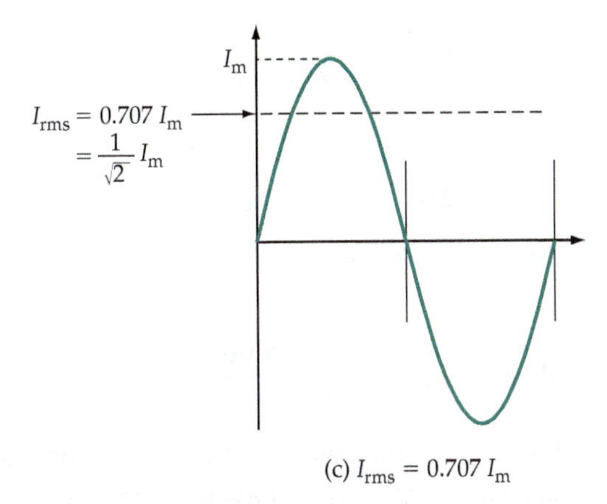

(c) $I_{rms} = 0.707\, I_m$

Figure 17-15 The rms level of a sinusoidal current is the dc-equivalent level; that is, the level that produces the same power dissipation in a resistor as a given direct current.

It is seen that I^2 (the dc current squared) is equal to the average value of the instantaneous *ac current squared* or the *ac mean-squared value.* So, the *effective value* (or *dc-equivalent value*) of the alternating current is,

$$I = \sqrt{\frac{i_1^2 + i_2^2 + \cdots + i_n^2}{n}}$$

Here, I is the *root of the mean-squared value*, or *rms value*, of the alternating current. So, the rms value of an alternating current is the dc-equivalent value, as far as power dissipation is concerned [Figure 17-15(b)].

Like the average value, the relationship between the rms value and the peak value of an *ac* waveform can be determined by calculus. However, it can also be worked out quite simply. From Equation 17-12, the ac power dissipation is,

$$P = 0.5\, P_m = 0.5\, I_m^2 R$$

and from Equation 3-5 for *dc* power dissipation,

$$P = I^2 R$$

So,
$$I^2 R = 0.5\, I_m^2 R$$

or,
$$I^2 = 0.5\, I_m^2$$

$$I = \frac{1}{\sqrt{2}} I_m = 0.707\, I_m \tag{17-15}$$

Alternatively,
$$I = \frac{1}{1.414} I_m$$

This is illustrated in Figure 17-15(c). Similarly, for an alternating voltage, the *effective value* is the rms value of the waveform, and where the waveform is sinusoidal,

$$E = \frac{1}{\sqrt{2}} E_m = 0.707\, E_m \tag{17-16}$$

Alternatively,
$$E = \frac{1}{1.414} E_m$$

The relationship between peak and rms quantities is termed the *peak factor* or *crest factor*.

$$\text{Peak factor} = \frac{\text{peak value}}{\text{rms value}}$$

The *form factor* is another term sometimes used in the study of ac waveforms. This is simply the relationship between rms and average values.

$$\text{Form factor} = \frac{\text{rms value}}{\text{average value}}$$

With alternating current and voltage, the rms values are the normally quoted values. A domestic alternating voltage supply of 115 V, for example, has a peak value of,

$$E_m = 1.414 \times 115\ \text{V} \approx 163\ \text{V}$$

The rms quantities are normally used in all Ohm's law calculations involving alternating current and voltage.

It is important to note that the relationships stated in Equations 17-13 through 17-16 *apply only to pure sine waves.* In the case of other waveforms, the peak, average, and rms quantities are related by other (different) factors.

Example 17-6

A 300 V sinusoidal ac supply is applied to a 50 Ω resistor. Determine the peak, rms, and average values of the current through the resistor. Also, calculate the power dissipated in the resistor.

Solution

From Eq. 17-16, the peak voltage is,

$$E_m = \sqrt{2}\, E = 1.414 \times 300\ \text{V}$$
$$\approx 424\ \text{V}$$

Peak current, $I_m = \frac{E_m}{R} = \frac{424\text{ V}}{50\ \Omega}$

$= 8.48\text{ A}$

Rms current, $I = \frac{E}{R} = \frac{300\text{ V}}{50\ \Omega}$

$= 6\text{ A}$

From Eq. 17-13, the average current for one half-cycle is,

$$I_{av} = 0.636\, I_m = 0.636 \times 8.48\text{ A}$$

$$\approx 5.4\text{ A}$$

The average current for whole cycle $= 0$

power dissipation, $P = I^2 \times R = (6\text{ A})^2 \times 50\ \Omega$

$= 1.8\text{ kW}$

Practice Problems

17-5.1 The power dissipated in a 560 Ω resistor is measured as 250 mW. Calculate the peak, average, and rms values of the voltage applied to the resistor.

17-5.2 A 220 Ω resistor has a 33 V ac supply. Determine the resistor power dissipation and peak current.

17-6 OSCILLOSCOPES

Basic Oscilloscope

The *oscilloscope* is the basic instrument for the study of waveforms. Typically, an oscilloscope has two input for connecting to two points in a circuit under investigation. Waveforms are displayed on a small computer-type screen, as shown in Figure 17-16. Various controls on the front panel of the instrument facilitate measurement of waveform voltage, frequency, time period, and other quantities. As in the case of other electronic instruments, both digital and analog oscilloscopes are available. Analog oscilloscopes accept an input voltage and process it for display by analog means. Digital instruments use analog-to-digital converters (see Section 13-4) to convert the analog input voltages to digital form. The signals are then processed digitally for display and measurement.

Oscilloscope Probes

Input signals are normally applied to an oscilloscope via *coaxial cables* with *probes* on their ends, (see Figure 17-17). These are essentially just convenient-to-use

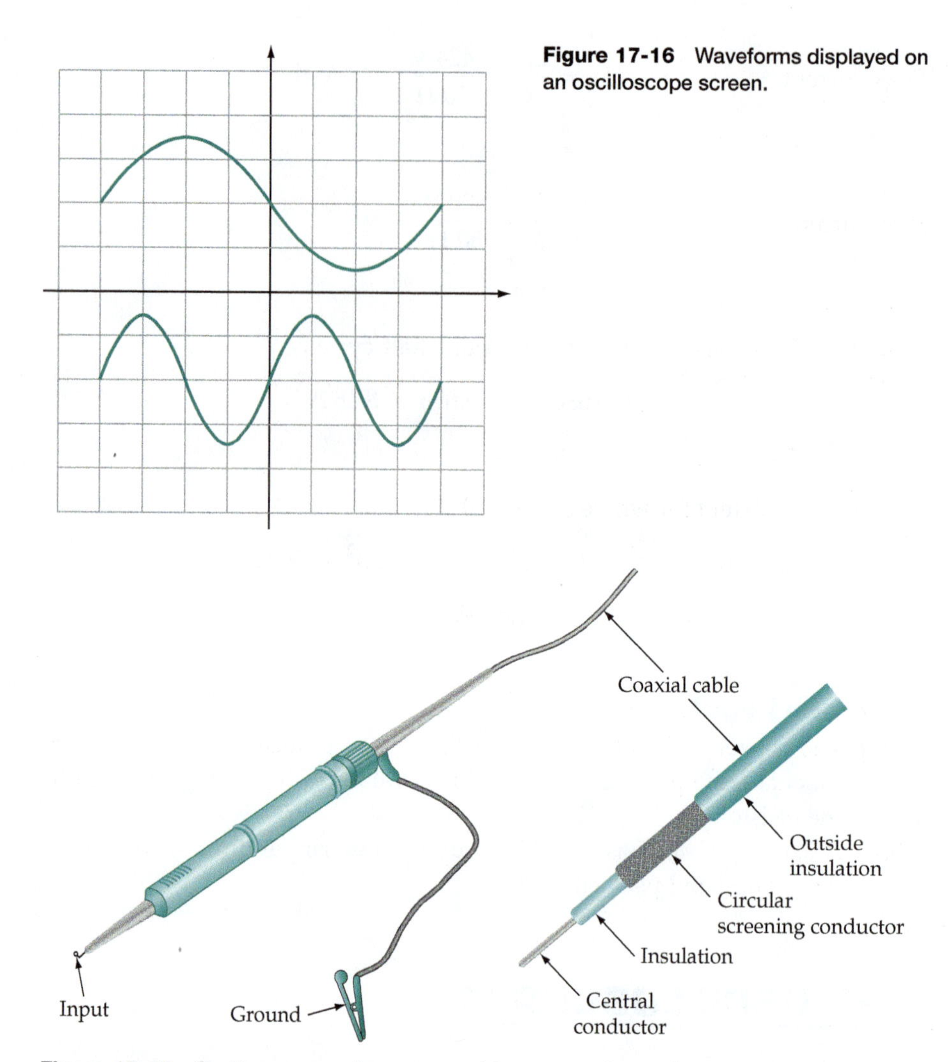

Figure 17-16 Waveforms displayed on an oscilloscope screen.

Figure 17-17 Oscilloscope probes are used to connect input signals to an oscilloscope. Coaxial cable is employed to screen-out unwanted signals. Some probes contain adjustable capacitors to eliminate distortion in displayed waveforms.

insulated connecting clips. Each probe has two connections, as illustrated, one input and one ground. It is important to connect both ground terminals (from the two input probes) to the same (grounded) point in the circuit under investigation; otherwise, there will be more than one grounded point, and the circuit may not function correctly. As illustrated in Figure 17-17, a coaxial cable consists of a central conductor with its insulation surrounded by another, braided circular, conductor. The circular conductor is grounded, and it acts as a *screen*, which helps to prevent unwanted radio frequency signals being picked up by the oscilloscope input.

Some probes, termed *attenuator probes*, have resistors inside them to increase the input resistance from the normal 1 MΩ input resistance of the oscilloscope to 10 MΩ. These probes also have the effect of reducing the voltage applied to the oscilloscope by a factor of 10. So, they are usually referred to as *10:1 probes* (or *10 times probes*), and the ordinary (non-attenuator) probes are termed *1:1 probes.*

Oscilloscope Controls

The font panels of representative analog and digital oscilloscopes are shown in Figures 17-18 and 17-19, respectively. It is seen that both instruments have display screens that are vertically and horizontally calibrated. Both also have *Channel 1* and *Channel 2* inputs, to facilitate the display of two waveforms at the same time. A detailed explanation of all of the control functions is beyond the scope of this book, however, the most important controls are discussed below.

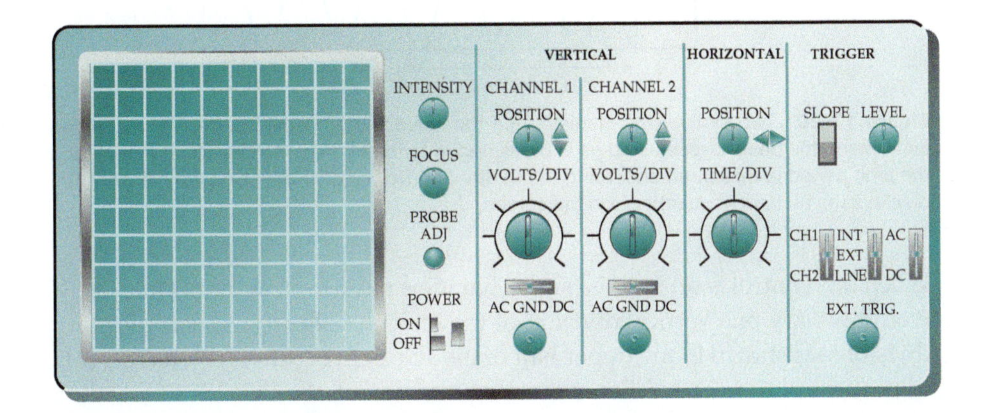

Figure 17-18 Front panel and controls of a typical analog oscilloscope.

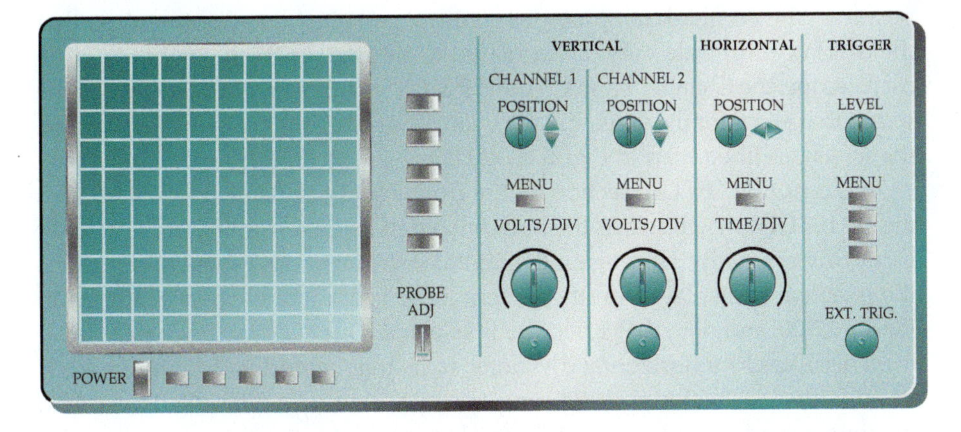

Figure 17-19 Front panel and controls of a typical digital oscilloscope.

Digital and analog oscilloscopes both have VERTICAL, HORIZONTAL, and TRIGGER sections on the front panel, (see Figures 17-18 and 17-19). Each VERTICAL section has Channel 1 and Channel 2 POSITION and VOLTS/DIV knobs. The vertical position controls are used to move each waveform up or down the screen to the best position for viewing. The VOLTS/DIV switch for each channel selects the vertical sensitivity of the display (see Figure 17-20).

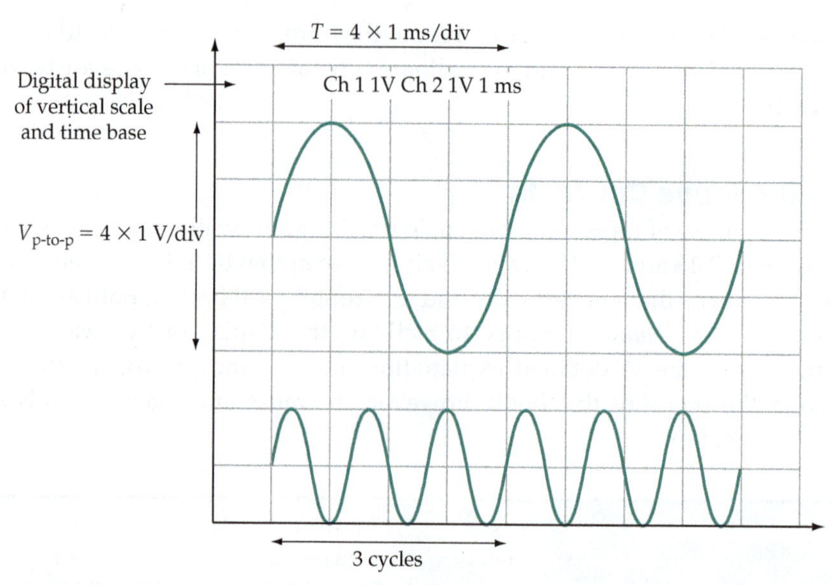

Figure 17-20 Two sine waves displayed on the screen of an oscilloscope. For an analog instrument, the peak-to-peak voltage is measured in terms of the volts/(vertical division), and the time period is measured according to the time/(horizontal division). A digital oscilloscope displays the numerical quantities on the screen.

When this control is set to 1 V, a signal having a peak-to-peak amplitude of 1 V would occupy one vertical division of the screen. A signal that occupies four divisions, as shown in the upper half of the screen in Figure 17-20, has a peak-to-peak amplitude of 4 V. For the analog instrument, the VOLTS/DIV setting is correct only when the *vernier* knob at its center is in the *CAL* (calibrated) position. The vernier knob provides continuous volts/div adjustment, so that the display amplitude may be decreased as desired. In the case of the digital oscilloscope, the Channel 1 and Channel 2 MENU buttons produce menus on the screen that allow for the volts/div selection. The selections are then displayed on the screen, as illustrated in Figure 17-20 [Ch 1 1V Ch 2 1V 1 ms].

In the HORIZONTAL section on the front panels of the analog instrument, the TIME/DIV switch selects the horizontal deflection sensitivity of the display in ms/div or μs/div. Here again the digital instrument has a menu button, and the selection is displayed on the screen. Refer again to the two waveforms in Figure 17-20, and note that 1 ms/div is selected. One cycle of the upper waveform occupies four horizontal divisions, so its time period is $(4 \times 1 \text{ ms}) = 4$ ms. For the lower waveform, three complete cycles occupy 4 ms. Consequently, the time period is (4/3) ms, or 1.3 ms. Here again, for the analog oscilloscope the setting is correct only when the vernier knob at the center of the time/div control is in its CAL position. This vernier knob provides continuous time/division adjustment, so that a cycle of displayed waveform may be widened up to ten times the horizontal time/div setting. The horizontal POSITION knob performs a similar function to the vertical position control; the displayed waveforms may be moved horizontally across the screen as desired.

In order to provide a stable display, the waveform must be made to commence (at the left hand side of the screen) exactly when one of the input waveforms is at its zero position. So, the time-base circuits (which produce the horizontal movement of the point that traces out the waveform) are said to be *triggered* at this instant. The TRIGGER SOURCE switch on the analog instrument, and the TRIGGER MENU on the digital oscilloscope, allow for the selection of INT (internal), EXT (external), or LINE as triggering sources for the time base. With INT selected, the time base is triggered from one of the input waveforms, and this is the most commonly used triggering source.

Triggering may be *positive* (+) or *negative* (−), which simply means that the displayed waveform may be made to commence either when it's instantaneous value is increasing in a positive direction or when it is going in a negative direction [see Figure 17-21(a) and (b)]. The TRIGGER LEVEL knob continuously adjusts the instant in time at which the waveform display commences. The display may be made to commence exactly when the signal goes through its zero position or at some time shortly before or after this instant [Figure 17-21(c)].

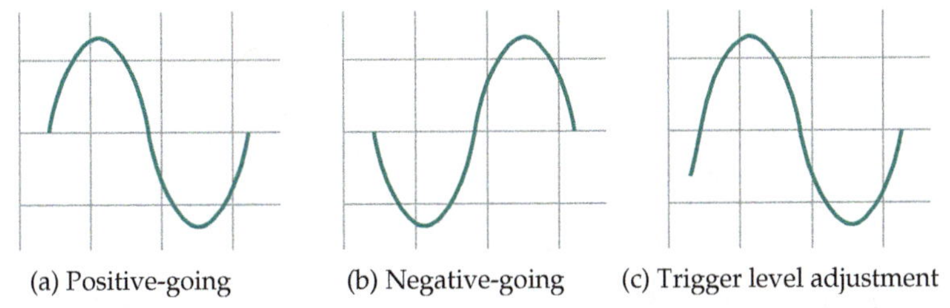

Figure 17-21 By use of the trigger controls, a displayed waveform may be made to commence positive-going or negative-going, and the starting point can be adjusted.

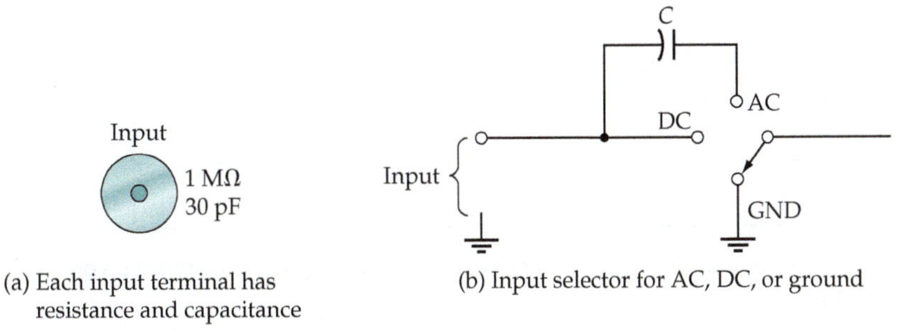

Figure 17-22 Oscilloscope input resistance and capacitance is usually marked at each input terminal. The *AC-DC-GND* switch facilitates selection of ac or dc inputs, or grounded input.

Each oscilloscope input terminal typically has a 1 MΩ input resistance and a 30 pF input capacitance, [see Figure 17-22(a)]. The AC-GND-DC switches immediately above each input terminal on the analog oscilloscope, and a menu selection on the digital instrument, facilitate ac, ground, or dc connection of the input terminals [Figure 17-22(b)]. AC is selected when it is necessary to display an alternating voltage and block any dc component of the input waveform.

DC is choosen when the dc component of the input is to be measured. The GND (ground) selection disconnects the input signals and grounds the input terminals, to permit each trace to be set to a convenient zero position on the screen.

Measuring Voltage and Time Period

Figure 17-23 illustrates the method that should be used to achieve the most accurate measurement of voltage and time period with an analog oscilloscope. The appropriate VERTICAL POSITION control is used to shift the unwanted trace off the screen. The VOLTS/DIV and TIME/DIV knobs are then adjusted to give the largest possible display of one cycle of the input waveform. Note that the VOLTS/DIV and TIME/DIV (center) vernier knobs must be in their CAL positions. From Figure 17-23, the time period of the waveform is (number of horizontal divisions) × (time/division), and the measured amplitude is (number of vertical divisions) × (volts/division).

The above procedure is unnecessary with a digital oscilloscope, because the quantities are normally measured automatically and printed on the screen.

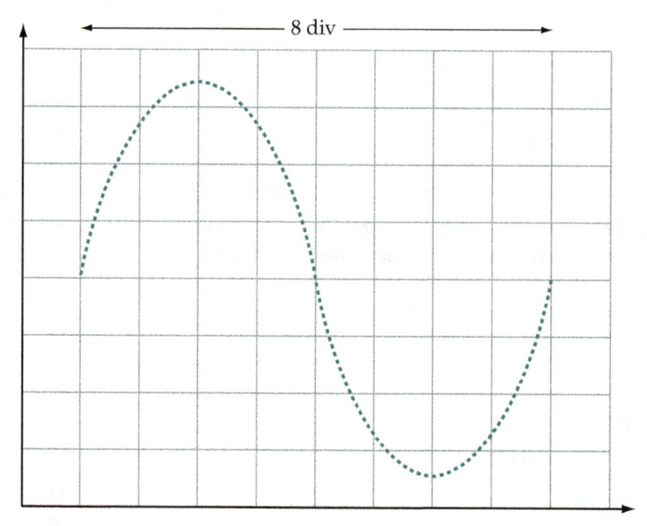

Figure 17-23 Accurate measurement of waveform time period and amplitude on an analog oscilloscope.

Phase Difference Measurement

Figure 17-24 shows how the phase difference between two waveforms can be determined. From a measurement of the time period and the fact that one cycle represents 360°, the degrees/(horizontal division) can be calculated. The difference (in horizontal divisions) between commencement of each waveform is measured. This is then converted into degrees of phase shift using the calculated degrees/division. Once again, the desired quantity can be automatically determined by a digital oscilloscope and printed on the screen.

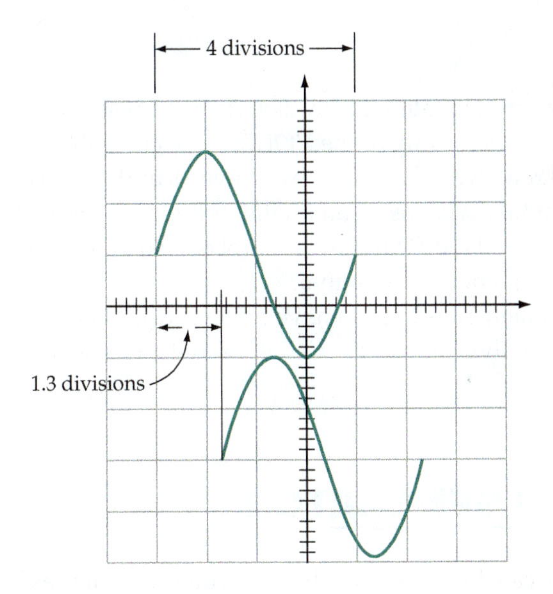

Figure 17-24 Measurement of the phase difference between two waveforms. Because one complete cycle represents 360°, the degrees/horizontal division can be determined, and the phase difference in divisions can be converted into degrees.

Example 17-7

Determine the frequency and amplitude of the upper waveform in Figure 17-20 if the oscilloscope controls are set at 2 V/div and 5 ms/div. Also, determine the phase difference between the two waveforms shown in Figure 17-24.

Solution

For the upper waveform in Figure 17-20,

$$T = (4 \text{ horizontal divisions}) \times 5 \text{ ms/div}$$
$$= 20 \text{ ms}$$
$$f = 1/T = 1/20 \text{ ms}$$
$$= 50 \text{ Hz}$$

Peak amplitude,

$$V_\text{p} = (2 \text{ vertical divisions}) \times 2 \text{ V/div}$$
$$= 4 \text{ V}$$

Peak-to-peak amplitude,

$$V_\text{p-to-p} = (4 \text{ vertical divisions}) \times 2 \text{ V/div}$$
$$= 8 \text{ V}$$

For Figure 17-24,

$$T = 4 \text{ divisions} = 360°$$
$$1 \text{ div} = 360°/4 = 90°$$
$$\text{Phase difference} = (1.3 \text{ div}) \times 90°/\text{div}$$
$$= 117°$$

Practice Problems

17-6.1 A 15 V sine wave occupies approximately 4.2 divisions peak to peak on an oscilloscope screen. What is the setting of the VOLTS/DIV control?

17-6.2 Two cycles of a sinusoidal ac waveform occupy 6.5 vertical divisions peak to peak and 8 horizontal divisions on an oscilloscope screen. If the VOLTS/DIV setting is 0.1 V, and the TIME/DIV control is at 50 μs, determine the rms value and frequency of the signal.

17-6.3 Two waveforms as in Problem 17-6.2 are to have a phase difference of 63°. Estimate the horizontal divisions difference between the two when displayed on the oscilloscope.

17-7 AC FUNCTION GENERATOR

Alternating voltage waveforms can be generated by electronic circuits, as well as by the method described in Section 17-1. The laboratory-type *Function Generator* shown in Figure 17-25(a) can produce a sine wave output with

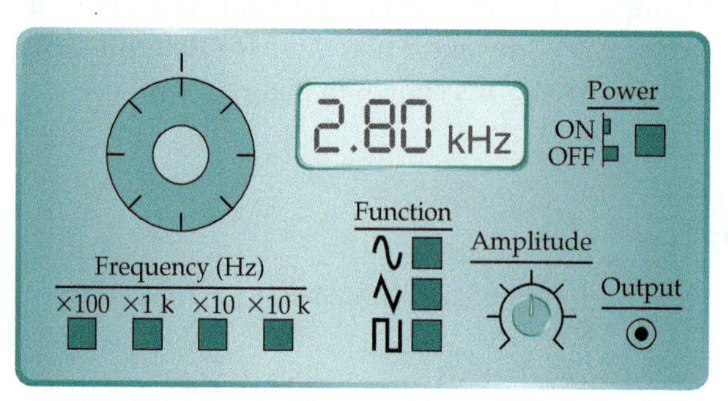

(a) Function generator

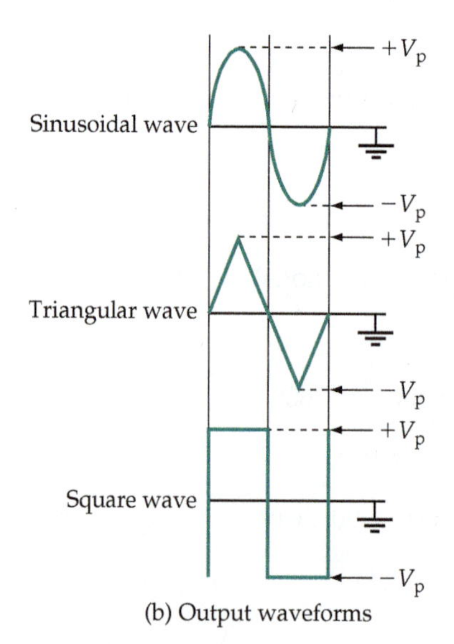

(b) Output waveforms

Figure 17-25 A typical function generator produces sinusoidal, triangular, and square waves with frequencies ranging from 10 Hz to 10 kHz.

amplitude adjustable from 1 V to 10 V, and with frequencies ranging from 10 Hz to 100 kHz. The instrument also produces two non-sinusoidal waveforms; a triangular wave, and a square wave. Figure 17-25(b) shows that in a triangular wave the instantaneous voltage increases or decreases at a constant rate with time until it reaches a (positive or negative) peak. At the peak level, the voltage reverses from positive-going to negative-going, or vice versa. In a square wave, as illustrated, the instantaneous voltage level is a constant positive value for one half of the waveform, and a constant negative quantity for the other half. The change between positive and negative levels appears to occur instantaneously, however, there are measurable *rise* and *fall times.*

17-8 BASIC AC VOLTMETERS

PMMC Instrument on AC

The permanent magnet moving-coil (PMMC) instrument is *polarized,* (see Section 11-6 and Chapter 13). This means that the terminals of the instrument are identified as plus (+) and minus (–), and that for correct deflection the meter must be connected with that polarity. When incorrectly connected, the pointer attempts to deflect to the left of zero (i.e., off-scale). Now consider what occurs when a PMMC instrument is connected directly to an alternating current source, as shown in Figure 17-26(a). For the illustration showing instantaneous pointer position [Figure 17-26(b)], it is assumed that the frequency of the alternating current is very low; around 0.1 Hz. As the current level increases positively

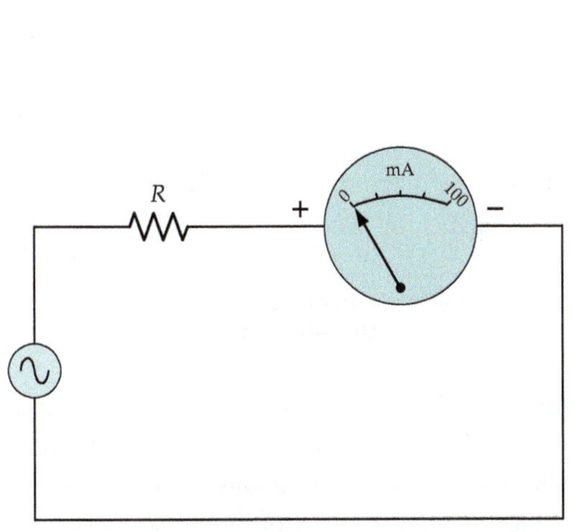

(a) PMMC ammeter connected to alternating current source

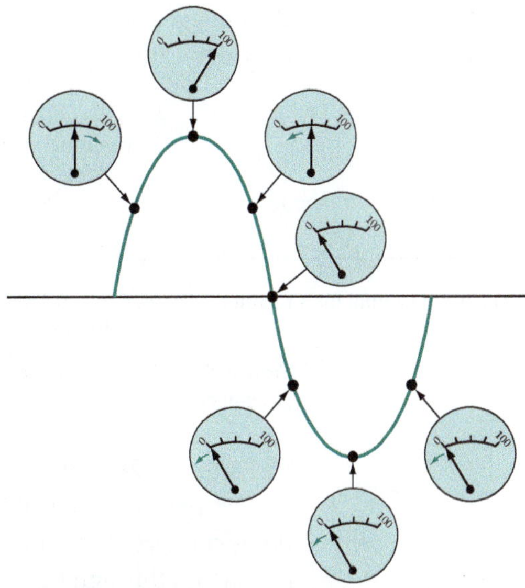

(b) Behavior of PMMC ammeter when measuring a very low frequency alternating current

Figure 17-26 A permanent magnet moving-coil (PMMC) instrument used directly on sinusoidal alternating current normally indicates zero. On very low frequency ac, the pointer would move up and down the scale.

from zero, the meter pointer moves to the right until it indicates peak current. Then the pointer falls back toward zero again as the current level drops. When the current direction reverses during the second half-cycle of the waveform, the meter attempts to deflect to the left, which, of course, is not possible. Consequently, the pointer remains at (or just below) the zero mark during the negative half-cycle of the current.

Most ac frequencies are much greater than 0.1 Hz; 60 Hz being the normal ac supply frequency in North America. At this frequency, the meter pointer would have to rise and fall 60 times in every second. The damping mechanism of the instrument, as well as the inertia of the moving system, prevents the moving coil and pointer from moving that fast. Consequently, the instrument pointer settles at the average level of the alternating current passing through the windings. For normal sinusoidal alternating current, the average level is zero. So, a PMMC instrument used directly to measure 60 Hz alternating current indicates zero.

Rectification

A *rectifier* (or semiconductor diode) has two terminals, *anode* and *cathode*, as illustrated in Figure 17-27(a). Current can flow through the device in only one direction. When the anode terminal is positive with respect to the cathode, current flows (in the conventional direction) from anode to cathode [see Figure 17-27(b)]. Current will not flow through the rectifier when the cathode is positive with respect to the anode [Figure 17-27(c)]. When *forward-biased* (i.e., conducting), there is a small volt drop from anode to cathode. This is typically 0.7 V or less, depending on the semiconductor material employed in manufacturing the rectifier. The conventional direction of current flow through the device is indicated by the arrowhead in the circuit symbol, pointing from anode to cathode.

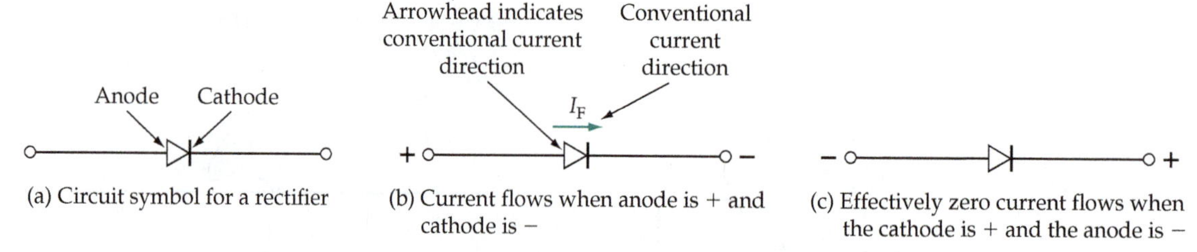

Figure 17-27 The semiconductor rectifier used in ac instruments is a one-way device; it passes current in one direction and blocks current flow in the other direction.

Figure 17-28 shows how a rectifier is used to convert alternating current into a series of pulses that all have the same polarity. In Figure 17-28(a) a single rectifier is connected in series with a resistor. Only the positive half-cycles of current are passed through the rectifier, so that the negative portion of the ac waveform is cut off. The waveform of the current that passes through the rectifier and the resistor to the output terminals is a continuous series of positive pulses with intervening spaces. This waveform is termed *half-wave-rectified.*

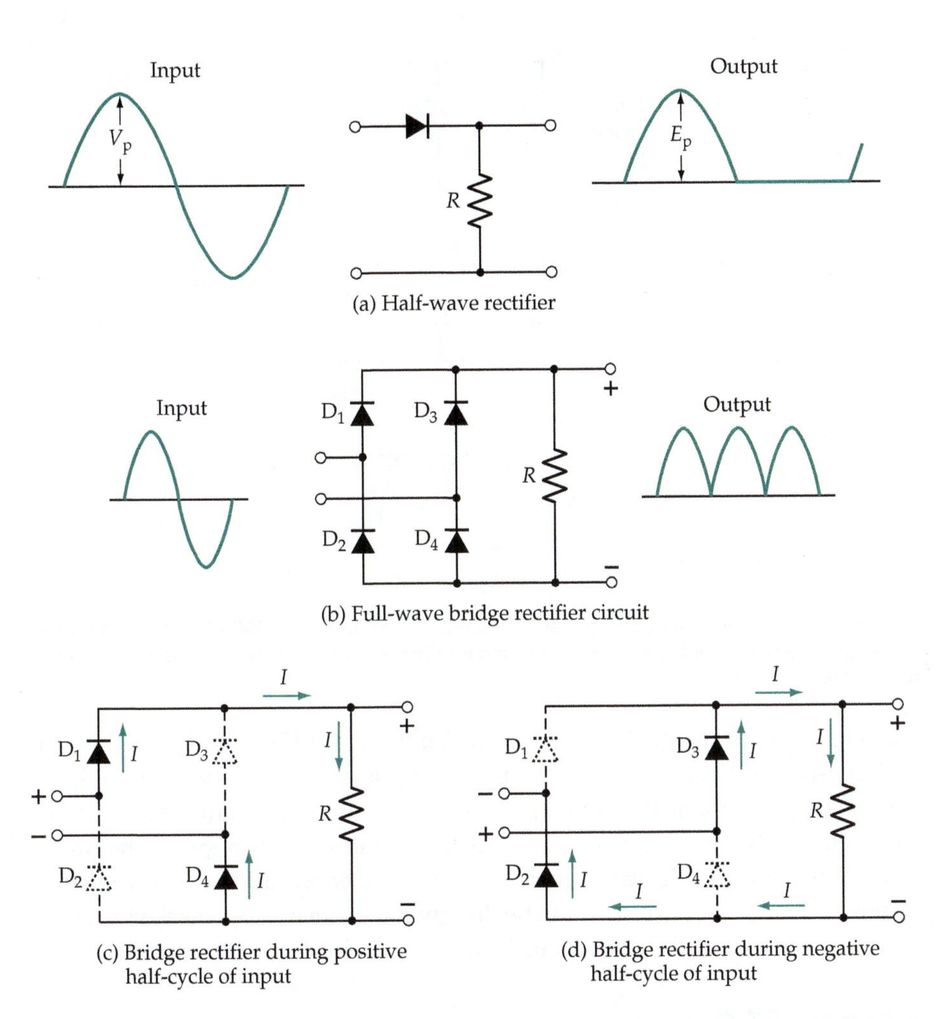

Figure 17-28 Half-wave rectifiers circuits pass only one half of each cycle of sine wave input. Full-wave rectifier circuits convert all half-cycles to the same polarity and pass them to the output.

The circuit shown in Figure 17-28(b) employs four rectifiers and is known as a *bridge rectifier circuit*. In this case, when the input waveform is positive [Figure 17-28(c)], rectifiers D_1, and D_4 are forward-biased, while D_2 and D_3 are reverse-biased. So, current flows through D_1, resistor R, and D_4, as illustrated. Conversely, when the input waveform is negative [Figure 17-28(d)], D_2 and D_3 are forward-biased, and D_1 and D_4 are reverse-biased. Current now flows through D_3, resistor R, and D_2, and it is seen that the current direction through the resistor is the same as before. The result of this is that the output waveform [Figure 17-28(b)] is a continuous series of unidirectional pulses, and the process is termed *full-wave rectification*.

Rectifier Voltmeter

The circuit shown in Figure 17-29(a) is that of an ac voltmeter using a PMMC instrument and a full-wave rectifier circuit. As in the case of a dc voltmeter, a *multiplier* resistance must be included in the circuit to limit the current through the instrument. The actual current that flows through the deflection instrument

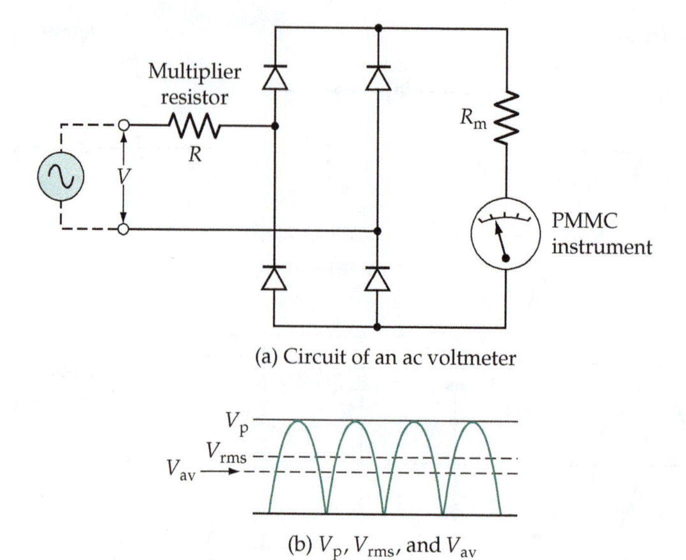

Figure 17-29 AC voltmeter using a full-wave rectifier circuit and a PMMC meter. The meter deflection is proportional to the average of the rectified waveform, but the scale is calibrated to indicate rms values.

has the full-wave-rectified waveform shown in Figure 17-29(b). The deflection of the instrument is proportional to the average level of the current through its coil, so the pointer tends to indicate 0.636 of V_p. However, the effective value, or rms value, of the alternating waveform is the quantity that is normally required in any measurement. Because there is a direct relationship between rms and average values, the instrument can be designed to have its scale marked to indicate rms volts. The following example shows how this is accomplished.

Example 17-8

An ac voltmeter to indicate 100 V rms maximum is to be constructed using a deflection instrument that has full-scale deflection (FDS) of 500 μA. The coil resistance (R_m) is 1 kΩ, and the rectifiers used each have a 0.7 V forward voltage drop. Using the full-wave rectifier circuit shown in Figure 17-29(a), determine the required multiplier resistance.

Solution

At FSD the average current through the deflection instrument is,

$$I_{av} = 500\ \mu A$$

For each half-cycle of sinusoidal waveform:

From Eq. 17-13,

Peak current,
$$I_p = \frac{\pi}{2} I_{av} = \frac{\pi}{2} \times 500\ \mu A$$

$$\approx 785\ \mu A$$

peak voltage, $V_P = 1.414V_{rms} = 1.414 \times 100\text{ V}$

$= 141.4\text{ V}$

and $$I_P = \frac{V_P - (2V_F)}{R + R_m}$$

Giving, $$R = \frac{V_P - (2V_F)}{I_P} - R_m$$

$$= \frac{141.4\text{ V} - (2 \times 0.7\text{ V})}{785\ \mu\text{A}} - 1\text{ k}\Omega$$

$$= 177\text{ k}\Omega$$

It is important to note that rectifier voltmeters designed for sine-wave operation can be used only where pure sine waves are involved. Where the voltage to be measured has any other waveform, the voltmeter will *not* correctly indicate rms voltage. This is because the 1.414 relationship between rms and peak current levels applies only to pure sine waves. For multirange ac voltmeters, a rotary switch is used to select one of several values of multiplier resistor, exactly as in the case of dc voltmeters.

Practice Problem

17-8.1 A full-wave rectifier voltmeter circuit uses a 75 μA meter with a coil resistance of 2.98 kΩ and diodes that have a forward voltage drop of 0.3 V. If the multiplier resistance is 352 kΩ, determine the full-scale rms voltage.

17-9 AC ELECTRONIC VOLTMETERS

Voltmeter Circuit

The circuit of one type of ac electronic voltmeter illustrated in Figure 17-30 is quite similar to the dc electronic voltmeter circuit discussed in Section 13-4. As in the case of the dc instrument, the input attenuator functions as a range selector, and the amplifier usually functions as a (high Z_{in}, low Z_{out}) *buffer stage* between the attenuator and the voltage measuring circuit.

A major difference between the ac voltmeter circuit in Figure 17-30 and the dc electronic instrument is that a full-wave bridge rectifier is included with the amplifier. The combination of amplifier and rectifiers is known as a *precision rectifier,* and it produces a rectified version of the input waveform without any rectifier voltage drop. The amplifier can also have a voltage gain greater than 1, so that small-amplitude waveforms can be amplified to measurable levels, and again, the diode voltage drop is eliminated.

As illustrated, the rectifier-amplifier stage is followed by a *peak detector.* This means that the output waveform to the voltage measuring stage is a constant

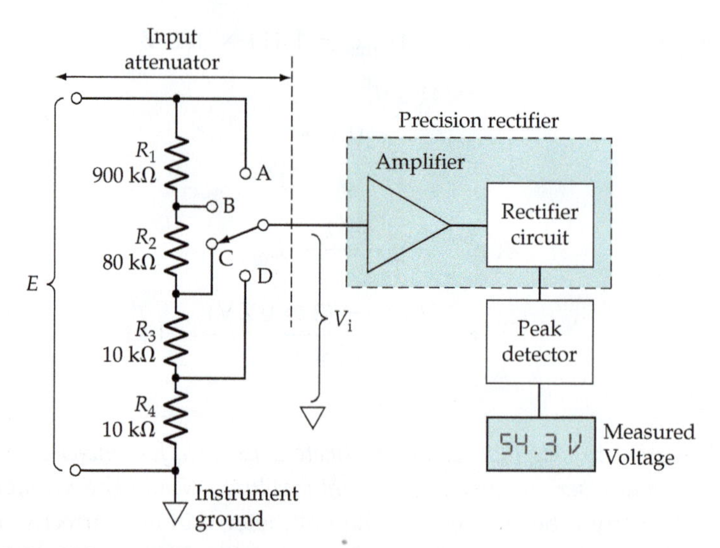

Figure 17-30 AC electronic voltmeter using a *precision rectifier* and a *peak detector*. The dc output from the peak detector can be made directly equal to the rms value of the input. The voltage measuring stage can be either analog or digital.

dc voltage equal to the peak value of the input waveform. If the amplifier has a voltage gain of 0.707 (instead of a gain of 1), the dc output voltage (from the peak detector) equals the rms value of the input ($V_{i(rms)}$) instead of the peak value. Similarly, for low-amplitude inputs, the amplifier voltage gain could be 7.07 to give a dc output equal to 10 $V_{i(rms)}$. Because the voltage from the peak detector stage is a dc quantity, it can be measured directly by either an analog dc voltmeter or by a digital dc meter, as illustrated. The indicated (dc) voltage is the rms value of the input.

As for all ac instruments using rectifiers, it is important to note that measurements made by this type of ac electronic voltmeter are accurate only for purely sinusoidal voltages. Special *true rms* instruments are available for measuring the rms value of voltages with nonsinusoidal waveforms.

Frequency Response of AC Electronic Instruments

The frequency response of ac voltmeters can be an important quantity. For example, if an instrument with an upper frequency limit of 500 Hz is used to measure a voltage that has a 1 kHz frequency, the measurement is completely unreliable. Figure 17-31 illustrates the typical frequency responses of several different types of analog and digital voltmeters. As shown, most handheld instruments have a frequency range from 40 Hz to 500 Hz. The bandwidth of high-performance portable instruments is typically 40 Hz to 30 kHz. Many bench-type digital instruments have a frequency range of 20 Hz to 100 kHz, while similar analog instruments can operate satisfactorily to 1 MHz. Special instruments are available for higher frequency ranges.

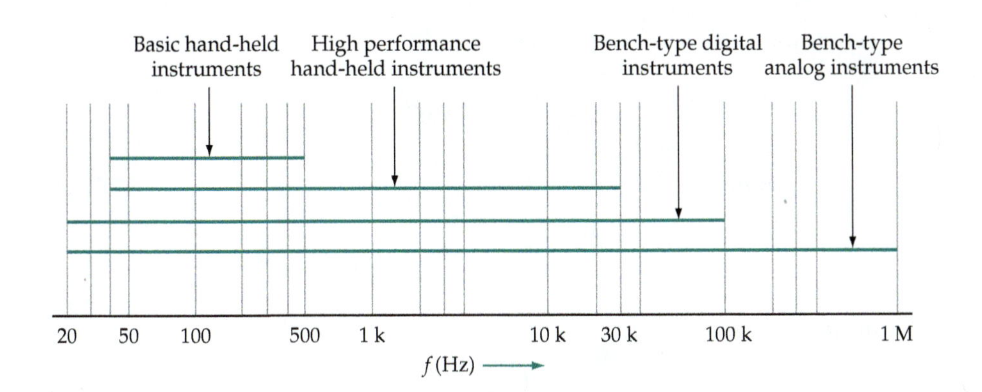

Figure 17-31 Typical frequency ranges for several types of ac instruments.

Summary of Formulas

- ***Peak voltage generated:***

$$E_m = B\ell v$$

- ***Frequency:***

$$f = \frac{1}{T}$$

- ***Wavelength:***

$$\lambda = \frac{c}{f}$$

- ***Instantaneous ac voltage:***

$$e = E_m \sin \omega t$$

- ***Instantaneous ac current:***

$$i = I_m \sin \omega t$$

- ***Instantaneous ac power:***

$$p = P_m \sin^2 \omega t$$

- ***Average ac power:***

$$P = 0.5P_m$$

- ***Average (of half-cycle) ac current:***

$$I_{av} = \frac{2}{\pi} I_m \approx 0.636 I_m$$

- ***Average (of half-cycle) ac voltage:***

$$E_{av} = \frac{2}{\pi} E_m \approx 0.636 E_m$$

- ***Rms ac current:***

$$I = \frac{1}{\sqrt{2}} I_m = 0.707 I_m$$

- ***Rms ac voltage:***

$$E = \frac{1}{\sqrt{2}} E_m = 0.707 E_m$$

Review Questions

Section 17-1

17-1 Draw sketches to illustrate and explain how an alternating current is generated in a conducting loop rotated in a magnetic field.

17-2 Derive an expression for the maximum voltage induced in a conducting loop rotated in a magnetic field. Also, derive an expression for the instantaneous value of the voltage.

Section 17-2

17-3 Define ac current, ac voltage, sinusoidal waveform, sine wave, peak value, phase angle, and amplitude.

Section 17-3

17-4 Sketch two sine waves that are: (a) in-phase, (b) in antiphase, (c) have a phase difference of $\pi/2$ radians.

Section 17-4

17-5 Derive equations for the instantaneous level of current through a resistor and for the instantaneous power dissipated when a sinusoidal ac voltage is applied to the resistor.

17-6 Draw a graph of instantaneous power versus time for the power dissipated in a resistor when a sinusoidal ac voltage is applied. Also, show that the average power dissipated is half the maximum instantaneous power.

Section 17-5

17-7 Explain how the average value of a sinusoidal voltage may be calculated, and write an expression for the average value in terms of the maximum value for: (a) a half-cycle of the waveform, (b) a full cycle of the waveform.

17-8 Define the rms value of an alternating current, and write an expression for the rms value in terms of the maximum value. Show that the rms value is the equivalent dc value or effective value.

Section 17-6

17-9 Referring to Figure 17-18 and 17-19 identify each of the major controls on the front panels of analog and digital oscilloscopes, and explain the function of each control.

17-10 Using illustrations, explain how an oscilloscope is used to measure voltage, frequency, and phase difference.

Section 17-8

17-11 Sketch half-wave and bridge rectifier circuits. Show the input and output waveform, and explain briefly the operation of each circuit.

17-12 Sketch the circuit of a rectifier voltmeter that uses a bridge rectifier circuit. Explain how the instrument operates, and discuss its limitations.

17-13 Sketch the basic circuit/block diagram of an electronic *ac* voltmeter. Show the waveforms at various points in the diagram, and explain the instrument operation.

17-14 Draw a diagram to show the typical frequency range of various *ac* voltmeters. Briefly explain.

Problems

Section 17-2

17-1 A conducting loop rotated in a magnetic field has an axial length of 30 cm and a width of 8 cm. The magnetic field flux density is 0.25 T, and the loop is rotated at 140 rpm. Calculate the loop output frequency and the peak output voltage if the loop has: (a) one turn, (b) 10 turns.

17-2 The generator described in Problem 17-1(a) is to be modified to produce a peak output of 100 mV. The alternatives are: (a) increase the conductor length, (b) increase the speed of rotation, (c) increase the flux density. Determine the required new quantity in each case.

17-3 Calculate the instantaneous levels of output voltage from the generator described in Problem 17-1(b) at $\pi/4$, $3\pi/4$, and $5\pi/4$ radians from $e = 0$.

17-4 For the generator described in Problems 17-1 and 17-2, calculate the angles of the conducting loop at which the instantaneous output voltage is 5 mV, 10 mV, 12 mV, and 19 mV.

17-5 A 25 turn coil of insulated copper wire with an axial length of 5 cm and a width of 4 cm is rotated at 2400 rpm in a magnetic field with a flux density of 0.13 T. Calculate the peak output voltage.

17-6 Determine the angles of the rotating coil in Problem 17-5 at the instants that the output voltage is 50 mV, 100 mV, 200 mV, 400 mV, and 800 mV.

Section 17-3

17-7 For the generator described in Problem 17-1(b), calculate the instantaneous levels of output voltage at 5 ms, 10 ms, 25 ms, and 30 ms from the instant of zero output.

17-8 For the generator described in Problem 17-5, determine the times from zero output until the instantaneous output voltages are 70 mV, 140 mV, 280 mV, and 560 mV.

17-9 An ac waveform has a peak value of 9 V and a frequency of 150 kHz. Determine the instantaneous voltage levels at 1.1 μs, 5 μs, and 29 μs from the instant of zero voltage.

17-10 Calculate the frequencies and wavelengths of the outputs from the generators in: (a) Example 17-1, (b) Problem 17-1, (c) Problem 17-5.

17-11 Calculate the frequency of an ac voltage with a wavelength of 300 km. Also, calculate the wavelengths of 120 kHz and 12 MHz ac waveforms.

17-12 Calculate the peak value of a 60 Hz waveform if its instantaneous level is measured as 110 V at 2 ms from the instant of zero output.

Section 17-4

17-13 A 120 Hz ac supply with a peak level of 100 V is applied to a 27 Ω resistor. Calculate the instantaneous levels of power dissipated at phase angles of $\pi/3$, $2\pi/3$, $5\pi/3$ and $3\pi/2$ radians.

17-14 The ac voltage in Problem 17-12 is applied to a 332 Ω resistor. Calculate the instantaneous power dissipations at 0.5 ms, 1 ms, 2 ms, and 4 ms from the instant of zero output. Also, determine the average power dissipated in the resistor.

17-15 A sinusoidal waveform with an rms value of 6 V and a frequency of 30 Hz is applied to a 120 Ω resistor. Determine the instantaneous power dissipated in the resistor at 5.5 ms from the instant of zero voltage. Also, calculate the peak instantaneous power and the average power dissipated in the resistor.

17-16 A resistor connected to the generator described in Problem 17-1(b) is to have a peak instantaneous power dissipation of 352 mW. Calculate the required resistance value. Also, determine the instantaneous power dissipation that occurs at 100 ms, 200 ms, and 400 ms from the instant of zero output voltage.

17-17 An instantaneous power of 410 μW is to be dissipated in a resistor connected to the generator described in Problem 17-5 at a time of 2.1 ms from zero output. Calculate the resistor value and determine the peak instantaneous power and the average power dissipated in the resistor.

17-18 A 400 Hz supply produces a peak power dissipation of 10 W in a 500 Ω resistor. Calculate the instantaneous power dissipated by the supply in a 3.3 kΩ resistor at angles of $\pi/4$, $\pi/2$, and $\pi/3$ radians from the instant of zero output.

Section 17-5

17-19 For Problem 17-13, calculate the rms and average levels of current in the resistor. Also, calculate the average power dissipation.

17-20 For Problem 17-14, determine the rms and average levels of current in the resistor and the average power dissipation.

17-21 A 297 V sine wave is applied to a 3.9 kΩ resistor. Calculate the peak, average, and rms levels of current flowing in the resistor. Also, determine the resistor power dissipation.

17-22 Calculate the average and rms output voltages from the generator in Problem 17-1(b). Determine the power dissipation in a 1 kΩ resistor connected to the output terminals.

17-23 Determine the rms and average output voltages from the generator in Problem 17-5. Calculate the value of a resistor that will dissipate 1 mW when connected to the generator output terminals.

17-24 A sine wave with a 166 V peak value dissipates 1 kW in a certain resistor. Calculate the resistor value and the average and rms levels of the current.

Section 17-6

17-25 Determine the peak-to-peak, average, and rms values, and the frequency of the waveform shown in Figure 17-32. Calculate the instantaneous amplitude of the wave at a phase angle of 120° and determine the amplitude at that angle from the display in Figure 17-32.

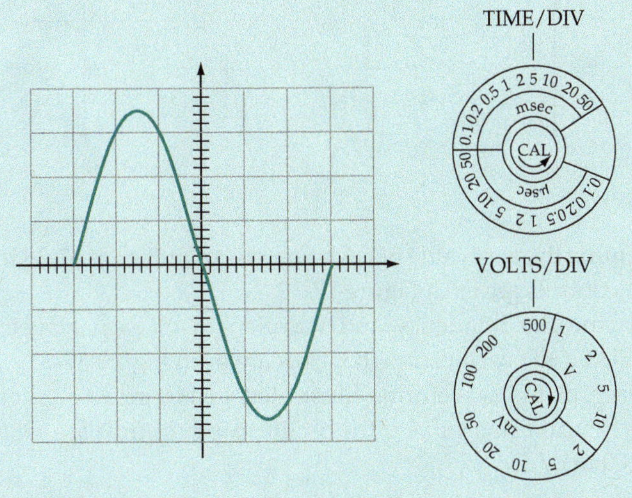

Figure 17-32

17-26 A voltage waveform with an amplitude of 1 V and a frequency of 25 Hz is to be displayed on an oscilloscope with the controls set as in Figure 17-32. Determine the number of vertical and horizontal divisions that would be occupied by the waveform.

17-27 (a) If the T1ME/DIV control in Figure 17-32 is changed to 10 ms, how many cycles of the waveform shown will be displayed on the oscilloscope screen? (b) How many cycles of a 500 Hz waveform could be displayed on an oscilloscope with the controls set as in Figure 17-32?

17-28 Calculate the rms value of the largest amplitude sinusoidal waveform that could be displayed on an oscilloscope with the controls set as in Figure 17-32.

17-29 Determine the number of vertical divisions on an oscilloscope screen that will be occupied by a waveform with a 0.9 V rms value when the controls are as illustrated in Figure 17-32, and the probe used is: (a) 1:1, (b) 10: 1.

17-30 Calculate the minimum frequency of a sine wave that can be displayed as one cycle on an oscilloscope with the TIME/DIV control set (a) as in Figure 17-32, (b) to 0.1 ms/div.

17-31 Determine the phase difference between the two waveforms shown in Figure 17-33(a), and between the two in Figure 17-33(b).

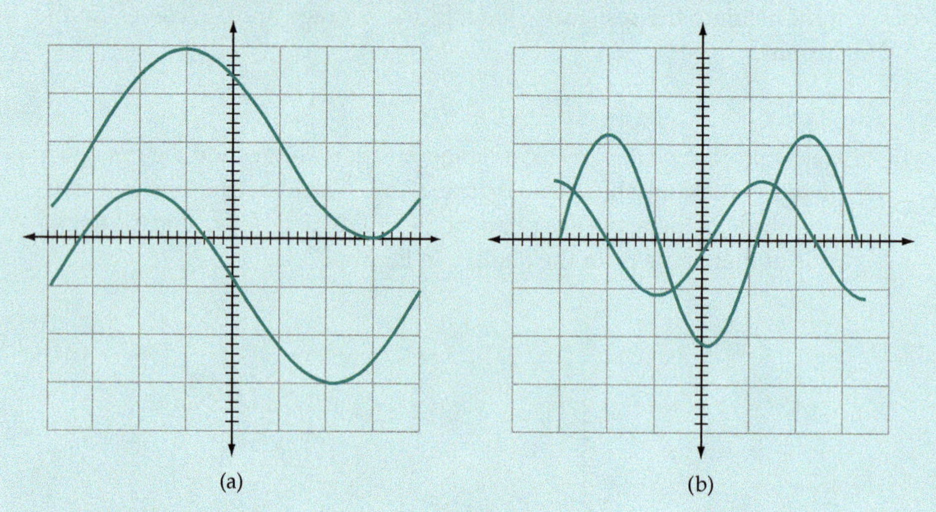

Figure 17-33

17-32 Determine the rms value of each waveform in Figure 17-33 if the oscilloscope controls are as in Figure 17-32.

17-33 Determine the frequency and wavelength of each waveform in Figure 17-33 if the oscilloscope controls are as in Figure 17-32.

17-34 For each of the waveforms illustrated in Figure 17-33, determine the frequency and rms value if the oscilloscope controls are set to 2 ms/div and 100 mV/div.

Section 17-8

17-35 A PMMC voltmeter with a 100 V FDS is used to measure an alternating voltage with an rms value of 70.7 V. Determine the meter reading when the frequency is (a) 60 Hz, (b) 0.05 Hz.

17-36 A 200 μA PMMC instrument with a coil resistance of 75 Ω is to be connected to a bridge rectifier circuit to construct an ac voltmeter with a range of 100 V. If the rectifiers each have a forward volts drop of 0.3 V, calculate the required multiplier resistance value.

17-37 The voltmeter in Problem 17-36 is to have additional ranges of 250 V and 50 V. Determine the multiplier resistance for each range.

17-38 A PMMC instrument with 50 μA FSD and a coil resistance of 850 Ω is to be used with a bridge rectifier to construct an ac voltmeter. The rectifiers each have a 0.7 V forward voltage drop. If the voltmeter is to have FSD at 150 V, calculate the value of the required multiplier resistance.

17-39 The ac voltmeter in Figure 17-34 consists of a 150 μA PMMC meter with a 100 Ω coil resistance, three series-connected resistors, and a bridge rectifier with rectifier voltage drops of 0.7 V. The resistor values are $R_1 = 174$ kΩ, $R_2 = 719$ kΩ, and $R_3 = 1.8$ MΩ. Determine the voltmeter range when the multiplier resistance is: (a) R_1 alone, (b) $R_1 + R_2$, (c) $R_1 + R_2 + R_3$.

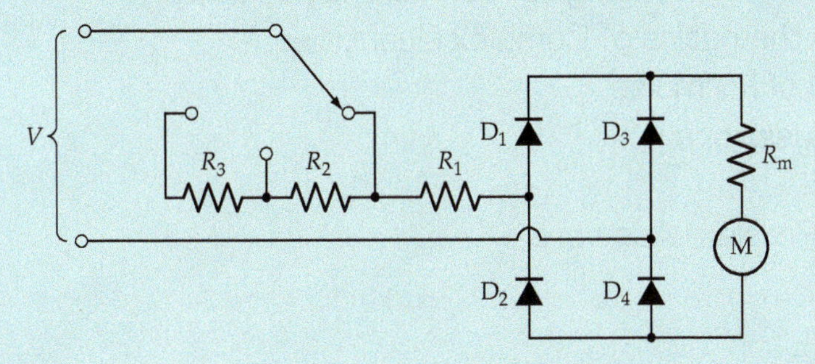

Figure 17-34

17-40 For the voltmeter in Problem 17-39, determine the voltmeter ranges if two rectifiers become open-circuited so that bridge rectifier functions as a half-wave rectifier circuit with two rectifiers connected in series.

Practice Problem Answers

17-2.1 796 rpm
17-2.2 41.3°, 45°, 58.2°
17-3.1 48.5 V, −48.5 V
17-3.2 405 m, 250 m
17-4.1 5.2 V, 5.2 V
17-5.1 16.73 V, 10.65 V, 11.83 V
17-5.2 4.95 W, 212 mA
17-6.1 10 V/div
17-6.2 230 mV, 5 kHz
17-6.3 0.7 div
17-8.1 30 V

CHAPTER 18
Phasors and Complex Numbers

CONTENTS

Objectives

You will be able to:

1. Show how the instantaneous value of a sinusoidal waveform can be represented by a phasor.
2. Show how two sinusoidal waveforms can be represented by two phasors with a phase angle between them.
3. Determine the resultant of two sinusoidal waveforms by addition of instantaneous levels of the waveforms and by phasor addition.
4. Perform phasor addition and subtraction by resolving each phasor into horizontal and vertical components.
5. Express sinusoidal quantities in polar form and in rectangular form, and convert from one form into the other.
6. Solve problems involving addition, subtraction, multiplication, and division of complex quantities.

INTRODUCTION

A sinusoidal alternating current or voltage can be graphically represented by a line of fixed length rotating about one of its ends. Such a line is termed a *phasor*, and the instantaneous value of a sinusoidal quantity can be determined from the length of the phasor and its angle with respect to the horizontal at the particular instant. A phasor can be mathematically represented by its length and angle or by its vertical and horizontal components. The vertical and horizontal components of a phasor can be written in the form of a complex number. Provided that certain rules are followed, phasors can be added, subtracted, multiplied, or divided. The same is true for complex numbers.

18-1 PHASOR REPRESENTATION OF ALTERNATING VOLTAGE

Scalars and Vectors

Quantities that can be completely represented by a number of units are termed *scalar* quantities. Some examples of scalar quantities are temperature, volume, and resistance. Other quantities, such as force and velocity, must have a direction indicated as well as the number of units stated if they are to be completely specified. Such quantities can be graphically represented by a line that has a length proportional to the number of units, with the line drawn at an angle with respect to some reference direction. This line is termed a *vector*, and the quantity it represents is classified as a *vector quantity*.

Phasors

The instantaneous levels of alternating current and voltage are vector quantities, but because the instantaneous levels are continuously changing, an ac waveform must be represented by a *rotating vector*, or *phasor*. A phasor is a vector that is rotating at a constant angular velocity. The sinusoidal output voltage produced by the simple generator described in Section 17-1 can be represented by the *phasor diagram* shown in Figure 18-1. Here *OA* is a rotating vector (or phasor) with a constant angular velocity and a length that represents the peak output voltage E_m. The arrowhead identifies the end of the phasor that moves, and the other end is the axis of rotation. By convention, the direction of phasor rotation is taken as counterclockwise.

The instantaneous value of the generated voltage represented by the phasor depends on the angle of the phasor with respect to the zero level. At time $t = 0$ on the waveform in Figure 18-1, the angle is $\theta = 0$ and the instantaneous voltage

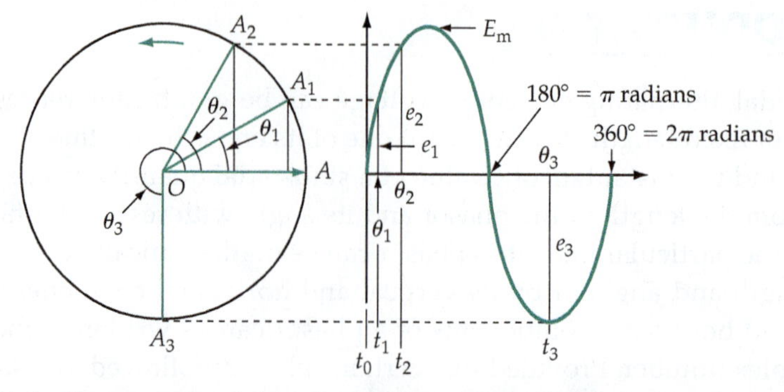

Figure 18-1 Phasor representation of a sine wave. Phasor *OA* rotates in a counterclockwise direction, and its length equals the peak voltage E_m. At any angle θ, *OA* sin θ equals the instantaneous value of the sine wave.

level is zero. At t_1, the angle is θ_1 and the instantaneous level is e_1, as illustrated in the figure.

At t_1,
$$\begin{aligned} e_1 &= (OA)\sin\theta_1 \\ &= E_m \sin\theta_1 \end{aligned}$$

At t_2,
$$e_2 = E_m \sin\theta_2$$

and at t_3,
$$e_3 = E_m \sin\theta_3$$

At each of these instants, the voltage level is represented by the *stopped phasor* (or vector) OA_1 at an angle θ_1, OA_2 at angle θ_2, and OA_3 at angle θ_3. Any sinusoidal ac voltage or current can be represented by a phasor, and instead of degrees, the angles may be expressed in radians, as explained in Section 17-3.

Example 18-1

Draw the phasor diagrams for the following voltages:

(a) $V_1 = 15$ V sin 35°, (b) $V_2 = 10$ V sin 60°, (c) $V_3 = 12$ V sin 135°.

Solution

See Figure 18-2.

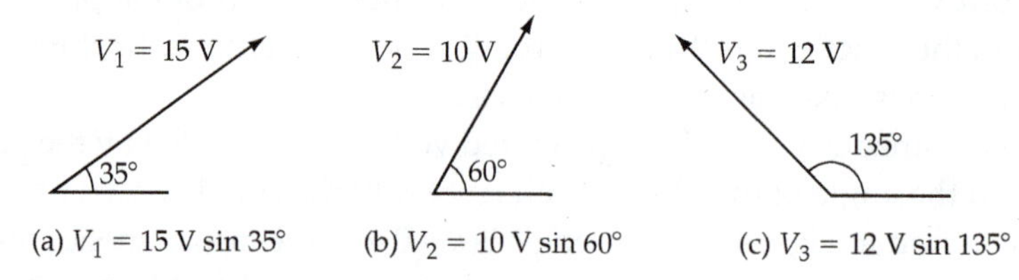

Figure 18-2 Phasor diagrams for Example 18-1.

Practice Problem

18-1.1 Draw phasor diagrams for $i_1 = 10$ mA sin 30° and $i_2 = 8$ mA sin 100°.

18-2 ADDITION AND SUBTRACTION OF PHASORS

Phasor Addition

In Figure 18-3 two sinusoidal waveforms (A and B) are shown together with the phasor representing each. It is seen that waveform B has a larger amplitude than waveform A. So, phasor *OB* is longer than phasor *OA*. Also, waveform B starts to grow positively from its zero level $\phi°$ after the beginning of waveform A. Consequently, waveform B lags waveform A by $\phi°$, and so the phasor *OB* is shown $\phi°$ behind *OA*. It can also be said that waveform A leads waveform B by $\phi°$.

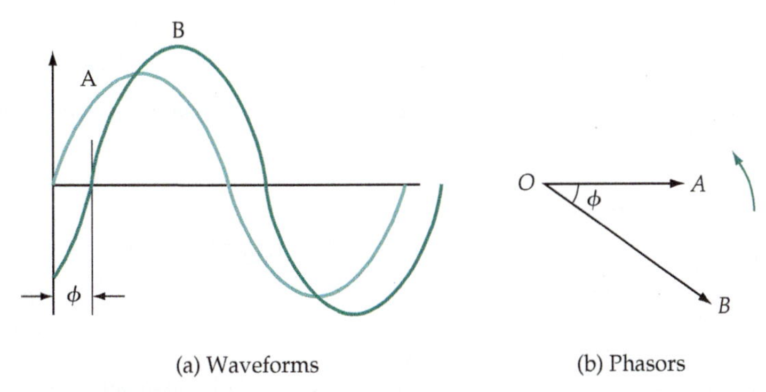

Figure 18-3 Two waveforms and their phasors. *OB* is larger than *OA* because waveform B has a higher peak value than waveform A. Waveform A leads waveform B by the angle ϕ; consequently, phasor *OA* leads phasor *OB* by ϕ.

The instantaneous level of waveform A can be written as,

$$e_A = OA \sin \theta$$

and the instantaneous level of waveform B is,

$$e_B = OB \sin(\theta - \phi)$$

Now consider Figure 18-4(a), in which the two waveforms are shown to be outputs from each of two series-connected generators. The resultant output of the two in series is the *phasor sum* of the two waveforms. Figure 18-4(b) shows how the instantaneous levels of A and B may be added to give the resultant waveform C. At t_1, $a_1 + b_1$ gives point 1 on waveform C. At t_2, $a_2 + b_2$ gives point 2 on waveform C, and so on. It is seen that waveform C lags waveform A by an angle $\alpha°$. So, waveform C is represented by phasor *OC* lagging $\alpha°$ behind *OA* [see Figure 18-4(c)].

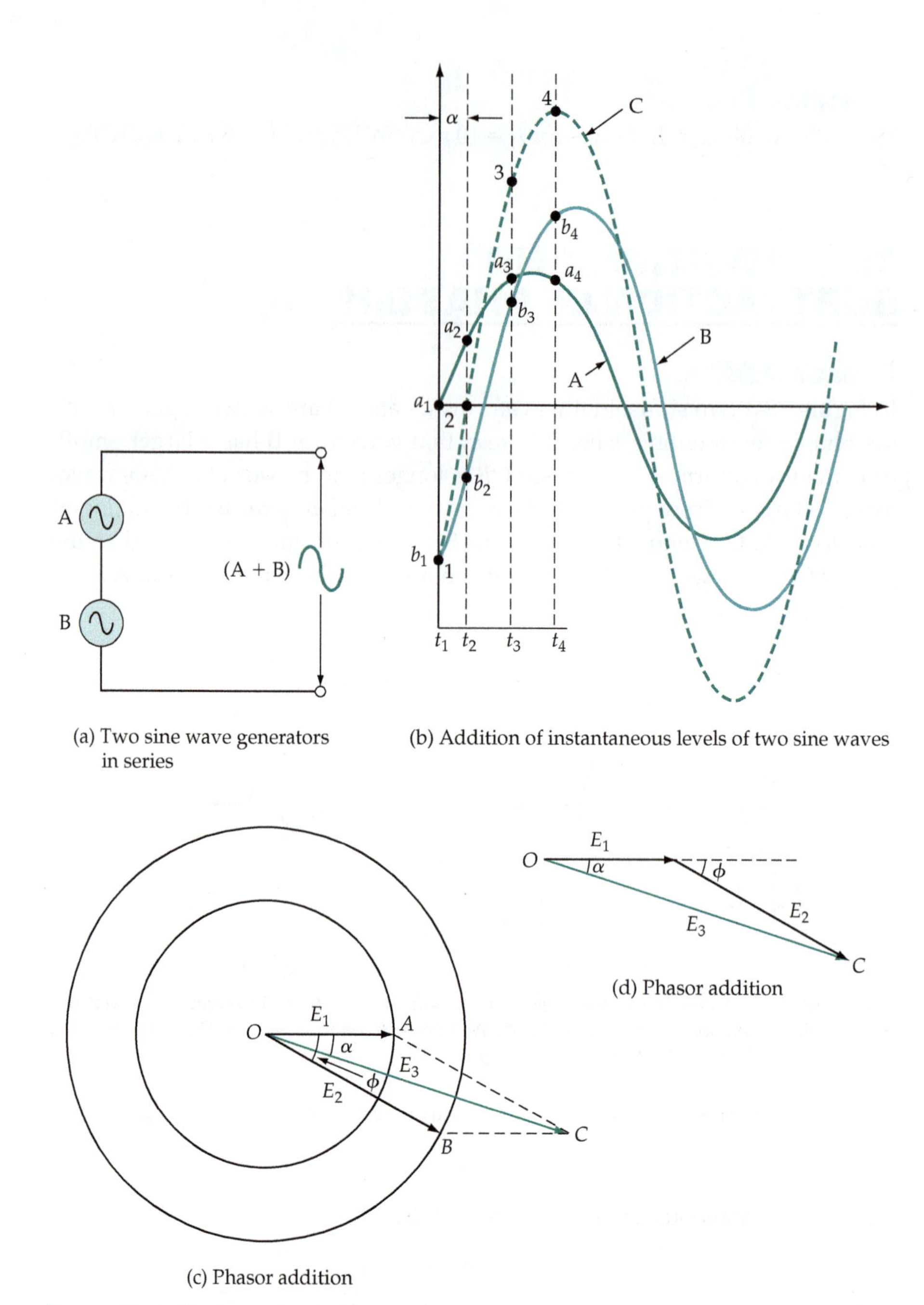

(a) Two sine wave generators in series

(b) Addition of instantaneous levels of two sine waves

(c) Phasor addition

(d) Phasor addition

Figure 18-4 **To determine the resultant output from two series-connected ac generators, the instantaneous levels of the waveforms can be added together. Alternatively, phasors *OA* and *OB*, representing each voltage, can be added graphically to obtain the resultant *OC*.**

Another method of obtaining the resultant of waveforms A and B is illustrated in Figure 18-4(c). A parallelogram is constructed, with *AC* drawn equal to and parallel with *OB*, and *BC* equal to and parallel with *OA*. The resultant diagonal of the parallelogram (*OC*) represents the amplitude of waveform C, and the angle $\alpha°$ is the angle by which waveform C lags waveform *A*. Another approach to obtain the same result is shown in Figure 18-4(d). Here, OA, identified as E_1, is first drawn horizontally. Then E_2 (which is OB) is drawn from the end of E_1 and at angle ϕ with respect to E_1. The resultant (E_3 or OC) is found by drawing a line from O to the end of E_2. The angle of E_3 with respect to E_1 is again seen to be $-\alpha°$.

Instead of graphically adding the two phasors, the resultant may be obtained mathematically. Referring to Figure 18-5, each phasor is resolved into horizontal and vertical components (by trigonometry). The arithmetical sum of the horizontal components is then the horizontal component of the resultant. Similarly, the arithmetical sum of the vertical components is the vertical component of the resultant.

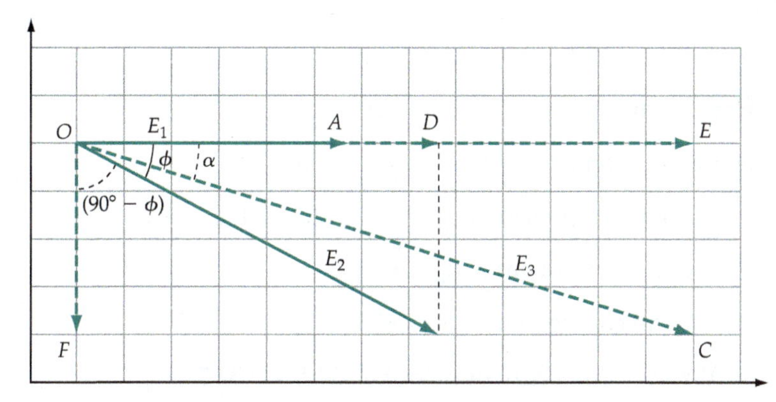

Figure 18-5 The resultant of two phasors can be determined by first resolving each into its horizontal and vertical components. The two horizontal quantities are added together, and the two vertical quantities are added, to give the horizontal and vertical components of the resultant.

$$\begin{aligned}
\text{Horizontal component of } E_1 &= OA \\
\text{Horizontal component of } E_2 &= OD = E_2 \cos \phi \\
\text{Total of horizontal components} &= OE = OA + OD \\
&= E_1 + E_2 \cos \phi \\
\text{Vertical component of } E_1 &= 0 \\
\text{Vertical component of } E_2 &= OF = -E_2 \cos(90° - \phi) = -E_2 \sin \phi \\
\text{Total of vertical components} &= 0 + OF = -E_2 \sin \phi
\end{aligned}$$

The negative sign here indicates that the vertical component is measured *down* from the horizontal, rather than up.

Resultant, $$E_3 = OC = \sqrt{(OE)^2 + (OF)^2}$$

Angle, $$\alpha = \tan^{-1}(OF/OE)$$

Example 18-2

Two sinusoidal quantities are given as $v_1 = 120 \sin \theta$ and $v_2 = 75 \sin(\theta - 30°)$. Determine the resultant obtained when v_1 and v_2 are added.

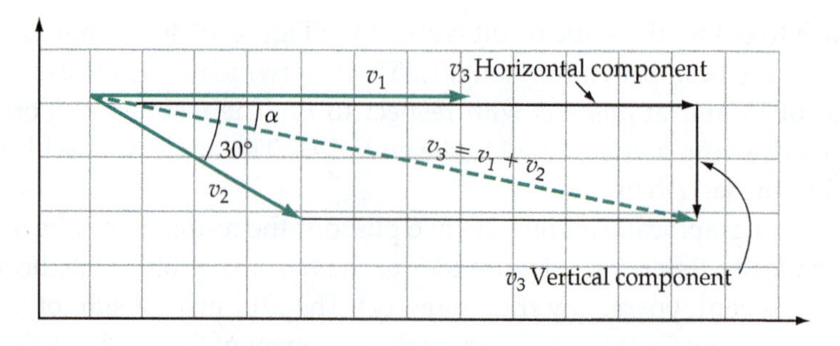

Figure 18-6 Sum of two phasors, $v_3 = v_1 + v_2$.

Solution

Referring to Figure 18-6:

$$\text{Horizontal component of } v_1 = 120$$

$$\text{Horizontal component of } v_2 = 75 \cos 30° \approx 65$$

$$\text{Horizontal component of } v_3 = 120 + 65 = 185$$

$$\text{Vertical component of } v_1 = 0$$

$$\text{Vertical component of } v_2 = -75 \sin 30° = -37.5$$

$$\text{Vertical component of } v_3 = 0 - 37.5 = -37.5$$

$$v_3 = \sqrt{185^2 + 37.5^2}$$

$$= 188.8$$

$$\text{Angle } \alpha = \tan^{-1}\left[\frac{v_3 \text{ vertical component}}{v_3 \text{ horizontal component}}\right]$$

$$= \tan^{-1} \frac{-37.5}{185}$$

$$= -11.5°$$

$$v_3 = 188.8 \sin(\theta - 11.5°)$$

Phasor Subtraction

Sometimes instead of phasor addition, *phasor subtraction* is required. An example of this occurs with the circuit shown in Figure 18-7(a), where

$$E_3 = E_1 - E_2$$

Figure 18-7(b) shows the process of phasor subtraction. E_2 is shown as $\theta°$ ahead of E_1. For subtraction, E_2 is reversed to become $-E_2$, as illustrated. The resultant (E_3) is then the phasor sum of E_1 and $-E_2$. Again, the resultant may be found mathematically or by either of the two graphical methods described.

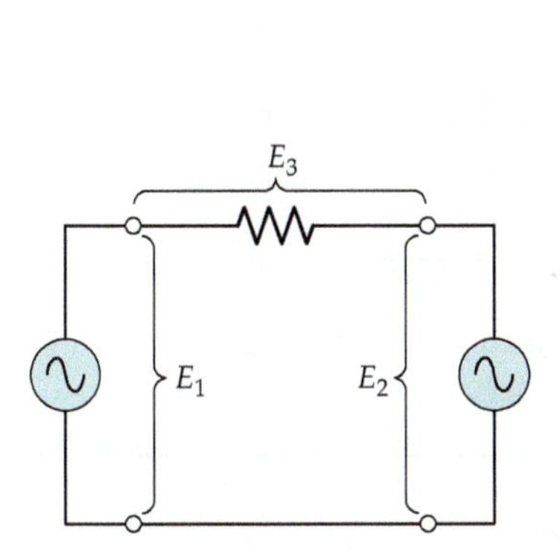

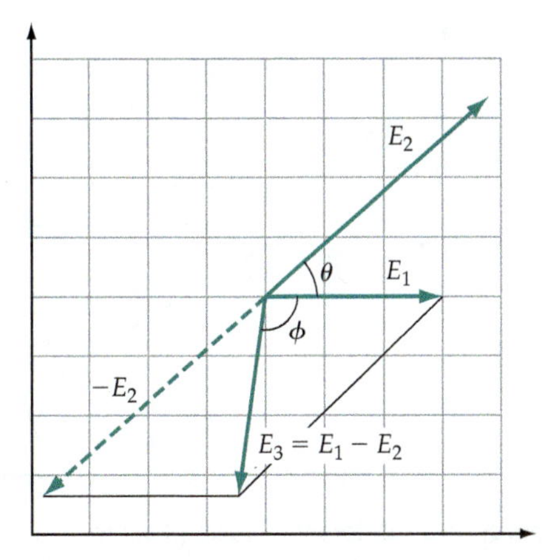

(a) E_3 is the phasor difference of E_1 and E_2 (b) Determination of E_3

Figure 18-7 The phasor difference of two phasors, E_1 and E_2, can be determined by summing E_1 and $-E_2$.

Phasor addition and subtraction are not limited to two phasor quantities. The resultant of any number of phasor quantities can be found; some of these may require phasor addition, while others may have to be subtracted.

Example 18-3

Determine $v_1 - v_2$ for the two quantities specified in Example 18-2.

Solution

Referring to Figure 18-8:

$$\text{Horizontal component of } v_1 = 120$$

$$\text{Horizontal component of } -v_2 = 75\cos(180^\circ - 30^\circ) \approx -65$$

$$\text{Horizontal component of } v_4 = 120 - 65 = 55$$

$$\text{Vertical component of } v_1 = 0$$

$$\text{Vertical component of } -v_2 = 75\sin(180^\circ - 30^\circ) = 37.5$$

$$\text{Vertical component of } v_4 = 0 + 37.5 = 37.5$$

$$v_4 = \sqrt{55^2 + 37.5^2}$$

$$= 66.6$$

$$\text{Angle } \phi = \tan^{-1}\frac{-37.5}{55}$$

$$= 34.3^\circ$$

$$v_4 = 66.6\sin(\theta + 34.3^\circ)$$

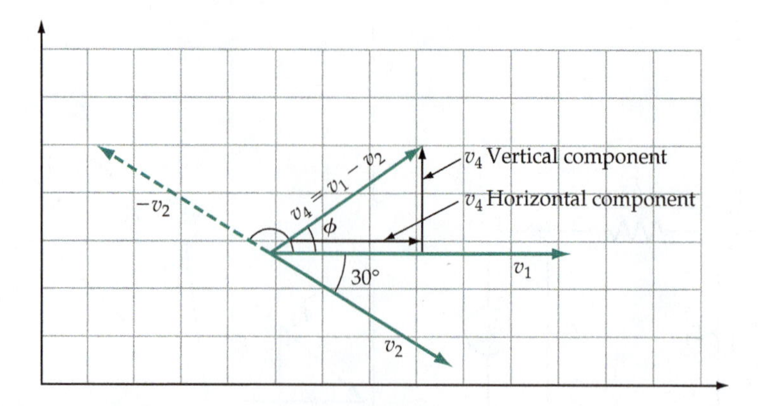

Figure 18-8 Difference of two phasors, $v_4 = v_1 - v_2$.

Example 18-4

Graphically determine the resultant of $v_1 + v_2 - v_3$, where $v_1 = 50 \sin \theta$, $v_2 = 30 \sin(\theta + 25°)$, and $v_3 = 25 \sin(\theta - 90°)$.

Solution

1. The phasor diagram for v_1, v_2, and v_3 is drawn in Figure 18-9(a).
2. In Figure 18-9(b), v_1 is drawn to scale.

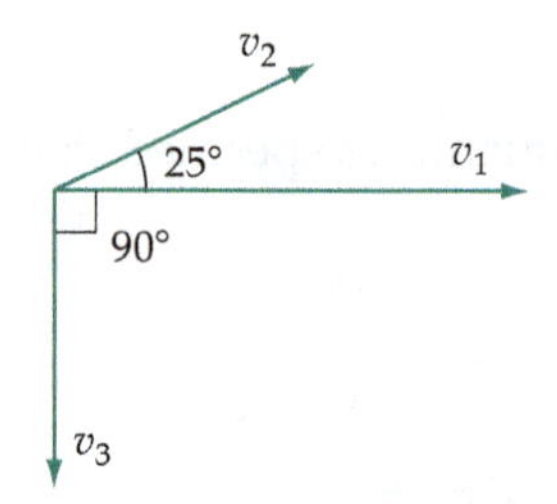

(a) Phasors v_1, v_2, and v_3

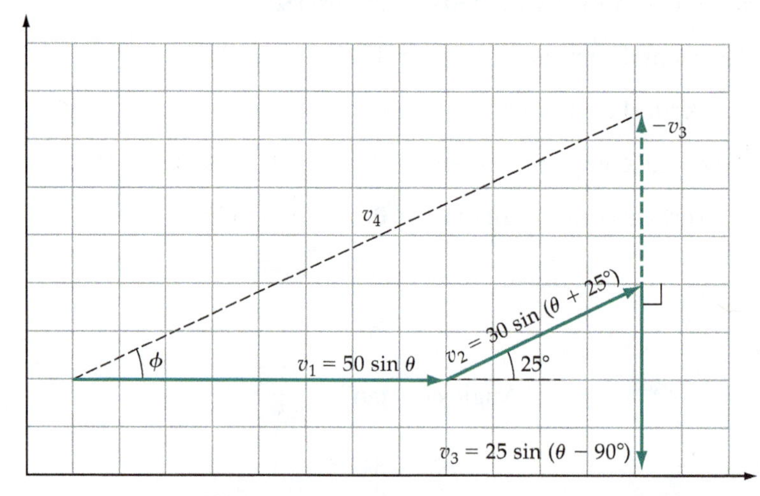

(b) Graphical solution of $v_1 + v_2 - v_3$

Figure 18-9 Graphical determination of $v_1 + v_2 - v_3$.

3. Phasor v_2 is drawn to scale from the end of v_1 and at an angle of +25° with respect to v_1.
4. Phasor v_3 is drawn to scale from the end of v_2 and at an angle of −90° with respect to v_1.
5. Phasor $-v_3$ is drawn opposite v_3, as illustrated.
6. Phasor v_4 is drawn by joining the origin of v_1 to the end of $-v_3$.
7. Phasor v_4 and angle ϕ are measured as 86 and 26°.

$$v_4 = 86 \sin(\theta + 26°)$$

Practice Problems

18-2.1 Two ac signal generators with identical frequencies have peak amplitudes of $E_1 = 100$ mV and $E_2 = 75$ mV. If E_2 lags E_1 by 25°, determine the resultant when the two outputs are connected in series.

18-2.2 For the quantities specified in Example 18-4, graphically determine $v_1 - v_2 + v_3$.

18-3 POLAR AND RECTANGULAR FORMS, THE *j* OPERATOR

Polar Form

Another way of mathematically expressing a sinusoidal quantity, termed *polar form,* leaves out *sin* and substitutes ∠ for angle. So,

$$e_1 = E_m \sin\theta \qquad \text{becomes } e_1 = E_m\angle\theta$$

and,

$$i_1 = I_m \sin(\theta + \alpha) \quad \text{becomes } i_1 = I_m\angle\theta + \alpha$$

Polar form is very convenient for phasor quantities. The terms *modulus* and *argument* are sometimes used for the magnitude and angle, respectively, of a polar quantity. For e_1, the modulus is E_m, and θ is the argument.

Because there is a direct (0.707) relationship between the rms and the peak values of a sinusoidal waveform, rms values are normally employed when using polar form. In fact, *with alternating voltages and currents, all quantities are assumed to be rms quantities unless otherwise indicated.* Phasor addition and subtraction using rms quantities follows exactly the same graphical or mathematical procedure as when using peak quantities. The resultant obtained is, of course, an rms quantity. Rms values and peak values must not be mixed together in calculations. Where different types of quantities are to be combined, *all must be converted into one type* before phasor addition or subtraction.

As already explained, phasors must be resolved into horizontal and vertical components for addition and subtraction. So, in Figure 18-10(a), the polar form,

$$e = E\angle\theta$$

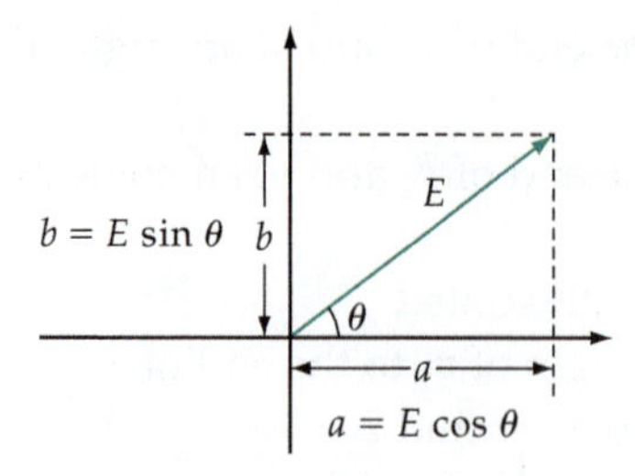

(a) Resolution of a phasor into vertical and horizontal components

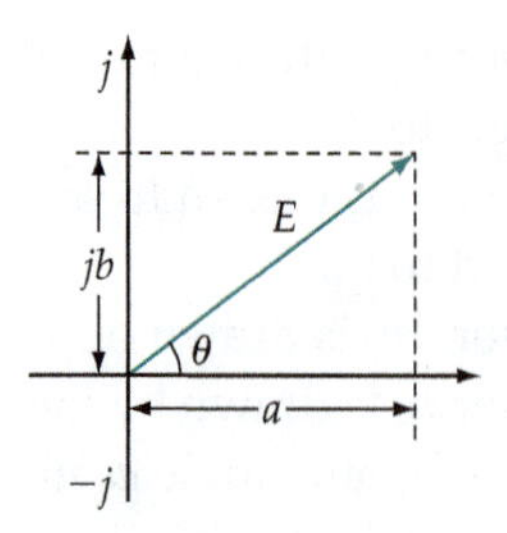

(b) Use of the *j* operator

Figure 18-10 The *j* operator can be used to identify the vertical component of a phasor. Vertical component *b* is written as *jb*, and horizontal component *a* is simply written as *a*. So, phasor $E\angle\theta$ is represented as $a + jb$.

becomes,

$$\text{Horizontal component, } a = E\cos\theta$$

$$\text{Vertical component, } b = E\sin\theta$$

Rectangular Form, the *j* Operator

For convenience in writing vertical and horizontal components the *j operator* is employed, as illustrated in Figure 18-10(b). All positive vertical components are given the prefix *j*, and all negative vertical components are prefixed with $-j$. The quantity $e = E\angle\theta$ can now be written as,

$$E\angle\theta = a + jb$$

or,

$$E\angle\theta = E\cos\theta + jE\sin\theta \tag{18-1}$$

A phasor stated in this way is expressed in *rectangular form*. Conversion from polar to rectangular form is simply a matter of resolving the phasor into its horizontal and vertical components. Conversion from rectangular to polar form is the reverse of that process, as follows:

$$E = \sqrt{a^2 + b^2}$$

and,

$$\theta = \tan^{-1}(b/a)$$

or

$$E\angle\theta = \sqrt{a^2 + b^2}\angle\tan^{-1}(b/a) \tag{18-2}$$

The *j* operator is not merely a prefix employed for convenience in identifying the vertical component of a phasor. Expressions such as Equation 18-1, using the *j* operator, are termed *complex numbers*, and the mathematics of complex numbers require that certain rules be followed. Consider phasor $E\angle 0°$ in Figure 18-11. When multiplied by *j*, the phasor is rotated counterclockwise by 90° or $\pi/2$ radians. So, it becomes,

$$E\angle 90° = jE$$

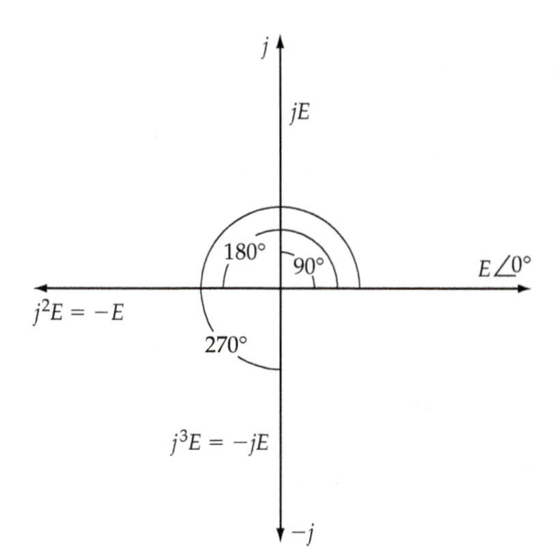

Figure 18-11 Multiplying $E\angle 0$ by j gives jE. Multiplying by j^2 gives $-E$. Multiplying by j^3 produces $-jE$.

When jE is multiplied by j, the phasor is,

$$E\angle 180^\circ = j^2E = -E$$

So,

$$j^2 = -1 \qquad (18\text{-}3)$$

or,

$$j = \sqrt{-1}$$

The actual mathematical square root of +1 is 1, but the square root of −1 cannot be determined. For this reason j is sometimes referred to as an *imaginary number*. For the complex number $a + jb$, a is sometimes termed the *real part*, while jb is called the *imaginary part*.

Returning to Figure 18-11, when $-E$ is once more multiplied by j, the phasor is rotated counterclockwise by a further 90°, giving,

$$E\angle 270^\circ = -jE$$

and when multiplied by a fourth j, the quantity is,

$$E\angle 0^\circ = j^2E = -(-1)E$$
$$= E$$

Example 18-5

Convert the following quantities into rectangular form: (a) $12\angle 30^\circ$, (b) $270\angle 1.7\pi$, (c) $40\angle 105^\circ$.

Solution

(a) Eq. 18-1,

$$E\angle\theta = E\cos\theta + jE\sin\theta$$
$$12\angle 30^\circ = 12\cos 30^\circ + j12\sin 30^\circ$$
$$= 10.4 + j6 \quad \text{[see Figure 18-12(a)]}$$

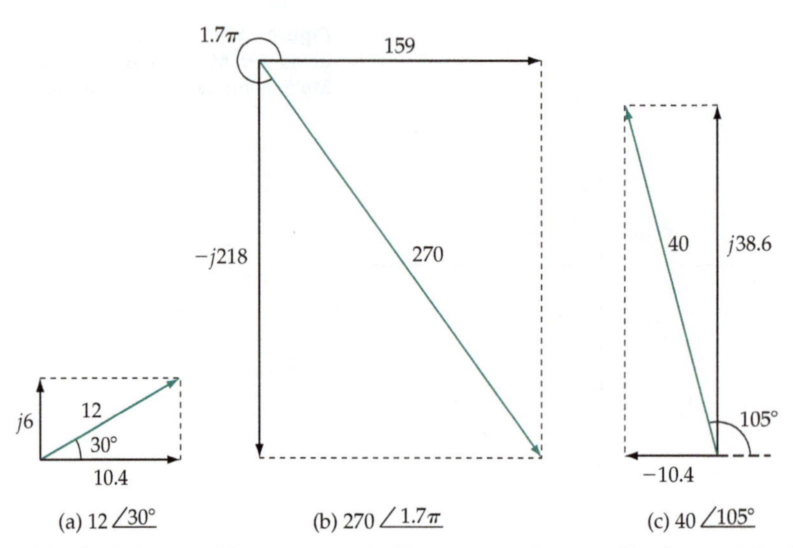

Figure 18-12 Polar quantities are converted into rectangular quantities by resolving them into their horizontal and vertical components. $E\angle\theta$ is resolved into $E\cos\theta$ and $jE\sin\theta$.

(b)
$$270\angle 1.7\pi = 270\cos(1.7\pi) + j270\sin(1.7\pi)$$
$$= 159 - j218 \quad \text{[see Figure 18-12(b)]}$$

(c)
$$40\angle 105° = 40\cos 105° + j40\sin 105°$$
$$= -10.4 + j38.6 \quad \text{[see Figure 18-12(c)]}$$

Practice Problems

18-3.1 Taking $\theta = 0$, convert each of the quantities in Example 18-4 into rectangular form.

18-3.2 Express the following quantities in polar form: $a = 3 + j4$, $b = 77 - j220$, $c = -53 - j47$.

18-4 MATHEMATICS OF COMPLEX QUANTITIES

Addition and Subtraction

When mathematically processing complex quantities in rectangular form, the usual rules of algebra apply as long as it is remembered that $j^2 = -1$. For addition or subtraction, the real parts of the numbers are added or subtracted, and the imaginary parts are added or subtracted separately. This is, of course, simply adding or subtracting the horizontal and vertical components of a phasor.

$$(a + jb) + (c + jd) = (a + c) + j(b + d) \tag{18-4}$$

$$(a + jb) - (c + jd) = (a - c) + j(b - d) \tag{18-5}$$

Multiplication

Multiplication of complex quantities in rectangular form follows the normal procedure for multiplication of algebraic expressions, as shown below.

$(a + jb) \times (c + jd)$:

$$\begin{array}{r} (a + jb) \\ \times (c + jd) \\ \hline \end{array}$$

Multiplying the top line by c, $\quad ac + jbc$

Multiplying the top line by jd, $\quad jad + j^2bd$

Adding, $\quad ac + jbc + jad - bd$

or, $\quad (ac - bd) + j(bc + ad)$

So,

$$(a + jb) \times (c + jd) = (ac - bd) + j(bc + ad) \tag{18-6}$$

Division

Division of complex quantities in rectangular form requires the use of the *conjugate* of the denominator. The conjugate of a complex number is the same number with the sign changed on the j component. So, the conjugate of $(a + jb)$ is $(a - jb)$.

$$\frac{a + jb}{c + jd} = \frac{a + jb}{c + jd} \times \frac{c - jd}{c - jd}$$

Multiplying by $(c - jd)/(c - jd)$ is the same as multiplying by 1. Nothing has been changed, and so,

$$\frac{a + jb}{c + jd} \times \frac{c - jd}{c - jd} = \frac{ac + jbc - jad - j^2bd}{c^2 + jcd - jcd - j^2d^2}$$

$$= \frac{(ac + bd) + j(bc - ad)}{c^2 + d^2}$$

So,

$$\frac{a + jb}{c + jd} = \frac{ac + bd}{c^2 + d^2} + j\frac{bc - ad}{c^2 + d^2} \tag{18-7}$$

Another important consideration in processing j quantities is the reciprocal of a j quantity as shown below:

$$\frac{1}{jx} = \frac{1}{jx} \times \left(\frac{-j}{-j}\right) = \frac{-j}{-j^2x} = \frac{-j}{-(-1)x}$$

$$= \frac{-j}{x}$$

So,

$$\frac{1}{jx} = -j\frac{1}{x}$$

or,

$$\frac{1}{j} = -j \tag{18-8}$$

Example 18-6

Determine: (a) $(75 - j50) \times (25 + j5)$, (b) $(75 - j50)/(25 + j5)$.

Solution

(a) $(75 - j50) \times (25 + j5)$:

$$\begin{array}{l} \quad (75 - j50) \\ \underline{\times (25 + j5)} \\ 1875 - j1250 \\ \underline{\qquad + j375 - j^2 250} \\ 1875 - j1250 + j375 + 250 \\ = 2125 - j875 \end{array}$$

(b) $(75 - j50)/(25 + j5)$:

$$\frac{75 - j50}{25 + j5} \times \frac{25 - j5}{25 - j5} = \frac{1875 - j1250 - j375 + j^2 250}{625 + j125 - j125 - j^2 25}$$

$$= \frac{1625 - j1625}{650}$$

$$= 2.5 - j2.5$$

Polar Form Mathematics

Multiplication and division in polar form are much simpler than in rectangular form:

$$E_1\angle\theta_1 \times E_2\angle\theta_2 = E_1E_2\angle\theta_1 + \theta_2 \tag{18-9}$$

and,

$$\frac{E_1\angle\theta_1}{E_2\angle\theta_2} = \frac{E_1}{E_2}\angle\theta_1 - \theta_2 \tag{18-10}$$

For addition and subtraction, quantities stated in polar form must first be converted into rectangular form.

Example 18-7

Determine: (a) $(90\angle -33.7°) \times (25.5\angle 11.3°)$, (a) $(90\angle -33.7°)/(25.5\angle 11.3°)$.

Solution

(a) $$(90\angle -33.7°) \times (25.5\angle 11.3°) = (90 \times 25.5)\angle -33.7° + 11.3°$$
$$= 2295\angle -22.4°$$

(b) $$(90\angle -33.7°)/(25.5\angle 11.3°) = (90/25.5)\angle -33.7° - 11.3°$$
$$= 3.5\angle -45°$$

Note that $90\angle{-33.7°}$ and $25.5\angle{11.3°}$ are the polar forms of the quantities used in Example 18-6. When the answers in Example 18-7 are checked, they are found to be the polar forms of the answers in Example 18-6.

Practice Problems

18-4.1 Using the quantities in Problem 18-3.2, determine $a + b - c$.

18-4.2 Using the quantities in Problem 18-3.2, determine $a \times b/c$.

18-4.3 Solve $(100 + j85)/(37 - j29)$ in rectangular form. Then convert each quantity into polar form and find the result in polar form.

Summary of Formulas

- ***Polar to rectangular conversion:***

$$E\angle\theta = E\cos\theta + jE\sin\theta$$

- ***Rectangular to polar conversion:***

$$E\angle\theta = \sqrt{a^2 + b^2}\angle\tan^{-1}(b/a)$$

- ***j^2 and $1/j$:***

$$j^2 = -1$$

$$1/j = -j$$

- ***Addition:*** $(a + jb) + (c + jd) = (a + c) + j(b + d)$
- ***Subtraction:*** $(a + jb) - (c + jd) = (a - c) + j(b - d)$
- ***Multiplication:*** $(a + jb) \times (c + jd) = (ac - bd) + j(bc + ad)$
- ***Division:***

$$\text{For } \frac{a + jb}{c + jd} \text{ multiply by } \frac{c - jd}{c - jd}$$

- ***Multiplication:*** $E_1\angle\theta_1 \times E_2\angle\theta_2 = E_1E_2\angle\theta_1 + \theta_2$
- ***Division:***

$$\frac{E_1\angle\theta_1}{E_2\angle\theta_2} = \frac{E_1}{E_2}\angle\theta_1 - \theta_2$$

Review Questions

Section 18-1

18-1 Sketch a sinusoidal waveform and show how its instantaneous value may be represented by a phasor.

18-2 Show how two sinusoidal waveforms that have a phase difference may be represented by phasors. Briefly explain.

Section 18-2

18-3 Sketch the waveform from two series-connected ac generators that have ac voltages that differ in amplitude and phase. Show how the resultant of the two waveforms may be obtained by adding the instantaneous voltage values. Also show how the resultant may be obtained by phasor addition.

18-4 Sketch two phasors with different amplitudes and phase angles. Show how the resultant of the two can be obtained by resolving each into horizontal and vertical components.

18-5 Draw the diagram of a circuit that involves phasor subtraction. Sketch two phasors differing in amplitude and phase, and show how one should be subtracted from the other.

Section 18-3

18-6 Sketch a phasor diagram to show the relationship between polar and rectangular forms of phasor representation. Show the effect of multiplying $E\underline{/0}$ by j, j^2, and j^3. Explain briefly.

Problems

Section 18-2

18-1 Graphically determine $v_1 + v_2$, where $v_1 = 47 \sin \phi$ and $v_2 = 33 \sin (\phi + 20°)$.

18-2 For the quantities given in Problem 18-1, graphically determine $v_1 - v_2$.

18-3 Graphically determine the resultant of $v_1 + v_2 + v_3 - v_4$, where $v_1 = 10 \sin \phi$, $v_2 = 15 \sin(\phi - 15°)$, $v_3 = 20 \sin(\phi + 10°)$, and $v_4 = 18 \sin (\phi + 25°)$.

18-4 For the quantities given in Problem 18-3, graphically determine $v_1 - v_2 - v_3 + v_4$.

18-5 Graphically determine $a - 2b$, where $a = 150\underline{/22°}$ and $b = 85\underline{/2.5\pi}$.

18-6 Graphically determine $b + c$, where $b = 85\underline{/2.5\pi}$ and $c = 64\underline{/72°}$.

18-7 Draw the phasor diagram for $i_1 + i_2 - i_3$, where $i_1 = 12\underline{/125°}$, $i_2 = 10\underline{/0}$ and $i_3 = 15\underline{/86°}$.

18-8 For the quantities given in Problem 18-7, graphically determine $i_1 - i_2 + i_3$.

18-9 Graphically determine the resultant of $v_1 + v_2 - v_3$, where $v_1 = 45\angle 30°$, $v_2 = 27\angle 21°$, and $v_3 = 30\angle 42°$.

18-10 For the quantities given in Problem 18-9, graphically determine $v_1 - v_2 - v_3$.

18-11 For $a = 150\angle 22°$, $b = 85\angle 2.5\pi$, and $c = 64\angle 72°$, graphically determine $a + b - c$.

18-12 For the quantities given in Problem 18-11, graphically determine $2a + b - 2c$.

Section 18-3

18-13 Convert $(75 - j50)$ and $(25 + j5)$ into polar form.

18-14 Convert $i_1 = 12\angle 125°$, $i_2 = 10\angle 0$, and $i_3 = 15\angle 86°$ into rectangular form.

18-15 Convert $a = 150\angle 22°$, $b = 85\angle 2.5\pi$, and $c = 64\angle 72°$ into rectangular form.

18-16 Convert the following quantities into polar form: $3 + j4$, $120 - j75$, $680 + j170$, $-35 + j84$.

18-17 Convert the quantities stated in Problem 18-9 into rectangular form.

18-18 Convert the following quantities into polar form: $-23 + j23$, $-333 - j270$, $80 + j17$, $-75 + j56$.

Section 18-4

18-19 Algebraically solve Problem 18-1.

18-20 Algebraically solve Problem 18-2.

18-21 For the quantities given in Problem 18-9, determine $(v_1 \times v_2)/v_3$.

18-22 Solve $(16 + j12) \times (22 - j18)/(14 + j25)$.

18-23 Algebraically solve Problem 18-11.

18-24 For the quantities given in Problems 18-5 and 18-6, determine $(a - 2b)/(b + c)$.

18-25 Solve $[(25 + j15) + (45 - j50)] \times (33 - j29)/[(62 + j70) - (32 + j100)]$.

18-26 For the quantities given in Problem 18-1, determine $(v_1 \times v_2)/[2(v_1 + v_2)]$.

18-27 Solve $[(47\angle 35° - 22\angle 60°) \times 54\angle 30°]/(76 + j29)$.

18-28 Add $6.4\angle 22°$, $3.9\angle 45°$, and $4.1\angle 68°$.

18-29 Algebraically solve Problem 18-7; $i_1 + i_2 - i_3$, where $i_1 - 12\angle 125°$, $i_2 = 10\angle 0°$, and $i_3 = 15\angle 86°$.

18-30 Solve $Z = (v_1 + v_2)/i_1$, where $v_1 = 115\angle 0°$, $v_2 = 115\angle 120°$, and $i_1 = 22\angle 73°$.

18-31 Find $(a \times c)/b$, where $a = 29 - j73$, $b = 64 + j55$, and $c = 49 - j22$.

18-32 Algebraically solve Problem 18-9; $v_1 + v_2 - v_3$, where $v_1 = 45\angle 30°$, $v_2 = 27\angle 21°$, and $v_3 = 30\angle 42°$.

18-33 Algebraically solve Problem 18-8.

18-34 Algebraically solve Problem 18-4.

18-35 Algebraically solve Problem 18-12.

Practice Problem Answers

18-2.1 171 mV$\angle 10.7°$
18-2.2 44 sin(0 − 59°)
18-3.1 (50 + *j*0), (27.2 + *j*12.7), (0 − *j*25)
18-3.2 5$\angle 53.1°$, 233$\angle 70.7°$, 70.8$\angle 222°$
18-4.1 133 − *j*169
18-4.2 16.5$\angle -239°$
18-4.3 (0.559 + *j*2.74), 2.8$\angle 78.4°$

CHAPTER 19
Inductance and Capacitance in AC Circuits

CONTENTS

Objectives

You will be able to:

1 Sketch the waveforms of supply voltage and current in a purely inductive circuit and explain the phase relationships between the waveforms.

2 Calculate the inductive reactance of a given inductor, and the current that flows when the inductor is connected to an ac supply.

3 Sketch the waveforms of supply voltage and current in a purely capacitive circuit and explain the phase relationships between the waveforms.

4 Calculate the capacitive reactance of a given capacitor and the current that flows when the capacitor is connected to an ac supply.

5 Sketch and explain all voltage and current waveforms in series *RL* and series *RC* circuits.

6 Calculate all current levels, voltage levels, and phase relationships in series-connected and parallel-connected *RC* circuits. Sketch phasor diagrams, impedance diagrams, and admittance diagrams where appropriate.

7 Calculate all current levels, voltage levels, and phase relationships in series-connected and parallel-connected *RL* circuits. Sketch phasor diagrams, impedance diagrams, and admittance diagrams where appropriate.

8 Calculate all current levels, voltage levels, and phase relationships in series-connected and parallel-conneced *RLC* circuits. Sketch phasor diagrams, impedance diagrams, and admittance diagrams where appropriate.

INTRODUCTION

Alternating current flow in an inductor depends on the applied voltage and the inductive reactance of the inductor. The inductive reactance is proportional to the inductance value and the frequency of the alternating supply voltage. Similarly, the alternating current flow in a capacitor depends on the supply voltage and the capacitive reactance, which is related to the capacitance value and the supply frequency.

When an alternating voltage is applied to a pure inductance, the current that flows through the inductance lags its terminal voltage by 90°. Conversely, the alternating current through a capacitor leads the capacitor terminal voltage by 90°. For *RL* and *RC* series circuits, the phase angle of current with respect to supply voltage is less than 90°.

A series *RLC* circuit may behave as a resistive-inductive circuit or as a resistive-capacitive circuit, depending on the component values and supply frequency. The same can be said of a parallel *RLC* circuit.

19-1 ALTERNATING CURRENT AND VOLTAGE IN AN INDUCTIVE CIRCUIT

In Section 17-4, it is explained that when an alternating voltage is applied to a purely resistive circuit the resultant alternating current through the resistor is in phase with the applied voltage. This is illustrated by the voltage and current waveforms in Figure 19-1. Because of the presence of the counter-emf in an inductor, the current and voltage phase relationships in an inductive circuit are considerably different from those in a purely resistive circuit.

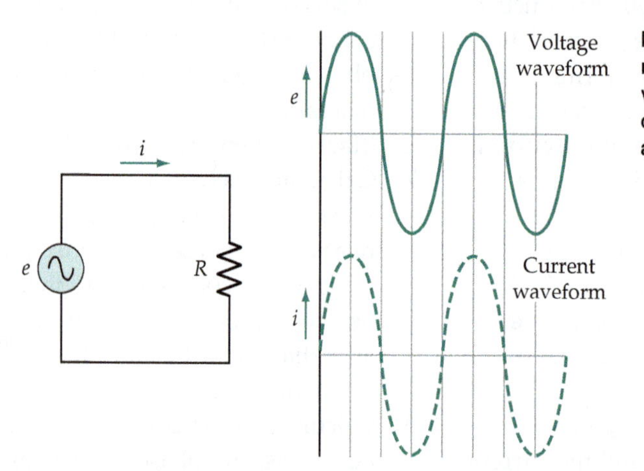

Figure 19-1 In a purely resistive circuit with an ac voltage source, the circuit current is in phase with the applied voltage.

Figure 19-2(a) shows the circuit of an inductor (L) connected to an ac voltage source, and the resultant current and voltage waveforms are illustrated in Figure 19-2(b). For the waveforms shown, it is assumed that the coil resistance is very much smaller than its inductance, so that the circuit is considered to

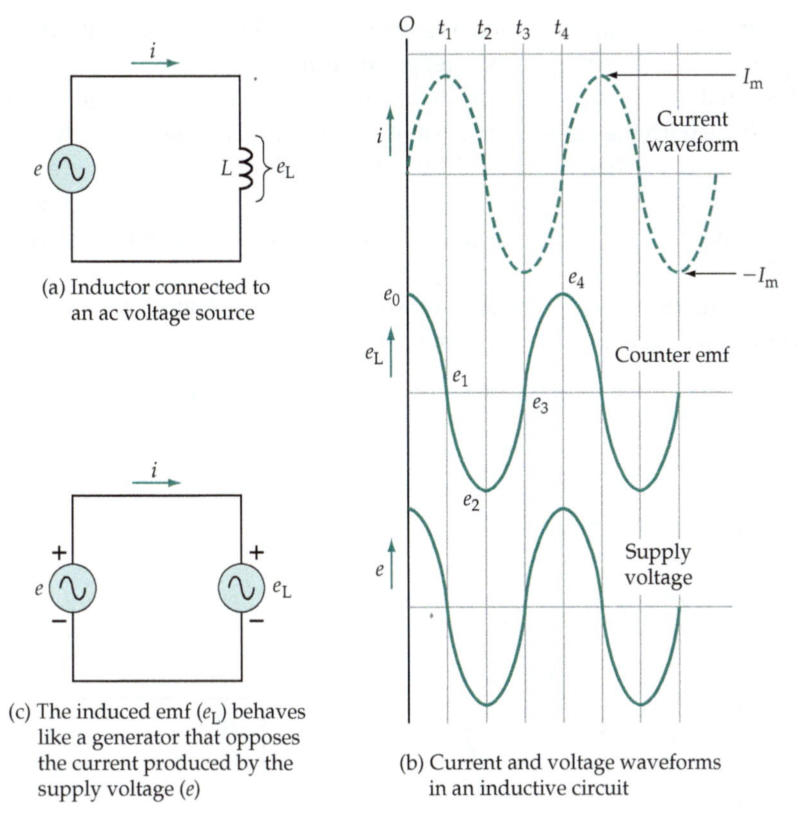

(a) Inductor connected to an ac voltage source

(c) The induced emf (e_L) behaves like a generator that opposes the current produced by the supply voltage (e)

(b) Current and voltage waveforms in an inductive circuit

Figure 19-2 **For a purely inductive circuit, $\Delta i/\Delta t$ is a maximum positive quantity when the current (i) crosses zero in a positive-going direction. The counter emf (e_L) opposes Δi and is proportional to $\Delta i/\Delta t$. So, e_L is a maximum positive quantity when $\Delta i/\Delta t$ is at its positive maximum.**

be purely inductive. Because the counter-emf is proportional to the rate of change of current through the inductance, it is appropriate to start with the current waveform.

Figure 19-2(b) shows the sinusoidal current waveform with a zero level at time $t = 0$, positive peak (I_m) at t_1, zero again at t_2, and so on. At time t_1, the current has stopped increasing positively and has not yet started to decrease. Therefore, the rate-of-change of current ($\Delta i/\Delta t$) is zero at t_1. Now recall from Equation 14-3 that the counter-emf in an inductor is,

$$e_L = L\frac{\Delta i}{\Delta t}$$

So, because the counter-emf is directly proportional to the rate-of-change of current, e_L is zero when $\Delta i/\Delta t$ is zero at time t_1. This gives *point* e_1 on the waveform representing counter-emf.

At time t_2, the current is decreasing at its maximum rate, which means that it has a maximum negative rate of change. Thus, *point* e_2 (at t_2) on the e_L waveform is a peak negative value. At t_3, the current has a zero rate-of-change once again, and the e_L value is plotted as zero at *point* e_3. Finally, at t_4, the current has a maximum positive rate of change, giving *point* e_4 on the e_L waveform.

The final waveform in Figure 19-2(b) shows the supply voltage waveform (e) in-phase with the counter-emf wave. The relationship between the e and e_L waveforms is further illustrated by Figure 19-2(c), which shows a voltage generator representing the counter-emf connected in phase opposition with the supply voltage source. Thus, e_L always opposes the current flow produced by e.

In Figure 19-3(a), the supply voltage waveform (e) and inductor current waveform (i) are drawn on the same zero level. It is seen that,

the current in a purely inductive circuit lags the voltage by 90°, or $\pi/2$ radians.

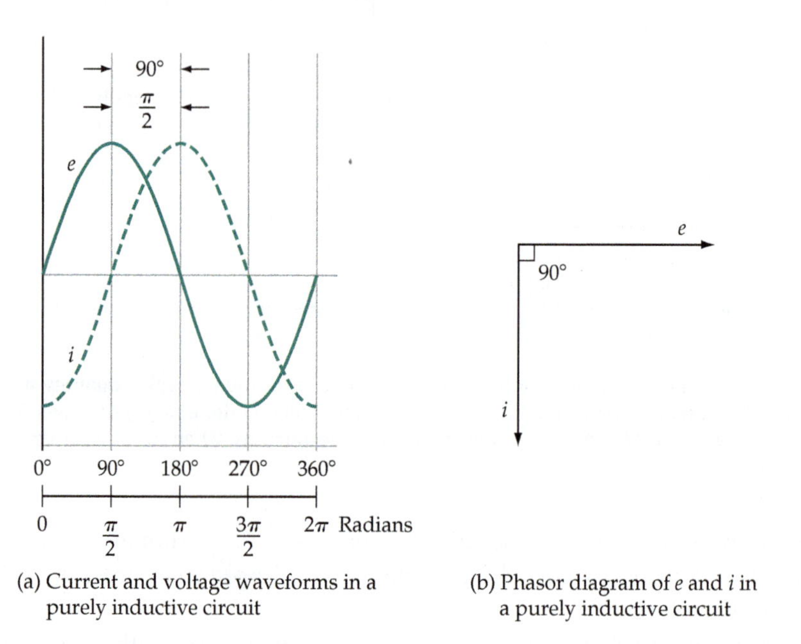

(a) Current and voltage waveforms in a purely inductive circuit

(b) Phasor diagram of e and i in a purely inductive circuit

Figure 19-3 In a purely inductive circuit, the current lags the voltage by 90° ($\pi/2$ radians).

Alternatively, it may be stated that in a purely inductive circuit the supply voltage leads the current by 90°. The phasor diagram of voltage and current in an inductive circuit [Figure 19-3(b)] uses the voltage as a reference and shows the current phasor 90° behind the voltage. When the inductor terminal voltage is,

$$e = E_m \sin \omega t$$

the inductor current is represented as,

$$i = I_m \sin\left(\omega t - \frac{\pi}{2}\right) \tag{19-1}$$

19-2 INDUCTIVE REACTANCE AND SUSCEPTANCE

Inductive Reactance

Referring once again to the current waveform in Figure 19-2, it is seen that the current grows from zero to its maximum value I_m in one quarter of a cycle. The time period of one cycle is,

$$T = \frac{1}{f}$$

Therefore, the time taken for the current to grow from zero to I_m is,

$$t = \frac{1}{4f}$$

The average rate of change *of* current during time t is,

$$\frac{\Delta i}{\Delta t} = \frac{I_m}{1/(4f)} = 4fI_m$$

Substituting for $\Delta i/\Delta t$ from Equation 14-3 [$e_L = L(\Delta i/\Delta t)$], the average counter-emf induced is,

$$e_L = 4fLI_m$$

The counter-emf is equal to the supply voltage in a purely inductive circuit, so the average supply voltage is,

$$E_{av} = 4fLI_m$$

Actually, there is always a slight difference between the counter-emf and the supply voltage; otherwise, no current would flow and no counter-emf would be generated. From Equation 17-14,

$$E_{av} = \frac{2}{\pi}E_m$$

Therefore,

$$4fLI_m = \frac{2}{\pi}E_m$$

Giving,

$$\frac{E_m}{I_m} = 2\pi fL$$

Substituting rms quantities for maximum quantities,

$$\frac{1.414E}{1.414I} = 2\pi fL$$

So, for an inductor,

$$\frac{E}{I} = 2\pi fL \qquad (19\text{-}2)$$

where E and I are rms quantities.

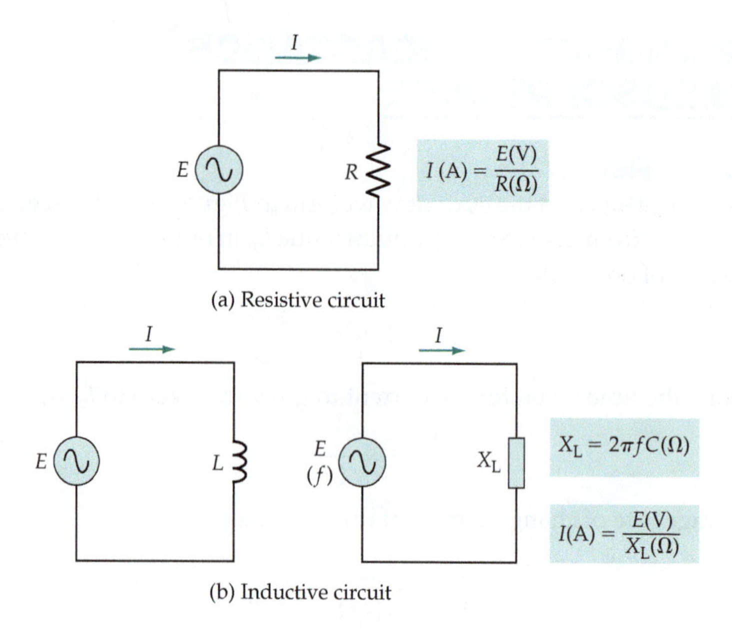

Figure 19-4 For a resistive circuit, $I = E/R$. For an inductive circuit, $I = E/X_L$, where $X_L = 2\pi fL$.

From Ohm's law, (E volts)/(I amps) gives resistance in ohms [Figure 19-4(a)]. In a pure inductance there is zero resistance; however, $2\pi fL$ does represent an opposition to current flow. The name given to this quantity is *inductive reactance,* and like resistance, inductive reactance is measured in ohms. The symbol X_L is employed for inductive reactance [see Figure 19-4(b)].

Because $2\pi f = \omega$ (the angular velocity in radians per second), Equation 19-2 can be rewritten as

$$\frac{E}{I} = X_L = \omega L$$

Inductive reactance,

$$X_L = \omega L = 2\pi fL \qquad (19\text{-}3)$$

When f is in hertz and L is in henrys, X_L is given in ohms. Note that X_L is directly proportional to frequency and inductance.

Substituting for I_m in Equation 19-1, the expression for the instantaneous current in an inductive circuit becomes,

$$i = \frac{E_m}{\omega L} \sin\left(\omega t - \frac{\pi}{2}\right) \qquad (19\text{-}4)$$

Inductive Susceptance

The reciprocal of resistance (R) is conductance (G), which is the measure of the ability of a resistive circuit to pass current. Similarly, inductive reactance (X_L) has its reciprocal in *inductive susceptance* for which the symbol is B_L. Inductive susceptance is a measure of the ability of a purely inductive circuit to pass current, and like conductance its unit is the siemens (S).

Inductive susceptance,

$$B_L = \frac{1}{X_L} \quad (19\text{-}5)$$

When X_L is in ohms, B_L is in siemens (see Figure 19-5).

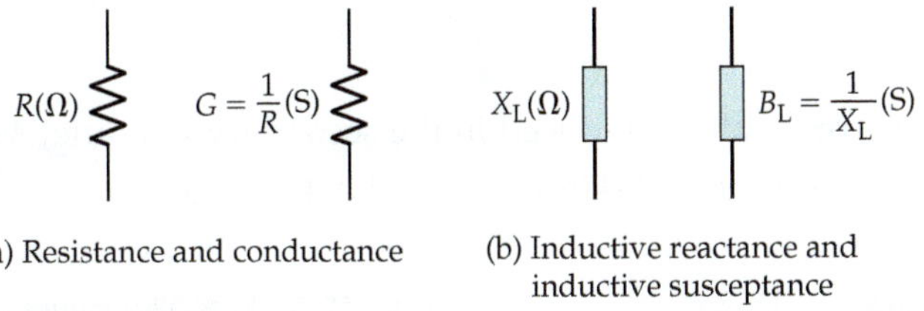

$R(\Omega)$ $G = \frac{1}{R}(\text{S})$ $X_L(\Omega)$ $B_L = \frac{1}{X_L}(\text{S})$

(a) Resistance and conductance

(b) Inductive reactance and inductive susceptance

Figure 19-5 Inductive susceptance (B_L) is the reciprocal of inductive reactance (X_L), just like the relationship between conductance (G) and resistance (R).

Example 19-1

A 500 mH inductor is supplied from a 115 V source with a 60 Hz frequency (see Figure 19-6). Calculate the inductive reactance and the current that flows in the circuit.

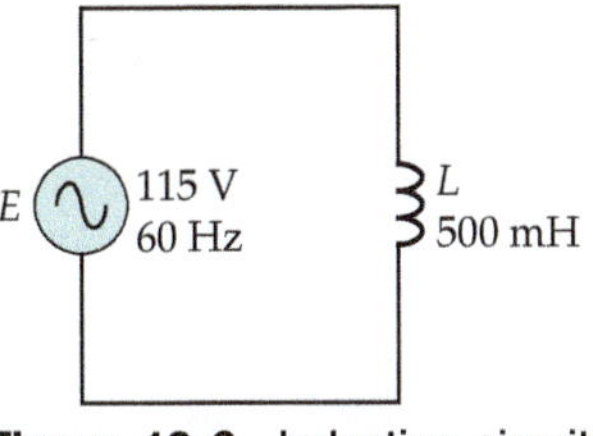

Figure 19-6 Inductive circuit for Example 19-1.

Solution

Eq. 19-3,

$$X_L = 2\pi f L = 2\pi \times 60 \text{ Hz} \times 500 \text{ mH}$$

$$\approx 188.5 \ \Omega$$

$$I = \frac{E}{X_L} = \frac{115 \text{ V}}{188.5 \ \Omega}$$

$$= 610 \text{ mA}$$

Example 19-2

A current of 7.96 mA is measured in inductor which has a 50 V supply with a 10 kHz frequency (see Figure 19-7). Determine the inductance of the inductor.

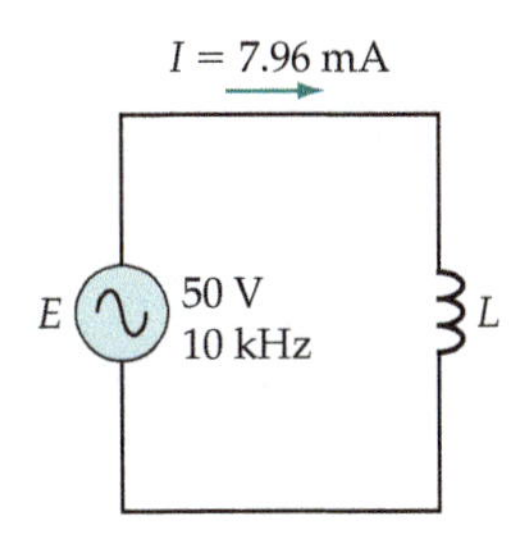

Figure 19-7 Inductive circuit for Example 19-2.

Solution

$$X_L = \frac{E}{I} = \frac{50 \text{ V}}{7.96 \text{ mA}}$$

$$= 6.28 \text{ k}\Omega$$

From Eq. 19-3,

$$L = \frac{X_L}{2\pi f} = \frac{6.28 \text{ k}\Omega}{2\pi \times 10 \text{ kHz}}$$

$$\approx 100 \text{ mH}$$

Effect of Mutual Inductance

Figure 19-8 shows two coupled coils with a mutual inductance (M) (see Section 14-4). By substituting Equation 14-7 for mutual inductance instead of Equation 14-3 for self-inductance in the derivation of Equation 19-2, an expression is obtained for the alternating voltage induced in the secondary of two mutually coupled coils. When an alternating current (I_1) flows in the primary,

$$E_2 = \omega M I_1 \quad (19\text{-}6)$$

where E_2 is the rms voltage induced in the secondary winding, M is the mutual inductance, and I_1 is the rms current in the primary.

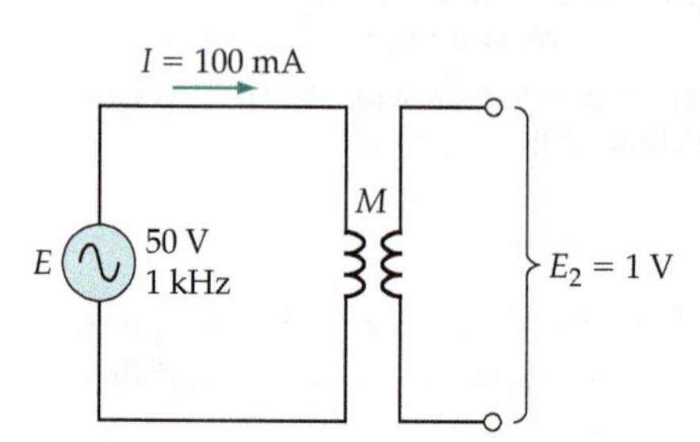

Figure 19-8 Mutually inductive circuit for Example 19-3.

Example 19-3

A 100 mA alternating current with a frequency of 1 kHz flowing in the primary of two coupled coils produces a 1 V secondary output. Calculate the mutual inductance between the coils.

Solution

From Eq. 19-6,

$$M = \frac{E_2}{\omega I_1} = \frac{1\ \text{V}}{2\pi \times 1\ \text{kHz} \times 100\ \text{mA}}$$

$$= 1.59\ \text{mH}$$

Practice Problems

19-2.1 An inductor passes a current of 25 mA when connected to a 40 mV, 200 Hz supply. Calculate the inductance.

19-2.2 Determine the current that flows in a 50 μH inductor supplied from a 35 V, 1 MHz source.

19-2.3 A secondary coil is wound on the inductor referred to in Problem 19-2.2. The mutual inductance between the two coils is 3.3 μH. Determine the emf induced in the secondary winding.

19-3 ALTERNATING CURRENT AND VOLTAGE IN A CAPACITIVE CIRCUIT

For the circuit and waveforms shown in Figure 19-9, it is assumed that C is a pure capacitance with a dielectric having infinite resistance, and that the connecting wires have zero resistance. The first waveform illustrated is that of the instantaneous level of supply voltage (e). This is also the voltage (e_C) that appears at the capacitor terminals. The waveform representing the capacitor current is drawn by considering the supply voltage waveform.

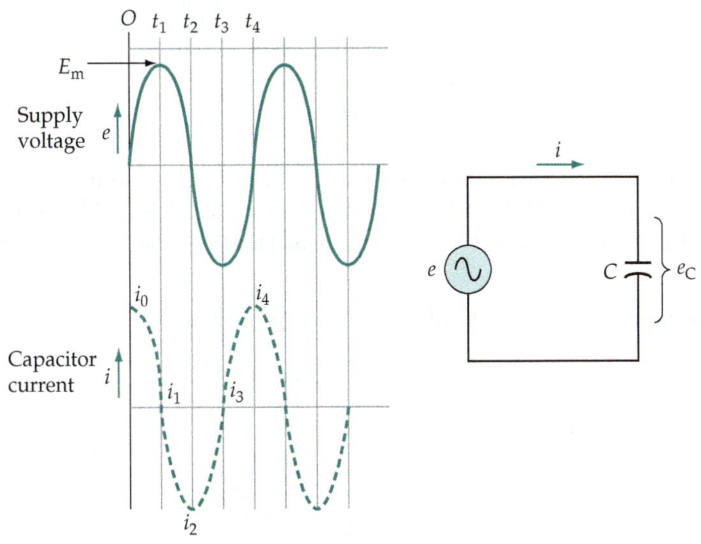

Figure 19-9 For a purely capacitive circuit, $\Delta v/\Delta t$ is a maximum positive quantity when the supply voltage (e) crosses zero in a positive-going direction. Consequently, the instantaneous current is a maximum positive quantity at this time.

For the capacitor terminal voltage (e_C) to follow the supply voltage (e), a charge and discharge current (i) must flow, as required. At time $t = 0$, e is changing rapidly in a positive direction; so i is a maximum, (i_o on Figure 19-9). At time t_1 e_C is not changing, consequently $i = 0$. The instantaneous current i_1 is plotted as zero on the current waveform. At time t_2, e_C is changing rapidly in a negative direction; so, the instantaneous current at t_2 is a maximum negative quantity i_2, as illustrated. At t_3, e_C is constant once again, giving i_3 equal to zero. When the rate of change of voltage is a maximum positive quantity at t_4, the instantaneous current is also a positive maximum; plotted as i_4 on the current waveform.

In Figure 19-10(a) the waveforms of supply voltage and capacitor current are shown together. The waveforms and phasor diagram in Figure 19-10(b), show that,

the current in a purely capacitive circuit leads the voltage by 90° or $\pi/2$ radians.

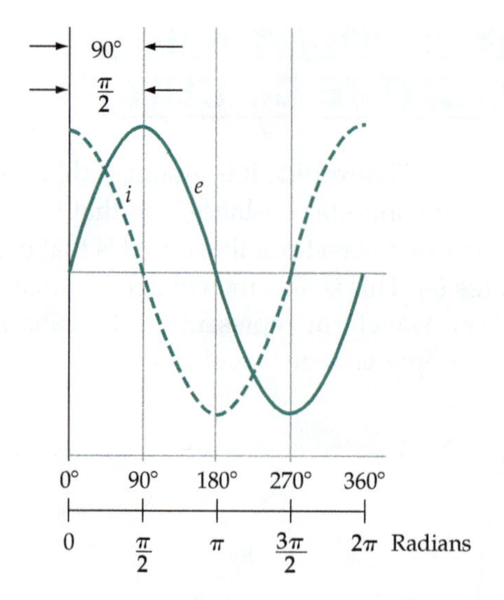

(a) Current and voltage waveforms in a purely capacitive circuit

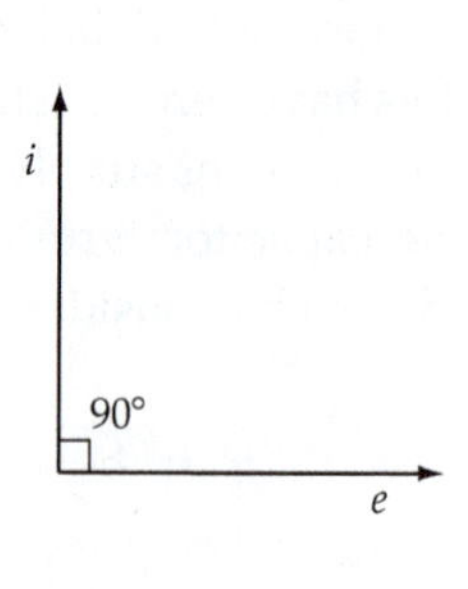

(b) Phasor diagram of e and i in a purely capacitive circuit

Figure 19-10 In a purely capacitive circuit, the current leads the voltage by 90° (π/2 radians).

When the capacitor terminal voltage is represented as,

$$e = E_m \sin \omega t$$

the capacitor current is,

$$i = I_m \sin\left(\omega t + \frac{\pi}{2}\right) \tag{19-7}$$

19-4 CAPACITIVE REACTANCE AND SUSCEPTANCE

Capacitive Reactance

As already explained for the inductive circuit, the capacitor current grows from zero to its maximum level I_m in a time of $t = 1/(4f)$.

From Eq. 17-13, $$I_{av} = \frac{2}{\pi} I_m$$

The capacitor charge is,

$$Q = I_{av} \times t = C\Delta V = CE_m$$

so, $$CE_m = I_{av} \times t = \frac{2}{\pi} I_m \times \frac{1}{4f}$$

Giving, $$\frac{E_m}{I_m} = \frac{1}{2\pi f C}$$

Substitution of rms quantities for maximum voltage and current gives the same result. So, for a capacitor,

$$\frac{E}{I} = \frac{1}{2\pi f C} \qquad (19\text{-}8)$$

where E and I are rms quantities.

As in the case of the pure inductor, a pure capacitor has no resistive component, but $1/(2\pi fC)$ represents an opposition to current flow. Termed *capacitive reactance,* this quantity is measured in ohms and is given the symbol X_C, (see Figure 19-11).

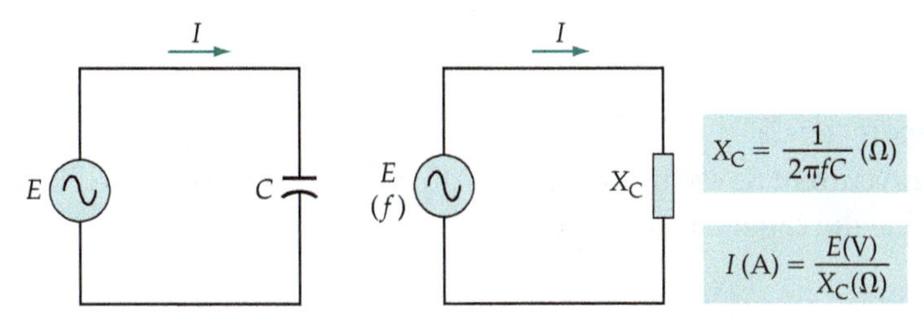

Figure 19-11 For a capacitive circuit, $I = E/X_C$, where $X_C = 1/(2\pi fC)$.

Capacitive reactance,

$$X_C = \frac{1}{\omega C} = \frac{1}{2\pi f C} \qquad (19\text{-}9)$$

When f is in hertz and C is in farads, Equation 19-9 gives X_C in ohms. Note that X_C is inversely proportional to frequency and capacitance.

Substituting for I_m in Equation 19-7, the instantaneous current level in a capacitive circuit is given by,

$$i = \frac{E_m}{X_C} \sin\left(\omega t + \frac{\pi}{2}\right) \qquad (19\text{-}10)$$

Capacitive Susceptance

The reciprocal of capacitive reactance (X_C) is *capacitive susceptance* (B_C), which is a measure of the ability of a purely capacitive circuit to pass current (see Figure 19-12). Like inductive susceptance, capacitive susceptance is comparable to the conductance of a resistive circuit.

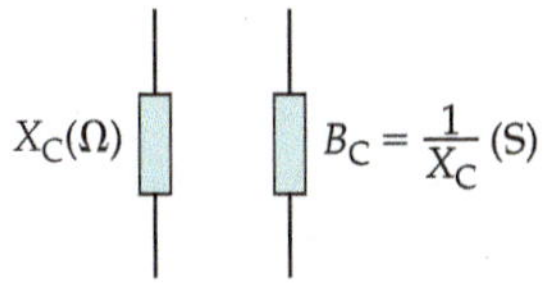

Figure 19-12 Capacitive susceptance is the reciprocal of capacitive reactance.

Capacitive susceptance,

$$B_C = \frac{1}{X_C} \qquad (19\text{-}11)$$

When X_C is in ohms, B_C is in siemens (see Figure 19-12).

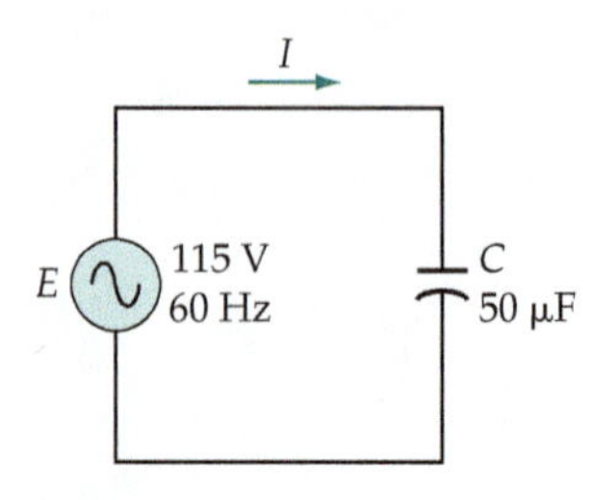

Figure 19-13 Capacitive circuit for Example 19-4.

Example 19-4

Figure 19-13 shows a 50 μF capacitor supplied from a 115 V, 60 Hz source. Determine the capacitive reactance, and calculate the current that flows in the circuit.

Solution

Eq 19-9,

$$X_C = \frac{1}{2\pi f C} = \frac{1}{2\pi \times 60\text{ Hz} \times 50\ \mu\text{F}}$$

$$\approx 53.1\ \Omega$$

$$I = \frac{E}{X_C} = \frac{115\text{ V}}{53.1\ \Omega}$$

$$\approx 2.2\text{ A}$$

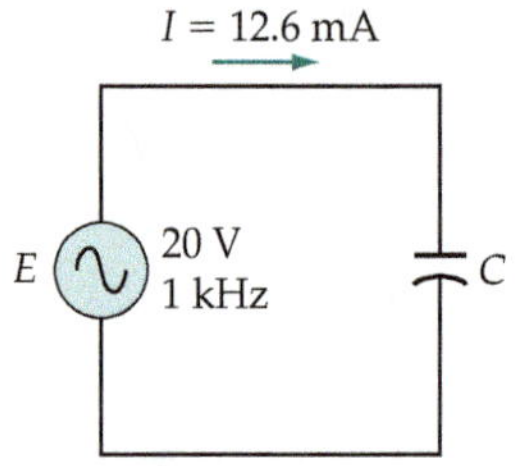

Figure 19-14 Capacitive circuit for Example 19-5.

Example 19-5

A 12.6 mA current flows in a capacitor with 20 V, 1 kHz supply (see Figure 19-14). Determine the capacitance value.

Solution

$$X_C = \frac{E}{I} = \frac{20\text{ V}}{12.6\text{ mA}}$$

$$\approx 1.59\text{ k}\Omega$$

From Eq. 19-9,

$$C = \frac{1}{2\pi f X_C} = \frac{1}{2\pi \times 1\text{ kHz} \times 1.59\text{ k}\Omega}$$

$$= 0.1\ \mu\text{F}$$

Practice Problems

19-4.1 A 1000 pF capacitor has a 1.5 V terminal voltage when passing a 6.6 mA current. Determine the frequency of the supply.

19-4.2 The supply current for the circuit in Problem 19-4.1 changes to 6.93 mA when an oscilloscope is connected across the 1000 pF capacitor. Calculate the input capacitance of the oscilloscope.

19-5 SERIES *RL* CIRCUITS

Current and Voltage Waveforms

A series circuit consisting of inductance (L) and resistance (R) is shown in Figure 19-15(a), and the waveforms and phasor diagram for the circuit are illustrated in Figure 19-15(b) and (c), respectively. Referring to the circuit

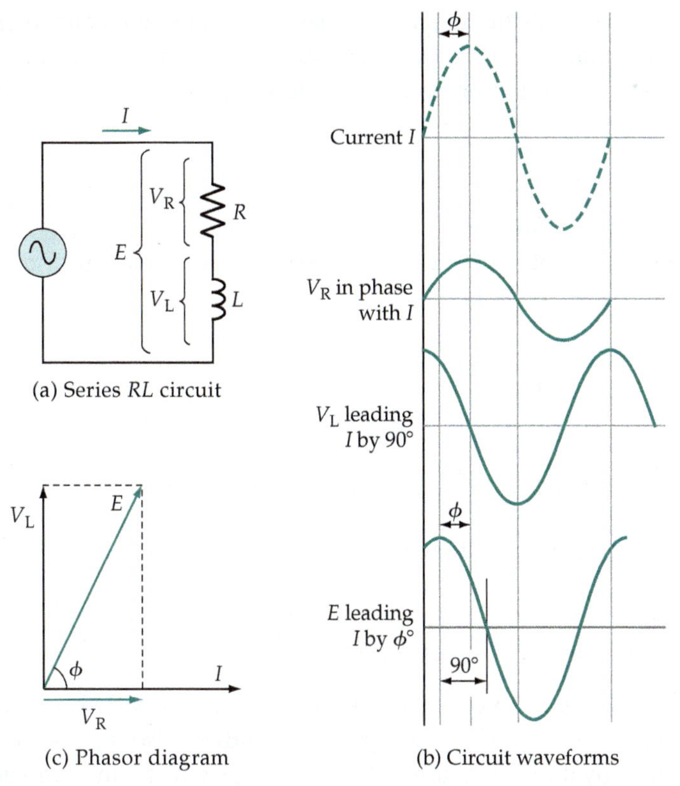

Figure 19-15 In a series-connected *RL* circuit, the current lags the inductor voltage (V_L) by 90, and lags the supply voltage (*E*) by an angle less than 90°.

diagram, it is seen that (as for all series circuits) the current *I* is common to both *R* and *L*. So the circuit waveforms are drawn starting with the waveform of current.

The voltage (V_R) across the resistance is always in phase with the current through the resistance. Thus, the waveform of V_R in Figure 19-15(b) is drawn in phase with the current waveform. The inductor voltage (V_L) leads the current through the inductor by 90°, (see Section 19-1). So, the waveform of V_L is drawn in Figure 19-15(b) leading *I* by 90°. The applied voltage *E* is the resultant of the two component voltages V_R and V_L, and its waveform can be obtained simply by summing the instantaneous levels of V_R and V_L. It is seen that *E* leads *I* by an angle (ϕ), which is less than 90°.

Phasor Diagram

The phasor diagram for the series *RL* circuit is drawn by starting with the current phasor, because the current is the common quantity in a series circuit. A horizontal line is drawn to scale representing current *I* [Figure 19-15(c)]. Because V_R is in phase with *I*, another horizontal line is drawn alongside *I* to represent V_R. The inductor voltage (V_L) is 90° ahead of *I*; so, the phasor for V_L is drawn vertically at an angle of 90° with respect to *I*. The phasor addition of

V_L and V_R gives a resultant that represents the applied voltage (E). Once again, it is seen that the applied voltage leads the circuit current by an angle ϕ, which is less than the 90° angle that would exist between E and I in a pure inductive circuit.

Circuit Equations

As explained in Chapter 18, E can be expressed in rectangular or polar form. Referring to Figure 19-15(c),

$$E = V_R + jV_L \tag{19-12}$$

From Eq. 18-2,

$$E = \sqrt{V_R^2 + V_L^2}\,\angle \tan^{-1}(V_L/V_R) \tag{19-13}$$

Dividing Equation 19-12 by I gives,

$$\frac{E}{I} = \frac{V_R}{I} + j\frac{V_L}{I}$$

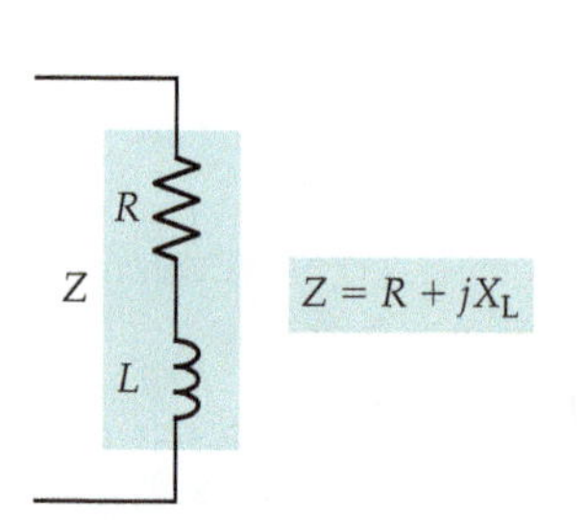

Figure 19-16 The impedance of a circuit consisting of an inductor and resistor connected in series is ($Z = R + jX_L$).

The quantity V_R/I is the voltage across the resistance divided by the current through the resistance. So V_R/I can be replaced with R. Also, V_L/I is the inductor voltage divided by the current through the inductor, which can be replaced with the inductive reactance X_L. The quantity E/I is neither resistance nor inductive reactance, because it has both as component parts. In this case E/I is termed *impedance* and is given the symbol Z. Therefore, as illustrated in Figure 19-16, the above equation can be restated.

Impedance,

$$Z = R + jX_L \tag{19-14}$$

When R and X_L are expressed in ohms, the units of Z are also ohms.

The numerical value or modulus of Z (i.e., not including the angle) is written as $|Z|$.

So,

$$|Z| = \sqrt{R^2 + X_L^2} \tag{19-15}$$

The phase angle of Z is,

$$\phi = \tan^{-1}\left(\frac{X_L}{R}\right) \tag{19-16}$$

Combining Equations 19-15 and 19-16,

$$Z = \sqrt{R^2 + X_L^2}\,\angle \tan^{-1}(X_L/R) \tag{19-17}$$

Equation 19-17 represents conversion from rectangular form ($Z = R + jX$) to polar form ($Z\angle\phi$). For conversion from polar to rectangular form, Equation 18-1 is rewritten using impedances,

$$Z\angle\phi = Z\cos\phi + jZ\sin\phi$$

Impedance Diagram

The quantities Z, R, and X_L can be represented on a vector diagram. They are not phasor quantites because they have fixed values. Unlike ac voltage and current quanties, Z, R, and X_L do not have continuously changing instantaneous values. To distinguish the vector diagram of impedance, reactance, and resistance from a phasor diagram, it is usually drawn in triangular form, and is referred to as an *impedance diagram.* Figure 19-17 shows the impedance diagram for the series RL circuit of Figure 19-15(a). A horizontal line is first drawn to represent the resistive component R. The $+j$ component (X_L) is then drawn at +90° with respect to R, as illustrated. The Z component is the hypotenuse of the triangle, and the angle ϕ is the phase angle of the impedance Z with respect to resistance R.

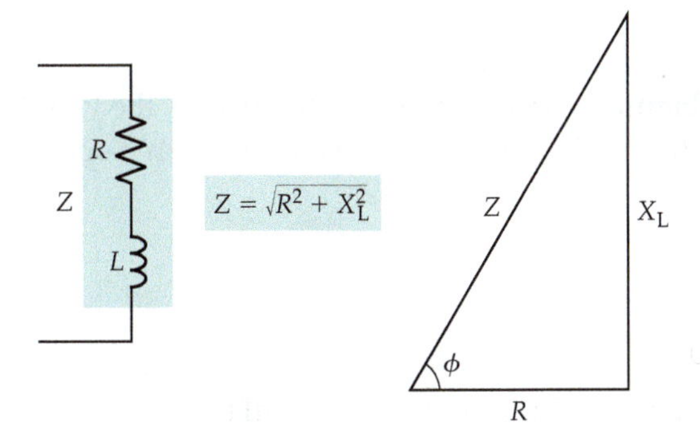

Figure 19-17 Impedance diagram (or impedance triangle) for a series RL circuit. Inductive reactance vector X_L is drawn (up) from the resistance vector at an angle of 90°. The impedance (Z) is the resultant of R and X_L.

Admittance

As in the case of conductance (the reciprocal of resistance) and susceptance (the reciprocal of reactance), the reciprocal of impedance is an important quantity that can be usefully applied in circuit analysis. The *admittance* (symbol Y) is the reciprocal of impedance (Z) and its unit is the siemens (S).

Admittance,
$$Y = \frac{1}{Z} \qquad (19\text{-}18)$$

When Z is expressed in ohms in Equation 19-18, Y is in siemens.

Practical Inductors

All inductors are coils that have some winding resistance, and as already explained, practical inductors can be represented by a pure inductance in series with the winding resistance. So, the diagrams in Figures 19-15 and 19-17 can apply to the case of an inductor connected directly to an ac supply. Where an external resistance is connected in series with the inductor, the winding resistance should also be shown as a series component in the equivalent circuit.

Analysis Procedure for a Series *RL* Circuit

1. ***Calculate the inductive reactance;*** $X_L + (2\pi fL)$
2. ***If there is more than one resistive component, calculate the total resistance;***

$$R = R_1 + R_2 + \cdots$$

3. ***Calculate the circuit impedance;***

$$|Z| = \sqrt{R^2 + X_L^2}$$

4. ***Calculate the phase angle;*** $\phi = \tan^{-1}(X_L/R)$
5. ***Calculate the current;*** $I = E/Z$
6. ***Determine the resistive voltage;*** $V_R = IR$ ***or*** $V_R = E\cos\phi$
7. ***Determine the inductive voltage;*** $V_L = IX_L$ ***or*** $V_L = E\sin\phi$

Example 19-6

Analyze the series *RL* circuit in Figure 19-18 to determine the current, the voltage across *R*, the voltage across *L*, and the phase angle of current with respect to the supply voltage.

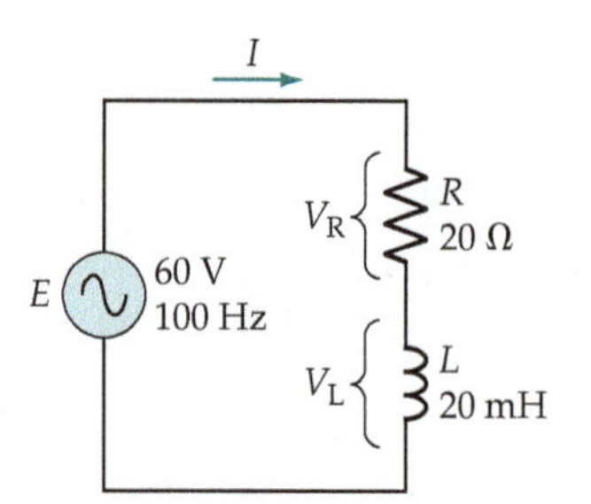

Figure 19-18 Series-connected *RL* circuit for Example 19-6.

Solution

Inductor reactance,

Eq. 19-3,
$$X_L = 2\pi fL = 2\pi \times 100\text{ Hz} \times 20\text{ mH} \approx 12.57\ \Omega$$

Circuit impedance,

Eq. 19-15,
$$|Z| = \sqrt{R^2 + X_L^2} = \sqrt{(20\ \Omega)^2 + (12.57\ \Omega)^2} = 23.6\ \Omega$$

Phase angle,

Eq. 19-16,
$$\phi = \tan^{-1}\left(\frac{X_L}{R}\right) = \tan^{-1}\left(\frac{12.57\ \Omega}{20\ \Omega}\right) = 32.1^\circ$$

Circuit current,
$$I = \frac{E}{Z} = \frac{60\text{ V}}{23.6\ \Omega\angle 32.1^\circ} \approx 2.54\text{ A}\angle -32.1^\circ$$

Referring to the series *RL* circuit phasor diagram Figure 19-15(c):

Resistor voltage, $V_R = E\cos\phi = 60\text{ V}\cos 32.1°$

$= 50.8\text{ V } (= I \times R)$

Inductor voltage, $V_L = E\sin\phi = 60\text{ V}\sin 32.1°$

$= 31.9\text{ V } (= I \times X_L)$

Example 19-7

Figure 19-19 shows 64 mH inductor with a 700 Ω winding resistance connected in series with a 3.3 kΩ resistor. A 10 V, 5 kHz supply is applied to the circuit. Calculate the circuit current and the inductor terminal voltage.

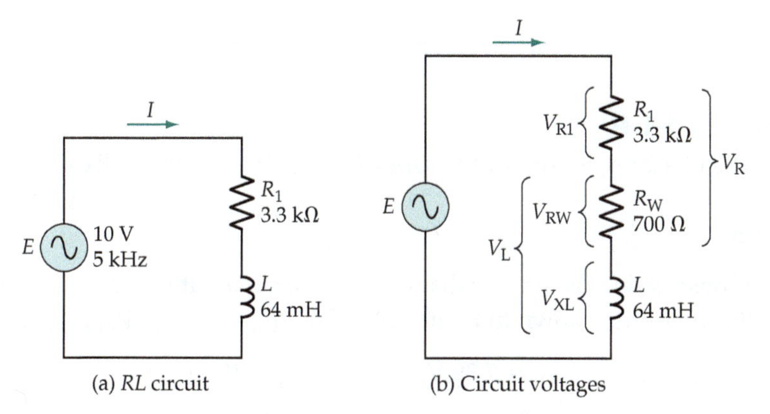

Figure 19-19 Series-connected *RL* circuit for Example 19-7.

Solution

Inductor reactance,

Eq. 19-3, $X_L = 2\pi fL = 2\pi \times 5\text{ kHz} \times 64\text{ mH}$

$\approx 2\text{ k}\Omega$

Total resistance, $R = R_1 + R_w = 3.3\text{ k}\Omega + 700\ \Omega$

$= 4\text{ k}\Omega$

Circuit impedance,

Eq. 19-15, $|Z| = \sqrt{R^2 + X_L^2} = \sqrt{(4\text{ k}\Omega)^2 + (2\text{ k}\Omega)^2}$

$= 4.47\text{ k}\Omega$

Phase angle,

Eq. 19-16, $\phi = \tan^{-1}\left(\frac{X_L}{R}\right) = \tan^{-1}\left(\frac{2\text{ k}\Omega}{4\text{ k}\Omega}\right)$

$= 26.6°$

Circuit current, $I = \frac{E}{Z} = \frac{10\ \text{V}}{4.46\ \text{k}\Omega\angle 26.6°}$

$= 2.24\ \text{mA}\angle -26.6°$

Resistor voltage, $V_{R1} = I \times R_1 = 2.24\ \text{mA} \times 3.3\ \text{k}\Omega$

$= 7.39\ \text{V}$

Inductor impedance,

Eq. 19-15, $|Z_L| = \sqrt{R_W^2 + X_L^2} = \sqrt{(700\ \Omega)^2 + (2\ \text{k}\Omega)^2}$

$= 2.12\ \text{k}\Omega$

Inductor voltage, $V_L = I_L \times Z_L = 2.24\ \text{mA} \times 2.12\ \text{k}\Omega$

$= 4.75\ \text{V}$

Example 19-8

Draw the phasor diagram for the series *RL* circuit analyzed in Example 19-7.

Solution

Before constructing the phasor diagram for the circuit, it is necessary to determine the quantities shown in Figure 19-19(b): V_{R1}, V_{RW}, V_R, V_{XL}, V_L, and θ.

Resistor voltage, $V_{R1} = 7.39$ V (from Example 19-7)

Resistive component of inductor voltage,

$$V_{RW} = I \times R_W = 2.24\ \text{mA} \times 700\ \Omega$$

$$= 1.57\ \text{V}$$

Total resistor voltage, $V_R = V_{R1} + V_{RW} = 7.39\ \text{V} + 1.57\ \text{V}$

$= 8.95\ \text{V}$

Reactive component of inductor voltage,

$$V_{XL} = I \times X_L = 2.24\ \text{mA} \times 2\ \text{k}\Omega$$

$$= 4.48\ \text{V}$$

Inductor voltage, $V_L = \sqrt{V_{RW}^2 + V_{XL}^2} = \sqrt{(1.57\ \text{V})^2 + (4.48\ \text{V})^2}$

$= 4.75\ \text{V}$

Inductor voltage phase angle,

$$\theta = \tan^{-1}\left(\frac{V_{XL}}{V_{RW}}\right) = \tan^{-1}\left(\frac{4.48\ \text{V}}{1.57\ \text{V}}\right)$$

$$\approx 70.7°$$

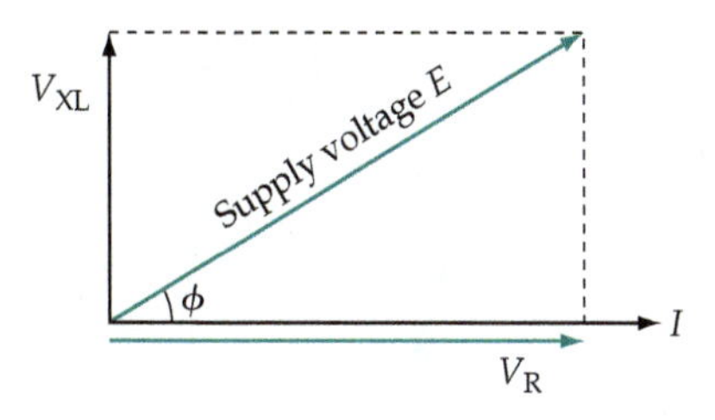

(a) Phasor diagram showing relationship between supply voltage, current, and the inductive and resistive voltage components

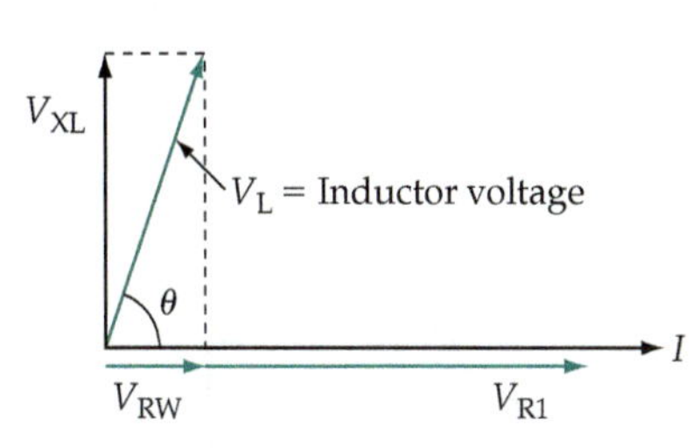

(b) Phasor components of inductor voltage

Figure 19-20 **Phasor diagrams for the series-connected *RL* circuit in Example 19-7.**

- The phasor for the current (I) is first drawn horizontally [see Figure 19-20(a)].
- V_R is drawn in phase with the current.
- V_{XL} is drawn leading the current by 90°.
- E leads the current by ϕ, and is also the resultant of V_{XL} and V_R.
- V_R is divided into its component parts, V_{R1} and V_{RW} [see Figure 19-20(b)].
- V_L is drawn leading the current by θ, and is also the resultant of V_{RW} and V_{XL}.

Practice Problems

19-5.1 A series circuit has $R_1 = 120\ \Omega$, $L = 30$ mH, and the coil resistance of the inductor is $R_W = 40\ \Omega$. If the circuit is connected to a 24 V, 400 Hz supply, calculate the circuit current and its phase angle with respect to the supply.

19-5.2 An inductor in series with a 2.7 kΩ resistor is connected to a 100 mV, 250 kHz supply. The resistor voltage is measured as 40.5 mV. Calculate the inductance.

19-6 SERIES *RC* CIRCUITS

Current and Voltage Waveforms

Figure 19-21(a) shows a series-connected capacitor and resistor with an ac supply voltage. The circuit waveforms are illustrated in Figure 19-21(b) and the phasor diagram for the circuit is drawn in Figure 19-21(c).

The waveform of current (I) is drawn first because it is common to both series-connected components (R and C) [Figure 19-21(b)]. The resistor voltage is always in phase with its current, so the waveform of V_R is drawn in phase with the current waveform. The current through the capacitor leads the capacitor terminal voltage (V_C) by 90° (see Section 19-3), consequently, the V_C waveform is drawn 90° lagging the current wave. The applied voltage (E) is the vector sum of V_R and V_C, and it is seen that E lags I by an angle ϕ, which is less than 90°.

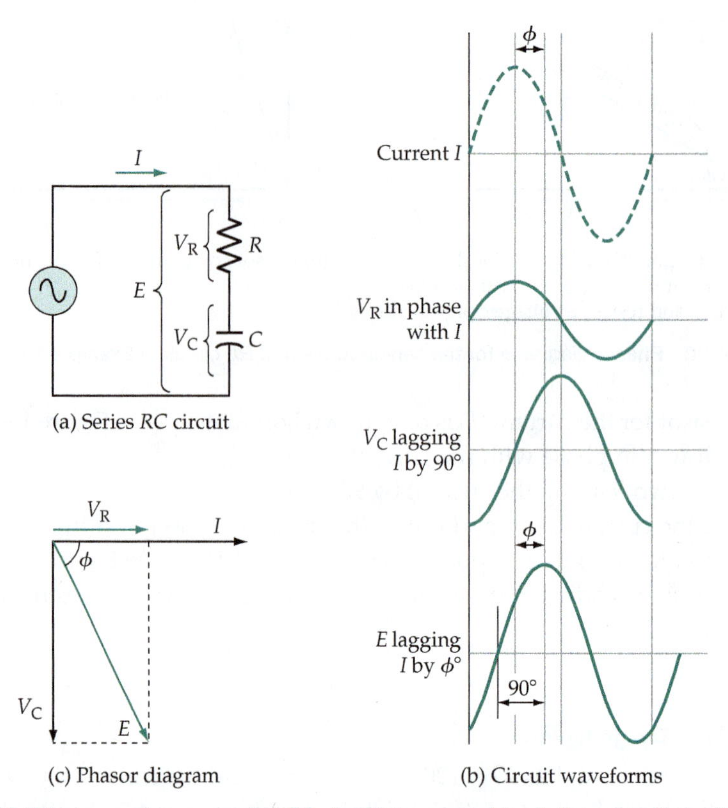

Figure 19-21 In a series-connected *RC* circuit, the current leads the capacitor voltage (V_C) by 90°, and leads the supply voltage (*E*) by an angle less than 90°.

Phasor Diagram

The phasor diagram for the series *RC* circuit is drawn by starting with the current phasor; again because the current is the common quantity in a series circuit. A horizontal line is drawn to scale representing current (I), [Figure 19-21(c)]. Because the resistor voltage is in phase with I, V_R is represented by another horizontal line drawn alongside I. The capacitor voltage lags I by 90°; so the V_C phasor is drawn vertically at an angle of −90° with respect to I. The resultant of the phasor addition of V_C and V_R represents the applied voltage (E). Once again, it is seen that E lags I by an angle ϕ, which is less than the 90° angle that would exist between E and I in a pure capacitive circuit.

Circuit Equations

From Chapter 18 and Figure 19-21(c), the rectangular form expression for E in the series *RC* circuit is,

$$E = V_R - jV_C \tag{19-19}$$

From Eq. 18-2,

$$E = \sqrt{V_R^2 + V_C^2}\ \angle \tan^{-1}(V_C/V_R) \tag{19-20}$$

Divide Equation 19-19 by I,

$$\frac{E}{I} = \frac{V_R}{I} - j\frac{V_C}{I}$$

giving impedance,

$$Z = R - jX_C \qquad (19\text{-}21)$$

In Equation 19-21, R is resistance in ohms, X_C is capacitive reactance in ohms, and Z is impedance in ohms. This is illustrated in Figure 19-22.

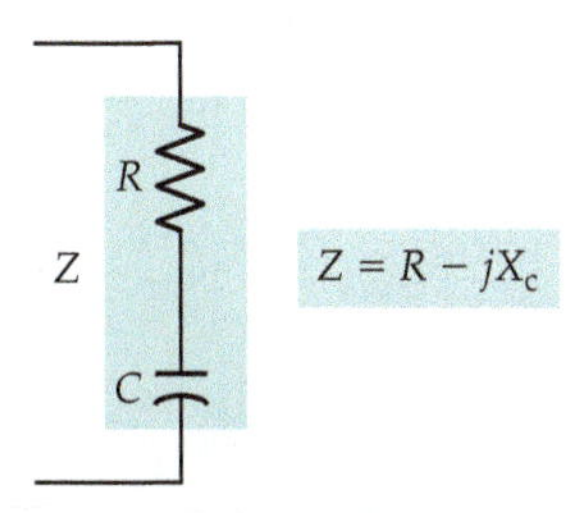

Figure 19-22 The impedance of a circuit consisting of a resistor and capacitor connected in series is ($Z = R - jX_C$).

Also,

$$|Z| = \sqrt{R^2 + X_C^2} \qquad (19\text{-}22)$$

and,

$$\phi = \tan^{-1}\left(\frac{-X_C}{R}\right) \qquad (19\text{-}23)$$

Combining Equations 19-22 and 19-23,

$$Z = \sqrt{R^2 + X_C^2}\,\underline{/\tan^{-1}(-X_C/R)} \qquad (19\text{-}24)$$

Equation 19-24 converts from rectangular to Polar form. Conversion from polar to rectangular form is performed using Equation 18-1 rewritten for impedances,

$$Z\underline{/\phi} = Z\cos\phi + jZ\sin\phi$$

As in the case of an RL circuit, the admittance of an RC circuit is calculated as $Y = 1/Z$.

Impedance Diagram

As explained for the series RL circuit the quantities Z, R, and X_C are not phasor quantites because they have fixed values. To distinguish the vector diagram of impedance, reactance, and resistance from a phasor diagram, it is usually drawn in triangular form and is referred to as an *impedance diagram.* The impedance diagram for a series RC circuit is shown in Figure 19-23. A horizontal

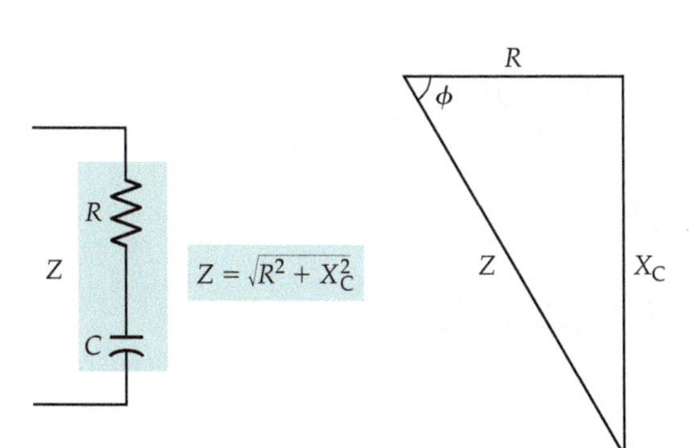

Figure 19-23 Impedance diagram (or impedance triangle) for a series-connected RC circuit. Capacitive reactance vector X_C is drawn (down) from the resistance vector at a −90° angle. The impedance vector Z is the resultant of R and X_C.

line is first drawn to scale to represent R, then the $-j$ component (X_C) is drawn at $-90°$ with respect to R. Z is the hypotenuse of the impedance triangle, and the phase angle of Z with respect to R is ϕ.

Analysis Procedure for Series *RC* Circuit

1. *Calculate the capacitive reactance;* $X_C = 1/(2\pi fC)$
2. *If there is more than one resistive component, calculate the total resistance;*

$$R = R_1 + R_2 + \cdots.$$

3. *Calculate the impedance;*

$$|Z| = \sqrt{R^2 + X_C^2}$$

4. *Calculate the phase angle;* $\phi = \tan^{-1}(X_C/R)$
5. *Calculate the current;* $I = E/Z$
6. *Determine the resistive voltage;* $V_R = IR$ *or* $V_R = E\cos\phi$
7. *Determine the capacitive voltage;* $V_C = IX_C$ *or* $V_C = E\sin\phi$

Example 19-9

Analyze the series *RC* circuit in Figure 19-24 to determine the current, the voltage across *R*, the voltage across *C*, and the phase angle of the current with respect to the supply voltage.

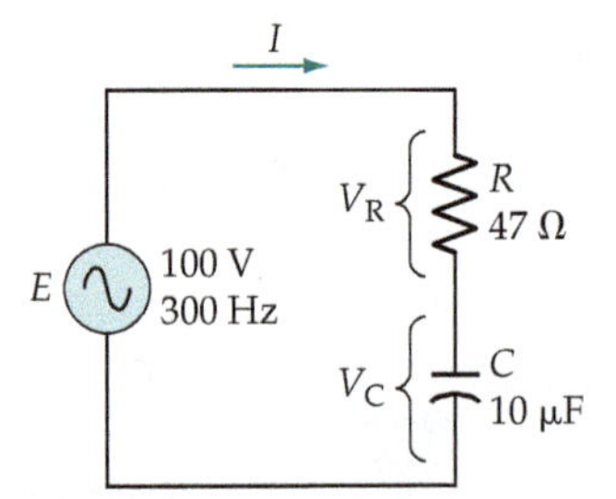

Figure 19-24 Series-connected *RC* circuit for Example 19-9.

Solution

Capacitive reactance,

Eq. 19-9,

$$X_C = \frac{1}{2\pi fC} = \frac{1}{2\pi \times 300\text{ Hz} \times 10\ \mu\text{F}}$$

$$\approx 53.1\ \Omega$$

Circuit impedance,

Eq. 19-22,

$$|Z| = \sqrt{R^2 + X_C^2} = \sqrt{(47\ \Omega)^2 + (53.1\ \Omega)^2}$$

$$= 70.9\ \Omega$$

Phase angle,

Eq 19-23,

$$\phi = \tan^{-1}\left(\frac{-X_C}{R}\right) = \tan^{-1}\left(\frac{-53.1\ \Omega}{47\ \Omega}\right)$$

$$\approx -48.5°$$

Circuit current,

$$I = \frac{E}{Z} = \frac{100\text{ V}}{70.9\ \Omega\angle -48.5°}$$

$$\approx 1.41\text{ A}\angle 48.5°$$

Referring to the series *RC* circuit phasor diagram Figure 19-21(c):

Resistor voltage, $V_R = E \cos \phi = 100 \text{ V} \cos(-48.5°)$

$\approx 66.3 \text{ V} (= I \times R)$

Capacitor voltage, $V_C = E \sin \phi = 100 \text{ V} \sin(-48.5°)$

$\approx -74.7 \text{ V} (= I \times X_C)$

Example 19-10

Draw the phasor diagram for the series *RC* circuit analyzed in Example 19-9.

Solution

- The phasor for the current (I) is first drawn horizontally (see Figure 19-25).
- V_R is drawn in phase with the current.
- V_C is drawn lagging the current by 90°.
- E lags the current by ϕ, and is also the resultant of V_C and V_R.

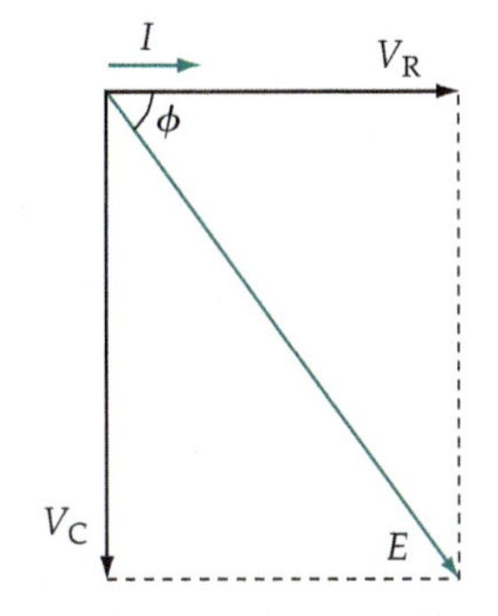

Figure 19-25 Phasor diagram for the series *RC* circuit in Example 19-9.

Practice Problems

19-6.1 A series circuit with $R = 470\ \Omega$ and $C = 0.1\ \mu\text{F}$ is connected to a 50 V, 1.8 kHz supply. Calculate V_C, V_R, and the phase angle of the current with respect to the supply.

19-6.2 A 33 V signal is applied to a 10 kΩ resistor in series with a 4000 pF capacitor. Calculate the capacitor voltage when the signal frequency is (a) 10 kHz, and (b) 55 kHz.

19-7 SERIES *RLC* CIRCUITS

Circuit

A circuit consisting of a resistor, inductor, and capacitor connected in series is shown in Figure 19-26(a). This circuit will behave either as a series *RL* circuit or as a series *RC* circuit, depending on which of the two reactances (X_L or X_C) is the larger. The special case where X_L and X_C are equal is considered in Chapter 23. As illustrated, the current (I) is common to all three components, and the component voltage drops (V_R, V_L, and V_C) add up vectorially to the supply voltage (E).

Phasor Diagram

In the *RLC* series circuit shown in Figure 19-26(a), because the current is common to all series-connected components, the phasor diagram is drawn by starting with the horizontal current phasor (I) [see Figure 19-26(b)]. The

Figure 19-26 Series-connected *RLC* circuit and phasor diagrams. Where $X_L > X_C$, the current lags the supply voltage. With $X_C > X_L$, the current leads the supply voltage.

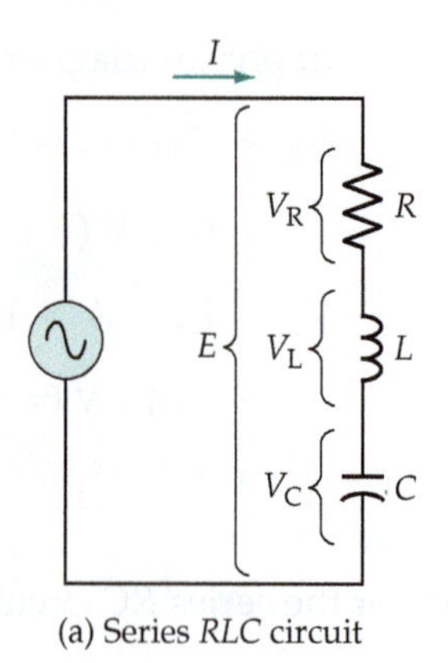

(a) Series *RLC* circuit

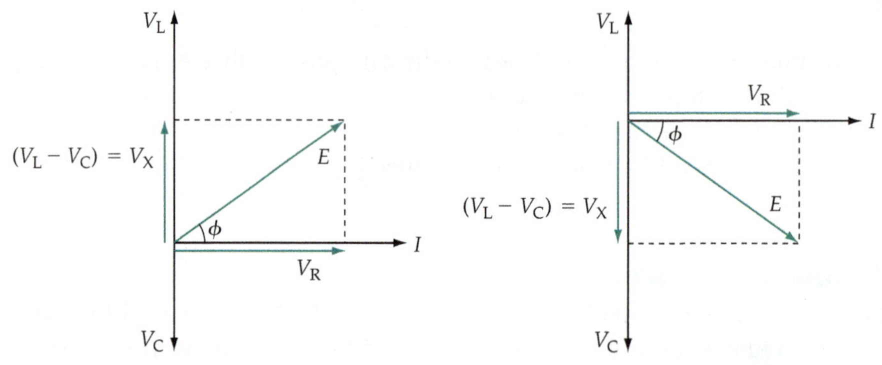

(b) Phasor diagram for series *RLC* circuit in which $X_L > X_C$

(c) Phasor diagram for series *RLC* circuit in which $X_C > X_L$

resistor voltage phasor is in phase with the current; so, the phasor representing V_R is drawn in phase with the current phasor, as shown. Because the current lags the voltage across the inductor, the inductor voltage phasor (V_L) is drawn 90° leading the current phasor. Similarly, the capacitor voltage phasor (V_C) is shown 90° lagging the current phasor, because the current leads the capacitor voltage by 90°.

The phasor diagram in Figure 19-26(b) is for a circuit where the inductive reactance is greater than the capacitive reactance. As illustrated, V_L is greater than V_C, and two reactive phasors (V_L and V_C) are added vectorially to find the resultant reactive phasor (V_X). In this case, the circuit behaves as an *RL* series circuit. The resultant of V_X and V_R is the supply voltage *E*, and ϕ is its phase angle with respect to the current. In Figure 19-26(c), the phasor diagram illustrates the case where X_C is greater than X_L, giving a capacitor voltage V_C larger than the inductor voltage V_L. With V_X capacitive, *E* lags *I* by ϕ, and the circuit behaves as a *RC* series circuit.

Circuit Equations

The rectangular expression for the voltage in a series *RLC* circuit is,

$$E = V_R + j(V_L - V_C) \tag{19-25}$$

or, $$E = V_R + jV_X \tag{19-26}$$

where, $$V_X = V_L - V_C$$

Converting from rectangular form,

$$E = \sqrt{V_R^2 + V_X^2} \angle \tan^{-1}(V_X/V_R) \tag{19-27}$$

Recall from Equation 18-1 that to convert from polar form to rectangular form,

$$E\angle\phi = E\cos\theta + jE\sin\theta$$

Dividing Equation 19-25 through by I,

$$\frac{E}{I} = \frac{V_R}{I} + j\left(\frac{V_L}{I} - \frac{V_C}{I}\right)$$

Impedance, $$Z = R + j(X_L - X_C) \tag{19-28}$$

Also, $$|Z| = \sqrt{R^2 + X_{eq}^2} \tag{19-29}$$

where, $$X_{eq} = X_L - X_C$$

and, $$\phi = \tan^{-1}\left(\frac{X_{eq}}{R}\right) \tag{19-30}$$

Combining Equations 19-29 and 19-30,

$$Z = \sqrt{R^2 + X_{eq}^2} \angle \tan^{-1}(X_{eq}/R) \tag{19-31}$$

As already noted, for conversion from polar to rectangular form, Equation 18-1 is rewritten using impedances,

$$Z\angle\phi = Z\cos\phi + jZ\sin\phi$$

Impedance Diagrams

The impedance diagrams shown in Figures 19-27(a) and (b) relate to the phasor diagrams in Figures 19-26(b) and (c) respectively, and are self-explanatory.

Analysis Procedure for Series *RLC* Circuit

1. *Calculate the inductive reactance;* $X_L = (2\pi fL)$
2. *Calculate the capacitive reactance;* $X_C = 1/(2\pi fC)$
3. *Determine the equivalent reactance;* $X_{eq} = X_L - X_C$
4. *If there is more than one resistive component, calculate the total resistance;*

$$R = R_1 + R_2 + \cdots$$

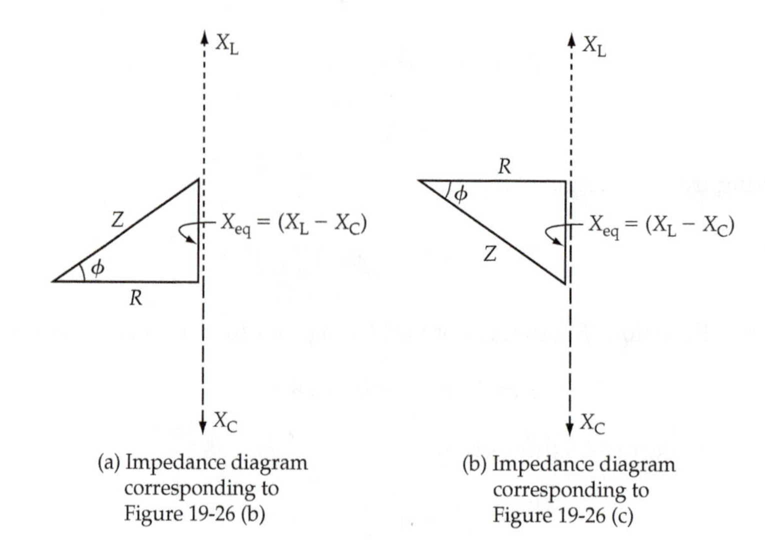

Figure 19-27 Impedance diagrams for a series-connected *RLC* circuit. If the circuit is largely inductive, the impedance triangle is that of an *RL* circuit. When the circuit is largely capacitive an *RC* circuit impedance triangle results.

5. *Calculate the impedance;*

$$|Z| = \sqrt{R^2 + X_{eq}^2}$$

6. *Calculate the phase angle:* $\phi = \tan^{-1}(X_{eq}/R)$
7. *Calculate the current;* $I = E/Z$
8. *Determine the resistive voltage;* $V_R = IR$ *or* $V_R = E\cos\phi$
9. *Determine the reactive voltage;* $V_X = IX_{eq}$ *or* $V_X = E\sin\phi$
10. *Determine the inductive voltage;* $V_L = IX_L$
11. *Determine the capacitive voltage;* $V_C = IX_C$

Example 19-11

Analyze the series *RLC* circuit in Figure 19-28 to determine: I, V_R, V_L, V_C, and the phase angle of the current with respect to the supply voltage.

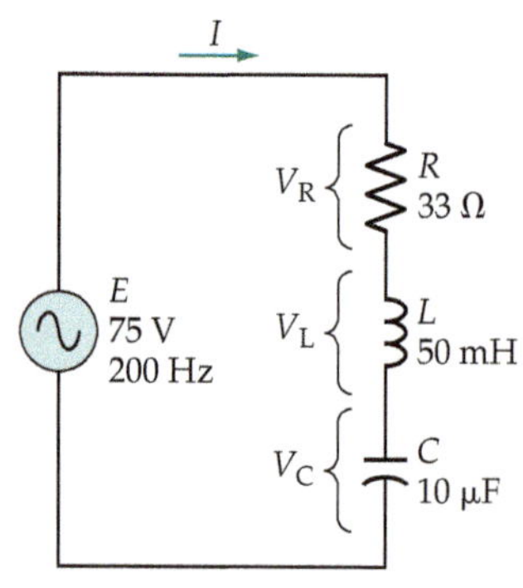

Figure 19-28 Series-connected *RLC* circuit for Example 19-11.

Solution

Eq. 19-3,
$$X_L = 2\pi fL = 2\pi \times 200\text{ Hz} \times 50\text{ mH} = 62.8\ \Omega$$

Eq. 19-9,
$$X_C = \frac{1}{2\pi fC} = \frac{1}{2\pi \times 200\text{ Hz} \times 10\ \mu\text{F}} = 79.6\ \Omega$$

Eq. 19-28,
$$Z = R + j(X_L - X_C) = 33\ \Omega + j(62.8\ \Omega - 79.6\ \Omega) = 33\ \Omega - j16.8\ \Omega$$

$$X_{eq} = X_L - X_C = 62.8\ \Omega - 79.6\ \Omega = -16.8\ \Omega \text{ (capacitive because } X_C > X_L)$$

Eq. 19-29, $|Z| = \sqrt{R^2 + X_{eq}^2} = \sqrt{(33\ \Omega)^2 + (16.8\ \Omega)^2}$

$\approx 37\ \Omega$

Eq. 19-30, $\phi = \tan^{-1}\left(\frac{X_{eq}}{R}\right) = \tan^{-1}\left(\frac{-16.8\ \Omega}{33\ \Omega}\right)$

$= -27°$

$$I = \frac{E}{Z} = \frac{75\ \text{V}}{37\ \Omega\angle -27°}$$

$$= 2.03\ \text{A}\angle 27°$$

Because X_{eq} is a capacitive reactance, the circuit current I leads the supply voltage E.

$$V_R = IR = 2.03\ \text{A} \times 33\ \Omega$$

$$\approx 67\ \text{V}$$

$$V_L = IX_L = 2.03\ \text{A} \times 62.8\ \Omega$$

$$= 127\ \text{V}$$

$$V_C = IX_C = 2.03\ \text{A} \times 79.6\ \Omega$$

$$= 162\ \text{V}$$

Example 19-12

Draw the phasor diagram for the series *RLC* circuit analyzed in Example 19-11.

Solution

- The phasor for the current (I) is first drawn horizontally, see Figure 19-29.
- V_R is drawn in phase with the current.
- V_C is drawn lagging the current by 90°.
- V_L is drawn leading the current by 90°.
- $(V_L - V_C)$ is identified.
- E lags the current by ϕ, and is also the resultant of $(V_L - V_C)$ and V_R.

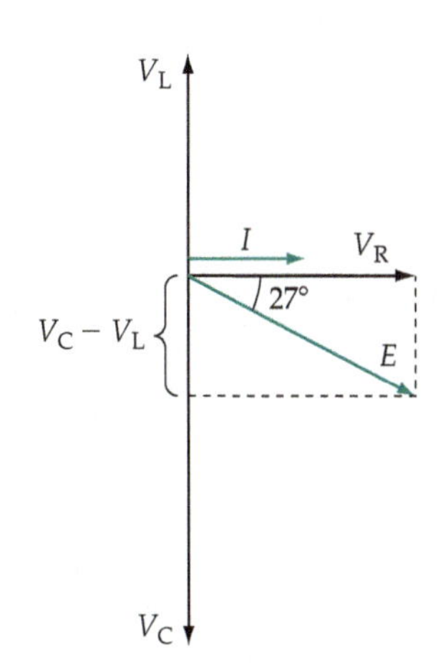

Figure 19-29 Phasor diagram for the series-connected *RLC* circuit in Example 19-11.

Practice Problems

19-7.1 For the circuit in Example 19-11, recalculate the quantities when the supply frequency is 300 Hz.

19-7.2 A series circuit consists of a 10 mH inductor, 0.1 μF capacitor, and 1 kΩ resistor. If the resistor voltage is measured as 1 V with a frequency of 10 kHz, calculate the supply voltage and its phase angle with respect to the current.

19-8 PARALLEL *RL* CIRCUITS

Circuit and Phasor Diagrams

For all cases of parallel-connected components, the supply voltage appears across the terminals of every component. In the parallel *RL* circuit in Figure 19-30(a), the supply voltage (E) is common to both components. The branch current in the circuit (i_R and i_L) are phasor quantities, so the supply current (I) is the vector sum of i_R and i_L. The phasor diagram is commenced by first drawing the phasor for E horizontally, because E is common to both R and L [see Figure 19-30(b)]. The resistive current phasor (i_R) is drawn in phase with E. The inductive current i_L lags the applied voltage, so the i_L phasor is drawn 90° lagging E. The supply current (I) is the resultant of the phasor sum of i_R and i_L, and the phase angle of I with respect to E is ϕ, as shown in the figure.

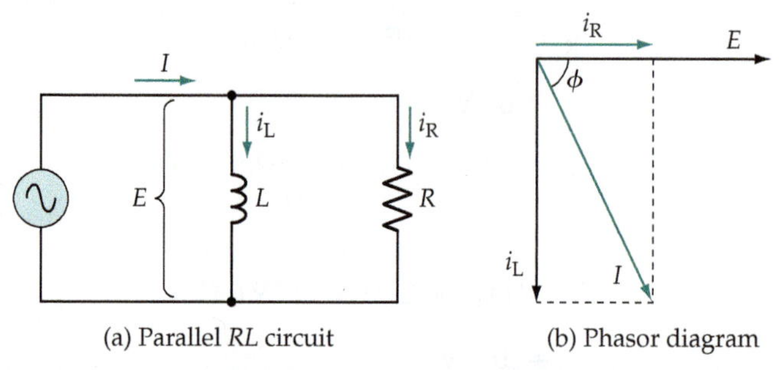

Figure 19-30 In a parallel-connected *RL* circuit, the inductor current (i_L) lags the supply voltage (E) by 90°. The resistor current (i_R) is in phase with the supply voltage. The supply current (I) is the resultant of i_L and i_R.

Circuit Equations

The supply current equation for the parallel-connected *RL* circuit in Figure 19-30 is,

$$I = i_R - ji_L \qquad (19\text{-}32)$$

The polar form expression for the current is,

$$I = \sqrt{i_R^2 + i_L^2}\angle \tan^{-1}(-i_L/i_R) \qquad (19\text{-}33)$$

Dividing Equation 19-32 by E gives,

$$\frac{I}{E} = \frac{i_R}{E} - j\left(\frac{i_L}{E}\right)$$

which can be rewritten as,

$$\frac{1}{Z} = \frac{1}{R} - j\left(\frac{1}{X_L}\right) \qquad (19\text{-}34)$$

or,

$$Y = G - jB_L \quad (19\text{-}35)$$

In Equation 19-35, Y is admittance, G is conductance, and B_L is inductive susceptance, all of which are measured in siemens (S).

Also,

$$|Y| = \sqrt{G^2 + B_L^2} \quad (19\text{-}36)$$

and,

$$\phi = \tan^{-1}\left(\frac{-B_L}{G}\right) \quad (19\text{-}37)$$

Combining Equations 19-36 and 19-37,

$$Y = \sqrt{G^2 + B_L^2}\,\angle \tan^{-1}(-B_L/G) \quad (19\text{-}38)$$

Admittance Diagram

It is impossible to draw an impedance diagram for a parallel *RL* circuit; instead, an *admittance diagram* is prepared, as shown in Figure 19-31. A horizontal line is first drawn to scale to represent the conductance (G). The inductive susceptance vector (B_L) is next drawn vertically at an angle of −90° with respect to the conductance vector. The resultant of B_L and G gives the admittance (Y).

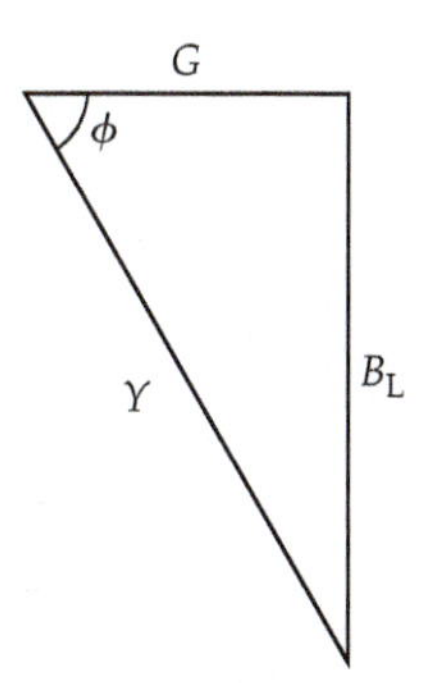

Figure 19-31 Admittance diagram for a parallel-connected *RL* circuit. The conductance vector (*G*) is drawn horizontally. Inductive susceptance vector B_L is drawn (down) from the conductance vector at a −90° angle. The admittance (*Y*) is the resultant of *G* and B_L.

Analysis Procedure for Parallel *RL* Circuits

1. *Calculate the inductive reactance;* $X_L = 2\pi fL$.
2. *Determine the currents through each component;* $i_R = E/R$, $i_L = E/X_L$
3. *Calculate the supply current;*

$$I = \sqrt{i_R^2 + i_L^2}$$

4. *Calculate the phase angle;* $\phi = \tan^{-1}(-i_L/i_R)$.

Example 19-13

Analyze the parallel *RL* circuit in Figure 19-32 to determine i_R, i_L, I, and the phase angle of I with respect to the supply voltage. Also, determine the circuit admittance and calculate the current directly using the admittance.

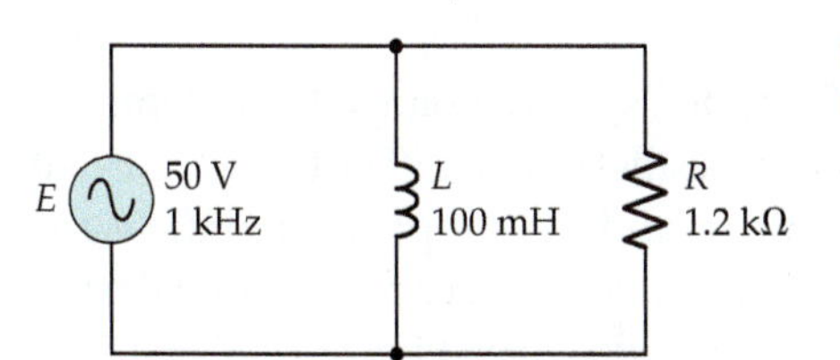

Figure 19-32 Parallel-connected *RL* circuit for Example 19-13.

Solution

$$X_L = 2\pi f L = 2\pi \times 1\text{ kHz} \times 100\text{ mH}$$
$$= 628\ \Omega$$
$$i_R = \frac{E}{R} = \frac{50\text{ V}}{1.2\text{ k}\Omega} = 41.7\text{ mA}$$
$$i_L = \frac{E}{X_L} = \frac{50\text{ V}}{628\ \Omega} = 79.6\text{ mA}$$
$$|I| = \sqrt{i_R^2 + i_L^2} = \sqrt{(41.7\text{ mA})^2 + (79.6\text{ mA})^2}$$
$$= 89.9\text{ mA}$$
$$\phi = \tan^{-1}(-i_L/i_R) = \tan^{-1}(-79.6\text{ mA}/41.7\text{ mA})$$
$$= -62.3^\circ$$
$$B_L = \frac{1}{X_L} = \frac{1}{62.8\ \Omega} = 15.9\text{ mS}$$
$$G = \frac{1}{R} = \frac{1}{1.2\text{ k}\Omega} = 0.833\text{ mS}$$
$$Y = \sqrt{G^2 + B_L^2} = \sqrt{(0.833\text{ mS})^2 + (15.9\text{ mS})^2}$$
$$= 1.79\text{ mS}$$
$$I = EY = 50\text{ V} \times 1.79\text{ mS}$$
$$= 89.5\text{ mA}$$

Practice Problems

19-8.1 Calculate B_L, G, Y, i_R, i_L, I, and ϕ for the circuit in Figure 19-32 when the supply frequency is changed to 500 Hz.

19-8.2 Draw phasor and admittance diagrams for Problem 19-8.1.

19-9 PARALLEL RC CIRCUITS

Circuit and Phasor Diagrams

In the parallel *RC* circuit in Figure 19-33(a), the supply voltage (E) is common to both components, (as in the case of other parallel circuits). The branch current in the circuit (i_R and i_C) are phasor quantities, and the supply current (I) is the vector sum of i_R and i_C. For the phasor diagram, the phasor for E is first drawn horizontally, because E is common to both R and C, [see Figure 19-33(b)]. The

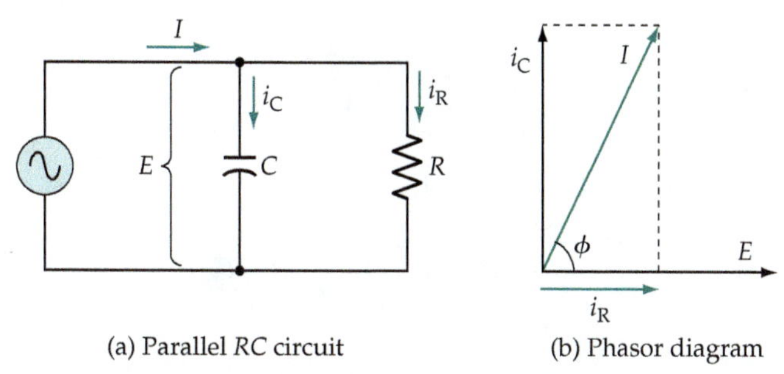

(a) Parallel *RC* circuit (b) Phasor diagram

Figure 19-33 In a parallel-connected *RC* circuit, the capacitor current (i_C) leads the supply voltage (E) by 90°. The resistor current (i_R) is in phase with the supply voltage. The supply current (I) is the resultant of i_C and i_R.

resistive current phasor (i_R) is drawn in phase with E. The capacitive current i_C leads the applied voltage, so the i_C phasor is drawn 90° leading E. The supply current (I) is the resultant of the phasor sum of i_R and i_C, and the phase angle of I with respect to E is ϕ, as illustrated.

Circuit Equations

The supply current equation for the parallel-connected *RC* circuit in Figure 19-33 is,

$$I = i_R + ji_C \tag{19-39}$$

The polar form expression for the current is,

$$I = \sqrt{i_R^2 + i_C^2}\angle \tan^{-1}(i_C/i_R) \tag{19-40}$$

Dividing Equation 19-39 by E gives,

$$\frac{I}{E} = \frac{i_R}{E} + j\left(\frac{i_C}{E}\right)$$

which can be rewritten as,

$$\frac{1}{Z} = \frac{1}{R} + j\left(\frac{1}{X_C}\right) \tag{19-41}$$

or,

$$Y = G + jB_C \tag{19-42}$$

In Equation 19-42, Y is admittance, G is conductance, and B_C is capacitive susceptance, all of which are measured in siemens.

Also,

$$|Y| = \sqrt{G^2 + B_C^2} \tag{19-43}$$

and,

$$\phi = \tan^{-1}\left(\frac{B_C}{G}\right) \tag{19-44}$$

Combining Equations 19-43 and 19-44,

$$Y = \sqrt{G^2 + B_C^2}\angle\tan^{-1}(B_C/G) \tag{19-45}$$

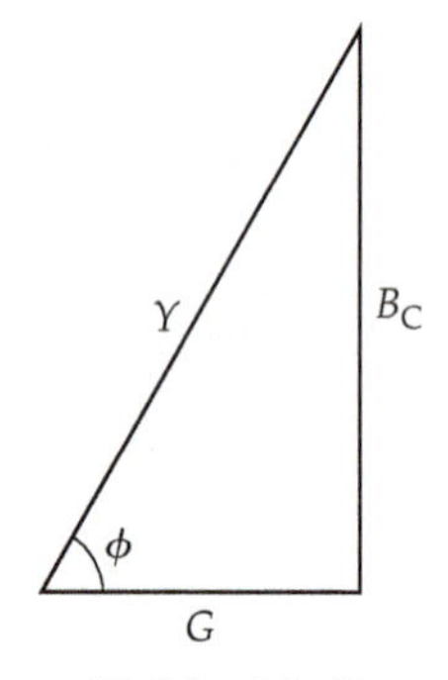

Figure 19-34 Admittance diagram for a parallel-connected *RC* circuit. The capacitive susceptance vector B_C is drawn (up) from the (horizontal) conductance vector at a 90° angle. The admittance (Y) is the resultant of G and B_C.

Admittance Diagram

As discussed for the parallel *RL* circuit, it is impossible to draw an impedance diagram for a parallel *RC* circuit. An *admittance diagram* is constructed instead, as shown in Figure 19-34. A horizontal line is first drawn to scale to represent the conductance (G). The capacitive susceptance vector (B_C) is next drawn vertically at an angle of 90° with respect to the conductance vector. The resultant of B_C and G gives the admittance (Y).

Analysis Procedure for Parallel *RC* Circuits

1. *Calculate the capacitive reactance;* $X_C = 1/(2\pi fC)$
2. *Determine the currents through each component;* $i_R = E/R$, $i_C = E/X_C$
3. *Calculate the supply current;*

$$I = \sqrt{i_R^2 + i_C^2}$$

4. *Calculate the phase angle;* $\phi = \tan^{-1}(i_C/i_R)$

Example 19-14

Analyze the parallel *RC* circuit in Figure 19-35 to determine i_R, i_C, I, and the phase angle of I with respect to the supply voltage.

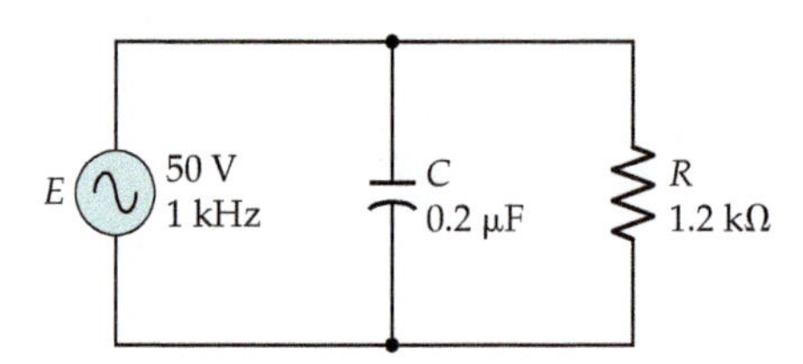

Figure 19-35 Parallel-connected *RC* circuit for Example 19-14.

Solution

$$X_C = 1/(2\pi fC) = 1/(2\pi \times 1\text{ kHz} \times 0.2\text{ μF})$$
$$= 796\ \Omega$$

$$i_R = \frac{E}{R} = \frac{50\text{ V}}{1.2\text{ k}\Omega} = 41.7\text{ mA}$$

$$i_C = \frac{E}{X_C} = \frac{50\text{ V}}{796\ \Omega} = 62.8\text{ mA}$$

$$|I| = \sqrt{i_R^2 + i_C^2} = \sqrt{(41.7\ \text{mA})^2 + (62.8\ \text{mA})^2}$$
$$= 75.4\ \text{mA}$$
$$\phi = \angle \tan^{-1}(i_C/i_R) = \angle \tan^{-1}(62.8\ \text{mA}/41.7\ \text{mA})$$
$$= 56.4°$$

Practice Problems

19-9.1 If the supply current for the circuit analyzed in Example 19-14 changes to 89 mA when the capacitor is replaced, determine the new capacitance value.

19-9.2 Draw phasor and admittance diagrams for the circuit in Example 19-14.

19-10 PARALLEL *RLC* CIRCUITS

Phasor Diagrams

A resistor, an inductor, and a capacitor are shown connected in parallel in Figure 19-36(a), and the phasor diagram for the circuit is illustrated in Figure 19-36(b). Because the supply voltage (E) is common to all components, the phasor diagram is commenced by first drawing the phasor for E horizontally. The i_R phasor is next drawn in phase with E. The inductive current (i_L) lags the applied voltage, so the i_L phasor is drawn at −90° with respect to E. Similarly, the capacitive current leads the applied voltage, so its phasor is drawn 90° leading E. The phasors of i_L and i_C are in antiphase, so, as illustrated, they are added to produce the equivalent reactance phasor ($i_X = i_C - i_L$). The supply

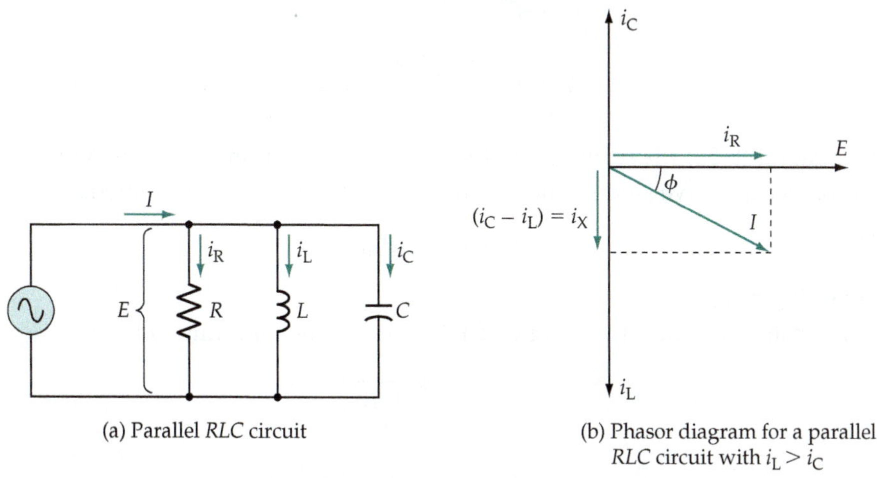

(a) Parallel *RLC* circuit

(b) Phasor diagram for a parallel *RLC* circuit with $i_L > i_C$

Figure 19-36 Parallel *RLC* circuit and phasor diagrams. If $i_L > i_C$, the supply current lags the supply voltage. With $i_C > i_L$, the supply current leads the supply voltage.

current (I) is the resultant of the phasor sum of i_X and i_R, and the phase angle of I with respect to E is ϕ, as shown.

In the phasor diagram shown in Figure 19-36(b), i_L is greater than i_C; so i_X is inductive, and the complete circuit behaves as a resistive-inductive parallel combination. It is also possible for i_C to be greater than i_L, giving a circuit that performs as a resistor and capacitor in parallel. The special case where i_L and i_C are equal is considered in Chapter 23.

Circuit Equations

Consideration of the circuit and phasor diagram in Figure 19-36 produces the equation for the total supply current as,

$$I = i_R + j(i_C - i_L) \tag{19-46}$$

or,

$$I = i_R + j i_X \tag{19-47}$$

where $i_X = i_C - i_L$.

The expression for the current in polar form is,

$$I = \sqrt{i_R^2 + i_X^2} \angle \tan^{-1}(i_X / i_R) \tag{19-48}$$

Dividing Equation 19-46 by E,

$$\frac{I}{E} = \frac{i_R}{E} + j\left(\frac{i_C}{E} - \frac{i_L}{E}\right)$$

which gives,

$$\frac{1}{Z} = \frac{1}{R} + j\left(\frac{1}{X_C} - \frac{1}{X_L}\right) \tag{19-49}$$

or,

$$Y = G + j(B_C - B_L) \tag{19-50}$$

In Equation 19-50, Y is admittance, G is conductance, B_L is inductive susceptance, and B_C is capacitive susceptance, all of which are measured in siemens.

Also,

$$Y = G + jB_{eq}$$

where $B_{eq} = B_C - B_L$.

The modulus and phase angle of Y can readily be determined from,

$$|Y| = \sqrt{G^2 + B_{eq}^2} \tag{19-51}$$

and,

$$\phi = \tan^{-1}\left(\frac{B_{eq}}{G}\right) \tag{19-52}$$

Combining Equations 19-51 and 19-52,

$$Y = \sqrt{G^2 + B_{eq}^2}\,\underline{/\tan^{-1}(B_{eq}/G)} \qquad \textbf{(19-53)}$$

Equation 19-53 converts from the rectangular $G + jB$ to the polar $Y\underline{/\phi}$. Equation 18-1 can be rewritten for conversion of admittance from polar to rectangular form,

$$Y\underline{/\phi} = Y\cos\phi + jY\sin\phi.$$

Admittance Diagram

As already discussed for parallel *RL* and *RC* circuits, is impossible to draw an impedance diagram for a parallel *RLC* circuit; instead, an *admittance diagram* is prepared, as shown in Figure 19-37. A horizontal line is first drawn to scale to represent the conductance (G). The capacitive susceptance vector (B_C) is next drawn vertically at an angle of 90° with respect to the conductance vector. Then, the inductive susceptance vector (B_L) is drawn at −90° to G. The resultant susceptance vector (B_{eq}) is found by vectorially summing B_C and B_L. Finally, the resultant of B_{eq} and G is established, to determine the admittance Y.

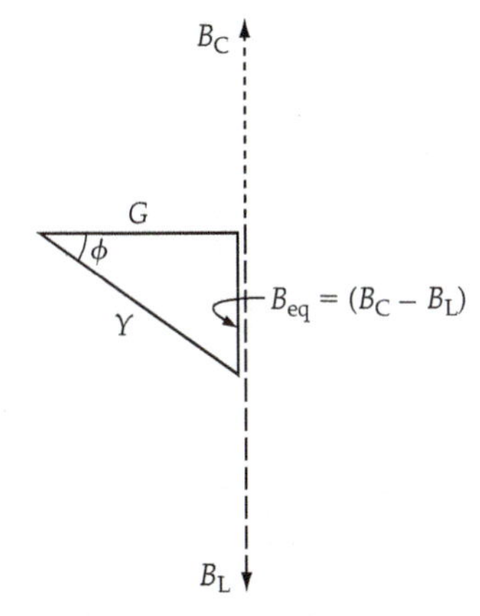

Figure 19-37 Admittance diagram for a parallel *RLC* circuit. Capacitive susceptance vector B_C is drawn (up) from the conductance vector G at an angle of 90°. Inductive susceptance vector B_L is drawn (down) from G at an angle of −90°. The admittance (*Y*) is the resultant of G and ($B_C - B_L$).

Parallel *RLC* Circuit Analysis Procedure

1. *Calculate the inductive reactance;* $X_L = 2\pi fL$
2. *Calculate the capacitive reactance;* $X_C = 1/(2\pi fC)$
3. *Determine the currents through each component;*

$$i_R = E/R,\ i_L = E/X_L,\ i_C = E/X_C$$

4. *Calculate the reactive current;* $i_X = i_C - i_L$
5. *Calculate the supply current;*

$$I = \sqrt{i_R^2 + i_X^2}$$

6. *Calculate the phase angle;* $\phi = \tan^{-1}(i_X/i_R)$

Alternatively:

1. *Calculate the inductive reactance;* $X_L = 2\pi fL$
2. *Calculate the capacitive reactance;* $X_C = 1/(2\pi fC)$
3. *Determine the conductance and susceptances;*

$$G = 1/R,\ B_L = 1/X_L,\ B_C = 1/X_C$$

4. *Determine the currents through each component;*

$$i_R = EG,\ i_L = EB_L,\ i_C = EB_C$$

5. *Calculate the reactive current;* $i_X = i_C - i_L$
6. *Calculate the supply current;*

$$I = \sqrt{i_R^2 + i_X^2}$$

7. *Calculate the phase angle;* $\phi = \tan^{-1}(i_X/i_R)$

Example 19-15

Analyze the parallel *RLC* circuit in Figure 19-38 to determine; i_R, i_L, i_C, I, and the phase angle of I with respect to the supply voltage.

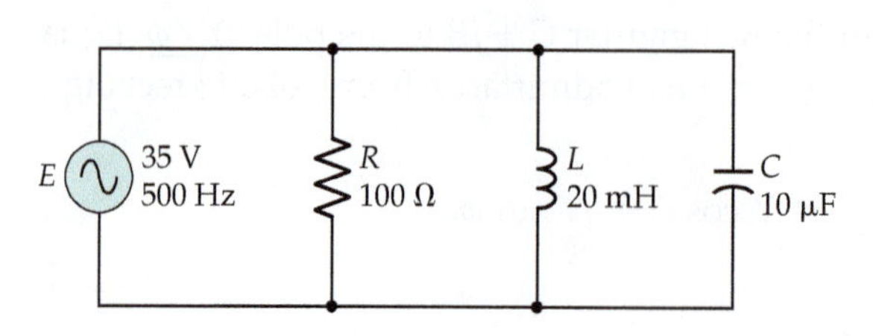

Figure 19-38 Parallel-connected *RLC* circuit for Example 19-15.

Solution

$$X_L = 2\pi fL = 2\pi \times 500 \text{ Hz} \times 20 \text{ mH}$$
$$= 62.8\ \Omega$$

$$X_C = \frac{1}{2\pi fC} = \frac{1}{2\pi \times 500 \text{ Hz} \times 10\ \mu\text{F}}$$
$$= 31.8\ \Omega$$

$$i_R = \frac{E}{R} = \frac{35 \text{ V}}{100\ \Omega} = 350 \text{ mA}$$

$$i_L = \frac{E}{X_L} = \frac{35 \text{ V}}{62.8\ \Omega} = 557 \text{ mA}$$

$$i_C = \frac{E}{X_C} = \frac{35 \text{ V}}{31.8\ \Omega} = 1.1 \text{ A}$$

$$i_X = i_C - i_L = 1.1 \text{ A} - 557 \text{ mA}$$
$$= 543 \text{ mA}$$

Eq. 19-47,
$$I = i_R + ji_X = 350 \text{ mA} + j543 \text{ mA}$$

From Eq. 19-48,
$$|I| = \sqrt{i_R^2 + i_X^2} = \sqrt{(350 \text{ mA})^2 + (543 \text{ mA})^2}$$
$$= 646 \text{ mA}$$

and,
$$\phi = \tan^{-1}\left(\frac{i_X}{i_R}\right) = \tan^{-1}\left(\frac{543 \text{ mA}}{350 \text{ mA}}\right)$$
$$= 57.2^\circ$$

Example 19-16

Draw a phasor diagram and an admittance diagram for the circuit analyzed in Example 19-15. Determine the circuit impedance.

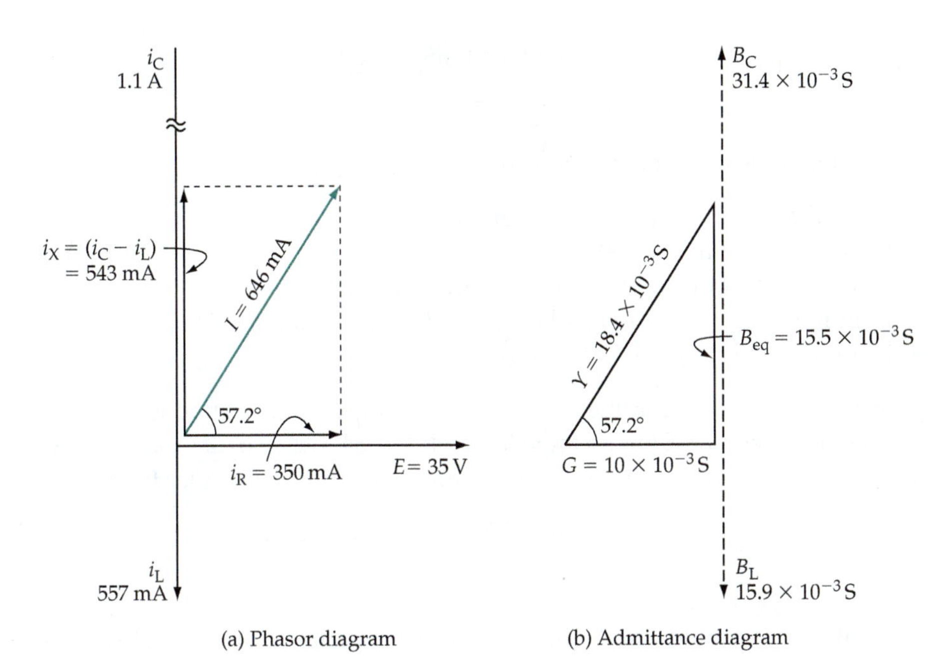

Figure 19-39 Phasor diagram and admittance diagram for the parallel *RLC* circuit in Example 19-15.

Solution

Phasor diagram—Figure 19-39(a):

- Start by drawing the $E = 35$ V phasor horizontally.
- Draw $i_R = 350$ mA in phase with E.
- Draw $i_C = 1.1$ A at $+90°$ with respect to E.
- Draw $i_L = 557$ mA at $-90°$ with respect to E.
- Determine $i_X = (i_C - i_L)$ and draw the i_X phasor alongside i_C.
- Draw I as the resultant of i_R and $(i_C - i_L)$.

$$G = \frac{1}{R} = \frac{1}{100\ \Omega} = 10\text{ mS}$$

$$B_L = \frac{1}{X_L} = \frac{1}{62.8\ \Omega} = 15.9\text{ mS}$$

$$B_C = \frac{1}{X_C} = \frac{1}{31.8\ \Omega} = 31.4\text{ mS}$$

$$B_{eq} = B_C - B_L = (31.4 \times 10^{-3}) - (15.9 \times 10^{-3})$$

$$= 15.5\text{ mS}$$

Eq. 19-51,

$$|Y| = \sqrt{G^2 + B_{eq}^2} = \sqrt{(10\text{ mS})^2 + (15.5\text{ mS})^2}$$

$$= 18.4\text{ mS}$$

Eq. 19-51,

$$\phi = \tan^{-1}\left(\frac{B_{eq}}{G}\right) = \tan^{-1}\left(\frac{15.5\text{ mS}}{10\text{ mS}}\right)$$

$$= 57.2°$$

Eq. 19-19, $Z = 1/Y = 1/(18.4\ \text{mS}\angle 57.2^\circ)$

$$= 54.3\ \Omega\angle -57.2^\circ$$

Admittance diagram—Figure 19-39(b):

- Draw the $G = 10$ mS phasor horizontally.
- Draw $B_C = 31.4$ mS at $+90^\circ$ with respect to G.
- Draw $B_L = 15.9$ mS at -90° with respect to G.
- Determine $B_{eq} = (B_C - B_L)$ and draw the B_{eq} phasor alongside B_C.
- Draw Y as the resultant of G and $(B_C - B_L)$.

Example 19-16 shows that the parallel *RLC* circuit with the component values and frequency as stated has an impedance of $Z = 54.3\ \Omega\angle -57.2^\circ$. Using Equation 18-1, the impedance can be resolved from polar into rectangular form:

$$Z = 54.3 \cos(-57.2^\circ) + j54.3 \sin(-57.2^\circ)$$

$$= 29.4\ \Omega - j45.6\ \Omega$$

This value of Z would also be obtained with a resistance of $R = 29.4\ \Omega$ connected *in series* with a capacitive reactance of $X_C = 45.6\ \Omega$. Thus, it is seen that this particular parallel *RLC* circuit has an equivalent series *RC* circuit. Depending on the component values, the equivalent series circuit of a parallel *RLC* circuit could be an *RL* circuit.

Practice Problems

19-10.1 Recalculate the quantities for the circuit in Example 19-15 when the supply frequency is changed to 200 Hz.

19-10.2 Draw phasor and admittance diagrams for Problem 19-10.1, and determine the total circuit impedance.

Summary of Formulas

- *Inductive reactance:*

$$X_L = \omega L = 2\pi f L$$

- *Inductive susceptance:*

$$B_L = 1/X_L$$

- *Capacitive reactance:*

$$X_C = \frac{1}{\omega C} = \frac{1}{2\pi f C}$$

- ***Capacitive susceptance:***

$$B_C = 1/X_C$$

- ***For a series RL circuit:***

$$E = \sqrt{V_R^2 + V_L^2}\,\angle\tan^{-1}(V_L/V_R)$$

$$Z = \sqrt{R^2 + X_L^2}\,\angle\tan^{-1}(X_L/R)$$

- ***For a series RC circuit:***

$$E = \sqrt{V_R^2 + V_C^2}\,\angle\tan^{-1}(-V_C/V_R)$$

$$Z = \sqrt{R^2 + X_C^2}\,\angle\tan^{-1}(-X_C/R)$$

- ***For a series RLC circuit:***

$$E = \sqrt{V_R^2 + V_X^2}\,\angle\tan^{-1}(-V_X/V_R)$$

$$Z = \sqrt{R^2 + X_{eq}^2}\,\angle\tan^{-1}(X_{eq}/R)$$

- ***For a parallel RL circuit:***

$$I = \sqrt{i_R^2 + i_L^2}\,\angle\tan^{-1}(-i_L/i_R)$$

$$Y = \sqrt{G^2 + B_L^2}\,\angle\tan^{-1}(-B_L/G)$$

- ***For a parallel RC circuit:***

$$I = \sqrt{i_R^2 + i_C^2}\,\angle\tan^{-1}(i_C/i_R)$$

$$Y = \sqrt{G^2 + B_C^2}\,\angle\tan^{-1}(B_C/G)$$

- ***For a parallel RLC circuit:***

$$I = \sqrt{i_R^2 + i_X^2}\,\angle\tan^{-1}(i_X/i_R)$$

$$Y = G + j(B_C - B_L)$$

$$Y = \sqrt{G^2 + B_{eq}^2}\,\angle\tan^{-1}(B_{eq}/G)$$

Review Questions

Section 19-1

19-1 Sketch typical waveforms of alternating voltage and current in a purely resistive circuit with an ac supply. Briefly explain.

19-2 Sketch typical waveforms of alternating current, counter emf, and supply voltage for a purely inductive circuit with an ac supply. Show the phase relationships between the waveforms, and carefully explain.

Section 19-2

19-3 Define inductive reactance and inductive susceptance, and write equations for each quantity.

Section 19-3

19-4 Sketch typical waveforms of alternating current and supply voltage for a purely capacitive circuit with an ac supply. Show the phase relationships between the waveforms, and carefully explain.

Section 19-4

19-5 Define capacitive reactance and capacitive susceptance, and write equations for each quantity.

Section 19-5

19-6 For a series-connected *RL* circuit with an ac supply, sketch typical waveforms of current, resistor voltage, inductor voltage, and supply voltage. Explain the phase relationships between the waveforms.

19-7 Write rectangular-form and polar-form equations for applied voltage, impedance, and phase angle for a series-connected *RL* circuit.

19-8 Sketch typical phasor and impedance diagrams for a series-connected *RL* circuit. Explain each diagram.

Section 19-6

19-9 For a series-connected *RC* circuit with an ac supply, sketch typical waveforms of current, resistor voltage, capacitor voltage, and supply voltage. Explain the phase relationships between the waveforms.

19-10 Write rectangular-form and polar-form equations for applied voltage, impedance, and phase angle for a series-connected *RC* circuit.

19-11 Sketch typical phasor and impedance diagrams for a series-connected *RC* circuit. Explain each diagram.

Section 19-7

19-12 Sketch typical phasor diagrams for a series-connected *RLC* circuit for the case of (a) $X_C > X_L$, and (b) $X_L > X_C$. Explain each diagram.

19-13 Sketch typical impedance diagrams for a series-connected *RLC* circuit for the case of (a) $X_C > X_L$, and (b) $X_L > X_C$. Explain each diagram.

19-14 For a series-connected *RLC* circuit, write polar-form and rectangular-form equations for the supply voltage and the circuit impedance.

Section 19-8

19-15 Sketch typical phasor and admittance diagrams for a parallel-connected *RL* circuit. Explain each diagram.

19-16 Write rectangular-form and polar-form equations for current, admittance, and phase angle for a parallel-connected *RL* circuit.

Section 19-9

19-17 Sketch typical phasor and admittance diagrams for a parallel-connected *RC* circuit. Explain each diagram.

19-18 Write rectangular-form and polar-form equations for current, admittance, and phase angle for a parallel-connected *RC* circuit.

Section 19-10

19-19 Write rectangular-form and polar-form equations for current, admittance, and phase angle for a parallel-connected *RLC* circuit.

19-20 Sketch typical phasor diagrams for a parallel *RLC* circuit for the case of (a) $X_C > X_L$, and (b) $X_L > X_C$. Explain each diagram.

19-21 Sketch typical admittance diagrams for a parallel *RLC* circuit for the case of (a) $X_C > X_L$, and (b) $X_L > X_C$. Explain each diagram.

Problems

Section 19-2

19-1 Calculate the current that flows in a 300 mH inductor connected to a 100 V, 250 Hz supply. Sketch the voltage and current waveforms and draw the phasor diagram for the circuit.

19-2 A 50 mH inductor passes a current of 398 mA when connected to a 400 Hz supply. Calculate the supply voltage.

19-3 A 100 μH inductor passes a current of 10 mA when its terminal voltage is 6.3 V. Calculate the frequency of the ac supply.

19-4 A 29 mA current flows in an inductor connected to a 15 V, 1 kHz ac supply. Determine the inductance value.

Section 19-4

19-5 A 100 V, 250 Hz supply is applied to a 30 μF capacitor. Calculate the capacitive reactance and the current that flows. Sketch the voltage and current waveforms and draw the phasor diagram for the circut.

19-6 A 4.5 μF capacitor passes a current of 34 mA when connected to a 100 Hz supply. Calculate the supply voltage.

19-7 A 100 μF capacitor passes a current of 0.25 mA when its terminal voltage is 4 V. Determine the supply frequency.

19-8 A current of 270 mA flows in a capacitor connected to a 25 V, 300 Hz ac supply. Determine the capacitance value.

Section 19-5

19-9 A series-connected *RL* circuit with $R = 68\ \Omega$ and $L = 10$ mH is connected to a 33 V, 1 kHz supply, as in Figure 19-40. Determine the circuit current, the resistor voltage, the inductor voltage, and the phase angle of the current with respect to the supply voltage. Sketch a phasor diagram and an impedance diagram for the circuit.

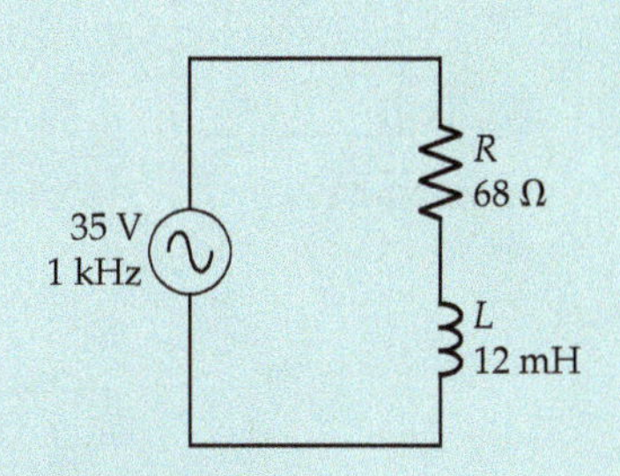

Figure 19-40

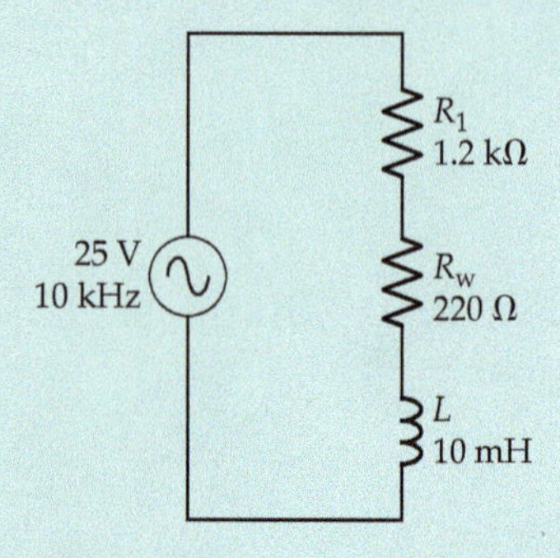

Figure 19-41

19-10 In the circuit described in Problem 19-9, the phase angle of the current with respect to the supply voltage is to be altered to 35°. Calculate a suitable new resistance value to effect this change, and determine the new current level.

19-11 A 25 V, 10 kHz supply is applied to a 10 mH inductor in series with a 1.2 kΩ resistor (see Figure 19-41). The inductor has a 220 Ω winding resistance. Calculate the circuit current and the inductor terminal voltage, and draw a phasor diagram and an impedance diagram for the circuit.

19-12 In the circuit described in Problem 19-11 the circuit current is to be changed to 10 mA by adjusting the supply frequency. Calculate the appropriate frequency and determine the new level of inductor terminal voltage.

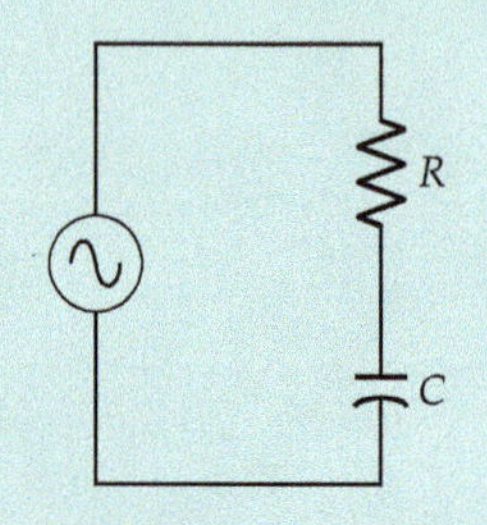

Figure 19-42

Section 19-6

19-13 A series *RC* circuit with $R = 120\ \Omega$ and $C = 3.3\ \mu F$ is connected to a 12 V, 1 kHz supply (see Figure 19-42). Determine the circuit current, the resistor voltage, the capacitor voltage, and the phase angle of the current with respect to the supply voltage. Sketch a phasor diagram and an impedance diagram for the circuit.

19-14 In the circuit described in Problem 19-13 the circuit current is to be changed to 50 mA by adjusting the capacitor. Determine the required new capacitance and the new capacitor terminal voltage.

19-15 A 35 V, 500 Hz supply applied to a capacitor in series with a 220 Ω resistor produces a 66 mA circuit current. Calculate the capacitance of the capacitor.

19-16 In the circuit described in Problem 19-15, the phase angle of the current is to be altered to −35° by adjusting the supply frequency. Determine the appropriate frequency and the new current level in the circuit.

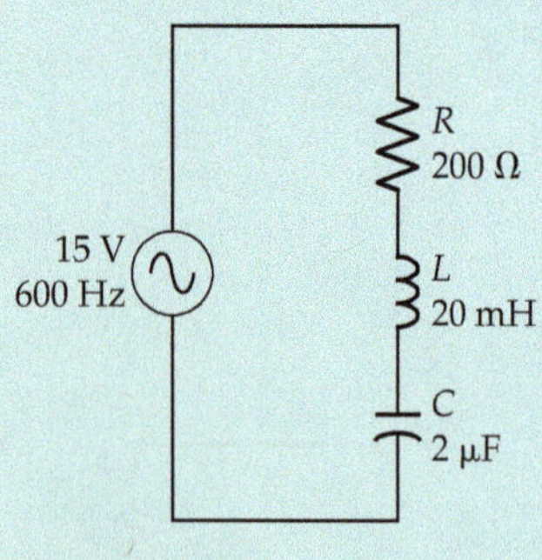

Figure 19-43

Section 19-7

19-17 A 20 mH inductor is connected in series with a 2 μF capacitor and a 200 Ω resistor, as shown in Figure 19-43. The supply voltage is 15 V with a frequency of 600 Hz. Calculate the circuit current and its phase angle with respect to the supply voltage. Also calculate the terminal voltage of each component.

19-18 For the circuit described in Problem 19-17, accurately sketch phasor and impedance diagrams.

19-19 The resistor in the circuit described in Problem 19-17 is to be adjusted to give an inductor terminal voltage of 7.5 V. Calculate the appropriate resistor value.

19-20 The supply frequency is altered to 900 Hz for the circuit described in Problem 19-17 Recalculate all of the quantities.

19-21 A coil with an inductance of 390 mH is connected in series with a 620 Ω resistor and a 6 μF capacitor. The supply is 45 V. Determine the frequency at which $X_C = 300\ \Omega$, and at this frequency find the circuit current and $(V_L + V_R)$. Draw the phasor and impedance diagrams.

19-22 Recalculate the quantities listed in Problem 19-21 for the case of $X_L = 300\ \Omega$.

Section 19-8

19-23 A parallel *RL* circuit consists of a 68 Ω resistor and a 10 mH inductor with a 5 V, 1 kHz supply, as illustrated in Figure 19-44. Calculate the branch currents, supply current, and the phase angle of the supply current with respect to the supply voltage.

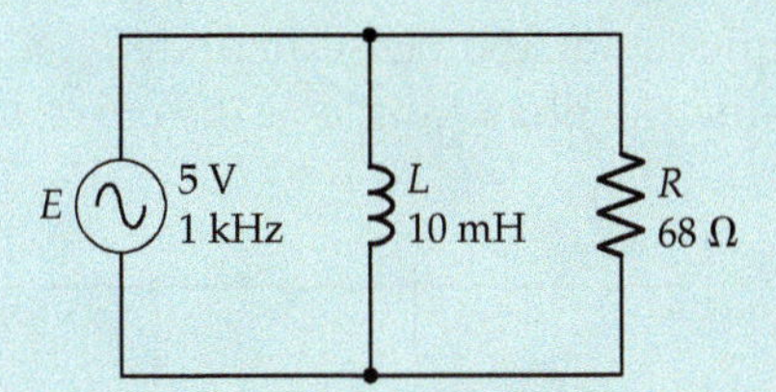

Figure 19-44

19-24 The circuit for Problem 19-23 is to have the frequency altered to adjust the current phase angle to −35°. Determine the required frequency.

19-25 A 5 mH inductor with a 220 Ω coil resistance is connected in parallel with a 1.2 kΩ resistor, and a 25 V, 10 kHz supply is applied to the two components. Calculate the supply current and its phase angle with respect to the supply voltage.

19-26 A 10 V, 3.3 kHz supply is applied to a 2.7 kΩ resistor and a 50 mH inductor connected in parallel. Calculate the branch currents and the supply current.

Section 19-9

19-27 A parallel *RC* circuit consists of a 68 Ω resistor and a 0.015 μF capacitor with a 3 V, 3.3 kHz supply (see Figure 19-45). Calculate the branch currents, the supply current, and the phase angle of the supply current with respect to the supply voltage.

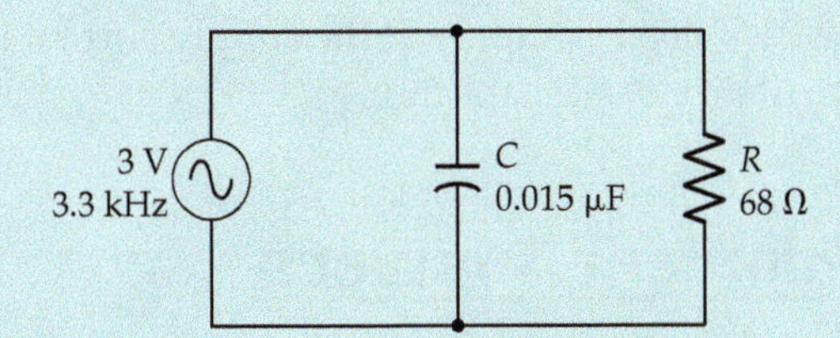

Figure 19-45

19-28 The circuit for Problem 19-27 is to have the resistor altered to adjust the supply current phase angle to 25°. Determine required resistance.

19-29 A 5000 pF capacitor is connected in parallel with a 5.6 kΩ resistor, and a 5 V, 6 kHz supply is applied to the two components. Calculate the supply current and its phase angle with respect to the supply voltage.

19-30 A 12 V, 9 kHz supply is applied to a 2.7 kΩ resistor and a 0.015 μF capacitor connected in parallel. Calculate the branch currents and the supply current.

Section 19-10

19-31 A 220 mH inductor is connected in parallel with an 8 μF capacitor and a 330 Ω resistor, as illustrated in Figure 19-46. The supply is 10 V with a frequency of 100 Hz. Calculate i_R, i_L, i_C, and the supply current. Determine the phase angle of the current with respect to the supply voltage.

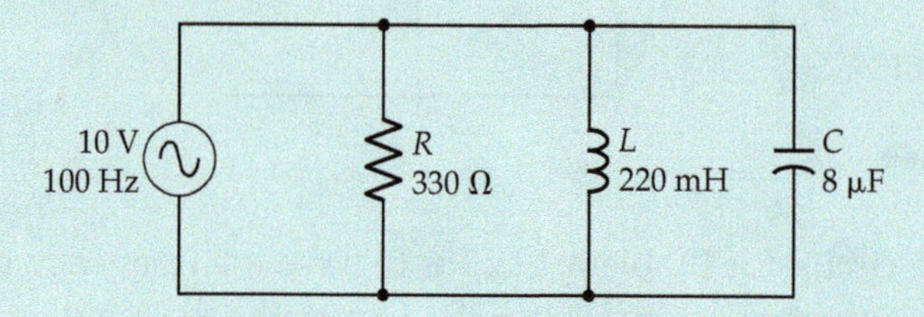

Figure 19-46

19-32 The supply voltage for the circuit in Problem 19-31 is to be altered to give $i_C = 61$ mA. Calculate the required new level of supply voltage, and recalculate the total supply current and its phase angle with respect to the voltage.

19-33 Accurately sketch phasor and admittance diagrams for the circuit described in Problem 19-31.

19-34 A 20 mH inductor, a 2 μF capacitor, and a 200 Ω resistor are connected in parallel to a 15 V, 600 Hz source. Calculate the total supply current and its phase angle with respect to the voltage.

19-35 A 390 mH inductor, a 620 Ω resistor, and a 6 μF capacitor are connected in parallel to a 120 V, 60 Hz source. Determine the supply current and $(i_R + i_L)$. Sketch the phasor and admittance diagrams.

19-36 A 200 mH inductor is connected in parallel with a 1.5 kΩ resistor, a capacitor, and a 33 V, 400 Hz supply. If the supply current is 40 mA, determine the capacitance of the capacitor.

Practice Problem Answers

19-2.1 1.27 mH
19-2.2 111 mA
19-2.3 2.3 V
19-4.1 700 kHz

19-4.2 50 pF
19-5.1 136 mA $\angle -25.2°$
19-5.2 3.88 mH
19-6.1 44.2 V, 23.5 V, 62°
19-6.2 12.2 V, 2.38 V
19-7.1 1.42 A $\angle -51.2°$, 47 V, 134 V, 75.4 V
19-7.2 1.1 V $\angle 25°$
19-8.1 159 mA, 41.7 mA, 164 mA, –75.3°
19-9.1 0.25 μF
19-10.1 350 mA, 1.39 A, 440 mA, 1.01 A $\angle -70°$
19-10.2 34.5 Ω $\angle 70°$

CHAPTER 20
Series and Parallel AC Circuits

CONTENTS

Objectives

You will be able to:

1. Analyze series circuits consisting of several impedances each of which is an *RL* or *RC* series combination. Determine the circuit current and component voltage drops, and the phase angle of each quantity.
2. Draw complete phasor diagrams for series circuits consisting of several impedances.
3. Apply the voltage-divider theorem to series impedance networks to determine component voltages and phase angles.
4. Analyze parallel circuits consisting of several impedances each of which is an *RL* or *RC* series or parallel combination. Determine all current levels, voltage drops, and phase angles.
5. Apply the current-divider rule to parallel impedance networks to determine branch currents and phase angles.
6. Analyze series-parallel impedance networks to determine all current levels, voltage drops, and phase angles.
7. Convert series *RL* and *RC* circuits into equivalent parallel networks, and vice versa.

INTRODUCTION

AC series impedance circuits can be analyzed in a similar way to dc series resistance circuits, as long as the phase angle of each impedance is taken into consideration. For addition, impedances must be in rectangular form, and for division or multiplication they should be in polar form. The ac voltage divider can be treated in a similar way to the dc voltage divider as long as the phase angles are correctly considered. Parallel impedances should be converted into admittances and added together; then the result may be inverted to determine the equivalent total impedance. In dealing with series-parallel impedance combinations, each parallel group of impedances should be resolved into a single equivalent impedance, so that the entire configuration reduces to a simple series impedance circuit. For every simple series impedance circuit there is an equivalent parallel impedance circuit, and vice versa.

20-1 SERIES-CONNECTED IMPEDANCES

Impedances in series in an ac circuit can be treated similarly to resistances in series in a dc circuit. The equivalent total impedance is determined, and using the applied voltage, the current through the circuit can be calculated. However, series impedances cannot be added directly but must be resolved into rectangular form; then the resistive and reactive components are added separately. Figure 20-1 illustrates the process. The three series impedances in Figure 20-1(a) are each converted to their rectangular form, Figure 20-1(b). The total values of R and X are calculated separately and then converted back to polar form, to give the equivalent circuit in Figure 20-1(c). The phasor diagram for the series impedance circuit is drawn as explained in Section 19-7.

Analysis Procedure for Series Impedance Circuits:

1. ***Convert each impedance into rectangular form.***
2. ***Sum the individual resistive components to determine the total equivalent resistance.***
3. ***Sum the individual reactive components to determine the total equivalent reactance.***
4. ***Convert the equivalent resistance and reactance values into the polar form equivalent impedance.***
5. ***Divide the equivalent impedance into the applied voltage to determine the supply current.***

Example 20-1

Analyze the circuit in Figure 20-2(a) to determine the current taken from the supply. Draw a phasor diagram of supply voltage and current.

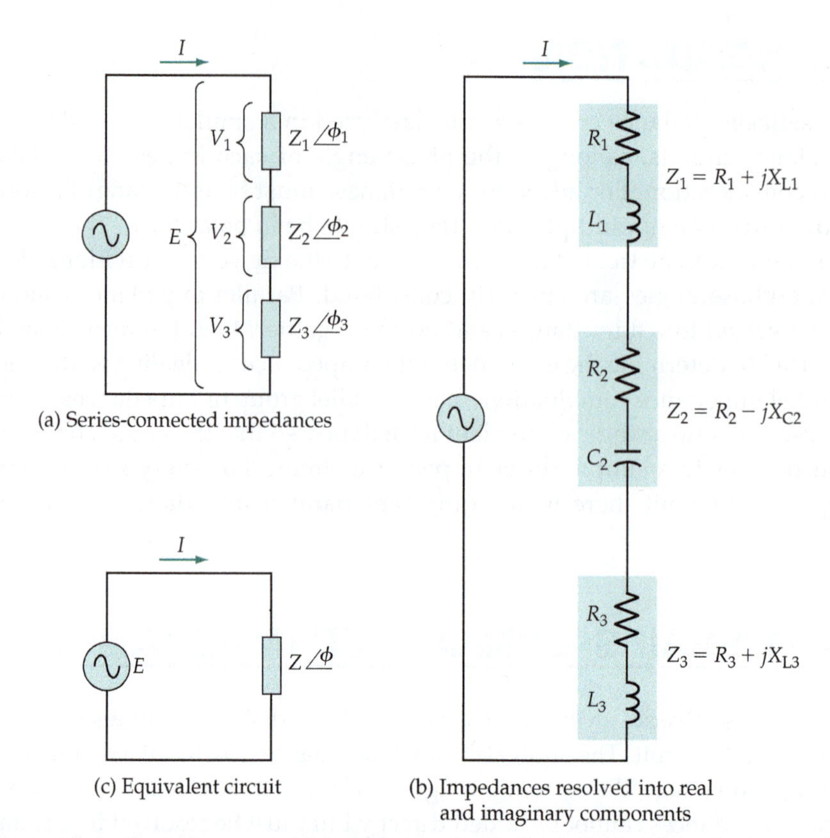

Figure 20-1 To analyze a series impedance circuit, the individual impedances are first converted into rectangular form. All the resistive components are summed and all the reactive components are summed, to give the equivalent circuit impedance.

Solution

Converting to rectangular form.

From Eq. 18-1,
$$Z_1 = (70.7\ \Omega \cos 45°) + j(70.7\ \Omega \sin 45°)$$
$$= 50\ \Omega + j\,50\ \Omega$$
$$Z_2 = (92.4\ \Omega \cos 330°) + j(92.4\ \Omega \sin 330°)$$
$$= 80\ \Omega - j\,46.2\ \Omega$$
$$Z_3 = (67\ \Omega \cos 60°) + j(67\ \Omega \sin 60°)$$
$$= 33.5\ \Omega + j\,58\ \Omega$$

Adding impedances in series.

$$Z = Z_1 + Z_2 + Z_3$$
$$= (50 + 80 + 33.5)\ \Omega + j(50 - 46.2 + 58)\ \Omega$$
$$= 163.5\ \Omega + j\,61.8\ \Omega$$

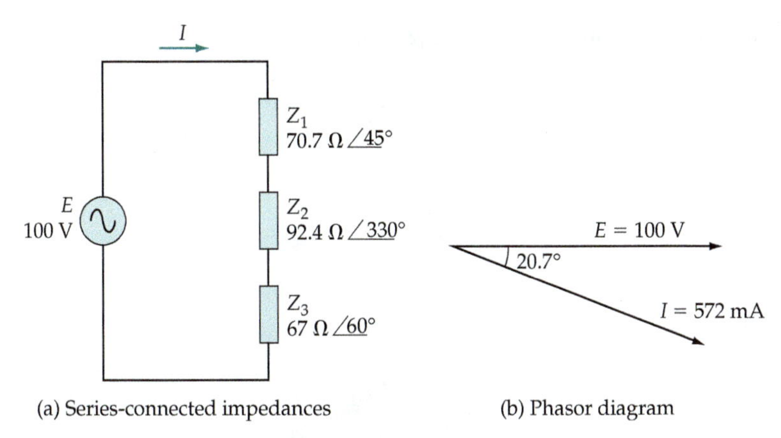

Figure 20-2 Circuit and phasor diagram for Example 20-1.

Converting to polar form.

From Eq. 19-17, $Z = \sqrt{(163.5\ \Omega)^2 + (61.8\ \Omega)^2}\angle\tan^{-1}(61.8\ \Omega/163.5\ \Omega)$

$$= 174.8\ \Omega\angle 20.7^\circ$$

Determining the current.

$$I = \frac{E}{Z} = \frac{100\text{ V}}{174.8\ \Omega\angle 20.7^\circ}$$

$$= 572\text{ mA}\angle -20.7^\circ$$

The current is 572 mA, lagging the supply voltage by 20.7° [see Figure 20-2(b)].

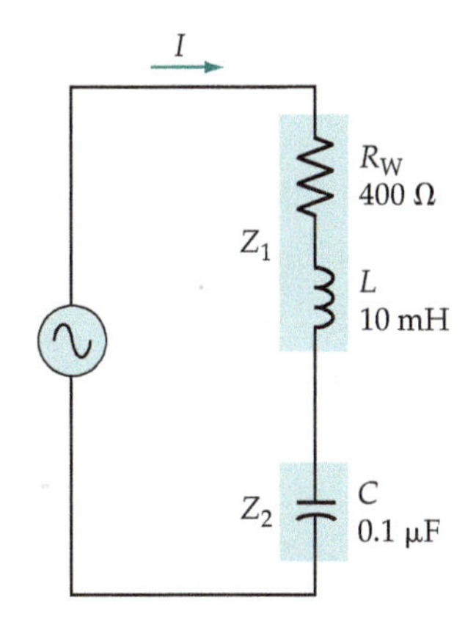

Figure 20-3 Circuit for Problem 20-1.1.

Practice Problems

20-1.1 A 0.1 μF capacitor is connected in series with a 10 mH inductor that has a coil resistance of 400 Ω, as illustrated in Figure 20-3. Calculate the circuit current when the supply is a 1 V, 3.18 kHz source.

20-1.2 A fourth impedance, $Z_4 = 26\ \Omega\angle 30^\circ$, is connected in series with the three impedances shown in Figure 20-2(a). Calculate the new current level.

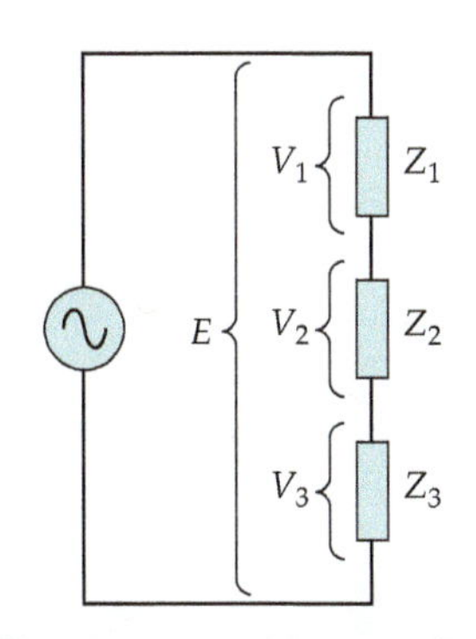

Figure 20-4 An ac voltage divider using impedances is similar to a dc resistive voltage divider. The phase angles of the impedances must be taken into account.

20-2 AC VOLTAGE DIVIDER

Multi-Component Voltage Divider

The voltage-divider rule as applied to series-connected resistors can also be applied to series-connected impedances. As always in an ac circuit, the phase angles must be taken into consideration. Referring to Figure 20-4, it is seen that the voltage drop across Z_3 is,

$$V_3 = IZ_3$$

Also, the current through Z_1, Z_2, and Z_3, is,

$$I = \frac{E}{Z_1 + Z_2 + Z_3}$$

So,

$$V_3 = \frac{EZ_3}{Z_1 + Z_2 + Z_3} \tag{20-1}$$

Similarly,

$$V_2 = \frac{EZ_2}{Z_1 + Z_2 + Z_3}$$

and,

$$V_1 = \frac{EZ_1}{Z_1 + Z_2 + Z_3}$$

Example 20-2

Use the voltage-divider equation to determine the values of V_1, V_2, and V_3 for the series impedance circuit described in Example 20-1. Also, draw a complete phasor diagram for the circuit voltages. The component values are $Z_1 = 70.7\ \Omega\angle 45°$, $Z_2 = 92.4\ \Omega\angle 330°$, $Z_3 = 67\ \Omega\angle 60°$, and the supply is 100 V.

Solution

From Example 20-1,

$$Z_1 + Z_2 + Z_3 = 174.8\ \Omega\angle 20.7°$$

Eq. 20-1,

$$V_3 = \frac{EZ_3}{Z_1 + Z_2 + Z_3} = \frac{100\text{ V} \times (67\ \Omega\angle 60°)}{174.8\ \Omega\angle 20.7°}$$

$$= 39.3\text{ V}\angle 39.3°$$

$$V_2 = \frac{100\text{ V } (92.4\ \Omega\angle 330°)}{174.8\ \Omega\angle 20.7°}$$

$$= 52.9\text{ V}\angle 309.3°$$

$$V_1 = \frac{100\text{ V } (70.7\ \Omega\angle 45°)}{174.8\ \Omega\angle 20.7°}$$

$$= 40.4\text{ V}\angle 24.3°$$

The voltage phasor diagram is as shown in Figure 20-5. Note that the phasor sum of V_1, V_2, and V_3 gives the supply voltage E.

Compensated Oscilloscope Probe

The circuit of an oscilloscope attenuator probe is shown in Figure 20-6(a). The probe is simply a voltage divider consisting of the 9 MΩ probe resistor (R_1) and a 1 MΩ resistor (R_2) representing the oscilloscope input resistance. As discussed in Section 17-6, this arrangement attenuates the signal input

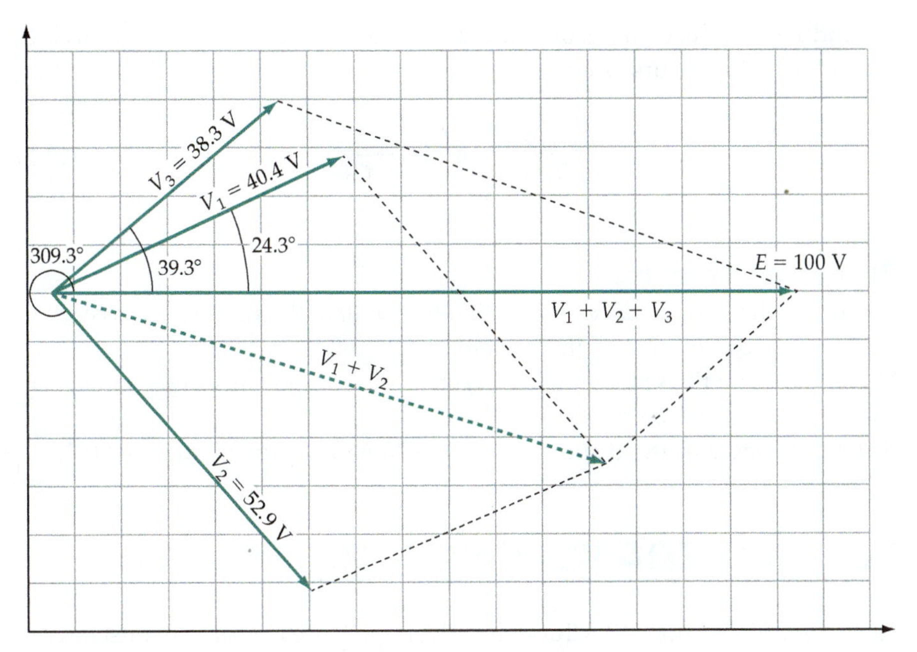

Figure 20-5 Phasor diagram of supply voltage and individual impedance voltages for the circuit in Figure 20-1.

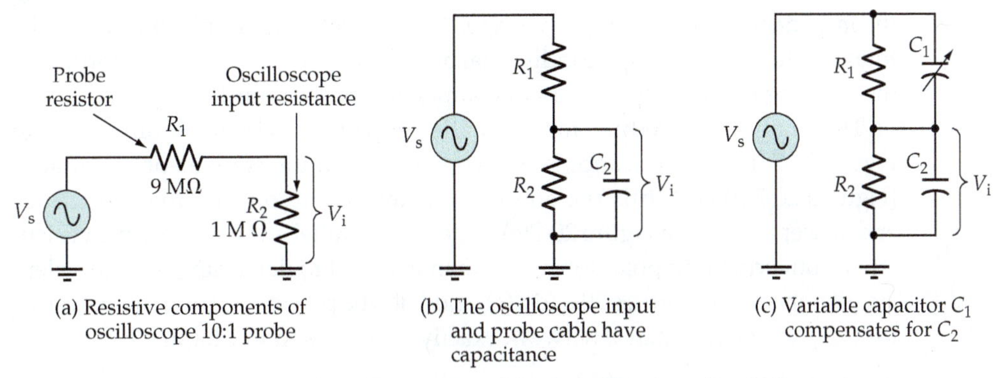

Figure 20-6 A compensated oscilloscope probe uses a variable capacitor (C_1) to compensate for the changes in X_{C2} when the signal frequency changes.

voltage (v_s) by a factor of 10 to produce the input voltage (v_i) for display on the oscilloscope.

Figure 20-6(b) illustrates the fact that the probe cable and the oscilloscope input have capacitance (C_2) which appears in parallel with the oscilloscope input resistance. These typically add up to approximately 130 pF. As the signal frequency increases, the capacitive impedance decreases; consequently, high-frequency signals suffer greater attenuation than low-frequency signals. To counter this effect, a variable capacitor (C_1) is provided across resistor R_1, as shown in Figure 20-6(c). Complete compensation is achieved (all signal frequencies attenuated by the same amount) when the capacitive voltage divider

(C_1 and C_2) produces the same attenuation as the resistive voltage divider (R_1 and R_2). This occurs when,

$$\frac{R_2}{R_1 + R_2} = \frac{C_1}{C_1 + C_2} \tag{20-2}$$

Practice Problems

20-2.1 Use the voltage-divider equation to calculate the voltage across each of the two components in Problem 20-1.1.

20-2.2 Calculate the voltages across each of the components in the circuit described in Problem 20-1.2.

20-2.3 Analyze the circuit shown in Figure 20-6(c) to prove Equation 20-2.

20-3 IMPEDANCES IN PARALLEL

As discussed in Sections 19-8 through 19-10, impedances in parallel must be converted into admittances to determine the total circuit admittance. This can then be inverted to give the total impedance of the circuit. Once the total impedance is known, the current taken from the supply can be determined simply by dividing the impedance into the supply voltage. Alternatively, the supply current can be found by first calculating the individual branch currents in rectangular form, and then adding the real and imaginary components of the branch currents.

The analysis procedure for a parallel impedance circuit is illustrated in Figure 20-7. Each impedance is first converted into the equivalent admittance [Figures 20-7(a) and (b)]. The admittances are then resolved into conductance and susceptance, [see Figure 20-7(c)]. The total parallel admittance is next calculated, converted into polar form, and then inverted to determine the equivalent circuit impedance [Figures 20-7 (d), (e), and (f)]. The phasor diagram can be drawn for the parallel impedance network exactly as discussed in Chapter 19.

Analysis Procedure for Parallel Impedance Circuit:

1. ***Operating in polar form, each individual impedance is converted into an admittance.***

   Eq. 19-18, $$Y = \frac{1}{Z}$$

2. ***The individual admittances are resolved into rectangular form.***

$$Y = G + j\,B_{eq}$$

where, $$B_{eq} = B_C - B_L$$

From Eq. 18-1, $$Y\angle\theta = Y\cos\phi + jY\sin\phi \tag{20-3}$$

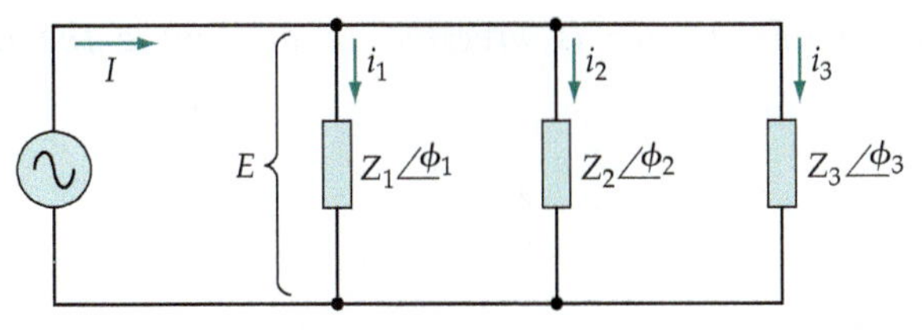

(a) Parallel-connected impedances

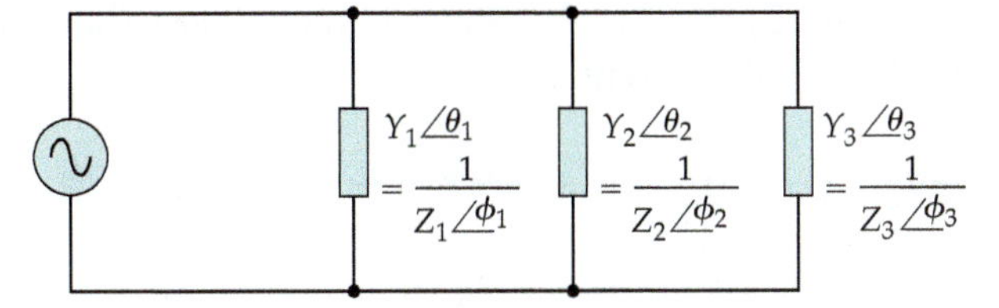

(b) Impedances converted into admittances

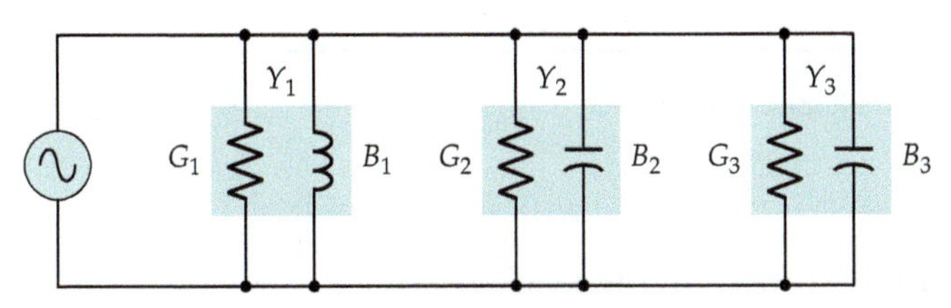

(c) Admittances resolved into conductances and susceptances

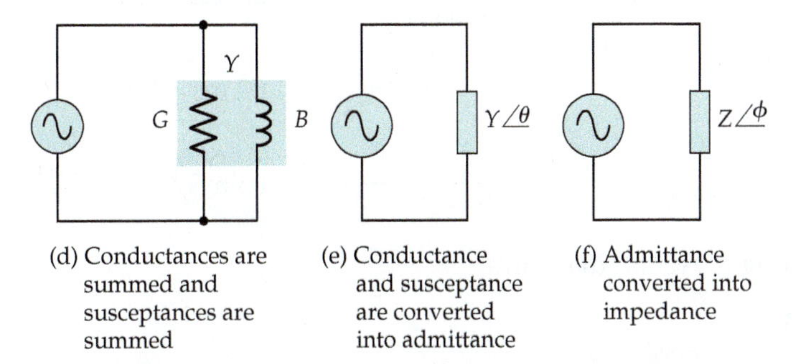

(d) Conductances are summed and susceptances are summed

(e) Conductance and susceptance are converted into admittance

(f) Admittance converted into impedance

Figure 20-7 To analyze a parallel impedance circuit, the individual impedances are first converted into admittances that are then resolved into conductances and susceptances. All the conductances are summed and all the susceptances are summed, to give the equivalent circuit admittance.

3. ***The total parallel admittance is found as,***

$$Y = Y_1 + Y_2 + Y_3$$

where the admittances are in rectangular form.

4. ***The total admittance is converted back to polar form.***

From Eq. 19-53,

$$Y = \sqrt{G^2 + B_{eq}^2} \angle \tan^{-1}(B_{eq}/G) \quad \textbf{(20-4)}$$

5. *The polar form admittance is inverted to determine the total circuit impedance.*

$$Z = \frac{1}{Y}$$

Example 20-3

Three impedances in parallel with a 33 V supply consist of: $Z_1 = 1606\ \Omega\angle 51^\circ$, $Z_2 = 977\ \Omega\angle -33^\circ$, and $Z_3 = 953\ \Omega\angle -19^\circ$, (see Figure 20-8). Calculate the total impedance and the supply current.

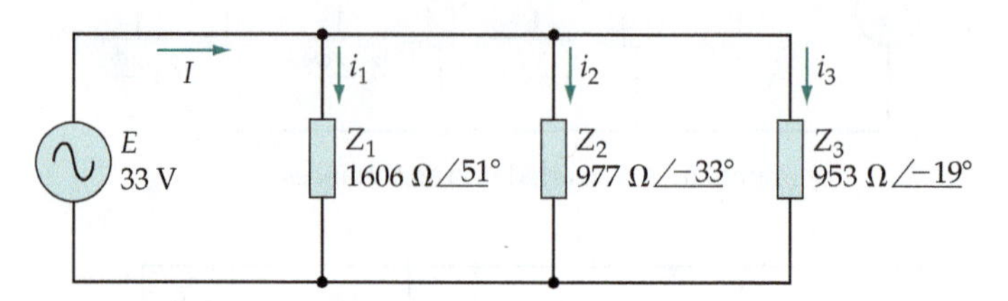

Figure 20-8 Circuit for Example 20-3.

Solution

Determining individual admittances,

$$Y_1 = \frac{1}{Z_1} = \frac{1}{1606\ \Omega\angle 51^\circ} = 622.7\ \mu\text{S}\angle -51^\circ$$

$$Y_2 = \frac{1}{Z_2} = \frac{1}{977\ \Omega\angle -33^\circ} = 1.02\ \text{mS}\angle 33^\circ$$

$$Y_3 = \frac{1}{Z_3} = \frac{1}{953\ \Omega\angle -19^\circ} = 1.05\ \text{mS}\angle 19^\circ$$

Converting to rectangular form,

Eq. 20-3

$$\begin{aligned}
Y_1\angle\phi_1 &= Y_1 \cos\phi_1 + jY_1 \sin\phi_1 \\
&= 622.7\ \mu\text{S}\,[\cos(-51^\circ) + j\sin(-51^\circ)] \\
&= 392\ \mu\text{S} - j484\ \mu\text{S} \\
Y_2\angle\phi_2 &= Y_2 \cos\phi_2 + jY_2 \sin\phi_2 \\
&= 1.02\ \text{mS}\,[\cos 33^\circ + j\sin 33^\circ] \\
&= 855\ \mu\text{S} + j556\ \mu\text{S} \\
Y_3\angle\phi_3 &= Y_3 \cos\phi_3 + jY_3 \sin\phi_3 \\
&= 1.05\ \text{mS}\,[\cos 19^\circ + j\sin 19^\circ] \\
&= 993\ \mu\text{S} + j342\ \mu\text{S}
\end{aligned}$$

Adding admittances,

$$Y = Y_1 + Y_2 + Y_3$$

$$= (392 + 855 + 993)\ \mu S + j(-484 + 556 + 342)\ \mu S$$

$$= 2.24\ mS + j0.414\ mS$$

Converting back to polar form,

Eq. 20-4, $$Y = \sqrt{G^2 + B_{eq}^2} \angle \tan^{-1}(B_{eq}/G)$$

$$= \sqrt{(2.24\ mS)^2 + (0.414\ mS)^2} \angle \tan^{-1}(0.414/2.24)$$

$$= 2.28\ mS \angle 10.5°$$

Total impedance,

$$Z = \frac{1}{Y} = \frac{1}{2.28\ mS \angle 10.5°}$$

$$= 439\ \Omega \angle -10.5°$$

Supply current,

$$I = \frac{E}{Z} = \frac{33\ V}{439\ \Omega \angle -10.5°}$$

$$= 75.2\ mA \angle 10.5°$$

Example 20-4

Determine the individual branch currents and the total supply current for the parallel impedance circuit in Example 20-3. Draw the phasor diagram for the circuit.

Solution

Refer to Figure 20-8.

Branch currents,

$$i_1 = \frac{E}{Z_1} = \frac{33\ V}{1606\ \Omega \angle 51°} = 20.5\ mA \angle -51°$$

$$i_2 = \frac{E}{Z_2} = \frac{33\ V}{977\ \Omega \angle -33°} = 33.8\ mA \angle 33°$$

$$i_3 = \frac{E}{Z_3} = \frac{33\ V}{953\ \Omega \angle -19°} = 34.6\ mA \angle 19°$$

Converting to rectangular form,

From Eq. 20-3,

$$
\begin{aligned}
i_1\angle\underline{\phi_1} &= i_1 \cos \phi_1 + j i_1 \sin \phi_1 \\
&= 20.5 \text{ mA} [\cos (-51°) + j\sin (-51°)] \\
&= 12.9 \text{ mA} - j15.9 \text{ mA} \\
i_2\angle\underline{\phi_2} &= i_2 \cos \phi_2 + j i_2 \sin \phi_2 \\
&= 33.8 \text{ mA} [\cos (33°) + j\sin (33°)] \\
&= 28.3 \text{ mA} + j18.4 \text{ mA} \\
i_3\angle\underline{\phi_3} &= i_3 \cos \phi_3 + j i_3 \sin \phi_3 \\
&= 34.6 \text{ mA} [\cos (19°) + j\sin (19°)] \\
&= 32.7 \text{ mA} + j11.3 \text{ mA}
\end{aligned}
$$

Adding the currents,

$$
\begin{aligned}
I &= i_1 + i_2 + i_3 \\
&= (12.9 + 28.3 + 32.7) \text{ mA} + j(-15.9 + 18.4 + 11.3) \text{ mA} \\
&= 73.9 \text{ mA} + j13.8 \text{ mA}
\end{aligned}
$$

Converting back to polar form,

From Eq. 20-4, $I = \sqrt{(73.9 \text{ mA})^2 + (13.8 \text{ mA})^2}\angle\underline{\tan^{-1}(13.8/73.9)}$

$$= 75.2 \text{ mA}\angle\underline{10.6°}$$

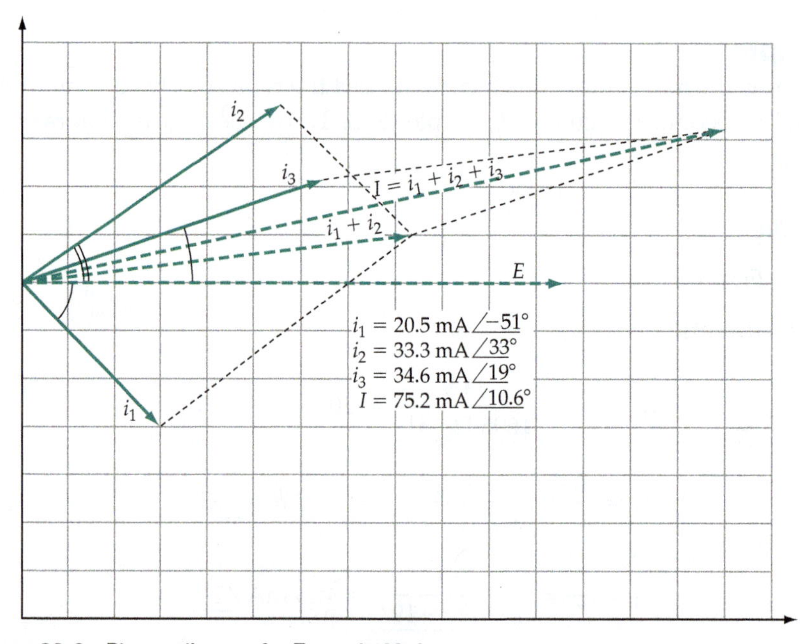

Figure 20-9 Phasor diagram for Example 20-4.

Practice Problems

20-3.1 A parallel impedance circuit with a 12 V supply has $Z_1 = 5.6\ \text{k}\Omega\angle 30^\circ$, $Z_2 = 2.2\ \text{k}\Omega\angle -20^\circ$, and $Z_3 = 3.3\ \text{k}\Omega\angle 45^\circ$. Determine the total circuit admittance and the supply current.

20-3.2 Calculate the branch currents and the total supply current for the circuit in Problem 20-3.1.

20-4 AC CURRENT DIVIDER

The current-divider rule can be applied to impedances in parallel as well as to parallel resistors. Referring to Figure 20-10, Equations 6-8 and 6-9 for a two-branch parallel circuit (as developed in Section 6-4) are rewritten as follows:

$$i_1 = I\frac{Z_2}{Z_1 + Z_2} \qquad \textbf{(20-5)}$$

$$i_2 = I\frac{Z_1}{Z_1 + Z_2} \qquad \textbf{(20-6)}$$

In applying these equations, it is necessary to convert the denominator impedances (Z_1 and Z_2) into rectangular form to add them together. The denominator should then be returned to polar form so that it may be easily divided into the numerator.

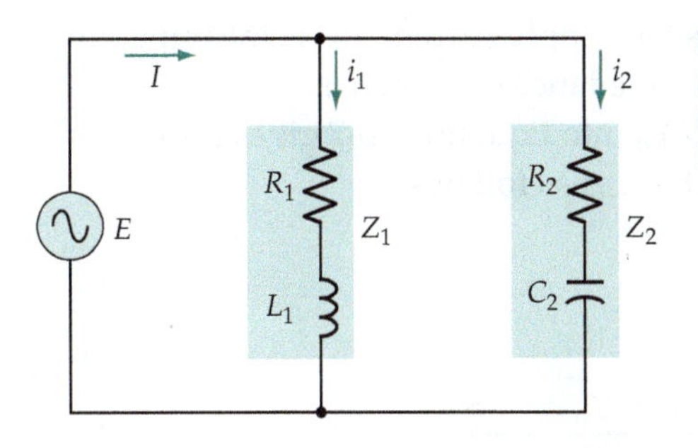

Figure 20-10 Alternating current divides between two parallel-connected impedances in a similar way to direct current dividing between two parallel-connected resistors.

Example 20-5

The two impedances in the circuit shown in Figure 20-10 are $Z_1 = (1\ \text{k}\Omega + j\,2.7\ \text{k}\Omega)$ and $Z_2 = (790\ \Omega - j\,1.6\ \text{k}\Omega)$. The total current taken from the supply is 15 mA. Use the current-divider rule to calculate the two branch currents.

Solution

$$Z_1 = \sqrt{(1\ \text{k}\Omega)^2 + (2.7\ \text{k}\Omega)^2}\angle \tan^{-1}(2.7/1)$$

$$= 2.88\ \text{k}\Omega\angle 69.7^\circ$$

$$Z_2 = \sqrt{(790\ \Omega)^2 + (1.6\ \text{k}\Omega)^2}\underline{/\tan^{-1}(1.6\ \text{k}\Omega/790\ \Omega)}$$

$$= 1.78\ \text{k}\Omega\underline{/-63.7^\circ}$$

$$Z_1 + Z_2 = (1\ \text{k}\Omega + j2.7\ \text{k}\Omega) + (790\ \Omega - j1.6\ \text{k}\Omega)$$

$$= 1.79\ \text{k}\Omega + j1.1\ \text{k}\Omega$$

$$= \sqrt{(1.79\ \text{k}\Omega)^2 + (1.1\ \text{k}\Omega)^2}\underline{/\tan^{-1}(1.1\ \text{k}\Omega/1.79\ \text{k}\Omega)}$$

$$= 2.1\ \text{k}\Omega\underline{/31.6^\circ}$$

Eq. 20-5,

$$i_1 = I\frac{Z_2}{Z_1 + Z_2} = (15\ \text{mA}) \times \frac{1.78\ \text{k}\Omega\underline{/-63.7^\circ}}{2.1\ \text{k}\Omega\underline{/31.6^\circ}}$$

$$= 12.7\ \text{mA}\underline{/-95.3^\circ}$$

Eq. 20-6,

$$i_2 = I\frac{Z_1}{Z_1 + Z_2} = 15\ \text{mA} \times \frac{2.88\ \text{k}\Omega\underline{/69.7^\circ}}{2.1\ \text{k}\Omega\underline{/31.6^\circ}}$$

$$= 20.6\ \text{mA}\underline{/38.1^\circ}$$

When a phasor diagram is drawn for i_1 and i_2, it is seen that their resultant (the supply current) is $I = 15$ mA.

Equations 20-5 and 20-6 apply only to a two-branch parallel circuit. For a multibranch current-divider circuit, Equations 6-10 and 6-11 can be rewritten in the form of impedances and admittances. Rewriting Equations 6-10,

$$I_n = I\left[\frac{Z_{eq}}{Z_n}\right] \tag{20-7}$$

where I_n is the current in branch n, I is the supply current, Z_{eq} is the impedance of the complete circuit, and Z_n is the impedance of branch n.

Substituting $Z = 1/Y_{eq}$ and $Z_n = 1/Y_n$ into Equation 20-7 gives the current-divider equation in the form of admittances as follows:

$$I_n = I\left[\frac{Y_n}{Y_{eq}}\right]$$

or,

$$I_n = \frac{I \times Y_n}{Y_1 + Y_2 + Y_3 + \cdots} \tag{20-8}$$

Equation 20-8 is comparable to voltage-divider Equation 20-1.

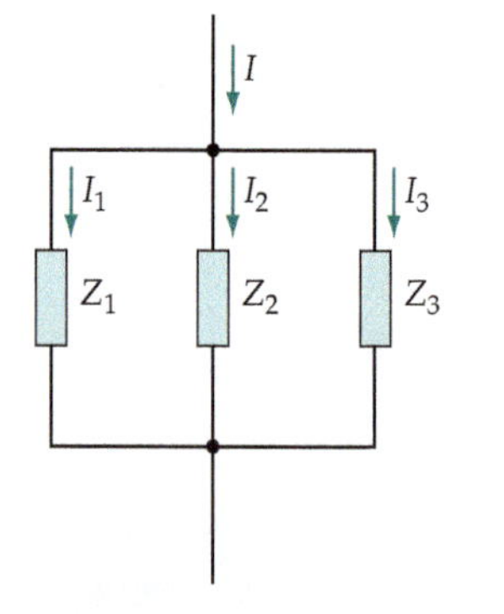

Figure 20-11 Alternating current division in a multi-impedance parallel circuit. The current through any branch is $I_n = I(Z/Z_n)$, where Z is the complete circuit impedance.

Example 20-6

The three impedances in Figure 20-11 are $Z_1 = 1606\ \Omega\underline{/51^\circ}$, $Z_2 = 977\ \Omega\underline{/-33^\circ}$, and $Z_3 = 953\ \Omega\underline{/-19^\circ}$. (These are the same component values as used in Examples 20-3 and 20-4.) If the supply current is $I = 75.2\ \text{mA}\underline{/10.5^\circ}$, determine I_1, I_2, and I_3.

Solution

From Example 20-3, $Y_1 = \frac{1}{Z_1} = 622.7\ \mu S\angle{-51°}$

$= 392\ \mu S - j484\ \mu S$

$Y_2 = \frac{1}{Z_2} = 1.02\ \mu S\angle{33°}$

$= 855\ \mu S + j556\ \mu S$

$Y_3 = \frac{1}{Z_3} = 1.05\ \mu S\angle{19°}$

$= 993\ \mu S + j343\ \mu S$

$Y_{eq} = Y_1 + Y_2 + Y_3$

$= (392 + 855 + 993)\ \mu S + j(-484 + 556 + 342)\ \mu S$

$= 2.24\ mS + j0.414\ mS$

$= 2.28\ mS\angle{10.5°}$

Eq. 20-8, $I_1 = I\left[\frac{Y_1}{Y_{eq}}\right] = 75.2\ mA\angle{10.5°}\left[\frac{622.7\ \mu S\angle{-51°}}{2.28\ mS\angle{10.5°}}\right]$

$= 20.5\ mA\angle{-51°}$

$I_2 = I\left[\frac{Y_2}{Y_{eq}}\right] = 75.2\ mA\angle{10.5°}\left[\frac{1.02\ mS\angle{33°}}{2.28\ mS\angle{10.5°}}\right]$

$= 33.6\ mA\angle{33°}$

$I_3 = I\left[\frac{Y_3}{Y_{eq}}\right] = 75.2\ mA\angle{10.5°}\left[\frac{1.05\ mS\angle{19°}}{2.28\ mS\angle{10.5°}}\right]$

$= 34.6\ mA\angle{19°}$

The above results are similar to those obtained in Example 20-4.

Practice Problems

20-4.1 Use the current-divider equation to determine the branch currents in a parallel circuit consisting of $Z_1 = 470\ \Omega\angle{39°}$ and $Z_2 = 390\ \Omega\angle{-75°}$. The supply current is $51.7\ mA\angle{26.2°}$.

20-4.2 Determine the new level of supply current if an additional parallel impedance $Z_3 = 560\ \Omega\angle{15°}$ is included in the circuit of Problem 20-4.1.

20-5 SERIES-PARALLEL IMPEDANCES

Analysis Procedure for Series-Parallel Impedance Circuits:

1. *Resolve all series-connected impedances into a single equivalent impedance.*
2. *Resolve all parallel-connected impedances into a single equivalent impedance.*
3. *Repeat procedures 1 and 2 until a single equivalent impedance is determined for the whole circuit.*
4. *Calculate the supply current.*
5. *Use the current-divider rule to determine the individual branch currents.*

Figure 20-12(a) shows a series-parallel impedance circuit, and Figures 20-12(b) and (c) shows the stages of circuit simplification for analysis. Impedances Z_3 and Z_4 are added together (in rectangular form) to give a single equivalent impedance. Then converting Z_2 and $(Z_3 + Z_4)$ into admittances, the equivalent impedance of Z_2 in parallel with $(Z_3 + Z_4)$ is found. At this stage the circuit has become two impedances in series, and these may be added in rectangular form to give the equivalent impedance for the whole circuit.

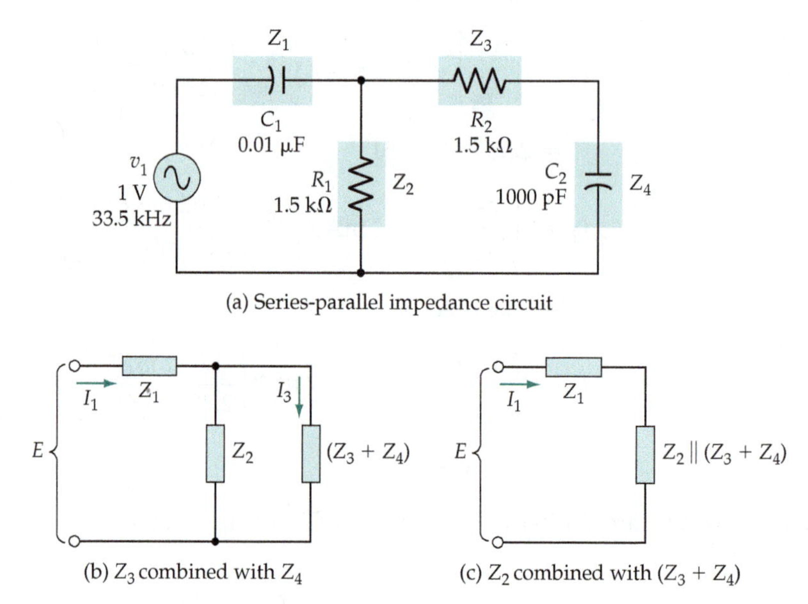

Figure 20-12 To analyze a series-parallel impedance circuit, all series-connected impedances are resolved into a single impedance and all parallel-connected impedances are resolved into a single impedance.

Example 20-7

Analyze the circuit in Figure 20-12(a) to determine the current through capacitor C_2.

Solution

$$I_1 = \frac{v_i}{Z_1 + [Z_2 \| (Z_3 + Z_4)]}$$

$$I_3 = \frac{I_1 Z_2}{Z_2 + (Z_3 + Z_4)}$$

$$X_{C1} = \frac{1}{2\pi f C_1} = \frac{1}{2 \times \pi \times 33.5\ \text{kHz} \times 0.01\ \mu\text{F}}$$

$$= 475\ \Omega$$

$$X_{C2} = \frac{1}{2\pi f C_2} = \frac{1}{2 \times \pi \times 33.5\ \text{kHz} \times 1000\ \text{pF}}$$

$$= 4.75\ \text{k}\Omega$$

$$Z_1 = -j475\ \Omega, Z_2 = 1.5\ \text{k}\Omega, Z_3 = 1.5\ \text{k}\Omega, Z_4 = -j4.75\ \text{k}\Omega$$

$$Z_3 + Z_4 = 1.5\ \text{k}\Omega - j4.75\ \text{k}\Omega = 4.98\ \text{k}\Omega \angle -72.5°$$

$$Y_{3,4} = \frac{1}{Z_3 + Z_4} = \frac{1}{4.98\ \text{k}\Omega \angle -72.5°}$$

$$= 201\ \mu\text{S} \angle +72.5° = 201\ \mu\text{S}\,[\cos(72.5°) + j\sin(-72.5°)]$$

$$= 60.4\ \mu\text{S} + j192\ \mu\text{S}$$

$$Y_2 = \frac{1}{Z_2} = \frac{1}{1.5\ \text{k}\Omega \angle 0°} = 667\ \mu\text{S} \angle 0°$$

$$= 667\ \mu\text{S} + j0$$

$$Y_2 + Y_{3,4} = (667\ \mu\text{S} + j0) + (60.4\ \mu\text{S} + j192\ \mu\text{S})$$

$$= 727\ \mu\text{S} + j192\ \mu\text{S}$$

$$= 752\ \mu\text{S} \angle 14.8°$$

$$Z_2 \| (Z_3 + Z_4) = \frac{1}{Y_2 + Y_{3,4}} = \frac{1}{752\ \mu\text{S} \angle 14.8°}$$

$$= 1.33\ \text{k}\Omega \angle -14.8°$$

$$= 1.29\ \text{k}\Omega - j340\ \Omega$$

$$Z_1 + [Z_2 \| (Z_3 + Z_4)] = -j475\ \Omega + (1.29\ \text{k}\Omega - j340\ \Omega)$$

$$= 1.29\ \text{k}\Omega - j815\ \Omega$$

$$= 1.53\ \text{k}\Omega \angle -32°$$

$$Z_2 + (Z_3 + Z_4) = 1.5\ \text{k}\Omega + (1.5\ \text{k}\Omega - j4.75\ \text{k}\Omega)$$

$$= 5.62\ \text{k}\Omega \angle -57.7°$$

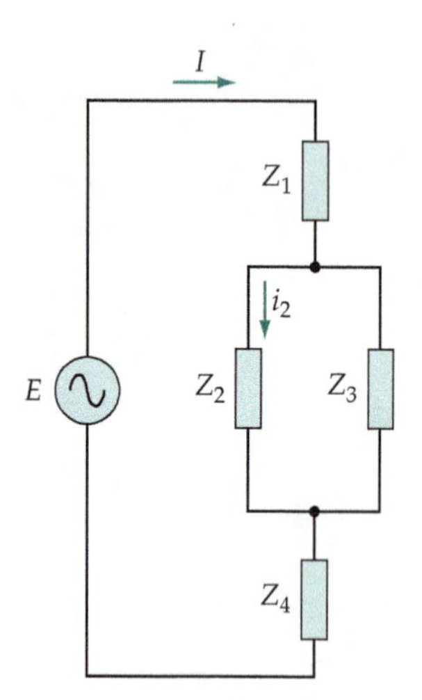

Figure 20-13 Circuit for Problems 20-5.1 and 20-5.2.

$$I_1 = \frac{v_i}{Z_1 + [Z_2 \| (Z_3 + Z_4)]} = \frac{1\ \text{V}}{1.53\ \text{k}\Omega\angle -32^\circ}$$

$$= 654\ \mu\text{A}\angle 32^\circ$$

$$I_3 = \frac{I_1 Z_2}{Z_2 + (Z_3 + Z_4)} = \frac{654\ \mu\text{A}\angle 32^\circ \times 1.5\ \text{k}\Omega\angle 0^\circ}{5.62\ \text{k}\Omega\angle -57.7^\circ}$$

$$= 174.6\ \mu\text{A}\angle 89.7^\circ$$

Practice Problems

20-5.1 The impedances in the circuit shown in Figure 20-13 are $Z_1 = 1.46\ \text{k}\Omega\angle 68.3^\circ$, $Z_2 = 200\ \Omega\angle 53.1^\circ$, $Z_3 = 707\ \Omega\angle 45^\circ$, and $Z_4 = 566\ \Omega\angle 58^\circ$. Determine the total impedance presented to the supply.

20-5.2 If a supply of 18 V is applied to the circuit in Problem 20-5.1, determine the current through Z_2.

20-6 SERIES AND PARALLEL EQUIVALENT CIRCUITS

Sometimes it is convenient to replace a series *RL* circuit with a parallel *RL* circuit that offers exactly the same impedance characteristics, and vice versa. In this case, each circuit is the equivalent circuit of the other, and if each were contained in a sealed box with only the terminals showing, it might be difficult to distinguish between them. Figure 20-14 shows a series *RL* circuit consisting of R_S and X_S, and its equivalent parallel circuit, R_P and X_P.

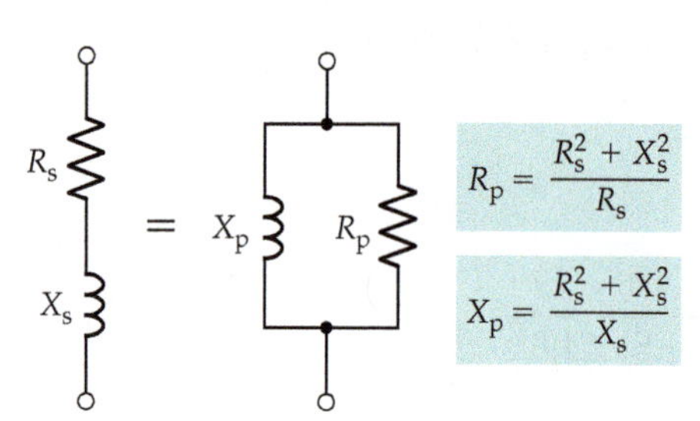

Figure 20-14 An *RL* series circuit may be replaced with an equivalent *RL* parallel circuit that has a similar terminal impedance. The circuit terminal resistances differ.

From Figure 20-14(a), $$Z_S = R_S + jX_S$$

and from Figure 20-14(b),

$$Y_P = \frac{1}{R_P} - j\frac{1}{X_P}$$

Also, $$Z_S = Z_P$$

or, $$\frac{1}{Z_S} = Y_P$$

So,

$$\frac{1}{R_S + jX_S} = \frac{1}{R_P} - j\frac{1}{X_P}$$

$$\frac{1}{R_S + jX_S} \times \frac{R_S - jX_S}{R_S - jX_S} = \frac{1}{R_P} - j\frac{1}{X_P}$$

$$\frac{R_S - jX_S}{R_S^2 + X_S^2} = \frac{1}{R_P} - j\frac{1}{X_P}$$

Equating the resistive terms,

$$\frac{R_S}{R_S^2 + X_S^2} = \frac{1}{R_P}$$

or,

$$R_P = \frac{R_S^2 + X_S^2}{R_S} \qquad \text{(20-9)}$$

Equating the reactive terms,

$$\frac{X_S}{R_S^2 + X_S^2} = \frac{1}{X_P}$$

or,

$$X_P = \frac{R_S^2 + X_S^2}{X_S} \qquad \text{(20-10)}$$

Equations 20-9 and 20-10 also apply to the series *RC* circuit and its parallel equivalent shown in Figure 20-15. These equations enable the component values of the parallel *RC* or *RL* equivalent circuit to be calculated for any given series *RC* or *RL* circuit.

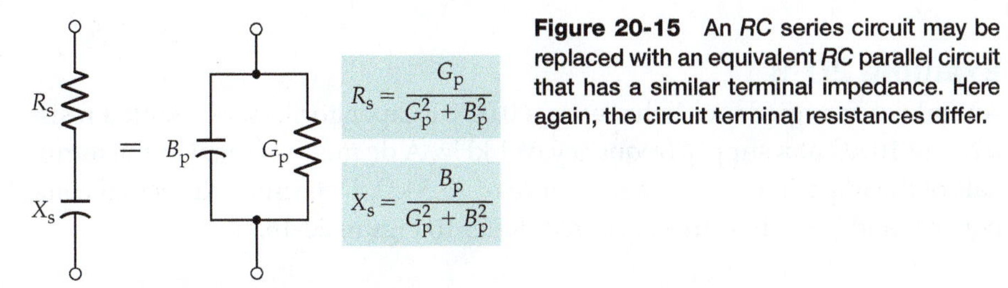

Figure 20-15 An *RC* series circuit may be replaced with an equivalent *RC* parallel circuit that has a similar terminal impedance. Here again, the circuit terminal resistances differ.

To derive equations for calculating the series *RC* or *RL* equivalent circuit for a given parallel circuit, Y_P is written in terms of a conductance and susceptance. Z_S is then equated to the reciprocal of Y_P. For the circuits in Figure 20-15,

$$Z_S = R_S - jX_S$$

and,

$$Y_p = G_p + jB_P$$

$$Z_S = \frac{1}{Y_P}$$

So,

$$R_S - jX_S = \frac{1}{G_P + jB_P}$$

$$R_S - jX_S = \frac{1}{G_P + jB_P} \times \frac{G_P - jB_P}{G_P - jB_P}$$

$$R_S - jX_S = \frac{G_P - jB_P}{G_P^2 + B_P^2}$$

Equating the resistive terms and the reactive terms,

$$R_S = \frac{G_P}{G_P^2 + B_P^2} \tag{20-11}$$

and,

$$X_S = \frac{B_P}{G_P^2 + B_P^2} \tag{20-12}$$

Because it may be more convenient to use R_p and X_p rather than G_p and B_p, the equations can be rewritten as follows:

From Eq. 20-11,

$$R_S = \frac{R_P X_P^2}{X_P^2 + R_P^2} \tag{20-13}$$

and from Eq. 20-12,

$$X_S = \frac{X_P R_P^2}{X_P^2 + R_P^2} \tag{20-14}$$

Example 20-8

An unknown impedance behaves as a 0.01 μF capacitor in series with a resistance of 10 kΩ at a supply frequency of 1 kHz. A dc measurement at the terminals of the impedance gives a resistance of 35.3 kΩ. Determine the actual components and how they are connected. Refer to Figure 20-16.

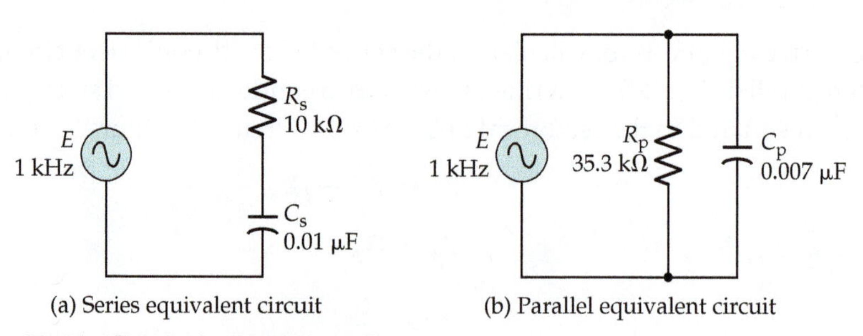

Figure 20-16 Circuits for Example 20-8.

Solution

$$X_S = \frac{1}{2\pi fC} = \frac{1}{2\pi \times 1\ \text{kHz} \times 0.01\ \mu\text{F}}$$

$$= 15.9\ \text{k}\Omega$$

$$R_S = 10\ \text{k}\Omega$$

Eq. 20-9,

$$R_P = \frac{R_S^2 + X_S^2}{R_S} = \frac{(10\ \text{k}\Omega)^2 + (15.9\ \text{k}\Omega)^2}{10\ \text{k}\Omega}$$

$$= 35.3\ \text{k}\Omega$$

Eq. 20-10,

$$X_P = \frac{R_S^2 + X_S^2}{X_S} = \frac{(10\ \text{k}\Omega)^2 + (15.9\ \text{k}\Omega)^2}{15.9\ \text{k}\Omega}$$

$$= 22.2\ \text{k}\Omega$$

$$C_p = \frac{1}{2\pi f X_p} = \frac{1}{2\pi \times 1\text{kHz} \times 22.2\ \text{k}\Omega}$$

$$= 7000\ \text{pF}$$

If the actual impedance consisted of a series-connected resistor and capacitor, the dc resistance measurement would give a value greater than R_S; it would equal R_S + (the capacitor dielectric resistance). Because the measured resistance corresponds with the value of R_p, the actual impedance consists of a 35.3 kΩ resistor connected in parallel with a 0.007 μF capacitor.

Practice Problem

20-6.1 Two impedances are $Z_1 = 1\ \text{k}\Omega + j27\ \text{k}\Omega$ and $Z_2 = 790\ \Omega = j1.6\ \text{k}\Omega$. Resolve each into its parallel equivalent circuit.

Summary of Formulas

- *AC voltage divider:*

$$V_3 = \frac{EZ_3}{Z_1 + Z_2 + Z_3}$$

- *Resolving polar form admittances into rectangular form:*

$$Y\angle\theta = Y\cos\theta + jY\sin\theta$$

- ***AC current divider:***

$$i_1 = I\frac{Z_2}{Z_1 + Z_2},\ i_2 = I\frac{Z_1}{Z_1 + Z_2}$$

$$I_n = I\left(\frac{Z}{Z_n}\right) = I\left(\frac{Y_n}{Y_1 + Y_2 + Y_3 + \cdots}\right)$$

- ***Series-parallel equivalent circuit:***

$$R_P = \frac{R_S^2 + X_S^2}{R_S}$$

$$X_P = \frac{R_S^2 + X_S^2}{X_S}$$

$$R_S = \frac{G_P}{G_P^2 + B_P^2} = \frac{R_P X_P^2}{X_P^2 + R_P^2}$$

$$X_S = \frac{B_P}{G_P^2 + B_P^2} = \frac{X_P R_P^2}{X_P^2 + R_P^2}$$

Problems

Section 20-1

20-1 A coil with a resistance of 100 Ω and an inductance of 300 mH is connected in series with a 10 μF capacitor, another inductor of 150 mH, and a resistance of 180 Ω. The supply voltage is 115 V with a frequency of 60 Hz. Calculate the supply current.

20-2 Two impedances, $Z_1 = 120\ \Omega\angle 33°$ and $Z_2 = 56\ \Omega\angle 60°$, are connected in series to a 15 V ac supply. Calculate the circuit current and the terminal voltage of each impedance.

20-3 For the circuit described in Problem 20-1, calculate the terminal voltages of the capacitor and inductor. Draw a phasor diagram showing the supply voltage and current and the voltages across the capacitor and inductor.

20-4 A 3 V, 12 kHz voltage is applied to a series circuit consisting of $R = 4.7$ kΩ, $L = 56$ mH, and $C = 0.001$ μF. The inductor has a coil resistance of 800 Ω. Calculate the terminal voltage of each component and the phase angle of each voltage with respect to the supply.

Section 20-2

20-5 A 55 V supply is applied to two series-connected impedances: (3.3 kΩ + j3.9 kΩ) and (6.8 kΩ − j5.6 kΩ). Use the voltage-divider rule to determine the voltage across each impedance.

20-6 The circuit shown in Figure 20-6(c) has $R_1 = 9$ MΩ, $R_2 = 1$ MΩ, $C_1 = 16.67$ pF, $C_2 = 150$ pF, and $v_S = 10$ V. Determine v_i when C_1 and C_2 are absent, and when R_1 and R_2 are absent (a) when the signal frequency is $f = 1$ kHz, and (b) when $f = 33$ kHz.

Section 20-3

20-7 Three impedances connected in parallel are $Z_1 = (5.6\ \text{k}\Omega - j3.3\ \text{k}\Omega)$, $Z_2 = (1.8\ \text{k}\Omega + j2.2\ \text{k}\Omega)$, and $Z_3 = (8.2\ \text{k}\Omega - j6.2\ \text{k}\Omega)$. Determine the total circuit impedance and the current taken from a 19 V supply.

20-8 A 56 mH inductor, a 0.001 μF capacitor, and a 4.7 kΩ resistor are connected in parallel to a 3 V, 12 kHz supply. Calculate the total circuit impedance and the supply current.

Section 20-4

20-9 Two impedances, $Z_1 = (3.3\ \text{k}\Omega + j3.9\ \text{k}\Omega)$ and $Z_2 = (6.8\ \text{k}\Omega - j5.6\ \text{k}\Omega)$, are connected in parallel to an ac supply. If the supply current is 22 mA, use the current-divider rule to calculate the current through each impedance.

20-10 Two impedances, $Z_1 = 120\ \Omega\angle 33°$, and $Z_2 = 56\ \Omega\angle 60°$, are connected in parallel to an ac supply. If the current through Z_1 is $83\ \text{mA}\angle -33°$, use the current-divider rule to determine the current through Z_2.

20-11 For the circuit described in Problem 20-7, use the calculated supply current and the appropriate current-divider equation to determine each of the branch currents.

20-12 A 0.1 μF capacitor in series with a 470 Ω resistor is connected in parallel with a 75 mH inductor that has a 330 Ω coil resistance. A 7.5 V, 1 kHz supply is applied to the parallel circuit. Calculate the total supply current, and then use the current-divider rule to determine the branch currents.

Section 20-5

20-13 Calculate the equivalent impedance of Z_3 and Z_4 in parallel in the circuit shown in Figure 20-17.

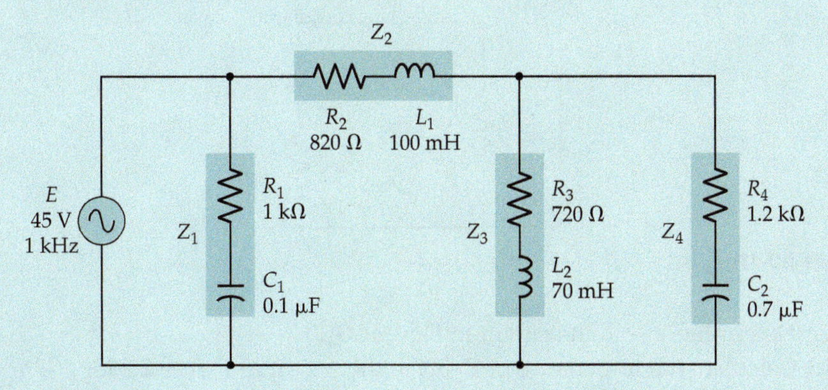

Figure 20-17

20-14 Calculate the total equivalent impedance of the circuit shown in Figure 20-17.

20-15 For the circuit shown in Figure 20-17, determine the voltage drops across L_2 and C_2.

20-16 In the circuit shown in Figure 20-17, L_1 is changed to 200 mH, and C_1 is altered to 0.3 μF. Calculate the new supply current.

20-17 Determine the total equivalent impedance for the circuit shown in Figure 20-18.

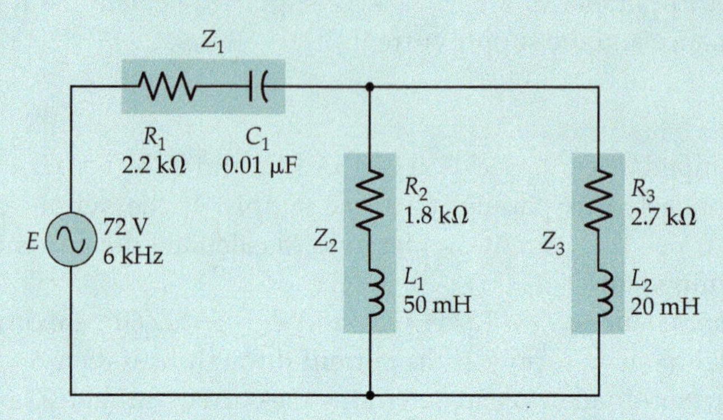

Figure 20-18

20-18 Calculate the current flowing in R_3 and L_2 in the circuit in Figure 20-18.

20-19 Calculate the current levels in R_2 and L_1 in the circuit in Figure 20-18.

20-20 A 6.8 kΩ resistor is connected in parallel with Z_1 in the circuit shown in Figure 20-18. Calculate the new supply current.

20-21 Determine the total impedance of the circuit in Figure 20-19.

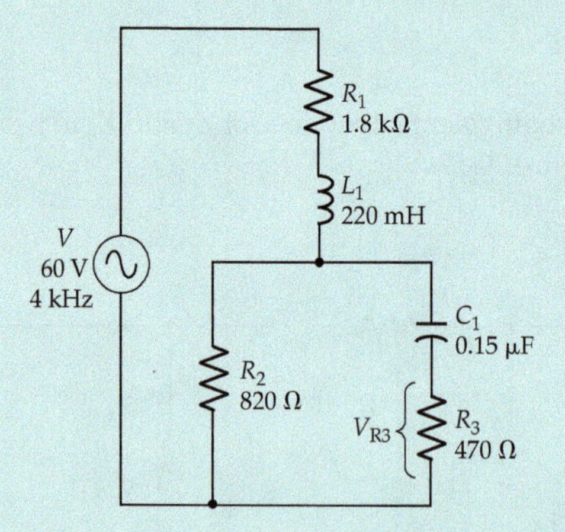

Figure 20-19

20-22 Determine V_{R3} in the circuit in Figure 20-19.

20-23 Calculate the impedance of $Z_2 \| Z_3$ in the circuit in Figure 20-20.

20-24 Calculate the total impedance of the circuit in Figure 20-20.

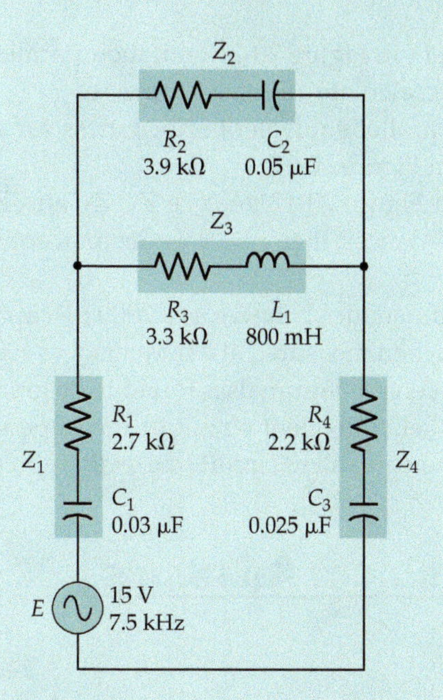

Figure 20-20

20-25 For the circuit in Figure 20-20, calculate the voltages across R_3 and C_2.

20-26 An 8.2 kΩ resistor is connected in parallel with Z_3 in Figure 20-18. Calculate the new level of supply current.

Section 20-6

20-27 Derive the equations for converting a series-connected inductance and resistance into a parallel-connected inductance and resistance.

20-28 Derive the equations for converting a parallel-connected capacitance and resistance into a series-connected capacitance and resistance.

20-29 An unknown impedance measured at a frequency of 5 kHz behaves as a 500 mH inductor in series with a 10 kΩ resistor. A dc measurement gives a terminal resistance of 10 kΩ. Determine the parallel equivalent circuit of the impedance, and decide the actual component values.

20-30 Calculate i_1, i_2, and I in the circuit shown in Figure 20-21.

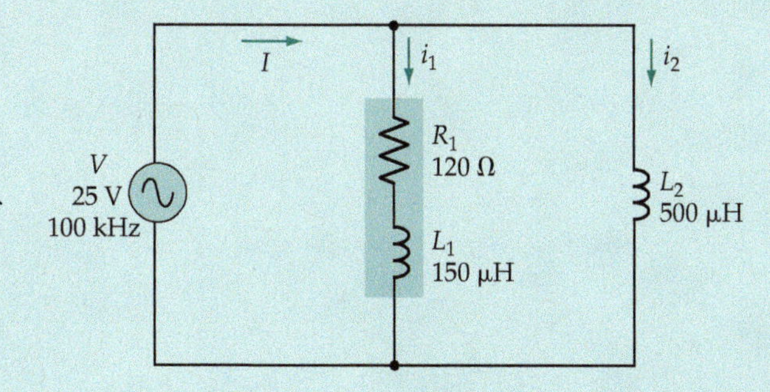

Figure 20-21

20-31 Replace R_1 and L_1 in Figure 20-21 with their parallel equivalent circuits, then once again calculate all branch currents.

20-32 Calculate the parallel equivalent impedances for Z_1, Z_2, and Z_4 in the circuit shown in Figure 20-17.

20-33 In the circuit in Figure 20-17, replace Z_1, Z_2, and Z_4 with their parallel equivalent impedances, then calculate the total equivalent impedance of the circuit.

20-34 An unknown impedance behaves as a 2000 pF capacitor in series with a 1.5 kΩ resistor when measured at a frequency of 100 kHz. The resistance measured between the terminals is found to be greater than 10 MΩ. Calculate the parallel equivalent circuit of the impedance, and determine which of the two equivalent circuits correctly represents the impedance.

Practice Problem Answers

20-1.1 2 mA$\angle 37°$

20-1.2 499mA$\angle -21.9°$

20-2.1 1 V$\angle -53°$, 0.89 V$\angle 63.6°$

20-2.2 35.3 V$\angle 23.1°$, 46.1 V$\angle 308.1°$, 33.4 V$\angle 38.1°$, 13 V$\angle 8.1°$

20-3.1 811 μS$\angle -10.6°$, 9.73 mA$\angle -10.6°$

20-3.2 2.14 mA$\angle -30°$, 5.45 mA$\angle 20°$, 3.63 mA$\angle -45°$, 9.73 mA$\angle 10.6°$

20-4.1 42.6 mA$\angle 39°$, 51.4 mA$\angle 75°$

20-4.2 82 mA$\angle 9.5°$

20-5.1 1.414 kΩ$\angle 45°$

20-5.2 12.73 mA$\angle -61.2°$

20-6.1 8.29 kΩ, 3.07 kΩ, 4.03 kΩ, 2 kΩ

CHAPTER 21
Power in AC Circuits

CONTENTS

Objectives

You will be able to:

1. Sketch and explain the waveforms of current, voltage, and power for a pure resistance connected in an ac circuit.
2. Solve problems involving current voltage and power dissipation in a purely resistive ac circuit.
3. Sketch and explain the waveforms of current, voltage, and power for a pure inductance and for a pure capacitance in ac circuits.
4. Sketch and explain the waveforms of current, voltage, and power for series *RL* and series *RC* ac circuits.
5. Calculate the true power and reactive power in purely resistive, purely inductive, and purely capacitive ac circuits.
6. Calculate the apparent power, true power, and reactive power in series *RL* and series *RC* circuits.
7. Define power factor and discuss power factor correction.
8. Sketch power triangles for series *RL* and *RC* circuits and calculate power factor correction capacitor values for inductive loads.
9. Show how a dynamometer wattmeter measures true power in an ac circuit.

INTRODUCTION

In an alternating-current circuit, power is dissipated in a resistor, but not in a pure inductor or capacitor. Because the current in an *RL* circuit lags the supply voltage by an angle ϕ, the amount of useful power supplied to the circuit is proportional to cos ϕ. Similarly, in an *RC* circuit the useful power is proportional to the angle by which the current leads the voltage. The presence of the reactive components causes reactive power to be supplied to the circuit. This gives an apparent power, which is greater than the true power utilized in the circuit. The ratio of true power to apparent power is termed the power factor of the circuit. The unit of true power is the watt (W), the unit of apparent power is the volt-amp (VA), and the reactive power unit is the volt-amp-reactive (var). These quantities all refer to ac circuits which typically have supply frequencies ranging from 25 Hz to 400 Hz.

21-1 POWER DISSIPATED IN A RESISTANCE

In Section 17-4 it was shown that when an alternating current flows in a resistance, the power dissipated can be calculated as follows:

Eq. 17-12, $$P = 0.5\,P_m$$

where, $$P_m = E_m I_m$$

So, $$P = 0.5\,E_m I_m$$

$$= \frac{E_m}{\sqrt{2}} \times \frac{I_m}{\sqrt{2}}$$

or, $$P = EI \qquad \textbf{(21-1)}$$

where *E* and *I* are rms values.

Equation 21-1 is exactly the same as for Equation 3-3, which was derived for a dc circuit. As long as rms quantities are used for voltage and current, the power calculations for a resistor in an ac circuit are exactly the same as those for a dc circuit. So, from Equations 3-4 and 3-5,

$$P = \frac{E^2}{R} \qquad \textbf{(21-2)}$$

and, $$P = I^2 R \qquad \textbf{(21-3)}$$

where *E* and *I* are again rms values.

Figure 21-1, reproduced from Figure 17-12, shows that the instantaneous power dissipated in a resistance alternates between zero and a peak level as

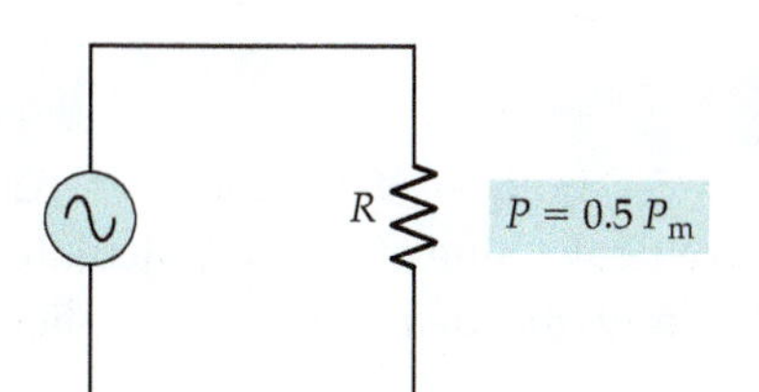

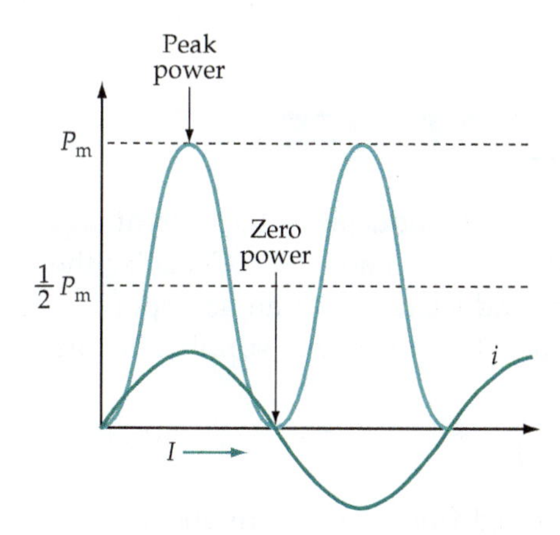

Figure 21-1 The average ac power dissipated in a resistor is half the peak power.

the current rises from zero to its positive and negative peak values. If the resistance were the tungsten filament of a lamp, and if the frequency of the alternating current were 0.1 Hz, the filament could clearly be seen to go bright and dim at a frequency of 0.2 Hz. The lamp would be bright when the current is $+I_m$, dim when the current is zero, and bright again when the current is $-I_m$. The waveform of power in Figure 21-1 shows that two positive peaks of power occur during each cycle of current. As the normal domestic and industrial power frequency is 60 Hz (in North America), any fluctuations in the brightness of an electric lamp are much too fast for the human eye to detect.

Example 21-1

A 100 W electric lamp is supplied from a 115 V, 60 Hz source, as illustrated in Figure 21-2. Calculate (a) the level of current that flows, (b) the resistance of the filament, (c) the peak instantaneous power dissipated in the filament.

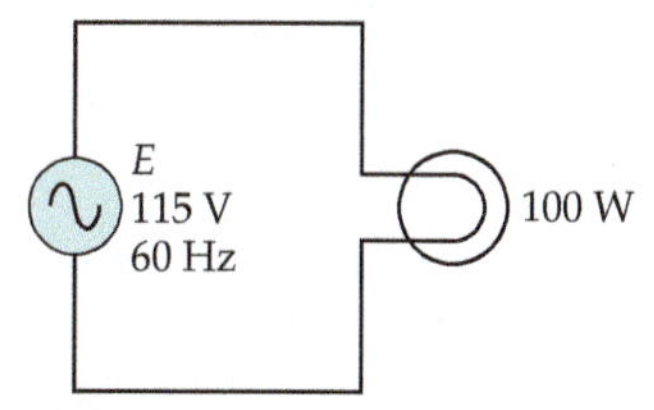

Figure 21-2 The power dissipated in a lamp with an ac supply depends on the rms levels of the voltage and current.

Solution

(a) From Eq. 21-1,

$$I = \frac{P}{E} = \frac{100 \text{ W}}{115 \text{ V}}$$

$$\approx 870 \text{ mA}$$

(b) From Eq. 21-2,

$$R = \frac{E^2}{P} = \frac{(115 \text{ V})^2}{100 \text{ W}}$$

$$= 132\ \Omega$$

(c) From Eq. 17-12,

$$P_m = 2P = 2 \times 100 \text{ W}$$

$$= 200 \text{ W}$$

Practice Problem

21-1.1 Three series-connected resistors, $R_1 = 500\ \Omega, R_2 = 680\ \Omega$, and $R_3 = 820\ \Omega$, are supplied from an ac source with a peak value of 39.6 V. Calculate the power dissipation in each resistor and the peak power supplied to the circuit.

21-2 POWER IN AN INDUCTANCE

When an alternating voltage is applied to a pure inductance, the current lags the applied voltage by 90°. This is discussed in Section 19-1. Consider the waveforms of current and voltage for an inductance with an ac supply, as reproduced in Figure 21-3. Because the instantaneous power supplied to any component is calculated as,

$$p = e \times i$$

the waveform of power can easily be derived from the current and voltage waves.

At time t_1 on Figure 21-3,

$$i = 0 \quad \text{and} \quad e = E_{\text{m}}$$

so,

$$p = i \times E_{\text{m}} = 0 \times E_{\text{m}}$$
$$= 0$$

At t_2,

$$i = I_{\text{m}} \sin 45° = 0.707 I_{\text{m}}$$
$$= I \text{ (rms value)}$$

and,

$$e = E_{\text{m}} \sin 135° = 0.707 E_{\text{m}}$$
$$= E \text{ (rms value)}$$

So,

$$p = P_{\text{m}} = IE$$

At t_3,

$$i = I_{\text{m}} \quad \text{and} \quad e = 0$$

so,

$$p = I_{\text{m}} \times 0$$
$$= 0$$

At t_4,

$$i = 0.707 I_{\text{m}} = I$$

and,

$$e = -0.707 E_{\text{m}} = -E$$

Giving,

$$p = -P_{\text{m}} = -IE$$

At this point (when $p = -IE$) the power supplied is a negative quantity, which means that the inductance is not absorbing power but supplying power.

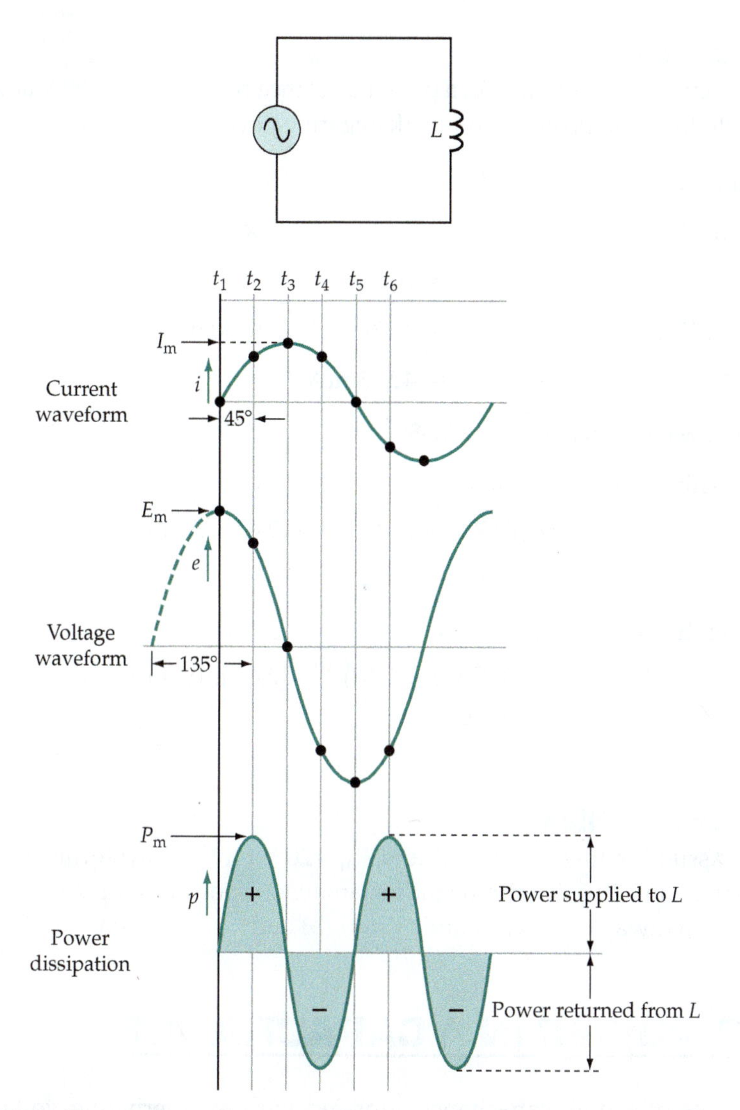

Figure 21-3 The waveform of power supplied from an ac source to a pure inductance can be derived from the voltage and current waveforms (using instantaneous levels, $p = ei$). The average power absorbed by the inductance is zero.

Continuing the process of calculating and plotting the instantaneous power levels in Figure 21-3, it is seen that the frequency of the power waveform is twice the voltage and current frequency. Also, because the negative half-cycles of power supplied are equal to the positive half-cycles, the average power supplied to the inductance is zero.

In Chapter 14 it is shown that energy can be stored in an inductor. So, referring to Figure 21-3 again, it can be said that the energy supplied to the inductor is stored during the time that energy input is a positive quantity, and that the stored energy is returned to the source of supply when the energy input is negative.

Example 21-2

A 30 mA current is measured in a pure inductance connected to a 50 V ac source. Calculate the peak positive and peak negative power dissipation.

Solution

From Eq. 17-16, $E_m = \sqrt{2}E = \sqrt{2} \times 50 \text{ V}$

$= 70.7 \text{ V}$

From Eq. 17-15, $I_m = \sqrt{2}I = \sqrt{2} \times 30 \text{ mA}$

$= 42.43 \text{ mA}$

From the waveforms in Figure 21-3:

Peak positive power dissipation,

$$P_+ = (70.7 \text{ V} \sin 135°) \times (42.43 \text{ mA} \sin 45°)$$
$$= 1.5 \text{ W}$$

Peak negative power dissipation,

$$P_- = (70.7 \text{ V} \sin 225°) \times (42.43 \text{ mA} \sin 135°)$$
$$= -1.5 \text{ W}$$

Practice Problem

21-2.1 Assuming that $I_m = 15$ mA and $E_m = 20$ V for the waveforms illustrated in Figure 21-3, determine the instantaneous power dissipations at points (a) halfway between t_1 and t_2, and (b) halfway between t_3 and t_4.

21-3 POWER IN A CAPACITANCE

In the case of a pure capacitance supplied with an alternating voltage, the current leads the voltage by 90°, (see Section 19-3). Figure 21-4 shows the current and voltage waveforms for a capacitance, and the waveform of power supplied as derived from the current and voltage waves.

At time t_1 on Figure 21-4,

$$i = 0 \quad \text{and} \quad e = E_m$$

So, $p = i \times E_m = 0$

At t_2, $i = I_m \sin 45° = 0.707 I_m$

$= I$ (rms value)

and, $e = E_m \sin 315° = -0.707 E_m$

$= -E$ (rms value)

So, $p = -P_m = -IE$

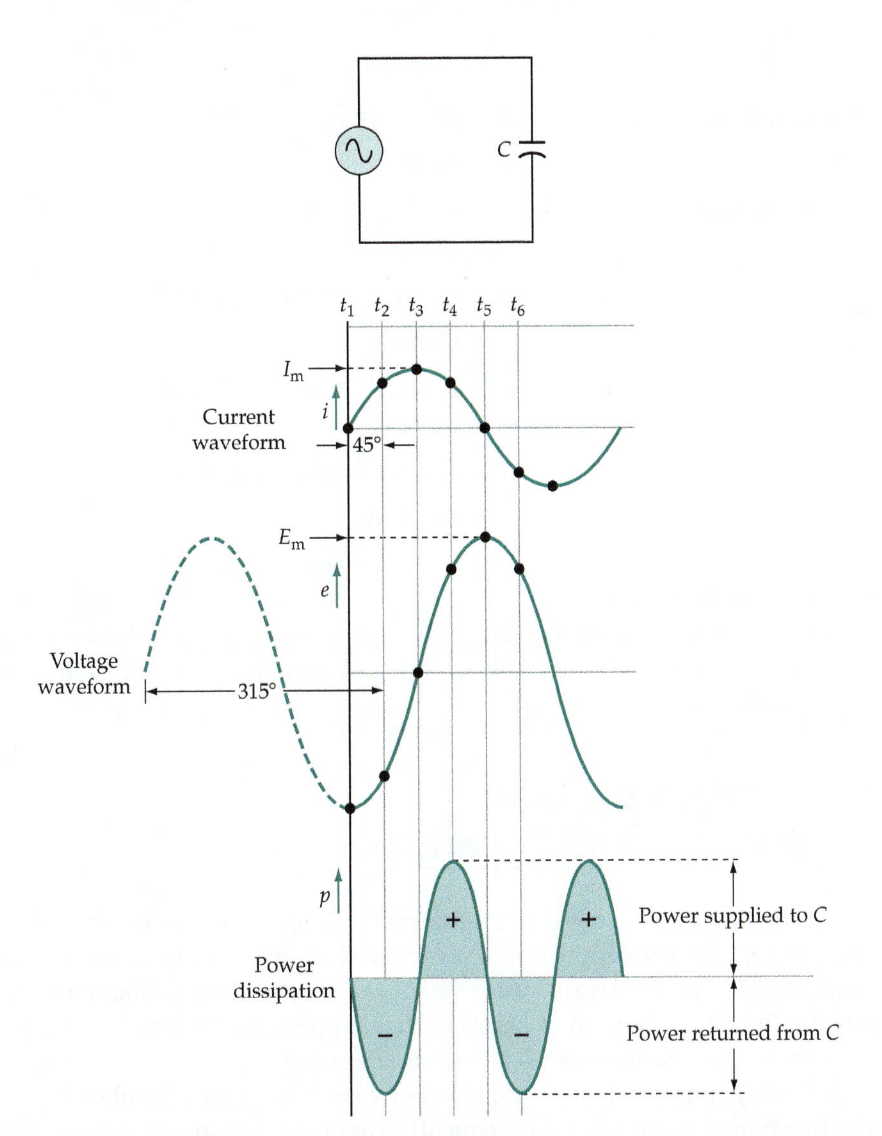

Figure 21-4 The waveform of power supplied from an ac source to a pure capacitance can be derived from the voltage and current waveforms (using instantaneous levels, $p = ei$). The average power absorbed by the capacitance is zero.

Continuing the process, exactly as was done for the inductive circuit, it is seen that the frequency of the waveform of power supplied to a capacitor is twice that of the voltage and current frequency. It is also seen that the power supplied to the capacitor is alternatively positive and negative, meaning that the capacitor stores the power supplied to it then returns the stored power to the source of supply. The result of this is that the average power supplied to the capacitance is zero.

Example 21-3

For the circuit and the waveforms illustrated in Figure 21-4, $I_m = 100$ mA and $E_m = 25$ V. Calculate the instantaneous power supplied to the capacitor at a time halfway between t_2 and t_3.

Solution

Phase angle for i, $\theta = 45° + 45°/2$

$= 67.5°$

Phase angle for e, $\phi = 315° + 45°/2$

$= 337.5°$

$$i = I_m \sin\theta = 100\text{ mA} \sin 67.5°$$
$$= 92.4\text{ mA}$$
$$e = E_m \sin\phi = 25\text{ V} \sin 337.5°$$
$$= -9.6\text{ V}$$
$$p = e \times i = -9.6\text{ V} \times 92.4\text{ mA}$$
$$= -884\text{ mW}$$

Practice Problem

21-3.1 Calculate the instantaneous power supplied to a capacitive circuit at $e = E_m \sin 30°$ if the peak voltage and current are 10 V and 15 mA, respectively.

21-4 TRUE POWER AND REACTIVE POWER

The power supplied to a resistance is sometimes referred to as *resistive power.* Similarly, the power supplied to a reactance (inductive or capacitive) can be termed *reactive power.* Also, because the power supplied to a resistance is actually dissipated in the form of heat, but that supplied to a reactance averages out to zero, the *resistive power* is known as *true power.*

It is very important to distinguish between true power (symbol P) and reactive power (symbol Q). Consequently, true power is always measured in *watts* (W), and the reactive power unit is given the name *volt-amp-reactive* (*var*). For a pure inductance or capacitance, the reactive power is calculated from,

$$Q_L = E_L I_L, \quad \text{or} \quad Q_C = E_C I_C \tag{21-4}$$

where Q is in var and E_L, I_L, E_C, and I_C are the component voltage and current levels.

Reactive power can also be calculated as,

$$Q_L = I_L^2 X_L, \quad \text{or} \quad Q_C = I_C^2 X_C \tag{21-5}$$

and,

$$Q_L = \frac{E_L^2}{X_L} \quad \text{or} \quad Q_C = \frac{E_C^2}{X_C} \tag{21-6}$$

Example 21-4

Calculate the power supplied when a 120 V, 60 Hz source is connected (as in Figure 21-5) to: (a) a 60 Ω resistor, (b) a 50 mH inductor, (c) a 33 μF capacitor.

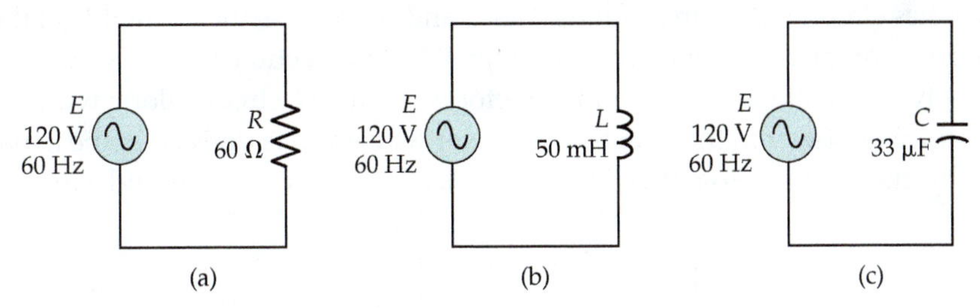

Figure 21-5 Circuits for Example 21-4.

Solution

(a)

$$P = \frac{E^2}{R} = \frac{(120\text{ V})^2}{60\ \Omega}$$

$$= 240\text{ W (true power)}$$

(b)

$$X_L = 2\pi fL = 2\pi \times 60\text{ Hz} \times 50\text{ mH}$$

$$\approx 18.8\ \Omega$$

From Eq. 21-6,

$$Q_L = \frac{E_L^2}{X_L} = \frac{(120\text{ V})^2}{18.8\ \Omega}$$

$$= 766\text{ var (reactive power)}$$

(c)

$$X_C = \frac{1}{2\pi fC} = \frac{1}{2\pi \times 60\text{ Hz} \times 33\ \mu\text{F}}$$

$$\approx 80.4\ \Omega$$

From Eq. 21-6,

$$Q_C = \frac{E_C^2}{X_C} = \frac{(120\text{ V})^2}{80.4\ \Omega}$$

$$= 179\text{ var (reactive power)}$$

Practice Problem

21-4.1 A 0.1 μF capacitor, 100 mH inductor, and a 1 kΩ resistor connected in parallel are supplied from a 400 Hz ac source, (see Figure 21-6). The power dissipated in the resistor is measured as 100 mW. Calculate the reactive power supplied to the capacitor and the reactive power supplied to the inductor.

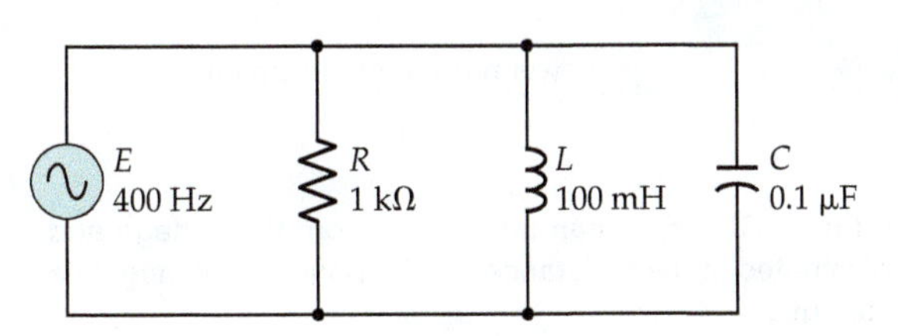

Figure 21-6 Circuit for Problem 21-4.1.

21-5 POWER IN *RL* AND *RC* CIRCUITS

RL Circuit Power

In a series circuit consisting of inductance and resistance, the current lags the voltage by an angle (ϕ) which is less than 90°, (see Section 19-5). Figure 21-7 shows typical voltage and current waveforms for an *RL* circuit, along with the waveform of power in the circuit. The power waveform is derived in the usual way by multiplying together instantaneous values of voltage and current.

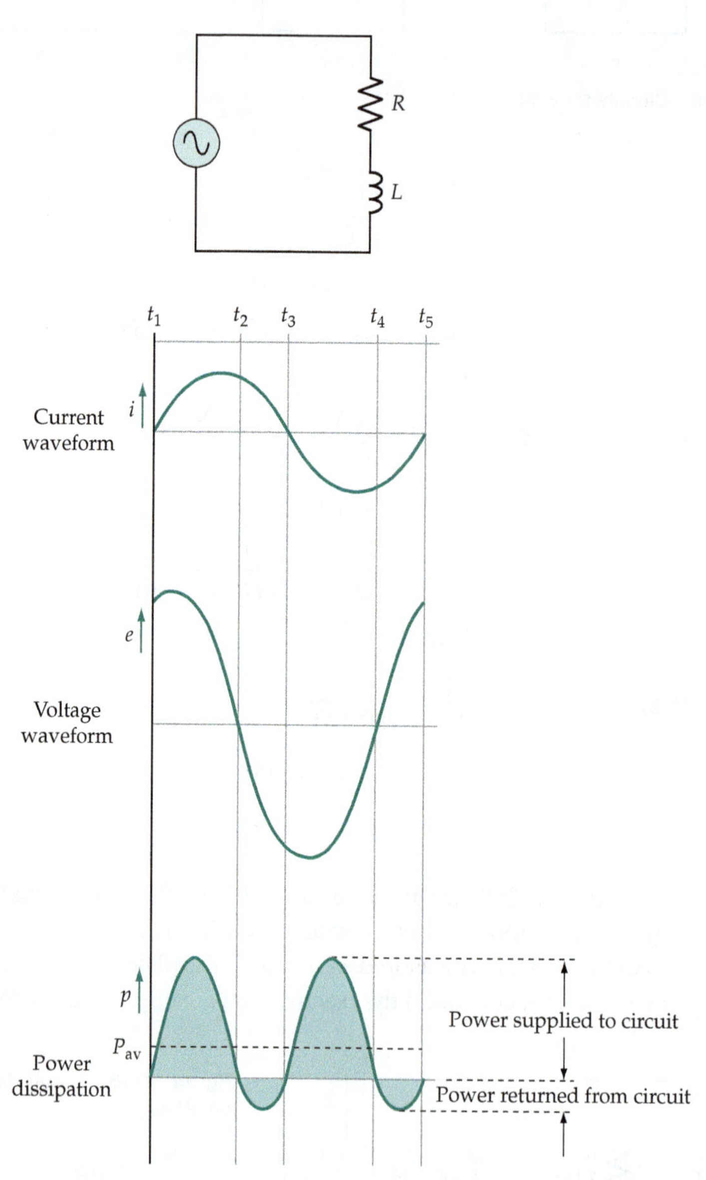

Figure 21-7 The power waveform for an *RL* circuit can be derived from the voltage and current waveforms. Some power is dissipated in the resistance, so the positive and negative portions of the power waveform are unequal.

During the period from t_1 to t_2 both i and e are positive quantities; so, power p remains positive throughout that time. From t_2 to t_3, i is positive and e is a negative quantity; consequently, the product of i and e is negative, and the power waveform is below the zero line. After t_3, the current and voltage are both negative until t_4. Because $(-e) \times (-i)$ is a positive quantity, the power is positive again from t_3 to t_4.

The power waveform clearly shows that more positive power than negative power is supplied to the circuit. This is to be expected, of course, because power is dissipated in the resistance, but the average power supplied to the inductance remains at zero. The average power supplied to the circuit can be represented by the dashed line shown on the power waveform.

Now consider the phasor diagram for a series *RL* circuit, as reproduced in Figure 21-8. The current I is shown lagging the applied voltage E by the angle ϕ. The voltage across the resistance (E_R) is in phase with I, so E_R also lags E by angle ϕ. The voltage across the inductance is E_L, and it leads the current by an angle of 90°. The phasor sum of E_R and E_L gives the supply voltage E, as illustrated.

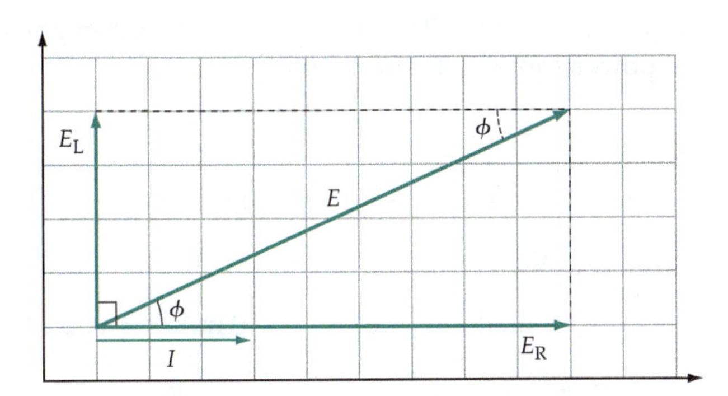

Figure 21-8 Phasor diagram for an *RL* series circuit. Because power is dissipated only in the resistive component, the circuit power is $P = I \times$ (voltage in phase with I), or $P = EI \cos \phi$.

The true power dissipated in the *RL* circuit is, of course, the power dissipated in the resistor. So, true power is,

$$P = E_R \times I$$

and from Figure 21-8,

$$E_R = E \cos \phi$$

Giving,

$$P = (E \cos \phi) \times I$$

or, true power is,

$$P = EI \cos \phi \qquad (21\text{-}7)$$

The reactive power is the product of the inductive voltage and current, E_L and I_L.

From Figure 21-8,

$$E_L = E \sin \phi$$

and,

$$I_L = I$$

So, $$Q = (E \sin \phi) \times I$$

or, reactive power is, $$Q = EI \sin \phi \text{ (var)} \quad (21\text{-}8)$$

If the supply voltage E and the measured current I in an RL circuit are multiplied together, the product is neither the true power nor the reactive power. However, it does give a quantity that appears to be the power supplied to the circuit. The term *apparent power* (symbol S) is applied to this quantity, and units of apparent power are *volts-amps* (VA):

apparent power, $$S = EI \quad (21\text{-}9)$$

where S is the apparent power and E and I are rms values of supply voltage and current.

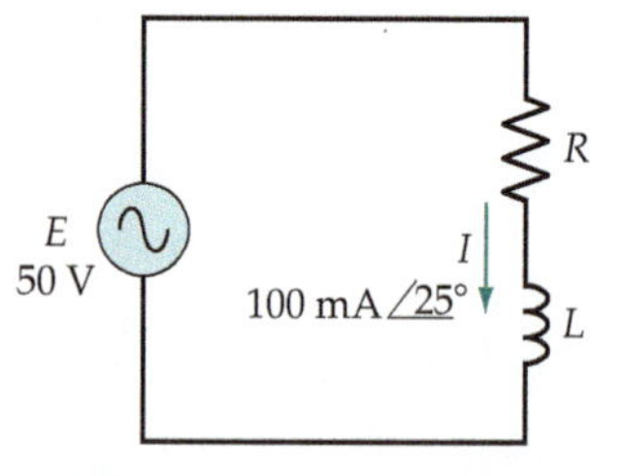

Figure 21-9 Circuit for Example 21-5.

Example 21-5

In a series RL circuit supplied with 50 V, the current is measured as 100 mA with a phase angle of 25°, (Figure 21-9). Calculate the apparent power, reactive power, and true power supplied to the circuit.

Solution

Eq. 21-9, $$S = EI \text{ volt-amp} = 50 \text{ V} \times 100 \text{ mA}$$
$$= 5 \text{ VA}$$

Eq. 21-8, $$Q = EI \sin \phi \text{ var} = 50 \text{ V} \times 100 \text{ mA} \times (\sin 25°)$$
$$\approx 2.1 \text{ var}$$

Eq. 21-7, $$P = EI \cos \phi \text{ watts} = 50 \text{ V} \times 100 \text{ mA} \times (\cos 25°)$$
$$\approx 4.5 \text{ W}$$

RC Circuit Power

When the current and voltage waveforms in a series RC circuit are employed to derive the waveform of power supplied (Figure 21-10), it is seen that a similar result is obtained as in the case of the RL circuit. The average power supplied to the circuit is a positive quantity, and this represents the power dissipated in the resistance. The phasor diagram for the RC circuit, as shown in Figure 21-11, gives the same equations for true power, reactive power, and apparent power as those derived for the RL circuit.

The power supplied to parallel RC and RL circuits can also be resolved into true power, reactive power, and apparent power components. Using the supply voltage and current, and the current-voltage phase angle, Equations 21-7, 21-8, and 21-9 are just as valid for parallel circuits as for series

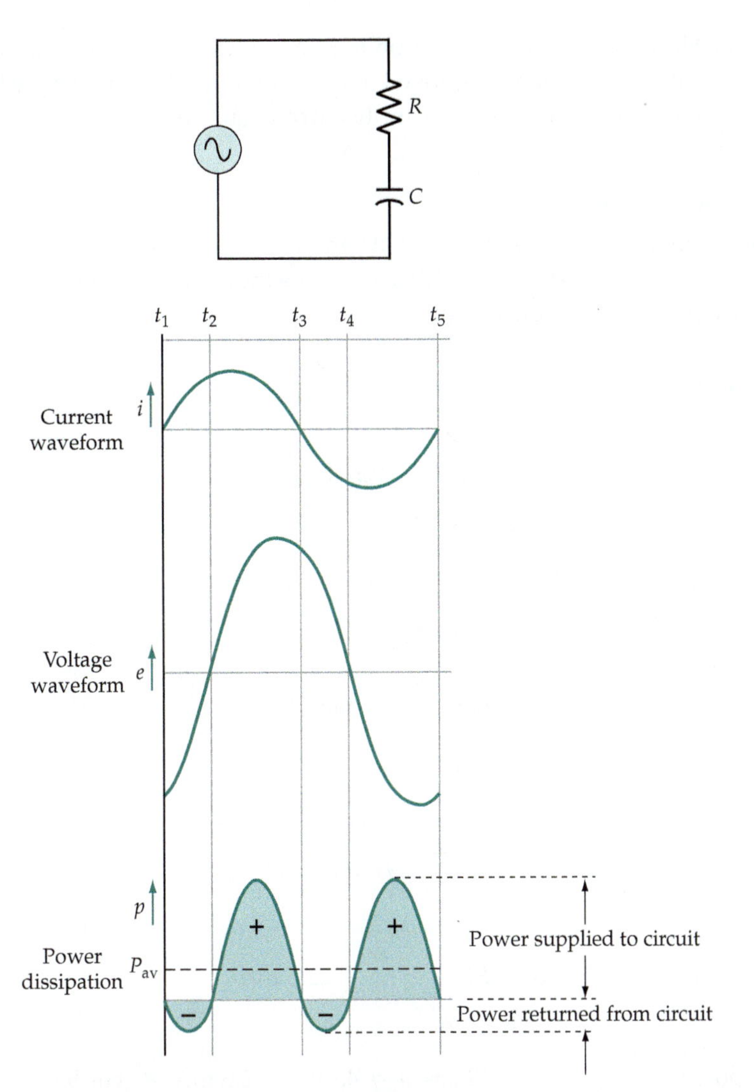

Figure 21-10 The power waveform for an *RC* circuit can be derived from the voltage and current waveforms. The positive and negative portions of the power waveform are unequal because some power is dissipated in the resistance.

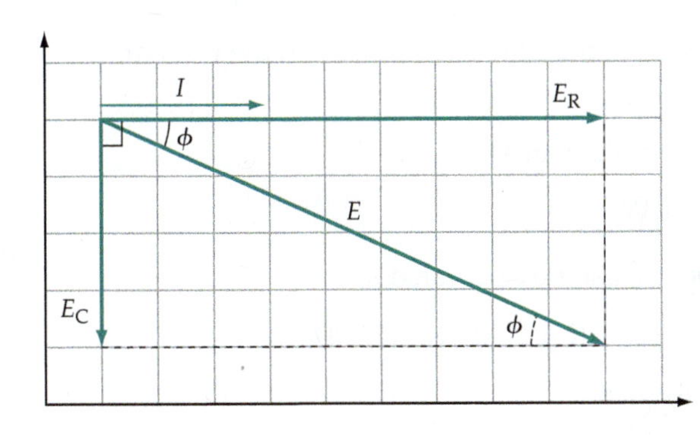

Figure 21-11 Phasor diagram for an *RC* circuit. Because power is dissipated only in the resistive component, the circuit power is $P = I \times$ (voltage in phase with I), or $P = EI \cos \phi$.

circuits. However, in the next section it is shown that capacitors connected in parallel with a load can be employed in correcting problems that arise because of the phase difference between supply current and voltage.

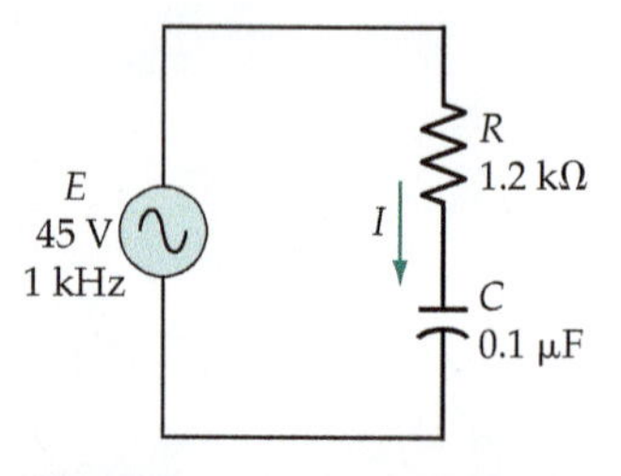

Figure 21-12 Circuit for Example 21-6.

Example 21-6

A series circuit consisting of $R = 1.2\ \text{k}\Omega$ and $C = 0.1\ \mu\text{F}$ is supplied with 45 V at a frequency of 1 kHz (Figure 21-12). Determine the apparent power, true power, and reactive power in the circuit.

Solution

$$X_C = \frac{1}{2\pi fC} = \frac{1}{2\pi \times 1\ \text{kHz} \times 0.1\ \mu\text{F}}$$

$$= 1.59\ \text{k}\Omega$$

$$|Z| = \sqrt{R^2 + X_C^2} = \sqrt{(1.2\ \text{k}\Omega)^2 + (1.59\ \text{k}\Omega)^2}$$

$$= 1.99\ \text{k}\Omega$$

$$\phi = \tan^{-1}\frac{X_C}{R} = \tan^{-1}\left(\frac{1.59\ \text{k}\Omega}{1.2\ \text{k}\Omega}\right)$$

$$\approx 53^\circ$$

$$|I| = \frac{E}{|Z|} = \frac{45\ \text{V}}{1.99\ \text{k}\Omega}$$

$$= 22.6\ \text{mA}$$

Apparent power, $S = EI = 45\ \text{V} \times 22.6\ \text{mA}$

$$\approx 1\ \text{VA}$$

True power, $P = EI\cos\phi = 45\ \text{V} \times 22.6\ \text{mA} \times (\cos 53^\circ)$

$$= 0.61\ \text{W}$$

Reactive power, $Q = EI\sin\phi = 45\ \text{V} \times 22.6\ \text{mA} \times (\sin 53^\circ)$

$$= 0.81\ \text{var}$$

Note that because the true power is the power dissipated in the resistor, it can also be calculated as I^2R.

Eq. 21-3, $P = I^2R = (22.6\ \text{mA})^2 \times 1.2\ \text{k}\Omega$

$$= 0.61\ \text{W}$$

Also, note that the reactive power can be calculated as I^2X_C.

Eq. 21-5, $Q = I^2X_C = (22.6\ \text{mA})^2 \times 1.59\ \text{k}\Omega$

$$= 0.81\ \text{var}$$

Example 21-7

A parallel impedance circuit with a 33 V supply has a current of $75.2\ \text{mA}\underline{/10.5^\circ}$. Calculate the apparent power, true power, and reactive power.

Solution

Eq. 21-9, $S = EI = 33\ \text{V} \times 75.2\ \text{mA}$

$\approx 2.5\ \text{VA}$

Eq. 21-7, $P = EI \cos \phi = 33\ \text{V} \times 75.2\ \text{mA} \times \cos 10.5^\circ$

$= 2.44\ \text{W}$

Eq. 21-8, $Q = EI \sin \phi = 33\ \text{V} \times 75.2\ \text{mA} \times \sin 10.5^\circ$

$= 0.45\ \text{var}$

Practice Problems

21-5.1 A series RL circuit with a 3.3 V, 250 Hz supply has $R = 1.8\ \text{k}\Omega$. The circuit current is measured as 1.5 mA lagging the input voltage by 35.1°. Determine the inductance value.

21-5.2 Calculate the apparent power, reactive power, and true power supplied to the circuit in Problem 21-5.1.

21-5.3 A series RC circuit with a 24 V, 400 Hz supply has $R = 1.2\ \text{k}\Omega$ and $C = 0.2\ \mu\text{F}$. Calculate the apparent power, reactive power, and true power in the circuit.

21-6 POWER FACTOR

Power Factor Equation and Power Triangle

Ideally, all the supply voltage and current should be converted into true power in a load. When this is not the case, a certain kind of inefficiency occurs. The ratio of *true power* to *apparent power* is termed the *power factor* of the load,

$$\text{Power factor} = \frac{\text{True power}}{\text{Apparent power}} = \frac{P}{EI}$$

As already discussed, the true power in a load is calculated from Equation 21-7 as,

$$P = EI \cos \phi$$

where E and I are the supply voltage and current, and ϕ is the phase angle between them. Equation 21-7 can be rewritten to determine $\cos \phi$,

$$\cos \phi = \frac{P}{EI} = \frac{\text{True power}}{\text{Apparent power}}$$

So,

$$\text{Power factor} = \frac{P}{EI} = \cos\phi \qquad \textbf{(21-10)}$$

If the phase angle is zero (I in phase with E) the power factor is 1. So, all of the apparent power ($E \times I$) from the supply is dissipated as true power in the load. Where ϕ is greater than zero, $\cos\phi$ is less than 1, and only a portion of the apparent power is converted into useful power. The power factor always has a maximum value of 1, and is normally less than 1. The power factor can be expressed as a ratio or as a percentage, and is also usually defined as *leading* or *lagging*. A 60% *lagging power factor* implies an inductive load in which the supply current lags the voltage by an angle with a cosine of 0.6, (i.e., by approximately 53°). A 90% *leading power factor* would indicate a capacitive load in which the current leads the voltage by an angle with a cosine of 0.9, or approximately 26°.

A *power triangle* can be drawn to represent load conditions, as illustrated in Figure 21-13. The true power (P) is represented by the horizontal vector, and the vector for apparent power (S) is drawn lagging the true power by the phase angle (ϕ). The reactive power vector (Q_L) completes the triangle. If ϕ is close to zero, as in Figure 21-13(a), then the power factor ($\cos\phi$) is approximately equal to 1. In this case, the true power is almost equal to the apparent power, and the reactive power is very small. The power triangle in Figure 21-13(b) represents a situation where there is a large lagging phase angle, which gives $\cos\phi$ very much less than 1. The true power ($P = EI\cos\phi$) is now substantially smaller than the apparent power, and the reactive power is relatively large.

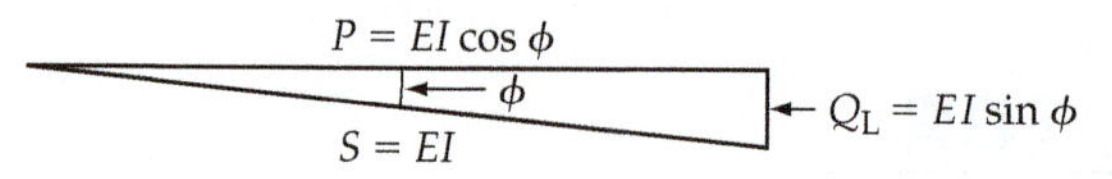

(a) Power triangle for load with $\cos\phi \approx 1$

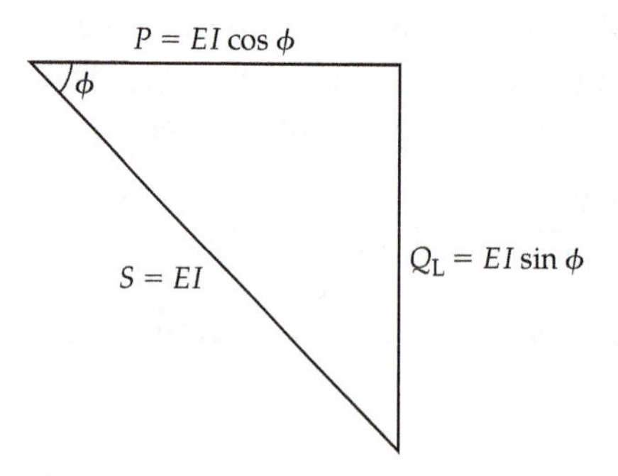

(b) Power triangle for load with $\cos\phi << 1$

Figure 21-13 Power triangles can be drawn for ac loads. When ϕ is small, the true power is close to the level of apparent power. When ϕ is large, the true power is considerably smaller than the apparent power.

Power Factor Correction

Most industrial loads consist of electric motors, which are inductive and have lagging power factors. Consider the situation illustrated in Figure 21-14(a),

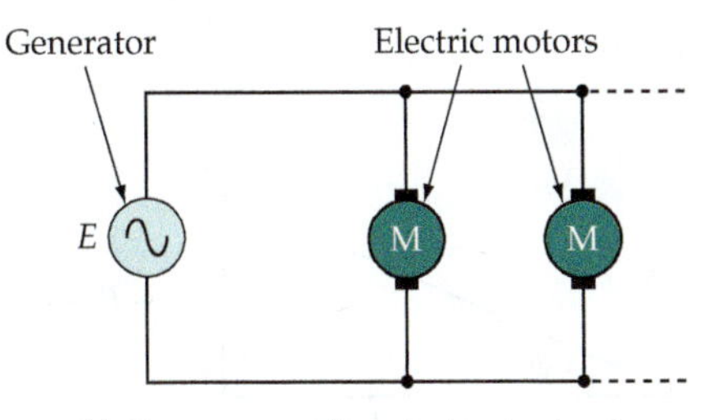

(a) Generator with an inductive load

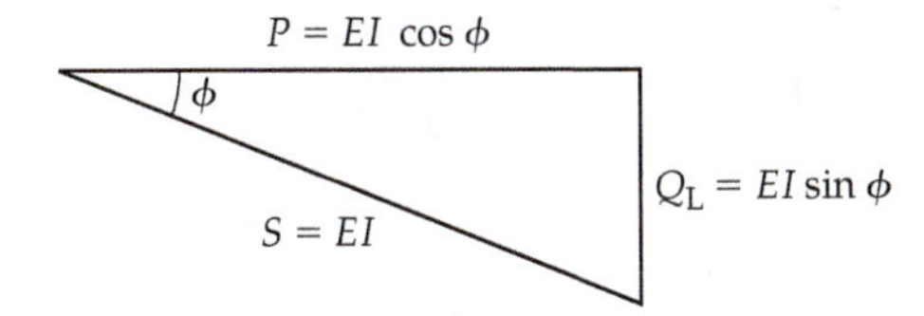

(b) Power triangle for an inductive load

Figure 21-14 Because electric motors are inductive, the motor supply current lags the supply voltage; so, the load has a lagging power factor. The power triangle shows the relationship between true power (P), apparent power (S), and reactive power (Q_L).

and assume that the supply voltage and current are measured as $E = 120$ V and $I = 100$ A, with the current lagging the voltage by an angle of 33.5°. The power triangle for the load is drawn in Figure 21-14(b). Using the appropriate equations, the apparent power, true power, and reactive power are calculated as:

Apparent power, $$S = EI = 120\text{ V} \times 100\text{ A}$$
$$= 12\text{ kVA}$$

True power, $$P = EI\cos\phi = 12\text{ kVA} \times (\cos 33.5°)$$
$$= 10\text{ kW}$$

Inductive reactive power,
$$Q_L = EI\sin\phi = 12\text{ kVA} \times (\sin 33.5°)$$
$$= 6.6\text{ kvar}$$

If the power factor was 1 ($\phi = 0$), then to supply the 10 kW true power (calculated above) to the load would require a current of,

$$I = \frac{P}{E} = \frac{10\text{ kW}}{120\text{ V}}$$
$$\approx 83\text{ A}$$

This shows that if the power factor could be increased to unity (ϕ reduced to zero), the generators supplying the load would have to produce a current of only 83 A instead of 100 A while still supplying the required amount of true power. Also, the conducting cables could be selected to carry 83 A instead of 100 A; and consequently, they would be less expensive. So, it is always best to have a power factor as near unity as possible, and in fact the power factor of an inductive load can be adjusted toward unity by a process known as *power factor correction*.

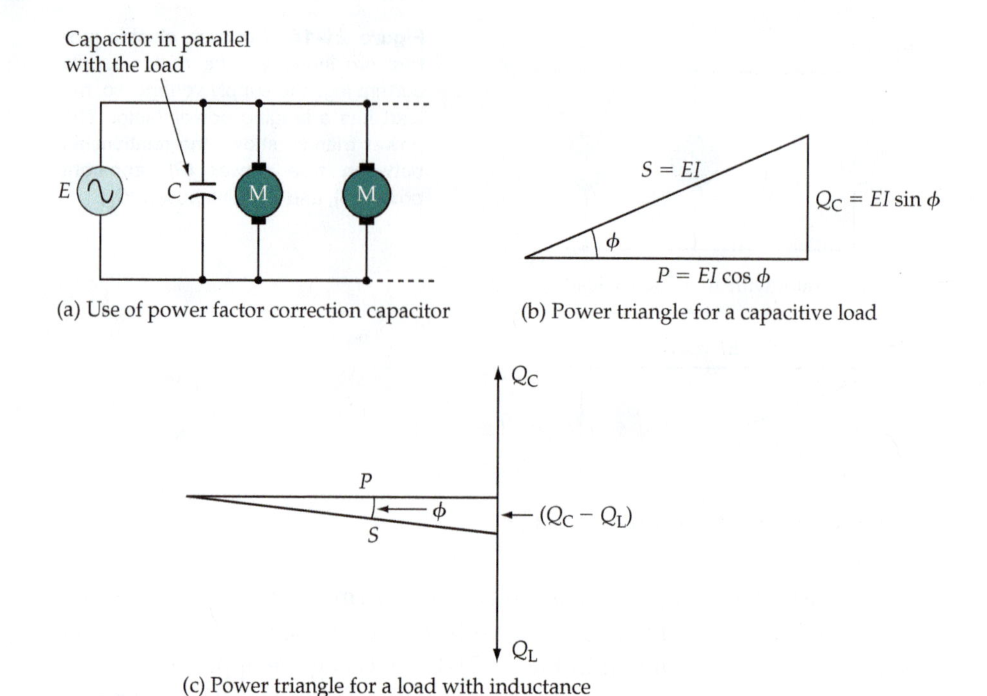

Figure 21-15 A capacitive load has a leading power factor; its reactive power vector Q_C is drawn vertically up from the vector for the true power. Capacitance can be used to partially correct a lagging power factor.

Power factor correction for an inductive load consists simply of connecting capacitance in parallel with the load, [Figure 21-15(a)]. Suppose that the 100 A load discussed above is capacitive instead of inductive. The current would then lead the supply voltage by the phase angle, and the power triangle would be drawn as illustrated in Figure 21-15(b). The apparent power vector S is shown leading the true power vector P by the angle ϕ, and the capacitive reactive power vector Q_C is drawn vertically up from the true power vector, as illustrated.

When capacitance is connected in parallel with an inductive load, the power triangle has a *capacitive reactive power* component as well as *inductive reactive power.* The diagram in Figure 21-15(c) illustrates the situation. The capacitive reactive power is represented by the vector Q_C drawn vertically *up*, while Q_L drawn *down* represents the inductive reactive power. The net reactive power is the difference between Q_L and Q_C, and in the diagram it is so small that it gives a very small phase angle which results in a near-unity power factor. Unity power factor would, of course, be achieved when Q_C and Q_L are equal. *This would create a state of resonance,* (see Chapter 23) *which should be avoided in a power factor correction situation.*

Example 21-8

The current taken from a 115 V, 60 Hz supply to a circuit is measured as 20 A with a lagging power factor of 75%, (see Figure 21-16). Calculate the apparent

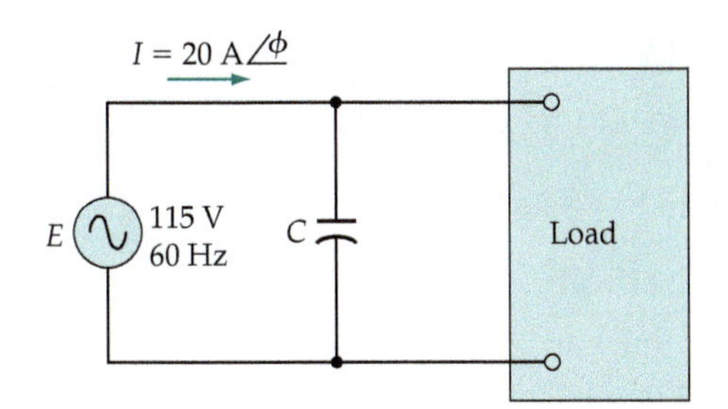

Figure 21-16 Circuit for Example 21-8.

power, true power, and reactive power. Also, determine the amount of capacitance that must be connected in parallel with the load to correct the power factors to 95% lagging. Draw the power triangle for the circuit.

Solution

$$\text{Power factor} = 75\%$$

or, $\cos\phi = 0.75$

So, $\phi = \cos^{-1} 0.75$

$= 41.4°$

Apparent power, $S = EI = 115\text{ V} \times 20\text{ A}$

$= 2.3\text{ kVA}$

True power, $P = EI\cos\phi = 2.3\text{ kVA} \times 0.75$

$= 1.725\text{ kW}$

Reactive power, $Q_L = EI\sin\phi = 2300\text{ VA} \times (\sin 41.4°)$

$= 1.52\text{ kvar}$

For a 95% power factor,

$$\cos\phi_2 = 0.95$$

and, $\phi_2 = \cos^{-1} 0.95$

$= 18.2°$

True power remains, $P = EI\cos\phi = 1.725\text{ kW}$

New value of apparent power,

$$S_2 = EI = \frac{P}{\cos\phi} = \frac{1.725\text{ k}\Omega}{0.95}$$

$= 1.82\text{ kVA}$

and new value of reactive power,

$$Q = EI\sin\phi = 1.82\text{ kVA} \times (\sin 18.2°)$$

$= 568\text{ var}$

and, $$Q = Q_L - Q_C$$

or, $$Q_C = Q_L - Q = 1.52\ \text{kvar} - 568\ \text{var}$$
$$= 952\ \text{var}$$

From Eq. 21-6, $$X_C = \frac{E^2}{Q_C} = \frac{(115\ \text{V})^2}{952\ \text{var}}$$
$$= 13.9\ \Omega$$

and, $$C = \frac{1}{2\pi f X_C} = \frac{1}{2\pi \times 60\ \text{Hz} \times 13.9\ \Omega}$$
$$= 191\ \mu\text{F}$$

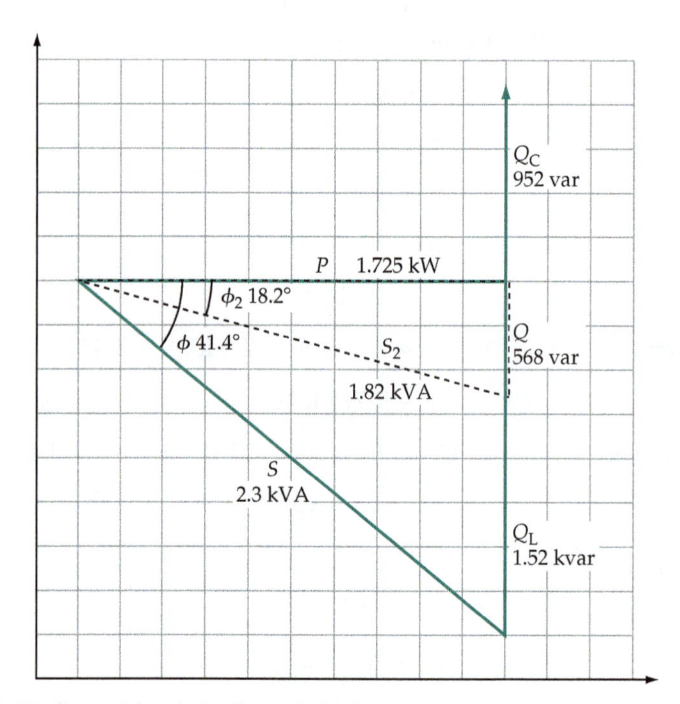

Figure 21-17 Power triangle for Example 21-8.

Practice Problems

21-6.1 An industrial load that uses a 120 V, 60 Hz supply has a 1000 μF power factor correction capacitor. The measured power is 23 kW with a power factor of 0.93. Calculate the power factor with the capacitor removed.

21-6.2 For Problem 21-6.1, determine the supply current with the capacitor connected and without the capacitor connected.

21-7 AC WATTMETER

As mentioned in Section 13-8, the dynamometer instrument can be employed as a voltmeter or ammeter, or in its major application as a wattmeter. Consider the illustrations in Figure 21-18 that show the field coils and moving coil of a dynamometer instrument connected in series. With current flowing in the direction shown in Figure 21-18(a), the field coils set up fluxes that have their *N* poles on the right-hand side and their *S* poles on the left-hand side. Also, the moving-coil flux has the polarity illustrated, *S* at the bottom and *N* at the top. The *N* pole of the moving-coil flux is repelled from the adjacent *N* pole of the field coil flux, and the two adjacent *S* poles also repel each other. The result of this is that the moving coil is deflected clockwise, causing the pointer to move over the scale from left to right.

Now consider the effect of reversing the direction of the current through the coils. As illustrated in Figure 21-18(b), the field coil fluxes are now *S* at the right-hand side and *N* at the left-hand side. The moving-coil flux is also reversed, having *S* at the top and *N* at the bottom. Once again, like poles are adjacent to each other, and the pointer moves from left to right over the scale.

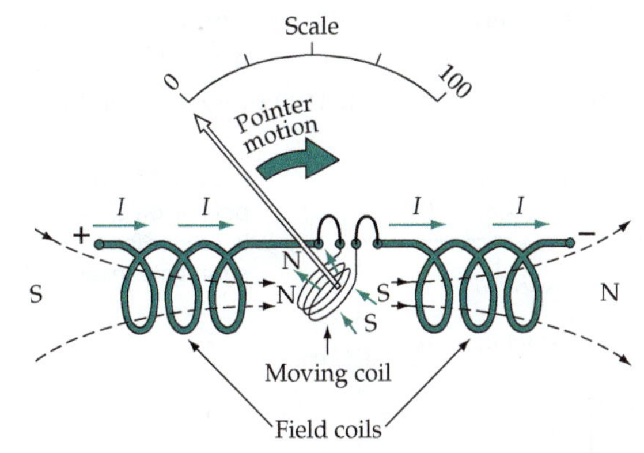

(a) Current flowing from left to right produces positive deflection

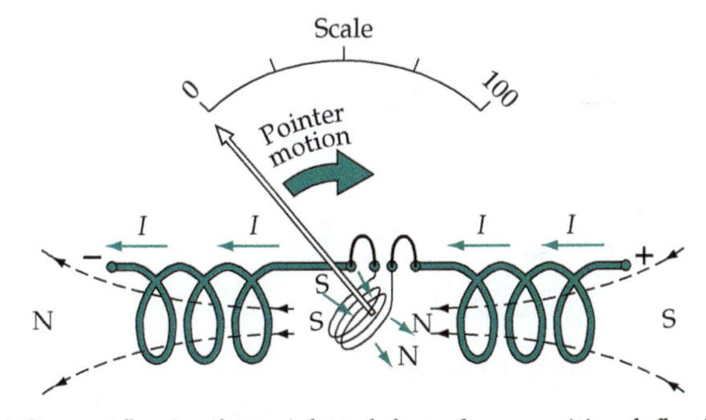

(b) Current flowing from right to left produces positive deflection

Figure 21-18 When the current flow through a dynamometer instrument reverses, both the field flux and the moving-coil flux reverse directions. The pointer deflection is unaffected. This makes the instrument suitable for application to ac measurements, as well as dc measurements.

It is seen that the dynamometer instrument has a positive deflection, regardless of the direction of the current through the meter. The meter terminals are not marked positive (+) and negative (−); that is, the dynamometer instrument is *unpolarized*. So, the instrument gives a positive deflection when either direct or alternating current flows through the coils. As an ammeter, the scale of the instrument can be read as direct current when measuring dc, and as the rms value of alternating current when measuring ac. Similarly, as a voltmeter, the dynamometer instrument indicates dc volts or rms ac volts. The scale of the instrument can be conveniently calibrated on dc and then used to measure ac.

For ac as well as dc, the major application of the dynamometer instrument is as a wattmeter. As illustrated in Figure 21-19(a), the connection for measuring ac power is similar to that for measuring power in a dc circuit. In Chapter 13 it is explained that the input terminals of the voltage and current coils are identified with a ±, ↓, or * sign. The marked terminal of the current coil should be connected to the supply, and the voltage coil marked terminal connected to the load side of the current coil. In alternating-current applications, the load current could lead or lag the load voltage by a phase angle (ϕ). The deflection of the instrument is proportional to the in-phase components of current and voltage. Thus, as shown in Figure 21-19(b), the instrument indication is proportional to $EI \cos \phi$. As already discussed, the true power dissipated in a load with an ac supply is, $P = EI \cos \phi$. So, when used as an ac wattmeter, the dynamometer instrument measures the true power supplied to the load.

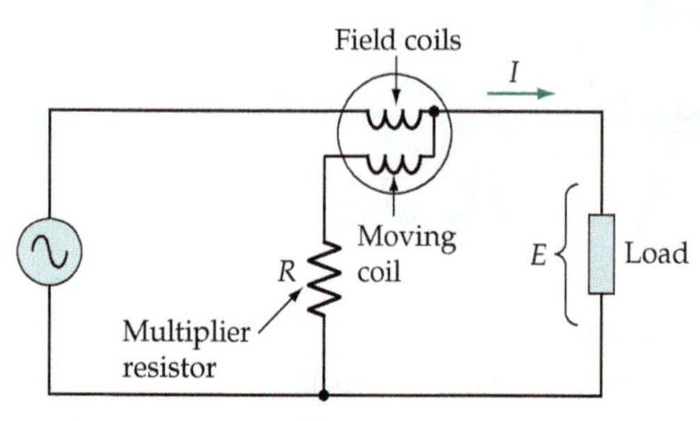

(a) Dynamometer instrument as an ac wattmeter

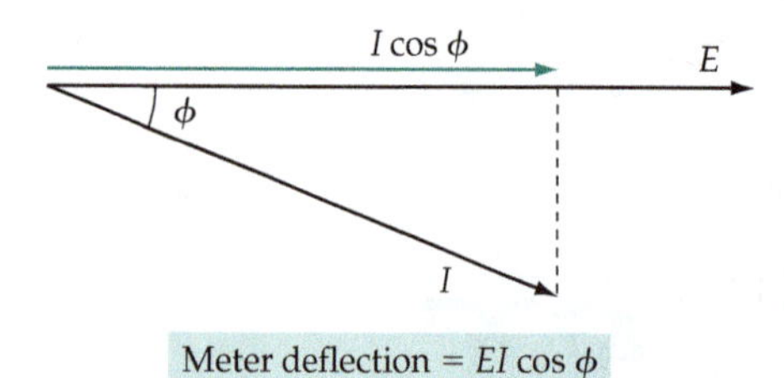

(b) Phasor diagram of load current and voltage

Figure 21-19 The deflection of the pointer on a dynamometer wattmeter is proportional to $EI \cos \phi$; so the instrument measures true ac power.

Example 21-9

A wattmeter measures the ac power delivered to a load as 100 W, and an ammeter and voltmeter indicate that the load current and voltage are 1.5 A and 100 V, respectively. Calculate the phase angle between the current and voltage.

Solution

$$\text{True power} = P = EI\cos\phi$$

So,

$$\cos\phi = \frac{P}{EI} = \frac{100\ \text{W}}{100\ \text{V} \times 1.5\ \text{A}}$$

$$= 0.667$$

and,

$$\phi = \cos^{-1} 0.667$$

$$= 48.2°$$

Practice Problem

21-7.1 A 115 V source supplies 700 mA to a load. If the power supplied to the load is measured as 75 W, determine the phase angle between the load current and the supply voltage.

Summary of Formulas

- *True power:*

$$P = EI,\ P = \frac{E^2}{R},\ P = I^2R$$

- *Reactive power:*

$$Q_L = E_L I_L \quad \text{or} \quad Q_C = E_C I_C$$

$$Q_L = I_L^2 X_L \quad \text{or} \quad Q_C = I_C^2 X_C$$

$$Q_L = \frac{E_L^2}{X_L} \quad \text{or} \quad Q_C = \frac{E_C^2}{X_C}$$

- *True power:*

$$P = EI\cos\phi$$

- *Reactive power:*

$$Q = EI\sin\phi$$

- *Apparent power:*

$$S = EI$$

- *Power factor:*

$$\cos\phi = \frac{P}{EI}$$

Review Questions

Section 21-1

21-1 Sketch typical waveforms of current and voltage for a pure resistance connected in an alternating current circuit. From the voltage and current waveforms, derive the waveform of instantaneous power dissipation. Explain briefly.

Section 21-2

21-2 Sketch typical waveforms of current and voltage for a pure inductance with an ac supply. Derive the waveform for instantaneous power in the inductance.

Section 21-3

21-3 Sketch typical waveforms of current and voltage for a pure capacitance with an ac supply. Derive the waveform for instantaneous power in the capacitance.

Section 21-4

21-4 Define reactive power and true power, and write equations for each.

Section 21-5

21-5 Sketch typical current and voltage waveforms for a series *RL* circuit with an ac supply, and derive the waveform for instantaneous power.

21-6 Sketch typical current and voltage waveforms for a series *RC* circuit with an ac supply, and derive the waveform for instantaneous power.

21-7 Define apparent power, and explain its relationship to true power and reactive power.

Section 21-6

21-8 Define power factor, and write the power factor equation.

21-9 Sketch typical power triangles for a series *RL* circuit and for a series *RC* circuit. Show the effect that a parallel capacitor can have on the power triangle for a series *RL* circuit.

Section 21-7

21-10 Using illustrations, explain why a dynamometer instrument gives a positive deflection for current flow through the meter in either direction. What changes are necessary to produce a negative deflection on a dynamometer instrument?

21-11 Show how a dynamometer wattmeter is connected to measure the power delivered to a load, and explain why the instrument measures the true power in an ac circuit.

Problems

Section 21-1

21-1 A 2 kW heating element is supplied from a 220 V ac source. Determine (a) the resistance of the element, (b) the current level, (c) the peak power dissipated in the element.

21-2 Determine the resistance of a 60 W lamp that is supplied from a 50 V ac source. Also, calculate the current and the peak power dissipated in lamp.

Section 21-2

21-3 Calculate the peak instantaneous power dissipated in a 300 mH pure inductance connected to a 115 V, 60 Hz source.

21-4 A 56 mH pure inductance has a 60 mA, 400 Hz current. Calculate the peak instantaneous power dissipated in the inductor.

Section 21-3

21-5 Determine the peak instantaneous power dissipated in a 0.1 μF pure capacitance connected to a 115 V, 60 Hz source.

21-6 A 100 pF pure capacitance has a supply of 0.9 V at a frequency of 1 MHz. Calculate the maximum instantaneous power dissipated in the capacitor.

Section 21-4

21-7 Calculate the true power and reactive power supplied by a 240 V, 100 Hz source connected to: (a) a 200 mH inductor, (b) a 100 Ω resistor, (c) a 100 μF capacitor.

21-8 Three components, $L = 400\ \mu H$, $C = 300$ pF, and $R = 1.8$ kΩ, are connected in parallel to a 3 V, 200 kHz supply. Calculate the power (true and reactive) supplied to each component.

Section 21-5

21-9 Determine the apparent power, true power, and reactive power dissipated in a series *RL* circuit when the supply voltage is 75 V, the current is 1 A, and the phase angle is 30° lagging. Also, draw a power triangle for the circuit.

21-10 Determine the apparent power, true power, and reactive power for the circuit in Figure 21-20.

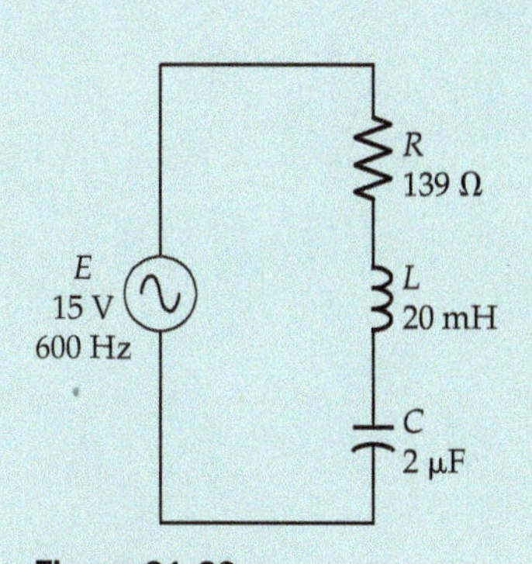

Figure 21-20

21-11 A 200 mH inductor is connected in series with a 600 Ω resistor. The ac supply voltage is 15 V with a frequency of 1 kHz. Determine the apparent power, true power, and reactive power in the circuit. Draw the power triangle.

21-12 The supply frequency for the circuit in Problem 21-11 is altered until the true power is measured as 50 mW. Calculate the new frequency, and determine the new apparent power and reactive power.

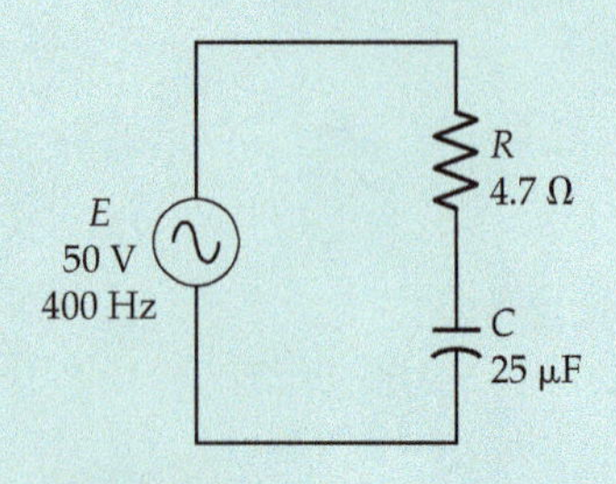

Figure 21-21

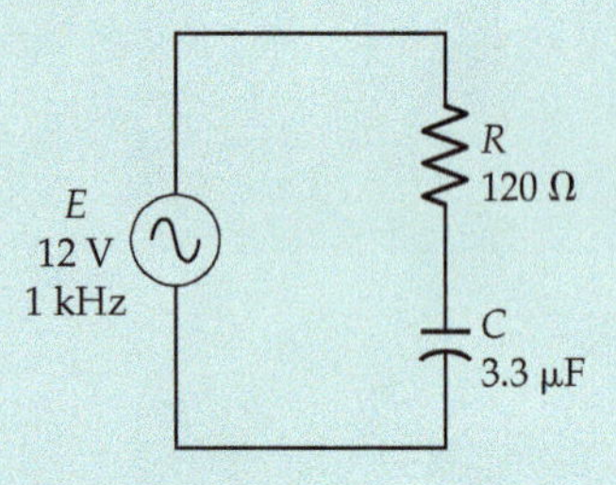

Figure 21-22

21-13 A 50 V, 400 Hz supply is applied to a load consisting of a 25 μF capacitor in series with a 4.7 Ω resistor, as in Figure 21-21. Determine the apparent power, true power, and reactive power in the circuit. Draw the power triangle.

21-14 The supply frequency for the circuit in Problem 21-13 is altered until the true power is measured as 30 W. Calculate the new frequency, and determine the new apparent power and reactive power.

21-15 Determine the apparent power, true power, and reactive power for the circuit in Figure 21-22.

21-16 Determine the apparent power, true power, and reactive power for the circuit in Figure 21-23.

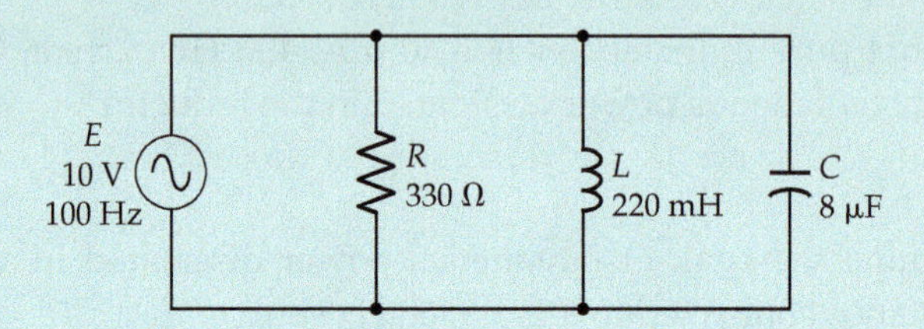

Figure 21-23

Section 21-6

21-17 The current taken by a certain load connected to a 220 V, 100 Hz supply is measured as 4 A with a power factor of 69% lagging. Determine the apparent power, true power, and reactive power. Draw the power triangle for the circuit.

21-18 For the load in Problem 21-17, calculate the parallel capacitance required to correct the power factor to 97% lagging. Draw the new power triangle for the circuit.

21-19 An inductive circuit with a 250 μF parallel-connected capacitor has a power factor of 93% lagging. The supply voltage is 120 V with a frequency of 60 Hz, and the supply current is 15 A. Determine the apparent power, true power, and reactive power.

21-20 For the circuit in Problem 21-19, determine the apparent power, true power, and reactive power when the capacitor is disconnected.

21-21 A 24 V, 400 Hz supply is connected to a 4 kW load with a 65% lagging power factor. Calculate the current that must be carried by the conductors. Also, calculate the new level of conductor current when the power factor is corrected to 85% lagging, and determine the capacitance required for power factor correction.

21-22 For Problem 21-21, calculate the capacitance required to reduce the conductor current to 185 A.

21-23 A 120 mH inductor is connected in series with a resistance of 70 Ω to a 50 V, 400 Hz supply. Determine the power dissipated in the circuit. Also, calculate the load current when a 1 μF capacitor is connected in parallel with the load.

21-24 For Problem 21-23, calculate the capacitor required to reduce the power factor to 0.87. Determine the new level of load current.

21-25 A 230 V, 120 Hz supply is connected to a 12 kW load with an 80% lagging power factor via conductors that can carry a maximum current of 58 A. Calculate the actual current level in the conductors if the load remains unaltered. Determine how the load should be changed to reduce the conductor current to 58 A. Draw the power triangle for the circuit.

21-26 For Problem 21-25, determine the conductor current that flows when a 75 μF power factor correction capacitor is employed.

Section 21-7

21-27 A wattmeter measures the ac power delivered to a load as 250 W. The supply voltage is 115 V, and the load is known to have a phase angle of $\phi = 33°$. Calculate the load current.

21-28 A wattmeter with a field coil resistance of 0.9 Ω and a moving coil resistance of 100 kΩ is designed to give FSD when the field coil current is 1 A and the moving coil current is 1 mA. (a) Calculate the maximum power that can be measured by the instrument. (b) Determine the voltage applied to the moving-coil circuit if the instrument indicates full scale when measuring the power dissipated in a 1 A load with a phase angle of 45°.

21-29 The wattmeter in Problem 21-28 has its moving coil contact moved to the supply side of the field coils, (see Figure 21-19). Determine the measurement error that results when the load current and moving coil current are 1 A and 1 mA respectively.

Practice Problem Answers

21-1.1 98 mW, 133 mW, 161 mW, 784 mW
21-2.1 106 mW, −106 mW
21-3.1 65 mW
21-4.1 25.1 mvar, 398 mvar
21-5.1 809 mH
21-5.2 4.95 mVA, 2.85 mvar, 4.05 mW
21-5.3 247 mVA, 212 mvar, 127 mW
21-6.1 0.845
21-6.2 206 A, 227 A
21-7.1 21.3°

CHAPTER 22
AC Network Analysis and Theorems

CONTENTS

Objectives

You will be able to:

1 Convert independent and dependent ac voltage sources into current sources and vice versa.
2 State Kirchhoff's voltage and current laws for impedance networks, and apply them to ac network analysis.
3 Analyze complex ac networks by use of loop equations.
4 State the superposition theorem as applied to ac networks, and use it for ac network analysis.
5 Analyze complex ac networks by the use of nodal analysis.
6 State Thévenin's theorem as applied to ac networks, and use it for ac network analysis.
7 State Norton's theorem as applied to ac networks, and use it for ac network analysis.
8 State the maximum power transfer theorem as applied to ac networks, and use it to determine optimum loads for ac sources.
9 Apply delta-wye transformations to simplify the analysis of ac networks.

INTRODUCTION

The analysis techniques and theorems employed to solve dc network problems in Chapters 8 and 9 can also be applied to ac networks. However, all impedances, voltages, and currents must be treated as phasor quantities. Consequently, there are a lot more calculations involved in analyzing an ac impedance network than for a similar dc resistance circuit. As in the case of purely resistive circuits, the calculations can be greatly simplified by application of the various network theorems.

22-1 AC VOLTAGE AND CURRENT SOURCES

AC Sources

In analyzing an ac network it is necessary to know the phase angle of each source as well as the amplitude of the voltage or current generated at each source. Also, although alternating voltages are continuously reversing polarity, positive and negative terminals must be identified at each generator. This is necessary because if a generator is connected in reverse the phase angle of the generator output is altered by 180°.

Figure 22-1(a) shows a circuit that has two voltage sources; $E_1 = 6\ \text{V}\angle 0$ and $E_2 = 12\ \text{V}\angle -20°$. The waveforms in the illustration show the phase relationship

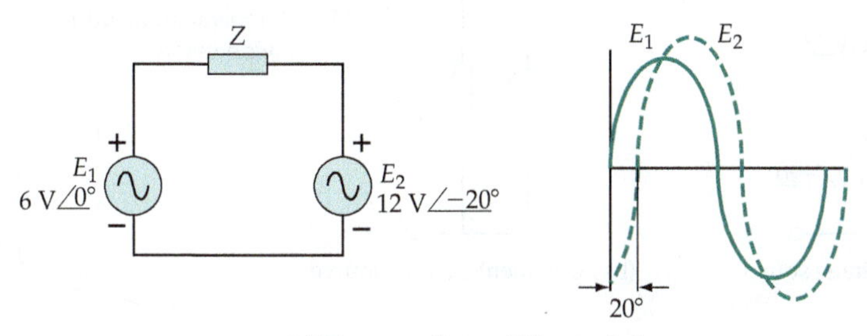

(a) Two ac voltages 20° out of phase

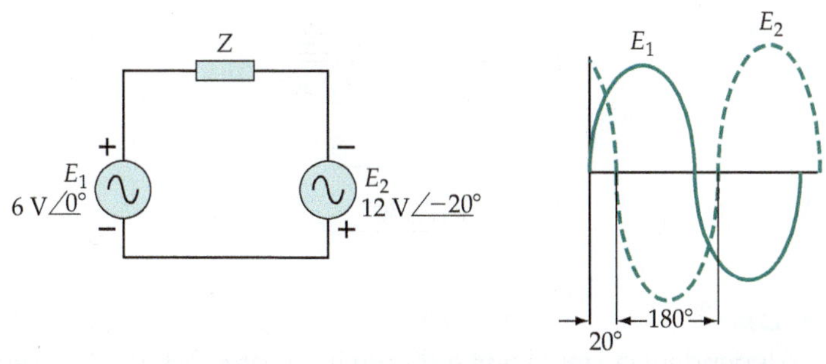

(b) E_2 connected in reverse adds a further 180° phase shift

Figure 22-1 All ac sources must have a terminal polarity identification to indicate whether the output is initially increasing in a positive direction or negative direction. This is necessary to define the phase relationships between two or more sources.

between these two sources. Because the phase angle of E_2 is $-20°$ and that of E_1 is 0°, the zero level of E_2 occurs 20° after the zero level of E_1, as illustrated. Now assume that the generator that produces E_2 is reconnected in reverse, as shown in Figure 22-1(b). E_2 still has its zero level 20° after E_1, but instead of increasing positively from zero, its output increases in the *negative* direction, as illustrated. Thus, the reversal of the output terminals has added a further 180° to the phase of E_2 with respect to E_1. It is for this reason that the terminals of all ac sources in a circuit diagram must have the polarity identified.

AC Source Conversion

The procedure for convertion between ac voltage and current sources is essentially the same as for dc sources (see Chapter 8) except that complex quantities are involved. Convertion between externally-dependent ac sources also involves the same procedures. More care is necessary in converting circuit-dependent sources, and this is further explored in Section 22-2.

Example 22-1

Convert the ac voltage source in Figure 22-2(a) into the current source in Figure 22-2(b).

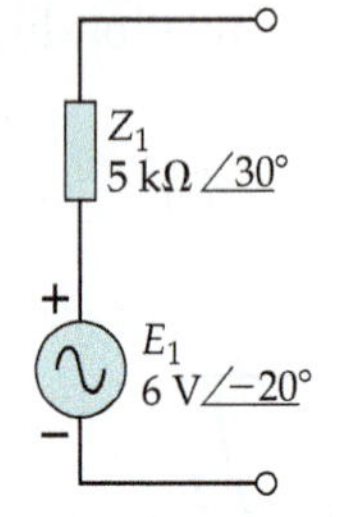

(a) AC voltage source

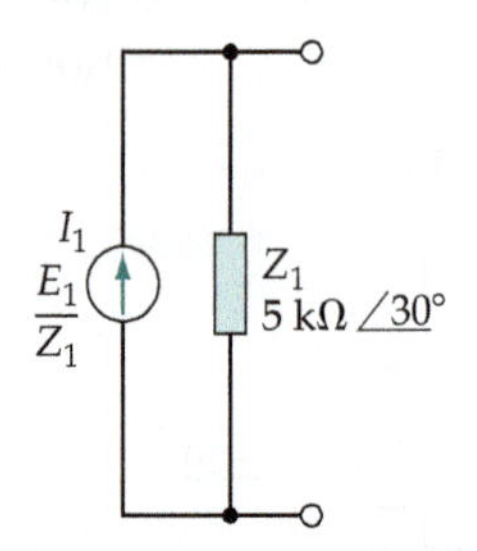

(b) Equivalent current source

Figure 22-2 An ac voltage source can be converted into an equivalent ac current source, and vice versa.

Solution

$$I_1 = \frac{E_1}{Z_1} = \frac{6\ \text{V}\angle -20°}{5\ \text{k}\Omega\angle 30°}$$

$$= 1.2\ \text{mA}\angle -50°$$

$$Z_1 = 5\ \text{k}\Omega\angle 30°$$

Example 22-2

Convert the dependent current source shown in Figure 22-3(a) into its equivalent voltage source [Figure 22-3(b)]. Assume $K = 6 \times 10^{-3}$ and $V_2 = 1.5\ \text{mV}\angle 10°$. Calculate the current that flows in a load of $Z_L = 10\ \text{k}\Omega\angle 15°$ connected to the voltage source, as illustrated in Figure 22-4.

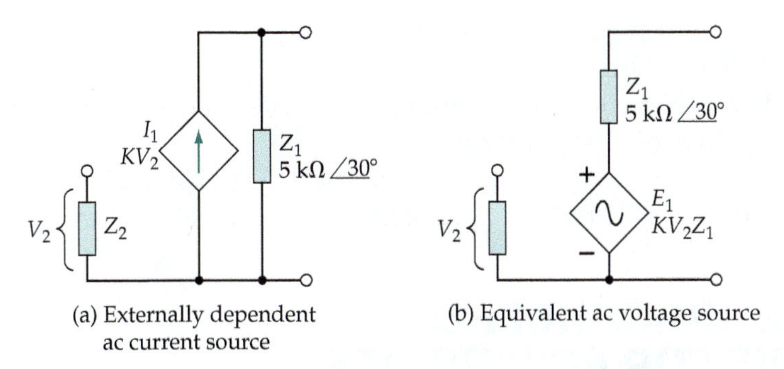

Figure 22-3 Externally dependent sources have the voltage or current controlled by a quantity outside the circuit.

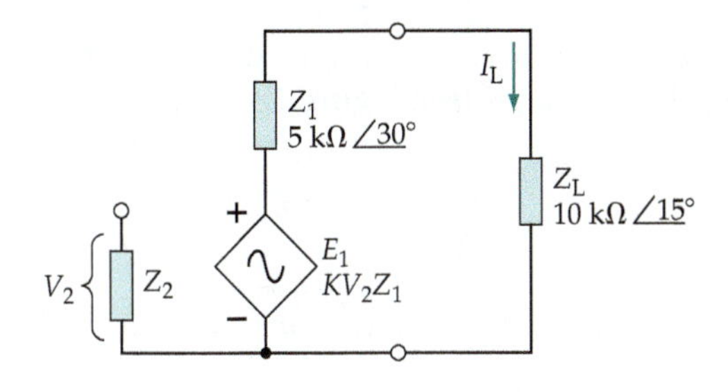

Figure 22-4 Dependent voltage source with a load impedance Z_L.

Solution

For the voltage source,

$$E_1 = I_1 Z_1 = KV_2 Z_1$$

and $Z_1 = Z_1$ for the current source [see Figure 22-3(b)]

$$E_1 = KV_2Z_1 = (6 \times 10^{-3}) \times (5\ \text{k}\Omega\angle 30^\circ) \times V_2$$

$$= 30\angle 30^\circ \times V_2 = 30\angle 30^\circ \times 1.5\ \text{mV}\angle 10^\circ$$

$$= 45\ \text{mV}\angle 40^\circ$$

$$Z_1 = 5\ \text{k}\Omega\angle 30^\circ$$

$$I_L = \frac{E_1}{Z_1 + Z_L} = \frac{45\ \text{mV}\angle 40^\circ}{5\ \text{k}\Omega\angle 30^\circ + 10\ \text{k}\Omega\angle 15^\circ}$$

$$Z_1 + Z_L = 5\ \text{k}\Omega(\cos 30^\circ + j\sin 30^\circ) + 10\ \text{k}\Omega(\cos 15^\circ + j\sin 15^\circ)$$

$$= (4.33\ \text{k}\Omega + j2.5\ \text{k}\Omega) + (9.66\ \text{k}\Omega + j2.59\ \text{k}\Omega)$$

$$= 14\ \text{k}\Omega + j5.1\ \text{k}\Omega$$

$$= 14.9\ \text{k}\Omega\angle 20^\circ$$

$$I_L = \frac{E_1}{Z_1 + Z_L} = \frac{45\ \text{mV}\angle 40^\circ}{14.9\ \text{k}\Omega\angle 20^\circ}$$

$$= 3.02\ \mu\text{A}\angle 20^\circ$$

Practice Problems

22-1.1 For the current source in Example 22-2, calculate the current that flows in a load of $Z_L = 10\ \text{k}\Omega\angle 15^\circ$.

22-1.2 A dependent voltage source consists of $Z_S = 33\ \text{k}\Omega\angle 20^\circ$ and $E_S = 5\ V_2$, where $V_2 = 150\ \text{mV}\angle -45^\circ$. Determine the equivalent current source.

22-2 KIRCHHOFF'S LAWS FOR AC CIRCUITS

Kirchhoff's Voltage Law for AC Circuits

In any closed electric circuit, the phasor sum of the voltage drops must equal the phasor sum of the applied voltages [see Figure 22-5(a)].

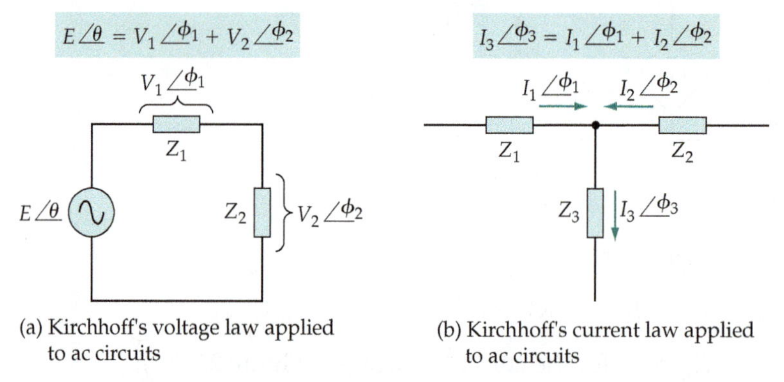

(a) Kirchhoff's voltage law applied to ac circuits

(b) Kirchhoff's current law applied to ac circuits

Figure 22-5 Application of Kirchhoff's voltage and current laws to ac circuits.

The statement differs from that made for dc networks (see Section 5-2) only in that it refers to the *phasor sum* of the voltages instead of the *algebraic sum*. The same remarks apply to the currents referred to in Kirchhoff's Current Law for ac circuits (Section 6-1 for dc).

Kirchhoff's Current Law for AC Circuits

The phasor sum of the currents entering a point in an electric circuit must equal the phasor sum of the currents leaving that point [see Figure 22-5(b)].

It is obvious that when solving ac networks, the phase angles of all impedances and of all voltages and currents must be carefully considered. When two quantities are to be multiplied or divided, they should be stated in polar form. When they are to be added or subtracted, they must be converted into rectangular form.

Example 22-3

Use Kirchhoff's laws to derive an equation for the voltage V_3 in the circuit shown in Figure 22-6.

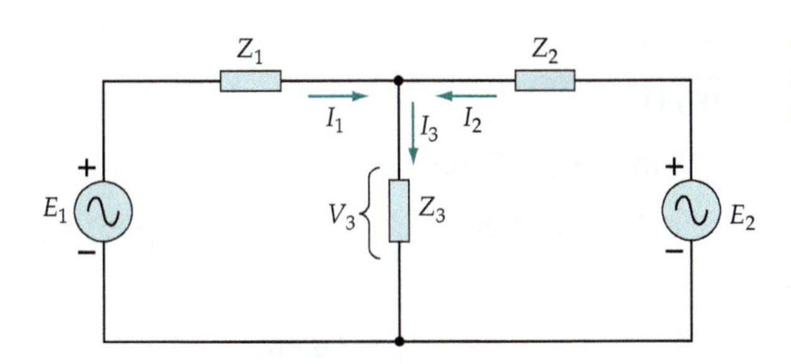

Figure 22-6 Circuit with independent voltage sources, analyzed in Examples 22-3 and 22-4.

Solution

$$I_3 = I_1 + I_2$$

$$\frac{V_3}{Z_3} = \frac{E_1 - V_3}{Z_1} + \frac{E_2 - V_3}{Z_1}$$

$$\frac{V_3}{Z_3} = \frac{E_1}{Z_1} - \frac{V_3}{Z_1} + \frac{E_2}{Z_2} - \frac{V_3}{Z_2}$$

$$V_3\left[\frac{1}{Z_1} + \frac{1}{Z_2} + \frac{1}{Z_3}\right] = \frac{E_1}{Z_1} + \frac{E_2}{Z_2}$$

$$V_3 = \frac{E_1/Z_1 + E_2/Z_2}{1/Z_1 + 1/Z_2 + 1/Z_3}$$

Example 22-4

Substitute the voltages and component values into the equation derived in Example 22-3 to calculate V_3 in Figure 22-6. The quantities are $Z_1 = 100\ \Omega\angle 25°$, $Z_2 = 180\ \Omega\angle 60°$, $Z_3 = 120\ \Omega\angle -30°$, $E_1 = 6\ \text{V}\angle 0°$, and $E_2 = 12\ \text{V}\angle -20°$.

Solution

$$1/Z_1 = 1/(100\ \Omega\angle 25°) = 10\ \text{mS}\angle -25°$$

$$= 9\ \text{mS} - j4.2\ \text{mS}$$

$$1/Z_2 = 1/(180\ \Omega\angle 60°) = 5.6\ \text{mS}\angle -60°$$

$$= 2.8\ \text{mS} - j4.8\ \text{mS}$$

$$1/Z_3 = 1/(120\ \Omega\angle -30°) = 8.3\ \text{mS}\angle 30°$$

$$= 7.2\ \text{mS} + j4.2\ \text{mS}$$

$$1/Z_1 + 1/Z_2 + 1/Z_3 = 19\ \text{mS} - j4.8\ \text{mS}$$

$$= 19.6\ \text{mS}\angle -14.2°$$

$$\frac{E_1}{Z_1} = \frac{6\ \text{V}\angle 0°}{100\ \Omega\angle 25°} = 60\ \text{mA}\angle -25°$$

$$= 54.4\ \text{mS} - j25.4\ \text{mA}$$

$$\frac{E_2}{Z_2} = \frac{12\text{ V}\angle -20^\circ}{180\ \Omega\angle 60^\circ} = 66.7\text{ mA}\angle -80^\circ$$
$$= 11.6\text{ mA} - j65.7\text{ mA}$$
$$E_1/Z_1 + E_2/Z_2 = 66\text{ mA} - j91\text{ mA}$$
$$= 112.4\text{ mA}\angle -54^\circ$$
$$V_3 = \frac{E_1/Z_1 + E_2/Z_2}{1/Z_1 + 1/Z_2 + 1/Z_3} = \frac{112.4\text{ mA}\angle -54^\circ}{19.6\text{ mA}\angle -14.2^\circ}$$
$$= 5.7\text{ V}\angle -39.8^\circ$$

The circuit in Figure 22-6 (analyzed in Examples 22-3 and 22-4) uses independent sources. If one of the sources is an externally dependent source (KV_4), as illustrated in Figure 22-7(a), then KV_4 is simply substituted in place of E_1 in the equation derived for V_3. This cannot be done when E_1 is an internally dependent source (dependent, for example, on the circuit voltage V_3) as shown in Figure 22-7(b). Instead, KV_3 must be used when deriving an equation for any current or voltage within the circuit. The process is demonstrated in Example 22-5.

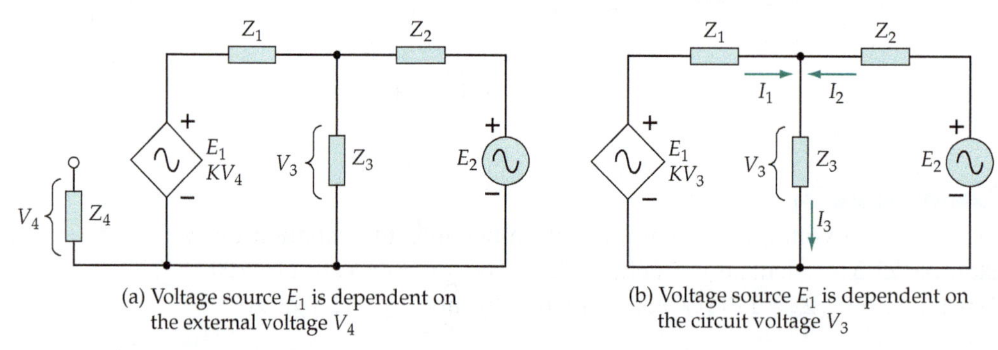

(a) Voltage source E_1 is dependent on the external voltage V_4

(b) Voltage source E_1 is dependent on the circuit voltage V_3

Figure 22-7 Circuits with externally dependent and circuit-dependent sources.

Example 22-5

Use Kirchhoff's laws to derive an equation for the voltage V_3 in the circuit in Figure 22-7(b). Note that E_1 is a dependent voltage source.

Solution

$$I_3 = I_1 + I_2$$
$$\frac{V_3}{Z_3} = \frac{KV_3 - V_3}{Z_1} + \frac{E_2 - V_3}{Z_1}$$
$$\frac{V_3}{Z_3} = \frac{KV_3}{Z_1} - \frac{V_3}{Z_1} + \frac{E_2}{Z_2} - \frac{V_3}{Z_2}$$
$$V_3\left[\frac{1}{Z_1} + \frac{1}{Z_2} + \frac{1}{Z_3} - \frac{K}{Z_1}\right] = \frac{E_2}{Z_2}$$
$$V_3 = \frac{E_2/Z_2}{1/Z_1 + 1/Z_2 + 1/Z_3 - K/Z_1}$$

Practice Problems

22-2.1 The voltage source E_1 and series impedance Z_1 in Figure 22-6 are replaced with the equivalent current source I_1 and Z_1. Derive an equation for the voltage V_3.

22-2.2 If I_1 in Problem 22-2.1 is a dependent source KI_{Z3}, derive the new equation for V_3.

22-2.3 The quantities for the circuit in problem 22-2.2 are $Z_1 = 150\ \Omega\angle 35°$, $Z_2 = 140\ \Omega\angle -50°$, $Z_3 = 95\ \Omega\angle 45°$, $E_2 = 10\ \text{V}\angle 33°$, and $K = 1.5$. Calculate V_3.

22-3 AC CIRCUIT LOOP EQUATIONS (MESH EQUATIONS)

The procedure for analysis of ac networks by loop equations (or mesh equations) is exactly the same as that for dc networks (see Chapter 8), with the exception that impedances are being dealt with instead of resistances. Loop currents are first identified as for dc circuits, usually in a clockwise direction (see Figure 22-8). When the analysis is complete, those branch currents that turn out as positive quantities are (instantaneously) in the same direction as that selected for the loop currents. The branch currents that have negative signs have an additional 180° phase shift in relation to the loop currents.

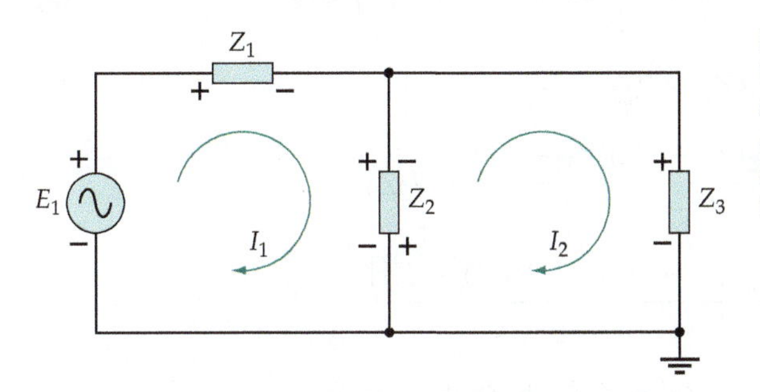

Figure 22-8 In the loop equation (or mesh equation) method of network analysis, the voltage drops around each circuit loop are equated to zero. For loop 1, $0 = I_1Z_1 + I_1Z_3 - I_2Z_3 - E_1$.

Procedure for AC Network Analysis by Loop Equations

1. *Convert all current sources into voltage sources.*
2. *Draw all loop currents in a clockwise direction and identify them by number.*
3. *Identify all impedance voltage drops as + to − in the direction of the loop current.*
4. *Identify all voltage sources with their correct polarity.*
5. *Write the equations for the voltage drops around each loop.*
6. *Solve the equations to find an equation for the required branch current.*
7. *Reduce the equations to the simplest possible form before substituting the impedance quantities into each equation.*

Because of the necessity of converting from polar to rectangular form, and vice versa, during calculations, there is much more work involved in analyzing an ac network than for a similar dc network. Consequently, the possibility of calculation errors is increased many times. To minimize the calculations, the equations for the unknown quantities should be reduced to their very simplest state. The following examples illustrate the application of loop equations to ac network analysis.

Example 22-6

Use the loop equation method to derive an equation for the current through impedance Z_3 in the circuit shown in Figure 22-8.

Solution

For loop 1: $$0 = I_1Z_1 + I_1Z_2 - I_2Z_2 - E_1$$

or, $$E_1 = I_1(Z_1 + Z_2) - I_2Z_2 \quad \textbf{(1)}$$

For loop 2: $$0 = I_2Z_3 + I_2Z_2 - I_1Z_2$$

or, $$0 = I_2(Z_2 + Z_3) - I_1Z_2 \quad \textbf{(2)}$$

Current I_2 is to be determined. So, Equations (1) and (2) should be manipulated to eliminate I_1.

Equation (2) $\times$ $(Z_1 + Z_2)/Z_2$:

$$0 = \frac{I_2(Z_2 + Z_3)(Z_1 + Z_2)}{Z_2} - I_1(Z_1 + Z_2) \quad \textbf{(3)}$$

Eq. (1) $$E_1 = I_1(Z_1 + Z_2) - I_2Z_2$$

(1) + (3), $$E_1 = \frac{I_2(Z_2 + Z_3)(Z_1 + Z_2)}{Z_2} - I_2Z_2$$

$$= \frac{I_2[(Z_2 + Z_3)(Z_1 + Z_2) - Z_2^2]}{Z_2}$$

$$= \frac{I_2[Z_1Z_2 + Z_1Z_3 + Z_2^2 + Z_2Z_3 - Z_2^2]}{Z_2}$$

$$= \frac{I_2[Z_1Z_2 + Z_1Z_3 + Z_2Z_3]}{Z_2}$$

$$I_{Z3} = I_2 = \frac{E_1Z_2}{Z_1Z_2 + Z_1Z_3 + Z_2Z_3}$$

Example 22-7

The quantities for the circuit in Example 22-6 are $E_1 = 10\ \text{V}\angle 0°$, $Z_1 = 100\ \Omega\angle 15°$, $Z_2 = 70\ \Omega\angle 30°$, $Z_3 = 50\ \Omega\angle 45°$. Calculate the level of the current I_{Z3}.

Solution

$$Z_1Z_2 = 100\ \Omega\angle 15^\circ \times 70\ \Omega\angle 30^\circ$$
$$= 7\ \text{k}\angle 45^\circ = 4.95\ \text{k} + j4.95\ \text{k}$$
$$Z_1Z_3 = 100\ \Omega\angle 15^\circ \times 50\Omega\angle 45^\circ$$
$$= 5\ \text{k}\angle 60^\circ = 2.5\ \text{k} + j4.33\ \text{k}$$
$$Z_2Z_3 = 70\ \Omega\angle 30^\circ \times 50\ \Omega\angle 45^\circ$$
$$= 3.5\ \text{k}\angle 75^\circ = 906 + j3.38\ \text{k}$$
$$Z = Z_1Z_2 + Z_1Z_3 + Z_2Z_3$$
$$= (4.95\ \text{k} + 2.5\ \text{k} + 906) + j(4.95\ \text{k} + 4.33\ \text{k} + 3.38\ \text{k})$$
$$= 8.36\ \text{k} + j12.66\ \text{k} = 15.17\ \text{k}\angle 56.6^\circ$$
$$I_{Z3} = \frac{E_1Z_2}{Z_1Z_2 + Z_1Z_3 + Z_2Z_3} = \frac{10\ \text{V}\angle 0 \times 70\angle 30^\circ}{15.17\ \text{k}\angle 56.6^\circ}$$
$$= 41.4\ \text{mA}\angle -26.6^\circ$$

Treatment of ac network analysis by computer is offered in Section 28-6.

Example 22-8

Using loop equations, analyze the impedance network shown in Figure 22-9 to derive an equation for the current through Z_3.

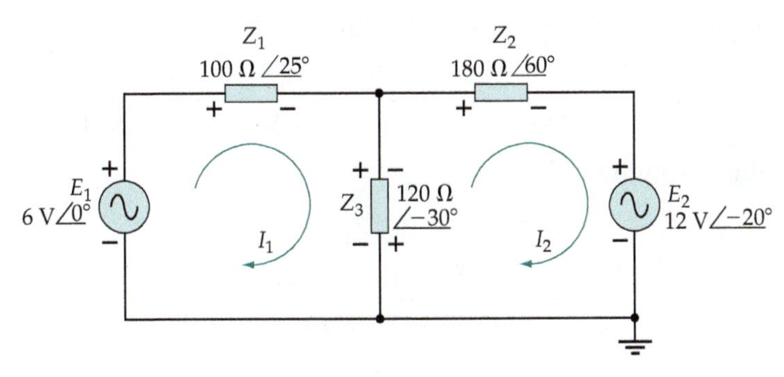

Figure 22-9 Circuit for Examples 22-8 and 22-9.

Solution

Loop currents I_1 and I_2 are drawn clockwise, as shown in the circuit diagram, and voltage drops are identified as + to − in the direction of the loop current.

For loop 1: $$0 = I_1Z_1 + I_1Z_3 - I_2Z_3 - E_1$$
$$E_1 = I_1(Z_1 + Z_3) - I_2Z_3 \qquad \textbf{(1)}$$

For loop 2: $$0 = I_2Z_2 + E_2 + I_2Z_3 - I_1Z_3$$

Giving, $$-E_2 = I_2(Z_2 + Z_3) - I_1Z_3 \qquad \textbf{(2)}$$

To make the multiple of I_2 in Equation (1) the same as in Equation (2), Equation (1) $\times$ $(Z_2 + Z_3)/Z_3$ gives

$$\frac{E_1(Z_2 + Z_3)}{Z_3} = \frac{I_1(Z_1 + Z_3)(Z_2 + Z_3)}{Z_3} - I_2(Z_2 + Z_3) \tag{3}$$

Adding Equation (2) to Equation (3) to eliminate I_2 gives

$$\frac{E_1(Z_2 + Z_3)}{Z_3} - E_2 = I_1\left[\frac{(Z_1 + Z_3)(Z_2 + Z_3)}{Z_3} - Z_3\right] \tag{4}$$

Multiplying through by Z_3

$$E_1(Z_2 + Z_3) - E_2Z_3 = I_1[(Z_1 + Z_3)(Z_2 + Z_3) - Z_3^2]$$

and

$$I_1 = \frac{E_1(Z_2 + Z_3) - E_2Z_3}{(Z_1 + Z_3)(Z_2 + Z_3) - Z_3^2}$$

$$= \frac{E_1Z_2 + E_1Z_3 - E_2Z_3}{Z_1Z_2 + Z_1Z_3 + Z_2Z_3 + Z_3^2 - Z_3^2}$$

$$= \frac{E_1Z_2 + E_1Z_3 - E_2Z_3}{Z_1Z_2 + Z_1Z_3 + Z_2Z_3} \tag{5}$$

To make the multiple of I_1 in Equation (2) the same as in Equation (1), Equation (2) $\times$ $(Z_1 + Z_3)/Z_3$ gives

$$\frac{-E_2(Z_1 + Z_3)}{Z_3} = \frac{I_2(Z_2 + Z_3)(Z_1 + Z_3)}{Z_3} - I_1(Z_1 + Z_3) \tag{6}$$

Adding Equation (1) to Equation (6) to eliminate I_1 gives,

$$E_1 - \frac{E_2(Z_1 + Z_3)}{Z_3} = I_2\left[\frac{(Z_2 + Z_3)(Z_1 + Z_3)}{Z_3} - Z_3\right]$$

Multiplying through by Z_3:

$$E_1Z_3 - E_2(Z_1 + Z_3) = I_2[(Z_2 + Z_3)(Z_1 + Z_3) - Z_3^2]$$

and,

$$I_2 = \frac{E_1Z_3 - E_2(Z_1 + Z_3)}{(Z_2 + Z_3)(Z_1 + Z_3) - Z_3^2}$$

$$= \frac{E_1Z_3 - E_2Z_1 - E_2Z_3}{Z_1Z_2 + Z_2Z_3 + Z_1Z_3 + Z_3^2 - Z_3^2}$$

$$= \frac{E_1Z_3 - E_2Z_1 - E_2Z_3}{Z_1Z_2 + Z_1Z_3 + Z_2Z_3} \tag{7}$$

Determining I_3, $I_3 = I_1 - I_2$

$$= \frac{E_1Z_2 + E_1Z_3 - E_2Z_3 - E_1Z_3 + E_2Z_1 + E_2Z_3}{Z_1Z_2 + Z_1Z_3 + Z_2Z_3}$$

$$= \frac{E_1Z_2 + E_2Z_1}{Z_1Z_2 + Z_1Z_3 + Z_2Z_3} \tag{8}$$

The equation for the current through Z_3 is stated in its simplest possible form. The component values for the equation can now be calculated separately, and then used for determination of I_3.

Example 22-9

Substitute the component values into the equation derived in Example 22-8 to calculate the current through Z_3 in Figure 22-9.

Solution

$$Z_1 = 100\ \Omega\angle 25^\circ = 90.6\ \Omega + j42.3\ \Omega$$

$$Z_2 = 180\ \Omega\angle 60^\circ = 90\ \Omega + j156\ \Omega$$

$$Z_3 = 120\ \Omega\angle -30^\circ = 104\ \Omega - j60\ \Omega$$

$$E_1 = 6\ \text{V}\angle 0^\circ = 6\ \text{V} + j0$$

$$E_2 = 12\ \text{V}\angle 20^\circ = 11.3\ \text{V} - j4.1\ \text{V}$$

$$Z_1Z_2 = 18\ \text{k}\angle 85^\circ = 1.57\ \text{k} + j17.9\ \text{k}$$

$$Z_1Z_3 = 12\ \text{k}\angle -5^\circ = 11.95\ \text{k} - j1.05\ \text{k}$$

$$Z_2Z_3 = 21.6\ \text{k}\angle 30^\circ = 18.7\ \text{k} + j10.8\ \text{k}$$

$$Z_1Z_2 + Z_1Z_3 + Z_2Z_3 = 32.22\ \text{k} + j27.65\ \text{k}$$

$$= 42.46\ \text{k}\angle 40.6^\circ$$

$$E_1Z_2 = 1080\angle 60^\circ = 540 + j935$$

$$E_2Z_1 = 1200\angle 5^\circ = 1195 + j105$$

$$E_1Z_2 + E_2Z_1 = 1735 + j1040$$

$$= 2023\angle 30.9^\circ$$

Substituting into Equation (8),

$$I_3 = \frac{2023\angle 30.9^\circ}{42.46 \times 10^3\angle 40.6^\circ}$$

$$= 47.6\ \text{mA}\angle -9.7^\circ$$

Although the circuit in Figure 22-9 is not an extremely complex network, the analysis of it (in Examples 22-8 and 22-9) is quite lengthy. In following sections, it is shown that the analysis can be simplified by application of one of the circuit theorems. In general, the solution of more complex networks requires use of the circuit theorems.

Practice Problems

22-3.1 For the circuit in Figure 22-9, use loop equations to determine I_2 when $E_2 = 0$.

22-3.2 For the circuit in Figure 22-9, use loop equations to determine I_1 when $E_1 = 0$.

22-4 SUPERPOSITION THEOREM APPLIED TO AC NETWORKS

The superposition theorem allows a complex network involving several voltage and/or current sources to be reduced to several simpler networks each of which has only one voltage or current source.

Superposition Theorem for AC Circuits

> **In a network containing more than one source of voltage or current, the current through any branch is the phasor sum of the currents produced by each source acting independently.**

As illustrated in Figure 22-10, the current (I_3) through impedance Z_3 is determined by first calculating I_c for E_1 acting alone [Figure 22-10(b)]. Then I_d is calculated for E_2 acting alone [Figure 22-10(c)]. The current through Z_3 is the phasor sum of I_c and I_d.

Procedure for Determining AC Network Branch Currents by Use of the Superposition Theorem

1. ***Select one source and replace all other sources with their internal impedances.***
2. ***Determine the amplitude and phase angle of the current that flows through the desired branch as a result of the single source acting alone.***
3. ***Repeat steps 1 and 2 using each source in turn until the branch current components have been calculated for all sources.***
4. ***Determine the phasor sum of the current components to obtain the actual branch current.***

Example 22-10

Using the superposition theorem, analyze the impedance network in Figure 22-10(a) (reproduced from Figure 22-9) to determine the current through Z_3.

Solution

From Figure 22-10(b),

$$I_a = \frac{E_1}{Z_1 + Z_2 \| Z_3} = \frac{E_1}{Z_1 + (Z_2 Z_3)/(Z_2 + Z_3)}$$

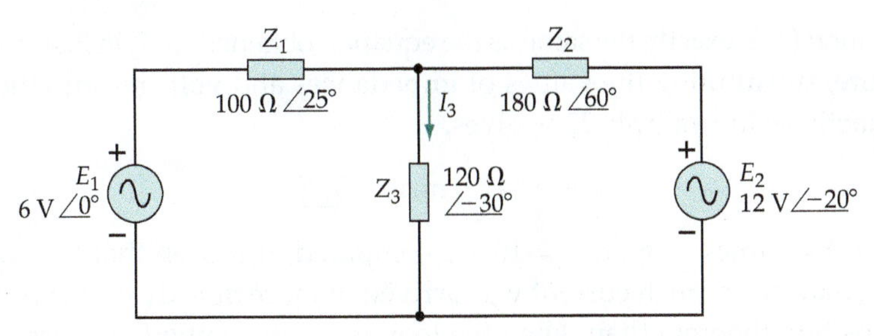

(a) Circuit to be analyzed (reproduced from Figure 22-9)

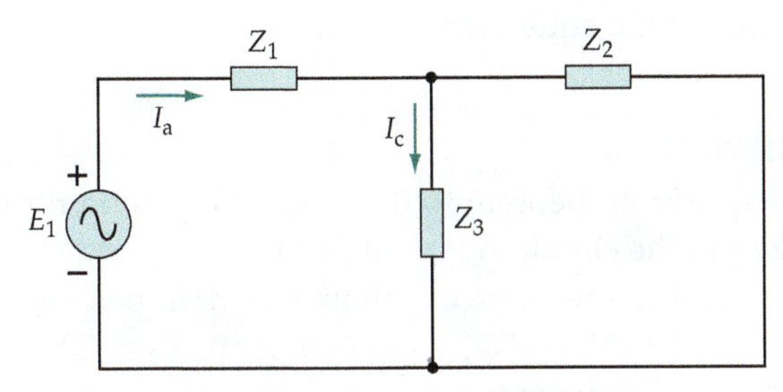

(b) Source E_1 operating alone

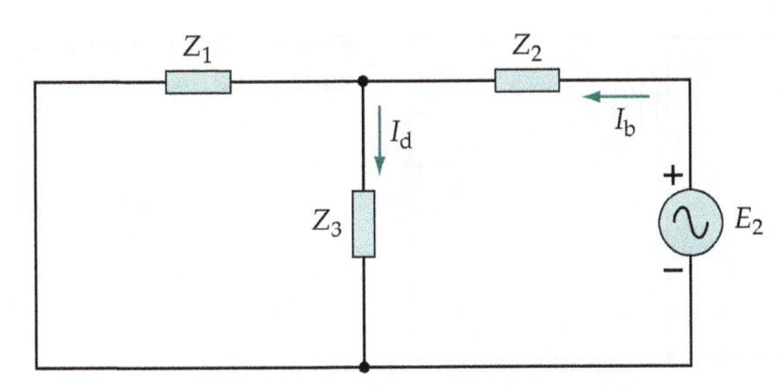

(c) Source E_2 operating alone

Figure 22-10 **Application of the superposition theorem to determine the current through impedance Z_3. Current levels (I_c and I_d) are determined for each source operating alone, and then the levels are added in rectangular form.**

By the current-divider rule,

$$I_c = I_a \frac{Z_2}{Z_2 + Z_3} = \frac{Z_2 E_1}{(Z_2 + Z_3)[Z_1 + (Z_2 Z_3)/(Z_2 + Z_3)]}$$

$$= \frac{E_1 Z_2}{Z_1 Z_2 + Z_1 Z_3 + Z_2 Z_3} \tag{1}$$

From Figure 22-10(c),

$$I_b = \frac{E_2}{Z_2 + (Z_1 Z_3)/(Z_1 + Z_3)}$$

$$I_d = I_b \frac{Z_1}{Z_1 + Z_3} = \frac{E_2 Z_1}{Z_1 Z_2 + Z_1 Z_3 + Z_2 Z_3} \tag{2}$$

$$I_3 = I_c + I_d$$

$$I_3 = \frac{E_1 Z_2 + E_2 Z_1}{Z_1 Z_2 + Z_1 Z_3 + Z_2 Z_3} \tag{3}$$

Equation (3) is exactly the same as the equation obtained for I_3 in Example 22-8. Therefore, substituting the values of impedances and voltages into the equation, exactly as in Example 22-9, gives,

$$I_3 = 47.6 \text{ mA}\angle{-9.7°}$$

When Examples 22-8 and 22-10 are compared, it is seen that the equation for the unknown branch current was arrived at more quickly by the use of the superposition theorem than when the loop currents method was used. However, in the case of a more complex circuit, the superposition theorem may not simplify the derivation of the equations.

Practice Problems

22-4.1 Use the superposition theorem to develop an equation for the current from source E_2 in the circuit in Figure 22-11.

22-4.2 For Problem 22-4.1, the source voltages and impedances are $E_1 = 100 \text{ mV}\angle 0$, $E_2 = 70 \text{ mV}\angle{-38°}$, $Z_1 = 600\ \Omega\angle 0$, $Z_2 = 200\ \Omega\angle 90°$, and $Z_L = 238\ \Omega\angle{-45°}$. Calculate the current from source E_2.

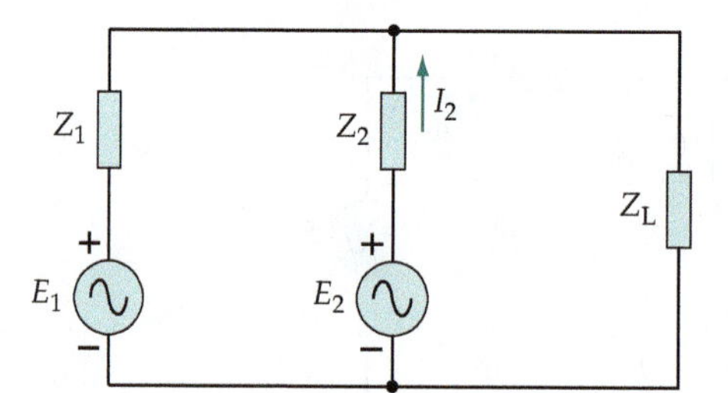

Figure 22-11 Circuit for Problems 22-4.1 and 22-4.2.

22-5 NODAL ANALYSIS FOR AC CIRCUITS

The procedure for nodal analysis of an impedance network is exactly the same as that for a dc resistance network, with the exception that, once again, the phase angle of every quantity must be taken into consideration (see Figure 22-12).

Procedure for Nodal Analysis of AC Networks

1. *Convert all voltage sources into current sources and redraw the circuit.*
2. *Identify all nodes and choose a reference node.*
3. *Write the equations for the currents flowing into and out of each node, with the exception of the reference node.*
4. *Solve the equations to determine the node voltage and the required branch currents.*

Example 22-11

Using nodal analysis on the impedance network shown in Figure 22-12(a), determine the current through Z_3.

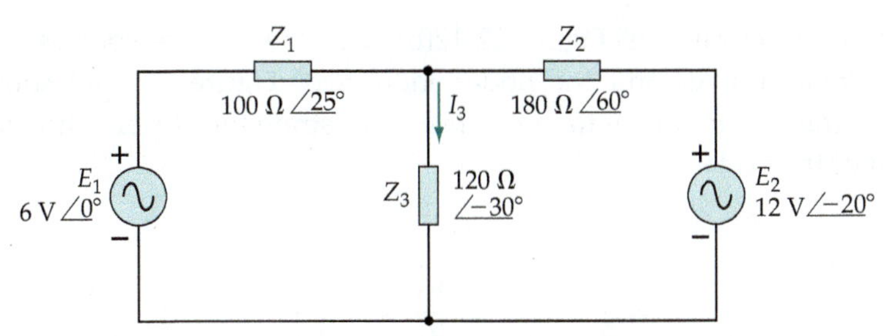

(a) Circuit to be analyzed (reproduced from Figure 22-9)

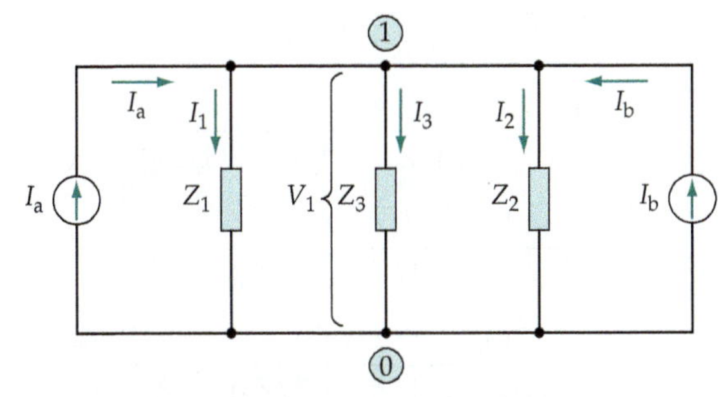

(b) Voltage sources replaced by current sources

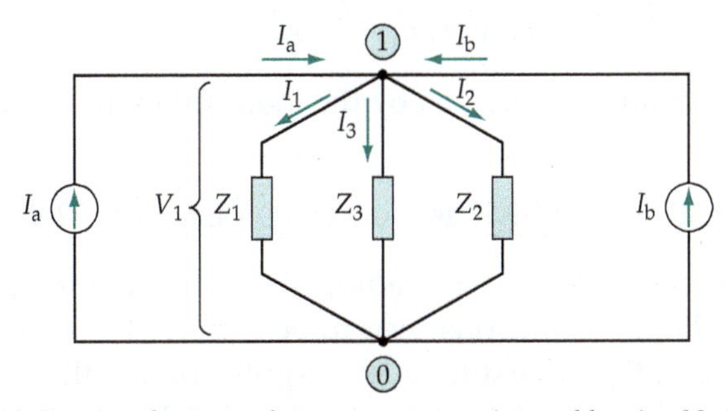

(c) Circuit redrawn to show currents entering and leaving Node 1

Figure 22-12 Nodal analysis of an impedance network. All voltage sources are first replaced with current sources, and the voltage at each node is identified with respect to ground (node zero). Circuit equations are written to determine the branch currents and node voltages.

Solution

Converting the voltage sources to current sources

$$I_a = \frac{E_1}{Z_1} = \frac{6\text{ V}\angle 0^\circ}{100\ \Omega\angle 25^\circ}$$

$$= 60\text{ mA}\angle -25^\circ = 54.4\text{ mA} - j25.4\text{ mA}$$

$$I_b = \frac{E_2}{Z_2} = \frac{12\text{ V}\angle -20^\circ}{180\ \Omega\angle 60^\circ}$$

$$= 66.7\text{ mA}\angle -80^\circ = 11.58\text{ mA} - j65.69\text{ mA}$$

The circuit is redrawn in Figure 22-12(b) with the voltage sources replaced with current sources and the nodes identified. Figure 22-12(c) shows the circuit redrawn once again to more clearly illustrate the currents flowing into and out of the nodes.

$$I_a + I_b = I_1 + I_2 + I_3 \quad (1)$$

$$= \frac{V_1}{Z_1} + \frac{V_1}{V_2} + \frac{V_1}{Z_3} = V_1\left(\frac{1}{Z_1} + \frac{1}{Z_2} + \frac{1}{Z_3}\right)$$

or,

$$V_1 = \frac{I_a + I_b}{1/Z_1 + 1/Z_2 + 1/Z_3} \quad (2)$$

and,

$$I_3 = \frac{V_1}{Z_3} = \frac{I_a + I_b}{Z_3(1/Z_1 + 1/Z_2 + 1/Z_3)} \quad (3)$$

$$= \frac{E_1/Z_1 + E_2/Z_2}{Z_3(1/Z_1 + 1/Z_2 + 1/Z_3)}$$

$$= \frac{E_1/Z_1 + E_2/Z_2}{Z_3(1/Z_1 + 1/Z_2 + 1/Z_3)} \times \frac{Z_1Z_2Z_3}{Z_1Z_2Z_3}$$

Giving,

$$I_3 = \frac{E_1Z_2 + E_2Z_1}{Z_1Z_2 + Z_1Z_3 + Z_2Z_3}$$

This is the same equation as derived by the loop equation method in Example 22-8.

So, $I_3 = 47.6\text{ mA}\angle -9.7^\circ$ as calculated in Example 22-9.

The nodal analysis method of finding the current through Z_3 in the circuit shown in Figure 22-12(a) is seen to be simpler than the use of loop currents and around the same difficulty as application of the superposition theorem. Again, this method may not be the simplest for more complex circuits.

Practice Problem

22-5.1 Use nodal analysis to develop an equation for the current from source E_1 in the circuit in Figure 22-13.

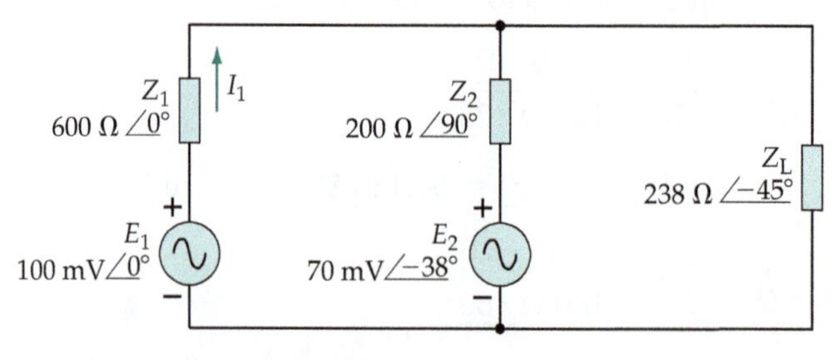

Figure 22-13 Circuit for Problem 22-5.1.

22-6 THÉVENIN'S THEOREM APPLIED TO AC CIRCUITS

Thévenin's theorem allows any single impedance in a network to be isolated. The rest of the network is replaced by a single impedance and a single voltage source. In this way the entire network is reduced to one voltage source in series with two impedances. Calculation of the current level through the desired impedance is then quite simple. Also, as in the case of dc circuit analysis by Thévenin's theorem (see Chapter 9), when the value of the impedance involved is changed there is no need for a complete new circuit analysis.

Thévenin's Theorem for AC Circuits

Any two-terminal network containing impedances and voltage and/or current sources may be replaced by a single voltage source in series with a single impedance. The emf of the voltage source is the open-circuit emf at the network terminals, and the series impedance is the impedance between the network terminals when all sources are replaced by their internal impedances.

Procedure for Thévenizing an AC Network

1. *Calculate the open-circuit terminal voltage (V_{th}) of the network.*
2. *Redraw the network with each voltage source replaced by a short-circuit in series with its internal impedance, and each current source replaced by an open-circuit in parallel with its internal impedance.*
3. *Calculate the impedance (Z_{th}) of the redrawn network as seen from the output terminals.*

To apply Thévenin's theorem to the problem of finding the current through impedance Z_3 in the circuit in Figure 22-14(a), Z_3 is first removed from the circuit, as shown in Figure 22-14(b). The open-circuit terminal voltage (V_{th}) is then calculated. Next, the circuit is redrawn with the voltage sources replaced by their internal impedances [Figure 22-14(c)], and the impedance 'looking into' the output terminals is calculated. Finally, as illustrated in Figure 22-14(d), the Thévenin's equivalent circuit is drawn in series with the impedance Z_3, and the output current is calculated.

Example 22-12

Using Thévenin's theorem, determine the current through Z_3 in the circuit shown in Figure 22-14(a).

Solution

Referring to Figure 22-14(b),

$$I = \frac{E_2 - E_1}{Z_1 + Z_2}$$

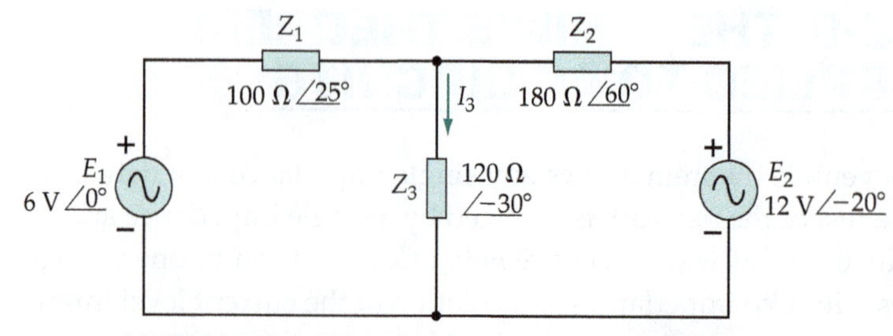

(a) Circuit to be analyzed (reproduced from Figure 22-9)

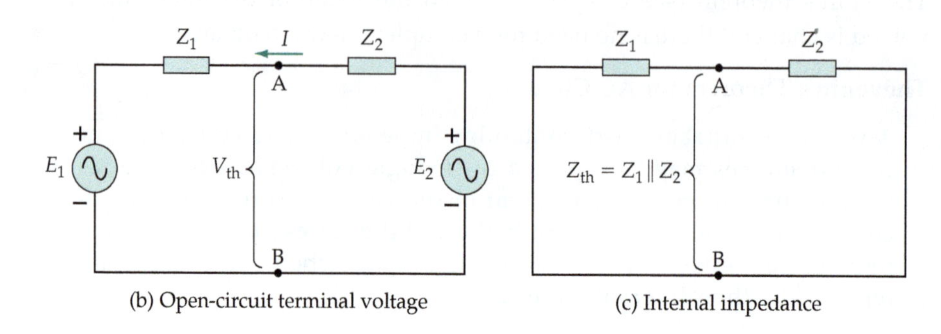

(b) Open-circuit terminal voltage

(c) Internal impedance

Thévenin circuit

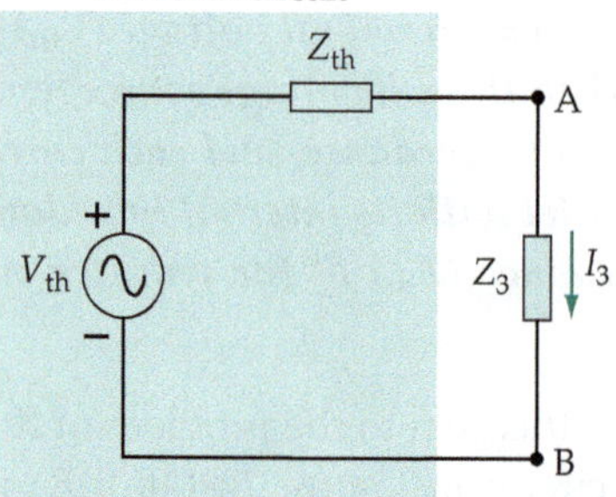

(d) Thévenin equivalent circuit and load

Figure 22-14 To Thévenize an ac network, the open-circuit terminal voltage of the network is first determined. Then, with all sources replaced with their impedances, the network internal impedance is calculated.

and,

$$V_{th} = E_1 + IZ_1 = E_1 + \frac{Z_1(E_2 - E_1)}{Z_1 + Z_2}$$

$$= \frac{E_1(Z_1 + Z_2) + Z_1(E_2 - E_1)}{Z_1 + Z_2}$$

$$= \frac{E_1 Z_2 + E_2 Z_1}{Z_1 + Z_2} \qquad \textbf{(1)}$$

From Figure 22-14(c):

$$Z_{th} = Z_1 \| Z_2 = \frac{Z_1 Z_2}{Z_1 + Z_2} \qquad \textbf{(2)}$$

and from Figure 22-14(d):

$$I_3 = \frac{V_{th}}{Z_3 + Z_{th}} \tag{3}$$

$$Z_1 = 100\ \Omega\angle 25^\circ = 90.6\ \Omega + j42.3\ \Omega$$

$$Z_2 = 180\ \Omega\angle 60^\circ = 90\ \Omega + j156\ \Omega$$

$$Z_3 = 120\ \Omega\angle -30^\circ = 104\ \Omega - j60\ \Omega$$

$$E_1 = 6\ \text{V}\angle 0 = 6\ \text{V} + j0$$

$$E_2 = 12\ \text{V}\angle -20^\circ = 11.3\ \text{V} - j4.1\ \text{V}$$

$$Z_1Z_2 = 18\ \text{k}\angle 85^\circ = 1.57\ \text{k} + j17.9\ \text{k}$$

$$Z_1 + Z_2 = 180.6\ \Omega + j193.3\ \Omega = 268.2\ \Omega\angle 47.7^\circ$$

$$E_1Z_2 = 1080\angle 60^\circ = 540 + j935$$

$$E_2Z_1 = 1200\angle 5^\circ = 1195 + j105$$

$$E_1Z_2 + E_2Z_1 = 1735 + j1040 = 2023\angle 30.9^\circ$$

Eq. (1),
$$V_{th} = \frac{E_1Z_2 + E_2Z_1}{Z_1 + Z_2} = \frac{2023\angle 30.9^\circ}{268.2\angle 47.7^\circ}$$

$$= 7.54\ \text{V}\angle -16.8^\circ$$

$$Z_{th} = \frac{Z_1Z_2}{Z_1 + Z_2} = \frac{100\ \Omega\angle 25^\circ \times 180\ \Omega\angle 60^\circ}{268.2\ \Omega\angle 47.7^\circ}$$

$$= 67.1\ \Omega\angle 37.3^\circ = 53.4\ \Omega + j40.7\ \Omega$$

$$Z_3 + Z_{th} = 157.4\ \Omega - j19.3\ \Omega = 158.6\ \Omega\angle -7^\circ$$

Eq. (3),
$$I_3 = \frac{V_{th}}{Z_3 + Z_{th}} = \frac{7.54\ \text{V}\angle -16.8^\circ}{158.6\ \Omega\angle -7^\circ}$$

$$= 47.5\ \text{mA}\angle -9.8^\circ$$

Comparing Example 22-12 to Example 22-10, it is seen that the solution of this particular problem (analysis of Figure 22-9) by Thévenin's theorem is approximately as simple (or as complex) as the use of the superposition theorem. The major advantage of the Thévenin's theorem method is that, once the Thévenin equivalent circuit is determined, new calculations can be very quickly made for new impedance values.

Practice Problem

22-6.1 Use the Thévenin's theorem to develop an equation for the load current in the circuit in Figure 22-15.

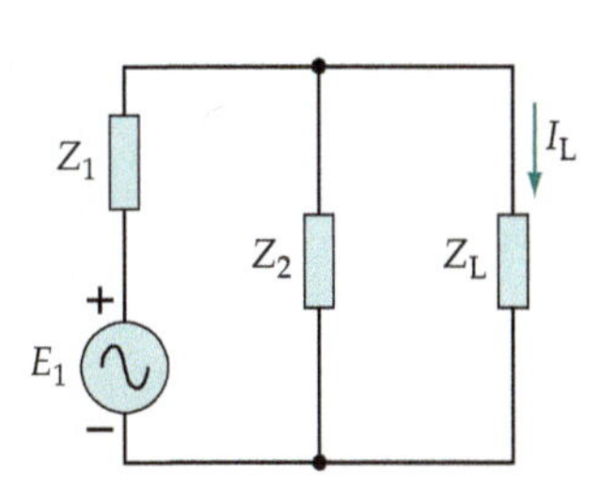

Figure 22-15 Circuit for Problem 22-6.1.

22-7 NORTON'S THEOREM APPLIED TO AC CIRCUITS

By application of Norton's theorem, any impedance in a network can be isolated and the rest of the network replaced by a single current source in parallel with a single impedance.

Norton's Theorem for AC Circuits

> **Any two-terminal network containing impedances and voltage sources, and/or current sources, may be replaced by a single current source in parallel with a single impedance. The output from the current source is the short-circuit current at the network terminals, and the parallel impedance is the impedance between the network terminals when all sources are replaced by their internal impedances.**

Procedure for Nortonizing an AC Network

1. ***Calculate the short-circuit current at the network terminals.***
2. ***Redraw the network with each voltage source replaced by a short-circuit in series with its internal impedance, and each current source replaced by an open-circuit in parallel with its internal impedance.***
3. ***Calculate the impedance of the redrawn network as seen from the output terminals.***

Example 22-13

Using Norton's theorem, determine the current through Z_3 in the circuit shown in Figure 22-16(a).

Solution

From Figure 22-16(b),

$$I_{SC} = I_1 + I_2 = \frac{E_1}{Z_1} + \frac{E_2}{Z_2}$$

From Figure 22-16(c),

$$Z_o = Z_1 \| Z_2 = \frac{Z_1 Z_2}{Z_1 + Z_2}$$

In Figure 22-16(d),

$$\begin{aligned} I_N = I_{SC} &= \frac{E_1}{Z_1} + \frac{E_2}{Z_2} = \frac{6\text{ V}\underline{/0}}{10\ \Omega\underline{/25^\circ}} + \frac{12\text{ V}\underline{/-20^\circ}}{180\ \Omega\underline{/60^\circ}} \\ &= 60\text{ mA}\underline{/-25^\circ} + 66.7\text{ mA}\underline{/-80^\circ} \\ &= (54.3\text{ mA} - j25.4\text{ mA}) + (11.58\text{ mA} - j65.7\text{ mA}) \\ &= 65.96\text{ mA} - j91.1\text{ mA} = 112.47\text{ mA}\underline{/-54.1^\circ} \end{aligned}$$

$$\begin{aligned} Z_N = Z_o &= \frac{Z_1 Z_2}{Z_1 + Z_2} \\ &= 67.1\ \Omega\underline{/37.3^\circ} = 53.4\ \Omega + j40.7\ \Omega\ (= Z_{th}\text{ from Ex. 22-12}) \end{aligned}$$

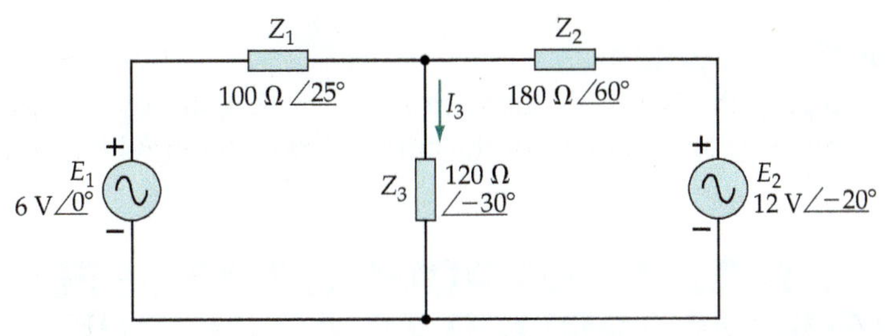

(a) Circuit to be analyzed (reproduced from Figure 22-9)

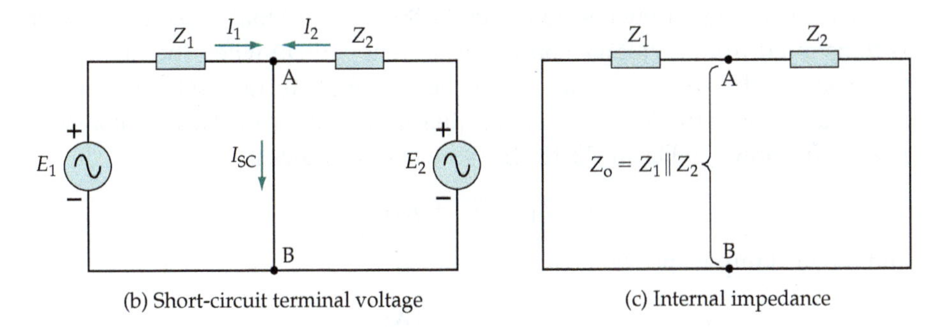

(b) Short-circuit terminal voltage

(c) Internal impedance

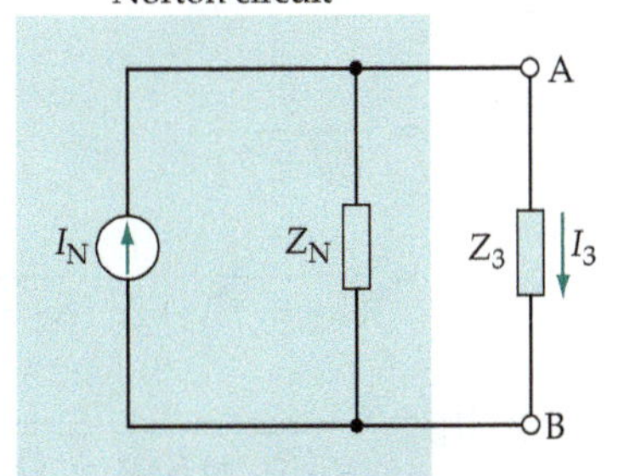

(d) Norton equivalent circuit and load

Figure 22-16 To Nortonize an ac network, the short-circuit output current is first determined. Then, with all sources replaced with their impedances, the network internal impedance is calculated.

Using the current-divider rule,

$$I_3 = \frac{I_N Z_N}{Z_3 + Z_N}$$

$$Z_3 + Z_N = 158.6\ \Omega\angle{-7^\circ}\ (= Z_3 + Z_{th} \text{ from Ex. 22-12})$$

$$I_3 = \frac{112.47\ \text{mA}\angle{-54.1^\circ} \times 67.1\ \Omega\angle{37.3^\circ}}{158.6\ \Omega\angle{-7^\circ}}$$

$$= 47.6\ \text{mA}\angle{-9.8^\circ}$$

The use of Norton's theorem for determining the current through a given impedance in a network is seen to be very similar to the Thévenin's theorem approach. Both methods produce an equivalent circuit that simplifies the calculations when the impedance involved in changed in value.

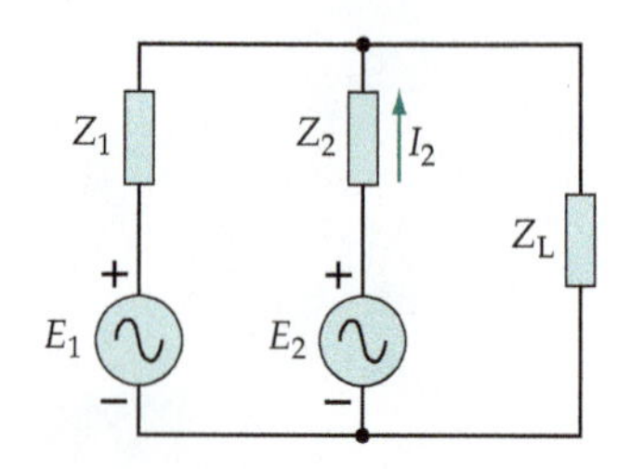

Figure 22-17 Circuit for Problem 22-7.1.

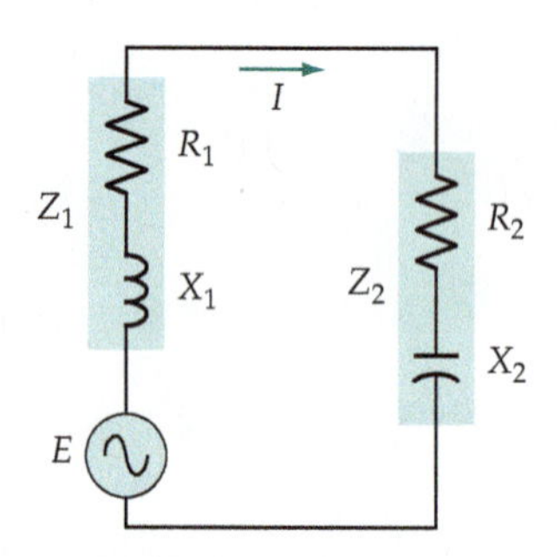

Figure 22-18 Maximum power transfer from an ac source to a load occurs when the load impedance is the conjugate of the network output impedance. That is, when $X_2 = -X_1$ and $R_2 = R_1$.

Practice Problem

22-7.1 Apply the Norton's theorem to the circuit shown in Figure 22-17 to develop an equation for the current through impedance Z_2.

22-8 MAXIMUM POWER TRANSFER THEOREM APPLIED TO AC CIRCUITS

When considering purely resistive circuits, it was found that maximum power is transferred from a voltage source (or current source) when the load resistance equals the source resistance (see Section 9-5). To understand how the maximum power transfer theorem applies to impedance networks, consider the circuit shown in Figure 22-18. The source impedance is,

$$Z_1 = R_1 + jX_1$$

and the load impedance is,

$$Z_2 = R_2 - jX_2$$

The output current is,

$$I = \frac{E}{Z_1 + Z_2}$$

$$= \frac{E}{(R_1 + jX_1) + (R_2 - jX_2)}$$

or,

$$|I| = \frac{E}{\sqrt{(R_1 + R_2)^2 + (X_1 - X_2)^2}}$$

The power dissipation in the load impedance occurs in the resistive portion of the load (R_2).

Thus,

$$P_o = I^2R_2$$

or,

$$P_o = \frac{E^2R_2}{(R_1 + R_2)^2 + (X_1 - X_2)^2} \quad \textbf{(22-1)}$$

Maximum power output occurs when the denominator in Equation 22-1 has its minimum value (i.e., when $|X_1| = |X_2|$), which gives,

$$P_{o(max)} = \frac{E^2R_2}{(R_1 + R_2)^2} \quad \textbf{(22-2)}$$

When this condition is obtained, the circuit behaves as if there were no reactive components, and so $R_1 = R_2$ is required for maximum power output.

It is seen that maximum power output is derived from an ac source when the resistive components of the source impedance and load impedance are equal, and when the reactive components of the source impedance and load impedance are equal in magnitude but opposite in sign. Thus, for maximum power transfer, an *RL* source must have an *RC* load in which the capacitive reactance has the same magnitude as the inductive reactance. Similarly, an *RC* source must have an *RL* load, with $|X_L| = |X_C|$ Another way of stating this requirement is that the load impedance must be the *conjugate* of the source impedance.

Maximum Power Transfer Theorem for AC Circuits

Maximum output power is obtained from a network or source when the load impedance is the conjugate of the output impedance of the network or source, as seen from the terminals of the load.

Example 22-14

A voltage source has an equivalent circuit consisting of $R_1 = 100\ \Omega$ in series with $L_1 = 20\ \mu\text{H}$, as in Figure 22-19. Calculate the optimum load for maximum output power at a frequency of 500 kHz.

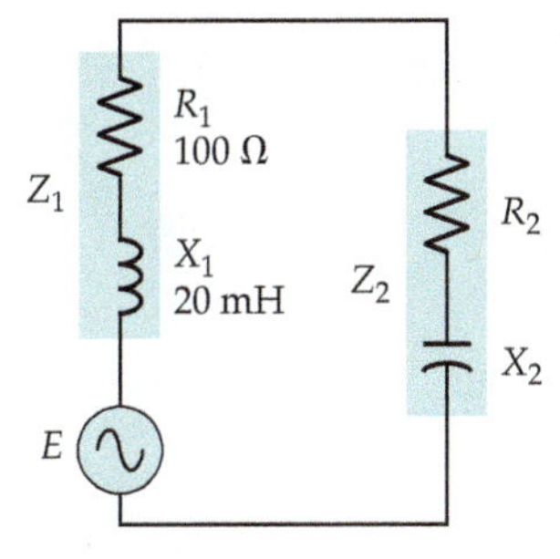

Figure 22-19 Circuit for Example 22-14.

Solution

Reactive component of source impedance:

$$X_1 = 2\pi f L_1 = 2\pi \times 500\ \text{kHz} \times 20\ \mu\text{H}$$
$$= 62.83\ \Omega$$

Reactive component of load impedance:

$$X_2 = -X_1 = -62.83\ \Omega\ \text{(capacitive)}$$

Therefore,
$$\frac{1}{2\pi f C_2} = 2\pi f L_1$$

and,
$$C_2 = \frac{1}{(2\pi f)^2 L_1} = \frac{1}{(2\pi \times 500\ \text{kHz})^2 \times 20\ \mu\text{H}}$$
$$= 5000\ \text{pF}$$

Resistive component of load impedance:

$$R_2 = R_1 = 100\ \Omega$$

Example 22-15

A parallel *RC* circuit is to be used as the load for the voltage source described in Example 22-14, as in Figure 22-20. Determine the required component values for optimum load.

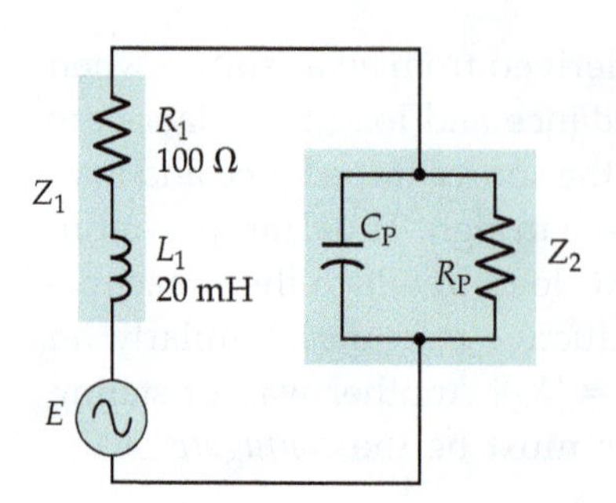

Figure 22-20 Circuit for Example 22-15.

Solution

From Example 22-14, a parallel *RC* equivalent circuit is required for the series connected $R_2 = 100\ \Omega$ and $C_2 = 0.005\ \mu\text{F}$. The series circuit equation is,

$$R_S - jX = R_2 - jX_2$$

$$= 100\ \Omega - j62.83\ \Omega$$

Equation are derived for equivalent series and parallel circuits in Section 20-6.

Eq. 20-9,
$$R_P = \frac{R_S^2 + X_S^2}{R_S} = \frac{(100\ \Omega)^2 + (62.83\ \Omega)^2}{100\ \Omega}$$

$$= 139.5\ \Omega$$

Eq. 20-10,
$$X_P = \frac{R_S^2 + X_S^2}{X_S} = \frac{(100\ \Omega)^2 + (62.83\ \Omega)^2}{62.83\ \Omega}$$

$$= 222\ \Omega$$

$$C_P = \frac{1}{2\pi f X_P} = \frac{1}{2\pi \times 500\ \text{kHz} \times 222\ \Omega}$$

$$= 1400\ \text{pF}$$

Practice Problem

22-8.1 Determine an optimum load (Z_L) for maximum power output for the circuit in Figure 22-21.

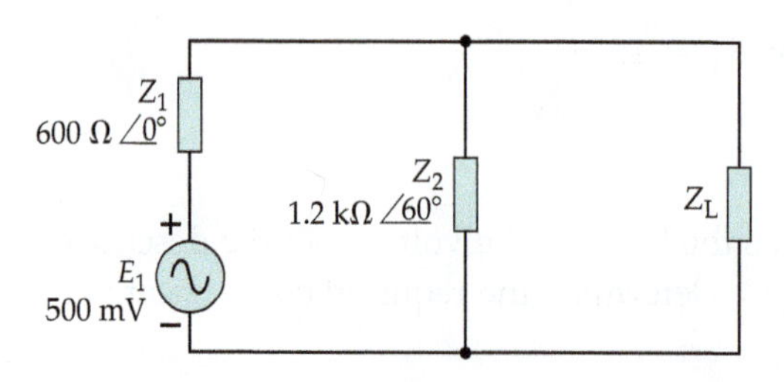

Figure 22-21 Circuit for Problem 22-8.1.

22-9 DELTA-WYE TRANSFORMATIONS FOR AC NETWORKS

Conversion between *delta* (Δ) and *wye* (Y) impedance networks can be performed using the equations derived in Section 8-5 for delta-wye conversion of resistance networks. Figure 22-22 shows Δ and Y impedance networks, and the conversion equations for the networks are listed below. As always, the impedances must be treated as vector quantities.

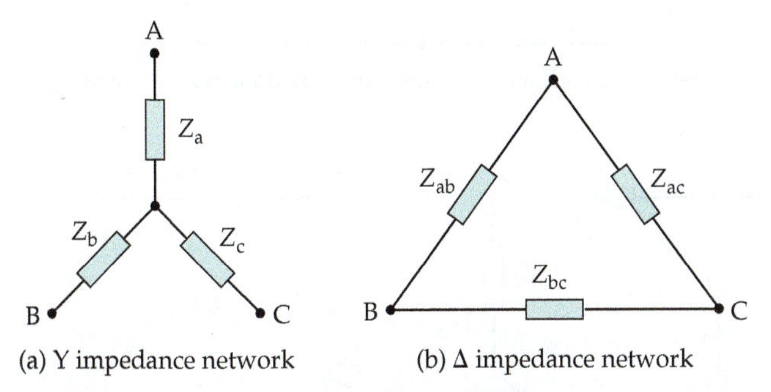

Figure 22-22 Wye and delta impedance networks identified for convenience in converting between circuit types. In the wye network, each impedance is named for the terminal it is connected to: Z_a, Z_b, or Z_c. For the delta network, each impedance is named for the two terminals it is connected across: Z_{ab}, Z_{bc}, or Z_{ac}.

Equations 8-5, 8-6, and 8-7 are rewritten in the form of impedances.

Converting from Δ to Y:

$$Z_a = \frac{Z_{ab}Z_{ac}}{Z_{ab} + Z_{ac} + Z_{bc}} \tag{22-3}$$

$$Z_b = \frac{Z_{ab}Z_{bc}}{Z_{ab} + Z_{ac} + Z_{bc}} \tag{22-4}$$

$$Z_c = \frac{Z_{ac}Z_{bc}}{Z_{ab} + Z_{ac} + Z_{bc}} \tag{22-5}$$

Equations 8-8, 8-9, and 8-10 are rewritten in the form of impedances.

Converting from Y to Δ:

$$Z_{ab} = \frac{Z_aZ_b + Z_aZ_c + Z_bZ_c}{Z_c} \tag{22-6}$$

$$Z_{ac} = \frac{Z_aZ_b + Z_aZ_c + Z_bZ_c}{Z_b} \tag{22-7}$$

$$Z_{bc} = \frac{Z_aZ_b + Z_aZ_c + Z_bZ_c}{Z_a} \tag{22-8}$$

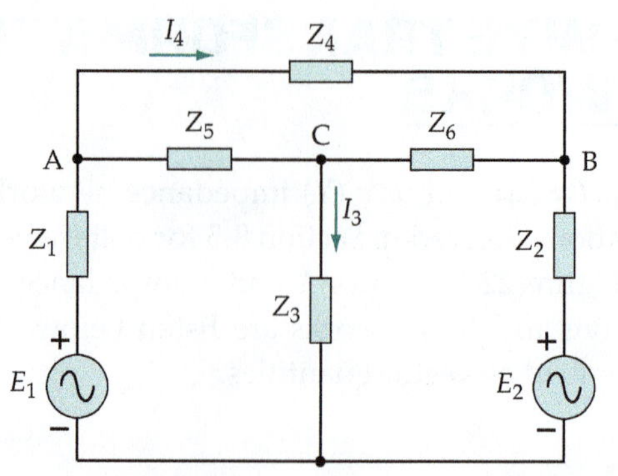

(a) Impedance network with Δ-connected impedances Z_4, Z_5, and Z_6

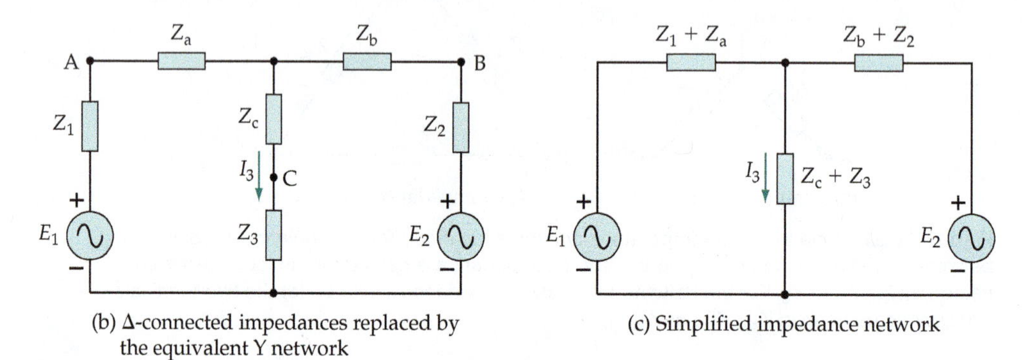

(b) Δ-connected impedances replaced by the equivalent Y network

(c) Simplified impedance network

Figure 22-23 Use of delta-wye transformation to simplify a complex network. Delta network Z_4, Z_5, Z_6 is replaced by its equivalent wye network Z_a, Z_b, Z_c.

A typical application of Δ-Y conversion is illustrated in Figure 22-23. The current through impedance Z_3 is to be determined in the circuit shown in Figure 22-23(a). To simplify the calculations, the delta-connected impedances Z_4, Z_5, and Z_6 are converted to the Y network Z_a, Z_b, and Z_c shown in Figure 22-23(b). Note that the value of Z_3 is unaffected by the conversion. Figure 22-23(c) shows that the network has been very much simplified by the Δ to Y conversion.

The procedure for Δ-Y and Y-Δ transformations of impedance networks is exactly the same as that listed for resistance networks in Section 8-5. As always, impedances must be in rectangular form for addition and subtraction and in polar form for multiplication and division.

Example 22-16

Convert the delta-connected impedances Z_4, Z_5, and Z_6, in Figure 22-23(a), into the wye network Z_a, Z_b, and Z_c shown in Figure 22-23(b). The impedance values are: $Z_4 = 100\ \Omega\angle 30°$, $Z_5 = 95\ \Omega\angle 40°$, $Z_6 = 60\ \Omega\angle 20°$.

Solution

$$Z_{ab} = Z_4 = 100\ \Omega\angle 30° = 86.6\ \Omega + j50\ \Omega$$

$$Z_{ac} = Z_5 = 95\ \Omega\angle 40° = 72.8\ \Omega + j61.1\ \Omega$$

$$Z_{bc} = Z_6 = 60\ \Omega\angle 20° = 56.4\ \Omega + j20.5\ \Omega$$

$$Z_{ab} + Z_{ac} + Z_{bc} = 215.8\ \Omega + j131.6\ \Omega$$

$$= 253\ \Omega\angle 31.4^\circ$$

From Equation 22-3,

$$Z_a = \frac{100\ \Omega\angle 30^\circ \times 95\ \Omega\angle 40^\circ}{253\ \Omega\angle 31.4^\circ}$$

$$= 37.5\ \Omega\angle 38.6^\circ$$

From Equation 22-4,

$$Z_b = \frac{100\ \Omega\angle 30^\circ \times 60\ \Omega\angle 20^\circ}{253\ \Omega\angle 31.4^\circ}$$

$$= 23.7\ \Omega\angle 18.6^\circ$$

From Equation 22-5,

$$Z_c = \frac{95\ \Omega\angle 40^\circ \times 60\ \Omega\angle 20^\circ}{253\ \Omega\angle 31.4^\circ}$$

$$= 22.5\ \Omega\angle 28.6^\circ$$

Practice Problem

22-9.1 Convert the impedances in Figure 22-24 into the equivalent delta-network impedances.

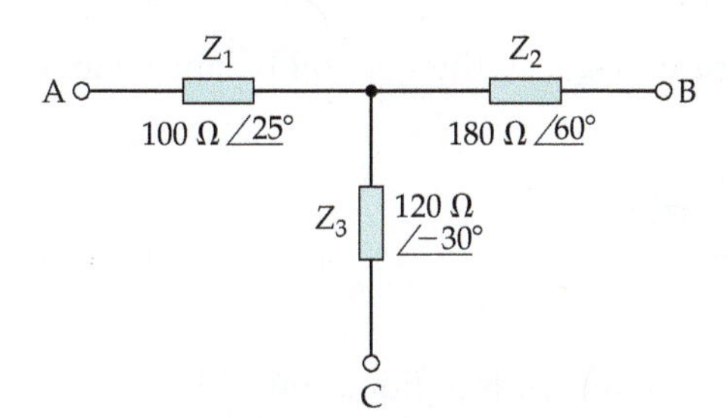

Figure 22-24 Circuit for Problem 22-9.1.

22-10 AC BRIDGES

Simple Capacitance Bridge

AC bridges are impedance networks that can be used for measurement of inductance and capacitance. All ac bridge circuits are based on the Wheatstone bridge (see Section 13-9). In the simple capacitance bridge in Figure 22-25(a), C_S is a precise standard capacitor, C_x is an unknown capacitance, and Q and P are standard resistors, one or both of which is adjustable. An ac supply is applied as shown, and the null detector (D) must be an ac instrument. A low-current rectifier ammeter (a microammeter) is frequently employed as null detector. Variable resistor Q is adjusted until the null detector indicates zero, and when this is obtained the bridge is said to be balanced.

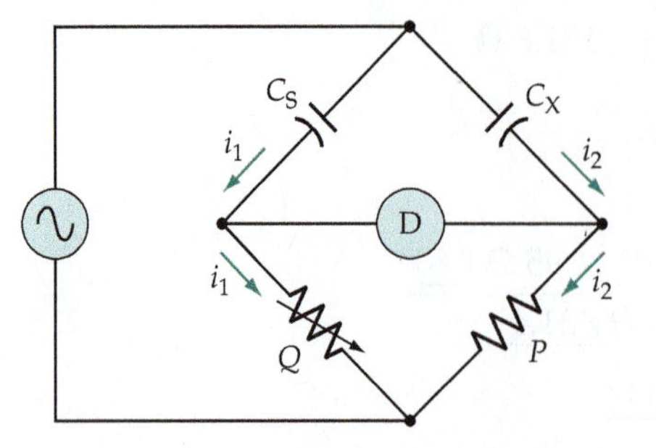

(a) Circuit of a simple capacitance bridge

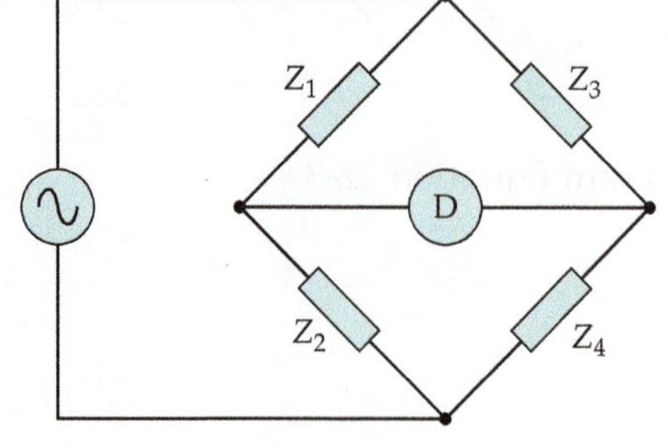

(b) General circuit diagram for an ac bridge

Figure 22-25 A simple capacitance bridge is similar to a Wheatstone bridge, except that two capacitors are used instead of two resistors, the bridge is supplied from an ac source, and the null detector is an ac instrument.

When the detector indicates null, the voltage drop across C_s must equal that across C_X, and similarly, the voltage across Q must be equal to the voltage across P. Therefore,

$$V_{CS} = V_{CX}$$

or,

$$i_1 X_{CS} = i_2 X_{CX} \quad (1)$$

and,

$$V_Q = V_P$$

giving,

$$i_1 Q = i_2 P \quad (2)$$

Dividing Equation (1) by Equation (2):

$$\frac{X_{CS}}{Q} = \frac{X_{CX}}{P} \quad (22\text{-}9)$$

Referring to Equation 22-9 and to Figure 22-25(b), the general balance equation for all ac bridges can be written as,

$$\frac{Z_1}{Z_2} = \frac{Z_3}{Z_4} \quad (22\text{-}10)$$

Substituting $1/(\omega C_S)$ for X_{CS}, and $1/(\omega C_X)$ for X_{CS} into Equation 22-9,

$$\frac{1}{\omega C_S Q} = \frac{1}{\omega C_X P}$$

Giving,

$$C_X = \frac{QC_S}{P} \quad (22\text{-}11)$$

It is seen that the unknown capacitance C_X can now be calculated from the known values of Q, C_S, and P.

Series-Resistance Capacitance Bridge

One disadvantage of the simple capacitance bridge is that perfect balance of the bridge is obtained only when C_S and C_x are both pure capacitances, (i.e., they have virtually no resistive component). In general, this occurs only with capacitors that have air or mica dielectrics. Capacitors with other types of dielectric have a leakage current, and consequently the equivalent circuits for the capacitors have resistive components that must be included in the bridge circuit.

The circuit of the series resistance capacitance bridge shown in Figure 22-26 eliminates the balance problems that can occur with the simple capacitance bridge. Resistance r_X in series with the unknown capacitance represents the resistive component of the capacitor equivalent circuit. The standard capacitor C_S normally has mica dielectric, and thus has a very small resistive component. Consequently, the adjustable resistance S must be included in the circuit to balance the effect of r_X.

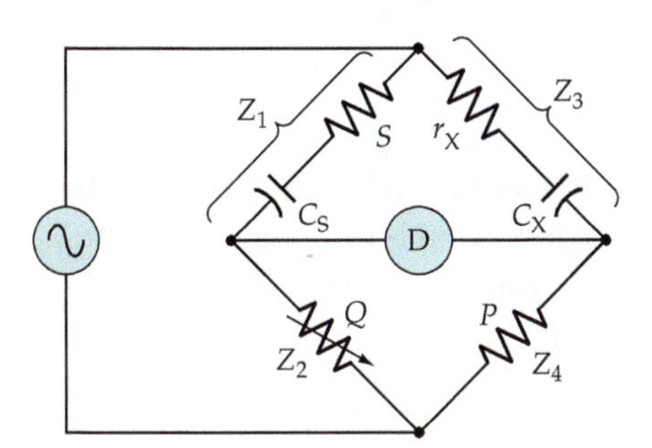

Figure 22-26 Series-resistance capacitance bridge circuit. The inclusion of resistors in series with the capacitors makes bridge balance easier to achieve and allows the capacitor resistive component to be measured.

The balance equations for the series resistance capacitance bridge are derived as follows:

Eq. 22-10,
$$\frac{Z_1}{Z_2} = \frac{Z_3}{Z_4}$$

Therefore,
$$\frac{S - j(1/\omega C_S)}{Q} = \frac{r_X - j(1/\omega C_X)}{P}$$

Giving,
$$\frac{S}{Q} - j\frac{1}{\omega C_S Q} = \frac{r_X}{P} - j\frac{1}{\omega C_X P} \qquad \text{(22-12)}$$

For Equation 22-12 to be correct, the real parts on each side must be equal, and the imaginary parts on each side must be equal. Equating the real parts,

$$\frac{S}{Q} = \frac{r_X}{P}$$

so,
$$r_X = \frac{PS}{Q} \qquad \text{(22-13)}$$

Equating the imaginary parts,

$$\frac{1}{\omega C_S Q} = \frac{1}{\omega C_X P}$$

so,
$$C_X = \frac{QC_S}{P} \qquad \text{(22-14)}$$

The resistive and capacitive components of the unknown capacitor can now be calculated by means of Equations 22-13 and 22-14. Note that neither the supply voltage nor the frequency of the ac supply is involved in the balance equations for the bridge.

Because of the need to balance the real and imaginary components of the bridge impedances, the process of obtaining balance in an ac bridge is a little more complicated than with the Wheatstone bridge. One of the adjustable components (Q or S in Figure 22-26) is first altered to obtain the lowest possible indication on the null meter. Then the other adjustable component is varied to obtain a lower reading. The process is repeated until further adjustment of either component cannot produce a lower reading on the null meter. At this point the bridge is balanced.

Example 22-17

The capacitance bridge shown in Figure 22-26 has a 0.1 μF standard capacitor (C_S) and a standard resistor of $P = 1\ \text{k}\Omega$. Zero deflection is obtained on the null detector when $Q = 10.25\ \text{k}\Omega$ and $S = 2.25\ \text{k}\Omega$. Calculate the value of the unknown capacitance and its resistive component.

Solution

Eq. 22-14,
$$C_X = \frac{QC_S}{P} = \frac{10.25\ \text{k}\Omega \times 0.1\ \mu\text{F}}{1\ \text{k}\Omega}$$
$$= 1.025\ \mu\text{F}$$

Eq. 22-13,
$$r_X = \frac{PS}{Q} = \frac{1\ \text{k}\Omega \times 2.25\ \text{k}\Omega}{10.25\ \text{k}\Omega}$$
$$= 219.5\ \Omega$$

Inductance Bridges

For measurement of inductance, the *Maxwell bridge* shown in Figure 22-27(a) can be employed. It is seen that the circuit of the Maxwell bridge is simply a repeat of the series resistance capacitance bridge, with the capacitors replaced by inductors. A disadvantage of this bridge is that standard inductors are larger and more difficult to manufacture than standard capacitors. Consequently, a variation of this circuit, known as the *Maxwell-Wein bridge*, is most often employed for inductance measurement.

The circuit of the Maxwell-Wein bridge is shown in Figure 22-27(b). L_X is the unknown inductance to be measured, and r_X is the resistance of its windings. C_S is again a precise standard capacitor, and P is a standard resistor. Q and S are accurate adjustable resistors. At balance, Equation 22-10 again applies.

$$\frac{Z_1}{Z_2} = \frac{Z_3}{Z_4}$$

so,
$$\frac{S}{1/(1/Q + j\omega C_S)} = \frac{r_X + j\omega L_X}{P} \qquad \textbf{(22-15)}$$

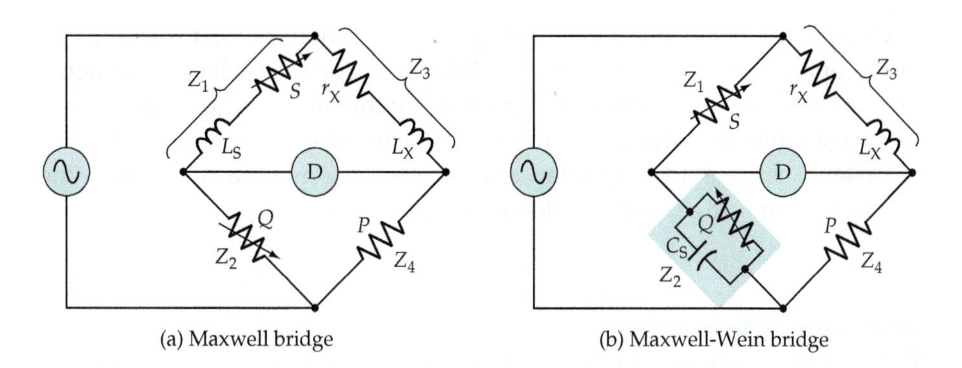

Figure 22-27 A Maxwell bridge is an inductance bridge arranged similarly to a series-resistance capacitance bridge. A Maxwell-Wein bridge uses resistors and standard capacitors to measure inductance.

Which gives,

$$\frac{S}{Q} + j\omega C_S S = \frac{r_X}{P} + j\frac{\omega L_X}{P} \qquad (22\text{-}16)$$

Equating the real terms and the imaginary terms of Equation 22-16:

$$\frac{S}{Q} = \frac{r_X}{P}$$

Giving,

$$r_X = \frac{SP}{Q} \qquad (22\text{-}17)$$

and,

$$\omega C_S S = \frac{\omega L_X}{P}$$

$$L_X = PC_S S \qquad (22\text{-}18)$$

Once again it is seen that the supply voltage and frequency are not involved in the balance equations for the bridge. This is not always the case with ac bridges; indeed, one particular bridge can be used to measure the frequency of the supply in terms of the bridge component values at balance.

Example 22-18

The Maxwell-Wein bridge shown in Figure 22-27(b) uses a 0.1 μF standard capacitor (C_S) and a standard resistor of $Q = 1$ kΩ. Zero deflection of the null detector is obtained when $P = 1.33$ kΩ and $S = 870$ Ω. Calculate the inductance and resistance of Z_3.

Solution

Eq. 22-17,

$$r_X = \frac{SP}{Q} = \frac{870\ \Omega \times 1.33\ \text{k}\Omega}{1\ \text{k}\Omega}$$

$$= 1.16\ \text{k}\Omega$$

Eq. 22-18,

$$L_X = PC_S S = 1.33\ \text{k}\Omega \times 0.1\ \mu\text{F} \times 870\ \Omega$$

$$= 115.7\ \text{mH}$$

Comparing the circuits shown in Figure 22-26 and 22-27(b), it is seen that a series capacitance bridge and a Maxwell-Wein bridge can each be constructed from the same set of components. In fact, many commercial ac bridges use one set of components, which are connected in the form of a series resistance capacitance bridge for capacitance measurement, and switched into the Maxwell-Wein bridge configuration for inductance measurement.

Practice Problems

22-10.1 In the capacitance bridge in Example 22-17, resistor Q is variable from 500 Ω to 10 kΩ and S ranges from 1 kΩ to 3 kΩ. Calculate the range of measurement of C_x and r_x.

22-10.2 An ac bridge has the following components: $C_s = 0.2\ \mu F$, $S = 500\ \Omega$ to 1.6 kΩ, $Q = 500\ \Omega$ to 15 kΩ, and $P = 100\ \Omega$ to 1 MΩ. If the bridge can be either a series resistance capacitance circuit or a Maxwell-Wein circuit, determine the range of unknown capacitance and inductance that can be measured.

Summary of Formulas

- ***Power output from an ac source to an impedance:***

$$P_o = \frac{E^2 R_2}{(R_1 + R_2)^2 + (X_1 - X_2)^2}$$

- ***Δ-to-Y network conversion:***

$$Z_a = \frac{Z_{ab} Z_{ac}}{Z_{ab} + Z_{ac} + Z_{bc}}$$

$$Z_b = \frac{Z_{ab} Z_{bc}}{Z_{ab} + Z_{ac} + Z_{bc}}$$

$$Z_c = \frac{Z_{ac} Z_{bc}}{Z_{ab} + Z_{ac} + Z_{bc}}$$

- ***Y-to-Δ network conversion:***

$$Z_{ab} = \frac{Z_a Z_b + Z_a Z_c + Z_b Z_c}{Z_c}$$

$$Z_{ac} = \frac{Z_a Z_b + Z_a Z_c + Z_b Z_c}{Z_b}$$

$$Z_{bc} = \frac{Z_a Z_b + Z_a Z_c + Z_b Z_c}{Z_a}$$

- *AC bridge general equation:*

$$\frac{Z_1}{Z_2} = \frac{Z_3}{Z_4}$$

- *Series capacitance bridge:*

$$r_X = \frac{PS}{Q}$$

$$C_X = \frac{QC_S}{P}$$

- *Maxwell Bridge:*

$$r_X = \frac{PS}{Q}$$

$$L_x = PC_SS$$

Review Questions

Section 22-1

22-1 Sketch circuit diagrams for ac voltage and current sources with impedance loads. Write equations for the load current in each case.

22-2 Write the equations for converting (a) an ac voltage source into a current source, and (b) a current source into a voltage source.

22-3 Explain why ac voltage and current sources in circuit diagrams should have their terminals identified as positive and negative.

22-4 Sketch a circuit showing an ac voltage source that is dependent on a voltage within the circuit containing the source. Briefly explain.

22-5 Sketch a circuit showing an ac current source that is dependent on a current within the circuit containing the source. Briefly explain.

Section 22-2

22-6 State Kirchhoff's voltage law as applied to ac impedance networks.

22-7 State Kirchhoff's current law as applied to ac impedance networks.

Section 22-3

22-8 List the procedure for ac network analysis by loop equations.

Section 22-4

22-9 State the superposition theorem as applied to ac impedance networks.

22-10 List the procedure for ac network analysis by use of the superposition theorem.

Section 22-5

22-11 List the procedure for using nodal analysis for determining branch currents and node voltages in an ac network.

Section 22-6

22-12 State Thévenin's theorem as applied to ac impedance networks.

22-13 List the procedure for Thévenizing an ac network.

Section 22-7

22-14 State Norton's theorem as applied to ac impedance networks.

22-15 List the procedure for Nortonizing an ac network.

Section 22-8

22-16 State the maximum power transfer theorem as applied to ac sources. Sketch the circuit of an ac source and load, and derive the appropriate equation to show the truth of the theorem.

Section 22-9

22-17 Sketch delta and wye networks and show how the terminals and branches of the two circuits should be identified for calculating equivalent impedance values.

Section 22-10

22-18 Sketch the circuit of a simple capacitance bridge. Explain how it operates and derive an expression for the unknown capacitance.

22-19 Sketch the circuit of a Maxwell bridge. Explain the steps involved in balancing the bridge, and derive the balance equations.

22-20 Sketch the circuits of two ac bridges that may be constructed from a standard capacitor and three variable standard resistors. Identify each bridge by name and derive the balance equations for the bridge.

Problems

Section 22-1

22-1 Determine the equivalent current source components for an ac voltage source that has an rms value of $5\ \text{V}\angle 0°$ with a 1 kHz frequency, and an impedance consisting of a 1 kΩ resistance in series with a 200 mH inductance.

22-2 A dependent current source consists of $Z_s = 20\ \text{k}\Omega\angle -10°$ and $I_s = 50\ I_1$, where $I_1 = 20\ \mu\text{A}\angle 45°$. Determine the components of the equivalent voltage source.

Section 22-2

22-3 Apply Kirchhoff's laws to derive an equation for the current through impedance Z_3 in the circuit shown in Figure 22-28.

22-4 Substitute the component values into the equation derived for Problem 22-3 to determine the current through Z_3.

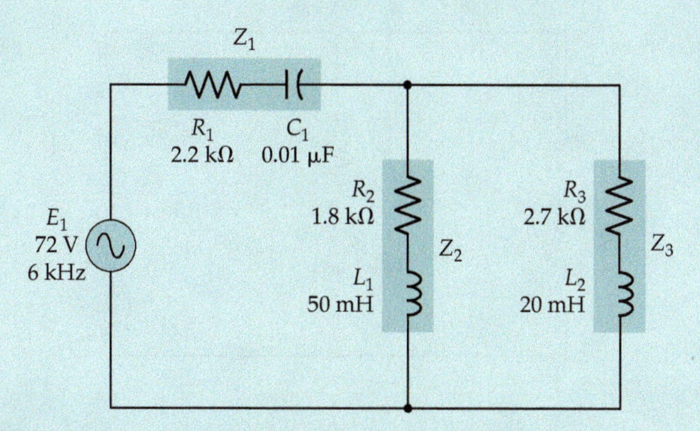

Figure 22-28

22-5 Apply Kirchhoff's laws to derive an equation for the current through impedance Z_4 in the circuit shown in Figure 22-29.

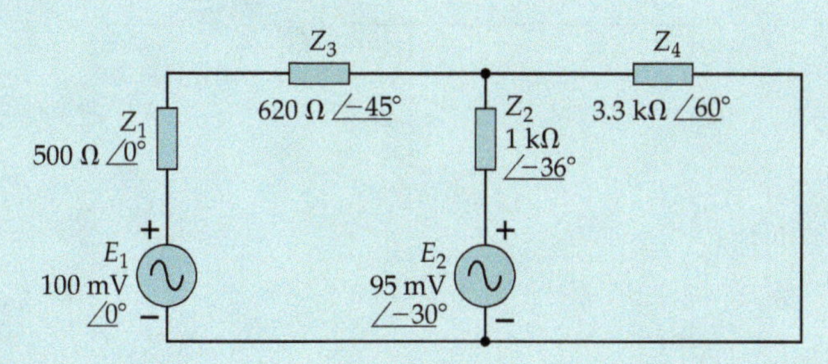

Figure 22-29

22-6 An additional voltage source $E_3 = 120\ \text{mV}\angle 20^\circ$ is included in series with Z_4 in the circuit in Figure 22-29. Derive an equation for the voltage across Z_2.

22-7 If E_2 in Figure 22-29 is replaced with a circuit-dependent voltage source KV_{Z4}, derive the equation for I_{Z4}.

22-8 Substitute the component values into the equation derived for Problem 22-7 to determine the current through Z_4. Assume $K = 3$.

Section 22-3

22-9 Use loop equations to determine the current through impedance Z_3 in the circuit shown in Figure 22-28.

22-10 Using loop equations, derive an equation for the current through Z_4 in the circuit shown in Figure 22-29.

22-11 Substitute the component values into the equation derived for Problem 22-10 to determine the current through Z_4.

22-12 Use loop equations to derive an equation for the voltage across R_2 in the circuit shown in Figure 22-30.

22-13 Substitute the component values into the equation derived for Problem 22-12 to determine the voltage across R_2.

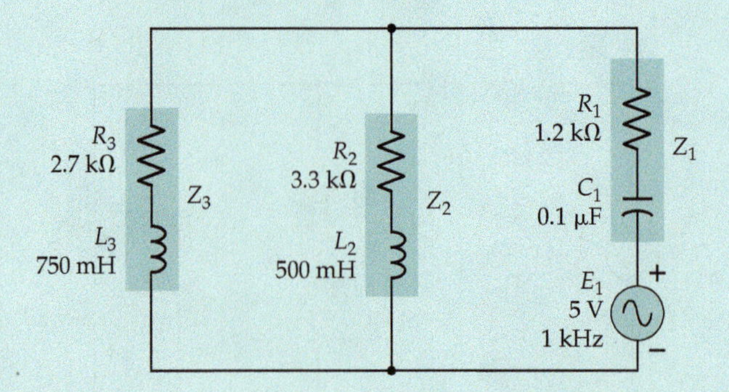

Figure 22-30

22-14 Use the loop equation method to derive an equation for the current supplied by voltage source E_1 in the circuit shown in Figure 22-31.

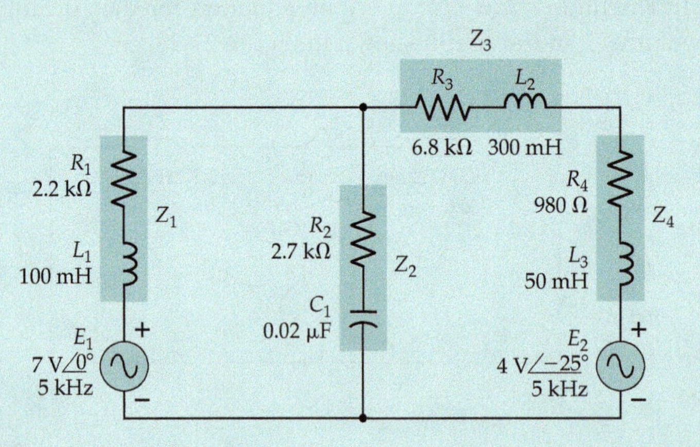

Figure 22-31

22-15 Substitute the component values into the equation derived for Problem 22-14 to determine the current supplied by E_1.

22-16 Use loop equations to determine the current in impedance Z_2 in circuit in Figure 20-28.

Section 22-4

22-17 Using the superposition theorem, derive an equation for the current through Z_4 in the circuit in Figure 22-29.

22-18 Use the superposition theorem to derive an equation for the current through Z_3 in the circuit in Figure 22-29.

22-19 Substitute the component values into the equation derived for Problem 22-18 to determine the current through Z_3.

22-20 Use the superposition theorem to derive an equation for the current through Z_2 in Figure 22-31.

22-21 Substitute the component values into the equation derived for Problem 22-20 to determine the current through Z_2.

22-22 Apply the superposition theorem to derive an equation for the current through Z_4 in the circuit shown in Figure 22-32.

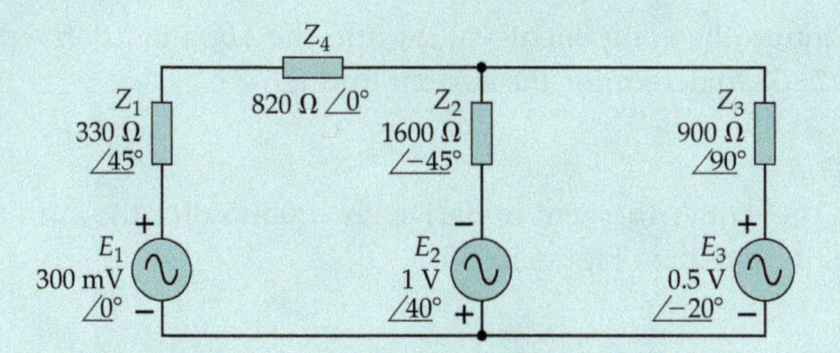

Figure 22-32

22-23 Substitute the component values into the equation derived for Problem 22-22 to determine the current through Z_4.

22-24 Use the superposition theorem to derive an equation for the voltage drop across Z_4 in the circuit shown in Figure 22-33.

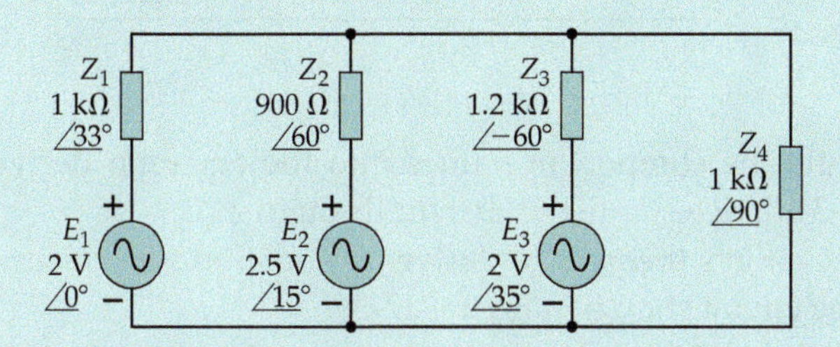

Figure 22-33

22-25 Substitute the component values into the equation derived for Problem 22-24 to determine the voltage across Z_4.

22-26 Using the superposition theorem, derive an equation for the current supplied by voltage source E_1 in the circuit shown in Figure 22-31.

Section 22-5

22-27 Use nodal analysis to determine I_{Z3} in Figure 22-28.

22-28 Use nodal analysis to derive an equation for V_{Z4} in Figure 22-33.

22-29 Use nodal analysis to derive an equation for I_{Z4} in Figure 22-29.

22-30 Use nodal analysis to derive an equation for V_{R2} in Figure 22-30.

22-31 Use nodal analysis to derive an equation for I_{Z1} in Figure 22-31.

22-32 Use nodal analysis to derive an equation for the current through Z_3 in the circuit in Figure 22-34.

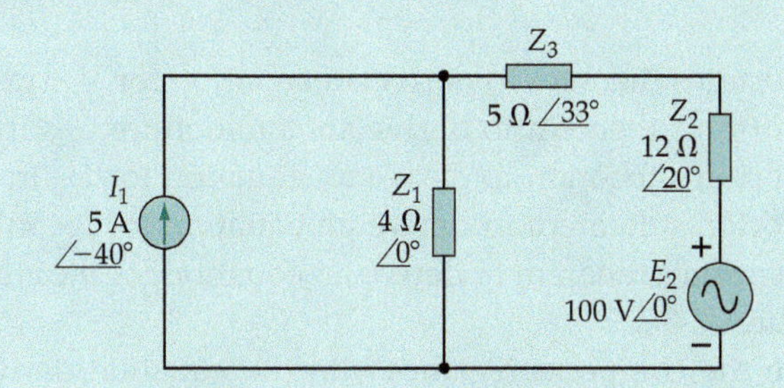

Figure 22-34

22-33 Substitute the component values into the equation derived for Problem 22-32 to determine the current through Z_3.

Section 22-6

22-34 Use Thévenin's theorem to derive an equation for the current through Z_5 in the circuit in Figure 22-35.

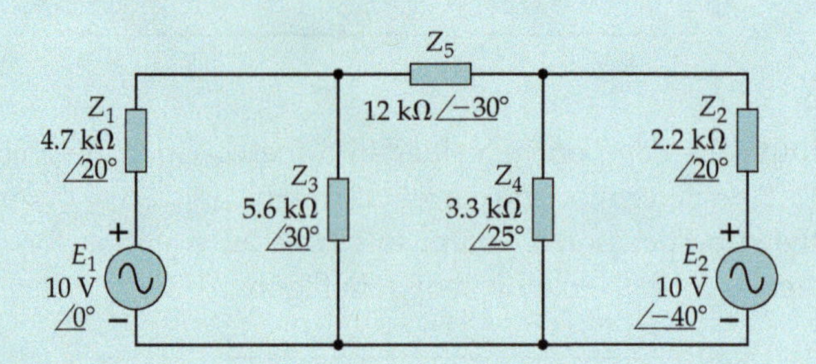

Figure 22-35

22-35 Substitute the component values into the equation derived for Problem 22-34 to determine the current through Z_5.

22-36 Use Thévenin's theorem to derive an equation for the current through R_3 in the circuit shown in Figure 22-36.

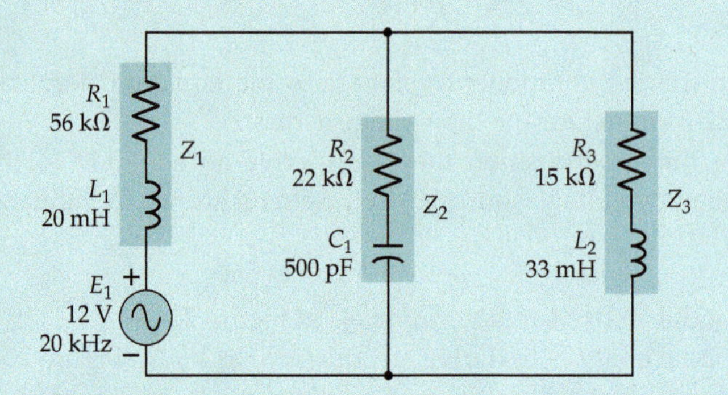

Figure 22-36

22-37 Substitute the component values into the equation derived for Problem 22-36 to determine the current through R_3.

Section 22-7

22-38 Apply Norton's theorem to derive an equation for V_{Z4} in Figure 22-33.

22-39 Apply Norton's theorem to derive an equation for I_{Z3} in Figure 22-34.

22-40 Apply Norton's theorem to derive an equation for I_{Z3} in Figure 22-28.

22-41 Apply Norton's theorem to derive an equation for I_{R3} in Figure 22-36.

22-42 Use the Norton's theorem to derive an equation for the current through R_2 in Figure 22-36.

22-43 Substitute the component values into the equation derived for Problem 22-42 to determine the current through R_2.

Section 22-8

22-44 A voltage source has an equivalent circuit consisting of $R = 500\ \Omega$ in series with $C = 0.01\ \mu F$. Calculate the required component values for a series *RL* circuit that will draw maximum power from the source when the signal frequency is 1 MHz.

22-45 For Problem 22-44, calculate the required component values for a parallel *RL* circuit that will draw maximum power from the source.

22-46 For the circuit in Figure 22-33, calculate a suitable impedance that may be substituted in place of Z_4 to give maximum power dissipation.

22-47 For the circuit in Figure 22-36, calculate suitable series-connected components that may be substituted in place of R_3 and L_2 for maximum power dissipation.

Section 22-9

22-48 Apply delta-wye transformation and nodal analysis to derive an equation for the current through Z_4 in Figure 22-37.

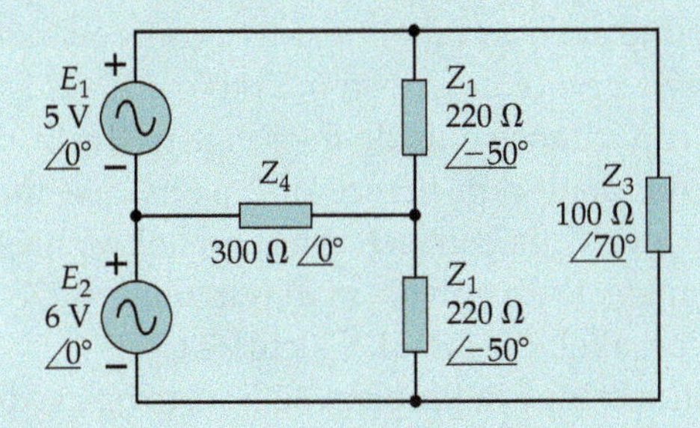

Figure 22-37

22-49 Substitute the component values into the equation derived for Problem 22-48 to determine the current through Z_4.

22-50 Apply delta-wye transformation to derive an equation for the current supplied by the voltage source in Figure 22-38.

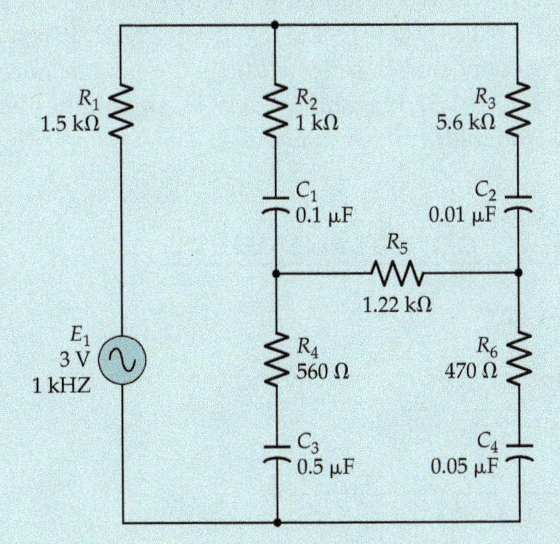

Figure 22-38

22-51 Substitute the component values into the equation derived for Problem 22-50 to determine the source current.

Section 22-10

22-52 A series-connected capacitance bridge (Figure 22-26) uses a 0.15 μF standard capacitor, a series resistance S = 1 kΩ to 1.8 kΩ, and a standard resistor Q = 12 kΩ. If the capacitance to be measured is known to be between 2 μF and 2.5 μF, calculate the required range of adjustment for resistor P. Determine the range of measurement for the resistive component of the unknown capacitor.

22-53 A simple capacitance bridge, as in Figure 22-25, has C_S = 0.2 μF, P = 10 kΩ, and Q adjustable from 3 kΩ to 15 kΩ. Calculate the range of unknown capacitance C_x that can be measured with the bridge.

22-54 A 1 μF capacitor and three-decade resistors that may be adjusted from 100 Ω to 5 kΩ are available for construction of the two bridges referred to in Review Question 22-20. Calculate the measurable ranges of capacitance and inductance and their resistive components.

22-55 The capacitance bridge described in Problem 22-52 is to be used to measure a small capacitance outside the range of the bridge. A capacitor C_1 with an approximate capacitance of 2 μF is first measured. Then, the small capacitance C_2 is connected in parallel with C_1, and a new measurement is made. With only C_1 in the circuit, P = 875 Ω at balance. With C_1 and C_2 in parallel, P = 869 Ω. Calculate C_2.

22-56 The capacitance bridge in Problem 22-52 is to be used to measure a large capacitance outside the range of the bridge. A capacitor C_1 with an approximate capacitance of 2 μF is first measured. Then, the large capacitance C_2 is connected in series with C_1, and a new measurement is made. With only C_1 in the circuit, P = 875 Ω at balance. If C_2 has a capacitance of 56.5 μF, determine the expected setting for P at balance.

22-57 An inductance L_1 measured on the bridge in Problem 22-54 gives balance with P = 1 kΩ, Q = 1.98 kΩ, and S = 122 Ω. When another inductance L_2 is connected in series with L_1, the new balance is found with P = 852 Ω, Q = 1.31 kΩ, and S = 149 Ω. Determine the inductance of L_1 and L_2 and their coil resistances.

Practice Problem Answers

22-1.1 3.02 $\mu A\angle 20°$

22-1.2 22.7 $\mu A\angle -65°$, 33 $k\Omega\angle 20°$

22-2.1 $\dfrac{E_2/Z_2 + E_1/Z_1}{1/Z_1 + 1/Z_2 + 1/Z_3}$

22-2.2 $\dfrac{E_2/Z_2}{1/Z_1 + 1/Z_2 + 1/Z_3 - K/Z_2}$

22-2.3 $8.6 \text{ V}\angle 42.2^\circ$
22-3.1 $17 \text{ mA}\angle -70.6^\circ$
22-3.2 $-34 \text{ mA}\angle -90.6^\circ$
22-4.1 $(E_2Z_1 + E_2Z_L - E_1Z_L)/(Z_1Z_2 + Z_2Z_L + Z_1Z_L)$
22-4.2 $181 \ \mu\text{A}\angle -72.8^\circ$
22-5.1 $\dfrac{E_1}{Z_1} - \dfrac{1}{Z_1}\left[\dfrac{E_1/Z_1 + E_2/Z_2}{1/Z_1 + 1/Z_2 + 1/Z_3}\right]$
22-6.1 $\dfrac{E_1Z_2}{Z_2Z_1 + Z_2Z_L + Z_1Z_L}$
22-7.1 $\dfrac{E_1Z_L + Z_L(E_1 - E_2)}{Z_1Z_2 + Z_2Z_L + Z_1Z_L}$
22-8.1 $429\ \Omega - j148\ \Omega$
22-9.1 $236\ \Omega\angle -19.3^\circ$, $354\ \Omega\angle 70.7^\circ$, $425\ \Omega\angle 15.7^\circ$
22-10.1 $0.05\ \mu\text{F}$ to $1\ \mu\text{F}$, $100\ \Omega$ to $6\ \text{k}\Omega$
22-10.2 $100\ \text{pF}$ to $30\ \mu\text{F}$, $10\ \text{mH}$ to $320\ \text{H}$

CHAPTER 23
Resonance

CONTENTS

Objectives

You will be able to:

1 Write the equations for the impedance of series and parallel *RLC* circuits and explain what occurs at the resonance frequency.

2 Draw graphs for inductive reactance, capacitive reactance, and total impedance versus frequency for a series *RLC* circuit.

3 Analyze series-connected and parallel-connected *RLC* circuits to determine the resonance frequency, and the component voltages and current levels at various frequencies.

4 Sketch phasor diagrams and component waveforms for series and parallel *RLC* circuits at resonance.

5 Define and calculate the circuit *Q* factor, half-power points, and bandwidth for series and parallel resonance circuits.

6 Sketch the graphs of primary current and secondary current versus frequency for coupled coils tuned to resonance. Calculate the critical value of the coefficient of coupling for the coils and determine the levels of current in each coil.

INTRODUCTION

When a series-connected *RLC* circuit has the frequency of its alternating supply voltage varied, it is found that $X_L = X_C$ at a particular frequency. The two reactances cancel, and the result is that the series impedance of the circuit has a minimum value of $Z = R$. The circuit is said to be in a state of *resonance*, and the frequency at which this occurs is termed the *resonance frequency*. Because the impedance is a minimum, the series current is a maximum, and there is a rise in the voltage developed across *L* and *C*.

A parallel *LC* circuit can also be in a state of resonance, and again this occurs when $X_L = X_C$. In this case, the circuit impedance has a maximum value, and the circuit current is a minimum, but there is a rise in the *L* and *C* current levels. An *LC* circuit can be tuned to resonate over a range of frequencies by making *L* or *C* adjustable.

23-1 SERIES RESONANCE

Frequency Effect on Circuit Impedance

The series-connected *RLC* circuit shown in Figure 23-1(a) has an impedance of,

$$Z = R + j(X_L - X_C) \tag{23-1}$$

or

$$Z = \sqrt{R^2 + X^2}\,\underline{/\tan^{-1}(X/R)}$$

Because $X_L = 2\pi fL$ and $X_C = 1/(2\pi fC)$, the actual impedance of the circuit depends on the frequency of the alternating supply, as illustrated in Figure 23-1(b). Figure 23-2 shows the values of jX_L and $-jX_C$ plotted versus supply frequency. It is seen that, because X_L is directly proportional to

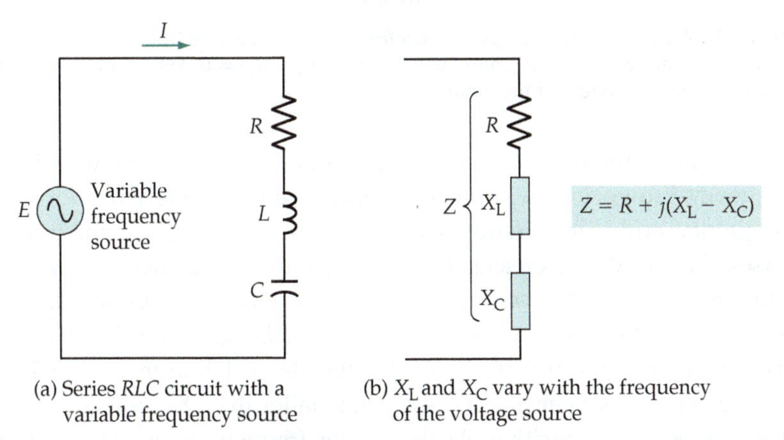

(a) Series *RLC* circuit with a variable frequency source

(b) X_L and X_C vary with the frequency of the voltage source

Figure 23-1 In a series *RLC* circuit with a variable frequency source, X_L becomes equal to X_C at a particular frequency, known as the resonance frequency. The circuit impedance is then equal to *R*.

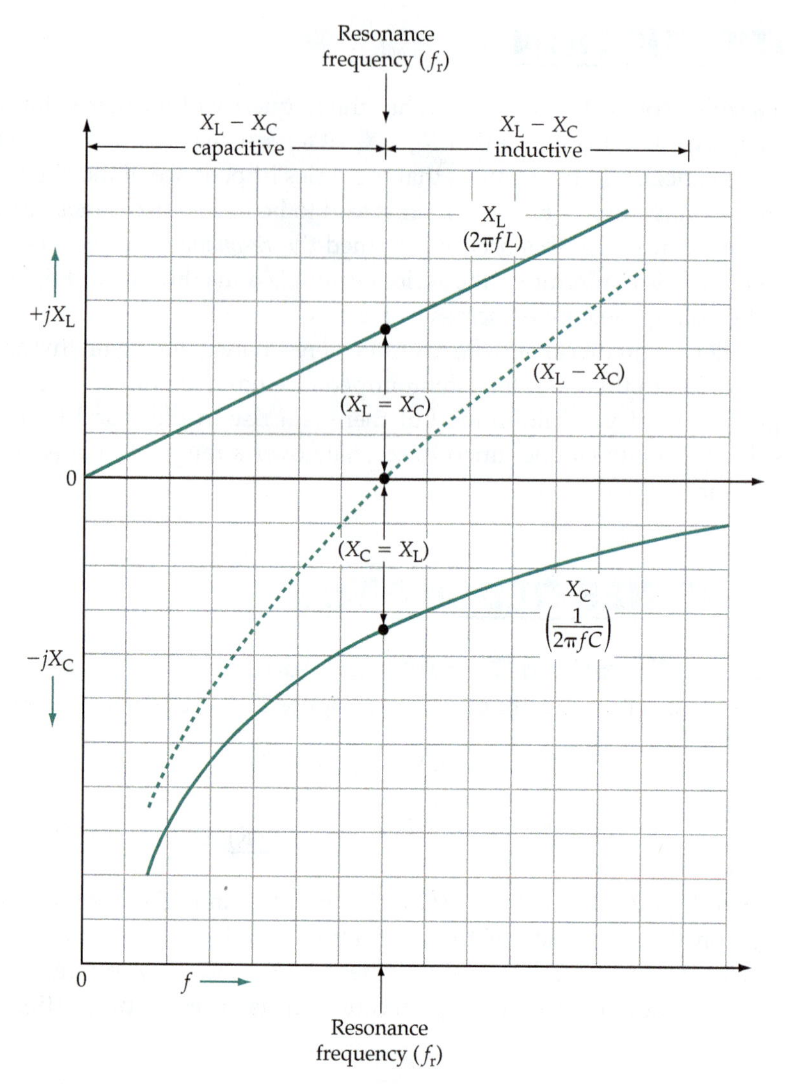

Figure 23-2 A plot of $+jX_L$ and $-jX_C$ versus frequency for a series *RLC* circuit shows that at a particular frequency (f_r), $X_L = X_C$, and consequently $(X_L - X_C) = 0$. The frequency f_r is known as the resonance frequency for the circuit.

frequency, the inductive reactance increases linearly from zero with increase in frequency. However, the capacitive reactance is inversely proportional to the frequency, and at low frequencies X_C is infinitely large. As the frequency increases, X_C rapidly decreases from infinity but never becomes zero. The resulting graph of $(-jX_C)$ versus frequency is curved, as illustrated.

The total reactance of the series-connected *RLC* circuit is $(X_L - X_C)$, and the graph of this quantity is shown by the dashed line in Figure 23-2. At low frequencies, it is clear that X_L is much smaller than X_C, and thus the total reactance is largely capacitive. At the higher frequencies, where X_L is much larger than X_C, the total reactance becomes largely inductive. At one particular frequency, identified as f_r on the graph, X_L and X_C are numerically equal. Consequently, the impedance of the series *RLC* circuit at f_r becomes,

$$Z = R + j(X_L - X_C)$$
$$= R + j(0)$$

or

$$Z = R$$

When this occurs, the circuit is said to be in a state of *electrical resonance*, and the frequency at which X_L equals X_C is known as the *resonance frequency* (f_r).

The resonance frequency for a series *RLC* circuit is defined as the frequency at which $X_L = X_C$.

Referring again to Figure 23-2, it is seen that at frequencies above and below resonance, the reactance ($X = X_L - X_C$) has a large value, which is either inductive or capacitive. So, when the circuit impedance (Equation 23-1) is plotted versus frequency, it is found that the impedance has a high value above and below the resonance frequency, and at resonance it dips to its minimum value of $Z = R$ [see Figure 23-3(a)].

A series *RLC* circuit has minimum impedance at the resonance frequency.

Current at Resonance

The current in the series *RLC* circuit is determined from,

$$I = \frac{E}{R + j(X_L - X_C)} \quad (23\text{-}2)$$

or

$$|I| = \frac{E}{\sqrt{R^2 + (X_L - X_C)^2}}$$

where $|I|$ is the numerical value of the current, without reference to its phase angle. Thus, at resonance, when $X_L = X_C$, the current equation becomes

$$I = \frac{E}{R}$$

A typical graph of current versus frequency for the series *RLC* circuit is shown in Figure 23-3(b). The current is a minimum at frequencies above and below resonance but peaks sharply at resonance as the impedance falls to a minimum. The current/frequency graph is sometimes referred to as the *frequency response curve* of the circuit.

Phase Angle

As noted from Figure 23-2, the impedance of the series *RLC* circuit is largely capacitive at frequencies well below resonance. This means that the circuit

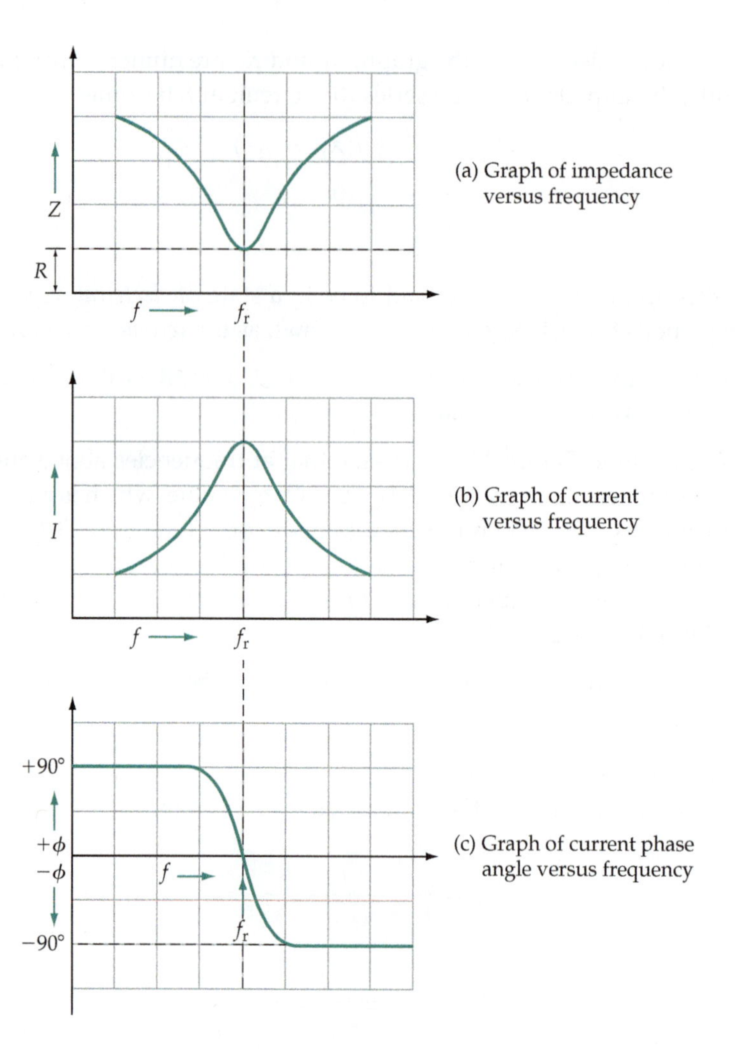

Figure 23-3 Graphs of impedance, current, and phase angle versus frequency for a series *RLC* circuit. Because $Z = R + j(X_L - X_C)$, the impedance dips to R at f_r and the current peaks. The current phase angle is zero at f_r, becomes leading below f_r, and is lagging above f_r.

current leads the applied voltage by a phase angle of approximately 90°. Conversely, because the impedance is largely inductive at frequencies much greater than the resonance frequency, the phase angle of the current above resonance is approximately −90°. The graph of phase angle versus frequency for the series *RLC* circuit [Figure 23-3(c)] shows the 90° leading phase angle at low frequencies, changing to 0° at the resonance frequency, and moving to a 90° lagging phase angle above f_r.

Resonance Frequency

The frequency at which resonance occurs is easily calculated by equating X_L and X_C:

$$X_L = 2\pi f_r L$$

$$X_C = \frac{1}{2\pi f_r C}$$

So, $$2\pi f_r L = \frac{1}{2\pi f_r C}$$

Giving, $$f_r = \frac{1}{2\pi\sqrt{LC}} \tag{23-3}$$

When L and C are in henrys and farads, respectively, Equation 23-2 gives f_r in hertz.

Example 23-1

Determine the resonance frequency for the series RLC circuit in Figure 23-4. Also, calculate the circuit currents at $0.25f_r$, $0.5f_r$, $0.8f_r$, f_r, $1.25f_r$, $2f_r$, and $4f_r$. Plot a graph of current versus frequency to a logarithmic base.

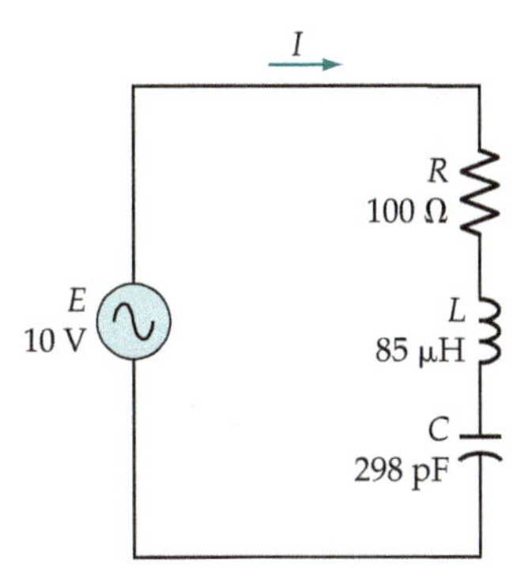

Figure 23-4 Circuit for Example 23-1.

Solution

Eq. 23-3, $$f_r = \frac{1}{2\pi\sqrt{LC}} = \frac{1}{2\pi\sqrt{85\ \mu\text{H} \times 298\ \text{pF}}}$$

$$= 1\ \text{MHz}$$

Eq. 23-2, $$|I| = \frac{E}{\sqrt{R^2 + (X_L - X_C)^2}}$$

The values of the various quantities are tabulated below for each frequency, and the graph of current versus frequency is plotted in Figure 23-5.

f (Hz)	X_L (Ω)	X_C (Ω)	$(X_L - X_C)$ (Ω)	|Z| (Ω)	*I* (mA)
250 k	134	2136	$-j2$ k	≈2 k	5
500 k	267	1068	$-j801$	≈807	12.4
800 k	427	668	$-j241$	≈261	38.3
1 M	534	534	0	100	100
1.25 M	668	427	$+j241$	≈261	38.3
2 M	1068	267	$+j801$	≈807	12.4
4 M	2136	134	$+j2$ k	≈2 k	5

Computer analysis of the above problem is offered in Section 28-7.

The graph of current versus frequency in Figure 23-5 clearly shows that the current rises to a maximum at the resonance frequency and falls off sharply at frequencies above and below resonance. Note from the table that when the frequency is doubled from 250 kHz to 500 kHz, the current increases from 5 mA to 12.4 mA. Also note that above resonance the current decreases from 12.4 mA to 5 mA when the frequency is doubled from 2 MHz to 4 MHz. The same sort of

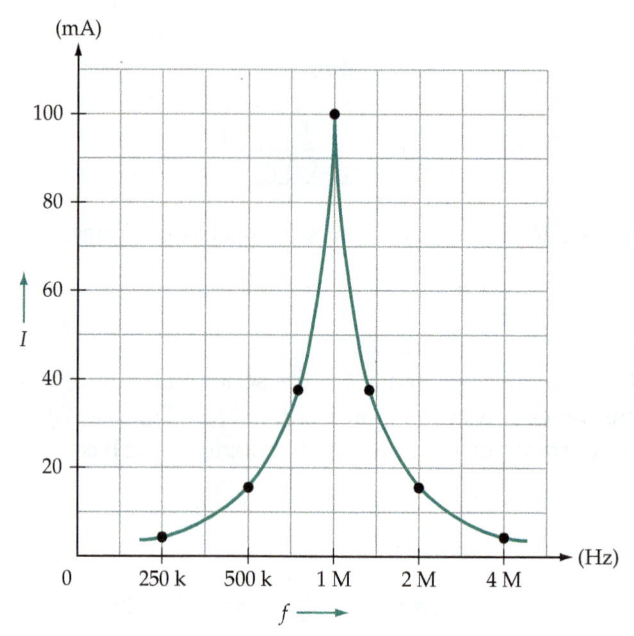

Figure 23-5 Plot of current versus frequency (to a logarithmic base) for a series *RLC* circuit. The current peaks sharply at the resonance frequency.

corresponding current ratios are evident for other frequency changes above and below the resonance frequency. It is seen that the ratio of current to frequency is logarithmic, and so a logarithmic frequency scale is used on the graph.

Resonance Rise in Voltage

The voltage across the resistance in a series *RLC* circuit is $V_R = IR$ and at resonance V_R equals the supply voltage [see Figure 23-6(a)]. As illustrated, the capacitor and inductor voltages are

$$V_C = IX_C \qquad (23\text{-}4)$$

and

$$V_L = IX_L \qquad (23\text{-}5)$$

When the capacitor voltage levels and inductor voltage levels are calculated for various frequencies and plotted to a logarithmic frequency scale, it is found that the V_C and V_L *frequency response curves* are similar in shape to the circuit current-versus-frequency graph. It is also found that

V_C and V_L at resonance can be many times greater than the supply voltage.

This effect is termed the *resonant rise in voltage.*

Figure 23-6(b) shows that the waveforms of V_L and V_C are in phase opposition at resonance. Thus, the instantaneous levels of V_L and V_C cancel each other, and all of the supply voltage appears across *R*.

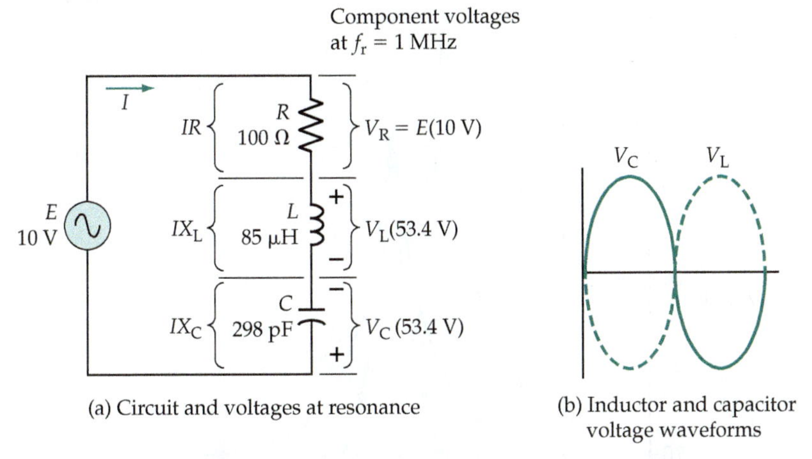

(a) Circuit and voltages at resonance

(b) Inductor and capacitor voltage waveforms

Figure 23-6 In a series *RLC* circuit the resonance frequency $V_R = E$, and V_L and V_C are larger than *E*. V_L and V_C are in phase opposition.

Example 23-2

For the series *RLC* circuit in Figure 23-6 (as in Example 23-1), determine the capacitor voltage and inductor voltage at each frequency. Also plot the graphs of V_R, V_C, and V_L versus frequency.

Solution

$$V_C = IX_C,\ V_L = IX_L, \text{ and } V_R = IR$$

The quantities are tabulated for each frequency, and the graphs are plotted in Figure 23-7. Note from the graph that $V_C > V_L$ below resonance, and $V_L > V_C$ above resonance.

f (Hz)	V_R (V)	V_L (V)	V_C (V)
250 k	0.5	0.67	10.7
500 k	1.24	3.3	13.2
800 k	3.83	16.4	25.5
1 M	10	53.4	53.4
1.25 M	3.83	25.5	16.4
2 M	1.24	13.2	3.3
4 M	0.5	10.7	0.67

Example 23-2 demonstrates that at resonance, the voltages across the capacitor and inductor become much greater than the supply voltage. The voltage across the resistance (V_R) is, of course, in phase with the circuit current, while V_L leads the current by 90° and V_C lags the current by 90°. This is illustrated in the phasor diagram in Figure 23-8(a). From the phasor diagram, it is clear that V_L and V_C cancel at resonance and the supply voltage is applied across the resistance.

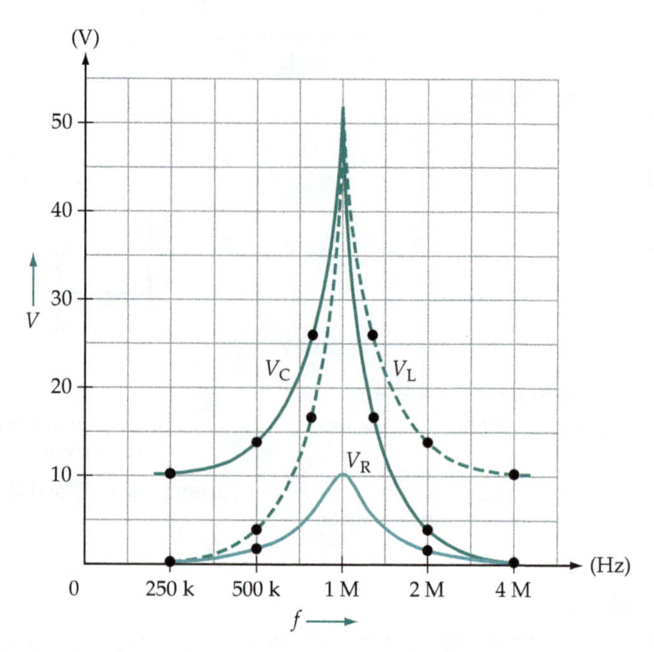

Figure 23-7 Graphs of capacitor voltage V_C, inductor voltage V_L, and resistor voltage V_R versus frequency (to a logarithmic base) for a series *RLC* circuit. At the resonance frequency, V_C and V_L become much greater than the supply voltage.

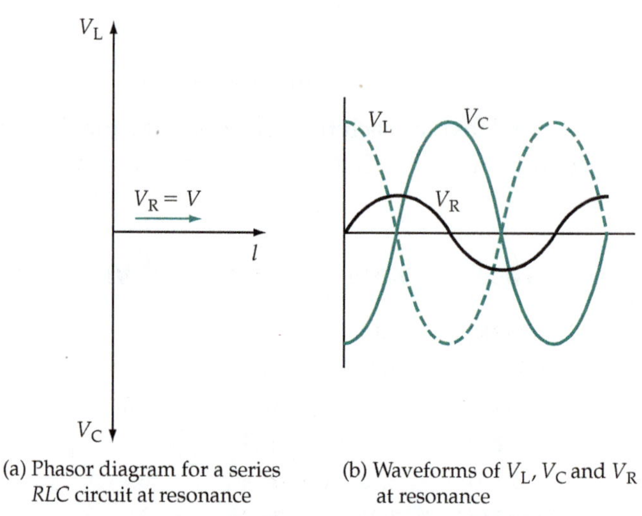

(a) Phasor diagram for a series *RLC* circuit at resonance

(b) Waveforms of V_L, V_C and V_R at resonance

Figure 23-8 Voltage phasor diagram and waveforms for a series *RLC* circuit at resonance. V_C and V_L cancel each other, V_R equals the supply voltage, and the current is in phase with the supply voltage.

Energy Transfer between *L* and *C*

The waveforms of V_L, V_R, and V_C at the resonance frequency are shown in Figure 23-8(b). It is evident that as V_L grows positively, V_C becomes more negative, and vice versa. The positive peak of V_L occurs at the same instant as the negative peak of V_C, and the positive peak of V_C corresponds in time with the negative peak of V_L. Also note that the positive and negative peaks of voltage across the resistance coincide with the instant when V_L and V_C are zero. The information to be derived from these waveforms is that there is energy stored in the resonant circuit. The energy is continuously being transferred from the inductor to the capacitor and back again at the resonance frequency. The resonance frequency is the only frequency at which this transfer of energy can take place for the particular values of capacitor and inductor involved. Note, however, that *the energy storage and energy transfer between reactive components occurs only when there is an energy input at the resonance frequency.*

Practice Problems

23-1.1 Determine the new resonance frequency for the series *RLC* circuit in Example 23-1 when an oscilloscope with an input capacitance of 40 pF is connected in parallel with the capacitor, as in Figure 23-9.

23-1.2 Calculate the new levels of current, inductor voltage, and capacitor voltage at $0.5f_r$ and $0.8f_r$ for the circuit in Figure 23-9.

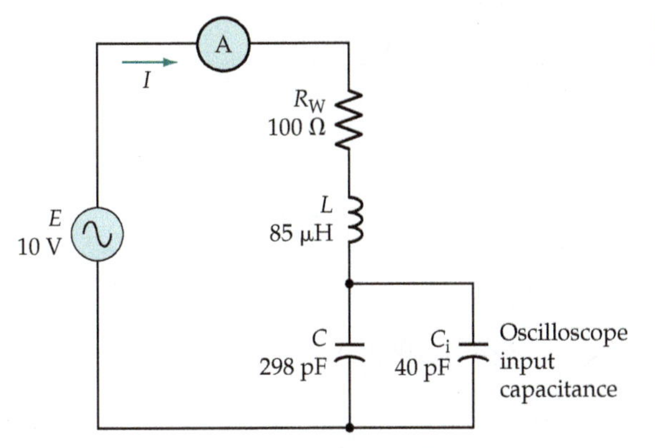

Figure 23-9 Circuit for Problems 23-1.1 and 23-1.2.

23-2 TUNING FOR RESONANCE

Refer again to the equation for calculating the resonance frequency,

Eq. 23-3, $$f_r = \frac{1}{2\pi\sqrt{LC}}$$

As discussed, when L and C have fixed values, the frequency of the supply must be made equal to f_r to achieve resonance. Suppose, instead, that f had a fixed value; then either L or C (or perhaps both) could be adjusted until the circuit resonates at the supply frequency. In fact, this is exactly what is done in a radio receiver. Figure 23-10 shows a circuit in which v_S represents a small voltage induced from the atmosphere into the inductance (L). Resistance R_W represents the resistive component of the inductor winding, and capacitor C is variable. The electromagnetic signals sent into the atmosphere by different

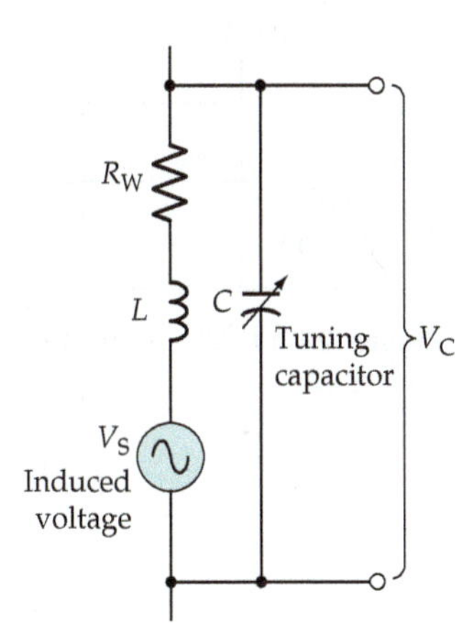

Figure 23-10 Series resonance circuit with a tuning capacitor. Capacitor C can be adjusted to give $X_C = X_L$ at any desired frequency, and so the resonance frequency is selected by adjustment of C.

radio stations have different frequencies. So, by adjusting C, the circuit may be made to resonate at any one of a wide range of radio frequencies. Adjusting the value of C is termed *tuning* the circuit, and C is referred to as a *tuning capacitor*. The resonance frequency causes a *resonance rise in voltage* across the tuning capacitor. The signal developed across the capacitor can then be passed on to the other circuits of the radio receiver.

Example 23-3

The circuit shown in Figure 23-10 has $L = 100\ \mu\text{H}$ and is to be tuned to resonate at frequencies ranging from 500 kHz to 1 MHz. Calculate the range of adjustment of the tuning capacitor.

Solution

Eq. 23-3, $$f_r = \frac{1}{2\pi\sqrt{LC}}$$

Which gives, $$C = \frac{1}{4\pi^2 f_r^2 L}$$

At $f_r = 500$ kHz, $$C = \frac{1}{4\pi^2 \times (500\ \text{kHz})^2 \times 100\ \mu\text{H}} \approx 1000\ \text{pF}$$

At $f_r = 1$ MHz, $$C = \frac{1}{4\pi^2 \times (1\ \text{MHz})^2 \times 100\ \mu\text{H}} \approx 253\ \text{pF}$$

The tuning capacitor range is from 253 pF to 1000 pF.

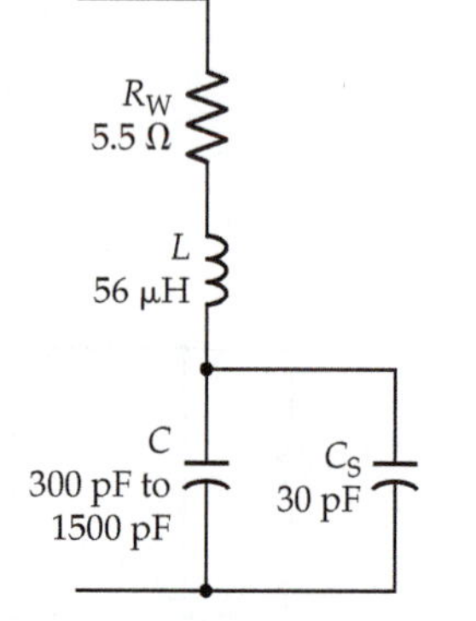

Figure 23-11 Circuit for Problem 23-2.1.

Practice Problem

23-2.1 A series *RLC* circuit (as in Figure 23-11) has $R_W = 5.5\ \Omega$, $L_1 = 56\ \mu\text{H}$, C_1 adjustable from 300 pF to 1500 pF, and there is stray capacitance of 30 pF in parallel with C_1. Determine the range of resonance frequencies for the circuit.

23-3 *Q* FACTOR OF A SERIES RESONANT CIRCUIT

Reconsider the equations for current, inductor voltage, and capacitor voltage at resonance, and refer to Figure 23-12.

$$V_L = I\,X_L \quad \text{and} \quad I = E/R$$

Therefore, $$V_L = \frac{E}{R} \times X_L$$

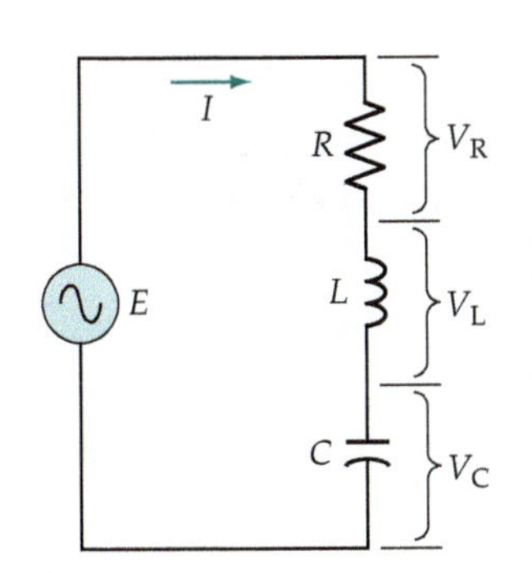

$$Q = \frac{V_L}{V_R} = \frac{V_C}{V_R}$$

$$Q = \frac{\omega L}{R} = \frac{1}{\omega CR}$$

Figure 23-12 The voltage magnification factor or Q factor for a series resonant circuit is V_L/V_R, or V_C/V_R.

or

$$\frac{V_L}{E} = \frac{X_L}{R} \tag{23-6}$$

Similarly,

$$\frac{V_C}{E} = \frac{X_C}{R} \tag{23-7}$$

The ratio of the capacitor voltage, or inductor voltage, at resonance to the supply voltage is a measure of the *quality* of a resonance circuit. This is termed the *Q factor* of the circuit, also known as the *voltage magnification factor*.

From Equations 23-6 and 23-7, the equations for the Q factor are,

$$Q = \frac{X_L}{R} \quad \text{and} \quad Q = \frac{X_C}{R}$$

or

$$Q = \frac{2\pi f_r L}{R} \tag{23-8}$$

and

$$Q = \frac{1}{2\pi f_r CR} \tag{23-9}$$

Because the inductor winding resistance is often the only resistance in a series resonance circuit, the Q is sometimes referred to as the *Q factor of the inductor*, or the *figure of merit* for the inductor. Substituting for f_r from Equation 23-3,

$$Q = \frac{2\pi L\left(\dfrac{1}{2\pi\sqrt{LC}}\right)}{R}$$

Which reduces to

$$Q = \frac{1}{R}\sqrt{\frac{L}{C}} \tag{23-10}$$

Equation 23-10 shows that the Q factor of a series resonance circuit may be increased by reducing R, or by increasing the L/C ratio. The Q factor can also be defined in terms of the ratio of the reactive power to the power dissipated in the circuit resistance. Using this definition, the equations for Q come out exactly as derived above.

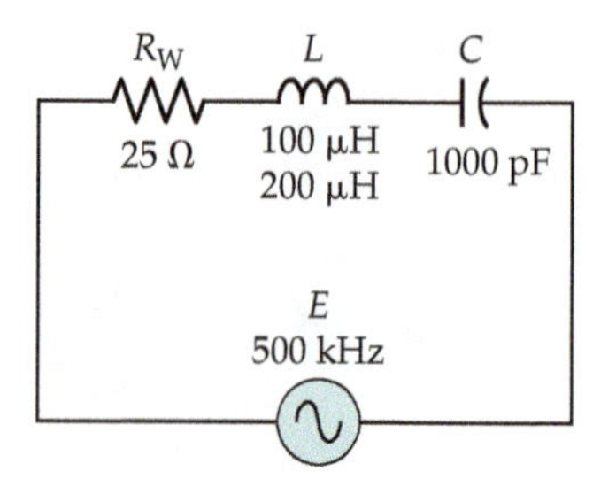

Figure 23-13 Circuit for Example 23-4.

Example 23-4

Determine the Q factor for the series RLC circuit in Figure 23-13. Also determine the new capacitance value required for resonance at 500 kHz when the inductance is doubled, and calculate the new Q factor.

Solution

Eq. 23-10,
$$Q_1 = \frac{1}{R}\sqrt{\frac{L}{C}} = \frac{1}{25\ \Omega}\sqrt{\frac{100\ \mu\text{H}}{1000\ \text{pF}}}$$
$$\approx 12.6$$

From Eq. 23-3,
$$C = \frac{1}{4\pi^2 f_r^2 L} = \frac{1}{4\pi^2 \times (500\ \text{kHz})^2 \times 200\ \mu\text{H}}$$
$$\approx 500\ \text{pF}$$

Eq. 23-8,
$$Q_1 = \frac{2\pi f_r L}{R} = \frac{2\pi \times 500\ \text{kHz} \times 200\ \mu\text{H}}{25\ \Omega}$$
$$\approx 25$$

Practice Problem

23-3.1 Calculate the Q factor of the circuit in Problem 23-2.1 at the maximum and minimum resonance frequencies.

23-4 BANDWIDTH OF A SERIES RESONANT CIRCUIT

Consider the current-versus-frequency response graph for a series resonant circuit once again, as reproduced in Figure 23-14. It is clear from the graph that the current reaches a maximum level (I_m) at resonance. It is also clear that at frequencies close to resonance, the current level is only a little below its maximum value. So, the resonant circuit is said to select a *band* of frequencies rather than just one frequency. In Figure 23-14, the lowest and highest frequencies of the band are identified as f_1 and f_2, respectively, and the *bandwidth* (B or Δf) is,

$$\Delta f = f_2 - f_1 \qquad \textbf{(23-11)}$$

The resonant circuit frequencies f_1 and f_2 are arbitrarily defined as those frequencies at which the power delivered to the circuit is half the power delivered at resonance.

So, f_1 and f_2 are termed the *half-power points*. As illustrated in Figure 23-15, the power levels at these frequencies can also be shown to be 3 dB below the

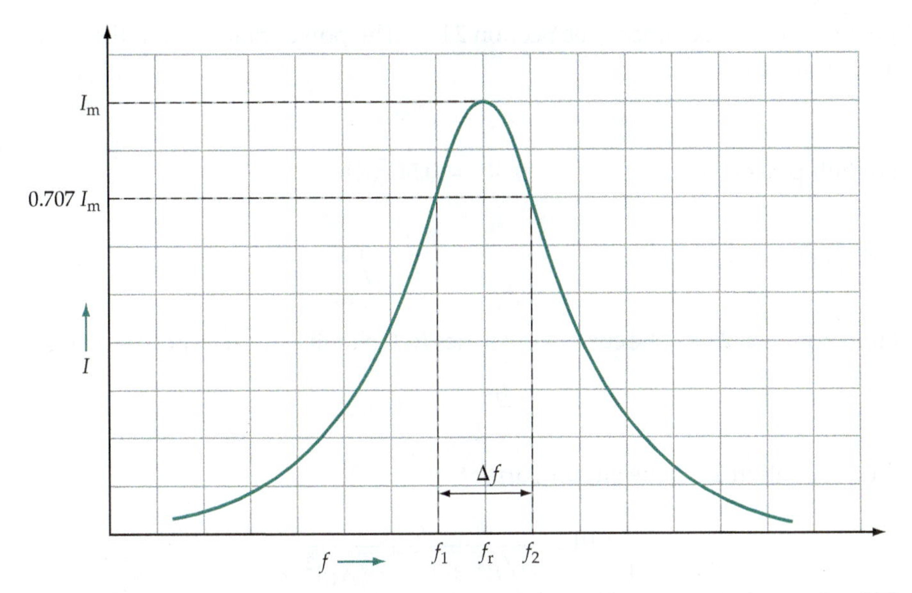

Figure 23-14 The lowest and highest frequencies for maximum current in a series *RLC* circuit are defined as those frequencies (f_1 and f_2) at which the current is 0.707 of its peak level. The bandwidth of the circuit is ($f_2 - f_1$).

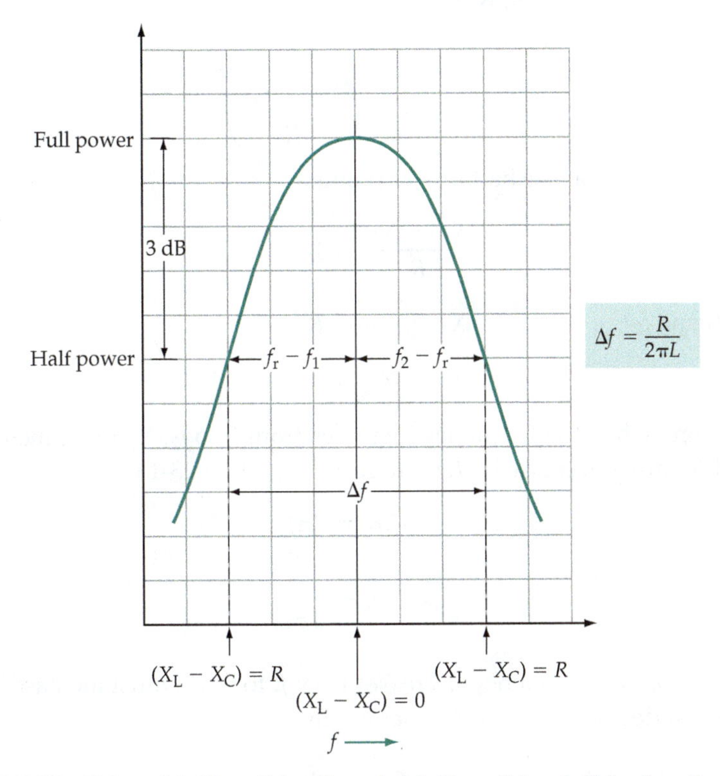

Figure 23-15 At f_1 and f_2 $(X_L - X_C) = R$, and at f_r $(X_L - X_C) = 0$. This gives the resonant circuit bandwidth as $\Delta f = R/(2\pi L)$.

power level at resonance, (see Section 24-2). The power delivered to the circuit at resonance is,

$$P = I_m^2 R$$

For half power, $$0.5P = 0.5(I_m^2 R)$$

Which gives, $$0.5P = \left(\frac{I_m}{\sqrt{2}}\right)^2 R \quad \text{(23-12)}$$

From Equation 23-12, the current in a series *RLC* circuit at the half-power points is,

$$I = \frac{I_m}{\sqrt{2}} = 0.707 I_m$$

From Equation 23-2, the circuit current is,

$$|I| = \frac{E}{\sqrt{R^2 + (X_L - X_C)^2}}$$

or $$|I| = \frac{E}{R\sqrt{1 + \left(\frac{X_L - X_C}{R}\right)^2}}$$

and $$E/R = I_m$$

So, $$|I| = \frac{I_m}{\sqrt{1 + \left(\frac{X_L - X_C}{R}\right)^2}}$$

To give $I = I_m/\sqrt{2}$ at the half power points,

$$\frac{X_L - X_C}{R} = 1$$

and (at f_1 and f_2), $$X_L - X_C = R \quad \text{(23-13)}$$

At f_r, $$X_L - X_C = 0$$

Therefore, when the frequency increases from f_r to f_2, X_L must increase by $R/2$ and X_C must decrease by $R/2$, to fulfill Equation 23-13.

So, $$0.5R = 2\pi f_2 L - 2\pi f_r L$$

Which gives $$f_2 - f_r = \frac{R}{4\pi L} \quad \text{(23-14)}$$

Similarly, when the frequency decreases from f_r to f_1, X_C must increase by $R/2$ and X_L must decrease by $R/2$. Consequently,

$$f_r - f_1 = \frac{R}{4\pi L} \quad \text{(23-15)}$$

Adding Equations 23-14 and 23-15, the circuit bandwidth is,

$$\Delta f = \frac{R}{2\pi L} \tag{23-16}$$

Multiplying the right-hand side of Equation 23-16 by f_r/f_r gives,

$$\Delta f = \frac{Rf_r}{2\pi f_r L}$$

or

$$\Delta f = \frac{f_r}{Q} \tag{23-17}$$

The ability of a resonant circuit to select one particular frequency and to discriminate against other frequencies is termed the *selectivity* of the circuit. Circuits with the narrowest bandwidths obviously have the greatest selectivity. Equation 23-17 shows that the largest Q factors produce the narrowest bandwidths. So, circuits with large Q factors have better selectivity than those with small Q factors.

Example 23-5

Calculate the half-power frequencies and the bandwidths for the circuit in Example 23-4 (reproduced in Figure 23-16) (a) for $L = 100\ \mu\text{H}$ and (b) for $L = 200\ \mu\text{H}$.

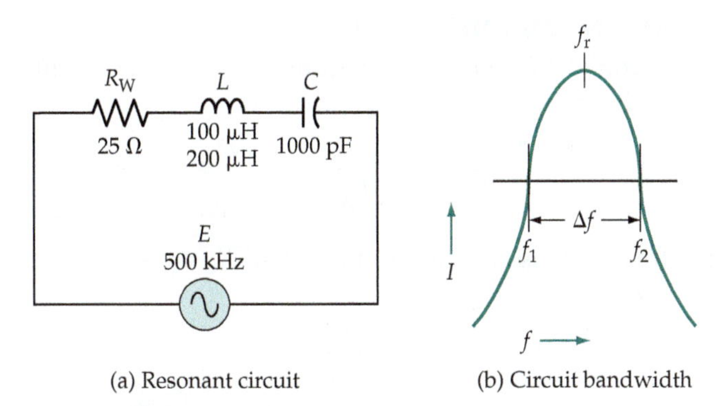

Figure 23-16 Illustration for Example 23-5.

Solution

(a) for $L = 100\ \mu\text{H}$.

From Eq. 23-14, $$f_2 = \frac{R}{4\pi L} + f_r = \frac{25\ \Omega}{4 \times \pi \times 100\ \mu\text{H}} + 500\ \text{kHz}$$
$$\approx 520\ \text{kHz}$$

From Eq. 23-15, $$f_1 = f_r - \frac{R}{4\pi L} = 500\ \text{kHz} - \frac{25\ \Omega}{4 \times \pi \times 100\ \mu\text{H}}$$
$$\approx 480\ \text{kHz}$$

Eq. 23-11, $$\Delta f = f_2 - f_1 = 520 \text{ kHz} - 480 \text{ kHz}$$
$$\approx 40 \text{ kHz}$$

Alternatively, from Example 23-4, $Q = 12.6$ when $L = 100\ \mu\text{H}$.

Eq. 23-17, $$\Delta f = \frac{f_r}{Q} = \frac{500 \text{ kHz}}{12.6}$$
$$= 40 \text{ kHz}$$

(b) For $L = 200\ \mu\text{H}$.

From Example 23-4, $Q = 25$ when $L = 200\ \mu\text{H}$.

Eq. 23-17, $$\Delta f = \frac{f_r}{Q} = \frac{500 \text{ kHz}}{25}$$
$$= 20 \text{ kHz}$$

Practice Problem

23-4.1 Calculate the bandwidth of the circuit in Problem 23-2.1.

23-5 PARALLEL RESONANCE

Ideal Parallel Resonant Circuit

Consider the parallel *RLC* circuit in Figure 23-17(a). The admittance of the circuit is seen to be

$$Y = \frac{1}{R} - j\frac{1}{X_L} + j\frac{1}{X_C}$$

If the supply frequency is adjusted until X_L and X_C are equal, the admittance becomes

$$Y = \frac{1}{R}$$

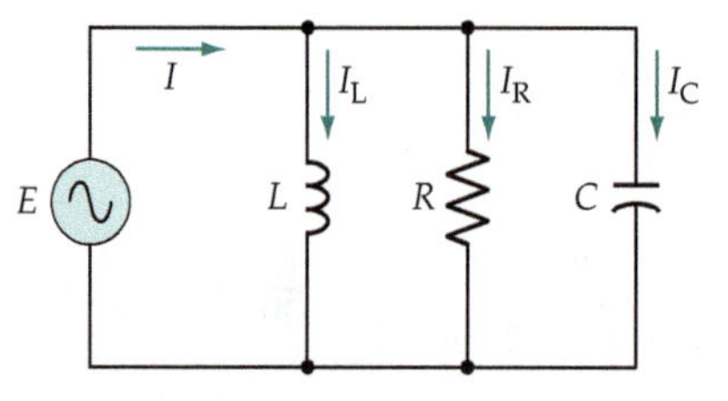

(a) Parallel *RCL* circuit with a variable frequency source

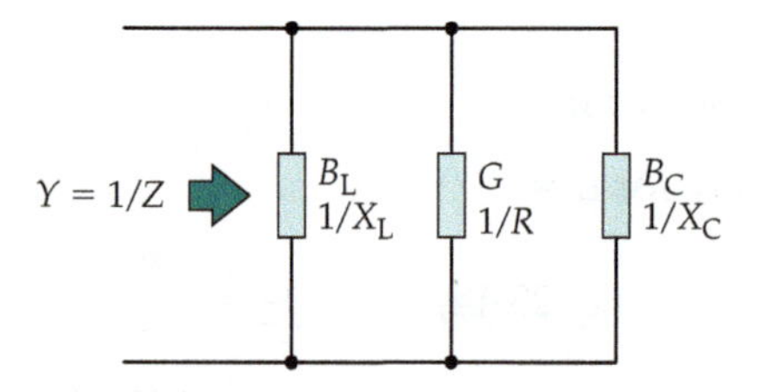

(b) B_L and B_C vary with the frequency of the source

Figure 23-17 Ideal parallel resonance circuit. At resonance, $X_L = X_C$, $I = E/R$, and I_C and I_L are much larger than I.

and the circuit impedance is

$$Z = R$$

Consequently, the current taken from the supply source is

$$I = \frac{E}{R}$$

and the currents through the inductor and capacitor are

$$I_L = \frac{E}{X_L} \quad \text{and} \quad I_C = \frac{E}{X_C}$$

As will be explained, I_L and I_C are normally much larger than the supply current at resonance.

The resistor current (in the circuit in Figure 23-17) is in phase with the supply voltage, the inductor current lags the supply voltage by 90°, and the capacitor current leads the supply voltage by 90°. This is illustrated by the phasor diagram in Figure 23-18(a) and by the current waveforms in Figure 23-18(b). When X_L and X_C are equal, the inductive and capacitive currents are equal and opposite, as illustrated in the phasor diagram. Thus, the total current supplied by the voltage source is I_R. Currents I_C and I_L are then the result of the energy stored in the circuit being continuously transferred from the inductor to the capacitor and back again.

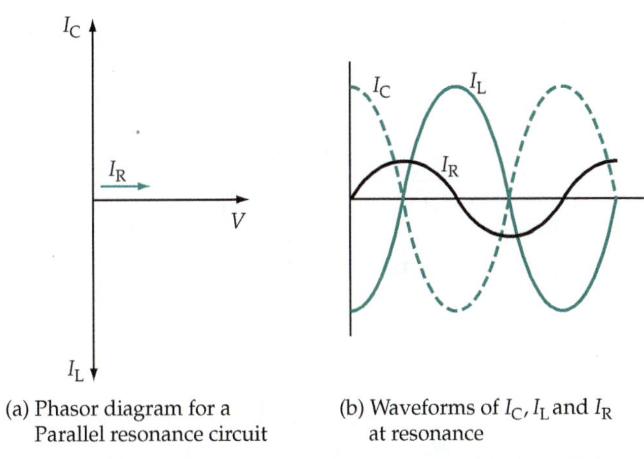

Figure 23-18 Current phasor diagram and waveforms for an ideal parallel resonance circuit. I_C and I_L cancel each other, and the total supply current is I_R, which is in phase with the supply voltage.

Practical Parallel Resonant Circuit

A practical inductor is not purely reactive but has a winding resistance, and this must be represented as a resistance R_W in series with an inductance L. Some capacitors must also be shown as having a resistive component; however, in general, capacitors can be assumed to be purely reactive. The circuit in

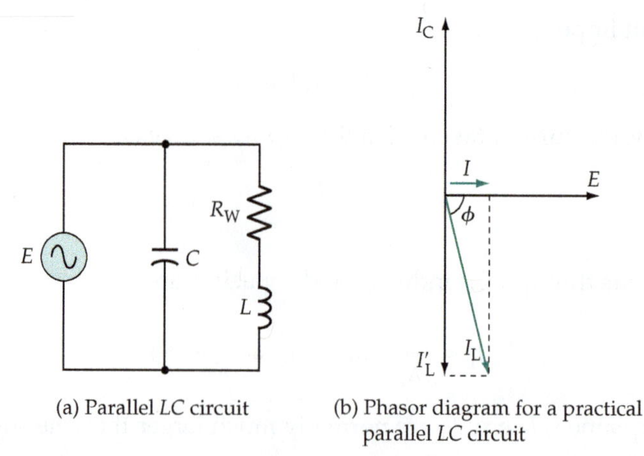

(a) Parallel *LC* circuit (b) Phasor diagram for a practical parallel *LC* circuit

Figure 23-19 In a practical parallel resonance circuit, the presence of a resistive component in the inductance gives an inductor current phase angle less than 90° at resonance.

Figure 23-19(a) is that of a practical parallel *LC* circuit, and the phasor diagram in Figure 23-19(b) represents the circuit conditions at resonance. It is seen that the presence of the resistive component in the inductive branch makes the I_L phase angle (ϕ) slightly less than 90°. Resonance is achieved when the vertical component of I_L [I'_L in Figure 23-19(b)] is equal to the capacitor current (I_C).

The admittance of the parallel circuit in Figure 23-19(a) is,

$$Y = \frac{1}{R_W + jX_L} + j\frac{1}{X_C}$$

Multiplying $\frac{1}{R_W + jX_L}$ by $\frac{R_W - jX_L}{R_W - jX_L}$

gives,

$$Y = \frac{R_W}{R_W^2 + X_L^2} - j\frac{X_L}{R_W^2 + X_L^2} + j\frac{1}{X_C} \tag{23-18}$$

For the circuit to become purely resistive at resonance,

$$\frac{1}{X_C} = \frac{X_L}{R_W^2 + X_L^2}$$

or

$$X_C = \frac{R_W^2 + X_L^2}{X_L} \tag{23-19}$$

and Equation 23-18 gives the admittance at resonance as,

$$Y = \frac{R_W}{R_W^2 + X_L^2}$$

So, the circuit impedance at resonance is,

$$Z = \frac{R_W^2 + X_L^2}{R_W}$$

From Eq. 23-19, $$X_L X_C = R_W^2 + X_L^2$$

Therefore, $$Z = \frac{X_L X_C}{R_W} = \frac{2\pi f_r L}{2\pi f_r C R_W}$$

So, at resonance, $$Z = \frac{L}{CR_W} \tag{23-20}$$

and this is a resistive quantity.

When the impedance of the parallel *LC* circuit (as represented by the inverse of Equation 23-18) is plotted to a logarithmic frequency base, it is found that the impedance peaks at the resonance frequency [see Figure 23-20(a)]. Therefore,

a parallel LC circuit has a maximum impedance at the resonance frequency,

Recall that a series resonant circuit has a *minimum* impedance at the resonance frequency. So the impedances of series and parallel *LC* circuits at resonance are opposites. As a consequence of the peak in the impedance value of a parallel resonant circuit, there is a dip in the current taken from the supply at the resonance frequency. This is illustrated in Figure 23-20(b). Once again, this is the opposite of the case with series resonance.

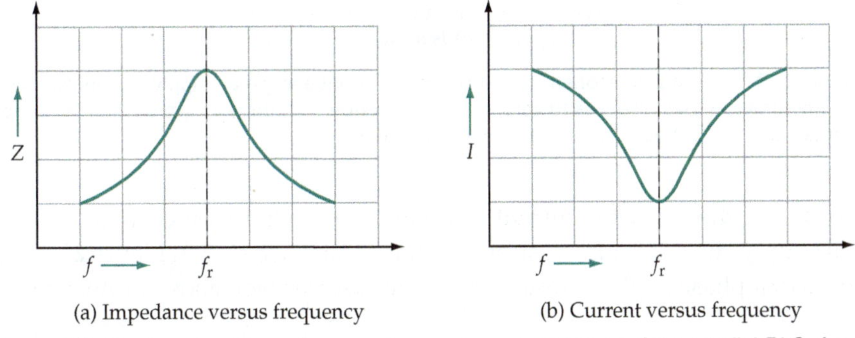

Figure 23-20 Graphs of impedance and current versus frequency for a parallel *RLC* circuit. Because X_L and X_C cancel each other at f_r, the impedance peaks to $Z = L/(CR_W)$ and the current dips.

Phase Angle

As discussed, the impedance of a parallel *LC* circuit is resistive at the resonance frequency. At supply frequencies below resonance, the inductive reactance is smaller than the capacitive reactance; $[(2\pi f L) < 1/(2\pi f C)]$, see Figure 23-21(a). Thus, the inductive current is greater than the capacitive current, and the total supply current lags the supply voltage. This effect becomes most pronounced at frequencies well below f_r, so that, as shown in Figure 23-21(c), the current phase angle is close to $-90°$.

When the supply frequency is above f_r, the capacitive reactance becomes smaller than the inductive reactance; $[1/(2\pi f C) < (2\pi f L)]$, see Figure 23-21(b).

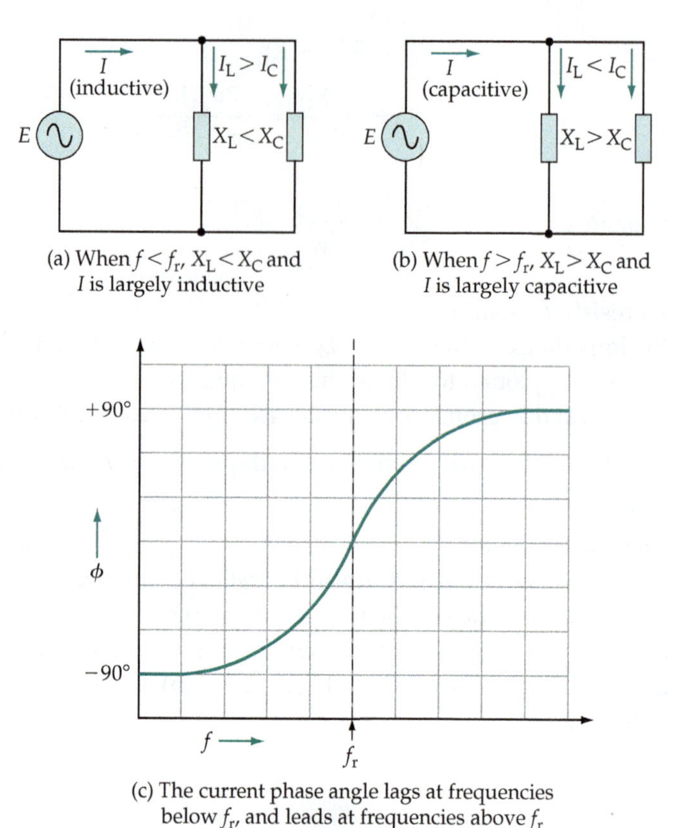

Figure 23-21 A parallel resonant circuit has a −90°phase angle at supply frequencies well below resonance. This changes to zero as the resonance frequency is approached, and then approaches +90° at frequencies well above resonance.

In this case, the capacitor current is larger than the inductor current, and the total supply current leads the supply voltage. As illustrated in Figure 23-21(c), the current phase angle is close to 90° at frequencies well above resonance.

Example 23-6

A parallel LC circuit, as in Figure 23-19(a), has $E = 100$ mV, $L = 150$ μH, $R_W = 15\ \Omega$, and $C = 750$ pF. Calculate the supply current at resonance.

Solution

Eq. 23-20,

$$Z = \frac{L}{CR_W} = \frac{150\ \mu\text{H}}{750\ \text{pF} \times 15\ \Omega}$$

$$= 13.3\ \text{k}\Omega$$

$$I = \frac{E}{Z} = \frac{100\ \text{mV}}{13.3\ \text{k}\Omega}$$

$$= 7.52\ \mu\text{A}$$

Practice Problem

23-5.1 Determine the maximum and minimum impedance at resonance for the circuit in Figure 23-22.

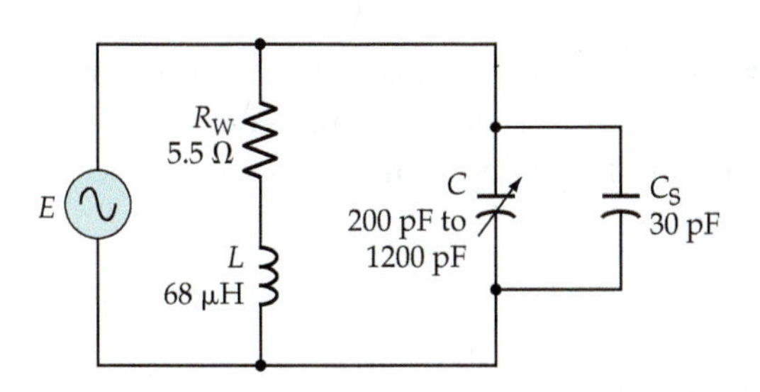

Figure 23-22 Circuit for Problem 23-5.1.

23-6 *Q* FACTOR FOR PARALLEL *LC* CIRCUITS

Referring again to Figure 23-19(b), it is seen that the total current drawn from the supply at resonance (I), is in phase with the supply voltage (E). Also, I is much smaller than the inductive and capacitive currents (I_C and I_L). Thus, in the parallel resonant LC circuit, there is a *current magnification*, which is analogous to the voltage magnification that occurs in a series resonant circuit. The Q factor of the parallel resonant circuit is the ratio of I_L (or I_C) to I.

As illustrated in Figure 23-23,

$$I = \frac{E}{Z}$$

From Eq. 23-20,

$$Z = \frac{L}{CR_W}$$

So,

$$I = \frac{E}{L/(CR_W)}$$

Also,

$$I_L = I_C = \frac{E}{X_L} = E \times \omega C$$

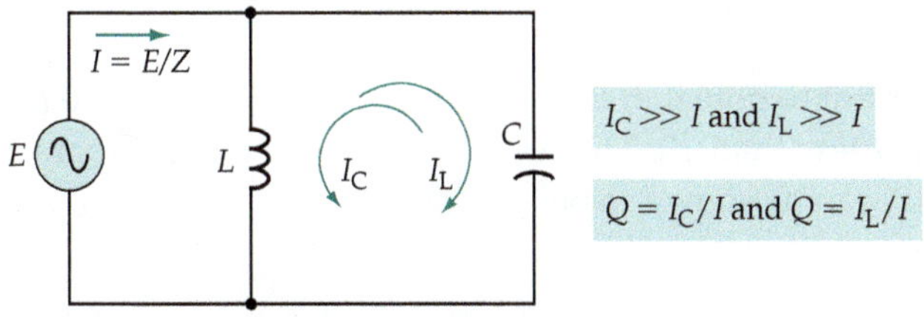

$I = E/Z$

E L I_C I_L C

$I_C >> I$ and $I_L >> I$

$Q = I_C/I$ and $Q = I_L/I$

Figure 23-23 In a parallel resonance circuit, I_L and I_C are much larger than the supply current (I). The circuit Q factor is the ratio I_L/I or I_C/I.

and $$Q = \frac{I_C}{I} = E \times \omega C \times \frac{L/(CR_W)}{E}$$

Giving, $$Q = \frac{\omega L}{R_W} \tag{23-21}$$

Equation 23-21 is exactly the same as the Q factor equation for a series resonant circuit, (Equation 23-8). If R_W is the only resistance in the series LC circuit, then the Q factor for both series and parallel resonant circuits is the Q factor of the inductor.

Practice Problem

23-6.1 Calculate the Q factor of the circuit in Figure 23-22 when the adjustable capacitor is set to 700 pF.

23-7 RESONANCE FREQUENCY FOR PARALLEL *LC* CIRCUITS

Reconsider Equation 23-19, which gives the value of X_C at resonance.

Eq. 23-19, $$X_C = \frac{R_W^2 + X_L^2}{X_L}$$

Eq. 23-21, $$Q = \frac{\omega L}{R_W} = \frac{X_L}{R_W}$$

When $Q > 10$, $$X_L^2 \gg R_W^2$$

So, $$X_C = \frac{R_W^2 + X_L^2}{X_L} \approx \frac{X_L^2}{X_L}$$

Giving, $$X_C = X_L \text{ at resonance}$$

This means that the resonance frequency for a parallel LC circuit with $Q > 10$ is the same as that for a series LC circuit,

$$f_r = \frac{1}{2\pi\sqrt{LC}} \tag{23-22}$$

It is important to note that Equation 23-22 does not apply to parallel LC circuits which Q less than 10. The resonance frequency for a parallel LC circuit with $Q < 10$, can be shown to be,

$$f_r = \frac{1}{2\pi\sqrt{LC}}\sqrt{1 - \frac{CR_W^2}{L}} \tag{23-23}$$

The bandwidth of a parallel resonant circuit is determined in exactly the same way as that for a series resonant circuit (Equation 23-17).

$$\Delta f = \frac{f_r}{Q} \tag{23-24}$$

Example 23-7

Calculate the resonance frequency, Q factor, and inductor current for the parallel LC circuit in Example 23-6.

Solution

Eq. 23-22,
$$f_r = \frac{1}{2\pi\sqrt{LC}} = \frac{1}{2\pi \times \sqrt{150\ \mu\text{H} \times 750\ \text{pF}}}$$
$$= 474.5\ \text{kHz}$$

Eq. 23-21,
$$Q = \frac{\omega L}{R_W} = \frac{2\pi \times 474.5\ \text{kHz} \times 150\ \mu\text{H}}{15\ \Omega}$$
$$= 29.8$$

$$I_L = \frac{E}{2\pi f_r L} = \frac{100\ \text{mV}}{2\pi \times 474.5\ \text{kHz} \times 150\ \mu\text{H}}$$
$$= 223.6\ \mu\text{A}$$

Example 23-8

A parallel LC circuit [as in Figure 23-24(a)] has a 100 μH inductance with a 12 Ω coil resistance. The capacitor is adjustable over the range 200 pF to 300 pF. Determine the maximum and minimum resonance frequencies for the circuit. Also, calculate the Q factor and bandwidth of the circuit at the two resonance frequency extremes [see Figure 23-24(b)].

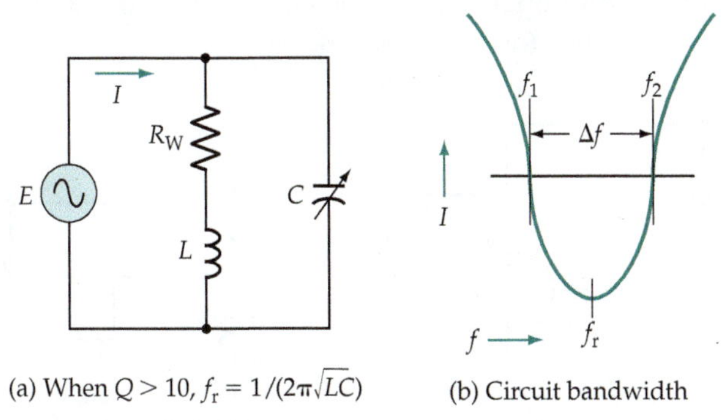

(a) When $Q > 10$, $f_r = 1/(2\pi\sqrt{LC})$ (b) Circuit bandwidth

Figure 23-24 In a parallel resonance circuit with Q > 10, the resonance frequency and bandwidth are calculated as in a series resonance circuit.

Solution

When $C = 200$ pF,

Eq. 23-22, $$f_r = \frac{1}{2\pi\sqrt{LC}} = \frac{1}{2\pi \times \sqrt{100\ \mu\text{H} \times 200\ \text{pF}}}$$
$$= 1.13\ \text{MHz}$$

When $C = 300$ pF, $$f_r = \frac{1}{2\pi\sqrt{LC}} = \frac{1}{2\pi \times \sqrt{100\ \mu\text{H} \times 300\ \text{pF}}}$$
$$= 919\ \text{kHz}$$

At $f_r = 1.13$ MHz,

Eq. 23-21, $$Q = \frac{\omega L}{R_W} = \frac{2\pi \times 1.13\ \text{MHz} \times 100\ \mu\text{H}}{12\ \Omega}$$
$$= 59.2$$

At $f_r = 919$ kHz, $$Q = \frac{\omega L}{R_W} = \frac{2\pi \times 919\ \text{kHz} \times 100\ \mu\text{H}}{12\ \Omega}$$
$$= 48$$

Eq. 23-16, $$\Delta f = \frac{R_W}{2\pi L} = \frac{12\ \Omega}{2\pi \times 100\ \mu\text{H}}$$
$$= 19.1\ \text{kHz}$$

Practice Problems

23-7.1 For the circuit in Problem 23-5.1 [reproduced in Figure 23-25(a)], determine the maximum and minimum resonance frequencies, the circuit bandwidth, and the Q factors at each resonance frequency.

23-7.2 The highest resonance frequency for the circuit in Figure 23-25(a) is found to be 10 kHz below the calculated frequency. As illustrated in Figure 23-25(b), the effect could be due to stray inductance in series with L or to additional stray capacitance in parallel with C. Calculate the stray quantities for each case.

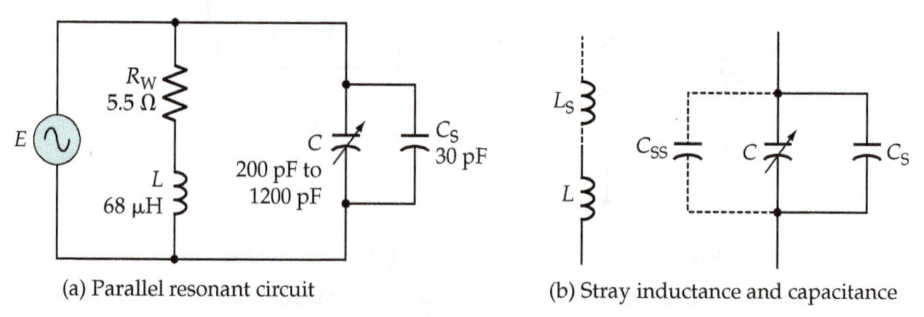

Figure 23-25 Illustrations for Problems 23-7.1 and 23-7.2.

23-8 RESISTANCE DAMPING OF PARALLEL *LC* CIRCUITS

Figure 23-26 shows a parallel *LC* circuit with a resistor (R_D), connected in parallel with the capacitor and inductor. Note the circuit currents identified as the *LC* circuit current (at resonance) (I_Z), the parallel resistor current (I_{RD}), and the total supply current (I_T). The impedance of the parallel *LC* circuit is a maximum at resonance, and the presence of R_D reduces the total impedance of the circuit. As will be shown, it also reduces the circuit *Q* factor.

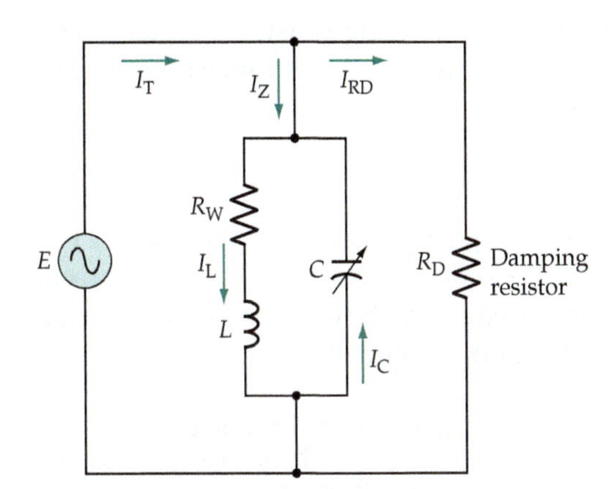

Figure 23-26 A damping resistor may be connected to a parallel *LC* circuit to reduce the circuit *Q* factor, and thus adjust the bandwidth.

The *Q* factor for the parallel resonance *RLC* circuit is the ratio of the inductor current (or the capacitor current) at resonance to the total supply current. So the resonant circuit current is

$$I_Z = \frac{I_L}{Q_L}$$

where Q_L is the *Q* factor of the inductor.

The total supply current is, $I_T = I_Z + I_{RD}$

So, the *Q* factor of the parallel combination of *LC* and R_D is,

$$Q_P = \frac{I_L}{I_T} = \frac{I_L}{I_Z + I_{RD}} = \frac{I_L}{(I_L/Q_L) + I_{RD}}$$

$$= \frac{E/(\omega L)}{\{[E/(\omega L)]/Q_L\} + (E/R_D)}$$

$$= \frac{1}{(1/Q_L) + (\omega L/R_D)}$$

or

$$Q_P = Q_L \| [R_D/(\omega L)] \qquad (23\text{-}25)$$

When $Q_L \gg R_D/(\omega L)$,

$$Q_P \approx \frac{R_D}{\omega L} \qquad (23\text{-}26)$$

Equation 23-26 should be used only when $Q_L \gg (R_D/\omega L)$; typically, Q_L should be ten times $(R_D/\omega L)$ or greater, otherwise Equation 23-25 should be used.

Note that the approximate equation for the Q factor of a parallel RLC circuit is inverted by comparison with the equation for the Q factor of the inductance ($Q_L = \omega L/R_L$). It is seen that when R_D is made smaller, the circuit Q is reduced. The effect is known as *damping*, and so R_D is referred to as a *damping resistor*. The reduced Q factor results in a broader circuit bandwidth:

Eq. 23-24,
$$\Delta f = \frac{f_r}{Q}$$

Bandwidth adjustment in one purpose of a damping resistor. Damping may also be employed where resonance of an LC circuit is undesirable; for example, to avoid oscillations in audio circuits.

Example 23-9

A parallel resonant circuit with a damping resistor, as in Figure 23-27, has $L = 150\ \mu\text{H}$, $R_W = 5\ \Omega$, and $C = 100$ pF. The damping resistor is adjustable from 15 kΩ to 20 kΩ. Calculate the resonance frequency, the Q factor of the inductor, and the maximum and minimum Q factor of the circuit.

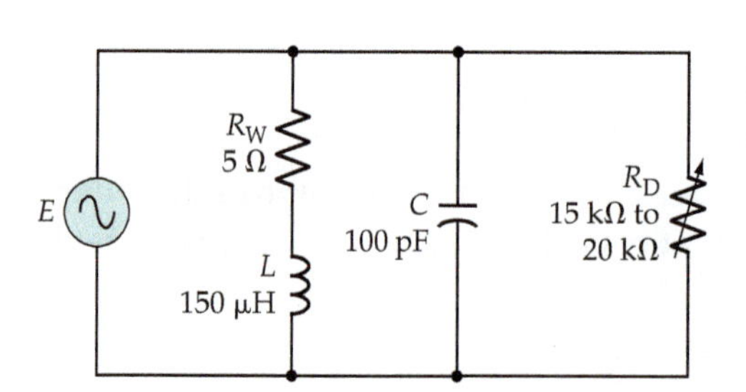

Figure 23-27 Parallel resonant circuit with an adjustable damping resistor; analyzed in Example 23-9.

Solution

Eq. 23-22,
$$f_r = \frac{1}{2\pi\sqrt{LC}} = \frac{1}{2\pi \times \sqrt{150\ \mu\text{H} \times 100\ \text{pF}}}$$
$$\approx 1.3\ \text{MHz}$$

Eq. 23-8,
$$Q_L = \frac{\omega L}{R_W} = \frac{2\pi \times 1.3\ \text{MHz} \times 150\ \mu\text{H}}{5\ \Omega}$$
$$\approx 245$$

Eq. 23-26,
$$Q_{P(\text{min})} \approx \frac{R_{D(\text{min})}}{\omega L} = \frac{15\ \text{k}\Omega}{2\pi \times 1.3\ \text{MHz} \times 150\ \mu\text{H}}$$
$$\approx 12.2$$

and
$$Q_{P(\text{max})} \approx \frac{R_{D(\text{max})}}{\omega L} = \frac{20\ \text{k}\Omega}{2\pi \times 1.3\ \text{MHz} \times 150\ \mu\text{H}}$$
$$\approx 16.3$$

More precisely from Eq. 23-25,

$$Q_P = Q_L \| (R_D/\omega L)$$
$$Q_{P(min)} = 245 \| 12.2 = 11.6$$
$$Q_{P(max)} = 245 \| 16.3 = 15.3$$

Practice Problem

23-8.1 The circuit in Figure 23-28 is to be damped to have a maximum Q factor of 6. Calculate the resonance frequency, the required damping resistance, and circuit bandwidth.

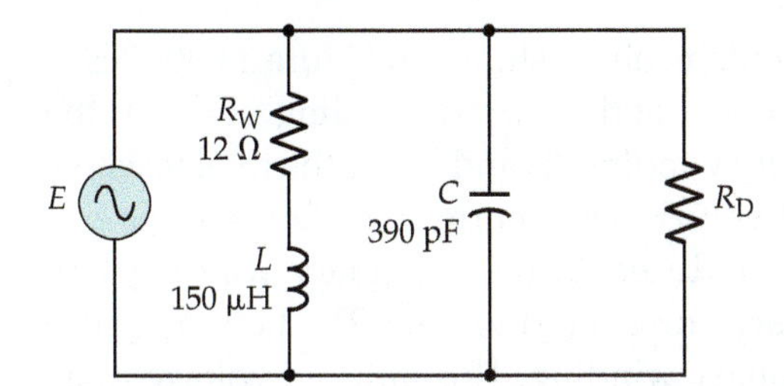

Figure 23-28 Circuit for Problem 23-8.1.

23-9 TUNED COUPLED COILS

Coupled Coils

In Chapter 14 it is shown that mutual inductance exists between adjacent coils, and that when a current flows in one coil, a terminal voltage may be induced in a mutally coupled coil. Figure 23-29(a) shows a typical arrangement of two mutually coupled coils wound on a nonmagnetic core. The coefficient of coupling between the coils (see Section 14-4) can be adjusted simply by altering the spacing between them. The circuit diagram for coupled coils is shown in Figure 23-29(b). Note the dots at one end of each coil. These are employed to show the phase relationship between input and output voltages. The dots simply identify the in-phase ends of each coil. Thus, when an input voltage applied to the left-hand coil has the instantaneous polarity shown (positive at the dotted end of the coil), the output voltage from the secondary (right-hand) coil is also positive at the dotted end of the coil.

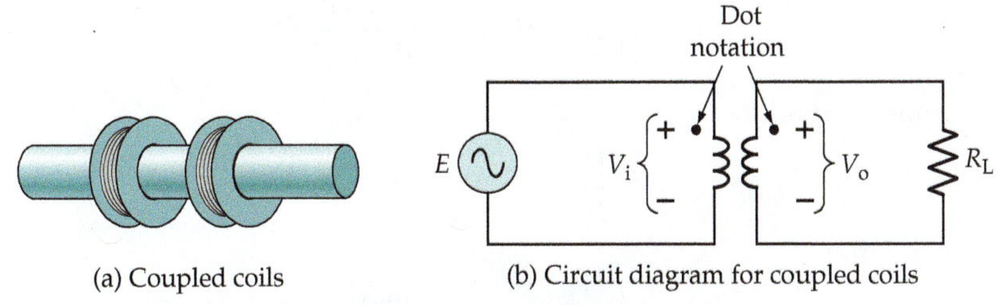

Figure 23-29 Coupled coils and their circuit diagram. The dot notation on the circuit diagram denotes the relative polarity of the coil input and output voltages.

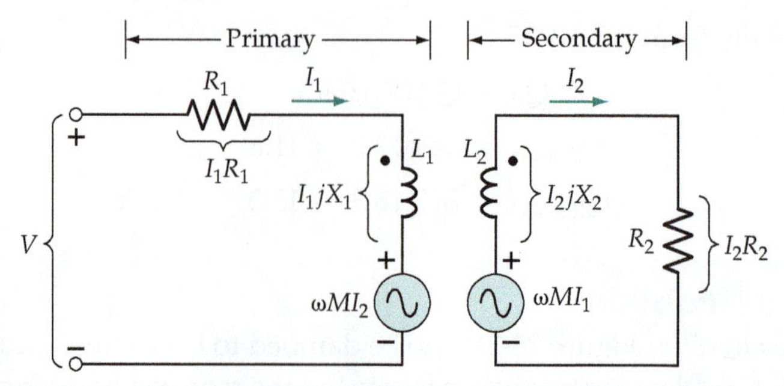

Figure 23-30 Equivalent circuit for coupled coils. Each coil has resistive and inductive components (*R* and *L*), and each has a voltage (ωMI) induced in it by the current flow in the other coil.

The equivalent circuit for the coupled coils is shown in Figure 23-30. Resistor R_1 represents the resistive component of the *primary* winding, and L_1 is the inductive component. In the *secondary* winding R_2 and L_2 are the total resistive and inductive components, respectively. When a current (I_1) flows in the primary winding, a voltage (ωMI_1) is induced in the secondary winding [see Equation (19-6)], causing a secondary current (I_2) to flow. The flow of I_2 also induces a voltage (ωMI_2) in the primary winding. This induced voltage must be shown as having a polarity that opposes the primary current. Two voltage generator symbols are included in the equivalent circuit to represent the voltages induced in each winding.

Coupled Impedance

The equation for the voltage drops around the primary circuit may be written

$$V = I_1R_1 + jI_1X_1 + \omega MI_2 \tag{23-27}$$

and the equation for the voltage drops around the secondary circuit is,

$$\omega MI_1 = I_2R_2 + jI_2X_2$$

Which gives,

$$I_2 = \frac{\omega MI_1}{R_2 + jX_2} \tag{23-28}$$

Substituting for I_2 in Equation 23-27,

$$V = I_1R_1 + jI_1X_1 + \frac{(\omega M)^2 I_1}{R_2 + jX_2}$$

The primary impedance is,

$$Z_1 = \frac{V}{I_1}$$

So,

$$Z_1 = R_1 + jX_1 + \frac{(\omega M)^2}{R_2 + jX_2} \tag{23-29}$$

The primary circuit impedance is seen to consist of a *self-impedance* Z and a *coupled impedance* Z_2' (coupled from the secondary into the primary) where the self-impedance is,

$$Z = R_1 + jX_1$$

and the coupled impedance is,

$$Z_2' = \frac{(\omega M)^2}{R_2 + jX_2} \qquad \text{(23-30)}$$

The coupled impedance can be resolved into resistive and reactive components by multiplying it by the conjugate of the denominator (see Section 18-4).

$$Z_2' = \frac{(\omega M)^2}{R_2 + jX_2} \times \frac{R_2 - jX_2}{R_2 - jX_2}$$

$$= \frac{(\omega M)^2 R^2 - j(\omega M)^2 X_2}{R_2^2 + X_2^2}$$

$$Z_2' = \frac{(\omega M)^2 R_2}{R_2^2 + X_2^2} - j\frac{(\omega M)^2 X_2}{R_2^2 + X_2^2} \qquad \text{(23-31)}$$

It is seen that the coupled impedance is made up of a coupled resistance R_2' and a coupled reactance X_2'.

Coupled resistance, $$R_2' = \frac{(\omega M)^2 R_2}{R_2^2 + X_2^2} \qquad \text{(23-32)}$$

Coupled reactance, $$X_2' = -j\frac{(\omega M)^2 R_2}{R_2^2 + X_2^2} \qquad \text{(23-33)}$$

Coefficient of Coupling

Now consider the coupled circuits shown in Figure 23-31(a), in which the primary and secondary circuits include capacitors. The circuits are tuned to resonate at the same frequency, and so the series reactance in each circuit is zero at the resonance frequency. When X_2 is zero, only the resistive component of the impedance is coupled from the secondary into the primary. Also, because of the presence of X_2 in the denominator of the expression for the coupled resistance [Equation 23-32], the coupled resistance is a maximum when X_2 becomes zero (i.e., at the resonance frequency). This (maximum) coupled resistance adds to the coil resistance of the primary and thus has the effect of reducing the primary current at resonance.

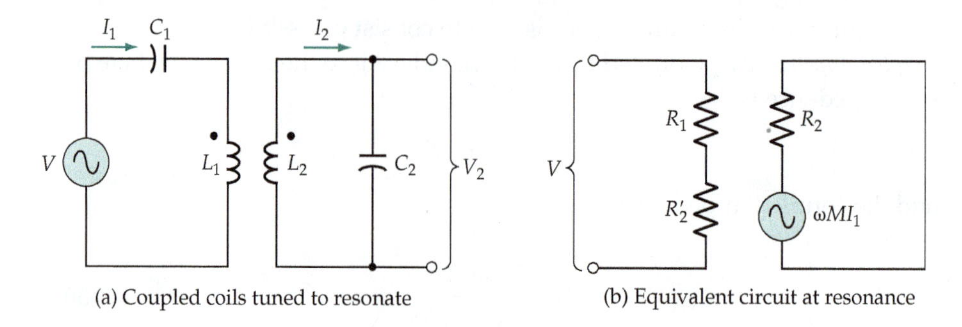

Figure 23-31 Coupled coils can be tuned to resonate at the same frequency by including appropriate capacitors in the primary and secondary circuits. At resonance, X_{C1} and X_{C2} cancel X_{Ll} and X_{L2}, respectively, and maximum primary and secondary currents flow.

Figure 23-31(b) shows the equivalent circuit of the coupled circuits at resonance. R_1 and R_2 are the primary and secondary coil resistances, respectively, and R_2' is the secondary resistance coupled into the primary. At resonance (with $X_2 = 0$), Equation 23-32 for R_2' becomes,

$$R_2' = \frac{(\omega M)^2}{R_2} \tag{23-34}$$

Viewed from the input terminals, the primary circuit (operating alone) is a series resonant circuit with a maximum impedance at resonance,

$$Z = R_1$$

The secondary circuit behaves as a series resonant circuit with a supply voltage of (ωMI_1). The coupled resistance R_2' causes the maximum (resonance) impedance of the primary circuit to become,

$$Z_1 = R_1 + R_2' \tag{23-35}$$

It is clear that the effect of R_2' is to increase Z_1 and consequently to reduce the current flowing through the primary winding. The degree to which the primary current is affected by the secondary depends on how tightly the coils are coupled; that is, on the coefficient of coupling k.

The frequency response curves of I_1 and I_2 for *loosely coupled coils* are shown in Figure 23-32(a). The shape of the I_1 response curve is largely unaffected by the coupled impedance, and the I_2 curve is seen to be very much smaller than I_1 but with the same shape as the I_1 curve.

With *tight coupling* [Figure 23-32(b)], the coupled resistance has a considerable effect at the resonance frequency. The total primary impedance is increased and the primary current is significantly reduced. Also, the primary response curve is distorted by the occurrence of two *humps*; one above and one below the resonance frequency (f_r). These humps result from the fact that the secondary impedance is capacitive below the f_r and inductive above f_r. When coupled into the primary, the coupled impedances are inductive below f_r and capacitive above f_r. These result in minimum primary impedances, and

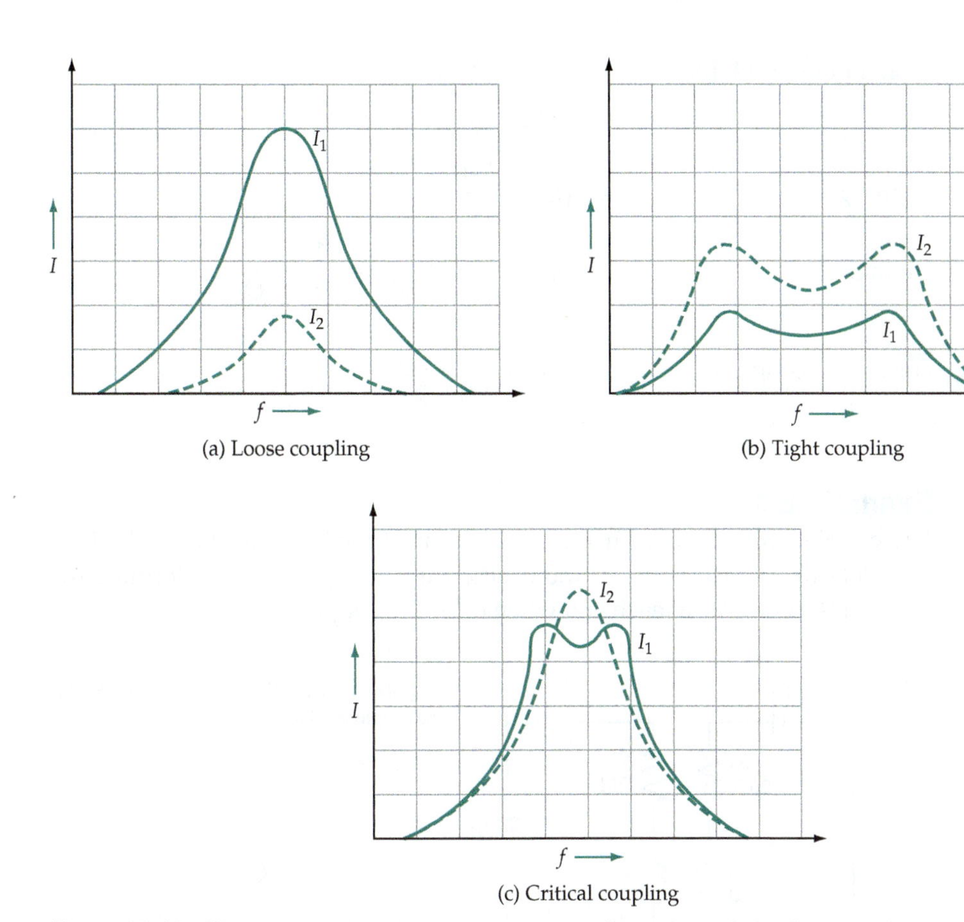

Figure 23-32 The current-versus frequency response of tuned coupled coils depends on the degree of coupling. Loose coupling gives low secondary current, tight coupling produces distortion, and critical coupling affords maximum power transfer from primary to secondary.

therefore maximum currents at two frequencies above and below resonance. Because of the tight coupling, the response curve of the secondary current is simply a slightly amplified version of the primary.

With *critical coupling* [Figure 23-32(c)], maximum power transfer occurs from the primary to the secondary, and thus the secondary current is greater than in either of the other two cases. The condition for maximum power transfer is that the source impedance and load impedance be equal (see Section 22-8). The two humps still occur in the primary response curve but are absent in the secondary. For maximum power transfer and critical coupling:

$$R_1 = R_2' = \frac{(\omega M)^2}{R_2}$$

or

$$M^2 = \frac{R_1 R_2}{\omega^2} \tag{23-36}$$

From Equation 14-13,

$$M^2 = k^2L_1L_2$$

Therefore, $$k^2L_1L_2 = \frac{R_1R_2}{\omega^2}$$

Giving, $$k^2 = \frac{R_1}{\omega L_1} \times \frac{R_2}{\omega L_2} = \frac{1}{Q_1} \times \frac{1}{Q_2}$$

For critical coupling, $$k = \frac{1}{\sqrt{Q_1Q_2}} \qquad (23\text{-}37)$$

Example 23-10

The coupled coils shown in Figure 23-33 are tuned to resonate at 1 MHz. Calculate the critical value of the coefficient of coupling and determine the levels of the currents in each coil when critically coupled.

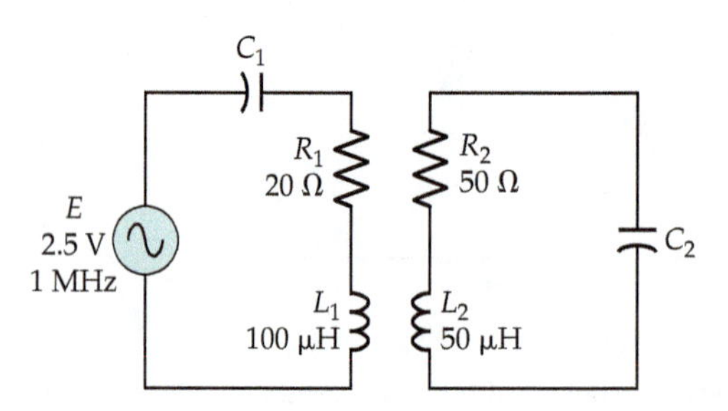

Figure 23-33 Coupled coil circuit for Example 23-10.

Solution

$$Q_1 = \frac{\omega L_1}{R_1} = \frac{2\pi \times 1\text{ MHz} \times 100\ \mu\text{H}}{20\ \Omega}$$

$$= 31.4$$

$$Q_2 = \frac{\omega L_2}{R_2} = \frac{2\pi \times 1\text{ MHz} \times 50\ \mu\text{H}}{50\ \Omega}$$

$$= 6.3$$

Eq. 23-37, $$k = \frac{1}{\sqrt{Q_1Q_2}} = \frac{1}{\sqrt{31.4 \times 6.3}}$$

$$= 0.071$$

When critically coupled, $$R_1 = R_2'$$

and, $$I_1 = \frac{E}{R_1 + R_2'} = \frac{2.5\text{ V}}{20\ \Omega + 20\ \Omega}$$

$$= 62.5\text{ mA}$$

From Eq. 23-36,

$$\omega M = \sqrt{R_1 R_2} = \sqrt{20\ \Omega \times 50\ \Omega}$$

$$= 31.6$$

Secondary voltage,

$$e_2 = \omega M I_1$$

Secondary current,

$$I_2 = \frac{e_2}{R_2} = \frac{\omega M I_1}{R_2} = \frac{31.6 \times 62.5\ \text{mA}}{50\ \Omega}$$

$$= 39.5\ \text{mA}$$

Example 23-11

Calculate the required secondary capacitance (C_2) in the circuit for Example 23-10. Also calculate the secondary capacitor voltage.

Solution

From Eq. 23-3,

$$C_2 = \frac{1}{(2\pi f_r)^2 L_2} = \frac{1}{(2\pi \times 1\ \text{MHz})^2 \times 50\ \mu\text{H}}$$

$$= 507\ \text{pF}$$

$$X_{C2} = \frac{1}{2\pi f C_2} = \frac{1}{2\pi \times 1\ \text{MHz} \times 507\ \text{pF}}$$

$$= 314\ \Omega$$

$$v_{C2} = I_2 \times X_{C2} = 39.5\ \text{mA} \times 314\ \Omega$$

$$= 12.4\ \text{V}$$

Practice Problem

23-9.1 The coils in Figure 23-34 are in resonance at a supply frequency of 800 kHz, and the coefficient of coupling between the coils is $k = 0.04$. Calculate the resistance (R_L) that must be connected in series with the secondary coil to obtain critical coupling.

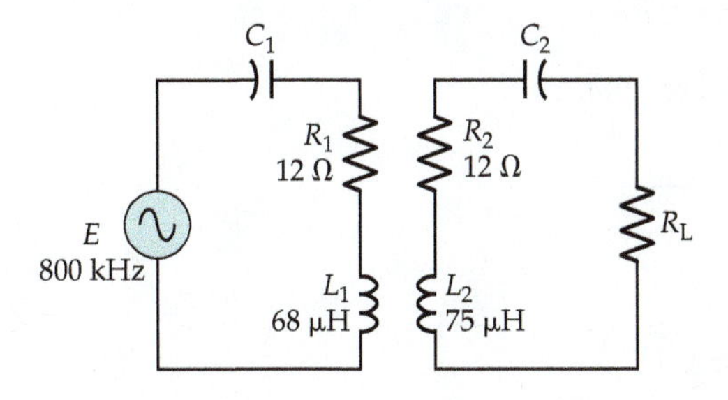

Figure 23-34 Circuit for Problem 23-9.1.

Summary of Formulas

- ***Impedance of series RLC circuit:***

$$Z = R + j(X_L - X_C)$$

- ***Resonance frequency:***

$$f_r = \frac{1}{2\pi\sqrt{LC}}$$

- ***Q factor for series resonance:***

$$Q = \frac{2\pi f_r L}{R}$$

$$Q = \frac{1}{2\pi f_r CR}$$

$$Q = \frac{1}{R}\sqrt{\frac{L}{C}}$$

- ***Bandwidth:***

$$B = \Delta f = f_2 - f_1$$

$$\Delta f = \frac{f_r}{Q}$$

- ***For parallel LC circuit:***

$$Y = \frac{1}{R_W + jX_L} + j\frac{1}{X_C}$$

- ***At resonance:***

$$Z = \frac{L}{CR_W}$$

- ***Q factor for parallel resonance:***

$$Q = \frac{2\pi f_r L}{R_W}$$

$$Q = \frac{1}{2\pi f_r CR_W}$$

- ***Resonance frequency for parallel resonance:***

When Q > 10, $$f_r = \frac{1}{2\pi\sqrt{LC}}$$

When Q < 10, $$f_r = \frac{1}{2\pi\sqrt{LC}}\sqrt{1 - \frac{CR_W^2}{L}}$$

- ***For parallel RLC circuit:***

$$Q_P = Q_L \| (R_D/\omega L)$$

$$Q'_P \approx \frac{R_D}{\omega L}$$

- ***For coupled coils:***

Secondary current:

$$I_2 = \frac{\omega M I_1}{R_2 + jX_2}$$

- ***Coupled impedance:***

$$Z'_2 = \frac{(\omega M)^2}{R_2 + jX_2}$$

- ***Coupled resistance:***

$$R'_2 = \frac{(\omega M)^2 R_2}{R_2^2 + X_2^2}$$

- ***At resonance:***

$$R_1 = R'_2 = \frac{(\omega M)^2}{R_2}$$

- ***Coupled reactance:***

$$X'_2 = -j\frac{(\omega M)^2 R_2}{R_2^2 + X_2^2}$$

- ***For critical coupling:***

$$M^2 = \frac{R_1 R_2}{\omega^2}$$

$$k = \frac{1}{\sqrt{Q_1 Q_2}}$$

Review Questions

Section 23-1

23-1 Sketch a series *RLC* circuit with an *ac* supply. Write the equation for the impedance of the circuit, and explain what occurs at the resonance frequency.

23-2 Draw a sketch to show the graphs of inductive reactance and capacitive reactance plotted versus frequency. Also, show the graph of total impedance for a series *RLC* circuit. Briefly explain.

23-3 Sketch a phasor diagram for a series *RLC* circuit at resonance. Also, draw typical graphs showing the variation of impedance, current, and the current phase angle, above and below resonance.

23-4 Derive an equation for the resonance frequency of a series *RLC* circuit.

Section 23-2

23-5 Show how a series *RLC* circuit can be tuned to resonate at a range of frequencies.

Section 23-3

23-6 Define Q factor with respect to a series resonance circuit. Derive an expression for the Q factor of the inductance coil in a series resonance circuit.

23-7 Derive an equation for the circuit Q factor of a series resonant circuit in terms of R, L, and C.

Section 23-4

23-8 Using illustrations, define half-power points, bandwidth, and selectivity, as applied to a resonance circuit.

23-9 Derive an equation for the bandwidth of a resonance circuit in terms of the resonance frequency and the circuit Q factor.

Section 23-5

23-10 Sketch a parallel *LC* circuit with inductor coil resistance and an *ac* supply. Write the equation for the circuit admittance, and explain what occurs at resonance.

23-11 Derive an expression for the impedance of a parallel *LC* circuit at resonance.

23-12 Sketch the phasor diagram for the component currents in a parallel *LC* circuit at resonance. Briefly explain.

23-13 For a parallel *LC* circuit, sketch the typical graphs of circuit impedance, inductor current, and supply current plotted versus frequency. Explain the shape of the graphs.

Section 23-6

23-14 Derive equations for the Q factor and resonance frequency of a parallel *LC* circuit. Discuss any approximations made.

Section 23-8

23-15 Explain the process resistance damping of a parallel resonant circuit, and discuss its effects.

23-16 Derive accurate and approximate equations for the Q factor of a parallel resonant circuit with resistance damping.

Section 23-9

23-17 A coil supplied with an alternating voltage (v) is inductively coupled to another coil with a load resistance R_2. Sketch the circuit diagram for the coils, and sketch the equivalent circuit showing the voltage drops around the primary and secondary circuits.

23-18 For coupled coils, write the equations for the resistance and reactance coupled from the secondary into the primary.

23-19 Two coupled coils each have a capacitor connected in series with them and are tuned to resonate at the same frequency. Explain the effect of resonance on the coupled reactance and resistance.

23-20 For tuned couples coils, write the equation for the total primary circuit resistance at resonance.

23-21 Define the term critical coupling as applied to two coupled coils that are tuned to resonate at the same frequency. Sketch the graphs of primary current and secondary current versus frequency for various degrees of coupling between the coils. Explain the shapes of the graphs. Write the equation for the critical coupling factor.

Problems

Section 23-1

23-1 Calculate the resonance frequency for the series *RLC* circuit in Figure 23-35, and determine the supply current at resonance.

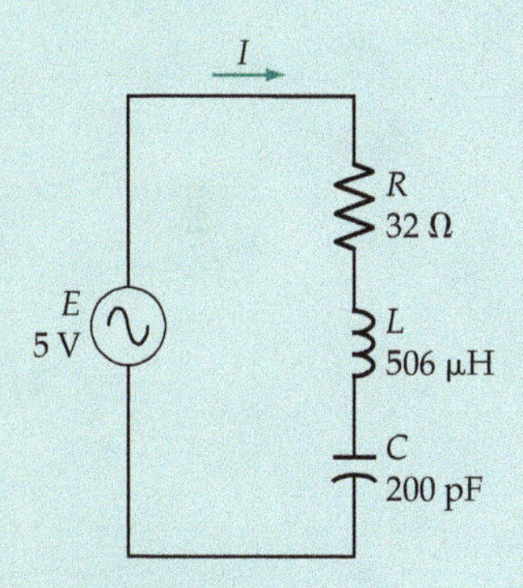

Figure 23-35

23-2 For the resonance circuit in Figure 23-35, calculate the new capacitance value that will produce resonance at a frequency of 450 kHz.

23-3 For the resonance circuit in Figure 23-35, calculate the levels of supply current at frequencies of $0.25f_r$, $0.5f_r$, $0.8f_r$, f_r, $1.25f_r$, $2f_r$, and $4f_r$. Plot a graph of current to a logarithmic frequency base.

23-4 A series *RLC* circuit is to resonate at a frequency of 198 kHz. If $C = 320$ pF and $R = 390\ \Omega$, determine the required inductance. Also calculate the circuit current at resonance when the signal voltage is 500 mV.

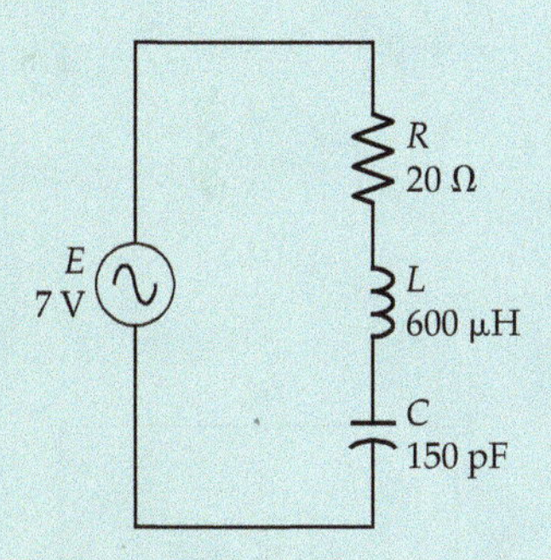

Figure 23-36

23-5 For the resonance circuit in Figure 23-35, calculate the component terminal voltages at frequencies of $0.25f_r$, $0.5f_r$, $0.8f_r$, f_r, $1.25f_r$, $2f_r$, and $4f_r$. Plot a graph of these voltages to a logarithmic frequency base.

23-6 For the resonance circuit in Problem 23-4, calculate the component terminal voltages at frequencies of $0.25f_r$, $0.5f_r$, $0.8f_r$, f_r, and $1.25f_r$.

23-7 Determine the resonance frequency for the series *RLC* circuit in Figure 23-36, and plot the graph of circuit current at frequencies of $0.25f_r$, $0.5f_r$, $0.8f_r$, f_r, $1.25f_r$, $2f_r$, and $4f_r$.

23-8 For the resonance circuit in Problem 23-7, calculate the component terminal voltages at frequencies of $0.25f_r$, $0.5f_r$, $0.8f_r$, f_r, $1.25f_r$, $2f_r$, and $4f_r$. Plot a graph of these voltages to a logarithmic frequency base.

Section 23-2

23-9 The circuit in Figure 23-37 is to be tuned to resonate at frequencies ranging from 100 kHz to 900 kHz. Determine the required range of adjustment of the capacitor.

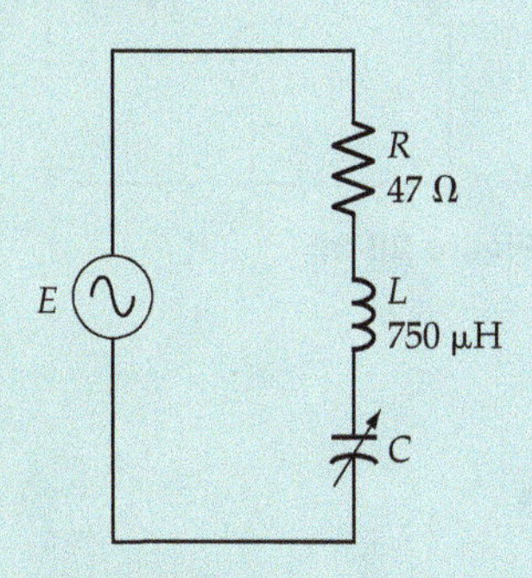

Figure 23-37

23-10 Determine the range of adjustment of f_r for the circuit in Problem 23-9 if the capacitor is variable from 100 pF to 2000 pF, and a 10 pF stray capacitance is present in parallel with the capacitor.

23-11 The voltage across the capacitor in Problem 23-7 is monitored on an oscilloscope. If the oscilloscope has an input capacitance of 40 pF, determine its effect on the resonance frequency of the circuit.

23-12 If the 200 pF capacitor in Figure 23-35 is coupled into the resonance circuit by means of a series-connected 0.01 μF capacitor, calculate the circuit resonance frequency.

Section 23-3

23-13 Calculate the Q factors for the circuits in Problems 23-1 and 23-7. Also determine the new Q factor for the circuit in Problem 23-7 when an oscilloscope is connected to the circuit as explained in Problem 23-11.

23-14 Calculate the Q factor of the circuit in Problem 23-4. Also determine the Q factors of the circuit in Problem 23-9 at the maximum and minimum resonance frequencies.

Section 23-4

23-15 Determine the half-power frequencies and the bandwidth for the circuit described in Problem 23-1.

23-16 Determine the half-power frequencies and the bandwidth for the circuit described in Problem 23-4.

23-17 Calculate the half-power frequencies and the bandwidth for the circuit in Problem 23-7. Determine the new bandwidth when an oscilloscope is connected to the circuit as described in Problem 23-11.

23-18 The bandwidth in the circuit in Problem 23-1 is to be altered to 13 kHz. Determine how this can be done without affecting the resonance frequency of the circuit.

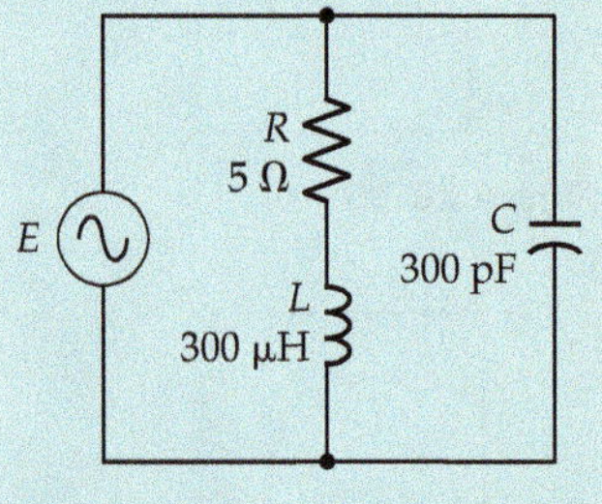

Figure 23-38

Section 23-5

23-19 An inductance with $L = 300$ μH and $R = 5\ \Omega$ is connected in parallel with a capacitor having $C = 300$ pF, as in Figure 23-38. Determine the circuit impedance at resonance.

23-20 The components in Figure 23-38 are changed to $R = 32\ \Omega$, $L = 506$ μH, and $C = 200$ pF. Determine the circuit impedance at resonance, and calculate the effect of a 40 pF oscilloscope capacitance in parallel with the capacitor.

Section 23-6

23-21 Calculate the resonance frequency and Q factor for the circuit in Problem 23-19.

23-22 Calculate the Q factor for the circuit in Problem 23-20 (a) without the 40 pF capacitance, and (b) with the 40 pF capacitance.

Section 23-7

23-23 Determine the bandwidth of the circuit in Problem 23-19.

23-24 Calculate the upper and lower bandwidth frequencies for the circuit in Problem 23-20 (a) without the 40 pF capacitance, and (b) with the 40 pF capacitance.

23-25 An instrument with an unknown input capacitance is connected across the capacitor in Problem 23-19. The resonance frequency is found to be 505.8 kHz. Calculate the instrument input capacitance.

23-26 A parallel *LC* circuit that is to resonate at 270 kHz has a 600 μH inductance with a 37 Ω coil resistance. Calculate the required capacitance and determine the circuit *Q* factor and bandwidth.

Section 23-8

23-27 Calculate the damping resistance (R_D) required to give a *Q* of 15 to the circuit described in Problem 23-19, as reproduced in Figure 23-39. Determine the precise Q factor with R_D connected.

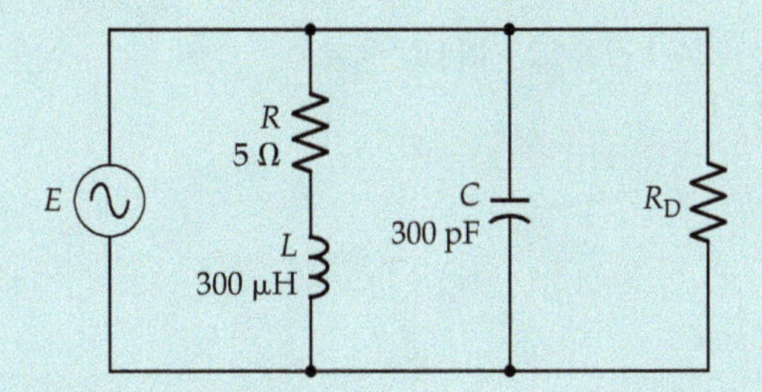

Figure 23-39

23-28 The parallel resonance circuit in Problem 23-26 is to be resistance-damped to obtain a *Q* factor of 2. Calculate the required damping resistance, and determine the precise Q factor with R_D connected to the circuit.

Section 23-9

23-29 A 300 μH coil with a resistance of $R_1 = 15\ \Omega$ is connected in series with a 3.3 V supply and a 300 pF capacitor. A second coil with $L = 100$ μH and $R_2 = 10\ \Omega$ is inductively coupled to the first coil, and tuned to resonate at the same frequency as the first (see Figure 23-40). Determine the critical

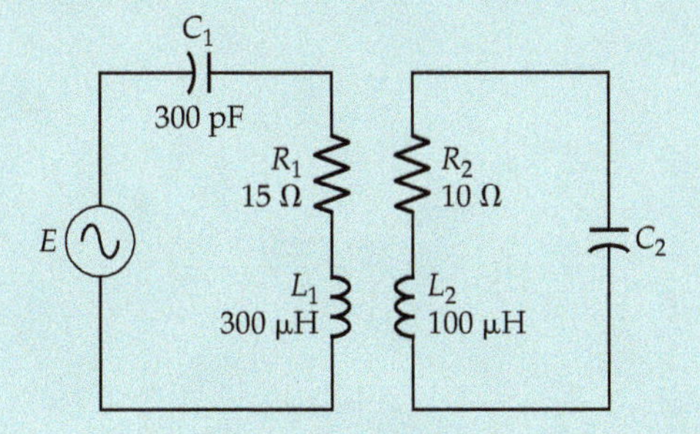

Figure 23-40

value of the coefficient of coupling for the coils, and calculate the current in each coil when the coils are critically coupled.

23-30 The resistance in the primary coil circuit in Problem 23-29 is to be adjusted to set the secondary current at 100 mA when the coils are critically coupled. Determine the required resistance for R_1 and the new coefficient of coupling.

Practice Problem Answers

23-1.1 939 kHz
23-1.2 13.2 mA, 3.3 V, 13.2 V, 40.5 mA, 16.2 V, 25.4 V
23-2.1 544 kHz, 1.17 MHz
23-3.1 35, 75
23-4.1 15.6 kHz
23-5.1 53.8 kΩ, 10.1 kΩ
23-6.1 55.5
23-7.1 1.27 MHz, 550 kHz, 12.9 kHz, 99, 43
23-7.2 1.4 μH, 5 pF
23-8.1 658 kHz, 3.7 kΩ, 110 kHz
23-9.1 5.2 Ω

CHAPTER 24
Filters

CONTENTS

Objectives

You will be able to:

1. Explain the function of low-pass, high-pass, bandpass, and band-stop filters, and draw graphs showing typical filter gain/frequency and phase/frequency response.
2. Calculate power level changes in decibels from power or voltage measurements.
3. Sketch the circuits of *RC* and *RL* low-pass and high-pass filters, explain the filter operation, and determine the cutoff frequency.
4. Write the transfer functions for various low-pass and high-pass filters, and plot the gain/frequency and phase/frequency response graphs.
5. Draw straight-line approximations for the gain/frequency and phase/frequency response graphs for various low-pass and high-pass filters.
6. Show how bandpass and notch filters can be constructed by combining low-pass and high-pass circuits.
7. Sketch the circuits and gain/frequency response graphs for bandpass and notch filters that use series and parallel resonance, and explain the operation of each circuit.
8. Analyze bandpass and notch filter circuits to determine the center frequency and bandwidth.

INTRODUCTION

A filter blocks or attenuates unwanted signals or noise voltages, and passes wanted signals with little or no attenuation. Filters are usually classified according to their frequency characteristics; a low-pass filter passes low frequency signals and attenuates all frequencies above a selected cutoff frequency. Similarly, a high-pass filter passes only those inputs with frequencies above the selected cutoff frequency. A bandpass filter passes signals within a selected frequency band, and a notch (or band-stop) filter blocks frequencies within its band.

24-1 BASIC FILTER TYPES

The functions of various types of filters are illustrated by the waveforms and typical frequency response graphs in Figures 24-1 and 24-2. A *low-pass filter* passes low frequency signals and attenuates (or blocks) signals with frequencies above a given *cutoff frequency* (f_c), (also known as the *critical frequency*). The input waveform for the low-pass filter in Figure 24-1(a) is composed of a low-frequency signal with an unwanted high-frequency input superimposed. The low-frequency input is reproduced at the output with very little attenuation, while the high-frequency noise voltage is severely attenuated.

The gain/frequency response graph for the filter is a graph of v_o/v_i plotted versus frequency (f). As illustrated in Figure 24-1(a), the filter has a low-frequency *pass-band* where the output voltage approximately equals the input, ($v_o/v_i \approx 1$). The pass-band applies to all signal frequencies from zero up to the

Figure 24-1 A low-pass filter passes low-frequency input voltages and blocks high-frequency inputs. A high-pass filter passes high-frequency inputs and blocks low frequencies.

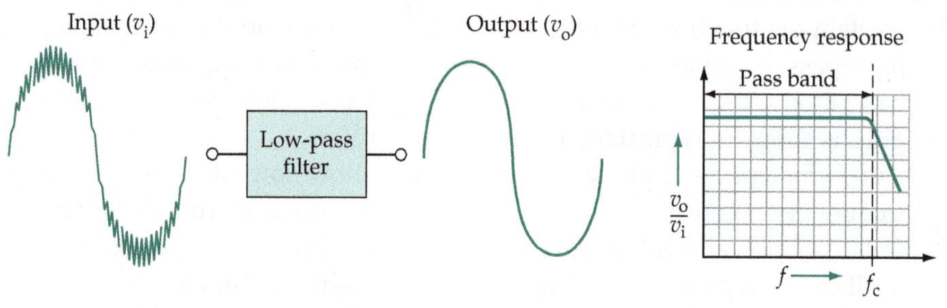

(a) Waveforms and frequency response for a low-pass filter

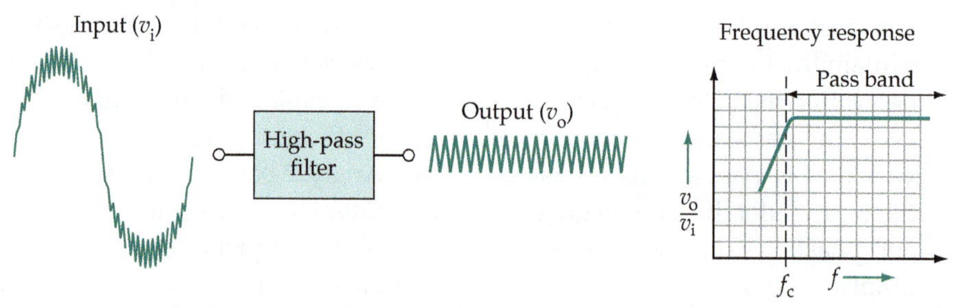

(b) Waveforms and frequency response for a high-pass filter

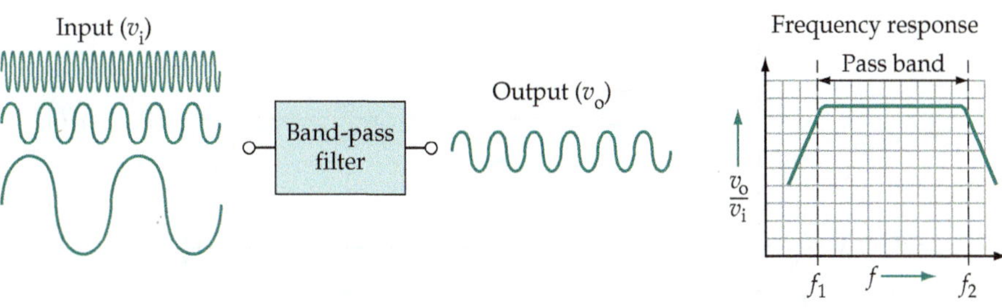

(a) Waveforms and frequency response for a band-pass filter

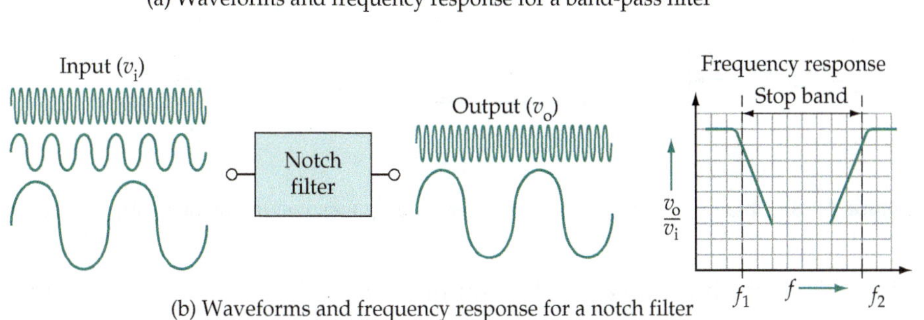

(b) Waveforms and frequency response for a notch filter

Figure 24-2 A band-pass filter passes a band of signal frequencies and blocks lower and higher frequencies outside the band. A notch filter blocks a band of signal frequencies and passes lower and higher frequency inputs.

cutoff frequency. For input frequencies above the cutoff frequency, v_o/v_i is less than one and rapidly falls-off to lower values. So, as shown by the graph, signals with frequencies above f_c are attenuated.

The *high-pass* filter in Figure 24-1(b) passes high-frequency input voltages and attenuates low-frequency inputs. In this case, the pass-band is above the cutoff frequency (f_c) and the attenuation band is below f_c. So, low-frequency inputs are attenuated while high-frequency signals are passed to the output terminals.

A *bandpass* filter application is illustrated in Figure 24-2(a), and Figure 24-2(b) shows the action of a *notch*, or *band-stop*, filter. In each case, there is a band of frequencies affected between two cutoff frequencies (f_1 and f_2). As illustrated, this is the pass-band for the bandpass filter, and the stop-band (attenuation band) for the notch circuit. The bandpass filter passes a middle range of signal frequencies within its pass-band, and blocks signals frequencies above and below the pass-band. The band-stop filter blocks the middle range of signal frequencies, and passes signals with frequencies above and below the stop-band.

24-2 POWER MEASUREMENT IN DECIBELS

The *change* in output from a filter, (or from an amplifier, signal generator, or other electronics equipment) is measured in *decibels* (dB). The decibel is one-tenth of the unit known as the *Bel*[1], which is inconveniently large. Consider

[1]Named in honor of the Scottish-Canadian inventor of the telephone, Alexander Graham Bell (1847–1922).

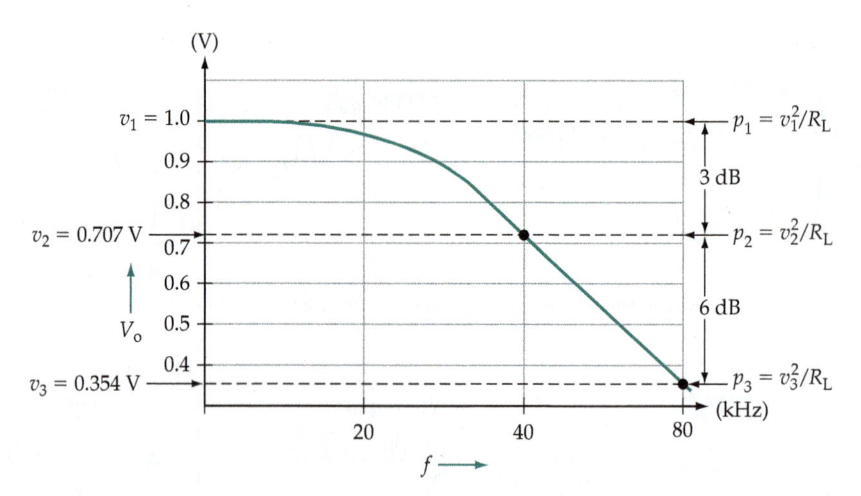

Figure 24-3 **Changes in filter output power levels are measured in decibels (dB).**

Figure 24-3 which shows a graph of output voltage (v_o) versus frequency (f). When the output voltage changes from v_1 to v_2, the output power level changes from p_1 to p_2, as illustrated. The change in power level (Δp) is expressed as the log of the ratio of p_2 to p_1.

$$\Delta p = \log (p_2/p_1) \text{ Bel}$$

or

$$\Delta p = 10 \log (p_2/p_1) \text{ dB} \tag{24-1}$$

If p_1 and p_2 are dissipated in a load resistor (R_L), the measured output voltages or currents can be employed to calculate the power change.

$$p_1 = v_1^2/R_L, \quad \text{and} \quad p_2 = v_2^2/R_L$$

or

$$p_1 = i_1^2 R_L, \quad \text{and} \quad p_2 = i_2^2 R_L$$

Substituting the voltages into Equation 24-1,

$$\Delta p = 10 \log \left[\frac{v_2^2/R_L}{v_1^2/R_L}\right] \text{ dB}$$

$$= 10 \log (v_2^2/v_1^2) \text{ dB}$$

or

$$\Delta p = 20 \log (v_2/v_1) \text{ dB} \tag{24-2}$$

Substituting the currents into Equation 24-1,

$$\Delta p = 20 \log (i_2/i_1) \text{ dB} \tag{24-3}$$

When power levels p_1 and p_2 are directly measured, the output power change in decibels is calculated by the use of Equation 24-1. When voltage or current measurements are made, Equation 24-2 or 24-3 applies.

Example 24-1

Two output voltage levels from a filter are measured as $v_1 = 1$ V and $v_2 = 0.707$ V, as illustrated in Figure 24-3. Calculate the output power change in decibels, and the ratio of p_2 to p_1. Also determine the output power change from p_2 to p_3 in Figure 24-3.

Solution

Eq. 24-2,
$$\Delta p = 20 \log (v_2/v_1) = 20 \log (0.707\ \text{V}/1\ \text{V})$$
$$= -3\ \text{dB}$$
$$p_1 = \frac{(1\ \text{V})^2}{R_L} = \frac{1}{R_L}$$
$$p_2 = \frac{(0.707\ \text{V})^2}{R_L} = \frac{0.5}{R_L}$$
$$= 0.5 p_1$$

From p_2 to p_3,
$$\Delta p = 20 \log (v_3/v_2) = 20 \log (0.354\ \text{V}/0.707\ \text{V})$$
$$= -6\ \text{dB}$$

When the absolute output power of a device is to be measured (rather than a change in output power), the reference level employed is 1 mW dissipated in a resistance of 600 Ω. The unit of absolute power level for this situation is known as the *decibel-milliwatt* (dBm). A signal generator that produces an output of 100 mW into a 600 Ω load has an absolute output power of,

$$10 \log (100\ \text{mW}/1\ \text{mW}) = 20\ \text{dBm}.$$

Practice Problem

24-2.1 The output voltage levels from a low-pass filter are measured as: $v_{o1} = 100$ mV at f_c, and $v_{o2} = 10$ mV at $10 f_c$. Calculate the power change in decibels. Also, determine the expected output voltage level (v_{o3}) for a 10 dB decrease from v_{o1}.

24-3 *RC* LOW-PASS FILTERS

RC Low-Pass Filter Circuit

Figure 24-4(a) shows a basic *RC* low-pass filter circuit (R_1 and C_1) together with a signal voltage (v_i) and a load resistor (R_L). Resistance R_1 and capacitive reactance X_{C1} constitute a voltage divider, as illustrated in Figure 24-4(b). When the filter input voltage (v_i) has a low frequency (f), the capacitor impedance [$X_{C1} = 1/(2\pi f C_1$] is much larger than the resistance of R_1. In this case there is very little voltage division, and the output voltage (v_o) approximately equals the input. When v_i has a high frequency, X_{C1} becomes much smaller than R_1, and

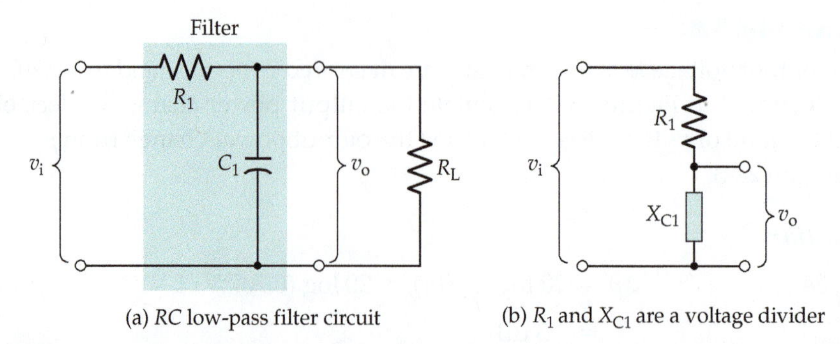

Figure 24-4 In a low-pass *RC* filter, input voltages are divided across the capacitor impedance (X_{C1}) and the resistor (R_1). Because X_{C1} is large at low frequencies, low-frequency inputs are passed to the output with very little attenuation. High-frequency signals are substantially attenuated as X_{C1} becomes smaller with increasing frequency.

so v_o becomes much smaller than v_i. It is seen that low frequency input voltages are passed to the output, and high frequency inputs are attenuated.

Gain/Frequency Response

The typical graph of output-to-input voltage ratio (v_o/v_i in decibels) plotted versus input frequency (f) for a low-pass filter is shown in Figure 24-5(a). This is the filter *gain/frequency response* graph (also known as a *Bode plot*). It is seen that the output remains substantially equal to the input ($v_o/v_i \approx 1 \approx 0$ dB) for all frequencies up to the *cutoff frequency* (f_c). This is upper limit of the *pass-band* for the filter, and all frequencies above f_c are in the *attenuation-band*, as illustrated. The cutoff frequency is defined as the frequency at which the output voltage falls by 3 dB from its normal (low-frequency) level. It is also referred to as the *3 dB frequency*, and as the *critical frequency*. The cutoff frequency can be expected to occur when X_{C1} becomes equal to the resistance of R_1.

$$X_{C1} = \frac{1}{2\pi f_c C_1} = R_1$$

Giving,

$$f_c = \frac{1}{2\pi R_1 C_1} \tag{24-4}$$

Equation 24-4 is correct only when the resistance of R_1 is very much smaller than the load resistance, ($R_1 \ll R_L$ in Figure 24-4). When R_1 is not much smaller than R_L, then f_c occur when X_{C1} equals $R_1 \| R_L$.

An equation for the ratio of output-to-input voltage for the *RC* low-pass filter is easily derived from the voltage divider in Figure 24-4(b) by the methods already covered, (see Section 19-6).

$$\frac{v_o}{v_i} = \frac{X_{C1}}{\sqrt{R_1^2 + X_{C1}^2}} \tag{24-5}$$

When $X_{C1} = R_1$,

$$\frac{v_o}{v_i} = \frac{R_1}{\sqrt{R_1^2 + R_1^2}} = \frac{1}{\sqrt{2}} = 0.707$$

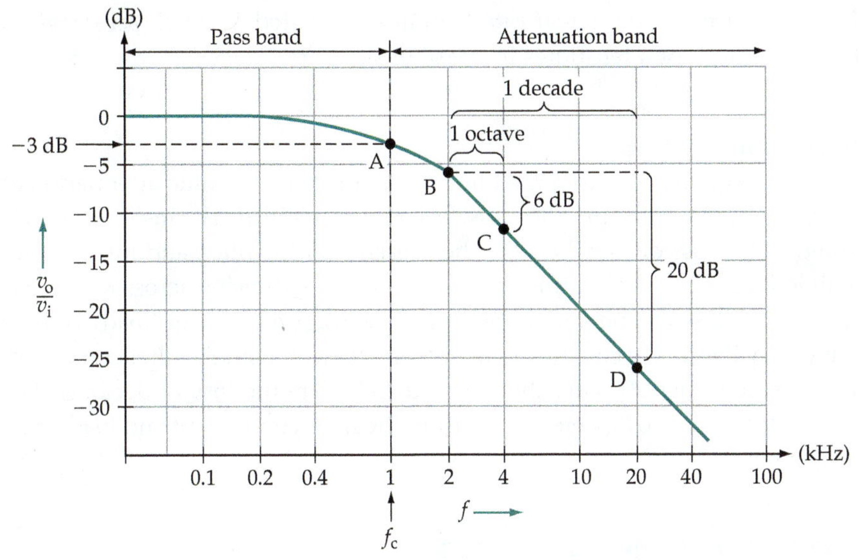

(a) Gain/frequency response for low-pass filter

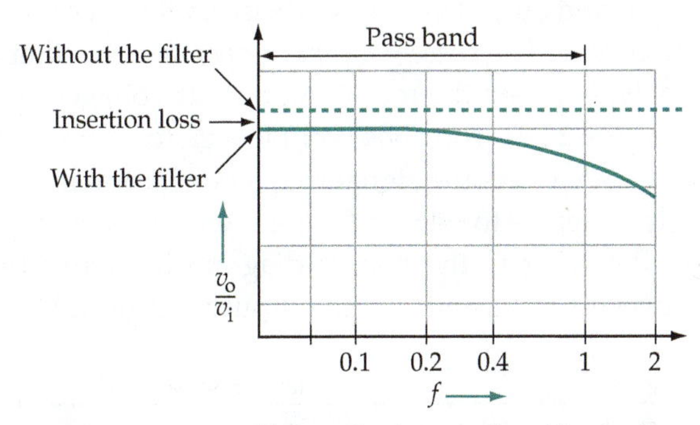

(b) Insertion loss produced by the filter

Figure 24-5 Typical gain/frequency response graph for a low-pass filter. At the cutoff frequency (f_c), the output level is 3 db below the normal low-frequency output. At frequencies higher than f_c, the gain falls-off at 6 dB/octave, which is the same as 20 dB/decade. The presence of the filter introduces an insertion loss between the input and output.

So, $X_{C1} = R_1$ at f_c, and this occurs when $v_o/v_i = 0.707$ (or $1/\sqrt{2}$), which is also −3 dB below its normal (low-frequency) level, (see Example 24-1). Example 24-1 also demonstrates that the output power is halved at the −3 db point, and this gives *the half-power point* as one more name for the cutoff frequency.

Ideally the gain/frequency response graph for the low-pass filter should be perfectly flat below the cutoff frequency and perfectly vertical above f_c, to produce zero attenuation of signals with frequencies within the pass-band and infinite attenuation of those with frequencies in the attenuation band. This is impossible with practical circuits, however, the steepness of the frequency response graph within the atenuation band can be an important consideration for a filter. Figure 24-5(a) shows that for signal frequencies above f_c, the output voltage decreases at a rate of *6 dB per octave;* a 6 dB reduction each time the frequency is doubled. As illustrated, this fall-off rate (also referred to as the *roll-off* rate) can also be stated as *20 dB per decade;* a 20 dB reduction for a ten-times increase in signal frequency.

Insertion Loss

For signal frequencies within the filter pass-band there is some attenuation of the input, known as the *insertion loss*, [see Figure 24-5(b)]. The insertion loss is defined as the loss of signal due to the presence of the filter, and for the filter circuit in Figure 24-4 it is simply the result of voltage division across R_1 and R_L. The insertion loss can be determined by measuring the output (load) voltage without the filter in the circuit, and measuring it again with the filter connected. The measured quantities are then used to calculate the loss in decibels. The insertion loss can also be determined from the filter component and load resistance values.

Phase/Frequency Response

Because the low-pass filter output is the capacitor voltage, there is a phase shift between the input and output voltages. (Phase shift for a series *RC* circuit is discussed in Section 19-6.) The *phase/frequency response* of the filter is a plot of the phase difference (θ) between the input and output voltages versus signal frequency (f), (see Figure 24-6). The phase shift is seen to be zero at low signal frequencies where $X_{C1} \gg R_1$. As the signal frequency increases, the (lagging) phase shift gradually increases to $-45°$ at f_c, and then continues beyond f_c to a maximum of $-90°$. Referring to the phasor diagram in Figure 19-21(c), the phase angle (ϕ) between the current and the input voltage can be calculated

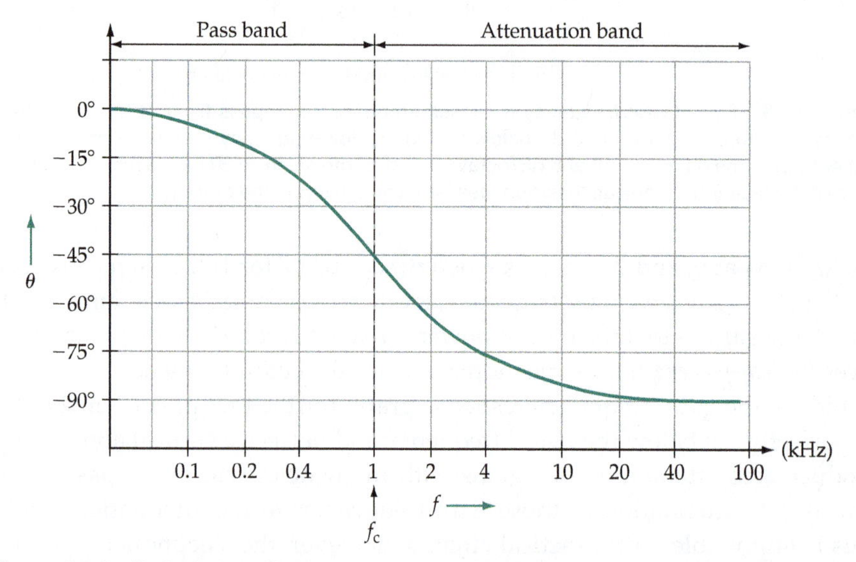

Figure 24-6 Typical phase/frequency response graph for a low-pass filter. At the cutoff frequency (f_c), the output voltage is phase shifted by $-45°$ with respect to the input. At frequencies higher and lower than f_c, the phase shift changes at a rate of approximately 45° per decade.

from Equation 19-24. The phase angle (θ) between the input and capacitor (filter output) voltages is ($90° - \phi$), and it can also be determined directly as,

$$\theta = \tan^{-1}\left[\frac{-R_1}{X_{C1}}\right] \qquad (24\text{-}6)$$

At f_c, where $X_{C1} = R_1$, θ is calculated from Equation 24-6 as $-45°$ (see Example 24-2).

Example 24-2

Determine the cutoff frequency (f_c), and the phase shift at f_c for the low-pass filter circuit in Figure 24-7.

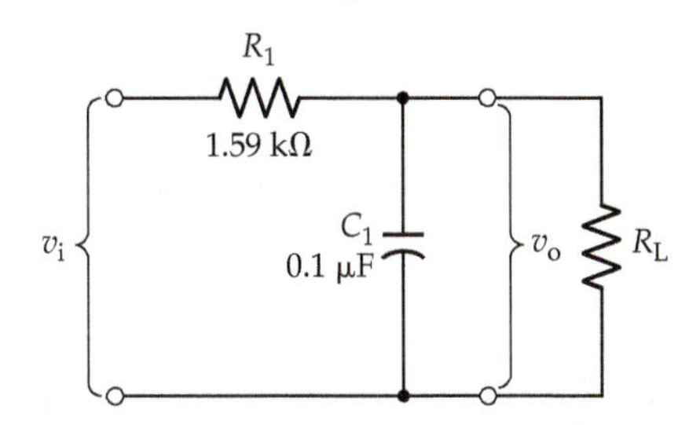

Figure 24-7 Low-pass filter circuit for Examples 24-2 and 24-3.

Solution

Eq. 24-4,
$$f_c = \frac{1}{2\pi R_1 C_1} = \frac{1}{2\pi \times 1.59\ \text{k}\Omega \times 0.1\ \mu\text{F}}$$
$$= 1\ \text{kHz}$$

$X_{C1} = R_1$ at f_c,

Eq. 24-6,
$$\theta = \tan^{-1}\left[\frac{-R_1}{X_{C1}}\right] = \tan^{-1}(-1)$$
$$= -45°$$

Example 24-3

If the filter in Figure 24-7 has a 100 kΩ load resistor (R_L), calculate the insertion loss. Also determine the effect of R_L on the circuit cutoff frequency.

Solution

$$\frac{v_o}{v_i} = \frac{R_L}{R_1 + R_L} = \frac{100\ \text{k}\Omega}{1.59\ \text{k}\Omega + 100\ \text{k}\Omega}$$
$$= 0.98$$

Eq. 24-2, insertion loss $= 20 \log (v_o/v_i) = 20 \log 0.98$
$$= -0.18\ \text{dB}$$

Eq. 24-4,
$$f_c = \frac{1}{2\pi (R_1 \| R_L) C_1} = \frac{1}{2\pi \times (1.59\ \text{k}\Omega \| 100\ \text{k}\Omega) \times 0.1\ \mu\text{F}}$$
$$= 1.02\ \text{kHz}$$

Note from the above example that, as discussed, an insertion loss is produced by the presence of the load resistor (R_L) connected at filter output. The insertion loss causes the low-frequency output to be reduced from $v_o = v_i$ to $v_o = 0.98\ v_i$. The new filter cutoff frequency calculated with R_L connected is the frequency at which the output is 3 db down from the attenuated output level ($v_o = 0.98\ v_i$).

Low-Pass Filter Transfer Function

Combination of equations 24-5 and 24-6 gives the *transfer function* for the filter, which is the phasor input-output relationship.

$$\frac{v_o}{v_i} = \frac{X_{C1}}{\sqrt{R_1^2 + X_{C1}^2}} \angle \tan^{-1}(-R_1/X_{C1}) \qquad \textbf{(24-7)}$$

Substituting $R_1 = 1/(2\pi f_c C_1)$, and $X_{C1} = 1/(2\pi f C_1)$, Equation 24-7 can be simplified to,

$$\frac{v_o}{v_i} = \frac{1}{\sqrt{(f/f_c)^2 + 1}} \angle \tan^{-1}(-f/f_c) \qquad \textbf{(24-8)}$$

Expressing v_o/v_i in decibels,

$$\frac{v_o}{v_i} = 20 \log\left(\frac{1}{\sqrt{(f/f_c)^2 + 1}}\right) \angle \tan^{-1}(-f/f_c) \qquad \textbf{(24-9)}$$

Once f_c is calculated for a given *RC* low-pass filter, Equation 24-9 can be used to quickly determine values of v_o/v_i and θ at various frequencies (multiples of f_c), for plotting the gain/frequency and phase/frequency responses.

Example 24-4

Calculate the voltage gain in dB and the phase angle for the low-pass filter circuit in Example 24-2 at frequencies of: 100 Hz, 500 Hz, 1 kHz, 2 kHz, and 10 kHz.

Solution

at $f = 100$ Hz, $\qquad f/f_c = 100\ \text{Hz}/1\ \text{kHz} = 0.1$

Substituting into Eq. 24-9,

$$\frac{v_o}{v_i} = 20 \log\left(\frac{1}{\sqrt{(0.1)^2 + 1}}\right) \angle \tan^{-1}(-0.1)$$

$$\approx 0\ \text{dB} \angle -5.7^\circ$$

at $f = 500$ Hz, $f/f_c = 500\text{ Hz}/1\text{ kHz} = 0.5$

$$\frac{v_o}{v_i} = 20\log\left(\frac{1}{\sqrt{(0.5)^2 + 1}}\right)\underline{/\tan^{-1}(-0.5)}$$

$$= -1\text{ dB}\underline{/-26.6°}$$

at $f = 1$ kHz, $f/f_c = 1\text{ kHz}/1\text{ kHz} = 1$

$$\frac{v_o}{v_i} = 20\log\left(\frac{1}{\sqrt{(1)^2 + 1}}\right)\underline{/\tan^{-1}(-1)}$$

$$= -3\text{ dB}\underline{/-45°}$$

at $f = 2$ kHz, $f/f_c = 2\text{ kHz}/1\text{ kHz} = 2$

$$\frac{v_o}{v_i} = 20\log\left(\frac{1}{\sqrt{(2)^2 + 1}}\right)\underline{/\tan^{-1}(-2)}$$

$$= -7\text{ dB}\underline{/-63.4°}$$

at $f = 10$ kHz, $f/f_c = 10\text{ kHz}/1\text{ kHz} = 10$

$$\frac{v_o}{v_i} = 20\log\left(\frac{1}{\sqrt{(10)^2 + 1}}\right)\underline{/\tan^{-1}(-10)}$$

$$= -20\text{ dB}\underline{/-84.3°}$$

Computer analysis of the above type of circuit is offered in Section 28-8.

Practice Problems

24-3.1 A low-pass *RC* filter circuit has $R_1 = 4.8\text{ k}\Omega$ and $C_1 = 1000$ pF. Calculate the filter cutoff frequency, and determine the attenuation at $20f_c$.

24-3.2 An low-pass *RC* filter circuit has $R_1 = 2\text{ k}\Omega$ and $C_1 = 5000$ pF. Calculate the filter gain and phase shift at a 25 kHz signal frequency.

24-3.3 Calculate the insertion loss for the filter circuit in Problem 24-3.2 if $R_L = 80\text{ k}\Omega$. Also, determine the effect of R_L on the circuit cutoff frequency.

24-4 *RC* HIGH-PASS FILTERS

RC High-Pass Filter Circuit

The basic *RC* high-pass filter circuit in Figure 24-8(a) is seen to be similar to the *RC* low-pass circuit in Figure 24-4 except that the capacitor and resistor are interchanged. Here again all input voltages are divided across R_1 and X_{C1}, [Figure 24-8(b)], and because the capacitive reactance is very high at low signal frequencies, low-frequency input voltages are heavily attenuated. As

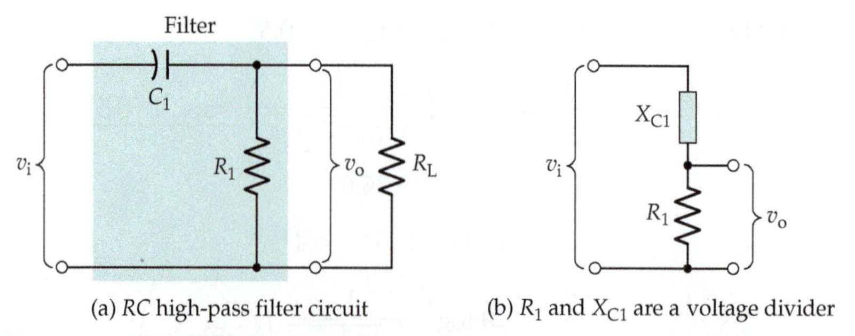

Figure 24-8 In a high-pass *RC* filter, input voltages are divided across the resistor (R_1) and the capacitor impedance (X_{C1}). Because X_{C1} is small at high frequencies, high-frequency inputs are passed to the output with very little attenuation. Low-frequency signals are substantially attenuated as X_{C1} becomes larger with decreasing signal frequency.

the signal frequency increases, X_{C1} becomes progressively smaller, so that high-frequency input voltages are only slightly attenuated.

Frequency Response Graphs

A typical high-pass filter gain/frequency response graphs is shown in Figures 24-9. It is seen that the output remains substantially equal to the input ($v_o/v_i \approx 0$ dB) for all frequencies above the cutoff frequency (f_c). In this case f_c is the lower frequency at which the output voltage falls by 3 dB from its normal (high-frequency) level. For signal frequencies below f_c, the output voltage decreases at a rate of 6 dB per octave. For the high-pass circuit this is a 6 dB reduction each time the frequency is halved. As already stated, 6 dB per octave is equivalent to 20 dB per decade, meaning that the filter gain is reduced by

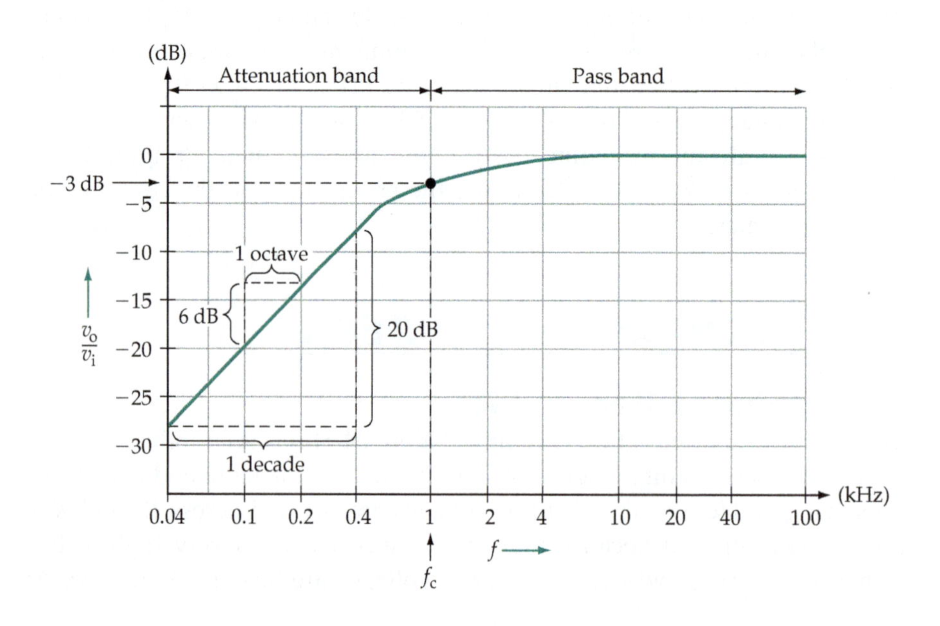

Figure 24-9 Typical gain/frequency response graph for a high-pass filter. At the cutoff frequency (f_c), the output level is 3 dB below the normal high-frequency output. At frequencies lower than f_c, the gain falls-off at 6 dB/octave, or 20 dB/decade.

20 dB whenever the signal frequency is reduced by a factor of 10. As in the case of the *RC* low-pass circuit, the cutoff frequency occurs when $X_{C1} = R_1$.

Eq. 24-4,
$$f_c = \frac{1}{2\pi R_1 C_1}$$

The high-pass filter phase/frequency response graph in Figure 24-10 shows that the output is substantially in phase with the input at high signal frequencies, and increasingly leads the input as the signal frequency decreases. At the cutoff frequency, the phase lead is 45°, and below f_c the lead increases to a maximum of 90°. Referring to the phasor diagram for a series *RC* circuit in Figure 19-21(c), it is seen that the resistor voltage leads the circuit input voltage by angle ϕ, and that,

$$\phi = \tan^{-1}\left[\frac{X_{C1}}{R_1}\right] \qquad (24\text{-}10)$$

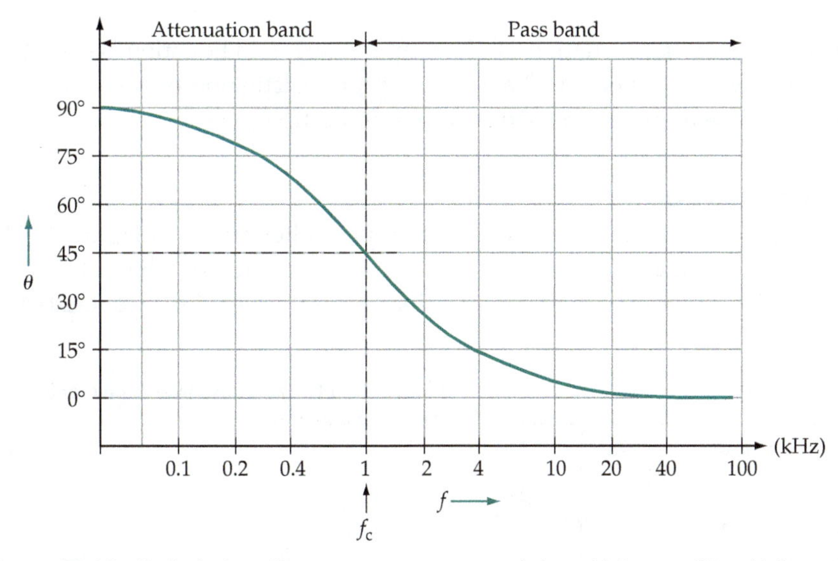

Figure 24-10 Typical phase/frequency response graph for a high-pass filter. At the cutoff frequency (f_c), the output voltage is phase shifted by +45° with respect to the input. At frequencies higher and lower than f_c, the phase shift changes at a rate of approximately 45° per decade.

Insertion Loss

Insertion loss (see Section 24-3) in the pass-band of a high-pass filter is the result of the filter resistor effectively reducing the load resistance presented to the signal source. Refer to the circuit in Figure 24-8, and note that, without the filter in the circuit, the load on the signal source is resistor R_L. The signal voltage (v_i) is divided across (source resistance) R_S and R_L to produce the output (v_o). Normally, $R_L \gg R_S$, so that v_o effectively equals v_i. With the filter inserted, the capacitor offers a very low impedance in the circuit pass-band, and the load on the source becomes $R_1 \| R_L$. The input is now divided across R_S and $R_1 \| R_L$, (see Example 24-5).

High-Pass Filter Transfer Function

The transfer function, or input/output phasor relationship, for an *RC* high-pass filter is easily derived from the voltage divider in Figure 24-8(b), (see Section 19-6).

$$\frac{v_o}{v_i} = \frac{R_1}{\sqrt{R_1^2 + X_{C1}^2}} \angle \tan^{-1}(X_{C1}/R_1) \qquad \textbf{(24-11)}$$

Substituting $R_1 = 1/(2\pi f_c C_1)$, and $X_{C1} = 1/(2\pi f C_1)$, Equation 24-11 becomes,

$$\frac{v_o}{v_i} = \frac{1}{\sqrt{(f_c/f)^2 + 1}} \angle \tan^{-1}(f_c/f) \qquad \textbf{(24-12)}$$

Expressing v_o/v_i in decibels,

$$\frac{v_o}{v_i} = 20 \log\left(\frac{1}{\sqrt{(f_c/f)^2 + 1}}\right) \angle \tan^{-1}(f_c/f) \qquad \textbf{(24-13)}$$

Note the factor f_c/f in Equation 24-13, compared to f/f_c in Equation 24-9 for the low-pass filter. Equation 24-13 allows for rapid calculation of points on the gain and phase frequency response graphs at multiples of f_c.

Example 24-5

Calculate the cutoff frequency for the high-pass filter circuit in Figure 24-11, and determine the effect of a 100 kΩ load resistance on the cutoff frequency. Also, determine the filter insertion loss if the signal source resistance is $R_S = 100\ \Omega$.

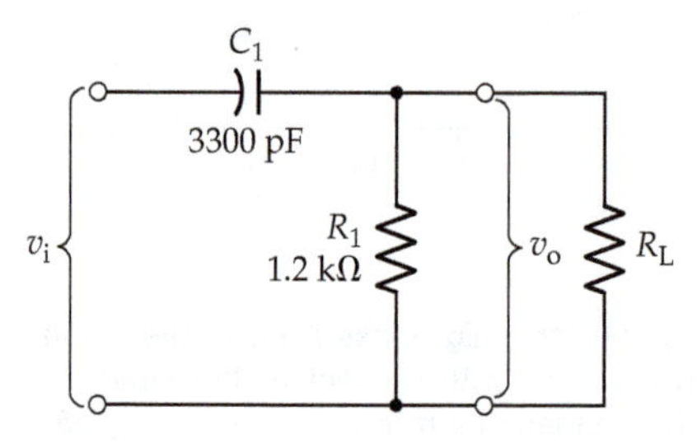

Figure 24-11 High-pass filter circuit for Examples 24-5 and 24-6.

Solution

Eq. 24-4,

$$f_c = \frac{1}{2\pi R_1 C_1} = \frac{1}{2\pi \times 1.2\ \text{k}\Omega \times 3300\ \text{pF}}$$

$$\approx 40\ \text{kHz}$$

for $R_L = 100\ \text{k}\Omega$,

Eq. 24-4,

$$f_c = \frac{1}{2\pi(R_1 \| R_L)C_1} = \frac{1}{2\pi \times (1.2\ \text{k}\Omega \| 100\ \text{k}\Omega) \times 3300\ \text{pF}}$$

$$\approx 40.7\ \text{kHz}$$

Without the filter in the circuit,

For $v_i = 1$ V, $$v_{o1} = \frac{v_i \times R_L}{R_S + R_L} = \frac{1\text{ V} \times 100\text{ k}\Omega}{100\ \Omega + 100\text{ k}\Omega}$$

$$\approx 1\text{ V}$$

With the filter in the circuit, and the signal frequency in the pass band,

For $v_i = 1$ V, $$v_{o2} = \frac{v_i \times (R_1 \| R_L)}{R_S + (R_1 \| R_L)} = \frac{1\text{ V} \times (1.2\text{ k}\Omega \| 100\text{ k}\Omega)}{100\ \Omega + (1.2\text{ k}\Omega \| 100\text{ k}\Omega)}$$

$$\approx 0.9\text{ V}$$

Eq. 24-2, insertion loss $= 20\log(v_{o2}/v_{o1}) = 20\log(0.9\text{ V}/1\text{ V})$

$$\approx -0.9\text{ dB}$$

Example 24-6

For the high-pass filter circuit in Figure 24-11, determine the gain and phase shift at $0.1f_c$, $0.5f_c$, f_c, $2f_c$, and $10f_c$.

Solution

at $f = 0.1f_c$, $f_c/f = 10$

Substitution into Eq. 24-13,

$$\frac{v_o}{v_i} = 20\log\left(\frac{1}{\sqrt{10^2 + 1}}\right)\angle\tan^{-1} 10$$

$$= -20\text{ dB}\angle 84.3°$$

at $f = 0.5f_c$, $f_c/f = 2$

$$\frac{v_o}{v_i} = 20\log\left(\frac{1}{\sqrt{2^2 + 1}}\right)\angle\tan^{-1} 2$$

$$= -7\text{ dB}\angle 63.4°$$

at $f = f_c$, $f_c/f = 1$

$$\frac{v_o}{v_i} = -3\text{ dB}\angle 45°$$

at $f = 2f_c$, $f_c/f = 0.5$

$$\frac{v_o}{v_i} = -1\text{ dB}\angle 26.6°$$

at $f = 10f_c$, $f_c/f = 0.1$

$$\frac{v_o}{v_i} = 0\text{ dB}\angle 5.7°$$

Practice Problems

24-4.1 Determine the cutoff frequency for a high-pass RC filter which has $R_1 = 1.5\ k\Omega$ and $C_1 = 4000\ pF$. Also, calculate the circuit gain and phase shift at $0.2f_c$ and $0.4f_c$.

24-4.2 Determine the effect of a 82 kΩ load resistance on the filter in Problem 24-4.1. Also calculate the filter insertion loss if the signal source resistance is 150 Ω.

24-5 FILTER FREQUENCY RESPONSE GRAPHS

Low-Pass Straight-Line Approximations

A straight-line approximation of the low-pass filter gain/frequency response graph is shown in Figure 24-12(a). The horizontal line drawn at 0 dB from the lowest frequency to f_c represents the pass band. The attenuation band is represented by the line drawn from f_c at a 20 dB/decade slope. The phase/frequency response can be similarly represented, as illustrated in Figure 24-12(b). In this case, a line is drawn through the point at $\theta = -45°$ and $f = f_c$ at a slope of 45°/decade. The approximate graphs can be used as a convenient way to prepare the actual frequency response graphs for a filter. The straight-line graphs are drawn first, then the most important points are plotted on the actual graphs, as demonstrated in Example 24-5.

Drawing the Actual Low-Pass Frequency Response Graphs

Example 24-7

Plot points on the actual gain/frequency and phase/frequency response graphs for the low-pass filter in Examples 24-2 and 24-4 at frequencies of $0.1f_c$, $0.5f_c$, f_c, $2f_c$, and $10f_c$.

Solution

From Example 24-4,

at $f = 100$ Hz, $v_o/v_i = 0$ dB, and $\theta = -5.7°$

Plot Point 1 on each graph in Figures 24-13 and 24-14.

at $f = 500$ Hz, $v_o/v_i = -1$ dB, and $\theta = -26.6°$ Point 2 on each graph.

at $f = 1$ kHz $v_o/v_i = -3$ dB, and $\theta = -45°$ Point 3 on each graph.

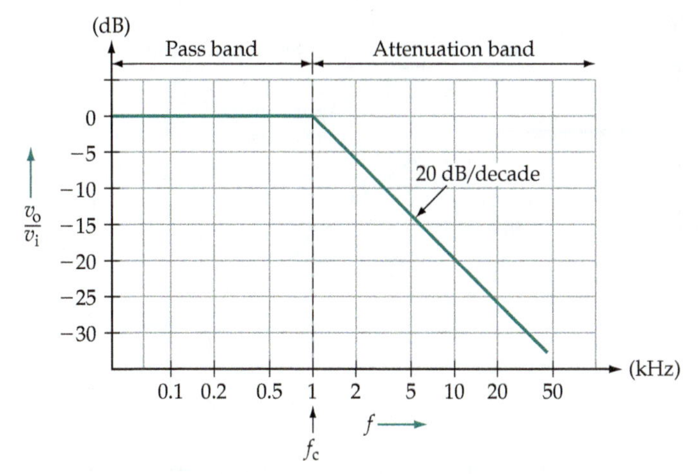

(a) Straight-line approximation of low-pass filter gain/frequency response

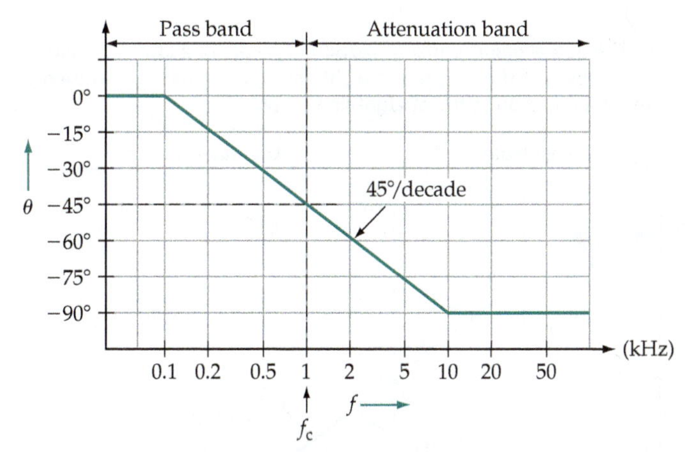

(b) Straight-line approximation of low-pass filter phase/frequency response

Figure 24-12 **The gain/frequency response of a low-pass filter can be represented approximately by a horizontal line from zero to f_c, and a sloping line (at −20 db/decade) above f_c. The phase/frequency response is approximately represented by a line sloping at −45°/decade from 0.1f_c to 10f_c and horizontal lines above and below those frequencies.**

at f = 2 kHz v_o/v_i = −7 dB, and θ = −63.4° Point 4 on each graph.

at f = 10 kHz v_o/v_i = −20 dB, and θ = −84.3° Point 5 on each graph.

Consider the points plotted on the actual gain/frequency response graph in Figure 24-13 (from the values listed in Example 24-7) as compared to the straight-line graph.

- At f_c, v_o/v_i is always 3 db below the low-frequency level. So, the −3 db point on the actual gain/frequency point can be plotted at f_c *without any additional calculations*.

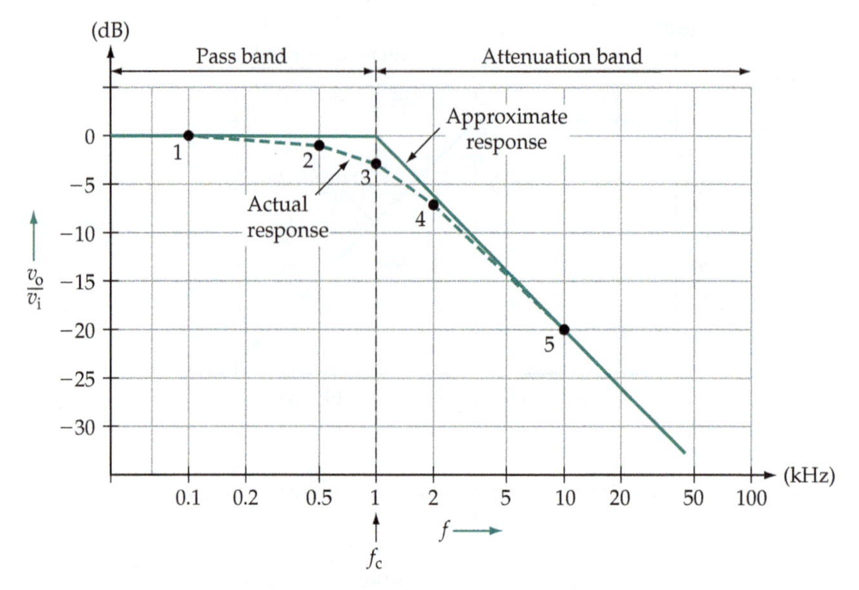

Figure 24-13 The actual gain/frequency response graph for a low-pass filter is most conveniently constructed by first drawing the straight-line approximation and then plotting points on the actual graph in relation to the straight-line graph.

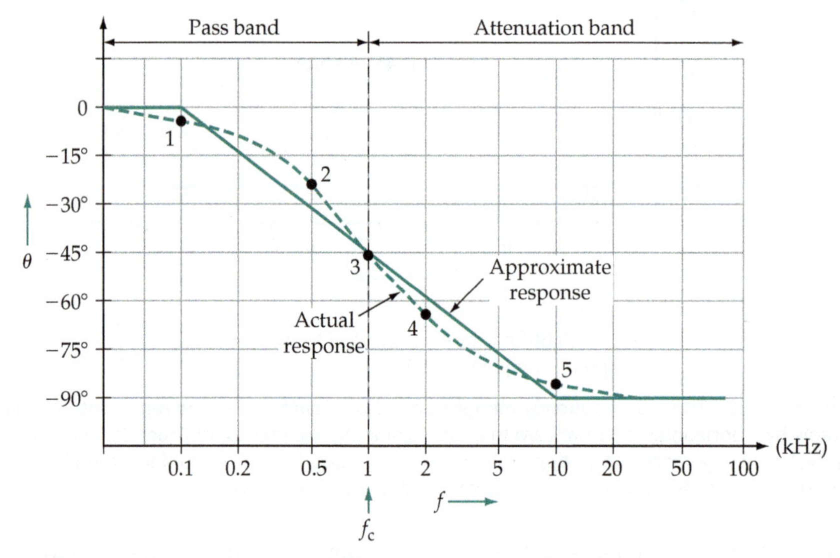

Figure 24-14 The actual phase/frequency response graphs for filters are most conveniently constructed by first drawing a straight-line approximation and then plotting points on the actual graph in relation to the straight-line graph.

- At $f/f_c = 0.1$, the actual v_o/v_i level is approximately 0 dB. So, 0 db can be plotted at $f = 0.1f_c$ right on the straight-line graph.
- At $f = 0.5f_c$, the actual v_o/v_i level is −1 dB, which is 1 db below the level at $f = 0.5f_c$ on the straight-line graph.
- At $f = 2f_c$, the actual v_o/v_i level is −7 dB, which again is 1 dB below the straight-line graph level at $f = 2f_c$
- At $f = 10f_c$, the actual v_o/v_i level is −20 dB, right on the straight-line graph.

It is seen that, after the straight-line graph is drawn, the points on the actual gain/frequency response graph can be plotted (at $0.1f_c$, $0.5f_c$, f_c, $2f_c$, and $10f_c$) for *any* RC low-pass filter with *any* cutoff frequency *without any calculations*. The actual gain/frequency response graph can then be drawn through these plotted points.

Now consider the points plotted for the actual phase/frequency response graph in Figure 24-14, as compared to the straight-line approximate graph.

- At f_c, $\theta = -45°$, which coincides with the $-45°$ point on the straight-line graph.
- At $f/f_c = 0.1$, $\theta = -5.7°$. This is 5.7° below the 0° point at $f = 0.1f_c$ on the straight-line graph.
- At $f/f_c = 0.5$, $\theta = -26.5°$. Approximately 5° above the $f = 0.5f_c$ point on the straight-line graph.
- At $f/f_c = 2$, $\theta = -63.4°$. Approximately 5° below the $f = 2f_c$ point on the straight-line graph.
- At $f/f_c = 10$, $\theta = -84.3°$. This is 5.7° above the $f = 10f_c$ point on the straight-line graph.

As in the case of the gain/frequency graph, points on the actual phase/frequency response graph can be plotted (at $0.1f_c$, $0.5f_c$, f_c, $2f_c$, and $10f_c$) *without any calculations*.

Selecting f_c on the Frequency Response Graphs

The frequency response graphs in Figures 24-13 and 24-14 are plotted for a circuit with a 1 kHz cutoff frequency. So, it makes sense to use one of the decade points on the frequency scale for f_c, and then the other points on the scale are easily selected as multiples of f_c. This is illustrated in Figure 24-15(a). When the cutoff frequency is not a multiple of 10, there are two possible approaches to selection of the f_c point on the horizontal scale. Suppose $f_c = 33$ kHz. A '33' point could be identified between '3' and '4' points on the log scale, as shown in Figure 24-15(b). Then, the 330 kHz and 3.3 kHz points on the scale are located one decade above and below f_c. The position of these points is obviously not very precise. A better approach is to always select one of the decade points as f_c, regardless of the actual f_c value. Points on the frequency scale are then easily identified at $0.1f_c$, $0.5f_c$, etc., as shown in Figure 24-15(c).

High-Pass Straight-Line Approximations

Straight-line approximations of gain/frequency and phase/frequency response graphs can be constructed for a high-pass filter in a similar manner to those for a low-pass circuit [see Figure 24-16(a) and (b)]. Here again, the approximate graphs can be used as a convenient way to prepare the actual filter response graphs. The straight-line graphs are drawn first, and then the most important points are plotted on the actual graphs.

Figure 24-15 Careful selection of the critical frequency point on a logarithmic scale can simplify the process of plotting a response graph.

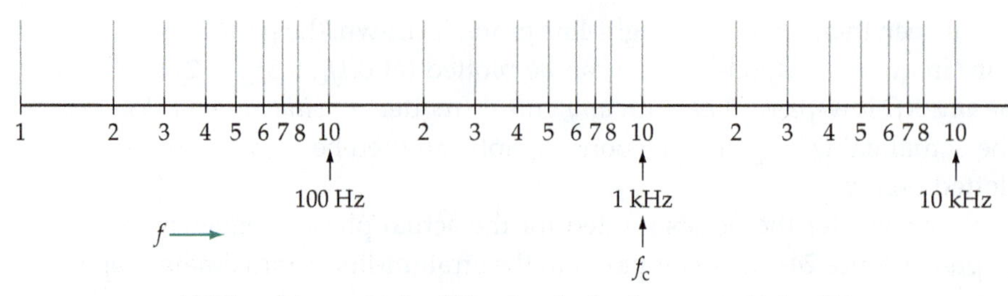

(a) When f_c is a multiple of 10, select the f_c point at a multiple of 10

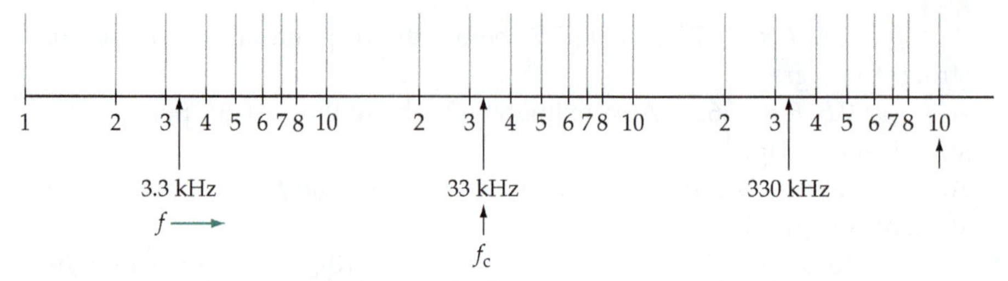

(b) When f_c is not a multiple of 10, the f_c point can be selected directly on the scale

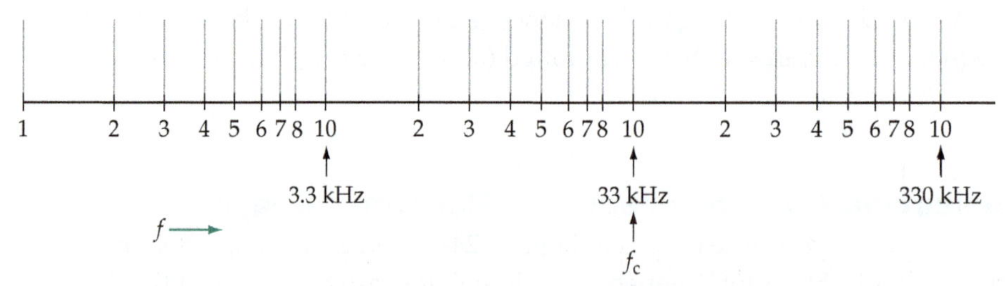

(c) Always selecting the f_c point at multiple of 10 can be most convenient

Example 24-8

Draw the straight-line approximate gain/frequency and phase/frequency response graphs for the high-pass filter circuit in Examples 24-5 and 24-6.

Solution

- The gain/frequency response is started by first identifying f_c at 40 kHz at one of the decade points on the horizontal axis of the graph. In Figure 24-17, f_c is marked at the '1' point.
- One decade below f_c is marked as 4 kHz, and a further decade below 4 kHz would be marked as 400 Hz. In a similar manner, one decade above f_c is marked as 400 kHz, and the next one decade higher becomes 4 MHz.
- The straight-line approximate gain/frequency graph is now drawn horizontally at 0 dB from f_c at 40 kHz to the higher frequencies, and then at a slope of 20 dB/decade from f_c down to the lower frequencies.
- The horizontal axis for the phase/frequency graph is marked exactly as for the gain/frequency graph (Figure 24-18).

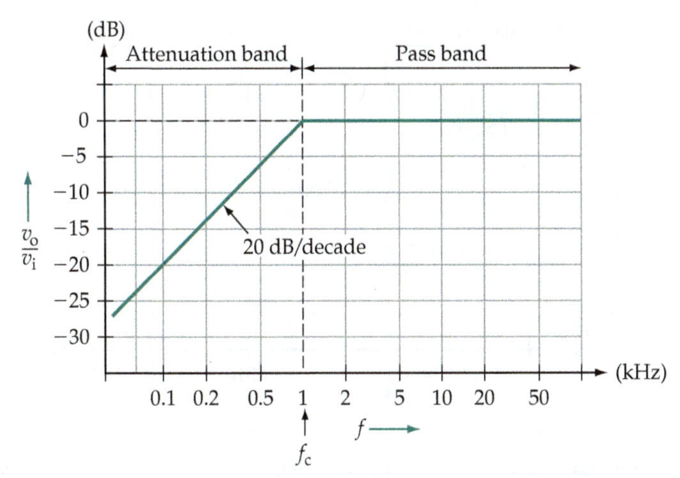

(a) Straight-line approximation of high-pass filter gain/frequency response

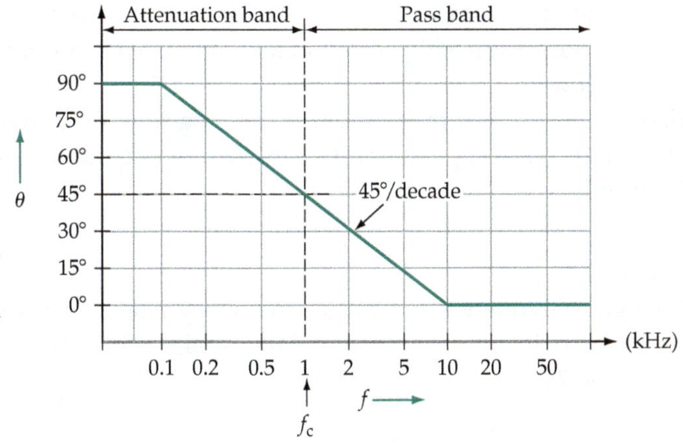

(b) Straight-line approximation of high-pass filter phase/frequency response

Figure 24-16 Straight-line approximations of gain/frequency and phase/frequency response graphs for a high-pass filter. The gain is represented by a horizontal line from f_c to higher frequencies and a line sloping at −20 dB/decade below f_c. The phase/frequency response is represented by a line sloping at +45°/decade from $10f_c$ to $0.1f_c$ and horizontal lines above and below those frequencies.

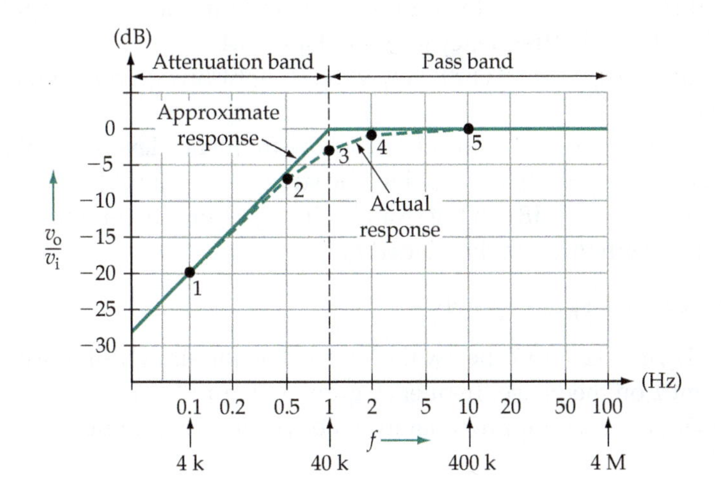

Figure 24-17 The actual gain/frequency response graph for a high-pass filter is constructed by plotting points on the actual graph in relation to the straight-line approximation.

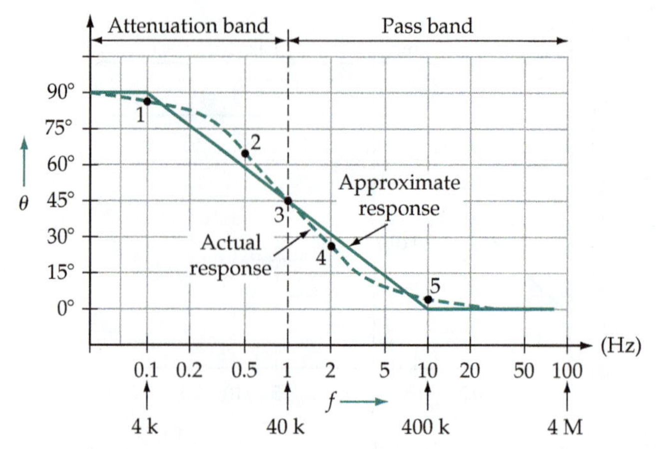

Figure 24-18 The actual phase/frequency response graph for a high-pass filter is constructed by plotting points on the actual graph in relation to the straight-line approximation.

- The straight-line approximate phase/frequency graph is now drawn at a slope of 45°/decade, from a high of 90° at low frequencies through the +45° point at f_c, to a low of 0°.
- The horizontal portions of the phase/frequency graph are drawn from 0° toward the high-frequency end, and from 90° toward the low-frequency end of the graph.

Example 24-9

Plot the actual gain/frequency and phase/frequency response graphs, using the points calculated in Example 24-6, on the straight-line approximations drawn for Example 24-8.

Solution

Gain/frequency response graph:

- At $f = 0.1f_c$, $v_o/v_i = -20$ dB (right on the straight-line v_o/v_i graph)—Plot Point 1 on the gain/frequency graph in Figure 24-17.
- At $f = 0.5f_c$, $v_o/v_i = -7$ dB (1 dB below the straight-line graph)—Plot Point 2 on the graph.
- At $f = f_c$, v_o/v_i is always 3 db below the low-frequency level—Point 3.
- At $f = 2f_c$, $v_o/v_i = -1$ dB (1 dB below the straight-line graph)—Point 4.
- At $f = 10f_c$, $v_o/v_i = 0$ dB (on the straight-line v_o/v_i graph)—Point 5.
- Draw the graph through the plotted points.

Phase/frequency response graph:

- At $f = 0.1f_c$, $\phi = 84.3°$ (5.7° below the 90° level on the straight-line ϕ graph)—Plot Point 1 on the phase/frequency graph in Figure 24-18.
- At $f = 0.5f_c$, $\phi = 63.4°$ (approximately 5° above the straight-line graph)—Plot Point 2.

- At $f = f_c$, $\phi = 45°$ at f_c, which coincides with the +45° point on the phase/frequency straight-line graph—Point 3.
- At $f = 2f_c$, $\phi = 26.6°$ (approximately 5° below the straight-line graph)—Point 4.
- At $f = 10f_c$, $\phi = 5.7°$ (5.7° above the straight-line graph)—Point 5.
- Draw the graph through the plotted points.

Drawing the Actual High-Pass Frequency Response Graphs

Example 24-9 demonstrates that, as in the case of a low-pass *RC* filter, the actual gain/frequency and phase/frequency responses for a high-pass *RC* filter can be can readily be constructed after the straight-line approximate graphs are drawn. Points on the actual graphs can be plotted (at $0.1f_c$, $0.5f_c$, f_c, $2f_c$, and $10f_c$) for *any RC* high-pass filter with *any* cutoff frequency *without any calculations*. The actual gain/frequency response graphs are then drawn through the plotted points.

Practice Problems

24-5.1 Prepare straight-line approximations of the gain/frequency and phase/frequency response graphs for a low-pass filter with $R_1 = 4.8\ \text{k}\Omega$, $C_1 = 1000\ \text{pF}$, (as in Problem 24-3.1). Also, draw the actual response curves by plotting points at $0.1f_c$, $0.5f_c$, f_c, $2f_c$, and $10f_c$.

24-5.2 Prepare gain/frequency and phase/frequency straight-line and actual graphs for the high-pass filter in Problem 24-4.1.

24-6 *RL* LOW-PASS AND HIGH-PASS FILTERS

RL Low-Pass Filter

Because inductors are usually larger and more expensive than capacitors, *RC* filters are normally preferable. However, *RL* filters can be constructed, and their performance is similar to the performance of *RC* filters.

The low-pass filter circuit in Figure 24-19(a) uses an inductor and a resistor instead of a capacitor and resistor. As in the case of the *RC* circuit, the two components form a voltage divider which attenuates the input, [Figure 24-19(b)]. The inductive reactance ($2\pi fL$) is very small at low signal frequencies, so very little voltage is dropped across the inductor, and v_o (across the resistor) is substantially equal to v_i. As the inductive reactance becomes progressively larger with increasing frequency, the output is increasingly attenuated.

The gain/frequency and phase/frequency responses for the *RL* low-pass circuit are similar to those of the *RC* low-pass filter, (see Figures 24-5 and 24-6). A question arises about how it can be that the *RL* low-pass filter has a lagging

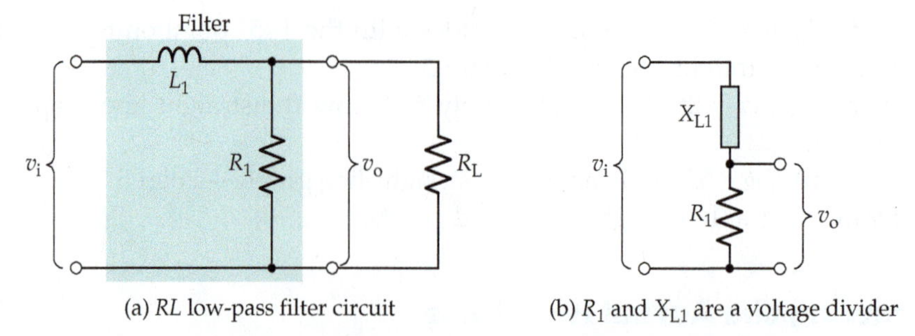

(a) *RL* low-pass filter circuit (b) R_1 and X_{L1} are a voltage divider

Figure 24-19 In a low-pass *RL* filter, input voltages are divided across the inductor impedance (X_{L1}) and the resistor (R_1). Because X_{L1} is low at low frequencies, low-frequency inputs are passed to the output with very little attenuation. High-frequency signals are substantially attenuated as X_{L1} becomes larger with increasing frequency.

phase/frequency response, when it has already been shown that the phase angle is lagging for the *RC* low-pass filter. Refer to the phasor diagram for a series *RL* circuit in Figure 19-20(a), and note that V_R lags the supply voltage. The *RL* low-pass filter output is taken across *R* (in the *RC* circuit it is taken across *C*), so the *RL* filter output lags the input. The cutoff frequency for an *RL* low-pass or high-pass filter occurs when $X_L = R$. So, the equations for the *RL* low-pass filter are:

$$f_c = \frac{1}{2\pi(L_1/R_1)} \tag{24-14}$$

$$\theta = \tan^{-1}\left[\frac{-X_{L1}}{R_1}\right] \tag{24-15}$$

The transfer function for the *RL* low-pass filter is,

$$\frac{v_o}{v_i} = \frac{R_1}{\sqrt{R_1^2 + X_{L1}^2}}\angle\tan^{-1}(-X_{L1}/R_1) \tag{24-16}$$

Substituting $R_1 = 2\pi f_c L_1$, and $X_{L1} = 2\pi f L_1$, Equation 24-16 simplifies to the same equation as used for the *RC* low-pass filter,

Eq. 24-9, $$\frac{v_o}{v_i} = 20\log\left(\frac{1}{\sqrt{(f/f_c)^2 + 1}}\right)\angle\tan^{-1}(-f/f_c)$$

RL High-Pass Filter

A high-pass *RL* filter circuit is shown in Figure 24-20. As for the low-pass *RL* filter, the frequency response graphs are similar to those for the equivalent *RC* circuit (Figures 24-9 and 24-10). The input voltage applied to the *RL* high-pass circuit is divided across X_{L1} and R_1. At high frequencies where $X_{L1} \gg R_1$ there is very little attenuation of the input. At low frequencies where $X_{L1} \ll R_1$ input

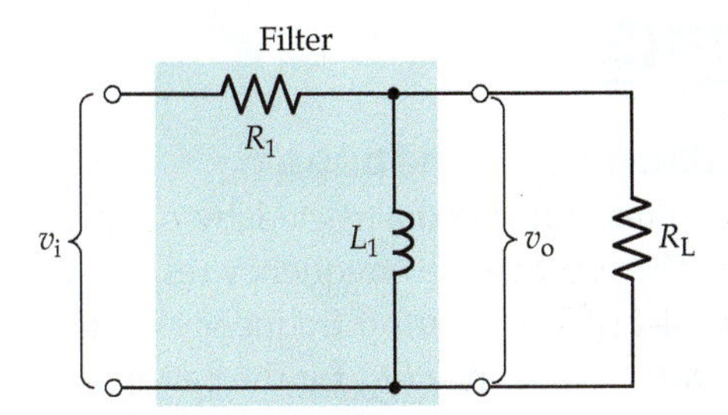

Figure 24-20 In a high-pass *RL* filter, input voltages are divided across the inductor impedance (X_{L1}) and the resistor (R_1). Because X_{L1} is large at high frequencies, high-frequency inputs are passed to the output with very little attenuation. Low-frequency signals are substantially attenuated as X_{C1} becomes smaller with decreasing signal frequency.

voltages are severely attenuated. Once again, the cutoff frequency is calculated from,

Eq. 24-14, $$f_c = \frac{1}{2\pi(L_1/R_1)}$$

and the transfer function is the same as for an *RC* high-pass filter,

Eq. 24-13, $$\frac{v_o}{v_i} = 20\log\left(\frac{1}{\sqrt{(f_c/f)^2 + 1}}\right)\underline{/\tan^{-1}(f_c/f)}$$

Example 24-10

An *RL* low-pass filter has $R_1 = 8.2\ \text{k}\Omega$ and $L_1 = 50$ mH. Calculate the cutoff frequency, and determine the signal attenuation and phase shift at 50 kHz.

Solution

Eq. 24-14, $$f_c = \frac{1}{2\pi(L_1/R_1)} = \frac{1}{2\pi(50\ \text{mH}/8.2\ \text{k}\Omega)}$$

$$= 26\ \text{kHz}$$

Eq. 24-9, $$\frac{v_o}{v_i} = 20\log\left(\frac{1}{\sqrt{(f/f_c)^2 + 1}}\right)\underline{/\tan^{-1}(-f/f_c)}$$

$$= 20\log\left(\frac{1}{\sqrt{(50\ \text{kHz}/26\ \text{kHz})^2 + 1}}\right)\underline{/\tan^{-1}(-50\ \text{kHz}/26\ \text{kHz})}$$

$$= -6.7\ \text{dB}\underline{/-62.5^\circ}$$

Practice Problems

24-6.1 Determine the cutoff frequency for an *RL* low-pass filter with $R_1 = 1\ \text{k}\Omega$ and $L_1 = 3.18$ mH. Also, calculate the filter gain and phase shift at $3f_c$.

24-6.2 The circuit in Problem 24-6.1 has $R_L = 27\ \text{k}\Omega$, and the winding resistance for L_1 is $R_W = 22\ \Omega$. Calculate the insertion loss.

24-6.3 Prepare straight-line approximations of the gain/frequency and phase/frequency response graphs for the circuit in Problem 24-6.1.

24-7 BANDPASS FILTERS

High-Pass Low-Pass Combination for Bandpass

Figure 24-21(a) shows how a bandpass filter can be constructed by simply cascading high-pass and low-pass stages. The typical gain/frequency response for this arrangement is shown in Figure 24-21(b). The cutoff frequency of the high-pass circuit ($f_{c(H)}$) becomes the low cutoff frequency (f_1) for the bandpass filter, and the cutoff frequency for the low-pass circuit ($f_{c(L)}$) becomes the high cutoff frequency (f_2), as illustrated. For example, if $f_{c(H)}$ is 10 kHz and $f_{c(L)}$ is 100 kHz, the high-pass circuit blocks signal frequencies below 10 kHz, and the low-pass filter blocks frequencies above 100 kHz. Consequently, only signals with frequencies between 10 kHz and 100 kHz will pass through both filters. A bandpass filter constructed in this way should have at least a decade difference between f_1 and f_2. With a smaller difference, the second stage tends to load the first stage and thus affect its cutoff frequency. So, cascading of high-pass and low-pass filters is suitable only for constructing *wide-band* bandpass filters.

As shown in Figure 24-21(b), the *bandwidth* (B) of the circuit is,

$$B = f_2 - f_1 \qquad \textbf{(24-17)}$$

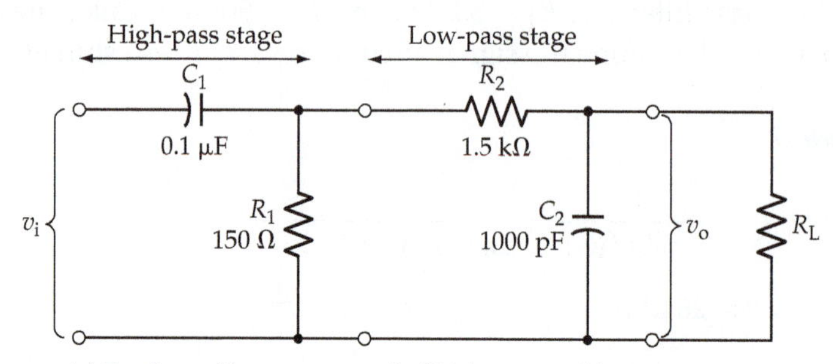

(a) Bandpass filter constructed of high-pass and low-pass stages

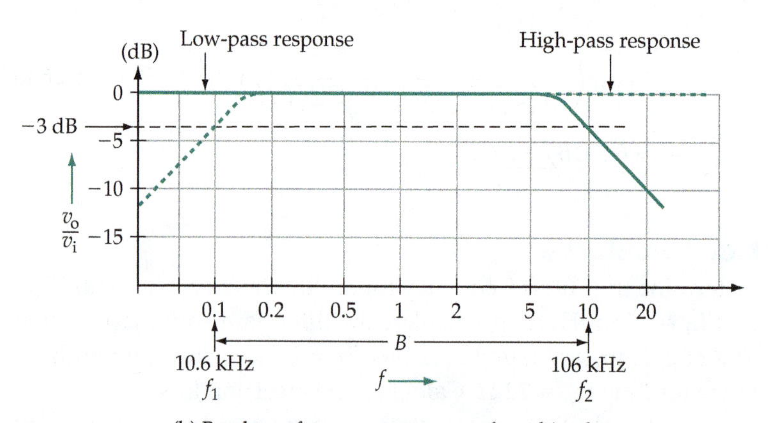

(b) Bandpass frequency response of combined stages

Figure 24-21 Bandpass filter consisting of cascaded high-pass and low-pass stages. The high-pass stage blocks signal frequencies up to f_1 and the low-pass stage blocks frequencies above f_2, thus passing the band of frequencies between f_1 and f_1.

The centre frequency (f_o) for a bandpass filter is defined as the *geometric mean*,

$$f_o = \sqrt{f_1 \times f_2} \tag{24-18}$$

Example 24-11

Determine the cutoff frequencies for the bandpass filter shown in Figure 24-21(a). Also calculate the pass-band centre frequency (f_o).

Solution

High-pass stage.

Eq. 24-4,
$$f_{cH} = \frac{1}{2\pi R_1 C_1} = \frac{1}{2\pi \times 150\ \Omega \times 0.01\ \mu\text{F}}$$
$$= 10.6\ \text{kHz}$$

Low-pass stage.

Eq. 24-4,
$$f_{cL} = \frac{1}{2\pi R_1 C_1} = \frac{1}{2\pi \times 1.5\ \text{k}\Omega \times 1000\ \text{pF}}$$
$$= 106\ \text{kHz}$$

Eq. 24-18,
$$f_o = \sqrt{f_1 \times f_2} = \sqrt{f_{cH} \times f_{cL}}$$
$$= \sqrt{10.6\ \text{kHz} \times 106\ \text{kHz}}$$
$$= 33.5\ \text{kHz}$$

Series Resonant Bandpass Filter

A bandpass filter using a series resonant circuit is shown in Figure 24-22(a), and its typical gain/frequency response is illustrated in Figure 24-23. The total impedance of the series *LCR* circuit is,

$$Z = (R_1 + R_w) + j(X_L - X_C)$$

where R_w is the resistance of the inductor winding. As shown by the filter equivalent circuit in Figure 24-22(b), the input voltage (v_i) is divided across the

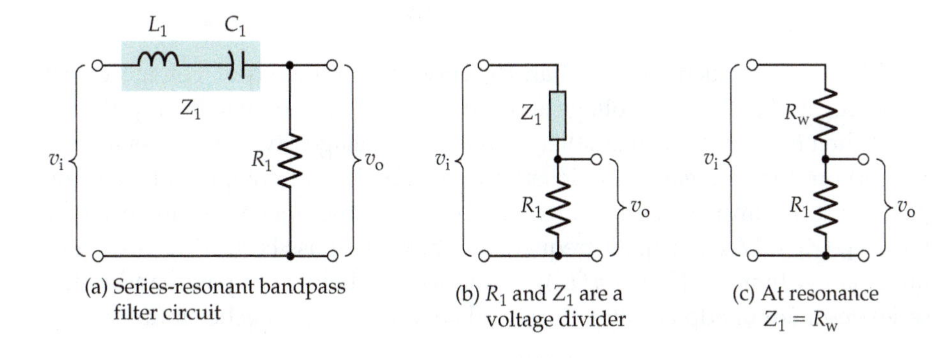

(a) Series-resonant bandpass filter circuit

(b) R_1 and Z_1 are a voltage divider

(c) At resonance $Z_1 = R_w$

Figure 24-22 In a series-resonant bandpass filter, input voltages are divided across the series *LC* circuit impedance (Z_1) and the resistor (R_1). When Z_1 is a minimum at the resonance frequency, inputs voltages are passed to the output with very little attenuation. Because Z_1 becomes quite large at frequencies above and below resonance, inputs with off-resonance frequencies are severely attenuated.

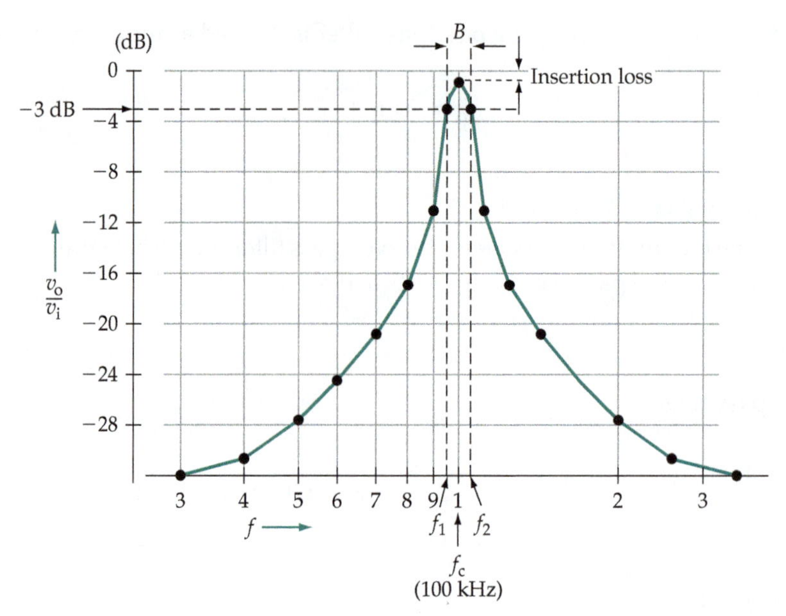

Figure 24-23 Gain/frequency response graph for the series-resonant bandpass filter in Example 24-12.

impedance of the resonant circuit (Z_1) and the resistor (R_1) to produce the output voltage (v_o). So, the voltage gain (attenuation) is

$$\frac{v_o}{v_i} = \frac{R_1}{R_1 + Z_1} \tag{24-19}$$

The impedance of the LC circuit is a minimum at resonance, and consequently, v_o/v_i is a maximum. At frequencies above and below resonance, Z_1 is much greater than R_1, and so the output voltage is very much smaller than the input. The graph of v_o/v_i plotted versus frequency shows that the output voltage remains high for the pass-band between the upper and lower cutoff frequencies (f_1 and f_2), and that there is increasing signal attenuation above and below these frequencies. At the resonant frequency, $X_L = X_C$, and the impedance of the series LC circuit becomes, $Z = R_W$, [see Figure 24-22(c)]. The voltage gain at resonance is

$$\frac{v_o}{v_i} = \frac{R_1}{R_1 + R_W} \tag{24-20}$$

If R_W is very much smaller than R_1, very little of the input voltage is lost across R_W, and the output voltage at resonance is almost equal to the input voltage. When R_W is not much smaller than R_1, the voltage division across R_1 and R_W can produce a measurable insertion loss. At signal frequencies below the pass-band the impedance of the series LC circuit becomes very much larger than R_1, $[Z_1 \approx 1/(2\pi f C_1)]$. At frequencies above the pass band Z_1 is also very much larger than R_1, $[Z_1 \approx 2\pi f L_1)]$. So, above and below the pass-band of the series-resonant bandpass filter, the signal attenuation approaches infinity.

Example 24-12

The series-resonant bandpass filter circuit in Figure 24-22(a) has $L = 2.53$ mH, $C = 1000$ pF, $R_1 = 100\ \Omega$, and $R_W = 10\ \Omega$. Determine the resonance frequency, bandwidth, and insertion loss at resonance.

Solution

Eq. 23-3,
$$f_r = \frac{1}{2\pi\sqrt{LC}} = \frac{1}{2\pi\sqrt{2.53\text{ mH} \times 1000\text{ pF}}}$$
$$= 100\text{ kHz}$$

Eq. 23-8,
$$Q = \frac{2\pi f_r L}{R_1 + R_W} = \frac{2\pi \times 100\text{ kHz} \times 2.53\text{ mH}}{100\ \Omega + 10\ \Omega}$$
$$= 14.45$$

Eq. 23-17,
$$B = \Delta f = \frac{f_r}{Q} = \frac{100\text{ kHz}}{14.45}$$
$$= 6.92\text{ kHz}$$

Eq. 24-20,
$$\frac{v_o}{v_i} = \frac{R_1}{R_1 + R_W} = \frac{100\ \Omega}{100\ \Omega + 10\ \Omega}$$
$$\approx 0.91$$

$$\text{Insertion loss} = 20\log(v_o/v_i) = 20\log 0.91$$
$$\approx -0.82\text{ dB}$$

The gain/frequency response graph for the series-resonant bandpass filter in Example 24-12 is plotted in Figure 24-23. It is seen that the filter bandwidth appears to be very narrow, and that the gain falls off rapidly close to resonance. As discussed in Chapter 23, the bandwidth depends upon the circuit Q factor; the largest Q values produce the narrowest bandwidths.

Parallel Resonant Bandpass Filter

A bandpass filter using a parallel resonant circuit is shown in Figure 24-24. In this case the output voltage is developed across the resonant circuit.

$$\frac{v_o}{v_i} = \frac{Z_1}{R_1 + Z_1} \tag{24-21}$$

At frequencies above and below resonance, Z_1 becomes much smaller than R_1, so that the output is much smaller than the input. For a parallel LC circuit, Z_1 is a maximum at the resonance frequency (much larger than R_1); so v_o is just a little smaller than v_i. The shape of frequency response for the parallel-resonant bandpass filter is very similar to that for the series-resonant filter. The centre frequency is calculated exactly as for the series-resonant bandpass filter, but the bandwidth must be determined using the parallel circuit Q factor, which is

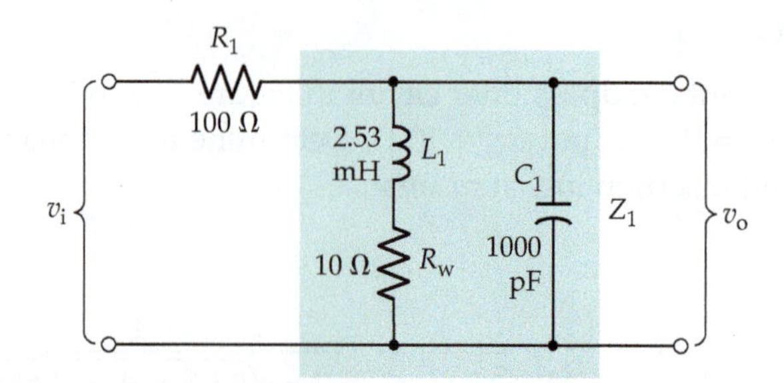

Figure 24-24 Parallel-resonant bandpass filter circuit. Input voltages are divided across the resistor (R_1) and the parallel *LC* circuit impedance (Z_1). Z_1 is a maximum at the resonance frequency, and so inputs voltages are passed to the output with very little attenuation. Z_1 becomes quite small at frequencies above and below resonance, causing inputs with off-resonance frequencies to be severely attenuated.

the Q factor of the inductor. There are also differences (between series and parallel resonant bandpass filters) in the circuit attenuation at the resonance frequency and at frequencies outside the pass band.

Example 24-13

The parallel resonant bandpass circuit in Figure 24-24 uses the same component values as the series-resonant filter in Example 24-12. Determine the resonance frequency, bandwith, and insertion loss at resonance.

Solution

From Example 24-12, $f_r = 100 \text{ kHz}$

Eq. 23-21, $$Q_p = \frac{2\pi f_r L_1}{R_W} = \frac{2\pi \times 100 \text{ kHz} \times 2.53 \text{ mH}}{10 \ \Omega}$$
$$= 159$$

Eq. 23-24, $$B = \Delta f = \frac{f_r}{Q_p} = \frac{100 \text{ kHz}}{159}$$
$$= 629 \text{ Hz}$$

Eq. 23-20, $$Z_1 = \frac{L_1}{C_1 R_W} = \frac{2.53 \text{ mH}}{1000 \text{ pF} \times 10 \ \Omega}$$
$$= 253 \text{ k}\Omega$$

Eq. 24-21, $$\frac{v_o}{v_i} = \frac{Z_1}{R_1 + Z_1} = \frac{253 \text{ k}\Omega}{253 \text{ k}\Omega + 10 \ \Omega}$$
$$= 0.9999$$

$$\text{Insertion loss} = 20 \log (v_o/v_i) = 20 \log 0.9999$$
$$\approx -0.87 \times 10^{-3} \text{ dB}$$

Comparing the results obtained in Examples 24-12 and 24-13, it is seen that the bandwidth of the parallel-resonant bandpass filter is much smaller than that of the series-resonant filter using the same components. There is virtually zero insertion loss with the parallel-resonant filter, while the series-resonant filter has a loss approaching 1 dB. As already discussed, the series-resonant bandpass filter effectively has infinite attenuation at frequencies well above and below the pass-band. For the parallel-resonance circuit at very low frequencies, X_{C1} becomes very large, X_{L1} is quite small, and $Z_1 \approx R_W$. Thus, at low frequencies the signal attenuation is limited to a minimum of,

$$\frac{v_o}{v_i} = \frac{R_W}{R_1 + R_W} \tag{24-22}$$

At frequencies above the pass-band, X_{L1} increases and X_{C1} becomes smaller as the signal frequency increases. Consequently, the attenuation approaches infinity, just as in the case of the series-resonant bandpass filter.

Practice Problems

24-7.1 A bandpass filter using a series resonance circuit (as in Figure 24-22) is to be designed to have a centre frequency of 50 kHz and a bandwidth of 15 kHz. A 3.3 mH inductor with a 14 Ω coil resistance is to be used. Determine a suitable value for capacitor C and resistor R_1, and calculate the filter insertion loss at the resonance frequency.

24-7.2 Determine the insertion loss at the resonance frequency when the components used in Problem 24-7.1 are reconnected to function as a parallel-resonant bandpass filter.

24-8 NOTCH FILTERS

High-Pass Low-Pass Combination for Notch

A *notch filter* (also known as a *band-stop filter* and as a *band-reject filter*) can be constructed by parallel-connecting low-pass and high-pass filters, as shown in Figure 24-25(a). The low-pass circuit passes signal frequencies up to f_1 while the high-pass circuit passes frequencies above f_2. Consequently, as illustrated by the gain/frequency graph in Figure 24-25(b), the band of frequencies between f_1 and f_2 is rejected. Note the use of a *summing circuit* at the output terminals of the low-pass and high-pass filters. This is an electronic circuit that combines the two outputs while avoiding an overload on either filter circuit. The outputs of the low-pass and high-pass circuits *cannot be directly connected in parallel*, because at least one filter would overload the other circuit and thus affect the output amplitude and cutoff frequencies.

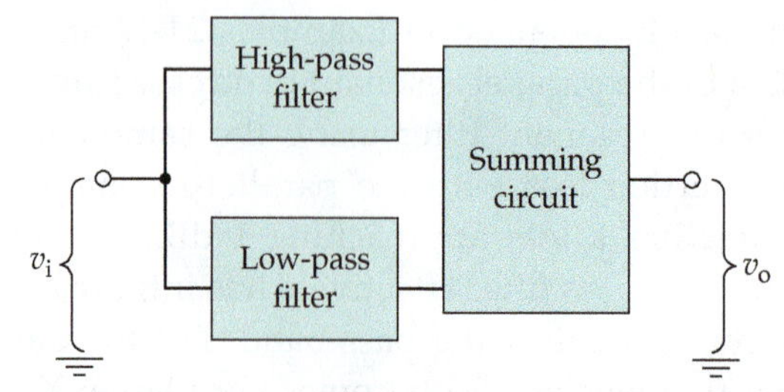

(a) Notch filter constructed of high-pass and low-pass stages

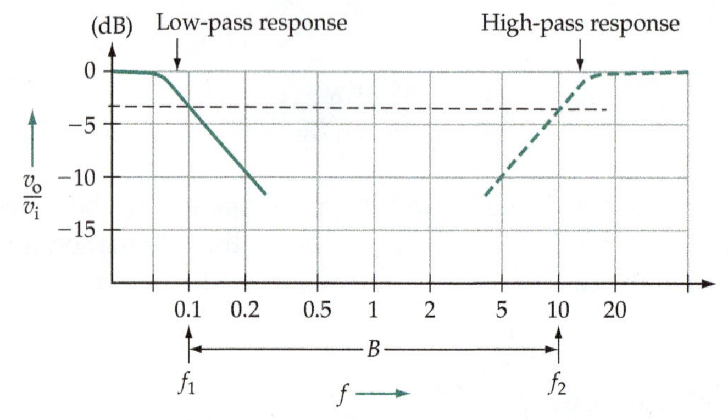

(b) Frequency response of combined stages

Figure 24-25 Notch filter consisting of high-pass and low-pass stages with the outputs connected via a summing circuit to avoid loading. The low-pass stage passes signal frequencies up to f_1, and the high-pass stage passes frequencies above f_2, thus blocking the band of frequencies between f_1 and f_2.

Notch Filters Using Resonant Circuits

Two notch filters using resonant circuits and their typical notch gain/frequency response graph are shown in Figure 24-26. The circuit in 24-26(a) has its output developed across a series *LC* circuit. So, because the impedance of the series *LC* circuit is a minimum at resonance (much smaller than R_1), the input voltage is severely attenuated (to produce v_o) for a band of signal frequencies close to resonance. At frequencies away from resonance, the impedance of the series *LC* circuit is much larger than R_1, and so there is minimum attenuation of v_i at off-resonance frequencies [see Figure 24-26(c)].

The parallel *LC* circuit in Figure 24-26(b) has an impedance that is much larger than R_1 at the resonance frequency and smaller than R_1 at frequencies off resonance. Because the output is developed across R_1, v_o is much smaller than v_i over a band of frequencies close to resonance. The gain/frequency response graphs for series-resonant and parallel-resonant notch filters using the same components are similar in shape, and in centre and cutoff frequencies. However, as in the case of similar bandpass filters, there are also bandwidth and signal attenuation differences.

For the series-resonant notch filter circuit [Figure 24-26(a)],

$$\frac{v_o}{v_i} = \frac{Z_1}{R_1 + Z_1}$$

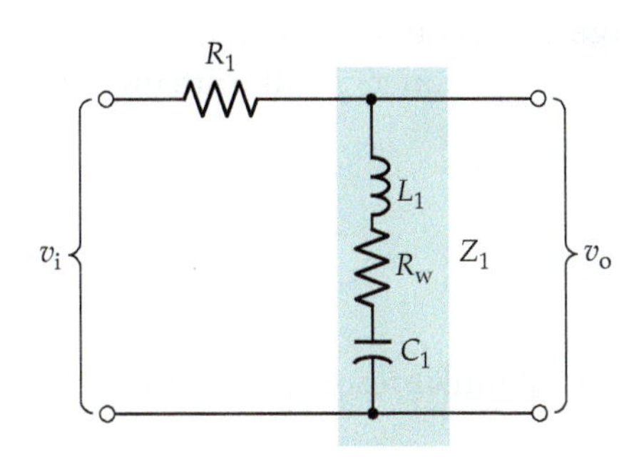

(a) Series-resonant notch filter circuit

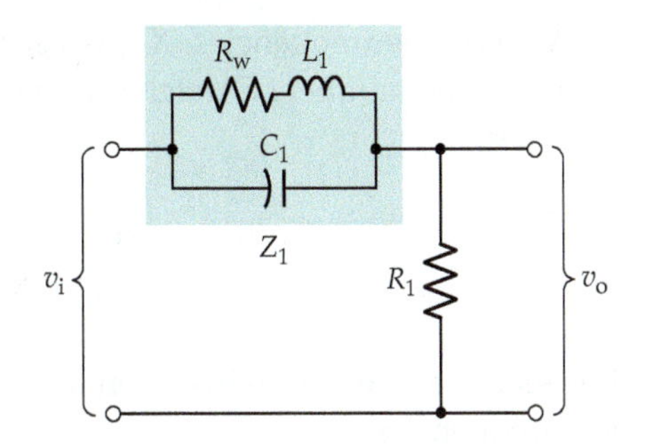

(b) Parallel-resonant notch filter circuit

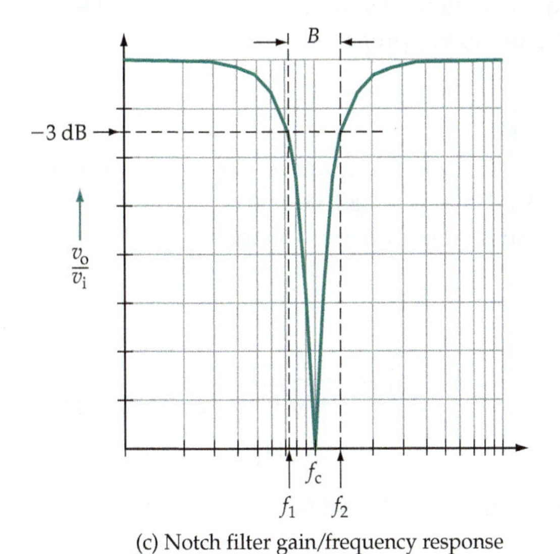

(c) Notch filter gain/frequency response

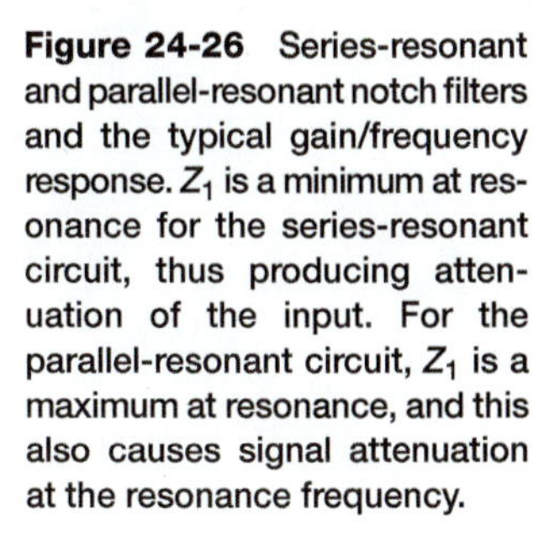
Figure 24-26 Series-resonant and parallel-resonant notch filters and the typical gain/frequency response. Z_1 is a minimum at resonance for the series-resonant circuit, thus producing attenuation of the input. For the parallel-resonant circuit, Z_1 is a maximum at resonance, and this also causes signal attenuation at the resonance frequency.

At the resonance frequency this becomes,

$$\frac{v_o}{v_i} = \frac{R_W}{R_1 + R_W} \qquad \textbf{(24-23)}$$

For signal frequencies above and below the pass-band, Z_1 is much larger than R_1 that there is virtual zero attenuation.

For the parallel-resonant notch filter [Figure 24-26(b)],

$$\frac{v_o}{v_i} = \frac{R_1}{R_1 + Z_1}$$

From Equation 23-20, Z_1 increases to $L_1/(C_1R_W)$ at the resonance frequency, so that the filter attenuation equation at f_r becomes,

$$\frac{v_o}{v_i} = \frac{R_1}{R_1 + L/(CR_W)} \qquad \textbf{(24-24)}$$

At very low frequencies, X_{C1} is very large, X_{L1} is quite small, and $Z_1 \approx R_W$. Thus, for the parallel LC notch filter at low signal frequencies, the attenuation becomes,

$$\frac{v_o}{v_i} = \frac{R_1}{R_1 + R_W} \tag{24-25}$$

It is seen that there is an insertion loss with the parallel-resonant notch filter at low frequencies.

At frequencies above the pass-band, X_{L1} increases and X_{C1} becomes smaller as the signal frequency increases. In this case, Z_1 becomes much smaller than R_1, and so the attenuation approaches zero.

Example 24-14

The two notch filter circuits in Figure 23-26 use the following components: $L = 200\ \mu H$, $C = 3000$ pF, $R_1 = 180\ \Omega$, and $R_w = 15\ \Omega$. Determine the centre frequency and signal attenuation at resonance for both filters.

Solution

Eq. 23-3,

$$f_r = \frac{1}{2\pi\sqrt{LC}} = \frac{1}{2\pi\sqrt{200\ \mu H \times 3000\ pF}}$$

$$= 205\ kHz$$

For the series-resonant notch filter at f_r,

From Eq. 24-23,

$$\frac{v_o}{v_i} = \frac{R_W}{R_1 + R_W} = \frac{15\ \Omega}{180\ \Omega + 15\ \Omega}$$

$$\approx 0.077$$

$$\text{Attenuation} = 20 \log (v_o/v_i) = 20 \log 0.077$$

$$\approx -22\ dB$$

For the parallel-resonant notch filter at f_r,

Eq. 23-20,

$$Z_1 = \frac{L_1}{C_1 R_W} = \frac{200\ \mu H}{3000\ pF \times 15\ \Omega}$$

$$= 4.44\ k\Omega$$

Eq. 24-24,

$$\frac{v_o}{v_i} = \frac{R_1}{R_1 + L/(CR_W)} = \frac{15\ \Omega}{180\ \Omega + 4.44\ k\Omega}$$

$$\approx 3.25 \times 10^{-3}$$

$$\text{Attenuation} = 20 \log (v_o/v_i) = 20 \log (3.25 \times 10^{-3})$$

$$\approx -50\ dB$$

Practice Problems

24-8.1 A series-resonant notch filter uses the following components: $R_1 = 33\ \Omega$, $L_1 = 100\ \mu H$, $R_w = 5\ \Omega$, $C_1 = 1000$ pF. Calculate the resonance frequency, cutoff frequencies, and filter attenuation at resonance.

24-8.2 If the components used in Problem 24-8.1 are rearranged in the form of a parallel-resonant notch filter, determine the filter attenuation at resonance, and the low-frequency insertion loss.

Summary of Formulas

- ***Power change in decibels:***

$$\Delta p = 10 \log (p_2/p_1) \text{ dB}$$

$$\Delta p = 20 \log (v_2/v_1) \text{ dB}$$

$$\Delta p = 20 \log (i_2/i_1) \text{ dB}$$

- ***RC filter low-pass and high-pass cutoff frequencies:***

$$f_c = \frac{1}{2\pi R_1 C_1}$$

- ***RC filter low-pass transfer function:***

$$\frac{v_o}{v_i} = \frac{X_{C1}}{\sqrt{R_1^2 + X_{C1}^2}} \angle \tan^{-1}(-R_1/X_{C1})$$

$$\frac{v_o}{v_i} = \frac{1}{\sqrt{(f/f_c)^2 + 1}} \angle \tan^{-1}(-f/f_c)$$

- ***RC filter high-pass transfer function:***

$$\frac{v_o}{v_i} = \frac{R_1}{\sqrt{R_1^2 + X_{C1}^2}} \angle \tan^{-1}(X_{C1}/R_1)$$

$$\frac{v_o}{v_i} = \frac{1}{\sqrt{(f_c/f)^2 + 1}} \angle \tan^{-1}(f_c/f)$$

- ***RL filter low-pass and high-pass cutoff frequencies:***

$$f_c = \frac{1}{2\pi(L_1/R_1)}$$

- ***RL filter low-pass transfer function:***

$$\frac{v_o}{v_i} = \frac{R_1}{\sqrt{R_1^2 + X_{L1}^2}} \angle \tan^{-1}(-X_{L1}/R_1)$$

- ***Bandpass filter:***

$$B = f_2 - f_1$$

$$f_o = \sqrt{f_1 \times f_2}$$

Review Questions

Section 24-1

24-1 Explain the function of low-pass and high-pass filters and draw typical gain/frequency graphs for each type of filter.

24-2 Explain the function of bandpass and band-stop filters and draw typical gain/frequency graphs for each type of filter.

Section 24-2

24-3 Write equations for the power level change in decibels between two power levels (p_1 and p_2). Also write the equations for the power level changes when the two voltage levels (v_1 and v_2) and two current levels (i_1 and i_2) are measured.

Section 24-3

24-4 Sketch the circuit of a low-pass *RC* filter, and explain the filter operation. Write the equation for the low-pass filter cutoff frequency.

24-5 Sketch the typical gain/frequency response graph for a low-pass *RC* filter. Identify the filter cutoff frequency and the rate of change of gain from the cutoff frequency.

24-6 Sketch the typical phase/frequency response graph for a low-pass *RC* filter. Identify the approximate rate of change of phase from the cutoff frequency.

24-7 Write the transfer function equation for a low-pass *RC* filter.

24-8 Define insertion loss and explain how it can occur with low-pass *RC* filters.

Section 24-4

24-9 Sketch the circuit of a high-pass *RC* filter and explain the filter operation. Write the equation for the high-pass filter cutoff frequency.

24-10 Sketch the typical gain/frequency response graph for a high-pass *RC* filter. Identify the filter cutoff frequency and the rate of change of gain from the cutoff frequency.

24-11 Sketch the typical phase/frequency response graph for a high-pass *RC* filter. Identify the approximate rate of change of phase from the cutoff frequency.

24-12 Write the transfer function equation for a high-pass *RC* filter.

24-13 Explain how insertion loss can occur with high-pass *RC* filters.

Section 24-5

24-14 Draw straight-line approximations of the typical gain/frequency and phase/frequency responses for a low-pass filter. Identify the filter cutoff frequency and the rate of change of gain and phase from the cutoff frequency.

24-15 Draw straight-line approximations of the typical gain/frequency and phase/frequency responses for a high-pass filter. Identify the filter cutoff frequency and the rate of change of gain and phase from the cutoff frequency.

Section 24-6

24-16 Sketch the circuit of a low-pass filter that uses a resistor and inductor and explain the filter operation. Write the equation for the filter cutoff frequency.

24-17 Sketch typical gain/frequency and phase/frequency response graphs for a low-pass *RL* filter. Identify the filter cutoff frequency and the rate of change of gain and phase from the cutoff frequency.

24-18 Sketch the circuit of a high-pass filter that uses a resistor and inductor. Explain the filter operation and write the equation for the filter cutoff frequency.

24-19 Sketch typical gain/frequency and phase/frequency response graphs for a high-pass *RL* filter. Identify the filter cutoff frequency and the rate of change of gain and phase from the cutoff frequency.

Section 24-7

24-20 Draw the circuit diagram of the combination of low-pass and high-pass filters that functions as a bandpass filter. Explain how the filter operates.

24-21 Sketch the typical gain/frequency response graph for the bandpass filter in Question 24-20. Write equations for the filter cutoff frequencies, centre frequency, and bandwidth.

24-22 Draw the circuit diagram of a series-resonant bandpass filter. Explain how the circuit operates and write the equation for the filter insertion loss at the centre frequency.

24-23 Sketch a parallel-resonant bandpass filter circuit, and explain its operation. Write the equation for the filter insertion loss at the centre frequency.

24-24 A bandpass filter circuit can be constructed either in series-resonant form or in parallel-resonant form. Which of the two arrangements produces the lowest insertion loss at the centre frequency? Explain.

Section 24-8

24-25 Draw the circuit diagram of the combination of low-pass and high-pass filters that functions as a notch filter. Explain how the filter operates.

24-26 Sketch the typical gain/frequency response graph for the notch filter in Question 24-25. Write equations for the filter cutoff frequencies, centre frequency, and bandwidth.

24-27 Draw the circuit diagram of a notch filter that uses series-resonance. Explain how the circuit operates and write the equation for the filter insertion loss at the centre frequency.

24-28 Sketch the circuit of a notch filter that uses parallel-resonance. Explain how the circuit operates and write the equation for the filter insertion loss at the centre frequency.

24-29 A notch filter circuit can be constructed either in series-resonant form or in parallel-resonant form. Which of the two arrangements produces the greatest attenuation at the centre frequency? Explain.

Problems

Section 24-2

24-1 The output voltage from a filter is measured as 1 V at a frequency f_1 and 0.5 V at f_2. Calculate the decibel change in the output power level.

24-2 A 3.3 mV output is produced when a 100 mV signal is applied to a filter. Calculate the attenuation in decibels.

24-3 The output from a filter decreases by 6 dB as the frequency increases from f_1 to f_2. If the output voltage at f_2 is 80 mV, calculate the voltage level at f_1.

Section 24-3

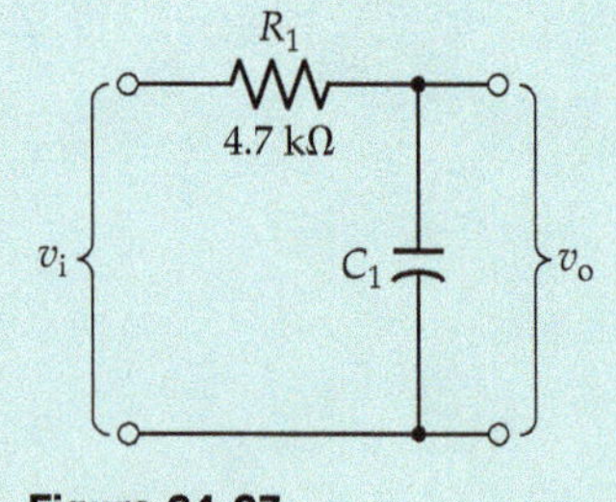

Figure 24-27

24-4 A low-pass *RC* filter (as in Figure 24-27) is to be used to attenuate an unwanted 120 Hz input by a factor of approximately 100. If $R_1 = 4.7\ \text{k}\Omega$, determine a suitable capacitor value.

24-5 If the circuit in Problem 24-4 has a 47 kΩ load resistor, calculate the insertion loss and the effect of R_L on the cutoff frequency.

24-6 A low-pass *RC* filter has $R_1 = 56\ \text{k}\Omega$ and $C_1 = 1000$ pF. Determine the filter cutoff frequency. Also calculate the insertion loss when a 330 kΩ load resistor is connected, and determine the effect of the load resistor on the filter cutoff frequency.

24-7 Determine the voltage gain and phase shift for the filter in Example 24-2 for the frequencies of 5 kHz, 10 kHz, and 15 kHz.

24-8 A low-pass *RC* filter is to be designed to attenuate a 12 kHz input signal by 12 dB. Determine a suitable resistance value if a 1000 pF capacitor is to be used. Calculate the filter cutoff frequency.

Section 24-4

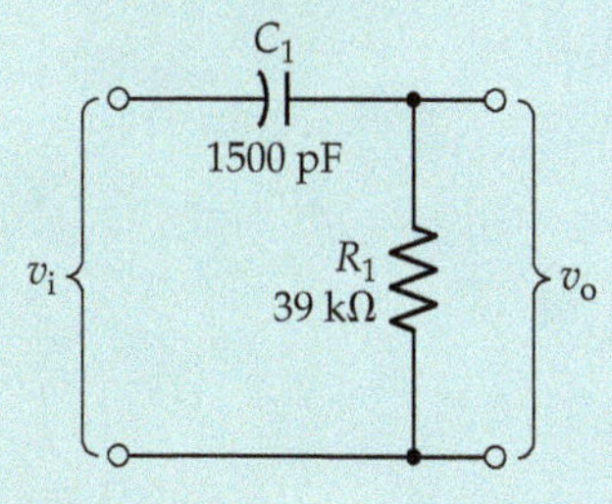

Figure 24-28

24-9 Calculate the cutoff frequency for the high-pass filter in Figure 24-28. Determine the filter attenuation and phase shift at signal frequencies of 1 kHz and 500 Hz.

24-10 A high-pass filter is to attenuate a 1 kHz signal by approximately 25 dB. Determine a suitable resistor value if a 0.01 μF capacitor is to be used.

24-11 Calculate the cutoff frequency for the high-pass filter in Problem 24-10. Determine the filter attenuation and phase shift at frequencies of $0.3f_c$ and $0.15f_c$.

24-12 A high-pass *RC* filter has $R_1 = 27$ kΩ and $C_1 = 1200$ pF. Determine the filter cutoff frequency. Also calculate the effect of a 220 kΩ load resistor on the cutoff frequency.

24-13 A high-pass filter is to attenuate a 5 kHz input by a factor of 10. Determine a suitable resistance value if a 0.01 μF capacitor is to be used. Calculate the filter cutoff frequency.

Section 24-5

24-14 Determine corresponding gain and frequency quantities for plotting the straight-line and actual gain/frequency response graphs for the low-pass filter in Problem 24-6.

24-15 Repeat Problem 24-14 for the straight-line and actual phase/frequency response graphs.

24-16 For the high-pass filter in Problem 24-9, determine corresponding gain and frequency quantities for plotting the straight-line and actual gain/frequency response graphs.

24-17 Repeat Problem 24-16 for the straight-line and actual phase/frequency response graphs.

Section 24-6

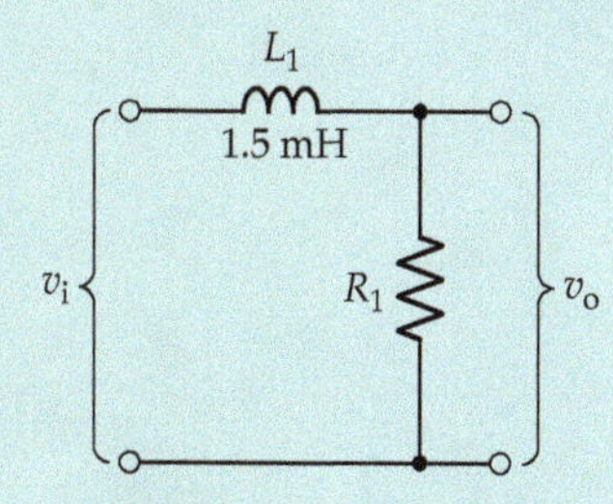

Figure 24-29

24-18 A low-pass *RL* filter (as in Figure 24-29) is to have a 4 kHz cutoff frequency. Determine a suitable resistance value if a 1.5 mH inductor is to be used. Calculate the filter attenuation at 18 kHz.

24-19 The inductor in the *RL* filter in Example 24-10 has a resistance of 15 Ω. Calculate the minimum load resistance value if the insertion loss is not to exceed 1 dB.

24-20 Calculate the cutoff frequency for an *RL* high-pass filter which has $L_1 = 10$ mH and $R_1 = 390$ Ω. Determine the filter attenuation and phase shift at 3 kHz.

24-21 A high-pass *RL* filter has $R_1 = 27$ kΩ and $L_1 = 130$ mH. The inductor winding resistance is $R_W = 25$ Ω, and the filter load resistance is 220 kΩ. Determine the cutoff frequency, insertion loss, and maximum signal attenuation at low frequencies.

Section 24-7

24-22 A bandpass filter constructed of cascaded high-pass and low-pass stages (as in Figure 24-30) is to have a 500 Hz lower cutoff frequency and a 12 kHz bandwidth. Determine suitable resistor values if 0.01 μF capacitors are to be used.

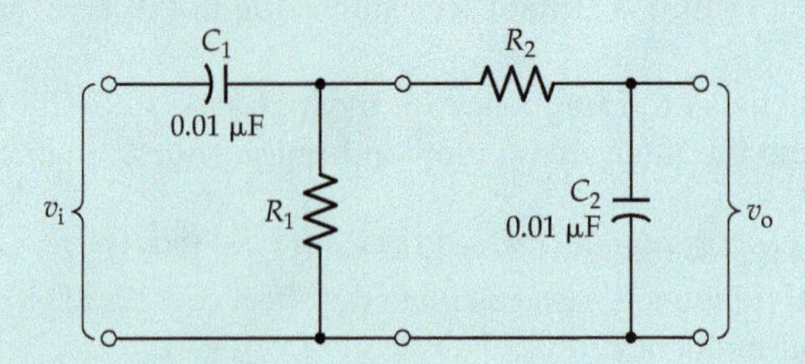

Figure 24-30

24-23 Determine the effect of a 68 kΩ load resistor on the bandwidth of the filter in Problem 24-22.

24-24 Calculate the pass-band frequencies and insertion loss for a series resonant bandpass filter circuit with the following component values: $R_1 = 120\ \Omega$, $L_1 = 1.5$ mH, $R_W = 20\ \Omega$, and $C_1 = 1500$ pF.

24-25 Determine the insertion loss for a parallel-resonant bandpass filter that uses the same component values as in Problem 24-24.

24-26 A bandpass filter using a series resonant circuit is to be designed to pass frequencies within the range 28 kHz to 32 kHz. A 100 μH inductor with a 13 Ω winding resistance is to be used, and the filter insertion loss at the centre frequency is not to exceed 1 dB. Determine suitable capacitor and resistor values.

24-27 A parallel-resonant bandpass filter is to pass a 90 kHz input with a 9 kHz bandwidth. The circuit is to use a 1500 pF capacitor and a 100 Ω resistor. Determine a suitable inductance value and the maximum coil resistance for the inductor.

Section 24-8

24-28 A notch filter constructed of high-pass and low-pass stages (as in Figure 24-31) is to have a 900 Hz lower cutoff frequency and a 15 kHz upper cutoff frequency. Determine suitable capacitor values if 39 kΩ resistors are to be used.

24-29 Determine the centre frequency and bandwidth for a series resonant notch filter that uses the following component values: $R_1 = 68\ \Omega$, $L_1 = 1.2$ mH, $R_W = 12\ \Omega$, and $C_1 = 2700$ pF. Calculate the filter attenuation at resonance.

24-30 Calculate the attenuation at resonance for a parallel-resonant notch filter that uses the same component values as in Problem 24-29.

24-31 A series-resonant notch filter is to block frequencies within the range 110 kHz to 130 kHz. A 1 mH inductor with a 9 Ω winding resistance is to

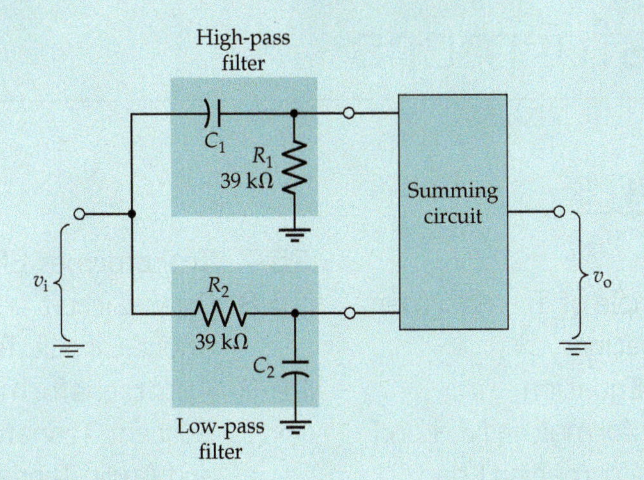

Figure 24-31

be used. Determine suitable capacitor and resistor values, and calculate the filter centre frequency and its attenuation at resonance.

24-32 A parallel-resonant notch filter is to attenuate a 90 kHz input by 33 dB. The circuit is to use a 5000 pF capacitor and a 220 Ω resistor. Determine a suitable inductance value and the coil resistance for the inductor, and calculate the filter low-frequency insertion loss.

Practice Problem Answers

24-2.1 −20 dB, 30.1 mV
24-3.1 33.2 kHz, 29 dB
24-3.2 −8.2 dB $\angle -57.5°$
24-3.3 −0.21 dB, 16.3 kHz
24-4.1 26.5 kHz, −7.8 dB $\angle 78.7°$, −5.4 dB $\angle 68.2°$
24-4.2 27 kHz, −0.81 dB
24-6.1 50 kHz, −10 dB $\angle -71.5°$
24-6.2 −0.2 dB
24-7.1 300 Ω, 3070 pF, −0.4 dB
24-7.2 −0.035 dB
24-8.1 503 kHz, 472.7 kHz, 533.3 kHz, −17.6 dB
24-8.2 −55.7 dB, −1.2 dB

CHAPTER 25
Transformers

CONTENTS

Objectives

You will be able to:

1. Sketch the basic construction of a transformer and explain its operation.
2. Sketch the circuit diagram for a transformer and derive equations relating the number of turns on each winding to the primary and secondary voltages and currents.
3. Given input voltage and secondary load, calculate the secondary voltages and the primary and secondary currents of a transformer, and determine the level of flux in a transformer core.
4. Sketch and explain the equivalent circuit and phasor diagrams for a transformer under no-load conditions.
5. Sketch and explain the complete equivalent circuit and the phasor diagrams for the primary and secondary of a loaded transformer.
6. Given the parameters of a transformer and the load at the output terminals, determine the amplitudes and phase angles of the supply voltage and current.
7. Define referred resistance and referred reactance, and show how the transformer equivalent circuit may be simplified by referring the secondary circuit components to the primary.
8. Solve problems involving transformer efficiency and voltage regulation.
9. Analyze the results of open-circuit and short-circuit tests on transformers to determine the component values of the transformer equivalent circuit.
10. Explain autotransformers, current transformers, and audio transformers.

INTRODUCTION

A transformer basically consists of two coils wound on a single iron core. When an alternating voltage is applied to one of the coils, the mutual inductance causes an alternating voltage to be induced in the other coil. The ratio of the input and output voltage amplitudes depends on the number of turns on each coil. A transformer may be used either to increase or decrease an applied voltage, or to increase or decrease a current.

The emf induced in each winding of a transformer can be calculated from the transformer emf equation. The behavior of the transformer under no-load and full-load conditions is best understood by drawing the appropriate phasor diagrams and by studying the transformer equivalent circuit. Simplification of the equivalent circuit is possible by the process of referring the secondary resistance and reactance to the primary winding.

The performance of a transformer is described in terms of its voltage regulation and efficiency, and can be predicted from the results of two tests, known as the *open-circuit test* and the *short-circuit test*.

25-1 PRINCIPLE OF TRANSFORMER OPERATION

Transformer Construction

The transformer is an application of mutual inductance. An alternating voltage applied to a *primary* winding generates an alternating magnetic flux that links with a *secondary* winding and induces an alternating emf in the secondary.

Figure 25-1(a) illustrates the basic construction of a transformer. The iron core forms a closed magnetic circuit, and the windings are simply coils of insulated wire wrapped around the core. The presence of the iron core causes virtually 100% of the magnetic flux generated by the primary to be linked with the secondary. The transformer is therefore similar to two *coupled coils* (see Section 14-4) in which the coefficient of coupling is 1. In fact, air-cored coupled coils (Section 23-9) are sometimes referred to as an *air-cored transformer.* The circuit symbol for the transformer is shown in Figure 25-1(b). The two lines between the coils indicate the presence of an iron core.

Primary and Secondary Voltages

The alternating magnetic flux in the iron core of the transformer links with both the primary and secondary windings (see Figure 25-1). The flux linking with the primary winding generates a counter-emf in the primary. As the counter-emf is essentially equal in magnitude to the supply voltage (see Sections 19-1 and 19-2), the counter-emf equation can be used to calculate the transformer primary voltage.

From Eq. 14-2, $E_P = N_P(\Delta\Phi/\Delta t)$

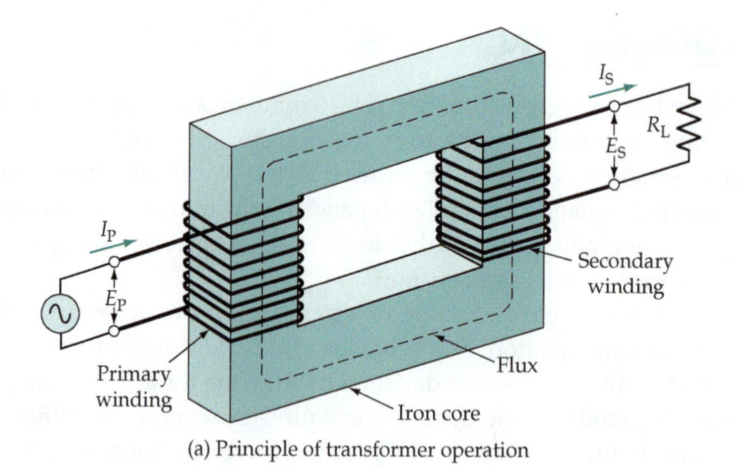

(a) Principle of transformer operation

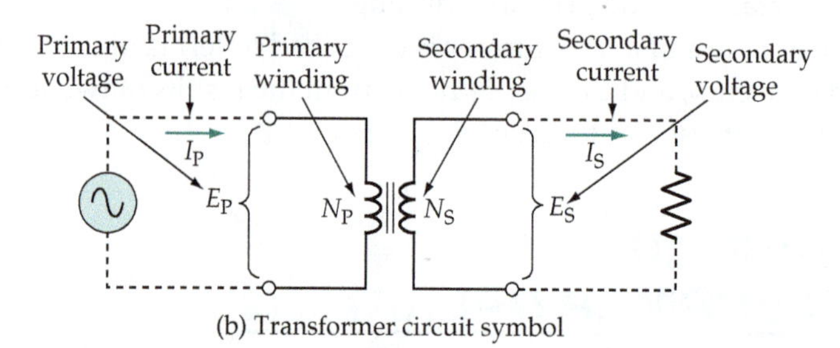

(b) Transformer circuit symbol

Figure 25-1 A transformer consists basically of two coils wound on an iron core. When an ac voltage is applied to the primary winding, the resultant core flux links with the secondary winding and induces a voltage in the secondary.

where E_P is the primary induced voltage and N_P is the number of turns on the primary winding. Also,

$$E_S = N_S(\Delta\Phi/\Delta t)$$

where E_S and N_S are the secondary induced voltage and the secondary turns, respectively. Combining the two equations,

$$\frac{E_P}{E_S} = \frac{\Delta\Phi/\Delta t}{\Delta\Phi/\Delta t} \times \frac{N_P}{N_S}$$

So,

$$\frac{E_P}{E_S} = \frac{N_P}{N_S}$$

It is seen that the ratio of primary voltage to secondary voltage is the same as the ratio of primary coil turns to secondary turns. When the primary (input) voltage and the turns ratio are known, the secondary (output) voltage can be calculated from,

$$E_S = E_P \times \frac{N_S}{N_P} \qquad \textbf{(25-1)}$$

Step-Up, Step-Down and Multi-Output Transformers

When the primary and secondary windings have an equal number of turns, the output voltage is equal to the input voltage, and the transformer is referred to as a *1:1 transformer.* If the secondary has more turns than the primary. E_S is greater than E_P, and the device is termed a *step-up transformer*. This means that the transformer steps up the input voltage to produce a higher output level. It is also possible to have a secondary winding with fewer turns than the primary. In this case, the output voltage is less than the input, and the device is a *step-down transformer*.

Figure 25-2 shows the circuit diagram of a transformer with three separate secondary windings. The output voltage from each winding depends on the primary voltage, the number of primary turns, and the number of turns on the secondary winding.

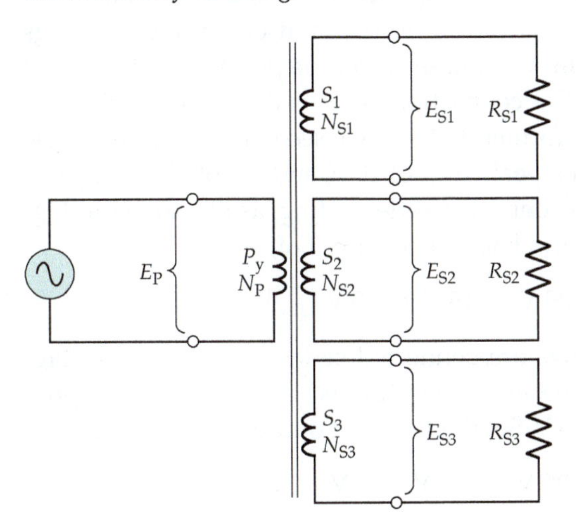

Figure 25-2 A transformer may have multiple secondary windings. The output voltage from each secondary depends on the primary voltage and the ratio of primary to secondary turns.

Example 25-1

The transformer in Figure 25-2 has an alternating input of 100 V. Determine the output voltage from each secondary winding. Also determine the total output voltage if all three secondary windings are connected series-aiding. The numbers of turns on each winding are $N_P = 375$, $N_{S1} = 750$, $N_{S2} = 500$, and $N_{S3} = 75$.

Solution

Eq. 25-1,

$$E_{S1} = E_P \times \frac{N_{S1}}{N_P} = 100\text{ V} \times \frac{750}{375}$$

$$= 200\text{ V}$$

$$E_{S2} = E_P \times \frac{N_{S2}}{N_P} = 100\text{ V} \times \frac{500}{375}$$

$$= 133\text{ V}$$

$$E_{S3} = E_P \times \frac{N_{S3}}{N_P} = 100\text{ V} \times \frac{75}{375}$$

$$= 20\text{ V}$$

Total secondary turns, $N_S = N_{S1} + N_{S2} + N_{S3} = 750 + 500 + 75$

$$= 1325$$

Total secondary voltage, $E_S = E_P \times \dfrac{N_S}{N_P} = 100\text{ V} \times \dfrac{1325}{375}$

$$= 353\text{ V}$$

or $E_S = E_{S1} + E_{S2} + E_{S3} = 200\text{ V} + 133\text{ V} + 20\text{ V}$

$$= 353\text{ V}$$

It is important to note that the total output voltage calculation in Example 25-1 assumes that the secondary windings are connected *series-aiding*; that is, such that all the secondary voltages are in phase. If one of the windings is connected with its voltage in anti-phase to the others, that voltage will subtract from the others, and the connection is termed *series-opposing*. Figures 25-3(a) and (b) show a transformer that has two secondary windings. One of these gives a waveform with a peak level of 20 V, and the other has a peak level of 15 V. When the two are connected series-aiding, as illustrated in Figure 25-3(a), the resultant waveform has a peak output of,

$$V_o = 20\text{ V} + 15\text{ V} = 35\text{ V}$$

When the secondary windings are connected series-opposing, as in Figure 25-3(b), the peak level of the lower winding is −15 V when the upper winding has a peak value of +20 V. So, the resultant output peak level is,

$$V_o = 20\text{ V} - 15\text{ V} = 5\text{ V}$$

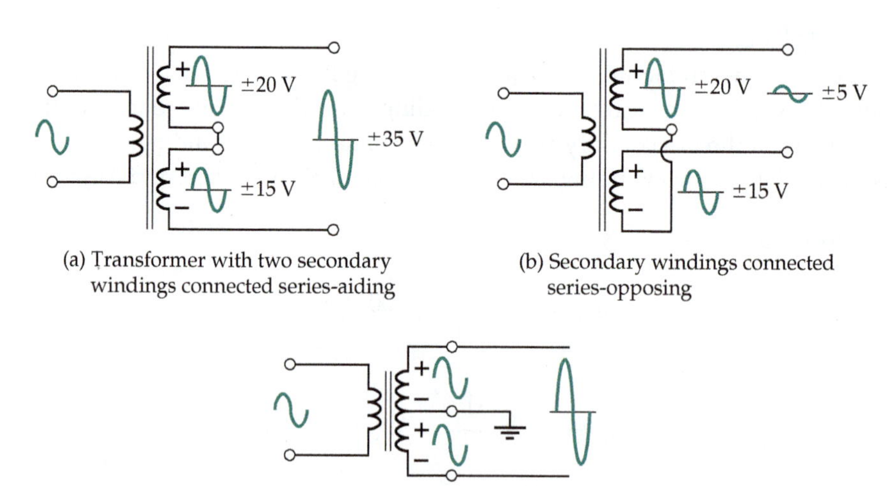

(a) Transformer with two secondary windings connected series-aiding

(b) Secondary windings connected series-opposing

(c) Transformer with centre-tapped secondary winding

Figure 25-3 A transformer with two secondary windings should have the secondaries connected series-aiding to produce the largest output voltage. A transformer with a center-tapped secondary produces two equal outputs.

The transformer shown diagrammatically in Figure 25-3(c) has a secondary that is said to be *center-tapped.* This means that the secondary winding consists of two equal-turn windings connected series-aiding. If the center tap (connection) is grounded, as shown, the output from one terminal is positive-going with respect to ground when the output from the other terminal is negative-going and vice versa. The total output voltage from the two windings is the sum of the outputs from each.

Primary and Secondary Currents

Refer to Figure 25-4 and recall that Equation 21-7 gives the transformer input power (in watts) as,

$$P_P = E_P I_P \cos \Phi_P$$

And the output power is,

$$P_S = E_S I_S \cos \Phi_S$$

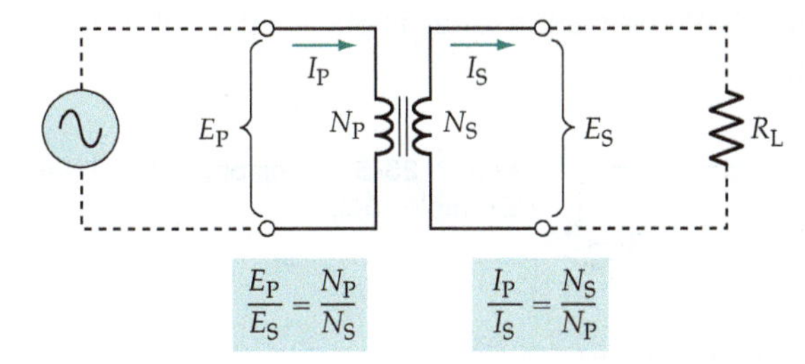

Figure 25-4 The ratio of transformer primary and secondary voltages is the same as the turns ratio. The ratio of primary and secondary currents is the inverse of the turns ratio.

The primary and secondary circuit phase angles can be assumed to be equal, and as an approximation the transformer efficiency can be taken as 100%. Therefore,

$$P_P \approx P_S$$

or,

$$E_P I_P = E_S I_S$$

which gives,

$$\frac{I_P}{I_S} = \frac{E_S}{E_P} = \frac{N_S}{N_P}$$

so,

$$I_S = I_P \times \frac{N_P}{N_S} \qquad (25\text{-}2)$$

Comparing Equations 25-2 and 25-1, it is seen that the ratio of primary and secondary currents in terms of the turns ratio is the inverse of the ratio of primary and secondary voltages. This shows that if a transformer steps up the voltage from the primary, it may be said to step down the current, and vice versa.

Example 25-2

A transformer is to produce a 15 V, 300 mA output from a 115 V supply. Determine the ratio of primary-to-secondary turns and calculate the supply current.

Solution

Eq. 25-1,

$$\frac{N_P}{N_S} = \frac{E_P}{E_S} = \frac{115\text{ V}}{15\text{ V}}$$

$$\approx 7.7$$

Eq. 25-2,

$$I_P = I_S \times \frac{N_S}{N_P} = 300\text{ mA} \times \frac{1}{7.7}$$

$$\approx 39\text{ mA}$$

Example 25-3

The three secondary windings on the transformer described in Example 25-1 have the following loads: $R_{S1} = 1\text{ k}\Omega$, $R_{S2} = 500\ \Omega$, and $R_{S3} = 100\ \Omega$, as illustrated in Figure 25-5. Assuming that the transformer is 100% efficient, determine the total primary current.

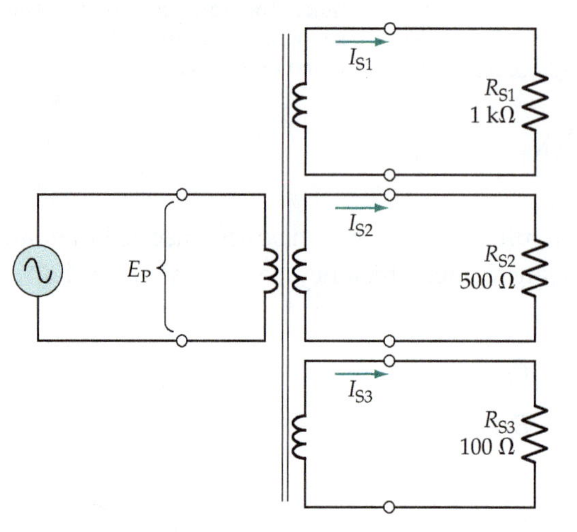

Figure 25-5 Transformer circuit for Example 25-3.

Solution

For secondary S_1,

$$I_{S1} = \frac{E_{S1}}{R_{S1}} = \frac{200\text{ V}}{1\text{ k}\Omega}$$

$$= 200\text{ mA}$$

Eq. 25-2,

$$I_{P1} = I_{S1} \times \frac{N_{S1}}{N_P} = 200\text{ mA} \times \frac{750}{375}$$

$$= 400\text{ mA}$$

For secondary S_2,

$$I_{S2} = \frac{E_{S2}}{R_{S2}} = \frac{133\text{ V}}{500\ \Omega}$$

$$= 266\text{ mA}$$

$$I_{P2} = I_{S2} \times \frac{N_{S2}}{N_P} = 266\text{ mA} \times \frac{500}{375}$$

$$= 355\text{ mA}$$

For secondary S_3:

$$I_{S3} = \frac{E_{S3}}{R_{S3}} = \frac{20\text{ V}}{100\ \Omega}$$

$$= 200\text{ mA}$$

$$I_{P3} = I_{S3} \times \frac{N_{S3}}{N_P} = 200\text{ mA} \times \frac{75}{375}$$

$$= 40\text{ mA}$$

Total primary current:

$$I_P = I_{P1} + I_{P2} + I_{P3}$$

$$= 400\text{ mA} + 355\text{ mA} + 40\text{ mA}$$

$$= 795\text{ mA}$$

Practice Problems

25-1.1 The transformer in Figure 25-6, is to produce peak outputs of 15 V and 22 V. The primary winding has 700 turns, and the input voltage is 115 V rms. Calculate the required number of turns for each of the secondary windings.

25-1.2 The transformer in Problem 25-1.1 has 470 Ω loads on each of the secondary windings. Determine the supply current.

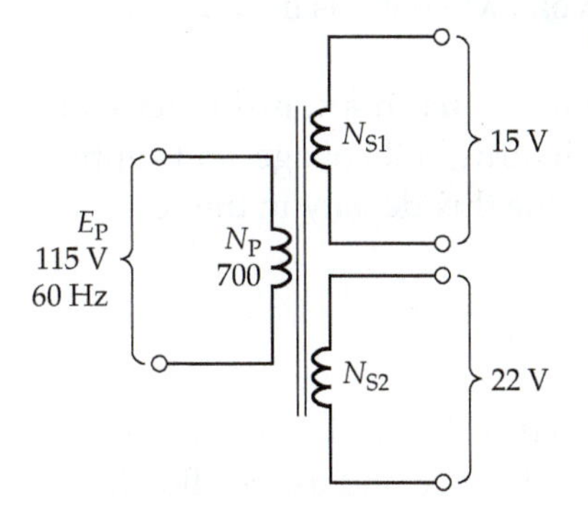

Figure 25-6 Transformer circuit for Problem 25-1.1.

25-2 EMF EQUATION

As discussed in the preceding section, the emfs induced in the primary and secondary windings are:

$$E_P = N_P(\Delta\Phi/\Delta t), \quad \text{and} \quad E_S = N_S(\Delta\Phi/\Delta t)$$

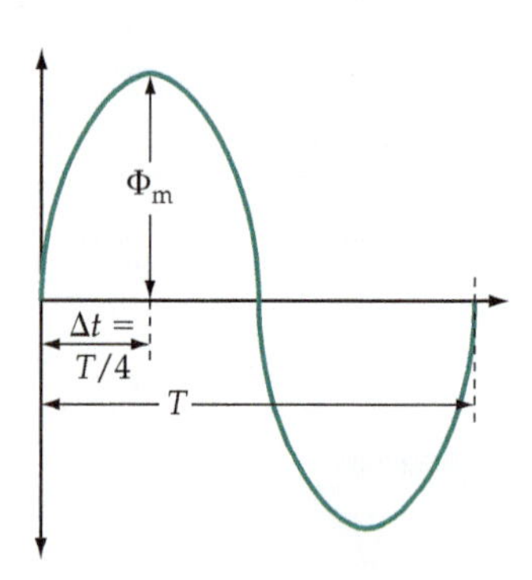

Figure 25-7 The flux in a transformer core has a sinusoidal waveform, so it changes from zero to maximum in a time of $T/4$ or $1/(4f)$. The voltage induced in each winding is $(4.44\ \Phi f N)$.

The flux in the transformer core has a sinusoidal waveform, because the (primary) current producing it is sinusoidal. So, as illustrated in Figure 25-7, the flux increases from zero to its maximum value Φ_m in a time period of

$$\Delta t = 0.25T$$

or

$$\Delta t = 1/(4f)$$

where f is the frequency of the flux waveform (i.e., the frequency of the primary voltage and current). Consequently, the average induced voltage is

$$E_{av} = \frac{\Phi_m}{1/(4f)} N$$

or

$$E_{av} = 4\Phi_m f N$$

For a sine wave,

$$E_{rms} = E_{av} \times \frac{0.707}{0.637} = 1.11 E_{av} \qquad \text{(see Section 17-5)}$$

So

$$E_{rms} = 1.11(4\Phi_m f N)$$
$$= 4.44\Phi_m f N$$

This gives the primary rms voltage as

$$E_P = 4.44\Phi_m f N_P \qquad \textbf{(25-3)}$$

and the secondary as

$$E_S = 4.44\Phi_m f N_S \qquad \textbf{(25-4)}$$

In Equations 25-3 and 25-4, E_P and E_S are in volts when Φ_m is in webers and f is in hertz.

Using Equations 25-3 and 25-4, the maximum flux in a transformer core may be calculated from a knowledge of the coil turns, coil voltage, and supply frequency. Once the maximum flux is known, the flux density in the core can be determined using the core dimensions.

Example 25-4

The input voltage to the transformer described in Example 25-1 has a frequency of 400 Hz. Determine the peak value of the flux, and use the flux level to calculate E_{S1}. The transformer quantities are: $E_P = 100$ V, $N_P = 375$, $N_{S1} = 750$, $N_{S2} = 500$, and $N_{S3} = 75$.

Solution

Eq. 25-3,
$$\Phi_m = \frac{E_P}{4.44fN_P} = \frac{100\text{ V}}{4.44 \times 400\text{ Hz} \times 375}$$
$$= 150\ \mu\text{Wb}$$

Eq. 25-4,
$$E_{S1} = 4.44\Phi_m fN_{S1}$$
$$= 4.44 \times 150\ \mu\text{Wb} \times 400\text{ Hz} \times 750$$
$$= 200\text{ V}$$

Practice Problems

25-2.1 Determine the peak flux in the core of the transformer described in Problem 25-1.1 (reproduced in Figure 25-8) if the supply frequency is 60 Hz.

25-2.2 A transformer, designed to operate from a 50 V, 400 Hz supply, has 390 primary turns. If the transformer is used on a 60 Hz supply, determine the required level of primary voltage for the core flux to remain constant.

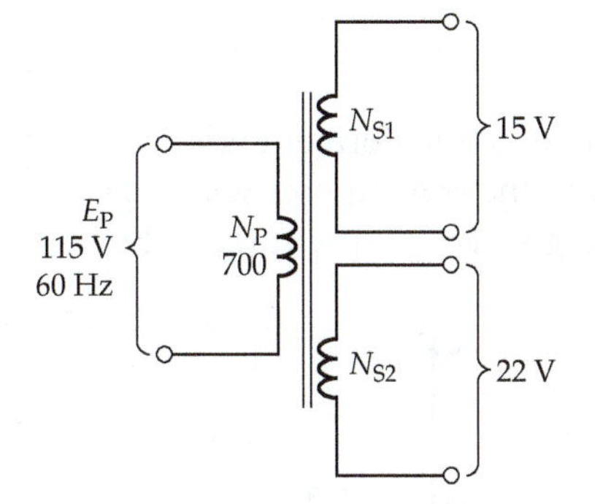

Figure 25-8 Transformer for Problem 25-2.1.

25-3 TRANSFORMER ON NO-LOAD

Core Losses

A transformer is said to be on *no-load* when the output (secondary) terminals are open-circuited. In this condition there is no current flowing in the secondary windings. However, with an alternating voltage applied to the input (primary) terminals, a small primary current flows in order to create the magnetic flux in the core. This is termed the *magnetizing current.*

In Section 12-7 it is explained that when the magnetic flux in a core is continuously increasing to a peak level in one direction and then reversing to a peak in the opposite direction, energy is absorbed by the core, due to *hysteresis* loss.

The alternating magnetic flux also induces *eddy currents* in the transformer core (see Section 12-7). The eddy currents cause additional energy to be dissipated in the core. Hysteresis loss is kept to a minimum by the use of a magnetic material with a narrow hysteresis loop, and eddy current loss is minimized by constructing the core of *laminations.* Because of the laminated construction, there are air gaps in the core, as illustrated in Figure 25-9. The magnetizing current must create a flux in the air gaps as well as in the core. Power dissipated in the transformer primary windings (the I^2R loss) is additional energy loss that must be supplied. The total primary *no-load current* is composed of the magnetizing current and the current required to supply the core losses.

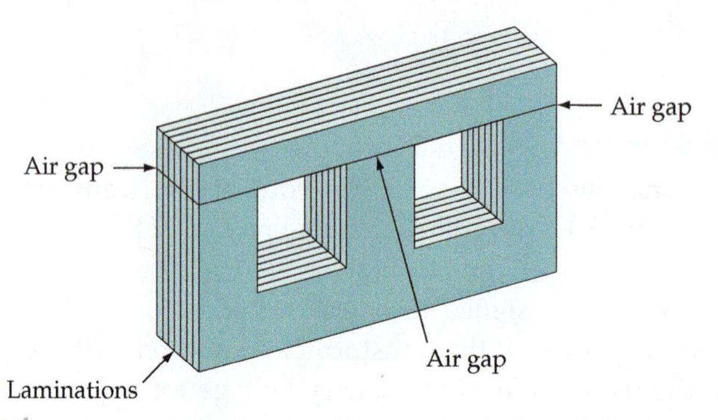

Figure 25-9 Energy losses occur in a transformer core because of eddy currents and air gaps. The primary no-load current must supply these core losses as well as create a magnetic flux in the core.

No-Load Phasor Diagram

A circuit and an approximate phasor diagram for a transformer under no-load conditions are shown in Figure 25-10. Because the transformer windings are inductive, the input voltage V_p leads the magnetizing current (I_{mag}) by 90°.

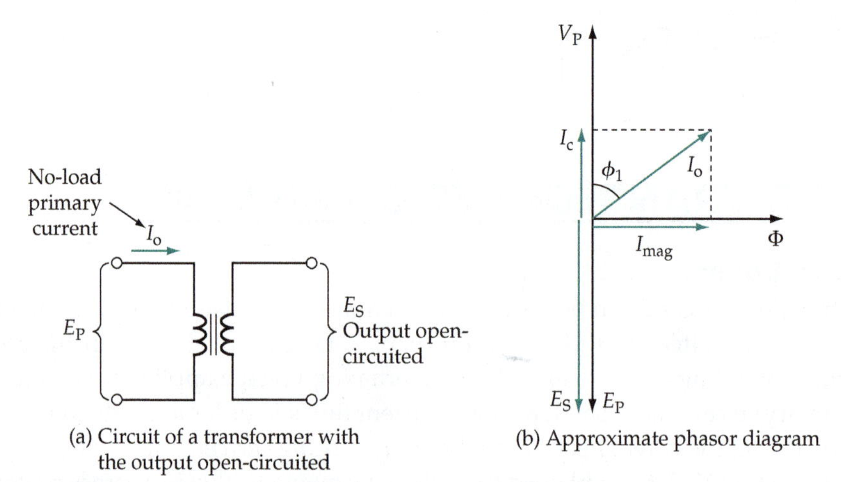

(a) Circuit of a transformer with the output open-circuited

(b) Approximate phasor diagram

Figure 25-10 Circuit and phasor diagram for a transformer on no-load. The core flux (Φ) is in phase with the magnetizing current (I_{mag}). The windings are inductive, so I_{mag} lags the primary voltage (V_P) by 90°.

The flux (Φ) increases and decreases as the magnetizing current rises and falls; consequently, the flux phasor is shown in phase with I_{mag}. The core loss current (I_c) is the component of the primary current that supplies the core losses and the small power loss in the primary winding. The total power losses are equal to ($I_c \times V_P$), and so I_c is in phase with V_P. The no-load primary current (I_o) is the phasor sum of I_c and I_{mag}. Because I_c is normally much smaller than I_{mag}, the no-load power factor ($\cos \Phi_1$) is very small. The voltages induced in the secondary and primary windings, E_S and E_P, respectively, lag the flux by 90°, and thus the E_S and E_P phasors are drawn opposite to the V_P phasor. In Figure 25-10, E_S and E_P are shown as equal voltages (i.e., assuming a 1:1 transformer). An approximation in the phasor diagram of Figure 25-10 occurs because E_P has been taken as exactly equal and opposite to V_P. In fact, E_P is equal and opposite to the phasor sum of V_P, and the winding voltage drops due to I_o. This becomes more evident when a transformer *on-load* is considered.

No-Load Equivalent Circuit

The no-load equivalent circuit for the transformer is shown in Figure 25-11. The transformer is replaced with an ideal (no-loss) transformer, with a resistance R_o, and an inductive reactance X_o in parallel with its primary. R_o represents the core losses, and so the current I_c that supplies the core losses is shown passing through R_o. The inductive reactance X_o represents a loss-free coil that passes the magnetizing current I_{mag}. Thus, the combination of R_a, X_o, and the ideal transformer simulates the actual transformer under no-load conditions.

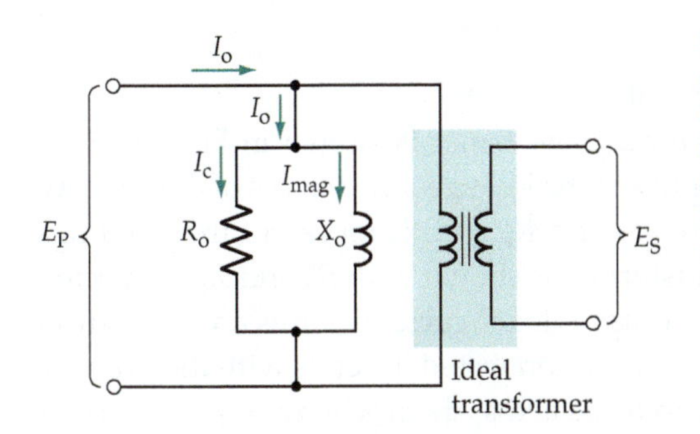

Figure 25-11 The no-load equivalent circuit of a transformer is made up of an ideal (no loss) transformer with its primary in parallel with a resistance (R_o) representing core losses and an inductive reactance (X_o) to represent the winding inductance.

25-4 TRANSFORMER ON LOAD

Leakage Inductance

When a load is connected to a transformer secondary, a secondary winding (load) current flows. As illustrated in Figure 25-12, the secondary current tends to generate a flux Φ_2 in the transformer core. To supply the secondary current, current must flow in the primary winding. The primary current generates a flux Φ_1 which is exactly equal and opposite to Φ_2. Thus, Φ_1 and Φ_2 cancel each other out, and the core flux remains at the level set up by the magnetizing current.

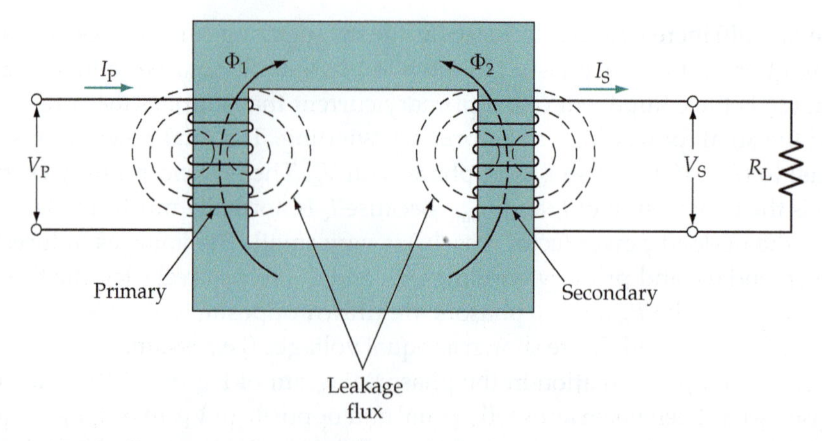

Figure 25-12 The currents flowing in the primary and secondary windings of a transformer set up core fluxes (Φ_1 and Φ_2). Leakage flux also occurs around each winding.

Figure 25-12 also shows that when primary and secondary currents flow, not all the flux set up by the currents passes through the iron core. Instead, there is some *leakage flux* passing through the air surrounding each coil. Because the magnetic path through the iron core has a very much smaller reluctance than the air path around each coil, the leakage flux is normally quite small. However, the leakage flux links with the winding turns in each coil and sets up emfs that oppose the flow of current through each coil. Thus, the leakage flux produces the same effect as an unwanted inductance connected in series with each winding. The effect is termed the *leakage inductance.*

Complete Equivalent Circuit

The complete equivalent circuit for a transformer is shown in Figure 25-13. Inductive reactances X_P and X_S represent the leakage inductances of the primary and secondary windings, respectively, and R_P and R_S represent the windings resistances. An ideal (no-loss) transformer is shown with R_S and X_S connected in series with the secondary, so an output (load) current causes a voltage drop across R_S and X_S. Similarly, R_P and X_P are connected in series with the primary windings, and voltage drops are produced across them when a primary current

Figure 25-13 The complete equivalent circuit for a transformer is simply the no-load equivalent circuit (Figure 25-11) with components R_P, X_P, R_S, and X_S included to represent the winding resistances and leakage inductances.

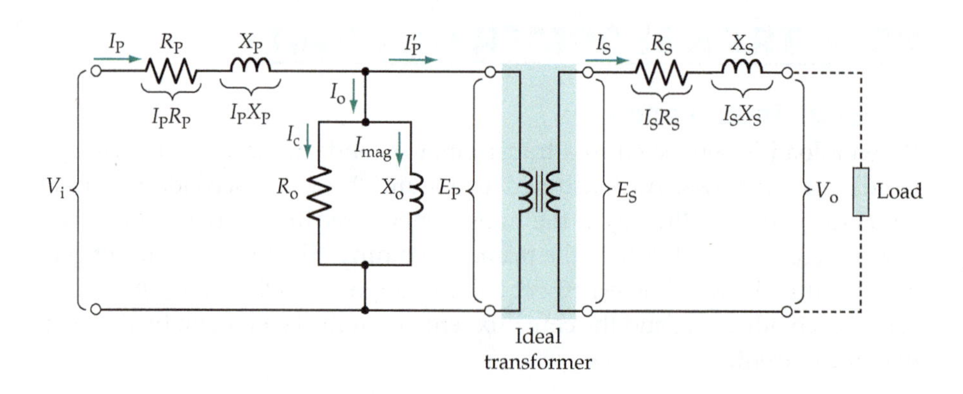

flows. As before, R_o and X_o are shown in parallel with the primary, to simulate the no-load losses and the magnetizing current.

Phasor Diagram for Secondary

The phasor diagram for the secondary circuit of a transformer under load is shown in Figure 25-14. Referring to Figures 25-13 and 25-14(a), V_o is the voltage at the transformer output terminals, and I_S is the secondary (load) current. For a load with a lagging phase angle (ϕ_o), the I_S phasor is shown lagging the V_o phasor by angle ϕ_o. The secondary current flows through R_S and X_S and produces voltage drops: I_SR_S and I_SX_S. The resistive voltage drop (I_SR_S) is in phase with I_S, and the inductive voltage drop (I_SX_S) leads the current by 90°, as shown in Figure 25-14(a).

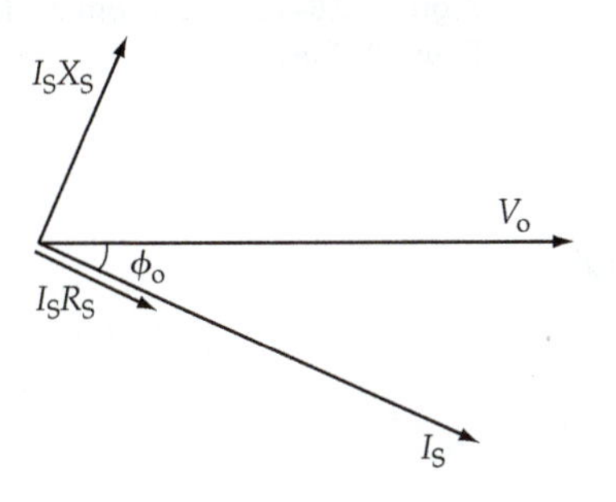

(a) Partial phasor diagram for secondary circuit with an inductive (lagging ϕ) load

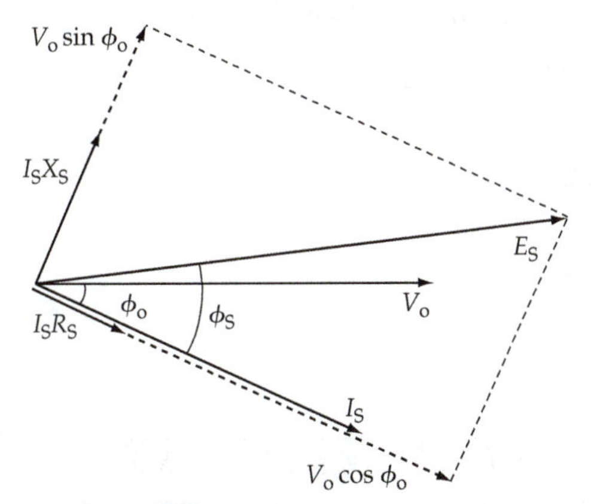

(b) E_S is the phasor sum of V_o, I_SX_S, and I_SR_S

Figure 25-14 Phasor diagram for the secondary of a transformer under load. Secondary current I_S lags the output voltage V_o by the load phase angle ϕ_o. (I_SX_S) and (I_SR_S) are the voltage drops across X_S and R_S. The secondary induced voltage E_S is the phasor sum of V_o, (I_SX_S), and (I_SR_S).

In Figure 25-14(b), the phasors of V_o, I_S, I_SX_S, and I_SR_S are reproduced from Figure 25-14(a). The secondary induced voltage E_S (also see Figure 25-13) is the phasor sum of V_o, I_SR_S, and I_SX_S. From the discussion of phasor addition in Chapter 18:

$$E_S = \sqrt{(V_o \cos \phi_o + I_SR_S)^2 + (V_o \sin \phi_o + I_SX_S)^2} \qquad \text{(25-5)}$$

and the phase angle between E_S and I_S is,

$$\phi_S = \tan^{-1}\left(\frac{V_o \sin \phi_o + I_S X_S}{V_o \cos \phi_o + I_S R_S}\right) \qquad (25\text{-}6)$$

Since the secondary induced voltage always lags 90° behind the core flux Φ, the core flux phasor may be drawn 90° ahead of E_S.

Example 25-5

A transformer with $R_S = 0.25\ \Omega$, $X_S = 1\ \Omega$, and $N_P/N_S = 2$ has a 50 V output and a load of $Z_L = 25\ \Omega\angle 30°$. Determine the secondary current, the secondary induced voltage, and the angle between the current and voltage.

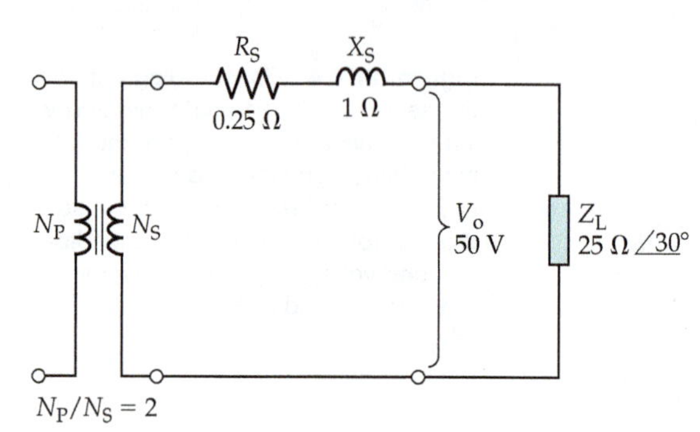

Figure 25-15 Transformer for Example 25-5.

Solution

Secondary circuit calculations (refer to Figures 25-14 and 25-15):

$$I_S = V_o/Z_L = 50\text{ V}/(25\ \Omega\angle 30°)$$
$$= 2\text{ A}\angle -30°$$

R_S volts drop, $I_S R_S = 2\text{ A} \times 0.25\ \Omega$
$$= 0.5\text{ V}$$

X_S volts drop, $I_S X_S = 2\text{ A} \times 1\ \Omega$
$$= 2\text{ V}$$

Phasor sum of V_o, $I_S R_S$, and $I_S X_S$,

Eq. 25-5,
$$E_S = \sqrt{(V_o \cos \phi_o + I_S R_S)^2 + (V_o \sin \phi_o + I_S X_S)^2}$$
$$= \sqrt{(50\text{ V} \cos 30° + 0.5\text{ V})^2 + (50\text{ V} \sin 30° + 2\text{ V})^2}$$
$$= 51.5\text{ V}$$

Eq. 25-6,
$$\phi_S = \tan^{-1}\left(\frac{V_o \sin \phi_o + I_S X_S}{V_o \cos \phi_o + I_S R_S}\right)$$
$$= \tan^{-1}\left(\frac{50\text{ V} \sin 30° + 2\text{ V}}{50\text{ V} \cos 30° + 0.5\text{ V}}\right)$$
$$= 31.7°$$

Phasor Diagram for Primary

The phasor diagram for the transformer primary may be constructed in a similar way to that just discussed for the secondary. In this case, it is necessary to commence with the voltage E_P and current I'_P right at the primary winding, (see Figure 25-13). E_P and I'_P may be calculated from knowledge of the turns ratio and the secondary current and voltage. The phase angle between E_P and I'_P is ϕ_S, the phase angle of the secondary circuit.

Referring to Figure 25-16(a), the $-E_P$ phasor is first drawn horizontally, and the I'_P current phasor is drawn at an angle of ϕ_S lagging $-E_P$. Note that because E_P is the voltage induced in the primary winding by the changing core flux, it is equal and opposite to the component of the applied voltage at the (ideal) transformer primary winding. Consequently, the applied voltage component is $-E_P$.

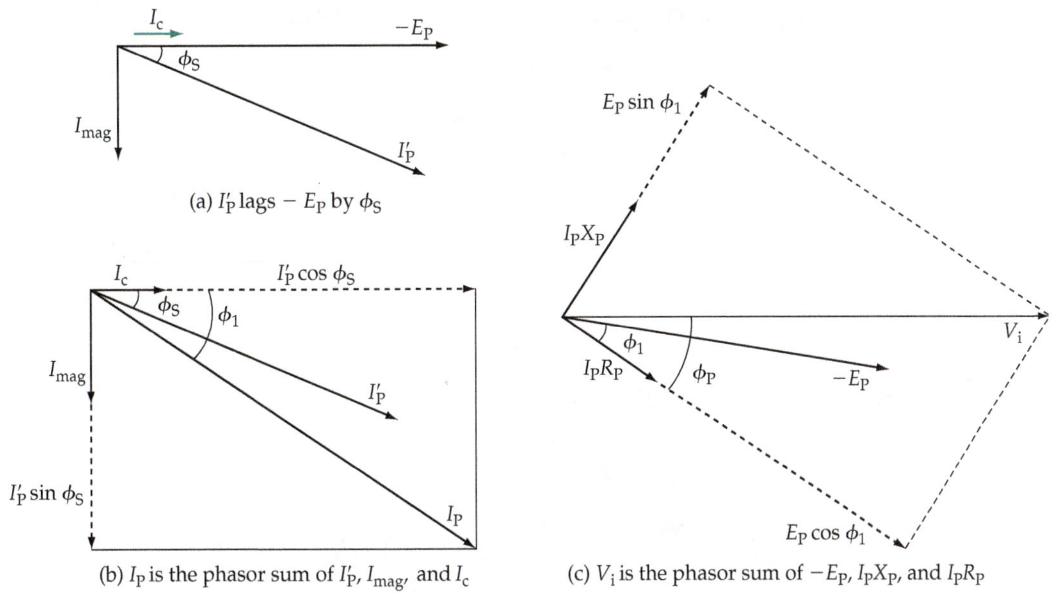

Figure 25-16 Phasor diagram for the primary of a transformer under load. The primary current (I'_P) lags the primary voltage ($-E_P$) by the secondary phase angle (ϕ_S). No-load currents I_c and I_{mag} are also included, and the total primary current (I_P) is the phasor sum of I'_P, I_c and I_{mag}. The supply voltage (V_i) is the phasor sum of $-E_P$, ($I_P X_P$), and ($I_P R_P$).

As well as I'_P, the no-load current I_o (composed of I_c and I_{mag}) must be supplied. The actual current drawn from the supply is I_P, which is the phasor sum of I'_P, I_c and I_{mag} [see Figures 25-13 and 25-16(b)].

$$I_P = \sqrt{(I'_P \cos \phi_S + I_c)^2 + (I'_P \sin \phi_S + I_{mag})^2} \qquad (25\text{-}7)$$

and

$$\phi_1 = \tan^{-1}\left(\frac{I'_P \sin \phi_S + I_{mag}}{I'_P \cos \phi_S + I_c}\right) \qquad (25\text{-}8)$$

The primary current (I_P) produces a voltage drop along R_P and X_P, (Figure 25-13). The $I_P R_P$ phasor is in phase with I_P, and the $I_P X_P$ phasor leads I_P by 90°, as shown in Figure 25-16(c). The phasor sum of $I_P R_P$, $I_P X_P$, and $-E_P$ gives the supply voltage V_i. The primary input phase angle is then ϕ_P, which is the angle between V_i and I_P, as illustrated.

Given the secondary load and the parameters of the transformer equivalent circuit, the primary input voltage and current can be calculated by means of the phasor diagram.

Example 25-6

For the transformer in Example 25-5, $R_P = 2.5\ \Omega$, $X_P = 6\ \Omega$, $R_o = 5\ \text{k}\Omega$, and $X_o = 2\ \text{k}\Omega$ (see Figure 25-17). Determine the supply voltage and current.

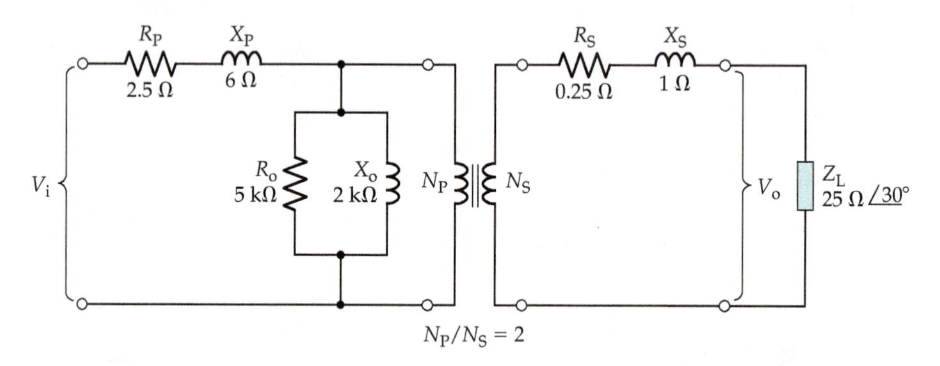

Figure 25-17 Transformer for Example 25-6.

Solution

Referring to Figures 25-16 and 25-17:

$$I'_P = \frac{N_S}{N_P} \times I_S = \frac{1}{2} \times 2\ \text{A} = 1\ \text{A}$$

$$E_P = \frac{N_P}{N_S} \times E_S = \frac{2}{1} \times 51.5\ \text{V} = 103\ \text{V}$$

$$I_c = \frac{E_P}{R_o} = \frac{103\ \text{V}}{5\ \text{k}\Omega} = 20.6\ \text{mA}$$

$$I_{mag} = \frac{E_P}{X_o} = \frac{103\ \text{V}}{2\ \text{k}\Omega} = 51.5\ \text{mA}$$

Eq. 25-7, $$I_P = \sqrt{(I'_P \cos\phi_S + I_c)^2 + (I'_P \sin\phi_S + I_{mag})^2}$$

$$= \sqrt{(1\ \text{A} \cos 31.7° + 21.6\ \text{mA})^2 + (1\ \text{A} \sin 31.7° + 51.5\ \text{mA})^2}$$

$$= 1.05\ \text{A}$$

Eq. 25-8, $$\phi_1 = \tan^{-1}\left(\frac{I'_P \sin\phi_S + I_{mag}}{I'_P \cos\phi_S + I_c}\right) = \tan^{-1}\left(\frac{1\ \text{A}\ \sin 31.7° + 51.5\ \text{mA}}{1\ \text{A}\ \cos 31.7° + 21.6\ \text{mA}}\right)$$

$$= 33.5°$$

$$I_P R_P = 1.05 \text{ A} \times 2.5 \ \Omega = 2.63 \text{ V}$$

$$I_P X_P = 1.05 \text{ A} \times 6 \ \Omega = 6.3 \text{ V}$$

$$V_i = \sqrt{(E_P \cos \phi_1 + I_P R_P)^2 + (E_P \sin \phi_1 + I_P X_P)^2}$$

$$= \sqrt{(103 \text{ V} \cos 33.5° + 2.63 \text{ V})^2 + (103 \text{ V} \sin 33.5° + 6.3 \text{ V})^2}$$

$$= 108.7 \text{ V}$$

$$\phi_P = \tan^{-1}\left(\frac{E_P \sin \phi_1 + I_P X_P}{E_P \cos \phi_1 + I_P R_P}\right) = \tan^{-1}\left(\frac{103 \text{ V} \sin 33.5° + 6.3 \text{ V}}{103 \text{ V} \cos 33.5° + 2.63 \text{ V}}\right)$$

$$= 35.5°$$

Practice Problems

25-4.1 A transformer (as in Figure 25-17) has an output of 35 V, $R_S = 0.3 \ \Omega$, $X_S = 0.8 \ \Omega$, and $Z_L = 144 \ \Omega\angle 45°$. Calculate the secondary current and the secondary induced voltage.

25-4.2 The transformer in Problem 25-4.1 has $N_P/N_S = 4/3$, $R_P = 3.3 \ \Omega$, $X_P = 4.2 \ \Omega$, $R_o = 14 \text{ k}\Omega$, and $X_o = 3.6 \text{ k}\Omega$. Determine the input current and voltage.

25-5 REFERRED RESISTANCE AND REACTANCE

Referred Quantities

The equivalent circuit of a transformer can be considerably simplified by replacing the secondary circuit resistive and reactive components by primary circuit components that have the same effect. Consider the transformer circuit in Figure 25-18(a). The secondary load resistance (R_L) can obviously be written as,

$$R_L = E_S/I_S$$

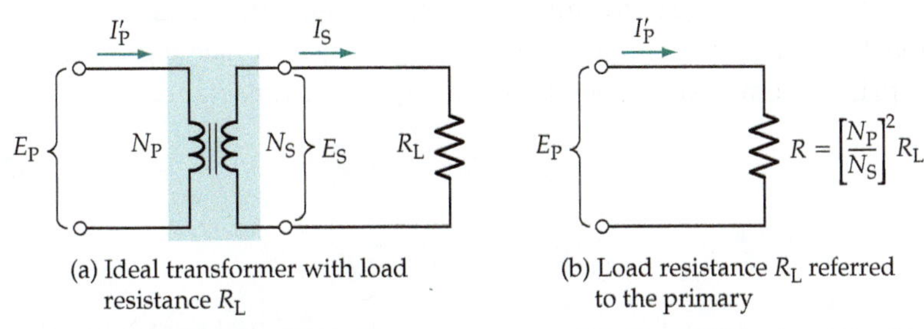

(a) Ideal transformer with load resistance R_L

(b) Load resistance R_L referred to the primary

Figure 25-18 A load connected to the secondary of a transformer may be *referred* (or *reflected*) to the primary to simplify circuitry and calculations.

Also, the primary circuit *sees* a resistance of

$$R = E_P/I'_P$$

However, $$E_P = \frac{N_P}{N_S} \times E_S \text{ and } I'_P = \frac{N_S}{N_P} \times I_S$$

So, $$R = \frac{E_S(N_P/N_S)}{I_S(N_S/N_P)} = R_L\left(\frac{N_P}{N_S}\right)^2$$

Writing $$N_P/N_S = a$$

gives $$R = a^2 R_L \quad (25\text{-}9)$$

The derivation of Equation 25-9 shows that a primary resistance of a^2R_L would have the same effect on the transformer primary circuit as a secondary resistance of R_L, [Figure 25-18(b)]. The primary equivalent of the secondary resistance calculated in this way is termed the *referred resistance* or the *reflected resistance* (i.e., the resistance of the secondary is said to be *referred to* or *reflected into* the primary). The secondary resistance, reactance, and load impedance may all be referred to the primary:

$$\text{Referred reactance} = a^2 X_S \quad (25\text{-}10)$$

$$\text{Referred load} = a^2 Z_L \quad (25\text{-}11)$$

Simplification of Equivalent Circuit

Figure 25-19 illustrates the simplification of the transformer equivalent circuit by the technique of referring everything to the primary. R_S, X_S, and Z_L become a^2R_S, a^2X_S, and a^2Z_L, respectively, when referred to the primary circuit, [see Figures 25-19(a) and (b)]. Note that these referred components are *seen looking into* the primary winding of the ideal transformer. Consequently, the three of them (in series) must be shown in parallel with R_o and X_o in the primary equivalent circuit.

As an approximation to further simplify the equivalent circuit of a transformer on load, R_o and X_o may be omitted. This is because I_c and I_{mag} are each a very small percentage of the total load current (I_P). The total primary resistive and reactive components may now be added together [see Figure 25-19(c)] to give a primary equivalent resistance and equivalent reactance:

$$R_e = R_P + a^2 R_S \quad (25\text{-}12)$$

and $$X_e = X_P + a^2 X_S \quad (25\text{-}13)$$

The referred load impedance remains a^2Z_L; however, it may also be taken care of by adding its resistive and reactive components to R_e and X_e.

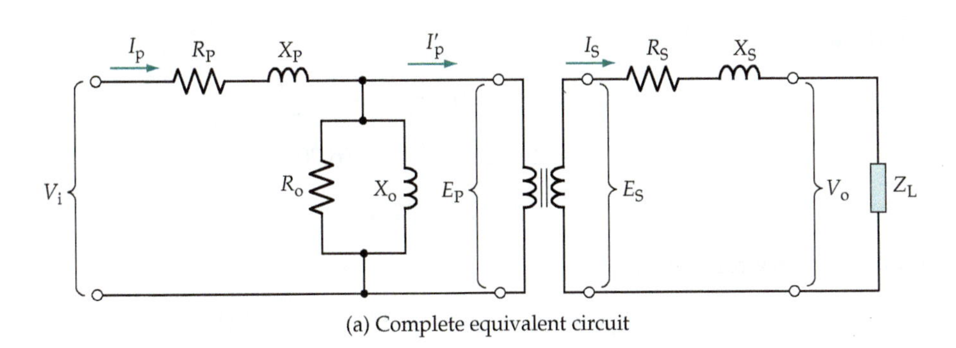

(a) Complete equivalent circuit

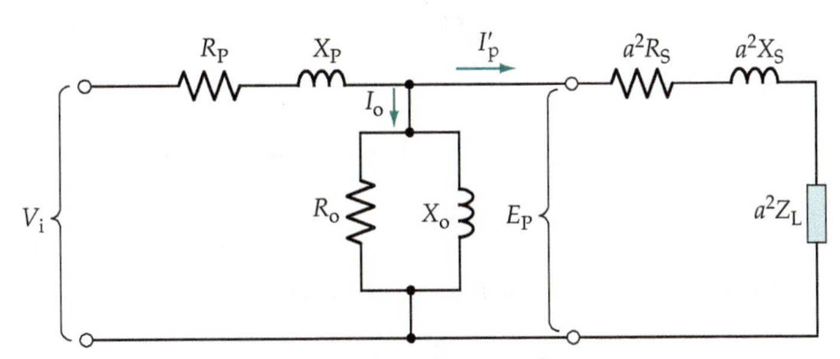

(b) Secondary components referred to the primary

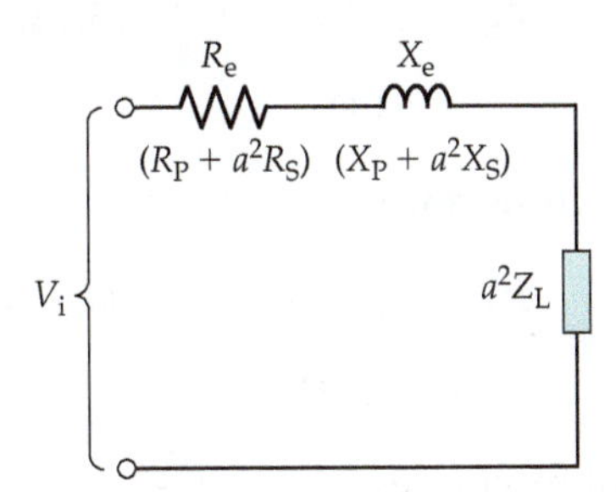

(c) Equivalent circuit simplified by neglecting R_o and X_o

Figure 25-19 The complete equivalent circuit of a transformer may be greatly simplified by referring all of the secondary circuit components to the primary.

Example 25-7

For the transformer circuit described in Examples 25-5 and 25-6, refer all secondary components to the primary. Neglecting R_o and X_o, calculate I_P and ϕ_P when V_i is 108.7 V. The transformer circuit is reproduced in Figure 25-20.

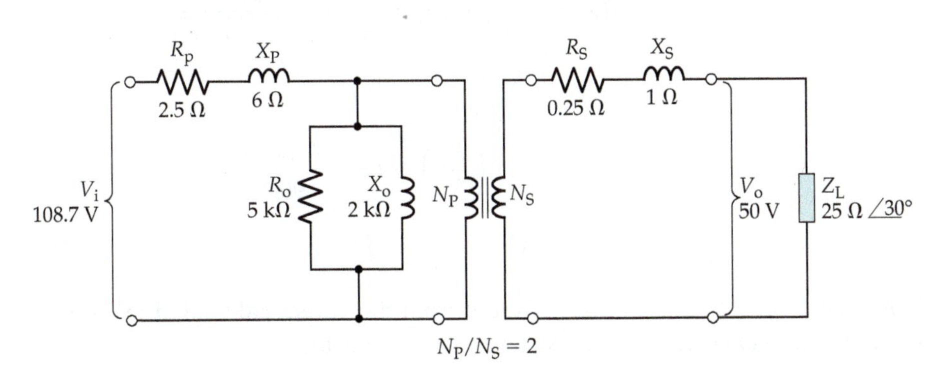

Figure 25-20 Transformer circuit for Example 25-7.

Solution

$$Z_L = 25\ \Omega\angle 30^\circ = R_L + jX_L$$
$$= 25\cos 30^\circ + j25\sin 30^\circ$$
$$= 21.65\ \Omega + j12.5\ \Omega$$

Total secondary resistance,

$$R_{SY} = R_S + R_L = 0.25\ \Omega + 21.65\ \Omega$$
$$= 21.9\ \Omega$$

Total secondary reactance,

$$X_{SY} = X_S + X_L = 1\ \Omega + 12.5\ \Omega$$
$$= 13.5\ \Omega$$

R_{SY} referred to primary, $R_S' = a^2R_{SY} = 2^2 \times 21.9\ \Omega$

$$= 87.6\ \Omega$$

X_{SY} referred to primary, $X_S' = a^2X_{SY} = 2^2 \times 13.5\ \Omega$

$$= 54\ \Omega$$

Total primary equivalent resistance,

$$R_e = R_P + R_S' = 2.5\ \Omega + 87.6\ \Omega$$
$$= 90.1\ \Omega$$

Total primary equivalent reactance,

$$X_e = X_P + X_S' = 6\ \Omega + 54\ \Omega$$
$$= 60\ \Omega$$

$$R_e + jX_e = 90.1\ \Omega + j60\ \Omega$$

$$|I_P| = \frac{V_i}{\sqrt{R_e^2 + X_e^2}} = \frac{108.7\ \text{V}}{\sqrt{(90.1\ \Omega)^2 + (60\ \Omega)^2}}$$
$$= 1\ \text{A}$$

$$\phi_P = \tan^{-1}\left(\frac{X_e}{R_e}\right) = \tan^{-1}\left(\frac{60\ \Omega}{90.1\ \Omega}\right)$$
$$= 33.7^\circ$$

Note that I_P and ϕ_P are slightly different from the values calculated in Example 25-6. This is because R_o and X_o have been neglected.

Practice Problems

25-5.1 Refer all components in the secondary of the transformer described in Problems 25-4.1 and 25-4.2 to the primary. Neglecting the no-load circuit quantities, determine the primary equivalent circuit components.

25-5.2 A transformer is to supply 500 mW to a 12 Ω load. The input is 25 V peak-to-peak. Neglecting all core losses, winding resistance, and reactance, calculate the required turns ratio.

25-6 TRANSFORMER VOLTAGE REGULATION

It is clear from the transformer equivalent circuit in Figure 25-19(a) that the secondary current (I_S) produces voltage drops ($I_S R_X$) and ($I_S X_S$) across the resistive and reactive components. Also, the primary current (I_P) produces the primary circuit voltage drops ($I_P R_P$) and ($I_P X_P$). Consequently, the effective primary voltage (E_P) is less than the input (V_i), and the output voltage (V_o) is less than the calculated value of the secondary voltage (E_S).

When there is no load connected to the output terminals of the transformer, no secondary current flows, and therefore, no voltage drops occur across R_S and X_S. With zero secondary current, the primary current drops to the no-load current (I_o) and the voltage drops across R_P and X_P become very small. So, in the no-load situation, E_P is almost equal to V_i and V_o equals E_S.

It appears that the transformer output voltage is greatest on no-load, and that under loaded conditions the voltage drops across the resistive and reactive components of the equivalent circuit cause V_o to drop below its no-load level. (Note that, depending on the power factor of the load, the output full-load voltage may actually be larger than the no-load voltage.) The percentage change in output voltage from no-load to full load is termed the *voltage regulation* of the transformer. Ideally, there should be no change in V_o from no-load to full-load (i.e., regulation = 0%). For the best possible performance, the transformer should have the lowest possible regulation.

$$\text{Voltage regulation} = \frac{V_{o(NL)} - V_{o(FL)}}{V_{o(FL)}} \times 100\% \qquad \textbf{(25-14)}$$

In Equation 25-14, $V_{o(NL)}$ is the transformer no-load output voltage, and $V_{o(FL)}$ is the full-load output voltage. Voltage regulation for a transformer is illustrated in Figure 25-21.

Example 25-8

Neglecting the no-load current, calculate the voltage regulation for the transformer described in Examples 25-5 and 25-6.

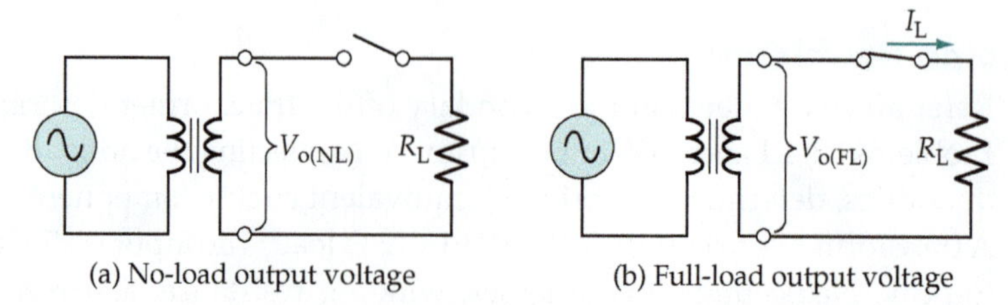

Figure 25-21 The change in transformer output voltage from no-load to full-load condition is defined by the transformer voltage regulation.

Solution

Full-load output voltage,

$$V_{o(FL)} = 50\ \text{V}$$

Input voltage, $V_i = 108.7\ \text{V}$

At no-load, $E_P \approx V_i$

No-load output,

$$V_{o(NL)} = E_S = E_P \times \frac{N_S}{N_P} = 108.7\ \text{V} \times \frac{1}{2}$$

$$= 54.35\ \text{V}$$

Eq. 25-14,

$$\text{Voltage regulation} = \frac{V_{o(NL)} - V_{o(FL)}}{V_{o(FL)}} \times 100\%$$

$$= \frac{54.35\ \text{V} - 50\ \text{V}}{50\ \text{V}} \times 100\%$$

$$= 8.7\%$$

Practice Problem

25-6.1 A transformer with a 115 V input, $N_P = 600$, and $N_S = 329$ has a specified voltage regulation of 7%. Calculate the full-load output voltage.

25-7 TRANSFORMER EFFICIENCY

The efficiency of a transformer, like any other piece of equipment, is the output power expressed as a percentage of the input power,

Efficiency,

$$\eta = \frac{P_o}{P_i} \times 100\% \qquad \textbf{(25-15)}$$

Because $P_i = (P_o + \text{losses})$,

$$\eta = \frac{P_o}{P_o + \text{losses}} \times 100\% \qquad \textbf{(25-16)}$$

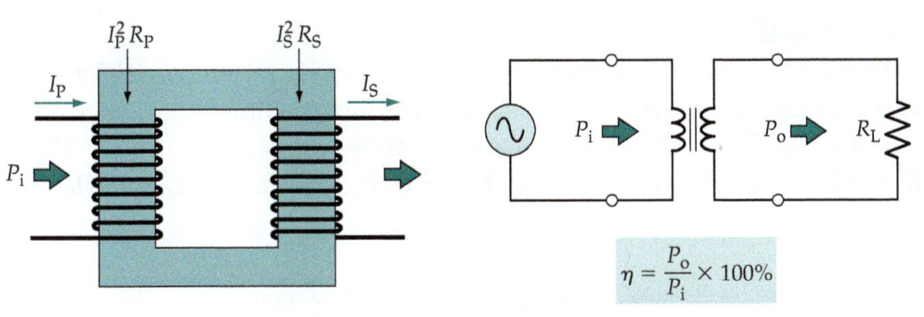

(a) Core losses occur in a transformer

(b) Transformer efficiency is calculated in terms of P_i and P_o

Figure 25-22 Power losses that occur in the windings and core of a transformer determine its efficiency.

The power losses in a transformer consist of core losses due to hysteresis and eddy currents, and copper losses due to the currents flowing in the primary and secondary windings. As long as the supply frequency remains constant, the core losses tend to be a constant quantity. As illustrated in Figure 25-22, the copper losses have two components:

$$\text{Primary winding copper loss} = I_P^2 R_P$$

$$\text{Secondary winding copper loss} = I_S^2 R_S$$

The two copper losses can be lumped into one power loss proportional to the load current if R_P is referred to the *secondary* winding and added to R_S. Thus, when R_{es} is the equivalent of $[(R_P \text{ referred}) + R_S]$,

$$\text{Total copper losses in both windings} = I_S^2 R_{es}$$

Rewriting Equation 25-16,

$$\eta = \frac{V_o I_S \cos \phi_o}{V_o I_S \cos \phi_o + I_S^2 R_{es} + P_c} \times 100\%$$

where $(V_o I_S \cos \phi_S)$ is the output power, $(I_S^2 R_{es})$ is the total copper losses, and P_c is the core loss. Dividing the numerator and denominator by I_S,

$$\eta = \frac{V_o \cos \phi_o}{V_o \cos \phi_o + I_S R_{es} + (P_c / I_S)} \times 100\%$$

The output voltage (V_o) remains substantially constant (within the limits of the regulation), so the maximum value of η is obtained from the equation above when the denominator has its minimum value. If I_S were zero, $I_S R_{es}$ becomes zero, but P_c / I_S becomes infinity. Similarly, if I_S were made very large, P_c / I_S might be made very small, but $I_S R_{es}$ would become very large. So, neither $I_S = 0$ nor $I_S =$ (a very large current) gives maximum efficiency. It can be shown by differential calculus, or by substituting practical values into $I_S R_{es}$ and P / I_S, that maximum transformer efficiency is obtained when

$$I_S R_{es} = P_c / I_S$$

or

$$I_S^2 R_{es} = P_c \qquad (25\text{-}17)$$

So, for maximum efficiency,

$$\text{Winding copper losses} = \text{Core loss}$$

Example 25-9

Determine the efficiency of the transformer described in Examples 25-5 and 25-6. Also, calculate the level of output current at which maximum efficiency occurs. Referring to Figure 25-20, the circuit quantities are $V_o = 50$ V, $Z_L = 25\angle 30°$, $R_S = 0.25\ \Omega$, $N_P/N_S = 2$, $R_P = 2.5\ \Omega$, $X_P = 6\ \Omega$, $R_o = 5$ kΩ, and $X_o = 2$ kΩ.

Solution

$$P_o = V_o I_S \cos\phi_o = 50 \text{ V} \times 2 \text{ A} \times \cos 30°$$
$$= 86.6 \text{ W}$$

$$P_i = V_i I_P \cos\phi_P = 108.7 \text{ V} \times 1.05 \text{ A} \times \cos 35.5°$$
$$= 92.9 \text{ W}$$

Eq. 25-15,
$$\eta = \frac{P_o}{P_i} \times 100\% = \frac{86.6 \text{ W}}{92.9 \text{ W}} \times 100\%$$
$$\approx 93\%$$

Core losses,
$$P_c = \frac{E_P^2}{R_{es}} = \frac{(103 \text{ V})^2}{5 \text{ k}\Omega}$$
$$= 2.12 \text{ W}$$

Referring R_P to the secondary, the equivalent secondary resistance is

$$R_{es} = R_S + R_P\left(\frac{N_S}{N_P}\right)^2 = 0.25\ \Omega + 2.5\ \Omega(1/2)^2$$
$$= 0.875\ \Omega$$

For maximum efficiency:

From Eq. 24-17,
$$I_S = \sqrt{\frac{P_c}{R_{es}}} = \sqrt{\frac{2.12 \text{ W}}{0.875\ \Omega}}$$
$$\approx 1.56 \text{ A}$$

Practice Problems

25-7.1 The transformer in Figure 25-23 has a 30 V output when supplying a load of $Z_L = 68\ \Omega\angle 29°$. The input voltage and current are measured as 115 V and 135 mA, with a phase angle of 33°. Calculate the transformer efficiency.

25-7.2 For the transformer described in Problems 25-4.1 and 25-4.2, determine the level of output current at which maximum efficiency occurs.

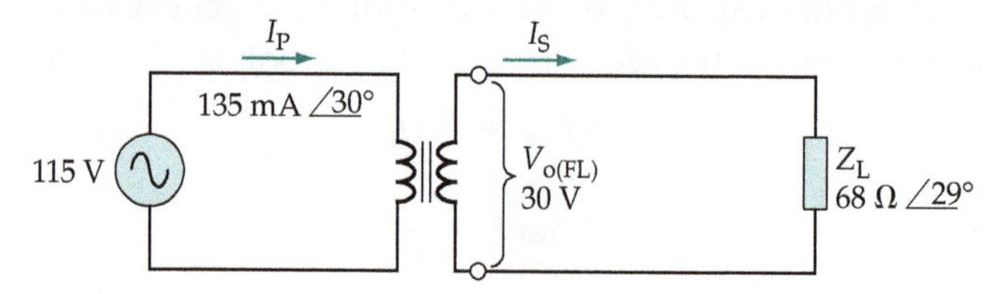

Figure 25-23 Transformer for Problem 25-7.1.

25-8 OPEN-CIRCUIT AND SHORT-CIRCUIT TESTS

Open-Circuit Test

A transformer could be tested under no-load and full-load conditions to determine its turns ratio, regulation, and efficiency. However, without fully loading the transformer it is possible to perform two tests from which all the important data can be derived.

Figure 25-24 shows the circuit for the transformer *open-circuit test*. The alternating input voltage is set to the normal primary level for the transformer, and the voltage at the open-circuited output terminals is monitored on a voltmeter, as illustrated. The input power is measured by the wattmeter (see Section 21-7), and the ammeter measures the primary current. Because the secondary is open-circuited, the primary current is very small, and the voltage drops across the ammeter and wattmeter can be assumed to be negligible. In this case, the input voltage can be taken as the transformer primary voltage, and thus the ratio of the voltmeter readings gives the turns ratio.

$$\frac{E_P}{E_S} = \frac{N_P}{N_S}$$

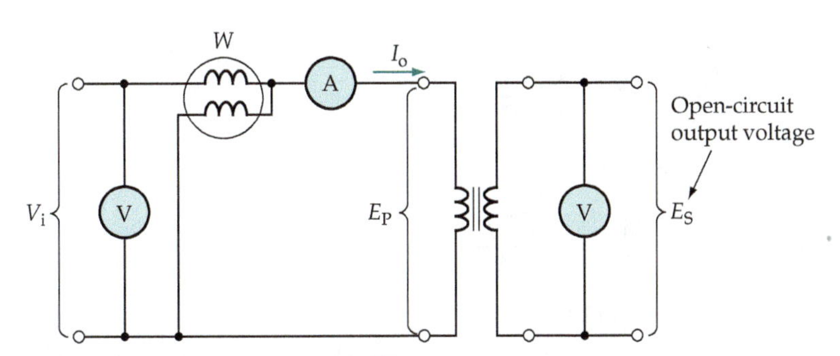

Figure 25-24 A transformer open-circuit test is performed by measuring the (no-load) secondary voltage, the primary current, and the input voltage and power.

With a very small primary current, and near-zero secondary current (i.e., the voltmeter current), the copper loss in the windings can be assumed negligible. The input power measured on the wattmeter is then the total transformer core losses, and the ammeter indicates the no-load primary current (I_o) [see Figure 25-19(b)]. From the measured values of input voltage, current, and power, the components of the no-load equivalent circuit can be determined.

True power, $$P = \frac{E_P^2}{R_o}$$

or $$R_o = \frac{E_P^2}{P} \qquad (25\text{-}18)$$

Apparent power, $S = E_P I_o$

and $S = \sqrt{(\text{true power})^2 + (\text{reactive power})^2}$

$= \sqrt{P^2 + Q^2}$ (see Section 21-4)

Giving $Q = \sqrt{S^2 - P^2}$

or $$Q = \sqrt{(E_P I_o)^2 - P^2} \quad (25\text{-}19)$$

From Eq. 21-6, $Q = \dfrac{E_P^2}{X_o}$

So $$X_o = \frac{E_P^2}{Q} \quad (25\text{-}20)$$

Example 25-10

An open-circuit test on a transformer produced the following measurements: $E_P = 115$ V, $E_S = 57.5$ V, $P = 9.5$ W, and $I_o = 180$ mA. Determine the transformer turns ratio and the values of R_o and X_o.

Solution

$$\frac{N_S}{N_P} = \frac{E_S}{E_P} = \frac{57.5\ \text{V}}{115\ \text{V}}$$

$$= 1/2$$

Eq. 25-18, $$R_o = \frac{E_P^2}{P} = \frac{(115\ \text{V})^2}{9.5\ \text{W}}$$

$$= 1.39\ \text{k}\Omega$$

Eq. 25-19, $$Q = \sqrt{(E_P I_o)^2 - P^2} = \sqrt{(115\ \text{V} \times 180\ \text{mA})^2 - (9.5\ \text{W})^2}$$

$$= 18.39\ \text{vars}$$

Eq. 25-20, $$X_o = \frac{E_P^2}{Q} = \frac{(115\ \text{V})^2}{18.39\ \text{vars}}$$

$$= 719\ \Omega$$

Short-Circuit Test

The transformer *short-circuit test* is performed with the secondary terminals short-circuited, as illustrated in Figure 25-25. Note that the primary voltage (E_P) is measured right at the transformer primary terminals to avoid error due to the voltage drops across the ammeter and wattmeter. The input voltage is increased from zero until the ammeter in the primary circuit indicates normal full-load primary current. When this occurs, the normal full-load secondary current is

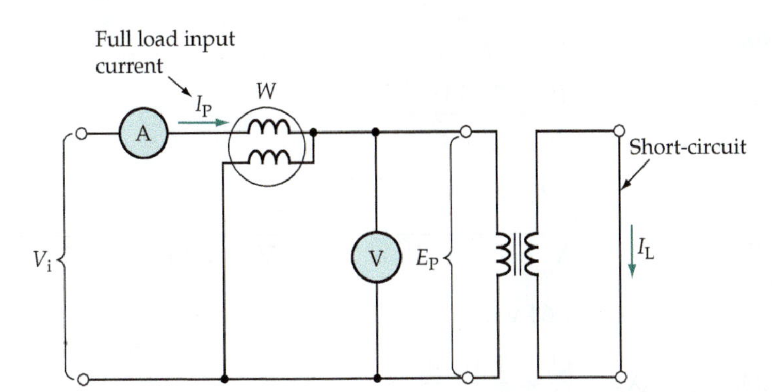

Figure 25-25 To perform a transformer short-circuit test, the secondary is shorted, and the primary current is adjusted to the normal full-load level. The primary voltage and input power are measured.

circulating in the secondary winding. Because the secondary terminals are short-circuited, the input voltage required to produce full-load primary and secondary currents is around 3% of the normal input voltage level. With such a low level of input voltage, the core losses are so small that they can be neglected. [Referring to Figure 25-19(b), it is seen that a low input voltage will produce a small core loss current (I_o).] However, the windings are carrying normal full-load current, and so the input is supplying the normal full-load copper losses.

The output power (to the short-circuit) is zero, so the wattmeter measuring true input power indicates the full-load copper losses. The product of the ammeter and voltmeter readings gives the apparent input power. From these quantities, calculation may be made of the resistive and reactive components of the full-load equivalent circuit referred to the primary, [see Figure 25-19(c)].

True power $$P = I_P^2 R_e$$

So $$R_e = \frac{P}{I_P^2} \tag{25-21}$$

Apparent power, $$S = E_P I_P$$

Reactive power, $$Q = \sqrt{(E_P I_P)^2 - P^2} \tag{25-22}$$

and $$Q = I_P^2 X_e$$

So $$X_e = \frac{Q}{I_P^2} \tag{25-23}$$

Example 25-11

Determine R_e and X_e for the transformer in Example 25-10 when the following measurements were made on a short-circuit test: $E_{P(SC)} = 5.5$ V, $I_P = 1$ A, and $P = 5.25$ W.

Solution

Eq. 25-21, $R_e = \frac{P}{I_P^2} = \frac{5.25\ \text{W}}{(1\ \text{A})^2}$

$= 5.25\ \Omega$

Eq. 25-22, $Q = \sqrt{(E_P I_P)^2 - P^2} = \sqrt{(5.5\ \text{V} \times 1\ \text{A})^2 - (5.25\ \text{W})^2}$

$= 1.64\ \text{var}$

Eq. 25-23, $X_e = \frac{Q}{I_P^2} = \frac{1.64\ \text{var}}{(1\ \text{A})^2}$

$= 1.64\ \Omega$

Example 25-12

From the open-circuit and short-circuit test results given in Examples 25-10 and 25-11, determine the transformer regulation.

Solution

$V_{o(NL)} = 57.5\ \text{V}$ (from the OC test)

and $E_{p(FL)} = E_{P(NL)} - [\text{volts drop due to } I_L]$

$= E_{P(NL)} - [E_{P(SC)} \text{ from the SC test}]$

$= 115\ \text{V} - 5.5\ \text{V}$

$= 109.5\ \text{V}$

$V_{o(FL)} = E_{P(FL)} \times \frac{N_S}{N_P} = 109.5\ \text{V} \times \frac{1}{2}$

$= 54.75\ \text{V}$

Eq. 25-14, Voltage regulation $= \frac{V_{o(NL)} - V_{o(FL)}}{V_{o(FL)}} \times 100\%$

$= \frac{57.5\ \text{V} - 54.75\ \text{V}}{54.75\ \text{V}} \times 100\%$

$= 5\%$

Example 25-13

From the open-circuit and short-circuit test results given in Examples 25-10, 25-11, and 25-12, calculate the transformer efficiency on full-load with a load phase angle of $\phi_S = 10°$ lagging.

Solution

Full-load output current,

$$I_{S(FL)} = I_{P(FL)} \times \frac{N_P}{N_S} = 1\text{ A} \times \frac{2}{1}$$
$$= 2\text{ A}$$

Power output, $$P_o = V_{o(FL)} \times I_{S(FL)} \times \cos\phi_S = 54.75\text{ V} \times 2\text{ A} \times \cos 10^\circ$$
$$= 107.8\text{ W}$$

Core losses = 9.5 W (from open-circuit test)

Full-load copper losses = 5.25 W (from short-circuit test)

Eq. 25-15, $$\eta = \frac{P_o}{P_i} \times 100\% = \frac{P_o}{P_o + \text{losses}}$$
$$= \frac{107.8\text{ W}}{107.8\text{ W} + 9.5\text{ W} + 5.25\text{ W}} \times 100\%$$
$$= 88\%$$

Practice Problems

25-8.1 A transformer with its secondary open-circuited has $E_P = 40$ V and $E_S = 9$ V. The input current is measured as 5 mA, and the input power is 120 mW. Determine the transformer turns ratio and the no-load equivalent circuit components.

25-8.2 When the transformer in Problem 25-8.1 is short-circuited, the input voltage required to maintain full load primary current is found to be 3 V. The primary current and input power are 125 mA and 250 mW. Determine the primary equivalent resistance and reactance.

25-8.3 Using the results of the open-circuit and short-circuit tests given in Problems 25-8.1 and 25-8.2, determine the voltage regulation and efficiency of the transformer when supplying full load to a pure resistance.

25-9 AUTOTRANSFORMER, CURRENT TRANSFORMER, AND AUDIO TRANSFORMER

Autotransformer

An *autotransformer* (also known as a *variable voltage transformer*, and as a *variac*) has a single winding on an iron core. One of the coil terminals is common to both input and output, and the other output terminal is movable so that it

Figure 25-26 An autotransformer produces a secondary voltage (V_S) that can be adjusted from zero to a maximum level of V_P. The output is not dc isolated from the input.

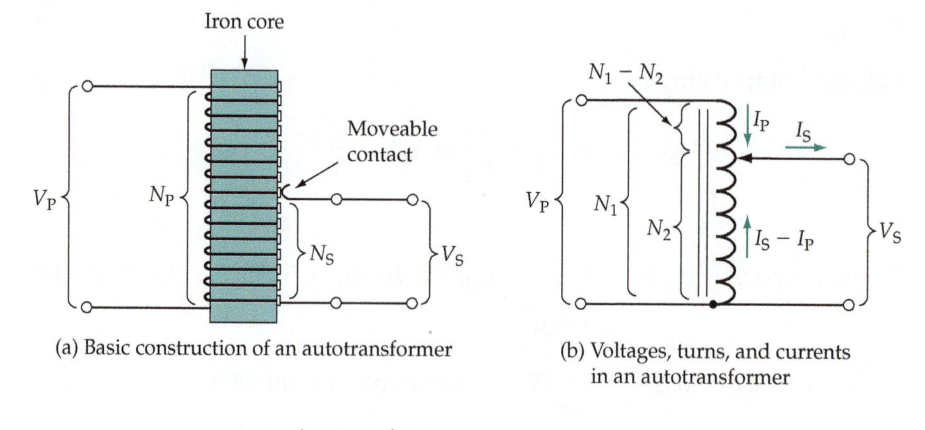

(a) Basic construction of an autotransformer

(b) Voltages, turns, and currents in an autotransformer

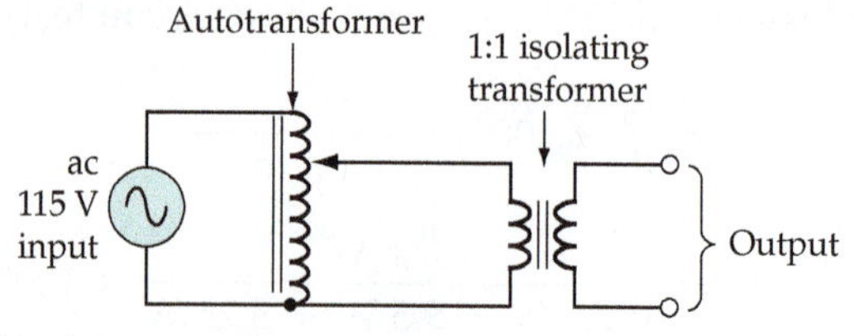

(c) Isolating transformer used with an autotransformer

can make contact with any turn on the winding. The principle is illustrated in Figure 25-26(a). Several fixed output terminals (or *taps*) are sometimes used instead of a continuously variable output. As in the case of a transformer with two windings,

$$\frac{V_S}{V_P} = \frac{N_S}{N_P}$$

The current identifications in Figure 25-26(b) show that the input current (I_P) passes through turns ($N_1 - N_2$), and then flows to the transformer output terminal. With the voltage step-down from primary to secondary, the secondary current (I_S) is larger than I_P. So, the remaining part of the output ($I_S - I_P$) is supplied from the N_2 turns section, as illustrated.

The obvious advantage of this kind of autotransformer is the facility for adjusting the output voltage to any desired level. The major disadvantage is that

an autotransformer output is NOT dc isolated from the input.

Figure 25-26(c) shows a typical laboratory-situation use of an autotransformer. An isolating transformer is included in the circuit to ensure that there is no dc contact between the ac supply and the output. The isolating transformer has the same number of primary and secondary turns; a 1:1 ratio. So, its ac output is the same as the autotransformer output.

Autotransformers can also be constructed to step-up the primary voltage to a higher secondary level, (see Figure 25-27). Here again the two portions of the transformer winding carry different current levels. The advantage of this over a two-coil transformer is that, because the N_1 section of the winding has a lower current level than the $N_2 - N_1$ section, the N_1 portion of the winding can be constructed of thinner copper wire than the $N_1 - N_2$ section. For a fixed-output autotransformer, the reduced thickness of the copper windings can result in a significant saving of copper compared to a double-wound transformer designed to do the same job.

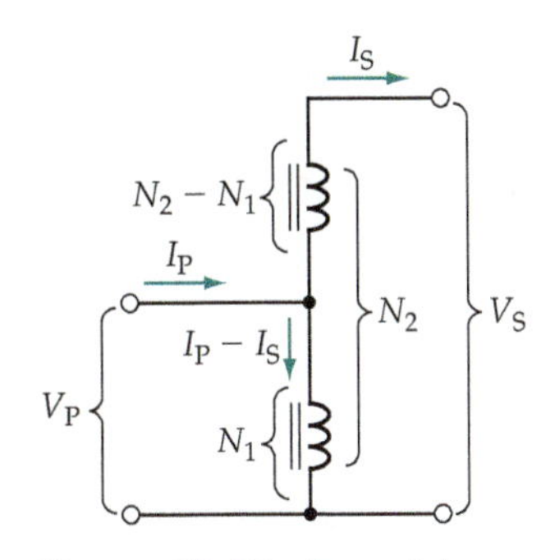

Figure 25-27 An autotransformer can be used to step-up the input voltage to a higher-level output.

Current Transformer

A high-level alternating current can be most easily measured by accurately transforming the current to a much lower level. Also, because conductors carrying large currents are frequently at high voltage levels, a measuring instrument directly connected to the conductor would have to be very well insulated. Another problem that arises when rectifier ammeters are to be used (see Section 17-8) is the relatively large voltage drop across each rectifier. The dc isolation afforded by a transformer solves the insulation and rectifier voltage problems.

The use of a current transformer for measuring high levels of alternating current is illustrated in Figure 25-28. A conductor carrying a large current passes through a circular iron core. The conductor constitutes a one-turn primary winding. The secondary winding consists of a number of turns of fine wire wrapped around the core, as illustrated. The secondary current is given by,

Eq. 25-2,
$$I_S = I_P \times \frac{N_P}{N_S}$$

So, with N_S being several times N_P, the secondary current is much smaller than the primary current. An additional resistor (R_1) is frequently included to divert some of the secondary current from the meter, as illustrated.

It is very important to note that a current transformer must never be operated with its secondary winding open-circuited. Serious overheating of the core can occur when there is no secondary current to oppose the core flux generated by the primary. It is also possible that the secondary open-circuit voltage could reach a dangerously high level.

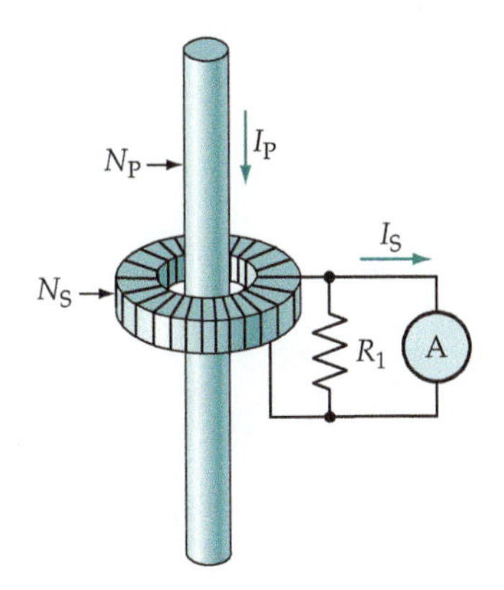

Figure 25-28 A current transformer converts a high primary current level to a low secondary level. This allows very high currents to be measured on low-current instruments.

Example 25-14

The current transformer circuit shown in Figure 25-28 has 909 secondary turns. The meter used has a 100 kΩ resistance and measures a maximum of 100 μA (rms). If the resistance of R_1 is 10 kΩ, calculate the current levels in the secondary circuit when the primary current is 1 A, and the meter indicates 100 μA.

Solution

$$I_S = I_P \times \frac{N_P}{N_S} = 1\text{ A} \times \frac{1}{909}$$

$$= 1.1\text{ mA}$$

$$I_m = 100\ \mu\text{A}$$

and,

$$V_m = I_m \times R_m = 100\ \mu\text{A} \times 100\text{ k}\Omega$$

$$= 10\text{ V}$$

$$I_{R1} = I_S - I_m = 1.1\text{ mA} - 100\ \mu\text{A}$$

$$= 1\text{ mA}$$

or,

$$I_{R1} = \frac{V_m}{R_1} = \frac{10\text{ V}}{10\text{ k}\Omega}$$

$$= 1\text{ mA}$$

Audio Transformer

Audio transformers are designed to operate over the range of audible frequencies, and are usually employed in coupling an audio amplifier to the speaker of a sound system, (see Figure 25-29). The audio range is approximately from 50 Hz to 12 kHz; however, because other factors reduce the frequency range of the system, most audio amplifiers are designed to amplify frequencies ranging from about 20 Hz to 20 kHz or higher.

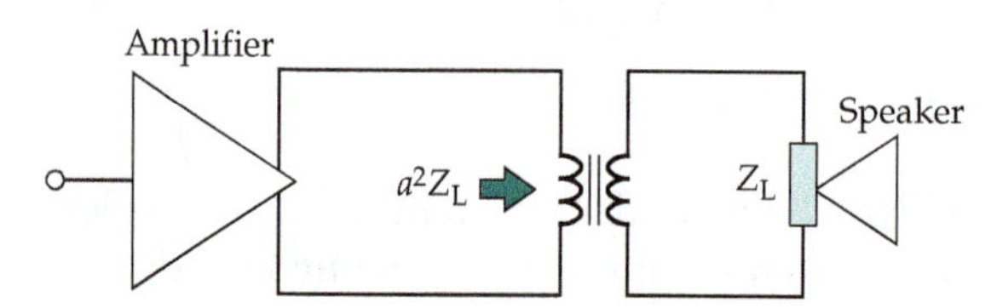

Figure 25-29 The output of an audio amplifier can be transformer-coupled to a speaker.

As discussed in Section 25-5, a load (Z_L) connected to the secondary appears as an impedance of (a^2Z_L) in series with the primary input terminals. Most speakers used in sound reproduction systems have impedance values ranging from 8 Ω to 16 Ω. By careful selection of the transformer turns ratio, the low speaker impedance can be made to appear in the transformer primary as a larger impedance, more suitable as an amplifier load. The process is referred to as *impedance matching*.

Example 25-15

An audio transformer designed to deliver a maximum of 4 W to a 16 Ω load has a 400 Ω referred load resistance and a 90% efficiency. Calculate the required peak input voltage when delivering full load. Also determine the transformer turns ratio.

Solution

$$P_{in} = \frac{P_{out}}{\eta} = \frac{4\text{ W}}{0.9}$$

$$= 4.44\text{ W}$$

$$P_{in} = \frac{V_{in}^2}{a^2 R_L}$$

so,

$$V_{in} = \sqrt{P_{in} \times a^2 R_L} = \sqrt{4.44\text{ W} \times 400\ \Omega}$$

$$= 42.14\text{ V (rms)}$$

$$V_{in(peak)} = 1.414 V_{in} = 1.414 \times 42.14\text{ V}$$

$$= 59.6\text{ V}$$

Turns ratio,

$$a = \sqrt{a^2 R_L / R_L} = \sqrt{400\ \Omega / 16\ \Omega}$$

$$= 5$$

Practice Problems

25-9.1 A step-down auto transformer with a 115 V input is adjusted to produce a 50 V output to a 1 kΩ load. Determine the turns ratio and calculate the current levels in each portion of the winding.

25-9.2 An ac ammeter uses a current transformer, as illustrated in Figure 25-28. The meter indicates a maximum of 1 mA, and its resistance is 15 kΩ. The transformer has 1000 secondary turns, and the primary current is 100 A. Calculate a suitable resistance for R_1.

25-9.3 An audio transformer with a 33 V peak output is to dissipate 10 W in an 8 Ω load. Assuming a 90% efficiency, determine the transformer input resistance and turns ratio. Also, calculate the primary and secondary current levels.

Summary of Formulas

- ***Primary/secondary voltage ratio:***

$$\frac{E_S}{E_P} = \frac{N_S}{N_P}$$

- ***Primary/secondary current ratio:***

$$\frac{I_S}{I_P} = \frac{N_P}{N_S}$$

- ***Transformer emf equations:***

$$E_P = 4.44\Phi_m f N_P$$
$$E_S = 4.44\Phi_m f N_S$$

- ***Secondary induced voltage:***

$$E_S = \sqrt{(V_o \cos\phi_o + I_S R_S)^2 + (V_o \sin\phi_o + I_S X_S)^2}$$

- ***Phase angle between E_S and I_S:***

$$\phi_S = \tan^{-1}\left(\frac{V_o \sin\phi_o + I_S X_S}{V_o \cos\phi_o + I_S R_S}\right)$$

- ***Primary current:***

$$I_P = \sqrt{(I_P' \cos\phi_S + I_c)^2 + (I_P' \sin\phi_S + I_{mag})^2}$$

- ***Phase angle between I_P and E_P:***

$$\phi_1 = \tan^{-1}\left(\frac{I_P' \sin\phi_S + I_{mag}}{I_P' \cos\phi_S + I_c}\right)$$

- ***Referred resistance:***

$$R = \left(\frac{N_P}{N_S}\right)^2 R_L = a^2 R_L$$

- ***Referred reactance:***

$$X = a^2 X_S$$

- ***Referred impedance:***

$$Z = a^2 Z_L$$

- ***Primary equivalent resistance:***

$$R_e = R_P + a^2 R_S$$

- ***Primary equivalent reactance:***

$$X_e = X_P + a^2 X_S$$

- ***Voltage regulation:***

$$\frac{V_{o(NL)} - V_{o(FL)}}{V_{o(FL)}} \times 100\%$$

- ***Efficiency:***

$$\eta = \frac{P_o}{P_i} \times 100\%$$

- ***From maximum efficiency:***

$$I_S^2 R_{es} = P_c$$

- ***From open-circuit test:***

$$R_o = \frac{E_P^2}{P}$$

$$Q = \sqrt{(E_P I_o)^2 - P^2}$$

$$X_o = \frac{E_P^2}{Q}$$

- ***From short-circuit test:***

$$R_e = \frac{P}{I_P^2}$$

$$Q = \sqrt{(E_P I_P)^2 - P^2}$$

$$X_e = \frac{Q}{I_P^2}$$

Review Questions

Section 25-1

25-1 Draw a sketch to show the basic construction of a transformer and explain how it functions. Sketch the circuit diagram for a simple two-winding transformer, and derive the approximate relationship between the numbers of coil turns and the primary and secondary voltages.

25-2 Define primary winding, secondary winding, turns ratio, 1:1 transformer, step-up transformer, and step-down transformer.

25-3 Derive an expression relating the number of primary and secondary turns on a transformer to the primary and secondary currents.

Section 25-2

25-4 Derive the emf equation for a transformer relating primary or secondary voltage to the core flux, the supply frequency, and the number of turns on the winding.

25-5 Discuss the power losses that occur in a transformer, and explain measures that may be taken to minimize the losses.

Section 25-3

25-6 Sketch and explain the approximate phasor diagram for transformer primary voltage and currents under no-load conditions.

25-7 Draw the no-load equivalent circuit for a transformer. Briefly explain.

Section 25-4

25-8 Explain the terms leakage flux and leakage inductance. Sketch the complete equivalent circuit for a transformer under load. Identify all components of the circuit and all voltages and currents. Explain the origin of each quantity.

25-9 Sketch a complete phasor diagram for the secondary circuit of a transformer under load. Briefly explain the diagram.

25-10 Sketch a complete phasor diagram for the primary circuit of a transformer under load. Briefly explain the diagram.

Section 25-5

25-11 Explain the principle of referred resistance and referred reactance. Show how the transformer equivalent circuit may be simplified by referring the secondary circuit components to the primary.

Section 25-6

25-12 Explain what is meant by the voltage regulation of a transformer, and write the equation for calculating the voltage regulation.

Section 25-7

25-13 Discuss the efficiency of transformers, and derive an equation for transformer efficiency in terms of the output voltage and current and core losses.

Section 25-8

25-14 Sketch the circuit diagrams for performing open- and short-circuit tests on a transformer.

25-15 Explain the open-circuit and short-circuit testing procedure, and develop the necessary equation for calculating R_o, X_o, R_e, and X_e from the test results.

Section 25-9

25-16 Sketch circuit diagrams for two types of autotransformer. Explain its operation and discuss the advantages and disadvantages of autotransformers.

25-17 Describe a current transformer. Explain its application, and discuss any precautions necessary when using a current transformer.

25-18 Explain the purpose of an audio transformer, and show that its input resistance is a^2R_L.

Problems

Section 25-1

25-1 A transformer with a primary winding of 250 turns has an input of 115 V rms and two secondary windings, one with 65 turns and the other with 80 turns, (see Figure 25-30). Determine the output voltage from each

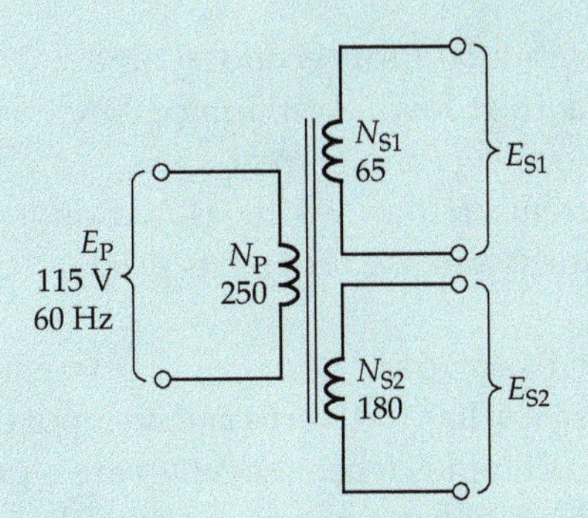

Figure 25-30

winding. Also calculate the combined output voltage when (a) the secondary windings are connected series-aiding, (b) the secondary windings are connected series-opposing.

25-2 A transformer with a 115 V, 60 Hz primary voltage is to have 35 V, 15 V, and 5 V secondaries with loads of 100 Ω, 500 Ω, and 25 Ω respectively. If the primary winding has 1000 turns, determine the required number of turns for each secondary. Calculate the primary current, assuming 100% efficiency.

25-3 Assuming that the transformer in Figure 25-31 is 100% efficient, determine the output voltage from each secondary winding and calculate the total primary current.

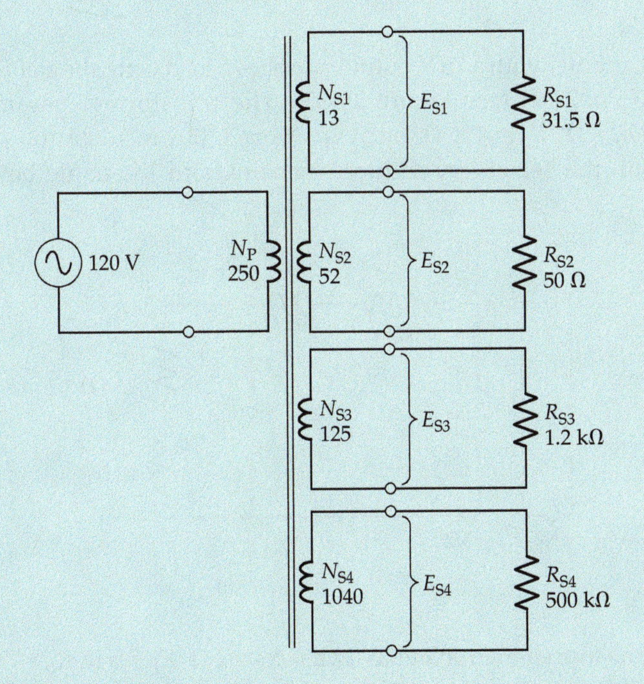

Figure 25-31

25-4 A transformer with a 50 turn secondary and a 270 turn primary is to deliver 5 W to a 16 Ω load. Assuming 100% efficiency, calculate the required primary voltage and current.

25-5 A secondary current of 10 mA is measured on a transformer that has 5000 secondary turns and one turn on its primary. Calculate the primary current.

25-6 The center-tapped secondary winding of a transformer has a total of 200 turns. The primary has 1000 turns and an input of 5 V (rms). Calculate the secondary peak outputs when the center tap is grounded. If the whole of the secondary is accidentally connected to the input voltage supply, calculate the output voltage from the primary.

Section 25-2

25-7 The transformer described in Problem 25-3 has a supply frequency of 60 Hz. Calculate the peak core flux.

25-8 Determine the peak core flux for the transformer in Problem 25-2.

25-9 The input frequency for the transformer in Problem 25-4 ranges from 75 Hz to 10 kHz. Calculate the maximum and minimum levels of peak core flux.

25-10 Determine the peak core flux for the transformer in Problem 25-5 if the supply frequency is 60 Hz and the secondary output is 18 V

25-11 Calculate the peak core flux for the transformer in Problem 25-6 when correctly connected, and when the secondary is connected as a primary. The minimum frequency of the input voltage is 50 Hz.

Section 25-4

25-12 A transformer with a 75 V output supplies a load consisting of $R_L = 33.5\ \Omega$ and $X_L = 22\ \Omega$ (see Figure 25-32). The transformer parameters are $R_S = 0.25\ \Omega$, $X_S = 1.2\ \Omega$, and $N_P/N_S = 3/2$. Calculate the secondary current, the secondary induced voltage, and the angle between the two.

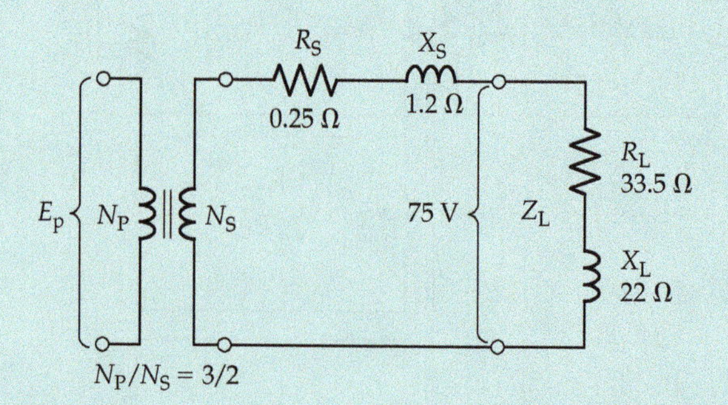

Figure 25-32

25-13 The transformer in Problem 25-12 has $R_P = 2\ \Omega$, $X_P = 5\ \Omega$, $R_o = 7.5\ \text{k}\Omega$, and $X_o = 3\ \text{k}\Omega$ (see Figure 25-33). Determine the supply voltage and current and the phase angle between the two.

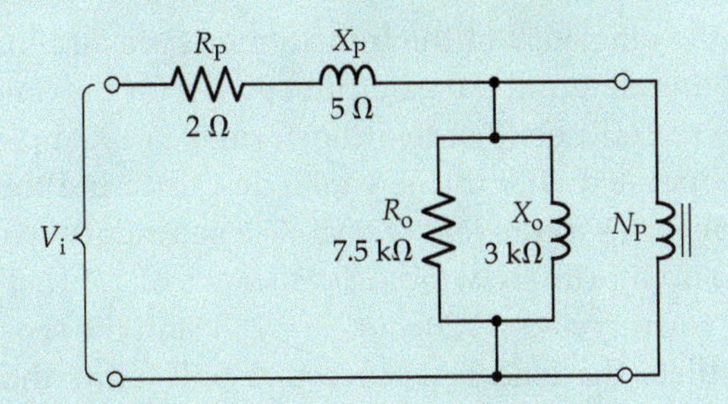

Figure 25-33

25-14 The load connected to the secondary of the transformer in Example 25-5 is changed to $Z_L = 40\ \Omega\angle 20°$. If the output remains at 50 V, calculate the new secondary current, induced voltage, and phase angle.

25-15 For the load change described in Problem 25-14, determine the new primary current, voltage, and phase angle for the transformer in Examples 25-5 and 25-6.

25-16 The transformer described in Problems 25-12 and 25-13 has the inductive part of its load doubled to 44 Ω. If the output remains at 75 V, calculate the new secondary current, induced voltage, and phase angle.

25-17 Determine the new primary current, voltage, and phase angle for the transformer in Problems 25-12 and 25-13 when the load is changed as described in Problem 25-16.

Section 25-5

25-18 For the transformer in Examples 25-5 and 25-6 with its load changed as described in Problem 25-14, refer all secondary components to the primary. Then, neglecting R_o and X_o, calculate I_P and ϕ_P when $V_i = 108.7$ V.

25-19 Refer all the secondary components of the transformer described in Problems 25-12 and 25-13 to the primary winding. Neglecting the components representing the core loss, calculate the primary currents and phase angles with respect to the input voltage when the input is 120 V.

25-20 For the transformer in Problems 25-12 and 25-13 with its load changed as described in Problem 25-16, refer all secondary components to the primary. Then, neglecting R_o and X_o calculate I_P and ϕ_P when $V_i = 120.85$ V.

Section 25-6

25-21 Neglecting the core losses determine the voltage regulation for the transformer specified in Problems 25-12 and 25-13.

25-22 The transformer in Problem 25-4 has an output of 9.5 V when the load is disconnected. Calculate its regulation.

25-23 A transformer has a primary input of 115 V, a regulation of 4%, and a turns ratio of 4.6:1. Calculate its no-load and full-load output voltages.

Section 25-7

25-24 The input power to a transformer is measured as 22 W when the output voltage is 31.6 V and the secondary load is 50 Ω. Calculate the transformer efficiency.

25-25 Calculate the efficiency of the transformer specified in Problems 25-12 and 25-13, and determine the output current for maximum efficiency.

25-26 Determine the efficiency of the transformer in Examples 25-5 and 25-6 when it is operated with the new load described in Problem 25-14.

25-27 A transformer has a core loss of 0.95 W, a secondary winding resistance of 0.18 Ω, and a primary winding resistance of 1.3 Ω. The primary and secondary turns are $N_P = 720$ and $N_S = 288$, and the secondary output is 50 V. Calculate the load resistance that will allow the transformer to operate at maximum efficiency.

Section 25-8

25-28 An open-circuit test on a transformer gave the following results: $E_P = 120$ V, $E_S = 35$ V, $P = 5$ W, and $I_o = 125$ mA. A short-circuit test on the same transformer produced $E_P = 4$ V, $I_P = 0.8$ A, and $P = 3$ W. Calculate R_o, X_o, N_P/N_S, R_e, and X_e.

25-29 Calculate the transformer full-load regulation from the open-circuit and short-circuit test results given in Problem 25-28, when the load phase angle is 15° lagging.

25-30 Using the open-circuit and short-circuit test results in Problem 25-28, calculate the transformer efficiency under full-load with a load phase angle of 15° lagging.

Section 25-9

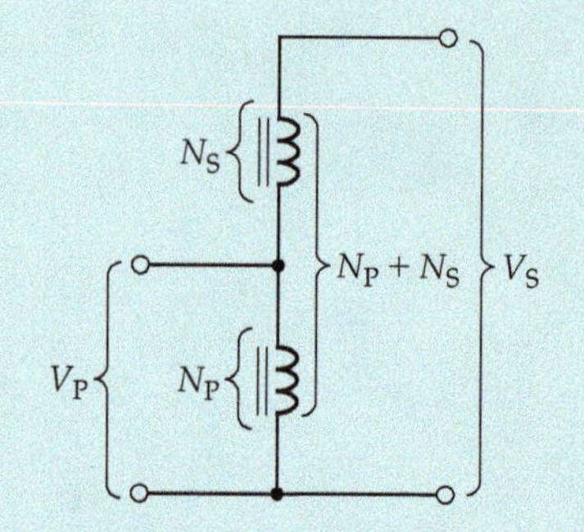

Figure 25-34

25-31 A two coil transformer is connected to function as a step-up autotransformer with the input applied to N_P and the output taken across $N_P + N_S$, as illustrated in Figure 25-34. The coil turns are: $N_P = 200$ and $N_S = 400$. The input voltage is 50 V and the output load is 100 Ω. Calculate the output voltage and the current levels in each winding.

25-32 An ac ammeter that uses a current transformer has 500 μA, 20 kΩ meter with shunt resistance of $R_1 = 2.2$ kΩ. The transformer primary and secondary turns are: $N_1 = 5$ and $N_2 = 200$. Calculate the primary current for maximum meter reading.

25-33 An audio amplifier applies a 35 V peak voltage to a transformer which has a 12 Ω secondary load. The transformer has a 5:1 turns ratio, and 95% efficiency. Determine the power delivered to the load, and calculate the primary and secondary current levels.

Practice Problem Answers

25-1.1 65, 95
25-1.2 6.6 mA
25-2.1 617 μWb
25-2.2 7.5 V
25-4.1 243 mA, 35.2 V

25-4.2 194 mA, 47.9 V
25-5.1 184.8 Ω, 186.6 Ω
25-5.2 3.6/1
25-6.1 58.6 V
25-7.1 89%
25-7.2 274 mA
25-8.1 4.4:1, 13.3 kΩ, 10 kΩ
25-8.2 16 Ω, 18 Ω
25-8.3 8.4%, 92%
25-9.1 2.3, 21.7 mA, 28.3 mA
25-9.2 151.5 Ω
25-9.3 48.9 Ω, 2.47, 476 mA, 1.12 A

CHAPTER 26
Three-Phase AC Systems

CONTENTS

Objectives

You will be able to:

1. Explain the generation of three-phase alternating voltages, sketch the output voltage waveforms, and draw a phasor diagram of the voltages.
2. Draw circuit diagrams of Y-connected and Δ-connected generators, and draw phasor diagrams for generator line and phase voltages, and line and phase currents.
3. Analyze Y-connected and Δ-connected four-wire and three-wire loads to determine load and line voltages, load and line currents, neutral current, and the phase relationships between the various quantities.
4. Explain the effects of phase sequence reversal and analyze circuits with different phase sequences.
5. Explain the process of power factor correction in three-phase loads and calculate the power factor correction components.
6. Describe various methods of measuring the power dissipated in a three-phase load and calculate wattmeter measurements.

INTRODUCTION

A three-phase generator produces three separate sinusoidal alternating voltages from three output terminals. The terminals are identified as *A*, *B*, and *C*; sometimes a neutral terminal (*N*) is also available. The three waveforms have 120° phase differences, with a particular phase sequence depending on the generator terminal connections. The generator output voltages can be measured as phase voltage (the output at the terminals of a single generator in a three-phase system), and line voltage (the voltage measured at the terminals of two interconnected generators). The three generator coils may be connected in one of two possible configurations; Y-connection and Δ-connection. Similarly, the load on a three-phase generator may be connected in Y or Δ fashion.

26-1 GENERATION OF THREE-PHASE VOLTAGES

The output from the simple rotating-loop ac generator described in Section 17-1 is one cycle of sine wave continuously repeating. In Figures 26-1(a) and (b) three similar conducting loops are shown situated between two magnetic poles. Each loop generates a sinusoidal alternating voltage as it rotates in the magnetic field, exactly as described in Section 17-1. Three separate sinusoidal voltages are produced at the three pairs of output terminals. Because the loops are set at an angle to each other, the individual output peak voltages occur at different times. This means that there are phase differences between the three output sine waves.

Referring to Figures 26-1(a) and (b), note that the two sides (or rotating conductors) of each loop are identified as *A* and *a*, *B* and *b*, and *C* and *c*. The loops are rotated in a counterclockwise direction. With conductor *A* moving past the N pole and conductor *a* moving past the S pole, as illustrated, the instantaneous emf (e_{Aa}) generated in *a* and *A* is a maximum. Also, terminal *A* is positive with respect to terminal *a*. (This is explained in Section 17-1, where conductors 1 and 2 correspond to conductors *A* and *a*, respectively.)

At the instant when the output from *A* and *a* is a positive maximum, conductor *B* is moving away from the S pole and *b* is leaving the N pole. Consequently, the output voltage from terminals *B* and *b* (e_{Bb}) is past its negative peak value and is moving toward zero (terminal *B* being negative with respect to *b*).

Loop *Bb* in Figure 26-1 is 120° behind loop *Aa*. The output voltage e_{Bb} can be written with respect to voltage e_{Aa}. From Equation 17-4,

$$e_{Aa} = E_m \sin \alpha$$

and,

$$e_{Bb} = E_m \sin(\alpha - 120°)$$

Now consider loop *Cc* in Figure 26-1(a). At the instant illustrated, conductors *C* and *c* are seen to be approaching the S and N poles, respectively. Thus,

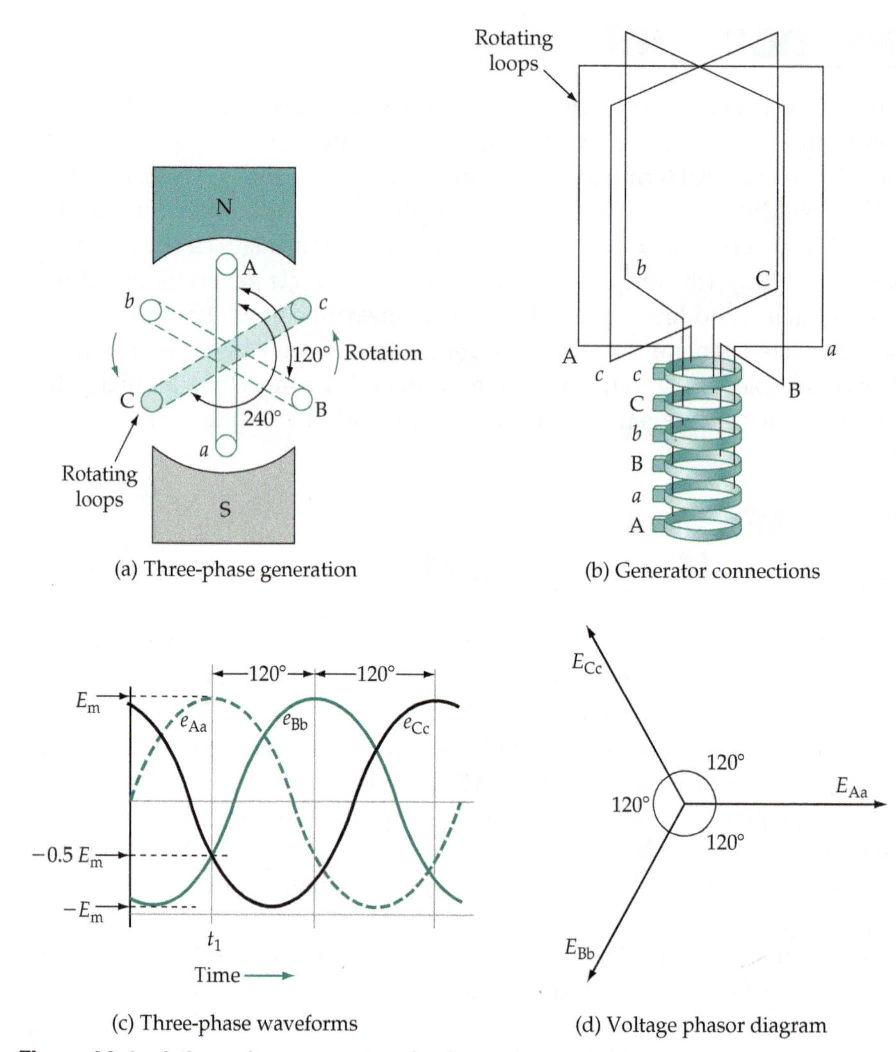

Figure 26-1 **A three-phase generator circuit consists basically of three conducting loops rotated in a magnetic field. Because the loops are 120° apart, the three output voltage waveforms have 120° phase angle differences.**

the output from terminals C and *c* is a negative voltage (*C* being negative with respect to *c*) growing toward its negative peak level. Loop *Cc* is 120° behind loop *Bb* and 240° behind loop *Aa*. This is the same as 120° ahead of loop *Aa*.

So,

$$e_{Cc} = E_m \sin(\alpha - 240°)$$
$$= E_m \sin(\alpha + 120°)$$

The three output voltage waveforms are shown in Figure 26-1(c) with the correct phase relationships. Waveform e_{Bb} reaches its positive peak 120°, or one-third of a cycle, after e_{Aa} peaks. The positive peak level of e_{Cc} occurs 240°, or two-thirds of a cycle, after the peak of e_{Aa}. The output voltages at the instant illustrated in Figure 26-1(a) are shown at t_1 in Figure 26-1(c). As discussed, at this point e_{Aa} is a positive maximum, e_{Bb} is negative moving toward zero, and e_{Cc} is negative moving in the direction of its negative peak level. Figure 26-1(c) shows that the waveforms peak with the sequence e_{Aa}, e_{Bb}, and e_{Cc}. The voltages are said to have a *phase sequence* of *ABC*.

An ac generator that produces a single sine wave output voltage is referred to as a *single-phase* generator. A generator that produces three separate sine wave output voltages with 120° phase differences is called a *three-phase* generator. (*Three-phase* is sometimes written as 3ϕ.)

As in the case of single-phase ac circuits, rms values of voltage and current are normally used in three-phase calculations rather than instantaneous or peak quantities. The phasor diagram of the three-phase rms output voltages is drawn in Figure 26-1(d). E_{Aa} is the rms output voltage from loop *Aa*. E_{Bb} is 120° behind E_{Aa}, and E_{Cc} is 240° behind E_{Aa}. It is seen that the phasors are arranged symmetrically with phase differences of 120° between them.

26-2 Y-CONNECTED GENERATOR

Phase Voltages and Currents

The output voltage levels produced by a three-phase generator depend on the connection arrangement of the three individual generator loops (or coils). The two possible connection methods are known as *wye* (Y), and *delta* (Δ) because of circuit resemblances to those symbols. Figure 26-2 represents a Y-connected three-phase generator, in which the coil terminals *a*, *b*, and *c* are connected to a single *neutral terminal* (*N*). There are four output terminals; *A*, *B*, *C*, and *N*.

The system illustrated in Figure 26-2 is termed a *three-phase four-wire supply*. In some cases the neutral conductor may not be connected, and the system becomes a *three-phase three-wire supply*.

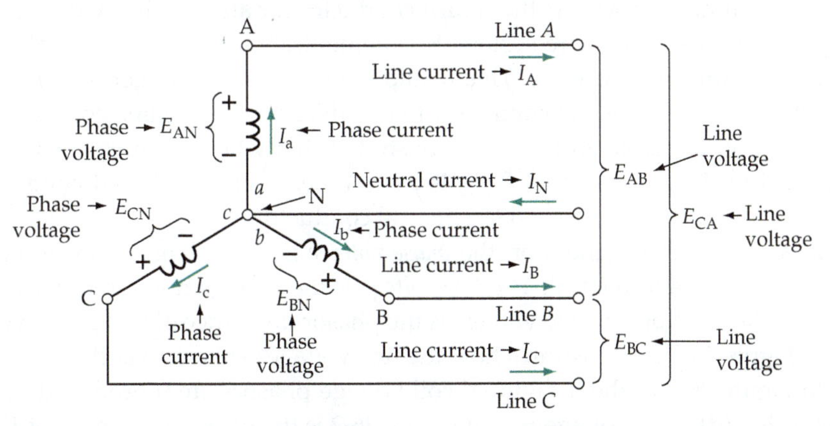

Figure 26-2 For a Y-connected three-phase generator, the phase voltage and phase current are the output voltage and current from each coil. The line voltage is the voltage measured between any two output lines (conductors). The line current is the current flowing in a line.

All three coils in a three-phase generator are identical, and all three produce the same level of rms output voltage. This is termed the *phase voltage* (V_p). Figure 26-2 shows E_{Aa}, E_{Bb}, and E_{Cc} now identified as E_{AN}, E_{BN}, and E_{CN}, respectively. All three are equal to V_p.

The *phase current* (I_p) is the current level supplied by each of the three generator coils. The directions of the phase currents (I_a, I_b, and I_c) shown in Figure 26-2 are those that occur when each coil has a positive output voltage. Thus, the phase current for coil *Aa* is shown flowing out of terminal *A* and into *a*.

Double Subscript Notation

Double subscript notation is used with the voltage symbols to indicate the direction of current flow when each coil (or loop) output is positive. Subscript Aa used with voltage E_{Aa} identifies terminal *A* as positive with respect to terminal *a* during the time that this coil output is positive. E_{Aa} is rewritten as E_{AN} in Figure 26-2, and this indicates that terminal *A* is positive with respect to the neutral terminal while the coil output is positive. Similarly, when the output of coil *Bb* is a positive quantity (see Figure 26-1), terminal *B* is positive with respect to *b*, and the internal current direction is from *b* to *B*. The voltage is identified as E_{Bb}, or E_{BN} in Figure 26-2. The voltage labeled E_{Cc} or E_{CN} shows that terminal C is positive with respect to *c* when coil *Cc* has a positive output. In a different subscript system, the order of the double subscripts indicates the current direction instead of the voltage polarity.

Line Currents and Voltages

The conductors that connect a three-phase generator to a load are referred to as *lines*, and the current carried by each of these conductors is known as the *line current* (I_L). Figure 26-2 shows that in a Y-connected generator the phase currents and line currents are the same quantities.

The neutral conductor is the return conductor for all three individual coils in a Y-connected, four wire system. So, the neutral current (I_N) is the phasor sum of all three line currents. This would seem to make I_N larger than I_L and therefore require a neutral conductor that is thicker than the line conductors. However, in Example 26-1 it is demonstrated that when all three coils have identical loads (*a balanced load*), I_N is zero. When a balanced load condition does not exist, current flows in the neutral conductor.

In a Y-connected generator, the *phase voltage* is the voltage measured between any line and neutral. The *line voltage* is the voltage measured between any two lines. Thus, the line voltage is the phasor difference of two phase voltages. Figure 26-3 demonstrates how the line voltages are calculated.

In Figure 26-3(a) the individual coil voltage phasors are reproduced from Figure 26-1(d). Line voltage E_{AB} in Figure 26-2 is the phasor difference of E_{AN} and E_{BN}. To determine E_{AB}, $-E_{BN}$ is first drawn equal and opposite to E_{BN} [see Figure 26-3(b)]. Because of the 120° phase difference between E_{AN} and E_{BN},

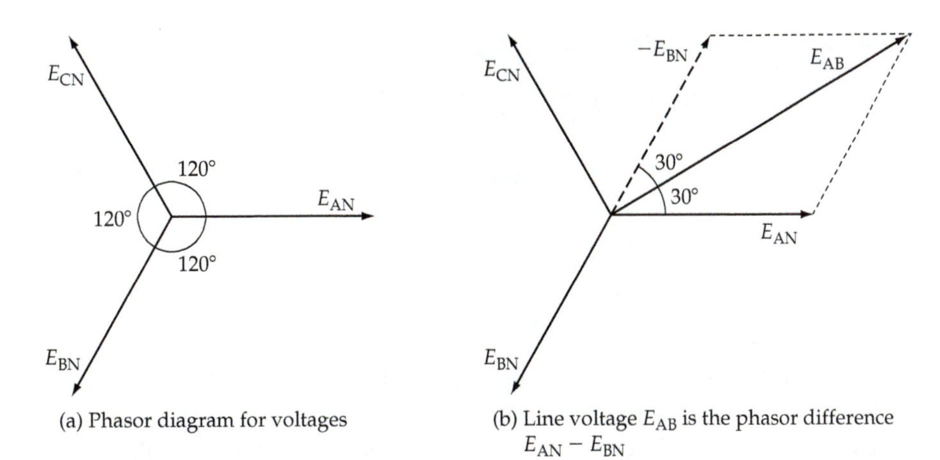

(a) Phasor diagram for voltages

(b) Line voltage E_{AB} is the phasor difference $E_{AN} - E_{BN}$

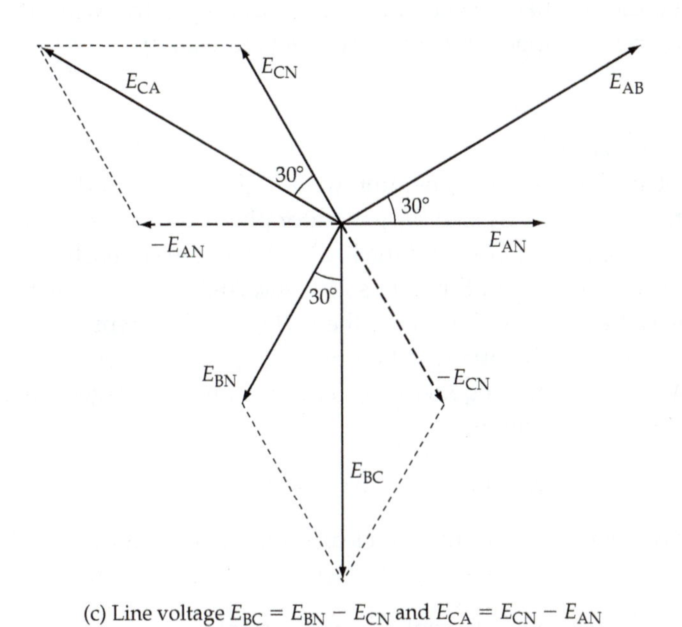

(c) Line voltage $E_{BC} = E_{BN} - E_{CN}$ and $E_{CA} = E_{CN} - E_{AN}$

Figure 26-3 For a Y-connected three-phase generator, the line voltages are the phasor sums of pairs of phase voltages. This makes each line voltage equal to $\sqrt{3}$ × (phase voltage).

there is a 60° angle between E_{AN} and $-E_{BN}$. Voltages E_{AN} and E_{BN} are equal in magnitude; therefore, phasors E_{AN} and $-E_{BN}$ are equal in length. Because of this equality, phasor E_{AB} (representing the sum of E_{AN} and $-E_{BN}$) is situated 30° from E_{AN} and 30° from $-E_{BN}$, as illustrated in Figure 26-3(b).

This gives $$E_{AB} = E_{AN} \cos 30° + (-E_{BN} \cos 30°)$$

Which is the same as, $$E_{AB} = V_p \cos 30° + V_p \cos 30°$$
$$= 2(V_p \cos 30°)$$
$$= 1.732V_p$$

or $$E_{AB} = \sqrt{3}\, V_p$$

So, for a Y-connected generator,

$$\text{Line voltage} = \sqrt{3}\,(\text{Phase voltage})$$

That is,

$$V_L = \sqrt{3}\,V_p \tag{26-1}$$

Also,

$$\text{Line current} = \text{Phase current}$$

or,

$$I_L = I_p \tag{26-2}$$

Figure 26-3(b) also shows that line voltage E_{AB} leads phase voltage E_{AN} by 30°.

Figure 26-3(c) shows the determination of E_{BC} and E_{CA} as the phasor differences $E_{BN} - E_{CN}$ and $E_{CN} - E_{AN}$, respectively. Note that for line voltage E_{AB}, the subscript indicates that the generator terminal *A* is positive with respect to terminal *B* when E_{AB} is a positive quantity (see Figure 26-2).

Y-Connected Loads

The circuit of a Y-connected generator with Y-connected load resistors is shown in Figure 26-4(a). Note that the generator phase voltages are once again identified as E_{AN}, E_{BN}, and E_{CN}, and the load voltages are similarly identified. The reasoning that developed Equations 26-1 and 26-2 for a Y-connected generator can also be applied to determine the voltage and current relationships for the Y-connected *load*. Referring to Figure 26-4(a), the voltages across the individual loads are obviously equal to the generator phase voltages. Therefore, Equation 26-1 can be rewritten as

$$\text{Line voltage} = \sqrt{3}\,(\text{Load voltage})$$

Also, the individual load currents are clearly the same currents that flow in the lines.

$$\text{Line current} = \text{Load current}$$

Example 26-1

Load resistors, R_1 R_2, and R_3 in Figure 26-4(a) are each 100 Ω, and the phase voltage is $V_p = 100$ V. Determine: (a) the line current, (b) the neutral current, and (c) the line voltage.

Solution

(a)

$$E_{AN} = E_{BN} = E_{CN} = V_p = 100\text{ V}$$

$$I_L = I_1 = I_2 = I_3$$

$$I_L = \frac{V_P}{R_1} = \frac{100\text{ V}}{100\ \Omega} = 1\text{ A}$$

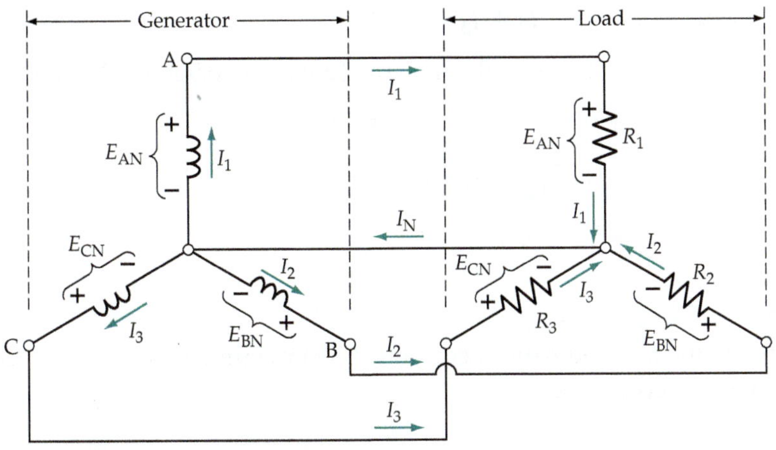

(a) Y-connected three-phase generator with Y-connected resistive loads

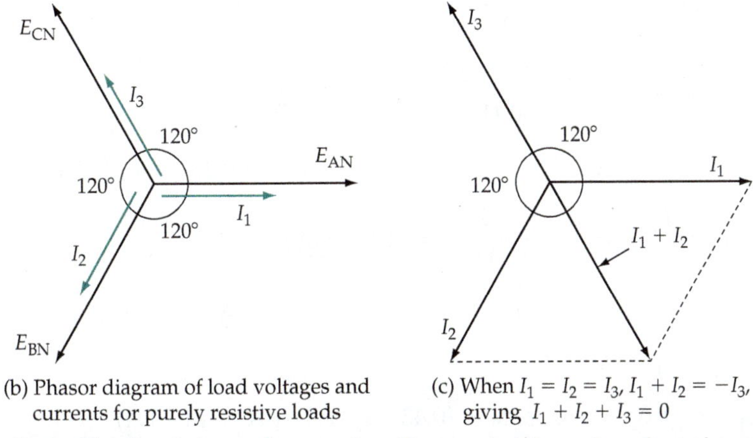

(b) Phasor diagram of load voltages and currents for purely resistive loads

(c) When $I_1 = I_2 = I_3$, $I_1 + I_2 = -I_3$, giving $I_1 + I_2 + I_3 = 0$

Figure 26-4 Circuit and phasor diagrams for a Y-connected three-phase four-wire generator with a Y-connected balanced load. In this case, the load voltages and currents are equal to the generator phase voltage and phase current.

(b) The phasor diagram for the currents is drawn in Figure 26-4(b). Because the loads are resistive, each load current is in phase with the phase voltages. Figure 26-4(c) shows that the phasor sum of I_1 and I_2 equals $-I_3$.

So, $$I_N = I_1 + I_2 + I_3 = 0$$

Alternatively, $$I_1 = \frac{E_{AN}}{R_1} = \frac{100\text{ V}\angle 0°}{100\ \Omega} = 1\text{ A}\angle 0° = 1\text{ A} + j0\text{ A}$$

$$I_2 = \frac{E_{BN}}{R_2} = \frac{100\text{ V}\angle -120°}{100\ \Omega} = 1\text{ A}\angle -120°$$

$$= 1\text{ A}[\cos(-120°) + j\sin(-120°)] = -0.5\text{ A} - j0.866\text{ A}$$

$$I_3 = \frac{E_{CN}}{R_3} = \frac{100\text{ V}\angle -240°}{100\ \Omega} = 1\text{ A}\angle -240°$$

$$= 1\text{ A}[\cos(-240°) + j\sin(-240°)] = -0.5\text{ A} + j0.866\text{ A}$$

$$I_N = I_1 + I_2 + I_3$$
$$= 1\text{ A} + (-0.5\text{ A} - j0.866\text{ A}) + (-0.5\text{ A} + j0.866\text{ A})$$
$$= 0$$

(c) Eq. 26-1, $V_L = \sqrt{3}\,V_p = \sqrt{3} \times 100\text{ V}$

$$= 173.2\text{ V}$$

Example 26-2

Recalculate the line and neutral currents if R_2 in Figure 26-4(a) is changed to 200 Ω, and R_3 becomes 50 Ω.

Solution

$$I_1 = \frac{E_{AN}}{R_1} = \frac{100\text{ V}\angle 0^\circ}{100\ \Omega} = 1\text{ A}\angle 0^\circ = 1\text{ A} + j0\text{ A}$$

$$I_2 = \frac{E_{BN}}{R_2} = \frac{100\text{ V}\angle -120^\circ}{200\ \Omega} = 0.5\text{ A}\angle -120^\circ$$
$$= 0.5\text{ A}[\cos(-120^\circ) + j\sin(-120^\circ)] = -0.25\text{ A} - j0.433\text{ A}$$

$$I_3 = \frac{E_{CN}}{R_3} = \frac{100\text{ V}\angle -240^\circ}{50\ \Omega} = 2\text{ A}\angle -240^\circ$$
$$= 2\text{ A}[\cos(-240^\circ) + j\sin(-240^\circ)] = -1\text{ A} + j1.73\text{ A}$$

$$I_N = I_1 + I_2 + I_3$$
$$= 1\text{ A} + (-0.25\text{ A} - j0.433\text{ A}) + (-1\text{ A} + j1.73\text{ A})$$
$$= -0.25\text{ A} + j1.3\text{ A}$$
$$= 1.32\text{ A}\angle 100.9^\circ$$

Example 26-3

In the circuit shown in Figure 26-5, the generator phase voltage is $V_p = 100$ V, and its frequency is 60 Hz. The load impedance components are: $R_1 = R_2 = R_3 = 100\ \Omega$, $C_2 = 66.3\ \mu\text{F}$, and $L_3 = 159.2$ mH. Calculate the three line currents and the neutral current.

Solution

$$I_1 = \frac{E_{AN}}{Z_1} = \frac{100\text{ V}}{100\ \Omega} = 1\text{ A}\angle 0^\circ$$

$$X_{C2} = \frac{1}{2\pi f C} = \frac{1}{2\pi \times 60\text{ Hz} \times 66.3\ \mu\text{F}}$$
$$= 40\ \Omega$$

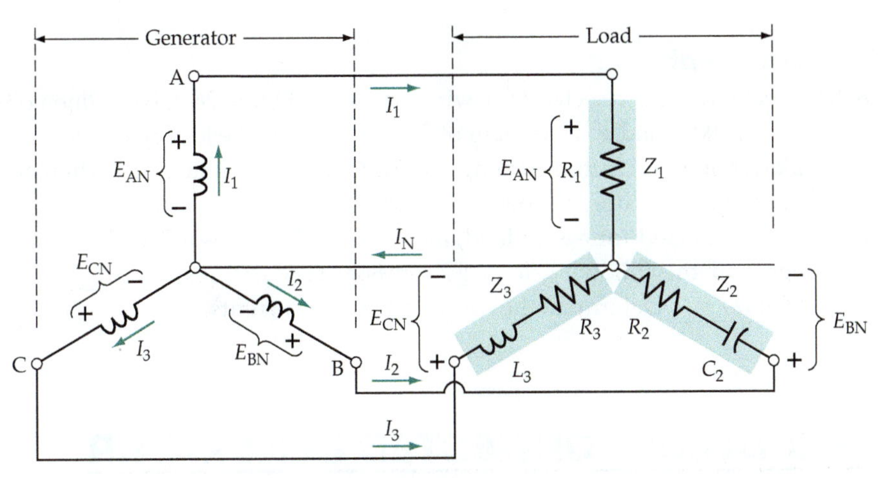

Figure 26-5 Circuit of a Y-connected three-phase four-wire generator with a Y-connected unbalanced load. The load voltages and currents, once again, equal the generator phase voltages and individual phase currents, but the neutral current is not zero.

$$Z_2 = R_2 - jX_{C2} = 100\ \Omega - j40\ \Omega$$
$$= 107.7\ \Omega\angle -21.8°$$
$$I_2 = \frac{E_{BN}}{Z_2} = \frac{100\ \text{V}\angle -120°}{107.7\ \Omega\angle -21.8°} = 0.929\ \text{A}\angle -98.2°$$
$$= 0.929\ \text{A}\,[\cos(-92.9°) + j\sin(-92.9°)]$$
$$= -0.133\ \text{A} - j0.92\ \text{A}$$
$$X_{L3} = 2\pi f L_3 = 2\pi \times 60\ \text{Hz} \times 159.2\ \text{mH}$$
$$= 60\ \Omega$$
$$Z_3 = R_3 + jX_{L3} = 100\ \Omega + j60\ \Omega$$
$$= 116.6\ \Omega\angle 31°$$
$$I_3 = \frac{E_{CN}}{Z_3} = \frac{100\ \text{V}\angle -240°}{116.6\ \Omega\angle 31°} = 0.858\ \text{A}\angle -271°$$
$$= 0.858\ \text{A}\,[\cos(-271°) + j\sin(-271°)]$$
$$= 15\ \text{mA} + j858\ \text{mA}$$
$$I_N = I_1 + I_2 + I_3$$
$$= 1\ \text{A} + (-133\ \text{mA} - j920\ \text{mA}) + (15\ \text{mA} + j858\ \text{mA})$$
$$= 882\ \text{mA} - j62\ \text{mA}$$
$$= 884\ \text{mA}\angle -4°$$

Example 26-1 shows that for a balanced load, the neutral current is zero. Examples 26-2 and 26-3 show that when the load is unbalanced, the neutral current can be substantial.

Practice Problems

26-2.1 A balanced Y-connected four-wire load, as in Figure 26-4, has a line voltage of 200 V and a line current of 3.5 A. Calculate the load resistances.

26-2.2 Determine the neutral current in the circuit of Problem 26-2.1 when one of the load resistors is open-circuited.

26-2.3 A Y-connected four-wire load, as in Figure 26-5, has a 300 V line voltage and the following loads: $Z_1 = 86.6\ \Omega$, $Z_2 = 86.6\ \Omega\angle{-30^\circ}$, $Z_3 = 86.6\ \Omega\angle 60^\circ$. Determine the line and neutral currents.

26-3 DELTA-CONNECTED GENERATOR

Voltages and Currents

As an alternative to Y connection, the individual coils of a three-phase generator may be connected in Δ (*delta*) configuration. In the Δ-connected generator shown in Figure 26-6, the coil terminals are connected *A* to *c*, *B* to *a*, and *C* to *b*. The three output lines are identified as *A*, *B*, and *C*. No neutral conductor is involved. The line voltage (V_L) is the voltage measured as E_{AB}, E_{BC}, or E_{CA}. Once again, the order of the subscripts indicates the terminal polarity when each line voltage is positive.

Recall that the phase voltage (V_p) is the rms voltage measured across the output terminal of each of the three coils. It is obvious from Figure 26-6 that the line voltage and phase voltage are the same quantities in a Δ-connected generator.

For a Δ-connected generator

$$\text{Line voltage} = \text{Phase voltage}$$

That is,

$$V_L = V_p \qquad (26\text{-}3)$$

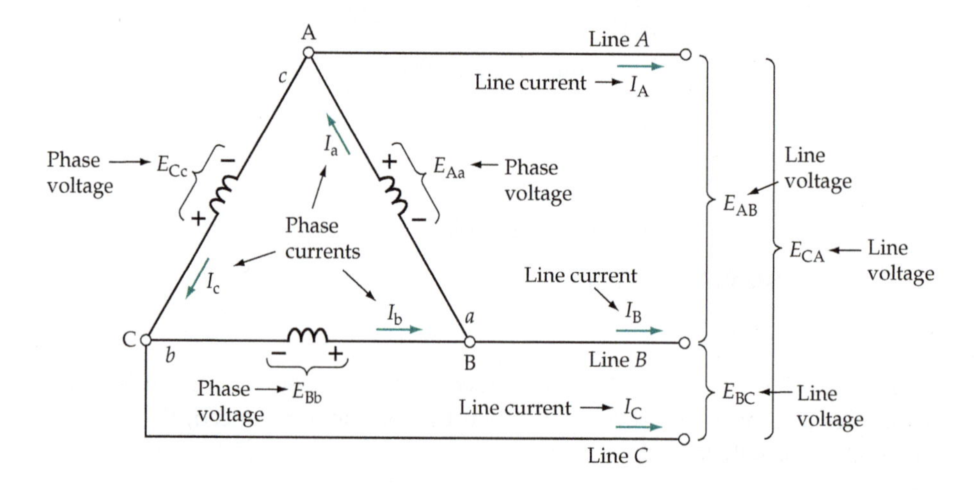

Figure 26-6 For a Δ-connected three-phase generator, the phase voltage and phase current are the output voltage and current from each coil. The line voltage is the voltage measured between any two output lines, and the line current is the current flowing in a line.

The line current (I_L) is the current level measured in each of the line conductors. The direction indicated for I_L in each line is the direction in which current flows when the line voltage is positive. When E_{AB} is positive, line A is positive with respect to line B. The direction of line current I_A is then out of terminal A, as indicated in Figure 26-6. Similarly, when B is positive with respect to C, I_B flows out of terminal B, and when C is more positive than A, the direction of I_C is away from terminal C.

The phase current (I_p) is the current level produced in each of the three generator coils. Once again, the current directions indicated in Figure 26-6 are those that occur when each individual coil output is positive. When E_{Aa} is positive, I_a flows out of the terminal A and into terminal a. When E_{Bb} is positive, the I_b direction within the coil is from b to B. Similarly, for E_{Cc}, I_c flows from c to C so that the output current is out of terminal C and into c.

With the phase currents having the directions indicated in Figure 26-6, it appears that a circulating current might flow around the coils of a Δ-connected generator. This would not be acceptable, because it would produce unnecessary and wasteful power dissipation in the generator coils. Consider the waveforms in Figure 26-1(c) once again, and note that at t_1:

$$e_{Aa} = E_m$$

and

$$e_{Bb} = -0.5E_m$$

$$e_{Cc} = -0.5E_m$$

These instantaneous voltage levels are easily confirmed by calculation as follows:

$$e_{Aa} = E_m \sin \alpha = E_m \sin 90°$$
$$= E_m$$
$$e_{Bb} = E_m \sin(\alpha - 120°) = E_m \sin (90° - 120°)$$
$$= -0.5E_m$$
$$e_{Cc} = E_m \sin (90° - 240°)$$
$$= -0.5E_m$$

The instantaneous level of voltage producing a circulating current in illustrated in Figure 26-6 is

$$e = e_{Aa} + e_{Bb} + e_{Cc}$$

At time t_1 in Figure 26-1(c)

$$e = E_m - 0.5E_m - 0.5E_m$$
$$= 0$$

It is seen that, at time t_1 there is no voltage acting around the closed loop formed by the coils. Consequently, there is no current circulating in the coils.

By taking any other instant in the three-phase waveforms illustrated in Figure 26-1(c), it can be demonstrated that the circulating current is always zero in a Δ-connected generator.

Referring again to Figure 26-6, note that I_a flows toward terminal A and that both I_c and I_A flow away from A.

This gives, $$I_a = I_A + I_c$$

or, the line current is, $$I_A = I_a - I_c$$

Because the currents are phasor quantities, this is a phasor difference.

Figure 26-7(a) shows the phasor diagram for the phase currents that might be produced by a three-phase generator with a *balanced load.* As explained, when a balanced load occurs, all three line currents are equal and have the same phase relationship to their respective line voltages. If load consists of three equal Δ-connected resistors, then a balanced load condition exists.

In Figure 26-7(a), the phasors for the three equal phase currents (I_a, I_b, and I_c) are drawn with 120° phase angle separation. Line current I_A is shown in Figure 26-7(a) as ($I_a - I_c$) (as determined in the preceding equation). It is seen that I_a and $-I_c$ are separated by 60°, and because they are equal in magnitude

$$I_A = I_a \cos 30° + I_c \cos 30° = 2(I_p \cos 30°)$$
$$= 1.732 I_p$$
$$= \sqrt{3}\, I_p$$

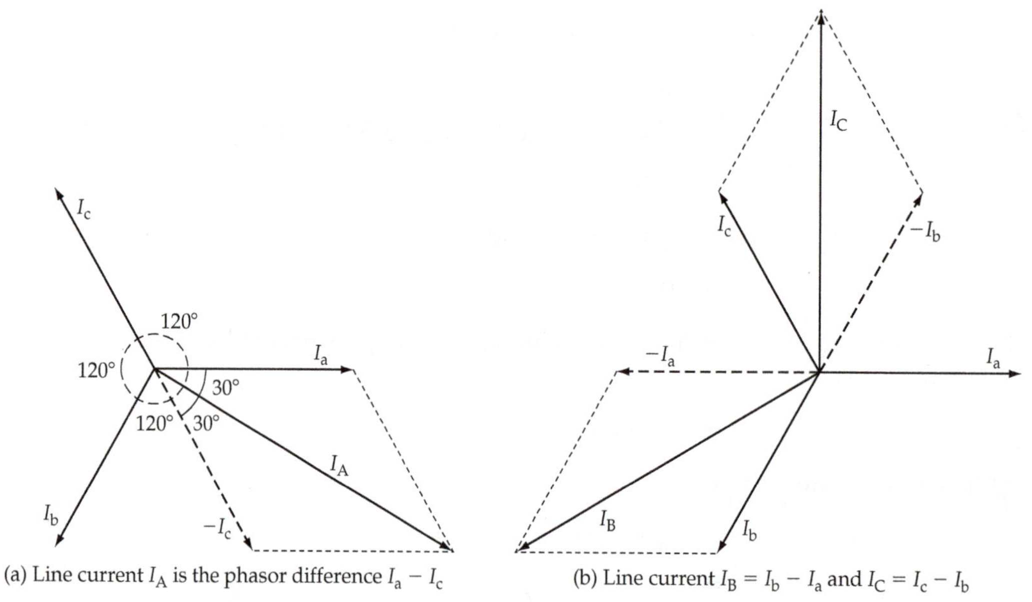

(a) Line current I_A is the phasor difference $I_a - I_c$ (b) Line current $I_B = I_b - I_a$ and $I_C = I_c - I_b$

Figure 26-7 For a Δ-connected three-phase generator, the line currents are the phasor sums of pairs of phase currents. With a balanced load, each line current is equal to $\sqrt{3}$ × (phase current).

So, with a balanced load,

$$\text{Line current} = \sqrt{3}\,(\text{phase current})$$

$$I_L = \sqrt{3}I_p \tag{26-4}$$

Figure 26-7(a) also shows that (for a balanced load) the phase current (I_a) leads the line current (I_A) by an angle 30°. In Figure 26-7(b) the determinations of I_B and I_C are shown as the phasor differences ($I_b - I_a$) and ($I_c - I_b$), respectively.

Delta-Connected Loads

The reasoning used in deriving Equations 26-3 and 26-4 for a Δ-connected generator can also be applied to a Δ-connected load. For the circuit in Figure 26-8(a), the individual Δ-connected load voltages are obviously equal to the line voltages.

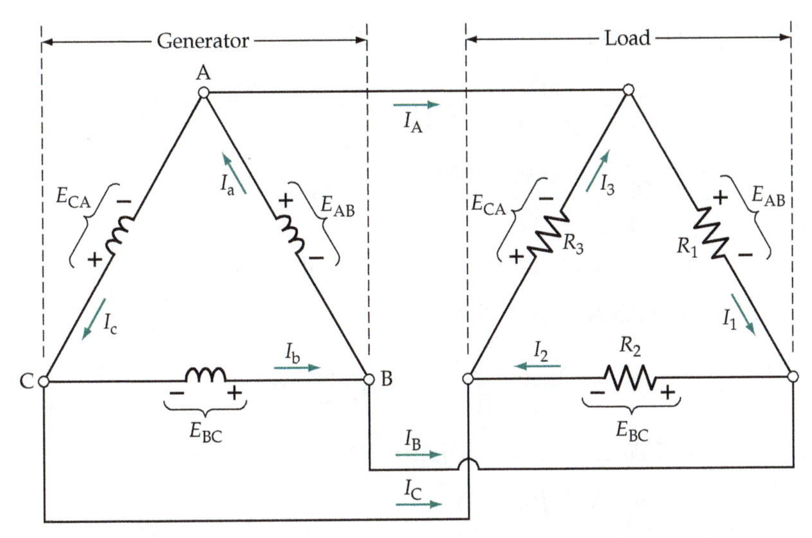

(a) Δ-connected three-phase generator with Δ-connected resistive loads

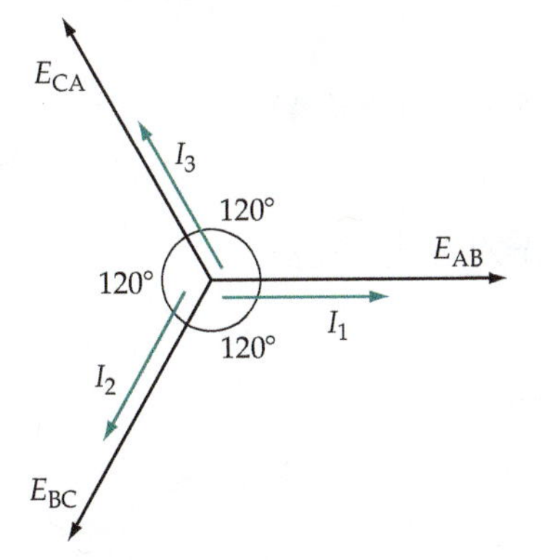

(b) Phasor diagram of load voltages and currents for purely resistive loads

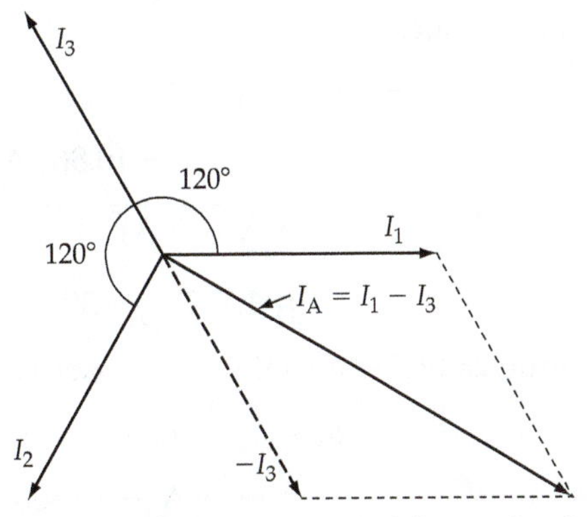

(c) Line current I_A is the phasor difference $I_1 - I_3$

Figure 26-8 Circuit and phasor diagrams for a Δ-connected three-phase generator with a Δ-connected balanced load. The load voltage equals the generator phase voltage, and the line current is $\sqrt{3}$ × (load current).

For a Δ-connected load,

$$\text{Load voltage} = \text{Line voltage}$$

Also, where all three load impedances are identical, giving a balanced load, the individual load currents combine to produce,

$$\text{Line current} = \sqrt{3}\,(\text{Load current})$$

Example 26-4

In the circuit shown in Figure 26-8(a), $R_1 = R_2 = R_3 = 100\ \Omega$, and $V_p = 100$ V. Calculate the load currents, and the line currents.

Solution

From Eq. 26-3, $$V_L = V_P = 100\ \text{V}$$

Taking voltage E_{AB} as the reference voltage for all phase angles,

$$I_1 = \frac{E_{AB}}{R_1} = \frac{100\ \text{V}}{100\ \Omega} = 1\ \text{A}\underline{/0°}$$

$$I_2 = \frac{E_{BC}}{R_2} = \frac{100\ \text{V}\underline{/-120°}}{100\ \Omega} = 1\ \text{A}\underline{/-120°}$$

$$= 1\ \text{A}\,[\cos(-120°) + j\sin(-120°)]$$

$$= -0.5\ \text{A} - j0.866\ \text{A}$$

$$I_3 = \frac{E_{CA}}{R_3} = \frac{100\ \text{V}\underline{/-240°}}{100\ \Omega} = 1\ \text{A}\underline{/-240°}$$

$$= 1\ \text{A}\,[\cos(-240°) + j\sin(-240°)]$$

$$= -0.5\ \text{A} + j0.866\ \text{A}$$

The phasor diagram for load voltages and currents is drawn in Figure 26-8(b). Because the loads are purely resistive, the load currents are in phase with the load voltages.

$$I_A = I_1 - I_3 = 1\ \text{A} - (-0.5\ \text{A} + j0.866\ \text{A})$$

$$= 1.5\ \text{A} - j0.866\ \text{A}$$

$$= \sqrt{(1.5\ \text{A})^2 + (0.866\ \text{A})^2}\,\underline{/\tan^{-1}(-0.866/1.5)}$$

$$= 1.732\ \text{A}\underline{/-30°}$$

Figure 26-8(c) shows the phasor derivation of I_A as $I_1 - I_3$.

$$I_B = I_2 - I_1 = (-0.5\ \text{A} - j0.866\ \text{A}) - 1\ \text{A}$$

$$= -1.5\ \text{A} - j0.866\ \text{A}$$

$$= 1.732\ \text{A}\underline{/-150°}$$

$$I_C = I_3 - I_2 = (-0.5\text{ A} + j0.866\text{ A}) - (-0.5\text{ A} - j0.866\text{ A})$$
$$= 0 + j1.732\text{ A}$$
$$= 1.732\text{ A}\underline{/-270°}$$

Example 26-5

Recalculate the line currents in the circuit of Figure 26-8(a) if R_2 is changed to 200 Ω and R_3 becomes 50 Ω.

Solution

$$I_1 = \frac{E_{AB}}{R_1} = \frac{100\text{ V}}{100\ \Omega} = 1\text{ A}\underline{/0°}$$

$$I_2 = \frac{E_{BC}}{R_2} = \frac{100\text{ V}\underline{/-120°}}{200\ \Omega} = 0.5\text{ A}\underline{/-120°}$$
$$= 0.5\text{ A}\,[\cos(-120°) + j\sin(-120°)]$$
$$= -0.25\text{ A} - j0.433\text{ A}$$

$$I_3 = \frac{E_{CA}}{R_3} = \frac{100\text{ V}\underline{/-240°}}{50\ \Omega} = 2\text{ A}\underline{/-240°}$$
$$= 2\text{ A}\,[\cos(-240°) + j\sin(-240°)]$$
$$= -1\text{ A} + j1.732\text{ A}$$

$$I_A = I_1 - I_3 = 1\text{ A} - (-1\text{ A} + j1.732\text{ A})$$
$$= 2\text{ A} - j1.732\text{ A}$$
$$= \sqrt{(2\text{ A})^2 + (1.732\text{ A})^2}\underline{/\tan^{-1}(-1.732/2)}$$
$$= 2.65\text{ A}\underline{/-40.9°}$$

$$I_B = I_2 - I_1 = (-0.25\text{ A} - j0.433\text{ A}) - 1\text{ A}$$
$$= -1.25\text{ A} - j0.433\text{ A}$$
$$= \sqrt{(1.25\text{ A})^2 + (0.433\text{ A})^2}\underline{/\tan^{-1}(-0.433/-1.25)}$$
$$= 1.32\text{ A}\underline{/-161°}$$

$$I_C = I_3 - I_2 = (-1\text{ A} + j1.732\text{ A}) - (-0.25\text{ A} - j0.433\text{ A})$$
$$= -0.75\text{ A} + j2.165\text{ A}$$
$$= \sqrt{(0.75\text{ A})^2 + (2.165\text{ A})^2}\underline{/\tan^{-1}(2.165/-0.75)}$$
$$= 2.29\text{ A}\underline{/-251°}$$

Example 26-6

The circuit in Figure 26-9 has $R_1 = R_2 = R_3 = 200\ \Omega$, $C_2 = 10\ \mu F$, $L_3 = 400$ mH. The line voltage is $V_L = 250$ V, and the supply frequency is 60 Hz. Calculate the load currents and the line currents.

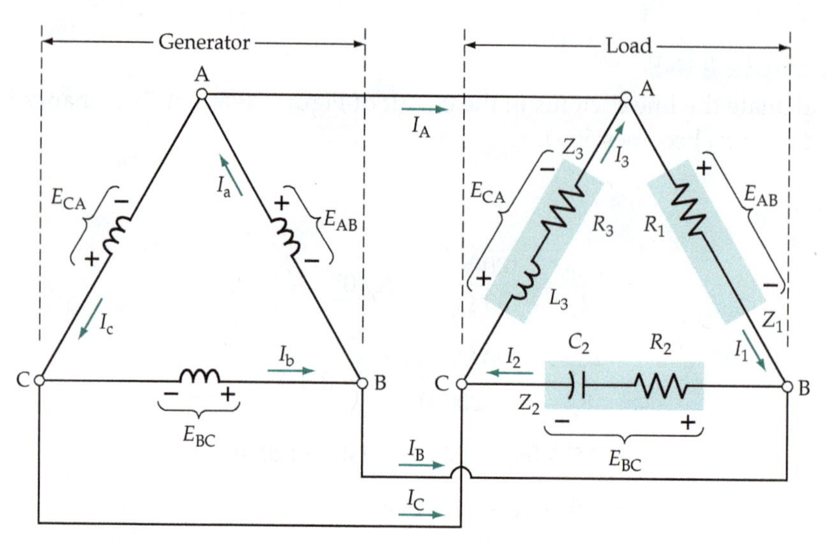

Figure 26-9 Circuit of a Δ-connected three-phase generator with a Δ-connected unbalanced load. The load voltage, once again, equals the generator phase voltage, but the individual line currents depend on the loads.

Solution

Taking E_{AB} as reference for all phase angles,

$$I_1 = \frac{E_{AB}}{Z_1} = \frac{250\ \text{V}}{200\ \Omega} = 1.25\ \text{A}\angle 0^\circ$$

$$X_{C2} = \frac{1}{2\pi f C} = \frac{1}{2\pi \times 60\ \text{Hz} \times 10\ \mu\text{F}}$$

$$= 265.3\ \Omega$$

$$Z_2 = R_2 - jX_{C2} = 200\ \Omega - j265.3\ \Omega$$

$$= \sqrt{(200\ \Omega)^2 + (265.3\ \Omega)^2}\angle \tan^{-1}(-265.3/200)$$

$$= 332.2\ \Omega\angle -53^\circ$$

$$I_2 = \frac{E_{BC}}{Z_2} = \frac{250\ \text{V}\angle -120^\circ}{332.2\ \Omega\angle -53^\circ} = 0.753\ \text{A}\angle -67^\circ$$

$$= 0.753\ \text{A}\,[\cos(-67^\circ) + j\sin(-67^\circ)]$$

$$= -0.295\ \text{A} - j0.693\ \text{A}$$

$$X_{L3} = 2\pi f L_3 = 2\pi \times 60\ \text{Hz} \times 400\ \text{mH}$$

$$= 151\ \Omega$$

$$Z_3 = R_3 + jX_{L3} = 200\ \Omega + j151\ \Omega$$
$$= \sqrt{(200\ \Omega)^2 + (151\ \Omega)^2}\underline{/\tan^{-1}(151/200)}$$
$$= 251\ \Omega\underline{/37.1^\circ}$$
$$I_3 = \frac{E_{CA}}{Z_3} = \frac{250\ \text{V}\underline{/-240^\circ}}{251\ \Omega\underline{/37.1^\circ}} \approx 0.996\ \text{A}\underline{/-277^\circ}$$
$$= 0.996\ \text{A}\,[\cos(-277^\circ) + j\sin(-277^\circ)]$$
$$= 0.123\ \text{A} + j0.988\ \text{A}$$
$$I_A = I_1 - I_3 = 1.25\ \text{A} - (0.123\ \text{A} + j0.988\ \text{A})$$
$$= 1.127\ \text{A} - j0.988\ \text{A}$$
$$= \sqrt{(1.127\ \text{A})^2 + (0.988\ \text{A})^2}\underline{/\tan^{-1}(-0.988/1.127)}$$
$$= 1.5\ \text{A}\underline{/-41.2^\circ}$$
$$I_B = I_2 - I_1 = (0.294\ \text{A} - j0.693\ \text{A}) - 1.25\ \text{A}$$
$$= -0.956\ \text{A} - j0.693\ \text{A}$$
$$= \sqrt{(0.956\ \text{A})^2 + (0.693\ \text{A})^2}\underline{/\tan^{-1}(-0.693/-956)}$$
$$= 1.18\ \text{A}\underline{/-144^\circ}$$
$$I_C = I_3 - I_2 = (0.123\ \text{A} + j0.988\ \text{A}) - (0.294\ \text{A} - j0.693\ \text{A})$$
$$= -0.171\ \text{A} + j1.68\ \text{A}$$
$$= \sqrt{(0.171\ \text{A})^2 + (1.68\ \text{A})^2}\underline{/\tan^{-1}(1.68/-0.171)}$$
$$= 1.69\ \text{A}\underline{/95.8^\circ}$$

Practice Problems

26-3.1 A balanced Δ-connected load, as in Figure 26-8(a), has a 200 V line voltage and a 3.5 A line current. Calculate the load resistances.

26-3.2 Determine the line currents in the circuit of Problem 26-3.1 when the load resistors are: $R_1 = 80\ \Omega$, $R_2 = 60\ \Omega$, and $R_3 = 120\ \Omega$.

26-3.3 A Δ-connected load, as in Figure 26-9, has a line voltage of 150 V and the following loads: $Z_1 = 86.6\ \Omega$, $Z_2 = 86.6\ \Omega\underline{/-30^\circ}$, $Z_3 = 86.6\ \Omega\underline{/60^\circ}$. Determine the line currents.

26-4 Y-Δ AND Y-Y SYSTEMS

Y-Δ System

Electricity is usually distributed in a three-phase, four-wire, system from the Y-connected secondary of a transformer, (see Figure 26-10). Each of the three lines may be used along with the neutral as a single-phase supply for relatively

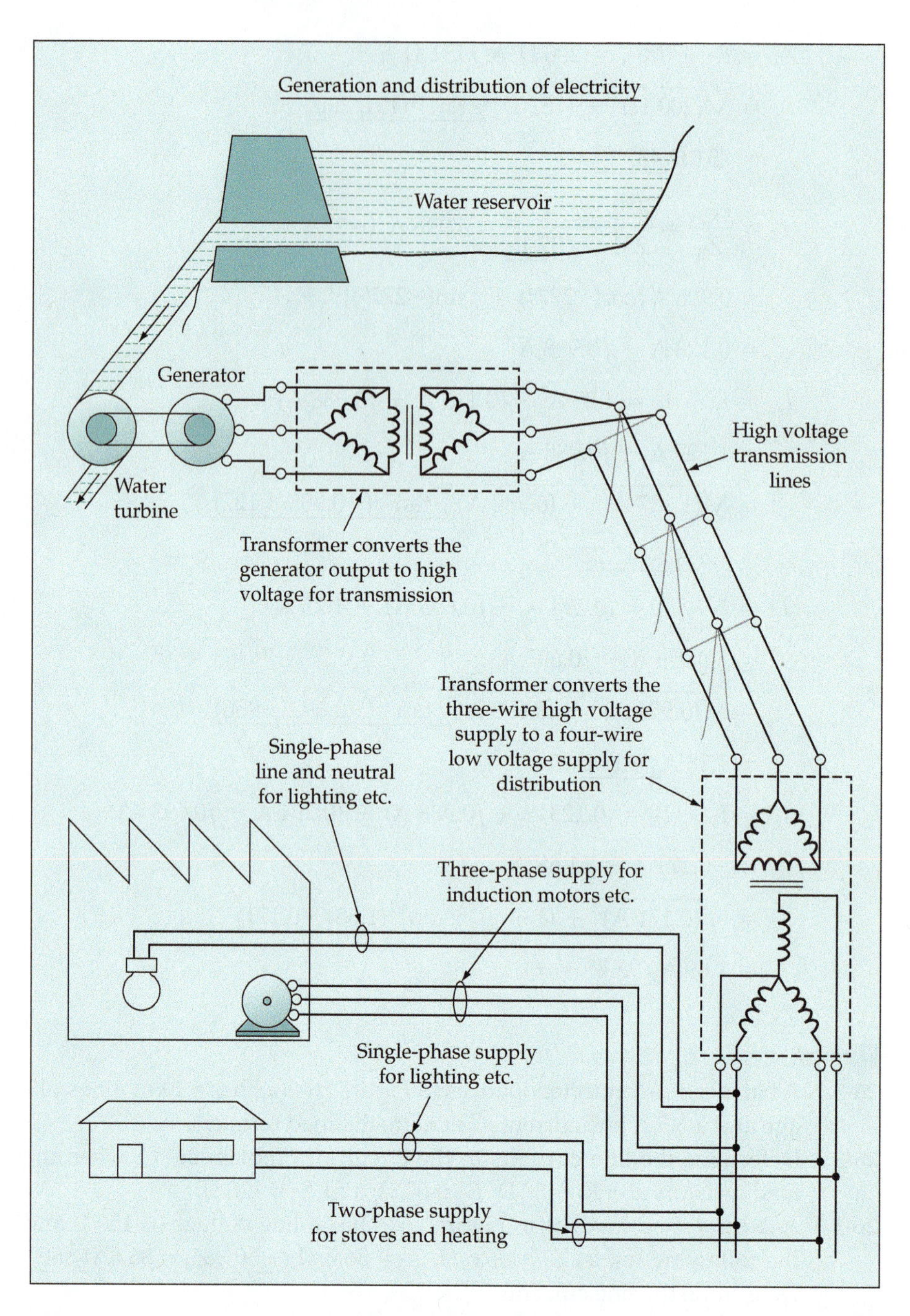

Figure 26-10 Electricity is usually produced in a three-phase generator, stepped-up to a high voltage for transmission, and then stepped-down again to a lower voltage level for distibution.

low power applications such as lighting. Individual homes normally have single-phase supplies, although two lines of a three-phase supply are sometimes provided for domestic cooking and heating purposes. Industrial consumers normally use a three-phase, four-wire supply, which is redistributed locally in single-phase form for low power loads, and in three-phase form for induction motors and high power loads. A typical generation and distribution system is illustrated in Figure 26-10.

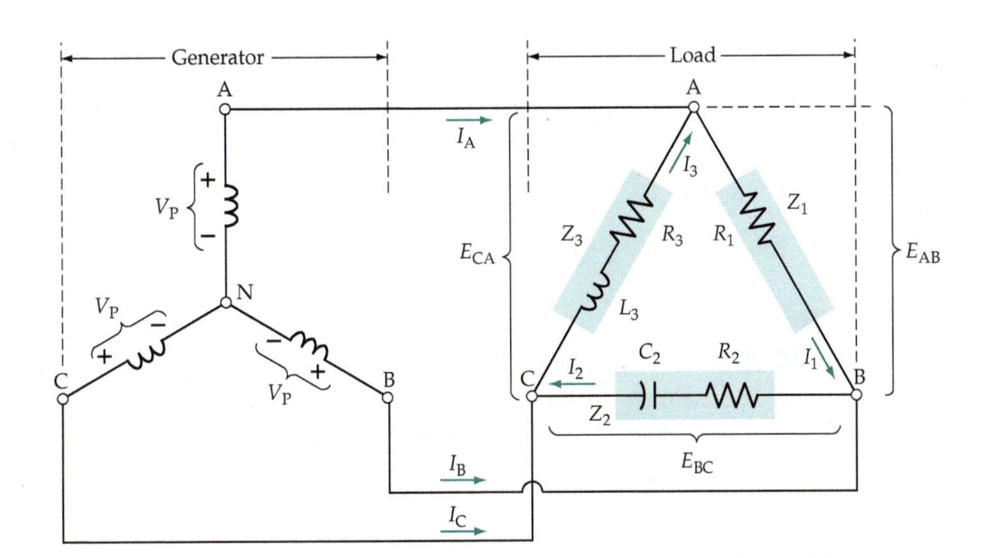

Figure 26-11 Balanced Δ-connected load supplied from a Y-connected generator. The load voltage is the generator line voltage, and the line current (for a balanced load) is $\sqrt{3}$ × (load current).

A three-phase generator may have several loads, some of which are Y-connected and some Δ-connected. In this case, each Y- or Δ-connected load may be analyzed separately to determine the required line currents at the generator. The superposition theorem can then be employed to calculate the total generator currents.

Figure 26-11 shows the circuit of a Y-connected generator with a balanced Δ-connected load. To determine the line currents, the load currents (I_1, I_2, and I_3) are first calculated. Because this is a balanced load, the load currents are equal ($I_1 = I_2 = I_3$), and only one calculation of I_L has to be made. Also, because the load is balanced, the line current is simply determined as $\sqrt{3} \times I_L$.

Example 26-7

A balanced Δ-connected load is supplied from a Y-connected generator, as illustrated in Figure 26-11. The load consists of $R_1 = R_2 = R_3 = 33.3\ \Omega$ and $L_1 = L_2 = L_3 = 523$ mH. The supply has a 115.5 V phase voltage and a 60 Hz frequency. Calculate the line current.

Solution

Eq. 26-1,

$$V_L = \sqrt{3}\,V_p = \sqrt{3} \times 115.5\ \text{V}$$
$$= 200\ \text{V}$$
$$X_L = 2\pi f L = 2\pi \times 60\ \text{Hz} \times 523\ \text{mH}$$
$$= 197.2\ \Omega$$
$$Z_1 = Z_2 = Z_3 = R + jX_L$$
$$= 33.3\ \Omega + j197.2\ \Omega$$

or

$$|Z| = \sqrt{(33.3\ \Omega)^2 + (197.2\ \Omega)^2} = 200\ \Omega$$

$$|I_P| = I_1 = I_2 = I_3$$

$$= \frac{V_L}{Z} = \frac{200\text{ V}}{200\ \Omega} = 1\text{ A}$$

Eq. 26-4,

$$I_L = \sqrt{3}\,I_p = \sqrt{3} \times 1\text{ A}$$

$$= 1.732\text{ A}$$

Example 26-8

Recalculate the line currents in the circuit of Figure 26-11 to determine the phase angle of each current with respect to phase voltage E_{AN}.

Solution

From Figure 26-3(b), E_{AB} leads E_{AN} by 30°

$$E_{AB} = \sqrt{3}\,E_{AN} = \sqrt{3} \times 115.5\text{ V}\angle 30^\circ$$

$$= 200\text{ V}\angle 30^\circ$$

and from Figure 26-3(c),

$$E_{BC} = 200\text{ V}\angle -90^\circ \text{ with respect to } E_{AN}$$

$$E_{CA} = 200\text{ V}\angle -210^\circ$$

$$Z_1 = Z_2 = Z_3 = 33.3\ \Omega + j197.2\ \Omega$$

$$= 200\ \Omega\angle 80.4^\circ$$

$$I_1 = \frac{E_{AB}}{Z_1} = \frac{200\text{ V}\angle 30^\circ}{200\ \Omega\angle 80.4^\circ} = 1\text{ A}\angle -50.4^\circ$$

$$= 0.637\text{ A} - j0.771\text{ A}$$

$$I_2 = \frac{E_{BC}}{Z_2} = \frac{200\text{ V}\angle -90^\circ}{200\ \Omega\angle 80.4^\circ} = 1\text{ A}\angle -170.4^\circ$$

$$= -0.986\text{ A} - j0.167\text{ A}$$

$$I_3 = \frac{E_{CA}}{Z_3} = \frac{200\text{ V}\angle -210^\circ}{200\ \Omega\angle 80.4^\circ} = 1\text{ A}\angle -290.4^\circ$$

$$= 0.349\text{ A} + j0.937\text{ A}$$

$$I_A = I_1 - I_3$$

$$= (0.637\text{ A} - j0.771\text{ A}) - (0.349\text{ A} + j0.973\text{ A})$$

$$= 0.288\text{ A} - j1.71\text{ A}$$

$$= 1.73\text{ A}\angle -80.4^\circ$$

$$
\begin{aligned}
I_B &= I_2 - I_1 \\
&= (-0.986\ \text{A} - j0.167\ \text{A}) - (0.637\ \text{A} - j0.771\ \text{A}) \\
&= -1.62\ \text{A} + j0.604\ \text{A} \\
&= \underline{1.73\ \text{A}\angle -200.4^\circ} \\
I_C &= I_3 - I_2 \\
&= (0.349\ \text{A} + j0.937\ \text{A}) - (-0.986\ \text{A} - j0.167\ \text{A}) \\
&= 1.34\ \text{A} + j1.1\ \text{A} \\
&= \underline{1.73\ \text{A}\angle -320.4^\circ}
\end{aligned}
$$

The phasor diagrams of voltages and currents for this circuit are drawn in Figure 26-12. Note that each line current is 80.4° behind each phase voltage.

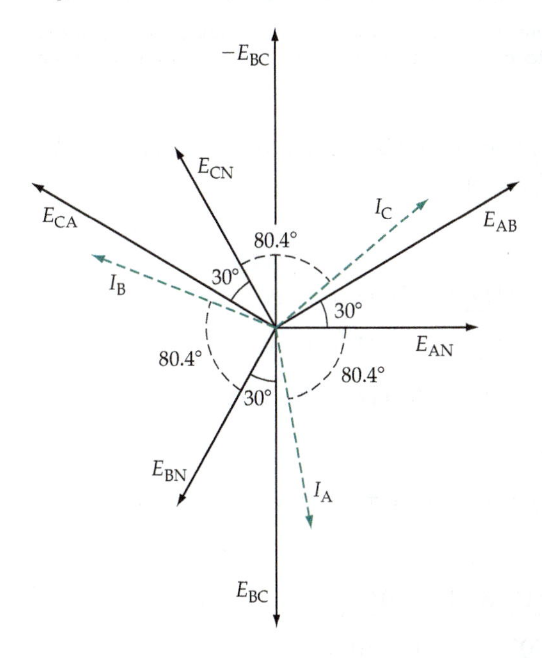

Figure 26-12 Phasor diagram of voltages and currents for a balanced Δ-connected load supplied from a Y-connected generator.

Y-Y System

Figure 26-13(a) shows a Y-connected load with a three-wire supply (no neutral conductor). To calculate the line currents for such a load, the Y-connected circuit must first be converted to its Δ-connected equivalent circuit, as in Figure 26-13(b). Equations 22-6, 22-7, and 22-8 for Y-to-Δ conversion are applied to determine the equivalent Δ-connected components. The phase currents for the Δ-connected load, I_1, I_2, and I_3 in Figure 26-13(b) are calculated next; then the line currents are determined as the phasor difference of the phase currents. The Y-connected load currents are the same as the line currents.

Example 26-9

An unbalanced Y-connected load has a three-wire supply, as in Figure 26-13(a). The load components are $R_1 = R_2 = R_3 = 50\ \Omega$, and $L_3 = 398$ mH. The supply

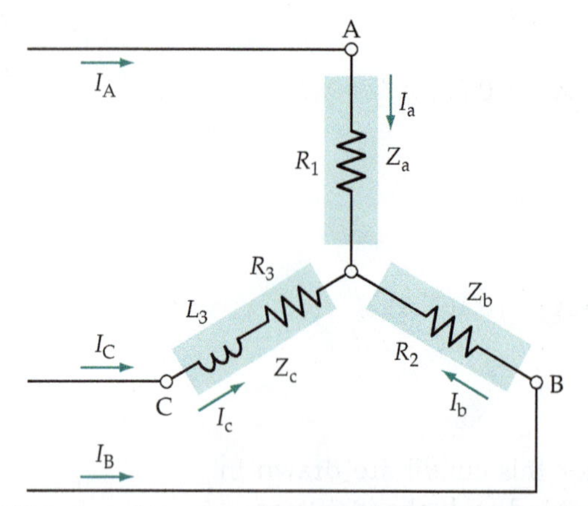

(a) Unbalanced Y-connected load with three-wire supply

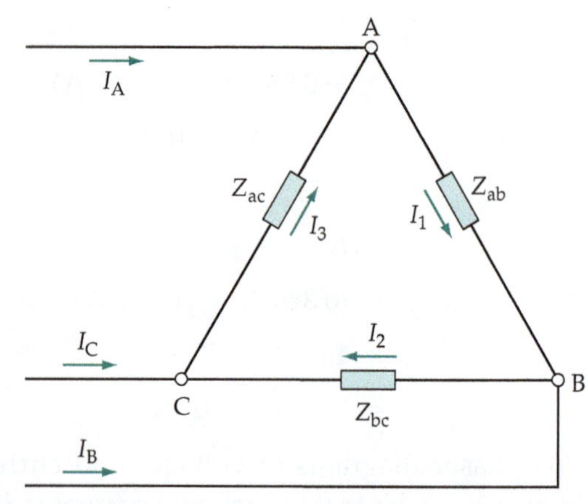

(b) Δ-connected equivalent circuit of Y-connected load

Figure 26-13 For analysis of a Y-connected load with a three-wire supply, the Y-Δ transformation equations should be used to convert the Y-connected load into an equivalent Δ-connected load.

frequency is 60 Hz, and the line voltage is 200 V. Calculate the line currents I_A, I_B, and I_C, and the load currents I_a, I_b, and I_c.

Solution

$$Z_a = R_1 = 50\ \Omega\angle 0 = 50\ \Omega + j0\ \text{A}$$

$$Z_b = R_2 = 50\ \Omega\angle 0 = 50\ \Omega + j0\ \text{A}$$

$$X_{L3} = 2\pi f L_3 = 2\pi \times 60\ \text{Hz} \times 398\ \text{mH}$$

$$= 150\ \Omega$$

$$Z_c = R_3 + jX_{L3} = 50\ \Omega + j150\ \Omega$$

$$= 158\ \Omega\angle 71.6^\circ$$

$$Z_aZ_b = 50\ \Omega\angle 0^\circ \times 50\ \Omega\angle 0^\circ$$

$$= 2500\angle 0^\circ = 2500 + j0\ \text{A}$$

$$Z_aZ_c = 50\ \Omega\angle 0^\circ \times 158\ \Omega\angle 71.6^\circ$$

$$= 7900\angle 71.6^\circ = 2494 + j7496$$

$$Z_bZ_c = 50\ \Omega\angle 0^\circ \times 158\ \Omega\angle 71.6^\circ$$

$$= 7900\angle 71.6^\circ = 2494 + j7496$$

$$Z_aZ_b + Z_aZ_c + Z_bZ_c = 7500 + j15\,000$$

$$\approx 16\,800\angle 63.4^\circ$$

Eq. 22-6,

$$Z_{ab} = \frac{Z_aZ_b + Z_aZ_c + Z_bZ_c}{Z_c} = \frac{16\,800\angle 63.4^\circ}{158\angle 71.6^\circ}$$

$$= 106\ \Omega\angle -8.2^\circ$$

Eq. 22-7,
$$Z_{ac} = \frac{Z_aZ_b + Z_aZ_c + Z_bZ_c}{Z_b} = \frac{16\,800\angle 63.4^\circ}{50\angle 0^\circ}$$
$$= 336\ \Omega\angle 63.4^\circ$$

Eq. 22-8,
$$Z_{bc} = \frac{Z_aZ_b + Z_aZ_c + Z_bZ_c}{Z_a} = \frac{16\,800\angle 63.4^\circ}{50\angle 0^\circ}$$
$$= 336\ \Omega\angle 63.4^\circ$$

$$I_1 = \frac{E_{AB}}{Z_{ab}} = \frac{200\ \text{V}}{106\ \Omega\angle -8.2^\circ}$$
$$= 1.88\ \text{A}\angle 8.2^\circ = 1.86\ \text{A} + j0.268\ \text{A}$$

$$I_2 = \frac{E_{BC}}{Z_{bc}} = \frac{200\ \text{V}\angle -120^\circ}{336\ \Omega\angle 63.4^\circ}$$
$$= 0.595\ \text{A}\angle -183.4^\circ = -594\ \text{mA} + j35.3\ \text{mA}$$

$$I_3 = \frac{E_{CA}}{Z_{ac}} = \frac{200\ \text{V}\angle -240^\circ}{336\ \Omega\angle 63.4^\circ}$$
$$= 0.595\ \text{A}\angle -303.4^\circ = 328\ \text{mA} + j497\ \text{mA}$$

$$I_A = I_1 - I_3 = 1.53\ \text{A} - j229\ \text{mA}$$
$$= 1.55\ \text{A}\angle -8.5^\circ$$

$$I_B = I_2 - I_1 = -2.45\ \text{A} - j233\ \text{mA}$$
$$= 2.46\ \text{A}\angle -174.6^\circ$$

$$I_C = I_3 - I_2 = 922\ \text{mA} + j461\ \text{mA}$$
$$= 1.03\ \text{A}\angle 26.6^\circ$$

$$I_a = I_A = 1.55\ \text{A}\angle -8.5^\circ$$
$$I_b = I_B = 2.46\ \text{A}\angle -174.6^\circ$$
$$I_c = I_C = 1.03\ \text{A}\angle 26.6^\circ$$

Practice Problems

26-4.1 A balanced Δ-connected load is supplied from a Y-connected generator, as in Figure 26-11. The generator phase voltage and phase current are: 115.5 V and 3.46 A. Calculate the load impedances.

26-4.2 A Y-connected load, as in Figure 26-13(a) has a three-wire supply with a 150 V line voltage. The loads are $R_a = 86.6\ \Omega$, $R_b = 173.2\ \Omega$, and $R_c = 43.3\ \Omega$. Determine the load currents.

26-5 PHASE SEQUENCE

Correct Phase Sequence

In Section 26-1 it was explained that in a three-phase system the three output waveforms are generated in a particular sequence. Refer to the waveforms in Figure 26-1(c) again, and recall that voltage e_{Bb} reaches its peak 120° or one-third of a cycle after the peak of e_{Aa} occurs. Also, e_{Cc} peaks 240° after e_{Aa} and 120° after E_{Bb}. The phase sequence of the voltages can be listed as e_{Aa} e_{Bb} e_{Cc}, or *ABC*. The phasor diagram in Figure 26-1(d) is drawn for a phase sequence of *ABC*. Because phasors are thought of as rotating in a counterclockwise direction, those illustrated would pass a stationary point with the sequence *ABC*. Reading the phasors in a clockwise direction is the same as thinking of them rotating in a counterclockwise direction. Either way, the phase sequence is *ABC*, and it can also be correctly stated as *BCA* or *CAB*, as the actual repeating sequence is *ABCABCABC*, and so on.

Figure 26-14(a) shows a Y-connected generator together with its voltage wave forms and phasor diagram. The sequence of the phase voltages is *ABC*, and the line voltages E_{AB}, E_{BC}, and E_{CA} also have the phase sequence *ABC*. This is easily verified from the phasor diagram by reading either the first or second of the double subscripts clockwise around the diagram. The first subscripts read *ABC*, and the second read *BCA*, which means that both give the phase sequence as *ABC*. The supply lines from the generator in Figure 26-14(a) are identified as *line A*, *line B*, and *line C*, and they are connected respectively to generator terminals *A*, *B*, and *C*.

Reversed Phase Sequence

Now refer to Figure 26-14(b), which shows a generator with the terminals incorrectly connected. The supply line identified as line *A* is connected to terminal C, and line C is connected to terminal *A*. This change in terminal connections has a significant effect on the waveforms and on the phasor diagram. The voltage measured between line *A* and neutral is identified as E_{AN} (e_{AN} in the waveforms), but is actually the output of terminal C, which reaches its peak 120° behind E_{BN}. Similarly, the voltage measured between line *C* and neutral is now 120° ahead of E_{BN}, because it is actual E_{AN} (the output of terminal *A*). So, the waveforms in Figure 26-14(b) show e_{CN} leading e_{BN} by 120°, and e_{AN} lagging e_{BN} by 120°. The phase sequence is *CBA*. This is, in fact, a reversal of the phase sequence *ABC*.

The phase sequence reversal is also apparent in the phasor diagram in Figure 26-14(b). Comparing the phase voltages to those in Figure 26-14(a), it is seen that E_{CN} (the output from terminals *N* and *C*) is now 120° behind E_{AN}, while previously it was 240° behind. Also, E_{BN} is now 240° behind E_{AN}, instead of 120°, as before.

The line voltage phasor E_{AB} in Figure 26-14(b), which is the phasor difference of E_{AN} and E_{BN}, is seen to be 30° *lagging* E_{AN}. The phasor diagram in

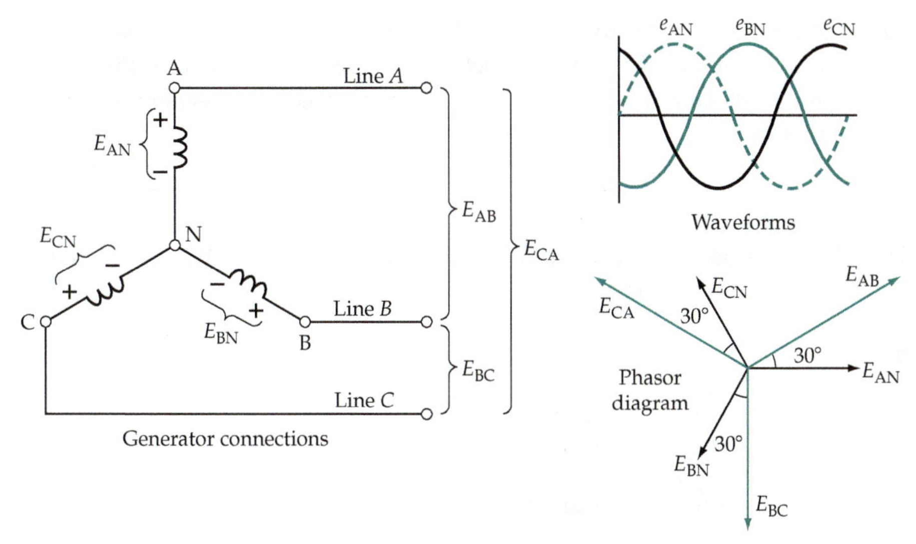

(a) Generator connections, waveforms, and phasor diagram for phase sequence *ABC*

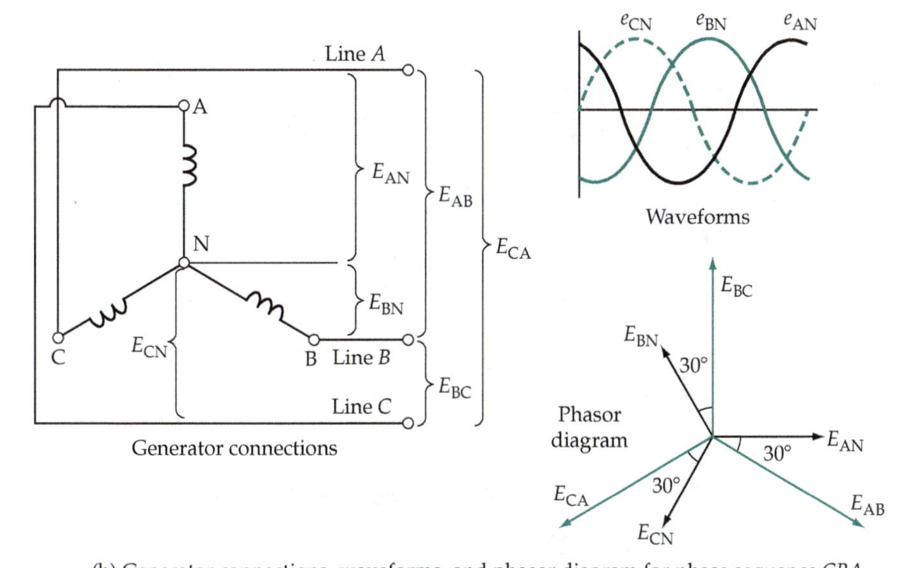

(b) Generator connections, waveforms, and phasor diagram for phase sequence *CBA*

Figure 26-14 When two line connections are interchanged on a three-phase generator, the phase sequence is reversed. This can reverse the direction of three-phase induction motors and have serious consequences for unbalanced loads.

Figure 26-14(a) shows that before the terminal connection change, E_{AB} led E_{AN} by 30°. Also, in Figure 26-14(b), E_{CA} and E_{BC} are now reversed in their phase relationships with respect to E_{AB}.

Phase Sequence Effects

Where balanced resistive loads are involved, the phase sequence is normally not important. However, with three-phase induction motors, any change in the phase sequence of a supply can reverse the direction of rotation. An induction motor is usually required to rotate in a particular direction. When initially connected, its direction of rotation can be checked and corrected by interchanging any two line connections to the motor. If the phase sequence of the supply is subsequently altered, the motor direction will be reversed again. Phase sequence reversal can also seriously affect an unbalanced load, as demonstrated in Example 26-10.

Example 26-10

The *ABC* phase sequence used in Example 26-9 is reversed to *ACB*. Recalculate the load currents I_a, I_b, and I_c.

Solution

$$I_1 = \frac{E_{AB}}{Z_{ab}} = \frac{200\text{ V}}{106\ \Omega\angle -8.2^\circ} = 1.88\text{ A}\angle 8.2^\circ$$

$$= 1.86\text{ A} + j0.268\text{ A}$$

$$I_2 = \frac{E_{BC}}{Z_{bc}} = \frac{200\text{ V}\angle -240^\circ}{336\ \Omega\angle 63.4^\circ} = 0.595\text{ A}\angle -303.4^\circ$$

$$= 328\text{ mA} + j497\text{ mA}$$

$$I_3 = \frac{E_{CA}}{Z_{ac}} = \frac{200\text{ V}\angle -120^\circ}{336\ \Omega\angle 63.4^\circ} = 0.595\text{ A}\angle -183.4^\circ$$

$$= -594\text{ mA} + j35.3\text{ mA}$$

$$I_a = I_A = I_1 - I_3$$

$$= (1.86\text{ A} + j0.268\text{ A}) - (-594\text{ mA} + j35.3\text{ mA})$$

$$= 2.45\text{ A} + j233\text{ mA}$$

$$= 2.46\text{ A}\angle 5.4^\circ$$

$$I_b = I_B = I_2 - I_1$$

$$= (328\text{ mA} + j497\text{ mA}) - (1.86\text{ A} + j0.268\text{ A})$$

$$= -1.53\text{ A} + j229\text{ mA}$$

$$= 1.55\text{ A}\angle -188.5^\circ$$

$$I_c = I_C = I_3 - I_2$$

$$= (-594\text{ mA} + j35.3\text{ mA}) - (328\text{ mA} + j497\text{ mA})$$

$$= -922\text{ mA} - j462\text{ mA}$$

$$= 1.03\angle -153.4^\circ$$

Phase Sequence Tester

Comparing the results of Example 26-10 to those obtained in Example 26-9, it is seen that the phase sequence reversal caused I_a to change from 1.55 A

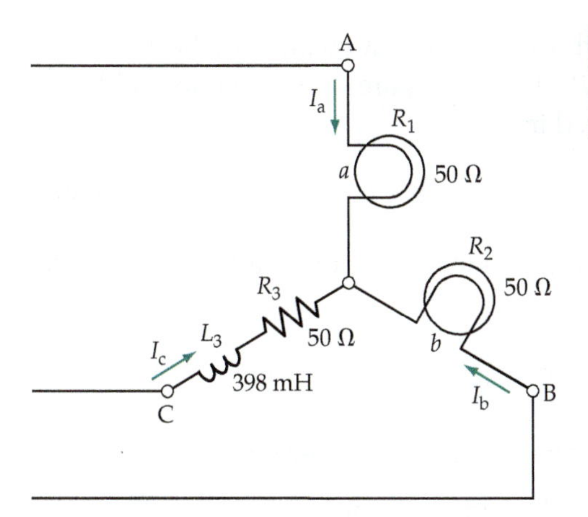

Figure 26-15 Phase sequence tester. With sequence *ABC*, lamp *b* is brighter than lamp *a*. When the sequence is *CBA*, lamp *a* is brightest.

to 2.46 A, I_b to change from 2.46 A to 1.55 A, and I_c to remain 1.03 A. These results can be employed to demonstrate the operation of a *phase sequence tester*.

The circuit of Figure 26-13(a) is redrawn in Figure 26-15 with resistors R_1 and R_2 replaced by lamps *a* and *b*. Each lamp has a filament resistance of 50 Ω, as for R_1 and R_3. So, the circuit is essentially the same as in Figure 26-13(a), and the analysis performed in Examples 26-9 and 26-10 applies. In Example 26-9, with phase sequence *ABC* it was found that $I_a = 1.55$ A and $I_b = 2.46$ A. The larger current through lamp *b* would cause it to glow more brightly than lamp *a*. When the phase sequence is changed to *CBA* (Example 26-10), I_a becomes 2.46 A and I_b is reduced to 1.55 A. Now lamp *a* is brighter than lamp *b*. Thus, the circuit in Figure 26-15 is a phase sequence tester. With phase sequence *ABC*, lamp *b* is brighter than lamp *a*. With sequence *CBA*, lamp *a* is brightest.

Practice Problem

26-5.1 Recalculate the load currents in the circuit in Problem 26-4.2 when the phase sequence is reversed.

26-6 POWER IN THREE-PHASE SYSTEMS

True Power

The power dissipated in each of the three loads or branches in a three-phase system is

$$P_p = V_p I_p \cos\phi \text{(watts)} \qquad (26\text{-}5)$$

where V_p and I_p are the phase voltage and current, and ϕ is the phase angle between V_p and I_p.

In the case of an unbalanced load, whether Y- or Δ-connected, the individual phase (load) power dissipations, P_1, P_2, and P_3 are calculated and added to determine the total power dissipated in the load.

$$P = P_1 + P_2 + P_3 \tag{26-6}$$

For a balanced load, the power dissipation is the same in all three branches. So, from Equation 26-5

$$P = 3V_pI_p \cos \phi \tag{26-7}$$

In the case of a Y-connected load, Equations 26-1 and 26-2 give

$$V_p = V_L/\sqrt{3} \quad \text{and} \quad I_p = I_L$$

For a Δ-connected load, Equations 26-3 and 26-4 give

$$V_p = V_L \quad \text{and} \quad I_p = I_L/\sqrt{3}$$

So, for both Y- and Δ-connected balanced loads, the total power is,

$$P = \frac{3V_LI_L}{\sqrt{3}} \cos \phi$$

or

$$P = \sqrt{3}\, V_LI_L \cos \phi \tag{26-8}$$

Equation 26-8 allows the total power (in watts) to be calculated from the measured line current and line voltage. Note that ϕ is still the phase angle between V_p and I_p, which is the phase angle of each individual load. It is *not* the phase angle between the line quantities V_L and I_L.

Reactive Power

The reactive power per phase (from Equation 21-8) is

$$Q_p = V_pI_p \sin \phi$$

or, total reactive power with a balanced load is

$$Q = 3V_pI_p \sin \phi$$

Substituting the line voltage and current into the equation, as before

$$Q = \sqrt{3}\, V_LI_L \sin \phi \tag{26-9}$$

Equation 26-9 gives the total reactive power (in var) for a balanced load. Once again, ϕ is the load phase angle.

Apparent Power

The apparent power per phase (Equation 21-9) is calculated as

$$S = V_pI_p \text{ (volt amp)}$$

and the total apparent power with a balanced load is

$$S = 3V_pI_p$$

Once again the line voltage and current may be substituted into the equation:

$$S = \sqrt{3}\,V_LI_L \tag{26-10}$$

Dividing Equation 26-8 by Equation 26-10,

$$\frac{P}{S} = \frac{\sqrt{3}\,V_LI_L\cos\phi}{\sqrt{3}\,V_LI_L}$$

Giving

$$\cos\phi = P/S \tag{26-11}$$

Example 26-11

Calculate the power dissipated in the load shown in Figure 26-5. As for Example 26-3, the generator phase voltage is $V_p = 100$ V, and its frequency is 60 Hz. The load components are $R_1 = R_2 = R_3 = 100\ \Omega$, $C_2 = 66.3\ \mu$F, and $L_3 = 159.2$ mH.

Solution

From Example 26-3,

$$Z_1 = 100\ \Omega\angle 0°,\ Z_2 = 107.7\ \Omega\angle -21.8°,\ Z_3 = 166.6\ \Omega\angle 31°,$$

$$I_1 = 1\text{ A},\ I_2 = 0.929\text{ A},\ I_3 = 0.858\text{ A}$$

Eq. 26-5, $P_1 = V_{p1}I_{p1}\cos\phi_1 = 100\text{ V} \times 1\text{ A} \times \cos 0$

$= 100$ W

$P_2 = V_{p2}I_{p2}\cos\phi_2 = 100\text{ V} \times 0.929\text{ A} \times \cos 21.8°$

≈ 86.3 W

$P_3 = V_{p3}I_{p3}\cos\phi_3 = 100\text{ V} \times 0.858\text{ A} \times \cos 31°$

$= 73.5$ W

Eq. 26-6, $P_T = P_1 + P_2 + P_3 = 100\text{ W} + 86.3\text{ W} + 73.5\text{ W}$

$= 259.8$ W

Example 26-12

For the balanced load shown in Figure 26-11, calculate the total power dissipated, the reactive power, and the apparent power. The circuit quantities are as listed in Examples 26-7 and 26-8.

Solution

From Examples 26-7 and 26-8:

$$V_L = 200\text{ V},\ I_L = 1.732\text{ A, and } \phi = 80.4°$$

Eq. 22-8,
$$P = \sqrt{3}\,V_L I_L \cos\phi$$
$$= \sqrt{3} \times 200\text{ V} \times 1.732\text{ A} \times \cos 80.4°$$
$$= 100\text{ W}$$

Eq. 26-9,
$$Q = \sqrt{3}\,V_L I_L \sin\phi$$
$$= \sqrt{3} \times 200\text{ V} \times 1.732\text{ A} \times \sin 80.4°$$
$$= 591.6\text{ var}$$

Eq. 26-10,
$$S = \sqrt{3}\,V_L I_L$$
$$= \sqrt{3} \times 200\text{ V} \times 1.732\text{ A}$$
$$= 600\text{ VA}$$

Practice Problems

26-6.1 Calculate the power dissipated in the circuit described in Problem 26-3.3.

26-6.2 Determine the apparent power, reactive power, and true power for a Y-connected load consisting of $Z_a = Z_b = Z_c = 47\ \Omega\angle 45°$. The line voltage is 122 V.

26-7 POWER FACTOR CORRECTION

In Section 21-6, it is explained that if the power factor ($\cos\phi$) of a circuit can be improved, or corrected, the current carried by the conductors can be reduced without any reduction of the power dissipated in the load. The discussion in Section 21-6 referred to a single-phase circuit, but it is also applicable to three-phase circuits.

Figure 26-16(a) shows a Y-connected circuit consisting of resistors and inductors, in parallel with a Δ-connected capacitor circuit. The *RL* circuit could represent an induction motor or other inductive circuit. Operating without the capacitors, the *RL* circuit would tend to have a large lagging power factor. Figure 26-16(b) shows the power triangle for the *RL* circuit. When the capacitors are included in the circuit, the capacitive reactive volt-amp (Q_C) subtracts from the inductive reactive volt-amp (Q_L) to give a reduced phase angle [Figure 26-16(c)]. As in the case of single-phase circuits, this results in an increased power factor and reduced line current. Example 26-13 demonstrates the effect of the capacitors.

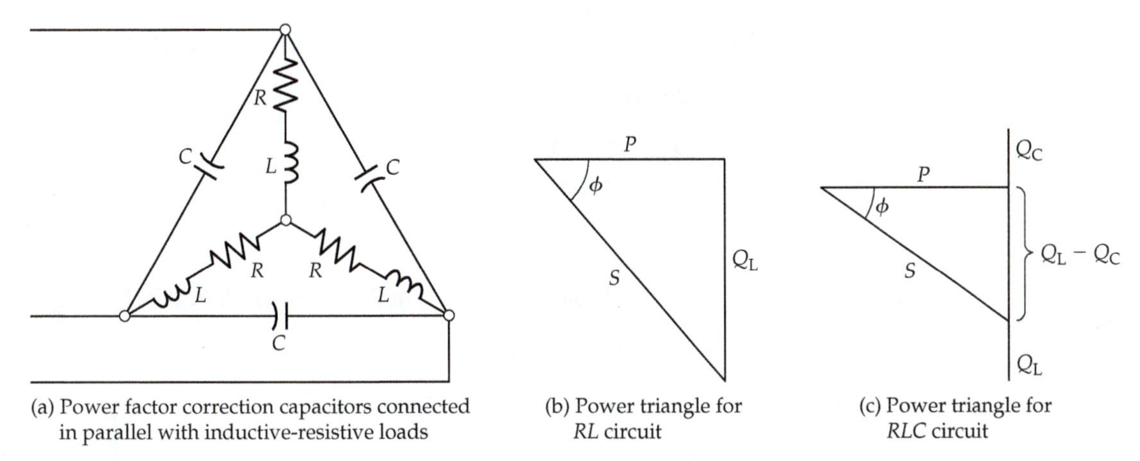

(a) Power factor correction capacitors connected in parallel with inductive-resistive loads

(b) Power triangle for *RL* circuit

(c) Power triangle for *RLC* circuit

Figure 26-16 Power factor correction capacitors for three-phase loads are usually Δ-connected. The effect of the capacitors is to increase the power factor and thus reduce the line current.

Instead of Δ-connecting the capacitors, a Y-connected arrangement could be used. However, it is found that larger capacitance values are required for the Y-connected circuit.

Example 26-13

The circuit shown in Figure 26-16(a) has $L = 530$ mH $R = 200\ \Omega$, $C = 1\ \mu$F, $V_L = 400$ V, and $f = 60$ Hz. Calculate P, I_L, and $\cos\phi$ with (a) only the inductors and resistors in the circuit, (b) only the capacitors in the circuit, and (c) all components present.

Solution

(a) Inductor and resistor circuit.

$$X_L = 2\pi f L = 2\pi \times 60\ \text{Hz} \times 530\ \text{mH}$$
$$= 200\ \Omega$$
$$Z = \sqrt{R^2 + X_L^2} = \sqrt{(200\ \Omega)^2 + (200\ \Omega)^2}$$
$$\approx 283\ \Omega$$
$$\cos\phi = \frac{R}{Z} = \frac{200\ \Omega}{283\ \Omega}$$
$$= 0.707$$

Eq. 26-1,
$$V_P = \frac{V_L}{\sqrt{3}} = \frac{400\ \text{V}}{\sqrt{3}}$$
$$\approx 231\ \text{V}$$
$$I_L = I_P = \frac{V_P}{Z} = \frac{231\ \text{V}}{283\ \Omega}$$
$$= 816\ \text{mA}$$

Eq. 26-8, $P_L = \sqrt{3}\,V_L I_L \cos\phi = \sqrt{3} \times 400\text{ V} \times 816\text{ mA} \times 0.707$

$= 400\text{ W}$

$$\sin\phi = \frac{X_L}{Z} = \frac{200\ \Omega}{238\ \Omega}$$

$= 0.707$

Eq. 26-9, $Q_L = \sqrt{3}\,V_L I_L \sin\phi = \sqrt{3} \times 400\text{ V} \times 816\text{ mA} \times 0.707$

$= 400\text{ var (inductive)}$

(b) Capacitor circuit.

$$X_C = \frac{1}{2\pi f C} = \frac{1}{2\pi \times 60\text{ Hz} \times 1\ \mu\text{F}}$$

$= 2.65\text{ k}\Omega$

$$I_p = \frac{V_L}{X_C} = \frac{400\text{ V}}{2.65\text{ k}\Omega}$$

$= 151\text{ mA}$

$$I_L = \sqrt{3}I_p = \sqrt{3} \times 151\text{ mA}$$

$= 261\text{ mA}$

$$\cos\phi = \frac{R}{Z} = \frac{0}{2.65\text{ k}\Omega}$$

$= 0$

Eq. 26-8, $P_C = \sqrt{3}\,V_L I_L \cos\phi = \sqrt{3}\,V_L I_L \times 0$

$= 0$

$$\sin\phi = \frac{X_L}{Z} = \frac{2.65\text{ k}\Omega}{2.65\text{ k}\Omega}$$

$= 1$

$$Q = \sqrt{3}\,V_L I_L \sin\phi = \sqrt{3} \times 400\text{ V} \times 216\text{ mA} \times 1$$

$= 181\text{ var (capacitive)}$

(c) All components in the circuit.

$$P = P_L + P_C = 400\text{ W} + 0$$

$= 400\text{ W}$

$$Q = Q_L - Q_C = 400\text{ var} - 181\text{ var}$$

$= 219\text{ var (inductive)}$

$$\phi = \tan^{-1}\left(\frac{Q}{P}\right) = \tan^{-1}\left(\frac{219}{400}\right)$$

$$= 28.7°$$

$$\cos\phi = \cos 28.7° = 0.877$$

From Eq. 26-8,
$$I_L = \frac{P}{\sqrt{3}\,V_L \cos\phi} = \frac{400\text{ W}}{\sqrt{3} \times 400\text{ V} \times 0.877}$$

$$= 658\text{ mA}$$

It is seen that the capacitors increase the power factor from 0.707 to 0.877, and this produced a reduction in line current from 816 mA to 658 mA without any reduction in load power dissipation.

Practice Problem

26-7.1 For the circuit in Problem 26-6.2, determine the size of power factor correction capacitors [connected as in Figure 26-16(a)] that will reduce the line current by 10%. The supply frequency is 60 Hz.

26-8 THREE-PHASE POWER MEASUREMENT

Wattmeter Use

Several methods are available for measuring the power delivered to a three-phase load. These depend on whether the load is Y- or Δ-connected, the accessibility of the load terminals, and whether the load is balanced or unbalanced. As with all instruments, some care must be taken to connect the wattmeter correctly. The current coil terminal identified as ± should be connected to the voltage source. (Sometimes a ↓ or * is used to identify this terminal.) The ± voltage coil terminal is connected to the load terminal of the current coil. If the instrument is not connected correctly, a negative deflection may occur.

Three-Wattmeter Methods

Figure 26-17 shows three wattmeters employed to measure the power in a Y-connected four-wire system with an unbalanced load. Each wattmeter measures the power dissipated in one phase of the load ($V_P I_P \cos\phi$), and the total power is then the sum of the three meter readings. When the load is balanced, all three meters in Figure 26-17 would read the same. So, only one wattmeter is needed, and the total power is three times the meter reading.

In Figure 26-18, a three-wattmeter arrangement is shown for measuring the power in a Δ-connected load. Here again, each meter indicates ($V_P I_P \cos\phi$) for each of the three load sections. The total power dissipated is the sum of the three

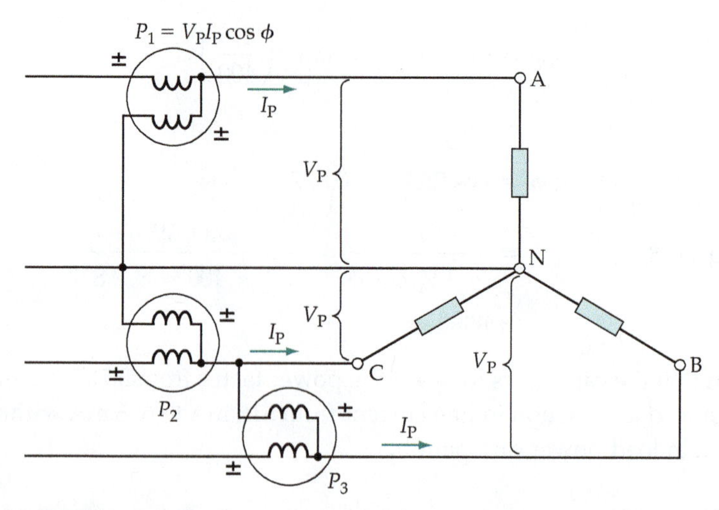

Figure 26-17 Three wattmeters connected to measure the power delivered to each section of a three-phase, four-wire load. The total load power is the sum of the three readings. For a balanced load, only one meter is required, and the load power is three times the meter reading.

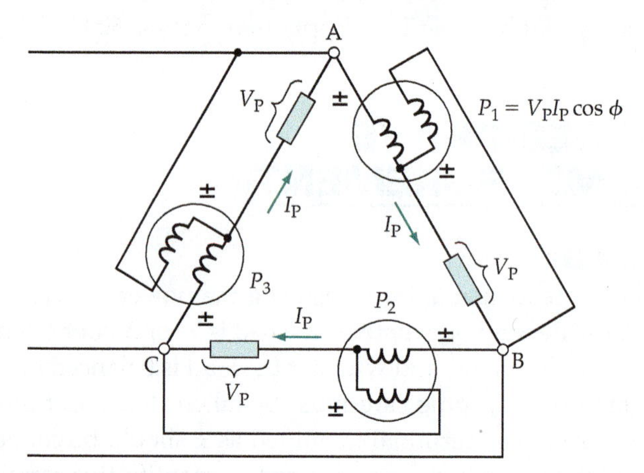

Figure 26-18 Three wattmeters connected to measure the power delivered to each section of a three-phase Δ-connected load. The total load power is the sum of the three readings. For a balanced load, only one meter is required.

meter readings. Only one wattmeter is required for a balanced system, and the measured quantity is then multiplied by three to determine the total power.

Single-Wattmeter Method

In a Y-connected three-wire system and in a Δ-connected system, no neutral conductor is available for connecting the common point of the wattmeter voltage coils. Figure 26-19 shows a circuit in which three equal resistors are Y-connected to provide a neutral point. The single wattmeter now measures the power (P_1) dissipated in one branch of the load, and so the total power is $P = 3P_1$. This circuit is suitable only for a balanced load.

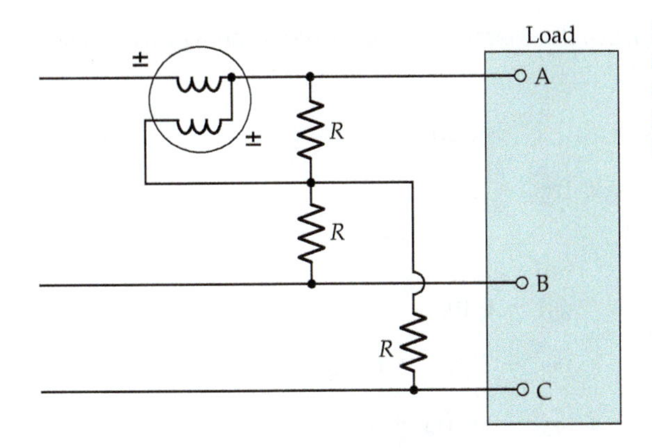

Figure 26-19 Single-wattmeter method of measuring the power delivered to a balanced three-wire load. The load power is three times the meter reading.

Two-Wattmeter Method

The most commonly used method of measuring power in a three-phase load is the *two-wattmeter method*. The circuit for this method is shown in Figure 26-20. The wattmeter current coils are each connected in series with one line, and the voltage coils are commoned at the remaining line, as illustrated. If the load is Y-connected, the instantaneous power dissipations in each phase can be written in terms of the instantaneous phase current and instantaneous phase voltage.

$$P_{\text{A}} = (\text{instantaneous phase current}) \times (\text{instantaneous phase voltage})$$

$$= [\text{line current } (i_{\text{A}})] \times [\text{line-to-neutral voltage } (v_{\text{AN}})]$$

$$= (i_{\text{A}})(v_{\text{AN}}) = i_{\text{A}}v_{\text{Aa}}$$

$$P_{\text{B}} = i_{\text{B}}v_{\text{BN}} = i_{\text{B}}v_{\text{Bb}}$$

$$P_{\text{C}} = i_{\text{C}}v_{\text{CN}} = i_{\text{C}}v_{\text{Cc}}$$

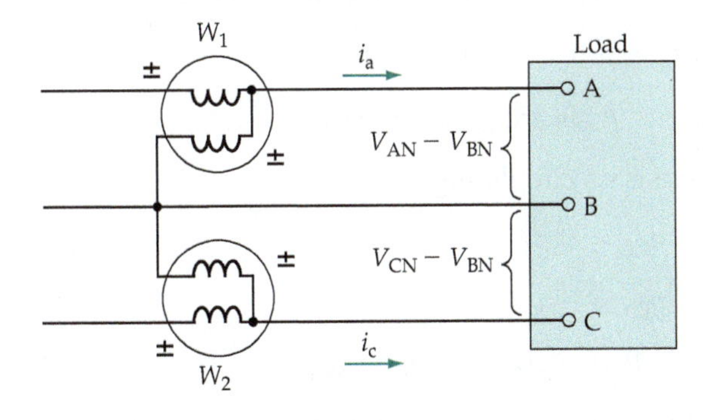

Figure 26-20 The two-wattmeter method of three-phase power measurement is most often used. It is suitable for Y-connected or Δ-connected, balanced or unbalanced loads. The load power is the sum of the meter readings.

The total instantaneous power is,

$$P = i_{\text{A}}v_{\text{Aa}} + i_{\text{B}}v_{\text{Bb}} + i_{\text{C}}v_{\text{Cc}} \tag{1}$$

The instantaneous current in meter W_1 is i_{A} and the instantaneous voltage across the voltage coil of W_1 is $(V_{\text{AN}} - V_{\text{BN}})$. Similarly, the instantaneous quantities for W_2 are i_c and $(V_{CN} - V_{BN})$. So, the instantaneous powers measured by each of these instruments are,

$$P_1 = i_A(v_{Aa} - v_{Bb})$$

And
$$P_2 = i_C(v_{Cc} - v_{Bb})$$

The sum of P_1 and P_2 is,

$$\begin{aligned} P &= i_A(v_{Aa} - v_{Bb}) + i_C(v_{Cc} - v_{Bb}) \\ &= i_A v_{Aa} - i_A v_{Bb} + i_C v_{Cc} - i_C v_{Bb} \\ &= i_A v_{Aa} + i_C v_{Cc} - v_{Bb}(i_A + i_C) \end{aligned} \quad \textbf{(2)}$$

With balanced or unbalanced loads, the algebraic sum of the currents at the neutral point of a Y-connected load must be zero (for a three-wire supply).

So,
$$i_A + i_B + i_C = 0$$

or
$$i_A + i_C = -i_B \quad \textbf{(3)}$$

Substituting from Equation (3) into Equation (2),

$$P = i_A v_{Aa} + i_C v_{Cc} + i_B v_{Bb} \quad \textbf{(4)}$$

Equation (4) is identical to Equation (1), showing that the sum of the instantaneous powers measured by the two wattmeters is equal to the total instantaneous power dissipated in the load. A wattmeter indicates the average of the instantaneous power dissipations in a load. Consequently, the sum of the two wattmeter readings is the average power dissipated in the three-phase load.

It can also be shown that for a Δ-connected load the two wattmeters indicate the total power dissipated. In this case the wattmeter current coils carry instantaneous currents of $(i_A - i_c)$ and $(i_c - i_B)$. The voltage coil instantaneous potential differences are $-v_{Aa}$ and $-v_{Bb}$. The sum of the meter readings is:

$$\begin{aligned} P &= v_{Aa}(i_A - i_C) - v_{Bb}(i_C - i_B) \\ &= v_{Aa} i_A + v_{Bb} i_B - i_C(v_{Aa} + v_{Bb}) \end{aligned}$$

For a Δ-connected system, there is no circulating current, and

$$v_{Aa} + v_{Bb} + v_{Cc} = 0$$

or
$$v_{Aa} + v_{Bb} = -v_{Cc}$$

Giving,
$$P = v_{Aa} i_A + v_{Bb} i_B + v_{Cc} i_C \quad \textbf{(5)}$$

Equation (5) is, once again, the same as Equation (1), showing that the sum of the meter readings gives the total load power. It is seen that the two-wattmeter method measures the power delivered to a three-phase load, whether the load is Y- or Δ-connected, balanced or unbalanced.

It should be noted that the phase angle between the voltage and current applied to the wattmeter may be greater than 90°. In this case $\cos\phi$ is a negative quantity, and the wattmeter reading will be a negative quantity. In order to obtain a positive deflection the terminal connections of the wattmeter voltage (or current) coil should be reversed. The meter reading must then be recorded as a negative quantity. Example 26-14 demonstrates this situation.

Example 26-14

The load described in Example 26-7 has its power dissipation measured by the two-wattmeter method illustrated in Figure 26-20. Calculate the load power and the power indicated by each wattmeter:

Solution

From Examples 26-7 and 26-8,

$$V_L = 200\text{ V}, I_p = 1\text{ A}, I_L = 1.73\text{ A, and } \phi = 80.4°$$

Power dissipation in each Δ-connected load,

$$P_p = V_p I_p \cos\phi = 200\text{ V} \times 1\text{ A} \times \cos 80.4°$$
$$= 33.35\text{ W}$$

Total load power,

$$P = 3P_p = 3 \times 33.35\text{ W}$$
$$= 100\text{ W}$$

Alternatively,

$$P = \sqrt{3}\, V_L I_L \cos\phi$$
$$= \sqrt{3} \times 200\text{ V} \times 1.73\text{ A} \times \cos 80.4°$$
$$= 100\text{ W}$$

Wattmeter readings,

$$P_1 = E_{AB} I_A \cos\phi_A$$

where ϕ_A is the phase angle between E_{AB} and I_A (see Figure 26-20).

From Figure 26-12,

$$\phi_A = (80.4° + 30°)$$

So,

$$P_1 = 200\text{ V} \times 1.73\text{ A} \times \cos(80.4° + 30°)$$
$$= -120.6\text{ W}$$

As noted, the meter terminal connections must be reversed in this case to obtain a positive reading,

$$P_2 = E_{CB} I_C \cos\phi_C$$

where ϕ_C is the phase angle between E_{CB} and I_C (see Figure 26-20).

$$E_{CB} = -E_{BC}$$

So, from Figure 26-12,

$$\phi_C = (80.4° - 30°)$$

and

$$P_2 = 200\text{ V} \times 1.73\text{ A} \times \cos(80.4° - 30°)$$
$$= 220.5\text{ W}$$

$$P_1 + P_2 = -120.6\text{ W} + 220.5\text{ W}$$
$$\approx 100\text{ W}$$

Power Factor Determination

When the two-wattmeter method is employed to measure the power dissipated in a balanced load, the power factor of the circuit can be calculated from the meter readings. Consider the phasor diagram for a balanced Y-connected load, as shown in Figure 26-21. The phase voltages E_{AN}, E_{BN}, and E_{CN} have 120° phase differences, and line voltages E_{AB}, E_{BC}, and E_{CA} are 30° ahead of E_{AN}, E_{BN}, and E_{CN}, respectively. (These phasors are explained in Section 26-2.) The line currents I_A, I_B, and I_C (which are also the phase currents in a Y-connected load) each lag the related phase voltage by phase angle ϕ. Now look at the circuit diagram in Figure 26-20 again. The current coil of wattmeter W_1 carries line current I_A, and the potential difference across its voltage coil is line voltage E_{AB}. The phase difference between E_{AB} and I_A is $(30° + \phi)$ (see Figure 26-21). Consequently, the power indicated by W_1 is

$$P_1 = E_{AB} I_A \cos(30° + \phi)$$

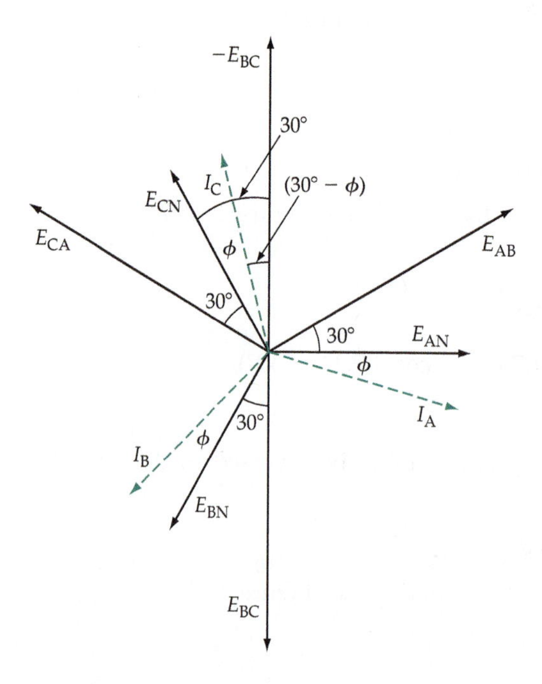

Figure 26-21 Phasor diagram for a two-wattmeter power measurement method with a balanced Y- connected load. The load power factor can also be determined from the meter readings.

Because E_{AB} is line voltage V_L, and I_A is line current I_L,

$$P_1 = V_L I_L \cos(30° + \phi) \quad \textbf{(26-12)}$$

In Figure 26-20, the current flowing in the current coil of wattmeter W_2 is line current I_C. The potential difference across its voltage coil is $-E_{BC}$. If the voltage coil of W_2 had been connected with its ± terminal to line B, the voltage would be $+E_{BC}$. But when connected as illustrated, the voltage coil potential difference is $-E_{BC}$. As shown in the phasor diagram in Figure 26-21, there is a 30° difference between $-E_{BC}$ and E_{CN}. Because I_C lags E_{CN} by angle ϕ, the phase difference between $-E_{BC}$ and I_C is $(30° - \phi)$. The power indicated by W_2 can now be written as

$$P_2 = E_{BC} I_C \cos(30° - \phi)$$

or

$$P_2 = V_L I_L \cos(30° - \phi) \quad \textbf{(26-13)}$$

$$\begin{aligned} P_2 + P_1 &= V_L I_L [\cos(30° - \phi) + \cos(30° + \phi)] \\ &= V_L I_L (2 \cos 30° \cos \phi) \\ &= \sqrt{3}\, V_L I_L \cos \phi \end{aligned}$$

and

$$\begin{aligned} P_2 - P_1 &= V_L I_L [\cos(30° - \phi) - \cos(30° + \phi)] \\ &= V_L I_L \sin \phi \end{aligned}$$

$$\frac{P_2 - P_1}{P_2 + P_1} = \frac{V_L I_L \sin \phi}{\sqrt{3}\, V_L I_L \cos \phi} = \frac{1}{\sqrt{3}} \tan \phi$$

Giving,

$$\phi = \tan^{-1} \sqrt{3} \left[\frac{P_2 - P_1}{P_2 + P_1} \right] \quad \textbf{(26-14)}$$

In deriving Equation 26-14, it was assumed that ϕ is less than 60°. When ϕ is greater than 60° $(30° + \phi)$ exceeds 90°, and cos $(30° + \phi)$ becomes a negative quantity, making the reading on wattmeter W_1 negative. As noted, to get a positive indication, the connections to either the current coil or the voltage coil must be reversed, and the power measured by the wattmeter must be recorded as a negative quantity.

The two-wattmeter method does not indicate whether the calculated power factor is leading or lagging. This must be determined by considering the load. A largely inductive load has a lagging power factor, and a predominantly capacitive load has a leading power factor.

Example 26-15

Calculate the load power factor from the wattmeter readings in Example 26-14.

Solution

Eq. 26-14,
$$\phi = \tan^{-1}\sqrt{3}\left[\frac{P_2 - P_1}{P_2 + P_1}\right]$$
$$= \tan^{-1}\left[\sqrt{3}\left(\frac{220.5\ \text{W} - (-120.6\ \text{W})}{220.5\ \text{W} + (-120.6\ \text{W})}\right)\right]$$
$$= 80.4°$$
$$\text{Power factor} = \cos\phi = \cos 80.4°$$
$$= 0.167$$

Practice Problems

26-8.1 The power supplied to the circuit described in Problem 26-6.2 is measured by the two-wattmeter method illustrated in Figure 26-20. Determine the power reading for each meter.

26-8.2 Calculate the load power factor from the wattmeter readings for the circuit in Problem 26-8.1.

Summary of Formulas

- ***For a Y-connected generator or load:***

$$V_L = \sqrt{3}\,V_p$$
$$I_L = I_p$$

- ***For a Δ-connected generator or load:***

$$V_L = V_p$$
$$I_L = \sqrt{3}\,I_p$$

- ***Power dissipated:***

$$P = \sqrt{3}\,V_L I_L \cos\phi$$

- ***Reactive power:***

$$Q = \sqrt{3}\,V_L I_L \sin\phi$$

- ***Apparent power:***

$$S = \sqrt{3}\,V_L I_L$$

- ***Power factor:***

$$\cos\phi = P/S$$

Review Questions

Section 26-1

26-1 Sketch illustrations to show how three-phase ac is generated. Also, sketch the output voltage waveforms and phasor diagrams. Briefly explain.

Section 26-2

26-2 Sketch the circuit of a Y-connected three-phase generator. Identify phase currents, phase voltages, line currents, and line voltages. Explain.

26-3 Sketch the phasor diagram for the phase voltages in a Y-connected generator. Show how the line voltage phasors are derived from the phase voltage phasors.

26-4 Sketch the circuit of a Y-connected generator providing a four-wire supply to a Y-connected load. Identify phase and line currents and voltages at the generator and at the load.

25-5 Draw the phasor diagram for the load voltages and currents for a purely resistive balanced Y-connected four-wire load. Show that the phasor sum of the load currents is zero.

26-6 Sketch the phasor diagram of load currents and voltages for a balanced Y-connected load with $\phi = 20°$ lagging. Show that the phasor sum of the load currents is zero.

Section 26-3

26-7 Sketch the circuit of a Δ-connected three-phase generator. Identify phase currents, phase voltages, line currents, and line voltages. Explain.

26-8 Sketch the phasor diagram for the phase currents in a Δ-connected generator. Show how the line current phasors are derived from the phase current phasors.

26-9 Sketch the circuit of a Δ-connected generator supplying a Δ-connected load. Identify phase and line currents and voltages at the generator and at the load.

26-10 For a Δ-connected purely resistive balanced load, sketch the phasor diagram for the load voltages and currents. Show that the phasor sum of the load currents is zero.

26-11 Sketch the phasor diagram of line voltages, phase currents, and line currents for a balanced Δ-connected load with $\phi = 40°$ lagging.

Section 26-4

26-12 Sketch the circuit of a Y-connected generator supplying a Δ-connected load. Identify phase and line currents and voltages at the generator and at the load.

26-13 Briefly discuss the procedure for analyzing Y-connected three-wire unbalanced loads and combinations of Y- and Δ-connected loads.

Section 26-5

26-14 Sketch the circuit waveforms and the line voltage phasor diagram for a Y-connected generator with phase sequence *ABC*. Repeat for a generator with sequence *CBA*. Explain why the phase sequence is important.

26-15 Sketch the circuit of a phase sequence tester, and briefly explain its operation.

Section 26-6

26-16 For a three-phase load, write equations for (a) power dissipated per phase, (b) total load power in an unbalanced load, (c) total balanced load power in terms of V_P and I_P, and (d) total balanced load power in terms of V_L and I_L. Briefly explain each equation.

26-17 For a three-phase balanced load, write equations for (a) reactive power per phase, (b) total load reactive power in terms of V_L and I_L, (c) apparent power per phase, (d) total apparent power in terms of V_P and I_P, (e) total apparent power in terms of V_L and I_L, and (f) $\cos \phi$. Briefly explain each equation.

Section 26-7

26-18 Sketch the circuit of a balanced Y-connected resistive-inductive load. Show how capacitors may be connected to increase the load power factor. Explain.

Section 26-8

26-19 Sketch the following wattmeter connection diagrams for three-phase power measurement in a Y-connected load: (a) three-wattmeter method for an unbalanced four-wire load, (b) single-wattmeter method for a balanced load. Briefly explain each diagram.

26-20 Sketch a three-wattmeter circuit for measuring power in an unbalanced Δ-connected load and a one-wattmeter circuit for measuring power in a balanced Δ-connected load. Explain.

26-21 Draw the circuit diagram of the two-wattmeter method for measuring power in a three-phase load. Show that the sum of the meter readings equals the total power dissipated in the load, whether the load is balanced or unbalanced, Y-connected or Δ-connected.

26-22 Show that for a balanced load the two-wattmeter method can be used to determine the load power factor.

Problems

Section 26-2

26-1 Three Y-connected 155 Ω resistors have a three-phase, four-wire supply with a line voltage of $V_L = 400$ V, as illustrated in Figure 26-22. Determine the phase voltage, the line current, the phase current, and the neutral current.

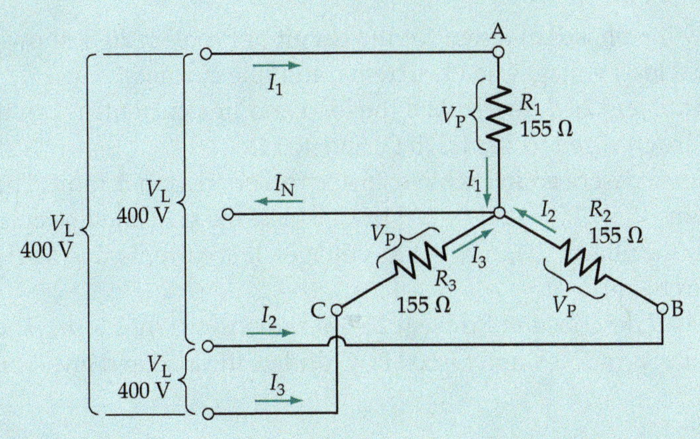

Figure 26-22

26-2 For Problem 26-1, draw the phasor diagram showing line and load voltages and line and load currents.

26-3 For Problem 26-1, recalculate the line and neutral currents when the resistors are changed to 300 Ω, 75 Ω, and 400 Ω.

26-4 A Y-connected load with a three-phase, four-wire supply consists of $Z_A = 180\ \Omega$, $Z_B = (180\ \Omega$ in series with $C = 33\ \mu F)$, and $Z_C = (180\ \Omega$ in series with $L = 300$ mH). The supply voltage has $V_L = 200$ V and $f = 60$ Hz. Calculate the line currents and the neutral current.

26-5 Draw a phasor diagram for the circuit in Problem 26-4.

26-6 Three impedances consist of $Z_A = R_1 + jX_{L1}$, $Z_B = R_2 - jX_{C2}$, and $Z_C = R_3$. The component values are $R_1 = 330\ \Omega$, $R_2 = 250\ \Omega$, $R_3 = 190\ \Omega$, $L_1 = 600$ mH, and $C_2 = 10\ \mu F$. The impedances are Y-connected and have a three-phase, four-wire supply. The supply voltage has $V_L = 150$ V and $f = 60$ Hz. Determine the line currents and the neutral current.

26-7 Draw a phasor diagram for the circuit described in Problem 26-6.

Section 26-3

26-8 Three Δ-connected 155 Ω resistors are supplied from a Δ-connected generator with a line voltage of $V_L = 400$ V, as shown in Figure 26-23. Calculate the current in each resistor and the line currents.

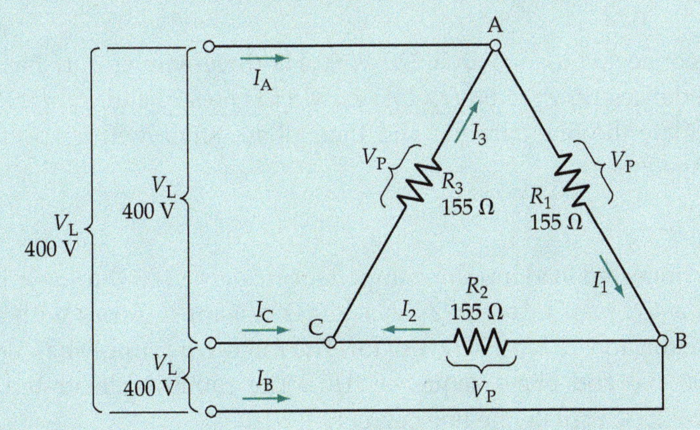

Figure 26-23

26-9 Draw the phasor diagram for the circuit of Problem 26-8, showing individual load voltages, load currents, and line currents.

26-10 For Problem 26-8, recalculate the load and line currents when the resistors are changed to 300 Ω, 75 Ω, and 400 Ω.

26-11 The load described in Problem 26-4 is reconnected in Δ form. The supply remains $V_L = 200$ V and $f = 60$ Hz. Calculate the load and line currents.

26-12 Draw the phasor diagram of line voltages, line currents, and load currents for Problem 26-11.

26-13 The load described in Problem 26-6 is reconnected in Δ form. The supply remains $V_L = 150$ V and $f = 60$ Hz. Calculate the load and line currents.

Section 26-4

26-14 A balanced Δ-connected load consists of three impedances, each of which have $R = 75$ Ω and $C = 2$ μF connected in series. The Y-connected supply and has a 50 V phase voltage and 400 Hz frequency (see Figure 26-24). Calculate the line currents, and the phase angle of each line current with respect to the line voltages.

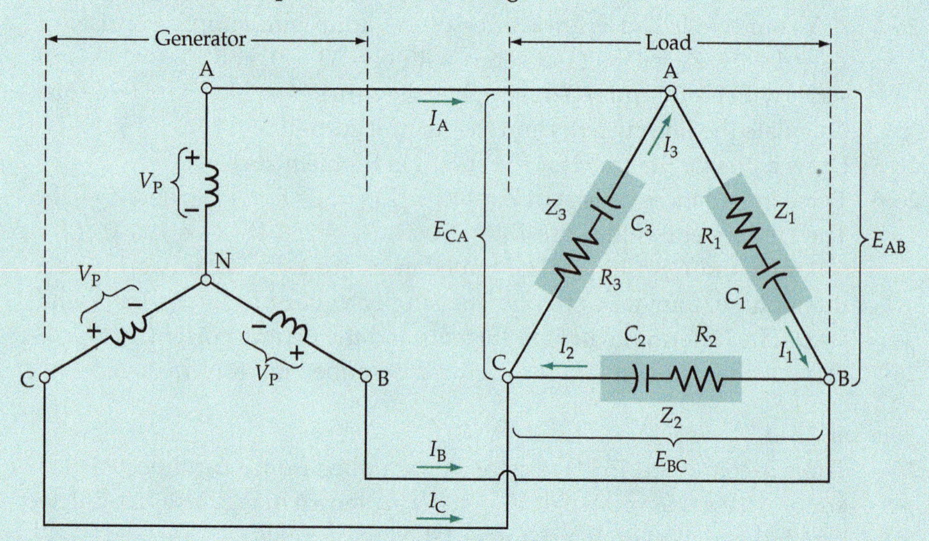

Figure 26-24

26-15 A Y-connected load with a 220 V, 60 Hz three-wire supply has branch impedances of: $Z_a = 66\ \Omega\angle 15°$, $Z_b = 85\ \Omega\angle -30°$, and $Z_c = 90\ \Omega\angle 20°$. Calculate the line currents and their phase angle with respect to line voltage E_{AB}.

Section 26-5

26-16 A Y-connected load has three impedances; Z_a is a 100 Ω resistor in series with a 400 mH inductor, Z_b is a 120 Ω resistor in series with a 10 μF capacitor, and Z_c is a 250 Ω resistor. The three-wire supply has $V_L = 75$ V, $f = 60$ Hz, and phase sequence *ABC*. The circuit is illustrated in Figure 26-25. Calculate the line currents.

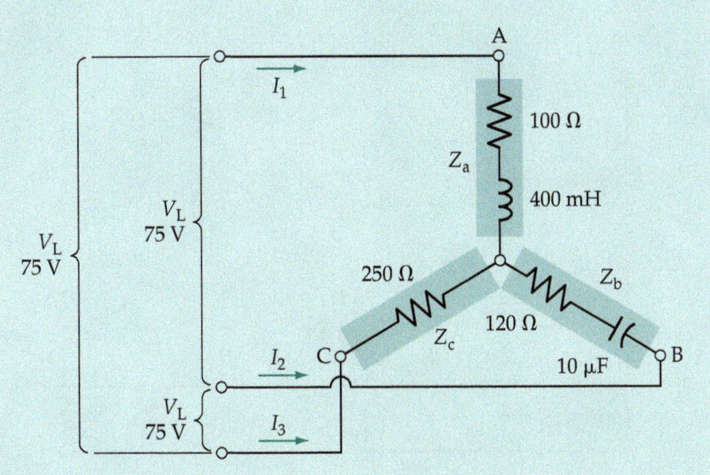

Figure 26-25

26-17 Recalculate the line currents for Problem 26-16 when the phase sequence is changed to *CBA*.

26-18 Recalculate the line currents in Problem 26-13 when the phase sequence is reversed.

26-19 A Y-connected load consists of $Z_a = Z_b =$ a 120 Ω resistor, and Z_c a 60 Ω resistor in series with a 200 mH inductor. The supply has $V_L = 220$ V and $f = 100$ Hz. Calculate the line currents (a) when the phase sequence is *ABC*, and (b) when the sequence is *CBA*.

26-20 Repeat Problem 26-19 with the 200 mH inductor replaced by a 15 μF capacitor.

Section 26-6

26-21 Calculate the total power dissipated in the load described in Problem 26-1.

26-22 Determine the total power dissipated in the circuit referred to in Problem 26-3.

26-23 Calculate the total power dissipated in the load described in Problem 26-4. Also, calculate the reactive power and the apparent power.

26-24 Calculate the total power dissipated in the load described in Problem 26-6. Also, calculate the reactive power and the apparent power.

26-25 For Problem 26-11, calculate the true power, apparent power, and reactive power.

26-26 For Problem 26-13, determine the true power, apparent power, and reactive power.

26-27 Calculate the apparent power, true power, and reactive power for the load described in Problem 26-14.

26-28 Calculate the apparent power, true power, and reactive power for the load described in Problem 26-19.

Section 26-7

26-29 A balanced Y-connected load consists of three impedances, each of which has $R = 75$ Ω and $L = 40$ mH (see Figure 26-26). The supply has

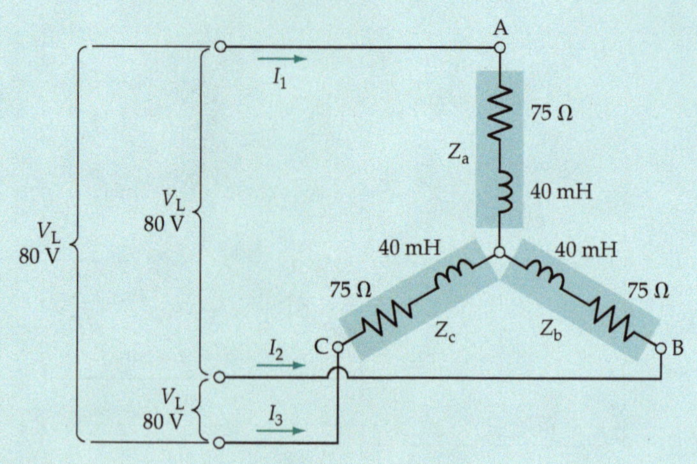

Figure 26-26

$V_L = 80$ V and $f = 400$ Hz. Calculate the line currents, the power factor, and the load power dissipation.

26-30 Three Δ-connected 0.5 μF capacitors are connected in parallel with the Y-connected components of Problem 26-29. The circuit is as illustrated in Figure 26-27 [reproduced from Figure 26-16(a)]. Recalculate the line currents, power factor, and load power.

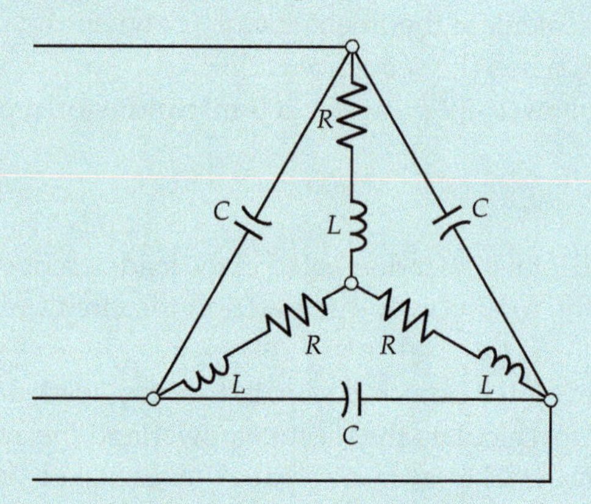

Figure 26-27

26-31 Three Δ-connected impedances each consist of a 180 Ω resistor in series with a 330 mH inductor. The supply voltage has $V_L = 300$ V and $f = 50$ Hz. Three Δ-connected 5 μF capacitors are connected in parallel with the load (see Figure 26-28). Calculate the line current, load power, and power factor (a) without the capacitors in the circuit, and (b) with the capacitors connected.

Section 26-8

26-32 Calculate each wattmeter indication when the two-wattmeter method is used to measure the power in the load described in Problem 26-29. Also, use the meter indications to determine the load power factor.

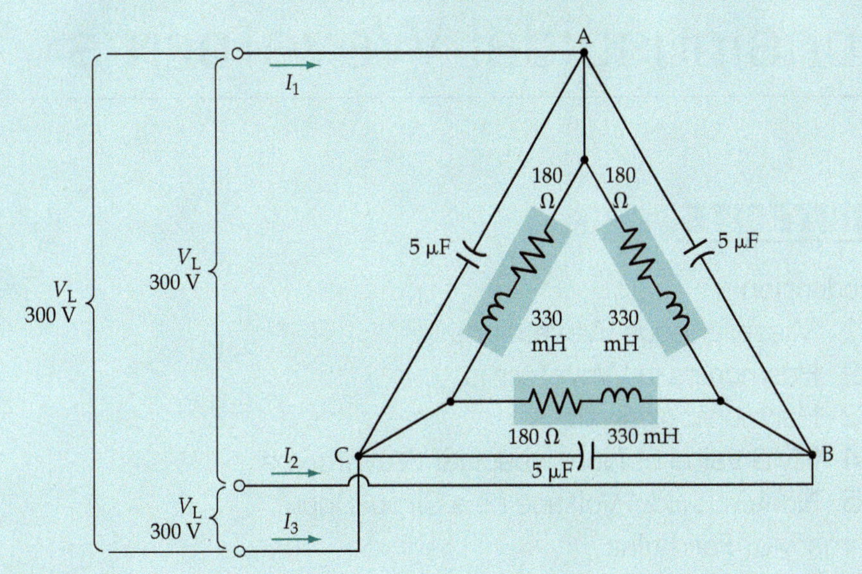

Figure 26-28

26-33 Calculate each wattmeter indication when the two-wattmeter method is used to measure the power delivered to the load described in Problem 26-16.

Practice Problem Answers

26-2.1 33 Ω
26-2.2 3.5 A
26-2.3 2 A, 3.01 A
26-3.1 99 Ω
26-3.2 $3.64\text{ A}\angle{-23.5°}$, $5.07\text{ A}\angle{-145°}$, $4.41\text{ A}\angle{79°}$
26-3.3 $1.732\text{ A}\angle{-60°}$, $2.45\text{ A}\angle{-135°}$, $3.35\text{ A}\angle{75°}$
26-4.1 100 Ω
26-4.2 $1.13\text{ A}\angle{-49°}$, $657\text{ mA}\angle{-139°}$, $1.31\text{ A}\angle{101°}$
26-5.1 $1.13\text{ A}\angle{49°}$, $657\text{ mA}\angle{139°}$, $1.31\text{ A}\angle{-101°}$
26-6.1 615 W
26-6.2 317 VA, 224 var, 224 W
26-7.1 2.8 μF
26-8.1 47 W, 177 W
26-8.2 0.707

CHAPTER 27
Nonsinusoidal Waveforms

CONTENTS

Objectives

You will be able to:

1. Sketch common nonsinusoidal waveforms and discuss the characteristics of each waveform.
2. Sketch waveforms to show that a square wave is made up of many sinusoidal harmonic components, and explain how waveform distortion can result when harmonics are not present in the correct proportion.
3. Use Fourier equations to determine the amplitudes of harmonic components in various nonsinusoidal waveforms.
4. Calculate rms values of nonsinusoidal waveforms.
5. Calculate the effect of a nonsinusoidal waveform applied as an input to *RC* and *RLC* circuits.

INTRODUCTION

Although preceding ac chapters refer exclusively to voltages and currents with sinusoidal waveforms, much electrical and electronics equipment involves waveforms that are nonsinusoidal. Some of these waveforms are deliberately created while others are the results of distortion added by equipment. All repetitive waveforms can be shown to be composed of combinations of many sinusoidal waves. Any waveform can be analyzed to determine the component quantities. These components may be used when investigating the response of an impedance network to a nonsinusoidal input.

27-1 MISCELLANEOUS WAVEFORMS

Figure 27-1 shows several common nonsinusoidal waveforms, all of which are *repetitive* waves; also termed *periodic.* This means that they are made up of identical cycles repeating over and over again. A waveform that does not have identical repeating cycles is termed *nonrepetitive,* or *aperiodic.*

Figures 27-1(a) and (b) show half-wave and full-wave rectified sine waves. These are discussed in Section 17-8. Although they are sinusoidal during half-cycles, rectified waves must be classified as nonsinusoidal. Many voltmeters are calibrated to indicate the rms value of voltages that have pure sinusoidal waveforms. When a rectified waveform is applied to such an instrument, the indicated voltage will *not* be the rms value.

A *square wave* and a *pulse wave* are illustrated in Figures 27-1(c) and (d), respectively. In both cases, the voltage rises rapidly from a low level to a high level, where it remains constant for a brief time. Then it returns rapidly to its low level, where it again remains constant for some time. The square wave has a high level for a time $T/2$ and a low level for an equal time ($T/2$). In the case of the pulse wave, the high-level and low-level times are unequal. Another way of stating this is, the *pulse width* and *space width* are different.

A square wave with severe distortion of its high and low levels is illustrated in Figure 27-1 (e). This type of distortion occurs most often when the wave is *capacitor coupled* from one point to another in an electronic circuit. The distortion results when the capacitor value is too small. When an extremely small capacitor is used in this situation, the spike waveform shown in Figure 27-1(f) might result.

The exponential waveform in Figure 27-1(g) occurs when a capacitor is charged and discharged via a resistor. It could also be produced at the output of a low-frequency amplifier when a square wave input is applied.

Figure 27-1(h) shows a *triangular wave.* This is produced for certain applications by some electronic signal generators. It usually results from a capacitor being charged and discharged using a constant current. The *sawtooth waveform* in Figure 27-1(i), which partially resembles the triangular wave, can be generated by charging a capacitor with a constant current and then rapidly discharging.

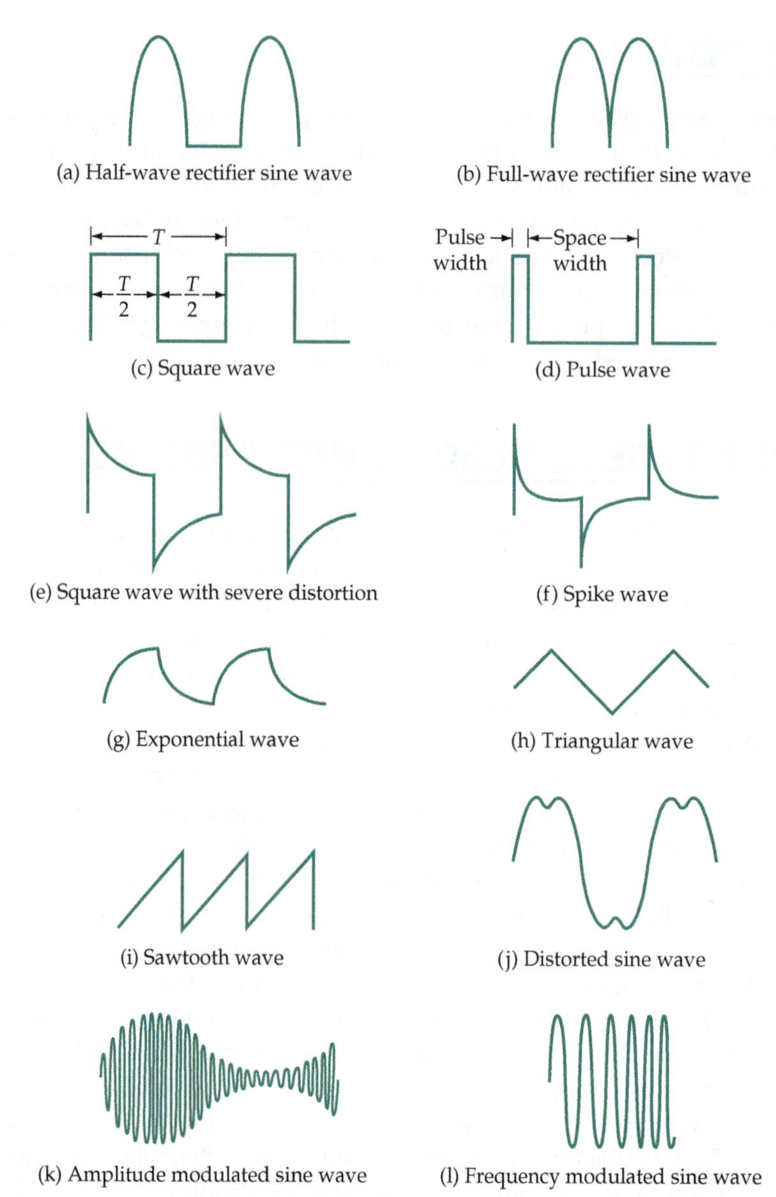

Figure 27-1 A wide variety of nonsinusoidal waveforms occur in electronics equipment. Many of these are deliberately created for useful purposes, while some are the result of unwanted distortion. All repetitive waveforms can be shown to be made up of combinations of sine waves.

The distorted sine wave shown in Figure 27-1(j) is the result of combining a low-amplitude sine wave with a larger-amplitude lower-frequency sine wave. This subject is further discussed in Section 27-2.

Figures 27-1(k) and (1) show the kind of output waveforms produced by some radio transmitters. A high-frequency *carrier* wave is generated for both *amplitude modulated* (*AM*) and *frequency modulated* (*FM*) transmissions. In the AM case, the low-frequency (audio) information is transmitted by increasing and decreasing the amplitude of the carrier. For FM radio, the frequency of the carrier is modulated by the audio signal, thus producing a sinusoidal wave with a changing time period.

The waveforms shown in Figure 27-1 are representative of many types of waves that occur in electronic equipment. Although many of those illustrated do not resemble sine waves, it can be shown that *they are all combinations of many sine waves* sometimes with a dc component.

27-2 HARMONICS IN WAVEFORMS

Frequency Synthesis and Harmonic Analysis

In Figure 27-2 two sine wave generators are shown connected in series. The output from generator B has a frequency which is three times the frequency of the generator A output. The amplitude of the output from B is also much less than that of A. The waveform produced by the two generators in series is the larger-amplitude (and lower-frequency) signal with the smaller-amplitude signal superimposed. It is seen that the combination approximately resembles a square wave with its peaks dented.

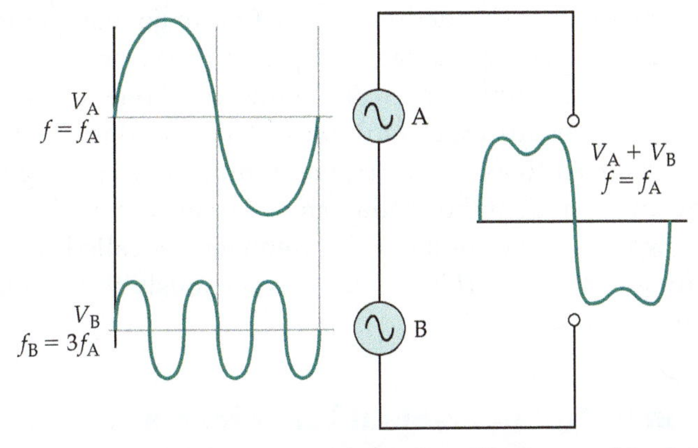

Figure 27-2 A sine wave with (fundamental) frequency f_A, when combined with a smaller-amplitude sine wave with (third harmonic) frequency $f_B = 3f_A$, produces an approximately square waveform.

Figure 27-3 shows a third generator (C) connected in series with A and B. The output from generator C is smaller in amplitude than that from B, and the frequency of the output from C is five times the frequency of the output from generator A. The waveform produced by the three generators in series now more closely resembles a square wave. If appropriate additional higher-frequency waveforms are included, the final wave becomes increasingly more square.

The process of constructing a waveform by combining several sine waves of different frequencies and amplitudes is referred to as *frequency synthesis.*

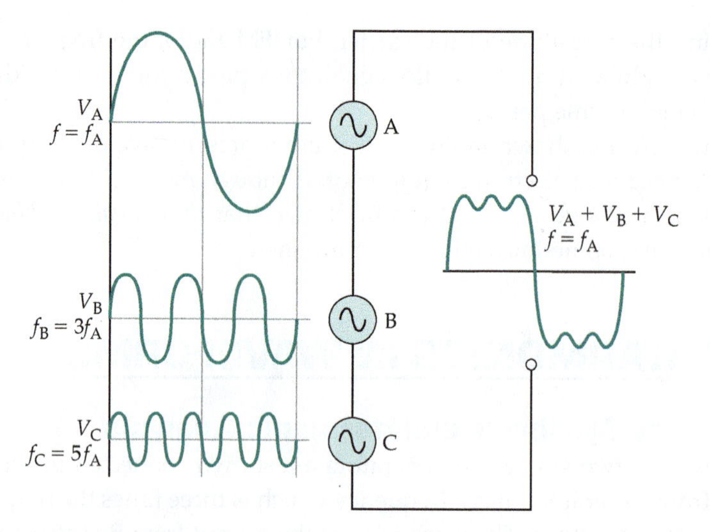

Figure 27-3 When a (fifth harmonic) sinewave with frequency $f_C = 5f_A$ is added to the two waveforms in Figure 27-2, the resultant output more closely resembles a square wave.

The converse of frequency synthesis is *harmonic analysis*; analyzing a waveform to discover its component sine waves. By harmonic analysis, periodic nonsinusoidal waveforms can be shown to consist of combinations of pure sine waves, sometimes with a dc component. One major component, a large-amplitude sine wave having the same frequency as the periodic wave being analyzed, is termed the *fundamental*. The other components are sine waves with frequencies that are exact multiples of the fundamental frequency. These waves, referred to as *harmonics*, are numbered according to the ratio of their frequency to that of the fundamental. For example, a harmonic with a frequency two times the fundamental frequency is called the *second harmonic*. The frequency of the *third harmonic* is obviously three times the fundamental frequency.

Generation of Nonsinusoidal Waveforms

Although it has been explained that nonsinusoidal waveforms are made up of sinusoidal components, generation of these waves does not normally-involve the addition of several sine waves. As discussed in Section 27-1, for example, a triangular wave can be produced by using a constant current to charge and discharge a capacitor. Square waves can be generated by causing an electronic device to switch its output from a low voltage level to a higher voltage level. Many nonsinusoidal waveforms are the result of distortion produced when an input signal (sinusoidal or nonsinusoidal) is processed through electronic circuitry.

When a sinusoidal input signal is applied to an amplifier, the output may be expected to be an amplified version of the input waveform. For this to occur, the signal amplitude must be within an acceptable range for the amplifier.

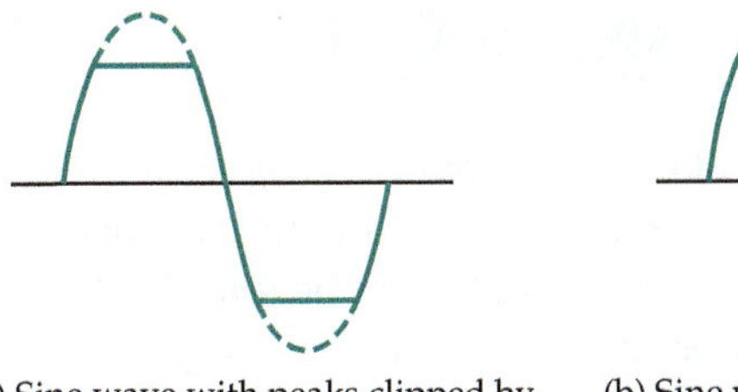

(a) Sine wave with peaks clipped by excessive amplification

(b) Sine wave with rounded negative peak due to second harmonic

Figure 27-4 A sine wave with clipped peaks has approximately the same harmonic content as a square wave. Second harmonic distortion produces a sine wave with one half-cycle slightly peaked and the other half-cycle flattened.

Also, the signal frequency must be within the frequency range (or bandwidth) of the amplifier.

If the input to an amplifier is much larger than it should be, the output is likely to look like the *clipped sine wave* illustrated in Figure 27-4(a). This output is fairly similar to a square wave, so its harmonic content is close to that of a square wave. When the sinusoidal input has an amplitude that is just a little greater than the acceptable maximum for the amplifier, the output might have the wave shape shown in Figure 27-4(b). In this case, the upper peak is almost sinusoidal, and the lower peak is rounded. The waveform can be constructed by adding a second harmonic to the fundamental. Consequently, this particular type of distortion is termed *second harmonic distortion.*

When a square wave is applied as input to a circuit that does not pass all the necessary frequency components, the resultant output is a distorted square wave. The type of distortion depends on whether the circuit has poor low-frequency response or poor high-frequency response. In Figure 27-5(a) the long rise and fall times of the square wave show that the high-frequency harmonics are attenuated. Thus, the circuit has poor high-frequency response. Figure 27-5(b) shows the output of a circuit that has good high-frequency response but poor low-frequency response. The *tilt* on the top and bottom of the waveform is the result of low-frequency components being attenuated by the circuit.

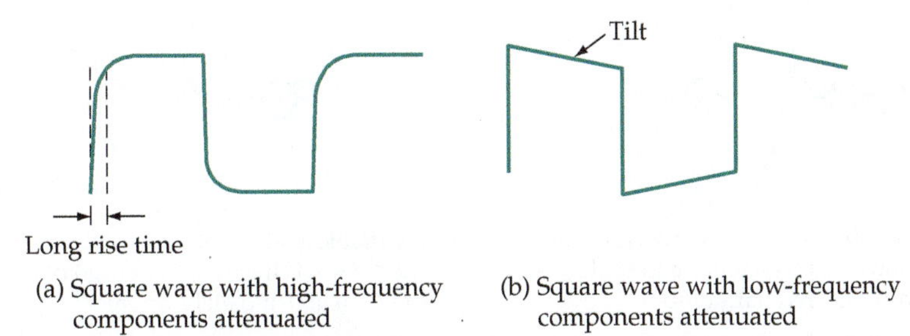

(a) Square wave with high-frequency components attenuated

(b) Square wave with low-frequency components attenuated

Figure 27-5 A square wave with excessive rise time and rounded corners has its high-frequency harmonic components attenuated. Attenuation of low-frequency components produces tilt on a square wave.

27-3 HARMONIC ANALYSIS

By the mathematical operation known as *Fourier analysis*[1], waveforms can be analyzed to determine their harmonic content. The amplitude of each harmonic and its phase relationship to the fundamental can be found. Also, the level of any dc component can be calculated.

Square Wave

A perfect square wave which is symmetrical above and below ground level [Figure 27-6(a)] can be shown by Fourier analysis to be represented by the following equation:

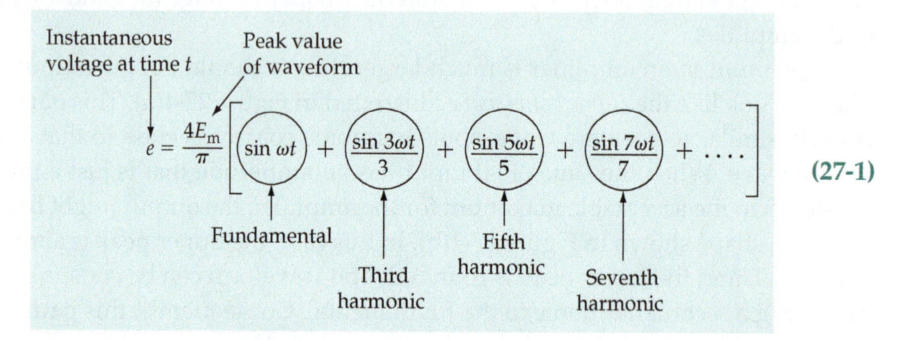

$$e = \frac{4E_m}{\pi}\left[\sin \omega t + \frac{\sin 3\omega t}{3} + \frac{\sin 5\omega t}{5} + \frac{\sin 7\omega t}{7} + \cdots\right] \qquad (27\text{-}1)$$

As shown by the equation, a symmetrical square wave can be said to be made up of a fundamental, odd-numbered harmonics, no even harmonics, and no dc component.

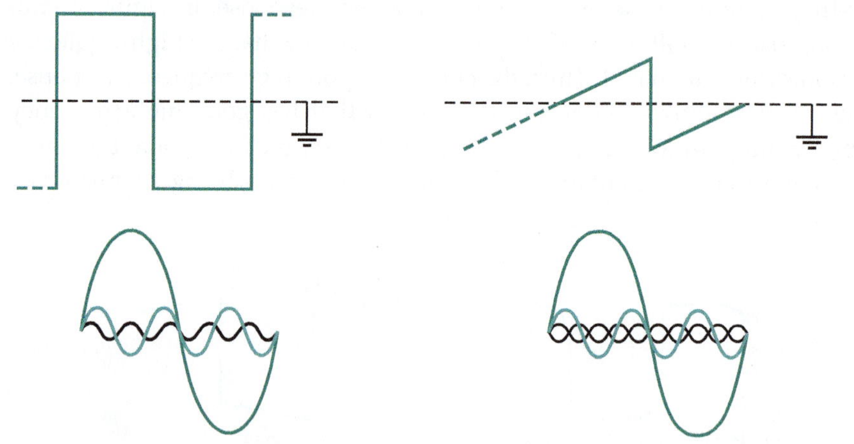

(a) Harmonic analysis of a symmetrical square wave shows that it contains a fundamental and odd-numbered harmonics

(b) Harmonic analysis of a symmetrical sawtooth wave reveals that it is composed of a fundamental and all harmonics

Figure 27-6 **Fourier analysis can be applied to all repetitive waveforms to determine their harmonic content.**

---

[1]Developed by French mathematician Jean Fourier (1768–1830).

Sawtooth Wave

The Fourier equation for the sawtooth waveform in Figure 27-6(b) is

$$e = \frac{2E_m}{\pi}\left[\sin \omega t - \frac{\sin 2\omega t}{2} + \frac{\sin 3\omega t}{3} - \frac{\sin 4\omega t}{4} + \frac{\sin 5\omega t}{5} \cdots\right] \quad (27\text{-}2)$$

In this case, all the harmonics are present, and once again, there is no dc component. In general, a waveform has no dc component when it is symmetrical above and below ground level.

Rectified Wave

The full-wave rectified sine wave in Figure 27-7(b) can be represented by

$$e = \frac{4E_m}{\pi}\left[\frac{1}{2} + \frac{\cos 2\omega t}{3} - \frac{\cos 4\omega t}{15} + \frac{\cos 6\omega t}{35} \cdots\right] \quad (27\text{-}3)$$

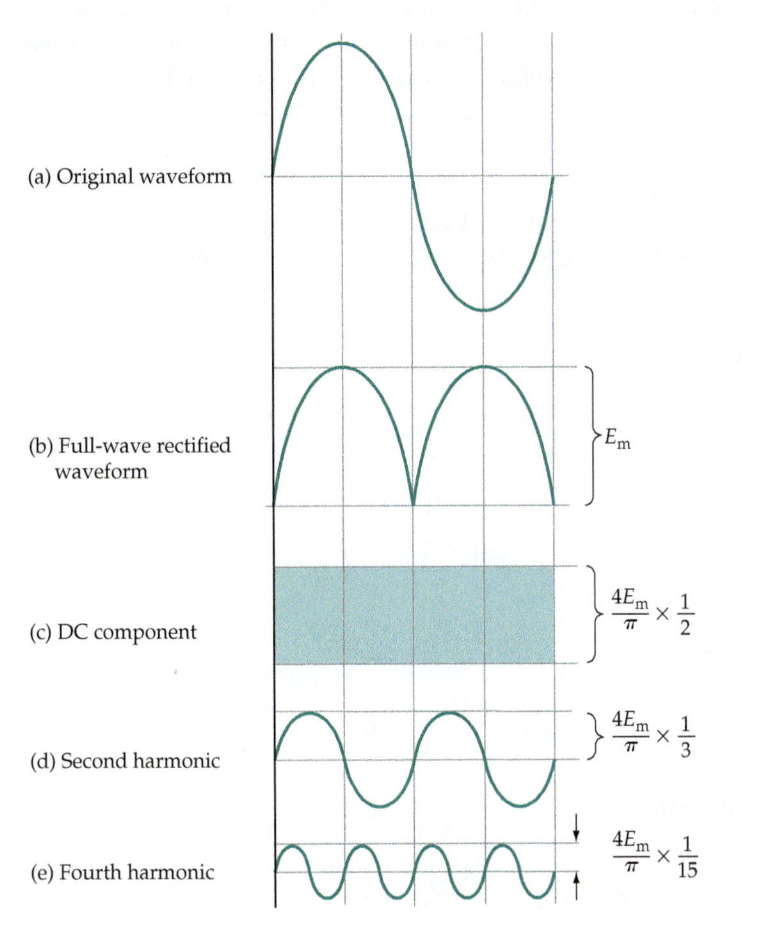

Figure 27-7 A full-wave rectified sine wave is made up of a dc component and even-numbered harmonics that decrease in amplitude with increasing harmonic number.

Equation 27-3 shows that the waveform has a dc component $4E_m/(2\pi)$ and even-numbered harmonics, $2\omega t$, $4\omega t$, $6\omega t$, and so on (see Figure 27-7). It would appear that there is no fundamental frequency component. However, in this case the fundamental frequency is taken as the input frequency (f) of the waveform prior to rectification. It could be argued that the fundamental frequency of the rectified waveform is actually $2f$. For example, a 60 Hz sine wave when full-wave rectified, produces a succession of sinusoidal half-cycles with a frequency of 120 Hz.

Amplitudes of Harmonics

Examination of the equations for the square, sawtooth, and rectified sine wave shows that in all cases the amplitudes of the harmonic components decrease as the harmonic frequency increases. Thus, the higher-order harmonics appear to have decreasing importance. This is certainly true in terms of the contribution of these components to the rms value of the waveform and to the power dissipated in a load. However, for good reproduction of the waveform, many of the higher-order harmonics must be present. For example, in the case of a square wave, all components up to the eleventh harmonic (or higher) may be required. For a pulse waveform, harmonics up to the one hundredth may have to be present to create a good output wave shape.

Example 27-1

A square wave with a 2 V peak-to-peak amplitude is symmetrical above and below ground level. Calculate the amplitudes of each component up to the seventh harmonic.

Solution

Eq. 27-1,

$$e = \frac{4E_m}{\pi}\left[\sin \omega t + \frac{\sin 3\omega t}{3} + \frac{\sin 5\omega t}{5} + \frac{\sin 7\omega t}{7}\right]$$

$$E_m = 1 \text{ V (peak)}$$

$$\frac{4E_m}{\pi} = \frac{4 \times 1 \text{ V}}{\pi} = 1.27 \text{ V}$$

Fundamental,

$$E_{m1} = \frac{4E_m}{\pi} = 1.27 \text{ V}$$

Third harmonic,

$$E_{m3} = \frac{4E_m}{\pi \times 3} = 0.42 \text{ V}$$

Fifth harmonic,

$$E_{m5} = \frac{4E_m}{\pi \times 5} = 0.25 \text{ V}$$

Seventh harmonic,

$$E_{m7} = \frac{4E_m}{\pi \times 7} = 0.18 \text{ V}$$

Note that the harmonic voltage components calculated are all peak values. Each must be multiplied by 0.707 to determine the rms values.

Example 27-2

A full-wave rectified sine wave has a peak amplitude of 30 V and a (pre-rectified) frequency of 60 Hz. Calculate the dc component and the rms values of the harmonic components up to the sixth harmonic. Also, determine the harmonic frequencies.

Solution

Eq. 27-3,
$$e = \frac{4E_m}{\pi}\left[\frac{1}{2} + \frac{\cos 2\omega t}{3} - \frac{\cos 4\omega t}{15} + \frac{\cos 6\omega t}{35}\dots\right]$$

$$\frac{4E_m}{\pi} = \frac{4 \times 30\text{ V}}{\pi} = 38.2\text{ V}$$

DC component,
$$E_{dc} = \frac{4E_m}{\pi \times 2} = 19.1\text{ V}$$

Fundamental,
$$E_{m1} = 0$$

Second harmonic,
$$E_{m2} = \frac{4E_m}{\pi \times 3} = 12.7\text{ V}$$

$$E_2 = 0.707E_{m2} = 9\text{ V}$$

$$f_2 = 2f_1 = 2 \times 60\text{ Hz} = 120\text{ Hz}$$

Fourth harmonic,
$$E_{m4} = \frac{4E_m}{\pi \times 15} = 2.5\text{ V}$$

$$E_4 = 0.707E_{m4} = 1.8\text{ V}$$

$$f_4 = 4f_1 = 240\text{ Hz}$$

Sixth harmonic,
$$E_{m6} = \frac{4E_m}{\pi \times 35} = 1.1\text{ V}$$

$$E_6 = 0.707E_{m6} = 0.78\text{ V}$$

$$f_6 = 6f_1 = 360\text{ Hz}$$

Practice Problems

27-3.1 A sawtooth wave, as in Figure 27-6(b), has a 3 V peak-to-peak amplitude. Determine the amplitude of all component waves up to the fifth harmonic.

27-3.2 The Fourier equation for a half-wave rectified sine wave is

$$e = \frac{2E_m}{\pi}\left[\frac{1}{2} + \frac{\pi \times \sin \omega t}{4} - \frac{\cos 2\omega t}{3} - \frac{\cos 4\omega t}{15}\dots\right]$$

Determine the amplitudes of all components up to the fourth harmonic for a wave with $E_m = 18$ V.

27-4 RMS VALUE OF NONSINUSOIDAL WAVEFORMS

Three ac voltage sources operating in series to produce a current (I) through a resistor (R) are shown in Figure 27-8. The total current can be determined by means of the superposition theorem. However, because these are ac quantities, phasor addition is involved, and the relative phase angles of the currents must be known. A calculation of the total power dissipated in the resistors can be made without any knowledge of the current phase angles.

Eq. 21-2,
$$P = \frac{E^2}{R}, \tag{1}$$

where E is the rms value of the E_1, E_2, and E_3 voltages combined.

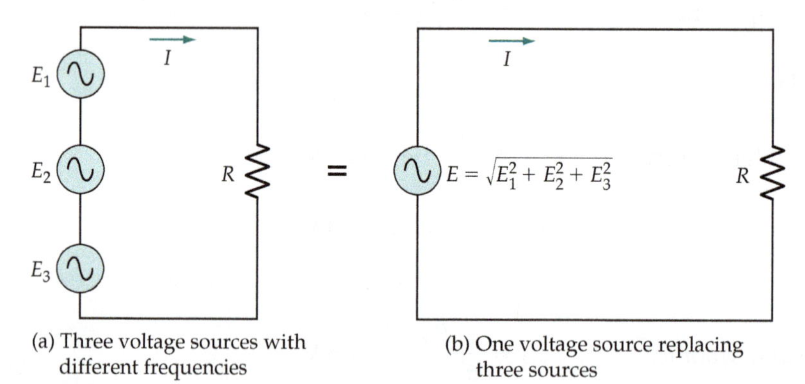

Figure 27-8 The rms value of nonsinusoidal waveforms can be determined from a knowledge of the harmonic components. The rms value of any waveform is the square root of the sum of the squared individual harmonic rms values.

The power dissipation due to each individual source can be determined as

$$P_1 = \frac{E_1^2}{R}, \quad P_2 = \frac{E_2^2}{R}, \quad \text{and} \quad P_3 = \frac{E_3^2}{R}$$

Total output power is,

$$P = \frac{E_1^2}{R} + \frac{E_2^2}{R} + \frac{E_3^2}{R} \tag{2}$$

$$P = \frac{1}{R}(E_1^2 + E_2^2 + E_3^2) \tag{3}$$

From Eqs. (1) and (3),

$$\frac{E^2}{R} = \frac{1}{R}(E_1^2 + E_2^2 + E_3^2)$$

So,
$$E^2 = E_1^2 + E_2^2 + E_3^2$$

Giving,
$$E = \sqrt{E_1^2 + E_2^2 + E_3^2} \tag{27-4}$$

A waveform composed of a fundamental and a number of harmonics can be treated as several series-connected voltage generators, as illustrated in Figure 27-8. The generators have voltages E_1, E_2, E_3, etc., that are the rms values of the fundamental and harmonics. Consequently, the rms value of a nonsinusoidal voltage wave is given by Equation 27-4. This reasoning can also be applied to the determination of the rms value of a nonsinusoidal current wave:

Eq. 21-3,
$$P = I^2R$$
$$= I_1^2R + I_2^2R + I_3^2R$$
$$= R(I_1^2 + I_2^2 + I_3^2)$$

So,
$$I^2 = I_1^2 + I_2^2 + I_3^2$$

or
$$I = \sqrt{I_1^2 + I_2^2 + I_3^2 + \cdots} \qquad (27\text{-}5)$$

Example 27-3

Determine the rms value of the square wave analyzed in Example 27-1.

Solution

The harmonic voltage components determined in Example 27-1 are peak values. They must be converted to rms values before substituting into Equation 27-4.

$$E_{1(\text{rms})} = 0.707 \times 1.27\text{ V} \approx 0.9\text{ V}$$
$$E_{3(\text{rms})} = 0.707 \times 0.42\text{ V} \approx 0.3\text{ V}$$
$$E_{5(\text{rms})} = 0.707 \times 0.25\text{ V} \approx 0.18\text{ V}$$
$$E_{7(\text{rms})} = 0.707 \times 0.18\text{ V} \approx 0.13\text{ V}$$

Eq. 27-4,
$$E = \sqrt{E_1^2 + E_2^2 + E_3^2}$$
$$= \sqrt{(0.9\text{ V})^2 + (0.3\text{ V})^2 + (0.18\text{ V})^2 + (0.13\text{ V})^2}$$
$$\approx 0.97\text{ V}$$

The actual rms value of a 2 V peak-to-peak square waveform is 1 V. The inaccuracy in the above calculation is due to the neglect of higher harmonic components.

Example 27-4

Determine the rms value of the full-wave rectified waveform analyzed in Example 27-2.

Solution

From Example 27-2, the component values of the waveform are

$$E_{dc} = 19.1\text{ V}$$

$$E_{2(rms)} = 9\text{ V}$$

$$E_{4(rms)} = 1.8\text{ V}$$

$$E_{6(rms)} = 0.78\text{ V}$$

Eq. 27-4,

$$E = \sqrt{E_{dc}^2 + E_2^2 + E_4^2 + E_6^2}$$

$$= \sqrt{(19.1\text{ V})^2 + (9\text{ V})^2 + (1.8\text{ V})^2 + (0.78\text{ V})^2}$$

$$= 21.2\text{ V}$$

Note that the rms value of a full-wave rectified sine wave should be equal to the rms value of the non-rectified waveform.

$$E_{rms} = 0.707E_p = 0.707 \times 30\text{ V}$$

$$= 21.2\text{ V}$$

Practice Problems

27-4.1 Calculate the rms value of the sawtooth wave analyzed for Problem 27-3.1

27-4.2 Calculate the rms value of the half-wave rectified sine wave in Problem 27-3.2.

27-5 NONSINUSOIDAL VOLTAGE AS A CIRCUIT INPUT

When a nonsinusoidal voltage is applied to a purely resistive circuit, the waveform of the resulting current is identical to the input voltage wave [see Figure 27-9(a)]. The peak level of the current can be calculated as $I_m = E_m/R$, and the instantaneous current at any time is $i = e/R$, where e is the instantaneous input voltage.

When the circuit is not purely resistive, the current waveform is a distorted version of the input voltage wave [Figures 27-9(b) and (c)]. This is because the circuit offers different impedances to the (different frequency) harmonic components of the input. Consequently, the various current components do not have the same proportional relationship as the input voltage components.

To determine the effect of a nonsinusoidal input voltage on an impedance network, the voltage must first be analyzed to find its harmonic components. The individual voltage components are then treated as separate inputs, and the (different) circuit impedances must be calculated for each harmonic frequency. When the current components have been determined, the superposition theorem

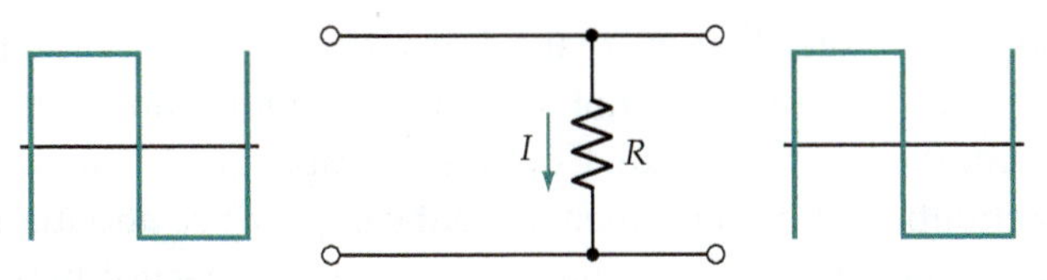

(a) A square wave input to a pure resistive circuit produces a square current waveform

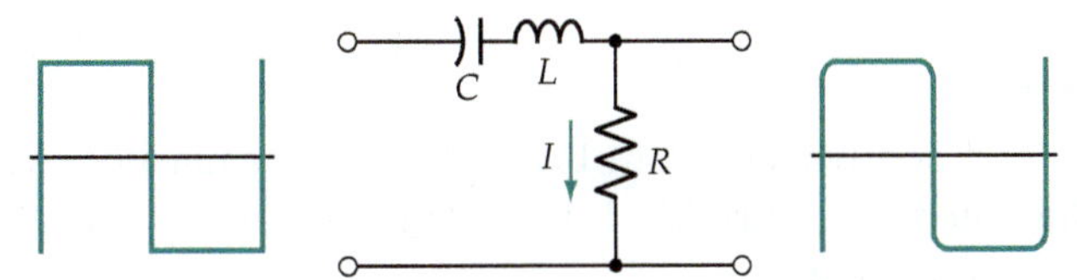

(b) A square wave input to an impedance circuit produces a distorted square current waveform

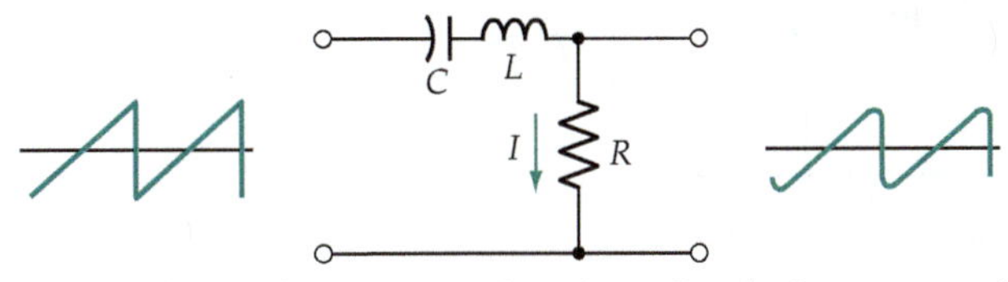

(c) A distorted current waveform is produced when a sawtooth input wave is applied to an impedance circuit

Figure 27-9 A purely resistive circuit has no effect on the harmonic components of a nonsinusoidal waveform. An impedance circuit attenuates different input frequencies by different amounts, thus producing distortion on the current waveform.

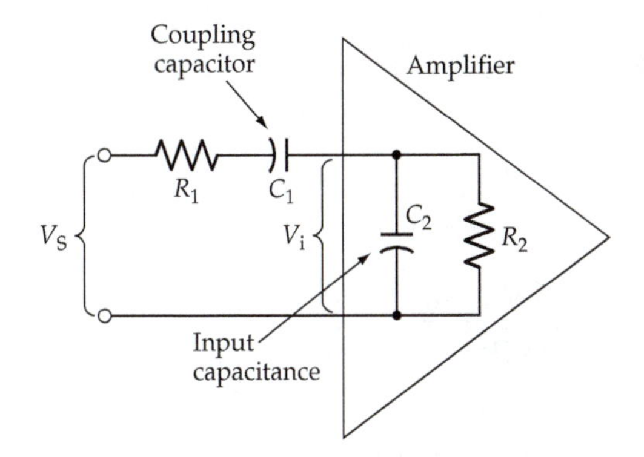

Figure 27-10 An *RC* network exists at the input of electronic amplifiers. R_1 is the signal source resistance, C_1 is a coupling capacitor, R_2 is the amplifier input resistance, and C_2 is the input capacitance.

may be applied to find the resultant current. Once again, note that because phasor quantities are involved, the component phase relationships must be taken into account. The rms value of the resultant current can be calculated using Equation 27-5, without reference to the phase angle of each component.

The circuit in Figure 27-10 illustrates the conditions at the input of many electronic amplifiers. The input resistance of the amplifier is R_2, but this is shunted by an input capacitance (C_2). Capacitor C_1 is a *coupling capacitor* for coupling the signal source to the amplifier input while maintaining dc isolation, and R_1 is the signal source resistance. At low frequencies, the impedance of C_2

is normally so much larger than R_2 that it has virtually no effect on the signal. At high frequencies, where X_{C2} is not very much larger than R_2, there can be significant attenuation of the signal. Conversely, capacitor C_1 causes insignificant signal attenuation at high frequencies but can produce attenuation at low frequencies. For a nonsinusoidal signal, the coupling capacitor may affect the fundamental frequency and perhaps the lower harmonics, while the amplifier input capacitance is most likely to attenuate the highfrequency harmonics.

Example 27-5

The full-wave rectified waveform in Example 27-2 is capacitor-coupled to a resistor as illustrated in Figure 27-11. Calculate the rms level of the voltage developed across the resistor.

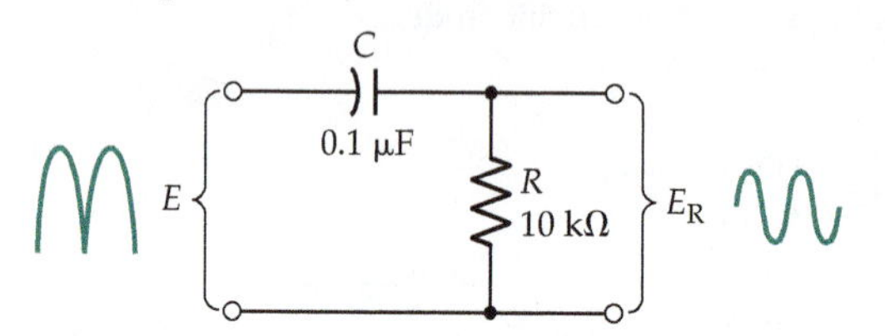

Figure 27-11 Illustration for Example 27-5.

Solution

From Example 27-2, the waveform components are (E_{dc} = 19.1 V), (E_2 = 9 V, 120 Hz), (E_4 = 1.8 V, 240 Hz), and (E_6 = 0.78 V, 360 Hz). The dc component is blocked by the capacitor, and the attenuation of each ac component depends upon the harmonic frequency.

At f_2,
$$X_{C2} = \frac{1}{2\pi f_2 C} = \frac{1}{2\pi \times 120\ \text{Hz} \times 0.1\ \mu\text{F}}$$
$$= 13.3\ \text{k}\Omega$$

At f_4,
$$X_{C4} = 6.63\ \text{k}\Omega$$

At f_6,
$$X_{C6} = 4.43\ \text{k}\Omega$$

For H_2,
$$|V_{R2}| = \frac{V_2 \times R}{\sqrt{R^2 + X_{C2}^2}} = \frac{9\ \text{V} \times 10\ \text{k}\Omega}{\sqrt{10\ \text{k}\Omega^2 + 13.3\ \text{k}\Omega^2}}$$
$$= 6.8\ \text{V}$$

For H_4,
$$|V_{R4}| = \frac{V_4 \times R}{\sqrt{R^2 + X_{C4}^2}} = \frac{1.8\ \text{V} \times 10\ \text{k}\Omega}{\sqrt{10\ \text{k}\Omega^2 + 6.63\ \text{k}\Omega^2}}$$
$$= 1.5\ \text{V}$$

For H_6,
$$|V_{R6}| = \frac{V_6 \times R}{\sqrt{R^2 + X_{C6}^2}} = \frac{0.78\ \text{V} \times 10\ \text{k}\Omega}{\sqrt{10\ \text{k}\Omega^2 + 4.43\ \text{k}\Omega^2}}$$
$$= 0.71\ \text{V}$$

Eq. 27-4,
$$E = \sqrt{E_2^2 + E_4^2 + E_6^2} = \sqrt{(6.8\ \text{V})^2 + (1.5\ \text{V})^2 + (0.71\ \text{V})^2}$$
$$= 7\ \text{V}$$

Practice Problems

27-5.1 The circuit in Figure 27-9(b) has $C = 1.6\ \mu F$, $L = 48$ mH, and $R = 2.2\ k\Omega$. The input is a square wave with a frequency of 1 kHz and an amplitude of 2 V (as analyzed in Example 27-1). Determine the rms value of the current.

27-5.2 The circuit in Figure 27-10 has: $C_1 = 1\ \mu F$, $C_2 = 300$ pF, $R_1 = 10\ k\Omega$, and $R_2 = 100\ k\Omega$. A 1 kHz square wave is applied as input signal. Determine the attenuation (from signal source to amplifier input) of the fundamental, and of the 15th harmonic of the signal.

Summary of Formulas

- *Fourier equation for a square wave:*

$$e = \frac{4E_m}{\pi}\left[\sin \omega t + \frac{\sin 3\omega t}{3} + \frac{\sin 5\omega t}{5} + \frac{\sin 7\omega t}{7}\right]$$

- *Fourier equation for a sawtooth wave:*

$$e = \frac{2E_m}{\pi}\left[\sin \omega t - \frac{\sin 2\omega t}{2} + \frac{\sin 3\omega t}{3} - \frac{\sin 4\omega t}{4} + \frac{\sin 5\omega t}{5} \cdots\right]$$

- *Fourier equation for a full-wave rectified sine wave:*

$$e = \frac{4E_m}{\pi}\left[\frac{1}{2} + \frac{\cos 2\omega t}{3} - \frac{\cos 4\omega t}{15} + \frac{\cos 6\omega t}{35} \cdots\right]$$

- *Fourier equation for a half-wave rectified sine wave:*

$$e = \frac{2E_m}{\pi}\left[\frac{1}{2} + \frac{\pi \times \sin \omega t}{4} - \frac{\cos 2\omega t}{3} - \frac{\cos 4\omega t}{15} \cdots\right]$$

- *Rms values of nonsinusoidal waveforms:*

$$E = \sqrt{E_1^2 + E_2^2 + E_3^2 + \cdots}$$

$$I = \sqrt{I_1^2 + I_2^2 + I_3^2 + \cdots}$$

Review Questions

Section 27-1

27-1 Sketch half-wave and full-wave rectified waveforms, and square and pulse waveforms. Briefly discuss the shape of each wave and how it is generated. Define the terms *periodic* and *aperiodic*.

27-2 Sketch exponential, triangular, and sawtooth waveforms. Briefly discuss the shape of each wave and how it is generated. Also, sketch amplitude modulated and frequency modulated sine waveforms.

Section 27-2

27-3 Sketch the circuit of two sine wave generators connected in series. Show how the waveforms combine to produce an approximation of a square wave. Explain the significance of this effect.

27-4 Draw sketches to show two kinds of distortion on a sine wave, and briefly explain the origin of the distortion in each case.

27-5 Draw sketches to show two kinds of distortion on a square wave, and briefly explain the origin of the distortion in each case.

Section 27-3

27-6 Refer to Fourier analysis Equations 27-1, 27-2, and 27-3 in the text. Briefly discuss each component of each equation in turn, and explain how the component quantities relate to the waveform represented.

Section 27-4

27-7 Derive the equations for the rms values of nonsinusoidal voltages and currents.

Section 27-5

27-8 Explain how an impedance network may be analyzed to determine its response to a nonsinusoidal input voltage.

Problems

Section 27-3

27-1 A sawtooth waveform symmetrical above and below ground, as in Figure 27-12, has a peak-to-peak amplitude of 20 V and a frequency of 1 kHz. Calculate the amplitude of each ac component up to the fifth harmonic.

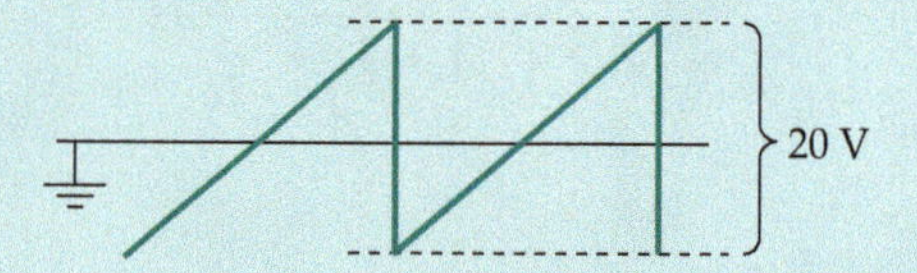

Figure 27-12

27-2 The amplitude and frequency of a full-wave rectified sinusoidal voltage is adjusted to give a fourth harmonic with a frequency of 18 kHz and an rms value of 150 mV. Determine the frequency and rms level of the second harmonic.

27-3 The Fourier equation for a half-wave rectified sinusoidal wave is

$$e = \frac{E_m}{\pi} + \frac{E_m \sin \omega t}{2} - \frac{2E_m}{\pi}\left[\frac{\cos 2\omega t}{3} + \frac{\cos 4\omega t}{15} + \frac{\cos 6\omega t}{35} + \cdots\right]$$

Determine the amplitude of each component of a 15 V, 400 Hz, half-wave rectified sine wave up to the sixth harmonic.

27-4 A square wave symmetrical above and below ground has a peak amplitude of 6 V, as illustrated in Figure 27-13. Determine the amplitude of each component up to the seventh harmonic.

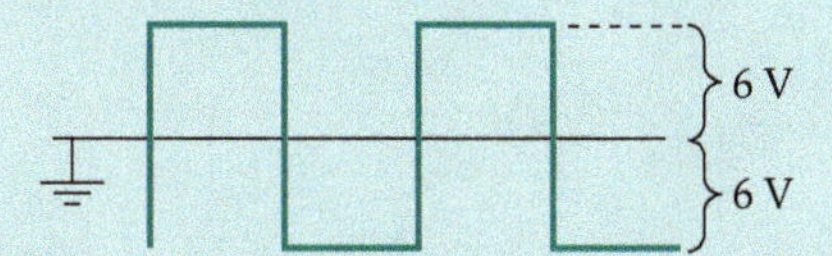

Figure 27-13

27-5 A 400 Hz sinusoidal voltage wave with an rms value of 10 V is half-wave rectified. Calculate the amplitude of each component of the wave up to the sixth harmonic.

27-6 A triangular wave symmetrical above and below ground, as in Figure 27-14, has the equation

$$e = \frac{8E_m}{\pi^2}\left[\cos \omega t + \frac{\cos 3\omega t}{3^2} + \frac{\cos 5\omega t}{5^2} + \frac{\cos 7\omega t}{7^2} + \cdots\right]$$

Calculate the peak values of all components up to the seventh harmonic of such a triangular waveform if its peak-to-peak amplitude is 25 V and its frequency is 1 kHz.

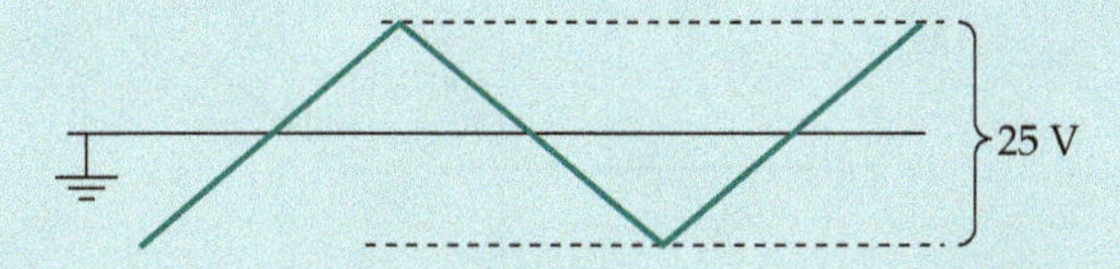

Figure 27-14

27-7 A 50 V, 400 Hz sine wave is full-wave rectified. Determine the dc component and the peak values of the harmonic components up to the sixth harmonic.

27-8 The fifth harmonic of a symmetrical triangular wave has an rms value of 162 mV. Calculate the peak value of the waveform.

Section 27-4

27-9 The sawtooth waveform in Problem 27-1 is applied to a 3.3 kΩ resistor. Determine the rms value of the current that flows.

27-10 Calculate the rms value of the half-wave rectified sine wave in Problem 27-3.

27-11 Calculate the rms value of the symmetrical square wave in Problem 27-4.

27-12 Calculate the rms value of the full-wave rectified waveform in Problem 27-7.

27-13 Determine the rms value of the symmetrical triangular wave in Problem 27-6.

Section 27-5

27-14 An impedance network consists of a resistor R_1 in series with parallel-connected components R_2 and C_1. The component values are $R_1 = 1.5\ \text{k}\Omega$, $R_2 = 15\ \text{k}\Omega$, and $C_1 = 10\ \mu\text{F}$ (see Figure 27-15). Calculate the dc and rms values of the capacitor voltage when the waveform in Problem 27-3 is applied as input.

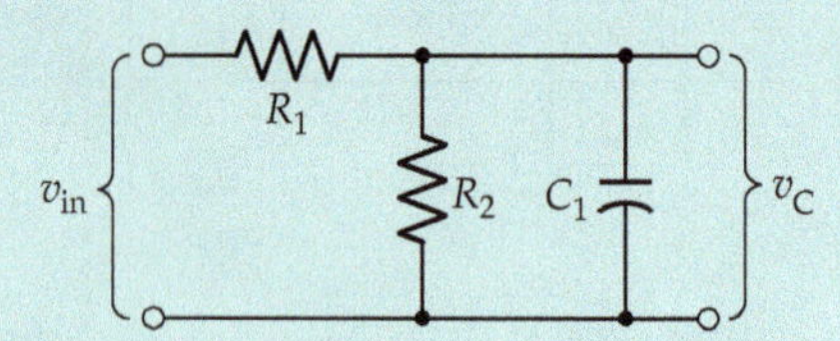

Figure 27-15

27-15 A symmetrical square wave with a 15 V peak-to-peak amplitude and 700 Hz frequency is applied to the network in Problem 27-14. Calculate the rms current level.

27-16 The waveform in Problem 27-6 is capacitor-coupled via a 0.01 μF capacitor to an amplifier with a 10 kΩ input resistance, as in Figure 27-16. Calculate the rms voltage level at the amplifier input terminals.

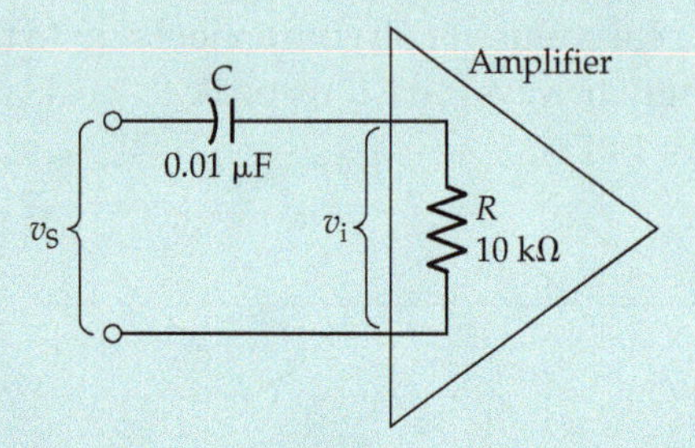

Figure 27-16

27-17 The sawtooth waveform in Problem 27-1 is investigated by means of an instrument that measures the dc level and the rms values of the harmonics. As shown by the circuit in Figure 27-17, the signal source resistance

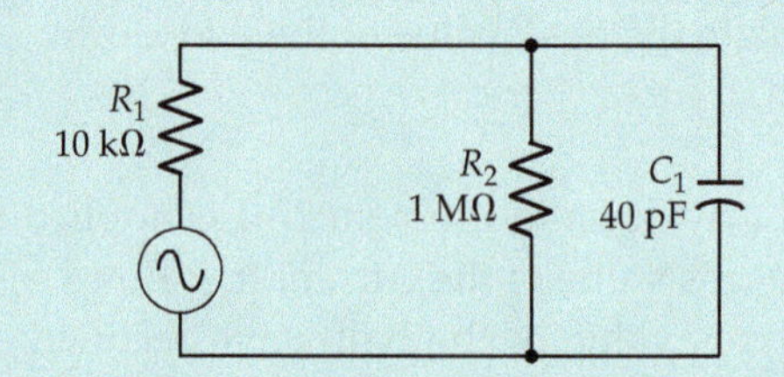

Figure 27-17

is 10 kΩ and the measuring instrument has an input resistance of 1 MΩ in parallel with a 40 pF input capacitance. Calculate the rms values of the fundamental and the fifth harmonic, and the measured value of each.

27-18 For Problem 27-17, recalculate the measured component values for a sawtooth waveform with a frequency of 100 kHz.

27-19 A full-wave rectified sine wave with an amplitude of 50 V and a (pre-rectified) frequency of 400 Hz is applied to an impedance network. The network consists of a resistor R_1 in series with parallel-connected components R_2 and C_1, as illustrated in Figure 27-18. If $R_1 = 1.5$ kΩ, $R_2 = 15$ kΩ, and $C_1 = 0.02$ μF, determine the dc and rms values of the voltage developed across R_2.

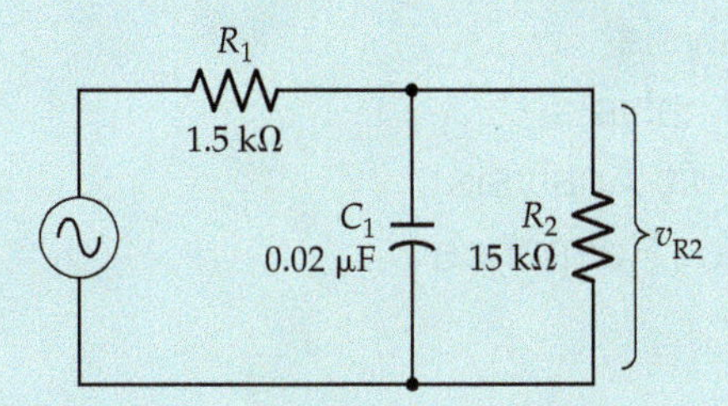

Figure 27-18

27-20 In the circuit in Problem 27-19, an inductor $L = 800$ mH with a coil resistance of $R_W = 18$ Ω is connected in series with R_1, and the capacitor is removed. Recalculate the dc and rms output voltages.

27-21 A half-wave rectified sine wave with a 50 V amplitude and a frequency of 400 Hz is applied to an impedance network. The network consists of a resistor $R_1 = 1.2$ kΩ in series with parallel-connected components $R_2 =$ 10 kΩ and $C_1 = 1000$ pF. Determine the dc and rms levels of the voltage developed across R_2.

27-22 An inductor $L = 1$ H with a coil resistance of $R_W = 10$ Ω is connected in series with R_1 in the circuit of Problem 27-21. The capacitor is disconnected. Recalculate the dc and rms output voltages.

Practice Problem Answers

27-3.1 0.955 V, 0.477 V, 0.318 V, 0.239 V, 0.191 V
27-3.2 5.73 V, 9 V, 3.82 V, 0.764 V
27-4.1 1.16 V
27-4.2 11.36 V
27-5.1 430 μA
27-5.2 0.91, 0.88

CHAPTER 28
Circuit Analysis by Computer

CONTENTS

Objectives

You will be able to use graphic analysis software to:

1. Analyze resistor networks to determine the various circuit voltage and current levels.
2. Analyze resistor networks that have several voltage and current sources and variable load resistors.
3. Perform transient analysis on *RC* and *RL* circuits to produce graphs of voltages and currents versus time.
4. Analyze impedance networks to determine the magnitude and phase of the circuit voltages and currents.
5. Determine the magnitude and phase of component voltage and currents in resonant circuits, and produce voltage/frequency and current/frequency graphs.
6. Analyze filter circuits to produce voltage/frequency and phase/frequency response graphs.

INTRODUCTION

Graphic analysis software packages allow users to draw a circuit on a computer screen by selecting graphic symbols for components, positioning them on the screen, and drawing lines to represent the connecting conductors. The computer then performs the circuit analysis and displays the requested voltages, currents, graphs, and so on. Several such packages are available, and there are many similarities in the procedures for using them. When a user becomes familiar with one system, it is not difficult to learn the procedures for another one. *Micro-Cap 9* (*MC9*) from *Spectrum Software* is the package mostly used with this presentation; however, application of *Electronic Workbench* from *Multisim*, and *PSpice* from *Orcad* is also demonstrated. The treatment is intended only as an introduction to the use of a computer for circuit analysis; detailed coverage is available in the on-line manuals provided by the software companies. A very simple program is offered first, and then progressively more interesting applications are treated.

28-1 SERIES RESISTOR CIRCUIT

Drawing the Circuit Using Micro-Cap (MC9)

Figure 28-1(a) shows the three-resistor series circuit from Examples 5-1 and 5-2. To draw the circuit on the computer screen (after the software is activated) proceed as follows:

- Select *Resistor* from the *Components* menu. Alternatively, the resistor symbol can be selected from the palette of frequently-used parts. Position the resistor at the desired location on the screen and click the mouse.

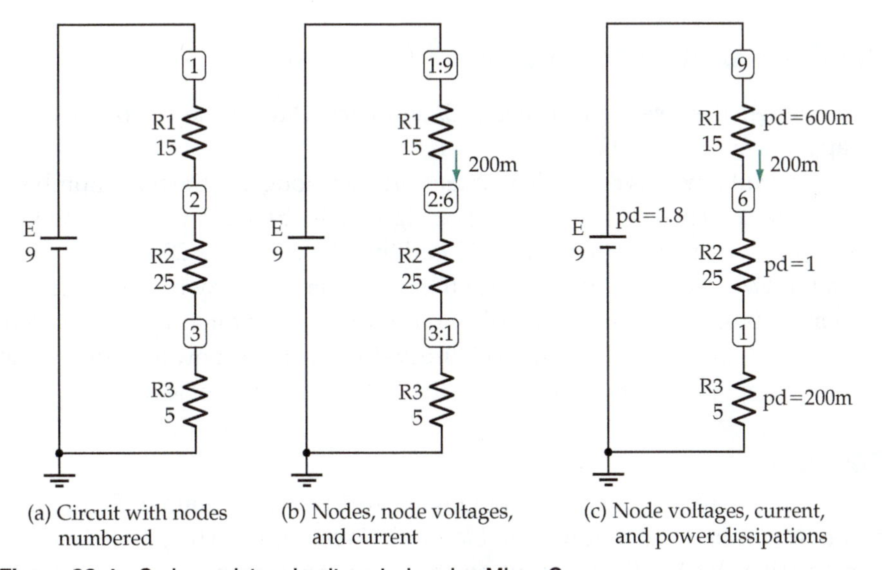

(a) Circuit with nodes numbered

(b) Nodes, node voltages, and current

(c) Node voltages, current, and power dissipations

Figure 28-1 Series resistor circuit analysis using Micro-Cap.

- Identify the resistor by name (*R1*, *R2*, etc.,) and resistance value in the dialog box that opens when the component is placed on the screen. Note that the resistance value does not use the ohms identification; *100* means 100 Ω, *100k* means 100 kΩ.
- Continue placing resistors and identifying them by name and value, as required.
- Select the pointer, and use it to position each resistor for drawing the circuit diagram. The *Rotation* tool may be used to rotate each component to the required orientation.
- From the components menu (or the parts pallet) select *Battery*, position it on the screen, and identify it by name and voltage in the dialog box. Once again, note that the voltage is identified only by quantity; *12* means 12 V, *12m* is 12 mV.
- The source voltage might not be displayed on the screen, although its value is recorded and may be checked by double-clicking on it to display the dialog box. Text (*T*) may be selected to type the voltage value beside the voltage source. Text may also be used to display comments on to the screen, to identify a circuit by number or name, for example.
- Select the *Wire Mode* symbol from the tool bar, and proceed to connect the circuit components by positioning the pointer on each component terminal, holding the mouse button, and dragging the line to the next component terminal or conductor junction.
- Select *Ground* (from the components menu or the parts pallet), and position it at the appropriate point in the circuit.
- If the file is not yet saved, go to *Save as*, name the file, and save it in an appropriate location.
- In the *View* menu (under *Options* in MC9) select *Node Numbers*, *Node Voltages/States*, and *Current*. The circuit should now look like Figure 28-1(a).

Analysis Using Micro-Cap

- Under *Analysis* select *Dynamic DC*, and click *O.K.* in the dialog box that appears on the screen.
- The circuit now has the node voltages printed alongside the node numbers, and the current level identified. See Figure 28-1(b). Note that the voltages for this circuit are 9 V, 6 V, and 1 V, and that the current is 200 mA.
- In the *View* menu (under *Options*) the node numbers may be deselected, to remove possible confusion with the node voltages. *Power* may also be selected. This gives the circuit diagram in Figure 28-1(c) where the power dissipation in each component is show along with the voltage and current levels.

Multisim

The three-resistor series circuit from Figure 28-1 is drawn again in Figure 28-2 using *Multisim* (also known as *Electronic Workbench*). The procedure for constructing the Multisim circuit on the computer screen is quite similar to

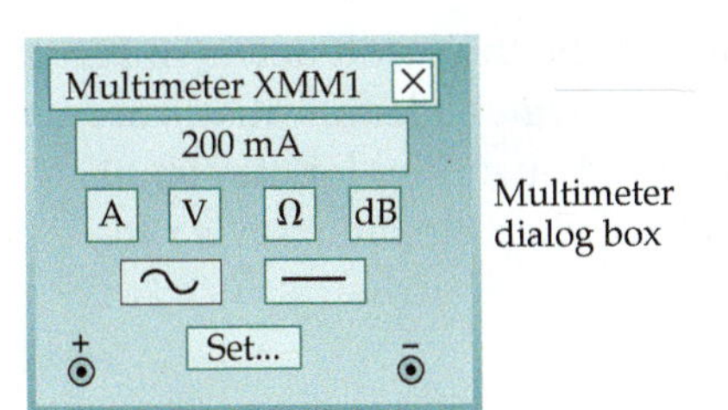

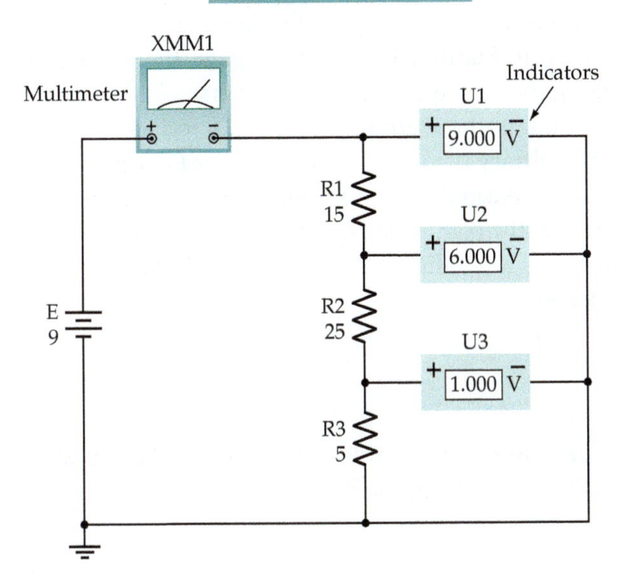

Figure 28-2 Multisim (Electronic Workbench) analysis of a series resistor circuit. Note the use of *Multimeters* and *Indicators*.

that for MC9. Components are selected either from menus or by clicking (or double-clicking) on graphic symbols. The components are then defined in their dialog boxes, and positioned on the screen. Connecting wires are drawn by placing the courser on a component terminal where it becomes cross-hairs. The mouse button is clicked and then the connecting conductor is traced to the next component terminal where the mouse is again clicked.

The Multisim circuit in Figure 28-2 appears different from the MC9 presentation because Multisim uses *Virtual Multimeters.* These are shown connected into the circuit as they would be in a laboratory situation. The ammeter (*XMM1*) is shown in series with the circuit dc supply, and the voltmeters (*U1, U2,* and *U3*) are connected between ground and the points where the voltage is to be measured.

Multimeters are selected from the instrument pallet (or from menus), and each one is identified as an ammeter or voltmeter (or as an ohmmeter or dB meter) by double-clicking on the multimeter symbol and selecting the appropriate button in the meter dialog box. With *A* and the horizontal line in the dialog box (at the top of the illustration) selected, meter *XMM1* is defined as a dc ammeter.

Multimeters (like *XMM1*) could also be used for the other three meters in the circuit, and each one would then have its associated dialog box on the screen to display the measured quantity. Instead of multimeters, the more

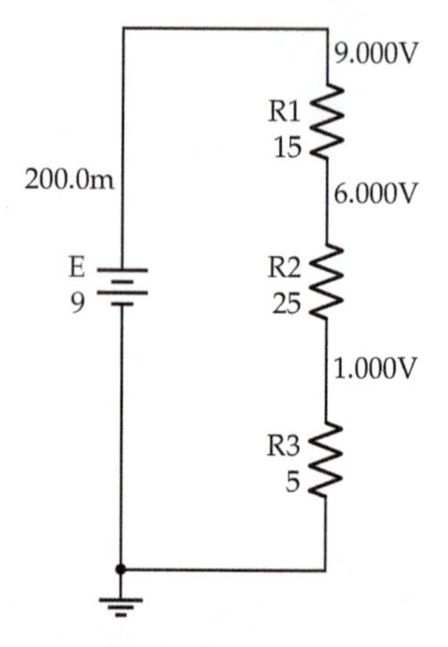

Figure 28-3 Series resistor circuit analysis by PSpice. The analyzed circuit is quite similar to that produced by the Micro-Cap analysis.

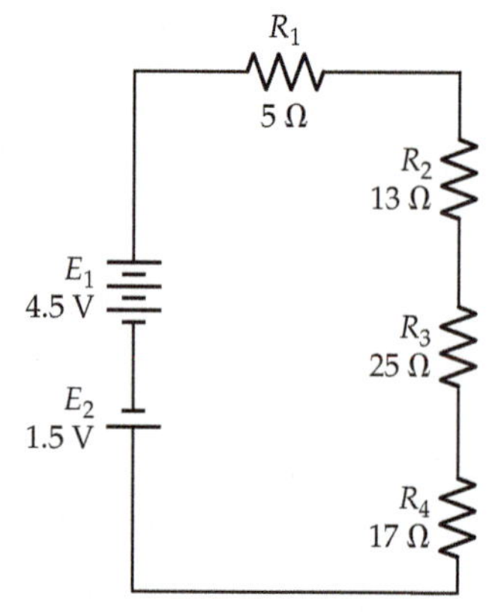

Figure 28-4 Series resistor circuit for Practice Problem 28-1.1.

compact *Indicators* are used for the three voltmeters in Figure 28-2. These are selected from the parts pallet (a *Figure 8* identification), or from the *Components* menu under *Place*. The measured quantities for the circuit are read from meter dialog boxes and/or indicators when the program is run.

PSpice

The PSpice version of the three-resistor series circuit shown in Figure 28-3 looks quite like the MC9 circuit in Figure 28-1(b). Once again, the procedure for constructing and analyzing the circuit using PSpice is fairly similar to that described for use of MC9, although there are important differences. Components are selected from a parts library using the *Get new part* or *Place part* command in the *Draw* menu. *Wire* is selected from the *Draw* menu, and the pencil icon that appears is used to draw the connecting conductors between the components. Once again, components are defined in dialog boxes that open when the component is double clicked.

Practice Problem

28-1.1 Use computer graphic software to analyze the circuit in Figure 28-4. Compare the results to those in Example 5-3.

28-2 SERIES-PARALLEL RESISTOR CIRCUITS

Micro-Cap

The procedure for drawing and analyzing a series-parallel resistor circuit, however complex, is precisely the same as for the simple series circuit discussed in the previous section. Figure 28-5 shows a series-parallel circuit (from Examples 7-3 and 7-5) analyzed using MC9, with the node voltages and

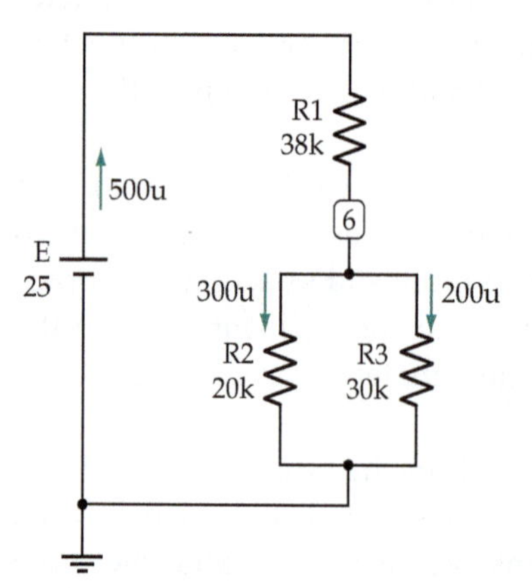

Figure 28-5 Series-parallel resistor circuit analysis by Micro-Cap.

current levels displayed. The resistor junction voltage is seen to be 6 V, and the branch currents are 500 μA, 300 μA, and 200 μA.

Multisim

The Multisim version of the series-parallel circuit in Figure 28-6 once again looks different from the MC9 version because of the *Indicators* used. Here too, regardless of how complex the resistor circuit, the procedure for construction and analysis is exactly the same as for the series resistor circuit. The displayed current levels differ slightly from those for the MC9 analysis.

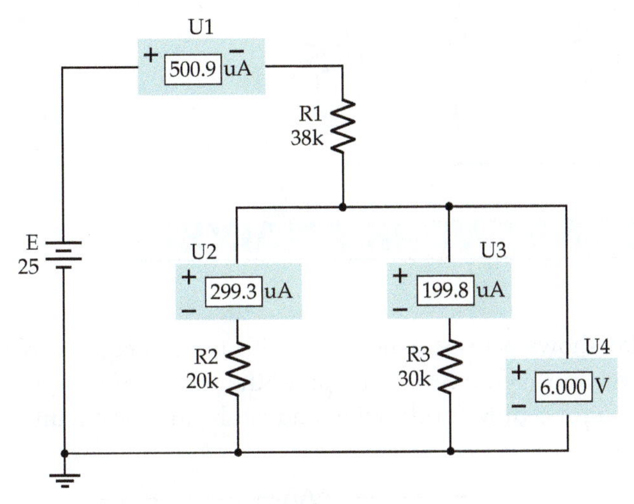

Figure 28-6 Series-parallel resistor circuit analysis by Multisim.

PSpice

The PSpice analysis of the series-parallel circuit shown in Figure 28-7 once more looks quite similar to the MC9 version in Figure 28-5. The only apparent difference is in how the node voltages are presented. Here again, the circuit construction and analysis procedure is the same as for the series circuit.

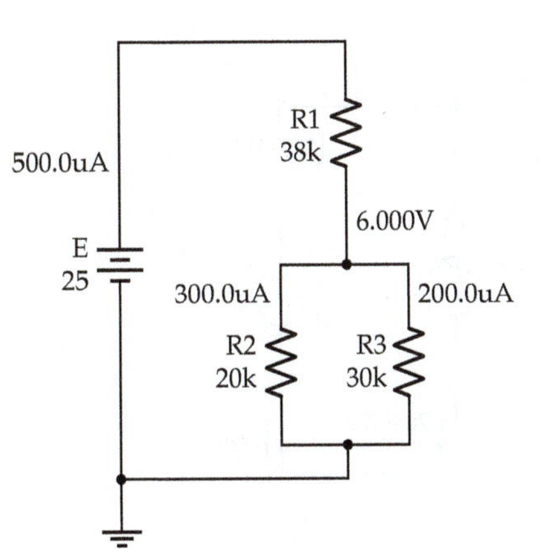

Figure 28-7 Series-parallel resistor circuit analysis by PSpice.

Practice Problem

28-2.1 Determine all current and voltage levels for the circuit in Figure 28-8. Compare the results to Example 7-11.

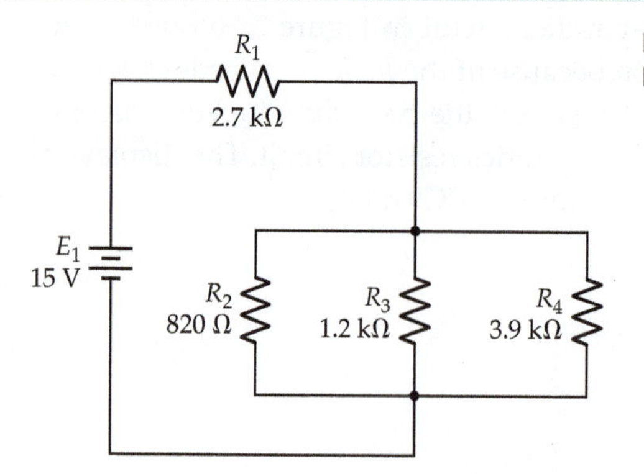

Figure 28-8 Circuit for Practice Problem 28-2-1.

28-3 DC NETWORK ANALYSIS

Micro-Cap

Figure 28-9(a) shows a resistor network with two voltage sources and one current source, and Figure 28-9(b) displays the results of computer analysis. Once again the procedure for drawing and analyzing the circuit is exactly as

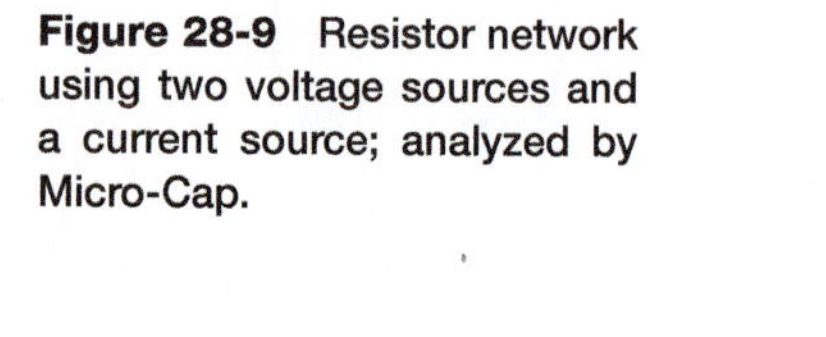

Figure 28-9 Resistor network using two voltage sources and a current source; analyzed by Micro-Cap.

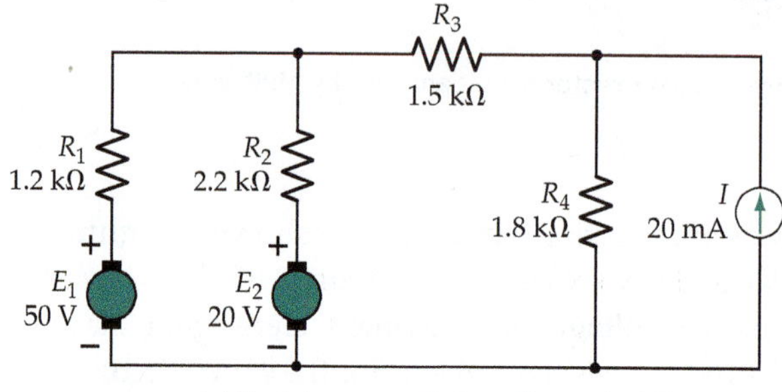

(a) Resistor network from Example 8-5

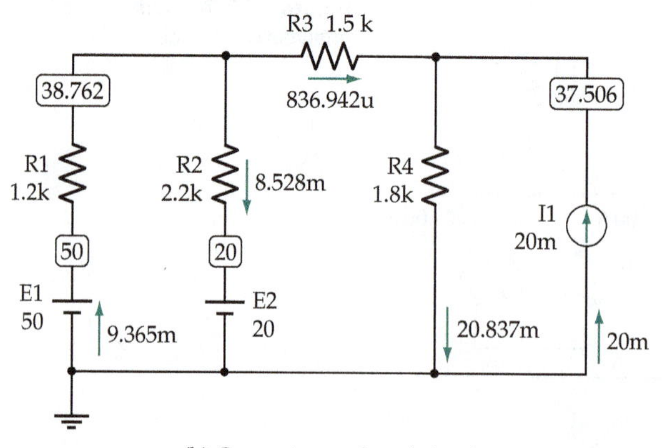

(b) Computer analyzed circuit

already discussed. Batteries are selected for voltage sources E_1 and E_2, as usual, and a current source is selected for I_1. All current and voltage levels are displayed in the final circuit, and the results verify the I_3 calculation (obtained with considerably more effort) in Example 8-5.

Multisim

The Multisim version of the analysis of the network from Figure 28-9(a) is shown in Figure 28-10. Once again the Multisim circuit appears more complex than the MC9 circuit because of the use of Indicators, and because of the need for additional wiring to connect the voltmeters.

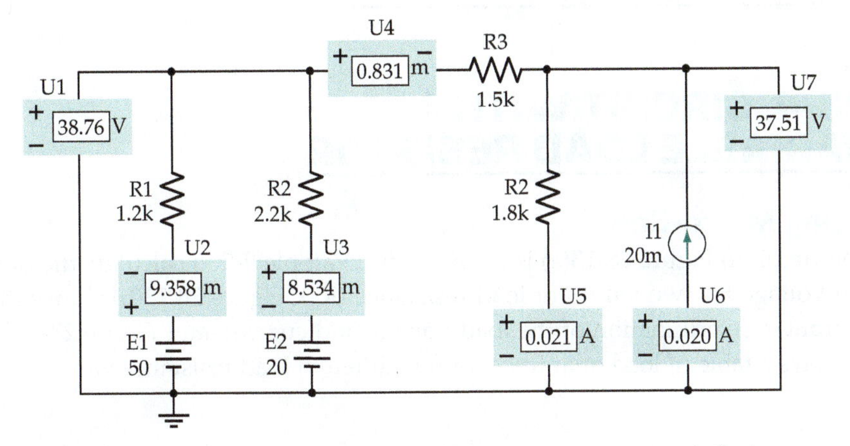

Figure 28-10 Resistor network analyzed by Multisim. Voltage or current *Indicators* are used wherever a quantity is to be measured.

PSpice

Figure 28-11 shows the PSpice version of the Figure 28-9(a) network analysis. As always, the similarities to the MC9 presentation are obvious.

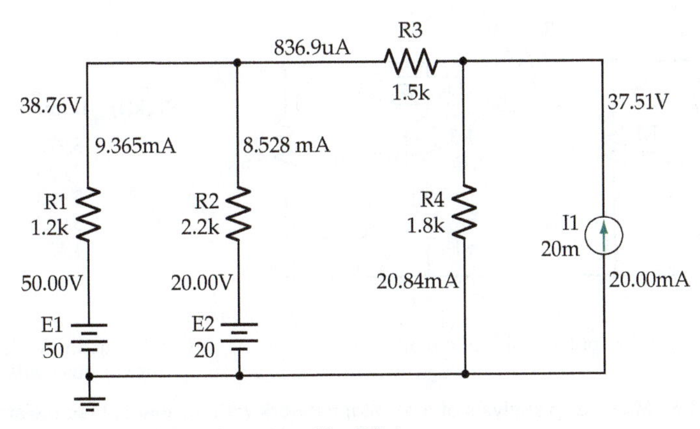

Figure 28-11 Resistor network analyzed by PSpice.

Practice Problem

28-3.1 Analyze the circuit in Figure 28-12 to determine the circuit current and voltage levels. Compare the results to Example 8-2.

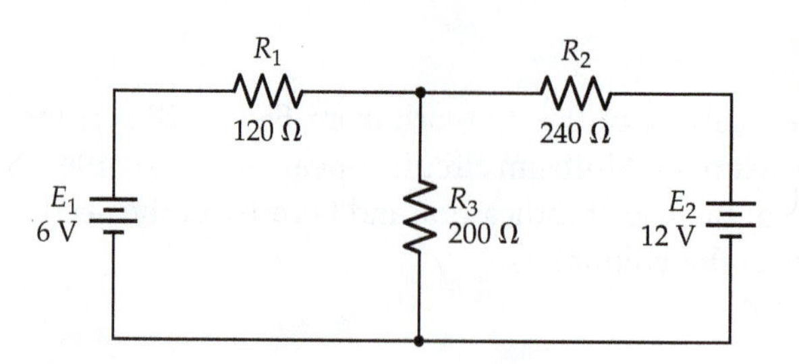

Figure 28-12 Resistor network circuit for Practice Problem 28-3-1.

28-4 CIRCUIT WITH A VARIABLE LOAD RESISTOR

Using Micro-Cap

The circuit in Figure 28-13(a) is analyzed in Example 9-5 to calculate the output voltage for two values of load resistance (R_L). Figure 28-13(b) shows the computer determination of the load voltage and current, and Figure 28-13(c) displays a table of load quantities for five different load resistance values.

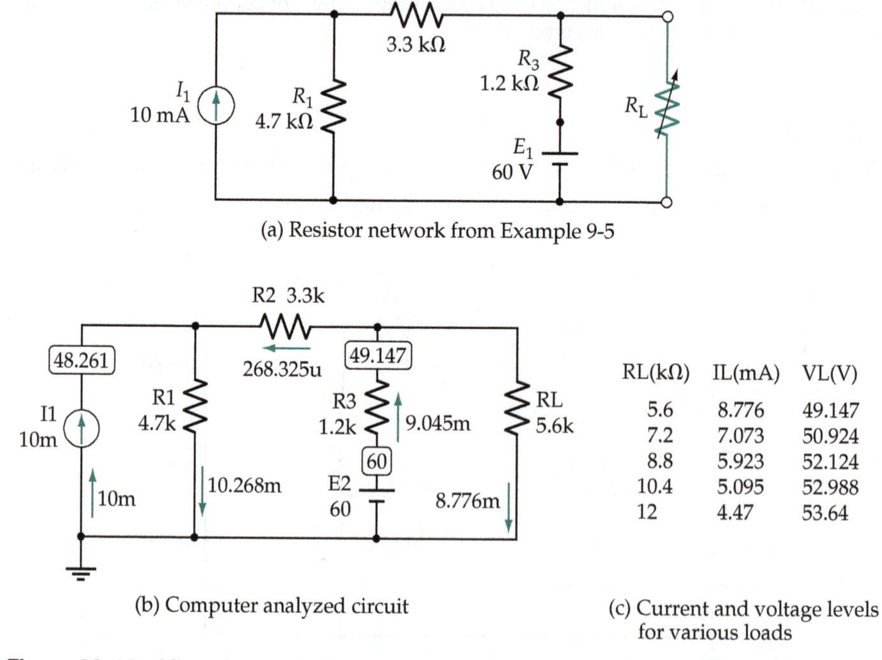

(a) Resistor network from Example 9-5

(b) Computer analyzed circuit

RL(kΩ)	IL(mA)	VL(V)
5.6	8.776	49.147
7.2	7.073	50.924
8.8	5.923	52.124
10.4	5.095	52.988
12	4.47	53.64

(c) Current and voltage levels for various loads

Figure 28-13 Micro-Cap analysis of a resistor network with a variable load resistor.

The circuit is drawn in the usual way, and the resistor to be changed is identified as *RL*. The analysis procedure is:

- Select *Stepping* from the *Dynamic DC* menu.
- Identify *RL* in the dialog box as the component to be stepped.
- Show that stepping is to occur from 5.6k to 12k in 1.6k steps.
- Click *Start*, and note the load voltage and current levels.
- Click *Next*, and again note the load voltage and current levels.
- Continue clicking *Next* and noting the quantities until all the load values have been worked through.

Practice Problem

28-4.1 Determine the load current and voltage levels for the circuit in Figure 28-14 for the following load resistor values: 5 Ω, 10 Ω, 25 Ω, 50 Ω, 100 Ω, 250 Ω, 500 Ω. Compare the computer analysis results to Example 9-9.

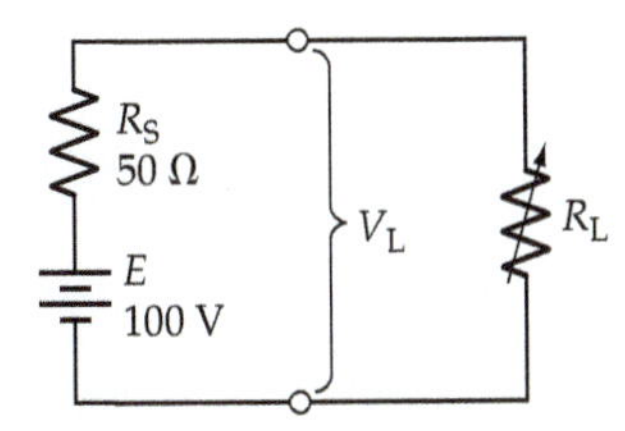

Figure 28-14 Variable load resistor circuit for Practice Problem 28-4-1.

28-5 TRANSIENT ANALYSIS

Using Micro-Cap

To investigate the voltage/time response of an *RC* circuit with a dc supply as in Figure 28-15(a), the circuit is drawn on the computer screen with a pulse source instead of a battery and switch [Figure 28-15(b)]. When *Pulse Source* is selected from the components menu a dialog box is opened for the source to be specified. The pulse characteristics must be defined in terms of the various times involved (time delay to zero level, time delay to trailing edge, etc.,), and the pulse maximum and minimum voltages must be listed. In the case of the circuit shown, the voltage levels are zero and 10 V, and the time period selected is 10 ms, which is 5CR (the time for a capacitor in an *RC* circuit to become charged to 99% of the dc input voltage). These quantities define the vertical and horizontal axis of the voltage/time graph.

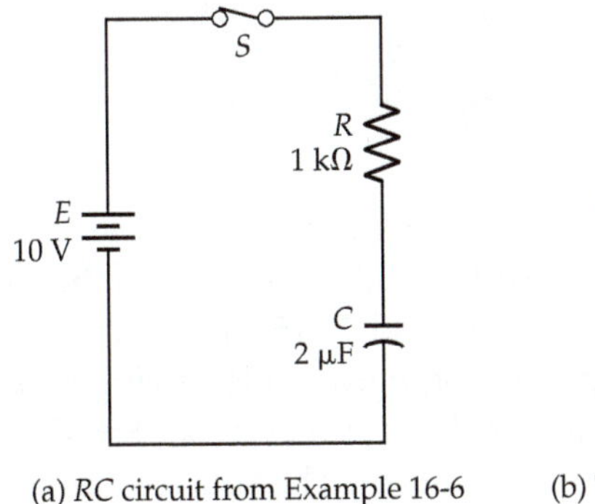

(a) *RC* circuit from Example 16-6

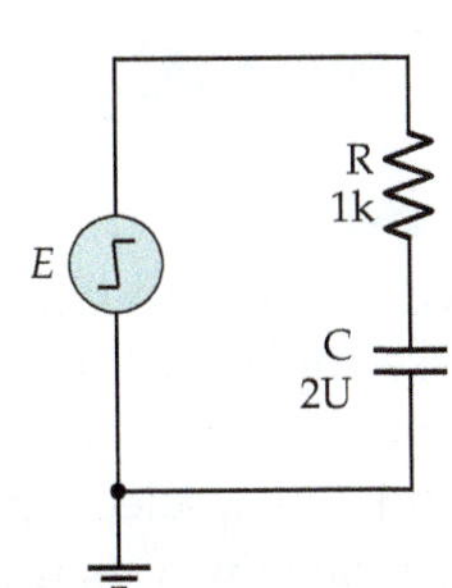

(b) The *RC* circuit drawn on the computer uses a pulse source

Figure 28-15 Micro-Cap analysis of an *RC* circuit for plotting the graph of capacitor voltage versus time.

When the pulse source dialog box is completed, *O.K.* is clicked, and *Transisent* is selected under *Analysis*. An *Analysis Limits* box is opened in which the time range is again specified as 10 ms and a maximum time step is defined. The maximum time step determines how smooth the voltage versus time graph will be (the number of points plotted on the graph). For this particular circuit the time step is listed as 100 ns. *Run* is clicked, and the graph of capacitor voltage versus time is immediately displayed, as illustrated in Figure 28-16. Note that this is the circuit analyzed in Example 16-6.

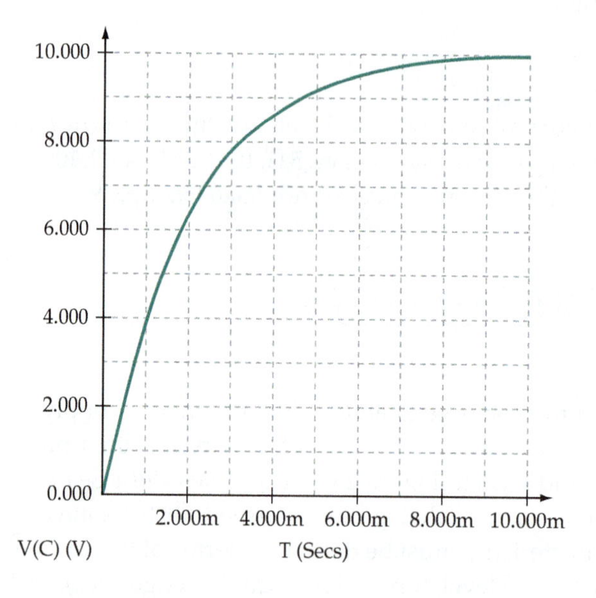

Figure 28-16 Capacitor voltage versus time graph for the *RC* circuit; analyzed by Micro-Cap.

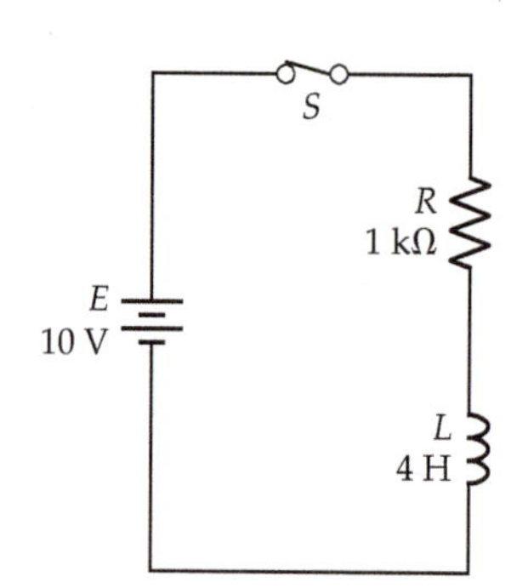

Figure 28-17 *RL* series circuit for Practice Problem 28-5-1.

Practice Problem

28-5.1 Analyze the *RL* circuit in Figure 28-17 to plot a graph of I_L versus time. Compare the results to Example 16-2 and Figure 16-4.

28-6 AC NETWORK ANALYSIS

Using Micro-Cap

The impedance network in Figure 28-18(a) has two 5 kHz voltage sources; one with a phase angle of 0°, and the other with −25°. (This is the circuit from Figure 22-31, analyzed in Problems 22-14 and 22-15.) The circuit is drawn on the screen as illustrated, selecting, positioning, and identifying inductors and capacitors in the same way as resistors. *Voltage Source* (not *Sine Source* which has a 1 V amplitude) is selected for E_1 and E_2. The sources are identified as *Sine* in the dialog box, and are defined in terms of voltage amplitude, frequency, and phase angle. *Dynamic AC analysis* is selected, producing a display

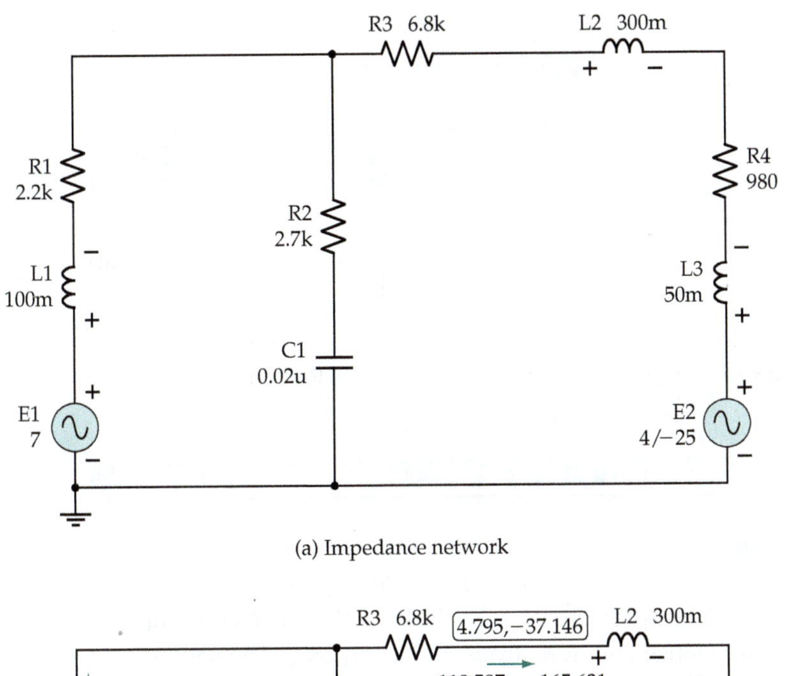

(a) Impedance network

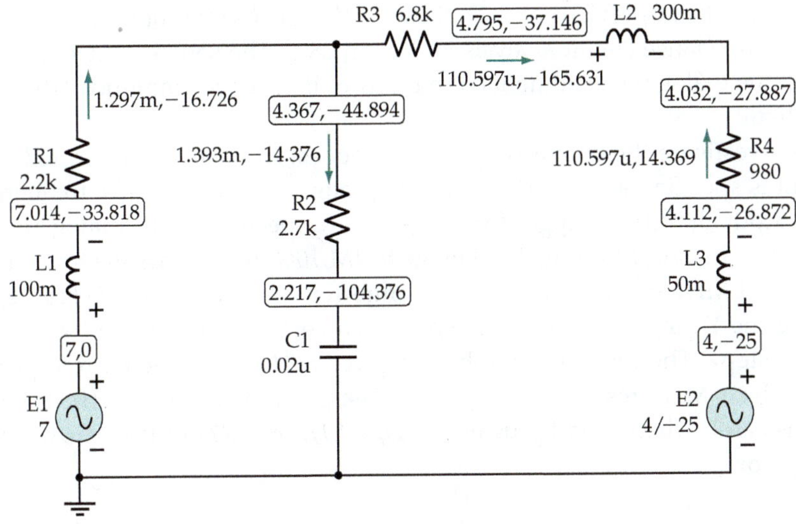

(b) Node voltages and currents

Figure 28-18 Impedance network analysis by Micro-Cap. Note that the voltage and current levels have their phase angles alongside.

of node voltage and current levels together with phase angles, as shown in Figure 28-18(b).

Practice Problem

28-6.1 Analyze the impedance network in Figure 28-19 to determine the component voltages and currents. Compare the computer output to the results for Problem 22-16.

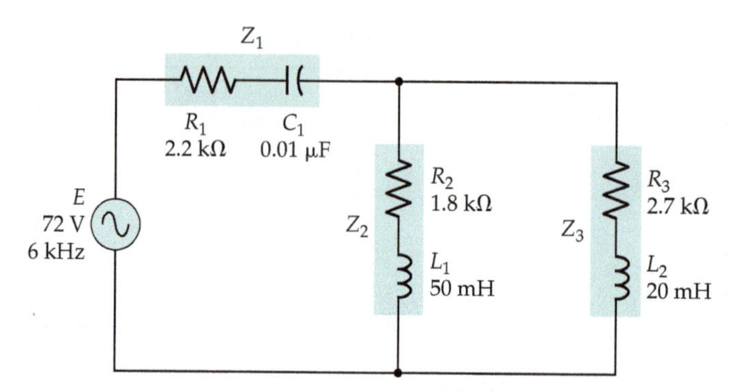

Figure 28-19 Impedance network circuit for Practice Problem 28-6.1.

28-7 RESONANT CIRCUIT ANALYSIS

Using Micro-Cap

The series resonant circuit in Figure 28-20 (reproduced from Figure 23-4 and analyzed in Examples 23-1 and 23-2) uses a *Voltage Source* (not *Sine Source*), as for the impedance circuit analysis in Section 28-6. The source is identified as *Sine* in the dialog box, and defined in terms of its voltage amplitude (10 V for this circuit).

To investigate the circuit frequency response, *AC analysis* is selected, the output is specified as *I(R1)* (the current through resistor *R1*). The frequency and output amplitude ranges for the graph are specified in the dialog box as *X range* and *Y range*. Setting the *X* range to *4M,100k,100* defines the frequency range as 4 MHz maximum, 100 kHz minimum, to be tested at 100 Hz steps. Setting the *Y* range to *100m,0,10m* specifies a vertical scale of 0 to 100 mA in 10 mA steps. The resulting graph in Figure 28-21 shows the current peaking at the 1 MHz resonance frequency. The component voltage/frequency graphs can be displayed by using *v(C1)*, *v(L1)*, or *v(R1)* in the *AC Analysis* dialog box.

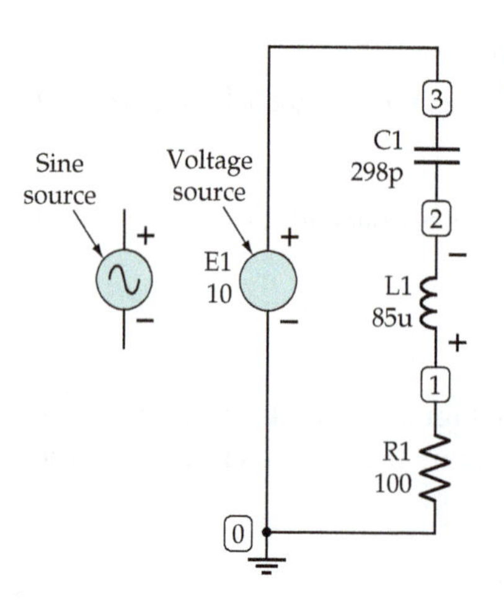

Figure 28-20 Series resonant circuit analysis using Micro-Cap.

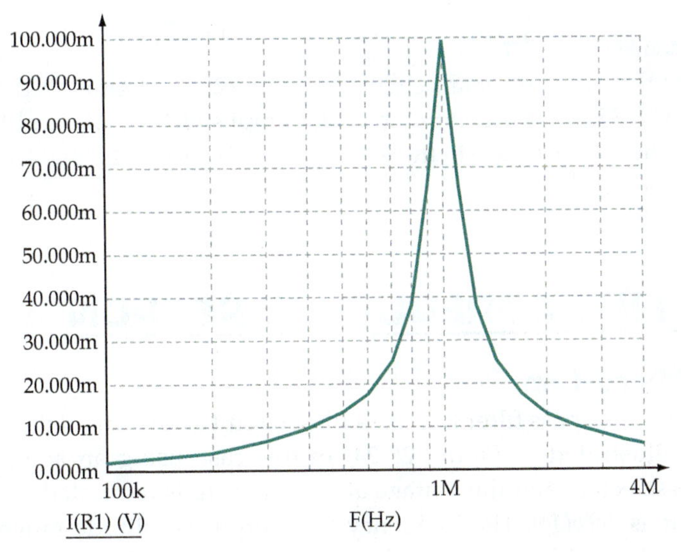

Figure 28-21 Micro-Cap current versus frequency graph for a series resonant circuit.

The decibel/frequency display of the response close to resonance in Figure 28-22 is obtained by substituting a *Sine Source* in place of the *Voltage Source*, selecting *v(R1)* as the output, resetting the *X* range to *2M,600k,100* (600 kHz to 2 MHz with 100 Hz steps), and resetting the *Y* range to *1,–15,3* (–15 dB to 1 dB with 3 dB steps). Note that the precise frequency and attenuation are displayed on the screen when the pointer is set at any point on the graph. This is convenient for determining the –3 dB frequencies.

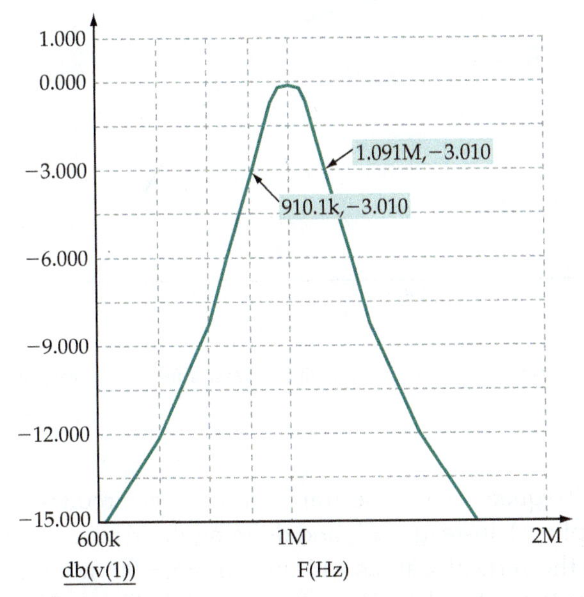

Figure 28-22 Micro-Cap graph of *node 1* voltage (in dB) versus frequency for the series resonant circuit close to the resonance frequency.

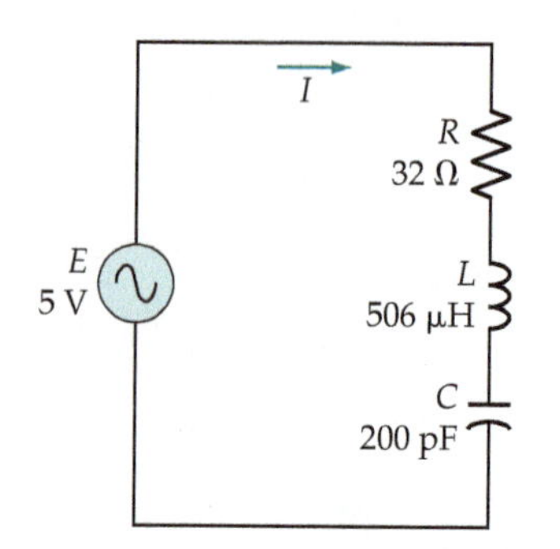

Figure 28-23 Series resonant circuit for Practice Problem 28-7.1.

Practice Problem

27-7.1 For the series resonant circuit in Figure 28-23 (reproduced from Figure 23-35) use the computer to plot graphs of I, V_R, V_L, and V_C versus frequency. Compare to the results for Problems 23-1, 23-2, and 23-3.

28-8 FILTER FREQUENCY RESPONSE

Using Micro-Cap

A simple *RC* low-pass filter low-pass filter circuit as drawn on the computer screen is illustrated in Figure 28-24, in this case using an ac source. *AC analysis* is selected, and the voltage at node 1 in dB is requested by defining the output as *db(v(1))*. The frequency and output amplitude ranges for the filter response graph are specified in the dialog box as *X range* and *Y range*. Setting the *X* range to *20k,10,200* defines the frequency range as 20 kHz maximum, 10 Hz minimum, tested at 200 Hz steps. Setting the *Y* range to *3,−15,3* specifies a vertical scale of +3 db maximum to −15 dB minimum in 3 dB steps. The resulting amplitude/frequency response graph is shown in Figure 28-25.

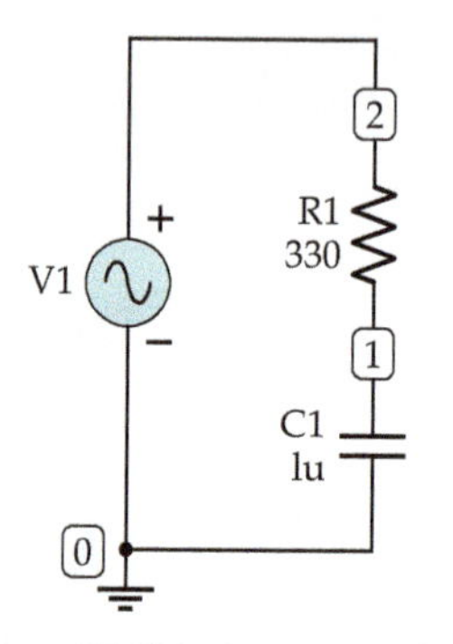

Figure 28-24 Low-pass filter circuit analysis using Micro-Cap.

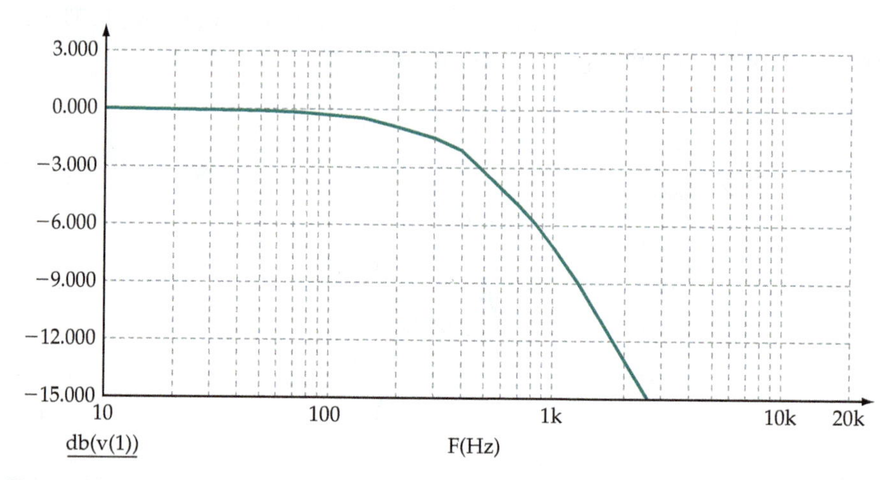

Figure 28-25 Low-pass filter output amplitude versus frequency response graph plotted using Micro-Cap.

The phase/frequency response graph for the filter circuit is plotted by specifying *ph* (for phase), instead of *db*, and defining the *Y* range in degrees. Using *10,−90,10* sets the vertical scale as +10° maximum, −90° minimum, with steps of 10°. This produces the phase/frequency graph in Figure 28-26.

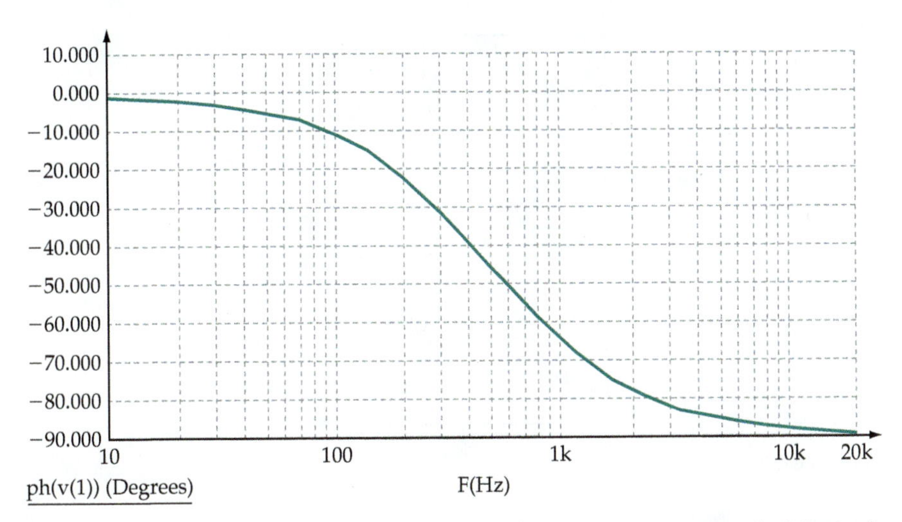

Figure 28-26 Low-pass filter output phase versus frequency response graph plotted using Micro-Cap.

Practice Problem

28-8.1 Analyze the high-pass filter circuit in Figure 28-27, to produce the amplitude/frequency and phase/frequency response graphs.

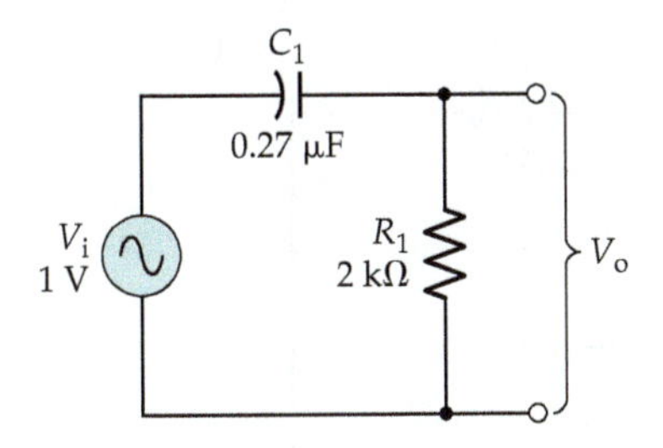

Figure 28-27 High-pass filter circuit for Practice Problem 28-8.1.

Problems

Section 28-1

28-1 Computer analyze the circuit in Figure 28-28 (reproduced from Figure 5-19) to determine the circuit current and the component voltage drops. Compare the results to those for Problem 5-20.

28-2 The circuit in Figure 28-29 is reproduced from Figure 5-17. Determine the node voltages for the circuit, and compare to the results for Problem 5-9.

28-3 Analyze the circuit in Figure 28-30 to determine the potential divider output voltages. Compare to the results for Problem 5-15.

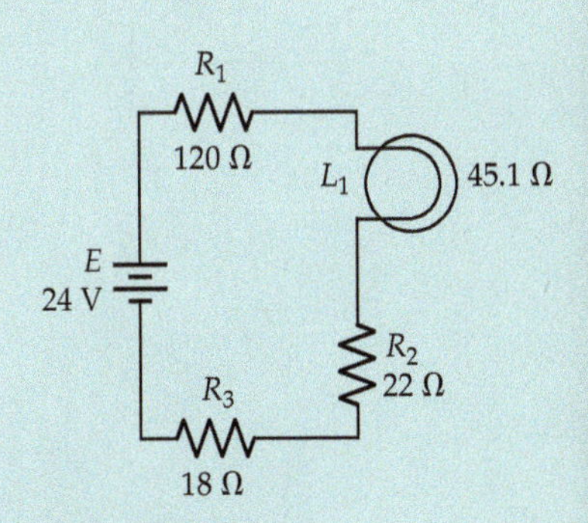

Figure 28-28

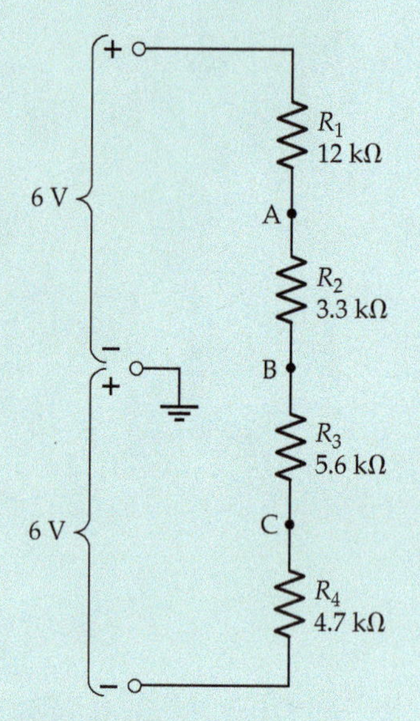

Figure 28-29

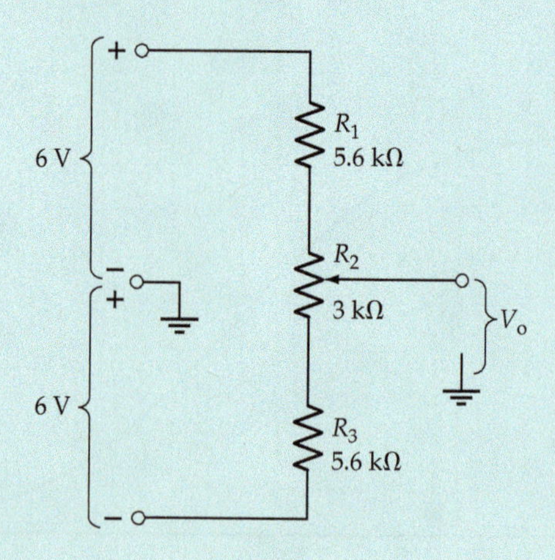

Figure 28-30

Section 28-2

28-4 Refer to Figure 28-31 [reproduced from Figure 7-10(a)]. Analyze the circuit to determine all component voltages and currents. Compare to the results for Example 7-10.

28-5 Determine all component voltages and currents for the circuit in Figure 28-32 (reproduced from Figure 7-19). Compare to the results for Problem 7-22.

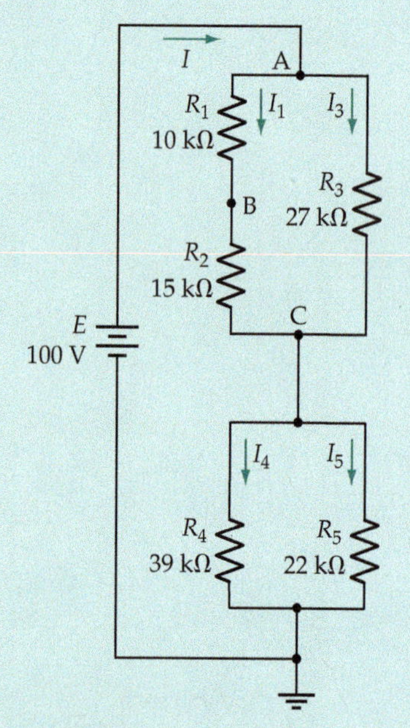

Figure 28-31

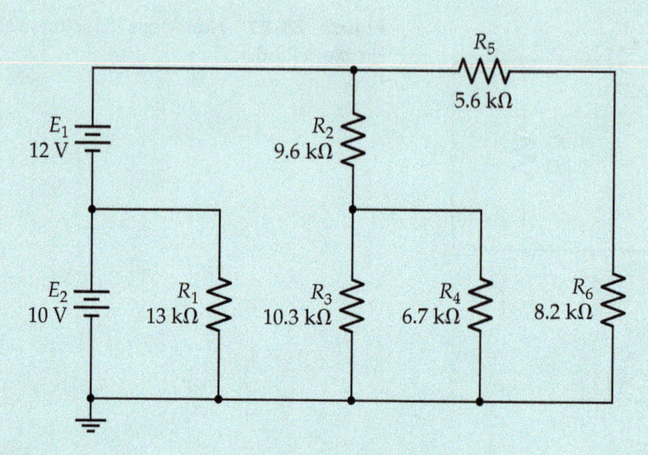

Figure 28-32

28-6 Determine the voltages at points *A*, *B*, and *C* in the circuit in Figure 28-33 (reproduced from Figure 7-20). Compare to the results of Problem 7-24.

Section 28-3

28-7 Use the computer to determine all component currents and voltages in Figure 28-34 (reproduced from Figure 8-25). Compare to the results of Problem 8-9.

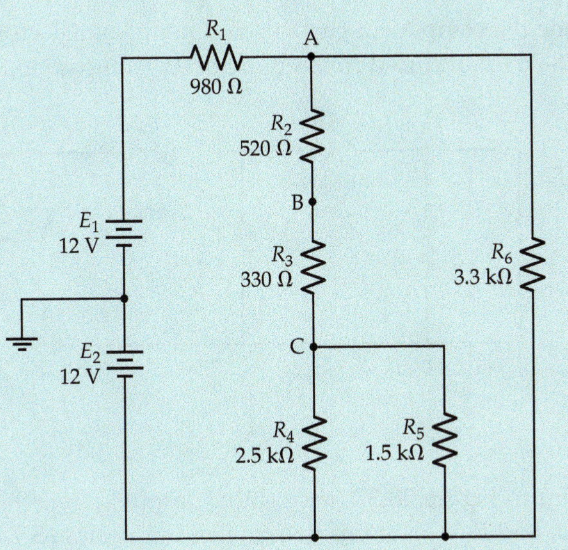

Figure 28-33

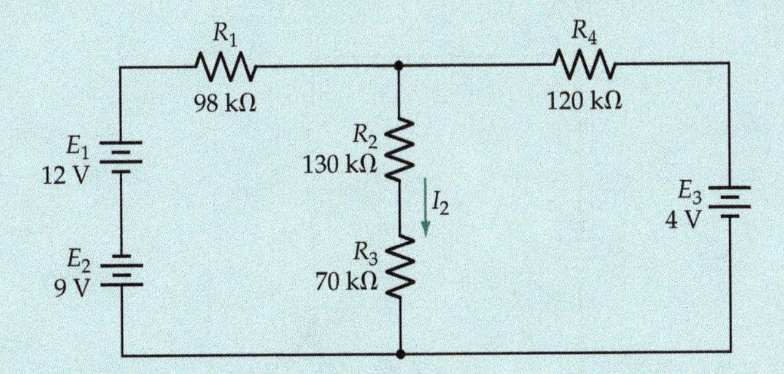

Figure 28-34

28-8 Use the computer to calculate the current through R_5 in Figure 28-35 (reproduced from Figure 8-28), as requested in Problem 8-12. Include the determination of all component currents and voltages.

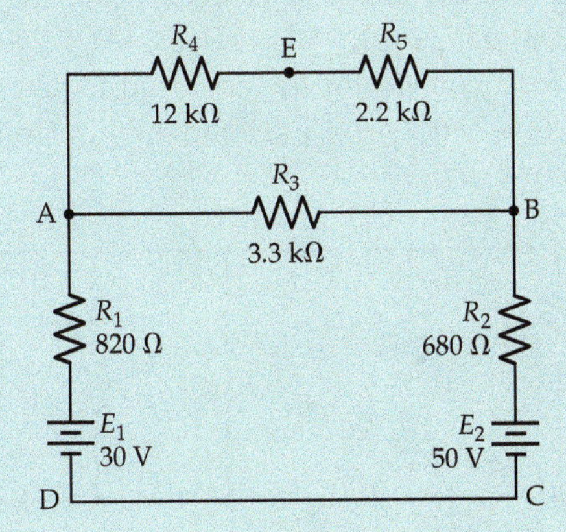

Figure 28-35

28-9 Determine the component currents and node voltages for the circuit in Figure 28-36 (reproduced from Figure 8-31). Compare to Problem 8-25.

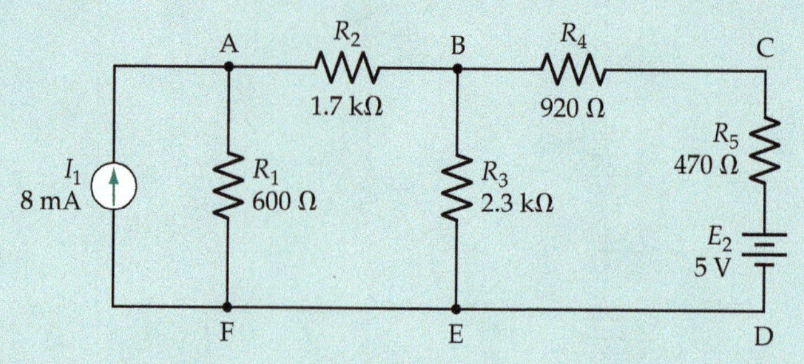

Figure 28-36

Section 28-4

28-10 The circuit in Figure 28-37 (reproduced from Figure 5-9) has $E = 16$ V, $R_1 = 1$ kΩ, and $R_2 = 560$ Ω, as determined in Example 5-7. Use the computer to calculate V_2 for the following load resistor values: 10 kΩ, 20 kΩ, 30 kΩ, 40 kΩ, 50 kΩ, 60 kΩ.

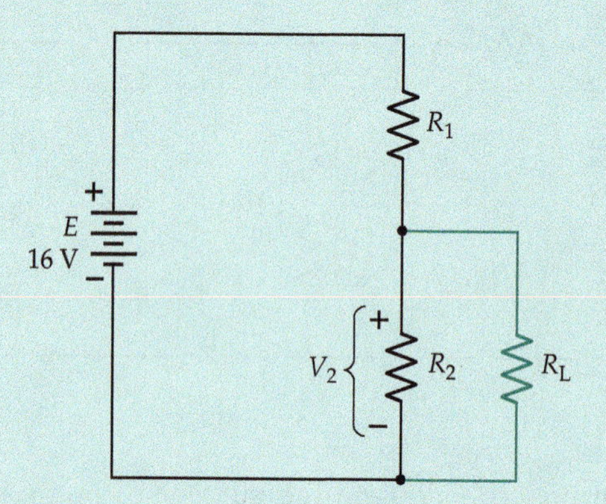

Figure 28-37

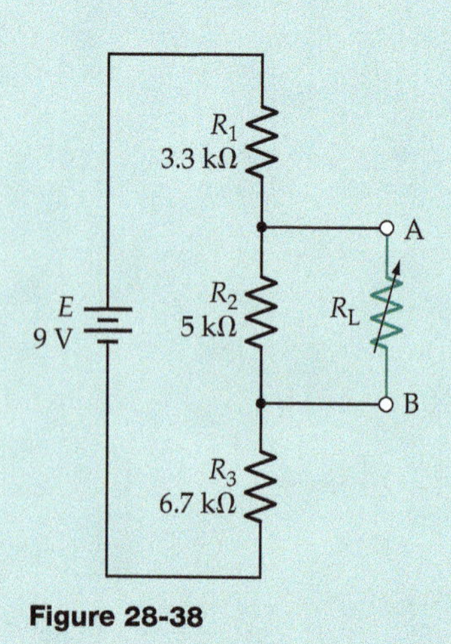

Figure 28-38

28-11 Example 9-6 analyzes the circuit in Figure 9-6(a) (reproduced in Figure 28-38) to determine the load current when $R_L = 5.6$ kΩ. Use the computer to calculate I_L for $R_L = 5.6$ kΩ, 6.8 kΩ, 8.2 kΩ, and 10 kΩ.

28-12 Determine the load current for the circuit in Figure 28-39 (reproduced from Figure 9-26) when $R_L = 3.3$ kΩ and 4.7 kΩ. Compare the reults to those for Problem 9-27.

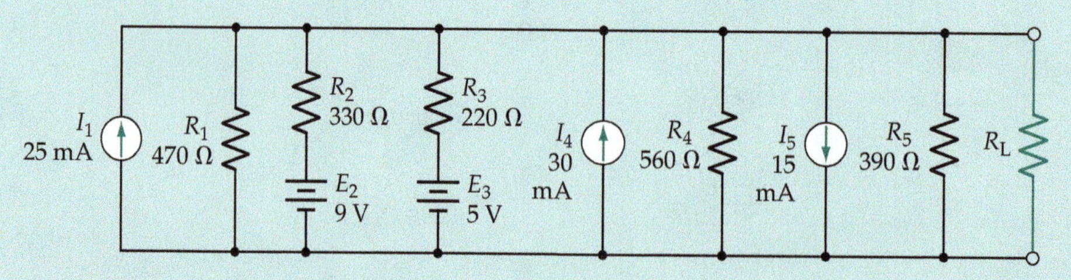

Figure 28-39

Section 28-5

28-13 A series *RC* circuit has $R = 3.3\ \text{k}\Omega$, $C = 10\ \mu\text{F}$, and $E = 12$ V, as in Figure 28-40 and in Practice Problem 16-6.1. Use the computer to draw the graphs of capacitor current and voltage from the instant of supply switch *on* until the capacitor voltage approaches maximum.

28-14 The *RC* circuit in Problem 28-13 has its supply switched *on* and *off* as illustrated in Figure 16-14. Use the computer to plot the graph of capacitor voltage versus time for $t_1 = t_2 = t_3 = 5CR$.

28-15 Modify Problem 28-14 to plot the v_C/t graph when $t_1 = t_2 = t_3 = CR$.

28-16 Plot the current versus time graph for the circuit in Figure 28-41 (reproduced from Figure 16-16). Compare to the results for Problem 16-7.

28-17 Modify the circuit for Problem 28-16 to investigate the effects of including a 5 kΩ resistor in parallel with the inductor.

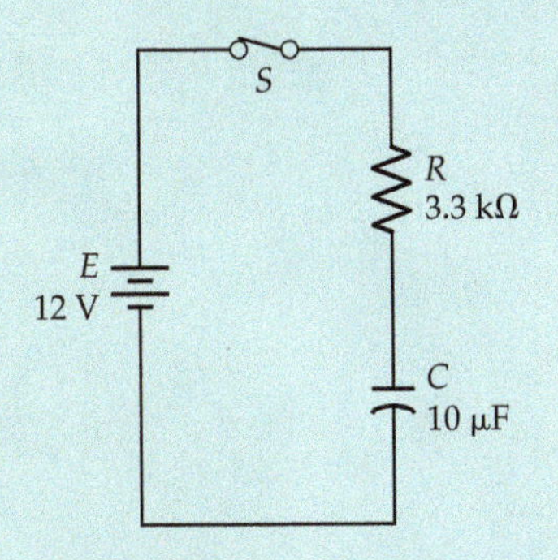

Figure 28-40

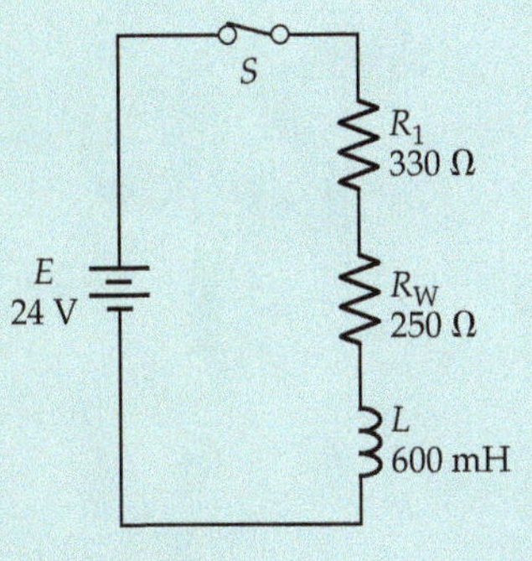

Figure 28-41

Section 28-6

28-18 Determine voltage drops across L_2 and C_2 in the impedance network in Figure 28-42 (from Figure 20-17). Compare to Problem 20-15.

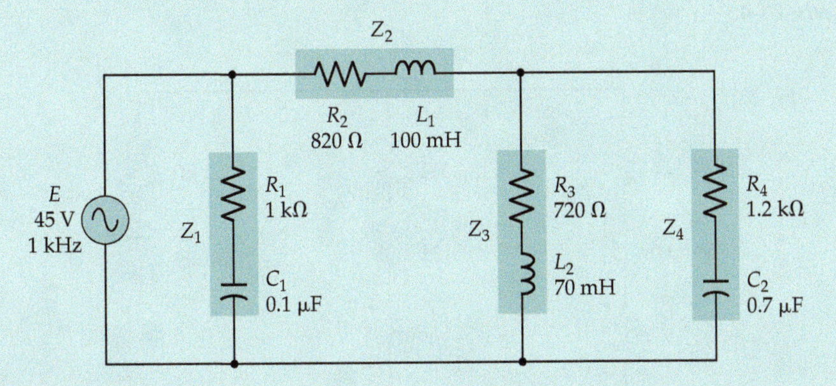

Figure 28-42

28-19 Determine the voltages across R_3 and C_2 in Figure 28-43 (from Figure 20-20). Compare to Problem 20-25.

28-20 Determine all component currents and voltages for the circuit in Figure 28-44.

Section 28-7

28-21 Analyze the resonant circuit in Figure 28-45 [from Figure 23-16(a)] for $L = 100\ \mu\text{H}$ and for $L = 200\ \mu\text{H}$, to determine the half-power frequencies and to prepare current-versus-frequency graphs for each case. Compare to Example 23-5.

28-22 The parallel resonant circuit in Figure 28-46 (from Figure 23-19) has $E = 100$ mV, $L = 150\ \mu\text{H}$, $R_W = 15\ \Omega$, and $C = 750$ pF. Analyzed the circuit to determine component currents and voltages and to draw current-versus-frequency graphs. Compare to Example 23-6.

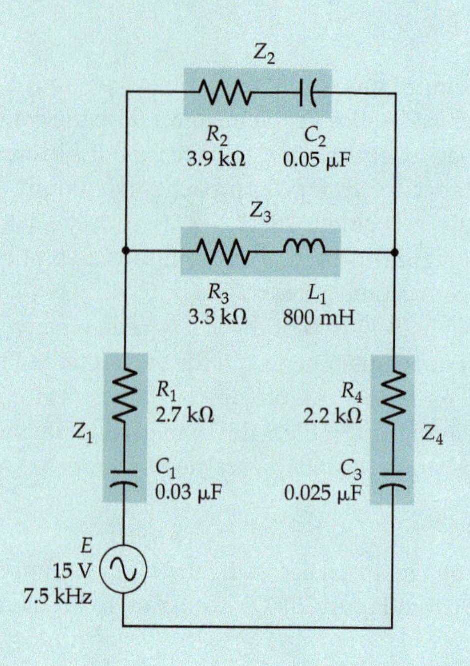

Figure 28-43

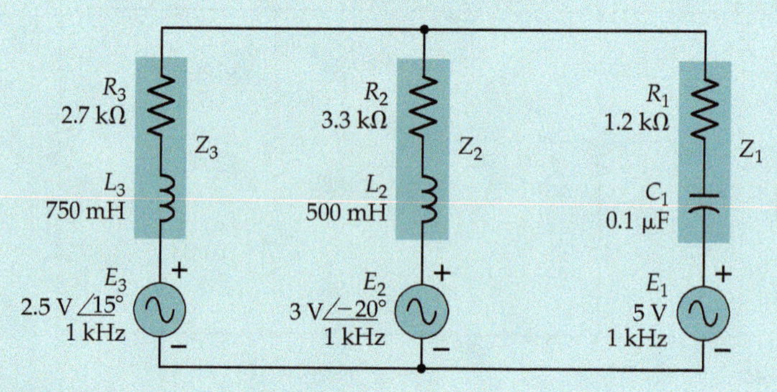

Figure 28-44

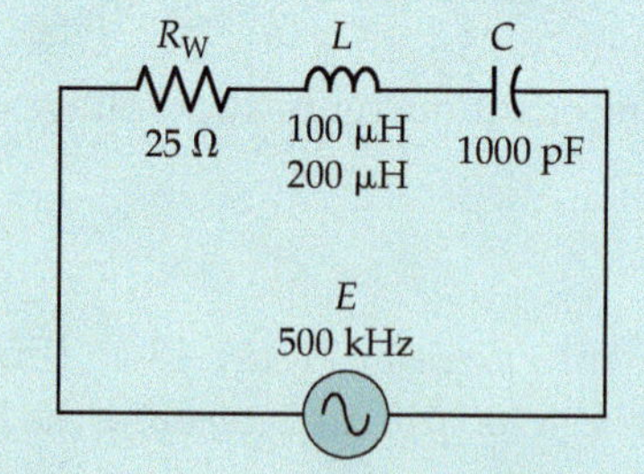

Figure 28-45

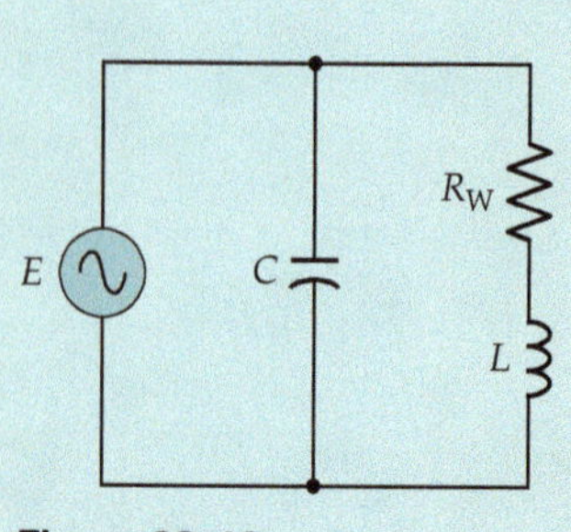

Figure 28-46

28-23 The series resonant circuit in Figure 28-47 (from 23-36) is to be analyzed for the two cases described in Problems 23-7 and 23-11. Prepare graphs of current versus frequency for each case.

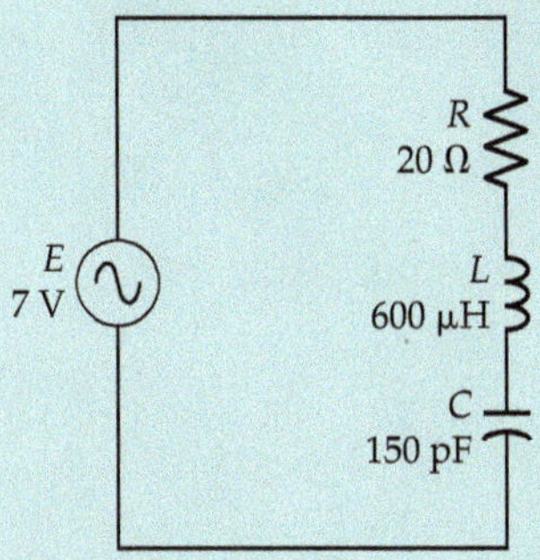

Figure 28-47

Section 28-8

28-24 Analyze the low-pass filter circuit in Figure 28-48 to prepare graphs of frequency and phase response. Compare to Problem 24-6.

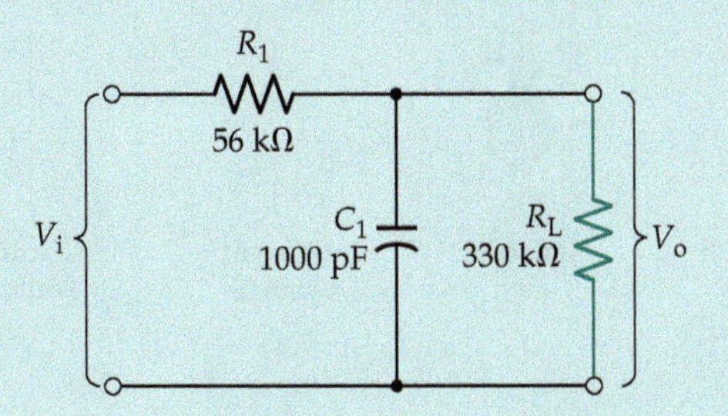

Figure 28-48

28-25 Analyze the high-pass filter circuit in Figure 28-49 (from Figure 24-11) to prepare graphs of frequency and phase response. Compare to Example 24-6.

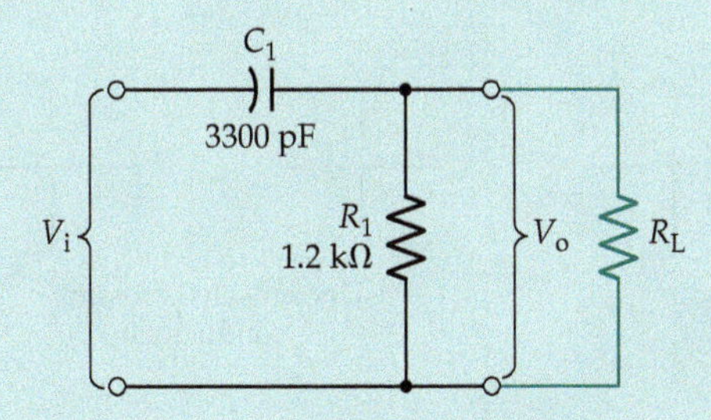

Figure 28-49

APPENDIX 1
Circuit Symbols

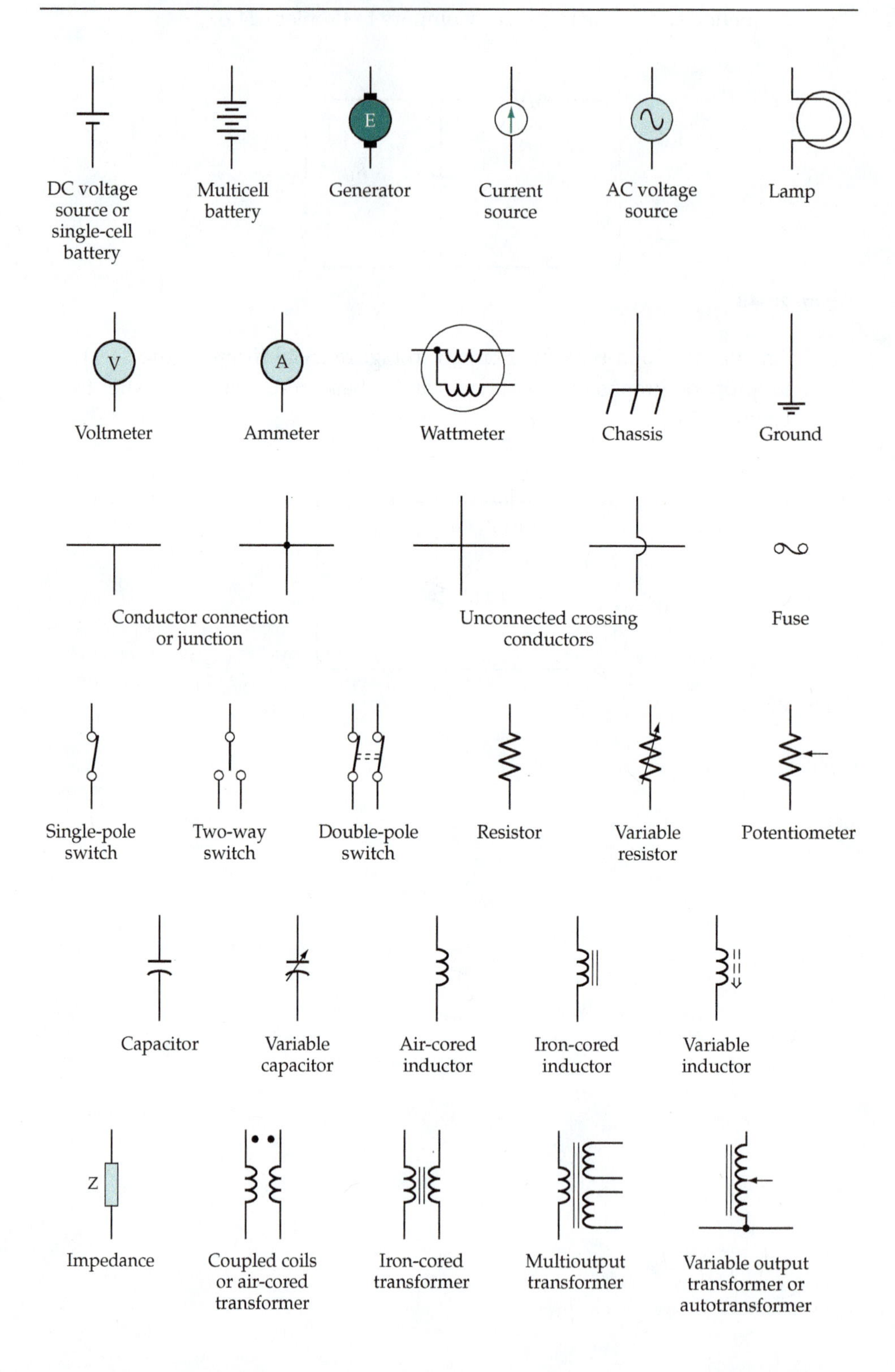

APPENDIX 2
The International System of Units (SI Units)

INTRODUCTION

Before standard systems of measurement were invented, many approximate units were used. A long distance was often measured by the number of *days* it would take to ride a horse over the distance; a horse's height was measured in *hands;* liquid was measured by the *bucket* or *barrel.*

With the development of science and engineering, more accurate units had to be devised. The English-speaking peoples adopted *the foot* and the *mile* for measuring distances, the *pound* for mass, and the *gallon* for liquid. Other nations followed the lead of the French in adopting a *metric system,* in which large and small units are very conveniently related by a factor of 10.

The increase of world trade and the exchange of scientific information between nations, made it necessary to establish a single system of units of measurement that would be acceptable internationally. After several world conferences on the matter, a metric system which uses the *meter, kilogram,* and *second* as fundamental units has now been generally adopted around the world. This is known, from the French term "systeme international," as the *SI,* or *international system.*

FUNDAMENTAL UNITS

In the SI system the three basic units are:

unit of length: the meter **(m)**[1]
unit of mass: the kilogram **(kg)**
unit of time: the second **(s)**

These are known as *fundamental units.* Other units derived from the fundamental units are termed *derived units.* Because of the use of the meter, kilogram, and second, the system is sometimes referred to as an *MKS system.*

The *meter* was originally defined as 1 ten-millionth of a meridian passing through Paris from the North Pole to the equator. The *kilogram* was defined as 1000 times the mass of 1 cubic centimeter of distilled water. The *liter*[2] is 1000 times

---

[1]Canadian and British spelling is *metre.*
[2]Canadian and British spelling is *litre.*

the volume of 1 cubic centimeter of liquid. Consequently, 1 liter of water has a mass of 1 kilogram. Because of the possibility of error in the original definitions, the meter was redefined in terms of atomic radiation. Also, the kilogram is now defined as the mass of a certain platinum-iridium standard bar kept at the International Bureau of Weights and Measures in France. The second is, of course, 1/(86 400) of a mean solar day, but it is also accurately defined by atomic radiation. Other fundamental SI units include units of electrical current and of temperature.

UNIT OF FORCE

The SI unit of force is the newton[3] (N), defined as that force which will give a mass of 1 kilogram an acceleration of 1 meter per second per second.

When a body is to be accelerated or decelerated, a force must be applied proportional to the desired rate of change of velocity, that is, proportional to the acceleration (or deceleration).

$$\text{Force} = \text{mass} \times \text{acceleration}$$

$$F = ma$$

When the mass is in kg and the acceleration is in m/s^2, the above equation gives the force in newtons. If the body is to be accelerated vertically from the earth's surface, the *acceleration due to gravity (g)* must be overcome before any vertical motion is possible. In SI units:

$$g = 9.81 \text{ m/s}^2$$

Thus, a mass of 1 kg has a gravitational force of 9.81 N.

WORK, POWER, AND ENERGY

Work

When a body is moved, a force is exerted to overcome the body's resistance to motion. The work done in moving a body is the product of the force and the distance through which the body is moved in the direction of the force.

$$\text{Work} = \text{force} \times \text{distance}$$

$$W = Fd$$

The SI unit of work is the joule[4] (J), defined as the amount of work done when a force of 1 newton acts through a distance of 1 meter. The *joule* may also be

[3]Named for the great English philosopher and mathematician Sir Isaac Newton (1642–1727).
[4]Named after the English physicist James P. Joule (1818–1899).

termed a *newton-meter.* For the equation $W = Fd$, work is expressed in joules when F is in newtons and d is in meters.

Power

Power is the time rate of doing work.

If a certain amount of work (W) is to be done in a time (t), the power required is,

$$\text{Power} = \frac{\text{work}}{\text{time}}$$

$$P = \frac{W}{t}$$

The SI unit of power is the watt[5] (W), defined as the power developed when 1 joule of work is done in 1 second. For $P = W/t$, P is in watts when W is in joules and t is in seconds.

Energy

Energy is defined as the capacity for doing work. Energy is measured in the same units as work; *joule* (J).

Energy Consumption

As discussed in Section 3-5, when 1 watt of power is expended for one hour, the energy consumed (or work done) is one *watt-hour* (1 Wh). When 1 kW is expended for 1 hour, the energy consumed is 1 *kilowatt-hour* (1 kWh). The kilowatt-hour is the usual unit applied to measure energy consumption.

Recall that *power is the time rate of doing work* and that a power of 1 W represents a work rate of *one joule per second* (1 J/s). Therefore, when 1 W of power is dissipated for 1 s, 1 J of work is done, or 1 J of energy is consumed. Similarly, when 1 kW of power is expended for 1 s, 1 kJ of energy is consumed. Also, when 1 kW of power is expended for 1 minute,

$$\begin{aligned}\text{Energy consumed} &= 1\ \text{kW minute} = 1\ \text{kJ} \times (60\ \text{s})\\ &= 60\ \text{kJ}\end{aligned}$$

When 1 kW of power is expended for 1 hour,

$$\begin{aligned}\text{Energy consumed} &= 1\ \text{kWh} = 60\ \text{kJ} \times (60\ \text{minutes})\\ &= 3600\ \text{kJ}\\ &= 3.6\ \text{MJ}\end{aligned}$$

The megajoule (MJ) is the SI unit of energy consumption.

[5]Named after the Scottish engineer and inventer James Watt (1736–1819).

TEMPERATURE AND HEAT

There are two SI temperature scales, the *Celsius scale*[6] and the *Kelvin scale.*[7] The Celsius scale has 100 equal divisions (or *degrees*) between the freezing temperature and the boiling temperature of water. At normal atmospheric pressure, water freezes at 0°C (*zero degrees Celsius*) and boils at 100°C.

The Kelvin temperature scale, also known as the *absolute scale,* commences at absolute zero of temperature, which corresponds to −273.15°C. Therefore, 0°C is equal to 273.15 K, and 100°C is the same temperature as 373.15 K. A temperature difference of 1 K is the same as a temperature difference of 1°C. With the *Fahrenheit* (non-SI) *scale,*[8] 32°F is the freezing temperature of water and 212°F is the boiling temperature.

Joules Equivalent

To raise 1 liter of water through 1°C requires an energy input of 4187 J. This is known as *Joules equivalent,* or the *mechanical equivalent of heat.* Using this figure, the energy required to raise any quantity of water through a given temperature change can be easily calculated. When water is heated, the container must also be raised to the same temperature as the water, so the container is usually defined as having a certain *water equivalent.* This is a quantity of water that would absorb the same amount of energy as the water container when heated through a given temperature change.

UNITS OF CURRENT AND CHARGE

As explained in Chapter 1, electric current (I) is a flow of charge carriers. Therefore, current could be defined in terms of the quantity of electricity (Q) that passes a given point in a conductor during a time of one second.

The coulomb[9] (C) is the unit of electrical charge or quantity of electricity.

The coulomb was originally selected as the fundamental electrical unit from which all other units were derived. However, since it is much easier to measure current accurately than it is to measure charge, the unit of *current* is now the *fundamental electrical unit* in the SI system. Thus, the coulomb is a *derived unit,* defined in terms of the unit of electric current.

The ampere[10] (A) is the unit of electric current.

---

[6]Invented by the Swedish astronomer and scientist Anders Celsius (1701–1744).
[7]Named for the Irish-born scientist and mathematician William Thomson, who became Lord Kelvin (1824–1907).
[8]Devised by German physicist Gabriel Daniel Fahrenheit (1686–1736).
[9]Named after the French physicist Charles Augustin de Coulomb (1736–1806).
[10]Named after the French physicist and mathematician André Marie Ampère (1775–1836).

The ampere is defined as that constant current which, when flowing in each of two infinitely long parallel conductors 1 meter apart, exerts a force of 2×10^{-7} newtons per meter of length on each conductor.

The coulomb is defined as that charge which passes a given point in a conductor each second, when a current of 1 ampere flows. Therfore, the coulomb could be termed an *ampere-second.* Conversely, the ampere can be described as a *coulomb per second.*

$$\text{Amperes} = \frac{\text{coulombs}}{\text{seconds}}$$

It has been established experimentally that *1 coulomb is equal to the total charge carried by 6.24×10^{18} electrons.* Therefore, the charge carried by one electron is,

$$Q = \frac{1}{6.24 \times 10^{18}} = 1.602 \times 10^{-19}\ \text{C}$$

EMF, POTENTIAL, DIFFERENCE, AND VOLTAGE

The volt[11] (V) is the unit of electromotive force (emf) and potential difference.

The volt (V) is defined as the potential difference between two points on a conductor carrying a constant current of 1 ampere when the power dissipated between these points is 1 watt.

As noted, the coulomb is the charge carried by 6.24×10^{18} electrons. One joule of work is done when 6.24×10^{18} electrons are moved through a potential difference of 1 V. One electron carries a charge of $1/(6.24 \times 10^{18})$ coulombs. If only one electron is moved through 1 V, the energy involved is one *electron-volt* (eV).

$$1\ \text{eV} = \frac{1}{6.24 \times 10^{18}}\ \text{J}$$

The electron-volt is frequently used in the case of the very small energy levels associated with electrons in orbit around the nucleus of an atom.

[11]Named in honor of the Italian physicist Count Alessandro Volta (1745–1827), inventor of the voltaic pile.

RESISTANCE AND CONDUCTANCE

The ohm[12] is the unit of resistance, and the symbol used for ohms is Ω, the Greek capital letter omega.

The ohm is defined as that resistance which permits a current flow of 1 ampere when a potential difference of 1 volt is applied to the resistance.

The term *conductance* (G) is applied to the reciprocal of resistance.

The siemens[13] (S) is the unit of conductance. The unit of conductance was previously the mho (ohm spelled backwards), and its symbol was an upside-down omega.

$$\text{conductance} = \frac{1}{\text{resistance}}$$

MAGNETIC FLUX AND FLUX DENSITY

The weber[14] (Wb) is the SI unit of magnetic flux.

The weber is defined as the magnetic flux which, linking a single-turn coil, produces an emf of 1 V when the flux is reduced to zero at a constant rate in one second.

The tesla[15] (T) is the SI unit of magnetic flux density.

The tesla is the flux density in a magnetic field when 1 Wb of flux occurs in a plane of 1 m^2. So, the tesla can be described as 1 Wb/m^2.

INDUCTANCE

The SI unit of inductance is the henry[16] (H).

The inductance of a circuit is 1 henry (1 H) when an emf of 1 V is induced by the current changing at the rate of 1 A/s.

CAPACITANCE

The farad[17] (F) is the SI unit of capacitance.

The farad is the capacitance of a capacitor that contains a charge of 1 coulomb when the potential difference between its terminals is 1 volt.

[12]Named after the German physicist Georg Simon Ohm (1787–1854), whose investigations led to his statement of "Ohm's law of resistance."

[13]Named after Sir William Siemens (1823–1883), a British engineer who was born Karl William von Siemens in Germany.

[14]Named after the German physicist Wilhelm Weber (1804–1890).

[15]Named for the Croatian-American researcher and inventor Nikola Tesla (1856–1943).

[16]Named for the American physicist Joseph Henry (1797–1878).

[17]Named for the English chemist and physicist Michael Farraday (1791–1867).

APPENDIX 3
Unit Conversion Factors

The following factors may be used for conversion between non-SI units and SI units.

To Convert	To	Multiply By
Area		
Acres	square meters (m^2)	4047
acres	hectares (ha)	0.4047
circular mils	square meters (m^2)	5.067×10^{-10}
square feet	square meters (m^2)	0.0929
square inches	square centimeters (cm^2)	6.452
square miles	hectares (ha)	259
square miles	square kilometers (km^2)	2.59
square yards	square meters (m^2)	0.8361
Electric and Magnetic		
amperes/inch	amperes/meter (A/m)	39.37
gauses	teslas (T)	10^{-4}
gilberts	ampere (turns) (A)	0.7958
lines/sq. inch	teslas (T)	1.55×10^{-5}
Maxwells	webers (Wb)	10^{-8}
mhos	Siemens (S)	1
Oersteds	amperes/meter	79.577
Energy and Work		
Btu	joules (J)	1054.8
Btu	kilowatt-hours (kWh)	2.928×10^{-4}
ergs	joules (J)	10^{-7}
ergs	kilowatt-hours (kWh)	0.2778×10^{-13}
foot-pounds	joules (J)	1.356
foot-pounds	kilogram meters (kgm)	0.1383
Force		
dynes	grams (g)	1.02×10^{-3}
dynes	newtons (N)	10^{-5}
pounds	newtons (N)	4.448
poundals	newtons (N)	0.1383
grams	newtons (N)	9.807×10^{-3}
Illumination		
foot-candles	lumens/cm^2	10.764

To Convert	To	Multiply By
Linear		
angstroms	meters (m)	1×10^{-10}
feet	meters (m)	0.3048
fathoms	meters (m)	1.8288
inches	centimeters (cm)	2.54
microns	meters (m)	10^{-6}
miles (nautical)	kilometers (km)	1.853
miles (statute)	kilometers (km)	1.609
mils	centimeters (cm)	2.54×10^{-3}
yards	meters (m)	0.9144
Power		
horsepower	watts (W)	745.7
Pressure		
atmospheres	kilograms/sq. meter (kg/m^2)	10 332
atmospheres	kilopascals (kPa)	101.325
bars	kilopascals (kPa)	100
bars	kilograms/sq. meter (kg/m^2)	1.02×10^{-4}
pounds/sq. foot	kilograms/sq. meter (kg/m^2)	4.882
pounds/sq. inch	kilograms/sq. meter (kg/m^2)	703
Temperature		
degrees Fahrenheit (°F)	degrees Celsius (°C)	(°F − 32)/1.8
degrees Fahrenheit (°F)	degrees kelvin (K)	273.15 + (°F − 32)/1.8
Velocity		
miles/hour (mph)	kilometers/hour (km/h)	1.609
knots	kilometers/hour (km/h)	1.853
Volume		
bushels	cubic meters (m^3)	0.035 24
cubic feet	cubic meters (m^3)	0.028 32
cubic inches	cubic centimeters (cm^3)	16.387
cubic inches	liters (l)	0.016 39
cubic yards	cubic meters (m^3)	0.7646
gallons (U.S.)	cubic meters (m^3)	3.7853×10^{-3}
gallons (imperial)	cubic meters (m^3)	4.546×10^{-3}
gallons (U.S.)	liters (l)	3.7853
gallons (imperial)	liters (l)	4.546
gills	liters (l)	0.1183
pints (U.S.)	liters (l)	0.4732
pints (imperial)	liters (l)	0.5683
quarts (U.S.)	liters (l)	0.9463
quarts (imperial)	liters (l)	1.137
Weight		
ounces	grams (g)	28.35
pounds	kilograms (kg)	0.453 59
tons (long)	kilograms (kg)	1016
tons (short)	kilograms (kg)	907.18

APPENDIX 4
American Wire Gauge Sizes and Metric Equivalents

Gauge	Diameter (mm)	Copper Wire Resistance (Ω/km)	Diameter (mil)	Copper Wire Resistance (Ω/1000 ft)
0000	11.68	0.160	460	0.049
000	10.40	0.203	409.6	0.062
00	9.266	0.255	364.8	0.078
0	8.252	0.316	324.9	0.098
1	7.348	0.406	289.3	0.124
2	6.543	0.511	257.6	0.156
3	5.827	0.645	229.4	0.197
4	5.189	0.813	204.3	0.248
5	4.620	1.026	181.9	0.313
6	4.115	1.29	162	0.395
7	3.665	1.63	144.3	0.498
8	3.264	2.06	128.5	0.628
9	2.906	2.59	114.4	0.792
10	2.588	3.27	101.9	0.999
11	2.30	4.10	90.7	1.26
12	2.05	5.20	80.8	1.59
13	1.83	6.55	72	2
14	1.63	8.26	64.1	2.52
15	1.45	10.4	57.1	3.18
16	1.20	13.1	50.8	4.02
17	1.15	16.6	45.3	5.06
18	1.02	21.0	40.3	6.39
19	0.912	26.3	35.9	8.05
20	0.813	33.2	32	10.1
21	0.723	41.9	28.5	12.8
22	0.644	52.8	25.3	16.1
23	0.573	66.7	22.6	20.3
24	0.511	83.9	20.1	25.7
25	0.455	106	17.9	32.4
26	0.405	134	15.9	41
27	0.361	168	14.2	51.4
28	0.321	213	12.6	64.9
29	0.286	267	11.3	81.4
30	0.255	337	10	103
31	0.227	425	8.9	130
32	0.202	537	8	164
33	0.180	676	7.1	206
34	0.160	855	6.3	261
35	0.143	1071	5.6	329
36	0.127	1360	5	415
37	0.113	1715	4.5	523
38	0.101	2147	4	655
39	0.090	2704	3.5	832
40	0.080	3422	3.1	1044

APPENDIX 5
Resistor Color Code

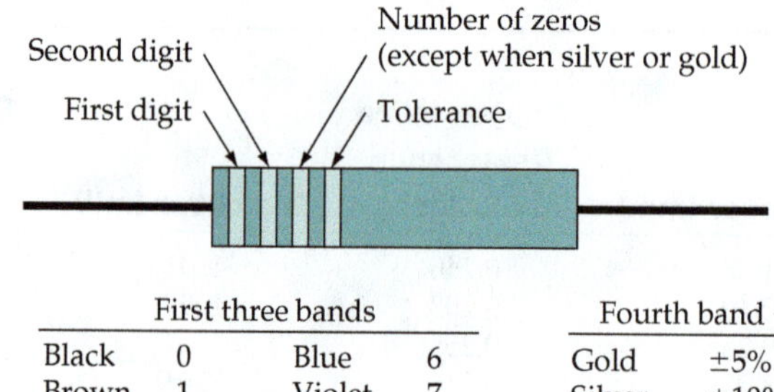

First three bands			
Black	0	Blue	6
Brown	1	Violet	7
Red	2	Grey	8
Orange	3	White	9
Yellow	4	Silver	0.01
Green	5	Gold	0.1

Fourth band	
Gold	±5%
Silver	±10%
None	±20%

COLOR CODE MEMORY AID

Black bruins relish ornery young greenhorns.
Blue violets growing wild smell good

Memory aid	**Color**	**Number**	**Tolerance**
Black	Black	0	
Bruins	Brown	1	
Relish	Red	2	
Ornery	Orange	3	
Young	Yellow	4	
Greenhorns	Green	5	
Blue	Blue	6	
Violets	Violet	7	
Growing	Grey	8	
Wild	White	9	
Smell	Silver	0.01	10%
Good	Gold	0.1	5%

APPENDIX 6
Resistor and Capacitor Values

Typical Standard Resistor Values (±10% Tolerance)							
Ω	Ω	Ω	kΩ	kΩ	kΩ	MΩ	MΩ
—	10	100	1	10	100	1	10
—	12	120	1.2	12	120	1.2	—
—	15	150	1.5	15	150	1.5	15
—	18	180	1.8	18	180	1.8	—
—	22	220	2.2	22	220	2.2	22
2.7	27	270	2.7	27	270	2.7	—
3.3	33	330	3.3	33	330	3.3	—
3.9	39	390	3.9	39	390	3.9	—
4.7	47	470	4.7	47	470	4.7	—
5.6	56	560	5.6	56	560	5.6	—
6.8	68	680	6.8	68	680	6.8	—
—	82	820	8.2	82	820	—	—

Typical Standard Capacitor Values

pF	pF	pF	pF	µF	µF	µF	µF	µF	µF	µF
5	50	500	5000		0.05	0.5	5	50	500	5000
—	51	510	5100		—	—	—	—	—	—
—	56	560	5600		0.056	0.56	5.6	56	—	5600
—	—	—	6000		0.06	—	6	—	—	6000
—	62	620	6200		—	—	—	—	—	—
—	68	680	6800		0.068	0.68	6.8	—	—	—
—	75	750	7500		—	—	—	75	—	—
—	—	—	8000		—	—	8	80	—	—
—	82	820	8200		0.082	0.82	8.2	82	—	—
—	91	910	9100		—	—	—	—	—	—
10	100	1000		0.01	0.1	1	10	100	1000	10 000
—	110	1100		—	—	—	—	—	—	—
12	120	1200		0.012	0.12	1.2	—	—	—	—
—	130	1300		—	—	—	—	—	—	—
15	150	1500		0.015	0.15	1.5	15	150	1500	
—	160	1600		—	—	—	—	—	—	
18	180	1800		0.018	0.18	1.8	18	180	—	
20	200	2000		0.02	0.2	2	20	200	2000	
22	220	2200		—	0.22	2.2	22	—	—	
24	240	2400		—	—	—	—	240	—	
—	250	2500		—	0.25	—	25	250	2500	
27	270	2700		0.027	0.27	2.7	27	270	—	
30	300	3000		0.03	0.3	3	30	300	3000	
33	330	3300		0.033	0.33	3.3	33	330	3300	
36	360	3600		—	—	—	—	—	—	
39	390	3900		0.039	0.39	3.9	39	—	—	
—	—	4000		0.04	—	4	—	400	—	
43	430	4300		—	—	—	—	—	—	
470	470	4700		0.047	0.47	4.7	47	—	—	

Answers to Odd-Numbered Problems

Chapter 2

2-1 1.5×10^{-2}, 1.6×10^{4}, 6.26×10^{3}, 7×10^{-4}, 9.89×10^{5}
2-3 7.8 kΩ, 60 μA, 19 mA, 91.5 kΩ, 50 mA
2-5 4.59 mA, 4.5 mA, 4 mA, 5 mA range
2-7 0.13 V, 13.6 V, over range, 19.99 V range
2-9 139 Ω, 0.13 kΩ, 0.1 kΩ, 200 Ω range
2-11 3 kΩ
2-13 7.217 to 7.263 mA, 1.886 to 1.9 mA, 4.221 to 4.249 mA
2-15 7.5 V to 7.7 V
2-17 22 kΩ to 28 kΩ

Chapter 3

3-1 15 μA, 150 A, 10 mA, 55.6 mA, 455 μA
3-3 33 V
3-5 0.8 Ω
3-7 5.45 mA
3-9 0.11 Ω
3-11 1 μS, 10 S, 667 μS, 3.7 mS, 30.3 μS
3-13 167 mS
3-15 180 W
3-17 10 W
3-19 870 mA, 132 Ω
3-21 0.2 A, 250 Ω
3-23 200 V, 2 kΩ
3-25 10.9 A, 6.25 kWh
3-27 8.7 A, 26.5 Ω, 750 Wh

Chapter 4

4-1 #15
4-3 118 V, 10.3 W
4-5 #14
4-7 108 kV
4-9 500 V/cm, 243 kV
4-11 60 V/cm, 140 kV, 60 kV
4-13 450 kV
4-15 2.19×10^{-2} Ω, 5.47×10^{-3} Ω, 2.43×10^{-3} Ω, 1.37×10^{-2} Ω, 8.76×10^{-4} Ω
4-17 0.06 Ω, 0.1 Ω
4-19 0.47 mm
4-21 0.516 mV
4-23 1.84 mm
4-25 13.12 Ω
4-27 440 Ω, 470 Ω, 530 Ω, 560 Ω, 590 Ω, 620 Ω, 650 Ω, 680 Ω, 710 Ω, 740 Ω
4-29 #10
4-31 17.46 mA, 14.32 V, 0.707 mA, 707 V
4-33 294 mW, 5.1 W, 893 mW, 6.1 W, 24 mW
4-35 2.64 Ω
4-37 56.5 kΩ, 827 Ω, 3.93 MΩ
4-39 Y-V-O, R-R-Bla, Br-Bla-Gr, Gr-R-Br, O-O-Br, Gr-Blu-Gr

Chapter 5

5-1 20 mA
5-3 1 A
5-5 3 V, 5 V, 2.5 V, 1.5 V
5-7 8.96 V
5-9 0.4 V, −1.2 V, −3.8 V
5-11 22.2 V, 52.8 V
5-13 25 kΩ
5-15 ±1.27 V
5-17 40.6 V
5-19 11.5 V, 5.16 V, 82 mW, 148 mW, 120 mW
5-21 0.55 Ω, 45.9 W
5-23 46 mW, 0.34 W
5-25 (a) 1.08 W, 1.98 W, 1.44 W, 4.5 W, (b) 43.2 mW, 79.2 mW, 57.6 mW, 180 mW
5-27 200 Ω, 45 mW
5-29 6.75 mA, 12.13 V, 4.58 V, 10.1 V, 3.17 V
5-31 2.31 A, 30.9 W
5-33 (a) −0.46 V, −0.46 V, −3.47 V, (b) 6 V, −6 V, −6 V

Chapter 6

6-1 115 mA, 250 mA, 365 mA
6-3 555 mA, 221 mA, 455 mA, 1.23 A
6-5 123.2 Ω, 365 mA
6-7 4.3 V
6-9 115 mA, 250 mA, 365 mA
6-11 252 Ω
6-13 95 Ω
6-15 115 mA, 250 mA
6-17 25.4 mA, 7.13 mA, 12.43 mA
6-19 5.19 W, 11.25 W, 16.43 W
6-21 870 mA, 348 mA, 522 mA, 217 mA, 225 W
6-23 44 mA, 31 mA, 27.28 V, 1.2 W, 0.85 W
6-25 271 mW, 76.3 mW, 133 mW, 480.3 mW
6-27 407 Ω
6-29 30.8 V
6-31 169.4 mA, 88 mA, 323.4 mA, 0

Chapter 7

7-1 958 Ω, 83.5 mA
7-3 6.8 kΩ, 1.32 mA
7-5 47.4 kΩ, 316 μA
7-7 83.5 mA, 52.5 mA, 31 mA
7-9 1.32 mA, 0.38 mA, 0.68 mA, 0.26 mA
7-11 183 μA, 133 μA, 72.3 μA, 60.7 μA, 133 μA, 79.1 μA, 53.9 μA
7-13 62.6 V, 17.4 V, 17.4 V
7-15 4.4 V, 4.6 V
7-17 15 V, 1.33 V, 3.39 V, 3.39V, 9 V, 1.19 V, 1.19 V
7-19 10 V, 2 V, 0.4 V
7-21 8 V, 8 V, 8 V
7-23 −770 μA, 146 μA, 58 μA, 88 μA, 145 μA, −10 V, 1.4 V, 0.6 V, 0.6 V, 0.8 V, 1.2 V
7-25 −12 V
7-27 3.5 V, −1.04 V, −3.9 V
7-29 726 μA, 406 μA, 329 μA, 1.45 mA, 6.9 V, 6.9 V, 26.1 V
7-31 27.1 kΩ
7-33 183 μA, 173 μA, 1.73 V, 0, 11.8 V, 1.5 V
7-35 3.73 mA
7-37 875 Ω

Chapter 8

8-1 75 V, 5 Ω, 62.5 V, 2.5 A
8-3 30.6 μA, 98 kΩ
8-5 4.8 V, 600 Ω
8-7 18.06 V, 42 kΩ
8-9 13.59 μA
8-11 775 μA
8-13 821 mV
8-15 1.55 mA
8-17 69 μA
8-19 13.59 μA
8-21 775 μA
8-23 821 mV
8-25 1.55 mA
8-27 −138 μA
8-29 13.59 μA
8-31 775 μA
8-33 823 mV
8-35 1.55 mA
8-37 69.5 μA
8-39 309 kΩ, 334 kΩ, 409 kΩ
8-41 499 μA, 481 μA
8-43 326 μA, 192 μA

Chapter 9

9-1 1.29 mA
9-3 374 μA
9-5 1.56 mA
9-7 1.41 mA (right to left)
9-9 543 μA
9-11 489 μA
9-13 902 μA
9-15 5.08 V
9-17 3.04 V
9-19 396 μA
9-21 1.29 mA
9-23 1.04 V
9-25 748 μA
9-27 1.9 mA, 1.3 mA
9-29 1.29 mA
9-31 6.3 kΩ
9-33 5.8 kΩ

Chapter 10

10-1 7.5 Ah, 30 h
10-3 5.3 h
10-5 1.4 V, 33 A
10-7 77 Ω

10-9 8 V, 7.5 V
10-11 8 cells, 0.4 Ω, 11.54 V
10-13 12 cells, 6 V
10-15 10 cells
10-17 1.7 V, 6 A, 72 Ah, 8.75×10^{-3} Ω
10-19 60 km
10-21 6664 Ah
10-23 16 cells
10-25 174
10-27 77 Ah
10-29 (a) 33.3 h, (b) 37.5 min, (c) 22.5 min
10-31 (a) 50 A, (b) 11.75 V

Chapter 11

11-1 3.3 mT, 0.2 mT
11-3 6.16 mWb
11-5 45 μWb
11-7 1.7 cm
11-9 1.8 mWb
11-11 562 μT
11-13 2100 A, 4200 A/m
11-15 858
11-17 60 mA
11-19 66.7 mA
11-21 0.9 N
11-23 12.5 N, 4.9 N
11-25 9.9 μJ
11-27 1.04 mA
11-29 18.2 mN
11-31 1.96 mA

Chapter 12

12-1 7.95×10^4 A/m, 0.1 T, 22.5 μWb
12-3 44.5 mA
12-5 28.4 A
12-7 45.07 mA
12-9 746 mA
12-11 191, 796
12-13 387 μWb
12-15 1.5×10^3 A/m, 0.4 T, 360 μWb
12-17 202 mA
12-19 2.26 A
12-21 222 mA
12-23 411×10^3
12-25 8.8 kg
12-27 73 kg
12-29 677 mA
12-31 92.1 mA

Chapter 13

13-1 0.27 Ω
13-3 (a) 0.027 Ω, (b) 0.0027 Ω
13-5 13.3 A
13-7 2.5 A, 3.3 A, 5 A, 10 A
13-9 2.5 MΩ, 1 MΩ, 500 kΩ
13-11 25 V, 75 V, 150 V
13-13 95 V, 142.5 V
13-15 11.85 V, 3.02 V
13-17 40 kΩ, 20 kΩ, 10 kΩ
13-19 70 kΩ, 30 kΩ, 16.7 kΩ
13-21 30 kΩ, 7.5 kΩ
13-23 1.933 MΩ, 53.34 kΩ, 13.33 kΩ, 6.67 kΩ
13-25 900 kΩ, 90 kΩ, 9 kΩ, 1 kΩ, 20
13-27 8
13-29 2 Ω, 1 V
13-31 3 mV, 30 mV
13-33 7
13-35 22 A
13-37 18 MΩ
13-39 1.5 kΩ to 150 kΩ

Chapter 14

14-1 40 V
14-3 6 V
14-5 1.13 A
14-7 24
14-9 10 V
14-11 712
14-13 1.76×10^{-8} Wb, 0.25 mV
14-15 4.74 mH
14-17 2.7 mH, 1.46 mH
14-19 17.35 mH, 17.35 mH
14-21 0.047
14-23 251 mH, 98.2 mH, 157 mH
14-25 14.2 J
14-27 3.8 H
14-29 563 mJ
14-31 (a) 1.15 H, (b) 733 μH
14-33 (a) 1.2 mH, (b) 800 μH
14-35 150 μH, 275 μH, 0.545

Chapter 15

15-1 40 000 V/m, 2 μC/m^2
15-3 50 V
15-5 (a) 88.5 pF, (b) 6.64 nF
15-7 37.7 m^2
15-9 0.27 μm
15-11 0.884 μF
15-13 0.796 μF, 9.55 μC
15-15 44.25 pF
15-17 67.9
15-19 0.62 mm
15-21 5.7
15-23 5.88 μF, 1.471 V, 2.94 V, 5.88 V, 14.7 V, 147 μC
15-25 21.4 μF
15-27 637 μJ, 319 μJ, 5.74 mJ
15-29 108 μJ, 216 μJ, 432 μJ, 1.08 mJ
15-31 1.59 nJ, 99 pJ

Chapter 16

16-1 63.2 mA, 86.5 mA, 95 mA, 98.2 mA, 99.3 mA
16-3 14.3 V, 4.1 V, 1.2 V, 0.34 V
16-5 92.9 H
16-7 1.6 ms, 0.36 ms
16-9 1.39 H
16-11 1.23 kΩ
16-13 12.5 kV, 24 kΩ
16-15 24.7 V, 37.2 V, 43.5 V, 46.7 V, 48.3 V, 49.2 V, 49.6 V
16-17 22.7 mA, 9.16 mA, 3.69 mA, 1.49 mA, 0.6 mA, 0.24 mA
16-19 492 Ω
16-21 128 ms
16-23 111 V
16-25 0.46 μF
16-27 34.6 s
16-29 5 kΩ, 1.175 s

Chapter 17

17-1 2.33 Hz, 88 mV, 880 mV
17-3 0.62 V, 0.62 V, –0.62 V
17-5 0.82 V
17-7 64.4 mV, 128.6 mV, 315.4 mV, 374.7 mV
17-9 7.7 V, –9 V, 7.28 V
17-11 1 kHz, 2.5 km, 25 m
17-13 277 W, 277 W, 277 W, 370 W
17-15 0.44 W, 0.6 W, 0.3 W
17-17 414 Ω, 1.62 mW, 813 μW
17-19 2.62 A, 2.35 A, 185 W
17-21 107.7 mA, 68.6 mA, 76.1 mA, 22.6 W
17-23 580 mV, 0, 336 Ω
17-25 3.15 V, 0, 1.11 V, 33.3 Hz, 1.5 V
17-27 2.7, 20
17-29 5 div, 0.5 div
17-31 45°, 83.7°
17-33 25 Hz, 12 000 km, 46.5 Hz, 6450 km
17-35 (a) 0, (b) 0 to 100 V peak
17-37 1.12 MΩ, 223 kΩ
17-39 30 V, 150 V, 450 V

Chapter 18

18-1 79 sin ($\theta + 8.2°$)
18-3 30 sin ($\phi + 15.3°$)
18-5 $180\angle{-39°}$
18-7 $5.5\angle{-68°}$
18-9 $44\angle{16°}$
18-11 $144\angle{34°}$
18-13 $90.1\angle{-33.7°}$, $25.5\angle{11.3°}$, $2298\angle{22.4°}$, $3.5\angle{-45°}$
18-15 ($139.1 + j56.2$), ($0 + j85$), ($19.8 + j60.9$)
18-17 ($39 + j22.5$), ($25.2 + j9.7$), ($22.3 + j20.1$)
18-19 77.8 sin($\phi + 8.2°$)
18-21 $40.5\angle{9°}$
18-23 $143.8\angle{33.9°}$
18-25 $81.07\angle{-22.9°}$
18-27 $19\angle{25.1°}$
18-29 $5.5\angle{-68°}$
18-31 $-34.2 - j36.4$
18-33 $29.4\angle{122°}$
18-35 $249.7\angle{17.6°}$

Chapter 19

19-1 471.2 Ω, 212 mA
19-3 1 MHz
19-5 21.2 Ω, 4.7 A
19-7 99.5 Hz
19-9 356 mA, 24.2 V, 22.4 V, 42.7°
19-11 16.1 mA, 10.7 V
19-13 92.8 mA, 11.1 V, 4.47 V, 21.9°

19-15 0.71 μF
19-17 72.1 mA, 16°, 14.4 V, 5.44 V, 9.56 V
19-19 139 Ω
19-21 88.42 Hz, 71.9 mA, $47.2\angle 19.3^\circ$
19-23 73.5 mA, 79.6 mA, 108.3 mA, −47.3°
19-25 79 mA$\angle -42.5^\circ$
19-27 933 μA, 441 μA, 1.03 mA, 64.7°
19-29 1.3 mA$\angle 46.5^\circ$
19-31 30.3 mA, 72.5 mA, 50.3 mA, 37.6 mA$\angle -36.2^\circ$
19-35 578 mA, 839 mA$\angle -76.9^\circ$

Chapter 20

20-1 389 mA$\angle 18.8^\circ$
20-3 103 V$\angle -71.2^\circ$, 58.7 V$\angle 67.3^\circ$
20-5 47.3 V$\angle -29.9^\circ$, 27.4 V$\angle 59.4^\circ$
20-7 2.2 kΩ$\angle -17.1^\circ$, 8.63 mA$\angle -17.1^\circ$
20-9 18.9 mA$\angle -29.9^\circ$, 10.96 mA$\angle 59.4^\circ$
20-11 2.93 mA$\angle 30.4^\circ$, 6.65 mA$\angle -50.4^\circ$, 18.49 mA $\angle 37.1^\circ$
20-13 534 Ω$\angle 14.4^\circ$
20-15 8.14 V$\angle 43.3^\circ$, 2.9 V$\angle -94.6^\circ$
20-17 3.9 kΩ$\angle -29.3^\circ$
20-19 9.94 mA$\angle 14.4^\circ$
20-21 5.83 kΩ$\angle 68.6^\circ$
20-23 3.9 kΩ$\angle -0.3^\circ$
20-25 0.574 V$\angle -75.2^\circ$, 0.71 V$\angle -74^\circ$
20-29 35 kΩ ‖ (14 kΩ + 703 mH)
20-31 79.6 mA$\angle -90^\circ$, 101 mA$\angle -90^\circ$, 128.9 mA$\angle 0^\circ$, 221.9 mA$\angle -54.5^\circ$
20-33 1.17 kΩ$\angle -6.8^\circ$

Chapter 21

21-1 (a) 24.2 Ω, (b) 9.09 A, (c) 4 kW
21-3 117 W
21-5 499 mW
21-7 (a) 457 var, (b) 576 W, (c) 3.6 kvar
21-9 75 VA, 37.5 var, 65 W
21-11 0.161 VA, 69.1 mW, 0.145 var
21-13 150.7 VA, 42.8 W, 144 var
21-15 1.11 VA, 1.03 W, 0.414 var
21-17 880 VA, 607.2 W, 637.3 var, 15.9 μF
21-19 1.8 kVA, 1.67 kW, 663 var
21-21 256.4 A, 196 A, 1530 μF
21-23 1.8 W, 0.047 A
21-25 65.2 A, 79.5 μF
21-27 2.59 A
21-29 0.91%

Chapter 22

22-1 3.1 mA$\angle -51.6^\circ$, 1.68 kΩ$\angle 51.6^\circ$
22-3 $E_1Z_2/(Z_1Z_2 + Z_1Z_3 + Z_2Z_3)$
22-5 $[E_1Z_2 + E_2(Z_1 + Z_3)]/[Z_2(Z_1 + Z_3 + Z_4) + Z_4(Z_1 + Z_3)]$
22-7 $E_1/\{Z_4(Z_1 + Z_3)[1/(Z_1 + Z_3) + 1/Z_2 + 1/Z_4 - K/Z_2]\}$
22-9 9.25 mA$\angle 45.1^\circ$
22-11 29 μA$\angle -66.2^\circ$
22-13 3.2 V$\angle 0.1^\circ$
22-15 1.3 mA$\angle -16.5^\circ$
22-17 $[E_1Z_2 + E_2(Z_1 + Z_3)]/[Z_2(Z_1 + Z_3 + Z_4) + Z_4(Z_1 + Z_3)]$
22-19 11.15 μA$\angle 88.8^\circ$
22-21 1.39 mA$\angle -14.3^\circ$
22-23 232.5 μA$\angle -27.9^\circ$
22-25 1.48 V$\angle 23^\circ$
22-27 9.21 mA$\angle 45.4^\circ$
22-29 $[E_1Z_2 + E_2(Z_1 + Z_3)]/[Z_2(Z_1 + Z_3 + Z_4) + Z_4(Z_1 + Z_3)]$
22-31 $[E_1(Z_2 + Z_3 + Z_4) - E_2Z_2]/(Z_1Z_2 + Z_1Z_3 + Z_1Z_4 + Z_2Z_3 + Z_2Z_4)$
22-33 4.2 A$\angle -10.5^\circ$
22-35 286.6 μA$\angle 98.2^\circ$
22-37 125.1 μA$\angle -20^\circ$
22-39 $(E_2 - I_1Z_1)/(Z_1 + Z_2 + Z_3)$
22-41 $E_1Z_2/(Z_1Z_2 + Z_1Z_3 + Z_2Z_3)$
22-43 72.3 μA$\angle 31.2^\circ$
22-45 500.5 Ω, 2.53 mH
22-47 17.5 kΩ, 60.5 mH
22-49 12.8 mA$\angle 91^\circ$
22-51 3.2 mA$\angle -124^\circ$
22-53 0.06 μF to 0.3 μF
22-55 0.024 μF
22-57 122 mH, 61.6 Ω, 5 mH; 35.3 Ω

Chapter 23

23-1 500 kHz, 156 mA
23-3 I(mA) 0.84 2.1 6.9 156 6.9 2.15 0.84

23-5

V_L(V)	0.33	1.67	8.76	248	13.7	6.71	5.34
V_C(V)	5.35	6.68	13.7	248	8.76	1.7	0.33
V_R(mV)	27	67	221	4990	221	69	27

23-7 530 kHz, I(mA) 0.93 2.3 7.8 350 7.8 2.3 0.93
23-9 40 pF to 3400 pF
23-11 471 kHz
23-13 49.7, 100, 88.9
23-15 505 kHz, 495 kHz, 10 kHz
23-17 527.3 kHz, 532.7 kHz, 468.4 kHz, 473.7 kHz
23-19 200 kΩ
23-21 530.5 kHz, 200
23-23 2.65 kHz
23-25 30 pF
23-27 15 kΩ, 13.9
23-29 0.02, 110 mA, 130 mA

Chapter 24

24-1 −6 dB
24-3 160 mV
24-5 −0.83 dB, 1.32 Hz
24-7 −14 dB$\angle -78.7^\circ$, −20 dB$\angle -84.3^\circ$, −23.5 dB$\angle -86.2^\circ$
24-9 2.72 kHz, −9.2 dB$\angle 69.8^\circ$, −14.9 dB$\angle 79.5^\circ$
24-11 18.25 kHz, −11.1 dB$\angle 73.2^\circ$, −16.5 dB$\angle 81.4^\circ$
24-13 320 Ω, 49.7 kHz
24-15 −5.7°, −26.6°, −45°, −63.4°, −84.3°
24-17 84.3°, 63.4°, 45°, 26.6°, 5.7°
24-19 123 Ω
24-21 29.4 kHz, −1 dB, −61 dB
24-23 11.27 kHz
24-25 0.02 dB
24-27 2.08 mH, 17.6 Ω
24-29 88.4 kHz, 10.65 kHz, −16.5 dB
24-31 1771 pF, 116 Ω, 119.6 kHz, −23 dB

Chapter 25

25-1 29.9 V, 36.8 V, 66.7 V, 6.9 V
25-3 6.24 V, 25 V, 60 V, 500 V, 143.5 mA
25-5 50 A
25-7 1.8 mWb
25-9 537 μWb, 4 μWb
25-11 22.5 μWb, 113 μWb
25-13 121 V, 1.28 A, 37.3°
25-15 663.5 mA, 104.5 V, 26.3°
25-17 906.7 mA, 119.7 V, 53.3°
25-19 1.24 A, 36.3°
25-21 7.6%
25-23 25 V, 24 V
25-25 93.8%, 1.24 A
25-27 32 Ω
25-29 3.6%
25-31 150 V, 4.5 A, 3 A
25-33 1.93 W, 82.5 mA, 401 mA

Chapter 26

26-1 231 V, 1.49 mA, 1.49 mA, 0
26-3 0.77 A, 3.08 A, 0.578 A, 2.42 A
26-11 1.11 A, 1.015 A, 0.941 A, 1.43 A, 1.58 A, 1.95 A
26-13 375 mA, 412 mA, 790 mA, 1.14 A, 264 mA, 1.19 A
26-15 1.24 A$\angle -34.7^\circ$, 1.72 A$\angle -139^\circ$, 1.85 A$\angle 81.5^\circ$
26-17 358 mA$\angle 16^\circ$, 286 mA$\angle -162^\circ$, 73.2 mA$\angle -170^\circ$
26-19 (a) 644 mA, 1.37 A, 1.1 A, (b) 1.37 A, 644 mA, 1.1 A
26-21 1.03 kW
26-23 189 W, 5.7 var, 204 VA
26-25 567 W, 613 VA, 17.1 var
26-27 106 VA, −99 var, 37.4 W
26-29 368 mA, 0.598, 30.5 W
26-31 (a) 2.5 A, 1.125 kW, 0.866; (b) 2.21 A, 1.125 kW, 0.98
26-33 7.61 W, 16.32 W

Chapter 27

27-1 6.37 V, 3.18 V, 2.21 V, 1.59 V, 1.27 V
27-3 4.77 V, 7.5 V, 3.18 V, 0.637 V, 0.27 V
27-5 4.5 V, 7.07 V, 3 V, 0.6 V, 0.26 V
27-7 45 V, 30 V, 6 V, 2.6 V
27-9 5.44 V
27-11 5.85 V
27-13 7.34 V
27-15 103 mV
27-17 4.5 V, 0.88 V, 4.45 V, 0.87 V
27-19 28.9 V, 32 V
27-21 14.2 V, 22.3 V

INDEX